British Plant Communities 2

British Plant Communities

VOLUME 2

MIRES AND HEATHS

J. S. Rodwell (editor)
C. D. Pigott, D. A. Ratcliffe
A. J. C. Malloch, H. J. B. Birks
M. C. F. Proctor, D. W. Shimwell
J. P. Huntley, E. Radford
M. J. Wigginton, P. Wilkins

for the
UK Joint Nature Conservation Committee

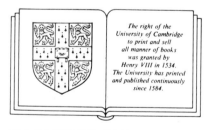

The right of the
University of Cambridge
to print and sell
all manner of books
was granted by
Henry VIII in 1534.
The University has printed
and published continuously
since 1584.

CAMBRIDGE UNIVERSITY PRESS
Cambridge
New York · Port Chester
Melbourne · Sydney

Published by the Press Syndicate of the University of Cambridge
The Pitt Building, Trumpington Street, Cambridge CB2 1RP
40 West 20th Street, New York, NY 10011–4211, USA
10 Stamford Road, Oakleigh, Victoria 3166, Australia

First published 1991

Printed in Great Britain at the University Press, Cambridge

British Library cataloguing in publication data
British plant communities.
Vol. 2 : Mires and heaths
1. Rodwell, J. S. (John S.)
581.5240941

Library of Congress cataloguing in publication data
British plant communities.

Includes bibliographical references.
Contents: v. 1. Woodlands and scrub—v. 2. Mires
and heaths.
1. Plant communities—Great Britain. 2. Vegetation
classification—Great Britain. I. Rodwell, J. S.
II. Nature Conservancy Council (Great Britain)
QK306.B857   1991   581.5'247'0941   90-1300

ISBN 0 521 39165 2 hardback

CONTENTS

Contents

FIGURES

PREFACE AND ACKNOWLEDGEMENTS

The formulation of our proposals for mires and heaths, which make up this second volume of *British Plant Communities*, occupied much of the middle years of the National Vegetation Classification project, when the enthusiasm of the research team could so easily have flagged and the financial support of the Nature Conservancy Council been put in jeopardy but for a continuing faith in the quality and value of the results.

As coordinator of the project and editor of the work, I owe a special and very substantial debt in the preparation of the accounts of the mire communities to Michael Proctor. With Paul Wilkins, the research assistant who worked under his supervision with great industry and wit, Michael undertook the analysis of a very large body of data collected by our own team together with numerous samples from existing work, and he prepared preliminary descriptions of most of the vegetation types. Both in these earlier stages and later, when together we shaped the proposals into a coherent whole and I completed the writing of the community accounts, I was enormously informed by Michael's experience of mire ecology and his always thoughtful comments on the developing scheme.

As the classification emerged, we were both helped greatly by the prodigious efforts of Dr Hilary Birks who, on secondment to the team for a brief period, confirmed the integrity of many of the vegetation types in a parallel analysis of upland data. Among other workers, we are especially grateful to Dr Bryan Wheeler, not just for his generosity in allowing us unhindered access to his numerous data from base-rich mires, but also for the open-hearted way in which he shared his understanding of mire vegetation and gave us enthusiastic critical comments.

As throughout our work, we are also pleased to acknowledge the assistance of Messrs Eric Birse and James Robertson of the Macaulay Institute in Aberdeen who supplied us with many data from Scotland. We are grateful, too, to Dr Peter Hulme also of the Macaulay, and to Drs Eric Bignal, Mike Clarke and Roger Meade,

Miss Ros Hattey, Mr John Ratcliffe and various staff of the England Field Unit, then under the leadership of Dr Tim Bines, for enabling us to draw upon NCC reports on mire vegetation. I have also been much helped in the writing by discussions with these workers and also with Drs Wanda Fojt, Henry Adams, Richard Jones and Martin Page, Miss Katherine Hearn and Messrs Stephen Evans, Richard Lindsay, Charles Pulteney and Derek Wells.

For the presentation of our heath data, responsibility among our team fell to David Shimwell who, with the diligent support of his research assistant Elaine Radford, came to grips nobly with large numbers of newly-collected samples, together with many existing data from maritime cliffs contributed by Andrew Malloch, and from the uplands, for which latter we were again fortunate to have a preliminary analysis from Hilary Birks. David always kept up his cheerful encouragement as I continued work on the preliminary community accounts that he prepared and I was spurred on by his enthusiasm for understanding even our most neglected kinds of heath vegetation.

Among other contributors of new data, we were particularly grateful to Miss Lynne Farrell, NCC's Lowland Heath Specialist, and Dr John Hopkins, also of the Chief Scientist Directorate. At various stages in the preparation of the material, I have been greatly helped too by discussions with Miss Alison Hobbs, Drs Chris Sydes and David Horsfield, Mr Alan Brown and other members of Dr Des Thompson's Uplands Team at NCC Edinburgh.

As our schemes for heaths and mires gained currency within the NCC and other organisations in these final years of the project, I have been especially encouraged by the spirited reception they received on the training courses which I ran with Michael Proctor in the New Forest and the Galloway hills. The kind of enthusiasm we encountered there from staff of the NCC, the National Trust, the Macaulay Institute and the Department of the Environment (Northern Ireland), individuals too numerous to name here, has helped carry through the work into these last arduous days and bodes well for the implementation of the proposals.

Among others who could easily go unsung, I want to thank Carol Barlow who, with great fortitude and cheerfulness, has continued to provide me with essential secretarial help, and, for Cambridge University Press, Jane Bulleid, the copy-editor whose patience and cool nerve have been invaluable in bringing these huge manuscripts that much closer to publication and use.

Finally, as with all sections of the work of the Natio-

nal Vegetation Classification, this has been part of a greater team effort and Michael Proctor, David Shimwell, Elaine Radford, Paul Wilkins and I are deeply indebted to our co-workers Andrew Malloch, John Birks, Jacqui Huntley and Martin Wigginton and the chairmen of the coordinating panel Donald Pigott and Derek Ratcliffe for their congenial effort, company and encouragement along the way.

John Rodwell

Lancaster

PREAMBLE

GENERAL INTRODUCTION

The background to the work

It is a tribute to the insight of our early ecologists that we can still return with profit to *Types of British Vegetation* which Tansley (1911) edited for the British Vegetation Committee as the first coordinated attempt to recognise and describe different kinds of plant community in this country. The contributors there wrote practically all they knew and a good deal that they guessed, as Tansley himself put it, but they were, on their own admission, far from comprehensive in their coverage. It was to provide this greater breadth, and much more detailed description of the structure and development of plant communities, that Tansley (1939) drew together the wealth of subsequent work in *The British Islands and their Vegetation*, and there must be few ecologists of the generations following who have not been inspired and challenged by the vision of this magisterial book.

Yet, partly because of its greater scope and the uneven understanding of different kinds of vegetation at the time, this is a less systematic work than *Types* in some respects: its narrative thread of explication is authoritative and engaging, but it lacks the light-handed framework of classification which made the earlier volume so very attractive, and within which the plant communities might be related one to another, and to the environmental variables which influence their composition and distribution. Indeed, for the most part, there is a rather self-conscious avoidance of the kind of rigorous taxonomy of vegetation types that had been developing for some time elsewhere in Europe, particularly under the leadership of Braun-Blanquet (1928) and Tüxen (1937). The difference in the scientific temperament of British ecologists that this reflected, their interest in how vegetation works, rather than in exactly what distinguishes plant communities from one another, though refreshing in itself, has been a lasting hindrance to the emergence in this country of any consensus as to how vegetation ought to be described, and whether it ought to be classified at all.

In fact, an impressive demonstration of the value of the traditional phytosociological approach to the description of plant communities in the British Isles was published in German after an international excursion to Ireland in 1949 (Braun-Blanquet & Tüxen 1952), but more immediately productive was a critical test of the techniques among a range of Scottish mountain vegetation by Poore (1955a, b, c). From this, it seemed that the really valuable element in the phytosociological method might be not so much the hierarchical definition of plant associations, as the meticulous sampling of homogeneous stands of vegetation on which this was based, and the possibility of using this to provide a multidimensional framework for the presentation and study of ecological problems. Poore & McVean's (1957) subsequent exercise in the description and mapping of communities defined using this more flexible approach then proved just a prelude to the survey of huge tracts of mountain vegetation by McVean & Ratcliffe (1962), work sponsored and published by the Nature Conservancy (as it then was) as *Plant Communities of the Scottish Highlands*. Here, for the first time, was the application of a systematised sampling technique across the vegetation cover of an extensive and varied landscape in mainland Britain, with assemblages defined in a standard fashion from full floristic data, and interpreted in relation to a complex of climatic, edaphic and biotic factors. The opportunity was taken, too, to relate the classification to other European traditions of vegetation description, particularly that developed in Scandinavia (Nordhagen 1943, Dahl 1956).

McVean & Ratcliffe's study was to prove a continual stimulus to the academic investigation of our mountain vegetation and of abiding value to the development of conservation policy, but their methods were not extended to other parts of the country in any ambitious sponsored surveys in the years immediately following. Despite renewed attempts to commend traditional phytosociology, too (Moore 1962), the attraction of this whole approach was overwhelmed for many by the heated debates that preoccupied British plant ecologists in the 1960s, on the issues of objectivity in the sampling and sorting of data, and the respective values of classification or ordination as analytical techniques. Others, though, found it perfectly possible to integrate multivariate analysis into phytosociological survey, and demonstrated the advantage of computers for the display and interpretation of ecological data, rather than the simple testing of methodologies (Ivimey-Cook & Proctor 1966). New generations of research students also began to draw inspiration from the Scottish and Irish initiatives by applying phytosociology to the solving of particular descriptive and interpretive problems, such as variation among British calcicolous grasslands (Shimwell 1968a), heaths (Bridgewater 1970), rich fens (Wheeler 1975) and salt-marshes (Adam 1976), the vegetation of Skye (Birks 1969), Cornish cliffs (Malloch 1970) and Upper Teesdale (Bradshaw & Jones 1976). Meanwhile, too, workers at the Macaulay Institute in Aberdeen had been extending the survey of Scottish vegetation to the lowlands and the Southern Uplands (Birse & Robertson 1976, Birse 1980, 1984).

With an accumulating volume of such data and the appearance of uncoordinated phytosociological perspectives on different kinds of British vegetation, the need for an overall framework of classification became ever more pressing. For some, it was also an increasingly urgent concern that it still proved impossible to integrate a wide variety of ecological research on plants within a generally accepted understanding of their vegetational context in this country. Dr Derek Ratcliffe, as Scientific Assessor of the Nature Conservancy's Reserves Review from the end of 1966, had encountered the problem of the lack of any comprehensive classification of British vegetation types on which to base a systematic selection of habitats for conservation. This same limitation was recognised by Professor Sir Harry Godwin, Professor Donald Pigott and Dr John Phillipson who, as members of the Nature Conservancy, had been asked to read and comment on the Reserves Review. The published version, A Nature Conservation Review (Ratcliffe 1977), was able to base the description of only the lowland and upland grasslands and heaths on a phytosociological treatment. In 1971, Dr Ratcliffe, then Deputy Director (Scientific) of the Nature Conservancy, in proposals for development of its research programme, drew attention to 'the need for a national and systematic phytosociological treatment of British vegetation, using standard methods in the field and in analysis/classification of the data'. The intention of setting up a group to examine the issue lapsed through the splitting of the Conservancy which was announced by the Government in 1972. Meanwhile, after discussions with Dr Ratcliffe, Professor Donald Pigott of the University of Lancaster proposed to the Nature Conservancy a programme of research to provide a systematic and comprehensive classification of British plant communities. The new

Nature Conservancy Council included it as a priority item within its proposed commissioned research programme. At its meeting on 24 March 1974, the Council of the British Ecological Society welcomed the proposal. Professor Pigott and Dr Andrew Malloch submitted specific plans for the project and a contract was awarded to Lancaster University, with sub-contractual arrangements with the Universities of Cambridge, Exeter and Manchester, with whom it was intended to share the early stages of the work. A coordinating panel was set up, jointly chaired by Professor Pigott and Dr Ratcliffe, and with research supervisors from the academic staff of the four universities, Drs John Birks, Michael Proctor and David Shimwell joining Dr Malloch. At a later stage, Dr Tim Bines replaced Dr Ratcliffe as nominated officer for the NCC, and Miss Lynne Farrell succeeded him in 1985.

With the appointment of Dr John Rodwell as full-time coordinator of the project, based at Lancaster, the National Vegetation Classification began its work officially in August 1975. Shortly afterwards, four full-time research assistants took up their posts, one based at each of the universities: Mr Martin Wigginton, Miss Jacqueline Paice (later Huntley), Mr Paul Wilkins and Dr Elaine Grindey (later Radford). These remained with the project until the close of the first stage of the work in 1980, sharing with the coordinator the tasks of data collection and analysis in different regions of the country, and beginning to prepare preliminary accounts of the major vegetation types. Drs Michael Lock and Hilary Birks and Miss Katherine Hearn were also able to join the research team for short periods of time. After the departure of the research assistants, the supervisors supplied Dr Rodwell with material for writing the final accounts of the plant communities and their integration within an overall framework. With the completion of this charge in 1989, the handover of the manuscript for publication by the Cambridge University Press began.

The scope and methods of data collection

The contract brief required the production of a classification with standardised descriptions of named and systematically arranged vegetation types and, from the beginning, this was conceived as something much more than an annotated list of interesting and unusual plant communities. It was to be comprehensive in its coverage, taking in the whole of Great Britain but not Northern Ireland, and including vegetation from all natural, semi-natural and major artificial habitats. Around the maritime fringe, interest was to extend to the start of the truly marine zone, and from there to the tops of our remotest mountains, covering virtually all terrestrial plant communities and those of brackish and fresh waters, except where non-vascular plants were the dominants. Only short-term leys were specifically excluded, and, though care was to be taken to sample more pristine and long-established kinds of vegetation, no undue attention was to be given to assemblages of rare plants or to especially rich and varied sites. Thus widespread and dull communities from improved pastures, plantations, run-down mires and neglected heaths were to be extensively sampled, together with the vegetation of paths, verges and recreational swards, walls, man-made waterways and industrial and urban wasteland.

For some vegetation types, we hoped that we might be able to make use, from early on, of existing studies, where these had produced data compatible in style and quality with the requirements of the project. The contract envisaged the abstraction and collation of such material from both published and unpublished sources,

and discussions with other workers involved in vegetation survey, so that we could ascertain the precise extent and character of existing coverage and plan our own sampling accordingly. Systematic searches of the literature and research reports revealed many data that we could use in some way and, with scarcely a single exception, the originators of such material allowed us unhindered access to it. Apart from the very few classic phytosociological accounts, the most important sources proved to be postgraduate theses, some of which had already amassed very comprehensive sets of samples of certain kinds of vegetation or from particular areas, and these we were generously permitted to incorporate directly.

Figure 1. Standard NVC sample card.

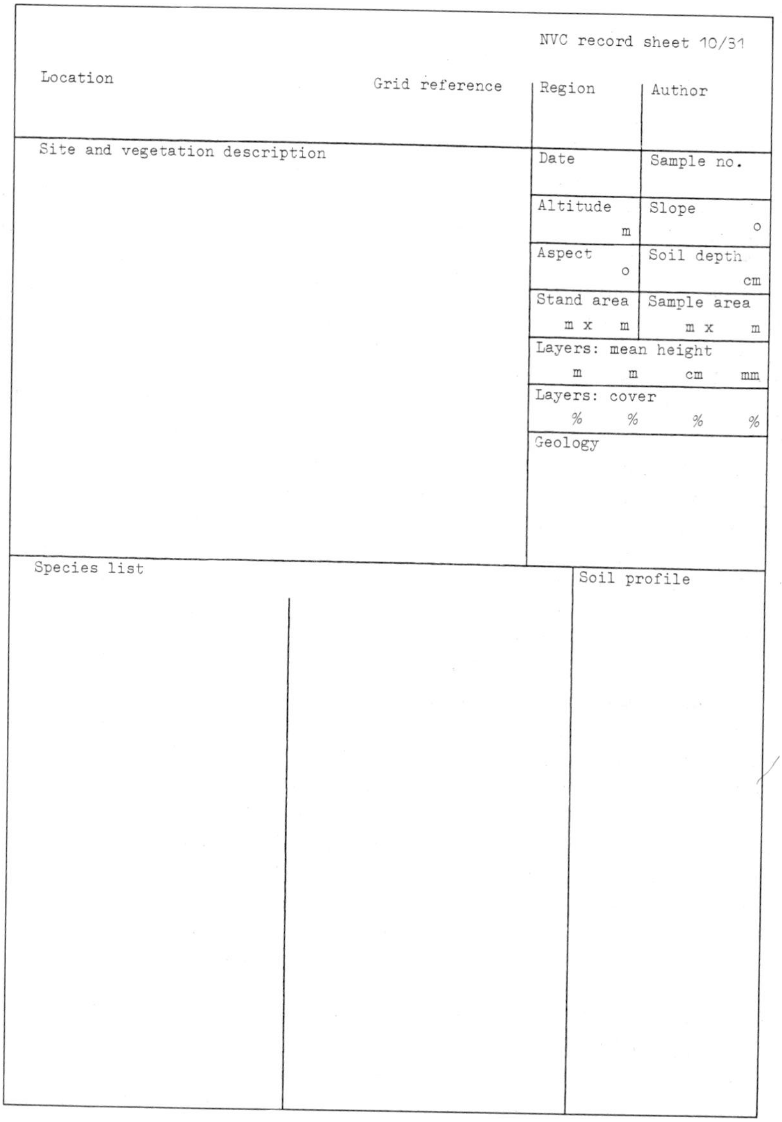

Then, from the NCC and some other government agencies, or from individuals who had been engaged in earlier contracts for them, there were some generally smaller bodies of data, occasionally from reports of extensive surveys, more usually from investigations of localised areas. Published papers on particular localities, vegetation types or individual species also provided small numbers of samples. In addition to these sources, the project was able to benefit from and influence ongoing studies by institutions and individuals, and itself to stimulate new work with a similar kind of approach among university researchers, NCC surveyors, local flora recorders and a few suitably qualified amateurs. An initial assessment and annual monitoring of floristic and geographical coverage were designed to ensure that the accumulating data were fairly evenly spread, fully representative of the range of British vegetation, and of a consistently high quality. Full details of the sources of the material, and our acknowledgements of help, are given in the preface and introduction to each volume.

Our own approach to data collection was simple and pragmatic, and a brief period of training at the outset ensured standardisation among the team of five staff who were to carry out the bulk of the sampling for the project in the field seasons of the first four years, 1976-9. The thrust of the approach was phytosociological in its emphasis on the systematic recording of floristic information from stands of vegetation, though these were chosen solely on the basis of their relative homogeneity in composition and structure. Such selection took a little practice, but it was not nearly so difficult as some critics of this approach imply, even in complex vegetation, and not at all mysterious. Thus, crucial guidelines were to avoid obvious vegetation boundaries or unrepresentative floristic or physiognomic features. No prior judgements were necessary about the identity of the vegetation type, nor were stands ever selected because of the presence of species thought characteristic for one reason or another, nor by virtue of any observed uniformity of the environmental context.

From within such homogeneous stands of vegetation, the data were recorded in quadrats, generally square unless the peculiar shape of stands dictated otherwise. A relatively small number of possible sample sizes was used, determined not by any calculation of minimal areas, but by the experienced assessment of their appropriateness to the range of structural scale found among our plant communities. Thus plots of 2 × 2 m were used for most short, herbaceous vegetation and dwarf-shrub heaths, 4 × 4 m for taller or more open herb communities, sub-shrub heaths and low woodland field layers, 10 × 10 m for species-poor or very tall herbaceous vegetation or woodland field layers and dense scrub, and 50 × 50 m for sparse scrub, and woodland canopy and understorey. Linear vegetation, like that in streams and ditches, on walls or from hedgerow field layers, was sampled in 10 m strips, with 30 m strips for hedgerow shrubs and trees. Quadrats of 1 × 1 m were rejected as being generally inadequate for representative sampling, although some bodies of existing data were used where this, or other sizes different from our own, had been employed. Stands smaller than the relevant sample size were recorded in their entirety, and mosaics were treated as a single vegetation type where they were repeatedly encountered in the same form, or where their scale made it quite impossible to sample their elements separately.

Samples from all different kinds of vegetation were recorded on identical sheets (Figure 1). Priority was always given to the accurate scoring of all vascular plants, bryophytes and macrolichens (sensu Dahl 1968), a task which often required assiduous searching in dense and complex vegetation, and the determination of difficult plants in the laboratory or with the help of referees. Critical taxa were treated in as much detail as possible though, with the urgency of sampling, certain groups, like the brambles, hawkweeds, eyebrights and dandelions, often defeated us, and some awkward bryophytes and crusts of lichen squamules had to be referred to just a genus. It is more than likely, too, that some very diminutive mosses and especially hepatics escaped notice in the field and, with much sampling taking place in summer, winter annuals and vernal perennials might have been missed on occasion. In general, nomenclature for vascular plants follows *Flora Europaea* (Tutin *et al.* 1964 *et seq.*) with Corley & Hill (1981) providing the authority for bryophytes and Dahl (1968) for lichens. Any exceptions to this, and details of any difficulties with sampling or identifying particular plants, are given in the introductions to each of the major vegetation types.

A quantitative measure of the abundance of every taxon was recorded using the Domin scale (sensu Dahl & Hadač 1941), cover being assessed by eye as a vertical projection on to the ground of all the live, above-ground parts of the plants in the quadrat. On this scale:

Cover of 91–100% is recorded as Domin	10
76–90%	9
51–75%	8
34–50%	7
26–33%	6
11–25%	5
4–10%	4
<4% with many individuals	3
with several individuals	2
with few individuals	1

In heaths, and more especially in woodlands, where the vegetation was obviously layered, the species in the different elements were listed separately as part of the same sample, and any different generations of seedlings or saplings distinguished. A record was made of the total cover and height of the layers, together with the cover of any bare soil, litter, bare rock or open water. Where existing data had been collected using percentage cover or the Braun-Blanquet scale (Braun-Blanquet 1928), it was possible to convert the abundance values to the Domin scale, but we had to reject all samples where DAFOR scoring had been used, because of the inherent confusion within this scale of abundance and frequency.

Each sample was numbered and its location noted using a site name and full grid reference. Altitude was estimated in metres from the Ordnance Survey 1:50 000 series maps, slope estimated by eye or measured using a hand level to the nearest degree, and aspect measured to the nearest degree using a compass. For terrestrial samples, soil depth was measured in centimetres using a probe, and in many cases a soil pit was dug sufficient to allocate the profile to a major soil group (*sensu* Avery 1980). From such profiles, a superficial soil sample was removed for pH determination as soon as possible thereafter using an electric meter on a 1:5 soil:water paste. With aquatic vegetation, water depth was measured in centimetres wherever possible, and some indication of the character of the bottom noted. Details of bedrock and superficial geology were obtained from Geological Survey maps and by field observation.

This basic information was supplemented by notes, with sketches and diagrams where appropriate, on any aspects of the vegetation and the habitat thought likely to help with interpretation of the data. In many cases, for example, the quantitative records for the species were filled out by details of the growth form and patterns of dominance among the plants, and an indication of how they related structurally one to another in finely organised layers, mosaics or phenological sequences within the vegetation. Then, there was often valuable information about the environment to be gained by simple observation of the gross landscape or microrelief, the drainage pattern, signs of erosion or deposition and patterning among rock outcrops, talus slopes or stony soils. Often, too, there were indications of biotic effects including treatments of the vegetation by man, with evidence of grazing or browsing, trampling, dunging, mowing, timber extraction or amenity use. Sometimes, it was possible to detect obvious signs of ongoing change in the vegetation, natural cycles of senescence and regeneration among the plants, or successional shifts consequent upon invasion or particular environmental impacts. In many cases, also, the spatial relationships between the stand and neighbouring vegetation types were highly informative and, where a number of samples was taken from an especially varied or complex site, it often proved useful to draw a map indicating how the various elements in the pattern were interrelated.

The approach to data analysis

At the close of the programme of data collection, we had assembled, through the efforts of the survey team and by the generosity of others, a total of about 35 000 samples of the same basic type, originating from more than 80% of the 10 × 10 km grid squares of the British mainland and many islands (Figure 2). Thereafter began a coordinated phase of data processing, with each of the four universities taking responsibility for producing preliminary analyses from data sets crudely separated into major vegetation types – mires, calcicolous grasslands, sand-dunes and so on – and liaising with the others where there was a shared interest. We were briefed in the contract to produce accounts of discrete plant communities which could be named and mapped, so our attention was naturally concentrated on techniques of multivariate classification, with the help of computers to sort the very numerous and often complex samples on the basis of their similarity. We were concerned to employ reputable methods of analysis, but the considerable experience of the team in this kind of work led us to resolve at the outset to concentrate on the ecological integrity of the results, rather than on the minutiae of mathematical technique. In fact, each centre was free to

Figure 2. Distribution of samples available for analysis.

Distribution of samples available for analysis

some extent to make its own contribution to the development of computer programs for the task, Exeter concentrating on Association and Information Analysis (Ivimey-Cook *et al.* 1975), Cambridge and Manchester on cluster analysis (Huntley *et al.* 1981), Lancaster on Indicator Species Analysis, later Twinspan (Hill *et al.* 1975, Hill 1979), a technique which came to form the core of the VESPAN package, designed, using the experience of the project, to be particularly appropriate for this kind of vegetation survey (Malloch 1988).

Throughout this phase of the work, however, we had some important guiding principles. First, this was to be a new classification, and not an attempt to employ computational analysis to fit groups of samples to some existing scheme, whether phytosociological or otherwise. Second, we were to produce a classification of vegetation types, not of habitats, so only the quantitative floristic records were used to test for similarity between the samples, and not any of the environmental information: this would be reserved, rather, to provide one valuable correlative check on the ecological meaning of the sample groups. Third, no samples were to be rejected at the outset because they appeared nondescript or troublesome, nor removed during the course of analysis or data presentation where they seemed to confuse an otherwise crisply-defined result. Fourth, though, there was to be no slavish adherence to the products of single analyses using arbitrary cut-off points when convenient numbers of end-groups had been produced. In fact, the whole scheme was to be the outcome of many rounds of sorting, with data being pooled and reanalysed repeatedly until optimum stability and sense were achieved within each of the major vegetation types. An important part of the coordination at this stage was to ensure roughly comparable scales of definition among the emerging classifications and to mesh together the work of the separate centres so as to avoid any omissions in the processing or wasteful overlaps.

With the departure from the team of the four research assistants in 1980, the academic supervisors were left to continue the preparation of the preliminary accounts of the vegetation types for the coordinator to bring to completion and integrate into a coherent whole. Throughout the periods of field work and data analysis, we had all been conscious of the charge in the contract that the whole project must gain wide support among ecologists with different attitudes to the descriptive analysis of vegetation. Great efforts were therefore made to establish a regular exchange of information and ideas through the production of progress reports, which gained a wide circulation in Britain and overseas, via contacts with NCC staff and those of other research agencies, and the giving of papers at scientific meetings. This meant that, as we approached the presentation of the results of the project, we were well informed about the needs of prospective users, and in a good position to offer that balance of concise terminology and broadly-based description that the NCC considered would commend the work, not only to their own personnel, but to others engaged in the assessment and management of vegetation, to plant and animal ecologists in universities and colleges, and to those concerned with land use and planning.

The style of presentation

The presentation of our results gives priority to the definition of the vegetation types, rather than to the construction of a hierarchical classification. We have striven to characterise the basic units of the scheme on roughly the same scale as a Braun-Blanquet association, but these have been ordered finally not by any rigid adherence to the higher phytosociological categories of alliance, order and class, but in sections akin to the formations long familiar to British ecologists. In some respects, this is a more untidy arrangement, and even those who find the general approach congenial may be surprised to discover what they have always considered to be, say, a heath, grouped here among the mires, or to search in vain for what they are used to calling 'marsh'. The five volumes of the work gather the major vegetation types into what seem like sensible combinations and provide introductions to the range of communities included: aquatic vegetation, swamps and tall-herb fens; grasslands and montane vegetation; heaths and mires; woodlands and scrub; salt-marsh, sand-dune and sea-cliff communities and weed vegetation. The order of appearance of the volumes, however, reflects more the exigencies of publishing than any ecological viewpoint.

The bulk of the material in the volumes comprises the descriptions of the vegetation types. After much consideration, we decided to call the basic units of the scheme by the rather non-committal term 'community', using 'sub-community' for the first-order sub-groups which could often be distinguished within these, and 'variant' in those very exceptional cases where we have defined a further tier of variation below this. We have also refrained from erecting any novel scheme of complicated nomenclature for the vegetation types, invoking existing names where there is an undisputed phytosociological synonym already in widespread use, but generally using the latin names of one, two or occasionally three of the most frequent species. Among the mesotrophic swards, for example, we have distinguished a *Centaurea nigra-Cynosurus cristatus* grassland, which is fairly obviously identical to what Braun-Blanquet & Tüxen (1952) called *Centaureo-Cynosuretum cristati*, and within which, from our data, we have characterised three sub-communities. For the convenience of shorthand description and mapping, every vegetation type has been given a code letter and number, so that *Centaurea-Cynosurus* grassland for example is MG5, MG referring to its place among the mesotrophic grasslands. The *Galium verum* sub-community of this vegetation type, the second to be distinguished within the description, is thus MG5b.

Vegetation being as variable as it is, it is sometimes expedient to allocate a sample to a community even though the name species are themselves absent. What defines a community as unique are rarely just the plants used to name it, but the particular combination of frequency and abundance values for all the species found in the samples. It is this information which is presented in summary form in the floristic tables for each of the communities in the scheme. Figure 3, for example, shows such a table for MG5 *Centaurea-Cynosurus* grassland. Like all the tables in the volumes, it includes such vascular plants, bryophytes and lichens as occur with a frequency of 5% or more in any one of the sub-communities (or, for vegetation types with no sub-communities, in the community as a whole). Early tests showed that records of species below this level of frequency could be largely considered as noise, but cutting off at any higher level meant that valuable floristic information was lost. The vascular species are not separated from the cryptogams on the table though, for woodlands and scrub, the vegetation is sufficiently complex for it to be sensible to tabulate the species in a way which reflects the layered structure.

Every table has the frequency and abundance values arranged in columns for the species. Here, 'frequency' refers to how often a plant is found on moving from one sample of the vegetation to the next, irrespective of how much of that species is present in each sample. This is summarised in the tables as classes denoted by the Roman numerals I to V: 1–20% frequency (that is, up to one sample in five) = I, 21–40% = II, 41–60% = III, 61–80% = IV and 81–100% = V. We have followed the usual phytosociological convention of referring to species of frequency classes IV and V in a particular community as its constants, and in the text usually refer to those of class III as common or frequent species, of class II as occasional and of class I as scarce. The term 'abundance', on the other hand, is used to describe how much of a plant is present in a sample, irrespective of how frequent or rare it is among the samples, and it is summarised on the tables as bracketed numbers for the Domin ranges, and denoted in the text using terms such as dominant, abundant, plentiful and sparse. Where there are sub-communities, as in this case, the data for these are listed first, with a final column summarising the records for the community as a whole.

The species are arranged in blocks according to their pattern of occurrence among the different sub-communities and within these blocks are generally ordered by decreasing frequency. The first group, *Festuca rubra* to *Trifolium pratense* in this case, is made up of the community constants, that is those species which have an overall frequency IV or V. Generally speaking, such plants tend to maintain their high frequency in each of the sub-communities, though there may be some measure of variation in their representation from one to the next: here, for example, *Plantago lanceolata* is somewhat less common in the last sub-community than the first two, with *Holcus lanatus* and a number of others showing the reverse pattern. More often, there are considerable differences in the abundance of these most frequent species: many of the constants can have very high covers, while others are more consistently sparse, and plants which are not constant can sometimes be numbered among the dominants.

The last group of species on a table, *Ranunculus acris* to *Festuca arundinacea* here, lists the general associates of the community, sometimes referred to as companions. These are plants which occur in the community as a whole with frequencies of III or less, though sometimes they rise to constancy in one or other of the sub-communities, as with *R. acris* in this vegetation. Certain of the companions are consistently common overall like *Rumex acetosa*, some are more occasional throughout as with *Rhinanthus minor*, some are always scarce, for example *Calliergon cuspidatum*. Others, though, are more unevenly represented, like *R. acris*, *Heracleum sphondylium* or *Poa trivialis*, though they do not show any marked affiliation to any particular sub-community. Again, there can be marked variation in the abundance of these associates: *Rumex acetosa*, for example, though quite frequent, is usually of low cover, while *Arrhenatherum elatius* and some of the bryophytes, though more occasional, can be patchily abundant; *Alchemilla xanthochlora* is both uncommon among the samples and sparse within them.

The intervening blocks comprise those species which are distinctly more frequent within one or more of the sub-communities than the others, plants which are referred to as preferential, or differential where their affiliation is more exclusive. For example, the group *Lolium perenne* to *Juncus inflexus* is particularly characteristic of the first sub-community of *Centaurea-Cynosurus* grassland, although some species, like *Leucanthemum vulgare* and, even more so, *Lathyrus pratensis*, are more strongly preferential than others, such as *Lolium*, which continues to be frequent in the second sub-community. Even uncommon plants can be good preferentials, as with *Festuca pratensis* here: it is not often found in *Centaurea-Cynosurus* grassland but, when it does occur, it is generally in this first sub-type.

The species group *Galium verum* to *Festuca ovina* helps to distinguish the second sub-community from the first, though again there is some variation in the strength

Floristic table MG5

	a	b	c	MG5
Festuca rubra	V (1–8)	V (2–8)	V (2–7)	V (1–8)
Cynosurus cristatus	V (1–8)	V (1–7)	V (1–7)	V (1–8)
Lotus corniculatus	V (1–7)	V (1–5)	V (2–4)	V (1–7)
Plantago lanceolata	V (1–7)	V (1–5)	IV (1–4)	V (1–7)
Holcus lanatus	IV (1–6)	IV (1–6)	V (1–5)	IV (1–6)
Dactylis glomerata	IV (1–7)	IV (1–6)	V (1–6)	IV (1–7)
Trifolium repens	IV (1–9)	IV (1–6)	V (1–4)	IV (1–9)
Centaurea nigra	IV (1–5)	IV (1–4)	V (2–4)	IV (1–5)
Agrostis capillaris	IV (1–7)	IV (1–7)	V (3–8)	IV (1–8)
Anthoxanthum odoratum	IV (1–7)	IV (1–8)	V (1–4)	IV (1–8)
Trifolium pratense	IV (1–5)	IV (1–4)	IV (1–3)	IV (1–5)
Lolium perenne	IV (1–8)	III (1–7)	I (2–3)	III (1–8)
Bellis perennis	III (1–7)	II (1–7)	I (4)	II (1–7)
Lathyrus pratensis	III (1–5)	I (1–3)	I (1)	II (1–5)
Leucanthemum vulgare	III (1–3)	I (1–3)	II (1–3)	II (1–3)
Festuca pratensis	II (1–5)	I (2–5)	I (1)	I (1–5)
Knautia arvensis	I (4)			I (4)
Juncus inflexus	I (3–5)			I (3–5)
Galium verum	I (1–6)	V (1–6)		II (1–6)
Trisetum flavescens	II (1–4)	IV (1–6)	II (1–3)	III (1–6)
Achillea millefolium	III (1–6)	V (1–4)	III (1–4)	III (1–6)
Carex flacca	I (1–4)	II (1–4)	I (1)	I (1–4)
Sanguisorba minor	I (4)	II (3–5)		I (3–5)
Koeleria macrantha	I (1)	II (1–6)		I (1–6)
Agrostis stolonifera	I (1–7)	II (1–6)	I (6)	I (1–7)
Festuca ovina		II (1–6)		I (1–6)
Prunella vulgaris	III (1–4)	III (1–4)	IV (1–3)	III (1–4)
Leontodon autumnalis	II (1–5)	II (1–3)	IV (1–4)	III (1–5)
Luzula campestris	II (1–4)	II (1–6)	IV (1–4)	III (1–6)
Danthonia decumbens	I (2–5)	I (1–3)	V (2–5)	I (1–5)
Potentilla erecta	I (1–4)	I (3)	V (1–4)	I (1–4)
Succisa pratensis	I (1–4)	I (1–5)	V (1–4)	I (1–5)
Pimpinella saxifraga	I (1–4)	I (1–4)	III (1–4)	I (1–4)
Stachys betonica	I (1–5)	I (1–4)	III (1–4)	I (1–5)
Carex caryophyllea	I (1–4)	I (1–3)	II (1–2)	I (1–4)
Conopodium majus	I (1–4)	I (1–5)	II (2–3)	I (1–5)
Ranunculus acris	IV (1–4)	II (1–4)	IV (2–4)	III (1–4)
Rumex acetosa	III (1–4)	III (1–4)	III (1–3)	III (1–4)
Hypochoeris radicata	III (1–5)	II (2–4)	III (1–4)	III (1–5)
Ranunculus bulbosus	III (1–7)	II (1–5)	III (1–2)	III (1–7)
Taraxacum officinale agg.	III (1–4)	III (1–4)	III (1–3)	III (1–4)
Brachythecium rutabulum	II (1–6)	III (1–4)	II (2)	III (1–6)
Cerastium fontanum	III (1–3)	II (1–3)	II (1–3)	II (1–3)
Leontodon hispidus	II (1–6)	III (2–4)	III (1–5)	II (1–6)
Rhinanthus minor	II (1–5)	II (1–4)	II (1–3)	II (1–5)
Briza media	II (1–6)	III (1–4)	III (2–3)	II (1–6)
Heracleum sphondylium	II (1–5)	II (1–3)	III (1–3)	II (1–5)
Trifolium dubium	II (1–8)	II (1–5)	I (2)	II (1–8)
Primula veris	II (1–4)	II (2–4)	I (2)	II (1–4)
Arrhenatherum elatius	II (1–6)	II (1–7)	I (3–4)	II (1–7)
Cirsium arvense	II (1–3)	II (1–4)	I (1)	II (1–4)
Eurhynchium praelongum	II (1–5)	II (1–4)	I (1–2)	II (1–5)
Rhytidiadelphus squarrosus	II (1–7)	II (1–5)	III (1–4)	II (1–7)
Poa pratensis	II (1–6)	II (2–5)		II (1–6)
Poa trivialis	II (1–8)	I (1–3)	I (1–2)	II (1–8)
Veronica chamaedrys	II (1–4)	I (1–4)	I (1)	II (1–4)
Alopecurus pratensis	I (1–6)	I (1–4)	I (1)	I (1–6)
Cardamine pratensis	I (1–3)	I (1)	I (3)	I (1–3)
Vicia cracca	I (1–4)	I (1–3)	I (1–2)	I (1–4)
Bromus hordeaceus hordeaceus	I (1–6)	I (2–3)	I (3)	I (1–6)
Phleum pratense pratense	I (1–6)	I (1–5)	I (1)	I (1–6)
Juncus effusus	I (2–3)	I (3)	I (1–2)	I (1–3)
Phleum pratense bertolonii	I (1–3)	I (1–3)	I (1)	I (1–3)
Calliergon cuspidatum	I (1–5)	I (2–4)	I (3)	I (1–5)
Ranunculus repens	II (1–7)	I (2)	II (1–4)	I (1–7)
Pseudoscleropodium purum	I (1–5)	I (3–4)	II (2)	I (1–5)
Ophioglossum vulgatum	I (1–5)	I (1)		I (1–5)
Silaum silaus	I (1–5)	I (1–3)		I (1–5)
Agrimonia eupatoria	I (1–5)	I (1–3)		I (1–5)
Avenula pubescens	I (1–3)	I (2–5)		I (1–5)
Plantago media	I (1–4)	I (1–4)		I (1–4)
Alchemilla glabra	I (2)	I (3)		I (2–3)
Alchemilla filicaulis vestita	I (1–3)	I (3)		I (1–3)
Alchemilla xanthochlora	I (1–3)	I (2)		I (1–3)
Carex panicea	I (1–4)	I (2–4)		I (1–4)
Colchicum autumnale	I (3–4)	I (1–3)		I (1–4)
Crepis capillaris	I (1–5)	I (3)		I (1–5)
Festuca arundinacea	I (1–5)	I (3–5)		I (1–5)

Figure 3. Floristic table for NVC community MG5 *Centaurea nigra-Cynosurus cristatus* grassland.

of association between these preferentials and the vegetation type, with *Achillea millefolium* being less markedly diagnostic than *Trisetum flavescens* and, particularly, *G. verum*. There are also important negative features, too, because, although some plants typical of the first and third sub-communities, such as *Lolium* and *Prunella vulgaris*, remain quite common here, the disappearance of others, like *Lathyrus pratensis*, *Danthonia decumbens*, *Potentilla erecta* and *Succisa pratensis* is strongly diagnostic. Similarly, with the third sub-community, there is that same mixture of positive and negative characteristics, and there is, among all the groups of preferentials, that same variation in abundance as is found among the constants and companions. Thus, some plants which can be very marked preferentials are always of rather low cover, as with *Prunella*, whereas others, like *Agrostis stolonifera*, though diagnostic at low frequency, can be locally plentiful.

For the naming of the sub-communities, we have generally used the most strongly preferential species, not necessarily those most frequent in the vegetation type. Sometimes, sub-communities are characterised by no floristic features over and above those of the community as a whole, in which case there will be no block of preferentials on the table. Usually, such vegetation types have been called Typical, although we have tried to avoid this epithet where the sub-community has a very restricted or eccentric distribution.

The tables organise and summarise the floristic variation which we encountered in the vegetation sampled: the text of the community accounts attempts to expound and interpret it in a standardised descriptive format. For each community, there is first a synonymy section which lists those names applied to that particular kind of vegetation where it has figured in some form or another in previous surveys, together with the name of the author and the date of ascription. The list is arranged chronologically, and it includes references to important unpublished studies and to accounts of Irish and Continental associations where these are obviously very similar. It is important to realise that very many synonyms are inexact, our communities corresponding to just part of a previously described vegetation type, in which case the initials *p.p.* (for *pro parte*) follow the name, or being subsumed within an older, more broadly-defined unit. Despite this complexity, however, we hope that this section, together with that on the affinities of the vegetation (see below), will help readers translate our scheme into terms with which they may have been long familiar. A special attempt has been made to indicate correspondence with popular existing schemes and to make sense of venerable but ill-defined terms like 'herb-rich meadow', 'oakwood' or 'general salt-marsh'.

There then follow a list of the constant species of the community, and a list of the rare vascular plants, bryophytes and lichens which have been encountered in the particular vegetation type, or which are reliably known to occur in it. In this context, 'rare' means, for vascular plants, an A rating in the *Atlas of the British Flora* (Perring & Walters 1962), where scarcity is measured by occurrence in vice-counties, or inclusion on lists compiled by the NCC of plants found in less than 100 10 × 10 km squares. For bryophytes, recorded presence in under 20 vice-counties has been used as a criterion (Corley & Hill 1981), with a necessarily more subjective estimate for lichens.

The first substantial section of text in each community description is an account of the physiognomy, which attempts to communicate the feel of the vegetation in a way which a tabulation of data can never do. Thus, the patterns of frequency and abundance of the different species which characterise the community are here filled out by details of the appearance and structure, variation in dominance and the growth form of the prominent elements of the vegetation, the physiognomic contribution of subordinate plants, and how all these components relate to one another. There is information, too, on important phenological changes that can affect the vegetation through the seasons and an indication of the structural and floristic implications of the progress of the life cycle of the dominants, any patterns of regeneration within the community or obvious signs of competitive interaction between plants. Much of this material is based on observations made during sampling, but it has often been possible to incorporate insights from previous studies, sometimes as brief interpretive notes, in other cases as extended treatments of, say, the biology of particular species such as *Phragmites australis* or *Ammophila arenaria*, the phenology of winter annuals or the demography of turf perennials. We trust that this will help demonstrate the value of this kind of descriptive classification as a framework for integrating all manner of autecological studies (Pigott 1984).

Some indication of the range of floristic and structural variation within each community is given in the discussion of general physiognomy, but where distinct sub-communities have been recognised these are each given a descriptive section of their own. The sub-community name is followed by any synonyms from previous studies, and by a text which concentrates on pointing up the particular features of composition and organisation which distinguish it from the other sub-communities.

Passing reference is often made in these portions of the community accounts to the ways in which the nature of the vegetation reflects the influence of environmental factors upon it, but extended treatment of this is reserved for a section devoted to the habitat. An opening paragraph here attempts to summarise the typical conditions which favour the development and maintenance of the vegetation types, and the major factors which control floristic and structural variation within it. This is followed by as much detail as we have at the present time about the impact of particular climatic, edaphic and biotic variables on the community, or as we suppose to be important to its essential character and distribution. With climate, for example, reference is very frequently made to the influence on the vegetation of the amount and disposition of rainfall through the year, the variation in temperature season by season, differences in cloud cover and sunshine, and how these factors interact in the maintenance of regimes of humidity, drought or frosts. Then, there can be notes of effects attributable to the extent and duration of snow-lie or to the direction and strength of winds, especially where these are icy or salt-laden. In each of these cases, we have tried to draw upon reputable sources of data for interpretation, and to be fully sensitive to the complex operation of topographic climates, where features like aspect and altitude can be of great importance, and of regional patterns, where concepts like continental, oceanic, montane and maritime climates can be of enormous help in understanding vegetation patterns.

Commonly, too, there are interactions between climate and geology that are best perceived in terms of variations in soils. Here again, we have tried to give full weight to the impact of the character of the landscape and its rocks and superficials, their lithology and the ways in which they weather and erode in the processes of pedogenesis. As far as possible, we have employed standardised terminology in the description of soils, trying at least to distinguish the major profile types with which each community is associated, and to draw attention to the influence on its floristics and structure of processes like leaching and podzolisation, gleying and waterlogging, parching, freeze-thaw and solifluction, and inundation by fresh- or salt-waters.

With very many of the communities we have distinguished, it is combinations of climatic and edaphic factors that determine the general character and possible range of the vegetation, but we have often also been able to discern biotic influences, such as the effects of wild herbivores or agents of dispersal, and there are very few instances where the impact of man cannot be seen in the present composition and distribution of the plant communities. Thus, there is frequent reference to the role which treatments such as grazing, mowing and burning have on the floristics and physiognomy of the vegetation, to the influence of manuring and other kinds of eutrophication, of draining and re-seeding for agriculture, of the cropping and planting of trees, of trampling or other disturbance, and of various kinds of recreation.

The amount and quality of the environmental information on which we have been able to draw for interpreting such effects has been very variable. Our own sampling provided just a spare outline of the physical and edaphic conditions at each location, data which we have summarised where appropriate at the foot of the floristic tables; existing sources of samples sometimes offered next to nothing, in other cases very full soil analyses or precise specifications of treatments. In general, we have used what we had, at the risk of great unevenness of understanding, but have tried to bring some shape to the accounts by dealing with the environmental variables in what seems to be their order of importance, irrespective of the amount of detail available, and by pointing up what can already be identified as environmental threats. We have also benefited by being able to draw on the substantial literature on the physiology and reproductive biology of individual species, on the taxonomy and demography of plants, on vegetation history and on farming and forestry techniques. Sometimes, this information provides little more than a provisional substantiation of what must remain for the moment an interpretive hunch. In other cases, it has enabled us to incorporate what amount to small essays on, for example, the past and present role of *Tilia cordata* in our woodlands with variation in climate, the diverse effects of dunging by rabbit, sheep and cattle on calcicolous swards, or the impact of burning on *Calluna-Arctostaphylos* heath on different soils in a boreal climate. Debts of this kind are always acknowledged in the text and, for our part, we hope that the accounts indicate the benefits of being able to locate experimental and historical studies on vegetation within the context of an understanding of plant communities (Pigott 1982).

Mention is often made in the discussion of the habitat of the ways in which stands of communities can show signs of variation in relation to spatial environmental differences, or the beginnings of a response to temporal changes in conditions. Fuller discussion of zonations to other vegetation types follows, with a detailed indication of how shifts in soil, microclimate or treatment affect the composition and structure of each community, and descriptions of the commonest patterns and particularly distinctive ecotones, mosaics and site types in which it and any sub-communities are found. It has also often been possible to give some fuller and more ordered account of the ways in which vegetation types can change through time, with invasion of newly available ground, the progression of communities to maturity, and their regeneration and replacement. Some attempt has been made to identify climax vegetation types and major lines of succession, but we have always been wary of the temptation to extrapolate from spatial patterns to temporal sequences. Once more, we have tried to incorporate the results of existing observational and experimental studies, including some of the classic accounts of patterns and processes among British vegetation, and to point up the great advantages of a reliable scheme of classification as a basis for the monitoring and management of plant communities (Pigott 1977).

Throughout the accounts, we have referred to particular sites and regions wherever we could, many of these visited and sampled by the team, some the location of previous surveys, the results of which we have now been able to redescribe in the terms of the classification we have erected. In this way, we hope that we have begun to make real a scheme which might otherwise remain abstract. We have also tried in the habitat section to provide some indications of how the overall ranges of the vegetation types are determined by environmental conditions. A separate paragraph on distribution summarises what we know of the ranges of the communities and sub-communities, then maps show the location, on the 10×10 km national grid, of the samples that are available to us for each. Much ground, of course, has been thinly covered, and sometimes a dense clustering of samples can reflect intensive sampling rather than locally high frequency of a vegetation type. However, we believe that all the maps we have included are accurate in their general indication of distributions, and we hope that this exercise might encourage the production of a comprehensive atlas of British plant communities.

The last section of each community description considers the floristic affinities of the vegetation types in the scheme, and expands on any particular problems of synonymy with previously described assemblages. Here, too, reference is often given to the equivalent or most closely-related association in continental phytosociological classifications and an attempt made to locate each community in an existing alliance. Where the fuller account of British vegetation that we have been able to provide necessitates a revision of the perspective on European plant communities as a whole, some suggestions are made as to how this might be achieved.

Meanwhile, each reader will bring his or her own needs and commitment to this scheme and perhaps be dismayed by its sheer size and apparent complexity. For those requiring some guidance as to the scope of each volume and the shape of that part of the classification with which it deals, the introductions to the major vegetation types will provide an outline of the variation and how it has been treated. The contents page will then give directions to the particular communities of interest. For readers less sure of the identity of the vegetation types with which they are dealing, a key is provided to each major group of communities which should enable a set of similar samples organised into a constancy table to be taken through a series of questions to a reasonably secure diagnosis. The keys, though, are not infallible short cuts to identification and must be used in conjunction with the floristic tables and community descriptions. An alternative entry to the scheme is provided by the species index which lists the occurrences of all taxa in the communities in which we have recorded them. There is also an index of synonyms which should help readers find the equivalents in our classification of vegetation types already familiar to them.

Finally, we hope that whatever the needs, commitments or even prejudices of those who open these volumes, there will be something here to inform and challenge everyone with an interest in vegetation. We never thought of this work as providing the last word on the classification of British plant communities: indeed, with the limited resources at our disposal, we knew it could offer little more than a first approximation. However, we do feel able to commend the scheme as essentially reliable. We hope that the broad outlines will find wide acceptance and stand the test of time, and that our approach will contribute to setting new standards of vegetation description. At the same time, we have tried to be honest about admitting deficiencies of coverage and recognising much unexplained floristic variation, attempting to make the accounts sufficiently open-textured that new data might be readily incorporated and ecological puzzles clearly seen and pursued. For the classification is meant to be not a static edifice, but a working tool for the description, assessment and study of vegetation. We hope that we have acquitted ourselves of the responsibilities of the contract brief and the expectations of all those who have encouraged us in the task, such that the work might be thought worthy of standing in the tradition of British ecology. Most of all, we trust that our efforts do justice to the vegetation which, for its own sake, deserves understanding and care.

MIRES

INTRODUCTION TO MIRES

The sampling of mire vegetation

In our survey of mires, the primary purpose has been to provide a classification of their vegetation according to its floristic composition, and not to categorise mires on the basis of their ecological development or hydrology, their situation in the landscape, or their gross morphology or fine surface patterning. Each of these other approaches to describing mires can yield benefits of its own, and we have tried to bring their various perspectives to bear in the interpretation of our results. Indeed, it was our hope that we might achieve some greater measure of understanding as to how the vegetation of mires relates to their form and function as perceived from these different standpoints. In the task of data collection, however, we were not concerned to dispose our samples according to any prior judgements about what did or did not constitute a particular kind of mire environment; rather, to ensure an adequate representation somewhere in our scheme of all vegetation types found in and around mires.

Most of that vegetation is described in this section of the work. Essentially, we are dealing here with communities made up of bryophytes, herbaceous plants and sub-shrubs sampled from a wide variety of bogs and wet heaths, fens, flushes, springs and soakways, where the ground is kept permanently or periodically waterlogged by high atmospheric humidity, a high ground water-table or lateral water flow. Many of the vegetation types occur on soils that are organic in character, but some are characteristic of mineral profiles, and the account presented here is thus not simply a classification of samples from peatlands, a category that is sometimes taken as synonymous with mires. On the other hand, there are certain kinds of vegetation which are often encountered in these different sorts of mires, indeed which frequently form an integral part of their mosaic of communities, but which are not included here. We did sample these, applying the same general principles of data collection throughout, but allowed the floristic distinctions among the vegetation types to emerge in the analyses, and have described certain communities in other sections of the work. Sometimes the reasons for this are more obvious, as with the various kinds of woodland that are associated with mires. In other cases, readers may be more perplexed as to where to search for an account of vegetation they have always themselves considered as some sort of mire. The assemblages of submerged and floating plants in peaty pools, for example, we have included among the aquatic communities, stands of emergent helophytes and large sedges in mire waters among the swamps and tall-herb fens, and certain wet swards with the mesotrophic grasslands. The contents lists and indices should direct readers aright here.

As always, the only criterion used to locate the samples within a particular stretch of mire vegetation was the floristic and structural homogeneity of a stand and we did not concentrate in any way on rarer or more species-rich assemblages. Indeed, we were particularly concerned to ensure as well the inclusion of more nondescript and neglected, though often widespread, kinds of mire vegetation. Sometimes, in these various situations, homogeneity was hard to discern, with the mire surface showing an intricate patterning among its plants and, in many cases, there was the additional problem of some sort of vertical layering, with tiers of bryophytes and small herbs, then tussocky or tall monocotyledons or sub-shrubs above. With practice, however, such complexities proved no great hindrance to sampling. As far as obvious mosaics were concerned, each element of a pattern was recorded in a separate sample only where the same vegetation could also be found elsewhere in more uniform tracts or in combination with different assemblages in other sorts of mosaic. A good example where such treatment was necessary was on mires with a differentiation of patterning over hummocks and hollows, with the vegetation in the latter occurring in more or less identical form among different hummock assemblages on other mires. By contrast, swards with strongly dominant *Schoenus nigricans* or *Molinia caerulea* among a network of intervening runnels were treated in their entirety, the individual tussocks of these species being taken as much more an integral part of unique kinds of vegetation. With vertical differentiation, a simpler solution was adopted: even where the cover of such dominants or of ericoids or *Myrica gale* attained considerable stature, the vegetation at a single location was never treated in separate layers.

Almost always, samples of 2 × 2 m or 4 × 4 m proved adequate for recording, with 10 × 10 m being only exceptionally used in very impoverished or grossly structured vegetation, as in some *Molinia* swards or *Eriophorum vaginatum* blanket mire. If stands were of an unusual shape, as was often the case with bog pools and hummocks, soakways, flushes or rills, then irregularly shaped samples of the same area were used. Where stands were smaller than the appropriate sample size, again frequently so with hummocks and pools, springs and rills, the whole stand was treated as the sample (Figure 4).

In the recording of floristic data, all vascular plants, bryophytes and macrolichens were scored as usual, with particular care being taken over the identification of cyperaceous plants and Sphagna, accurate information on which is quite vital for the diagnosis and understanding of many kinds of mire vegetation. With some taxa, however, we recorded to the aggregate, as with the occasional occurrences of *Taraxacum officinale* and *Euphrasia officinalis*, or, where there were especial difficulties, to the genus, as sometimes with *Utricularia*, *Cephaloziella* and *Cladonia*. *Lophocolea bidentata sensu lato* may also include some records for infertile *L.*

cuspidata. For the systematic treatment of Sphagna, we followed the proposals of Hill in Smith (1978) which particularly affect the understanding of *Sphagnum capillifolium*, *S. auriculatum* and *S. subsecundum*. We made every effort to search assiduously for smaller hepatics, but it is possible that fragments among dense bryophyte carpets were missed on occasion. Cryptogams growing on tussocks of plants like *Schoenus* and *Molinia* were recorded, though not those epiphytic on the twigs of sub-shrubs.

Very often, it was informative to supplement these numerical floristic data with details of the structural organisation of the vegetation: the role of any vascular dominants or especially prominent bryophytes, the differentiation of patterning or layering among the plants, the suggestion of phenological relationships or any impacts of the accumulation of litter or of the death and decay of important physiognomic components. Notes were also often made on the contribution of the vegetation being sampled to the overall fabric of a mire complex or landscape element, on particular zonations to neighbouring vegetation types and on any apparent successional processes that were in train (Figure 5).

In many kinds of mire, the association of different sorts of vegetation with distinctive environmental situations is rather obvious, even if the underlying reasons for this are complex and unclear: one thinks, for example, of the patterning over hummocks and hollows, on rand and lagg, alongside lines of soligenous influence and around springs and flushes. We tried, as far as possible, never to locate samples primarily by uniformity of habitat, always concentrating our attention initially on the vegetation itself. At the same time, we thought it very important to add to the rather few systematic environmental records for the sample – on altitude, slope, aspect, geology and soil – by noting details of relief and drainage in and around the sample, the structure of the mire complex of which it formed a part, the contribution of any weathering or erosive processes resulting from the action of water, wind or frost, and the impact of any biotic activities. Sometimes, these last concerned the predation of wild herbivores but, more often than not, they were anthropogenic, relating to processes such as grazing of stock, burning, draining, pollution and peat extraction, or natural responses to such activities, like the erosion of bog margins or flooding of turf ponds. With the heavy field work load, we were quite unable to research the treatment history of the sites we sampled, although some existing studies of mire vegetation were very informative in this respect, as sometimes about the character of the natural environment. Nevertheless, even field observations proved valuable in helping interpret the vegetational variation that emerged from our survey: in trying, for example, to assess the relative contributions of climate, soils and treatments to the

Figure 4. Sampling from bog hummock and hollow. Three homogeneous stands of mire vegetation (A–C) have been distinguished and a sample plot laid out in each, 2 × 2 m where possible on the hummock (a), with an identical area of different shape (b) in the hollow or the entire stand (c) in the small pool.

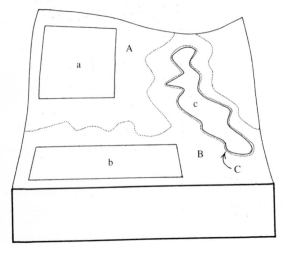

composition and distribution of fen-meadows and rush-pastures, to the maintenance of certain kinds of open flushed sward or to the degeneration of some sorts of bog and wet heath.

As well as having very substantial numbers of samples collected from mires by our own research team, we were most fortunate in being given full access to some important existing data sets. Along with published material from the Scottish Highlands in McVean & Ratcliffe (1962) and from the Southern Uplands and Scottish lowlands in work from the Macaulay Institute (Birse & Robertson 1976, Birse 1980, 1984), there 'were many unpublished samples from these sources. Then, studies such as those of Eddy et al. (1969) on Moor House, Birks (1973) on Skye, Jones (1973) in Upper Teesdale, Bignal & Curtis (1981) in Strathclyde, Ratcliffe & Hattey (1982) through the Welsh lowlands and Hulme & Blyth (1984) on the Hebrides provided valuable data from particular sites or regions. Continuing research by Wheeler (1975, 1978, 1980 a, b, c) has greatly enriched our coverage and understanding of calcareous mires throughout the British lowlands, and we have also benefited from work by Adam (1976), Page (1980) and Jones (1973) in dealing with transitions between mires and upper salt-marsh and mesotrophic grasslands.

Where samples from these sources, and from numer-

Figure 5. Three completed sample cards from mires.

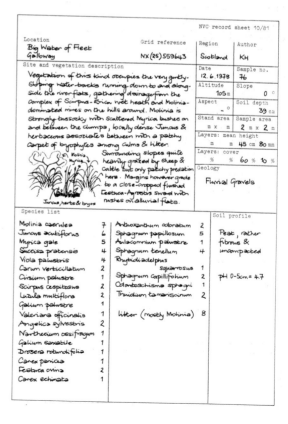

ous other smaller papers and NCC reports, were of compatible character to our own, we were able to include them in the data analyses, although some different recording conventions occasionally affected the way in which we had to name taxa. Thus, not all studies distinguished what is now known as *Agrostis vinealis*, so we retained the older convention of using *A. canina* ssp. *montana* where this taxon was separated from *A. canina* ssp. *canina*, with *A. canina* for records where such a distinction was not made. Similarly, *Festuca ovina* probably includes some *F. vivipara*, though we have scored this latter taxon separately where it was specifically recorded. Conversely, *J. bulbosus/kochii* has been used to accommodate records which claimed to discriminate between these two taxa. We were also sometimes aware of a possible confusion between *Juncus acutiflorus* and *J. articulatus*, rejecting doubtful data, and of likely underrecording of *J. conglomeratus*.

In total, over 3000 samples were available for analysis for this section of the work. Geographical coverage is good (Figure 6) and, apart from certain sorts of spring vegetation, the full range of mire communities seems to have been well sampled.

Data analysis and the description of mire communities

As usual in our data analysis, just the floristic records for the samples were employed to characterise the vegetation types. The quantitative scores for all vascular plants, bryophytes and lichens were used and no special weighting was given to rare or interesting species or supposed ecological or phytosociological indicators. In general, in the classification that emerged, the vegetation types were defined by assemblages of plants from all structural elements of the vegetation and from all taxonomic groups, with dominance alone rarely sufficient to make more than crude distinctions.

In all, 38 communities have been characterised: five kinds of vegetation from bog planes and two related wet heaths, four sorts of bog pools, four base-poor small-sedge and rush mires, six base-rich small-sedge and *Schoenus* mires, seven fen-meadows and rush-pastures, two sorts of soakway vegetation and eight kinds of spring and rill. Many of these communities are diverse enough internally to merit sub-division and, in certain cases, we have distinguished not just sub-communities but also variants within them. Every effort has been made to check the extensive floristic boundaries between the mire communities and vegetation types described elsewhere in the scheme and there seem to be no obvious areas of overlap in the descriptions nor any striking gaps. However, where there would be particular value in further sampling or analysis to augment or clarify our understanding, this is noted in the text.

Bog and wet heath communities

Five communities of bog vegetation are characterised by a prominent ground layer of peat-building Sphagna (most importantly *Sphagnum papillosum*, *S. magellanicum* and *S. capillifolium*), a low canopy of ericoid sub-shrubs (of which *Calluna vulgaris* and *Erica tetralix* are the most widespread) and some contribution from a limited range of graminoid monocotyledons (notably *Eriophorum vaginatum*, *Scirpus cespitosus* and *Molinia caerulea*). These are communities of damp to waterlogged, acid and oligotrophic peats, of greatest extent in the wetter north and west of Britain where ombrogenous deposits blanket much of the landscape, even on relatively steep slopes and down to sea level. However, essentially similar vegetation occurs on a more limited scale in drier parts of the country, in lowland raised mires and also in valley bogs where peat accumulation is dependent on the maintenance of a locally high ground water table by drainage impedence. In such mires, these communities constitute the bulk of the vegetation cover on the active plane and, on patterned surfaces, the lawn and hummock components, with hollows and pools occupied by different assemblages.

In their typical forms, the five communities are quite

Figure 6. Distribution of samples available from mires.

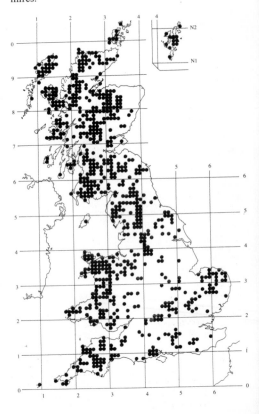

distinct and, indeed, they include extremes of mire vegetation that are very different from one another. Between these points, however, there is extensive inter-gradation: this is partly a reflection of the continuous nature of climatic variation across the country and of hydrological similarities between different kinds of mire, and partly the result of treatments such as draining, burning and grazing, which tend to produce a floristic convergence in these vegetation types. Even in their natural state, though, the communities are sufficiently close to one another to all fall fairly comfortably into the Erico-Sphagnion alliance of the class Oxycocco-Sphagnetea.

Most accounts of British ombrogenous bogs follow McVean & Ratcliffe (1962) in recognising two major communities, and expanded versions of both these figure in this scheme. The *Scirpus-Eriophorum* mire (M17) is the characteristic blanket bog of more oceanic parts of Britain, most extensive in areas with more than 2000 mm of rain (*Climatological Atlas* 1952) and over 200 wet days a year (Ratcliffe 1968), but largely restricted to sites below 500 m, where the annual temperature range is relatively small. On the widespread water-logged peats in this zone, *Scirpus cespitosus* and *Molinia* thrive but, though *E. vaginatum* is common, conditions are probably sub-optimal for it and Arctic-Alpine sub-shrubs are absent. *Calluna* and *E. tetralix* are the most

Figure 7. Bog and wet heath communities in relation to climate and soils.

frequent woody plants, though *Myrica gale* is locally prominent. In the extensive and varied *Sphagnum* carpet certain leafy hepatics, notably *Pleurozia purpurea*, can be abundant. Variation within the community is partly related to the differentiation of surface relief, with drier and wetter types occurring on hummocks and lawns respectively: very often, there are transitions from wet-ter lawns to *Sphagnum auriculatum* bog-pool vegetation (M1), the national distribution of which matches very closely that of the *Scirpus-Eriophorum* mire and which is a major component in the extensive pool complexes of many oceanic bogs. However, treatments, particularly burning and grazing, have also affected the floristic composition of the vegetation (Figure 7).

Over much of its range, the *Scirpus-Eriophorum* mire is separated altitudinally from the *Calluna-Eriophorum* mire (M19), the typical community of blanket peats on cold and wet, high-level plateaus throughout northern Britain. There the rainfall is somewhat less than in the oceanic zone, from 1200 to 2000 mm annually with 160–200 wet days a year (*Climatological Atlas* 1952, Ratcliffe 1968), but the summers are cool and cloudy and the winters harsh. Such conditions are optimal for *E. vagi-natum*, which can be an important peat-builder in the community, and favour the occurrence of montane sub-shrubs like *Empetrum nigrum* ssp. *nigrum* and ssp. *hermaphroditum*, *Vaccinium vitis-idaea*, *V. uliginosum*, *V. microcarpum*, *Arctostaphylos alpinus* and *Betula nana*, as well as the Arctic-Alpine herb *Rubus chamae-morus*. In this vegetation, our blanket mires make their

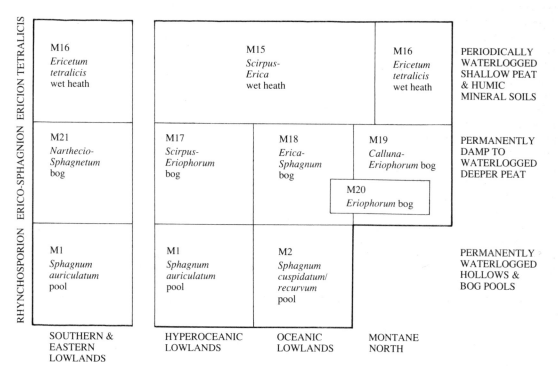

closest approach to the boreal Sphagnion fusci bogs of Scandinavia and, indeed, *S. fuscum* is preferential to the *Calluna-Eriophorum* mire, though very local. As well as such obviously montane vegetation, however, our community also includes much of the less rich Pennine blanket bog together with low-altitude stands where an unusually frequent occurrence of *Molinia* and *S. cespitosus* marks a climatic transition to more oceanic vegetation. In the *Calluna-Eriophorum* mire as a whole, though, the peats are often fairly dry superficially and hummock/hollow relief is typically absent: *S. capillifolium* is the only constant *Sphagnum* and it is hypnaceous mosses that provide greater cover in the ground carpet.

Grazing and burning can mediate a succession between *Calluna-Eriophorum* mire and a type of blanket mire vegetation from which sub-shrubs and bog mosses have been virtually eliminated, leaving *E. vaginatum* as the overwhelming dominant. Such *Eriophorum* mire (M20) is especially widespread in the southern Pennines where these treatments, combined with draining, pollution and erosion, have worked with especial ferocity.

As well as recognising the two major kinds of blanket bog, we have followed Moore (1968) in characterising a third community of virgin ombrogenous peatlands, comparable to vegetation described from the less oceanic lowlands of north-west Europe. This community, here termed the *Erica-Sphagnum papillosum* mire (M18), has much in common with the *Scirpus-Eriophorum* mire, but *Molinia* is much less frequent and, in the very extensive *Sphagnum* carpet, *S. magellanicum* is abundant and *S. imbricatum* a distinctive occasional (often now more abundant among the sub-fossil remains of the peat). Arctic-Alpine plants are absent but the Continental Northern *Vaccinium oxycoccos* and, more strikingly, *Andromeda polifolia* are characteristic associates. This community is concentrated in a zone to the east of the more oceanic *Scirpus-Eriophorum* mire and is essentially a vegetation type of raised bogs, although it can be found too where blanket peat extends over cols or deep sub-surface depressions. Surface patterning is often pronounced – it was from this kind of mire that the classic descriptions of hummock/hollow mosaics were made – but wetter depressions in this vegetation are typically occupied by the *Sphagnum cuspidatum/recurvum* bog pool community (M2).

Finally, among this group of bog types, there is the *Narthecium-Sphagnum papillosum* mire (M21, equivalent to the *Narthecio-Sphagnetum euatlanticum* of Continental schemes). This can be seen as continuing southwards the floristic trends seen in the *Scirpus-Eriophorum* mire, though both *S. cespitosus* and *E. vaginatum* are poorly represented and the community is characteristic, not of ombrogenous peats, but of valley bog deposits. With the *Sphagnum auriculatum* bog pool (M1), it occurs in generally shallow and ill-defined hummock/

hollow mosaics flanking the soligenous axes of the bogs, and is of local occurrence, being dependent on the maintenance of a high ground water table in more acidic rocks and superficials, and being highly susceptible to drainage operations.

The typical lowland English valley bog provides a good situation in which to compare the Erico-Sphagnion type of mire vegetation described above with our Ericion tetralicis wet heath, found on drier and usually shallower peat or humic mineral soils with only periodic waterlogging. On walking up on to the surrounds of such a bog, one finds that the peat-building Sphagna tend to fade in importance, being replaced by a patchier cover of *S. compactum* and *S. tenellum*, with a vascular canopy in which *E. tetralix*, *Calluna* and *Molinia* usually predominate, their proportions often much affected by burning and grazing. This *Erica-Sphagnum compactum* wet heath (M16, *Ericetum tetralicis*) in fact has a wider distribution than the *Narthecio-Sphagnetum*, extending to the north-east of Scotland where it occurs on thin ombrogenous peats, and in many sites it has been much degraded by treatments. However, it still remains a very distinctive feature of mire/heath zonations in southern England, providing an important locus for such plants as *Erica ciliaris*, *Gentiana pneumonanthe* and *Lepidotis inundata*.

Very similar vegetation, which we have gathered into a compendious *Scirpus-Erica* wet heath (M15), is of widespread occurrence in north-west Britain, on thinner ombrogenous peats that are too well drained to support Erico-Sphagnion bog, or on deeper deposits that have become dry because of climatic change or as a result of treatments. Among this vegetation, often because of particular local regimes of burning, grazing and draining, there is a very diverse pattern of dominance: the community subsumes vegetation previously separated on the basis of the varying abundance of *S. cespitosus*, *Calluna* and *Molinia*, but which is characterised throughout by the same suite of associates. This is typically a modest assemblage with *Potentilla erecta* and *Polygala serpyllifolia*, together with various bryophytes, the commonest members: in both kinds of wet heath, species richness is generally associated with soligenous influence, when there is some relief from the usually acidic and oligotrophic conditions and an appearance of moderately base-tolerant herbs and cryptogams.

Bog pool communities and poor fens

Although the pool and hollow vegetation dominated by the semi-aquatic Sphagna like *S. auriculatum*, *S. cuspidatum* and *S. recurvum* (M1 and M2) often occurs in the context of Erico-Sphagnion bog, it provides a convenient starting point for considering the more acidophilous of those mires that are dominated by small sedges or rushes, and which can be grouped in the Parvocaricetea (or Scheuchzerio-Caricetea). Among the bog pools of the Rhynchosporion alliance, stagnation of acidic and nutrient-poor waters is the rule: the vascular element in this vegetation is often sparse although, in the *Sphagnum auriculatum* and *S. cuspidatum/recurvum* communities, there can be occasional or local representation of plants such as *Rhynchospora alba*, *R. fusca*, *Drosera* and *Utricularia* spp., *Carex magellanica*, *Scheuchzeria palustris* and *Hammarbya paludosa* (Figure 8).

Two other common communities can be included among this group. The *Eriophorum angustifolium* pool (M3) is a widespead regenerating vegetation of bog hollows, channel bottoms and shallow peat-cuttings, in which *Drepanocladus fluitans* is the most frequent moss. The *Carex rostrata-Sphagnum recurvum* pool (M4) occurs in slightly more nutrient-rich waters and in seepage areas, and it has a mesophytic *Sphagnum* carpet with a variety of cyperaceous dominants, most often *C. rostrata*, but with local replacement by *C. nigra*, *C. curta*, *C. lasiocarpa*, *C. limosa* and the very rare *C. chordorrhiza*. Some of these sedges, particularly *C. rostrata*, maintain their abundance in a fifth community which lies on the junction of the Rhynchosporion and the alliance of small-sedge poor fens, the Caricion nigrae (formerly the Caricion curto-nigrae). Here, in the *Carex rostrata-Sphagnum squarrosum* mire (M5), a *Sphagnum* carpet is still prominent, but it is the moderately base-tolerant *S. squarrosum* and *S. teres* which tend to bulk largest, and these are often accompanied by such poor-

fen mosses as *Aulacomnium palustre* and *Calliergon stramineum*. Among the vascular element, too, there is some enrichment with plants such as *Succisa pratensis*, *Galium palustre*, *Viola palustris*, *Menyanthes trifoliata* and *Potentilla palustris*. This is a local community, characteristic of moderately-enriched soligenous zones around raised and basin mires or ombrotrophic nuclei on poor-fen rafts in open-water transitions.

Much more widespread and common than any of these kinds of vegetation, however, is the community we have called the *Carex echinata-Sphagnum* mire (M6). Essentially, this is a poor fen in which either small sedges (notably *C. echinata*, *C. nigra*, *C. panicea* or *C. demissa*) or rushes (usually *J. effusus* or *J. acutiflorus*) dominate over a carpet of oligotrophic and mesotrophic Sphagna (*S. auriculatum* in the more oceanic areas, *S. recurvum* in the less oceanic and more enriched sites, with *S. palustre* and *S. subnitens* also represented) and *Polytrichum commune*. Few herbs are constant throughout, but the suite of associates is nonetheless very characteristic: *Agrostis canina* ssp. *canina*, *Molinia*, *Anthoxanthum odoratum*, *Viola palustris*, *Galium palustre*, *G. saxatile*, *Potentilla erecta*, *Epilobium palustre*, *Succisa pratensis*, *Ranunculus flammula* and *Cardamine pratensis*. This is the major type of soligenous vegetation in Britain on peats and peaty gleys throughout the sub-montane zone where there is irrigation with more base-poor, though not excessively oligotrophic, waters. It has a distinctly north-western range, although this is partly a reflection of the prevalence of more acidic substrates in this region, and the frequency of flushing on gentle to moderately steep slopes. Stands are often small and frequently occur

Figure 8. Bog-pool and base-poor small-sedge mires of the Rhynchosporion and Caricion nigrae alliances.

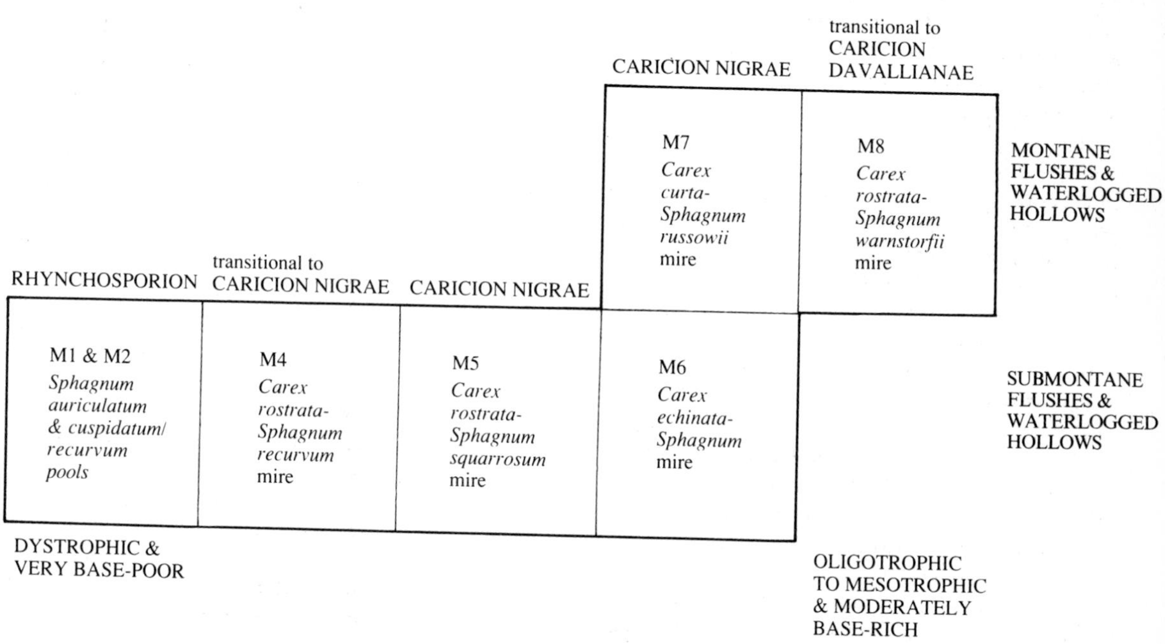

| | transitional to
CARICION | | |
CARICION NIGRAE	DAVALLIANAE		
M7 *Carex* *curta-* *Sphagnum* *russowii* mire	M8 *Carex* *rostrata-* *Sphagnum* *warnstorfii* mire		MONTANE FLUSHES & WATERLOGGED HOLLOWS

| | transitional to | | |
RHYNCHOSPORION	CARICION NIGRAE	CARICION NIGRAE	
M1 & M2 *Sphagnum* *auriculatum* *& cuspidatum/* *recurvum* *pools*	M4 *Carex* *rostrata-* *Sphagnum* *recurvum* mire	M5 *Carex* *rostrata-* *Sphagnum* *squarrosum* mire	M6 *Carex* *echinata-* *Sphagnum* mire	SUBMONTANE FLUSHES & WATERLOGGED HOLLOWS

DYSTROPHIC &
VERY BASE-POOR

OLIGOTROPHIC
TO MESOTROPHIC
& MODERATELY
BASE-RICH

within unenclosed pasture, so grazing often affects the floristics and physiognomy of the vegetation.

In essentially similar situations above 650 m and particularly in the Central Highlands of Scotland, the *Carex echinata-Sphagnum* mire is replaced by the *Carex curta-Sphagnum russowii* mire (M7). Here, there is a strong montane contingent among the flora, with *S. russowii* and *S. lindbergii* occurring with *S. riparium* in the ground carpet, *Carex curta, C. bigelowii, C. aquatilis, C. rariflora* and *Saxifraga stellaris* among the herbs, and a decline in the contribution from *Molinia, C. panicea* and the Junci.

Likewise at higher altitudes, though in distinctly more calcareous and stagnant conditions, is a final Caricion nigrae community, lying close to the boundary with the Caricion davallianae rich fens. This *Carex rostrata-Sphagnum warnstorfii* mire (M8) is restricted in its occurrence by the very local distribution of more base-rich peats in small hollows, but it is a very distinctive kind of vegetation with the poor-fen element enriched by the presence of *S. warnstorfii, S. contortum* and *S. subsecundum sensu stricto, Calliergon sarmentosum, Homalothecium nitens* and *Selaginella selaginoides.*

Sedge- and *Schoenus*-dominated rich fens

Along with the acidophilous small-sedge and rush-dominated communities, the classification recognises a series of more calcicolous mires, 'rich fen' in the sense of Sjörs (1948). These form an analogous suite of hydromorphological types with at least mildly base-rich and calcareous waters, and are therefore often of only local occurrence, their distribution reflecting the presence of lime-rich bedrocks and superficials. These communities, which can be grouped within the Caricion davallianae alliance, intergrade with the Caricion nigrae poor fens but, by and large, a *Sphagnum* carpet is more characteristic of the latter, together with such species as *Carex echinata, C. curta, Viola palustris* and *Calliergon stramineum*; in the former, the vegetation gains its distinctive character from plants like *Carex dioica, C. hostiana, C. lepidocarpa, Selaginella selaginoides, Pinguicula vulgaris, Parnassia palustris, Campylium stellatum, Bryum pseudotriquetrum* and *Scorpidium scorpioides.*

Among this group there is one community clearly associated with topogenous sites where soft, spongy peats are kept permanently moist by at least moderately base-rich but fairly nutrient-poor waters. This *Carex rostrata-Calliergon* mire (M9) is a very distinctive type of fen, though it shows considerable variation in floristics and physiognomy. It is the major community in Britain for *Carex diandra*, though this is often less abundant than *C. rostrata* and both can be accompanied by *C. lasiocarpa*, with *C. paniculata, C. appropinquata, Schoenus nigricans* and *Juncus subnodulosus* occasionally figuring as local dominants. *Phragmites australis*

or *Cladium mariscus* can also be patchily prominent in transitions to tall-herb fens. Associates such as *Potentilla palustris, Menyanthes trifoliata, Equisetum fluviatile* and *Eriophorum angustifolium* are also of widespread occurrence in related communities, but more especially distinctive here is an assemblage of smaller herbs like *Carex panicea, Valeriana dioica, Mentha aquatica, Epilobium palustre* and *Pedicularis palustris.* Also characteristic is a carpet of *Calliergon* spp. (usually with *C. cuspidatum* but also the more distinctive *C. cordifolium* and *C. giganteum*), larger Mniaceae, *Campylium stellatum, Scorpidium scorpioides, Drepanocladus revolvens* and *Bryum pseudotriquetrum.* The community is very local, though often a striking part of complex mosaics with poor fens and tall-herb fens, and there is sometimes a suspicion that it is associated with shallow peat-cuttings (Figure 9).

Much more widespread than the above kind of fen, and a calcicolous analogue to the *Carex echinata-Sphagnum* mire is the *Carex dioica-Pinguicula vulgaris* mire (M10, *Pinguiculo-Caricetum dioicae*), typical of calcareous surface-water gleys in flushes throughout the uplands. Small sedges such as *C. panicea, C. dioica, C. lepidocarpa, C. hostiana* and *C. pulicaris* are an important component here, although various aspects of the

Figure 9. Base-rich small-sedge mires of the Caricion davallianae alliance.

	M12 *Caricetum saxatilis* mire	HIGH MONTANE SOMETIMES WITH SNOW-LIE
	M11 *Carici-Saxifragetum* mire	
M9 *Carex rostrata-Calliergon* mire	M10 *Pinguiculo-Caricetum* mire	
TOPOGENOUS & SOLIGENOUS MIRES	SOLIGENOUS MIRES & FLUSHES	

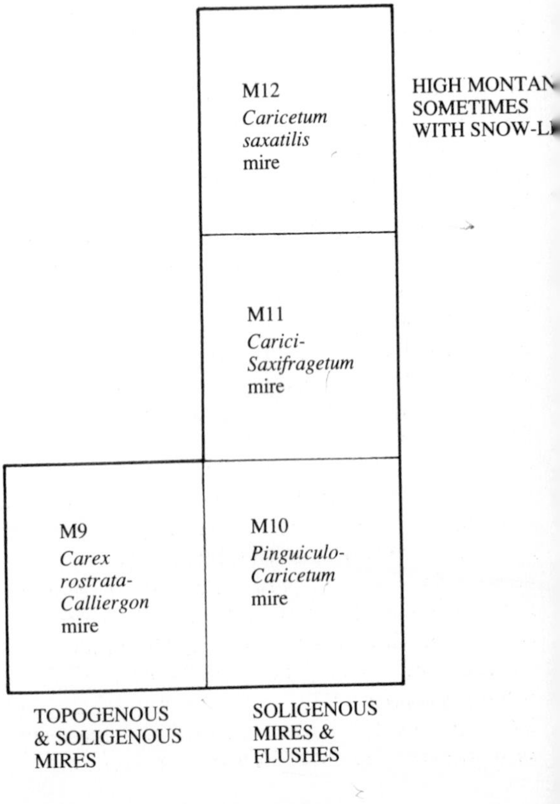

habitat contribute to keeping the vegetation rather open and diverse: the oligotrophic character of the soils and flushing waters, the scouring, wind erosion and freeze-thaw occurring at many of the sites, and the trampling and grazing activities of stock which often have unhindered access to the pastures where the flushes are typically found. Numerous small herbs find a place in the wet and frequently stony swards, among them *Pinguicula vulgaris*, *Selaginella*, *Linum catharticum*, *Primula farinosa*, *Parnassia palustris*, *Thymus praecox*, *Equisetum variegatum* and, more locally, *Bartsia alpina*, *Juncus alpinus*, *Kobresia simpliciuscula*, *Minuartia verna* and *M. stricta*. Bryophytes are typically very abundant, with *Campylium stellatum*, *Aneura pinguis*, *Drepanocladus revolvens*, *Ctenidium molluscum*, *Fissidens adianthoides*, *Cratoneuron commutatum* and *Gymnostomum recurvirostrum* occurring in crunchy tufaceous carpets and hummocks.

Continuing very much the same floristic trend at higher altitudes and largely confined to Scotland is the *Carex demissa-Saxifraga aizoides* mire (M11, *Carici-Saxifragetum aizoidis*), a community of stony montane flushes strongly irrigated with base-rich waters. More catholic Continental Northern plants such as *Carex dioica*, *C. pulicaris* and *Pinguicula* maintain their frequency here, but the lower temperatures favour the appearance of plants like *Saxifraga aizoides*, *Thalictrum minus* and *Juncus triglumis*, together with Arctic-Alpine rarities like *Carex atrofusca*, *C. microglochin*, *Juncus biglumis* and *J. castaneus* at higher levels. Among the bryophytes, many of the *Pinguiculo-Caricetum* species are well represented, but *Blindia acuta* and *Calliergon trifarium* are additionally distinctive.

At even higher altitudes, the margins of base-rich flushes, especially where there is prolonged snow-lie, can be distinguished by an abundance of *Carex saxatilis* among rich-fen associates. Such vegetation overlaps considerably with the high-level *Carici-Saxifragetum*, but some additional floristic features, like the high frequency of *Carex bigelowii* and certain poor-fen plants, argue for the separate recognition of a *Carex saxatilis* mire (M12, *Caricetum saxatilis*). Both this community and the *Carici-Saxifragetum* come close to Scandinavian Arctic-Alpine mires of the alliance Caricion bicolori-atrofuscae and they show a tendency to grade to tall-herb vegetation and willow scrub over dripping inaccessible banks and ledges.

A distinctive local feature in more low-altitude stands of rich-fen vegetation in parts of Britain with a relatively mild winter climate is an abundance of *Schoenus nigricans*: in the *Carex-Calliergon* mire, *Pinguiculo-Caricetum* and *Carici-Saxifragetum*, it can be locally dominant without disrupting the floristic integrity of the vegetation (Figure 10). However, in the *Schoenus-Juncus subnodulosus* mire (M13, *Schoenetum nigricantis*) *S.*

nigricans is consistently prominent and, though often accompanied by much *J. subnodulosus* and *Molinia*, it is *Schoenus* itself which exerts the strongest influence on the richness and organisation of the associated flora. In the runnels between the tussocks, for example, where the ground is flooded or at least kept moist for much of the year, it is Caricion davallianae herbs and bryophytes which predominate, although the Oceanic Southern *Anagallis tenella* is an additional distinctive associate, and many stands also have some tall dicotyledons like *Succisa pratensis*, *Angelica sylvestris*, *Cirsium palustre* and *Filipendula ulmaria*, and orchids, notably *Epipactis palustris* but also *Dactylorhiza majalis*, *D. incarnata*, *D. fuchsii*, *D. maculata* ssp. *ericetorum* and the rare *D. traunsteineri*. *Phragmites* and *Cladium* can be locally prominent too in stands which come close in their composition to tall-herb fen. The tussock tops, by contrast, providing niches far removed from the influence of the calcareous ground water, have such plants as *Potentilla erecta*, *Narthecium*, *Calluna*, *Erica tetralix* and *E. cinerea* with acidophilous bryophytes. On drier ground, an expansion of *Molinia*, together with *Festuca rubra* and *Holcus lanatus*, is often accompanied by a decline in the cover of *Schoenus* and a general

Figure 10. The changing community context of *Schoenus nigricans* in moving across Britain.

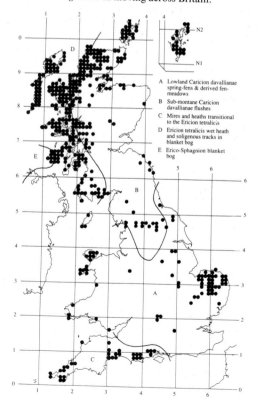

A   Lowland Caricion davallianae spring-fens & derived fen-meadows
B   Sub-montane Caricion davallianae flushes
C   Mires and heaths transitional to the Ericion tetralicis
D   Ericion tetralicis wet heath and soligenous tracks in blanket bog
E   Erico-Sphagnion blanket bog

floristic impoverishment, changes that have frequently been precipitated by draining.

Such treatment has reduced the *Schoenetum* to the status of a very local vegetation type over much of its range, although it still remains in some abundance on wet, oligotrophic soils below calcareous springs and seepage lines on lime-rich bedrocks and superficials in East Anglia and Anglesey. With a shift into areas with a more oceanic climate, however, *Schoenus* extends out of this narrow confinement to more calcicolous Caricion davallianae communities and related fen-meadows, as in north-west Scotland, and then, in Ireland, on to the bog plane proper. Part-way along this sequence is the kind of situation where *Schoenus* occurs in soligenous areas within lowland valley bogs, as in parts of the New Forest. Here, in a *Schoenus-Narthecium* mire (M14), *J. subnodulosus* is rare and there is a very sparse calcicolous element, even in the runnels, while *E. tetralix*, *Narthecium* and *Drosera rotundifolia* become common, together with *Sphagnum subnitens* and *S. auriculatum*.

Fen-meadows and rush-pastures

There are real reasons for this group being one of the most difficult to define and subdivide with satisfaction. In the first place, the vegetation included here is generally the product of some kind of agricultural treatment, frequently long-continued mowing and/or grazing. Stands thus often betray the influence of varying degrees of modification or neglect and can show site-specific features related to complex and unique management histories: indeed, many stands are in the course of change and show a local absence of species which, overall, can be considered characteristic. Second, such treatments have produced these kinds of vegetation from a wide variety of precursors in different parts of the country. The communities included here show a general association with soils that are kept moist for a substantial part of the year and are at least moderately base-rich, with a pH usually above 4.5, but otherwise edaphic conditions are very diverse, so remnant associated floras are likewise very varied, and they can show different phytogeographical affinities according to the particular location of the stand. Third, in the generally oceanic climate of the British Isles, several of the common Junci, *Molinia* and various other associates considered diagnostic of these vegetation types on the European mainland, occur widely in other communities, thus blurring the phytosociological boundaries.

Thus, although our fen-meadows and rush-pastures are readily recognisable in the field, and all fall fairly comfortably into the order Molinietalia of the class Molinio-Arrhenatheretea, it is hard to produce a list of reliable diagnostic species. On the negative side, however, plants of the Oxycocco-Sphagnetea and Parvocari-

cetea are poorly represented, while some of the following usually occur: *Achillea ptarmica*, *Angelica sylvestris*, *Cirsium palustre*, *Equisetum palustre*, *Filipendula ulmaria*, *Lathyrus palustris*, *Lychnis flos-cuculi*, *Sanguisorba officinalis*, *Juncus effusus*, *J. conglomeratus* and *J. acutiflorus*.

Five major kinds of vegetation can be recognised. The first comprises swards in which *J. subnodulosus* (occasionally *J. inflexus*) is very conspicuous among a ground of grasses and sedges (including *C. disticha*, *C. hirta*, *C. acutiformis*, and *C. panicea*) with smaller herbs (such as *Mentha aquatica* and *Lotus uliginosus*) and some emergent tall dicotyledons. Drier and swampier forms occur, but essentially this is vegetation of calcareous peaty gleys or fen peats below springs and seepage lines and in topogenous mires, where it has developed as a result of grazing and/or mowing. This *Juncus subnodulosus-Cirsium palustre* fen-meadow (M22) is largely confined to the central and eastern lowlands of Britain, matching the range of the *Schoenetum*, and it is similar to Continental Calthion communities, although *Caltha* itself and a number of other diagnostic species of this alliance in mainland Europe are more catholic in their behaviour in Britain (Figure 11).

With widespread draining and re-seeding of swards for agriculture, this kind of vegetation has become rather local and now a much more common Calthion community through lowland England and Wales is the *Holcus lanatus-Juncus effusus* rush-pasture (MG10, *Holco-Juncetum effusi*), the account of which will be found among the mesotrophic grasslands. Towards the west of Britain, where ill-drained ground is often extensive among relatively unimproved or reverted grazings, these communities give way to rushy swards which have much less in common with the Continental Calthion. The *Juncus-Galium palustre* rush-pasture (M23) in which *J. effusus* and *J. acutiflorus* bulk large, lacks the calcicoles and many of the mesophytic herbs typical to the south and east, being characterised rather by species such as *Ranunculus flammula*, *Agrostis canina* ssp. *canina*, *A. stolonifera*, *Potentilla erecta* and various associates with Caricion nigrae affinities. *Anagallis tenella*, *Carum verticillatum*, *Scutellaria minor* and *Wahlenbergia hederacea*, considered by some to be diagnostic of a Junction acutiflori alliance, can all occur but they are generally uncommon and not confined to the community.

Both the *Juncus-Cirsium* and *Juncus-Galium* communities show transitions to *Molinia*-dominated vegetation, the classification of which poses especially acute problems. *Molinia* is a natural member of a variety of vegetation types in Britain, including western ombrogenous bogs, wet heaths and some drier heaths and certain upland grasslands. In many of these it can become abundant, often overwhelmingly dominant, in vegetation which has few floristic affinities with the Molinietalia, as that order is understood on the Continent. However, there are two important kinds of vegetation with *Molinia* that are best considered as the British representatives of the Junco-Molinion alliance (Figure 12). The *Molinia-Cirsium dissectum* fen meadow (M24, *Cirsio-Molinietum caeruleae*) includes most of the anthropogenic *Molinia* swards on drier peats in southern and eastern England. It has much the same distribution as the *Juncus-Cirsium* community and could have been derived from it in some cases by selective nutrient depletion with long-continued mowing; in other cases, it has probably developed directly from tall-herb fen by repeated cutting. *Molinia* is generally dominant, though *J. subnodulosus* and *Schoenus* can both be locally abundant, and *Phragmites* or *Cladium* can persist patchily. Characteristic associates include *Cirsium dissectum*,

Carex panicea, *C. hostiana*, *Succisa*, *Potentilla erecta*, *Valeriana dioica* and *Centaurea nigra*. Very locally in northern England and southern Scotland, there are fairly similar *Molinia* swards which lack the Oceanic West European *C. dissectum*, but have such plants as the Continental Northern *Crepis paludosa* and Northern Montane *Trollius europaeus*. We have retained a separate *Molinia-Crepis* mire (M26) to include such vegetation and transitions to hay-meadows.

A second major kind of Junco-Molinion sward is much more widespread, but with a predominantly western distribution on a variety of moist but well-aerated, acid and oligotrophic peats and peaty mineral soils. Here, in the *Molinia-Potentilla* mire (M25), *Molinia* reaches a peak of abundance and vigour in vegetation which is consequently often species-poor. The frequent presence of *J. acutiflorus* makes for an indistinct boundary between the community and the western rush-pastures, although this plant is not so characteristic of this kind of vegetation as it is in Ireland, and many

Figure 11. Rush-dominated fen-meadow and pasture communities in relation to climate and soils.

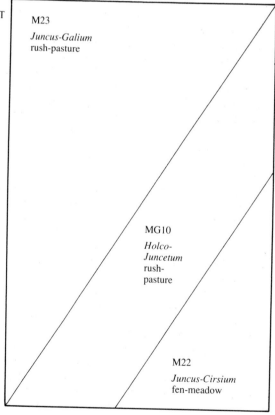

MOIST, ACID TO
NEUTRAL GLEYS,
OFTEN HUMIC ABOVE

COOL & WET
NORTH &
WEST OF
BRITAIN

M23

Juncus-Galium
rush-pasture

MG10

*Holco-
Juncetum*
rush-
pasture

M22

Juncus-Cirsium
fen-meadow

WARM & DRY
SOUTH-EAST
OF BRITAIN

MOIST & BASE-
RICH, MESOTROPHIC
PEATS & PEATY
GLEYS

stands look closer to wet heath with occasional bushes of *E. tetralix* and *Calluna*. Other tracts have rank mixtures of *Holcus lanatus*, *Anthoxanthum odoratum*, *Succisa*, *Lotus uliginosus* and *Vicia cracca*, or assemblages of tall dicotyledons or swamp helophytes. This community also provides one of the most common contexts for local dominance of *Myrica gale* over *Molinia*: the classification does not recognise a separate *Molinia-Myrica* mire, but treats such vegetation as a variant of a number of mire types.

Finally, among the Molinietalia, there are two communities with an abundance of *Filipendula ulmaria*. The bulk of this Filipendulion vegetation can be accommodated in a *Filipendula-Angelica sylvestris* fen (M27), a community which seems to be most clearly defined in northern Britain, where such species as *Valeriana officinalis*, *Rumex acetosa*, *Lychnis flos-cuculi*, *Geum rivale*

and *Caltha palustris* can be well represented. Often, though, the stands are species-poor and, particularly in the improved agricultural landscapes of the lowland south, show a marked susceptibility to invasion by rank, eutrophic weeds. A moist soil with a brisk nutrient turnover is generally characteristic of this kind of vegetation, together with freedom from grazing: it is typically developed on alluvial soils alongside rivers and streams and around ponds, and on more eutrophic gleys and peats. As in some other communities of wet ground, *Iris pseudacorus* tends to become locally dominant among this kind of fen in the more equable climate of western Britain, although the scheme has a separate *Filipendula-Iris* fen (M28, *Filipendulo-Iridetum pseudacori*) to include some striking vegetation of upper salt-marsh and raised beaches on the west coast. In this community, especially well-developed around the fringes of some Scottish sea-lochs, the two species occur together with Oceanic West European *Oenanthe crocata*, eutrophic weeds and a variety of maritime plants.

Figure 12. *Molinia*-dominated fen-meadows and mires in relation to climate and soils.

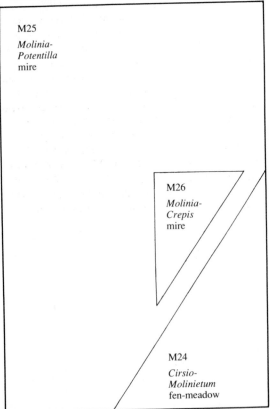

MOIST, BUT WELL-
AERATED PEATS &
PEATY MINERAL SOILS

COOL & WET
NORTH &
WEST OF
BRITAIN

M25

*Molinia-
Potentilla*
mire

M26

*Molinia-
Crepis*
mire

M24

*Cirsio-
Molinietum*
fen-meadow

WARM & DRY
SOUTH-EAST
OF BRITAIN

MOIST TO DRY,
CIRCUMNEUTRAL
PEATS & PEATY
MINERAL SOILS

Soakways and springs

Also included among the mires is a variety of vegetation types of generally very small extent around springs and soakways. One group is characteristic of still or slow-flowing waters, shallow but often fluctuating, acid to neutral and nutrient-poor, in bog soaks and pools and streams among heath and moorland. Such vegetation, which would be included in the Littorelletea class, is rather variable in its composition, with fragmented stands often showing site-specific differences, but further sampling is needed before a comprehensive account can be given. For the moment, we have distinguished a clearly-defined *Hypericum elodes-Potamogeton polygonifolius* soakway (M29, *Hyperico-Potametum polygonifolii*) with associates like *Juncus bulbosus*, *Ranunculus flammula*, *Eleocharis multicaulis*, and *Hydrocotyle vulgaris* among a carpet of *Sphagnum auriculatum*, a community of runnels and pools in mires and heaths in southern and western Britain; and an ill-characterised assortment of other Hydrocotylo-Baldellion vegetation (M30) with species like *Baldellia ranunculoides*, *Littorella uniflora* and *Lythrum portula*. These assemblages grade floristically to other Littorelletea vegetation included among the swamps and aquatics.

The small-scale communities of springs and rills with a constant flow of water and a probably fairly uniform temperature regime, have likewise been under-sampled in our survey but some major lines of variation can be discerned. As a whole, the vegetation types belong to the Montio-Cardaminetea class, and floristic divisions among them reflect such factors as water chemistry, climate and the amount of shade. Characteristic of spring-heads at moderate to high altitudes, with fairly slow irrigation by acid and oligotrophic waters, is the *Anthelia julacea-Sphagnum auriculatum* community (M31, *Sphagno-Anthelietum*). *Anthelia* dominates in carpets or thick cushions with but sparse shoots of *Sphagnum*, *Deschampsia cespitosa*, *Scapania undulata*

and *Marsupella* spp. with some Parvocaricetea plants. Here, the temperature regime is cold, but snow-lie is probably not prolonged.

Distinct from this vegetation, which is perhaps peculiar enough to warrant placing in a separate Anthelion alliance, is a group of spring communities of acid to near-neutral habitats, which could be grouped together in the Cardamino-Montion alliance. Upland stands of this kind are better represented in our data and four communities have been distinguished: the *Philonotis fontana-Saxifraga stellaris* spring (M32, *Philonoto-Saxifragetum*) from moderate to high altitudes in northern Britain where there is vigorous irrigation with calcium-poor waters, the *Ranunculus omiophyllus-Montia fontana* rill (M35) which largely replaces it on moorlands in the south-west, the *Pohlia wahlenbergii* var. *glacialis* spring (M33, *Pohlietum glacialis*) of late snow-beds at high altitudes, and the *Carex demissa-Koenigia islandica* flush (M34), restricted to moist stony ground on Skye.

Lowland stands of Cardamino-Montion vegetation (M36) are widespread, but have been very poorly covered as separate assemblages, although such characteristic species as *Chrysosplenium oppositifolium*, *Pellia epiphylla*, *Conocephalum conicum* and *Rhizomnium punctatum* have frequently been recorded from moist and shady banks in damper kinds of woodland.

Last among this group are the very striking Cratoneurion springs of calcareous waters in both the uplands, from where we have many samples, and the lowlands, where our coverage is very poor. *Cratoneuron commutatum* and *C. filicinum* are the typical dominants here in tufaceous masses, with associates like *Bryum pseudotriquetrum*, *Aneura pinguis* and *Saxifraga aizoides* and base-tolerant sedges. The *Cratoneuron-Festuca rubra* spring (M37) occurs in more inaccessible spring-heads, with the grazed *Cratoneuron-Carex nigra* spring (M38) representing a transition to the Caricion davallianae flushes of upland pastures.

KEY TO MIRES

With something as complex and variable as vegetation, no key can pretend to offer an infallible short cut to diagnosis. The following should thus be seen as simply a crude guide to identifying the types of mire vegetation in the scheme and must always be used in conjunction with the data tables and community descriptions. It relies on floristic (and, to a lesser extent, physiognomic) features of the vegetation and demands a knowledge of the British vascular flora and, in many cases here, of bryophytes. It does not make primary use of any habitat features, though these may provide a valuable confirmation of a diagnosis.

Because the major distinctions between the vegetation types in the classification are based on inter-stand frequency, the key works best when sufficient samples of similar composition are available to construct a constancy table. It is the frequency values in this (and, in some cases, the ranges of abundance) which are then subject to interrogation with the key. Most of the questions are dichotomous and notes are provided at particularly awkward choices or where confusing zonations are likely to be found.

Samples should always be taken from homogeneous stands and be 2 × 2 m or 4 × 4 m according to the scale of the vegetation or, where complex mosaics occur, of identical size but irregular shape. Very small springs or flushes can be sampled in their entirety.

---

1   At least some Sphagna constant and often forming an important structural element in the vegetation 2

Sphagna at most occasional and patchily prominent and often altogether absent 34

2   *Sphagnum auriculatum* constant and locally abundant in spring-head, soakway or pool vegetation with some of *Hypericum elodes, Potamogeton polygonifolius, Juncus bulbosus, Ranunculus flammula, Deschampsia cespitosa, Saxifraga stellaris, Ranunculus omiophyllus, Montia fontana, Anthelia julacea, Philonotis fontana, Scapania undulata, Marsupella emarginata, Dicranella palustris* 14

Sphagnum auriculatum can be frequent but not with the above 3

3   Sphagna occurring as an extensive carpet or in patches with graminoid monocotyledons (some of *Scirpus cespitosus, Eriophorum vaginatum, E. angustifolium, Molinia caerulea*) and sub-shrubs (some of *Erica tetralix, Calluna vulgaris, Myrica gale, Vaccinium* spp., *Empetrum* sspp.) 17

Sphagna extensive or patchy, often with some cyperaceous plants or rushes, but with sub-shrubs of only local importance 4

4   Semi-aquatic Sphagna (*S. auriculatum, S. cuspidatum, S. recurvum* or locally *S. pulchrum*) forming extensive floating masses or sodden lawns in and around bog pools with an often sparse cover of *Eriophorum angustifolium* and *Rhynchospora alba* or *R. fusca*, and *Drosera* spp.; sedges generally scarce though *C. limosa, C. curta* or *C. magellanica* may occur 5

Above Sphagna or others frequent and abundant in soakway or flush vegetation but usually without *Rhynchospora* spp. and with at least some sedges and a sometimes fairly rich associated flora 7

5   Mixtures of *S. auriculatum* and *S. cuspidatum* generally dominant with *S. recurvum* rare in rather open pools and lawns, with *Menyanthes trifoliata* constant but sparse, and *Utricularia* spp. and *Potamogeton polygonifolius* quite common in areas of open water

M1 *Sphagnum auriculatum* bog-pool community

S. cuspidatum and/or *S. recurvum* forming often extensive carpets, usually without *S. auriculatum*, but frequently with a little *Erica tetralix* and occasional *Andromeda polifolia*

> **M2** *Sphagnum cuspidatum/recurvum* bog-pool community   6

6   *S. recurvum* usually absent but *Rhynchospora alba* and *Andromeda* very frequent and *Drosera rotundifolia* occasionally accompanied by *D. anglica*

> **M2** *Sphagnum cuspidatum/recurvum* bog-pool community
> *Rhynchospora alba* sub-community

S. recurvum constant and often abundant with no *R. alba* and only occasional *Andromeda*, but frequent *Vaccinium oxycoccos* and patchily prominent *Calluna* and *Eriophorum vaginatum*

> **M2** *Sphagnum cuspidatum/recurvum* bog-pool community
> *Sphagnum recurvum* sub-community

7   Extensive sodden carpets of *S. recurvum* and *S. cuspidatum* or *S. auriculatum*, with occasional *S. palustre* and patches of *Polytrichum commune*; emergent, usually rather open, cover of sedges, most frequently *C. rostrata*, but locally accompanied by *C. curta*, *C. lasiocarpa*, *C. limosa* or *C. nigra* (the first two sometimes abundant) or rarely *C. chordorrhiza*; other vascular associates typically few and sparse but *Juncus acutiflorus*, *Molinia*, *Agrostis stolonifera* and *Potentilla erecta* occasional

> **M4** *Carex rostrata-Sphagnum recurvum* bog-pool community

C. rostrata, *C. nigra*, *C. lasiocarpa* or *C. curta* can be frequent and such an impoverished *Sphagnum* carpet can be extensive but not together   8

8   Extensive carpet or open patchwork of Sphagna with little or no *S. cuspidatum* or *S. auriculatum* and usually dominated by mixtures of *S. squarrosum*, *S. teres*, *S. recurvum* and *S. palustre*, with occasional *S. subnitens*; *Aulacomnium palustre* and *Calliergon stramineum* frequent; emergent sedges can be dense with *C. rostrata* the usual dominant, *C. nigra* also constant and sometimes abundant, *C. lasiocarpa* occasional and locally plentiful; occasional to frequent scattered individuals of *Potentilla palustris*, *Menyanthes*, *Viola palustris*, *Succisa pratensis* and *Galium palustre*

> **M5** *Carex rostrata-Sphagnum squarrosum* mire

Where this vegetation type occurs in complex mosaics in areas of soligenous influx, it can be difficult to separate richer stands from the *Carex-Calliergon* mire, but *C. diandra* becomes a common feature there, Sphagna apart from *S. contortum* are distinctly scarce, and some of *Campylium stellatum*, *Scorpidium scorpioides*, *Drepanocladus revolvens*, *Calliergon giganteum* and *C. cordifolium* are frequent.

Above combination of species not present   9

9   Usually extensive carpet of *S. recurvum* or *S. auriculatum* with *S. palustre*, *S. subnitens* and *S. papillosum* and patches of *P. commune*; vascular dominants are EITHER small sedges such as *C. echinata*, *C. nigra*, *C. panicea* and occasionally *C. demissa* with *C. rostrata* and *C. curta* of no more than local importance in swampier stands, OR bulkier rushes, usually *Juncus effusus*, occasionally *J. acutiflorus*; at least some of *Molinia*, *Agrostis canina* ssp. *canina*, *Anthoxanthum odoratum*, *Viola palustris*, *Galium palustre*, *G. saxatile* and *Potentilla erecta* present

> **M6** *Carex echinata-Sphagnum recurvum/auriculatum* mire   10

S. auriculatum and *S. recurvum* can be present but *S. papillosum* and *S. russowii* are constant, the former often in abundance, sometimes also with *S. lindbergii* and *S. subnitens*, with *Calliergon sarmentosum* or *C. stramineum* frequent; *C. rostrata* and bulkier Junci rare but *C. curta* constant and usually abundant, with sometimes plentiful *C. echinata*, and some of *C. bigelowii*, *C. aquatilis* and *C. rariflora*

> **M7** *Carex curta-Sphagnum russowii* mire   13

S. recurvum and *S. subsecundum sensu stricto* can be present, the former often in abundance, but commonest co-dominants are *S. warnstorfii* and *S. teres*; *C. sarmentosum* and *C. stramineum* frequent but often accompanied by *Aulacomnium palustre*, *Rhizomnium pseudopunctatum*, *Homalothecium nitens* and *Pellia endiviifolia*; *C. echinata*, *C. panicea* and *C. demissa* frequent but usual dominants are *C. rostrata* and *C. nigra*

> **M8** *Carex rostrata-Sphagnum warnstorfii* mire

10   Mixtures of *C. echinata*, *C. nigra* and *C. panicea* generally dominate among the vascular plants with sometimes abundant *E. angustifolium* but Junci at most sparse   11

Sedge cover less varied and extensive and *E. angustifolium* usually sparse but *J. effusus* or *J. acutiflorus* dominant   12

11   *C. echinata* generally the most abundant sedge with *C. panicea* and *C. nigra* subordinate, and *Nardus stricta* and *Juncus squarrosus* at most occasional in a fairly extensive and luxuriant *Sphagnum* carpet

> **M6** *Carex echinata-Sphagnum recurvum/auriculatum* mire
> *Carex echinata* sub-community

C. nigra and *C. panicea* as frequent as *C. echinata* and mixtures of these usually dominant, but *Nardus* and *J. squarrosus* very common, with occasional *Festuca ovina* and *Luzula multiflora* among a sometimes distinctly patchy *Sphagnum* cover

> **M6** *Carex echinata-Sphagnum recurvum/auriculatum* mire
> *Carex nigra-Nardus stricta* sub-community

Around flushes in calcifuge hill pastures, this kind of *Carex echinata-Sphagnum* mire can grade to damper forms of *Festuca-Agrostis-Galium* and *Nardus-Galium* grasslands, but there dominance among the vascular plants passes to mixtures of *F. ovina*, *F. rubra*, *Agrostis capillaris*, *A. canina*, *Anthoxanthum* and *Nardus* and Sphagna are usually totally extinguished.

---

12   *J. effusus* the more abundant rush among an extensive and luxuriant carpet usually dominated by *S. recurvum* with some *S. palustre*

> **M6** *Carex echinata-Sphagnum recurvum/auriculatum* mire
> *Juncus effusus* sub-community

J. acutiflorus usually totally replacing *J. effusus* with *Molinia* a frequent associate and a generally more varied *Sphagnum* carpet

> **M6** *Carex echinata-Sphagnum recurvum/auriculatum* mire
> *Juncus acutiflorus* sub-community

13   *S. lindbergii* very frequent and usually co-dominant with *S. papillosum*, and *S. subnitens*, *S. auriculatum*, *S. capillifolium*, *Calliergon sarmentosum* and *Drepanocladus exannulatus* all frequent; *C. echinata* usually co-dominant with *C. bigelowii* in the vascular tier with *Nardus* common

> **M7** *Carex curta-Sphagnum russowii* mire
> *Carex bigelowii-Sphagnum lindbergii* sub-community

S. lindbergii absent with *S. recurvum* abundant among frequent *S. papillosum* and *S. russowii*; *C. bigelowii* very

scarce and *C. echinata* usually dominant with *C. curta*, and with *C. aquatilis* and/or *C. rariflora* distinctive associates

> **M7** *Carex curta-Sphagnum russowii* mire
> *Carex aquatilis-Sphagnum recurvum* sub-community

14   *S. auriculatum* abundant, occasionally with other Sphagna, in pools and soakaways with a sometimes extensive cover of some of *Potamogeton polygonifolius*, *Ranunculus flammula*, *Juncus bulbosus*, *Agrostis stolonifera*, *Hypericum elodes*, *Ranunculus omiophyllus* and *Montia fontana*   15

S. auriculatum usually just patchily prominent among masses of other bryophytes around spring-heads and rills, with some of *Scapania undulata*, *Philonotis fontana*, *Dicranella palustris*, *Anthelia julacea* and *Marsupella emarginata* abundant and generally just scattered vascular plants   16

15   *Hypericum elodes*, *P. polygonifolius*, *J. bulbosus* and *R. flammula* constant with frequent *Hydrocotyle vulgaris*, *Eleocharis palustris*, *Anagallis tenella* and *Molinia*

> **M29** *Hypericum elodes-Potamogeton polygonifolius* soakway
> *Hyperico-Potametum polygonifolii* (Allorge 1921) Braun-Blanquet & Tüxen 1952

Impoverished examples of this kind of vegetation can be hard to distinguish from *Sphagnum auriculatum* bog pools and fragmentary stands lacking Sphagna are sometimes found around trampled pools and other seasonally-inundated hollows in heaths and moorlands in south-west Britain. Local abundance among such assemblages as these of species like *Eleocharis multicaulis*, *Deschampsia setacea*, *Scirpus fluitans* and *Baldellia ranunculoides* is best referred to M30 Hydrocotylo-Baldellion vegetation.

Often crowded semi-submerged masses of *R. omiophyllus*, *R. flammula*, *M. fontana*, *P. polygonifolius*, *J. bulbosus*, *J. articulatus* and *Myosotis secunda*

> **M35** *Ranunculus omiophyllus-Montia fontana* rill

16   *Anthelia julacea* (or, very locally, *A. juratzkana*) overwhelmingly dominant as swelling masses with *Marsupella emarginata* and *Scapania undulata* constant, *Pohlia ludwigii* locally prominent and small tufts of *Deschampsia cespitosa* and *Nardus stricta*; a very local community of montane areas

M31 *Anthelia julacea-Sphagnum auriculatum* spring
Sphagno auriculati-Anthelietum julaceae Shimwell 1972

Scapania undulata often present with *Sphagnum auriculatum*, but *Philonotis fontana* and *Dicranella palustris* also constant and abundant, together with scattered plants of *Saxifraga stellaris*

M32 *Philonotis fontana-Saxifraga stellaris* spring
Philonoto-Saxifragetum stellaris Nordhagen 1943
Sphagnum auriculatum sub-community

17   *Sphagnum* carpet often extensive and luxuriant with such species as *S. papillosum, S. magellanicum* and *S. capillifolium* usually dominant; *Eriophorum vaginatum, E. angustifolium, Scirpus cespitosus* and *Molinia* variously represented and sometimes structurally important; *Erica tetralix* and *Calluna*, sometimes with Empetra or Vaccinia, making up a shrub element that is often confined to drier hummocks   18

Sphagna often patchy in their cover with *S. papillosum* and *S. magellanicum* of restricted occurrence, *S. capillifolium, S. subnitens, S. compactum* and *S. tenellum* most frequent and abundant; vascular cover usually dominated by various mixtures of *Scirpus, Molinia, Calluna* and *E. tetralix* with *Eriophorum vaginatum* scarce   27

18   *E. vaginatum* usually at least moderately abundant and often a co-dominant, but with *Calluna, V. myrtillus* and *Empetrum nigrum* frequent and usually sufficiently abundant as to give a heathy aspect to the vegetation; *S. capillifolium* generally frequent but *Sphagnum* cover often species-poor and limited in extent with hypnaceous mosses like *Pleurozium schreberi, Rhytidiadelphus squarrosus, Hypnum cupressiforme* and *Plagiothecium undulatum* common; *Rubus chamaemorus* generally common except at lower altitudes

M19 *Calluna vulgaris-Eriophorum vaginatum* mire   19

E. vaginatum often of limited cover or altogether scarce and *V. myrtillus, Empetrum nigrum* and hypnaceous mosses not usually important elements in the vegetation but Sphagna generally abundant and varied, often dominant   21

19   *E. vaginatum* generally less abundant with *Erica tetralix* becoming common among the sub-shrubs; *S. cespitosus* frequent, *Molinia, Narthecium ossifragum* and *Drosera rotundifolia* occasional but *R. chamae-*

morus rare; *Sphagnum* cover often quite extensive with slight suggestion of hummock/hollow relief

M19 *Calluna vulgaris-Eriophorum vaginatum* mire
Erica tetralix sub-community

At lower altitudes through south-west Scotland and Wales, it can be difficult to partition samples between this vegetation type and drier kinds of *Scirpus-Eriophorum* and *Erica-Sphagnum* mires.

E. vaginatum usually abundant and *R. chamaemorus* frequent, but *E. tetralix, S. cespitosus* and *Molinia* rare and *Sphagnum* cover often quite patchy   20

20   *V. vitis-idaea* and *Empetrum nigrum* ssp. *hermaphroditum* constant and often co-dominant with *Calluna, V. myrtillus* and *E. vaginatum; V. uliginosum, J. squarrosus* and *C. bigelowii* occasional with *Hylocomium splendens, Racomitrium lanuginosum* and *Polytrichum alpestre* common in ground layer

M19 *Calluna vulgaris-Eriophorum vaginatum* mire
Vaccinium vitis-idaea-Hylocomium splendens sub-community

E. nigrum ssp. *hermaphroditum* and *V. vitis-idaea* occasional at most and rarely abundant; *V. uliginosum, J. squarrosus, C. bigelowii* and above bryophytes rare or absent

M19 *Calluna vulgaris-Eriophorum vaginatum* mire
Empetrum nigrum ssp. *nigrum* sub-community

On blanket and raised mires where grazing and burning have been severe, this latter vegetation type can grade to or be temporarily replaced by impoverished *Eriophorum vaginatum* mire.

21   *E. vaginatum* and *S. cespitosus* rare except in transitions to wet heath, and sub-shrub layer usually very open, with typically only *Erica tetralix* and *Calluna* frequent, and even the latter rather patchy in its cover; mixtures of *Eriophorum angustifolium, Molinia, Narthecium* and *Drosera rotundifolia* usually provide the constant vascular element among extensive *Sphagnum* lawns or gentle hummock/hollow with *S. papillosum* abundant and, especially in wetter areas, *S. auriculatum, S. cuspidatum, S. recurvum* and, very locally, *S. pulchrum* or *S. megallanicum*

M21 *Narthecium ossifragum-Sphagnum papillosum* mire

Narthecio-Sphagnetum euatlanticum Duvigneaud 1949   22

S. cespitosus and *E. vaginatum* both frequent, though not always abundant, with *S. capillifolium* or *S. magellanicum* often plentiful among extensive *Sphagnum* lawns or well-developed hummock/hollow   23

22   *Rhynchospora alba* very frequent and *Myrica gale* occasionally present, sometimes in abundance, among *Sphagnum* carpet usually dominated by mixtures of *S. papillosum* and *S. auriculatum*, with *S. recurvum* scarce

    M21 *Narthecio-Sphagnetum*
    Rhynchospora   alba-Sphagnum   auriculatum
    sub-community

In wetter hollows, this vegetation can grade to the *Sphagnum auriculatum* bog pool, often through a marginal zone of dense *R. alba*, but *S. auriculatum* becomes strongly dominant in the carpet there and plants such as *Menyanthes* and *Utricularia* spp. appear.

R. alba and *Myrica* rather infrequent, but *Vaccinium oxycoccos* patchily prominent and *Potentilla erecta* occasional in a carpet usually dominated by *S. recurvum* with some *S. papillosum* but little *S. auriculatum*

    M21 *Narthecio-Sphagnetum*
    Vaccinium oxycoccos-Sphagnum recurvum sub-
    community

23   *S. cespitosus* and *Molinia* very common and often abundant with *Potentilla erecta* and *Polygala serpyllifolia* frequent associates, but *Vaccinium oxycoccos* and *Andromeda polifolia* rare and *S. magellanicum* an insignificant component of the carpet

    M17 *Scirpus cespitosus-Eriophorum vaginatum*
    mire   24

S. cespitosus less frequent and *Molinia* only occasional and generally not very abundant; *Polygala serpyllifolia* and *Potentilla erecta* rare but *V. oxycoccos* and *Andromeda* common and *S. magellanicum* often important in the carpet

    M18 *Erica tetralix-Sphagnum papillosum* mire
                                                                          26

24   *E. tetralix* and *Molinia* usually well represented among the vascular dominants with *D. rotundifolia* very frequent and *Myrica* locally plentiful; extensive and varied *Sphagnum* carpet with *S. tenellum* and *S. subnitens* common; leafy hepatics often prominent with

Pleurozia purpurea and *Odontoschisma sphagni* very frequent

    M17 *Scirpus cespitosus-Eriophorum vaginatum*
    mire
    Drosera rotundifolia-Sphagnum sub-community

In wetter hollows, this vegetation can grade to the *Sphagnum auriculatum* bog pool, often through a fringe of *Rhynchospora alba*, but *S. auriculatum* and *S. cuspidatum* increase in abundance there, the sub-shrubs disappear and plants such as *Menyanthes* and *J. bulbosus* occur.

Calluna and *S. cespitosus* are the usual vascular dominants with *E. tetralix* and *Molinia* of reduced importance and *Myrica* rare; *D. rotundifolia* at most occasional and *Sphagnum* cover rather impoverished with *S. tenellum* and *S. subnitens* scarce; leafy hepatics infrequent.
                                                                          25

25   *Racomitrium lanuginosum* very common and often abundant with various *Cladonia* spp., particularly *C. impexa*, *C. arbuscula* and *C. uncialis*; *Erica cinerea* occasional and locally abundant among the sub-shrubs but *V. myrtillus* and *Empetrum nigrum* very scarce and *Nardus* and *J. squarrosus* at most occasional

    M17 *Scirpus cespitosus-Eriophorum vaginatum*
    mire
    Cladonia spp. sub-community

The *Cladonia* and *Drosera-Sphagnum* sub-communities of the *Scirpus-Eriophorum* mire are often found closely juxtaposed in mosaics over hummock/hollow relief on oceanic blanket bogs: this is usually well defined but gentler surface relief may present a hazier differentiation of the vegetation types.

J. squarrosus, *Nardus* and *Deschampsia flexuosa* frequent with small amounts of *V. myrtillus* and occasional *E. nigrum* but *Erica cinerea* rare; *R. lanuginosum* occasional but *Cladonia* spp. uncommon and most obvious feature among patchy Sphagna is frequency of *Hypnum jutlandicum*, *Rhytidiadelphus loreus*, *Pleurozium schreberi*, *Dicranum scoparium* and *Polytrichum commune*

    M17 *Scirpus cespitosus-Eriophorum vaginatum*
    mire
    Juncus squarrosus-Rhytidiadelphus loreus sub-
    community

Both the *Cladonia* and *Juncus-Rhytidiadelphus* sub-communities of the *Scirpus-Eriophorum* mire can grade to *Scirpus-Erica* wet heath on thin ombrogenous peats and drying blanket bogs, transitions with the former being especially obvious at lower altitudes in more oceanic regions, and with the latter on higher ground in drier climatic conditions. In eastern Scotland, too, the *Juncus-Rhytidiadelphus* sub-community can be hard to distinguish from the *Calluna-Eriophorum* mire.

---

26   Sphagna especially luxuriant with *S. magellanicum* frequent and abundant along with *S. papillosum* in wetter lawns; *Narthecium* and *D. rotundifolia* very common but *V. oxycoccos* and *Andromeda* especially distinctive; graminoids and sub-shrubs often showing rather poor growth

M18 *Erica tetralix-Sphagnum papillosum* mire *Sphagnum magellanicum-Andromeda polifolia* sub-community

Sphagna very abundant but with *S. capillifolium* usually dominant, *S. papillosum* frequent but usually subordinate and *S. magellanicum* only occasional; *Pleurozium*, *R. loreus*, *Cladonia impexa*, *C. uncialis* and *C. arbuscula* all frequent and locally abundant; *Narthecium* and *V. oxycoccos* occasional but *D. rotundifolia* and *Andromeda* very scarce; sub-shrubs often quite vigorous with *E. nigrum* ssp. *nigrum* frequent

M18 *Erica tetralix-Sphagnum papillosum* mire *Empetrum nigrum* ssp. *nigrum-Cladonia* spp. sub-community

---

These two kinds of *Erica-Sphagnum* mire are sometimes found closely juxtaposed in mosaics over hummock/hollow relief on raised bogs, but this is often rather ill-defined, when it is difficult to discern vegetation boundaries. In wetter hollows, the *Sphagnum-Andromeda* sub-community can also grade to the *Sphagnum cuspidatum/recurvum* bog pool, where *Andromeda* and *V. oxycoccos* may remain quite common, but where *S. recurvum* or *S. cuspidatum* becomes overwhelmingly abundant.

---

27   *S. capillifolium* and *S. subnitens* usually the most prominent members of an often patchy *Sphagnum* carpet with *Polygala serpyllifolia* and *Potentilla erecta* quite frequent among the herbaceous associates

M15 *Scirpus cespitosus-Erica tetralix* wet heath
28

S. compactum and *S. tenellum* generally the most prominent members of an often patchy *Sphagnum* carpet with *Polygala serpyllifolia* and *Potentilla erecta* patchily represented

M16 *Erica tetralix-Sphagnum compactum* wet heath
Ericetum tetralicis Schwickerath 1933   31

These two vegetation types are very similar in their general floristics, although the former is found mainly on and around blanket bogs to the north and west, the latter around valley bogs in the south and east; and the various sub-communities have additional distinctive features. In eastern Scotland, however, transitions remain problematical.

---

28   *Narthecium* and *E. angustifolium* frequent with *S. palustre* common and sometimes quite plentiful; *Myrica* often found, occasionally with local abundance   29

Narthecium occurs fairly commonly but these other species all scarce   30

---

29   Usually small stands, often in obvious soakways, with more extensive and varied *Sphagnum* carpet with frequent *S. recurvum* and *S. subnitens*; *D. rotundifolia* common with scattered tufts of small sedges such as *C. echinata*, *C. panicea* and *C. nigra* with occasional *Succisa pratensis*, *Juncus bulbosus* and *Viola palustris*

M15 *Scirpus cespitosus-Erica tetralix* wet heath *Carex panicea* sub-community

---

Sometimes this vegetation has a distinctly calcicolous element with locally common *C. pulicaris*, *Selaginella selaginoides*, *Pinguicula vulgaris*, *Drepanocladus revolvens*, *Aneura pinguis* and *Campylium stellatum* and *Schoenus nigricans* plentiful in some stands. It can then closely resemble certain types of *Pinguiculo-Caricetum*, although the continuing presence of at least some Sphagna and ericoid sub-shrubs should help effect a diagnosis.

D. rotundifolia and small sedges at most occasional and *Sphagnum* carpet generally patchy but *S. papillosum* is quite frequent and can be locally abundant, and there is generally a modest amount of *Eriophorum vaginatum* and *Odontoschisma sphagni*

M15 *Scirpus cespitosus-Erica tetralix* wet heath Typical sub-community

30 *Erica cinerea* frequent and sometimes abundant but *V. myrtillus* rare; *Racomitrium lanuginosum* common and *Cladonia* spp., particularly *C. impexa*, *C. uncialis* and *C. arbuscula*, often abundant

> **M15** *Scirpus cespitosus-Erica tetralix* wet heath *Cladonia* spp. sub-community

E. cinerea rare but *V. myrtillus* frequent and sometimes abundant, commonly with tussocks of *Nardus*, *J. squarrosus* and *Deschampsia flexuosa*; *R. lanuginosum* and *Cladonia* spp. scarce but *Dicranum scoparium*, *Pleurozium*, *Plagiothecium undulatum*, *Polytrichum commune* and *Rhytidiadelphus squarrosus* frequent

> **M15** *Scirpus cespitosus-Erica tetralix* wet heath *Vaccinium myrtillus* sub-community

31 *Molinia* tends to dominate among the vascular flora with *S. cespitosus* and *Narthecium* less common than usual; frequent scattered plants of *Potentilla erecta* and *Succisa*, with occasional *Polygala serpyllifolia Carex panicea*, *Salix repens*, *Cirsium dissectum* and *Serratula tinctoria*; bryophytes usually sparse but *S. auriculatum* is sometimes found

> **M16** *Ericetum tetralicis* *Succisa pratensis-Carex panicea* sub-community

---

On moist, but well-aerated, soils among heaths in south-west Britain, this vegetation type can grade to the *Cirsio-Molinietum* and is sometimes very difficult to distinguish from it, but the increased frequency of *Holcus lanatus*, *Anthoxanthum*, *Festuca rubra*, *Lotus uliginosus*, *Cirsium palustre*, *Angelica sylvestris*, *Carex pulicaris* and *C. hostiana* should help diagnose the latter.

Vegetation shows varying patterns of dominance with *S. cespitosus* and *Narthecium* generally frequent but other of listed associates usually uncommon 32

32 Sub-shrub cover generally rather patchy but *Sphagnum* carpet quite extensive between, with frequent *Kurzia pauciflora*; wetter hollows and runnels have *D. rotundifolia* and often *D. intermedia*, *Rhynchospora alba* and locally *R. fusca*, with sporadic appearance of *Lepidotis inundata* and crusts of *Zygogonium ericetorum* on bare patches of sandy peat

> **M16** *Ericetum tetralicis* *Rhynchospora alba-Drosera intermedia* sub-community

Sphagnum carpet often very patchy and sometimes altogether absent, and above associates hardly ever found 33

33 *Molinia* reduced in frequency and abundance and *Calluna* often exceeding *E. tetralix* with frequent tussocks of *S. cespitosus* and *J. squarrosus*; *Hypnum cupressiforme*, *D. scoparium*, *R. lanuginosum* and *Diplophyllum albicans* all common with *Cladonia impexa*, *C. uncialis* and *C. arbuscula* often occurring

> **M16** *Ericetum tetralicis* *Juncus squarrosus-Dicranum scoparium* sub-community

Very variable mixtures of *Calluna*, *E. tetralix* and *Molinia* generally dominate with occasional *S. cespitosus* and *J. squarrosus*, but other of above listed associates rare

> **M16** *Ericetum tetralicis* Typical sub-community

Very commonly throughout lowland Britain, this kind of vegetation has been severely degraded by burning, grazing and draining to impoverished patchworks of *Calluna*, *E. tetralix* and *Molinia* and little else. In the south and west of England, such stands are hard to distinguish from various kinds of drier heath.

---

34 *Carex rostrata* and *C. diandra* frequent with *C. lasiocarpa* occasional and various combinations of these sedges often dominant, with some stands showing a local abundance of *C. paniculata*, *C. appropinquata*, *Schoenus* or *J. subnodulosus*; *Eriophorum angustifolium*, *Menyanthes*, *Potentilla palustris*, *Equisetum fluviatile*, *E. palustre*, *Molinia*, *C. nigra*, *C. panicea*, *Galium palustre*, *Succisa* all common; bryophytes often conspicuous with some of *Calliergon cuspidatum*, *C. cordifolium*, *C. giganteum*, *Campylium stellatum*, *Scorpidium scorpioides*, *Plagiomnium rostratum* and *P. affine* frequent

> **M9** *Carex rostrata-Calliergon* spp. mire 35

Various of the above associates can occur frequently but not in combination with the listed dominants 36

35 *C. rostrata* usually dominant, sometimes with *C. lasiocarpa* or, more locally, *C. diandra* or *Schoenus*, with generally scattered plants of *C. limosa*, *C. echinata* and, more occasionally, *C. lepidocarpa*, *C. dioica* and *C. demissa*; *Calliergon cuspidatum* common but other *Calliergon* spp. and larger Mniaceae at most local and ground carpet usually dominated by mixtures of *Campylium stellatum*, *Scorpidium* and *D. revolvens* with occasional *Sphagnum contortum* and *S. subnitens*

> **M9** *Carex rostrata-Calliergon* spp. mire *Campylium stellatum-Scorpidium scorpioides* sub-community

C. rostrata and *C. diandra* constant with either or both dominant, with or without some *C. lasiocarpa* and very locally an abundance of *J. subnodulosus*; herbaceous associates often numerous and lush with frequent *Filipendula ulmaria*, *Angelica*, *Epilobium palustre*, *Lychnis flos-cuculi*, *Valeriana dioica*, *Caltha palustris*, *Cardamine pratensis* and *Mentha aquatica*; *Campylium stellatum* only occasional and *Scorpidium* and *D. revolvens* rare but *Calliergon cordifolium* and *C. giganteum* common, with *Plagiomnium rostratum*, *P. affine* or some other large Mniaceae common

> **M9** *Carex rostrata-Calliergon* spp. mire
> *Carex diandra-Calliergon giganteum* sub-community

36   Various mixtures of small sedges usually an important element in the swards with frequent records for at least some of *C. panicea*, *C. demissa*, *C. nigra*, *C. pulicaris*, *C. echinata* and *C. saxatilis*; *Pinguicula vulgaris* and *Selaginella* common among an often very rich associated flora; bryophyte carpet also frequently extensive and diverse with at least some of *Campylium stellatum*, *D. revolvens*, *A. pinguis*, *Scorpidium*, *Bryum pseudotriquetrum* and *Blindia acuta* common   37

Such species absent or forming just one element in vegetation with an additional dominant or other important structural components   42

Schoenus nigricans can be found throughout much of lowland Britain with mixtures of these species and most stands where it is a consistent dominant, often with *Juncus subnodulosus*, *Hydrocotyle*, *R. flammula*, *Anagallis* and *Epipactis palustris*, should follow 42. Particularly towards the north, however, *Schoenus* can be locally abundant in vegetation that should follow 37.

---

37   *C. saxatilis* constant and usually dominant with *C. bigelowii* frequent, especially in grassier flush surrounds; *Caltha* and *Viola palustris* common among scattered vascular associates; *Hylocomium splendens* and *Scapania undulata* often found in a patchy bryophyte carpet; distinctly local in Scottish Highlands

> **M12** *Carex saxatilis* mire
> *Caricetum saxatilis* McVean & Ratcliffe 1962

C. saxatilis at most a low-cover occasional and not occurring with the above associates   38

38   *Saxifraga aizoides* constant and quite often abundant with frequent *C. demissa*, but only occasional *C. lepidocarpa*, *C. hostiana*, *C. nigra* and *Molinia*, and frequent *Juncus triglumis* and *Thalictrum alpinum*; *Blindia acuta* constant

> **M11** *Carex demissa-Saxifraga aizoides* mire
> *Carici-Saxifragetum aizoidis* McVean & Ratcliffe 1962 *emend.*   41

S. aizoides only occasional and *J. triglumis* and *Thalictrum alpinum* rare but *C. lepidocarpa*, *C. hostiana*, *C. nigra* and *Molinia* common; *B. acuta* only locally present

> **M10** *Carex dioica-Pinguicula vulgaris* mire
> *Pinguiculo-Caricetum dioicae* Jones 1973 *emend.*   39

At lower altitudes in the upland north and west, it can be difficult to separate these vegetation types, particularly in northern England and Wales where *S. aizoides* and *Blindia* become scarce even in the *Carici-Saxifragetum*.

---

39   *C. hostiana*, *C. pulicaris* and *C. nigra* all frequent along with *E. angustifolium* and *Molinia*; *Potentilla erecta* and *Succisa* common with *Ctenidium molluscum* and *Fissidens adianthoides* often found among carpets or patches of bryophytes   40

Above species very scarce or absent and even *Selaginella* and *Campylium stellatum* rather infrequent; *Gymnostomum recurvirostrum* and, less commonly, *Catoscopium nigritum* forming prominent hummocks with *D. revolvens* and *Cratoneuron commutatum* often abundant between; *Plantago maritima*, *Sagina nodosa* and *Minuartia verna* frequent as scattered plants on bare ground among patches of eroding turf; rare vegetation type recorded only from Upper Teesdale

> **M10** *Pinguiculo-Caricetum*
> *Gymnostomum recurvirostrum* sub-community

40   *C. demissa* and *C. echinata* occasional to frequent, *C. lepidocarpa* and *C. flacca* scarce and *C. pulicaris* patchily represented; *J. bulbosus* and *E. tetralix* quite common; *Bryum pseudotriquetrum*, *F. adianthoides* and *C. molluscum* rather uncommon and *Campylium stellatum* and *Scorpidium* often more prominent

> **M10** *Pinguiculo-Caricetum*
> *Carex demissa-Juncus bulbosus* sub-community

This vegetation type shows considerable variation in the proportions of *C. hostiana* and *C. flacca* versus *Eleocharis quinqueflora*, of *Campylium* versus *Scorpidium* and in the prominence of the heathy element, with the local abundance of *Schoenus nigricans* (or, very rarely, *S. ferrugineus*) adding further diversity. Furthermore, around flushes in more calcifuge upland pastures, gradual transitions to *Festuca-Agrostis-Galium* and *Nardus-Galium* grasslands are commonplace.

C. demissa and *C. echinata* scarce, but *C. lepidocarpa, C. hostiana, C. pulicaris* and *C. flacca* frequent and, very locally, *Kobresia simpliciuscula; J. bulbosus* and *E. tetralix* of only local significance but *Briza media, Primula farinosa, Linum catharticum, Festuca ovina* and *Agrostis stolonifera* all very common, and *Thymus praecox, Equisetum variegatum, Parnassia palustris, Triglochin palustris* and *Leontodon autumnalis* occasional to frequent; *Scorpidium* rare but *C. commutatum* and *R. lanuginosum* often found

> **M10** *Pinguiculo-Caricetum*
> *Briza media-Primula farinosa* sub-community

This vegetation type shows great diversity with some stands approaching the *Gymnostomum* sub-community, others transitional to fen-meadows and others having a striking abundance of *Molinia, Eriophorum latifolium* and *Kobresia*, often with differentiation of fine patterning over and around hummocks in the sward. Around flushes in more calcicolous upland pastures in northern Britain, there are also very often gradual transitions to *Festuca-Agrostis-Thymus* and *Sesleria-Galium* grasslands.

---

41   *J. triglumis* and *T. alpinum* constant, but *Eleocharis quinqueflora* uncommon; *Deschampsia cespitosa, Nardus* and *Anthoxanthum* frequent and *Alchemilla alpina* occasional; *C. commutatum, Scorpidium* and *F. adianthiodes* all scarce

> **M11** *Carici-Saxifragetum*
> *Thalictrum alpinum-Juncus triglumis* sub-community

E. quinqueflora constant with *J. triglumis* and *T. alpinum* becoming frequent at higher altitudes; above grasses at most occasional; *C. commutatum* and *Scorpidium* very common and abundant, often with some *F. adianthoides*

> **M11** *Carici-Saxifragetum*
> *Cratoneuron commutatum-Eleocharis quinqueflora* sub-community

---

42   *Schoenus nigricans* a constant and abundant feature of the vegetation, often with *Anagallis tenella, Campylium stellatum* and *Aneura pinguis*   43

S. nigricans at most a locally prominent species   46

43   *S. nigricans* generally dominant, very often with some *J. subnodulosus* and *Molinia*, and sometimes other larger sedges and rushes; damper runnels between the tussocks have a usually distinctly calcicolous sward with some of *C. panicea, C. lepidocarpa, C. flacca, C. hostiana, C. pulicaris, Parnassia* and *Pinguicula vulgaris*; tall herbs and orchids also often well represented but a heathy element is only sporadically represented and usually confined to drier tussock tops; *D. revolvens, C. commutatum, F. adianthoides, B. pseudotriquetrum* and *Calliergon cuspidatum* all common

> **M13** *Schoenus nigricans-Juncus subnodulosus* mire
> *Schoenetum nigricantis*   Koch 1926   44

S. nigricans and *Molinia* generally making up the bulk of the cover with no *J. subnodulosus* and very few calcicolous small sedges or dicotyledons, tall herbs or orchids; *E. tetralix* very common with frequent *Narthecium* and *D. rotundifolia*; above bryophytes all scarce or absent but *Sphagnum auriculatum, S. subnitens* and *S. papillosum* occasional to frequent on tussocks and raised bryophyte mats

> **M14** *Schoenus nigricans-Narthecium ossifragum* mire

44   *J. subnodulosus* and *Molinia* often very abundant with *S. nigricans* markedly reduced in vigour; *Festuca rubra, Holcus lanatus, Agrostis canina* and *A. stolonifera* frequent and sometimes plentiful; runnel flora, tall herbs, orchids and bryophytes patchy and sometimes virtually overwhelmed in ranker stands

> **M13** *Schoenetum nigricantis*
> *Festuca rubra-Juncus acutiflorus* sub-community

In this widely-distributed vegetation type of drying fen remnants in southern lowland Britain, *S. nigricans* itself can be extinguished and then it is especially difficult to separate stands from the *Cirsio-Molinietum* or *Juncus-Cirsium* fen-meadow.

General floristic and structural features of the community well preserved with usually at least some of *Anagallis tenella, Pedicularis palustris, Angelica, Cirsium palustre, Mentha aquatica, Equisetum palustre, Epipactis palustris* and *Phragmites australis* frequent   45

45   *C. hostiana* and *C. pulicaris* very frequent among an abundant and diverse small herb flora in the runnels with *Briza media, Pinguicula vulgaris, Linum catharticum* and *Juncus articulatus* common; *Centaurea nigra, Serratula tinctoria, Oenanthe lachenalii* and orchids well represented

M13 *Schoenetum nigricantis*
Briza media-Pinguicula vulgaris sub-community

Many smaller runnel herbs sporadic, but *Caltha* and *Valeriana dioica* become common and taller dicotyledons are often prominent, with *Filipendula*, *Eupatorium cannabinum* and *Lychnis flos-cuculi* frequent, and *Galium uliginosum* often occurs; larger sedges such as *C. rostrata*, *C. diandra* and *C. elata* can be patchily abundant or *Phragmites* or *Cladium mariscus*, and there is sometimes a pool element among the flora with *Menyanthes*, *Equisetum fluviatile* and *Utricularia* spp.

M13 *Schoenetum nigricantis*
Caltha palustris-Galium uliginosum
sub-community

These two kinds of *Schoenetum* can sometimes be found in close association, the *Briza-Pinguicula* sub-community on moist ground, the *Caltha-Galium* sub-community on the wettest, with gradual transitions between. In swampy hollows on peat, the latter vegetation type can also pass to *Carex-Calliergon* mire where *S. nigricans* may remain quite prominent, but where dominance usually passes to larger sedges with distinctive herb and bryophyte elements.

---

46 Big rushes always a prominent feature of the vegetation with one or more of *Juncus subnodulosus*, *J. inflexus*, *J. articulatus*, *J. effusus* and *J. acutiflorus* constant and dominant; *Molinia* and *Filipendula* can be frequent but are subordinate in cover 47

Big rushes can be frequent but are typically not dominant and often at most locally abundant 52

This can be a difficult separation to make, especially where an abundance of *Festuca rubra*, *Holcus lanatus*, *Anthoxanthum*, *Cirsium palustre*, *Angelica*, *Mentha aquatica*, *Lotus uliginosus* and *Vicia cracca* masks differences in the proportions of the possible dominants.

---

47 *J. subnodulosus* constant and often dominant with *J. inflexus* and *J. articulatus* quite frequent but of patchy abundance and *J. effusus* and *J. acutiflorus* at most occasional and typically of low cover; *Molinia* rather uncommon and usually sparse but *C. acutiformis* and *C. disticha* frequent and sometimes very abundant

M22 *Juncus subnodulosus-Cirsium palustre* fen-
meadow 48

J. effusus and/or *J. acutiflorus* constant and usually dominant with *J. articulatus* only locally prominent and *J. inflexus* and *J. subnodulosus* hardly ever found; *Molinia* frequent, though generally not very abundant and larger sedges typically absent

M23 *Juncus effusus/acutiflorus-Galium palustre* rush-pasture 51

48 *Phragmites*, *Lythrum salicaria* and *Hydrocotyle vulgaris* frequent with *C. acutiformis* or *C. elata* common and sometimes abundant 49

Above species at most occasional and hardly ever abundant 50

---

49 *C. elata* constant and sometimes abundant among *J. subnodulosus* with frequent *Potentilla palustris*, *Epilobium palustre*, *Equisetum fluviatile*, *Dactylorhiza incarnata* and occasional *Pedicularis palustris*, *Berula erecta* and *Thelypteris palustris*

M22 *Juncus subnodulosus-Cirsium palustre* fen-
meadow
Carex elata sub-community

J. subnodulosus usually dominant with little or no *C. elata*, though with frequent and sometimes abundant *C. acutiformis*; *Galium palustre* common but other associates listed above rare; *Iris pseudacorus*, *R. flammula*, *Valeriana officinalis*, *Lysimachia vulgaris* and *Thalictrum flavum* all frequent; *Epilobium hirsutum*, *Phalaris arundinacea* and *Calystegia sepium* occasional and patchily abundant

M22 *Juncus subnodulosus-Cirsium palustre* fen-
meadow
Iris pseudacorus sub-community

In topogenous fens where there has been a history of mowing for litter, particularly in East Anglia, the *Iris* and, more locally, the *Carex* sub-community can be found grading to the *Peucedano-Phragmitetum* or *Phragmites-Eupatorium* fen. These are variously dominated by *Phragmites*, *Cladium* or *Glyceria maxima* and have a stronger contingent of tall dicotyledons, but transitions may be continuous and are now often further confused by neglect of traditional treatments.

---

50 *J. subnodulosus* often accompanied by *J. inflexus* and *J. articulatus*, although the rush tier is often not very dense and *C. disticha* and *Deschampsia cespitosa* can be patchily abundant; *Molinia*, *Briza media* and *Anthoxan-*

thum are common at low cover with frequent records for *Cardamine pratensis, Cerastium fontanum, Trifolium repens, T. pratense, Rumex acetosa, Plantago lanceolata* and *Epilobium parviflorum*

> **M22** *Juncus subnodulosus-Cirsium palustre* fen-meadow
> *Briza media-Trifolium* spp. sub-community

J. subnodulosus generally dominant in rather rank and impoverished vegetation with bulkier grasses abundant and *J. inflexus, J. articulatus, C. disticha, C. acutiformis, C. hirta* or *C. paniculata* occasional and sometimes locally prominent; *Molinia* and above-listed dicotyledons all infrequent at most

> **M22** *Juncus subnodulosus-Cirsium palustre* fen-meadow
> Typical sub-community

51   *J. effusus* very common but typically exceeded in cover by *J. acutiflorus* with *Molinia* and *Holcus lanatus* frequent and sometimes abundant; *Filipendula* occasional with some of *Ranunculus acris, Potentilla erecta, Achillea ptarmica* and *Equisetum palustre* and locally prominent tall-fen herbs like *Lythrum salicaria* and *Iris*

> **M23** *Juncus effusus/acutiflorus-Galium palustre* rush-pasture
> *Juncus acutiflorus* sub-community

Grazing-mediated transitions between this vegetation type and the *Filipendula-Angelica* mire on the one hand and Cynosurion swards on the other are very common in patchily-improved pastures in western Britain.

J. effusus constant and usually dominant with *J. acutiflorus* scarce; *H. lanatus* common but *Molinia* and above-listed dicotyledons all scarce

> **M23** *Juncus effusus/acutiflorus-Galium palustre* rush-pasture
> *Juncus effusus* sub-community

52   *Molinia* always a prominent feature of the vegetation and generally strongly dominant; *Filipendula* can be frequent but it is typically subordinate in cover   53

Molinia can be frequent but it is not typically dominant and often at most only locally abundant   59

53   *Carex nigra* constant and sometimes abundant with *Crepis paludosa* very frequent, *Trollius europaeus, Caltha palustris* and *Anemone nemorosa* common

> **M26** *Molinia caerulea-Crepis paludosa* fen   58

C. nigra occasional at most and rarely abundant, with *Crepis paludosa, Trollius, Caltha* and *Anemone* very rare or totally absent   54

54   *Molinia* always dominant but *J. subnodulosus, J. acutiflorus, J. conglomeratus* and, less commonly, *J. inflexus* and *J. effusus* represented in some stands, occasionally in abundance; other associates can be of low cover but *Cirsium dissectum, Succisa, Potentilla erecta, Lotus uliginosus* and *Carex panicea* are constant, and there is often some *Cirsium palustre, Angelica, Mentha aquatica* and *Carex hostiana*

> **M24** *Molinia caerulea-Cirsium dissectum* fen-meadow
> *Cirsio-Molinietum caeruleae*   Sissingh & de Vries 1942 *emend.*   55

Molinia always very abundant, often overwhelmingly dominant, with *J. acutiflorus* frequent and *J. effusus* occasional though both subordinate in cover, *J. subnodulosus, J. inflexus* and *J. conglomeratus* all rare; *P. erecta* constant and *Succisa, L. uliginosus* and *C. panicea* occasional to frequent but other associates often few and sparse and *Cirsium dissectum* occurs only towards its western limit

> **M25** *Molinia caerulea-Potentilla erecta* mire   57

This is a very variable vegetation type which, as well as taking in much of our most species-poor *Molinia* vegetation, also includes transitions to blanket mire, rush pasture and tall-herb fen.

---

55   *Molinia* generally dominant with *J. subnodulosus* common and with frequent records for some of *Valeriana dioica, Galium uliginosum, Centaurea nigra, Vicia cracca, Filipendula* and *Equisetum palustre*   56

J. subnodulosus absent but *J. acutiflorus* and *J. conglomeratus* common, *J. effusus* occasional, each sometimes with modest abundance among the *Molinia*; above-listed associates all scarce but *Erica tetralix, Calluna, Galium palustre* and *Dactylorhiza maculata* frequent

> **M24** *Cirsio-Molinietum*
> *Juncus acutiflorus-Erica tetralix* sub-community

56   *Phragmites* constant and *Cladium* quite common, but both typically subordinate in cover, with *Eupatorium cannabinum* and *Lythrum salicaria* frequent; bryophytes can be conspicuous with *Campylium stellatum* often occurring

> **M24** *Cirsio-Molinietum*
> *Eupatorium cannabinum* sub-community

Phragmites and *Eupatorium* only occasional and *Cladium*, *Lythrum* and *Campylium* usually absent, but *H. lanatus*, *Anthoxanthum*, *Briza media*, *C. pulicaris*, *Luzula multiflora* and *Hydrocotyle vulgaris* all frequent, the smaller herbs often extensive in fine-grained mosaics

M24 *Cirsio-Molinietum*
Typical sub-community

In topogenous fens where there has been a history of mowing for litter, particularly in eastern England, the *Eupatorium* and, less commonly, the Typical sub-community of the *Cirsio-Molinietum* can be found grading to the *Peucedano-Phragmitetum* or *Phragmites-Eupatorium* fen. These are variously dominated by *Phragmites* and *Cladium* and have a stronger contingent of tall herbs, but transitions may be continuous and are now often further confused by neglect of traditional treatments. Shifts to somewhat more enriched soils can also involve gradations to the *Juncus-Cirsium* fen-meadow, with a switch to rush-dominance.

---

57   *Erica tetralix* constant, *Calluna* and *Myrica* quite frequent, with *Eriophorum angustifolium* common and occasional *Narthecium*, *D. rotundifolia* and *V. oxycoccos*; *Aulacomnium palustre* and *Polytrichum commune* quite common with patchy *Sphagnum auriculatum*, *S. papillosum* and *S. palustre*

M25 *Molinia caerulea-Potentilla erecta* mire
Erica tetralix sub-community

J. acutiflorus and occasionally *J. effusus* patchily prominent in a grassy ground with frequent *H. lanatus*, *F. rubra*, *Anthoxanthum*, *Agrostis capillaris* (and in the south-west *A. curtisii*), *Danthonia decumbens*, *Luzula multiflora* and *L. campestris*; *E. tetralix* rare but *Calluna* occasional and *Ulex gallii* and *U. europaeus* sometimes found

M25 *Molinia caerulea-Potentilla erecta* mire
Anthoxanthum odoratum sub-community

Above-listed associates usually all occasional at most but tall herbs prominent among the *Molinia* and rush clumps with frequent *Angelica* and *Cirsium palustre*, and occasional *Epilobium palustre*, *Eupatorium*, *Filipendula*, *Galium palustre* and *Mentha aquatica*; *Schoenus* can be locally abundant

M25 *Molinia caerulea-Potentilla erecta* mire
Angelica sylvestris sub-community

58   *Molinia* and *C. nigra* often both abundant, with locally prominent *C. rostrata* or *C. appropinquata*, and

frequent *Sanguisorba officinalis*, *Angelica*, *Serratula*, *Galium palustre* and *G. uliginosum*; bryophytes patchy but *Campylium stellatum*, *C. elodes*, *Ctenidium molluscum*, *Aulacomnium palustre* and *Plagiochila asplenoides* all common

M26 *Molinia caerulea-Crepis paludosa* mire
Sanguisorba officinalis sub-community

C. nigra often subordinate in cover to *Molinia* in altogether grassier or rushier vegetation with frequent and sometimes abundant *F. rubra*, *F. ovina*, *H. lanatus*, *B. media*, *D. cespitosa*, *Anthoxanthum*, *J. acutiflorus* and *J. conglomeratus*; above-listed dicotyledonous associates and bryophytes all occasional at most but *Geum rivale*, *Centaurea nigra*, *Leontodon hispidus*, *Climacium dendroides* and *Plagiomnium undulatum* common

M26 *Molinia caerulea-Crepis paludosa* mire
Festuca rubra sub-community

59   *Filipendula ulmaria* constant and usually very abundant   60

Filipendula generally absent and never present at high cover   63

60   *Iris pseudacorus* and *Oenanthe crocata* constant, the former sometimes exceeding *Filipendula* in cover, with *Lycopus europaeus*, *Stellaria alsine*, *Rumex crispus* and *Scutellaria galericulata* common

M28 *Iris pseudacorus-Filipendula ulmaria* mire
Filipendulo-Iridetum pseudacori Adam 1976 emend.   62

Iris and *O. crocata* occasional at most and usually not abundant or with the above associates

M27 *Filipendula ulmaria-Angelica sylvestris* mire   61

Beyond the range of *O. crocata*, as along the north coast of Scotland, it can be difficult to separate these two vegetation types.

---

61   *Angelica*, *Valeriana officinalis*, *Rumex acetosa*, *Lychnis flos-cuculi*, *Succisa* and *Geum rivale* common among taller associates, with *Caltha*, *Ranunculus flammula*, *R. repens*, *R. acris*, *Cardamine flexuosa* and *C. pratensis* occasional to frequent below and *Galium palustre*, *G. uliginosum* and *Lathyrus pratensis* climbing and sprawling; nitrophilous herbs and Junci infrequent and of low cover

M27 *Filipendula ulmaria-Angelica sylvestris* mire
Valeriana officinalis-Rumex acetosa sub-community

Above associates at most occasional but sparse shoots of *Phragmites* common and there are often prominent clumps of *Urtica dioica*, *Eupatorium* and *Epilobium hirsutum*; *Cirsium arvense* occasional with *Galium aparine* and *Vicia cracca* the commonest sprawlers

> **M27** *Filipendula ulmaria-Angelica sylvestris* mire
> *Urtica dioica-Vicia cracca* sub-community

In and around open water transition mires and topogenous fens throughout the British lowlands, these two kinds of *Filipendula-Angelica* mire can be found in close association with *Phragmites-Eupatorium* and *Phragmites-Urtica* fens. *Filipendula* and other tall herbs can remain frequent there but are typically subordinate in cover to *Phragmites*, *Glyceria maxima* or *Carex paniculata*.

Above associates at most occasional but *J. effusus* and *H. lanatus* are constant, *J. acutiflorus* and *Molinia* occasional in ranker swards with *Anthoxanthum*, *Agrostis stolonifera*, *Mentha aquatica* and *Lotus uliginosus* quite common

> **M27** *Filipendula ulmaria-Angelica sylvestris* mire
> *Juncus effusus-Holcus lanatus* sub-community

62　　*J. effusus* and/or *J. acutiflorus* constant and patchily abundant with frequent *Rumex acetosa*, *Cirsium palustre*, *Epilobium palustre*, *Lychnis flos-cuculi*, *Ranunculus acris*, *Caltha*, *L. uliginosus* and *Galium palustre*

> **M28** *Filipendulo-Iridetum*
> *Juncus effusus/acutiflorus* sub-community

Above-listed species occasional at most, but *Urtica* and *Cirsium arvense* constant with sprawling *G. aparine*, these sometimes in abundance, and occasional to frequent *Elymus repens*, *Stellaria media*, *Arrhenatherum elatius* and *Dactylis glomerata*

> **M28** *Filipendulo-Iridetum*
> *Urtica dioica-Galium aparine* sub-community

Above groups of species and even *Filipendula* infrequent in often rather open vegetation, with *Atriplex prostrata* and *Samolus valerandi* common and sporadic records for maritime plants like *Oenanthe lachenalii*, *Glaux maritima*, *Triglochin maritima* and *Matricaria maritima*

> **M28** *Filipendulo-Iridetum*
> *Atriplex prostrata-Samolus valerandi*
> sub-community

63　　Spring-heads, rills or flushes dominated by bryophyte carpets with an abundance of some of *Philonotis*

fontana, *Dicranella palustris*, *Scapania undulata*, *Pohlia wahlenbergii* var. *glacialis* and scattered plants of *Saxifraga stellaris*, *Chrysosplenium oppositifolium*, *Deschampsia cespitosa* and *Stellaria alsine*　　64

Spring-heads, rills or flushes with *P. fontana*, *C. oppositifolium* and *D. cespitosa* occasional to frequent, but dominated by *Cratoneuron commutatum* or *C. filicinum*, often with some *Bryum pseudotriquetrum* and *Aneura pinguis*; *Carex panicea* and *C. dioica* occasional to frequent　　65

Above species at most occasional and of low cover　　66

64　　*P. fontana* usually dominant, often with an abundance of *D. palustris* and *S. undulata*, but little or no *Pohlia wahlenbergii* var. *glacialis*; *Saxifraga stellaris* and *Montia fontana* constant with *Anthoxanthum* and *Agrostis canina* ssp. *canina* common

> **M32** *Philonoto-Saxifragetum*
> *Montia fontana-Chrysosplenium oppositifolium*
> sub-community

P. wahlenbergii var. *glacialis* dominant, often with a little *P. ludwigii* and *Philonotis fontana* but only occasional *D. palustris* and *S. undulata*; *Montia fontana* and above grasses scarce but *Cerastium cerastoides* common; a very local community in parts of the Scottish Highlands

> **M33** *Pohlia wahlenbergii* var. *glacialis* spring
> *Pohlietum glacialis*　　McVean & Ratcliffe 1962

65　　*C. commutatum* and/or *C. filicinum* dominant with only occasional *P. fontana* and sometimes just scattered plants of *Saxifraga aizoides*, *Festuca rubra* and *Cardamine pratensis* and a usually insignificant small-sedge element

> **M37** *Cratoneuron commutatum-Festuca rubra* spring

One or both *Cratoneuron* spp. still abundant but *P. fontana* constant and of locally high cover and an often rich associated flora: *Carex panicea*, *C. nigra* and *C. demissa* constant and often quite abundant with frequent *Selaginella*, *Polygonum viviparum*, *Trifolium repens* and *Leontodon autumnalis*

> **M38** *Cratoneuron commutatum-Carex nigra* spring

Both these vegetation types, and particularly the latter, can grade almost imperceptibly to flushed calcicolous pastures through an intermediate zone of the *Pinguiculo-Caricetum* or *Carici-Saxifragetum*. In critical situations, the proportions of

Cratoneuron spp. to small sedges and dicotyledons should effect a separation.

---

66   *Eriophorum vaginatum* or *E. angustifolium* dominant in generally very species-poor vegetation   67

E. vaginatum and *E. angustifolium* absent or at most a very insignificant element in the vegetation   69

67   *E. angustifolium* abundant, though sometimes in distinct patches, with only occasional scattered tussocks of *E. vaginatum*; bryophytes very variable in their occurrence, but *Drepanocladus fluitans* is common and locally abundant

    M3 *Eriophorum angustifolium* bog pool community

E. angustifolium can be frequent but is typically sparse among dominant *E. vaginatum*; bryophytes variable in numbers and cover but *D. fluitans* is rare

    M20 *Eriophorum vaginatum* mire   68

68   *Calluna*, *V. myrtillus* and *Empetrum nigrum* ssp. *nigrum* occur frequently, the latter often in some abundance; *Campylopus paradoxus* and *Dicranum scoparium* frequent in small amounts with occasional patches of peat-encrusting lichens

    M20 *Eriophorum vaginatum* mire
    Calluna vulgaris-Cladonia spp. sub-community

Sub-shrubs at most occasional and sparse among very abundant *E. vaginatum* with an impoverished bryophyte and lichen flora, often just scattered tufts of *C. paradoxus*, *Orthodontium lineare* and *Calypogeia* spp.

    M20 *Eriophorum vaginatum* mire
    Species-poor sub-community

69   Open patchwork of *Scapania undulata*, *Calliergon sarmentosum* and *Blindia acuta* with areas of wet silty or stony ground between on which are scattered plants of *Carex demissa*, *Deschampsia cespitosa*, *Saxifraga stellaris*, *Juncus triglumis* and *Koenigia islandica*; very local on Skye

    M34 *Carex demissa-Koenigia islandica* flush

Patches of *Chrysosplenium oppositifolium* with some of *Hookeria lucens*, *Rhizomnium punctatum*, *Trichocolea tomentella*, *Pellia epiphylla* and *Conocephalum conicum* on damp, shady banks in the lowlands

    M36 Cardaminion vegetation

COMMUNITY DESCRIPTIONS

M1
Sphagnum auriculatum bog pool community

Synonymy

Sphagnetum Rankin 1911*b p.p.*; *Sphagnetum* regene-
ration complex Tansley 1939 *p.p.*; *Scheuchzeria palus-
tris* vegetation Sledge 1949; *Drosera intermedia-
Schoenus nigricans* Gesellschaft Braun-Blanquet &
Tüxen 1952 *p.p.*; *Sphagnum* Hummock Complex,
Sphagnum cuspidatum & *S. pulchrum* phases Rose
1953; Pool & furrow communities Pearsall 1956; Pool
communities Ratcliffe & Walker 1958, Boatman &
Armstrong 1968, Moore 1977, Boatman *et al.* 1981;
Rhynchospora-Sphagnum vegetation Newbould 1960;
Trichophoreto-Eriophoretum pool component
McVean & Ratcliffe 1962; *Eriophorum angustifolium-
Sphagnum cuspidatum* Association Birks 1973; *Rhyn-
chosporetum albae* Koch 1926 *sensu* Birse 1980 *p.p.*;
Caricetum limosae Dierssen 1982 *p.p.*, Osvald 1923
emend Dierssen 1978 *sensu* Birse 1984 *p.p.*; Pool communities 2.ii & 2.iii Hulme & Blythe 1984; Peatland noda
13–20 Lindsay *et al.* 1984; *Sphagnum* lawn bog hollows NCC New Forest Bogs Report 1984.

Constant species

*Eriophorum angustifolium, Menyanthes trifoliata,
Sphagnum auriculatum, S. cuspidatum.*

Rare species

*Hammarbya paludosa, Rhynchospora fusca, Scheuch-
zeria palustris, Utricularia intermedia, Sphagnum
pulchrum.*

Physiognomy

The *Sphagnum auriculatum* bog pool community typically consists of floating masses or soft wet carpets of
Sphagna with scattered vascular plants growing on or
through them or in areas of open water between. The
dominant Sphagna are *S. auriculatum* (including var.
inundatum, sometimes considered within *S. subsecundum*) and *S. cuspidatum*: the constancy and general
abundance of the former, together with the rarity here of
S. recurvum, provide a good contrast with the closely-
related *Sphagnum cuspidatum/recurvum* bog pools.

Locally, throughout the range of the community, the
bright orange-yellow *S. pulchrum* provides a conspicuous enrichment of this element of the vegetation,
growing semi-submerged with the other species (as on
Brishie bog, the only one of the Silver Flowe mires in
Dumfries and Galloway where Ratcliffe & Walker
(1958) found it) or as a somewhat firmer mat upon them
(as on some of the lowland English valley mires described by Rose (1953) who distinguished a separate *S.
pulchrum* phase). Then, where these species provide a
more substantial base, there can be some *S. papillosum*,
occasionally with *S. subnitens* or *S. tenellum*, and, in
patches raised above the water level, even a little *S.
magellanicum* or *S. capillifolium*. Generally, however,
these species are of low frequency and cover and, on
mires with a well-defined surface patterning, they represent a clear transition from the *Sphagnum auriculatum*
community to the lawns and hummocks of the pool
surrounds. Other bryophytes are generally scarce but
Cladopodiella fluitans is characteristic of this kind of
vegetation at low frequencies and there can also be some
Gymnocolea inflata.

Among the vascular plants, the commonest throughout are *Menyanthes trifoliata* (another good preferential
against the *Sphagnum cuspidatum* bog pools) and *Eriophorum angustifolium*. Together, these usually make up
a cover of less than 30% over the more intact carpets of
Sphagna, though they often extend out as a sparse but
quite tall cover into areas of deeper open water where the
semi-aquatic Sphagna are reduced to scattered, flaccid
and submerged shoots. Such barer patches tend to be
more common here than in the *Sphagnum cuspidatum/
recurvum* community and they quite frequently have
some *Utricularia* spp., usually *U. minor* or, more locally,
U. intermedia, occasional specimens of Littorelletea
species such as *Eleocharis multicaulis*, *Juncus bulbosus*
and *Potamogeton polygonifolius* and a thick algal
growth in which *Batrachospermum* spp. figure prominently. In some areas, *Carex limosa* is a frequent component of such stands (e.g. Sledge 1949, Hulme 1985, Birse
1984 who placed his samples in a *Caricetum limosae*),

though, being a somewhat shy flowerer, it can be easily overlooked. On many north-western mires on which the *Sphagnum auriculatum* community is represented, there is a continuous structural and floristic transition through these more open stands to what Sjörs (1948) called 'mud bottoms', deeper pools with little more than submerged accumulations of wind-blown *Molinia caerulea* litter over muddy peat (e.g. Ratcliffe & Walker 1958, Lindsay *et al.* 1984).

Also very characteristic of this community, though more strictly confined to shallower water around pool margins or extending across smaller mire pools, is *Rhynchospora alba*. Unlike *E. angustifolium* which is conspicuous throughout the year with its often red-tinged leaves with long triquetrous tips, *R. alba* dies back in winter to slender, green, spindle-shaped perennating structures hidden by the old leaf-bases and is then easily missed. In Hampshire and east Dorset, and at a few scattered localities through the range of the community, *Sphagnum auriculatum* bog pools also provide a major locus for the much rarer *R. fusca* (Newbould 1960, Moore 1977, British Ecological Society Mires Group unpub.). Then, among the *Sphagnum* carpet, there can be occasional plants of *Narthecium ossifragum*, *Drosera rotundifolia* and, more locally, *D. anglica* (and sometimes the hybrid between these two sundews, *D.* × *obovata*) or *D. intermedia*. *Pinguicula lusitanica* has been recorded from some stands and also the rare *Hammarbya paludosa*, an inconspicuous orchid typically found in the transition zone to surrounding hummocks. Finally among these species of more restricted national distribution, the only surviving plants of *Scheuchzeria palustris* in Britain, on Rannoch Moor on the Argyll/Perth border, occur in this community, in stands with frequent *Carex limosa* and *Drosera anglica* (Sledge 1949), an association characteristic of both its extinct English stations and its Eurasian localities (Katz 1926, Hegi 1965, Tallis & Birks 1965).

Also around the pool margins, *Molinia caerulea* can extend down from the mire surface, though its cover is generally low and it is typically not tussock-forming. Occasionally there may also be a little *Myrica gale*, *Scirpus cespitosus* or *Erica tetralix*.

Habitat

The *Sphagnum auriculatum* community is confined to pools and wetter hollows on ombrogenous and topogenous mires with base-poor and oligotrophic raw peat soils in the more oceanic parts of Britain. Where mires are grazed or burned, the wetness of the ground affords some protection to this vegetation but it has been reduced on many sites by draining and cutting of the peat.

Of our two major *Sphagnum*-dominated Rhynchosporion pool communities of bog peats, this is the more oceanic in its distribution. By and large, it is restricted to the far west of Britain, where annual precipitation exceeds 1200 mm (*Climatological Atlas* 1952) with more than 180 wet days yr^{-1} (Ratcliffe 1968), and to south-western valley mires, where local topography maintains a consistently high water-table in an area with an Atlantic climate. Within the former region its range coincides closely with that of the *Scirpus-Eriophorum* mire, in the latter with the distribution of the floristically similar *Narthecio-Sphagnetum* mire, and it typically comprises the pool and wet hollow component in these two communities.

A major part of the difference between the community and its analogue in the eastern lowlands of Britain, the *Sphagnum cuspidatum/recurvum* community, is the increased prominence here of *S. auriculatum*, a feature seen not only in Rhynchosporion vegetation and in the wetter flats of the mires in which these pools occur, but also in sedge- and rush-dominated *Sphagnum* carpets of oligotrophic soligenous areas in mires, heaths and grasslands in moving into the western parts of Britain. But other positive floristic features of the *Sphagnum auriculatum* community may be a coincidental reflection of the fact that the bog pool habitat has become very scarce outside the more oceanic parts of Britain because of the extensive draining of lowland mires. *Menyanthes*, for example, though much commoner here than in the *Sphagnum cuspidatum/recurvum* community, is widely recorded in other kinds of open water transition in the eastern lowlands and, of the other preferentials, only *Pinguicula lusitanica* has an oceanic distribution through Europe as a whole. Thus, *Rhynchospora fusca*, *Drosera anglica*, *D. intermedia* and *Carex limosa*, though more frequent here, are occasionally found in *Sphagnum cuspidatum* bog pools and have a Continental Northern range in Europe. *Sphagnum pulchrum*, likewise, is only western in Britain (Ratcliffe 1968). *Hammarbya*, though not recorded in *Sphagnum cuspidatum/recurvum* pools, has a number of old stations within their range and *Scheuchzeria palustris* a few (Sledge 1949). These species, then, are probably best regarded as plants which maintain their general affinities with Rhynchosporion vegetation in Britain but which have become increasingly confined to one particular kind of bog pool because of regional differences in the intensity of mire drainage.

Whether such real floristic differences as there are between the two communities are based directly on climate is unknown. Certainly, there seems little in the character of the peats and waters to separate the two types of bog pool: the pH range here, varying from 3 to 5, is only a little broader than in the *Sphagnum cuspidatum/recurvum* community and both are generally typical of acidic and impoverished substrates.

Throughout its range, the *Sphagnum auriculatum* community occupies wetter situations on the vegetated mire surface, comprising the shallow pool and wet hollow element in patterned systems. The distinctness of such patterning and the relative contribution of this component are very varied, but there is a clear tendency for pools and hollows to become proportionally more important in moving to the far north-west of Britain (e.g. Lindsay *et al*. 1984). In the lowland valley mires in which the community is represented, the hollows are generally small with a fairly smooth transition to the hummocks, whereas around the western sea board of Scotland, the blanket and raised mires can have complex and extensive systems of well-defined pools, locally so numerous as to give the surfaces a shining appearance (hence the name Silver Flowe: Ratcliffe & Walker 1958; see also photographs in Moore 1977, Ratcliffe 1977, Boatman *et al*. 1981). Even on the same stretch of mire, pools can vary considerably in size and shape. They may be but a few metres long to over 50 m in their longer dimension and, though deeper pools tend to be of the 'mud-bottom' type, the community can extend down in fragmentary form to depths of about 25 cm (Ratcliffe & Walker 1958, Lindsay *et al*. 1984). Pools can be straight, curved, simple, branched or irregular in shape though, at many sites, there is a strong tendency for them to be elongated and oriented along the contours of gently-sloping mire surfaces with more rounded, island-dotted pools on flatter areas. The overall effect is thus often of arcs or eccentric rings of pools with hummocks between.

Just how such pools form is a matter of considerable debate, though most authorities now discount Pearsall's (1956) suggestion that mire corrugations arise by wrinkling and tearing of the peat as it moves under gravitational flow (Ratcliffe & Walker 1958, Boatman & Armstrong 1968, Boatman & Tomlinson, 1973, Boatman *et al*. 1981). Neither do surface freeze-thaw effects (Moore & Bellamy 1973) nor wind erosion (Osvald 1949) nor a combination of the two seem adequate to account for the origin of pools, though Ratcliffe & Walker (1958) considered that deeper pools could only have developed from shallower ones by wind and water erosion. More likely seem to be proposals that pools form in areas where mire waters collect, either over depressions in the underlying mineral basins (Osvald 1923, Goode 1970, Birks 1973, Boatman & Tomlinson 1977, Boatman *et al*. 1981) or where waters are channelled within mires or from their surrounds (Ratcliffe & Walker 1958, Boatman & Tomlinson 1973), differentiating from initials of more aquatic Sphagna on the very wet and more or less level mire surface. Boatman (1977) has suggested that, in the climatic conditions characteristic of such mires as those of the Silver Flowe, such nuclei might be inherently unstable, tending to degenerate into self-maintaining pools, because of the sub-optimal growth of *S. cuspidatum*, an abundant species even in this community (see also Boatman & Tomlinson 1977, Boatman *et al*. 1981). It was also thought possible that colonisation of

pool surrounds by the more robust *S. papillosum* might be set back where the free-floating and flaccid shoots of *S. cuspidatum* were blown across the water and deposited as a smothering mass (Boatman & Tomlinson 1977). Such observations as these cast some doubt on the traditional view of regenerative alternations between pools and hollows on the one hand and flats and hummocks on the other (Osvald 1923, Tansley 1939) and stratigraphic evidence seems to support the view that, in sites like the Silver Flowe and the Claish/Kentra mosses in Argyll, the pools are quasi-permanent structures. The same is probably true of the larger pool systems in the New Forest and Dorset valley mires.

Once pools are established, it is relatively easy to see how they might take on the distinctive shapes and alignments seen in more extensive systems. Thus, larger pools could form by the flooding of low intervening strips of vegetation between smaller ones. Only where such initials lie on the same contour can water level be maintained evenly throughout the new, enlarged pool, so pools on flatter mire surfaces can enlarge in all directions, whereas those on gently-sloping areas tend to become elongated along the contours (Boatman & Armstrong 1968, Boatman & Tomlinson 1973, Boatman *et al*. 1981). In some situations, differential peat growth above and below long pools on slopes could pond back mire waters and allow fusion at right angles to the slope (Boatman *et al*. 1981).

Zonation and succession
On more featureless mires, the *Sphagnum auriculatum* community can contribute to ill-defined areas of wetter vegetation (e.g. Newbould 1960, Goode & Lindsay 1979, Boatman *et al*. 1981), but it is best developed as the pool and wet hollow component in certain patterned Erico-Sphagnion mires where it grades, with increasing height above the water-table, to flat and hummock vegetation. There is only limited evidence that such mosaics represent a full regenerative sequence of alternating pools and hummocks.

Throughout western Britain, the community is found most consistently within the *Scirpus-Eriophorum* mire, on domed or slightly sloping areas within blanket mire or in areas which bear a more obvious resemblance to raised mire. Where surface patterning is pronounced, it passes, round the margins of pools and wetter hollows to the *Drosera-Sphagnum* sub-community of the mire proper, with a switch in the *Sphagnum* carpet from *S. auriculatum* and *S. cuspidatum* to *S. papillosum*, *S. subnitens* and *S. tenellum* (and, locally, *S. magellanicum*) and thence to *S. capillifolium*. *Drosera rotundifolia* and, more locally, *D. anglica* run some way into this transitional vegetation, *Narthecium* and *Eriophorum angustifolium* much further; and there is a substantial increase in *Molinia* and *Erica tetralix* and an appearance of *Erio-*

phorum vaginatum and *Scirpus cespitosus*. Mixtures of these species continue on as vascular dominants of the *Cladonia* sub-community of the *Scirpus-Eriophorum* mire which typically makes up the vegetation cover of the hummock sides and tops in such systems. The vertical stratification of the major species in relation to water-table is well seen in profiles from the Silver Flowe mosses (Ratcliffe & Walker 1958) and Kentra Moss (British Ecological Society Mires Group unpub.). On degraded blanket mires in western Britain, fragmentary *Sphagnum auriculatum* pools may persist in the *Juncus-Rhytidiadelphus* sub-community of the *Scirpus-Erio-phorum* mire. On moving further east in Scotland, where the *Erica-Sphagnum* mire replaces the *Scirpus-Erio-phorum* mire on lowland raised bogs, the *Sphagnum auriculatum* bog pool is replaced by the *Sphagnum cuspidatum/recurvum* type, and intermediate stands may be encountered in the transitional zone.

In southern lowland valley bogs, the *Sphagnum auri-culatum* community occupies an analogous position in patterned *Narthecio-Sphagnetum* mire, though here hummock-hollow systems tend to be less well defined and there is the complication of soligenous influence along the axes and around the margins of the bogs. Also, *Sphagnum capillifolium* and *Eriophorum angustifolium* play a less prominent role in the drier elements of the vegetation than to the north-west. But, these things apart, the general pattern is the same: typically, the community gives way to the *Sphagnum-Rhynchospora* sub-community of *Narthecio-Sphagnetum* mire, *S. papillosum*, *S. tenellum* and *S. subnitens* increasing their cover and *Molinia*, *Erica tetralix* and *Calluna* becoming important. Such transitions are well described in the various phases of Rose's (1953) *Sphagnum* Hummock Complex and were included in the 'general bog commu-nities' characterised by Newbould (1960) from Cranes-moor in the New Forest.

It has sometimes been assumed that such zonations as these are a spatial expression of a cyclic alternation of hollows and hummocks in regenerative bog growth (e.g. Tansley 1939) but, as noted above, the evidence for this is limited. The *Sphagnum auriculatum* community may thus often be a fairly permanent self-maintaining feature of mire surfaces with rather limited switches from more to less aquatic Sphagna but no guaranteed progression to the vegetation of drier flats and hummocks.

Where mire drainage lowers the water-table, the community is readily damaged or destroyed and it has been widely lost where Erico-Sphagnion communities have been converted to Ericion heaths or their degraded derivatives. Shallow peat-digging can create flooded hollows which become suitable for recolonisation by Sphagna, *Rhynchospora alba* and *Drosera* ssp. but such locally reconstituted stands often lie in much modified mire contexts (Rose 1953).

Distribution

The community is a widespread component of the *Scirpus-Eriophorum* mire in western Scotland and parts of the Lake District, Wales and the South-West Penin-sula (and it continues into Ireland in essentially the same form, though with abundant *Schoenus nigricans*: Braun-Blanquet & Tüxen 1952).In the range of the *Narthecio-Sphagnetum* mire, it is best developed in the valley bogs of the New Forest with local occurrences elsewhere.

Floristic table M1

Sphagnum auriculatum	IV (1–10)
Eriophorum angustifolium	IV (1–7)
Menyanthes trifoliata	IV (1–9)
Sphagnum cuspidatum	IV (1–10)
Utricularia minor	II (1–5)
Rhynchospora alba	II (1–6)
Potamogeton polygonifolius	II (1–8)
Drosera rotundifolia	II (1–4)
Narthecium ossifragum	II (1–4)
Carex limosa	II (1–8)
Sphagnum papillosum	II (1–5)
Drosera anglica	II (1–4)
Molinia caerulea	I (1–7)
Erica tetralix	I (1–3)
Sphagnum subnitens	I (1–4)
Eleocharis multicaulis	I (2–6)
Sphagnum palustre	I (1–4)
Sphagnum magellanicum	I (1–8)
Juncus bulbosus/kochii	I (1–6)
Sphagnum pulchrum	I (2–9)
Myrica gale	I (1–4)
Gymnocolea inflata	I (1–2)
Drosera intermedia	I (1–3)
Utricularia intermedia	I (2–3)
Scirpus cespitosus	I (1–3)
Cladopodiella fluitans	I (1–3)
Sphagnum recurvum	I (8–9)
Odontoschisma sphagni	I (1)
Rhynchospora fusca	I (1–3)
Scheuchzeria palustris	I (1–3)
Number of samples	63
Number of species/sample	7 (2–15)
Vegetation height (cm)	20 (10–45)
Vegetation cover (%)	88 (30–100)
Altitude (m)	64 (6–280)
Soil pH	4.3 (3.2–5.1)

Affinities
Bog pool vegetation of this kind has sometimes been described as an integral part of more broadly-defined mire communities (e.g. Rankin 1911*b*, Tansley 1939, Rose 1953) but, though it is characteristically found within Erico-Sphagnion bogs, it is floristically distinct enough to be considered separately. Phytosociologically, it belongs to the Rhynchosporion albae alliance, which comprises the most calcifuge and oligotrophic of the communities dominated by Sphagna and cyperaceous plants within the Scheuchzerio-Caricetea nigrae (*sensu* Ellenberg 1978). Some stands show clear affinities with the Littorelletea vegetation of shallow acid bog pools and runnels, where such species as *Eleocharis multicaulis*, *Juncus bulbosus* and *Potamogeton polygonifolius* provide a link, or with the submerged aquatic vegetation with *Utricularia minor* and *U. intermedia* placed by some authorities in the Utricularietea (Oberdorfer 1977, Birse 1984) or a class Utricularietalia in the Lemnetea (Ellenberg 1978). Similar bog pool vegetation has been described from Ireland in the *Drosera intermedia-Schoenus nigricans* Gesellschaft (Braun-Blanquet & Tüxen 1952; see also Morrison 1959) and from Scandinavia (e.g. Nordhagen 1928, 1943, Sjörs 1948). It falls within the rather more broadly defined *Caricetum limosae* of Dierssen (1982).

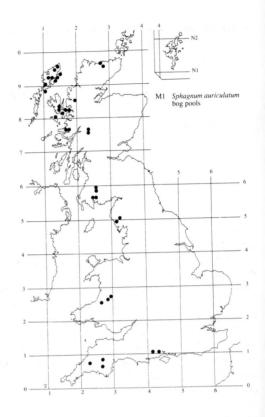

M1 *Sphagnum auriculatum* bog pools

M2
Sphagnum cuspidatum/recurvum bog pool community

Synonymy

Sphagnetum pools Rankin 1911*a*; *Sphagnetum* regeneration complex, Stages 1 & 2 Tansley 1939; *Sphagnum* lawn Poore & Walker 1958, Sinker 1962, Green & Pearson 1968; *Sphagnum cuspidatum-Eriophorum angustifolium, Sphagnum recurvum-Vaccinium oxycoccos* & *Sphagnum recurvum-Erica tetralix* Noda, Normal Series Tallis 1973; *Sphagnum flexuosum* noda 16-19 Daniels 1978; *Sphagnum cuspidatum* pool community Bignal & Curtis 1981 *p.p.*; *Sphagno tenelli-Rhynchosporetum albae* Dierssen 1982.

Constant species

Erica tetralix, Eriophorum angustifolium, Drosera rotundifolia, Sphagnum cuspidatum/recurvum.

Rare species

Andromeda polifolia, Carex magellanica, Sphagnum pulchrum.

Physiognomy

The *Sphagnum cuspidatum/recurvum* community is typically dominated by extensive soft wet carpets of *Sphagnum cuspidatum* and/or *S. recurvum* with, very locally, *S. pulchrum* (e.g. Sinker 1962, Ratcliffe 1977). In marked contrast to the bog pools of more oceanic parts of Britain, *S. auriculatum* is rare here. There is occasionally a little *S. tenellum, S. magellanicum* or *S. papillosum* and, where the community forms the pool and wet hollow component of patterned mire surfaces, these species generally represent a clear transition to the drier surrounds. Quite often, however, this kind of vegetation occurs as more extensive lawns where the differentiation of these structural elements is ill-defined. Other bryophytes are scarce but there can be occasional patches of *Polytrichum commune* or *Aulacomnium palustre* or scattered shoots of leafy hepatics like *Gymnocolea inflata, Odontoschisma sphagni* or *Mylia anomola* in the *Sphagnum* carpet.

Vascular plants typically occur as scattered indivi-

duals of low total cover but *Eriophorum angustifolium* and *Erica tetralix* are both constant throughout, the former often extending into deeper pools, the latter more confined to drier areas. *Drosera rotundifolia* is very frequent and *Narthecium ossifragum* occurs occasionally. As in the *Sphagnum auriculatum* bog pools, *Rhynchospora alba* can be quite abundant around pool margins but it is very much more common in one particular sub-community here. A little more evenly distributed and especially distinctive of this kind of Rhynchosporion vegetation is the Continental Northern *Andromeda polifolia*, the national distribution of which largely coincides with the range of this community and the type of Erico-Sphagnion mire within which it forms the wetter element.

There can also be some sedges but, though these may have some measure of local abundance, they do not occur as consistent physiognomic dominants. *Carex limosa* is sometimes found and it also seems best to include here bog pool vegetation where *C. curta* and/or *C. magellanica* occur in species-poor carpets of *Sphagnum cuspidatum* and *S. recurvum*. *C. rostrata* is very occasional and some stands may be transitional to the *Carex rostrata-Sphagnum* community where this sedge is dominant in a poor-fen assemblage in which *Polytrichum commune, Agrostis canina* ssp. *canina* and *Carex nigra* are characteristic associates. Some of the now extinct English stations of *Scheuchzeria palustris* seem to have been in stands of the *Sphagnum cuspidatum/recurvum* community (Sledge 1949, Sinker 1962, Green & Pearson 1977).

Sub-communities

***Rhynchospora alba* sub-community:** *Sphagnetum* pools Rankin 1911*a*; *Sphagnetum* regeneration complex, Stages 1 & 2 Tansley 1939; *Sphagnum cuspidatum* pool community Bignal & Curtis 1981 *p.p. Sphagnum cuspidatum* is the typical dominant in the carpet here or, very locally, *S. pulchrum*, with no *S. recurvum*. Among the vascular plants, *Rhynchospora alba* and *Andromeda* join *Eriophorum angustifolium* and *Erica tetralix* as constants on the pool surrounds and, with frequent *Drosera rotundifolia*, there can be occasional *D. anglica* or *D. intermedia*. *Myrica gale* occurs in some stands.

Sphagnum recurvum sub-community: *Sphagnum* lawn Poore & Walker 1958, Sinker 1962, Green & Pearson 1968; *Sphagnum cuspidatum-Eriophorum angustifolium, Sphagnum recurvum-Vaccinium oxycoccos* & *Sphagnum recurvum-Erica tetralix* Noda, Normal Series Tallis 1973; *Sphagnum flexuosum* noda 16–19 Daniels 1978. *Sphagnum recurvum* is a constant companion here to *S. cuspidatum* and is often the more abundant of the two species. *Rhynchospora alba* is typically absent and *Andromeda* is much reduced but *Eriophorum angustifolium, Erica tetralix* and *Drosera rotundifolia* maintain their high frequency and *Vaccinium oxycoccos* appears as a good preferential. In some cases, these species form the bulk of the cover in quite well-defined bog pools but elsewhere this sub-community occurs as extensive lawns with a somewhat enriched flora. Then, *S. recurvum* can be an overwhelming dominant in the carpet with *S. cuspidatum* largely confined to very wet depressions and *S. papillosum* marking out low hummocks. *Polytrichum commune* and *Aulacomnium palustre* can occur sporadically in the slightly drier areas, together with some *Eriophorum vaginatum* and *Calluna vulgaris*. *Molinia caerulea* may be locally prominent. This kind of patterning is well seen in the *Sphagnum* lawns described by Green & Pearson (1968) and among the wetter noda in Tallis's (1973*a*) Normal Series from Cheshire basin mires.

Habitat
The *Sphagnum cuspidatum/recurvum* community is typically found in pools and lawns on very wet and base-poor, though not always highly oligotrophic, raw peats on ombrogenous and topogenous mires in the less oceanic parts of Britain. It has been much reduced by the widespread drainage and cutting of such mires but it can readily colonise shallow flooded workings and seems to have expanded its coverage in sites where there has been some enrichment of the waters.

This kind of Rhynchosporion vegetation is characteristic of areas where the annual precipitation is generally between 800 and 1200 mm (*Climatological Atlas* 1952) with around 140–180 wet days yr^{-1} (Ratcliffe 1968). Its range coincides closely with that of the *Erica-Sphagnum* mire and it typically forms the pool and wet hollow and lawn element in that community on lowland raised bogs, on locally raised areas within low-altitude blanket mires and in base-poor basin mires, replacing the *Sphagnum auriculatum* pools of the *Scirpus-Eriophorum* mire in moving away from the very wet far west of Britain. The

less oceanic pair of communities is well represented on raised mires in mature river valleys running into Cardigan Bay and along the Welsh borders, in the Solway estuaries and through Dumfries & Galloway and Strathclyde, where there are widespread transitions to low-altitude blanket mire (Ratcliffe 1977, Bignal & Curtis 1981). Locally raised areas of *Erica-Sphagnum* bog with *Sphagnum cuspidatum/recurvum* pools also characterise the Border Mires in Northumberland. Fragments of raised mire also persist in the Shropshire–Cheshire Plain but, in this area, the communities are best seen in numerous small basin mires some of which have a *schwingmoor* structure (e.g. Lind 1949, Poore & Walker 1959, Sinker 1962, Green & Pearson 1968, 1977, Tallis 1973*a*). The *Sphagnum cuspidatum/recurvum* community can also be found marking out soligenous areas within the *Narthecio-Sphagnetum* mire of valley bogs, as at Dersingham Bog in Norfolk.

The floristic differences between the two kinds of Rhynchosporion pool are not very great. Some important species, like *Sphagnum cuspidatum, Eriophorum angustifolium, Rhynchospora alba* and *Drosera rotundifolia* are well represented in both, and other less common species, like *Carex limosa* and the rarer sundews, may owe their preferential survival in one or the other community to accidents of local destruction of the habitat with drainage and reclamation. But the great scarcity here of *Sphagnum auriculatum* and the occurrence of *Andromeda* through most of the range of the *Sphagnum cuspidatum/recurvum* community seem to provide real distinctions. With the somewhat lower rainfall here, there may be a greater tendency for the *Sphagnum* carpet to dry out in late summer, and the pH range of the substrates is a little more towards the acid end, at pH3–4, than in the *Sphagnum auriculatum* pools, but whether such differences are critical is uncertain.

On active, patterned surfaces of more undisturbed raised and basin mires, the community is typically represented by the *Rhynchospora* sub-community. As in blanket mire systems, the sharpness of such patterning is very varied. Some mires, like Cors Goch Glan Teifi (Tregaron Bog) in Dyfed, have pronounced hummock/hollow systems with the *Rhynchospora* sub-community occupying clearly-defined shallow pools and the margins of deeper ones (Godwin & Conway 1939, Tansley 1939); in other sites, as at Cors Fochno (Borth Bog), also in Dyfed, and at Glasson Moss, on the Solway (Ratcliffe 1977), the undulations are of lower amplitude and stands less well delineated. These pool and hollow systems have not been subject to the kind of developmental studies pursued in mires with the *Sphagnum auriculatum* community but they often show similar patterning and could presumably arise in the same way.

The *Sphagnum recurvum* sub-community can occur in similar situations to the *Rhynchospora* sub-community but it is more localised in its distribution and seems to be consistently associated with some measure of enrichment of the mire waters. It can thus pick out bog pools in which there is some natural soligenous influence and can be found in seepage areas in the laggs of raised mires and in some lowland valley bogs. But it is especially characteristic of and locally extensive in the numerous small basin mires of the Shropshire–Cheshire Plain (Lind 1949, Poore & Walker 1959, Sinker 1962, Green & Pearson 1968, Tallis 1973*a*) where it seems to have spread secondarily over solid and *schwingmoor* peats as a result of eutrophication of the waters (Sinker 1962, Green & Pearson 1968, 1977, Tallis 1973*a*). An abundance of *S. recurvum* in these mires (and possibly the reduction in *R. alba*) has been clearly correlated with raised total cation content, especially of potassium, and it is possible that this enrichment originates from fertiliser run-off or drift from the agricultural land which closely hems in these sites (Tallis 1973*a*, Green & Pearson 1977). Such lawns often have a fairly even or only gently-undulating surface, but in some sites seem to have extended from existing well-defined hollows or to have subsequently developed hummocks (Tallis 1973*a*).

Throughout its range, the extent of the mire vegetation in which this community occurs has been greatly reduced by reclamation or deep peat extraction, such that small and much-modified fragments often now remain within predominantly agricultural landscapes (Sinker 1962, Tallis 1973*a*, Ratcliffe 1977, Bignal & Curtis 1981). Frequently, and particularly in small and isolated sites, the *Sphagnum cuspidatum/recurvum* community persists in impoverished form. However, this kind of vegetation seems readily able to colonise shallow peat cuttings and, in some places, as on Whixall and Wem Moss in Shropshire, such situations provide the bulk of the wetter element of the mire surface (Sinker 1962, Ratcliffe 1977).

Zonation and succession
The *Sphagnum cuspidatum/recurvum* community is typically found as the pool, wet hollow or lawn component in the *Erica-Sphagnum* mire, grading to drier flat and hummock vegetation with increasing height above the water-table. There is some evidence that such patterns may represent cyclical regeneration complexes but such succession may be extremely slow and, on drained mires, the community may remain as a fragment of the previously active surface or in artificial pools among run-down wet heath and woodland.

A typical zonation on an active raised mire runs from the *Rhynchospora* sub-community through the *Sphagnum-Andromeda* sub-community of the *Erica-Sphagnum* mire on the flats to the *Empetrum-Cladonia* sub-community on the hummock sides and tops. In the *Sphagnum* carpet, there is a switch from *S. cuspidatum* to *S. papillosum*, *S. tenellum* and *S. magellanicum* and thence to *S. capillifolium*. Among the vascular plants, *Rhynchospora* continues only a little way on to the flats, forming a fringe to the pools, but *Drosera rotundifolia* and, especially noticeable on these mires, *Andromeda* maintain their frequency in the transition to the *Sphagnum-Andromeda* sub-community of the mire. *Eriophorum angustifolium* and *Erica tetralix* also remain very common and are joined by *Eriophorum vaginatum*, *Calluna* and *Scirpus cespitosus* and these five species contribute the bulk of the vascular cover on the flats and hummocks. The clarity of the sequence and the relative contributions of the different elements vary considerably according to the degree of structural patterning on the mire surface but the general transition is well seen in the series of noda characterised by Bignal & Curtis (1981) from Strathclyde mires and illustrated diagrammatically in Godwin & Conway's (1939) classic study of Cors Goch Glan Teifi.

Essentially similar sequences can be found where the community occurs within the *Erica-Sphagnum* mire on solid and *Schwingmoor* peats in basins, notably in the Shropshire–Cheshire Plain, though here it is now most often represented by the *Sphagnum recurvum* sub-community which tends to be the dominant element in extensive lawns with rather poor internal structural differentiation. A range of zonations with different degrees of clarity is well described in Tallis's (1973*a*) survey of these sites and the general predominance of *S. recurvum* throughout such sequences is clearly seen in published profiles of Clarepool Moss (Sinker 1962) and Wybunbury Moss (Green & Pearson 1968).

At least some of the hummock/hollow complexes in which this community is found, on active mire surfaces developed under undisturbed conditions, may represent regeneration complexes of the kind described by Osvald (1923; see also Godwin & Conway 1939, Tansley 1939), though stratigraphical evidence suggests that the cyclical pattern of replacement is probably very slow, proceeding over centuries. In such situations, the *Rhynchospora* sub-community is probably the more natural kind of *Sphagnum cuspidatum/recurvum* bog pool with the *Sphagnum recurvum* sub-community playing a minor role and particularly associated with areas of soligenous influence. With increased disturbance and modest eutrophication of mire surfaces, however, the latter sub-community has become more frequent, not only in scattered flooded peat workings, but also more extensively where mires have been subject to increased input of enriched waters from surrounding land. In the basin mires of Shropshire and Cheshire, this seems to have been a fairly recent process attendant upon agricultural improvement of the surrounding land and, in some cases, the decay of old drainage systems on the mires (Sinker 1962, Green & Pearson 1968, 1977, Tallis

1973a). At some sites, renewed growth of the *Sphagnum* carpet under such conditions has caused a reversion from the heath and woodland of the once-drained surfaces.

Where drainage has proceeded without this kind of interruption, the *Sphagnum cuspidatum/recurvum* bog pools have been progressively reduced with conversion of the mire surface to Ericion tetralicis wet heath. Fragments may remain within tracts of the *Scirpus-Erica* or *Ericetum tetralicis* wet heaths, generally dominated by mixtures of *Molinia*, *Scirpus* or ericoids (e.g. Sinker 1962, Tallis 1973a) or among developing woodland, usually birch- or pine-dominated stands of the *Betula-Molinia* woodland, on the drying peats (e.g. Sinker 1962, Green & Pearson 1968, Ratcliffe 1977).

Distribution

The *Sphagnum cuspidatum/recurvum* bog pools occur within the *Erica-Sphagnum* mire and its degraded derivatives throughout its range from Wales, up through the Borders and south-west Scotland with some far-flung localities in north-east Scotland. The *Rhynchospora* sub-community is more widely distributed on active, undisturbed raised mires, the *Sphagnum recurvum* sub-community more restricted to soligenous areas occurring also with local abundance, in disturbed basin mires.

Affinities

As with the *Sphagnum auriculatum* community, this kind of vegetation has sometimes been included as the pool component within more broadly-defined Erico-Sphagnion mires (e.g. Rankin 1911a, Tansley 1939, Godwin & Conway 1939) and to have attracted attention as a distinct unit mainly in the '*Sphagnum* lawns' of the *Sphagnum recurvum* sub-community (Poore & Walker 1959, Sinker 1962, Green & Pearson 1968). Its affinities with the Rhynchosporion are very clear and essentially the same vegetation occurs throughout northern Europe, from Norway down through Germany and The Netherlands into northern France (Westhoff & den Held 1969, Dierssen 1982). The *Drosera anglica-Rhynchospora fusca* Gesellschaft recorded from Ireland by Braun-Blanquet & Tüxen (1952) appears to be at most a local variant of this kind of vegetation.

Floristic table M2

	a	b	2
Eriophorum angustifolium	V (3–9)	V (1–9)	V (1–9)
Sphagnum cuspidatum	V (3–10)	III (2–7)	III (2–10)
Erica tetralix	IV (3–4)	III (1–6)	IV (1–6)
Drosera rotundifolia	III (1–3)	III (2–3)	III (1–3)
Rhynchospora alba	V (1–8)		II (1–8)
Andromeda polifolia	IV (1–4)	II (1–2)	II (1–4)
Drosera anglica	II (2–4)		I (2–4)
Sphagnum pulchrum	II (1–10)		I (1–10)
Myrica gale	II (1–7)		I (1–7)
Menyanthes trifoliata	I (3)		I (3)
Drosera intermedia	I (2)		I (2)
Sphagnum magellanicum	I (3)		I (3)
Cephalozia lunulifolia	I (2)		I (2)
Cephalozia connivens	I (1)		I (1)
Cladonia impexa	I (2)		I (2)
Cladonia uncialis	I (1)		I (1)
Sphagnum recurvum		V (2–10)	III (2–10)
Vaccinium oxycoccos		V (1–5)	III (1–5)
Calluna vulgaris	I (1–2)	III (1–6)	III (1–6)
Eriophorum vaginatum	I (1–4)	III (4–5)	III (1–5)
Sphagnum papillosum	I (3–7)	III (1–10)	II (1–10)
Polytrichum commune		II (3–4)	I (3–4)
Aulacomnium palustre		II (1–3)	I (1–3)
Empetrum nigrum nigrum		I (5)	I (5)

	a	b	2
Agrostis canina canina		I (3)	I (3)
Carex magellanica		I (2–5)	I (2–5)
Carex rostrata		I (2–8)	I (2–8)
Carex curta		I (3)	I (3)
Polytrichum alpestre		I (1)	I (1)
Deschampsia flexuosa		I (1)	I (1)
Molinia caerulea		I (3–5)	I (3–5)
Sphagnum palustre		I (2)	I (2)
Sphagnum tenellum	II (2–4)	I (1)	I (1–4)
Gymnocolea inflata	II (4–6)	I (2)	I (2–6)
Odontoschisma sphagni	I (1–5)	II (1–2)	I (1–5)
Mylia anomola	I (1–4)	II (1–3)	I (1–3)
Narthecium ossifragum	I (3–4)	I (1–4)	I (1–4)
Number of samples	11	21	32
Number of species/sample	8 (3–15)	8 (4–12)	8 (3–15)
Herb height (cm)	20 (10–25)	24 (12–60)	22 (10–60)
Herb cover (%)	52 (4–100)	45 (4–95)	50 (4–100)
Bryophyte cover (%)	89 (35–100)	94 (70–100)	91 (35–100)
Altitude (m)	70 (10–430)	77 (45–440)	74 (10–440)
Soil pH	3.6 (3.3–4.4)	3.3 (3.1–3.7)	3.5 (3.1–4.4)

a *Rhynchospora alba* sub-community

b *Sphagnum recurvum* sub-community

2 *Sphagnum cuspidatum/recurvum* bog pools (total)

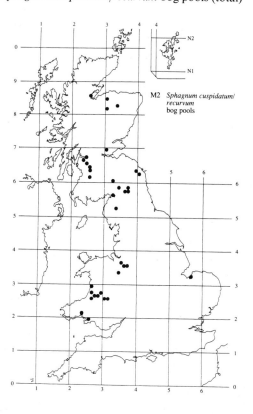

M2 *Sphagnum cuspidatum/recurvum* bog pools

M3
Eriophorum angustifolium bog pool community

Synonymy
Eriophoretum angustifolii   Lewis 1904, Lewis & Moss 1911, Watson 1932, Tansley 1939, all *p.p.*; *Eriophorum angustifolium* Gesellschaften   Dierssen 1982.

Constant species
Eriophorum angustifolium.

Physiognomy
Eriophorum angustifolium is a frequent and locally abundant plant in both kinds of *Sphagnum*-dominated bog pool community in Britain but here it is a consistent dominant in swards where other vascular species and Sphagna play a relatively minor role. The shoot density of the cotton-grass is very variable and can change through time as clumps spread and die from behind; and the sward height may be very short or up to half a metre or more, though usually the shoots reach about 30 cm. No other vascular plants attain anything more than very occasional frequency, though a variety of Oxycocco-Sphagnetea species occur sporadically and sometimes with local prominence: there can be scattered small tussocks of *Eriophorum vaginatum* or *Molinia caerulea* or sparse individuals of *Drosera rotundifolia*, *Erica tetralix* or *Empetrum nigrum* ssp. *nigrum*.

Bryophyte cover is likewise very variable and no species is constant but *Drepanocladus fluitans* is quite frequent and characteristic of this kind of vegetation, often growing submerged in flooded stands. There may also be some sparse shoots or small tufts of Sphagna, usually *S. cuspidatum*, but sometimes *S. recurvum* or *S. papillosum*, scattered *Polytrichum commune*, *Odontoschisma sphagni* or *Gymnocolea inflata* or, on drier patches, *Campylopus brevipilus*.

Habitat
The *Eriophorum angustifolium* community is typically found as small stands on barer exposures of acid raw peat soils in depressions, erosion channels or shallow peat cuttings on a wide range of mire types.

E. angustifolium is an important coloniser of bare peat, and, freed from the shading effect that its young shoots experience in thick *Sphagnum* carpets or among the vigorous growth of other vascular plants on mire flats and hummocks, is able to expand rapidly by vegetative growth of its rhizome system and become dominant on natural or artificial exposures (Phillips 1954). It shows a fair range of tolerance of soil moisture conditions and can dominate here in permanent shallow pools or in dried-up bottoms or in situations where there is some seasonal variation in water-level (e.g. Pearsall 1938, Phillips 1954): lowland stands in particular often show marked formation of ochre or iron concretions, suggestive of periods of surface oxidation. *E. angustifolium* is also tolerant of quite a wide range of pH, and the community can occur among poor-fen complexes, though the available stands are from uniformly acid substrates.

Suitable habitats for the development of the community can arise in various ways. It can sometimes be found in natural hollows on the surfaces of more or less intact mires but it is more common among erosion features like those found so extensively on the surface and margins of Pennine blanket mires where the peat has been worn down in gullies or redistributed (e.g. Bower 1961, Tallis 1985*b*). On lowland mires, it is often, though not invariably, associated with abandoned shallow peat-workings (e.g. Rose 1953).

Zonation and succession
The *Eriophorum angustifolium* community characteristically forms a minor component in the mosaics of vegetation on intact or modified surfaces of Erico-Sphagnion mires or Ericion tetralicis wet heaths. It may represent a seral stage in the redevelopment of active mire vegetation following disruption.

Exceptionally, the community can replace the *Sphagnum*-dominated bog pools in the wetter hollows of patterned Erico-Sphagnion mires, grading, with an increase in the *Sphagnum* carpet and an appearance of vascular associates, to flat and hummock vegetation. Much more commonly, it forms well-delineated stands on expanses of bare peat which may or may not be flooded, passing to unvegetated ground or open water or, rather sharply, to the mire or wet heath context around. Where it occurs on eroded blanket mire, for example, it occupies part or all of stretches of the flatter gullies between the haggs, giving way to the bare hagg sides or degraded vegetation of eroded peat surfaces. In shallow peat cuttings, the stands often conform to the sharp geometric shapes of the excavations, passing abruptly to the unaltered mire vegetation on the uncut surface around or to its degraded derivatives on drained and much-modified deposits.

The occurrence of scattered patches of *Sphagnum papillosum* and tussocks of *Eriophorum vaginatum* in some stands suggests that the *Eriophorum angustifolium* community may sometimes be a precursor to renewed establishment of Erico-Sphagnion mire but such successions have not been followed.

Distribution
The community is particularly associated with eroded blanket mire in the north-west of Britain, being a common feature in tracts of the *Calluna-Eriophorum* and *Eriophorum* mires, less so of the *Scirpus-Eriophorum* mire. It is also widespread but local in lowland Erico-Sphagnion mires and Ericion wet heaths.

Affinities
Early descriptive accounts of an *Eriophoretum angusti-folii* (Lewis 1904*a,b*, Lewis & Moss 1911, Watson 1932, Tansley 1939) were more broadly defined than the community characterised here, taking in vegetation of intact mire surfaces in which *E. angustifolium* continued to figure as a prominent component. In this scheme, the community is essentially a type of Rhynchosporion vegetation, closely related to *Sphagnum*-dominated bog pools. Similar communities have been described from Germany (Oberdorfer 1977), The Netherlands (West-hoff & den Held 1969), Norway and Iceland (Dierssen 1982).

Floristic table M3

Eriophorum angustifolium	V (3–9)
Drepanocladus fluitans	III (2–6)
Sphagnum cuspidatum	II (1–6)
Sphagnum papillosum	I (2–5)
Polytrichum commune	I (3–4)
Campylopus brevipilus	I (3–5)
Eriophorum vaginatum	I (5)
Erica tetralix	I (4)
Drosera rotundifolia	I (3)
Sphagnum recurvum	I (4)
Odontoschisma sphagni	I (2)
Gymnocolea inflata	I (2)
Molinia caerulea	I (4)
Empetrum nigrum nigrum	I (1)
Agrostis canina canina	I (3)
Number of samples	14
Number of species/sample	4 (1–10)
Herb height (cm)	27 (10–75)
Herb cover (%)	74 (10–100)
Bryophyte cover (%)	14 (1–35)
Altitude (m)	324 (5–770)
Soil pH	3.5 (3.3–4.5)

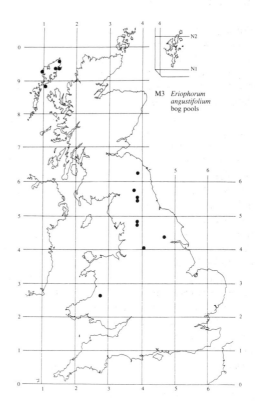

M3 *Eriophorum angustifolium* bog pools

M4
Carex rostrata-Sphagnum recurvum mire

Synonymy
Sphagneto-Juncetum effusi, Carex rostrata facies   Eddy et al. 1969; *Carex lasiocarpa-Menyanthes trifoliata* Association   Birks 1973 *p.p.*; *Carex rostrata-Carex limosa* nodum   Birks 1973 *p.p.*; *Sphagno-Caricetum curtae* Passarge 1964 *sensu* Birse 1980; *Caricetum chordorrhizae* Paul & Lutz 1941 *sensu* Birse 1980; *Caricetum rostratae* Dierssen 1982 *p.p.*; *Caricetum limosae* Dierssen 1982 *p.p.*; *Caricetum lasiocarpae* Dierssen 1982 *p.p.*

Constant species
Carex rostrata, Polytrichum commune, Sphagnum cuspidatum, S. recurvum.

Rare species
Carex chordorrhiza, Lysimachia thyrsiflora.

Physiognomy
The *Carex rostrata-Sphagnum recurvum* mire typically has a cover of sedges over a carpet of semi-aquatic Sphagna, with generally a very small contribution from vascular associates. *Carex rostrata* is the commonest sedge throughout, forming a usually rather open cover of shoots a few decimetres tall, but it can be accompanied by *C. curta, C. lasiocarpa, C. limosa* or *C. nigra* and the first two especially can be locally prominent, as in some stands in Birks' (1973) *Carex lasiocarpa-Menyanthes* Association and the *Sphagno-Caricetum curtae* (Passarge 1964) of Birse (1980). The kind of vegetation in which the very rare *C. chordorrhiza* occurs in Sutherland is also probably best considered as part of this community (Birse 1980, who placed it in a distinct *Caricetum chordorrhizae* Paul & Lutz 1941). Occasionally, this taller element is enriched by *Eriophorum angustifolium*, though this is typically less frequent here than in the Rhynchosporion bog pools, or by *Juncus effusus* or *J. acutiflorus*, though these do not show the regular pattern of dominance that they can exhibit in the *Carex echinata-Sphagnum* mire.

Beneath this cover, there is a generally extensive soft wet carpet of Sphagna, among which *Sphagnum recurvum* and *S. cuspidatum* are the most frequent and usually the most abundant. *S. auriculatum* (including var. *inundatum*) occurs quite commonly, too, and there is the suggestion among the available samples that it shows the same increase towards more oceanic regions that can be seen in the Rhynchosporion (e.g. Birks 1973, Birse 1980). *S. palustre* is occasional and there are sparse records for *S. subnitens* and *S. papillosum*, but more base-tolerant *S. squarrosum* and *S. teres* are characteristically rare, a good contrast with the *Carex rostrata-Sphagnum squarrosum* mire. Other bryophytes are few, but *Polytrichum commune* is very frequent and it can be abundant in the carpet as scattered patches. *Aulacomnium palustre* and *Calliergon stramineum* occur very sparsely, a further distinction between this community and the *Carex-Sphagnum squarrosum* and *Carex-Sphagnum warnstorfii* mires.

Scattered through this ground cover are individuals of a rather impoverished poor-fen herb flora. The commonest species throughout are *Agrostis canina* ssp. *canina* and *A. stolonifera*, which can be locally quite abundant as stoloniferous mats, *Molinia caerulea*, *Potentilla erecta, Galium palustre, Rumex acetosa, Viola palustris, Succisa pratensis* and *Stellaria alsine*, though usually only one or two of these are present in any particular stand. *Potentilla palustris, Menyanthes trifoliata* and *Equisetum fluviatile* occur occasionally and locally can be found together with the consistency typical of other *Carex rostrata*-dominated communities. In some stands there may be such Littorelletea species as *Potamogeton polygonifolius* and *Juncus bulbosus* (e.g. Birse's (1980) *Caricetum chordorrhizae*) and *Hypericum elodes* has been recorded from this vegetation.

Habitat
The *Carex rostrata-Sphagnum recurvum* mire is characteristic of pools and seepage areas on the raw peat soils of topogenous and soligenous mires where the waters

are fairly acid and only slightly enriched. The habitat is typically a little less oligotrophic than that of the Rhynchosporion communities, a feature reflected in the prominence of *Sphagnum recurvum* and *S. palustre* and the sparse occurrence of poor-fen herbs. However, though the community can occur in bog pools on the surface of basin (and sometimes raised) mires, it is more common in obviously soligenous areas, as in mire laggs and in the very wettest parts of water-tracks. Even here, however, the enrichment is only modest: the drainage is generally from acid soils or ombrogenous peat and the surface pH is typically around 4. No data are available on the calcium content of the irrigating waters but it is likely to be less than even the fairly small amounts characteristic of the *Carex rostrata-Sphagnum squarrosum* mire where *Carex nigra*, more base-tolerant Sphagna, *Aulacomnium palustre*, *Calliergon stramineum* and the poor-fen dicotyledons begin their ascendancy.

Zonation and succession

The community is typically found within or alongside Erico-Sphagnion mires, such as the *Erica-Sphagnum papillosum* raised or basin mire, or Ericion wet heaths, notably the *Scirpus-Erica* wet heath, either alone in fairly well-defined bog pools and soakways or in more complex mosaics with other poor-fen communities in more extensive soligenous areas. Most often, it marks out the wettest areas in the latter where water movement is quite diffuse, commonly passing to a fringe or water-track of the *Carex echinata-Sphagnum* mire on shal-lower and firmer peats where the throughput is more pronounced. In some places, Littorelletea communities occur alongside the *Carex rostrata-Sphagnum recurvum* mire in such complexes.

The place of the community in terrestrialising successions is obscure. As with the Rhynchosporion bog pools, this kind of vegetation may be very stable provided the high water-table and modest irrigation are maintained. The most common changes encountered now are probably those caused by drainage which produces a demise of the more aquatic Sphagna and perhaps a transition to the *Carex echinata-Sphagnum* mire with, where there is grazing, a spread of *Juncus* dominance.

Distribution

The community is of widespread but local occurrence throughout the north-west of Britain and probably remains as remnants in drained mire systems in the lowlands.

Affinities

The *Carex rostrata-Sphagnum recurvum* mire takes in stands dominated by various less common sedges (e.g. *C. lasiocarpa*: Birks 1973, Birse 1980, Dierssen 1982) but which are nevertheless similar in overall species composition and physiognomy. The community lies very close to the floristic boundary between the Caricion nigrae and the Rhynchosporion and on balance, it is probably best placed in the former as the most impoverished of the suite of poor fens.

Floristic table M4

Carex rostrata	V (3–9)	Rumex acetosa	I (1)
Sphagnum recurvum	IV (6–9)	Menyanthes trifoliata	I (5)
Polytrichum commune	IV (2–8)	Galium uliginosum	I (3)
Sphagnum cuspidatum	IV (3–10)	Deschampsia flexuosa	I (4)
Agrostis canina canina	III (2–8)	Aulacomnium palustre	I (3)
Carex nigra	II (2–5)	Viola palustris	I (2–3)
Carex curta	II (2–9)	Stellaria alsine	I (1)
Eriophorum angustifolium	II (2–4)	Succisa pratensis	I (4)
Juncus effusus	II (2–7)	Galium saxatile	I (4)
Sphagnum palustre	II (4–10)	Calliergon stramineum	I (4)
Agrostis stolonifera	II (2–4)	Lophocolea bidentata s.l.	I (3)
Sphagnum auriculatum	II (1–9)	Sphagnum subnitens	I (4)
Potentilla erecta	II (1–4)	Equisetum fluviatile	I (2)
Juncus acutiflorus	II (4–5)	Potamogeton polygonifolius	I (5)
Molinia caerulea	II (3–5)	Drepanocladus exannulatus	I (5)
Potentilla palustris	I (4–5)	Cardamine pratensis	I (2)
Hydrocotyle vulgaris	I (4)	Juncus bulbosus/kochii	I (2–3)
Galium palustre	I (3)	Carex echinata	I (2–3)
Sphagnum squarrosum	I (4–5)	Drosera rotundifolia	I (2)

Floristic table M4 *(cont.)*

Anthoxanthum odoratum	I (1–3)	Eriophorum vaginatum	I (4)

Hypnum jutlandicum	I (4)
Iris pseudacorus	I (6)
Epilobium anagallidifolium	I (2)

Number of samples	25
Number of species/sample	10 (2–15)

Montia fontana	I (1)
Saxifraga stellaris	I (1)
Carex limosa	I (4)
Lysimachia thyrsiflora	I (5)

Herb height (cm)	33 (12–50)
Herb cover (%)	67 (15–100)
Bryophyte height (mm)	40 (5–120)
Bryophyte cover (%)	88 (30–100)

Hypericum elodes	I (1)
Glyceria fluitans	I (1)
Sphagnum papillosum	I (7)

Altitude (m)	365 (107–730)
Soil pH	4.5 (3.5–5.8)

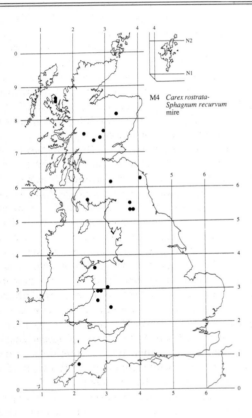

M4 *Carex rostrata-Sphagnum recurvum* mire

M5
Carex rostrata-Sphagnum squarrosum mire

Synonymy
Oxyphilous fen vegetation Pallis 1911, Godwin & Turner 1933, Tansley 1939 *p.p.*; *Carex rostrata-Acrocladium cordifolium/cuspidatum* sociation Spence 1964 *p.p.*; *Carex lasiocarpa-Myrica* & *Myrica-Carex lasiocarpa* sociations Spence 1964 *p.p.*; *Carex nigra-Acrocladium* sociation Spence 1964 *p.p.*; *Carex rostrata-Aulacomnium palustre* Association Birks 1973; Poor fen communities Proctor 1974 *p.p.*; General fen Adam *et al.* 1975 *p.p.*; *Carex curta-Carex rostrata* community Wheeler 1978; *Caricetum lasiocarpae* Koch 1926 *sensu* Birse 1980 *p.p.*

Constant species
Carex nigra, C. rostrata, Eriophorum angustifolium, Potentilla palustris, Succisa pratensis, Aulacomnium palustre, Sphagnum squarrosum.

Physiognomy
The *Carex rostrata-Sphagnum squarrosum* mire is a fairly heterogenous vegetation type characterised overall by the dominance of sedges with scattered poor-fen herbs over a patchy carpet of moderately base-tolerant Sphagna. The sedge cover is usually a few decimetres tall but quite variable in its abundance. The commonest species throughout are *Carex rostrata* and *C. nigra*, the former generally the more extensive and sometimes with a dense cover, the latter typically less abundant, though locally dominant, as in some stands of the *Carex nigra-Acrocladium* sociation described from Scottish lochs by Spence (1964) and in some of the Malham Tarn fens in North Yorkshire (Proctor 1974). *C. lasiocarpa* extends its occurrence among British mires into this kind of vegetation and it too can attain local prominence, so much so that stands which, on general grounds, clearly belong here have sometimes been included in distinct communities characterised by this sedge: e.g. the *Carex lasiocarpa* sociations of Spence (1964) and the *Caricetum lasiocarpae* (Koch 1926) of Birse (1980). *C. curta*, though generally more characteristic of oligotrophic

and base-poor Rhynchosporion pools, is occasionally found (Spence 1964, Birks 1973, Wheeler 1978 who diagnosed a *Carex curta-Carex rostrata* community from the Ant valley in Norfolk) but *C. limosa* is typically absent. Likewise, towards the other extreme, *C. diandra* does not penetrate into this community: it is a good diagnostic species for the closely-related *Carex rostrata-Calliergon* mire. Finally, there may be a little *C. echinata*, especially where the *Carex-Sphagnum squarrosum* mire extends on to firmer flushed peats.

Other vascular plants vary in their prominence and are often limited to scattered individuals but the most frequent species overall are *Potentilla palustris*, *Eriophorum angustifolium*, *Menyanthes trifoliata*, *Galium palustre* and such typical poor-fen herbs as *Succisa pratensis*, *Viola palustris*, *Ranunculus flammula*, *Epilobium palustre* and *Lychnis flos-cuculi* with, rather less commonly, *Equisetum palustre*, *Agrostis canina* ssp. *canina*, *Mentha aquatica*, *Myosotis scorpioides*, *Caltha palustris*, *Filipendula ulmaria* and *Cardamine pratensis*. Very much the same suite of species occurs in a variety of other mire types, notably the *Potentillo-Caricetum rostratae* and the *Carex-Calliergon* mire, in close spatial association with which the *Carex-Sphagnum squarrosum* community is often found. Transitional stands, where there is a strong continuity in this layer of the vegetation, can thus often be found.

The community also encompasses a number of particular lines of floristic variation in this vascular component which, with further sampling, might form the basis of distinct sub-communities but which, in the available data, present the appearance of a continuum within the overall definition. In some stands, for example, the representation of all the herbaceous associates is very poor, such that, with a cover of sedges, there may be just a few plants of *Potentilla palustris* and *Menyanthes*, when the vegetation takes on much of the character of the *Carex rostrata-Sphagnum recurvum* mire or even a *Caricetum rostratae* swamp: *Carex curta* seems to be particularly associated with this kind of transitional vegetation (e.g. Birse 1980). In other cases, *Juncus effusus* which is quite frequent here, though usually of low cover, can become prominent, suggestive of some kinds of rush-dominated *Carex echinata-Sphagnum* mire. Then, there are stands in which *Molinia caerulea* and/or *Myrica gale* have local abundance and it is in such situations that *Carex lasiocarpa* seems to be best represented (e.g. Spence 1964, Birse 1980). Finally, on rare occasions, *Phragmites australis* can be patchily abundant with a prominent contingent of poor-fen herbs in vegetation that resembles reed-dominated *Potentillo-Caricetum*.

Throughout the community, even in these transitional stands, the bryophyte carpet helps define the *Carex-Sphagnum squarrosum* mire against the closely-related vegetation types. Sphagna are usually at least patchily prominent but, though the more strictly aquatic and oligotrophic Rhynchosporion species, *S. cuspidatum* and *S. auriculatum*, can occasionally be found here, they are eclipsed by the more broadly-tolerant *S. recurvum*, the more mesotrophic *S. palustre* and, especially distinctive, the mildly base-tolerant *S. squarrosum* and *S. teres*. There can also be a little *S. subnitens* or *S. fimbriatum*. The species-balance in this assemblage, which forms discrete small domes over the ground or sometimes an extensive carpet, is thus quite different from that in the *Carex rostrata-Sphagnum recurvum* mire. Towards the other extreme, *S. contortum*, the most base-tolerant of the Sphagna, is typically rare here: this species has its major locus in the more calcicolous *Carex-Calliergon* mire. And, though *S. warnstorfii* is sometimes recorded where it descends to lower altitudes in north-west Scotland (as in Skye samples: Birks 1973), this species is much more diagnostic of the montane counterpart of the *Carex-Sphagnum squarrosum* mire, the *Carex-Sphagnum warnstorfii* mire.

Some other bryophytes occur commonly and together help confirm the distinctive character of this element of the vegetation. *Aulacomnium palustre*, though it has a fairly wide distribution through the Caricion nigrae mires, is a good diagnostic species of the particular range of variation encompassed by this community and the *Carex-Sphagnum warnstorfii* mire. *Calliergon stramineum* has a rather similar amplitude and here it is as common as, and much more characteristic than, the broadly-tolerant *C. cuspidatum*. Conditions may be sufficiently base-rich here for *C. trifarium* and *C. sarmentosum* though these are essentially species of montane flushes in Britain. More noticeable is the absence of *C. giganteum* and *C. cordifolium* which are such a distinctive feature of some kinds of *Carex-Calliergon* mire and the *Potentillo-Caricetum*. Other obvious absentees along this boundary of the *Carex-Sphagnum squarrosum* mire are larger Mniaceae, like *Plagiomnium rostratum* and *Rhizomnium pseudopunctatum*, and the 'brown mosses', *Campylium stellatum*, *Cratoneuron commutatum*, *Scorpidium scorpioides* and *Drepanocladus revolvens*, so characteristic of more calcicolous sedge-rich mires. Towards the transition to more oligotrophic and base-poor mires, it is noticeable that *Polytrichum commune* is of reduced frequency here.

Habitat

The *Carex-Sphagnum squarrosum* mire is typically found as a floating raft or on soft, spongy peats in topogenous mires and soligenous sites with midly acid, only moderately calcareous and rather nutrient-poor waters. In normal circumstances, the ground surface is kept very moist, though it can become periodically dry and this may permit access to grazing animals and damage by burning.

Among the range of vegetation types in which *Carex rostrata* plays a prominent role, differences in base-status, cation and nutrient content of the substrates and waters are probably of major importance in influencing which particular community develops once the limit of very swampy conditions has been passed. As yet, we do not know very much about how these variables work but what little information there is suggests that the controlling balances might be quite delicate, that the different factors operate in complex interactions and perhaps affect the various structural elements of the vegetation, the bryophyte carpet, the sedge canopy and the herbaceous associates, in only a crudely congruent fashion. It is thus very difficult to provide precise and absolute limits to the kind of mire environment in which the *Carex-Sphagnum squarrosum* community might be found: all that can be done is to indicate its relative position along continuous scales of variation in these factors with some crude thresholds. In general terms, its environmental preferences lie between those of the poor-fen *Carex rostrata-Sphagnum recurvum* mire on the one hand and the rich fen *Carex-Calliergon* mire and *Potentillo-Caricetum* on the other. In the former, where *C. rostrata* is dominant over a ground that has much in common with Rhynchosporion vegetation, the base-status is around pH4; in the latter two communities, where more calcicolous assemblages are characteristic, it is often above pH6, with the *Carex-Sphagnum squarrosum* mire spanning the range between. In this community, dissolved calcium levels are usually of the order $5-15$ mg l^{-1} (Proctor 1974) whereas, in the *C. rostrata* rich fens, they are generally higher, sometimes much higher, attaining 50 mg l^{-1} or more (Spence 1964, Proctor 1974, Wheeler 1983). In terms of trophic levels, the substrates and waters of the *Carex-Sphagnum squarrosum* mire are probably richer than those of the *Carex rostrata-Sphagnum recurvum* mire, though perhaps not substantially poorer than those of the *Carex-Calliergon* mire or some kinds of *Potentillo-Caricetum* (Proctor 1974).

As might be expected, the *Carex-Sphagnum squarrosum* mire is most commonly found in mire systems where the waters are of generally intermediate quality throughout or in situations where one or other of the extremes in environmental variation is locally ameliorated. Two particular habitat types are especially characteristic. Of the former kind are open water transitions or flood-plain mires around lakes or pools fed by only moderately oligotrophic or mesotrophic waters and where the base-status and cation content are fairly low. Such sites are particularly associated with catchments where slates, shales and some kinds of schist predominate and include many of the Scottish localities of the community described by Spence (1964), Birks (1973) and Birse (1980), where the *Carex-Sphagnum squarrosum* mire forms a fairly widespread component of swamp and fen sequences, often developed as floating mats. In some situations of this kind, fluctuations in the water-table may mediate supplies of bases and nutrients to the fen raft and it is possible that the community marks out areas which have become a little elevated above the limit of frequent inundation: this seems to be the case at Crag Lough in Northumberland where Lock & Rodwell (1981) described distinct discs of the *Carex-Sphagnum squarrosum* mire within a tract of the *Potentillo-Caricetum*. Similar oligotrophic nuclei have been reported from the rich-fen systems of lowland Britain (e.g. Pallis 1911, Godwin & Turner 1933, Tansley 1939) though this particular community is rather rare within *Phragmites*-dominated fens.

The other characteristic situation where the *Carex-Sphagnum squarrosum* mire is found is where soligenous influx ameliorates a more extreme mire environment, usually around raised or basin mires but sometimes along the margins of flood-plain mires. By and large, it is seepage of base-rich and calcareous waters which modifies a predominantly oligotrophic and base-poor mire system, a pattern very well seen at Malham Tarn, where a series of springs emerges from surrounding Carboniferous Limestone to feed a soligenous fen complex in the lagg of the raised Tarn Moss (Sinker 1960, Proctor 1974, Adam *et al.* 1975). In this and other analogous sites, the community forms part of the complex transition zone between ombrogenous bog and calcicolous rich-fen, the proportions of the different vegetation types depending on distance from water source and the pattern of drainage and flooding. Much less common is the situation where base-poor waters seep into more calcareous mires, but Wheeler (1978) reported the community from the Ant valley in Norfolk where there was some soligenous influence from nutrient-poor, decalcified brick earth along the margin of the flood-plain.

In both these kinds of habitat, the water-table is typically very close to the surface for much of the year, with perhaps some modest flooding of the bryophyte mat when heavy rain works its way through into basins (e.g Lock & Rodwell 1981) or, more frequently, in winter. Quite commonly, the vegetation occurs as a floating mat bound together below by the robust rhizomes of *Carex rostrata*, *Potentilla palustris* and *Menyanthes*, which can grow very vigorously in the aquatic environment, and then the raft may rise and fall with the fluctuations in water-level, a feature which may be of some significance where the *Carex-Sphagnum squarrosum* mire develops in more mesotrophic situations. Even where the substrates are continuous, the peat is often very soft, sometimes sloppy, which gives the vegetation a measure of protection against the trampling and grazing effects of larger herbivores where stands occur in pastoral landscapes. Where the community runs on to firmer peats around the margins of lakes and basins, it tends to pass to the *Carex echinata-Sphagnum* mire and, where herbivores have access to such transitions, their grazing may favour the spread of *Juncus effusus* throughout the transition.

Zonation and succession

The *Carex-Sphagnum squarrosum* mire is characteristically found in zonations and mosaics related to variation in water-level, base-status, cation and nutrient content of the waters and substrates. Such patterns can reflect a process of succession but many seem to be static at the present time.

The simplest sequences are seen in open-water transitions like those described from Scottish lakes by Spence (1964) where the community typically occurs behind a front of the *Caricetum rostratae* (sometimes with *C. lasiocarpa* locally prominent) or the *Equisetetum fluviatile*, with the *Phragmitetum* sometimes represented as a bank of reed beyond the other swamps or on the transition to the mineral soils around the basin. At scattered localities, the *Caricetum vesicariae* is part of such zonations and the whole sequence may be terminated to the landward side by the *Phalaridetum*. Sometimes, the *Carex-Sphagnum squarrosum* mire is an extensive component of the middle portion of these patterns, in other cases it is but a local patchy development within tracts of the *Potentillo-Caricetum*, perhaps representing more oligotrophic nuclei.

Much more complex mosaics are characteristic of those sites where the *Carex-Sphagnum squarrosum* mire occurs as part of soligenous sequences. Here, below springs and around tortuous seepage lines and streams, it can form part of a perplexing mixture of poor and rich fens, notably with the *Potentillo-Caricetum* and the *Carex-Calliergon* mires, with the *Pinguiculo-Caricetum* on more solid base-rich peats and peaty gleys around, and Filipendulion vegetation on mesotrophic alluvium, the whole closely hemmed in by Ericion tetralicis and Erico-Sphagnion vegetation on the ombrogenous deposits above the lagg. As Proctor (1974) showed at Malham Tarn, such patterns have a clear underlying structure in relation to the environmental variables discussed earlier, even though its expression at particular sites can present a superficial picture of great confusion.

Further studies of this kind are needed to give a precise indication of what conditions control the development of this or that kind of poor or rich fen and how the general process of terrestrialisation interacts with the differentiation of flushed and leached areas within mire complexes. We may then be able to see where the *Carex-Sphagnum squarrosum* community fits in successional sequences. In the kinds of mires in which it occurs, two lines of seral development seem possible. In more base-rich complexes, the community is often found in close juxtaposition with the *Salix pentandra-Carex rostrata* woodland, the field layer of which has much in common with this kind of mire (Proctor 1974, Adam *et al.* 1975, Lock & Rodwell 1981) and which, at Crag Lough, seems certainly to have developed from a mosaic of the *Carex-Sphagnum squarrosum* mire and the *Potentillo-Caricetum* with release from grazing (Lock & Rodwell 1981). Where the balance is tipped towards the more base-poor and oligotrophic side, it is possible that some kinds of *Betula-Molinia* woodland (the *Sphagnum* or *Juncus effusus* sub-communities, for example) are a natural successor. The *Carex-Sphagnum squarrosum* mire may itself play an important part in the differentiation of more oligotrophic nuclei that are preferentially invaded by *Betula pubescens*, though this line of development is probably greatly speeded where mires are drained but not subject to too much surface disturbance and eutrophication of the peats. Such successions can be seen locally in lowland flood-plain mires, where different mire and woodland types are involved (Wheeler 1978, 1980c) and seem to have been extensive in the past

(e.g. Walker 1970), though, at the present time, many zonations in which the *Carex-Sphagnum squarrosum* mire is found look to be more or less stable.

Distribution
The community has a widespread but fairly local distribution mainly in the north-western parts of Britain, probably a reflection more of the survival of suitable sites than of any strict association with the sub-montane environment. It was probably once much more widespread in the lowland south and east where relic stands are likely to occur more widely than the map suggests.

Affinities
The varied and transitional floristic character of the *Carex-Sphagnum squarrosum* mire means that it has sometimes been considered part of more broadly-defined communities (as in some of Spence's (1964) sociations and the 'general fen' of Adam *et al.* (1975)) or placed in fairly heterogeneous vegetation types based on the dominance of particular species, e.g. *Carex lasiocarpa* (Spence 1964, Birse 1980). It clearly belongs among the poor fens of the Caricion nigrae, though its affinities are diverse and multi-directional. First and foremost, it occupies a median position in the sequence of communities ranging from the oligotrophic and calcifuge vegetation of the *Carex-Sphagnum recurvum* mire to the calcicolous rich fens of the Caricion davallianae, like the *Carex-Calliergon* mire. Second, within this middle range, it is closely related to the *Carex-Sphagnum warnstorfii* mire, which can be seen essentially as its montane replacement. Third, it grades to the *Carex echinata-Sphagnum* mire, the typical Caricion nigrae community of shallower, firmer peats with pronounced lateral water movement, and also to the fens and swamps of the Phragmitetea where the *Potentillo-Caricetum* and the *Caricetum rostratae* continue the floristic line into open-water transitions. It seems to have no close equivalent among the mire types of Dierssen (1982).

Floristic table M5

Carex rostrata	V (4–8)	Sphagnum palustre	III (1–7)
Potentilla palustris	V (1–9)	Menyanthes trifoliata	III (1–6)
Aulacomnium palustre	IV (1–6)	Calliergon cuspidatum	III (2–4)
Carex nigra	IV (1–5)	Ranunculus flammula	III (1–4)
Eriophorum angustifolium	IV (1–4)	Epilobium palustre	III (2–4)
Succisa pratensis	IV (1–5)	Calliergon stramineum	III (2–5)
Sphagnum squarrosum	IV (7–9)	Juncus effusus	III (1–6)
		Phragmites australis	II (2–5)
Galium palustre	III (2–4)	Lychnis flos-cuculi	II (1–3)
Sphagnum recurvum	III (1–6)	Polytrichum commune	II (1–5)
Viola palustris	III (1–4)		

Sphagnum teres	II (2–9)	*Carex curta*	I (2)
Carex echinata	II (2–4)	*Hydrocotyle vulgaris*	I (4–8)
Equisetum palustre	II (1–5)	*Juncus acutiflorus*	I (1–3)
Molinia caerulea	II (1–4)	*Holcus lanatus*	I (3–4)
Agrostis canina canina	II (2–6)	*Chiloscyphus polyanthos*	I (1–2)
Mentha aquatica	II (1–4)	*Stellaria alsine*	I (1–2)
Myosotis scorpioides	II (1–2)	*Plagiomnium rostratum*	I (1–2)
Rumex acetosa	II (1–3)	*Luzula multiflora*	I (1–3)
Deschampsia flexuosa	II (1)	*Juncus articulatus*	I (1–3)
Caltha palustris	II (3–6)	*Rhytidiadelphus squarrosus*	I (2–3)
Filipendula ulmaria	II (2–5)		
Sphagnum subnitens	II (4–5)	Number of samples	22
Cardamine pratensis	II (1–3)	Number of species/sample	17 (10–26)
Sphagnum fimbriatum	I (5)		
Equisetum fluviatile	I (1–2)	Herb height (cm)	47 (20–80)
Rhizomnium pseudopunctatum	I (2–3)	Herb cover (%)	63 (40–80)
Juncus bulbosus/kochii	I (2)	Bryophyte height (mm)	55 (5–200)
Cirsium palustre	I (1–3)	Bryophyte cover (%)	77 (40–100)
Bryum pseudotriquetrum	I (1–2)		
Sphagnum contortum	I (2–3)	Altitude (m)	327 (61–823)
		Soil pH	4.8 (4.4–5.2)

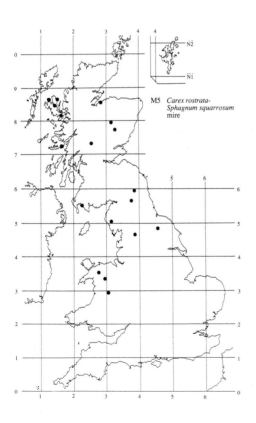

M5 *Carex rostrata-Sphagnum squarrosum* mire

M6
Carex echinata-Sphagnum recurvum/auriculatum mire

Synonymy

Juncus acutiflorus flush bog   Ratcliffe 1959*a*; *Juncus effusus* flush bog   Ratcliffe 1959*a*; *Sphagneto-Juncetum effusi*   McVean & Ratcliffe 1962, Eddy *et al.* 1969 *p.p.*, Birks 1973, Prentice & Prentice 1975; *Sphagneto-Caricetum sub-alpinum*   McVean & Ratcliffe 1962, Prentice & Prentice 1975, Evans *et al.* 1977; *Juncus effusus-Sphagnum* mire   Ratcliffe 1964, Ferreira 1978; Sub-alpine *Carex-Sphagnum* mire   Ratcliffe 1964, Ferreira 1978; *Juncus effusus-Sphagnum recurvum* sociation   Edgell 1969; *Carex-Sphagnum recurvum* nodum   Birks 1973; *Violo-Epilobietum sphagnetosum recurvae*   Jones 1973 *p.p.*; *Carex-Sphagnum-Polytrichum* nodum   Huntley 1979; *Caricetum echinato-paniceae* (Birse & Robertson 1976) Birse 1980 *p.p.*, *Juncus effusus-Sphagnum recurvum* Community (Birse & Robertson 1976) Birse 1980; *Sphagnum recurvum-Juncus effusus* mire   Ratcliffe & Hattey 1982, *Sphagnum-Carex nigra* mire   Ratcliffe & Hattey 1982; *Caricetum nigrae*   Dierssen 1982; *Juncus effusus* Gesellschaft   Dierssen 1982.

Constant species

Carex echinata, Polytrichum commune, Sphagnum auriculatum/recurvum (*Agrostis canina* ssp. *canina, Molinia caerulea, Potentilla erecta, Viola palustris*).

Physiognomy

The *Carex echinata-Sphagnum recurvum/auriculatum* mire has a quite distinct general character but shows a wide range of variation in its composition, more particularly in the proportional representation of its different structural components, something which can have a marked effect on the gross physiognomy of the vegetation. Essentially, this is a poor-fen in which either small sedges or rushes dominate over a carpet of more oligotrophic and base-intolerant Sphagna and it is in these two elements that most of the differences in the community can be seen: variation among the former helps characterise the four sub-communities; differences in the latter allow variants to be distinguished.

The constants of the *Carex echinata-Sphagnum* mire are very few. Among the vascular plants, only *Carex echinata* itself attains uniformly high frequency throughout but, of other sedges that can be found here, *C. nigra* and *C. panicea* are quite common and *C. demissa* occasional; and, although this combination can be found in other Caricion nigrae mires, taken together with the other floristic features of this community, it is quite diagnostic. Two further negative characteristics of the sedge component serve to sharpen up the definition of the vegetation. First, more calcicolous species, such as *C. dioica, C. pulicaris, C. lepidocarpa* and *C. flacca* are either very infrequent or totally absent; and second, though species like *C. rostrata* and *C. curta* can occur in more swampy stands, they are of no more than local significance. The rarity of the former sedges helps to separate the community from the Caricion davallianae rich fens; the infrequency of the latter marks off the vegetation from communities like the *Carex rostrata-Sphagnum* mire which comes close to the Rhynchosporion and in which the *Carex rostrata* facies of Eddy *et al.* (1969) is best placed.

The most frequent vascular associates of the sedges are grasses and poor-fen dicotyledons. Among the former, *Agrostis canina* ssp. *canina* and *Molinia caerulea* are the commonest species, though both show considerable variation in frequency from one sub-community to another and in abundance from stand to stand: when the latter is prominent, it may be difficult to separate this community from more species-poor kinds of Junco-Molinion vegetation. *Anthoxanthum odoratum* is also quite frequent throughout and in some kinds of *Carex echinata-Sphagnum* mire, *Nardus stricta* and *Festuca ovina* are well represented, together with *Juncus squarrosus*, when the community comes close in its floristics to flushed Nardetalia grasslands. The most common of the dicotyledons are *Viola palustris* and *Potentilla erecta* and each can be locally abundant, though generally both occur as scattered individuals. More occasionally, there can be some *Galium saxatile, G. palustre, Cirsium palustre, Epilobium palustre, Succisa pratensis, Ranunculus flammula* or *Cardamine pratensis* usually represented in fragmentary assortments rather than as a consistently rich suite. Sometimes, species such as *Narthecium ossifragum, Drosera rotundifolia* and *Erica tetralix* can be found, particularly where the community marks out flushed areas within Oxycocco-Sphagnetea mires and wet heaths.

In two of the sub-communities, mixtures of these vascular associates form a background to dominance of sedges or mixtures of sedges and grasses or, occasionally, of *Eriophorum angustifolium*, which is rather patchily represented in the community, but much more common among these kinds of *Carex echinata-Sphagnum* mire. Such vegetation types have usually been characterised as some form of *Sphagneto-Caricetum sub-alpinum* (McVean & Ratcliffe 1962, Ratcliffe 1964, Prentice & Prentice 1975, Evans *et al.* 1977) or an equivalent (Birks 1973, Birse & Robertson 1976, Birse 1980, Ratcliffe & Hattey 1982). But it also makes good sense to include in this community poor fens which have essentially the same floristic composition but which are dominated by the rushes, *Juncus effusus* or *J. acutiflorus* (usually not both together), vegetation which, in earlier schemes, has usually been designated as a distinct *Sphagneto-Juncetum* (McVean & Ratcliffe 1962, Ratcliffe 1964, Eddy *et al.* 1969, Birks 1973, Prentice & Prentice 1975) or an equivalent (Ratcliffe 1959*a*, Edgell 1969, Birse & Robertson 1976, Birse 1980, Ratcliffe & Hattey 1982). These rushes are sparsely represented in sedge-dominated kinds of *Carex echinata-Sphagnum* mire but their predominance helps characterise two further sub-communities.

Apart from the continuation throughout all these sub-communities of the vascular species already described, this kind of mire is also distinguished by the prominence of a ground carpet of Sphagna, the luxuriance of which provides a generally good separation from closely-related Nardetalia and Molinietalia communities. The most frequent species throughout are *Sphagnum recurvum* and *S. auriculatum* and generally one or the other of these predominates, this pattern of replacement enabling variants to be recognised, with varying degrees of clarity, in each of the four sub-communities. *S. subnitens* and *S. papillosum* which are occasional, and *S. capillifolium*, which is much scarcer, all tend to follow *S. auriculatum*; *S. palustre* is more evenly distributed throughout. Montane species, such as *S. russowii* and *S. lindbergii*, and more base-tolerant ones, like *S. squarrosum, S. teres, S. warnstorfii* and *S. contortum*, are very scarce or absent, features which help separate the community from related mires of higher altitudes or of more calcareous flushes.

Other bryophytes are few in number but *Polytrichum commune* is very frequent and often abundant, particu-

larly where the community extends on to gleyed mineral soils with a humic top rather than deeper peats. *Rhytidiadelphus squarrosus* is occasional and *Calliergon stramineum* and *Aulacomnium palustre* occur patchily throughout. In contrast to similar Molinietalia communities, species such as *Calliergon cuspidatum* and *Plagiomnium undulatum* are conspicuously rare.

Sub-communities

***Carex echinata* sub-community:** *Sphagneto-Caricetum sub-alpinum* McVean & Ratcliffe 1962 *p.p.*, Evans *et al.* 1977; *Carex-Sphagnum recurvum* nodum Birks 1973; *Violo-Epilobietum sphagnetosum recurvae, Sphagnum palustre* variant, typical sub-variant Jones 1973 *p.p.*; *Caricetum echinato-paniceae typicum*, typical variant (Birse & Robertson 1976) Birse 1980; *Sphagnum-Carex nigra* mire Ratcliffe & Hattey 1982 *p.p.* The vegetation here is usually dominated by mixtures of sedges in which *C. echinata* is generally the leading member with *C. nigra* and *C. panicea* rather less common, though locally abundant, and *C. demissa* occasional. *Eriophorum angustifolium* is quite frequent and it, too, may dominate but rushes are typically scarce and of low cover: *Juncus effusus* occurs occasionally but generally in small isolated tussocks. Apart from *Molinia* and *Agrostis canina* ssp. *canina*, both of which are very common and sometimes abundant, grasses are rather scarce: *Nardus* figures in some stands but does not occur with the consistency characteristic of the next sub-community.

The *Sphagnum* carpet is typically extensive and luxuriant. *S. palustre* occurs fairly often throughout but *S. recurvum* and *S. auriculatum* show a well-marked pattern of replacement, the latter tending to become more prominent in the more Atlantic climate of the far west of Britain, where it is sometimes accompanied by *S. subnitens* and *S. papillosum*. In this variant, too, there are some distinctive features of the vascular element of the vegetation because, though *Viola palustris* and *Potentilla erecta* remain frequent throughout, *Drosera rotundifolia, Narthecium ossifragum, Erica tetralix, Juncus bulbosus/kochii* and *Menyanthes trifoliata* show some preference for the *S. auriculatum* type of flush. Such stands come close to Oxycocco-Sphagnetea mires in their composition and are often found in contact with them.

***Carex nigra-Nardus stricta* sub-community:** *Sphagneto-Caricetum sub-alpinum* McVean & Ratcliffe 1962 *p.p.*, Prentice & Prentice 1975; Sub-alpine *Carex-Sphagnum* mire Ratcliffe 1964, Ferreira 1978. The general predominance, among the vascular element, of cyperaceous plants continues in this sub-community: indeed, *C. nigra* and *C. panicea* rival *C. echinata* in their high frequency and mixtures of these species often account for the bulk of the cover. *Eriophorum angustifolium* is very common, too, though usually not abundant, and *Luzula multiflora* can sometimes be found in small amounts. More specifically diagnostic, however, is the high frequency of *Nardus stricta* and the occasional occurrence of *Festuca ovina* which, with frequent *Anthoxanthum odoratum*, often give the vegetation the appearance of a flushed grassland. *Juncus squarrosus* is a further good preferential but *J. effusus* is very scarce and *J. acutiflorus* absent.

The *Sphagnum* cover is variable in its extent and, as in the previous sub-community, in its composition, features which again enable variants to be distinguished. In some stands, *S. recurvum* and *S. palustre* are the commonest species in a rather patchy carpet. In such vegetation, poor-fen herbs such as *Ranunculus flammula*, *Epilobium palustre* and *Cirsium palustre* tend to be a little more frequent and the sedge component can occasionally be enriched by *C. demissa*, *C. dioica* or *C. pulicaris*. *Rhytidiadelphus squarrosus* is very frequent and there can be some *Drepanocladus exannulatus* and *Calliergon cuspidatum*. It is here that the community most closely approaches the Nardetalia in its composition. By contrast, other stands preserve the general features of the sub-community but have a greater abundance of Sphagna, among which *S. auriculatum*, *S. subnitens* and *S. papillosum* are preferential, and quite often a small amount of *Molinia* among the other grasses.

***Juncus effusus* sub-community:** *Juncus effusus* flush bog Ratcliffe 1959; *Sphagneto-Juncetum effusi* McVean & Ratcliffe 1962, Eddy *et al.* 1969 *p.p.*, Birks 1973, Prentice & Prentice 1975; *Juncus effusus-Sphagnum recurvum* sociation Edgell 1969; *Violo-Epilobietum sphagnetosum recurvae*, *Sphagnum palustre* variant, typical sub-variant Jones 1973 *p.p.*; *Juncus effusus-Sphagnum* mire Ferreira 1978; *Juncus effusus-Sphagnum recurvum* Community (Birse & Robertson 1976) Birse 1980; *Sphagnum recurvum-Juncus effusus* mire Ratcliffe & Hattey 1982; *Sphagnum-Carex nigra* mire Ratcliffe & Hattey 1982 *p.p.*; *Juncus effusus* Gesellschaft Dierssen 1982. Sedges are less frequent and varied and much less abundant in this community and the vegetation is dominated physiognomically by *Juncus effusus*, occurring as prominent tussocks, sometimes of high total cover. *J. acutiflorus* can also occur occasionally though the two rushes are generally mutually exclusive in this community. Consistently frequent vascular associates are few and sometimes this vegetation is very impoverished but there is quite often some *Agrostis canina* ssp. *canina* and scattered plants of *Potentilla erecta* and, rather diagnostic here, *Galium saxatile*. *Carex echinata*, *Molinia caerulea* and *Viola palustris* are also fairly common, *Eriophorum angustifolium* and *C.*

nigra occasional and *Rumex acetosa* and *Lotus uliginosus* can occur as weak preferentials.

The *Sphagnum* carpet is generally extensive and luxuriant and almost always *S. recurvum* is the dominant species, occasionally with some *S. palustre*, but stands can be found where these are supplemented or more usually replaced by *S. auriculatum*, often with *S. subnitens* and occasionally *S. papillosum*. *Polytrichum commune* remains very frequent, and, on less deep peats and on peaty gleys, it can be very abundant.

***Juncus acutiflorus* sub-community:** *Juncus acutiflorus* flush bog Ratcliffe 1959a; *Caricetum echinato-paniceae*, *Juncus acutiflorus* Subassociation, typical variant (Birse & Robertson 1976) Birse 1980. The general physiognomic character of rush-dominance over an extensive *Sphagnum* carpet is maintained here, but *J. acutiflorus* almost totally replaces *J. effusus* as the leading species and, among the vascular associates, *Molinia* becomes more consistently frequent. Where this grass is abundant, this vegetation comes close in its floristics to the more impoverished of the Junco-Molinion communities, with which it also shares frequent records for such species as *Potentilla erecta*, *Viola palustris*, *Agrostis canina* ssp. *canina*, *Anthoxanthum odoratum*, *Carex echinata* and, less commonly, *C. nigra* and *C. panicea*. What helps separate the vegetation types is that here, in common with other kinds of *Carex echinata-Sphagnum* mire, the vascular element is generally not much richer than this sparse assemblage of poor-fen species. *Succisa pratensis* occurs fairly often but *Cirsium palustre*, *Angelica sylvestris* and *Eupatorium cannabinum* are either very sparse or absent. And, though *Molinia* can be locally prominent, it does not consistently share dominance with *J. acutiflorus*.

Also, here, the Sphagna are more generally abundant than in the Junco-Molinion with *S. palustre* common throughout and *S. recurvum* or *S. auriculatum* with *S. subnitens*, *S. papillosum* and *S. capillifolium* having dominance in the carpet. *Hypnum jutlandicum* is a frequent preferential in the *S. auriculatum* variant and quite commonly there can also be some bushes of *Myrica gale* and some *Erica tetralix* among the rushes.

Habitat

The *Carex echinata-Sphagnum* mire is the major soligenous community of peats and peaty gleys irrigated by rather base-poor but not excessively oligotrophic waters in the sub-montane zone in Britain. Its very widespread occurrence in the north-western parts of the country is in large measure a reflection of the prevalence there of more acidic substrates and the frequency of flushing on gentle to moderate slopes, though climate probably plays a direct part in limiting the distribution of one sub-community and of the variants. This kind of mire is commonly found within tracts of unenclosed pasture on the upland fringes and grazing has some influence on its floristics and physiognomy.

A high water-table and some amelioration of the very impoverished nutrient status characteristic of many waterlogged acid peats are essential pre-requisites for the development of this community and usually here both of these conditions are met by flushing from obvious springs or seepage lines or by the concentration of drainage waters into water-tracks or streams. Typically, the *Carex echinata-Sphagnum* mire marks out such sites on slopes of 1–10° within mire systems, where the soils are raw peats or, more often, in tracts of mineral soils where it is usually found over stagnohumic gleys or humic stagnopodzols, or sometimes over humic ground-water gleys on small alluvial terraces alongside upland streams (Ratcliffe 1959a, McVean & Ratcliffe 1962, Edgell 1969, Birks 1973, Birse & Robertson 1976, Birse 1980).

In such habitats as these, the through-put of waters gathered from the mire surfaces or from expanses of acidic soils or emerging from siliceous bedrocks or superficials provides sufficient enrichment to allow the development of the fairly diverse sedge or rush canopy and a modest poor-fen flora. Available data are very sparse but the studies of Birse & Robertson (1976) suggest that C/N ratios in vegetation of this kind are about 15 (compared with around 30 for ombrogenous peats) with levels of total phosphorus generally much higher than in the kind of habitat typical of the Rhynchosporion pools. The rise to prominence here of *Carex nigra*, *C. echinata*, *C. panicea*, *Juncus effusus*, *J. acutiflorus*, *Viola palustris* and *Potentilla erecta* provide a good indication of this environmental difference. The waters and soils, however, remain quite acid and poor in calcium, with a superficial pH usually between 4.5 and 5.5 (McVean & Ratcliffe 1962, Birks 1973, Birse & Robertson 1976), so that even moderately base-tolerant Sphagna, like *S. squarrosum* and *S. teres*, which mark the beginnings of a transition to the Caricion davallianae in the *Carex-Sphagnum squarrosum* mire, are poorly represented in the carpet.

The community is commonest at altitudes between 200 and 400 m and, though it can extend considerably higher, it is generally replaced in similar habitats above 650 m by the *Carex-Sphagnum russowii* mire, where species like *C. panicea*, the Junci and *Molinia* are scarce and where there is a distinct high-montane element among the sedges and Sphagna that is quite absent here. Stands can be encountered virtually at sea-level but the *Carex echinata-Sphagnum* mire is absent or no more than fragmentarily represented over most of the south-eastern lowlands of Britain, probably because suitable edaphic conditions are rare, rather than because of climatic exclusion. Within the community, however,

climate is probably involved in the association of the *Juncus acutiflorus* sub-community with sites at lower altitude along the more westerly oceanic fringe. In this kind of *Carex echinata-Sphagnum* mire, too, *Molinia* tends to be better represented such that the vegetation comes very close to the distinctly Atlantic Junco-Molinion communities. And, as among the Rhynchosporion bog pools, the prominence of *Sphagnum auriculatum* here, which enables variants to be recognised in each of the sub-communities, is probably also related in some way to increased oceanicity in moving to the far west.

The other floristic differences within the community are of more uncertain origin but are perhaps related to a combination of the degree of waterlogging and the extent of grazing, factors which are often related because of the inaccessibility of wetter stands to the stock and deer which commonly graze the upland fringes where this kind of mire occurs. The general prominence of Nardo-Galion species throughout British Caricion nigrae mires is quite marked (in comparison, say, with their equivalents in Scandinavia: see Nordhagen 1928, 1943, Dahl 1956) and is a measure of the great prevalence of pasturing throughout our uplands. Within the *Carex echinata-Sphagnum* mire, more intensive grazing on somewhat firmer, drier ground may account for the differences between the *Carex nigra-Nardus* sub-community which is more frequent among calcifugous hill pastures and the *Carex echinata* sub-community of wetter flushes and soligenous areas within mires. And it has also been suggested that the general switch from sedge- to rush-dominance seen among the sub-communities might be related to treatment, perhaps sometimes grazing alone or, on mire surfaces, combinations of grazing, burning and drainage.

Zonation and succession

The *Carex echinata-Sphagnum* mire typically occurs as small stands among other mire communities, grasslands and heaths and sometimes with swamps and spring vegetation, where zonations are related to the water regime and the base- and nutrient-status of the environment. Grazing, especially where combined with drainage, can convert the community to grasslands and exclusion of herbivores would be expected to permit progression to wet scrub and woodland, though this would, in many cases, probably be slow and patchy.

Where the community occurs within or alongside mires, it usually marks out linear tracks where there is an obvious channelling of soligenous waters on the sloping peat surface, as on blanket and raised mires, or around the margins, as in the lagg of raised and basin mires. It can be found in association with all the major Erico-Sphagnion communities. In more oceanic areas, it occurs with the *Scirpus-Eriophorum* blanket mire, where it is usually represented by the *Carex echinata* or *Juncus acutiflorus* sub-communities, typically their *Sphagnum auriculatum* variants (McVean & Ratcliffe 1962, Birks 1973). Further east, low-altitude *Calluna-Eriophorum* blanket mire provides a common context for the *Juncus effusus* sub-community, generally its *Sphagnum recurvum* variant (Eddy *et al.* 1969). The community is also found with the *Erica-Sphagnum papillosum* raised mire and occasionally as a strip alongside the central zone of more acidic valley bogs with the *Narthecio-Sphagnetum* mire (Rose 1953). In such situations, it may occur in close association with Littorelletea pools and pass laterally to the drier mire surface with changes in the *Sphagnum* carpet and a switch to dominance of the vascular plants of the active plane. Rather more complex patterns can be seen where the *Carex echinata-Sphagnum* mire occupies the wetter areas within tracts of degraded or disturbed peat, where it is often found with the *Scirpus-Erica* wet heath to the far-west (Figure 13) or the *Erica-Sphagnum compactum* wet heath further east,

Figure 13. Base-poor flushes among mire (*a*) and pasture (*b*).

M6b   *Carex echinata-Sphagnum* mire, *Carex-Nardus* sub-community

M6c   *Carex echinata-Sphagnum* mire, *Juncus effusus* sub-community

M15a   *Scirpus-Erica* wet heath, *Carex* sub-community

M15b   *Scirpus-Erica* wet heath, Typical sub-community

M17a   *Scirpus-Eriophorum* mire, *Drosera-Sphagnum* sub-community

M23a   *Juncus-Galium* rush-pasture, *Juncus acutiflorus* sub-community

M25a   *Molinia-Potentilla* mire, *Erica* sub-community

M32a   *Philonoto-Saxifragetum*, *Sphagnum* sub-community

U4a   *Festuca-Agrostis-Galium* grassland, Typical sub-community

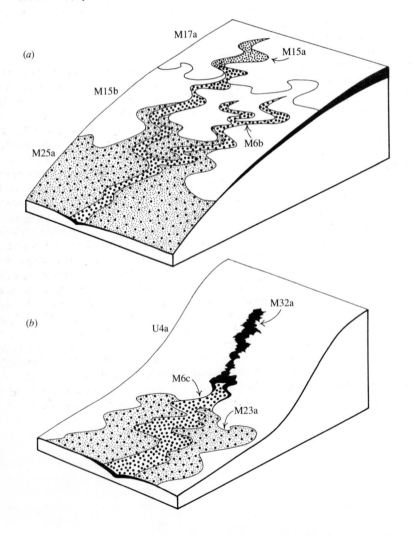

and fragmentary recolonising vegetation, in pools and over redistributed peat and on streamside gravels and alluvium.

On the gentler slopes of the upland fringes, the *Carex echinata-Sphagnum* mire picks out seepage areas and water tracks where there is local accumulation of peat or a peaty topsoil to gleys of various kinds. Here, the sharpness of the vegetation boundaries is very variable according to the degree of flushing. With diffuse irrigation, there may be a very gradual transition to the surrounding grassland or heath; where waters emerge from vigorous springs or are channelled into definite trickles and streams, the junctions are correspondingly sharp and stands can be clearly seen and mapped from a great distance (and from aerial photographs), particularly if the dominants are rushes or if the *Sphagnum* carpet is well developed. The way the distribution of the community can provide an effective indication of the drainage system is well seen on the published maps of Cader Idris in Dyfed (Edgell 1969) and of Moor House in Cumbria (Eddy *et al*. 1969).

Where waters originate from definite springs, the community may pass upstream, and fragmentarily along its length, to the *Philonoto-Saxifragetum*. Laterally, around its margins, there is usually a transition to Nardo-Galion or Juncion acutiflori grasslands, or to heaths, as the peat thins to a humose topsoil over less strongly-gleyed profiles. These communities, particularly the grasslands, have some species in common with the *Carex echinata-Sphagnum* mire and, where grazing stock have access, there may be a strong continuity among the sedges and dicotyledons, blurring the boundary. Where the community occurs on drift-smeared slopes, outlying patches of the mire may mark areas of local waterlogging in hollows and on gentle slopes, creating a mosaic with the grasslands and heaths and, where uncontrolled grazing has allowed a spread of rushes over generally poorly-drained ground, this may mask the patterning among the smaller herbs and bryophytes.

Long-continued grazing, especially on naturally drier soils or where there has been some drainage, probably converts the *Carex echinata-Sphagnum* mire to the *Juncus-Galium* mire or swards dominated by mixtures of *Nardus stricta* and *Juncus squarrosus*. And with a concerted attempt at improvement with drainage and lime and fertiliser application, lower-altitude stands can be reduced to narrow strips of wet ground within pastures of *Lolio-Cynosuretum*. At all stages, neglect and blockage of drains may permit a reversion.

A combination of grazing on drier fringes and pronounced waterlogging elsewhere helps prevent the establishment of woody plants in the community but where these get a hold, they would probably lead to the development of the *Salix-Galium* woodland or the wetter forms of the *Betula-Molinia* woodland. Under existing conditions, however, the *Carex echinata-Sphagnum* mire seems to be a substantially stable component of the sub-montane landscape.

Distribution

The community is virtually ubiquitous in the upland fringes of Britain and intensive survey (as in Wales on the distribution map, which includes data from Ratcliffe & Hattey 1982) is guaranteed to increase the density of known occurrences. All the sub-communities occur throughout the range, though the *Juncus acutiflorus* type is commoner in the far west (not well shown on the maps) and the variants show a clear relationship to oceanicity.

Affinities

As defined here, the *Carex echinata-Sphagnum* mire brings together a variety of poor fens which have generally been separated on the basis of physiognomic dominance by either small sedges or rushes. The community is the major Caricion nigrae mire of the sub-montane parts of Britain, occupying a central position on a floristic axis related to base-status and calcium content of the waters between the mires of the Rhynchosporion and those of the Caricion davallianae. It grades in both directions, its combination of poor-fen herbs and absence of more base-tolerant Sphagna or hypnoid mosses and calcicolous vascular plants providing good general characters. It also shows close affinities with Juncion and Junco-Molinion communities in the Molinietalia and with Nardo-Galion vegetation and, in these directions, the balance between Sphagna and poor-fen herbs and grasses helps define the limits of the vegetation types. Caricion nigrae mires of this kind are widespread in Europe and similar communities have been described from France, Germany (Tüxen 1937, Schwickerath 1940, Ellenberg 1978), The Netherlands (Westhoff & den Held 1969) and Belgium (LeBrun *et al*. 1949), Norway and Iceland (Dierssen 1982 who also included British stands in his *Caricetum nigrae*).

Floristic table M6

	a	b	c	d	6
Carex echinata	V (1–8)	V (3–8)	III (1–5)	III (1–5)	IV (1–8)
Polytrichum commune	IV (1–8)	II (2–5)	V (2–9)	III (1–7)	IV (1–9)
Agrostis canina canina	IV (1–6)	III (1–5)	III (5–9)	III (2–6)	III (1–9)
Potentilla erecta	III (1–7)	IV (1–5)	II (1–5)	V (1–6)	III (1–7)
Viola palustris	III (2–5)	IV (1–5)	II (2–5)	III (1–6)	III (1–6)
Sphagnum recurvum	III (3–10)	III (1–9)	III (2–10)	III (1–9)	III (1–10)
Molinia caerulea	V (2–9)	II (2–5)	III (2–7)	IV (1–8)	III (1–9)
Sphagnum auriculatum	III (2–10)	II (3–6)	II (1–8)	II (1–6)	II (1–10)
Sphagnum palustre	II (2–8)	II (2–8)	II (2–10)	III (1–8)	II (1–10)
Drosera rotundifolia	II (1–5)		II (1–4)	I (1–4)	I (1–5)
Potentilla palustris	I (1–5)		II (1–7)		I (1–5)
Menyanthes trifoliata	I (3–5)		I (3–4)		I (3–5)
Vaccinium oxycoccos	I (1–4)				I (1–4)
Eriophorum angustifolium	III (1–6)	V (2–7)	II (1–4)	I (1–3)	III (1–7)
Carex nigra	II (1–6)	V (3–8)	II (1–7)	II (1–4)	II (1–8)
Carex panicea	III (1–6)	IV (1–8)	I (3–4)	II (2–7)	II (1–8)
Nardus stricta	II (1–7)	IV (1–6)		I (2–3)	II (1–7)
Juncus squarrosus	II (2–7)	III (1–4)		I (1)	II (1–7)
Festuca ovina	I (1–5)	II (1–4)	I (2–6)	I (1–5)	I (1–6)
Ranunculus flammula	I (1–4)	II (1–4)		I (1–3)	I (1–4)
Luzula multiflora		II (1–2)		I (1–4)	I (1–4)
Carex dioica		II (1–2)			I (1–2)
Drepanocladus exannulatus		II (1–9)			I (1–9)
Agrostis capillaris		I (2–3)			I (2–3)
Cardamine pratensis		I (1–2)			I (1–2)
Juncus effusus	II (1–5)	I (1)	V (4–10)	I (1–3)	II (1–10)
Rumex acetosa			II (1–6)	I (2)	I (1–6)
Lotus uliginosus			I (1–5)		I (1–5)
Juncus acutiflorus		I (4)	I (4–5)	V (4–9)	II (4–9)
Sphagnum capillifolium	I (2–4)			II (1–7)	I (1–7)
Hypnum jutlandicum				II (1–5)	I (1–5)
Myrica gale				II (6–8)	I (6–8)
Anthoxanthum odoratum	II (2–4)	III (1–3)	II (1–5)	III (1–5)	II (1–5)
Sphagnum subnitens	II (4–6)	II (3–5)	II (2–6)	I (1–7)	II (1–7)

	a	b	c	d	6
Rhytidiadelphus squarrosus	I (1-4)	II (1-3)		II (1-5)	II (1-7)
Narthecium ossifragum	II (1-5)	II (1-5)		II (1-5)	II (1-5)
Sphagnum papillosum	II (3-8)	II (3-5)		I (1-8)	II (1-5)
Juncus bulbosus/kochii	II (1-5)	II (1-5)	I (1-9)	I (1)	II (1-9)
Carex demissa	II (1-6)	II (1-5)	I (1-3)	I (2-3)	I (1-5)
Aulacomnium palustre	I (1-6)	II (1-5)		I (1-3)	I (1-6)
Galium saxatile	I (1-5)	II (1-5)		I (1-4)	I (1-6)
Erica tetralix	II (1-4)	II (1-4)	II (1-4)	II (1-6)	I (1-5)
Calliergon stramineum	I (3-4)	I (5)	II (1-5)	I (1-3)	I (1-6)
Galium palustre	I (1-4)	I (1-7)	I (1-3)	I (1)	I (1-7)
Cirsium palustre	I (1-4)	I (1-3)	I (1)		I (1-4)
Succisa pratensis	I (1-4)	I (1-2)	I (1-3)	I (1-3)	I (1-4)
Epilobium palustre	I (2-3)	I (1-3)	I (1-6)	I (1-6)	I (1-6)
Carex curta	I (1-4)	I (1-3)	I (1-3)	I (1)	I (1-3)
Carex rostrata	I (4)	I (1-3)			I (1-4)
Juncus articulatus	I (1-6)	I (3-4)			I (3-4)
Eriophorum vaginatum	I (2-4)	I (2)			I (1-6)
Calliergon cuspidatum		I (4-5)			I (2-5)
Hydrocotyle vulgaris		I (1-4)	I (1-5)	I (1-5)	I (1-5)
Carex pulicaris		I (2-6)	I (2-3)	I (2-3)	I (2-6)
Holcus lanatus		I (1-3)	I (1-3)	I (1-3)	I (1-3)
Lophocolea bidentata s.l.		I (1-3)	I (1-2)	I (1-4)	I (1-4)
Number of samples	77	23	62	42	204
Number of species/sample	13 (6-24)	19 (11-26)	16 (2-32)	17 (2-28)	15 (2-32)
Herb height (cm)	23 (6-70)	22 (10-50)	52 (10-100)	53 (10-120)	36 (6-120)
Herb cover (%)	74 (15-100)	77 (40-95)	80 (30-100)	77 (40-100)	75 (15-100)
Bryophyte height (mm)	51 (20-150)	37 (30-50)	52 (10-150)	58 (10-150)	52 (10-150)
Bryophyte cover (%)	70 (5-100)	44 (2-95)	70 (1-100)	58 (3-100)	65 (1-100)
Altitude (m)	412 (7-1065)	419 (1-689)	389 (9-746)	203 (8-380)	371 (1-1065)
Slope (°)	7 (0-28)	6 (0-18)	3 (0-10)	4 (0-12)	5 (0-28)
Soil pH	4.9 (3.4-6.0)	3.9 (3.3-4.5)	4.3 (3.3-5.7)	4.3 (3.7-4.5)	4.6 (3.3-6.0)

a   *Carex echinata* sub-community
b   *Carex nigra-Nardus stricta* sub-community
c   *Juncus effusus* sub-community
d   *Juncus acutiflorus* sub-community
6   *Carex echinata-Sphagnum recurvum/auriculatum* mire (total)

Floristic table M6, variants

	ai	aii	bi	bii	ci	cii	di	dii
Carex echinata	IV (2-8)	V (1-8)	V (3-8)	V (4-8)	III (1-5)	III (1-5)	III (1-5)	IV (1-5)
Polytrichum commune	V (2-8)	III (1-6)	I (2-5)	III (2-5)	V (2-8)	V (5-9)	IV (1-7)	II (1-4)
Agrostis canina canina	IV (1-6)	III (3-6)	IV (1-5)	II (3)	III (2-7)	III (5-9)	IV (3-6)	II (2-5)
Potentilla erecta	III (1-7)	III (1-5)	IV (1-5)	V (2-4)	III (1-5)	I (3)	V (1-6)	V (2-5)
Viola palustris	IV (3-4)	IV (2-5)	IV (1-5)	III (3-5)	II (2-5)	II (4)	IV (1-6)	II (1-2)
Sphagnum recurvum	IV (3-10)	I (4-6)	III (1-8)	V (2-9)	V (2-10)		V (1-9)	II (1-9)
Molinia caerulea	V (2-9)	I (2-8)		III (2-5)	II (2-7)	III (4-5)	IV (1-8)	V (5-8)
Sphagnum auriculatum	I (2)	V (2-10)	II (3-6)	IV (1-9)	I (1)	V (4-8)	I (4-6)	IV (1-5)
Sphagnum palustre	III (2-8)	II (3-5)	II (2-6)	II (2-8)	II (2-10)	I (4)	III (1-8)	IV (1-8)
Drosera rotundifolia	I (4)	III (1-5)						II (1-4)
Potentilla palustris	I (3-5)	I (1-5)						
Menyanthes trifoliata	I (3)	II (3-5)						
Vaccinium oxycoccos	II (1-4)	I (4)						
Eriophorum angustifolium	II (1-4)	III (1-6)	V (2-7)	V (2-5)	II (1-4)	I (1)	I (1)	II (2-3)
Carex nigra	III (3-6)	III (1-6)	V (4-8)	V (3-8)	II (1-7)	I (4)	III (1-4)	I (3-4)
Carex panicea	II (2-6)	III (1-5)	IV (1-8)	III (2-4)	I (2-3)	II (3-4)	I (2-4)	IV (2-7)
Nardus stricta	I (3-4)	II (1-7)	IV (1-6)	IV (1-4)			I (1)	I (2-3)
Juncus squarrosus	I (3-6)	II (2-7)	II (1-4)	IV (1-3)				
Festuca ovina	I (3-5)	I (1-4)	I (1-3)	III (2-4)	II (2-6)		I (2-5)	I (1-3)
Ranunculus flammula		I (1-4)	III (1-4)	I (3)			I (1-3)	I (3)
Luzula multiflora			II (1-2)	II (1-2)			I (1-4)	
Carex dioica			II (1-2)	I (2)			I (1)	
Drepanocladus exannulatus			II (1-9)	I (1-3)				I (1)
Agrostis capillaris			I (3)	II (2-3)				
Cardamine pratensis			I (1-2)					
Juncus effusus	II (3-5)	II (1-5)	II (1)		V (4-10)	V (5-9)	I (1-3)	
Rumex acetosa					II (1-6)	I (4)	I (2)	
Lotus uliginosus					I (1-5)	II (2-4)		
Juncus acutiflorus	I (3-4)	I (2-4)		I (4)	I (4-5)		V (5-9)	V (4-7)
Sphagnum capillifolium							I (3)	III (1-7)
Hypnum jutlandicum							I (1)	IV (1-5)
Myrica gale								III (6-8)
Anthoxanthum odoratum	II (3-4)	II (2-4)	III (1-2)	III (3)	II (1-5)	II (4-5)	III (1-5)	II (1-4)
Sphagnum subnitens	II (4-6)	II (4-6)		III (3-5)	I (6)	IV (2-5)	I (2-4)	II (1-7)
Rhytidiadelphus squarrosus	II (2-4)	I (1-4)	III (1-3)	I (2)	II (1-6)	I (7)	II (1-5)	III (1-2)

	ai	aii	bi	bii	ci	cii	di	dii
Narthecium ossifragum	I (2–4)	II (1–5)	II (1–5)	III (2–3)	I (1–9)		I (5–8)	IV (2–5)
Sphagnum papillosum	I (4–8)	II (3–7)		III (3–5)		II (3–5)		II (1–7)
Juncus bulbosus/kochii	I (2–3)	III (1–5)	III (1–5)	II (2–3)		II (1–3)		II (1)
Carex demissa	I (6)	II (1–6)	II (1–5)	I (2)				I (2–3)
Aulacomnium palustre	II (1–6)	I (1–3)	II (1–4)	III (1–5)	II (1–4)	I (2)		I (2)
Galium saxatile	I (1–5)	I (1–3)	II (1–4)	II (2–3)	III (1–5)	I (2)		I (2)
Erica tetralix	I (1–4)	II (1–4)		I (5)	III (1–5)?			IV (1–6)
Calliergon stramineum	II (3–4)	I (3)	I (2–7)	II (1–3)	II (1–4)			
Galium palustre	I (1–4)	I (1–4)	I (1–3)		I (1–4)	I (1)		
Cirsium palustre	II (1–3)	I (1–4)	I (1–2)			I (1)		
Succisa pratensis	II (2–4)	I (1–4)	I (1)			I (1–3)		I (1)
Epilobium palustre	I (3)	I (2)	II (1–3)	I (3)		I (2–4)		III (1–6)
Carex curta	I (1–4)			I (1–3)		I (1)		
Carex rostrata	I (4)	I (4)	II (3–4)					
Juncus articulatus	II (1–6)	I (1–5)	II (2)					
Eriophorum vaginatum	I (2–4)	I (2)	II (4–5)					
Calliergon cuspidatum			II (1–4)	I (1)			I (5)	I (1–3)
Hydrocotyle vulgaris			II (2–6)				I (2–3)	
Carex pulicaris			II (1–3)				I (1)	I (1–3)
Holcus lanatus			I (1–3)				II (1)	I (1–2)
Lophocolea bidentata s.l.				I (2)	II (1–4)		I (4)	I (1–2)
Number of samples	34	43	14	9	57	5	27	15
Number of species/sample	12 (6–24)	14 (6–24)	18 (11–24)	21 (19–26)	16 (2–31)	17 (4–32)	16 (2–28)	19 (15–28)
Herb height (cm)	26 (6–70)	21 (10–60)	25 (12–50)	13 (10–15)	54 (10–100)	38 (10–80)	60 (35–120)	44 (10–75)
Herb cover (%)	81 (20–100)	69 (15–100)	78 (40–95)	70	82 (30–100)	73 (40–100)	74 (40–100)	81 (45–98)
Bryophyte height (mm)	66 (20–150)	41 (20–150)	37 (30–50)		54 (10–150)	44 (40–70)	58 (10–150)	60 (40–80)
Bryophyte cover (%)	66 (18–100)	72 (5–100)	42 (2–95)	55 (40–70)	68 (1–100)	76 (15–100)	64 (15–90)	50 (3–100)
Altitude (m)	412 (30–884)	413 (7–1065)	407 (1–689)	445 (390–540)	400 (9–746)	313 (140–440)	239 (8–380)	152 (40–290)
Slope (°)	6 (0–28)	8 (0–20)	5 (0–18)	8 (0–13)	3 (0–10)		6 (0–12)	2 (0–8)
Soil pH	4.9 (3.4–5.9)	4.8 (3.4–6.0)		3.9 (3.3–4.5)	4.4 (3.3–5.7)	4.0 (3.3–4.5)		4.3 (3.7–4.5)

ai   Carex echinata sub-community, Sphagnum recurvum variant
aii Carex echinata sub-community, Sphagnum auriculatum variant
bi   Carex nigra-Nardus stricta sub-community, Sphagnum recurvum variant
bii Carex nigra-Nardus stricta sub-community, Sphagnum auriculatum variant
ci   Juncus effusus sub-community, Sphagnum recurvum variant
cii Juncus effusus sub-community, Sphagnum auriculatum variant
di   Juncus acutiflorus sub-community, Sphagnum recurvum variant
dii Juncus acutiflorus sub-community, Sphagnum auriculatum variant

M6 Carex echinata-
Sphagnum recurvum/
auriculatum mire

M6 Carex echinata-
Sphagnum recurvum/
auriculatum mire

b Carex nigra-
Nardus stricta
sub-community

M6 Carex echinata-
Sphagnum recurvum/
auriculatum mire

a Carex echinata
sub-community

M6 Carex echinata-
Sphagnum recurvum/
auriculatum mire

c Juncus effusus
sub-community

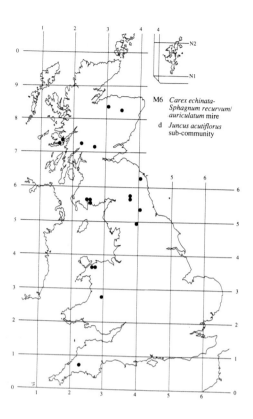

M6 *Carex echinata-*
    Sphagnum recurvum/
    auriculatum mire

d   *Juncus acutiflorus*
    sub-community

M7
Carex curta-Sphagnum russowii mire

Synonymy
Sphagneto-Caricetum alpinum McVean & Ratcliffe 1962, Eddy *et al.* 1969; Alpine *Carex-Sphagnum* mire Ratcliffe 1964; *Carex aquatilis-rariflora* nodum McVean & Ratcliffe 1962, Ratcliffe 1964; *Violo-Epilobietum sphagnetosum recurvae* Jones 1973 *p.p.*; *Caricetum nigrae* Dierssen 1982 *p.p.*; *Caricetum rariflorae* Dierssen 1982; *Drepanoclado exannulati-Caricetum aquatilis* Dierssen 1982 *p.p.*

Constant species
Carex curta, C. echinata, Eriophorum angustifolium, Viola palustris, Sphagnum papillosum, S. russowii.

Rare species
Carex aquatilis, C. rariflora, Sphagnum lindbergii, S. riparium.

Physiognomy
The *Carex curta-Sphagnum russowii* mire is a community whose prominent cyperaceous and *Sphagnum* components both have a distinct northern and montane character. Among the former element, *Eriophorum angustifolium* and *Carex echinata* are both very frequent and provide a floristic link with the lower-altitude *Carex echinata-Sphagnum* mire; the latter species can be abundant here and is sometimes co-dominant in the sedge canopy. But, in contrast to that community, the Continental Northern *Carex curta* is a constant in this kind of mire and is quite frequently of high cover; and it is often accompanied by either the Arctic-Alpine *C. bigelowii* or the Arctic-Subarctic *C. aquatilis* and *C. rariflora. C. nigra* can also occur, sometimes in abundance, but *C. rostrata*, which can accompany it in some wetter Caricion nigrae mires, is typically scarce and of low cover. *Scirpus cespitosus* is occasionally found and, particularly in transitions to blanket mire, there may be a little *Eriophorum vaginatum. Juncus bulbosus/kochii* and *J. squarrosus* occur at low frequencies but, in contrast to

the *Carex echinata-Sphagnum* mire, bulkier Junci, like *J. effusus* and *J. acutiflorus*, are very scarce here and do not function as alternative dominants over the *Sphagnum* carpet.

As in other Scheuchzerietalia mires, this carpet is typically very extensive and some of the prominent species are of wide distribution. Thus, *Sphagnum papillosum* is common throughout and it can be very abundant, especially where the community grades to surrounding blanket mire. Then, in the different sub-communities, there can be frequent records for *S. subnitens, S. auriculatum, S. capillifolium* or *S. recurvum.* Much more distinctive here is the constancy of the northern, high-altitude species *S. russowii* and, in one of the sub-communities, *S. lindbergii. S. girgensohnii* may also be found and *S. riparium* occurs in this community at some of its few Scottish stations. In contrast to more base-rich mires at high altitudes, *S. warnstorfii* and *S. contortum* are typically absent.

Among the Sphagna, there is frequently some *Polytrichum commune, Calliergon stramineum* or *C. sarmentosum* and, preferentially in one of the sub-communities, *Drepanocladus exannulatus, Polytrichum alpinum* and *P. alpestre. Lophozia ventricosa* occurs occasionally and there is sometimes a little *Scapania undulata* or *S. uliginosa.*

In this ground, grasses and dicotyledons play a relatively minor role, though, among the former, *Nardus stricta* is very common and *Agrostis canina* ssp. *canina* quite frequent, and both can be locally abundant. *A. stolonifera* and *Festuca vivipara* can also sometimes be found, but *Molinia caerulea*, a prominent species in the *Carex echinata-Sphagnum* mire, is typically absent. *Viola palustris* and *Galium saxatile* are the commonest dicotyledons of the community but are typically of low cover and not consistently accompanied by the richer suites of poor-fen herbs characteristic of lower-altitude mires. The Arctic-Alpine *Saxifraga stellaris* is fairly frequent in one sub-community.

Sub-communities

***Carex bigelowii-Sphagnum lindbergii* sub-community:**
Sphagneto-Caricetum alpinum McVean & Ratcliffe
1962, Eddy *et al.* 1969; Alpine *Carex-Sphagnum* mire
Ratcliffe 1964; *Violo-Epilobietum sphagnetosum recurvae, Sphagnum papillosum* variant Jones 1973 p.p.;
Caricetum nigrae Dierssen 1982. In this sub-community, *C. echinata* and the preferentially frequent *C. bigelowii* tend to dominate among the sedge canopy, with *C. curta* still common, though not usually of very high cover. Other sedges are rather scarce but *Nardus stricta* and *Agrostis canina* ssp. *canina* are frequent and, among the few dicotyledons, *Saxifraga stellaris* is preferential.

The other distinctive features of this kind of *Carex-Sphagnum russowii* mire are to be found in the bryophyte element. First, among the Sphagna, *S. lindbergii* is strongly diagnostic and, with *S. papillosum*, usually makes up the bulk of the cover, with *S. russowii* frequent, though usually less abundant. Then, there can be locally prominent patches of *S. subnitens, S. auriculatum* (both var. *auriculatum* and var. *inundatum*) or *S. compactum* with, less commonly, *S. magellanicum* or *S. cuspidatum*. Second, the associated bryophyte flora is a little richer than in the *Carex aquatilis-Sphagnum recurvum* sub-community. *Polytrichum commune* occurs frequently but more distinctive are *Calliergon sarmentosum, Drepanocladus exannulatus, Polytrichum alpinum, P. alpestre* and, more rarely, *Pellia epiphylla* and *Racomitrium lanuginosum*.

***Carex aquatilis-Sphagnum recurvum* sub-community:**
Carex aquatilis-rariflora nodum McVean & Ratcliffe 1962, Ratcliffe 1964; *Violo-Epilobietum sphagnetosum recurvae, Sphagnum papillosum* variant Jones 1973 p.p.; *Caricetum rariflorae* Dierssen 1982; *Drepanoclado exannulati-Caricetum aquatilis* Dierssen 1982 p.p. *Carex echinata* remains frequent and is often abundant here but, in contrast to the above, *C. curta* is often co-dominant and *C. bigelowii* very scarce. More striking is the occurrence in some stands of the rare *C. aquatilis*, here growing dwarfed in contrast to its appearance in the *Caricetum vesicariae* or *Caricetum rostratae* of open-water transitions and, especially good as a diagnostic species, *C. rariflora*, a very shy flowerer and thus easily missed or under-estimated. *C. nigra* is also more common here than in the other sub-community and, like the two rare sedges, it can be locally abundant.

There is also a little more diversity here among the vascular associates. *Nardus* is only occasional and sparse but *Festuca vivipara, Agrostis stolonifera, Deschampsia flexuosa, Luzula multiflora* and *Galium saxatile* are strongly preferential and there are occasional records for *Epilobium palustre* and *Trientalis europaea*.

The bryophyte element, too, is distinctive. *Sphagnum papillosum* and *S. russowii* both remain very frequent but the most abundant species is usually *S. recurvum*, a rather uncommon and low-cover member of the *Sphagnum* carpet in the other sub-community. *S. lindbergii* is absent and *S. subnitens, S. auriculatum, S. compactum* and *S. capillifolium* very scarce but *S. girgensohnii* and *S. riparium* are weakly preferential and there is occasionally some *S. teres*. Other bryophytes are rather few in number but *Polytrichum commune* tends to be better represented here and *Calliergon stramineum* replaces *C. sarmentosum*.

Habitat

The *Carex-Sphagnum russowii* mire is confined to high-altitude sites where peaty soils are irrigated by oligotrophic and base-poor waters, being most characteristic of hollows and drainage channels in blanket mire or flushes and seepage areas in tracts of montane moss-heaths.

Throughout its range, the community is found largely at altitudes of more than 650 m, extending up to more than 1100 m on the high summit plateaus of the Central Highlands of Scotland. In this region, which comprises the centre of distribution, the mean annual maximum temperature is generally less than 20 °C (Conolly & Dahl 1970), with conditions a little less extreme where the community extends on to the southern Pennines, at a few far-flung localities within a 25 °C isotherm (Eddy *et al.* 1969, Jones 1973, Bradshaw & Jones 1976). Winters in these areas are long and bitter and snow-lie may play some part in determining the distribution of stands, particularly those of the *Carex bigelowii-Sphagnum lindbergii* sub-community (McVean & Ratcliffe 1962).

This community is very much an altitudinal replacement for the *Carex echinata-Sphagnum* mire, a feature noted in McVean & Ratcliffe's (1962) designation of the two vegetation types as *Sphagneto-Caricetum alpinum* and *sub-alpinum* respectively. There is a fairly strong floristic overlap between the two communities but the pre-eminence here of montane plants like *Carex curta, C. bigelowii, C. aquatilis, C. rariflora, Saxifraga stellaris, Sphagnum russowii, S. lindbergii* and *S. riparium* and the scarcity of species such as *Carex panicea, Juncus effusus* and *Molinia* provide a generally good separation. However, the representation of some of the rare species is decidedly local and deliberate selection of stands in the data inherited from McVean & Ratcliffe (1962) may have produced an over-sharp picture of the differences. At lower altitudes, the two communities can grade into one another, a feature well seen in the *Sphagneto-Caricetum alpinum* which Eddy *et al.* (1969) described from Moor House in Cumbria. The upper altitudinal limit of the community may be partly set by scarcity of suitable sites, though *Sphagnum* growth, even among the more montane species, may be inhibited at extremely high levels (McVean & Ratcliffe 1962).

The *Carex-Sphagnum russowii* mire occupies very similar edaphic situations to those characteristic of its low-altitude counterpart, occurring on permanently moist peats fed by nutrient-poor waters collecting in hollows or percolating from granitic rocks, quartzose mica-schists or siliceous sedimentaries. The poverty of bases and cations here is a major difference between the habitat of this community and those of the *Carex-Sphagnum warnstorfii* mire, which is not so strikingly montane but which has a more base-tolerant *Sphagnum* component, and the *Caricetum saxatilis* which is found at comparable altitudes but which is much more obviously calcicolous.

The exact nature of the edaphic environment may play some part in determining the floristic differences between the two sub-communities, with the *Carex aquatilis-Sphagnum recurvum* type perhaps more characteristic of deeper peats with more stagnant ground-water (McVean & Ratcliffe 1962), but such a suggestion remains unconfirmed.

Zonation and succession
The community is typically found as small stands, most commonly in association with high-altitude *Calluna-Eriophorum* blanket mire, usually of the *Vaccinium-Hylocomium* sub-community (the *Empetreto-Eriophoretum* of McVean & Ratcliffe 1962), within which it can occupy hollows or drainage channels or form part of recolonising vegetation over eroded areas (McVean & Ratcliffe 1962, Eddy *et al.* 1969). Transitional zones around such stands of the *Carex-Sphagnum russowii* mire are often marked by an increase in *Eriophorum vaginatum* and a shift towards dominance of *S. papillosum* in the moss carpet. In other situations, the community can mark out spring or seepage lines within tracts of montane moss-heaths like the *Carex bigelowii-Polytrichum alpinum* or *Racomitrium-Carex bigelowii* communities, or high-level grasslands dominated by *Nardus stricta* or *Juncus squarrosus*, sometimes occurring in mosaics with floristically-related snow-bed vegetation. Where there is strong flushing, the community is often

found around the *Philonoto-Saxifragetum* spring and rill community.

Most of the occurrences of the *Carex-Sphagnum russowii* mire are close to or above the potential forest limit in the Scottish Highlands and the community is probably an essentially stable component of the vegetation pattern under present-day conditions; and it would probably remain so were grazing to be much reduced.

Distribution
 The community is largely confined to the higher reaches of the Central Highlands of Scotland. The *Carex aquatilis-Sphagnum recurvum* sub-community is the more local type, being concentrated around the Clova-Caenlochan area of the east-central Highlands, with the *Carex bigelowii-Sphagnum lindbergii* sub-community extending also to the hills of the Ben Alder and Creag Meagaidh massifs and to some isolated localities in the north-west Highlands. Essentially similar vegetation, with a poorer representation of the montane element, occurs at Moor House in Cumbria and on Widdybank Fell in Durham and perhaps also in Wales.

Affinities
As defined here, the *Carex-Sphagnum russowii* mire unites the two closely-related noda which McVean & Ratcliffe (1962) characterised from high-altitude oligotrophic flushes, retaining their distinction (with some minor re-allocation of samples) at sub-community level. The community is a high-montane counterpart of the *Carex echinata-Sphagnum* mire and clearly belongs among the poor-fens of the Caricion nigrae. Its nearest counterparts on the European mainland can be found among the series of mires described from Scandinavia by Nordhagen (1928, 1943), Dahl (1956) and Dierssen (1982) where *Sphagnum lindbergii* (and, to a lesser extent, *S. riparium*) occurs prominently in association with *Calliergon stramineum*, *Drepanocladus exannulatus* and a variety of sedges including *C. rariflora* and *C. aquatilis*. Such vegetation has usually been placed in a separate alliance, the Caricion lasiocarpae or Leuco-Scheuchzerion, within the Scheuchzerietalia.

Floristic table M7

	a	b	7
Eriophorum angustifolium	V (2–5)	V (2–4)	V (2–5)
Sphagnum papillosum	V (1–10)	IV (3–5)	V (1–10)
Carex curta	IV (1–5)	V (4–7)	IV (1–7)
Viola palustris	IV (2–3)	V (2–3)	IV (2–3)
Carex echinata	IV (2–7)	IV (3–6)	IV (2–7)
Sphagnum russowii	IV (1–3)	IV (2–3)	IV (1–3)

Sphagnum lindbergii	V (2–9)		III (2–9)
Nardus stricta	IV (1–5)	II (1–3)	II (1–5)
Carex bigelowii	IV (1–6)	I (3)	II (1–6)
Sphagnum subnitens	IV (1–5)	I (1)	II (1–5)
Sphagnum auriculatum auriculatum	IV (1–7)		II (1–7)
Agrostis canina canina	III (1–4)	I (1)	II (1–4)
Drepanocladus exannulatus	III (1–3)		II (1–3)
Calliergon sarmentosum	III (1–3)		II (1–3)
Sphagnum capillifolium	II (2–5)	I (5)	I (2–5)
Polytrichum alpestre	II (2–4)	I (3)	I (2–4)
Sphagnum compactum	II (2–5)	I (1)	I (1–5)
Saxifraga stellaris	II (1–2)	I (1)	I (1–2)
Polytrichum alpinum	II (2)	I (3)	I (2–3)
Juncus bulbosus/kochii	II (2–3)		I (2–3)
Sphagnum magellanicum	II (2)		I (2)
Pellia epiphylla	II (1–2)		I (1–2)
Rubus chamaemorus	I (1–2)		I (1–2)
Ptilidium ciliare	I (2)		I (2)
Hylocomium splendens	I (1–3)		I (1–3)
Racomitrium lanuginosum	I (1–2)		I (1–2)
Sphagnum cuspidatum	I (1–2)		I (1–2)
Dicranum scoparium	I (1–3)		I (1–3)
Philonotis fontana	I (1–2)		I (1–2)
Scapania uliginosa	I (3–5)		I (3–5)
Polytrichum commune	III (1–3)	V (1–4)	III (1–4)
Sphagnum recurvum	II (2–4)	V (5–9)	III (2–9)
Calliergon stramineum	II (1–3)	IV (1–2)	II (1–3)
Carex aquatilis	I (5)	IV (1–6)	II (1–6)
Festuca vivipara	I (1–2)	IV (1–2)	II (1–2)
Luzula multiflora	I (2)	IV (1)	II (1–2)
Carex nigra	I (2–5)	IV (4–6)	II (2–6)
Agrostis stolonifera		IV (3–4)	II (3–4)
Sphagnum girgensohnii	I (3)	III (5)	II (3–5)
Galium saxatile	I (1–2)	III (1–2)	II (1–2)
Carex rariflora	I (2)	III (3–5)	II (2–5)
Deschampsia flexuosa		III (1–3)	I (1–3)
Sphagnum riparium	I (6)	II (3–8)	I (3–8)
Sphagnum teres		II (3)	I (3)
Epilobium palustre		II (1–3)	I (1–3)
Trientalis europaea		II (1–3)	I (1–3)
Eriophorum vaginatum	III (1–3)	IV (1–4)	III (1–4)
Lophozia ventricosa	II (1–2)	II (1)	II (1–2)
Scirpus cespitosus	II (1–3)	II (1)	II (1–2)
Deschampsia cespitosa	I (1–3)	II (1)	II (1–3)
Vaccinium myrtillus	I (1–2)	II (1)	I (1–3)
Aulacomnium palustre	I (2–3)	I (1)	I (1–2)
Pohlia nutans	I (2–3)	I (1)	I (1–3)
Sphagnum auriculatum inundatum	I (2)	I (2)	I (2–3)
Cladonia arbuscula	I (2)	I (1)	I (1–2)

Floristic table M7 *(cont.)*

	a	b	7
Juncus squarrosus	I (1)	I (1)	I (1)
Anthoxanthum odoratum	I (1)	I (1)	I (1)
Scapania undulata	I (2)	I (1)	I (1–2)
Carex rostrata	I (1)	I (1)	I (1)
Epilobium anagallidifolium	I (1)	I (2)	I (1–2)
Number of samples	15	7	22
Number of species/sample	20 (11–27)	19 (14–23)	20 (11–27)

a   *Carex bigelowii-Sphagnum lindbergii* sub-community
b   *Carex aquatilis-Sphagnum recurvum* sub-community
7   *Carex curta-Sphagnum russowii* mire (total)

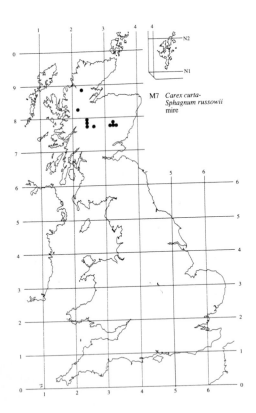

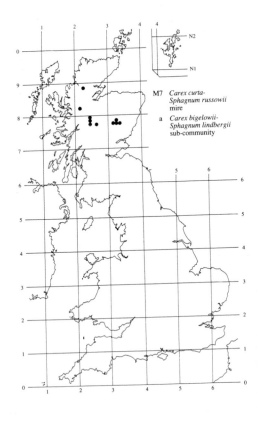

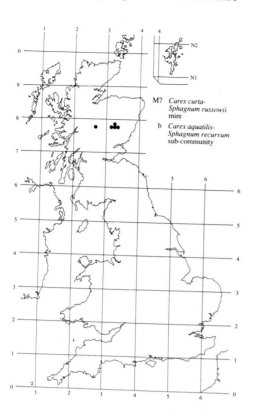

M7 *Carex curta-*
    Sphagnum russowii
    mire

b *Carex aquatilis-*
   Sphagnum recurvum
   sub-community

M8
Carex rostrata-Sphagnum warnstorfii mire

Synonymy
Carex rostrata-Sphagnum warnstorfianum nodum McVean & Ratcliffe 1962, Eddy *et al.* 1969; *Violo-Epilobietum sphagnetosum recurvae* Jones 1973 *p.p.*; *Menyantho-Sphagnetum teretis* Dierssen 1982.

Constant species
Carex nigra, C. rostrata, Epilobium palustre, Festuca ovina, Potentilla erecta, Selaginella selaginoides, Viola palustris, Aulacomnium palustre, Calliergon cuspidatum, Hylocomium splendens, Rhizomnium pseudopunctatum, Sphagnum teres, S. warnstorfii.

Rare species
Homalothecium nitens, Sphagnum subsecundum.

Physiognomy
The *Carex rostrata-Sphagnum warnstorfii* mire has a dominant cover of sedges over an extensive carpet of base-tolerant Sphagna and a fairly numerous and diverse assemblage of herbs. As in the lowland counterpart of this community, the *Carex-Sphagnum squarrosum* mire, *Carex rostrata* and *C. nigra* are the commonest sedges, the former generally the more abundant and often of high cover, the latter usually subordinate though locally dominant. Other poor-fen sedges, *C. panicea, C. echinata* and *C. demissa*, occur frequently and sometimes in abundance, and *C. pulicaris* is occasional, but the more calciolous *C. dioica* is rare and *C. curta*, an occasional associate of *C. rostrata* in oligotrophic mires, likewise scarce. There is frequently a little *Eriophorum angustifolium* among the sedge cover, much less often small amounts of one of the bulkier Junci, *Juncus articulatus* or, more rarely, *J. effusus* or *J. acutiflorus.*

The *Sphagnum* carpet is typically extensive and quite distinctive in the prominence, along with *S. recurvum*, of the base-tolerant *S. teres* and *S. warnstorfii*, the latter an especially good preferential for this kind of montane mire. *S. subsecundum sensu stricto* occurs more occasionally but is also very characteristic and the community provides one of the loci in Britain for *S. contortum*, though it is not so common here as in the *Carex-Calliergon* mire. *S. squarrosum* occurs occasionally but is not so consistent as in the *Carex-Sphagnum squarrosum* mire. Other low-frequency species include *S. palustre, S. girgensohnii, S. capillifolium, S. subnitens* and *S. papillosum*, all generally of low cover. *S. cuspidatum* and *S. auriculatum* are typically absent.

Other bryophytes are numerous with *Aulacomnium palustre* and *Rhizomnium pseudopunctatum* attaining higher frequency here than in any other of our montane mires. Also distinctive are *Calliergon cuspidatum, C. stramineum* and, less frequently, the montane *C. sarmentosum*, though not *C. trifarium* which is more typical of the *Caricetum saxatilis*. Quite common and more strictly diagnostic than any of these is *Homalothecium nitens*. Then there are frequent records for *Hylocomium splendens* and *Rhytidiadelphus squarrosus* and such indicators of some base-enrichment as *Pellia endiviifolia, Drepanocladus revolvens, Bryum pseudotriquetrum, Thuidium tamariscinum, Campylium stellatum* and *Fissidens adianthoides*. *Philonotis fontana, Polytrichum commune, Aneura pinguis, Lophocolea bidentata s.l., Calypogeia fissa, Scapania nemorosa* and *Chiloscyphus polyanthos* are occasional.

Scattered among the bryophytes is a quite rich mixture of herbaceous associates, though these are typically of low total cover. Some are poor-fen species like *Viola palustris* and *Epilobium palustre* and, more occasional, *Caltha palustris*; others, plants which are of broader affinities, but well represented in the Caricion nigrae, such as *Potentilla erecta* and *Galium saxatile*. But, along with these, is constant *Selaginella selaginoides*, a good preferential against other communities of the Scheuchzerietalia. Then, grasses can be quite numerous, with *Festuca ovina* (and *F. vivipara*), *Nardus stricta, Anthoxanthum odoratum* and *Agrostis stolonifera*, all generally present as scattered shoots or small tussocks; and there can be a few individuals of *Thalictrum alpinum, Polygonum viviparum, Leontodon autumnalis, Luzula multiflora* or, in wetter places, *Potentilla palustris* and *Juncus bulbosus.* Small bushes of *Erica tetralix* can be seen in some stands and there can be some seedlings of *Salix aurita,* though Arctic-Alpine willows, which are a prominent feature of similar vegetation in Scandinavia, are absent from British stands.

Habitat

The community is strictly confined to raw peat soils in waterlogged hollows in the montane zone of Britain where there is moderate base-enrichment by drainage from calcareous rocks.

Conditions suitable for the development of this kind of mire, which requires some degree of stagnation without the formation of base-poor and oligotrophic peat, are only rarely encountered: in the montane zone, small basins or perched flats generally develop more oligotrophic mires. The community is thus restricted to areas where ground waters drain from calcareous bedrocks and is especially characteristic of the Central Highlands of Scotland where Dalradian limestones, schists and epidiorites occur at high altitudes. It has also been recorded at Moor House in Cumbria (Eddy *et al.* 1969) and from Widdybank Fell in Durham (Jones 1973, Bradshaw & Jones 1976) where drainage is from Carboniferous Limestone; and it occurs in fragmentary form on more calcareous outcrops in the Moffat Hills in Dumfries and the Lake District (Eddy *et al.* 1969), though these localities have not been sampled.

The peat deposits on which the community is found are typically quite deep, usually more than 1 m (McVean & Ratcliffe 1962), with a high and stagnant water-table, features which help to mark off the habitat from that of montane flushed grasslands on the one hand (what McVean & Ratcliffe (1962) called *Hypno-Caricetum alpinum*) and, on the other, the *Caricetum saxatilis* mire, where there is a strong and constant through-put of water and greater calcium-enrichment. The base-status of the waters and peats here is usually in the range pH 5.5–6, a feature reflected in the intermediate character of the vegetation which is very similar to that of the *Carex-Sphagnum squarrosum* mire, found in analogous situations in the lowlands. Quite base-tolerant Sphagna predominate in the bryophyte carpet, with species such as *Aulacomnium* and *Calliergon stramineum,* and poor-fen sedges and dicotyledons are prominent above; and, although some moderately calcicolous bryophytes and herbs help mark off the community from the rest of the Caricion nigrae mires, these have not yet gained the ascendancy they show in the Caricion davallianae rich fens.

What helps separate this community from its lowland counterpart is a small but distinct montane element in the flora. This kind of mire is generally confined to altitudes between 400 and 800 m, where the mean annual maximum temperature is usually below 23 °C (Conolly & Dahl 1970), and the presence of *Sphagnum warnstorfii*

and the occasional *Calliergon sarmentosum,* together with Arctic-Alpine herbs such as *Thalictrum alpinum* and *Polygonum viviparum,* is quite diagnostic. In comparison with similar Scandinavian vegetation, however, this component is not well developed in Britain.

Zonation and succession

The *Carex-Sphagnum warnstorfii* mire typically occurs as small stands which can be sharply marked off from unirrigated surrounds or which can form part of swamp and mire complexes where water-depth and degree of base-richness influence the vegetation patterns. In the former situation, the community can occupy stagnant hollows in tracts of montane grasslands with a fringe of irrigated sward between, or mark out areas of base-enrichment below high-altitude ombrogenous bog, usually the *Calluna-Eriophorum* mire. In more extensive soligenous areas, it can be seen in zonations analogous to those in which the *Carex-Sphagnum squarrosum* mire occurs at lower altitudes, with the *Caricetum rostratae* replacing it towards open water and patches of the *Carex curta-Sphagnum russowii* and *Carex-Calliergon* mires indicating areas with a waning or increasing influence of base-richness.

The frequent presence of seedlings of *Salix aurita* in stands of the community may indicate a tendency towards the development of montane willow scrub but such successions have never been seen to progress further.

Distribution

Apart from a few far-flung (and sometimes fragmentary) stands in southern Scotland and northern England, the community is confined to the Central Highlands.

Affinities

The *Carex-Sphagnum warnstorfii* mire stands largely as originally defined by McVean & Ratcliffe (1962) and, though its relationships with other poor fens are very close, it seems better to retain it as a discrete community rather than unite it in a more broadly-defined unit like the *Violo-Epilobietum* of Jones (1973). It lies within the Caricion nigrae, though the presence of more basiphilous species place it close to the boundary with the rich-fens of the Caricion davallianae. It can be seen as the montane counterpart of the *Carex-Sphagnum squarrosum* mire and, among the upland mires, shares with the *Carex curta-Sphagnum russowii* mire a good representation of Nardetalia grassland species.

Although very local in Britain this kind of vegetation has clear affinities with Scandinavian mires like the *Salix lapponum-Carex rostrata-Sphagnum warnstorfii* sociation of Nordhagen (1943), the *Aulacomnieto-Sphagnum warnstorfii* of Dahl (1956) and the *Menyantho-Sphagnetum teretis* of Dierssen (1982).

Floristic table M8

Carex rostrata	V (6–9)	Drepanocladus fluitans	II (1–3)
Sphagnum warnstorfii	V (4–7)	Scapania nemorosa	II (1–3)
Rhizomnium pseudopunctatum	V (2–3)	Sphagnum palustre	II (1–4)
Viola palustris	V (2–4)	Chiloscyphus polyanthos	II (1–2)
Sphagnum teres	IV (2–6)	Juncus articulatus	II (1–2)
Aulacomnium palustre	IV (3–4)	Calliergon sarmentosum	II (2)
Hylocomium splendens	IV (1–4)	Climacium dendroides	I (1–2)
Carex nigra	IV (2–6)	Salix aurita	I (1)
Festuca ovina	IV (2–5)	Alchemilla vulgaris agg.	I (2)
Potentilla erecta	IV (1–3)	Ptilidium ciliare	I (1–2)
Calliergon cuspidatum	IV (1–3)	Tritomaria quinquedentata	I (1–3)
Epilobium palustre	IV (2)	Euphrasia officinalis agg.	I (3)
Selaginella selaginoides	IV (1–3)	Cerastium fontanum	I (1)
		Festuca vivipara	I (2)
Rhytidiadelphus squarrosus	III (1–3)	Pinguicula vulgaris	I (1–2)
Homalothecium nitens	III (3–6)	Rhytidiadelphus loreus	I (1–2)
Carex panicea	III (2–6)	Dicranum bonjeani	I (2)
Eriophorum angustifolium	III (2–4)	Sphagnum papillosum	I (1–7)
Pellia endiviifolia	III (1–3)	Sphagnum contortum	I (2–3)
Galium saxatile	III (1–3)	Pleurozium schreberi	I (1–2)
Philonotis fontana	III (1–4)	Pseudoscleropodium purum	I (1)
Carex echinata	III (4–6)	Equisetum palustre	I (3)
Nardus stricta	III (2–4)	Cirsium palustre	I (2)
Carex demissa	III (2–6)	Scapania undulata	I (3)
Drepanocladus revolvens	III (2–4)	Sphagnum squarrosum	I (1–9)
Bryum pseudotriquetrum	III (1–2)	Plagiomnium rostratum	I (2)
Calliergon stramineum	III (1–3)	Cardamine pratensis	I (1–2)
Sphagnum recurvum	III (2–6)	Jungermannia atrovirens	I (2)
Thalictrum alpinum	II (2–3)	Thuidium delicatulum	I (2)
Carex pulicaris	II (3)	Galium palustre	I (2)
Polygonum viviparum	II (2–3)	Dicranella palustris	I (1–3)
Luzula multiflora	II (1–2)	Brachythecium rutabulum	I (2–3)
Anthoxanthum odoratum	II (2–3)	Pedicularis palustris	I (1)
Leontodon autumnalis	II (2–3)	Carex lepidocarpa	I (2)
Thuidium tamariscinum	II (1–3)	Empetrum nigrum nigrum	I (3)
Caltha palustris	II (2–3)	Calluna vulgaris	I (1)
Campylium stellatum	II (3)	Alchemilla filicaulis vestita	I (1)
Juncus bulbosus/kochii	II (2–4)	Betula seedling	I (1)
Polytrichum commune	II (2)	Trientalis europaea	I (3)
Aneura pinguis	II (2–3)	Crepis paludosa	I (2)
Sphagnum girgensohnii	II (1–3)	Calypogeia trichomanis	I (1)
Lophocolea bidentata s.l.	II (1–3)	Empetrum nigrum hermaphroditum	I (1–2)
Agrostis stolonifera	II (1–4)	Carex bigelowii	I (4)
Fissidens adianthoides	II (2)	Deschampsia cespitosa	I (3)
Potentilla palustris	II (2–4)	Parnassia palustris	I (3)
Sphagnum subsecundum	II (3–8)	Ptilium crista-castrensis	I (2)
Erica tetralix	II (2–4)	Carex saxatilis	I (1)
Calypogeia fissa	II (1–2)		

Carex curta	I (2–3)	*Potamogeton polygonifolius*	I (1)
Harpanthus flotovianus	I (2)	*Luzula sylvatica*	I (1)
Ranunculus acris	I (4)	*Calliergon giganteum*	I (1)
Agrostis capillaris	I (4)	*Sphagnum subnitens*	I (1–4)
Festuca rubra	I (3)	*Hypnum jutlandicum*	I (3)
Linum catharticum	I (1)	*Ctenidium molluscum*	I (2)
Juncus effusus	I (2)	*Drosera rotundifolia*	I (1)
Carex dioica	I (1)	*Leiocolea bantriensis*	I (1)
Epilobium anagallidifolium	I (1)		

Number of samples	10
Number of species/sample	36 (22–46)

Riccardia multifida	I (2)
Alchemilla glabra	I (1)
Brachythecium rivulare	I (1)
Valeriana dioica	I (1)
Triglochin palustris	I (1)
Veronica scutellata	I (1)
Pellia epiphylla	I (1)
Juncus acutiflorus	I (2)

Vegetation height (cm)	32
Vegetation cover (%)	100
Altitude (m)	613 (427–833)
Slope (°)	5 (0–25)

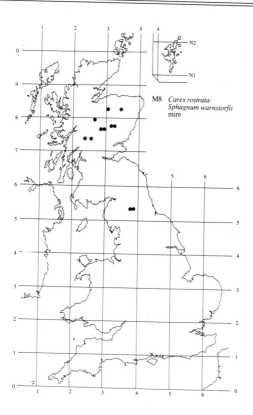

M8 *Carex rostrata-Sphagnum warnstorfii* mire

M9
Carex rostrata-Calliergon cuspidatum/giganteum mire

Synonymy

Carex rostrata 'reedswamp' Holdgate 1955*b p.p.*; Lower fens Holdgate 1955*b p.p.*; Mixed fen Holdgate 1955*b*; *Carex rostrata*-brown moss provisional nodum McVean & Ratcliffe 1962, Ferreira 1978; *Carex rostrata-Acrocladium* sociation Spence 1964 *p.p.*; *Carex nigra-Acrocladium* sociation Spence 1964 *p.p.*; *Potentilla-Acrocladium* sociation Spence 1964 *p.p.*; *Filipendula-Acrocladium* sociation Spence 1964 *p.p.*; *Carex rostrata-Scorpidium scorpioides* Association Birks 1973; *Potentilla palustris-Acrocladium* nodum & related fens Proctor 1974 *p.p.*; General fen Adam *et al.* 1975 *p.p.*; *Acrocladio-Caricetum diandrae* (Koch 1926) Wheeler 1975 *p.p.*; *Peucedano-Phragmitetum caricetosum* Wheeler 1978 *p.p.*; *Acrocladium cuspidatum-Carex diandra* mire Ratcliffe & Hattey 1982; *Caricetum diandrae* Dierssen 1982; *Caricetum rostratae* Dierssen 1982 *p.p.*

Constant species

Carex rostrata, Eriophorum angustifolium, Galium palustre, Menyanthes trifoliata, Potentilla palustris, Calliergon cuspidatum.

Rare species

Carex appropinquata, C. diandra, Cicuta virosa, Dactylorhiza traunsteineri, Liparis loeselii, Potamogeton coloratus, Pyrola rotundifolia, Sium latifolium, Utricularia intermedia, Cinclidium stygium.

Physiognomy

The *Carex rostrata-Calliergon cuspidatum* mire is a community of diverse composition and physiognomy, even within individual stands, but it is generally characterised by a fairly rich assemblage of vascular plants, among which sedges predominate, over a luxuriant carpet of bulky mosses, in which Sphagna are but locally represented. The commonest sedge overall is *Carex rostrata* and this is quite often abundant, sometimes a sole dominant, particularly in the *Campylium-Scorpi-*

dium sub-community. Very frequently, however, it is accompanied here by the nationally rare Continental Northern sedge *C. diandra*, which has its main locus in Britain in this community and which can likewise be very prominent, especially in the *Carex diandra-Calliergon giganteum* sub-community. In some stands, one or both of these species are accompanied by a third tall, rhizomatous sedge which can be locally abundant, *C. lasiocarpa*. When these plants are growing vigorously, they can make up a patchy canopy 60 cm or more high. In some localities, there are also prominent tussocks of *C. paniculata* or *C. appropinquata*, though these sedges are scarce in the community as a whole. Other possible local dominants of low frequency throughout are *Juncus subnodulosus* and *Schoenus nigricans* and, in a few places, the *Carex-Calliergon* mire occurs in some very striking mosaics with the *Peucedano-Phragmitetum* when a sparse or patchy cover of *Phragmites australis* or *Cladium mariscus* can occur (in what Wheeler (1978, 1980*a*) called *Peucedano-Phragmitetum caricetosum*).

Intermixed with these species or forming a fringe around their more extensive patches is a variety of associates of low to medium height. Some, like *Potentilla palustris* and *Menyanthes trifoliata*, are very common throughout the community and can attain high cover locally. These species probably play an important part in the establishment of the floating rafts in which form this vegetation is often found and, when they are abundant, they accentuate the similarity between the community and the closely-related *Potentillo-Caricetum rostratae*. Also very common are *Eriophorum angustifolium, Equisetum fluviatile, E. palustre, Succisa pratensis, Pedicularis palustris, Cirsium palustre* and *Ranunculus flammula* and, distributed somewhat more unevenly through the two sub-communities, *Mentha aquatica, Caltha palustris, Valeriana dioica, Angelica sylvestris, Epilobium palustre* and *Lychnis flos-cuculi.* For the most part, all these plants occur as scattered individuals but, when they are present in rich assortments, they give the vegetation a quite distinctive stamp. Other sedges apart from the bulkier species listed earlier can also make a contribution. *C. panicea* and *C. nigra* are both very frequent and the latter especially can be conspicuous in its tussock form (var. *tornata*, not the non-rhizomatous var. *juncea* (= *C. juncella*, which is Scandinavian: c.f. Holdgate 1955*b*)). In the *Campylium-Scorpidium* sub-community, there is further enrichment of the sedge component with records for a variety of species among which *C. limosa* and *C. echinata* are the most common. Grasses are few in number but *Molinia caerulea* is quite frequent and, in drier stands, it can be abundant, when the vegetation begins to approach the *Cirsio-Molinietum* in its composition. *Agrostis stolonifera* is also fairly common, though *A. canina* ssp. *canina* is very scarce in comparison with the Caricion nigrae poor fens. Apart from the occasional *Juncus subnodulosus*, rushes are characteristically uncommon, though there is some-times a little *J. articulatus* or *J. acutiflorus*.

Three other structural elements may attain some measure of prominence among the vascular cover. First, there is often some sprawling *Galium palustre* and, less commonly, *G. uliginosum*. Second, wetter areas, which may be sufficiently well defined as to form small com-plexes of pools, can have some submerged aquatic plants, notably *Utricularia* ssp., and emergents like *Ranunculus lingua* and the rare *Sium latifolium* and *Cicuta virosa*. And, third, there may be some small saplings of *Salix cinerea* (and, in northern Britain, *S. pentandra*) and *Betula pubescens*, when the vegetation may be transitional to the *Salix-Carex rostrata* woodland.

Bryophytes almost always form a conspicuous com-ponent of the vegetation, often attaining more than 50% cover in a luxuriant carpet. The most frequent species throughout are larger *Calliergon* ssp.: *C. cuspidatum* is a constant of the community and *C. giganteum* and *C. cordifolium* are frequent, though distinctly preferential to the *Carex diandra-Calliergon giganteum* sub-community. There is also quite often one or more of the larger Mniaceae, *Plagiomnium rostratum*, *P. affine*, *P. elatum*, *P. undulatum*, *P. ellipticum* or *Rhizomnium pseudopunctatum*. Members of these two groups of mosses can be quite common in the *Potentillo-Carice-tum rostratae*, but a further group of more or less calcicolous species helps define the community against both this fen and the Caricion nigrae mires. *Campylium stellatum* is the most frequent of these and it is often accompanied in the *Campylium-Scorpidium* sub-community by *Scorpidium scorpioides* and *Drepanocla-dus revolvens*; more occasional through the community as a whole are *Bryum pseudotriquetrum*, *Cratoneuron commutatum*, *C. filicinum* and *Ctenidium molluscum*. Other bryophytes recorded at low frequencies in this kind of mire are *Hylocomium splendens* and *Climacium dendroides*. In contrast to the Caricion nigrae communi-

ties, Sphagna, apart from the most base-tolerant of the genus, *S. contortum*, are distinctly uncommon, though there can be occasional patches of *S. subnitens*, *S. auriculatum*, *S. recurvum* and *S. warnstorfii*.

Sub-communities

***Campylium stellatum-Scorpidium scorpioides* sub-community:** *Carex rostrata* 'reedswamp' Holdgate 1955*b p.p.*; Lower fens Holdgate 1955*b p.p.*; *Carex rostrata*-brown moss provisional nodum McVean & Ratcliffe 1962, Ferreira 1978; *Carex rostrata-Scorpi-dium scorpioides* Association Birks 1973; *Acrocladio-Caricetum diandrae schoenetosum*, *sphagnetosum p.p.* and *juncetosum p.p.* (Koch 1926) Wheeler 1975. Although *Carex diandra* is quite common in this sub-community and can be locally abundant, the usual dominant is *C. rostrata*, sometimes accompanied by *C. lasiocarpa*. At some sites, *Schoenus nigricans* is a promi-nent associate in this taller stratum, as in the stands included in Wheeler's (1975, 1980*b*) *Acrocladio-Carice-tum schoenetosum*, but its overall frequency is low. Smaller sedges are much more numerous and varied than in the other sub-community, though they are gener-ally represented by scattered individuals: the commun-ity species *C. panicea* and *C. nigra* retain their high frequency here but, in addition, *C. limosa* and *C. echi-nata* are strongly preferential and there are occasional records for *C. lepidocarpa*, *C. dioica*, *C. serotina*, *C. demissa* and *C. hostiana*, though these latter species are never so structurally important here as in the *Pinguiculo-Caricetum dioicae*.

The representation of herbaceous associates is vari-able but generally not very rich and their total cover is typically quite low, giving a rather open look to the vascular component. *Menyanthes trifoliata* and *Poten-tilla palustris* can be patchily prominent but, for the most part, the community species occur as scattered plants. There are also but few preferentials among this element in this sub-community, though more aquatic plants of wet hollows and pools are a little more prominent with *Utricularia* ssp., *Potamogeton polygonifolius* and *Juncus bulbosus/kochii* giving the impression of a local develop-ment of Littorelletea vegetation. *Eleocharis quinqueflora* and *Pinguicula vulgaris* occur is some stands, *Viola palustris*, *Narthecium ossifragum* and *Drosera rotundifo-lia* in others and, sometimes, such assemblages can be closely juxtaposed in complex mosaics. The nationally rare *Pyrola rotundifolia* has been recorded in this kind of vegetation.

Characteristically, however, it is not these associates which comprise the prominent ground to sedge-domi-nance but the bryophytes. Apart from *Calliergon cuspi-datum*, the *Calliergon* spp. and Mniaceae are somewhat less frequent and conspicuous than in the other sub-community, though they can occur locally. But typically very extensive is a patchwork of the 'brown mosses' *Campylium stellatum*, *Scorpidium scorpioides* and *Drepanocladus revolvens* (sometimes determined to the more calcicolous var. *intermedius*: but see Smith 1978). There are also occasional clumps of Sphagna, notably the base-tolerant *S. contortum* but also, less commonly, *S. subnitens*, *S. auriculatum*, *S. recurvum* and *S. warnstorfii*. *Aneura pinguis* is also preferential at low frequency.

Carex diandra-Calliergon giganteum sub-community:

Mixed fen   Holdgate 1955*b p.p.*; *Carex rostrata-Acrocladium*, *Carex nigra-Acrocladium*, *Potentilla-Acrocladium* and *Filipendula-Acrocladium* sociations   Spence 1964 *p.p.*; *Potentilla palustris-Acrocladium* nodum & related fens   Proctor 1974 *p.p.*; Miscellaneous *Carex* fens   Proctor 1974 *p.p.*; General fen   Adam *et al.* 1975 *p.p.*; *Acrocladio-Caricetum diandrae typicum, cicutetosum, crepetosum* and *juncetosum subnodulosi*   Wheeler 1980*b*; *Peucedano-Phragmitetum caricetosum*   Wheeler 1978 *p.p.*; *Acrocladium cuspidatum-Carex diandra* mire   Ratcliffe & Hattey 1982. The pattern of dominance among the vascular plants is more variable here than in the *Campylium-Scorpidium* sub-community. Both *Carex rostrata* and *C. diandra* are very frequent and either or both may dominate, with or without *C. lasiocarpa*. Very locally, in eastern England, *Juncus subnodulosus* is abundant in this upper stratum of the vegetation (in what Wheeler (1980*b*) termed a *juncetosum*), and in parts of Broadland this sub-community occurs in intimate mosaics with the *Peucedano-Phragmitetum* when *Cladium* or *Phragmites* can be patchily represented.

Herbaceous associates, particularly taller dicotyledons, are also more numerous in this sub-community: together, they can constitute a quite lush and species-rich cover and some of them can attain local dominance along with the sedges, as in the series of fens defined by Spence (1964), Proctor (1974) and Adam *et al.* (1975). As well as *Potentilla palustris* and *Menyanthes*, *Filipendula ulmaria* can be prominent (especially along the sides of streams where there is some silting) and there are also frequent records for *Valeriana dioica*, *Epilobium palustre*, *Angelica sylvestris* and *Lychnis flos-cuculi*, occasionally for *Valeriana officinalis* and, more locally, in wetter places, for *Ranunculus lingua*, *Cicuta virosa* and *Sium latifolium* (as in the *cicutetosum* of Wheeler 1980*b*). Among the smaller herbs, too, there are some good preferentials with *Caltha palustris*, *Cardamine pratensis* and *Mentha aquatica* all common. *Crepis paludosa* has been recorded from some northern stands (as in Wheeler's (1980*b*) *crepetosum*) and, in Broadland, *Dactylorhiza traunsteineri* and the very rare *Liparis loeselii* occur in this vegetation (e.g. Wheeler 1978, 1980*a, b*). Occasional pools can have *Utricularia* ssp., *Potamogeton polygonifolius*, *P. coloratus* and *Hottonia palustris*.

As in the *Campylium-Scorpidium* sub-community, bryophytes are often extensive and it is here that *Calliergon giganteum* and *C. cordifolium*, *Plagiomnium rostratum* and *P. affine* show their maximum development, sometimes occurring in a variegated carpet, in other stands showing local replacement in a more well-defined mosaic (e.g. Proctor 1974). *Campylium stellatum* occurs occasionally but *Scorpidium*, *Drepanocladus revolvens* and Sphagna are very scarce.

Habitat

The *Carex-Calliergon* mire is characteristic of soft, spongy peats kept permanently moist by at least moderately base-rich and calcareous waters. It is commonest in the wetter parts of topogenous mires, in natural hollows or old peat-workings, but it can also occur in places with a strong soligenous influence, in small spring-head basins, along the margins of lagg streams around raised mires and in flushes on blanket mires. It is typically too wet to be grazed but in some areas it occurs within mowing-marsh that is still cropped.

The local distribution of this kind of mire is in part a reflection of the scarcity of the rather particular environmental conditions that it favours. To the north-west of Britain, it is generally found as part of the vegetation in basins receiving drainage waters from limestones or other highly calcareous rocks or superficials, as at Sunbiggin Tarn in Cumbria (Holdgate 1955*b*), Malham Tarn in North Yorkshire (Sinker 1960, Proctor 1974, Adam *et al.* 1975) and the Anglesey fens (Wheeler 1975, 1980*b*), all on Carboniferous Limestone, and at Newham Fen in Northumberland and the Whitlaw Mosses in southern Scotland (Ratcliffe 1977, Wheeler 1980*b*) developed in glacial drift. Here it occurs in open-water transitions or in more infilled basin mires, often with some marginal seepage from the surrounds, but in this region, too, it can also mark out smaller areas influenced by soligenous calcareous waters, within blanket mires (as on Skye: Birks 1973) or, at altitudes up to about 800 m, in slope flushes feeding little peat-filled hollows (McVean & Ratcliffe 1962). In the south-eastern lowlands of Britain, the *Carex-Calliergon* mire can again be found in basin mires, as on some of the north Norfolk commons (Wheeler 1980*b*), but it also occurs in wetter areas within some of the Broadland fens, though these stands do not appear to experience an orthodox flood-plain mire regime and may always occupy disused peat-workings (Wheeler 1978, Giller & Wheeler 1986*a*: see below). The distribution of these sites is such that this vegetation provides a locus for a number of Continental Northern species in Britain: *Carex diandra*, *C. lasiocarpa*, *Cicuta virosa*, *Crepis paludosa*, *Hottonia palustris*.

In all the various site types, the waters and substrates with which the community is associated are typically fairly base-rich, with a pH always above 5 and usually above 6, and with dissolved calcium levels in the range 15–50 mg l⁻¹ (Spence 1964, Proctor 1974, Wheeler 1983), conditions which are reflected in the shift away from the more calcifugous floristic elements prominent in the Caricion nigrae to the more calcicolous. In the community as a whole, the change is best seen among the bryophytes where, apart from the base-tolerant *Sphagnum contortum*, Sphagna are of rather patchy occurrence, *Calliergon cuspidatum*, *C. giganteum* and *C. cordifolium* replace *C. stramineum*, and the calcicolous 'brown mosses' become very prominent. But there are also occasionally records, too, for such typical Caricion davallianae herbs as *Eleocharis quinqueflora*, *Pinguicula vulgaris*, *Carex dioica* and *C. lepidocarpa*.

However, it is a very distinctive feature of some stands that pH varies quite widely within a small compass (e.g. Proctor 1974). This affects not only the sharpness with which the *Carex-Calliergon* mire is differentiated from the vegetational context in which it occurs but probably also plays some part in the development of the fine patterning often to be seen among the bryophytes and herbs, and is perhaps involved in the floristic differentiation of the two sub-communities. Such variations are sometimes related to complex patterns of seepage in sites with soligenous influence or, in topogenous fens, they may reflect the extent to which the vegetation is maintained in close contact with the ground waters. There may also be some autogenic differentiation of more acidic nuclei, around the patches of Sphagna best seen in the *Campylium-Scorpidium* sub-community, or other bryophytes like *Rhizomnium pseudopunctatum* (e.g. Proctor 1974: see also Clapham 1940).

Though not so impoverished as the habitats of the Caricion nigrae poor fens, those of the *Carex-Calliergon* mire are probably always relatively poor in available phosphorus and nitrogen (Proctor 1974, Giller & Wheeler 1986a, b). However, there may be some difference in the trophic state of the environments of the two sub-communities. The *Campylium-Scorpidium* sub-community is largely confined to the sub-montane north-west of Britain and often occurs there in situations which seem oligotrophic, like small flushes. The *Carex diandra-Calliergon giganteum* type is more consistently associated with topogenous mires where there may be some deposition of allochthonous material and, in the south-eastern lowlands, perhaps some influence from fertiliser run-off: its strong contingent of mesophytic tall herbs certainly bring it close in its floristics to some of the richer Phragmitetalia fens and Molinietalia fen-meadows.

The other important environmental feature associated with the development of the *Carex-Calliergon* mire is the maintenance of a more or less consistently high water-table in soft and often deep peats. Some sites

where the community occurs, notably basins in the wetter north-western uplands, do experience considerable fluctuations in water-level, from a few centimetres above the surface to up to 40 cm below (Holdgate 1955b, Spence 1964), but such changes are usually not seasonally maintained and probably do not leave the fen mat desiccated for long periods of time. Even where the *Carex-Calliergon* mire occurs within flood-plain mires, as in Broadland, the water-level fluctuations are somewhat irregular in the vicinity of the stands (Giller & Wheeler 1986a). Moreover, the vegetation often develops as a floating or semi-floating raft which can rise and fall with any change in water-level, as the typically unconsolidated deposits beneath expand and contract: this tends to reduce the amplitude of the variations relative to the surface and maintain a fairly consistent hydrological and redox environment. In the Ant valley fens in Broadland, Giller & Wheeler (1986a, b) found that it was this, rather than any consistent differences in cation concentration in the substrates or treatment of the vegetation, which distinguished the situations where the *Carex-Calliergon* mire developed, among a mosaic of Phragmitetalia swamps and fens, from the surrounding firm peat. Again, it may be possible that the floristic differences between the two sub-communities are related to fine variations in the height of the water-table. In comparable vegetation in The Netherlands, Segal (1966) noted that a *Scorpidium* phase was associated with wetter conditions, an *Acrocladium* (*Calliergon*) *cuspidatum* phase with drier conditions, local patches of more base-tolerant Sphagna developing in either. In some situations, this may represent a seral development with an upraising of the surface of the fen mat: certainly the establishment of an extensive bryophyte carpet seems to play an important role here in the colonisation by the community dicotyledons (Segal 1966, Giller & Wheeler 1986a).

In Broadland, the stretches of soft, unconsolidated peat in which the community develops are invariably associated with fairly shallow peat-cuttings, sometimes extensive turf-ponds 60–80 cm deep, excavated in the nineteenth century and now almost completely colonised (Wheeler 1978, Giller 1982, Giller & Wheeler 1986a). At some other sites, too, such as Malham Tarn (Proctor 1974), the *Carex-Calliergon* mire may mark out artificial hollows in which rather particular environmental conditions pertain, a feature of very considerable importance for the conservation of the community where terrestrialisation is proceeding apace. Giller & Wheeler (1986a) suggested that, in Broadland, maintenance of the striking species-rich mosaics in which the *Carex-Calliergon* mire occurs and which provide a locus for some notable rarities, would probably necessitate the excavation of new turf-ponds.

The generally wet and insubstantial nature of the

ground here usually preclude grazing of those stands which are open to stock. In the Ant valley fens, the stretches of the *Peucedano-Phragmitetum* in which the community is found are still summer-mown for sedge (*Cladium*) every three or four years but, though this helps maintain the richness of the mosaics (Wheeler & Giller 1982a), it does not explain why the *Carex-Callier-gon* component of the vegetation cover should be so very localised within the generally more species-poor tracts of fen which cover most of the turf-ponds (Giller & Wheeler 1986a).

Zonation and succession

The community is typically found among suites of swamps and mires whose distribution is related to the water regime of the environment and the base-status and calcium content of the waters and substrates. In some places, such zonations obviously reflect successions with increasing terrestrialisation but, in more oligotrophic situations, seral progression may be slow and, in isolated sites, woody seed-parents may be very remote. Under certain conditions, it is also possible that succession tends not to the establishment of mire forest but ultimately to ombrogenous bog.

The simplest sequences in which the *Carex-Calliergon* mire occurs are found in open-water transitions around lakes and pools with more base-rich waters, as in limestone and drift basins towards the upland north-west. Here, the community is often fronted by stands of the *Caricetum rostratae* or *Equisetetum fluviatile* swamps (or, in parts of Scotland, the *Caricetum vesicariae*) extending out into deeper open water (Holdgate 1955b, Spence 1964). More locally, in this kind of site, the *Cladietum marisci* may be represented in the zonation: it is especially well-developed in the Anglesey fens (Wheeler 1980a, b), though in the cooler conditions at higher latitudes and altitudes becomes increasingly sparse (e.g. Holdgate 1955b). Very commonly, the *Carex-Calliergon* mire forms a mosaic behind the swamp zone with the *Potentillo-Caricetum*, also often developed as a floating raft, within which there may also be a local occurrence of the *Carex-Sphagnum squarrosum* mire: through all of these, *Carex rostrata*, *Eriophorum angustifolium*, *Potentilla palustris* and *Menyanthes* may form a continuous backdrop to the differentiation of diagnostic suites of bryophytes and herbs in the various fens. Where there is any development of woody vegetation in such sequences, it is generally of the *Salix-Carex* type, characterised by rather diverse canopies of willows, in which *Salix pentandra* generally figures prominently, over a field layer that retains a strong floristic continuity with the surrounding fen. A more or less complete sequence of this general type is well seen at Malham Tarn (Proctor 1974, Adam et al. 1975).

At this site, too, it is possible to see some complications of this basic pattern. First, since base- and calcium-enrichment are strongly dependent on springs which debouch water from calcareous bedrocks (Carboniferous Limestone in this case), the *Carex-Calliergon* mire occurs closely juxtaposed to calcifugous mire vegetation more remote from soligenous influence or elevated above the ground water-table on ombrogenous peats. Directly comparable abrupt floristic contrasts can be seen where the community occurs locally around more base-rich flushes in stretches of blanket mire or wet heath in Scotland, where the basins are often insufficiently extensive for swamps to be represented, but where montane mires, like the *Carex-Sphagnum warnstorfii* community, and spring vegetation, like the *Carici-Saxifragetum aizoidis*, may also be represented (McVean & Ratcliffe 1962, Birks 1973).

Second, where calcareous conditions are maintained over shallower peats running away around the fringe of the basin, the *Carex-Calliergon* mire gives way at Malham, and around soligenous base-rich flushes on upland slopes, to the *Pinguiculo-Caricetum dioicae*, a community with which it shows quite strong floristic affinities but where species like *Carex rostrata*, *C. diandra*, *Potentilla palustris* and *Menyanthes* fade in importance and where there is a stronger representation of Caricion davallianae plants.

Third, on fairly moist and firm peats or peaty alluvium where there is a greater measure of nutrient enrichment, the *Carex-Calliergon* mire can pass to Molinietalia vegetation in which tall herbs become consistently prominent, together with plants like *Molinia caerulea* and/or *Juncus subnodulosus*. At Malham, such vegetation is represented by the rather particular *Molinia caerulea-Crepis paludosa* community, as well as the more widely distributed *Filipendula ulmaria-Angelica sylvestris* community (Proctor 1974, Adam et al. 1975). Similar transitions can be seen at Sunbiggin Tarn (Holdgate 1955b) and around some Scottish lochs (Spence 1964) and, where the *Carex-Calliergon* mire occurs very locally in fens in eastern England, it is generally in a context of Junco-Molinion fen-meadows (Wheeler 1980b, c).

The very distinctive occurrences of the community in Broadland preserve the general features of the swamp/fen transitions outlined above but the vegetation types represented are somewhat different, the mosaics very intricate and their origin, as noted above, rather special. In this area, the *Carex-Calliergon* mire occurs in intimate contact with various types of the *Peucedano-Phragmitetum* (notably the *Cicuta* sub-community) and was grouped within that association by Wheeler (1978, 1980a) as a distinct *caricetosum*. These mosaics, together with tracts of the *Cladietum marisci* and some species-rich types of *Phragmitetum*, occupy the turf-ponds and are surrounded by stretches of the *Peucedano-Phragmitetum*, *Schoenus* and *Myrica* sub-communities on the firmer, uncut peats. *Cladium* is generally the present fen-dominant (and is regularly mown) and seems to have figured prominently in the relatively short successional life of this quite striking vegetation in the disused cuttings (Giller & Wheeler 1986*a*).

Throughout its range, at least at lower altitudes and in less remote sites, the *Carex-Calliergon* mire is probably a successional stage to the development of Salicion cinereae mire forest, to the north the *Salix-Carex* woodland, to the south-east the *Salix-Betula-Phragmites* woodland. The maintenance of a generally high water-table or irregular fluctuations in water-level may hinder invasion by woody plants and, in Broadland, mowing for sedge can repeatedly set back colonisation but, where the fen mat begins to emerge permanently more than a few centimetres above the water level, progression to woodland is probably potentially quite rapid. There is also the possibility that the *Carex-Calliergon* mire is seral to the development of poor-fen and ombrogenous mire, through the local formation of *Sphagnum* nuclei, a progression to the *Carex-Sphagnum squarrosum* mire and selective invasion by *Betula pubescens*. Such a sere seems to be in operation in the Ant valley fens, where the *Sphagnum* sub-community of the *Salix-Betula-Phragmites* woodland forms striking islands in the fen, and it perhaps heralds the development of *Betula-Molinia* woodland and some kind of *Sphagnum*-dominated bog.

Distribution

The community has a widespread but rather local distribution, being limited by the fairly sparse occurrence of suitable natural situations and, in the lowland south-east, by wetland drainage and the cessation of shallow peat-digging. The *Campylium-Scorpidium* sub-community is largely north-western in its range, with scattered outliers in the south, and predominantly in smaller soligenous sites; the *Carex diandra-Calliergon giganteum* sub-community occurs throughout the range, mostly in topogenous mires.

Affinities

The *Carex-Calliergon* mire brings together a variety of previously-described vegetation types whose extremes are represented by the *Carex rostrata*-brown moss mires of McVean & Ratcliffe (1962), Birks (1973) and Ferreira (1978) and the communities dominated by mixtures of *C. rostrata*, *C. diandra* and *C. lasiocarpa* with a range of tall-fen herbs characterised by Holdgate (1955*b*), Spence (1964), Proctor (1974) and Adam *et al.* (1975) and subsumed by Wheeler (1975, 1980b) into his *Acrocladio-Caricetum diandrae*. Despite the considerable diversity in the pattern of dominance and in the composition of the bryophyte carpet, the community holds together well as the most calcicolous of the mires in which *C. rostrata* is well represented and provides the major British locus for *C. diandra*. It compares closely with *C. diandra*-dominated mires from other parts of north-west Europe such as Norway (Dierssen 1982), France (Wattez & Géhu 1972), Germany (Oberdorfer 1977, Ellenberg 1978), Belgium (Le Brun *et al.* 1949), The Netherlands (Segal 1966, Westhoff & Den Held 1969), Switzerland (Koch 1926) and Poland (Matuszkiewicz 1981) where the vegetation types have generally been related to the *Caricetum diandrae* (Jon. 1932) Oberdorfer 1957 or the *Scorpidio-Caricetum diandrae* Koch 1926. As in parts of Britain, such communities on the Continent are often associated with peat-cuttings (e.g. Duvigneaud 1949, Segal 1966). They have sometimes been placed in a Caricion lasiocarpae, more usually in the Caricion davallianae and, as here, provide a locus for such alliance indicators as *Liparis loeselii*, *Dactylorhiza traunsteineri*, *Sphagnum contortum* and *Cinclidium stygium*.

The *Carex-Calliergon* mire has close relationships with various other kinds of mire. It is fairly well marked off from the poor fens of the Caricion nigrae but grades floristically to other Caricion davallianae communities (through the smaller calciolous sedges and dicotyledons and some of the bryophytes), to Phragmitetalia fens (where tall herbs are prominent with bulky sedges and helophytes) and to Molinietalia fen-meadows (where tall herbs are associated with *Molinia* and *Juncus subnodulosus*).

Floristic table M9

	a	b	9
Carex rostrata	V (1–9)	V (1–6)	V (1–9)
Eriophorum angustifolium	V (1–4)	V (1–4)	V (1–4)
Menyanthes trifoliata	IV (1–7)	V (1–7)	V (1–7)
Potentilla palustris	IV (1–4)	IV (1–4)	IV (1–4)
Calliergon cuspidatum	IV (1–4)	IV (1–7)	IV (1–7)

Floristic table M9 *(cont.)*

	a	b	9
Galium palustre	III (1–3)	IV (1–4)	IV (1–4)
Campylium stellatum	V (1–5)	II (1–4)	III (1–5)
Scorpidium scorpioides	IV (1–6)	I (4)	II (1–6)
Drepanocladus revolvens	IV (1–4)	I (1–4)	II (1–4)
Carex limosa	III (1–4)	I (1–3)	II (1–4)
Utricularia spp.	III (1–3)	I (1)	II (1–3)
Sphagnum contortum	III (1–4)	I (1)	II (1–4)
Carex echinata	III (1–3)		II (1–3)
Viola palustris	II (1–3)	I (1–3)	I (1–3)
Carex lepidocarpa	II (1–4)	I (1–3)	I (1–4)
Potamogeton polygonifolius	II (1–2)	I (1–3)	I (1–3)
Sphagnum subnitens	II (1–3)	I (1)	I (1–3)
Eleocharis quinqueflora	II (2–3)	I (1)	I (1–3)
Potentilla erecta	II (1–2)	I (1–4)	I (1–4)
Aulacomnium palustre	II (1–3)	I (1–3)	I (1–3)
Triglochin palustris	II (1–2)	I (1)	I (1–2)
Aneura pinguis	II (1–3)	I (1–3)	I (1–3)
Carex dioica	II (1–4)		I (1–4)
Sphagnum auriculatum inundatum	II (2–3)		I (2–3)
Carex demissa	II (3–4)		I (3–4)
Pinguicula vulgaris	II (2–3)		I (2–3)
Carex serotina	I (3–4)		I (3–4)
Sphagnum recurvum	I (2–4)		I (2–4)
Juncus bulbosus/kochii	I (2)		I (2)
Carex hostiana	I (3–5)		I (3–5)
Sphagnum warnstorfii	I (1–4)		I (1–4)
Narthecium ossifragum	I (1–2)		I (1–2)
Pyrola rotundifolia	I (1–3)		I (1–3)
Dactylorhiza majalis purpurella	I (1–3)		I (1–3)
Drosera rotundifolia	I (1–3)		I (1–3)
Caltha palustris	II (1–3)	V (1–3)	III (1–3)
Cardamine pratensis	II (1–3)	V (1–3)	III (1–3)
Carex diandra	II (4–7)	IV (1–7)	III (1–7)
Calliergon giganteum	II (1–4)	IV (1–4)	III (1–4)
Valeriana dioica	II (1–4)	IV (1–4)	III (1–4)
Mentha aquatica	II (1–4)	IV (1–3)	III (1–4)
Epilobium palustre	I (1–3)	IV (1–4)	II (1–4)
Angelica sylvestris	II (1–3)	III (1–3)	II (1–3)
Lychnis flos-cuculi	II (1–3)	III (1–3)	II (1–3)
Filipendula ulmaria	I (1–3)	III (1–3)	II (1–3)
Galium uliginosum	I (1–3)	III (1–3)	II (1–3)
Calliergon cordifolium	I (1)	II (1–4)	II (1–4)
Ranunculus lingua	I (1–3)	II (1–3)	I (1–3)
Plagiomnium rostratum	I (1)	II (1–3)	I (1–3)
Plagiomnium affine	I (1)	II (1–3)	I (1–3)
Holcus lanatus	I (1)	II (1–3)	I (1–3)
Valeriana officinalis	I (1–3)	II (1–3)	I (1–3)

	a	b	9
Ranunculus acris	I (1)	II (1–3)	I (1–3)
Dactylorhiza fuchsii	I (1)	II (1–3)	I (1–3)
Juncus acutiflorus	I (1–3)	II (1–3)	I (1–3)
Juncus subnodulosus		II (1–7)	I (1–7)
Scutellaria galericulata		I (1–3)	I (1–3)
Geum rivale		I (1–2)	I (1–2)
Lotus uliginosus		I (1–3)	I (1–3)
Lythrum salicaria		I (1)	I (1)
Equisetum fluviatile	III (1–4)	III (1–4)	III (1–4)
Equisetum palustre	III (1–4)	III (1–3)	III (1–4)
Carex nigra	III (1–4)	III (1–3)	III (1–4)
Carex panicea	III (1–5)	III (1–3)	III (1–5)
Succisa pratensis	III (2–3)	III (1–3)	III (1–3)
Molinia caerulea	III (1–5)	II (1–4)	III (1–5)
Pedicularis palustris	III (1–4)	II (1–4)	III (1–4)
Cirsium palustre	II (1–3)	III (1–3)	III (1–3)
Agrostis stolonifera	II (1–3)	III (1–3)	III (1–3)
Ranunculus flammula	II (1–4)	II (1–3)	II (1–4)
Carex lasiocarpa	II (4–5)	II (1–7)	II (1–7)
Juncus articulatus	II (1–4)	II (1–3)	II (1–4)
Phragmites australis	II (1–4)	II (1–4)	II (1–4)
Bryum pseudotriquetrum	II (1–3)	II (1–3)	II (1–3)
Salix cinerea sapling	II (1–3)	II (1–5)	II (1–5)
Climacium dendroides	I (1–3)	II (1–4)	I (1–4)
Hylocomium splendens	I (1–3)	II (1–4)	I (1–4)
Hydrocotyle vulgaris	I (1–3)	I (1–3)	I (1–3)
Eupatorium cannabinum	I (1–3)	I (1–3)	I (1–3)
Rhizomnium pseudopunctatum	I (1–4)	I (1–3)	I (1–4)
Salix repens	I (1–3)	I (1–4)	I (1–4)
Cratoneuron commutatum	I (1–2)	I (1–3)	I (1–3)
Carex paniculata	I (1–4)	I (1–4)	I (1–4)
Cratoneuron filicinum	I (1–3)	I (1–3)	I (1–3)
Betula pubescens sapling	I (1–3)	I (1–3)	I (1–3)
Carex appropinquata	I (1–5)	I (1–6)	I (1–6)
Ctenidium molluscum	I (1–4)	I (1)	I (1–4)
Plagiomnium elatum	I (1–3)	I (1–3)	I (1–3)
Carex flacca	I (1)	I (1)	I (1)
Vicia cracca	I (2)	I (1–3)	I (1–3)
Plagiomnium undulatum	I (1)	I (1–3)	I (1–3)
Schoenus nigricans	I (3)	I (4)	I (3–4)
Crepis paludosa	I (1)	I (1–3)	I (1–3)
Dactylorhiza incarnata	I (1)	I (1–3)	I (1–3)
Cladium mariscus	I (4–5)	I (1–4)	I (1–5)
Parnassia palustris	I (1–3)	I (1)	I (1–3)
Number of samples	16	24	40
Number of species/sample	23 (16–33)	27 (12–35)	25 (12–35)

a   *Campylium stellatum-Scorpidium scorpioides* sub-community

b   *Carex diandra-Calliergon giganteum* sub-community

9   *Carex rostrata-Calliergon* spp. mire (total)

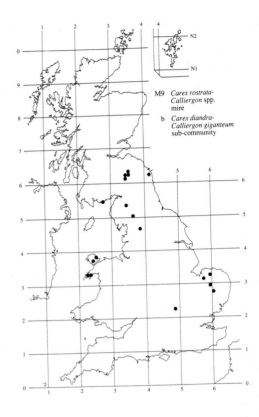

M9 *Carex rostrata-*
   Calliergon spp.
   mire

b *Carex diandra-*
   Calliergon giganteum
   sub-community

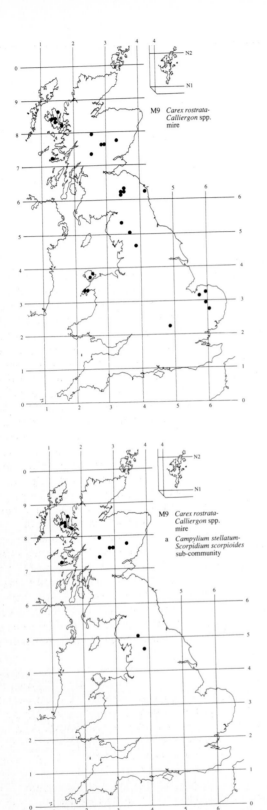

M9 *Carex rostrata-*
   Calliergon spp.
   mire

M9 *Carex rostrata-*
   Calliergon spp.
   mire

a *Campylium stellatum-*
   Scorpidium scorpioides
   sub-community

M10

Carex dioica-Pinguicula vulgaris mire
Pinguiculo-Caricetum dioicae   Jones 1973 *emend.*

Synonymy

Carex hostiana-C. demissa nodum Poore 1955*b*; *Carex demissa-C. panicea* nodum   Poore 1955*b p.p.*; Flush vegetation   Holdgate 1955*a p.p.*; Turfy marshes Pigott 1956*a*; Gravel flushes   Pigott 1956*a*; Calcareous marsh   Sinker 1960; *Carex panicea-Campylium stellatum* nodum   McVean & Ratcliffe 1962; *Schoenus nigricans* provisional nodum   McVean & Ratcliffe 1962 *p.p.*; *Kobresieto-Caricetum*   Ratcliffe 1965 *p.p.*; *Gymnostometo-Caricetum*   Ratcliffe 1965; *Caricetum lepidocarpae-hostianae*   Shimwell 1968*a*; *Carex panicea-Campylium stellatum* Association   Birks 1973; *Eriophorum latifolium-Carex hostiana* Association   Birks 1973; *Schoenus nigricans* Association   Birks 1973 *p.p.*; *Pinguiculo-Caricetum dioicae*   Jones 1973 *p.p.*; *Carex fens*   Proctor 1974 *p.p.*; *Pinguiculo-Caricetum dioicae* Jones 1973 *emend.* Wheeler 1975; *Schoenus nigricans* Community   Birse & Robertson 1976; *Caricetum hostiano-pulicaris* (Birse & Robertson 1976) Birse 1980; *Carici dioicae-Eleocharitetum quinqueflorae* (Birse & Robertson 1976) Birse 1980 *p.p.*; *Anagallis tenella-Equisetum variegatum* Association   Birse 1980; *Campylio-Caricetum dioicae*   Dierssen 1982; *Eleocharitetum quinqueflorae*   Dierssen 1982; *Schoenus ferrugineus* stands   Wheeler *et al.* 1983 *p.p.*

Constant species

Carex dioica, C. hostiana, C. lepidocarpa, C. panicea, C. pulicaris, Eriophorum angustifolium, Juncus articulatus, Pinguicula vulgaris, Aneura pinguis, Bryum pseudotriquetrum, Campylium stellatum, Ctenidium molluscum, Drepanocladus revolvens.

Rare species

Bartsia alpina, Carex capillaris, Equisetum variegatum, Juncus alpinus, Kobresia simpliciuscula, Minuartia stricta, M. verna, Primula farinosa, Schoenus ferrugineus, Sesleria albicans.

Physiognomy

The *Pinguiculo-Caricetum dioicae* includes some of the most distinctive of our more calcicolous flush vegetation in which small sedges, dicotyledons and bryophytes predominate, usually forming the bulk of a short sward, often hummocky and open. Although there is a strong contingent of constant species, the community is very variable in its composition, showing some marked differences in the proportions of its more frequent species and in the representation of groups of plants preferential to particular habitats or regions. And, where certain of these species are abundant, they can have a striking effect on the physiognomy of the vegetation. The habitat conditions, too, can have a marked influence on the appearance of the community. Most stands are grazed and the herbage is accordingly often cropped very short; where herbivore pressure is not so high, the vegetation can have a more luxuriant look, hemicryptophyte dicotyledons especially making an obvious contribution. Then, various factors contribute to the openness of the sward and there is often some deposition of tufa beneath the herbage which can produce a hummocky structure and give the turf a very characteristic crunchy texture. Frequently, then, there is a fine mosaic of micro-topographies within the vegetation which allows species of wetter and drier ground, and of more stable and disturbed areas, to grow closely juxtaposed.

The *Pinguiculo-Caricetum* is essentially a small-sedge mire and, among the more frequent Carices, *C. panicea, C. dioica, C. lepidocarpa, C. hostiana, C. pulicaris* can all occur in abundance, either together or in vegetation more obviously dominated by one or another species. *C. nigra* is also frequent in many stands and *C. demissa* and *C. echinata* or *C. flacca* are common in some sub-communities or their variants. Despite the very considerable variation among these commoner sedges, this particular combination of species is shared only by the closely-related *Carici-Saxifragetum aizoidis* and, even there, there are some obvious differences in the balance of those represented. Bulkier sedges, such as tend to predominate in swampier base-rich mires, are rarely present, though *C. rostrata* sometimes occurs. Finally, among this element, the community provides one locus for the rare *C. capillaris*.

Other cyperaceous plants can also be prominent. *Eriophorum angustifolium* is very frequent throughout and *Eleocharis quinqueflora* is common, especially in Scottish stands, and can be abundant. *Schoenus nigricans* also occurs here in some very distinctive flushes and so does the recently-rediscovered *S. ferrugineus*. *Kobresia simpliciuscula* can be a local dominant and *Eriophorum latifolium* occurs in some variants. *Scirpus cespitosus* is typically rare, though it can be well represented in transitional surrounds to the community where it occurs within blanket mire or wet heath.

Among other monocotyledons, certain rushes and grasses occur frequently. *Juncus articulatus* is a constant of the community, *J. bulbosus/kochii* is very typical of less markedly base-rich sites and, in some variants, *J. alpinus* and *J. triglumis* occur. *J. acutiflorus* and *J. squarrosus* occur very sparsely. *Molinia caerulea* is the commonest grass overall and locally it can dominate. *Festuca ovina* is rather more variable in its occurrence but, with species such as *Briza media*, *Sesleria albicans* and *Agrostis stolonifera*, it is very characteristic of the *Briza-Primula* sub-community which typically occurs within the context of Mesobromion grasslands. Other grasses found more occasionally include *Anthoxanthum odoratum*, *Festuca rubra* and *Nardus stricta*, all of which tend to increase where the community interrupts stretches of more calcifugous swards. *Triglochin palustris* and *Tofieldia pusilla* are typical of some variants of the community and *Narthecium ossifragum* of others.

The other herbs of the *Pinguiculo-Caricetum* generally occur as scattered plants and again they can show considerable variety, the frequency of particular species representing transitions in a number of directions. Very frequent throughout and distinctive of this kind of mire are *Pinguicula vulgaris* and *Selaginella selaginoides*, both of them plants which are rare within the Caricion nigrae poor fens. Then, in many stands, there is some *Potentilla erecta* and *Succisa pratensis* and, rather more unevenly represented, *Equisetum palustre*, and *Euphrasia officinalis* agg. Also characteristic, but more obviously preferential to particular sub-communities or their variants are *Linum catharticum*, *Primula farinosa*, *Prunella vulgaris*, *Parnassia palustris*, *Thymus praecox* and *Equisetum variegatum* (in the *Briza-Primula* sub-community) and *Drosera rotundifolia* and *Erica tetralix* (in the *Carex-Juncus* sub-community). Some of these species provide a strong continuity with the vegetation that typically surrounds the flushes of the community and many of them also occur in the closely related *Carici-Saxifrage-*

tum, though *S. aizoides* itself is typically of only local significance here. Apart from *Succisa pratensis*, taller fen herbs are generally uncommon, though in one variant there is a strong enrichment with Molinietalia dicotyledons. Among rarer herbs, this community provides a locus for *Bartsia alpina*, *Minuartia verna* and *M. stricta*. Bryophytes are almost always an obvious element in the sward and quite commonly comprise 50% or more of the ground cover, either in extensive carpets among the herbs or as discrete patches or hummocks. More calcicolous species such as *Campylium stellatum*, *Aneura pinguis*, *Drepanocladus revolvens* (sometimes determined to var. *intermedius*), *Ctenidium molluscum*, *Fissidens adianthoides* and *Cratoneuron commutatum* are among the more frequent species of the community. Also very common is *Bryum pseudotriquetrum* and in particular variants *Scorpidium scorpioides*, *Racomitrium lanuginosum*, *Pellia indiviifolia*, *Ditrichum flexicaule*, *Catascopium nigritum* and *Gymnostomun recurvirostrum*. *Blindia acuta* can also sometimes be found, though it is nothing like so frequent as in the *Carici-Saxifragetum*.

Sub-communities

***Carex demissa-Juncus-bulbosus/kochii* sub-community:** *Carex hostiana-C. demissa* nodum Poore 1955b; *Carex demissa-C. panicea* nodum Poore 1955b p.p.; Flush vegetation Holdgate 1955a p.p.; *Carex panicea-Campylium stellatum* nodum McVean & Ratcliffe 1962; *Schoenus nigricans* provisional nodum McVean & Ratcliffe 1962 p.p.; *Schoenetum nigricantis* Shimwell 1968a; *Carex panicea-Campylium stellatum* Association Birks 1973; *Eriophorum latifolium-Carex hostiana* Association Birks 1973; *Schoenus nigricans* Association Birks 1973 p.p.; *Pinguiculo-Caricetum eleocharetosum* Jones 1973 p.p.; *Schoenus nigricans* Community Birse & Robertson 1976; *Caricetum hostiano-pulicaris* (Birse & Robertson 1976) Birse 1980; *Carici dioicae-Eleocharitetum quinqueflorae* (Birse & Robertson 1976) Birse 1980 p.p.; *Schoenus ferrugineus* stands Wheeler et al. 1983 p.p. This sub-community comprises the less calcicolous kinds of *Pinguiculo-Caricetum*, a floristic shift that is visible in all the elements of the vegetation. Thus, among the sedges, though *Carex panicea*, *C. dioica*, *C. hostiana* and also *C. nigra* remain very frequent, *C. lepidocarpa* and *C. flacca* are only occasional or sparse, *C. pulicaris* is somewhat patchy and, on the positive side, *C. demissa* and *C. echinata* are preferentially frequent, the former especially being often prominent in the sward. Then, *Juncus bulbosus/kochii* and *Erica tetralix* are considerably more common here than in the other sub-communities and there are occasional records for *Narthecium ossifragum*, *Drosera rotundifolia* and *Scirpus cespitosus* (especially where the community occurs as flushed areas within ombrogenous mires) or for *Nardus stricta, Anthoxanthum odoratum* and *Festuca rubra* (where it interrupts stretches of calcifugous grassland). By contrast, more calcicolous herbs such as *Briza media, Primula farinosa, Linum catharticum* and *Sesleria albicans* are usually poorly represented. Bryophytes are generally a little less prominent in the sward than in the other sub-communities with species such as *Bryum pseudotriquetrum, Fissidens adianthoides* and *Ctenidium molluscum* of rather patchy occurrence, but a varied and locally extensive ground cover can occur and *Campylium stellatum* and *Scorpidium scorpioides* are often among the most obvious species. Three variants of this sub-community can be recognised, two of rather similar composition, one much more distinctive.

Eleocharis quinqueflora **variant:** *Carex hostiana-C. demissa* nodum Poore 1955b; *Carex demissa-C. panicea* nodum Poore 1955b p.p.; *Carex panicea-Campylium stellatum* nodum McVean & Ratcliffe 1962 p.p.; *Carex panicea-Campylium stellatum* Association Birks 1973 p.p.; *Carici dioicae-Eleocharitetum quinqueflorae* Typical subassociation (Birse & Robertson 1976) Birse 1980. In this variant, which is often very wet underfoot and closely associated with spring-heads, *Eleocharis quinqueflora* is preferentially frequent and quite commonly abundant, sharing dominance with the sedges or exceeding them in its cover. And, among the sedges, *C. hostiana* and *C. pulicaris* tend to be rather infrequent. *Erica tetralix*, too, is less common than usual in this sub-community. Associated herbs are relatively few in number (this is the most species-poor kind of *Pinguiculo-Caricetum*) and, though bryophytes can have a quite extensive ground cover, the frequent species are not numerous: apart from *Campylium stellatum* and *Scorpidium* there is usually just some *Aneura pinguis* and *Drepanocladus revolvens* and occasionally a little *Bryum pseudotriquetrum*.

Carex hostiana-Ctenidium molluscum **variant:** *Carex panicea-Campylium stellatum* nodum McVean & Ratcliffe 1962 p.p.; *Carex panicea-Campylium stellatum* Association Birks 1973 p.p.; *Caricetum hostiano-pulicaris* (Birse & Robertson 1976) Birse 1980. In its general floristics, this variant is very similar to the above, but it does show a noticeable shift in its sedge and bryophyte component. The former generally dominate among the vascular flora, *Eleocharis quinqueflora* being rather infrequent and usually of low cover, but *Carex hostiana* and *C. pulicaris* both show a rise in frequency here and, with *C. panicea*, they are usually the species with the highest cover. Plants of somewhat drier ground are also a little more common, with *Erica tetralix* frequent and locally fairly abundant and *Molinia caerulea, Festuca ovina, F. rubra* and *Nardus stricta* prominent in the context of flushed grasslands and wet heaths. Among the bryophytes, *Scorpidium* is much less common in this

variant than the former one but *Ctenidium molluscum* becomes common and often contributes a high proportion of the ground cover.

Schoenus nigricans **variant:** Flush vegetation Holdgate 1955b p.p.; *Schoenus nigricans* provisional nodum McVean & Ratcliffe 1962 p.p.; *Schoenetum nigricantis* Shimwell 1968a; *Eriophorum latifolium-Carex hostiana* Association Birks 1973; *Schoenus nigricans* Association Birks 1973; *Pinguiculo-Caricetum dioicae eleocharetosum* Jones 1973 p.p.; *Schoenus nigricans* Community Birse & Robertson 1976; *Schoenus ferrugineus* stands Wheeler *et al.* 1983 p.p. Through the *Pinguiculo-Caricetum* as a whole, *Schoenus nigricans* is only an occasional species but locally it attains high frequency and abundance in this kind of flush vegetation. On balance, this variant clearly belongs in the *Carex-Juncus* sub-community, with common records for such species as *Carex demissa, Juncus bulbosus/kochii, Erica tetralix* and *Scorpidium scorpioides*, but the prominence of *Schoenus* and the more or less consistent occurrence of some other species make this vegetation very distinct. Thus, *Eriophorum latifolium* is unusually frequent here (it is, generally speaking, a plant of the *Briza-Primula* sub-community) and *Drosera rotundifolia* is commonly accompanied by *D. anglica*, *Pinguicula vulgaris* occasionally by *P. lusitanica*. Then, along with common and sometimes abundant *Molinia*, there can be locally prominent *Myrica gale* and, among the bryophytes, there is frequently some *Blindia acuta*. As well as including many examples of sub-montane *S. nigricans* flushes, this variant also provides a locus for some of the occurrences of the very rare *S. ferrugineus*, recently rediscovered in what is apparently a natural locality in Perthshire (Wheeler *et al.* 1983).

Briza media-Primula farinosa **sub-community:** Flush vegetation Holdgate 1955a p.p.; Turfy marshes Pigott 1956; Calcareous marsh Sinker 1960; *Kobresieto-Caricetum* Ratcliffe 1965; *Caricetum lepidocarpae-hostianae* Shimwell 1968a; *Pinguiculo-Caricetum eleocharetosum p.p. & equisetosum p.p.* Jones 1973; *Carex* fens Proctor 1974 p.p.; *Pinguiculo-Caricetum dioicae* Jones 1973 *emend.* Wheeler 1975; *Anagallis tenella-Equisetum variegatum* Association Birse 1980. As in the *Carex-Juncus* sub-community, vascular plants still play a prominent role in the vegetation here, though many of the swards are open with extensive exposures of bare ground and some fine patterning of the different components over the micro-habitats. But in this sub-community calcicoles and more mesophytic herbs are much better represented. Thus, among the community species, *Carex lepidocarpa, C. hostiana* and *C. pulicaris* become more consistently frequent and are commonly accompanied by *C. flacca* and, in Upper Teesdale in Durham, by *Kobresia simpliciuscula. C. demissa* and *C.*

echinata, by contrast, are scarce. Then, among the preferentials, *Briza media*, *Primula farinosa*, *Linum catharticum*, *Sesleria albicans* and *Equisetum variegatum* occur frequently, *Eriophorum latifolium*, *Juncus alpinus* and *Tolfieldia pusilla* rather more occasionally. *Juncus bulbosus/kochii*, *Erica tetralix*, *Narthecium ossifragum* and *Drosera rotundifolia*, on the other hand, are much reduced in their occurrence and only of local significance. Also more common here than in the *Carex-Juncus* sub-community are *Festuca ovina*, *Agrostis stolonifera*, *Thymus praecox*, *Parnassia palustris*, *Prunella vulgaris*, *Euphrasia officinalis* agg. (including *E. confusa* and *E. micrantha* where species have been distinguished), *Leontodon autumnalis* and *Equisetum palustre*. Many of these preferential herbs provide a strong floristic continuity with the Mesobromion swards within which this sub-community typically occurs. The bryophyte element, which is generally prominent, especially in more open stands, also has some distinctive features: among the community species, *Aneura pinguis*, *Ctenidium molluscum* and *Fissidens adianthoides* are more consistently frequent and are often accompanied by *Cratoneuron commutatum*, less commonly by *C. filicinum*, *Calliergon cuspidatum*, *Ditrichum flexicaule* and *Racomitrium lanuginosum*. *Scorpidium scorpioides*, by contrast, is rare.

Three rather distinct variants can be recognised and further sampling may permit the characterisation of a fourth to contain stands which show a general affinity with this sub-community but which lie largely outside the range of such species as *Pinguicula vulgaris*, *Selaginella selaginoides*, *Carex dioica* and *Primula farinosa*. *Anagallis tenella*, generally speaking an uncommon plant in the community, is rather typical of this kind of vegetation which comes close in its floristics to our more base-rich dune slacks (e.g. the *Anagallis* variant of Wheeler's (1975, 1980*b*) *Pinguiculo-Caricetum* and the *Anagallis-Equisetum* Association of Birse (1980)).

Cirsium palustre **variant:** *Pinguiculo-Caricetum filipenduletosum* Wheeler 1975. The general features of the sub-community are maintained here, except that species such as *Sesleria albicans*, *Carex dioica* and *Primula farinosa*, whose range is more northern and montane, are of low frequency. More strikingly, the vegetation is consistently enriched by an assemblage of tall herbs most characteristic of Molinietalia fen-meadows. *Cirsium palustre*, *Valeriana dioica*, *Filipendula ulmaria*, *Angelica sylvestris* are all very frequent, with *Cardamine pratensis*, *Galium palustre* and *G. uliginosum* and, in wetter stands, *Carex rostrata*, *Menyanthes trifoliata* and *Caltha palustris*. Mixtures of these species, sometimes with much *Molinia*, may predominate over the small-sedge element in this vegetation, creating a quite different appearance from the usual kind of *Pinguiculo-Caricetum* flush; in other stands, *Juncus acutiflorus* is

locally abundant and, very occasionally, *Schoenus nigricans* can dominate. Among the bryophytes, *Calliergon cuspidatum* becomes constant and *Plagiomnium rostratum* and *Cratoneuron filicinum* are frequent.

Molinia caerulea-Eriophorum latifolium **variant:** Flush vegetation Holdgate 1955a *p.p.*; Turfy marshes Pigott 1956; Calcareous marsh Sinker 1960; *Caricetum lepidocarpae-hostianae molinietosum* Shimwell 1968a; *Pinguiculo-Caricetum eleocharetosum p.p.* & *equisetosum p.p.* Jones 1973; *Carex* fens Proctor 1974 *p.p.*; *Pinguiculo-Caricetum molinietosum* Wheeler 1975. This variant preserves the general features of the *Briza-Primula* sub-community much more obviously than the above with rich mixtures of small sedges and short herbaceous associates predominating in a usually close sward. *Carex flacca*, *Briza media*, *Festuca ovina*, *Sesleria caerulea*, *Primula farinosa*, *Linum catharticum*, *Equisetum palustre*, *Triglochin palustris* and *Parnassia palustris* all attain high frequencies along with the community constants. *Molinia* is well represented and it is in this variant that *Eriophorum latifolium* attains its highest frequency in the community. In Upper Teesdale, *Kobresia* is very common and there, too, *Bartsia alpina* and *Carex capillaris* are sometimes found. Among the bryophytes, all the community species are very frequent and, in addition, there is often some *Cratoneuron commutatum* and *Calliergon cuspidatum*.

Where this variant occurs in sites that are gently flushed, diverse mixtures of the above species can form a more or less intact and fairly smooth turf, sometimes dominated by sedges (or, in Upper Teesdale, by *Kobressia*), more rarely by *Molinia*. Grazing usually keeps the sward short but also very frequently the animals puddle the ground into a hummocky structure and wherever any of the factors tending to open up the sward becomes a predominant element in the environment, the vegetation cover (and often its shallow underlying soil) is disrupted, first to form a patchy carpet, then to leave fragments with extensive intervening areas of rock debris. The effects of this process on the floristics and physiognomy of this variant have been well described from Tarn Moor in Cumbria by Holdgate (1955a) and, in more detail, from Upper Teesdale by Pigott (1956a) and Jones (1973: see also Bradshaw & Jones 1976).

What happens in such situations is that, in addition to the fragmentation of the sward, which now appears as perched patches of varying size and shape, new microhabitats are created on the hummock sides, which are often eroded by undercutting with the flushing waters and by wind, and on the intervening very wet bare ground. The hummock sides provide an especially favourable situation for smaller rosette species such as *Pinguicula*, *Primula*, *Parnassia* and *Tofieldia*, as well as for occasional patches of *Saxifraga aizoides*, and also for bryophytes such as *Aneura pinguis*, *Ctenidium molluscum* and

Campylium stellatum: in such places, too, Pigott (1956a) also recorded occasional *Jungermannia atrovirens* and sparse *Leiocolea alpestris* and *Scapania nemorosa* var. *alata*. The gravelly flats in such mosaics in Teesdale are often very bare but where disturbance is less extreme the surface can have a cover of the sedges of the variant, often with *Triglochin palustre* and some *Juncus alpinus*, and locally abundant *Cratoneuron commutatum* and *Drepanocladus revolvens*. In this scheme, this kind of fine variation is all accommodated in this variant: in Jones (1973) and Bradshaw & Jones (1976), it was described and mapped under different sub-variants in the *Pinguiculo-Caricetum*.

Thymus praecox-Racomitrium lanuginosum variant: *Kobresieto-Caricetum* Ratcliffe 1965; *Caricetum lepidocarpae-hostianae kobresietosum* Shimwell 1968a; *Pinguiculo-Caricetum molinietosum* Wheeler 1975 *p.p.* This variant is floristically very similar to the last and in Upper Teesdale, where it is strongly centred, occurs in close association with it. But there are some obvious floristic differences. *Molinia* and *Eriophorum latifolium* are both very scarce and *Equisetum palustre*, *Triglochin palustre* and *Parnassia palustris* generally absent. *Sesleria* becomes rather uneven in its representation: it is absent from many stands (as in Ratcliffe's (1965) *Kobresieto-Caricetum*) but quite abundant in others (Jones 1973). The usual dominant, however, is *Kobresia*, with varying amounts of the sub-community sedges (though generally not much *C. dioica*) and, additionally here as a strong preferential, *C. capillaris*. More obviously, there is very frequently a little *Thymus praecox* and some scattered rosettes of *Plantago maritima*. *Thalictrum alpinum* is also preferential at lower frequency. Again, the ground is often rather hummocky in this variant and *Racomitrium lanuginosum* is often prominent on the raised areas. *Tortella tortuosa* and *Scapania aspera* have also been recorded from this vegetation.

Gymnostomum recurvirostrum sub-community: Gravel flushes Pigott 1956a; *Gymnostometo-Caricetum* Ratcliffe 1965; *Pinguiculo-Caricetum equisetosum* Jones 1973 *p.p.* This very striking vegetation is superficially very different from other kinds of *Pinguiculo-Caricetum* but almost all the constants of the community are well represented, with only *Carex hostiana, C. pulicaris, Eriophorum angustifolium* and *Ctenidium molluscum* markedly reduced. Floristically, it is quite close to the *Briza-Primula* sub-community, with frequent records for *Festuca ovina, Linum catharticum, Primula farinosa, Thymus praecox, Agrostis stolonifera* and *Cratoneuron commutatum*, though *Carex flacca, Briza media, Triglochin palustris, Sesleria albicans* and *Parnassia palustris* are uncommon. Typically, though, it is not these vascular species, which are as a rule of low individual and total cover, that give the vegetation its

character. Much more obvious is the moss *Gymnostomum recurvirostrum*, which occurs occasionally in the *Briza-Primula* sub-community as a coloniser of eroding turf fragments and on patches of other bryophytes, but which here forms prominent hummocks, up to 30 cm high and 60 cm or so across. *Catascopium nigritum* can also be found occasionally forming smaller cushions.

Between these, there is characteristically much bare ground: eroding patches of soil, exposed rock debris and patches of accumulating wind-borne material. On these, the vascular plants form a generally fragmentary cover, representing remnants of once more continuous turf or areas of recolonisation. And some good preferentials of this variant are especially associated with these areas. *Plantago maritima* is very common but more strictly confined to such situations are *Sagina nodosa, Minuartia verna* and, in its only British locus, the very rare *M. stricta*, like the other two species a perennial of open conditions, but perhaps behaving as an annual (Coombe & White 1951, Pigott 1956a). *Juncus alpinus* and *J. triglumis* also tend to be better represented in these open areas than anywhere else in the community. Among the bryophytes, *Drepanocladus revolvens* and, especially nearer to the sources of springs, *Cratoneuron commutatum*, can also be abundant between the *Gymnostomum* hummocks. Recolonisation of the moss hummocks themselves can also sometimes be observed, as these break down with age and wind erosion, when some of the small sedges and rosette dicotyledons may gain a hold, particularly on the sheltered lee side and where the mounds have not been hardened by tufa-deposition (Pigott 1956a).

Again, both the hummocks and the intervening areas are here treated as part of a single vegetation type: in Jones (1973) and Bradshaw & Jones (1976) they were described and mapped separately as sub-variants.

Habitat

The *Pinguiculo-Caricetum* is typically a soligenous mire of mineral soils and shallow peats kept very wet by base-rich, calcareous and oligotrophic waters. It is predominantly a community of north-western Britain and the cool, wet climate of the region has an obvious influence on the floristics of the vegetation and, in more extreme situations, on the structure of the sward. However, most stands are grazed and trampled by large herbivores and it is probably these factors, combined with nutrient impoverishment and the often strong and scouring effect of the irrigation, which play the major part in maintaining the community in its generally rich, varied and open state.

The pH of the flushing waters here is high, usually between 5.5 and 7.0, sometimes higher (Poore 1955a, Pigott 1956a, McVean & Ratcliffe 1962, Birks 1973, Wheeler 1983) and the amount of dissolved calcium large, up to 100 mg l^{-1} (Pigott 1956a, Wheeler 1983,

Wheeler *et al.* 1983), often with some re-precipitation as tufa among the herbage. These conditions are strongly reflected in the composition of the community which is one of the most calcicolous of our mires, much of its distinctive character being given by species such as *Carex dioica*, *C. hostiana*, *C. pulicaris*, *C. lepidocarpa*, *C. flacca*, *Sesleria albicans*, *Briza media*, *Eleocharis quinqueflora*, *Eriophorum latifolium*, *Linum catharticum* and *Tofieldia pusilla*, rarities like *Equisetum variegatum* and *Kobresia simpliciuscula*, and the bryophytes *Aneura pinguis*, *Drepanocladus revolvens*, *Campylium stellatum*, *Ctenidium molluscum*, *Cratoneuron commutatum*, *Gymnostomum recurvirostrum* and *Catascopium nigritum*. Such an assemblage provides a strong definition for the community against its counterparts in the Caricion nigrae and in flushed Oxycocco-Sphagnetea mires which occupy hydrologically similar, but more base-poor, situations.

Generally speaking, the source of base- and cation-enrichment here is calcareous bedrocks and superficials, from which ground water emerges where it strikes impervious substrates, running down over the ground surface and through the upper soil horizons to create a slope flush. But these materials are quite varied in their lithology and occur in diverse geological settings. They include sedimentary limestones, notably the Carboniferous Limestone, which crops out in massive deposits in the Pennines with interbedded limy partings also occurring in the Yoredale transitions to the Millstone Grit (Holdgate 1955a, b, Pigott 1956a, Sinker 1960, Shimwell 1968a) and the Cambrian/Ordovician Durness Limestone which is of more local importance on Skye (Birks 1973) and the north-west Scottish mainland (McVean & Ratcliffe 1962). Elsewhere, the community can mark out percolation from occasional calcareous strata in such deposits as the Ordovician/Silurian shales of the Southern Uplands (e.g. Ferreira 1978). Then, it can occur over metamorphosed calcareous deposits, as in Upper Teesdale, where it is part of the complex of vegetation found over the 'sugar-limestone' (Pigott 1956a, 1978b, Ratcliffe 1965, Jones 1973) and in the Scottish Highlands, where it occurs widely on the Dalradian metasediments and, less commonly, those of the Moine series (Poore 1955a, McVean & Ratcliffe 1962). In Scotland, too, and in the Lake District, certain igneous rocks provide local enrichment of flushing waters. Very often, throughout its range, glacial superficials, particularly fine-textured drift, make some contribution, ill-draining and calcareous, to the soil parent materials (Holdgate 1955a, Pigott 1956a, 1978b). The community can also be found on flushed shell-sand, as in some sites on the north Scottish coast (Ratcliffe 1977, Birse 1980), and on material redistributed by river erosion to which there has been a modest alluvial contribution (e.g. Pigott 1956a).

The soils which develop under the community in such diverse situations are often largely of mineral origin and, though the particular parent materials can be very different, the predominant influence of more or less continuous irrigation with lime-rich waters means that they can generally be classified as calcareous surface-water gleys (Soil Survey 1983). Typically, there is no stagnation and *in situ* formation of peat is very limited with usually no more than a humic topsoil, a feature which helps distinguish the habitat of the community from that of base-rich basins where it is replaced by vegetation like the *Carex-Calliergon* and *Carex-Sphagnum warnstorfii* mires in which swamp sedges play an important role, sometimes with more base-tolerant Sphagna, two groups of plants which are notably very scarce here. However, peat remnants can occur quite often, as where springs are eating back into tracts of blanket mire and, quite commonly, the community can be found on shallow peat surrounds to flushes, where irrigation occurs well within ombrogenous systems.

Variation in the degree of base- and cation-enrichment of the waters, the extent to which this effect is diffused as the waters emerge and the degree to which raised areas of the sward become removed from the immediate influence of irrigation, probably all play some part in the floristic differences between the various kinds of *Pinguiculo-Caricetum*, most obviously between the *Carex-Juncus* sub-community on the one hand and the *Briza-Primula* and *Gymnostomum* sub-communities on the other. The former, with its poorer representation of calcicoles and the preferential occurrence of species like *Carex demissa*, *Juncus bulbosus/kochii* and *Erica tetralix*, tends to occur more often on peaty gleys and shallow peats than the other types, and is predominantly a Scottish and Cumbrian vegetation type, marking out very local base-enrichment within the context of Nardo-Galion grasslands and Oxycocco-Sphagnetea wet heaths and mires on prevailingly acid soils. The latter types of *Pinguiculo-Caricetum*, with their more obvious calcicolous component, are usually found on surface-water gleys which are often little more than very wet lithomorphic soils, among tracts of calcicolous grasslands, most notably on the Pennine Carboniferous Limestone. But we need to know a great deal more about the ranges of edaphic variables here, about which are important and how their, perhaps quite subtle, control on the vegetation works (e.g. Poore 1955a, Clymo 1962, Wheeler *et al.* 1983).

Calcicolous flushes of the kind occupied by the *Pinguiculo-Caricetum* tend to occur more commonly towards the north-west of Britain, where suitable bedrocks and superficials are widely, though often locally, distributed on the slopes of the upland fringes in a region with a wet and cool climate. Annual precipitation through the range of the community is almost always in excess of 800 mm and usually more than 1200 mm (*Climatological Atlas* 1952) with typically more than 160 wet days yr^{-1} (Ratcliffe 1968), features which help maintain the constant flushing characteristic of the habitat. The generally low summer temperatures, with a mean annual maximum for the most part below 25 °C (Conolly & Dahl 1970), are also clearly reflected in the floristics of the community, with a good representation of Continental Northern plants like *Pinguicula vulgaris*, *Carex dioica*, *C. pulicaris* and *Eriophorum latifolium* and rarer species such as *Schoenus ferrugineus* and *Drosera anglica*, and the Northern Montane *Primula farinosa* and *Juncus alpinus*. In Scotland the community does not generally penetrate into high montane regions, but in the very harsh climate of Upper Teesdale, it provides a far-flung outpost for such Arctic-Alpine plants as *Kobresia*, *Tofieldia pusilla*, *Bartsia alpina*, *Carex capillaris*, *Juncus triglumis*, *Minuartia verna* and the only British locus for *M. stricta* (Pigott 1956a, Ratcliffe 1965). Towards the south-east of Britain, calcareous slope flushes are not unknown, but they are very local and, though similar vegetation can be found in this region, it lacks some of the most characteristic species of the community and, until further samples are available, its status must remain tentative. Certainly, at lower altitudes on flushed shell-sands, the *Pinguiculo-Caricetum* closely approaches certain kinds of *Salix repens-Campylium stellatum* dune-slack in its composition (e.g. Birse 1980). More generally in the lowland south-east, with its warmer and drier climate, the community is replaced by the *Schoenetum nigricantis*, though this vegetation type extends on to deeper base-rich peats of valley mires and some basin mires. The *Cirsium palustre* variant of the *Briza-Primula* sub-community shows some strong affinities with that community and with those Molinietalia fen-meadows that occupy fairly base-rich sites with a higher nutrient turnover than is usual in the habitat of the *Pinguiculo-Caricetum*, which is very typically of low productivity with little nitrogen and phosphorus in the profile, a feature of great importance in maintaining the open and often sedge-dominated character of the sward (Pigott 1956a, 1978b; Wheeler 1983, Wheeler *et al.* 1983).

Although bulkier dominants like *Schoenus* or *Molinia*, and a variety of taller fen herbs, can make some prominent contribution to the cover in this *Cirsium* variant and other kinds of *Pinguiculo-Caricetum*, the herbage of this kind of mire is much more commonly kept very short because of grazing. Throughout its range, the community typically occurs in unenclosed uplands to which, in summer, stock have free access: generally, sheep are the predominant grazers but in some sites cattle and ponies are important and, in Scotland especially, deer. The constant nibbling of the turf, which in early summer can yield a good bite

because of the irrigation, undoubtedly contributes to the richness of the sward and the great diversity in dominance in the community, by helping keep more vigorous species in check.

But there is a further effect of herbivores which is of great importance in controlling the physiognomy of the vegetation and the openness of the ground, and that is trampling (Pigott 1956a). On the generally very wet surfaces here, this tends to disrupt the sward and contributes to maintaining the turf in an irregular and discontinuous condition, with patches of exposed and redistributed mineral and organic matter and even bare bedrock fragments or drift. Flushing itself can play a large part in the erosion of the surface, particularly where the sward has already been broken. Sometimes the community marks out sites of rather diffuse irrigation below weak spring lines or where waters percolate some distance beneath a soil mantle; but more vigorous flushing, augmented in areas of harsher climate by snow-melt, and working more obviously on steeper slopes, can scour these flushes, washing away substantial amounts of soil and leaving systems of bare runnels and flats between irregular fragments of perched vegetation. Wind ablation of these blocks can undercut them and cause collapse and, where the winters are harsh, freeze-thaw can burst open the remnants of the sward and intervening areas of debris, causing further loosening of the surface. Strong flushes can also eat back upslope, enlarging the potential area of occupation of the community and feeding the process of redistribution of soil and rock. Such effects have been observed in many flushes of this kind (e.g. Holdgate 1955a, Birks 1973, Wheeler *et al.* 1983), and they contribute generally to a horizontal and vertical differentiation of intimately juxtaposed microhabitats. But they are especially well seen in Upper Teesdale, where a harsh climate and intensive trampling by stock are actively eroding flushed slopes of sugar-limestone, much of which has already been well disintegrated under a mantle of drift and peat, and creating a unique complex of variants of the *Pinguiculo-Caricetum*, pictured very clearly in the profiles of Pigott (1956a) and the maps of Bradshaw & Jones (1976). The *Briza-Primula* sub-community occupies most of the ground in these flushes (though there are areas which are close to the *Carex-Juncus* sub-community): the *Thymus-Racomitrium* variant is on the drier areas, the *Molinia-Eriophorum* on the wetter. And, in the most open places, there are some striking stands of the *Gymnostomum* sub-community which seems essentially to represent a regeneration vegetation, colonising fragments of existing sward, eroded soil material or accumulated aeolian detritus, offering a temporary site for invasion by vascular plants, and then declining as the hummocks open up and decay (Pigott 1956a, Ratcliffe 1965).

Zonation and succession

The *Pinguiculo-Caricetum* marks out areas of soligenous influence of base-rich and calcareous waters, often in association with spring and rill vegetation, within grasslands and more occasionally in ombrogenous mires and around topogenous mires. Stands are usually small and their definition in zonations depends on the strength of flushing and the degree of base-enrichment. Most stands

Figure 14. Base-rich flushes among pasture at high (*a*) and moderate (*b*) altitudes in northern Britain.
M10b   *Pinguiculo-Caricetum* mire, *Briza-Primula* sub-community
M11a   *Carici-Saxifragetum* mire, *Thalictrum-Juncus* sub-community
M12   *Caricetum saxatilis* mire
M37   *Cratoneuron-Festuca* spring
CG10a   *Festuca-Agrostis-Thymus* grassland, *Trifolium-Luzula* sub-community
CG10b   *Festuca-Agrostis-Thymus* grassland, *Carex* sub-community
CG11a   *Festuca-Agrostis-Alchemilla* grass-heath, Typical sub-community
CG11b   *Festuca-Agrostis-Alchemilla* grass-heath, *Carex* sub-community
CG12   *Festuca-Alchemilla-Silene* dwarf-herb community

are affected by herbivores and some would probably progress to Alno-Ulmion scrub or woodland if grazing were withdrawn; at higher altitudes, the vegetation may be a climatic climax, though with an internally dynamic character.

The most widespread situation in which the community occurs is within sub-montane pastures where it typically forms a zone below springs and flush-lines, often pear-shaped and elongated downslope below isolated sites of irrigation, in larger stands of more complex configuration where flushing is extensive. Towards the source of flushing it commonly gives way to Cratoneurion spring vegetation, usually the *Cratoneuron-Festuca rubra* community which can extend as a narrow strip down a central vigorous rill. In Scotland especially but also in the Lake District and more locally in the Pennines, *Saxifraga aizoides* rises to prominence in the continuously splashed zone to the centre of these flushes and there is frequently a clear intervening band of the *Carici-Saxifragetum* (McVean & Ratcliffe 1962, Birks 1973). Where the waters are somewhat less base-rich, the spring vegetation may be represented by the *Cratoneuron-Carex nigra* community, particularly towards higher altitudes in Scotland, where this vegetation type subsumes part of McVean & Ratcliffe's (1962) *Hypno-*

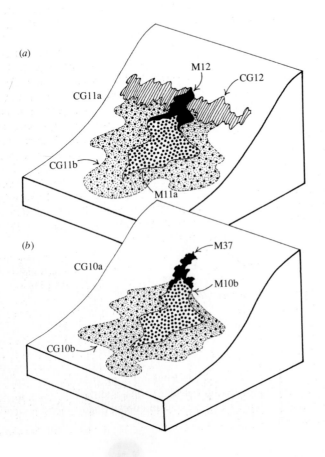

Caricetum alpinum, or even by some kind of *Philonotis* flush where the waters are only weakly calcareous (Holdgate 1955a, Pigott 1956a). Where flushing occurs on more level ground, complex mosaics of these communities can be seen around the springs (e.g. Poore 1955a, McVean & Ratcliffe 1962).

Towards the margins of these flushes, the *Pinguiculo-Caricetum* gives way, with varying degrees of abruptness, to the pasture vegetation on drier ground, often standing out in the smooth swards, by virtue of its uneven and broken surface. In Scotland, where the community is generally represented by the *Carex-Juncus* sub-community, a very widespread context for the *Pinguiculo-Caricetum* is the *Festuca-Agrostis-Thymus* grassland, the common plagioclimax sward of more base-rich brown earths associated with local calcareous outcrops in the Dalradian and, less so, the Moine Assemblages and the Durness Limestone of the far north-west (McVean & Ratcliffe 1962, Birks 1973, Ratcliffe 1977). In the Lake District, a similar zonation can be seen on more calcareous parts of the Borrowdale Volcanics. In Scotland, too, in situations where the surrounding soils are somewhat less base-rich, the *Pinguiculo-Caricetum* can occur within stretches of the *Festuca-Agrostis-Alchemilla* grass-heath. In both of these grassland types, a *Carex pulicaris-Carex panicea* sub-community often forms a transition zone on the mildly-irrigated surrounds to the flush (Figure 14) but more abrupt zonations can occur and where soligenous base-enrichment is very localised within stretches of more prevailingly pervious and acidic rocks and soils, the *Pinguiculo-Caricetum* may be very sharply marked off from calcifugous *Festuca-Agrostis* grassland.

Comparable zonations to the above can be found in the northern Pennines, though here the *Briza-Primula* sub-community is the usual kind of *Pinguiculo-Caricetum* and the pasture is generally the *Sesleria-Galium* grassland. With the more uniformly calcareous condition of the soils over the Carboniferous Limestones, there is often a very strong floristic continuity between the different vegetation types, the axis of variation being largely controlled by the amount of soil moisture. A common sequence is from the Typical sub-community of the grassland, through (again) a *Carex pulicaris-Carex panicea* sub-community and then to the *Molinia-Eriophorum* variant of the mire, species such as *Sesleria*, *Festuca ovina*, *Carex flacca*, *Briza media* and *Linum catharticum* running throughout with virtually undiminished frequency. In Upper Teesdale, the presence of species such as *Kobresia*, *Carex capillaris*, *Plantago maritima* and *Racomitrium uliginosum* enriches both the pasture (often represented by its *Carex-Kobresia* sub-community) and the drier parts of the mire (in the *Thymus-Racomitrium* variant) in this kind of zonation and the *Gymnostomum* sub-community appears on the

open gravels (Pigott 1956a, Ratcliffe 1965, Jones 1973, Bradshaw & Jones 1976). Sadly, some of the most intricate of these patterns were lost beneath the Cow Green reservoir. More locally in northern England and north Wales, the *Briza-Primula* sub-community can pick out flushes within the *Festuca-Avenula* grassland or, in Durham, the *Sesleria-Scabiosa* grassland.

Less common than these kinds of situation, though often very striking because of the close juxtaposition of rather different vegetation types, are those occurrences of the *Pinguiculo-Caricetum* around places of soligenous base-enrichment within ombrogenous mires. Fragments of perched blanket mire can sometimes be found where springs are biting back on to gentle peat-clad slopes above (as in upper Teesdale: Pigott 1956a) but, in Scotland, the *Carex-Juncus* sub-community can occur in prevailing ombrogenous systems, particularly at lower altitudes in the west, where it gives way directly to the *Scirpus-Eriophorum* mire or sometimes to an intervening zone of the *Carex panicea* sub-community of the *Scirpus-Erica* wet heath, into which *C. panicea*, *C. pulicaris*, *C. echinata*, *C. demissa*, *Pinguicula vulgaris*, *Selaginella selaginoides* and locally dominant *Schoenus nigricans* extend their cover (Poore 1955a, McVean & Ratcliffe 1962). In this kind of setting, and where the *Pinguiculo-Caricetum* occurs more locally in stretches of heath on thin, dry peats, the community typically stands out as fresh green patches.

Although the flushes with which this kind of mire is associated are often isolated (though repeated) on slopes, they sometimes feed topogenous fens in basins or channels and then the *Pinguiculo-Caricetum* can form a fringing zone around the central area, passing to communities like the *Carex-Calliergon* mire or, more exclusively at higher altitudes, the *Carex-Sphagnum warnstorfii* mire on the deeper peat of the more stagnant waters. Such transitions are well seen in small hollows in the central Highlands (McVean & Ratcliffe 1962), where the *Carex-Juncus* sub-community forms the margins, and at Sunbiggin Tarn (Holdgate 1955a) and Malham Tarn (Sinker 1960, Proctor 1974), where the *Briza-Primula* sub-community occurs.

Almost throughout its range, the *Pinguiculo-Caricetum* is strongly affected by grazing animals. Without such influence, many stands would probably be open to invasion by shrubs and trees though nutrient deficiency in the soils might be an important hindrance to some species. Likely colonisers are *Betula pubescens*, *Salix cinerea* and, in these northern localities, *S. pentandra*, *S. phylicifolia*, *S. nigricans* and *Prunus padus* with *Fraxinus excelsior* figuring on drier margins; *Alnus glutinosa* is more problematic because of the difficulty of fruit transport. Such mixtures are most characteristic of our Alno-Ulmion woodlands like the *Alnus-Fraxinus-Lysimachia* woodland and wetter stands of the *Fraxinus-*

Sorbus-Mercurialis woodland, beneath the canopies of which there is often a good representation of the kinds of tall herbs that figure in the *Cirsium palustre* variant here. In some localities, as in Teesdale and Weardale in Durham and in Swaledale in North Yorkshire, small patches of *Pinguiculo-Caricetum* persist in the wetter parts of traditionally-treated hay-meadows, whose vegetation has probably been secondarily derived from these kinds of forest (Pigott 1956*a*).

Distribution

The community is of widespread but local occurrence throughout northern England and Scotland with fragmentary and often rather impoverished stands in Wales and the Midlands. The *Carex-Juncus* sub-community is largely restricted to Scotland and the Lake District with outlying occurrences in north Wales and Upper Teesdale, the *Schoenus* variant confined to low altitudes, the other variants extending to higher levels. The *Briza-Primula* sub-community is the predominantly northern England type, being centred on the Pennines where the *Molinia-Eriophorum* variant is the common expression, but also including species-poor stands and the transitional vegetation of the *Cirsium* variant further south. The *Thymus-Racomitrium* variant and the *Gymnostomum* sub-community have been described only from Upper Teesdale.

Affinities

The *Pinguiculo-Caricetum* as defined here is an expanded version of the community given that name by Jones (1973: see also Bradshaw & Jones 1976) and amended by Wheeler (1975, 1980*b*), taking in a variety of other vegetation types with essentially similar floristics, most notably the Scottish flushes first described by Poore (1955*a*) and McVean & Ratcliffe (1962) and later by Birse (1980, 1984: see also Birse & Robertson 1976). Although quite diverse in its composition and structure, it is well defined against most of the closely-related mires and its floristic affinities are fairly straightforward.

Through the *Carex-Juncus* sub-community it grades to soligenous Caricion nigrae poor fens and some Oxycocco-Sphagnetea vegetation of more base-poor flushes, like the *Carex-Nardus* sub-community of the *Carex echinata-Sphagnum* mire and the *Carex panicea* sub-community of the *Scirpus-Erica* wet heath. *Molinia*, *Eriophorum angustifolium* and some of the more broadly-tolerant small sedges or plants like *Erica tetralix* and *Juncus bulbosus/kochii* provide links in this direction, but the great predominance of calcicoles here and the consistent replacement of Sphagna by brown mosses provide good diagnostic features. Through the *Briza-Primula* sub-community, the community shows a strong continuity with Mesobromion swards, notably the *Sesleria-Galium* grassland, and richer Nardo-Galion

communities like the *Festuca-Agrostis-Thymus* grassland, reflecting the impact of grazing over drier areas differentiated in the flushed sward: here the calcicolous or mesophytic element is maintained throughout the sequence, the proportion of hydrophilous species providing the major distinction. Affinities with Molinietalia communities can be seen locally but these are much more obvious in the *Schoenetum* which replaces this community, with some floristic transitions, towards the southern lowlands of Britain.

Transitions to spring and rill vegetation of the Cardamino-Montion and Cratoneurion alliances are a little more difficult to define, particularly in Scotland and the Lake District where *Saxifraga aizoides* provides a strong link between the two: the closest community to the *Pinguiculo-Caricetum* is the *Carici-Saxifragetum aizoidis* with which it shares many calcicolous herbs and bryophytes.

In phytosociological terms, this is our central Caricion davallianae small-sedge mire, showing clear affinities, particularly through the *Briza-Primula* sub-community, with communities like the *Caricetum davallianae* Dutoit 1924 *emend*. Görs 1963 described from Germany (Görs 1964, Oberdorfer 1977, Ellenberg 1978), Switzerland (Dutoit 1924) and Poland (Matuszkiewicz 1981) and the *Parnassio-Caricetum pulicaris* (Oberdorfer 1957) Görs 1963 and *Pinguiculo-Parnassietum* (Libbert 1928) Passarge 1964 described from Germany (Görs 1963, Passarge 1964) and The Netherlands (Westhoff & den Held 1969). Through higher-altitude stands in this sub-community and also through the *Carex-Juncus* type, there are also links with the Scandinavian base-rich mires, like the Braunmos-reiche *Carex panicea* Assoziation (Nordhagen 1928), generally placed in the Caricion bicolori-atrofuscae (Nordhagen 1928, 1943; see also Persson 1961, Svensson 1965, Hallberg 1971, Ellenberg 1978): such communities, however, often have a good representation of Arctic-Alpine willows, best seen in Britain in ungrazed wet tall-herb vegetation and high-altitude scrub. The *Schoenus* variant also includes vegetation which, on mainland Europe, has been placed in a variety of *Schoeneta*, like the *Orchio-Schoenetum nigricantis* Oberdorfer 1957 described from Germany (Oberdorfer 1957, 1977) and Poland (Matuszkiewicz 1981). *S. ferrugineus* figures much more extensively in these communities on the Continent, notably in the alpine foothills (as in the *Primulo-Schoenetum ferruginei* (Koch 1926), Oberdorfer (1957) 1962) and around the Baltic, and Nordhagen (1937) placed low-productivity, base-rich mires from low altitudes in an alliance Schoenion ferruginei (Wheeler *et al.* 1983: see also Görs 1964, Kloss 1965, Tyler 1979). British *Schoenus* flushes in the *Pinguiculo-Caricetum* have a distinctly oceanic character compared with European *Schoeneta* of this kind.

Floristic table M10

	a	b	c	10
Carex panicea	V (1–5)	V (1–6)	V (1–7)	V (1–7)
Pinguicula vulgaris	V (1–5)	IV (1–3)	IV (1–4)	V (1–5)
Carex dioica	III (1–6)	IV (1–5)	V (1–5)	IV (1–6)
Aneura pinguis	III (1–5)	IV (1–4)	IV (1–4)	IV (1–5)
Drepanocladus revolvens	III (1–9)	IV (1–7)	IV (2–7)	IV (1–9)
Juncus articulatus	III (1–5)	IV (1–4)	IV (1–3)	IV (1–5)
Carex lepidocarpa	II (1–6)	IV (1–6)	IV (2–5)	IV (1–6)
Bryum pseudotriquetrum	II (1–5)	IV (1–3)	IV (1–2)	IV (1–5)
Campylium stellatum	V (1–7)	IV (1–4)	II (2–4)	IV (1–7)
Selaginella selaginoides	III (1–5)	IV (1–4)	II (2–4)	III (1–5)
Carex hostiana	IV (1–7)	V (1–6)		IV (1–7)
Ctenidium molluscum	III (1–9)	V (1–8)	I (2)	IV (1–8)
Carex pulicaris	III (1–6)	IV (1–5)		IV (1–6)
Eriophorum angustifolium	III (1–7)	IV (1–4)		IV (1–7)
Potentilla erecta	III (1–5)	IV (1–6)		III (1–6)
Succisa pratensis	III (1–6)	IV (1–4)		III (1–6)
Molinia caerulea	III (1–7)	IV (1–6)		III (1–7)
Carex nigra	III (1–7)	III (1–4)	I (1–3)	III (1–7)
Fissidens adianthoides	II (1–5)	III (1–3)		II (1–5)
Anthoxanthum odoratum	II (1–7)	II (1–3)		II (1–7)
Ditrichum flexicaule	I (1–3)	I (1–3)		I (1–3)
Thalictrum alpinum	I (1–5)	I (1–3)		I (1–5)
Hylocomium splendens	I (1–7)	I (1–3)		I (1–7)
Sphagnum subnitens	I (1–3)	I (1)		I (1–3)
Ranunculus acris	I (1–4)	I (1–3)		I (1–4)
Taraxacum officinale agg.	I (1–3)	I (1–3)		I (1–3)
Festuca ovina	II (1–5)	III (1–7)	III (2–4)	II (1–7)
Linum catharticum	I (1–3)	IV (1–4)	III (1–2)	II (1–4)
Primula farinosa	I (1–5)	IV (1–4)	III (1–3)	II (1–4)
Cratoneuron commutatum	I (1–5)	III (1–6)	IV (1–4)	II (1–6)
Agrostis stolonifera	I (1)	III (1–3)	IV (1–4)	II (1–4)
Thymus praecox	I (1)	II (1–4)	III (1–2)	II (1–4)
Equisetum variegatum	I (1–3)	II (1–6)	II (1–3)	I (1–6)
Juncus alpinus	I (1–5)	II (1–3)	II (1–2)	I (1–5)
Cratoneuron filicinum		I (1–3)	I (2–3)	I (1–3)
Deschampsia cespitosa		I (1–4)	I (3)	I (1–4)
Eleocharis quinqueflora	III (1–7)	II (1–4)	II (1–4)	III (1–7)
Carex demissa	III (1–9)	I (1–3)	I (1–3)	III (1–9)
Juncus bulbosus/kochii	III (1–6)	I (1–3)		II (1–6)
Erica tetralix	III (1–7)	I (1–4)		II (1–7)
Scorpidium scorpioides	III (1–8)	I (1)		II (1–8)
Carex echinata	II (1–7)	I (1–3)		II (1–7)
Narthecium ossifragum	II (1–5)	I (1–3)		I (1–5)
Drosera rotundifolia	II (1–5)	I (1)		I (1–5)
Saxifraga aizoides	II (1–5)	I (2–5)		I (1–5)

Floristic table M10 *(cont.)*

	a	b	c	10
Schoenus nigricans	II (2–5)	I (1)		I (1–5)
Polygonum viviparum	I (1–4)			I (1–4)
Juncus squarrosus	I (1–5)			I (1–5)
Scirpus cespitosus	I (1–5)			I (1–5)
Pedicularis sylvatica	I (1–3)			I (1–3)
Drosera anglica	I (1–5)			I (1–5)
Blindia acuta	I (1–3)			I (1–3)
Schoenus ferrugineus	I (1–10)			I (1–10)
Myrica gale	I (1–5)			I (1–5)
Pinguicula lusitanica	I (1–5)			I (1–5)
Breutelia chrysocoma	I (1–3)			I (1–3)
Triglochin palustris	II (1–4)	IV (1–3)	II (1)	III (1–4)
Carex flacca	II (1–6)	IV (1–5)	I (3)	II (1–5)
Briza media	I (1–6)	IV (1–4)		II (1–4)
Racomitrium lanuginosum	I (1–4)	III (1–9)	II (1–2)	II (1–9)
Prunella vulgaris	I (1–6)	III (1–6)	I (2–4)	II (1–6)
Parnassia palustris	I (1–3)	III (1–3)	I (2)	II (1–5)
Equisetum palustre	II (1–5)	III (1–4)		II (1–5)
Eriophorum latifolium	II (1–8)	III (1–4)		II (1–8)
Kobresia simpliciuscula		III (2–8)	II (1–2)	II (1–8)
Leontodon autumnalis	I (1–5)	III (1–3)		I (1–5)
Tofieldia pusilla	I (1–3)	II (1–4)	I (1)	I (1–4)
Calliergon cuspidatum	I (1–8)	II (1–4)		I (1–8)
Pellia endiviifolia	I (1–5)	II (1–3)		I (1–5)
Pedicularis palustris	I (1–3)	II (1–4)		I (1–4)
Filipendula ulmaria	I (5)	II (1–3)		I (1–5)
Ranunculus flammula	I (1–4)	II (1–3)		I (1–4)
Cirsium palustre	I (1–3)	II (1–3)		I (1–3)
Cardamine pratensis	I (1)	II (1–3)		I (1–3)
Leontodon taraxacoides		II (1–4)		I (1–4)
Sesleria albicans		II (1–7)		I (1–7)
Valeriana dioica		II (1–3)		I (1–3)
Angelica sylvestris		II (1–3)		I (1–3)
Galium uliginosum		II (1–3)		I (1–3)
Plagiomnium rostratum		II (1–3)		I (1–3)
Tortella tortuosa		I (2–3)		I (2–3)
Scapania aspera		I (1–3)		I (1–3)
Bartsia alpina		I (2–3)		I (2–3)
Caltha palustris		I (1–3)		I (1–3)
Juncus acutiflorus		I (1–4)		I (1–4)
Luzula multiflora		I (1–3)		I (1–3)
Dactylorhiza fuchsii		I (1–3)		I (1–3)
Holcus lanatus		I (1–3)		I (1–3)
Galium palustre		I (1–3)		I (1–3)

	a	b	c	10
Plantago maritima	I (1–5)	II (1–4)	IV (1–3)	II (1–5)
Gymnostomum recurvirostrum		I (2–4)	V (1–10)	I (1–10)
Sagina nodosa		I (1–3)	III (1–3)	I (1–3)
Minuartia verna			III (2–4)	I (2–4)
Carex capillaris	I (1–3)	I (1–4)	II (1–4)	I (1–4)
Juncus triglumis	I (1–2)	I (1–3)	II (2–3)	I (1–3)
Catascopium nigritum			II (2–5)	I (2–5)
Minuartia stricta			I (1–2)	I (1–2)
Euphrasia officinalis agg.	II (1–5)	III (1–4)	II (1–2)	II (1–5)
Nardus stricta	I (1–7)	I (1–4)	I (2)	I (1–7)
Festuca rubra	I (1–5)	I (1–5)	I (1–2)	I (1–5)
Number of samples	143	55	19	217
Number of species/sample	23 (11–49)	37 (19–56)	21 (10–29)	25 (10–56)
Herb height (cm)	13 (4–35)			
Herb cover (%)	76 (20–100)	77 (50–100)	36 (20–50)	74 (20–100)
Bryophyte height (mm)	16 (2–30)			
Bryophyte cover (%)	45 (2–95)	58 (30–90)	65 (25–100)	49 (2–100)
Altitude (m)	269 (1–792)			
Slope (°)	8 (0–30)	6 (0–20)	2 (0–5)	7 (0–30)
Soil pH	6.6 (5.8–7.2)			

a   *Carex demissa-Juncus bulbosus/kochii* sub-community
b   *Briza media-Primula farinosa* sub-community
c   *Gymnostomum recurvirostrum* sub-community
10   *Pinguiculo-Caricetum dioicae* (total)

Floristic table M10, variants

	ai	aii	aiii	bi	bii	biii
Carex panicea	V (1–9)	V (1–7)	V (1–5)	V (1–4)	V (1–5)	V (1–6)
Pinguicula vulgaris	V (1–4)	IV (1–5)	V (1–5)	V (1–3)	IV (1–3)	III (1–3)
Campylium stellatum	V (1–7)	V (1–9)	V (1–7)	IV (1–4)	IV (1–4)	IV (1–4)
Carex dioica	III (1–6)	II (1–6)	III (1–5)	III (1–4)	V (1–4)	IV (1–5)
Aneura pinguis	III (1–5)	III (1–4)	II (1–3)	IV (1–3)	IV (1–3)	III (1–4)
Drepanocladus revolvens	III (1–9)	III (1–7)	III (1–5)	III (1–4)	IV (1–7)	IV (1–4)
Juncus articulatus	III (1–3)	III (1–5)	II (1–5)	IV (1–4)	IV (1–4)	IV (1–3)
Carex lepidocarpa	II (1–6)	II (1–6)	I (1–5)	IV (1–6)	IV (1–4)	IV (1–4)
Bryum pseudotriquetrum	II (1–5)	III (1–4)	I (1–3)	V (1–3)	IV (1–3)	III (1–3)
Selaginella selaginoides	III (1–3)	IV (1–3)	III (1–5)	II (1–3)	IV (1–3)	V (1–4)
Carex pulicaris	III (1–5)	III (1–6)	II (1–3)	IV (1–3)	IV (1–5)	IV (1–5)
Eriophorum angustifolium	V (1–7)	IV (1–5)	II (1–3)	III (1–3)	IV (1–4)	III (1–4)
Potentilla erecta	III (1–5)	III (1–5)	II (1–3)	III (1–3)	IV (1–6)	IV (1–4)
Carex nigra	III (1–7)	III (1–5)	III (1–5)	IV (1–3)	III (1–4)	II (1–3)
Molinia caerulea	II (1–5)	III (1–7)	IV (1–5)	V (1–6)	V (1–5)	I (1)
Succisa pratensis	II (1–3)	IV (1–6)	III (1–3)	V (1–3)	IV (1–4)	I (1–3)
Carex hostiana	II (1–6)	IV (1–7)	IV (1–5)	V (1–6)	V (1–5)	IV (1–5)
Ctenidium molluscum	I (1–5)	IV (1–9)	II (1–5)	IV (1–3)	V (1–6)	V (2–8)
Fissidens adianthoides	I (1–5)	IV (1–5)	I (1–3)	III (1–3)	IV (1–3)	III (1–3)
Eleocharis quinqueflora	IV (1–7)	II (1–7)	III (1–7)	III (1–4)	II (1–4)	II (1–4)
Carex demissa	III (1–7)	II (1–9)	III (1–5)		I (1–3)	I (1–3)
Juncus bulbosus/kochii	III (1–6)	III (1–5)	III (1–5)		II (1–3)	
Erica tetralix	II (1–5)	III (1–7)	IV (1–5)	II (1–3)	I (1–4)	I (1–2)
Scorpidium scorpioides	III (1–8)	II (1–8)	V (1–7)	I (1)		
Carex echinata	II (1–5)	II (1–7)	III (1–5)	II (1–3)		
Narthecium ossifragum	II (1–4)	I (1–5)	III (1–5)		I (1–2)	I (1–2)
Drosera rotundifolia	I (1–5)	II (1–3)	III (1–3)	I (1)	I (1–3)	
Saxifraga aizoides	I (1–3)	II (1–5)	II (1–5)		I (2–5)	I (2–5)
Polygonum viviparum	I (1–3)	II (1–4)				
Juncus squarrosus	I (1–3)	I (1–5)	I (1)			
Scirpus cespitosus	I (1–5)	I (1–3)	II (1–5)			
Pedicularis sylvatica	I (1–3)	I (1–3)	I (1–3)			

Species	1	2	3	4	5
Schoenus nigricans			V (2–5)	I (1)	
Drosera anglica			IV (1–5)		
Blindia acuta			III (1–3)		
Pinguicula lusitanica			II (1–5)		
Myrica gale			II (1–5)		
Schoenus ferrugineus			I (1–10)		
Breutelia chrysocoma			I (1–3)		
Briza media	I (1–6)	I (1–4)	II (1–5)	IV (1–3)	IV (1–4)
Carex flacca	I (1–5)	II (1–6)	I (1)	IV (1–4)	III (1–5)
Linum catharticum	I (1–3)	II (1–3)	I (1–3)	IV (1–4)	IV (1–3)
Cratoneuron commutatum	I (1–5)	II (1–5)	I (1)	III (1–5)	III (1–4)
Prunella vulgaris	I (1–4)	II (1–5)		III (1–3)	IV (1–6)
Primula farinosa				IV (1–3)	IV (1–4)
Euphrasia officinalis agg.	II (1–4)	III (1–5)	I (1–3)	III (1–4)	IV (1–3)
Leontodon autumnalis	I (1–3)	II (1–5)		III (1–3)	II (2–3)
Agrostis stolonifera	I (1)			III (1–3)	II (1–2)
Equisetum palustre	III (1–5)	II (1–3)	I (5)	IV (1–4)	
Triglochin palustris	II (1–3)	II (1–4)	II (1–3)	IV (1–3)	
Parnassia palustris	I (1)	II (1–3)		IV (1–3)	I (3)
Festuca ovina	I (1–5)	III (1–5)	I (1–5)	IV (1–3)	IV (2–7)
Equisetum variegatum	I (1–3)		I (1)	IV (1–6)	III (1–4)
Sesleria albicans			I (1)	III (1–6)	III (2–7)
Juncus alpinus	I (1–3)	I (1–5)	I (1–3)	III (1–4)	II (1–3)
Tofieldia pusilla	I (1–3)	I (1–3)	I (1)	II (1–2)	II (2–4)
Juncus triglumis	I (1)	I (2)		II (1–3)	I (1–3)
Calliergon cuspidatum	I (1–4)	II (1–8)	I (1)	V (1–4)	III (1–3)
Cirsium palustre		I (1–3)		V (1–3)	I (1)
Cardamine pratensis		I (1)		IV (1–3)	I (1–3)
Valeriana dioica				IV (1–3)	I (1–3)
Filipendula ulmaria			I (5)	IV (1–3)	II (1–3)
Angelica sylvestris				IV (1–3)	
Galium uliginosum				IV (1–3)	
Plagiomnium rostratum				IV (1–3)	
Cratoneuron filicinum				III (1–3)	I (1–3)
Menyanthes trifoliata			II (1–5)	III (1–4)	I (1–2)

Floristic table M10, variants (cont.)

	ai	aii	aiii	bi	bii	biii
Carex rostrata			II (1–7)	III (1–3)		
Luzula multiflora				III (1–3)		
Caltha palustris				III (1–3)		
Juncus acutiflorus				III (1–4)		
Dactylorhiza fuchsii				III (1–3)		
Holcus lanatus				III (1–3)		
Galium palustre				III (1–3)		
Eriophorum latifolium	I (2)	I (1–6)	III (1–8)	II (1–4)	IV (1–4)	I (1)
Leontodon taraxacoides				II (1–3)	III (1–4)	I (2)
Pellia endiviifolia	I (5)	I (1–3)	I (1)	II (1–3)	II (1–2)	I (1–3)
Scapania aspera					I (1–3)	I (2)
Bartsia alpina					I (2–3)	
Thymus praecox	I (1)		I (1)		II (1–4)	IV (1–4)
Racomitrium lanuginosum		I (1–4)			I (1–4)	IV (1–9)
Kobresia simpliciuscula					II (2–6)	IV (2–8)
Plantago maritima	I (1–2)	I (1–5)	I (1)		I (1–3)	III (1–4)
Carex capillaris	I (1–3)	I (1–5)			I (2–3)	III (1–4)
Thalictrum alpinum	I (1–5)	I (1–3)	I (1)		I (1)	II (1–3)
Ditrichum flexicaule					I (1)	II (1–3)
Tortella tortuosa						II (2–3)
Anthoxanthum odoratum	II (1–7)	II (1–4)		III (1–3)	I (1)	
Ranunculus flammula	II (1–4)	I (1–3)	I (1–3)	III (1–3)	I (1–3)	
Pedicularis palustris	I (1–3)	I (1–3)	I (1–3)	II (1–4)	I (1–4)	
Festuca rubra	I (1–3)	II (1–5)		II (1–3)	I (1–2)	I (2–5)
Nardus stricta	I (1–5)	II (1–7)				I (1–4)
Hylocomium splendens	I (1)	II (1–7)		II (1–3)	I (1–3)	
Ranunculus acris	I (1–3)	I (1–4)		II (1–3)		
Taraxacum officinale agg.	I (1–3)	I (1–3)		I (1–3)		
Sphagnum subnitens	I (1–3)		I (1–3)	I (1)		

Number of samples	51	56	36	9	20	26
Number of species/sample	21 (11–41)	29 (15–49)	22 (13–35)	41 (32–56)	34 (22–47)	32 (19–46)
Herb height (cm)	11 (4–30)	15 (3–30)	18 (10–35)			
Herb cover (%)	64 (20–100)	86 (65–100)	76 (50–100)		80 (60–90)	76 (50–100)
Bryophyte height (mm)	12 (2–20)	30	10			
Bryophyte cover (%)	41 (2–80)	49 (5–95)	45 (5–80)		53 (30–80)	61 (20–90)
Altitude (m)	364 (1–730)	310 (2–792)	80 (1–231)			
Slope (°)	8 (0–30)	10 (0–35)	5 (0–17)		10 (3–20)	2 (0–10)
Soil pH	6.5 (5.8–7.2)	6.6	6.8			

a   *Carex demissa-Juncus bulbosus/kochii* sub-community
ai   *Eleocharis quinqueflora* variant
aii   *Carex hostiana-Ctenidium molluscum* variant
aiii   *Schoenus nigricans* variant
b   *Briza media-Primula farinosa* sub-community
bi   *Cirsium palustre* variant
bii   *Molinia caerulea-Eriophorum latifolium* variant
biii   *Thymus praecox-Racomitrium lanuginosum* variant

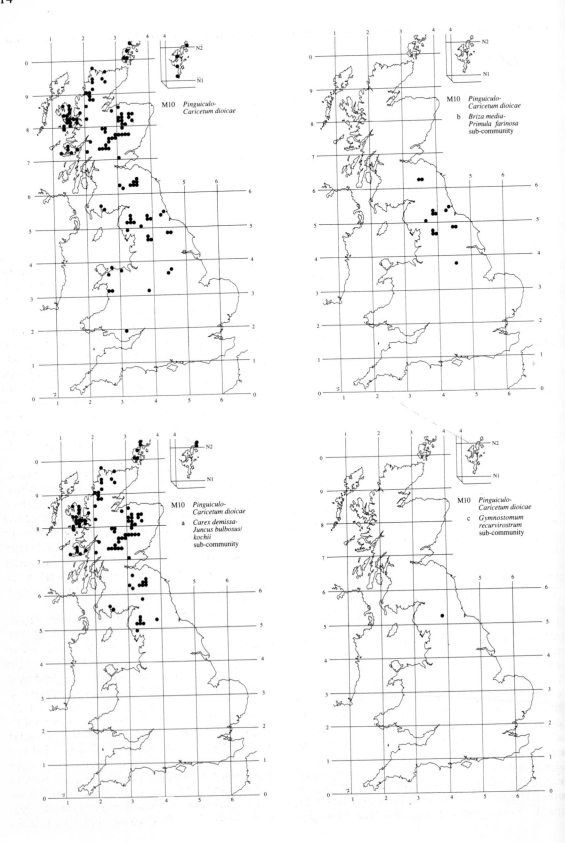

M10 *Pinguiculo-Caricetum dioicae*

M10 *Pinguiculo-Caricetum dioicae*
b *Briza media-Primula farinosa* sub-community

M10 *Pinguiculo-Caricetum dioicae*
a *Carex demissa-Juncus bulbosus/kochii* sub-community

M10 *Pinguiculo-Caricetum dioicae*
c *Gymnostomum recurvirostrum* sub-community

M11

Carex demissa-Saxifraga aizoides mire
Carici-Saxifragetum aizoidis   McVean & Ratcliffe
1962 emend.

Synonymy

Carex demissa-C. panicea nodum   Poore 1955*a p.p.*;
Cariceto-Saxifragetum aizoidis McVean & Ratcliffe
1962, Prentice & Prentice 1975; *Carex-Saxifraga
aizoides* nodum   Birks 1973, Huntley & Birks 1979;
Pinguiculo-Caricetum dioicae Jones 1973 *p.p.*; *Saxifraga aizoides-Juncus triglumis* nodum   Huntley 1979;
Caricetum atrofusco-vaginatae Dierssen 1982; *Schoenus ferrugineus* stands   Wheeler *et al.* 1983 *p.p.*.

Constant species

*Carex demissa, C. panicea, C. pulicaris, Juncus articulatus, Pinguicula vulgaris, Saxifraga aizoides, Aneura
pinguis, Blindia acuta, Bryum pseudotriquetrum, Campylium stellatum, Drepanocladus revolvens.*

Rare species

Alchemilla filicaulis ssp. *filicaulis, Carex atrofusca, C.
microglochin, C. vaginata, Equisetum variegatum, Juncus
alpinus, J. biglumis, J. castaneus, Kobresia simpliciuscula, Salix reticulata, Schoenus ferrugineus, Calliergon
trifarium, Meesia uliginosa.*

Physiognomy

The *Carici-Saxifragetum aizoidis* is typically an open
community in which rich mixtures of small sedges, other
herbs and bryophytes occur among water-scoured runnels with much exposed silt and rock debris on sometimes steeply-sloping ground. Typically, there is no
single vascular dominant, though sedges and other
monocotyledons almost always compose an important
element of the vegetation, providing a strong floristic
link with the *Pinguiculo-Caricetum*. Thus, *Carex
demissa, C. panicea* and *C. pulicaris* are very frequent
almost throughout, and each can attain moderately high
cover (usually less than 25%), and *C. flacca* and *C.
dioica* become common in particular variants, though
not generally with any abundance. *Juncus articulatus* is
also a constant, *Eriophorum angustifolium* occurs frequently and, at lower altitudes, *Eleocharis quinqueflora*

becomes very characteristic. In some such stands, the
local dominance of *Schoenus nigricans* (or, in some
Perthshire localities, *S. ferrugineus*), together with some
Eriophorum latifolium, can accentuate the similarities
between the two communities.

In comparison with the *Pinguiculo-Caricetum*, however, *Carex lepidocarpa* and *C. hostiana* are much less
common, and *C. nigra* and *C. echinata* are likewise
rather scarce. And, at higher altitudes, an Arctic-Alpine
component becomes prominent with *Juncus triglumis*
attaining constancy and *Tofieldia pusilla* increasing in
frequency. Here, too, the community provides an occasional locus for *Juncus biglumis, J. castaneus, Carex
atrofusca, C. microglochin* and, more unusually, *C.
vaginata*, and for *Kobresia simpliciuscula*.

Grasses are typically of low cover, though a variety of
species can be found. *Festuca ovina/vivipara* is common
throughout and *Agrostis stolonifera* occurs occasionally. Then, at higher altitudes, there is often some
Deschampsia cespitosa (presumably ssp. *alpina*), *Nardus
stricta, Anthoxanthum odoratum, Agrostis canina* ssp.
canina and *Festuca rubra*. In contrast to the *Pinguiculo-
Caricetum, Molinia caerulea* is only occasional in this
community and only rarely found as more than scattered shoots.

Amongst the other herbaceous species, there is again
considerable continuity between the two kinds of mire.
Pinguicula vulgaris is constant and *Selaginella selaginoides* very frequent and there are occasional records for
Linum catharticum, Euphrasia officinalis agg. (including
E. scottica) and, in some types of *Carici-Saxifragetum,
Leontodon autumnalis* and *Thymus praecox*. But the
montane character of the vegetation is emphasised by
the presence of *Saxifraga aizoides*, generally speaking
only an occasional in the *Pinguiculo-Caricetum*, but here
constant, often quite abundant and very striking with its
yellow summer flowers. This species is confined to
Scotland, the Lake District and isolated localities in the
northern Pennines, but flushes with very similar vegetation to the *Carici-Saxifragetum* can be found in the

Southern Uplands (Ferreira 1978) and in Snowdonia, and are probably best regarded as impoverished stands of the community. Other montane plants characteristic here are *Thalictrum alpinum*, which is very common at higher altitudes, *Saxifraga stellaris*, *S. oppositifolia* and *Alchemilla filicaulis* ssp. *filicaulis*, which occur more occasionally.

Typically, all these species occur in a rather short, uneven and broken sward, particularly in more strongly-eroded situations at higher altitudes, but at Caenlochan, Huntley (1979) recorded stands of the community which had some of the more luxuriant character associated with the *Saxifraga aizoides-Alchemilla glabra* tall-herb community (see below).

Bryophytes are a frequent and varied element of the *Carici-Saxifragetum*, though, again, their cover is typically discontinuous. Species such as *Aneura pinguis*, *Campylium stellatum*, *Drepanocladus revolvens*, *Bryum pseudotriquetrum* and, at lower altitudes, *Cratoneuron commutatum*, *Fissidens adianthoides*, *Ctenidium molluscum* and *Scorpidium scorpioides* are all common (another similarity with many kinds of *Pinguiculo-Caricetum*) and the brown mosses especially can be fairly abundant. However, a good preferential for this community is the montane moss, *Blindia acuta*, which can be especially prominent at higher altitudes. *Calliergon trifarium* also occurs locally (e.g. Poore 1955a) and *Amblyodon dealbatus*, *Catascopium nigritum*, *Meesia uliginosa* and *Orthothecium rufescens* have also been recorded (McVean & Ratcliffe 1962). Spring and rill species such as *Philonotis fontana*, *P. calcarea* and *Dicranella palustris* are occasionally found.

Sub-communities

Thalictrum alpinum-Juncus triglumis **sub-community:** *Cariceto-Saxifragetum*, typical and high-level facies McVean & Ratcliffe 1962; *Carex-Saxifraga aizoides* nodum Birks 1973 *p.p.*; *Pinguiculo-Caricetum thalictro-saxifragetosum* Jones 1973; *Saxifraga aizoides-Juncus triglumis* nodum Huntley 1979. This is the typical form of the *Carici-Saxifragetum* found at higher altitudes where a montane component is much more obvious in the vegetation, with constant records for *Juncus triglumis* and *Thalictrum alpinum* and correspondingly low frequency of *Eleocharis quinqueflora*. *Alchemilla alpina* also occurs occasionally and there are more commonly some grasses in the cover with *Festuca vivipara* and *Deschampsia cespitosa* frequent. Typically the vegetation is more open than at lower altitudes with scattered herbs and a patchy bryophyte cover disposed in stony flushes. *Saxifraga aizoides*, *Carex demissa* and *C. panicea* are usually the most abundant vascular plants with *Blindia*, *Campylium* or *Drepanocladus* predominating among the mosses. Bryophytes such as

Cratoneuron commutatum, *Fissidens adianthoides* and *Scorpidium* are characteristically scarce. Two variants can be distinguished.

Juncus bulbosus/kochii-Saxifraga stellaris **variant:** *Cariceto-Saxifragetum*, typical and high-level facies McVean & Ratcliffe 1962; *Pinguiculo-Caricetum thalictro-saxifragetosum* Jones 1973. In this most widespread form of the *Thalictrum-Juncus* sub-community, there is a further enrichment of the Arctic-Alpine element with the frequent presence of small quantities of *Saxifraga stellaris* and, more occasionally, *S. oppositifolia*. *Carex dioica* joins the constant sedges of the community and more open gravelly places provide a niche for such rarities as *C. atrofusca*, *C. microglochin*, *Juncus biglumis* and, especially in the south-western Highlands, *J. castaneus*, species which are largely confined to this vegetation type and the closely similar *Caricetum saxatilis* (McVean & Ratcliffe 1962; see also Raven & Walters 1956). *Kobresia simpliciuscula* also occurs occasionally. Then, there are frequent records for *Juncus bulbosus/kochii*, *Agrostis canina* ssp. *canina*, *Potentilla erecta* and *Ranunculus flammula*, which provide a link with the more base-poor mires of the Caricion nigrae and, for *Thymus drucei* and *Plantago maritima*, in somewhat drier areas of turf. Where this kind of flush occurs within a context of ombrogenous mire, quite a common occurrence, *Calluna vulgaris* and *Scirpus cespitosus* can occur; and where water splashes over the vegetation, *Philonotis fontana* can be present among the bryophytes.

Polygonum viviparum **variant:** *Carex-Saxifraga aizoides* nodum Birks 1973 *p.p.*; *Saxifraga aizoides-Juncus triglumis* nodum Huntley 1979. This variant, recorded largely from Caenlochan (Huntley 1979), preserves the general montane character of the *Thalictrum-Juncus* sub-community but lacks frequent records for many of the species of the *Juncus-Saxifraga* variant, notably *Juncus bulbosus/kochii* and *Saxifraga stellaris* themselves. *Carex flacca* becomes constant, largely replacing *C. dioica*, and the presence of plants such as *Leontodon autumnalis*, *Polygonum viviparum*, *Alchemilla glabra*, *Angelica sylvestris*, *Crepis paludosa* and the rare Arctic-Alpine dwarf willow, *Salix reticulata*, give something of the feel of a base-rich tall-herb ledge. *Carex vaginata* has been recorded from this variant.

Cratoneuron commutatum-Eleocharis quinqueflora **sub-community:** *Carex demissa-C. panicea* nodum Poore 1955a *p.p.*; *Cariceto-Saxifragetum*, low-level facies McVean & Ratcliffe 1962; *Carex-Saxifraga aizoides* nodum Birks 1973, Huntley & Birks 1979; *Schoenus ferrugineus* stands Wheeler et al. 1983 *p.p.* It is in this sub-community that the *Carici-Saxifragetum* comes closest to, and grades into, the *Pinguiculo-Caricetum*, with more extreme montane plants, apart from *Saxifraga aizoides* and *Blindia acuta*, much more poorly represented than above; in more southerly stands, even these species become rare. By contrast, *Eleocharis quinqueflora* becomes constant and, on occasion, it can be quite abundant, rivalling the sedges, among which *Carex hostiana* and, in wetter stands, *C. rostrata* are sometimes found. Vascular plant cover is typically more extensive than in the *Thalictrum-Juncus* sub-community, though only rarely approaching a continuous sward. In some stands, however, *Schoenus nigricans* or, much more locally, *S. ferrugineus*, can dominate and this gives the vegetation a distinctive stamp. The other positive features of this sub-community are to·be found among the bryophytes, where *Cratoneuron commutatum* and *Scorpidium* are very common and often the most abundant species with *Fissidens adianthoides* frequent and *Philonotis calcarea* occasional.

Habitat

The *Carici-Saxifragetum* is characteristic of open, stony flushes, strongly irrigated with moderately base-rich waters, on generally steep slopes in the sub-montane and montane parts of Britain.

Although the community can occur almost at sea-level in the far north-west of Scotland, it is generally confined to high altitudes, overlapping to some extent with the altitudinal range of the *Pinguiculo-Caricetum*, though having a considerably higher mean, at 510 m, and extending to levels never reached by that other kind of base-rich mire. Thus, although both communities have a representation of more catholic Continental Northern plants like *Pinguicula vulgaris, Carex pulicaris* and *C. dioica*, an Arctic-Alpine element is much more obvious here, reflecting the extreme character of the climate. Throughout the range of the *Carici-Saxifragetum* the mean annual maximum temperature is almost everywhere less than 23 °C (Conolly & Dahl 1970), with annual accumulated temperatures below 830 day-degrees (*Climatological Atlas* 1952). February minima are sometimes above freezing but winters are characteristically long and harsh, with much snow in the central Highlands, where the community is most common, and late frosts. Within this zone, the *Thalictrum-Juncus* sub-community occurs at the higher levels, with a mean altitude of 573 m and occasionally reaching more than 850 m. Here, mean annual maxima are typically less than 21 °C, the isotherm of which roughly corresponds with the distributions of *Saxifraga aizoides, Thalictrum alpinum* and *Juncus triglumis* and includes the stations of Arctic-Alpine rarities like *Carex atrofusca, C. microglochin, Juncus biglumis* and *J. castaneus* which are confined to this kind of *Carici-Saxifragetum*. Outside this high-montane area, the *Cratoneuron-Eleocharis* sub-community represents a transition to the *Pinguiculo-Caricetum*, extending down to considerably lower levels than the *Thalictrum-Juncus* sub-community, with a

mean altitude of 415 m; not only on the north-west seaboard of Scotland, where February minima are consistently above freezing, but also into the Southern Uplands, the Lake District, the north Pennines and north Wales, where mean annual maxima sometimes reach 25 °C (*Climatological Atlas* 1952, Conolly & Dahl 1970). Towards these limits, *Saxifraga aizoides, Thalictrum alpinum* and *Juncus triglumis* become increasingly rare and the community loses its integrity.

Within this climatic zone, the *Carici-Saxifragetum* is confined to soils irrigated with more base-rich waters and is thus consistently associated with calcareous bedrocks, the occurrence of which is, through much of the region, decidedly local. It is best developed over the Dalradian meta-sediments of the south and east-central Highlands, especially on the Breadalbane range and in the Clova-Caenlochan region (Poore 1955a, McVean & Ratcliffe 1962, Huntley 1979, Huntley & Birks 1979a) with more isolated, though still quite common, occurrences further to the north-west on the Moine meta-sediments and on such calcareous rocks as are present among Lewisian gneiss. It is also found on the Cambrian/Ordovician Durness Limestone along the Moine Thrust and on Skye (Birks 1973) and it occurs in the last site on Tertiary limestones and igneous rocks. Lavas and intrusions provide a local source of base-enrichment in the Borrowdale Volcanics of the Lake District and in Snowdonia (Ratcliffe 1977) and, in Upper Teesdale, there are stands on metamorphosed sugar-limestone (Pigott 1956a, Jones 1973, Bradshaw & Jones 1976). Flushing from such substrates as these generally maintains the soil at a pH of between 5.5 and 7.0, very much as in the *Pinguiculo-Caricetum*, though variations in base-richness may play some part in floristic differences in this community. More calcicolous species such as *Saxifraga aizoides, Selaginella selaginoides, Carex pulicaris, Thalictrum alpinum, Linum catharticum* and the bryophytes *Aneura pinguis, Campylium stellatum, Drepanocladus revolvens, Cratoneuron commutatum* and *Fissidens adianthoides* are well represented in one kind of *Carici-Saxifragetum* or another, but there is a somewhat less calcicolous character in the *Juncus-Saxifraga* variant and variations in the soil environment would repay investigation.

The other characteristic feature of the habitat of this mire is that flushing is vigorous. Rainfall or snowfall provide ample supplies of ground water and the community typically occupies more strongly sloping ground than the *Pinguiculo-Caricetum* and/or occurs closer to surface-flowing rills which rarely dry up. Erosion of the surface is therefore often pronounced (and, in some places, perhaps progressive: McVean & Ratcliffe 1962) and the soil cover little more than scoured accumulations of silt and organic matter, with much exposed rock debris, both smaller gravel and, where drift has contributed to the parent materials, larger boulders. The wet expanses of rotting mica-schist in flushes of this kind in Breadalbane are especially distinctive. The permanence and rate of surface water-flow are probably of particular importance to the extent of the moss cover of the community, which may largely control the humus content of the soil (McVean & Ratcliffe 1962).

Zonation and succession

The *Carici-Saxifragetum* is found around springs and flushes, often in association with other calcicolous soligenous mire vegetation, among sub-montane and high-altitude grasslands and dwarf herb communities and, more occasionally, within ombrogenous bogs. Typically, the stands are small and the definition of the vegetation sequences depends on the strength of irrigation and the amount of base-enrichment of the flush and its surrounds. Grazing may help maintain the open structure of the community and prevent the development of a woody cover, though colonisation by shrubs and trees on the wet and cold soils characteristic here would probably be slow and, at higher altitudes, this kind of mire is probably a climatic climax.

Where the *Carici-Saxifragetum* marks out flushes within stretches of upland pasture, the general pattern of vegetation is the same throughout its range, running, in its fullest development, from Cratoneurion communities around spring-heads, flush-lines and their associated rills, through the *Carici-Saxifragetum*, then through small-sedge soligenous mire on the less strongly-irrigated surrounds, to the swards of the drier soils beyond. Where vigorous flushes emerge locally on steeper slopes, this kind of pattern can be very well defined, with concentric zones of the different communities elongated downhill; where irrigation is more diffuse and more extensively spread, the zonation may be much less clear, often expressed as a complex mosaic, with the elements associated with stronger water-flow less well developed.

At low altitudes, generally below 500 m, where the *Carici-Saxifragetum* is typically represented by the *Cratoneuron-Eleocharis* sub-community, it is often surrounded by the *Pinguiculo-Caricetum*, usually some variant of the *Carex-Juncus* sub-community or, in the north Pennines where the *Carici-Saxifragetum* occurs very locally, some variant of the *Briza-Primula* sub-community. In such situations, the two kinds of mire can come very close floristically, particularly in those localities beyond the southern limit of *Saxifraga aizoides*, and the transition between the two is consequently very diffuse. Locally, too, an abundance throughout of *Schoenus nigricans* (or, in a few localities in Perthshire, *S. ferrugineus*: Wheeler *et al.* 1983) can give an overlying impression of structural uniformity.

Beyond this zone of more strongly-flushed, often unstable and sometimes heavily stock-poached soils, there is commonly a gradation to some type of calcicolous grassland, with a closing up of the turf, over drier profiles. Most often, this is the *Festuca-Agrostis-Thymus* grassland: *Saxifraga aizoides* and some other hydrophilous herbs can run some way into this pasture in its *Saxifraga-Ditrichum* sub-community (the *Saxifrageto-Agrosto-Festucetum* of McVean & Ratcliffe 1962) which can form a transitional zone to the unflushed sward. At somewhat higher altitudes, and perhaps on somewhat less base-rich soils, the *Festuca-Agrostis-Alchemilla* grass-heath can provide the context, with its *Carex pulicaris-C. panicea* sub-community immediately around the flushed areas. In the east-central Highlands, *Betula pubescens* and *Juniperus communis* ssp. *communis* are ready invaders of the less strongly-waterlogged soils of such sequences and the *Carici-Saxifragetum* forms an important part of the ground mosaic beneath open stands of birch and juniper at sites like Morrone (McVean & Ratcliffe 1962, Ratcliffe 1977, Huntley & Birks 1979*a*).

At higher altitudes throughout its range, the *Carici-Saxifragetum* is represented in such zonations by the *Thalictrum-Juncus* sub-community and, particularly in Breadalbane and more locally to the north-west of Scotland, there is a shift in all the components of the sequence to present a more strikingly montane character. Thus, the *Pinguiculo-Caricetum* in the surrounding zone is replaced by the *Caricetum saxatilis* and it is open stony areas in both this community and the *Carici-Saxifragetum* which provide the major locus for such Arctic-Alpine rarities as *Carex atrofusca, C. microglochin, Juncus biglumis* and *J. castaneus* and many of the Scottish stations for *Kobresia*. Then, at these high levels, the *Festuca-Alchemilla-Silene* dwarf-herb community often forms the surround to the springs on soils which are not so strongly flushed by the springs, though which are kept continually moist by high rainfall and snow-melt and maintained in a similar unstable state by solifluction and cryoturbation. Complexes of these vegetation types, forming a short, diverse and open cover over sparkling mica-schist soils, make a major contribution to the glory of sites like Ben Lawers (e.g. Poore 1955*b*, Raven & Walters 1956, McVean & Ratcliffe 1962).

The Cratoneurion communities represented in such sequences are typically low swards like the *Cratoneuron-Festuca* vegetation, in which *S. aizoides* often continues to be very prominent, or the less calcicolous *Cratoneuron-Carex nigra* community. However, where flush waters trickle down very steep rocky faces with ledges, the *Saxifraga-Alchemilla* tall-herb community can occur in close juxtaposition with the *Carici-Saxifragetum*. The two vegetation types are floristically quite similar and the *Polygonum* variant of the latter community, which Huntley (1979) described from the Clova-Caenlochan area, is floristically intermediate. *Salix reticulata* finds an occasional place in this kind of *Carici-Saxifragetum* and it is possible that these vegetation types represent a grazing-mediated succession on more inaccessible dripping crags that would culminate in the development of the *Salix-Luzula* Arctic-Alpine willow scrub. Even under favourable environmental conditions, however, this sere is, at most, only locally active because of the now very wide dispersal of potential parent bushes.

In zonations of the above types, the continuity of floristic variation, with a gradual lessening of the influence of irrigation and base-enrichment in moving away from the springs, is often very obvious. Where the *Carici-Saxifragetum* marks out flushed sites within ombrogenous mires and related wet-heaths, the soligenous areas are usually much more sharply delineated from their surrounds. The context of flushing can be the *Calluna-Eriophorum* blanket mire or, at lower altitudes towards the west, the *Scirpus-Eriophorum* blanket mire or degraded forms of this bog placed in the *Scirpus-Erica* wet-heath. The *Carici-Saxifragetum* can again be surrounded by a zone of the *Caricetum saxatilis* or, at lower levels, the *Pinguiculo-Caricetum*, but there is typically a very sharp switch from these mires on sloppy mineral soil in the soligenous tracks to the ombrogenous vegetation on the peat around, often with a low, steep bank between. Poore (1955*b*) and McVean & Ratcliffe (1962) both illustrated some very distinctive mosaics of this kind, with isolated hummocks of ombrogenous mire occurring detached from the surrounding bog within the soligenous zone. The latter authors noted how the flushes could isolate such fragments from the effects of burning of the main bog surface, sheltering a refuge for such species as *Sphagnum imbricatum* and *Dicranum bergeri*.

Distribution

The community is largely confined to Scotland, where it is especially common in the southern and central Highlands with more scattered localities further north-west, but it also occurs in the Lake District and, more locally, in the Southern Uplands, the northern Pennines and in north Wales. The *Thalictrum-Juncus* sub-community is confined to higher altitudes and virtually restricted to Scotland; the *Cratoneuron-Eleocharis* sub-community is also frequent there at lower altitudes and takes in most of the English and Welsh stands. Towards these southern limits of its range, the *Carici-Saxifragetum* grades into the *Pinguiculo-Caricetum*.

Affinities

The *Carici-Saxifragetum* is largely based on McVean & Ratcliffe's (1962) *Cariceto-Saxifragetum*, taking in additional data from a variety of sources (Poore 1955*a*, Birks 1973, Jones 1973, Prentice & Prentice 1975, Huntley & Birks 1979*a*, Huntley, 1979, Wheeler *et al.* 1983) and making a more well-defined division between high- and low-altitude types. Although there is a considerable continuity with the *Pinguiculo-Caricetum* (within which Jones (1973) placed her Teesdale stands), the community is worth recognising as a distinct vegetation type and, with its strong montane and Arctic-Alpine element, it comes closer to the northern European mires traditionally placed in the Caricion bicolori-atrofuscae (Nordhagen 1928, 1943: see also Coombe & White 1951, Persson 1961, Dierssen 1982). Such species as *Carex atrofusca*, *C. microglochin*, *Juncus biglumis*, *J. castaneus* and *Kobresia* are, however, of rather restricted distribution even within the higher-altitude stands of the community and its more general affinities are perhaps with the Caricion davallianae mires distributed throughout Europe (Poore 1955*b*, McVean & Ratcliffe 1962).

Floristic table M11

	ai	aii	a	b	11
Carex demissa	V (1–5)	V (1–7)	V (1–7)	IV (1–5)	V (1–7)
Saxifraga aizoides	V (1–7)	IV (1–8)	IV (1–8)	IV (1–8)	IV (1–8)
Pinguicula vulgaris	IV (1–3)	IV (1–5)	IV (1–5)	V (1–5)	IV (1–5)
Blindia acuta	IV (1–5)	IV (1–7)	IV (1–7)	IV (1–5)	IV (1–7)
Aneura pinguis	V (1–3)	IV (1–3)	IV (1–3)	III (1–3)	IV (1–3)
Carex pulicaris	V (1–5)	IV (1–5)	IV (1–5)	III (1–5)	IV (1–5)
Carex panicea	II (1–3)	V (1–5)	IV (1–5)	IV (1–5)	IV (1–5)
Drepanocladus revolvens	V (1–5)	III (1–5)	IV (1–5)	IV (1–5)	IV (1–5)
Campylium stellatum	IV (1–5)	III (1–5)	IV (1–5)	IV (1–5)	IV (1–5)
Juncus articulatus	IV (1–3)	III (1–5)	IV (1–5)	IV (1–5)	IV (1–5)
Bryum pseudotriquetrum	V (1–5)	III (1–3)	IV (1–5)	III (1–3)	IV (1–5)
Thalictrum alpinum	V (1–3)	IV (1–5)	IV (1–5)	II (1–3)	III (1–5)
Juncus triglumis	V (1–3)	IV (1–5)	IV (1–5)	II (1–3)	III (1–5)
Deschampsia cespitosa	IV (1–3)	III (1–5)	III (1–5)	I (1)	II (1–5)
Nardus stricta	III (1–3)	III (1–7)	III (1–7)		II (1–7)
Anthoxanthum odoratum	II (1–3)	III (1–5)	III (1–5)		II (1–5)
Alchemilla alpina	II (1–3)	II (1–3)	II (1–3)		I (1–3)
Caltha palustris	I (1–5)	I (5)	I (1–5)		I (1–5)
Carex flacca	V (1–5)	I (5)	II (1–5)	III (1–5)	II (1–5)
Leontodon autumnalis	V (1–3)	II (1–3)	II (1–3)	II (1–3)	II (1–3)
Polygonum viviparum	V (1–3)		II (1–3)		I (1–3)
Equisetum palustre	IV (1–3)	I (1–3)	II (1–3)	I (1–5)	I (1–5)
Alchemilla glabra	IV (1–5)	I (1–5)	II (1–5)		I (1–5)
Cerastium fontanum	II (1–3)	I (1–3)	I (1–3)		I (1–3)
Angelica sylvestris	II (1–3)	I (1)	I (1–3)		I (1–3)
Sagina procumbens	II (1–3)	I (1)	I (1–3)		I (1–3)
Campanula rotundifolia	II (1–3)	I (1–3)	I (1–3)		I (1–3)
Salix reticulata	II (1–3)		I (1–3)		I (1–3)
Hieracium spp.	II (1–3)		I (1–3)		I (1–3)
Carex dioica	II (1–3)	IV (1–5)	III (1–5)	II (1)	II (1–5)
Juncus bulbosus/kochii	I (1)	III (1–3)	III (1–3)	I (1)	II (1–3)
Saxifraga stellaris	I (1)	III (1–3)	III (1–3)		II (1–3)
	I (1)	III (1–3)	II (1–3)	I (1–3)	I (1–3)

Species					
Plantago maritima		III (1-5)	II (1-5)	I (1-5)	I (1-5)
Agrostis canina canina		III (1-3)	II (1-3)	I (1)	I (1-3)
Potentilla erecta		III (1-3)	II (1-3)	I (1)	I (1-3)
Festuca rubra		III (1-5)	II (1-5)	I (1)	I (1-3)
Ranunculus flammula		III (1-3)	I (1-3)	I (1-3)	I (1-5)
Saxifraga oppositifolia	I (1)	II (1-5)	I (1-5)	I (1)	I (1-3)
Philonotis fontana	I (1)	II (1-3)	I (1-3)	I (1)	I (1-5)
Scapania undulata	I (1-3)	II (1-5)	I (1-5)	I (1)	I (1-3)
Viola riviniana	I (1)	II (1-3)	I (1-3)		I (1-5)
Rhytidiadelphus triquetrus		II (1-3)	I (1-3)		I (1-3)
Calluna vulgaris		II (1-3)	I (1-3)	I (1)	I (1-3)
Juncus squarrosus		II (1-3)	I (1-3)	I (1)	I (1-3)
Scirpus cespitosus		III (1-3)	I (1-3)	I (2)	I (1-3)
Geum rivale		II (1-3)	I (1-3)	I (1)	I (1-3)
Carex echinata		II (1-3)	I (1-3)	I (1)	I (1-3)
Taraxacum officinale agg.		II (1-3)	I (1-3)	I (1)	I (1-3)
Hylocomium splendens		II (1-3)	I (1-3)		I (1-3)
Carex lepidocarpa		II (1-5)	I (1-5)		I (1-5)
Calliergon sarmentosum		I (1-5)	I (1-5)		I (1-5)
Cratoneuron commutatum	II (1-5)	II (1-9)	II (1-9)	V (1-8)	III (1-9)
Eleocharis quinqueflora	I (1)	I (1-3)	I (1-3)	IV (1-10)	II (1-10)
Fissidens adianthoides	II (1-3)	I (1-3)	I (1-3)	III (1-3)	II (1-3)
Scorpidium scorpioides		II (1-5)	I (1-5)	III (1-7)	II (1-7)
Carex hostiana		I (1-5)	I (1-5)	II (1-7)	I (1-7)
Triglochin palustris		I (1)	I (1-5)	II (1-3)	I (1-3)
Philonotis calcarea			I (1)	II (1-3)	I (1-3)
Carex rostrata				II (1-7)	I (1-7)
Schoenus nigricans				II (5-10)	I (5-10)
Selaginella selaginoides	III (1-3)	III (1-5)	III (1-5)	III (1-3)	III (1-5)
Tofieldia pusilla	III (1-3)	II (1-5)	II (1-5)	II (1-3)	II (1-5)
Linum catharticum	II (1-3)	II (1-3)	II (1-3)	III (1-3)	II (1-5)
Eriophorum angustifolium		III (1-5)	II (1-5)	III (1-3)	II (1-3)
Euphrasia officinalis agg.		III (1-3)	II (1-3)	II (1-3)	II (1-5)
Festuca vivipara		III (1-7)	II (1-7)	II (1-5)	II (1-3)
Ctenidium molluscum		II (1-3)	II (1-3)	II (1-3)	II (1-3)
Molinia caerulea		II (1-5)	II (1-5)	II (1-5)	II (1-5)

Floristic table M11 (cont.)

	ai	aii	a	b	11
Agrostis stolonifera	II (1–3)	I (1)	I (1–3)	II (1–3)	I (1–3)
Festuca ovina	I (1–3)	II (1–5)	I (1–5)	II (1–3)	I (1–5)
Tussilago farfara	II (1–3)	I (1)	I (1–3)	I (1–3)	I (1–3)
Dicranella palustris	I (1)	II (1–3)	I (1–3)	I (1)	I (1–3)
Alchemilla filicaulis filicaulis	I (1)	II (1–3)	I (1–3)	I (1–3)	I (1–3)
Calliergon cuspidatum	I (1)	I (1)	I (1)	I (1–3)	I (1–3)
Carex nigra	I (1–5)	I (1–3)	I (1–5)	I (1–5)	I (1–5)
Cardamine pratensis	II (1–3)		I (1–3)	I (1–3)	I (1–3)
Conocephalum conicum	II (1–3)		I (1–3)	I (1)	I (1–3)
Prunella vulgaris	I (1)	I (1)	I (1)	I (1–3)	I (1–3)
Juncus alpinus		I (1)	I (1)	I (1)	I (1–3)
Erica tetralix		I (1–3)	I (1–3)	I (1–3)	I (1–3)
Pellia endiviifolia		I (1–3)	I (1–3)	I (1)	I (1–3)
Narthecium ossifragum		I (1–3)	I (1–3)		I (1–3)
Succisa pratensis		I (1–3)	I (1–3)	I (1–3)	I (1–3)
Number of samples	9	26	35	24	59
Number of species/sample	31 (18–38)	29 (12–53)	29 (12–53)	24 (10–40)	26 (10–53)
Herb height (cm)	30 (20–40)	9 (5–20)	9 (5–20)	28 (10–50)	
Herb cover (%)		63 (20–100)	54 (5–100)	80 (15–100)	
Bryophyte height (mm)		30	30	31 (5–50)	
Bryophyte cover (%)		18 (10–25)	18 (10–25)	18 (5–40)	
Slope (°)	61 (30–80)	20 (1–50)	32 (1–80)	21 (0–85)	27 (0–85)
Altitude (m)	764 (740–820)	582 (213–853)	573 (213–853)	415 (30–883)	510 (30–883)
Soil pH		5.5	5.5	6.6	

a   *Thalictrum alpinum-Juncus triglumis* sub-community

ai *Polygonum viviparum* variant

aii *Juncus bulbosus/kochii-Saxifraga stellaris* variant

b   *Cratoneuron commutatum-Eleocharis quinqueflora* sub-community

11 *Carici-Saxifragetum aizoidis* (total)

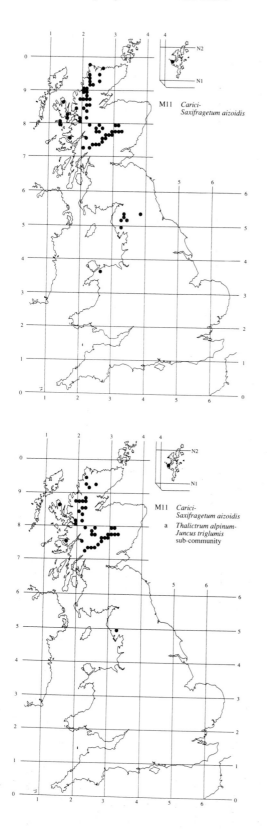
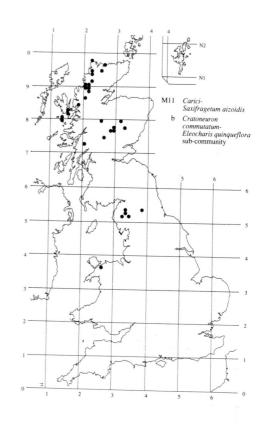

M11 *Carici-Saxifragetum aizoidis*

M11 *Carici-Saxifragetum aizoidis*
b *Cratoneuron commutatum-Eleocharis quinqueflora* sub-community

M11 *Carici-Saxifragetum aizoidis*
a *Thalictrum alpinum-Juncus triglumis* sub-community

M12
Carex saxatilis mire
Caricetum saxatilis   McVean & Ratcliffe 1962

Synonymy
Carex saxatilis sociation   Poore 1955*b*; *Carex saxatilis*
mire   McVean & Ratcliffe 1962: *Calliergono sarmen-*
tosi-Caricetum saxatilis   Dierssen 1982.

Constant species
Carex demissa, C. saxatilis, Eriophorum angustifolium,
Polygonum viviparum, Thalictrum alpinum, Aneura
pinguis, Drepanocladus revolvens, Hylocomium splen-
dens, Scapania undulata.

Rare species
Alchemilla filicaulis ssp. *filicaulis, Carex atrofusca, C.*
microglochin, C. saxatilis, C. vaginata, Juncus biglumis,
J. castaneus, Kobresia simpliciuscula.

Physiognomy
Carex saxatilis occurs at low frequency in a variety of
wetter vegetation types at high altitudes but here it is
typically dominant in a kind of montane mire which has
a distinctive assemblage of associates. The sward is
generally low, less than 20 cm tall, and often rather open
with patchy exposures of soil. Sedges as a group figure
quite prominently: apart from *C. saxatilis, C. demissa,*
C. echinata and *C. nigra* are very frequent and each can
be abundant and *C. dioica* and *C. pulicaris* also occur
occasionally. Such species provide considerable continuity with other more calcicolous mires like the *Pingui-*
culo-Caricetum and lower-altitude stands of the *Carici-*
Saxifragetum but, in contrast to those communities, *C.*
panicea is very scarce here and *C. bigelowii* is fairly
consistent, particularly in grassier transitions to the
surrounding swards. As in high-altitude stands of the
Carici-Saxifragetum, open stony patches in the vegetation provide a niche for the very rare Arctic-Alpine
sedges, *C. atrofusca* and *C. microglochin. C. vaginata* has
also been recorded in this community.

Apart from *Eriophorum angustifolium*, a frequent
species here which can attain covers of over 10%, almost
all the other herbs occur as sparse, scattered individuals.

Selaginella selaginoides and *Pinguicula vulgaris*, both of
which are very common, emphasise general similarities
with our other calcicolous flushes, but more characteristic are the Arctic-Alpines *Thalictrum alpinum, Polygo-*
num viviparum and *Juncus triglumis*. All these can occur
in more montane stands of the *Carici-Saxifragetum*,
together with *Saxifraga oppositifolia* and *S. stellaris*,
which are scarce to occasional here, but *S. aizoides* itself
is not frequent in the *Caricetum saxatilis* and, when it
does occur, it is typically of low cover. Also rather
distinctive is the presence of certain poor-fen herbs like
Viola palustris, Caltha palustris and *Agrostis canina* ssp.
canina. Other vascular species recorded occasionally
include *Leontodon autumnalis, Euphrasia officinalis* agg.
(including *E. scottica*), *Geum rivale, Huperzia selago,*
Ranunculus acris, Rumex acetosa, Alchemilla glabra and
the rare *A. filicaulis* ssp. *filicaulis*. Then, in some stands,
particularly on steeper slopes, the sward can have a
distinctly grassy look, with more frequent records and
slightly higher covers for *Deschampsia cespitosa, Nardus*
stricta, Festuca rubra and *F. vivipara*. Finally, the very
rare *Juncus castaneus* and *J. biglumis* can be found in
more open areas in this community.

Bryophytes compose an important element of the
vegetation though, apart from *Drepanocladus revolvens*,
the cover of individual species is usually low. This moss,
together with frequent *Aneura pinguis* and occasional
Bryum pseudotriquetrum, Blindia acuta and *Campylium*
stellatum, provides a further floristic link with the *Car-*
ici-Saxifragetum and *Pinguiculo-Caricetum* and, as
there, spring and rill bryophytes such as *Cratoneuron*
commutatum and *Philonotis fontana* can sometimes be
found. *Calliergon trifarium*, a montane moss which
otherwise occurs mainly in the *Carici-Saxifragetum*, is
also occasional here. More peculiar is the high frequency
of *Hylocomium splendens*, typically only as scattered
single shoots but nonetheless a rather unexpected moss
to find in this kind of vegetation, and of *Scapania*
undulata: other *Scapania* spp., like *S. uliginosa* and *S.*
irrigua, are also sometimes present. Then, there is occa-

sionally some *Rhytidiadelphus loreus*, *R. squarrosus*, *Racomitrium lanuginosum*, *Dicranum scoparium*, *Polytrichum alpinum* and poor-fen bryophytes such as *Polytrichum commune*, *Calliergon sarmentosum*, *C. stramineum* and *Drepanocladus exannulatus*. There can also be some small patches of Sphagna including *S. auriculatum*, *S. capillifolium*, *S. recurvum*, *S. subnitens*, *S. girgensohnii* and *S. warnstorfii*. The total cover of these species is never high but their presence, together with certain of the herbs, can make some stands look transitional to Caricion nigrae mires. Other bryophytes of restricted range recorded here are *Tayloria lingulata*, *Cinclidium stygium*, *Barbilophozia lycopodioides* and *Tritomaria polita*.

Habitat

The *Caricetum saxatilis* is strictly confined to the margins of high-montane flushes irrigated with base-rich and calcareous waters and perhaps influenced by long snow-lie.

C. saxatilis is an Arctic-Subarctic species whose British range falls almost entirely within the 21 °C mean annual maximum isotherm (Conolly & Dahl 1970), an area which takes in the higher peaks, generally over 750 m, in the southern and central Scottish Highlands, with more far-flung localities further to the north-west, and which also includes most of the British stations for the Arctic-Alpine *Thalictrum alpinum* and *Juncus triglumis* and, less exclusively, for *Polygonum viviparum*, all frequent here. Each of these species occurs in a variety of vegetation types in this region of inhospitable climate and they can sometimes be found together in other communities of springs and dripping banks, but only in the *Caricetum saxatilis* do they coincide consistently with the dominance of *C. saxatilis* and the other frequent species of this mire.

The distribution of the community is considerably less extensive than that of these Arctic-Alpine components, being further confined by the often local occurrence of flushes in more calcareous substrates in the high-montane region. *C. saxatilis* is, in fact, tolerant of a wide range of base-status in wetter soils (McVean & Ratcliffe 1962, Jermy *et al.* 1982) but the *Caricetum saxatilis* is strongly centred on the more lime-rich Dalradian meta-sediments, especially the calcareous mica-schists of Breadalbane, with its north-western stations on calcareous rocks of the Moine or Lewisian series. Even then, some sites where the community might be expected, like Caenlochan, have the sedge but not this kind of mire (McVean & Ratcliffe 1962, Huntley 1979).

Yet, though this is the most calcicolous of the montane mires, more so than the *Carex-Sphagnum warnstorfii* mire which sometimes penetrates to these altitudes, and much more so than the *Carex-Sphagnum russowii* mire, which has a very similar altitudinal range, some of its most frequent species are not calcicoles. *Thalictrum alpinum*, *Juncus triglumis* and *Selaginella selaginoides*, together with *Aneura pinguis*, *Drepanocladus revolvens* and the rare preferential *Calliergon trifarium*, are broadly calcicolous, but *Polygonum viviparum* not so obviously so above 800 m and exacting Arctic-Alpine calcicoles are much more characteristic of vegetation like the *Festuca-Alchemilla-Silene* and *Dryas-Silene* communities and the *Salix-Luzula* scrub. The soils here, though continuously irrigated, are not of especially high pH, ranging from 4.6 to 6.3 (Poore 1955b, McVean & Ratcliffe 1962). Direct snow-melt, rather than lateral flushing, may also provide much of the soil moisture: most of the stands encountered face north or east and Poore (1955b) noted that snow lay over the vegetation long into the spring. A chionophilous element is not prominent in the community but snow-melt may have an effect by diluting base-enrichment or even induce sufficient surface-leaching to allow the good representation of non-calcicolous species. With the combination of flushing, snow-melt, cryoturbation and solifluctional flow, even over gentler slopes, the profiles are typically unstructured, raw gleys, silty in texture or, where there is more organic matter, humic. Continual instability is important in maintaining open, stony areas where the rare Arctic-Alpine sedges and rushes find a niche.

Zonation and succession

The *Caricetum saxatilis* typically occurs as small stands bordering rills or more strongly-irrigated soligenous mires. It is possible that grazing prevents colonisation by Arctic-Alpine willows, though in the extreme environment characteristic here, the community is probably a climatic climax.

Towards the lower end of its altitudinal range, the *Caricetum saxatilis* overlaps considerably with the *Carici-Saxifragetum* and it can often be found as a fringing zone to this mire, giving way around to the *Festuca-Alchemilla-Silene* dwarf-herb community or, at its lowest stations, to the *Festuca-Agrostis-Alchemilla* or *Festuca-Agrostis-Thymus* grasslands. *Dryas-Silene* vegetation can also figure among these zonations on rocky banks and, on dripping crags close to the source of irrigation, the *Saxifraga-Alchemilla* community. Such sequences are very characteristic of the Breadalbane area and more isolated localities in the north-west Highlands.

At the very highest altitudes, the *Carici-Saxifragetum* may be absent from the flushes, when the *Caricetum saxatilis* gives way directly to the spring or rill vegetation, usually some kind of *Cratoneuron-Festuca* or *Cratoneuron-Carex nigra* flush. It can also occur in close association with some of the more calcicolous snow-bed communities.

As with the other kinds of calcicolous flushes, the

Caricetum saxatilis can sometimes be found within a markedly less calcicolous context, marking out very local areas of base-enrichment. Then, it can occur with the *Carex-Sphagnum russowii* mire, in water-tracks more remote from the source of flushing, and give way laterally to montane Nardo-Galion grasslands with *Nardus stricta*, *Juncus squarrosus* and *Deschampsia cespitosa*.

Distribution
The community is of fairly widespread, though distinctly local, occurrence through the southern and central Scottish Highlands, with scattered localities in north-west Scotland.

Affinities
No new samples have been added to those published by Poore (1955*b*), McVean & Ratcliffe (1962) and Birse (1980) and the definition of the community largely confirms McVean & Ratcliffe's (1962) diagnosis. The community in part replaces the *Pinguiculo-Caricetum* in mildly calcareous flush sequences at high altitudes though it shows some affinities with Caricion nigrae poor fens and is placed in that alliance by some authors (Dierssen 1982). With its striking Arctic-Alpine element, it is also close to the Caricion bicolori-atrofuscae mires described from Scandinavia, like the *Carex saxatilis-Drepanocladus intermedius* sosiadjon of Nordhagen (1943). There, however, *Carex vaginata* and/or *C. atrofusca* and *Juncus biglumis* and/or *J. castaneus* become very frequent and Arctic-Alpine willows occur commonly. On balance, it may be best to consider the *Caricetum saxatilis* as one of the most montane of our Caricion davallianae communities.

Floristic table M12

Carex saxatilis	V (5–9)	Carex dioica	II (1–4)
Drepanocladus revolvens	IV (2–6)	Philonotis fontana	II (1–2)
Hylocomium splendens	IV (1–3)	Dicranum scoparium	II (1)
Thalictrum alpinum	IV (1–4)	Festuca vivipara	II (1–3)
Polygonum viviparum	IV (1–3)	Racomitrium lanuginosum	II (1–3)
Scapania undulata	IV (1–4)	Leontodon autumnalis	II (1–3)
Aneura pinguis	IV (1–3)	Euphrasia officinalis agg.	II (1–3)
Eriophorum angustifolium	IV (3–5)	Polytrichum alpinum	II (1–2)
Carex demissa	IV (1–4)	Taraxacum officinale agg.	II (1–2)
		Huperzia selago	II (1–2)
Selaginella selaginoides	III (1–3)	Cratoneuron commutatum	II (3–6)
Carex echinata	III (1–6)	Campylium stellatum	II (1–3)
Juncus triglumis	III (2–3)	Juncus castaneus	II (1–3)
Festuca ovina	III (1–3)	Carex pulicaris	II (2–3)
Caltha palustris	III (1–4)	Geum rivale	I (1–2)
Viola palustris	III (1–2)	Alchemilla filicaulis filicaulis	I (1–2)
Nardus stricta	III (2–4)	Ranunculus acris	I (1–3)
Carex bigelowii	III (1–5)	Rhytidiadelphus squarrosus	I (1–2)
Agrostis canina canina	III (1–4)	Rumex acetosa	I (1–3)
Pinguicula vulgaris	III (1–3)	Fissidens osmundoides	I (1–3)
Carex nigra	III (1–5)	Equisetum palustre	I (3)
Deschampsia cespitosa	III (3–5)	Scorpidium scorpioides	I (1–3)
Rhytidiadelphus loreus	II (1–2)	Saxifraga oppositifolia	I (1–2)
Festuca rubra	II (1–4)	Scapania uliginosa	I (1–2)
Blindia acuta	II (1–3)	Drepanocladus exannulatus	I (2–3)
Bryum pseudotriquetrum	II (1–5)	Alchemilla glabra	I (1–2)
Calliergon sarmentosum	II (1–7)	Rhizomnium pseudopunctatum	I (1)
Polytrichum commune	II (1–3)	Sphagnum auriculatum	I (2–6)
Saxifraga aizoides	II (2–3)	Rhytidiadelphus triquetrus	I (1)
Calliergon trifarium	II (2–5)	Cinclidium stygium	I (1)
Agrostis capillaris	II (2–3)	Riccardia multifida	I (1)
Saxifraga stellaris	II (2)		

Salix herbacea	I (1)	*Carex panicea*	I (1–2)
Calliergon stramineum	I (1)	*Cladonia arbuscula*	I (1–2)
Armeria maritima	I (2)	*Sphagnum subnitens*	I (1–2)
Marsupella emarginata emarginata	I (1–6)	*Sphagnum warnstorfii*	I (1–4)
Epilobium anagallidifolium	I (1–2)	*Mnium hornum*	I (3)
Luzula multiflora	I (1)	*Alopecurus alpinus*	I (1)
Sphagnum capillifolium	I (1)	*Fissidens adianthoides*	I (1)
Sphagnum girgensohnii	I (1–4)	*Jungermannia obovata*	I (1–2)
Cerastium fontanum	I (1)	*Scapania irrigua*	I (2)
Sphagnum recurvum	I (1–2)	*Barbilophozia lycopodioides*	I (2)
Carex vaginata	I (3)	*Carex curta*	I (2)
Narthecium ossifragum	I (2)		

Potentilla erecta	I (1–2)	Number of samples	24
Anthelia julacea	I (2)	Number of species/sample	26 (9–42)
Juncus articulatus	I (1)		
Alchemilla alpina	I (1)	Vegetation height (cm)	16 (10–20)
Luzulu spicata	I (3)	Vegetation cover (%)	81 (20–100)
Tofieldia pusilla	I (1)		
Carex atrofusca	I (1–2)	Altitude (m)	966 (716–1052)
Pellia endiviifolia	I (2)	Slope (°)	12 (1–38)

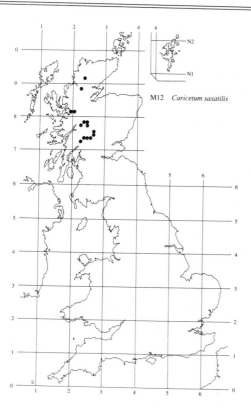

M12   *Caricetum saxatilis*

M13

Schoenus nigricans-Juncus subnodulosus mire
Schoenetum nigricantis   Koch 1926

Synonymy

Juncus-Schoenus community   Clapham 1940; Valley fen communities   Bellamy & Rose 1961 *p.p.*; *Cladium-Schoenus-Juncus* community   Haslam 1965; *Schoenus* edge   Haslam 1965; *Schoeno-Juncetum subnodulosi* (Allorge 1922) Wheeler 1975, 1980*b p.p.*; *Schoenus nigricans-Juncus subnodulosus* nodum   Ratcliffe & Hattey 1982; *Schoenetum nigricantis*   Dierssen 1982.

Constant species

Carex panicea, Juncus subnodulosus, Molinia caerulea, Potentilla erecta, Schoenus nigricans, Succisa pratensis, Calliergon cuspidatum, Campylium stellatum.

Rare species

Carex diandra, Dactylorhiza traunsteineri, Potamogeton coloratus.

Physiognomy

Some of the calcicolous mire vegetation in which *Schoenus nigricans* figures prominently is best included within the *Pinguiculo-Caricetum* and *Carici-Saxifragetum*, where *Schoenus* is overall only occasional and attains local dominance without a marked disruption of the floristic integrity of the communities. In the *Schoenetum nigricantis*, by contrast, *Schoenus* is a very frequent species and consistently associated with other distinctive floristic features. Although its cover is somewhat variable (it can even be absent from fragmentary stands), it is generally the dominant in this community, giving the vegetation a distinctive grey-green coloration through the year, with its semi-evergreen foliage. Very commonly, however, it occurs intermixed with at least some *Juncus subnodulosus*, a rush that is very rare in our submontane calcicolous mires, and sometimes this predominates, when stands have an olive-green hue in spring, turning reddish-brown with the death of the shoots in winter. *Molinia caerulea* is also constant and though usually not of very great abundance, it can be locally prominent, particularly in the more run-down kinds of

Schoenetum. By and large, it is these plants, accompanied in some sub-communities by other rushes and sedges of medium stature (e.g. *Juncus articulatus, J. acutiflorus, Carex elata, C. diandra, C. rostrata*), that form the major structural element of the community, creating the general impression of a rough sward, half a metre or so in height (Wheeler 1975).

On closer inspection, however, it can be seen that it is the size and spacing of the *Schoenus* plants (and, to a lesser extent of the *Molinia*) that exert the strongest influence on the richness and organisation of the associated flora. *Schoenus* is a strongly gregarious, caespitose hemicryptophyte, forming loose to dense tufts of shoots on a partially-buried or emergent rootstock (Sparling 1962*a*, 1968). The tussocks commonly attain a height of 40 cm or more, often having prominent fibrous stools: in some stands much larger individuals can be found; in others, the plants are of considerably smaller stature throughout. Some of this variation may be attributable to the effect of certain environmental factors (see below), but part probably reflects inherent differences in growth form (Wheeler 1975). Usual tussock densities are of the order of 4–5 plants m^{-2}, but values of 8–10 m^{-2} are quite frequent, when the canopy of leaves can be virtually continuous (Wheeler 1975). The development of the larger tussocks creates a variety of micro habitats: the depressions or runnels between, which are variously shaded and which can have standing or running water or can be no more than moist for much of the year, and the tussocks themselves, their sides and their tops, the latter far removed from the influence of the calcareous ground water. Such differentiation increases the floristic richness of the community, with 10×10 m samples commonly having more than 40 species, and is the basis of the frequent juxtaposition here of calcicoles and calcifuges and of aquatics and plants of drier habitats.

Among the associated vascular flora, smaller herbs are important throughout the community and, where the *Schoenus* tussocks are not too closely-spaced, they can comprise much of the runnel vegetation, forming a lower stratum 10–30 cm tall. Sedges are often a major element with *Carex panicea*, *C. lepidocarpa* and *C. flacca* occurring frequently, *C. nigra* more occasionally. *C. hostiana* and *C. pulicaris* can also be found, though they are strongly preferential for one particular kind of *Schoenetum*, and the community also provides a locus for some of the south-eastern occurrences of *C. dioica*. In stands where the summer water-table is close to or at the surface, mixtures of these sedges tend to form a rather open cover within which there is considerable enrichment by species such as *Equisetum palustre*, *Pedicularis palustris*, *Mentha aquatica*, *Valeriana dioica* and *Cardamine pratensis* and, in some cases, *Parnassia palustris*, *Pinguicula vulgaris* and *Eriophorum latifolium*. More calcicolous assemblages provide a clear floristic link with the *Pinguiculo-Caricetum*, although plants like *Eleocharis quinqueflora*, *Selaginella selaginoides* and *Triglochin palustris*, which are predominantly northern in their distribution, are at most occasional here.

More distinctively, there are frequent records in this component for *Hydrocotyle vulgaris*, *Ranunculus flammula* and the Oceanic Southern *Anagallis tenella*, and for a variety of orchids. *Epipactis palustris* is the commonest of these throughout the community as a whole and it is very characteristic of this kind of vegetation (its other major community being the floristically quite similar *Salix-Campylium* dune-slack) but *Dactylorhiza majalis* also occurs frequently, ssp. *praetermissa* to the south and east, ssp. *purpurella* to the north and west, and *D. fuchsii*, *D. incarnata* and *Gymnadenia conopsea* are also quite common. More locally, there can also be some *Dactylorhiza maculata* ssp. *ericetorum*, *Ophrys insectifera*, *Coeloglossum viride*, *Platanthera bifolia* and the nationally rare *D. traunsteineri*. Where numbers of these are present together and flowering between May and August, this vegetation can present a splendid sight, though variation within and hybridisation between the dactylorchids can make recording a vexatious task.

Taller herbs are generally not of high individual cover in the community, although they can be locally abundant and the emergent flowering shoots of the dicotyledons often make them conspicuous. Among these plants, *Succisa pratensis* is the most common, but also frequent are *Angelica sylvestris*, *Cirsium palustre*, *Filipendula ulmaria*, *Eupatorium cannabinum*, *Oenanthe lachenalli* with *Centaurea nigra*, *Serratula tinctoria*, *Lythrum salicaria*, *Cirsium dissectum* and *Valeriana officinalis* more occasional. Trailing among the vegetation, there can be some *Galium uliginosum* and, less commonly, *G. palustre*.

Phragmites australis is also a frequent member of the *Schoenetum*, particularly in ungrazed stands, but it is usually not very vigorous, occurring as sparse shoots or scattered clumps. *Cladium mariscus* occurs occasionally,

too, but it is likewise only locally abundant. Where these species are present, then, they typically form a very open upper tier to the vegetation, a metre or so high, but they can give the community a rather different superficial appearance and, particularly where tall-fen dicotyledons are also numerous, bring it close in floristics and structure to certain kinds of Phragmitetalia vegetation, notably the *Peucedano-Phragmitetum*.

In drier stands, where the summer water-table is well below the surface, much of this floristic richness is lost. Then, the smaller sedges, together with much *Molinia*, other grasses like *Holcus lanatus* and *Festuca rubra*, and often *Lotus uliginosus*, bulk large in the runnels and, especially where *Schoenus* itself is reduced, the vegetation approaches the communities of the Junco-Molinion: such a trend is best seen in the *Festuca-Juncus* subcommunity. To the other extreme, on very wet and flat sites, there can be pools of standing water between the tussocks with enrichment with a more obviously aquatic element, including *Menyanthes trifoliata*, *Equisetum fluviatile*, *Utricularia* ssp. and *Chara* ssp.: this is most noticeable in the *Caltha-Galium* sub-community which is floristically transitional to the *Carex-Calliergon* mire.

Even within the runnels of the *Schoenetum*, drier areas can have a scatter of less calcicolous plants, but more usually it is the tops of the *Schoenus* tussocks (and, sometimes, of the *Molinia*) which provide a niche for such species. *Potentilla erecta* and *Erica tetralix* are the most frequent members of this component but there can also be other ericoids like *Calluna vulgaris* and, to the west, *Erica cinerea*, and some *Drosera rotundifolia* or *Narthecium ossifragum*. *Molinia* can also be found growing epiphytically on the *Schoenus*.

Bryophytes vary considerably in their cover and variety in the community but they can be very extensive (with a total cover up to 70% or so) and, like the vascular plants, show a striking patterning over the various microhabitats, a feature first described by Clapham (1940) from stands of the community in Berkshire. The commonest mosses throughout are *Campylium stellatum* and, particularly where the shade is deeper, *Calliergon cuspidatum*. *Drepanocladus revolvens* (including var. *intermedius*), *Aneura pinguis*, *Cratoneuron commutatum* and *C. filicinum* are also quite frequent and extensive mats of all these species can colonise wetter ground and form a base, kept moist but rarely submerged, for the establishment of the vascular plants. These bryophytes can also grow over very young *Schoenus* tussocks but where larger stools have developed, they are very much confined to the area around their bases. Also common in some stands are *Bryum pseudotriquetrum*, *Plagiomnium elatum*, *P. rostratum* and *Rhizomnium punctatum* which Clapham (1940) noted as characteristic of compacted and rarely submerged peat and the sides of the *Schoenus* stools; this latter habitat also affords a niche for *Fissidens adian-thoides*. More typical of the tussock tops are *Pseudo-scleropodium purum*, *Brachythecium rutabulum*, *Aula-comnium palustre* and *Sphagnum subnitens* and, often spreading in considerable abundance over tops, sides and down into drier runnels, *Ctenidium molluscum*. *Lophocolea bidentata s.l.* and *Pellia endiviifolia* also occur occasionally in this community and more unusual species recorded in some stands are *Campylium elodes*, *Riccardia multifida*, *R. chamedryfolia* and *Drepanocla-dus lycopodioides*.

Sub-communities

***Festuca rubra-Juncus acutiflorus* sub-community:** *Schoeno-Juncetum typicum* (Allorge 1922) Wheeler 1980b *p.p.*; *Schoenus nigricans-Juncus subnodulosus* nodum Ratcliffe & Hattey 1982 *p.p.* This sub-community comprises the more impoverished stands of the *Schoenetum* where *Schoenus* itself is sometimes much reduced in vigour and cover or even totally absent from vegetation which preserves the general floristics of the community. *Juncus subnodulosus* and *Molinia* are often proportionately more important and mixtures of grasses can make a major contribution to the cover, with *Festuca rubra*, *Holcus lanatus*, *Anthoxanthum odoratum*, *Agrostis canina* ssp. *canina* and *A. stolonifera* well repre-sented. Where tussocks are strongly developed, it is these species, with mixtures of *Carex panicea*, *C. lepido-carpa* and *C. flacca*, which usually clothe much of the runnels. In some stands, *Juncus acutiflorus* is present in small amounts, emphasising similarities with Junco-Molinion vegetation.

   Various vascular plants generally characteristic of the *Schoenetum* are reduced here: *Anagallis tenella*, for example, is totally absent and *Pedicularis palustris*, *Epipactis palustris* and other orchids very scarce. The commonest herbs are *Succisa* and *Hydrocotyle* with occasional *Mentha aquatica*, *Equisetum palustre*, *Lotus uliginosus*, *Cardamine pratensis* and *Ranunculus acris* and scattered individuals of *Eupatorium*, *Angelica*, *Cir-sium palustre*, *Filipendula ulmaria* and *Valeriana dioica*. *Potentilla erecta* and occasional *Erica tetralix* occur on the tussock tops. *Phragmites* and *Cladium* are infrequent and of low cover but some stands have a patchy canopy of *Myrica gale*.

   Bryophytes, too, are relatively few in number here and generally sparse among the rank herbage. *Callier-gon cuspidatum* is the commonest species with *Campy-lium stellatum* somewhat reduced in frequency and other mosses distinctly scarce.

***Briza media-Pinguicula vulgaris* sub-community:** *Schoeno-Juncetum leontodetosum* (Allorge 1922) Wheeler 1975; *Schoeno-Juncetum serratuletosum* (Allorge 1922) Wheeler 1980b; *Schoenus nigricans-Jun-cus subnodulosus* nodum   Ratcliffe & Hattey 1982 *p.p.* Compared with the *Festuca-Juncus* sub-community, this kind of *Schoenetum* is often strikingly rich, sharing, with the next, frequent records for such characteristic herbs as *Anagallis tenella*, *Pedicularis palustris* and *Epipactis palustris* and a greater variety of bryophytes. Mixtures of *Schoenus*, *J. subnodulosus* and *Molinia* usually share dominance but the small-herb component of the runnel flora is especially distinctive. Here, *Carex hostiana* and *C. pulicaris* join the other sedges as strong preferentials and there is often some *Briza media*, *Pinguicula vulgaris*, *Parnassia palustris*, *Juncus articulatus* and *Linum cath-articum* and occasionally *Selaginella selaginoides*, *Trig-lochin palustris* and *Plantago maritima* can be found. Such assemblages emphasise the floristic links with the Northern sub-montane fen vegetation of the *Pinguiculo-Caricetum*. Among the sometimes rich mixtures of orchids with frequent *Gymnadenia conopsea* var. *densif-lora* and *Dactylorhiza fuchsii*, there is commonly some *D. majalis* ssp. *purpurella*, a further northern element. *Epipactis palustris*, on the other hand, and *Ophrys insectifera*, which occurs occasionally, are species of Continental distribution, as is *Serratula tinctoria*, another good preferential herb. This kind of floristic mix, with both northern and southern plants occurring together, is a striking feature of this sub-community. Other smaller herbs to be found in the runnels include *Polygala vulgaris* and *Trifolium pratense* and, in some stands, *Isolepis setacea*, *Sagina procumbens* and *S. nodosa* have been recorded together (the *Isolepis* variant of Wheeler 1980b).

   Along with *Succisa pratensis* and *Serratula*, taller herbs are represented by frequent *Angelica sylvestris*, *Cirsium palustre*, *Eupatorium cannabinum*, *Oenanthe lachenalii* and occasional *Filipendula ulmaria*, and bulk-ier grasses such as *Festuca rubra* and *Holcus lanatus* can be locally prominent too. *Phragmites* and *Cladium* are sometimes present, though typically they are not abun-dant. More calcifuge plants can be an important compo-nent on the tussock tops, with *Potentilla erecta* and *Erica tetralix* being sometimes joined by *Calluna vul-garis*, *E. cinerea*, *Danthonia decumbens* and *Luzula mul-tiflora*. Where these are prominent over much *Molinia* and with scattered bushes of *Myrica*, they can give this vegetation a rather different look (as in the *Myrica* variant of Wheeler 1980b).

   Bryophytes are quite numerous and sometimes of high cover with, in addition to *Campylium stellatum* and *Calliergon cuspidatum*, frequent *Drepanocladus revol-vens*, *Fissidens adianthoides*, *Cratoneuron commutatum* and occasional *C. filicinum*, *Ctenidium molluscum*, *Bryum pseudotriquetrum*, *Aneura pinguis* and *Pellia endiviifolia*.

***Caltha palustris-Galium uliginosum* sub-community:** Valley fen communities   Bellamy & Rose 1961 *p.p.* *Schoenus* and *Cladium-Schoenus-Juncus* communitie

Haslam 1965; *Schoeno-Juncetum caricetosum & cladietosum* (Allorge 1922) Wheeler 1975. *Schoenus, J. subnodulosus* and *Molinia* remain of structural importance here but this bulkier component of the vegetation is variously augmented by *Carex rostrata, C. diandra, C. elata* and, somewhat better represented in this sub-community than in other kinds of *Schoenetum*, *Cladium* and, to a lesser extent, *Phragmites*. The individual cover of these plants is generally low (although *Cladium* can be locally so abundant as to give the superficial impression of a sedge-bed) and usually they occur in a mosaic with the three constant monocotyledons of the community.

Runnels are often well developed but many of the smaller preferentials of the *Briza-Pinguicula* sub-community are reduced to occasional occurrences. The commonest species of this element here are *Carex panicea, C. lepidocarpa, Mentha aquatica, Hydrocotyle vulgaris* and, preferentially frequent, *Caltha palustris* and *Valeriana dioica*. In addition to *Epipactis palustris*, there is quite often some *Dactylorhiza incarnata, D. majalis* ssp. *praetermissa* and, in some sites, *D. traunsteineri*. Taller dicotyledons are rather more common than elsewhere in the community with, as well as the community species *Succisa, Angelica* and *Cirsium palustre*, preferentially frequent *Eupatorium, Filipendula ulmaria* and *Lynchnis flos-cuculi. Galium uliginosum* and, less commonly, *G. palustre* can be found sprawling among the vegetation and there is often some *Vicia cracca*.

Further distinctive enrichment occurs where, as is quite commonly the case here, there are pools of standing water in the runnels. The presence of *Carex rostrata* and *C. diandra* is especially associated with this kind of situation and these may be accompanied by *Menyanthes trifoliata, Equisetum fluviatile, Eriophorum angustifolium, Utricularia* spp. (including *U. vulgaris, U. neglecta* and the rare *U. intermedia*), the rare *Potamogeton coloratus* and *Chara* spp. (including *C. hispida* var. *hispida, C. globularis* var. *aspera* and *C. contraria*).

Bryophytes tend to be fewer and less abundant under dense vegetation of this kind but, where the cover is rather more open, and especially where there is a strong differentiation of *Schoenus* tussocks, diversity and cover can be high. All the species of the *Briza-Pinguicula* sub-community occur frequently here, sometimes with additional rarer plants in the wetter runnels, e.g. *Riccardia chamedryfolia, R. multifida, Drepanocladus lycopodioides*. But preferentially frequent are those species which seem to be most typical of the sides of the tussocks (*Plagiomnium rostratum, P. elatum* and *Rhizomnium punctatum*) or their tops (*Pseudoscleropodium purum*).

Habitat

The *Schoenetum* is confined to peat or mineral soils in and around lowland mires irrigated by base-rich, highly calcareous and oligotrophic waters. It is commonest in sites with a strongly soligenous character, being often found below springs and seepage lines and on the flushed margins of more fully-developed valley mires, but it also extends into topogenous basins provided there is close contact with waters draining from lime-rich substrates. Climate, particularly summer warmth, plays an important part in restricting the community to the southern parts of Britain, though this kind of vegetation provides an important locus for outlying occurrences of some northern mire plants and, towards the north and west, begins to grade to sub-montane Tofieldietalia communities. Grazing sometimes influences the structure and floristics of the *Schoenetum* and some stands have been affected by mowing and burning. Shallow peat-digging has been locally important in providing a suitable habitat for the community, but more drastic treatments of mires, particularly draining and eutrophication, have reduced its extent and eliminated it from some areas.

The representation of *Schoenus nigricans* in British vegetation types is a function of a complicated interaction between climatic and edaphic factors. It is generally restricted to lowland areas with a February minimum temperature at or above freezing (*Climatogical Atlas* 1952), a climatic relationship which reflects its general distribution through Europe and which may be related to frost-sensitivity (Sparling 1968). Within this zone, it occurs on wetter, oligotrophic soils, but these show a wide variation in base-status and are represented in a diversity of mire types. In moving across Britain into western Ireland, *Schoenus* becomes less exclusively a plant of base-rich soligenous habitats, extending in north-west Scotland into flushed areas within base-poor ombrogenous mires and then, in Ireland, on to the acidic peats of the bog-plane proper (e.g. Tansley 1911, 1939, Pearsall & Lind 1941, Osvald 1949, Boatman 1957). Sparling (1962b, 1967a, b) has related this trend to a complex of factors connected with increasing oceanicity to the west, and perhaps working through an amelioration in aluminium levels in the substrate, to which *Schoenus* is very sensitive.

Within this spectrum, *Schoenus* occurs as an occasional local dominant in base-rich soligenous mires around the upland fringes of north-west Britain, where the general character of the vegetation is very much determined by the cool sub-montane climate: this is the role it has in the low-altitude stands of the *Carici-Saxifragetum* and, more especially, in the *Pinguiculo-Caricetum*. The bulk of our more calcicolous *Schoenus* vegetation, however, belongs to the *Schoenetum* which is, by contrast, a community of southern Britain, occurring largely in an area bounded by the 25 °C mean annual maximum isotherm (Conolly & Dahl 1970). More Continental species, such as *Juncus subnodulosus, Epipactis palustris, Serratula tinctoria, Dactylorhiza majalis* ssp. *praetermissa* and, more occasionally, *D. traunsteineri* and *Ophrys insectifera*, which are very scarce in northern mires, are thus characteristic here. However, the ranges of such plants as these and more northerly mire species are by no means mutually exclusive. Arctic-Alpine plants with which *Schoenus* is sometimes associated in the *Carici-Saxifragetum* and the *Pinguiculo-Caricetum* are, of course, quite absent, but some Continental Northern species maintain quite a good representation in the *Schoenetum*. This is especially true towards the north-western limit of its range where summer temperatures are considerably cooler than they are in East Anglia. There, notably in the Anglesey fens, the floristic richness of the *Briza-Pinguicula* sub-community, which is mainly represented in this region, owes much to the mixing of the different phytogeographic elements, frequent *Carex pulicaris*, *Pinguicula vulgaris*, *Parnassia palustris* and *Dactylorhiza majalis* ssp. *purpurella* and occasional *Selaginella selaginoides* occurring with Continental species. Such plants thin out moving further to the south-east but, in East Anglia, the community still provides them (and others like *Carex dioica*, *C. diandra*, *Menyanthes* and *Utricularia* spp.) with important outposts.

The very local occurrence of the *Schoenetum* within the generally warmer parts of Britain is a strong reflection of its dependence upon continuous irrigation with base-rich and calcareous waters by seepage from lime-rich bedrocks or superficials. The community is best developed on the Chalk and chalky drift of East Anglia, where it occurs around the head-waters of tributaries to the Ouse and the Waveney (Bellamy & Rose 1961, Haslam 1965) and the Ant (Wheeler 1978, 1980*b*; see also Ratcliffe 1977), and in Anglesey, where it is found on the fringes of basin mires that have developed within shallow hollows in Carboniferous Limestone (Wheeler 1975, 1980*b*; Ratcliffe & Hattey 1982). Scattered sites occur elsewhere, notably over Corallian Limestone in the Cothill basin in Berkshire (Clapham 1940), on Oolite in Northamptonshire, on Magnesian Limestone in Nottinghamshire and Durham, and more distantly, on Carboniferous Limestone again in north Lancashire and Northumberland (Wheeler 1975, 1980*b*; Ratcliffe 1977).

Typically, in such situations as these, the flushing waters have a pH between 6.5 and 8.0, with dissolved calcium levels of 60–200 mg l^{-1} (Wheeler 1975, 1983), conditions which are reflected in the prominent calcicolous element among the smaller herbs and bryophytes of the runnels, where the effects of irrigation are felt most directly. The soil types, however, are quite variable. The community can occur on wetter mineral soils, including very ill-structured profiles which are little more than sloppy muds or marls, and on more well-developed surface water gleys, with or without a humic top, and also on peats proper which, on the margins of mires, can be moderately deep. Often, the sites are gently-sloping

with a more or less continuous through-put of water but the community can extend on to wet topogenous sites, so long as the base-richness of the substrate is maintained.

One further important characteristic of the soils is that they are very poor in major nutrients, probably especially in phosphorus (Haslam 1965, Wheeler 1983) and this is likely to be important in excluding the community from those parts of valley and flood-plain mires where there is enrichment by the regular deposition of allochthonous mineral material. In such situations, *Phragmites* and tall mesophytic herbs become progressively more important in vegetation like that of the *Peucedano-Phragmitetum* and the *Phragmites-Eupatorium* fen. The *Schoenetum* grades floristically into these communities, particularly through the *Caltha-Galium* sub-community, and, in some sites, it can form a fringe around them, but these elements are generally of low cover in the community itself, so that the vegetation is maintained in a moderately open condition with ample opportunity for the development of smaller, shade-sensitive species. *Schoenus* itself will not tolerate deep shade (Sparling 1968) and is also damaged by the deposition of silt over its shoots (Haslam 1965).

There is no doubt that for the optimal development of the *Schoenetum*, the soil must be maintained in a reasonably moist state throughout the year. *Schoenus* itself will stand some fluctuation in ground-water level (particularly towards the wetter west: Sparling 1962*a*, 1968; Haslam 1965) but it grows best where there is a fairly high and stable water-table and its seed germinates best when the soil surface is kept moist outside the winter months (Sparling 1968). Nonetheless, there is considerable variation in the height of the water-table here and this factor, often interacting with past or present treatments, plays some part in differentiating the sub-communities. The highest summer water-tables are found in the *Caltha-Galium* sub-community, where the *Schoenetum* extends on to flat sites which can have a few centimetres of standing water between the tussocks. This permits the development of an obvious aquatic element in the flora and can encourage an increase in the cover of *Phragmites* and *Cladium*, especially where stands have not been grazed or regularly summer-mown. Then, this vegetation takes on something of the character of a Phragmitetalia fen or of swampy Caricion davallianae vegetation like that in the *Carex-Calliergon* mire. In some situations, this sub-community may represent an early stage in primary hydrarch succession (e.g. Clapham 1940), though there is often the suspicion that it has developed secondarily in wet abandoned peat-cuttings, a common feature on lowland valley and basin mires. And it has probably sometimes been included with neighbouring Phragmitetalia vegetation in marsh crops harvested by mowing.

The *Briza-Pinguicula* sub-community, by contrast, is more typical of situations where the summer water-table is at, or not far below, the surface. In such places, this vegetation has often been open to grazing and trampling by stock and this probably plays some part in the differentiation of more pronounced tussock/runnel systems and in the maintenance of a low, open sward between the tussocks where smaller herbs and bryophytes sensitive to competition can flourish. Other stands of this kind have probably been mown for a mixed litter crop, traditionally cut annually between July and October (e.g. Lambert 1946). Such treatment would be expected to prevent the development of large tussocks but, in transitions to topogenous mires, could maintain the *Schoenetum* against the spread of *Phragmites* and, especially in the typically oligotrophic habitat here, of *Cladium* which, unlike *Schoenus*, cannot recover from annual summer mowing and cannot regenerate well from seed (Godwin 1941, Haslam 1965).

On drier soils still, the *Festuca-Juncus* sub-community is characteristic. In some cases, this kind of *Schoenetum* may represent a natural transition to Junco-Molinion vegetation on the surrounding mineral soils, or a secondary development where mowing and grazing have been relaxed on drier soils, but in many places it seems to have developed because of an artifical lowering of the water-table by drainage operations. Fragmentary stands of this sub-community can still provide relic localities for rarer mire species but they have little of the striking richness of the other kinds of *Schoenetum* (Wheeler 1975, Ratcliffe & Hattey 1982).

Zonation and succession

The *Schoenetum* occurs in zonations around springs and seepage lines where natural variation in the vegetation is related primarily to the height of the water-table in sequences of mineral and organic soils, the character and disposition of which are determined by geology and the pattern of flushing. But such zonations have been widely affected by a variety of treatments such as grazing, mowing and peat-cutting, so that some of their components are plagioclimax vegetation, now often further altered by successional changes attendant upon neglect. Other changes have been induced by drainage and eutrophication and widespread improvement of surrounding land has also left many stands isolated and much modified within intensive agricultural landscapes.

In more intact zonations developed over entirely calcareous substrates, the *Schoenetum* typically grades, on the gleyed mineral soils above seepage lines or around more localised springs, to richer kinds of Junco-Molinion vegetation like the *Cirsio-Molinietum*, in which both *Schoenus* and *J. subnodulosus* can persist but where dominance generally passes to *Molinia*. This, in turn, may grade to some kind of Mesobromion grass-

land on the dry, lithomorphic calcareous soils beyond: usually, within the range of the *Schoenetum*, this is the *Festuca-Avenula* grassland. Where flushing occurs more abruptly over limestone slopes, the *Schoenetum* can give way directly to such a calcicolous sward (Figure 15).

In the other direction, where the *Schoenetum* occurs above and sometimes extends on to deeper peats which have accumulated in valleys and basins under the close influence of calcareous and oligotrophic ground waters, the community can pass directly to some kind of Phragmitetalia vegetation, often the *Cladietum marisci*, on the waterlogged organic soils. Sometimes, in such sequences, the *Schoenetum* can itself be zoned, the *Festuca-Juncus* sub-community occupying the drier ground, the *Briza-Pinguicula* the moister soils and the *Caltha-Galium* the wettest, and the last type can also be found in hollows on peat with the *Carex-Calliergon* mire among a matrix of the *Cladietum*. Such patterns as these are especially characteristic of the Anglesey mires such as Cors Bodeilio, Cors y Farl, Cors Goch and, to a lesser extent, Cors Erddreiniog (Wheeler 1975, Ratcliffe & Hattey 1982 Site group IX).

In less consistently oligotrophic situations, the *Schoenetum* occurs in generally similar zonations, but in association with other communities. Where it occurs around valley mires, for example, in which the peats experience some deposition of allochthonous mineral material, it can give way below to the *Phragmitetum australis* swamp or some kind of Phragmitetalia tall-herb fen, generally the *Phragmites-Eupatorium* community. This is well seen in some of the East Anglian fens like Smallburgh, Scarning, Redgrave/South Lopham and Thelnetham/Blo' Norton (Bellamy & Rose 1961, Haslam 1965, Wheeler 1975, 1978, Ratcliffe 1977). And, where the gleyed soils around the flushes are somewhat richer than those typically supporting the *Cirsio-Molinietum*, the *Schoenetum* can grade above to the *Juncus-Cirsium* community or the *Holco-Juncetum* and thence, on the drier mineral profiles, to some kind of Cynosurion sward. Where irrigation with calcareous waters is a very local phenomenon within tracts of acidic substrates and soils, the community can be found among more calcifugous vegetation. On some of the Norfolk heaths, like Buxton Heath for example, where calcareous drift occupies the bottoms of some hollows and valleys the *Schoenetum* passes, above the seepage lines, to the *Ericetum tetralicis* on base-poor gleys. Other *Schoenus* vegetation in these 'mixed mires' is, however, perhaps better placed in the *Schoenus-Narthecium* community.

Traditional treatments have variously influenced zonations such as these. The middle and drier parts of the sequences were often subject to grazing by stock, the middle and wetter parts to mowing for hay, litter or marsh crops like sedge and reed. Such treatments had the general effect of holding back succession to woodland throughout, but they have probably also influenced the balance of the different components, increasing the extent of one community against another according to the particular regimes of grazing and mowing adopted, treatments which affect the vegetation directly but also modify the nutrient budget of the system. As far as the *Schoenetum* itself is concerned, the former effects have probably modified transitions to tall-herb fens and swamps by influencing the amounts of *Phragmites* and *Cladium* in the vegetation (e.g. Godwin 1941, Haslam 1965), and altered the balance within the community between the *Schoenus*, *J. subnodulosus* and *Molinia*. And alterations in nutrient budget perhaps play a part in the

balance between the *Schoenetum* and the *Cirsio-Molinietum* and *Juncus-Cirsium* community (Wheeler 1983). Treatment regimes have been very variable so actual zonations can be similarly diverse, added to which we now often see their influence blurred by the effects of long neglect, particularly over the surface of peats where mowing for marsh crops has fallen into virtually total disuse. A common pattern now is for the *Schoenetum* to form the wettest component in a fairly open herbaceous sequence, still maintained by grazing, and to give way below to very dense sedge- or reed-beds or to woodland that has colonised tall-herb fen and swamp.

Such transitions have become more frequent with the

Figure 15. *Schoenus* communities among fen (*a*) and bog (*b*) sequences.
M13b   *Schoenetum* mire, *Briza-Pinguicula* sub-community
M13c   *Schoenetum* mire, *Caltha-Galium* sub-community
M14   *Schoenus-Narthecium* mire
M16a   *Ericetum tetralicis* wet heath, Typical sub-community
M16b   *Ericetum tetralicis* wet heath, *Succisa-Carex* sub-community

M21   *Narthecio-Sphagnetum* valley bog
M24a   *Cirsio-Molinietum* fen-meadow, *Eupatorium* sub-community
M24b   *Cirsio-Molinietum* fen-meadow, Typical sub-community
M24c   *Cirsio-Molinietum* fen-meadow, *Juncus-Erica* sub-community
CG2   *Festuca-Avenula* grassland
S2   *Cladietum* swamp
S25   *Phragmites-Eupatorium* fen
H4   *Ulex minor-Agrostis* heath

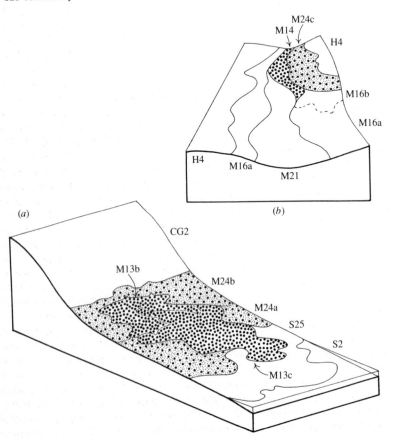

extensive drainage of impeded soils in the British low-lands, either locally or as a result of major works which have lowered the water-table in whole catchments (as on the Little Ouse where the Thelnetham and Blo' Norton fens have been affected: Ratcliffe 1977). Such treatment tends to speed colonisation of any unreclaimed mire fragments by woody vegetation and has probably des-troyed many stands of *Schoenetum* itself. Even where the community persists, it is often in the form of the impoverished *Festuca-Juncus* sub-community or transi-tions to rank Junco-Molinion vegetation. On its other boundary, where it grades on to drier soils, the commun-ity has suffered by assiduous agricultural improvement right up to spring and flush lines such that natural zonations are often truncated above, any remaining *Schoenetum* closely hemmed in by arable land or modi-fied pasture. Eutrophication by fertiliser run-off or drift is a common danger to the community in such situations and, with enrichment of peats by drainage and surface oxidation, the vegetation is open to conversion to Fili-pendulion tall-herb communities, with the *Phragmites-Urtica* fen on any fragments of wetter soils.

With such complex modifications of the habitat of the *Schoenetum*, it is difficult to tell what the natural seral development of the community might be. Scattered saplings of *Alnus glutinosa*, *Betula pubescens* and *Salix cinerea* can sometimes be found in ungrazed stands (Wheeler 1975) and it is possible that mixed canopies of such species could come to dominate in some kind of *Salix-Betula-Phragmites* woodland over topogenous peats (Wheeler 1980*c*). Local development of ombrotro-phic nuclei in such situations, or the formation of raised areas within soligenous mires, probably favours a spread of *Molinia* and preferential invasion by *B. pubes-cens* with the eventual formation of some type of *Betula-Molinia* woodland. *Schoenus* itself, of course, may play an important role in providing initial sites for the development of acidophilous elements in the flora but, although it can persist for a considerable time under shading branches, it quickly succumbs to the formation of a closed canopy.

Distribution

The *Schoenetum* is widespread but decidedly local throughout lowland England and Wales, being res-tricted by both the natural scarcity of suitable situations and extensive destruction of habitats. The *Caltha-Galium* sub-community is concentrated in East Anglia, the *Briza-Pinguicula* sub-community in Anglesey, where it represents a geographical transition to the *Pinguiculo-Caricetum* which replaces the *Schoenetum* in the sub-montane north-west. The *Festuca-Juncus* sub-commun-ity occurs throughout the range and in many much-modified sites is the only type of this mire represented.

Affinities

The *Schoenetum* as defined here includes most of the calcicolous mire vegetation from lowland Britain in which *Schoenus nigricans* plays a leading role, along with a variety of more Continental associates. It corres-ponds closely with Wheeler's (1975, 1980*b*) *Schoeno-Juncetum*, uniting his sub-associations into a smaller number of vegetation types and adding new data, par-ticularly from more impoverished stands, like many of those in Ratcliffe & Hattey's (1982) survey of Welsh mires. Although the community has a unique combi-nation of constants, many elements of its flora are species of wide ecological amplitude and the *Schoenetum* has strong affinities in a number of directions. Among other calcicolous mires, it grades through the *Briza-Pinguicula* sub-community to the *Pinguiculo-Caricetum*, and through the *Caltha-Galium* sub-community to the *Carex-Calliergon* mire; there are also clear overlaps, through the tall-herb element, with the Phragmitetalia fens. Even more obviously, the *Schoenetum* has a sub-stantial component of Molinietalia herbs and through-out, but especially in the *Festuca-Juncus* sub-commun-ity, grades to Holco-Juncion fen-meadows.

In a broader phytosociological context, the commun-ity forms part of a series of *Schoenus*-dominated vege-tation types that extends throughout the lowlands of western Europe and can be regarded as part of a compendious central and west European *Schoenetum nigricantis* (*sensu* Koch 1926, but not Tansley 1939). Within this association, regional variations are recognised as geographical races (Koch 1926, Duvig-neaud 1949), often upgraded as vicariants represented in Czechoslovakia (Klika 1929), Poland (Matuszkiewicz 1981), Germany (Tüxen 1937, Kloss 1965), The Nether-lands (Westhoff & den Held 1969), Belgium (Duvig-neaud & Vanden Berghen 1945, Vanden Berghen 1952), France (Allorge 1922, Lemée 1937) and Ireland (Braun-Blanquet & Tüxen 1952). In such a series, our *Schoene-tum* falls between the communities described from France, where Atlantic species like *Anagallis tenella*, *Cirsium dissectum* and *Oenanthe lachenalii* occur along with a strong contingent of Continental plants, and the *Cirsio-Schoenetum* characterised from Ireland which tends towards the *Schoenus*-dominated stands in our *Pinguiculo-Caricetum*. Allorge (1922) and Lemée (1937) also include in their community some vegetation where *Schoenus* itself is more poorly represented, but where *J. subnodulosus* dominates the characteristic flora of the community, a phenomenon seen here in certain stands of our *Festuca-Juncus* sub-community. In a sense, how-ever, the prominence of *J. subnodulosus* through much of the *Schoenus*-dominated vegetation of southern Bri-tain is a fortuitous reflection of the particular situation in which many stands are found, in or close to wetter fen.

We have not therefore followed Wheeler (1975, 1980*b*) in diagnosing this community as synonymous with Allorge's (1922) *Schoeno-Juncetum*. The general balance of species in the associated flora of the *Schoenetum* in Britain places the community firmly among the Caricion davallianae mires.

Floristic table M13

	a	b	c	13
Carex panicea	V (3–4)	V (1–5)	V (1–4)	V (1–5)
Juncus subnodulosus	IV (1–4)	V (1–7)	V (1–7)	V (1–7)
Schoenus nigricans	IV (1–4)	V (1–7)	V (1–8)	V (1–8)
Molinia caerulea	IV (1–4)	V (1–7)	V (1–7)	V (1–7)
Calliergon cuspidatum	IV (1–4)	V (1–3)	V (1–5)	V (1–5)
Succisa pratensis	IV (1–3)	V (1–4)	IV (1–3)	IV (1–4)
Potentilla erecta	IV (1–3)	V (1–3)	IV (1–4)	IV (1–4)
Campylium stellatum	III (1–3)	V (1–4)	V (1–4)	IV (1–4)
Angelica sylvestris	II (1–3)	IV (1–3)	IV (1–3)	III (1–3)
Cirsium palustre	II (3)	IV (1–3)	IV (1–3)	III (1–3)
Mentha aquatica	II (1–3)	IV (1–3)	IV (1–3)	III (1–3)
Epipactis palustris	I (1–3)	III (1–3)	IV (1–3)	III (1–3)
Phragmites australis	II (1–3)	III (1–3)	IV (1–7)	III (1–7)
Anagallis tenella		V (1–3)	IV (1–4)	III (1–4)
Equisetum palustre	II (1–3)	III (1–3)	IV (1–3)	III (1–3)
Pedicularis palustris	I (1–3)	IV (1–3)	III (1–3)	III (1–3)
Drepanocladus revolvens	I (1)	III (1–4)	III (1–5)	II (1–5)
Fissidens adianthoides		III (1–3)	III (1–4)	II (1–4)
Bryum pseudotriquetrum		II (1–3)	III (1–3)	II (1–3)
Aneura pinguis		II (1–3)	III (1–3)	II (1–3)
Sphagnum subnitens	I (1–3)	II (1–4)	II (1–3)	II (1–4)
Pellia endiviifolia		II (1–3)	II (1–3)	II (1–3)
Ctenidium molluscum		II (1–6)	II (1–4)	II (1–6)
Eriophorum latifolium		II (1–4)	II (1–4)	II (1–4)
Anthoxanthum odoratum	III (1–3)	III (1–3)	I (1–3)	III (1–3)
Holcus lanatus	III (1–3)	III (1–3)	I (1–3)	III (1–3)
Festuca rubra	III (3–4)	II (1–3)	I (1–3)	II (1–4)
Juncus acutiflorus	II (1–4)			I (1–4)
Agrostis canina canina	II (1–3)			I (1–3)
Carex flacca	III (2–4)	IV (1–3)	II (1–4)	III (1–4)
Briza media		V (1–4)	II (1–3)	II (1–4)
Dactylorhiza fuchsii		V (1–3)	II (1–3)	II (1–3)
Cratoneuron commutatum	I (1–3)	IV (1–4)	II (1–4)	II (1–4)
Centaurea nigra	I (1–3)	IV (1–3)	II (1–3)	II (1–3)
Gymnadenia conopsea		IV (1–3)	II (1–4)	II (1–4)
Parnassia palustris		IV (1–3)	II (1–3)	II (1–3)
Pinguicula vulgaris		IV (1–4)	II (1–4)	II (1–4)
Carex hostiana		IV (1–4)	I (1–3)	II (1–4)
Serratula tinctoria		IV (1–3)	I (1–3)	II (1–3)
Carex pulicaris		IV (1–3)	I (1–3)	II (1–3)
Juncus articulatus		IV (1–4)	I (1–3)	II (1–4)

Species				
Polygala vulgaris		IV (1–3)		II (1–3)
Agrostis stolonifera	II (3–4)	III (1–3)	I (1–3)	II (1–4)
Oenanthe lachenalii	I (1)	III (1–3)	II (1–3)	II (1–3)
Ranunculus flammula	I (1–3)	III (1–3)	II (1–3)	II (1–3)
Luzula multiflora	I (1–2)	III (1–3)	I (1–3)	II (1–3)
Danthonia decumbens	I (2–3)	III (1–3)		II (1–3)
Dactylorhiza majalis purpurella		III (1–3)	I (1–3)	II (1–3)
Linum catharticum		III (1–3)	I (1)	II (1–3)
Trifolium pratense		III (1–3)		I (1–3)
Myrica gale	II (1–4)	II (1–4)		I (1–4)
Calluna vulgaris		II (1–3)	I (1)	I (1–3)
Triglochin palustris		II (1–3)	I (1–3)	I (1–3)
Selaginella selaginoides		II (1–3)	I (1–3)	I (1–3)
Hypericum pulchrum		II (1–3)		I (1–3)
Plantago maritima		II (1–4)		I (1–4)
Ophrys insectifera		II (1–3)		I (1–3)
Isolepis setacea		I (1–3)		I (1–3)
Eupatorium cannabinum	II (1–3)	III (1–3)	V (1–4)	III (1–4)
Valeriana dioica	II (1–3)	II (1–3)	IV (1–3)	III (1–3)
Filipendula ulmaria	II (1–3)	II (1–3)	IV (1–3)	III (1–3)
Galium uliginosum	I (1–3)	II (1–3)	IV (1–3)	III (1–3)
Caltha palustris	I (1)	I (1–3)	IV (1–3)	II (1–3)
Lychnis flos-cuculi	I (2)	I (1)	IV (1–3)	II (1–3)
Pseudoscleropodium purum	I (1–3)	I (1–3)	III (1–3)	II (1–4)
Dactylorhiza incarnata		II (1–3)	III (1–4)	II (1–4)
Dactylorhiza majalis praetermissa		I (1)	III (1–4)	II (1–3)
Plagiomnium elatum		I (1)	III (1–3)	II (1–3)
Vicia cracca		I (1–3)	III (1–3)	II (1–3)
Menyanthes trifoliata	I (1–3)	I (1)	III (1–3)	II (1–3)
Eriophorum angustifolium		I (1–3)	II (1–4)	I (1–4)
Plagiomnium rostratum		I (1)	II (1–3)	I (1–3)
Carex elata		I (1)	II (1–3)	I (1–3)
Rhizomnium punctatum			II (1–4)	I (1–4)
Carex diandra			II (1–3)	I (1–3)
Riccardia chamedryfolia			II (1–4)	I (1–4)
Philonotis calcarea			II (1–3)	I (1–3)
Riccardia multifida			II (1–3)	I (1–3)
Carex rostrata			II (1–3)	I (1–3)
Potamogeton coloratus			I (1–4)	I (1–4)
Utricularia spp.			I (1–3)	I (1–3)
Carex lepidocarpa	III (2–3)	IV (1–4)	III (1–4)	III (1–4)
Hydrocotyle vulgaris	III (2–3)	II (1–3)	III (1–5)	III (1–5)
Lotus uliginosus	II (1–3)	III (1–3)	III (1–3)	III (1–3)
Carex nigra	II (1–3)	II (1–3)	II (1–4)	II (1–4)
Erica tetralix	II (1–3)	II (1–3)	I (1–3)	II (1–3)
Lophocolea bidentata s.l.	II (1–3)	I (1–3)	II (1–3)	II (1–3)
Cardamine pratensis	II (1–2)	I (1)	II (1–3)	II (1–3)
Cladium mariscus	II (1–4)	I (1–4)	II (1–3)	II (1–3)
Ranunculus acris	II (1–3)	II (1–3)	II (1–7)	II (1–7)
Lythrum salicaria	I (1)	II (1–3)	I (1–3)	II (1–3)

Floristic table M13 *(cont.)*

	a	b	c	13
Cratoneuron filicinum	I (1–3)	II (1–3)	II (1–4)	I (1–4)
Cirsium dissectum	I (1–3)	I (1–3)	II (1–3)	I (1–3)
Epilobium palustre	I (1–3)	I (1–3)	II (1–3)	I (1–3)
Equisetum fluviatile	I (1–4)	I (1)	II (1–3)	I (1–4)
Leontodon taraxacoides	I (1)	I (1–3)	I (1–3)	I (1–3)
Aulacomnium palustre	I (1–4)	I (1–3)	I (1–3)	I (1–4)
Galium palustre	I (1–2)	I (1)	I (1–3)	I (1–3)
Valeriana officinalis	I (2–3)	I (1)	I (1–3)	I (1–3)
Brachythecium rutabulum	I (1–3)	I (1)	I (1–3)	I (1–3)
Euphrasia officinalis agg.		II (1–3)	I (1)	I (1–3)
Lathyrus pratensis	I (1–2)		I (1–3)	I (1–3)
Trifolium repens	I (1–3)	I (1–3)		I (1–3)
Taraxacum officinale agg.		I (1–3)	I (1)	I (1–3)
Eleocharis quinqueflora		I (1–3)	I (1–3)	I (1–3)
Carex dioica		I (1)	I (1–3)	I (1–3)
Hypericum tetrapterum		I (1–3)	I (1–3)	I (1–3)
Dactylorhiza traunsteineri		I (1–3)	I (1–3)	I (1–3)
Drosera rotundifolia		I (1–3)	I (1–3)	I (1–3)
Number of samples	69	25	32	126
Number of species/sample	14 (7–27)	45 (26–57)	44 (30–65)	27 (7–65)

a   *Festuca rubra-Juncus acutiflorus* sub-community
b   *Briza media-Pinguicula vulgaris* sub-community
c   *Caltha palustris-Galium uliginosum* sub-community
13 *Schoenetum nigricantis* (total)

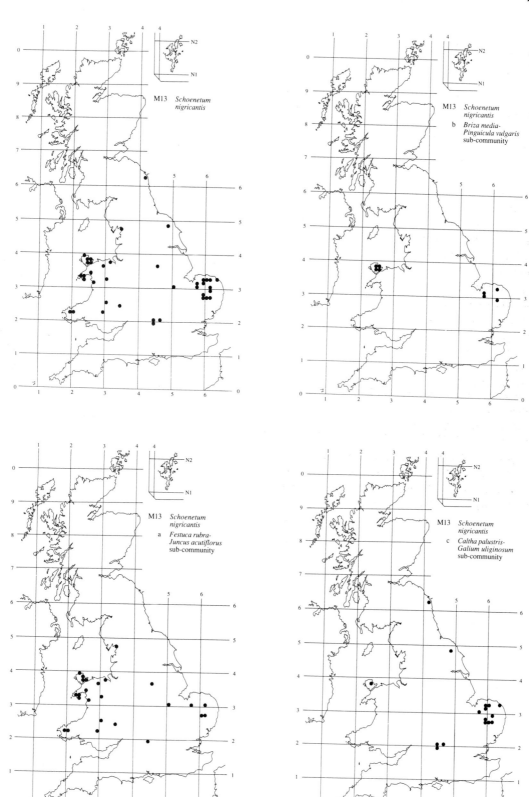

M14
Schoenus nigricans-Narthecium ossifragum mire

Synonymy
Schoenetum Newbould 1960; Roydon Common Zone
5a Daniels & Pearson 1974 *p.p.*; *Schoenetum nigrican-tis* Ivimey-Cook *et al.* 1975; *Schoeno-Juncetum subno-dulosi ericetosum* (Allorge 1922) Wheeler 1980*b p.p.*

Constant species
Anagallis tenella, Erica tetralix, Molinia caerulea, Narthecium ossifragum, Schoenus nigricans, Aneura pinguis, Campylium stellatum, Scorpidium scorpioides, Sphagnum auriculatum, S. subnitens.

Physiognomy
The *Schoenus nigricans-Narthecium ossifragum* mire includes mildly calcicole *Schoenus* vegetation of south-west lowland Britain which cannot readily be integrated in the *Schoenetum*. As in that community, *Schoenus* is usually a strong dominant here, and its grey-green, semi-evergreen foliage often enables stands to be picked out at a distance from their usual context of brown- or straw-coloured *Sphagnum*-dominated vegetation or heaths. And the tussock habit of the plant, with its robust rootstocks crowned by densely-caespitose clumps of shoots (Sparling 1962*a*, 1968), again gives the vegetation its distinctive structural character, with stools packed at varying densities and separated by systems of runnels. The tussock tops provide their characteristic niche for calcifuges but, in this community, the calcicolous element is confined to areas of close contact with the irrigating waters, other distinctive plants making an appearance around the bases of the tussocks.

Species occurring frequently in both communities are few in number and the most obvious is *Molinia caerulea*, which is generally abundant here, reinforcing the tussocky character of the vegetation. Typically, mixtures of *Schoenus* and *Molinia* cover the bulk of the ground without any *Juncus subnodulosus*, a species that is quite often co-dominant in the *Schoenetum*. Indeed, no other rushes or sedges of medium stature play an important role in this community: there is occasionally some *J. acutiflorus*, but it is not usually abundant. Other elements conspicuous by their absence are the various small sedges and other herbs that commonly line the runnels of the *Schoenetum* and provide much of its calcicolous stamp, and the taller mesophytic plants that give that community much of its colour: of these only *Carex panicea* and *C. demissa* are occasionally found, and sometimes *Dactylorhiza incarnata* ssp. *pulchella*.

Among other vascular plants represented here, only *Anagallis tenella* provides a floristic link between the two communities. More preferential to this vegetation type, and sometimes occurring in modest abundance, is *Narthecium ossifragum* and, less commonly, there can be some *Drosera rotundifolia* growing on *Sphagnum* cushions and sometimes, in wetter places, *D. intermedia*, *Eriophorum angustifolium*, *Rhynchospora alba*, *Eleocharis multicaulis* and *Pinguicula lusitanica*. Then, growing on the *Schoenus* or *Molinia*, there is very often some *Erica tetralix* and occasionally *Calluna vulgaris*. Some stands also have a local abundance of *Myrica gale* forming a low, bushy canopy.

Bryophytes are variable in their cover but this component, too, shows a shift towards a less calcicolous character. With the *Schoenetum*, the community shares frequent records for *Campylium stellatum* and *Aneura pinguis* and these, together with *Scorpidium scorpioides* and, less commonly, *Drepanocladus revolvens*, can form quite extensive mats in the wetter runnels, over very young *Schoenus* tussocks or around the bases of older, larger ones. By contrast, *Calliergon cuspidatum*, *Cratoneuron commutatum* and *C. filicinum*, which are a common feature in the runnels of the *Schoenetum*, and *Bryum pseudotriquetrum* and larger Mniaceae which figure often on the tussock sides, are either scarce or absent here. Certain Sphagna, on the other hand, which make but an infrequent appearance in the *Schoenetum*, become a consistent feature of this vegetation, growing on the tussocks or on top of a mat of other bryophytes raised a little above the level of the moister ground.

Sphagnum subnitens is the commonest species, but *S. auriculatum* is also frequent and there is occasionally some *S. papillosum*, *S. palustre*, *S. tenellum* or *S. recurvum*. *Hypnum jutlandicum* is also preferential to this community and there are sometimes patches of hepatics like *Kurzia pauciflora*, *Calypogeia* spp. (variously recorded as *C. muellerana*, *C. trichomanis* or *C. fissa*) and, less commonly, *Odontoschisma sphagni*, *Cephalozia connivens* and *C. bicuspidata*.

Habitat

The *Schoenus-Narthecium* mire is characteristic of peats and mineral soils irrigated by moderately base-rich and calcareous ground waters. It can be associated with more markedly soligenous zones within valley mires, but occurs more characteristically as isolated flushes among wet heath and moorland vegetation. It is largely a community of the oceanic south-west of Britain, its floristics betraying the influence of the relatively mild climate there.

By and large, *Schoenus* is a plant of base-rich habitats in the southern lowlands of Britain, not penetrating far into more acidic environments until the climate becomes markedly oceanic, first in the north-west of Scotland, where it can be found in flushed areas within blanket mires, and then in Ireland, where it occurs on the bog plane proper, a trend which Sparling (1962b, 1967a, b) has related to climatic amelioration of aluminium concentrations in the substrate. In the *Schoenetum*, in which most of the southerly stands of *Schoenus* mire fall, the vegetation thus takes most of its character from the influence of the calcareous substrate and the moderately Continental climate on the associated flora. The *Schoenus-Narthecium* mire, on the other hand, includes *Schoenus* vegetation from habitats which show a distinct shift towards the more base-poor and oceanic conditions prevailing further to the north and west, and it is the differences in these two edaphic and climatic factors that primarily determine the floristics of the community.

Typically, here, the pH of the flushing waters is in the range 5 to just over 7 (compared with 6.5–8 for the *Schoenetum*), with dissolved calcium levels rather variable, but mostly from 5 to 35 mg l^{-1} (compared with 60–200 mg l^{-1}) (Newbould 1960, Daniels & Pearson 1974, Wheeler 1983). This means that, even where the vegetation is in close contact with the water table, characteristically maintained at high levels throughout the year in the runnels, any calcicolous expression is limited. More basiphile species are best represented among the mat-forming bryophytes but, even there, are few in number and can themselves serve as an insulating layer upon which calcifuge plants can readily establish (cf. Clapham 1940). Thus, although there is the same structural diversity as in the *Schoenetum*, the edaphic character of the different microhabitats – runnels, tussock sides and

tops – is more uniform and the total flora less varied and rich. The positive response to the difference in soil conditions is , in fact, not so marked as is the absence of calcicoles, but nonetheless represents a decisive move towards the kind of vegetation in which *Schoenus* occurs in western ombrogenous mires, with the increase in Sphagna and bog herbs such as *Narthecium* and *Drosera rotundifolia*. Among the former, however, major peat-builders such as *S. papillosum* are not well represented. In the typical habitats of the *Schoenus-Narthecium* mire, an increase in such species in a luxuriant ground carpet marks a move to more acidic and stagnant conditions inimical to the vigour of *Schoenus* in southern Britain.

Continuous irrigation with moderately base-rich and calcareous waters is an uncommon phenomenon in lowland Britain and it usually occurs within tracts of largely lime-poor rocks and superficials, where flushing provides a local amelioration of prevailingly acidic soil conditions. In many cases, the specific geological contribution to base-enrichment is a modest one, increase in pH and calcium being mostly a reflection of the concentration of the flushing waters along lines of strong soligenous flow. Stands of the community can be found in such situations on Eocene clays, sands and gravels in south Dorset and Hampshire, on the Triassic pebble-beds of Devon and on the metamorphosed Killas shales in Cornwall.

In such situations, the *Schoenus-Narthecium* mire can be found on a variety of soil types, provided the general edaphic requirements are met. In slope flushes, the profile is usually some kind of wet mineral soil, on occasion very ill-structured, sloppy and gravelly, in other cases a better differentiated surface water gley, with or without a humic top. From such soils, there is a complete transition through to the moderately deep peats that have accumulated in elongated hollows under the influence of soligenous seepage to form valley mires. The profiles under the *Schoenus-Narthecium* mire are characteristically very poor in major nutrients, particularly phosphorus, and, if there is any eutrophication, as with the regular deposition of allochthonous silt, this vegetation is sharply replaced by some sort of Phragmitetalia fen, with its suite of helophytes, large sedges and tall dicotyledons.

The *Schoenus-Narthecium* mire is found on this range of soil types only within central southern and south-west England. Increased oceanicity in this region with its relatively high humidity may play some part in the extension of *Schoenus* into the moderately base-poor conditions typical here. Climate certainly plays some part in the occurrence of plants with a predominantly western distribution in Britain, like *Anagallis tenella*, and, also characteristic of *Schoenus* flushes in north-west Scottish blanket mires, *Drosera intermedia* and *Pinguicula lusitanica*. By contrast with the *Schoenetum*, more Continental species are poorly represented though, since some of these are calcicolous or mesophytic plants, this may be a coincidental effect of the differing edaphic conditions.

Even within this part of Britain, the community is very local, partly because of the natural scarcity of suitable habitats, but also because of the reduction in their extent by human activity. Occasional burning or light grazing, still of common occurrence over the tracts of heath in which this kind of mire usually occurs, are probably not very damaging: indeed, they may help maintain the vegetation by repeatedly setting back any invasion of woody plants. Stands on wetter ground within valley mires are protected to some extent from the worst effects of such treatments, though can be brought under their influence by draining. This is very damaging to flush and bog sites alike and has often been a prelude to agricultural improvement of areas of heathland in which the community may have been well represented.

Zonation and succession

The *Schoenus-Narthecium* mire is typically found in zonations and mosaics with valley bog and heath communities whose distribution reflects edaphic and hydrological variation. Some stands would probably progress to woodland in the absence of burning or grazing.

Where the community occurs in valley mires, it typically occupies the zones of moderate soligenous enrichment with bases and calcium, passing laterally to the *Narthecio-Sphagnetum* on more impoverished and acid peats with a more stagnant water-table, the transition being marked by the disappearance of *Schoenus* and the extension of the carpet of peat-building Sphagna. *Molinia* and *E. tetralix* continue to be well represented on the more raised areas and *Rhynchospora alba* becomes common, particularly around wetter hollows, where *Sphagnum auriculatum* pools can be found. Base-poor soligenous tracts in such sequences can have Littorelletea vegetation like the *Hyperico-Potametum*.

Characteristically, the *Narthecio-Sphagnetum* gives way around valley mires to the *Ericetum tetralicis* and, in some places, this community can be found in direct contact with the *Schoenus-Narthecium* mire, grading to it through the *Succisa-Carex* sub-community. Such vegetation may also form a surround to the community where it occurs as slope flushes within tracts of heath, but often it grades in such situations to some kind of Junco-Molinion grassland, in which *Molinia* is an overwhelming dominant and where *Schoenus* can be sporadically represented, a pattern well seen on the Aylesbeare Common heaths in Devon, described by Ivimey-Cook *et al.* (1975). Throughout such sequences as these, variations in base-richness and the amount of calcium are probably often quite small (Newbould 1960).

In contrast to related vegetation on north-western blanket bogs, the *Schoenus-Narthecium* mire is probably not a climax community. Succession has never been followed but might be expected to involve invasion by *Salix cinerea*, *Betula pubescens* and perhaps *Alnus glutinosa* and the eventual development of some sort of wet woodland.

Distribution

The community occurs very locally in Cornwall, east Devon, south-east Dorset and the New Forest. Some 'mixed mire' vegetation with *Schoenus* in west Norfolk could perhaps be accommodated here (e.g. Daniels & Pearson 1974); but is probably better regarded as leached fragments of the *Schoenetum*. It has not been sampled in Wales, though could well be found there, but is replaced in comparable situations on north-western blanket bogs by *Schoenus*-dominated stands of the *Scirpus-Erica* wet heath.

Affinities

Although the Schoenus-Narthecium mire shows some floristic continuity with the *Schoenetum*, certain stands being rather difficult to place, the two vegetation types are quite distinct in their characteristic expression and, in contrast to Wheeler (1975, 1980b), who included both within his *Schoeno-Juncetum*, are here considered as worth separating. Apart from Wheeler's account, this kind of *Schoenus* vegetation has figured only occasionally in descriptions of British vegetation, although Newbould (1960) and Ivimey-Cook *et al.* (1975) both recognised its essential character and place in the striking mire and heath zonations of southern Britain. Phytosociologically, it is difficult to place. Analogous vegetation on north-western blanket mires has been included in the *Scirpus-Erica* wet heath of the Ericion tetralicis, but the *Schoenus-Narthecium* community seems to occupy a more equivocal position on the borderline between that alliance and the Caricion davallianae, where the *Schoenetum* fairly obviously belongs.

Floristic table M14

Schoenus nigricans	V (6–8)
Molinia caerulea	V (6–8)
Erica tetralix	V (5–6)
Narthecium ossifragum	V (4–5)
Sphagnum subnitens	IV (4–5)
Anagallis tenella	IV (2–3)
Campylium stellatum	IV (4–5)
Aneura pinguis	IV (2–4)
Scorpidium scorpioides	IV (2–5)
Sphagnum auriculatum	IV (1–4)
Hypnum jutlandicum	III (1–3)
Kurzia pauciflora	III (1–2)
Drosera rotundifolia	III (1–2)
Juncus acutiflorus	II (1–4)
Calypogeia spp.	II (1–2)
Calluna vulgaris	II (1–4)
Sphagnum papillosum	II (3–4)
Eriophorum angustifolium	II (1–3)
Carex panicea	II (1–2)
Eleocharis multicaulis	II (1–3)
Drosera intermedia	II (1–2)
Riccardia multifida	II (1–2)
Pinguicula lusitanica	II (2–3)
Potentilla erecta	II (1–2)
Sphagnum palustre	II (2–4)
Rhynchospora alba	II (1–2)
Myrica gale	II (2–4)
Polygala serpyllifolia	I (1–2)
Odontoschisma sphagni	I (1–2)
Drepanocladus revolvens	I (1–2)
Sphagnum tenellum	I (1)
Ulex gallii	I (1)
Juncus bulbosus/kochii	I (1)
Ctenidium molluscum	I (1)
Pinus sylvestris seedling	I (1–2)
Pedicularis sylvatica	I (1)
Number of samples	15
Number of species/sample	17 (15–19)
Herb height (cm)	50 (30–90)
Herb cover (%)	81 (75–100)
Ground height (mm)	100
Ground cover (%)	15 (5–30)
Altitude (m)	55 (11–110)
Slope (°)	7 (5–10)
Soil pH	5.4 (5.0–5.9)

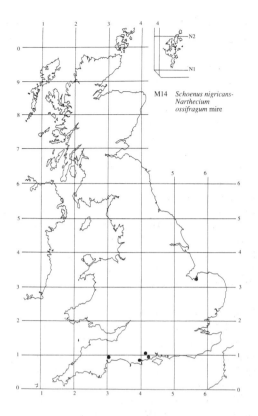

M14 *Schoenus nigricans-Narthecium ossifragum* mire

M15
Scirpus cespitosus-Erica tetralix wet heath

Synonymy
Scirpetum cespitosi Tansley 1939 *p.p.*; *Molinietum caeruleae* Tansley 1939 *p.p.*; *Trichophoreto-Eriophoretum caricetosum* McVean & Ratcliffe 1962; *Molinia-Myrica* mire McVean & Ratcliffe 1962, Birks 1973, Ratcliffe & Hattey 1982; *Molinieto-Callunetum* McVean & Ratcliffe 1962, Birks 1973, Prentice & Prentice 1975, Evans *et al.* 1977; *Trichophoreto-Callunetum* McVean & Ratcliffe 1962, Birks 1973, Prentice & Prentice 1975, Evans *et al.* 1977, Hill & Evans 1978; *Narthecio-Ericetum tetralicis* Moore (1964) 1968, *sensu* Birse 1980; *Erica tetralix-Molinia caerulea* sociation Edgell 1969; *Trichophorum cespitosum* nodum Edgell 1969; *Trichophorum cespitosum-Carex panicea* Association Birks 1973, Adam *et al.* 1977; *Narthecium-Campylopus atrovirens* nodum Prentice & Prentice 1975; Blanket bog communities Adam *et al.* 1977; *Calluno-Molinietum* Hill & Evans 1978; Mire noda 3 & 4 Daniels 1978 *p.p.*; *Trichophorum germanicum-Calluna vulgaris* Association McVean & Ratcliffe 1962 emend. Birse 1980; *Molinia caerulea-Potentilla erecta* community Bignal & Curtis 1981; *Trichophorum cespitosum-Eriophorum angustifolium* mire Ratcliffe & Hattey 1982; *Ericetum tetralicis* Dierssen 1982; Disturbed peatland 3.ii Hulme & Blyth 1984.

Constant species
Calluna vulgaris, Erica tetralix, Molinia caerulea, Potentilla erecta, Scirpus cespitosus.

Rare species
Campylopus atrovirens var. *falcatus, C. setifolius.*

Physiognomy
The *Scirpus cespitosus-Erica tetralix* wet heath is a compendious vegetation type with few constants and wide variation in the pattern of dominance and in the associated flora. *Molinia caerulea, Scirpus cespitosus, Erica tetralix* and *Calluna vulgaris* are all of high frequency throughout and, by and large, it is mixtures of these species that give the vegetation its general stamp.

But sometimes one of them, or occasionally even two, can be missing and their proportions are very diverse, varying not only with natural differences in climate and soils, but also very markedly with treatment. Of the four species, *Molinia* is the most consistent overall and it is often abundant, so some accounts have included the community within a *Molinietum* (e.g. Tansley 1939; see also McVean & Ratcliffe 1962, Edgell 1969, Bignall & Curtis 1981) or a *Molinieto-Callunetum*, where dominance is shared with *Calluna* (McVean & Ratcliffe 1962, Birks 1973, Prentice & Prentice 1975, Evans *et al.* 1977, Hill & Evans 1978). In other stands, *Scirpus* is very prominent, such that vegetation of this kind has been characterised as part of a *Scirpetum* (e.g. Tansley 1939, Edgell 1969) or, where *Calluna* is again also abundant, as communities like the *Trichophoreto-Callunetum* of McVean & Ratcliffe (1962) or its equivalents (Birks 1973, Prentice & Prentice 1975, Evans *et al.* 1977, Hill & Evans 1978, Birse 1980). Such units as these subsume a considerable amount of the variation in dominance within the community, but other combinations of species are quite common: *Molinia* sometimes dominates with *Scirpus*, or with both *Scirpus* and *Calluna*; and *E. tetralix*, which is generally subordinate in cover, can be abundant too, often with *Molinia*, as in many of the Welsh stands described by Ratcliffe & Hattey (1982).

Then, in the various sub-communities, some other plants make a contribution to the shrubby cover. *Erica cinerea* and *Vaccinium myrtillus* are of generally low frequency throughout but they are preferential to particular kinds of *Scirpus-Erica* wet heath and can attain local abundance, though usually with less than 25% cover. More striking is *Myrica gale* which, in other sub-communities, can be found as a low-cover occasional but which thickens up locally to dominate in a canopy 50 cm or so tall. Such vegetation has generally been characterised as a distinct *Molinia-Myrica* nodum (e.g. McVean & Ratcliffe 1962, Birks 1973), but, as where *Myrica* shows the same behaviour in other mire types, this variation can be readily subsumed within the community.

Typically, the dominants make up an extensive cover, though only 2–3 dm tall except where *Myrica* figures, and other vascular plants are often represented only as scattered individuals or in small clumps or tufts. Few of the associates are common throughout and one species notable for its scarcity here is *Eriophorum vaginatum*. In general appearance, this community often looks like a blanket mire, frequently occurs with such vegetation or replaces it, and grades floristically to it, but the low frequency of *E. vaginatum* overall, and its characteristically low cover when it does occur, provide one good separation between the *Scirpus-Erica* wet heath and the ombrogenous Sphagnetalia communities. Continuity, on the other hand, is stressed by the common occurrence here of *Potentilla erecta* and, particularly in the moister stands, of *Polygala serpyllifolia*, *Narthecium ossifragum* and *Eriophorum angustifolium*, all of which also characterise the *Scirpus-Eriophorum* blanket mire, the ombrogenous bog whose range coincides very closely with that of the *Scirpus-Erica* wet heath.

Other occasionals in the community include *Nardus stricta*, *Juncus squarrosus*, *Agrostis canina* ssp. *canina*, *Festuca ovina/vivipara* and, occurring more sparsely, *Blechnum spicant*, *Huperzia selago*, *Pedicularis sylvatica*, *Festuca rubra*, *Carex binervis*, *Juncus acutiflorus*, *J. effusus*, *J. conglomeratus* and *J. articulatus*. Enrichment among the vascular flora is of two kinds and helps define some of the sub-communities, involving an increase in Nardo-Callunetea species in the *Vaccinium myrtillus* sub-community or plants of the Caricion nigrae (or even the Caricion davallianae) in the *Carex panicea* sub-community. In general terms, however, this vegetation is of an impoverished character.

Bryophytes common throughout are likewise few in number. There are usually some Sphagna but they do not form the consistently luxuriant ground cover typical of the Sphagnetalia mires. The most frequent species overall are *Sphagnum capillifolium* and *S. subnitens* with *S. palustre*, *S. recurvum* and *S. auriculatum* becoming common in wetter stands, and each of these can attain some measure of abundance. *S. papillosum*, such an important component of the plane of Sphagnetalia bogs, is much more restricted here. *S. compactum* and *S. tenellum*, which are characteristic of the eastern, lowland counterpart of this community, the *Ericetum tetralicis*, are scarce but the *Scirpus-Erica* wet heath likewise provides an occasional locus for *S. molle*.

Other bryophytes found at moderate frequency include *Hypnum cupressiforme/jutlandicum*, *Hylocomium splendens*, *Aulacomnium palustre*, *Dicranum scoparium* and *Diplophyllum albicans* with, following the same general pattern as seen among the vascular plants, some enrichment in dry, heathy stands or those with some soligenous influence.

Lichens do not occur consistently throughout the *Scirpus-Erica* wet heath but they can be locally prominent, particularly in the *Cladonia* sub-community where *C. impexa*, *C. uncialis*, *C. arbuscula* and *C. pyxidata* are preferentially frequent.

Sub-communities

***Carex panicea* sub-community:** *Trichophoreto-Eriophoretum caricetosum* McVean & Ratcliffe 1962; *Molinia-Myrica* mire McVean & Ratcliffe 1962, Birks 1973, Ratcliffe & Hattey 1982; *Molinieto-Callunetum* McVean & Ratcliffe 1962 *p.p.*, Birks 1973 *p.p.*; *Narthecio-Ericetum tetralicis* Moore (1964) 1968, *sensu* Birse 1980 *p.p.*; *Erica tetralix-Molinia caerulea* sociation Edgell 1969; *Trichophorum cespitosum-Carex panicea* Association Birks 1973, Adam *et al.* 1977; *Narthecium-Campylopus atrovirens* nodum Prentice & Prentice 1975. This sub-community, which is typically found as small stands, often in obvious soakways or watertracks, is the richest and floristically most distinct kind of *Scirpus-Erica* wet heath. *Molinia* and *E. tetralix* retain high frequency throughout and the former especially is often abundant. *Scirpus*, on the other hand, is rather patchy and especially sparse on the more sloppy ground and *Calluna*, too, is typically of low cover and poor vigour. *Myrica* occurs frequently and quite often forms an extensive, low canopy but *E. cinerea* and *V. myrtillus* are hardly ever found.

Potentilla erecta and *Polygala serpyllifolia* are very commonly accompanied here by *Narthecium ossifragum* and *Eriophorum angustifolium*, but more obviously preferential are *Drosera rotundifolia* (and occasionally *D. anglica* and *D. intermedia*) and a variety of Caricion nigrae and Caricion davallianae species such as *Carex panicea*, *C. echinata*, *C. nigra*, *C. pulicaris*, *C. demissa*, *C. dioica*, *Selaginella selaginoides*, *Pinguicula vulgaris*, *Succisa pratensis*, *Viola palustris*, *Juncus bulbosus/kochii* and *Dactylorhiza maculata* ssp. *maculata*. *Schoenus nigricans* also occurs occasionally, sometimes with local abundance.

The bryophyte element, too, has distinctive features. Among the Sphagna, *S. capillifolium* is rather patchy, but *S. subnitens* occurs commonly though it is often exceeded in abundance by *S. palustre* and *S. recurvum*; *S. auriculatum* can also sometimes be found. More striking is the occasional occurrence of *Breutelia chrysocoma*, *Drepanocladus revolvens*, *Campylium stellatum*, *Scorpidium scorpioides*, *Aneura pinguis*, *Campylopus atrovirens* (including the rare var. *falcatus*), *C. setifolius* and *C. shawii*, the last three typically on shallow but very wet peat.

The local abundance of *Myrica* in this sub-community does not disrupt the overall floristic character of the vegetation, though under denser canopies the associates tend to thin out, such that in extreme cases there is little more beneath than patchy *Molinia* with scattered *E. tetralix*, *Potentilla erecta*, *Succisa pratensis*, a little

Carex panicea, C. echinata, Juncus acutiflorus or *J. effusus* and a carpet of *Sphagnum palustre*. Such vegetation could be distinguished as a variant.

Typical sub-community: *Trichophoreto-Callunetum, Sphagnum* facies McVean & Ratcliffe 1962, Birks 1973; *Molinieto-Callunetum* McVean & Ratcliffe 1962, Prentice & Prentice 1975, Evans *et al.* 1977, all *p.p.*; *Narthecio-Ericetum tetralicis* Moore (1964) 1968; *Trichophorum cespitosum* nodum Edgell 1969; *Trichophoreto-Callunetum*, Typical & *Eriophorum* facies Prentice & Prentice 1975; *Trichophoreto-Callunetum* Evans *et al.* 1978; *Calluno-Molinietum* Hill & Evans 1978; Mire nodum 3 Daniels 1978; *Trichophorum-Calluna* Association Birse 1980; *Molinia-Potentilla erecta* nodum Bignal & Curtis 1981 *p.p.*; *Trichophorum cespitosum-Eriophorum angustifolium* mire Ratcliffe & Hattey 1982. This sub-community shows virtually the complete range of possible permutations of the dominants from the kind of *Trichophoreto-Callunetum* described by McVean & Ratcliffe (1962) in which *Scirpus* and *Calluna* share dominance, with but little *Molinia* or *E. tetralix*, through stands in which *Molinia* and *Calluna* predominate (as in classic *Molinieto-Callunetum*) or *Molinia* and *E. tetralix*, to what is almost a pure *Molinietum*. *Myrica gale* is also quite common, though generally not abundant, and there is sometimes a little *E. cinerea*.

This kind of *Scirpus-Erica* wet heath shares with the previous sub-community frequent records for *Narthecium ossifragum* and *Eriophorum angustifolium* but the small sedges distinctive there are generally sparse, with only *C. panicea* and *C. echinata* occurring occasionally and the other poor- and rich-fen associates are very uncommon. In the other direction, some stands grade to the *Cladonia* or *Vaccinium* sub-communities, with *Nardus* and *Juncus squarrosus* showing local prominence. Positive features among the vascular element are rather weak, though it is here that *Eriophorum vaginatum* tends to occur, as a low-cover occasional.

A little better defined is the rise to fairly high frequency and local abundance of *Sphagnum papillosum* and, when this occurs with moderate amounts of *S. capillifolium* and *S. subnitens*, and a little *Odontoschisma sphagni*, the community makes its closest approach to the *Scirpus-Eriophorum* mire. Typically, however, the cover of these species, and of *S. palustre, S. recurvum* and *S. auriculatum*, is very variable and patchy. And, in some stands, Sphagna are very sparse and mosses such as *Racomitrium lanuginosum, Dicranum scoparium, Hypnum cupressiforme* and *Campylopus paradoxus* provide most of the cover.

Cladonia spp. sub-community: *Trichophoreto-Callunetum*, lichen facies McVean & Ratcliffe 1962; *Tricho-*

phoreto-Callunetum, Rhacomitrium facies Birks 1973, *Trichophoreto-Callunetum, Cladina* facies Prentice & Prentice 1975; Mire nodum 3 Daniels 1978; *Trichophorum-Calluna* Association Birse 1980 *p.p.*; *Narthecio-Ericetum* Moore (1964) 1968 *sensu* Birse 1980 *p.p.* Disturbed peatland 3.ii Hulme & Blyth 1984. All four of the possible dominants retain high frequency here but *Calluna* shows a fairly strong tendency to predominate and there is commonly a little *E. cinerea*. Among the associates, *Potentilla erecta* remains constant but *Polygala serpyllifolia* and, more so, *Narthecium* are reduced in frequency, and *Eriophorum angustifolium* and *Myrica* are very scarce.

Most of the distinctive features of the vegetation, however, are among the ground cover. Sphagna are but poorly represented, with only *S. capillifolium* attaining occasional records. *Racomitrium lanuginosum* and *Hypnum cupressiforme/jutlandicum* become frequent but more obvious is the variety and abundance of *Cladonia* spp., particularly *C. impexa* and *C. uncialis*, with *C. arbuscula, C. pyxidata, C. coccifera* and *C. gracilis* more weakly preferential.

Vaccinium myrtillus sub-community: *Juncus squarrosus* bog McVean & Ratcliffe 1962; *Trichophoreto-Callunetum, Juncus squarrosus* facies Prentice & Prentice 1975; *Trichophoreto-Callunetum* Hill & Evans 1978; *Molinia caerulea-Potentilla erecta* community Bignal & Curtis 1981 *p.p. Scirpus* and *E. tetralix* both become rather uneven in their occurrence here and rarely have high cover, mixtures of *Molinia* and *Calluna* generally dominating, quite often with some *Vaccinium myrtillus*. And, very commonly, there are small tussocks of *Nardus, Juncus squarrosus, Deschampsia flexuosa* and, more occasionally, some *Anthoxanthum odoratum, Festuca ovina/vivipara, F. rubra, Luzula multiflora* and *Carex pilulifera*, and *Galium saxatile* sometimes joins *Potentilla erecta*, so the overall impression is of vegetation intermediate between a blanket mire and a rough *Nardetalia* sward.

The bryophyte component, too, reflects this trend with most of the Sphagna infrequent and even *S. capillifolium* of low cover, their place being taken by *Hypnum cupressiforme/jutlandicum, Dicranum scoparium, Pleurozium schreberi, Plagiothecium undulatum, Polytrichum commune* and *Rhytidiadelphus loreus*. In this element, as well as among the vascular plants, this sub-community comes closest to the northern form of the *Ericetum tetralicis*, though the reduced role of *Molinia* there and the high frequency of *Sphagnum compactum* and *S. tenellum* will usually serve as distinguishing features. More problematic is the similarity of this *Vaccinium* sub-community to run-down forms of blanket mire vegetation, such as the *Juncus squarrosus-Rhytidiadelphus loreus* sub-community of the *Scirpus*

Eriophorum bog, but this is a very real reflection of the convergence of these vegetation types on drying peat which has been subject to particular treatments.

Habitat

The *Scirpus-Erica* wet heath is characteristic of moist and generally acid and oligotrophic peats and peaty mineral soils in the wetter western and northern parts of Britain, being especially associated with thinner or better-drained areas of ombrogenous peat, though also extending into places with soligenous influence. Grazing and burning have important effects on the floristics and structure of the vegetation, and draining and peat-cutting have extended its coverage on to once deeper and wetter peats.

Like its counterpart in the south and east of Britain, the *Ericetum tetralicis*, the *Scirpus-Erica* wet heath is typical of more acid soils (surface pH being generally between 4 and 5) that are too dry or freely drained for the development of Sphagnetalia mires and too wet for Calluno-Ulicetalia heaths. The community shares many of its major species with the bogs of the north and west of Britain, and grades to such vegetation on deeper peats with a high and stagnant water-table, but important peat-builders like *Eriophorum vaginatum*, *Sphagnum papillosum* and *S. magellanicum*, and their distinctive blanket mire associates such as *Pleurozia purpurea*, *Odontoschisma sphagni*, *Mylia anomala* and *M. taylori*, are of restricted occurrence here and help separate the communities. Conversely, the *Scirpus-Erica* wet heath extends on to drier ombrogenous peats and thin humic tops to podzolised soils, where the *Sphagnum* cover becomes attenuated and hypnoid mosses, lichens and Nardo-Callunetea herbs like *Nardus* and *Juncus squarrosus* increase their frequency, but *Erica cinerea* and *Vaccinium myrtillus* remain fairly confined and are generally of low cover. As in the *Ericetum*, this vegetation thus takes much of its basic character from species tolerant of the intermediate soil moisture regime and able to thrive with the reduced competition from plants better adapted to the extremes (Rutter 1955, Bannister 1966, Gimingham 1972): *Erica tetralix*, *Calluna*, *Molinia* and, more important here than in the wet heaths of the south and east, *Scirpus cespitosus*.

Environmental conditions favouring the development of the *Scirpus-Erica* wet heath are of widespread occurrence in the north and west, where high rainfall permits (or has, in the past, permitted) the accumulation of acid peat, not only in topogenous hollows, but over flatter plateaus and quite steep slopes, even on pervious and calcareous substrates and at low altitudes. This community is almost wholly confined to areas with a present annual precipitation in excess of 1200 mm, and generally with more than 1600 mm (*Climatological Atlas* 1952), at least 180 wet days yr^{-1} (Ratcliffe 1968) and

next to no potential water deficit. And within this zone, it is largely a community of the lowlands and sub-montane fringes, occurring almost at sea-level in the far north-west of Scotland and with a mean altitude of less than 250 m. In drier areas, it tends to be found at progressively higher altitudes, a feature well seen in traversing central Scotland where it eventually reaches over 500 m in the Cairngorms though, with this climatic shift, the *Scirpus-Erica* wet heath is largely replaced by the *Ericetum tetralicis* with its characteristic ground carpet of *Sphagnum compactum* and *S. tenellum*.

On the blanket of ombrogenous peat which clothes much of the ground in north-west Britain, the community typically occurs on well-humified deposits which are fairly dry or which at least show no stagnation. In areas with the highest rainfall, this generally means that it is found on sloping ground with a moderately thin cover of peat, giving way on flatter areas to blanket mire proper or, in regions where gentler slopes are rare, as in the west-central Highlands, comprising the bulk of the cover of ombrogenous mire vegetation. With decreasing precipitation, it occurs on progressively shallower slopes, with sometimes quite thick peat, which constitute the wettest ground locally. Again, this is best observed in moving across central Scotland: the contrast between Glen Torridon, in the west with more than 2000 mm of rain, and Glen Clova, in the east with about 1200 mm, is well illustrated in McVean & Ratcliffe (1962: Figure 20). This simple relationship is complicated by aspect since, on north-facing slopes, the community can run up on to steeper ground (of 30° or more at Glen Shiel, for example) and, on south-facing hills, is more restricted to gentler slopes. Overall, the *Scirpus-Erica* wet heath has geographical and altitudinal ranges roughly comparable to those of the *Scirpus-Eriophorum* mire, but occurs over a greater range of slopes (0–42°, mean 8°, for the former; 0–25°, mean 4°, for the latter) and, on more undisturbed ombrogenous peats, is found on shallow deposits (usually less than 2 m).

The community also extends, particularly in areas of higher rainfall or where there is local drainage impedence, on to peaty podzols with but a few centimetres of humic top, where blanket mire proper cannot develop but from which heaths are excluded by periodic water-logging (McVean & Ratcliffe 1962, Edgell 1969, Birse 1980). And it occurs widely on once deeper and wetter blanket peats and, in some areas, on raised mires which have become dry and eroded, or which have been drained, frequently burned, cut-over and disturbed (McVean & Ratcliffe 1962, Birks 1973, Birse 1980, Bignall & Curtis 1981, Hulme & Blyth 1984).

Much of the floristic and structural variation within and between the sub-communities can be accounted for in terms of differences in water relations in the generally intermediate kind of habitat characteristic of the

Scirpus-Erica wet heath. The Typical sub-community is generally found on deeper peats kept fairly moist by a moderately high water-table and it is here that the community makes its closest approach to the blanket mire vegetation typical of the region, with a substantial suite of Oxycocco-Sphagnetea species and quite vigorous *Sphagnum* growth, *Eriophorum vaginatum* figuring occasionally at low covers and *S. papillosum* showing local abundance. In such stands, *Calluna* is often poorly-grown or, quite often, totally absent, suffering from waterlogging and without the advantage of any differentiation of drier hummocks which maintain its frequency and abundance on blanket mires (Bannister 1964*b*, *c*, *d*; Gimingham 1972). *E. tetralix* is thus the commonest ericoid in such situations in mixtures with *Molinia* and *Scirpus*. Still within the Typical sub-community, but on somewhat drier and usually shallower peats, often on steeper slopes, the balance among the dominants shifts, with *Calluna* often gaining ascendance with *Scirpus*, and *Sphagnum* cover declines, *S. papillosum* in particular disappearing.

An increase in the vigour of *Calluna*, the appearance of *E. cinerea*, *Racomitrium lanuginosum* and lichens, marks the shift to the *Cladonia* sub-community on steeper, often convex as opposed to concave, slopes with a generally shallower and more sharply-draining cover of peat, or gentle slopes with peaty podzols or a much-eroded peat mantle. *E. cinerea* can perform moderately well on the drier surfaces here (Bannister 1964*a*, *d*, 1965), perhaps better than in comparable situations around valley mires, where higher levels of ferrous ions seem to restrict its transgression from the surrounding dry heaths (Jones 1971*a*, *b*).

The Typical sub-community also grades in another direction to the Carex sub-community with an increase in soligenous influence. Water movement through the peat is probably considerable within some stands of the Typical sub-community itself, particularly where these occur in slight hollows or flats that receive and shed surface and sub-surface water draining off surrounding slopes (McVean & Ratcliffe 1962, Birks 1973). Here, there is probably enhanced aeration and some modest nutrient-enrichment of the peat, with *Molinia* in particular being able to capitalise on this amelioration of the environment (Loach 1966, 1968*a*, *b*, Sheikh & Rutter 1969, Sheikh 1969*a*, *b*, 1970). Such conditions are much more pronounced in the Carex sub-community which typically occupies areas where downwash is concentrated or narrow soakways, often of gentle slope but with a continuous through-put. The soils in such situations are usually shallow peats over gleyed mineral profiles or undifferentiated stony hill-wash, with a surface pH noticeably higher than usual, generally between 5 and 6, and probably some increase in calcium and major nutrients (McVean & Ratcliffe 1962, Edgell 1969,

Birks 1973, Prentice & Prentice 1975). Here the *Scirpus-Erica* wet heath shows quite strong floristic affinities with Scheuchzerio-Caricetea poor fens, with an increase in semi-aquatic Sphagna and species such as *Carex panicea*, *C. echinata*, *C. nigra*, *Viola palustris* and *Juncus bulbosus/kochii*, and even provides a locus in the community for rich-fen plants like *Selaginella selaginoides* and *Pinguicula vulgaris*.

It is also in this kind of *Scirpus-Erica* wet heath that *Myrica gale* makes its most obvious contribution to the vegetation. It occurs occasionally in the Typical sub-community, usually with low cover (and also in the *Scirpus-Eriophorum* blanket mire) but reaches its greatest abundance on the better-aerated and more mineral-rich irrigated peats here, quite often with much strongly-tussocky *Molinia*, but with a consequent reduction in the variety of associates among the dense and shaded cover. Where streams flow from stretches of ombrogenous peat, this kind of vegetation often runs down on to the periodically-flooded flats where there may be modest deposition of fine mineral matter (McVean & Ratcliffe 1962, Birks 1973). In more nutrient-poor situations, *Myrica* may be of considerable importance in producing nitrogen-enrichment of the system by N-fixation in its root nodules (Bond 1951, 1967, Sprent *et al.* 1978).

Superimposed on these lines of variation related to natural edaphic differences are the diverse effects of treatments which influence the proportional contributions of the dominants and the composition of the associated flora. The natural surface drying of ombrogenous peats which has taken place with climatic changes in the recent post-Glacial has probably been greatly speeded by human activities such as deforestation and burning, which have been virtually universal in some parts of the range of the *Scirpus-Erica* wet heath, and more recently by direct drainage. Such activities have probably extended the cover of the community on to the wasting margins of once-wetter peats and, within the *Scirpus-Erica* wet heath itself, they tend to shift the composition of the vegetation away from the more *Sphagnum*-rich kind of the Typical sub-community through the more *Calluna*-rich to the distinctly heathy *Cladonia* sub-community. This latter vegetation is especially well represented on the deforested morainic country of the western Highlands, and, at higher altitudes, on ground that has probably been much burned. In extreme cases, *Calluna* and the other ericoids have been virtually eliminated and *Scirpus* reduced to scattered tussocks, with much of the cover made up of hummocks of *Racomitrium*, patches of *Cladonia* spp. and a surface crust of lichens over a thin and patchy mantle of humus with exposed gravel and stones between (McVean & Ratcliffe 1962). Where wastage has not been so extreme but where ericoids have been virtually lost, cessation of burning may favour a vigorous recovery of *Molinia* as against *Scirpus*: *Molinia*-dominated stands of the Typical sub-community are especially extensive in the western Highlands and in Galloway (McVean & Ratcliffe 1962). Other *Scirpus-Erica* wet heaths with an abundance of *Molinia* may have been derived from the *Carex* sub-community by elimination of *Myrica*. Where there has been frequent but more carefully-controlled burning, *Calluna* may be encouraged as a dominant, particularly where the community extends into less oceanic regions: in the east-central Highlands, for example, stands of the Typical and *Cladonia* sub-communities form part of poorer-quality grouse-moors (Birse 1980).

Grazing by sheep and deer, sometimes combined with burning, has also had some influence on the vegetation, particularly where it extends on to drier ground. Such a mixture of treatments may be partly responsible for the composition of the *Vaccinium* sub-community. The grazing-sensitive bilberry is present only at small covers but is often accompanied by mixtures of Nardetalia herbs and hypnoid mosses characteristic of sub-montane swards on base-poor and ill-drained ground where grazing is heavy.

Zonation and succession

Zonations within stands of the *Scirpus-Erica* wet heath and between the community and other vegetation types generally reflect variations in the soil moisture regime, sometimes with attendant differences in base- and nutrient-status. Such patterns are often overlain by the effects of treatments and, in some cases, the community may represent a seral intermediate between blanket mire and either dry heath or grassland. Without burning or grazing, less-damaged stands might be able to revert to blanket mire or progress to woodland. Much vegetation of this kind has been replaced by coniferous forest after draining of the ground.

The intermediate character of the *Scirpus-Erica* wet heath frequently finds spatial expression in its occurrence as a zone between Sphagnetalia bogs and Calluno-Ulicetalia heaths, the particular nature of the vegetation types and their proportions in the zonations varying with geographical locality, altitude and the character of the terrain. Most often, at lower altitudes in western Britain, where the community is best developed, it occurs in close association with the *Scirpus-Eriophorum* blanket mire, grading to it in hollows and over gently-undulating ground where there is accumulation of deeper peat with a high and stagnant water-table. More *Sphagnum*-rich stands of the Typical sub-community may form a more or less gradual transition to the blanket mire, being replaced, on more free-draining slopes, first by less *Sphagnum*-rich stands of this sub-community, then by the *Cladonia* sub-community over thinner, drier peats. This, in turn, can give way to dry

heath on sharply-draining podzols, generally some form of *Calluna-Erica cinerea* or *Calluna-Vaccinium* heath, though often dominated overwhelmingly by *Calluna*, a physiognomic feature which can extend a considerable way into the wet heath, where there has been more controlled burning, and which can therefore blur the upper part of the vegetation sequence.

In areas with lower rainfall, the whole zonation tends to move downslope, the *Sphagnum*-rich Typical sub-community, rather than the *Scirpus-Eriophorum* blanket mire, occupying the wettest ground in hollows and the *Cladonia* sub-community extending less far up the surrounding slopes before being replaced by drier heath. A similar effect on the upper part of the zonation can be seen where the communities occur over south-facing slopes where the drier heath component can be especially extensive; on northerly aspects, on the other hand, the *Scirpus-Erica* wet heath transgresses far on to steeper ground (McVean & Ratcliffe 1962).

Increased water movement through the peats is often marked by a rise in the proportion of *Molinia* through stands of the Typical sub-community and, where soligenous influence is most pronounced, with enhanced aeration and sometimes an increase in base-status and perhaps nutrient content, stretches of the *Carex* sub-community may clothe gentler slopes or occur within definite soakways, running through both the *Scirpus-Erica* wet heath and the *Scirpus-Eriophorum* mire and quite often along streams that drain from the peat through surrounding heaths and grasslands. Very frequently, such zones of water movement are marked most obviously by the presence of *Myrica*, a feature readily discerned from a distance on the ground or from aerial photographs.

Where the *Scirpus-Erica* wet heath extends to higher altitudes, as it does in Scotland and more locally in northern England and Wales, it can also be found in association with the *Calluna-Eriophorum* blanket mire, typically forming a zone below it where the peat thins out on steeper or convex slopes. In the north-west Highlands, the community often forms an altitudinal band separating the two kinds of blanket mire and grading in each direction to them. Further east in Scotland, where the *Scirpus-Eriophorum* bog becomes increasingly rare, the *Scirpus-Erica* wet heath can occur below the *Calluna-Eriophorum* mire in transitions to drier heath or grassland (McVean & Ratcliffe 1962, Birks 1973, Birse 1980). Generally, however, in this part of Britain, the community is replaced in such sequences by the *Ericetum tetralicis*.

Fragments of these kinds of zonations are of widespread occurrence where the climate or the terrain is unsuitable for the development of the full sequence and each of the sub-communities can be found alone in vegetation patterns. The Typical and *Cladonia* sub-communities, for example, often mark out the most ill-drained patches of ground in gently-undulating landscapes clothed with grasslands and heaths over podzols with varying degrees of drainage impedence and shallow peat accumulation. And the *Carex* sub-community is of widespread occurrence along fairly base-poor soakways in tracts of heaths and grasslands, well away from stands of other kinds of *Scirpus-Erica* wet heath or blanket mire (Edgell 1969). In such situations, it may pass, along the water-tracks, to different types of soligenous mire according to the degree of base- and nutrient-enrichment of the incoming waters.

Even where extensive zonations occur, however, they have been very widely affected by treatments. Less severe burning and/or grazing can impose a uniformity of dominance over much of the sequence, masking the transition from one vegetation type to another. More drastic or long-continued treatments, particularly where these accentuate the effects of natural degeneration of peatlands, have a more profound influence on the floristics. It is likely that some stands of the *Scirpus-Erica* wet heath have been derived from blanket mire as a result of a combination of climatic change, burning and grazing. Sphagnetalia communities affected by these factors tend to lose their luxuriant carpet of Sphagna and to show an increase in either *Racomitrium lanuginosum* and lichens or Nardetalia herbs and hypnoid mosses (McVean & Ratcliffe 1962), trends which can be seen, for example, in the *Cladonia* and *Juncus-Rhytidiadelphus* sub-communities of the *Scirpus-Eriophorum* blanket mire. Elimination of *Eriophorum vaginatum* from these vegetation types can produce assemblages which are indistinguishable from the *Cladonia* and *Vaccinium* sub-communities of the *Scirpus-Erica* wet heath and extensive mosaics of these vegetation types can be seen on many areas of degenerating ombrogenous peat. Deep sheet-erosion may lead to a replacement of the *Carex* sub-community in such patterns by the *Carex echinata-Sphagnum* mire in widespread systems of runnels (Birks 1973), and where the wasting peat is washed down over the slopes below, there can be transitions to Nardetalia grasslands with much *Nardus* and *Juncus squarrosus*.

Although in the far west, blanket mire seems to be able to regenerate in less-disturbed areas of wet ground (McVean & Ratcliffe 1962), it seems unlikely that progression to the *Scirpus-Erica* wet heath or beyond can be readily reversed. Cessation of burning, especially on peat that is naturally well-aerated or where there has been draining, may precipitate a vigorous expansion of *Molinia* and, in some areas, blanket mires and wet heath seem to have converged into vast tracts of *Molinia*-dominated grasslands in which Sphagnetalia or Sphagno-Ericetalia species play a relatively minor role, mosaics among them being shaded out or at least

masked by the dominant. Invasion of the *Scirpus-Erica* wet heath by woody species is probably theoretically possible over most, if not all, its altitudinal range but widespread deforestation has often removed potential seed-parents, and continued grazing by stock and deer and sporadic burning may be enough to set back succession continually. *Betula pubescens* and *Pinus sylvestris* are the most likely invaders of the characteristic soils here, especially on the drier ground of the *Cladonia* and *Vaccinium* sub-communities, but seedlings are rarely found. Coniferous plantations, on the other hand, have been established over extensive stretches of drained ombrogenous peat, some of which probably carried the *Scirpus-Erica* wet heath.

Distribution

The community occurs widely at lower altitudes in western and northern Britain, being particularly well represented in the western Highlands of Scotland, in south-west Scotland and Wales and, less extensively, in the Lake District and on Dartmoor and Exmoor. The distribution of the sub-communities is imperfectly known and, though the Typical and *Carex* types are especially common in western Scotland and the *Cladonia* and *Vaccinium* types in somewhat drier regions, all probably occur throughout the range of the community. In the eastern and southern lowlands of Britain, the *Scirpus-Erica* wet heath is replaced by the *Ericetum tetralicis* which, with increasing dryness of the climate, becomes progressively restricted to topogenous mires.

Affinities

The *Scirpus-Erica* wet heath brings together a variety of vegetation types on better-aerated ombrogenous peats in north-west Britain, uniting communities that have previously been distinguished to a great extent on the dominance of either *Scirpus*, *Calluna* or *Molinia*, very susceptible to variations in treatment (Tansley 1939, McVean & Ratcliffe 1962, Birks 1973, Prentice & Prentice 1975, Evans *et al.* 1977, Hill & Evans 1978), or on the degree of soligenous influence (McVean & Ratcliffe 1962, Edgell 1969, Birks 1973, Prentice & Prentice 1975). Although it embraces a fairly wide range of variation, particularly in including vegetation transitional to Caricion nigrae poor fens, the community holds together well as one of our two wet heaths, floristically intermediate between Sphagnetalia mires and Calluno-Ulicetalia heaths. With its south-eastern counterpart, the *Ericetum tetralicis*, floristic variation in which roughly parallels that found here, it can be placed unequivocally in the Sphagno-Ericetalia (= Ericetalia Moore (1964, 1968), with its single alliance, the Ericion tetralicis, comprising western European mires and wet heaths on shallower peats. The definition in this scheme retains some of the major divisions made among these vege ation types by McVean & Ratcliffe (1962) and fills out the description provided by Moore (1968) of what he termed the *Narthecio-Ericetum tetralicis*, the Ericion community characteristic of more Atlantic regions, like western Britain, Eire and Normandy (as in Duvigneaud 1949). Vegetation of this same general kind has also been recorded from Scandinavia by Nordhagen (1922), Böcher (1943), Tveitnes (1945), Knaben (1950), Faegri (1960), Skogen (1965) and Dierssen (1982).

Floristic table M15

	a	b	c	d	15
Molinia caerulea	V (4-8)	V (1-9)	V (4-7)	V (1-9)	V (1-9)
Potentilla erecta	V (1-4)	V (1-6)	V (1-4)	V (1-4)	V (1-6)
Erica tetralix	V (1-4)	V (1-6)	V (1-4)	IV (1-5)	V (1-6)
Calluna vulgaris	IV (1-4)	IV (1-8)	V (4-8)	V (1-9)	IV (1-9)
Scirpus cespitosus	III (1-8)	IV (1-8)	IV (1-6)	III (1-6)	IV (1-8)
Narthecium ossifragum	V (1-4)	V (1-6)	III (1-4)	II (1-4)	III (1-6)
Eriophorum angustifolium	IV (1-4)	III (1-6)	I (1-4)	II (1-4)	III (1-6)
Myrica gale	IV (4-8)	III (1-8)	I (4)	I (2-4)	III (1-8)
Sphagnum palustre	IV (1-6)	III (1-8)		I (1-6)	II (1-8)
Carex panicea	V (1-4)	II (1-4)	I (1-4)	I (1-4)	II (1-4)
Carex echinata	V (1-4)	II (1-4)		I (1)	II (1-4)
Drosera rotundifolia	IV (1-4)	II (1-4)		I (1)	II (1-4)
Pinguicula vulgaris	IV (1-4)	I (1)	I (1)	I (1)	I (1-4)
Sphagnum recurvum	III (1-8)	II (1-4)		I (1-4)	I (1-8)
Breutelia chrysocoma	III (1-4)	I (1-4)		I (1-4)	I (1-4)
Succisa pratensis	III (1-3)	I (1-4)	I (1-3)	I (1-4)	I (1-4)
Carex pulicaris	III (1-4)	I (1)	I (1-3)	I (1)	I (1-4)
Selaginella selaginoides	III (1-3)			I (1)	I (1-3)
Juncus bulbosus/kochii	II (1-4)	I (1-3)	I (1)	I (1-3)	I (1-4)
Carex nigra	II (1-6)	I (1-3)	I (1)	I (1-4)	I (1-6)
Viola palustris	II (1-3)	I (1-3)		I (1-3)	I (1-3)
Campylopus atrovirens	II (1-4)	I (1-3)	I (1-3)		I (1-4)
Euphrasia officinalis agg.	II (1-3)	I (1)		I (1)	I (1-3)
Schoenus nigricans	II (1-6)	I (2)			I (1-6)
Sphagnum auriculatum inundatum	II (1-4)	I (1)			I (1-4)
Carex demissa	II (1-4)	I (1)			I (1-4)
Drosera anglica	II (1-2)	I (2-4)			I (1-4)
Drepanocladus revolvens	II (1-3)	I (1-4)			I (1-4)
Aneura pinguis	II (1-3)	I (1)			I (1-3)
Campylium stellatum	II (1-4)				I (1-4)
Scorpidium scorpioides	II (1-3)				I (1-3)

Species					
Pleurozia purpurea	II (1-3)				I (1-3)
Dactylorhiza maculata maculata	II (1-3)				I (1-3)
Carex dioica	I (1-3)				I (1-3)
Menyanthes trifoliata	I (1-4)				I (1-4)
Sphagnum papillosum	I (4)	III (1-9)	I (1-6)	I (1-4)	II (1-9)
Juncus acutiflorus	I (1)	II (1-4)	I (4)	I (1)	I (1-4)
Odontoschisma sphagni	I (1)	II (1-3)	I (1-3)	I (1)	I (1-3)
Sphagnum auriculatum auriculatum	I (1)	II (1-4)	I (1)		I (1-4)
Eriophorum vaginatum		II (1-4)	I (1-3)	I (1-4)	I (1-4)
Campylopus paradoxus		II (1-3)	I (1-3)	I (1-2)	I (1-3)
Erica cinerea	I (1)	II (1-4)	V (1-6)	I (1-4)	II (1-6)
Hypnum cupressiforme	I (1-4)	II (1-3)	III (1-5)	II (1-8)	II (1-8)
Racomitrium lanuginosum	I (1-3)	II (1-4)	III (1-6)	I (1)	I (1-6)
Cladonia impexa	I (1)	I (1-3)	II (1-3)	I (1-4)	I (1-4)
Cladonia uncialis	I (1)	I (1)	II (1-3)	I (2)	I (1-3)
Vaccinium myrtillus		I (1-3)	I (1)	IV (1-6)	II (1-6)
Nardus stricta	II (1-4)	II (1-4)	II (1-6)	III (1-6)	II (1-6)
Juncus squarrosus	I (1)	II (1-4)	I (1-2)	III (1-4)	II (1-4)
Dicranum scoparium	I (1)	II (1-3)	I (1)	III (1-4)	II (1-4)
Hypnum jutlandicum	I (1)	I (1-4)	II (1-4)	III (1-9)	II (1-9)
Deschampsia flexuosa	I (1-3)	I (1)	I (1-4)	III (1-4)	I (1-4)
Pleurozium schreberi	I (1-4)	I (1-3)	I (1-3)	III (1-6)	I (1-6)
Plagiothecium undulatum	I (1)	I (1-3)	I (1-4)	II (1-4)	I (1-4)
Galium saxatile	I (1)	I (1)	I (1)	II (1-4)	I (1-4)
Anthoxanthum odoratum	I (1)	I (1-3)	I (1)	II (1-5)	I (1-5)
Polytrichum commune	I (1)	I (1-3)	I (1)	II (1-4)	I (1-4)
Festuca ovina	I (1-6)	I (1-4)	I (1)	II (1-6)	I (1-6)
Luzula multiflora	I (1)	I (1)		II (1-3)	I (1-3)
Carex pilulifera		I (1)		II (1-4)	I (1-4)
Rhytidiadelphus loreus		I (1-3)	I (1)	II (1-4)	I (1-4)
Cladonia chlorophaea	I (1-3)		I (1-3)	II (1-4)	I (1-4)
Polygala serpyllifolia	III (1-4)	III (1-4)	II (1-2)	II (1-4)	III (1-4)
Sphagnum capillifolium	III (1-4)	III (1-6)	II (1-5)	II (1-4)	III (1-6)
Sphagnum subnitens	III (1-4)	II (1-6)	I (1-3)	II (1-6)	II (1-6)

Floristic table M15 *(cont.)*

	a	b	c	d	15
Agrostis canina canina	I (1)	II (1–5)	II (1–3)	II (1–6)	II (1–6)
Hylocomium splendens	I (1–4)	I (1–3)	I (1–3)	I (1–4)	I (1–4)
Aulacomnium palustre	I (1–2)	I (1–3)	I (1)	I (1–4)	I (1–4)
Festuca vivipara	I (1–4)	I (1–2)	I (1–4)	I (1–3)	I (1–4)
Diplophyllum albicans	I (1)	I (1–4)	I (1)	I (1)	I (1–4)
Pedicularis sylvatica	I (1–3)	I (1–4)	I (1)	I (1–3)	I (1–4)
Empetrum nigrum nigrum	I (1)	I (1–3)	I (4)	I (1–3)	I (1–4)
Blechnum spicant	I (1)	I (1–2)	I (1–3)	I (1–4)	I (1–4)
Carex binervis	I (1)		I (1)	I (1–3)	I (1–3)
Huperzia selago	I (1)	I (1–3)	I (1–3)	I (1)	I (1–3)
Juncus conglomeratus	I (4)	I (1–3)	I (1)		I (1–4)
Rhytidiadelphus squarrosus	I (1–3)	I (1)		I (1–4)	I (1–4)
Festuca rubra	I (1–4)	I (1–3)		I (1–4)	I (1–4)
Sphagnum compactum	I (1–3)	I (1–6)		I (1–4)	I (1–6)
Juncus articulatus	I (1–4)	I (1)		I (1)	I (1–4)
Danthonia decumbens	I (1–4)	I (1–2)		I (1–4)	I (1–4)
Juncus effusus	I (1–4)	I (1–4)		I (1–4)	I (1–4)
Calypogeia muellerana	I (1)	I (1–3)		I (1–3)	I (1–3)
Leucobryum glaucum		I (1–4)	I (1)	I (1–3)	I (1–4)
Pteridium aquilinum		I (1–4)	I (1–3)	I (1–4)	I (1–4)
Agrostis stolonifera		I (1–3)	I (1)	I (1–4)	I (1–4)
Agrostis capillaris		I (1–3)	I (1)	I (1–4)	I (1–4)
Calypogeia fissa		I (1–3)	I (1)	I (1–3)	I (1–3)
Sphagnum tenellum		I (1–4)	I (1–3)	I (4)	I (1–4)
Cladonia arbuscula		I (1–2)	I (1–4)	I (1)	I (1–4)
Pedicularis palustris	I (1–3)	I (1)			I (1–3)
Hypericum pulchrum	I (1–3)	I (1)			I (1–3)
Triglochin palustris	I (1–2)	I (1)			I (1–2)
Trientalis europaea	I (1–4)			I (1)	I (1–4)
Lophozia ventricosa			I (1)	I (1–3)	I (1–3)
Pohlia nutans			I (1)	I (1–3)	I (1–3)
Cladonia coccifera			I (1–3)	I (1–3)	I (1–3)

	a	b	c	d	15
Number of samples	69	133	23	57	282
Number of species/sample	27 (10–45)	18 (8–41)	15 (7–27)	18 (6–57)	18 (6–57)
Herb height (cm)	27 (9–50)	21 (9–40)	23 (10–38)	26 (6–50)	24 (6–50)
Herb cover (%)	94 (60–100)	92 (85–100)	94 (70–100)	98 (85–100)	95 (60–100)
Bryophyte height (mm)	38 (10–100)	35 (10–80)	20 (10–40)	34 (10–100)	33 (10–100)
Bryophyte cover (%)	32 (10–70)	38 (2–60)	41 (2–80)	28 (2–90)	34 (2–90)
Altitude (m)	177 (45–533)	226 (43–472)	283 (57–470)	290 (35–550)	248 (35–550)
Slope (°)	4 (0–14)	7 (0–42)	16 (0–35)	8 (0–35)	8 (0–42)
Soil pH	5.6 (4.2–7.4)	4.6 (3.6–5.9)	4.7 (4.0–6.0)	4.2 (3.5–5.3)	4.6 (3.5–7.4)

a   *Carex panicea* sub-community
b   Typical sub-community
c   *Cladonia* sub-community
d   *Vaccinium myrtillus* sub-community
15   *Scirpus cespitosus-Erica tetralix* wet heath (total)

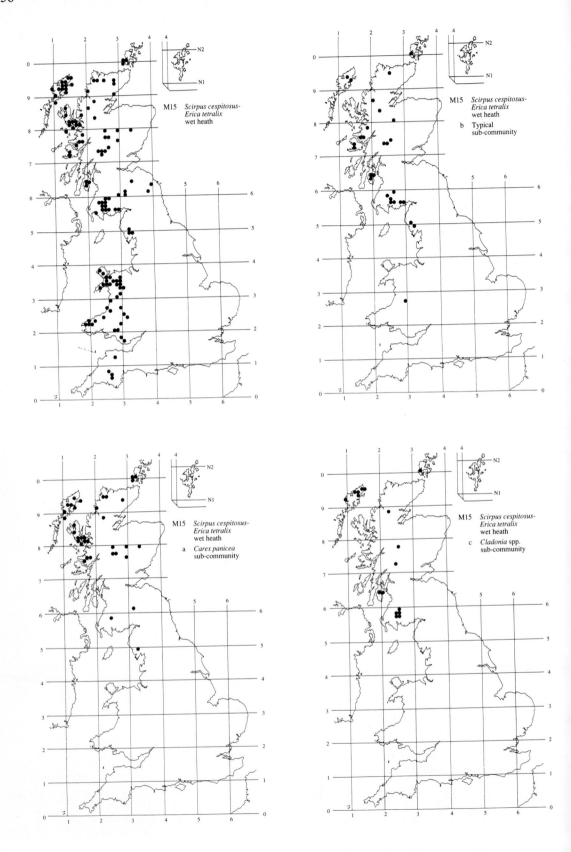

M15 *Scirpus cespitosus-Erica tetralix* wet heath

M15 *Scirpus cespitosus-Erica tetralix* wet heath

b Typical sub-community

M15 *Scirpus cespitosus-Erica tetralix* wet heath

a *Carex panicea* sub-community

M15 *Scirpus cespitosus-Erica tetralix* wet heath

c *Cladonia* spp. sub-community

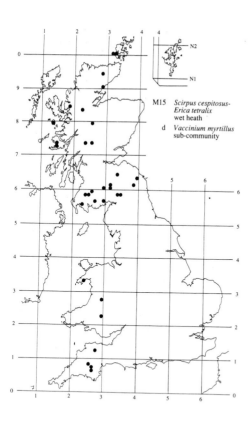

M15   *Scirpus cespitosus-*
      Erica tetralix
      wet heath

d   *Vaccinium myrtillus*
      sub-community

M16
Erica tetralix-Sphagnum compactum wet heath
Ericetum tetralicis   Schwickerath 1933

Synonymy

Molinietum caeruleae   Rankin 1911*b*, Tansley 1939 *p.p.*; *Molinia* consocies   Summerhayes & Williams 1926; Submoorland heath   Muir & Fraser 1940 *p.p.*; *Gentiana pneumonanthe* localities   Simmonds 1946; Damp heath   Rose 1953; Wet heath   Rose 1953, Rutter 1955, Newbould 1960, Ivimey-Cook & Proctor 1967; Three-species heath   Williams & Lambert 1959; *Calluna-Erica tetralix* wet heath   Gimingham 1964 *p.p.*, Bannister 1966 *p.p.*; *Ericetum tetralicis boreoatlanticum*   Ivimey-Cook *et al.* 1975; *Campylopo-Ericetum tetralicis*   Birse & Robertson 1976 *p.p.*; Mire noda 1–5   Daniels 1978 *p.p.*; *Narthecio-Ericetum tetralicis*   Moore (1964) 1968 *sensu* Birse 1980 *p.p.*

Constant species

Calluna vulgaris, Erica tetralix, Molinia caerulea, Sphagnum compactum.

Rare species

Erica ciliaris, Gentiana pneumonanthe, Lepidotis inundata, Rhynchospora fusca.

Physiognomy

The *Ericetum tetralicis* is characteristically dominated by mixtures of *Erica tetralix*, *Calluna vulgaris* and *Molinia caerulea*, but the proportions of these are very variable, being influenced by differences in the water regime and trophic state of the soils, and also by grazing and burning, the last factor being able to transform the appearance of particular stands over short periods of time and producing great structural diversity within a small compass (e.g. Williams & Lambert 1961). *E. tetralix* often grows very vigorously here, typically on these wetter soils adopting a semi-prostrate habit (Bannister 1966). *Calluna*, on the other hand, is often subordinate and of somewhat weak growth (Gimingham 1972), though it can become abundant in drier stands or where there has been deliberate encouragement by controlled burning. In other situations, *Molinia* is very

much the dominant, with the ericoids reduced to sparsely-scattered bushes among dense tussocks of the grass, such that some accounts of the community have grouped it within a broadly-defined *Molinietum* (Rankin 1911*b*, Summerhayes & Williams 1926, Tansley 1939).

Typically, however, no other sub-shrubs attain high frequency through the community as a whole, though some can be of local importance. Thus, *Erica cinerea* and *Ulex gallii* figure occasionally, and sometimes with local abundance, in transitions to drier heaths in south-west England, and *E. cinerea* and *U. minor* can occur in similar situations further east. More strikingly, around Poole Harbour and in Cornwall, the *Ericetum* provides one of the loci for the nationally-rare *E. ciliaris* and its hybrids with *E. tetralix* (= *E. × watsonii* Benth.: Chapman 1975). *U. europaeus* can also be found occasionally on less impoverished soils and can spread within the community in drier, disturbed situations.

As defined here, the *Ericetum tetralicis* includes vegetation in which there are no species other than the three vascular constants or very few sporadic associates. Such impoverished heaths are of wide distribution through lowland Britain and, though difficult to place on floristic grounds alone, have often been fairly obviously derived from this community by, for example, frequent uncontrolled burning and/or draining. However, although the *Ericetum* is at best only moderately species-rich, there are typically some additional plants. Most characteristic of these are the two Sphagna, *S. compactum*, which is a fairly strong preferential for the community and particularly associated with drier situations, and *S. tenellum*, less diagnostic and usually marking out wetter places. In some stands, these occur as scattered cushions; in others, they can be a virtually continuous carpet between the *Molinia* and ericoids, though they do not accumulate thick peat. *S. subnitens* and *S. cuspidatum* also occur very occasionally throughout and *S. molle*, though not common, shows fairly high fidelity to this community which provides its main locus in south-eastern Britain. Then, in different kinds of *Ericetum*, there can be some *S. auriculatum*, *S. papillosum* or *S. capillifolium*. The abundance of all these species is strongly influenced by burning but, in stands not recently fired, the balance between them can be of some value in setting the limits of the community: in transitions to the *Narthecio-Sphagnetum*, which is often surrounded by the *Ericetum* in lowland valley mires, there is generally an obvious switch to a luxuriant dominance of *S. papillosum* and *S. auriculatum* or *S. recurvum* (seen especially well in the tables of Newbould 1960 and Ivimey-Cook *et al*. 1975; see also Rose 1953).

No other bryophytes are as common throughout the community as the two characteristic Sphagna but there are occasional records for *Hypnum jutlandicum*, *Leucobryum glaucum*, *Campylopus paradoxus*, *C. pyriformis*, *Pohlia nutans*, *Pleurozium schreberi*, *Gymnocolea inflata*, *Cephalozia connivens*, *C. bicuspidata*, *Odontoschisma sphagni* and *O. denudatum*, in wetter situations for *Kurzia pauciflora* and *Campylopus paradoxus*, and in drier stands for *Dicranum scoparium* and *Racomitrium lanuginosum*. *Hypnum imponens* and *Dicranum spurium* are two mosses of rather local distribution in Britain which can sometimes be found here. Intermixed with bryophytes, there are also quite frequently some lichens, especially larger *Cladonia* spp. *C. impexa* and *C. uncialis* are the commonest of these but one or more of the following occur quite often: *C. arbuscula*, *C. floerkeana*, *C. squamosa*, *C. furcata* and *C. verticillata*.

The canopy of the three vascular constants, generally a few decimetres high at most and often rather open, and the ground cover of bryophytes and lichens, make up the more consistent structural elements of the vegetation, but there are frequently some herbaceous associates. *Scirpus cespitosus* is the commonest of these overall and the *Ericetum* provides the major locus for this plant in south-east Britain, as it does for *Eriophorum vaginatum*, though this is much scarcer in the community. *E. angustifolium* and *Narthecium ossifragum* also occur quite frequently and *Drosera rotundifolia* (sometimes accompanied by *D. intermedia*) can be found in wetter hollows, these species providing a strong floristic link with the *Narthecio-Sphagnetum*. Then, there is occasionally some *Polygala serpyllifolia*, *Potentilla erecta*, *Carex panicea*, *C. echinata*, *Juncus squarrosus*, *Pedicularis sylvatica* and *Dactylorhiza maculata* with, in some of the sub-communities, distinctive enrichment with preferential herbs. *Gentiana pneumonanthe*, a nationally rare species with a disjunct distribution through southern Britain, is almost always found in this community in this country (in contrast to the European mainland where it has more catholic affinities), seeming to prefer growing among bushes of *E. tetralix* and low *Calluna* and avoiding dense, tussocky *Molinia* (Simmonds 1946). *Rhynchospora fusca* and *Lepidotis inundata* are two

other rarities which occur in one particular kind of *Ericetum*.

Finally, the community can have some representation of taller woody species. *Myrica gale* occurs occasionally, sometimes with local abundance and, where *Molinia* is also prominent beneath, the vegetation can take on the appearance of a '*Molinia-Myrica* mire': some stands of this kind were perhaps included in the *Myricetum gale* of Wheeler (1980c) and the *Molinia/Myrica/Erica tetralix* bog of the NCC New Forest Bogs Report (1984). Then, there can also be some seedlings and saplings of *Betula* spp., usually *B. pubescens* but with *B. pendula* on drier substrates, and of *Pinus sylvestris*, often spreading from nearby plantations.

Sub-communities

Typical sub-community: *Molinietum caeruleae* Rankin 1911*b*, Tansley 1939 *p.p.*; *Molinia* consocies Summerhayes & Williams 1926; *Gentiana pneumonanthe* localities 1–7, 13, 18–19 Simmonds 1946; Damp heath Rose 1953; Wet heath Rose 1953, Rutter 1955, Newbould 1960, Ivimey-Cook & Proctor 1967 *p.p.*; Three-species heath Williams & Lambert 1959; *Ericetum tetralicis boreoatlanticum* Ivimey-Cook *et al.* 1975; Mire noda 1, 2 & 4 Daniels 1978 *p. p.* In this, the most widespread form of the *Ericetum tetralicis* throughout the southern part of the range of the community, all the variations in the proportions of *Molinia*, *E. tetralix* and *Calluna* can be found, so the gross appearance of the vegetation is very diverse (e.g. Rose 1953, Williams & Lambert 1959, 1961). Where stands have not been recently burned but have a fairly open cover of the dominants, *Sphagnum compactum* and *S. tenellum* can both be very frequent and often abundant, sometimes with a little *S. cuspidatum*, *S. auriculatum*, *S. papillosum* and *S. subnitens* and, more locally, *S. molle* in wetter places. *Hypnum jutlandicum* and *Kurzia pauciflora* are the commonest species among the other bryophytes though, in some areas, *Campylopus brevipilus* becomes very consistent (as in the samples from Aylesbeare Common in Devon included in Ivimey-Cook *et al.* 1975). Lichens, particularly *Cladonia impexa*, can also be frequent and, where the vegetation around lowland valley mires is strongly zoned, there is sometimes a tendency for this element to become more abundant and diverse in moving towards the drier heaths around the *Ericetum*, a feature seen in the data of Ivimey-Cook *et al.* (1975) and recognised in Rose's (1953) separation of 'wet' and 'damp' heaths.

In such transitions, there can also be a modest enrichment of the sub-shrub cover by scattered bushes of *Erica cinerea*, *Ulex gallii* and *U. minor* but, typically, the most frequent vascular associates of the dominants are *Scirpus cespitosus*, usually occurring as scattered tussocks and, becoming more frequent towards the wetter transitions to valley mire proper, *Narthecium ossifragum* and *Eriophorum angustifolium*. *E. vaginatum* is found very occasionally too, and there can be scattered plants of *Potentilla erecta* and *Juncus squarrosus* and sparse records for some of the preferentials of the next sub-community.

There is a continuous transition between this typical kind of *Ericetum* and the more impoverished wet heaths where bryophytes and lichens in particular are very much reduced, where herbaceous associates become very sporadic in their occurrence and where dominance sometimes passes very obviously to one or other of the vascular constants depending on the particular treatment history. Such stands could be included here, when they would markedly reduce the frequency of *Sphagnum compactum* and *S. tenellum*, or be treated as a separate species-poor sub-community.

Succisa pratensis-Carex panicea sub-community: *Gentiana pneumonanthe* localities 11–12 & 15–17 Simmonds 1946.

Although *E. tetralix* and *Calluna* retain high frequency here, there is a tendency for *Molinia* to predominate and the vascular component of the vegetation is considerably richer than in the Typical sub-community. *Potentilla erecta* and *Succisa pratensis* become constant and other fairly frequent preferentials are *Polygala serpyllifolia*, *Carex panicea*, *Danthonia decumbens*, *Salix repens*, *Cirsium dissectum* and *Serratula tinctoria*, with *Myrica gale* sometimes showing local abundance. More occasionally there can be small amounts of *Juncus acutiflorus* or *J. effusus* or some *Agrostis curtisii*, *Festuca rubra* and *Luzula multiflora*. Of scarcer occurrence, but still showing some preference to this sub-community, are *Carex hostiana*, *C. pulicaris*, *Dactylorhiza maculata* and *Platanthera bifolia*. By contrast, some of the community associates such as *Scirpus cespitosus*, *Narthecium ossifragum* and *Eriophorum angustifolium* are of rather reduced frequency here.

Bryophytes and lichens, too, tend to be less common among the more complete cover of angiosperms, with both *Sphagnum compactum* and *S. tenellum* occurring only occasionally. Unusually, however, *S. auriculatum* is more frequent here than in any other kind of *Ericetum* and it can be locally abundant.

Rhynchospora alba-Drosera intermedia sub-community: *Gentiana pneumonanthe* localities 8–10 Simmonds 1946; Wet heath Rose 1953, Newbould 1960 both *p.p.*

Molinia and *E. tetralix* both remain very frequent here but *Calluna* is a little less common than usual and, more obviously, the cover of all three is reduced. In the network of intervening open areas, there is typically an extensive cover of *Sphagnum compactum* and *S. tenellum* with frequent scattered plants of *Kurzia pauciflora*,

occasional *Hypnum jutlandicum*, *Odontoschisma sphagni*, *Campylopus brevipilus*, *Cephalozia* spp. and *Gymnocolea inflata* and locally prominent *Cladonia* spp., particularly *C. impexa*.

Scattered in this ground, there are frequent small tussocks of *Scirpus cespitosus* and often some *Narthecium* and sparse *Eriophorum angustifolium* and occasional *Eleocharis multicaulis*. More distinctive is the preferential occurrence here, often concentrated around wetter hollows and runnels, of *Drosera rotundifolia* and the rarer *D. intermedia*, *Rhynchospora alba* and, at some sites in Dorset and Hampshire and a few isolated stations elsewhere, *R. fusca*. Another rare species, very characteristic of this sub-community and typically occurring on bare peaty patches, is *Lepidotis inundata*, often associated with crusts of the purple alga *Zygogonium ericetorum s.l.* *Pinguicula lusitanica* can also be found in this vegetation at some sites in the south-west. Some of these species provide a strong floristic link between the *Ericetum* and the *Narthecio-Sphagnetum* and its typical *Sphagnum auriculatum* pools, but transitional stands can usually be separated by the scarcity here of *S. papillosum* and *S. auriculatum*.

Juncus squarrosus-Dicranum scoparium sub-community: Submoorland heath Muir & Fraser 1940 *p.p.*; *Calluna-Erica tetralix* wet heath Gimingham 1964 *p.p.*, Bannister 1966 *p.p.*; *Campylopo-Ericetum tetralicis* Birse & Robertson 1976; *Narthecio-Ericetum tetralicis* Moore (1964) 1968 *sensu* Birse 1980.

In this sub-community, which is the usual kind of *Ericetum* to the north and east of Britain, there are some noticeable shifts in the comparative frequency and abundance of the commonest species of the community. *Molinia*, for example, is rarely of high cover here and is sometimes totally absent and, though *E. tetralix* retains constancy and is often quite abundant, it generally yields dominance to *Calluna*. Then, *Scirpus cespitosus* is more frequent and of consistently higher cover in this sub-community than elsewhere.

The continuing high frequency of *Sphagnum compactum* and *S. tenellum*, and the relative scarcity of *S. capillifolium* and *S. subnitens*, emphasise the close relationships of this vegetation to the other kinds of *Ericetum*. *S. compactum*, in particular, is often abundant here and commonly accompanied by *Hypnum jutlandicum*, *Dicranum scoparium*, *Racomitrium lanuginosum*, *Diplophyllum albicans* and, less frequently, by *Pohlia nutans*, *Pleurozium schreberi* and *Campylopus paradoxus*. Lichens, too, often contribute prominently to the ground cover with *Cladonia impexa* and *C. uncialis* especially common, *C. arbuscula*, *C. floerkeana* and *C. squamosa* more occasional.

The general impression of the vegetation, then, is of a rather drier kind of heath than is usual in the other sub-communities, though it should be noted that the Oceanic *Erica cinerea*, which would be expected to accompany many of the above bryophytes and lichens in heaths to the west of Britain, is very scarce here. The only distinctive vascular plant is *Juncus squarrosus*, which is much more frequent than in other sub-communities, though rarely abundant. There is commonly some *Narthecium* and occasional plants of *Eriophorum angustifolium*, *Potentilla erecta* and *Polygala serpyllifolia*. Increased frequencies of the last three of these is one other feature of the floristic switch to the *Scirpus-Erica* wet heath in moving to the western uplands of Britain.

Habitat

The *Ericetum tetralicis* is a community of acid and oligotrophic mineral soils or shallow peats that are at least seasonally waterlogged. It is largely confined to the relatively dry lowlands of Britain, being particularly associated with the surrounds of valley mires maintained by a locally-high ground water-table, though in Scotland it extends on to thin ombrogenous peats at higher altitudes. Grazing and burning are important in maintaining the vegetation and, although draining has permitted its migration on to once-wetter peats in some places, many stands have been altered or destroyed by lowering the water-table.

The *Ericetum* occurs typically on acid soils (with a surface pH that is generally between 3.5 and 4.5) that are too dry for the development of Sphagnetalia magellanici bogs and too wet to sustain Calluno-Ulicetalia heaths. The soils here are maintained in a very moist state for much of the year but are not so thoroughly waterlogged to favour the luxuriant growth of the major peat-forming Sphagna. The frequent presence of such plants as *Scirpus cespitosus*, *Narthecium ossifragum* and *Eriophorum angustifolium*, and the occurrence in wetter depressions of *Drosera* spp. and *Rhynchospora* spp., provides some continuity with the vegetation of mire surfaces, but the scarcity of *Sphagnum papillosum* and *S. magellanicum* serves as a good marker of the boundary with the Sphagnetalia, typically a group of mire communities developed on deeper, wetter peats. The *Ericetum* can extend a little way on to the fringes of topogenous mires and, particularly in eastern Scotland, it is found on the thin margins of blanket peats, but mostly it occurs on wet mineral soils with sometimes but a thin humic top. Among the Sphagna which can maintain themselves on such surface-moist profiles are *Sphagnum tenellum* and, much more diagnostic of the community, *S. compactum* and the scarcer *S. molle*.

Towards the opposite extreme, the soil surface beneath the *Ericetum* can dry out intermittently in summer, but waterlogging is too sustained to permit the encroachment far into the community of Calluno-Ulicetalia species. *Erica cinerea*, for example, which is a common plant in heaths over free-draining acid soils throughout much of the range of the *Ericetum tetralicis*, cannot germinate well on the wet surfaces characteristic here (Bannister 1964a, 1965) and quickly fails if the water-table is raised (Bannister 1964d), perhaps because of the accumulation of ferrous ions (Jones & Etherington 1970, Jones 1971a, b): valley mires in particular seem to be rich in iron, which is often precipitated locally as ochre. Except in the more continental parts of East Anglia, where *E. cinerea* becomes increasingly scarce (Bannister 1965), the limit of this plant in heath sequences developed over progressively wetter soils provides a good indication of the switch to the *Ericetum*. Over south-eastern England, *Ulex minor*, and, further west, *U. gallii* and *Agrostis curtisii*, show the same behaviour, penetrating into the community only along its drier fringes or where increased through-put enhances aeration and nutrient status, as perhaps occurs in some stands of the *Succisa-Carex* sub-community.

The vascular plant which benefits most obviously from the intermediate wetness of the soils of the *Ericetum* is *E. tetralix* itself. This plant is widely represented on oligotrophic and poorly-aerated soils in Britain (Bannister 1966), extending on to the lawn and hummock component of Sphagnetalia mires and, in the more oceanic climate of south-west Britain, transgressing far into the Calluno-Ulicetalia heaths, but it tends to show a peak of abundance in the transitional zone occupied by the *Ericetum*. It may be able to avoid some of the effects of the waterlogged environment by virtue of its shallow root system (Sheikh 1970), but probably of greater importance is that here it can establish and grow free of some of the competition that limits its contribution on more free-draining and/or richer soils (Bannister 1966, Gimingham 1972). Of continuing importance through much of the *Ericetum*, however, are competitive interactions with *Calluna* and *Molinia*: the relationships between these species give rise to much of the more obvious variation within and between stands and help characterise some of the sub-communities (Rutter 1955).

The interactions with *Calluna* are probably largely related to waterlogging. *Calluna* can germinate and establish over a wider range of soils than *E. tetralix* (Bannister 1964c, Gimingham 1972) and, in particular, maintains its representation on substrates where there is some fluctuation in moisture content (Rutter 1955) or a consistently lower water-table, even in more continental areas where *Erica cinerea*, which it often accompanies in drier heaths, cannot thrive. On wetter, humic surfaces, however, *E. tetralix* can outstrip *Calluna* in its germination performance (Bannister 1964b) and maintain itself better in mixed populations (Smart in Gimingham 1972): consistent severe waterlogging induces reduced transpiration, chlorosis and failure of the root system in

Calluna (Bannister 1964*c*, *d*). The results of interactions of this kind can be seen in walking upslope through stands of the Typical sub-community around some valley mires, where shifts in dominance from *E. tetralix* to *Calluna* can occur, particularly outside south-western England, where there is no climatic complication of the amplitude of the former. And they are visible, too, in the poorer contribution of *Calluna* to the wet-hollow vegetation of the *Rhynchospora-Drosera* sub-community. The rather consistent way in which *Calluna* tends to exceed *E. tetralix* in the *Juncus-Dicranum* sub-community, on the other hand, is probably related to the treatment the *Ericetum* receives around the upland fringes, where it is often subject to controlled burning (see below).

The relationship between these two ericoids is complicated by interactions with *Molinia*, particularly towards the south-west of Britain where, like *E. tetralix*, it is represented not only through the Sphagnetalia mires but also in the Calluno-Ulicetalia heaths (as well, of course, as in a range of Molinietalia communities). In the *Ericetum*, *Molinia* fares less well on very badly-aerated profiles, but it is probably hindered more strongly by the generally poor trophic state of the soils (Loach 1966, 1968*a*, *b*, Sheikh & Rutter 1969, Sheikh 1969*a*, *b*, 1970). Thus, although *Molinia* maintains high frequency through much of the *Ericetum*, its competitive ability is often reduced, against *E. tetralix* on the more consistently waterlogged soils and against *Calluna* on somewhat drier, impoverished profiles. Where there is some amelioration of edaphic extremes, it can predominate: *Molinia*-dominated stands occur within the Typical sub-community and are especially common in the *Succisa-Carex* sub-community, where the appearance of species such as *Succisa*, *Serratula tinctoria*, *Cirsium dissectum*, *Juncus* spp. and *Myrica gale* is suggestive of a transition to Junco-Molinion vegetation. This kind of *Ericetum* can mark out areas of more base-rich soil (pH here sometimes exceeds 5) perhaps derived from less acidic parent materials or enriched with sub-surface water-flow.

Environmental conditions favouring the development of the *Ericetum* are of local occurrence in lowland Britain, depending as they do on the maintenance of a high water-table in acidic soils. In southern England, such conditions are generally associated with drainage-impedence over impervious parent materials, like non-calcareous clays, shales and superficials, which give rise to stagnogleys, or with the development of impervious horizons, such as an iron pan or strongly argillic B horizon, in podzolised profiles. These waterlogged, base-poor soils are concentrated on the more acidic components of the complex sequences of younger deposits that make up the landscape of south-eastern England. Here, the *Ericetum* is found over Eocene clays,

sands and gravels in south Dorset, Hampshire and Surrey, and over the clays and sands of the Lower Greensand and Hastings Beds in the Weald, and, more locally, in West Norfolk. Further west, in Devon and Cornwall, Triassic sandstones and pebble-beds carry the community and these were probably an important substrate for, now lost, wet heaths of this kind in the Cheshire Plain and east Yorkshire. In parts of the New Forest, the *Ericetum* occurs over Pleistocene gravels.

On such deposits, the community can pick out areas where a perched water-table occurs over stretches of flatter ground with drainage-impedence in podzols, a particularly distinctive habitat on ill-drained terrace fragments in the northern part of the New Forest (Fisher 1975*a*, *b*), or occupy the bottoms of hollows where erosion has cut down to impervious bedrocks, as in some of the Hampshire and Sussex sites described by Rose (1953). More strikingly, it occurs on slopes around more fully-developed mires in shallow valleys, where the water-table comes close to the surface with a switch to impervious strata, but is not yet so high as to encourage the development of Sphagnetalia vegetation, which supervenes below. In such situations, the hydrological regime is essentially topogenous, often with an artesian element maintaining a high-table and producing back-gleying in the mire surrounds. But the hollows are also fed by run-off from higher ground and tracks with more obvious soligenous inflow can sometimes be seen around the margins. The Typical sub-community often makes up the bulk of the cover in these sites, with the *Rhynchospora-Drosera* sub-community marking out wet and often sandy depressions: the latter vegetation can then be seen as transitional to the Rhynchosporion element in the hollows of the mire surface proper. The *Succisa-Carex* sub-community seems to replace the Typical sub-community over richer parent materials or where there is more marked sub-surface water-flow.

In a few remaining sites further north, the *Ericetum* can occur in this kind of setting: in some places on the North York Moors, for example, it fringes valley mire developed in drift-filled hollows cut into Jurassic rocks. But, increasingly in northern Britain, the community is found on gleys and gleyed podzols on drift-smeared slopes around the upland fringes and, particularly in eastern Scotland, it extends on to the drier peats on the edges of ombrogenous bog, extending to altitudes of 500 m or even more (Birse & Robertson 1976, Birse 1980). Here, the *Juncus-Dicranum* sub-community is the usual form of the *Ericetum*, *Molinia* becoming sparser in the continental climate and *Calluna* and Nardo-Callunetea associates increasing, perhaps in response to treatment. It is in Scotland, where the distribution of the community has not been too fragmented, that there is the most continuous geographical transition between the *Ericetum* and its north-western counterpart, the

Scirpus-Erica wet heath. The switch between the two communities occurs around the 1600 mm isohyet (*Climatological Atlas* 1952) or 180 wet days yr^{-1} line (Ratcliffe 1968), beyond which shallow ill-drained slopes often carry moderately thick ombrogenous peat with the *Scirpus-Erica* community.

Superimposed upon the natural patterns of variation in the *Ericetum*, developed in relation to the degree of waterlogging and, less so, the trophic state of the soil, are differences produced by biotic factors, most notably grazing, burning and draining. Throughout its range, the community lies within the forest zone and, without any grazing or burning, most stands would probably progress fairly quickly to some kind of woodland. Indeed, this has been the fate of some stands lying within tracts of heath on certain commons in south-east England where traditional management has fallen into disuse. Elsewhere, occasional burning, with grazing in some areas like the New Forest and around the upland fringes, continue to be of importance in maintaining the community. But both treatments, and especially burning, also have a marked effect on floristic variation within the community, particularly on the local pattern of dominance and the richness of the associated flora.

To some extent, the *Ericetum* is protected against the effects of very severe fires by the wetness of the ground, but more superficial fires commonly remove the bulk of the standing vegetation. *Calluna*, *E. tetralix* and *Molinia* are all well able to regenerate after fires, by sprouting from buried stools or basal shoots or, with the ericoids, by the rapid establishment of seedling populations. But the results of any particular burn are very variable, one or other of the species being able to take advantage according to such factors as the previous state of the vegetation, the time since the previous burning, the intensity of the fire and environmental conditions afterwards (Whittaker 1960, Gimingham 1972). Such opportunism creates great local diversity in the proportional contribution of the dominants in individual stands (e.g. Williams & Lambert 1959, 1961) and makes it difficult to discern consistent patterns of response among the species. In wetter habitats, such as are generally characteristic here, *E. tetralix* may be able to regenerate vegetatively better than *Calluna* because its semi-prostrate lower branches are protected to some extent by the *Sphagnum* carpet and litter (Fritsch & Salisbury 1915, Bannister 1966). But *Calluna* can re-establish in abundance from seed and, in somewhat drier situations and in areas with a cooler climate, it may have the edge. More controlled burning, deliberately to maximise the proportion of building-phase *Calluna*, may account for the consistent predominance of this ericoid in the *Juncus-Dicranum* sub-community in north-eastern Britain: this kind of *Ericetum* also has that abundance of *Cladonia* spp. typical of the middle years of regrowth in regularly-burned heaths. More severe burning of the community can destroy both ericoid stools and buried seed and probably gives *Molinia* an advantage in the long term (Tansley 1939, Rose 1953, Ratcliffe 1959a). This grass can also benefit by the burning off of choking litter and by the quick release of nutrients from the ash.

More deleterious to the community are the effects of burning on the associated flora, especially the typical bryophytes, which recover only very slowly and which can be virtually eliminated by frequent fires. Such treatment has probably contributed to the widespread impoverishment of the *Ericetum* on heaths throughout southern Britain, and has been especially destructive when combined, as it often has been, with draining. The survival of the community is ultimately dependent on the maintenance of a high water-table and, where this has been lowered, either by direct drainage or by such operations in the mires below, most of the associates are lost. Where the land is not reclaimed, mixtures of the dominants may survive for some considerable time, often with much *Calluna* on poorer soils or with a spread of *Molinia* where surface oxidation of humus has released nutrients. But the woeful effects of drainage on the *Ericetum* can be readily seen in the shrinking distributions of two of its most distinctive rare species, *Lepidotis inundata* and *Gentiana pneumonanthe*, and in losses in south-eastern Britain of *Scirpus cespitosus* and *Eriophorum vaginatum*, common enough to the north and west but very much confined to the community in this region.

Zonation and succession

The *Ericetum tetralicis* occurs as part of zonations related to the height of the water-table in sequences of acid and oligotrophic soils. Locally, these can be complicated by increased soligenous influence on slopes or by variations in the base-richness and trophic state of soil parent materials or waters. Burning can superimpose differences in floristics and structure and frequent firing and/or draining degrades the community. In the absence of grazing and burning, the *Ericetum* progresses to woodland.

When fully developed around lowland valley mires, the pattern of communities in which the *Ericetum* occurs is very striking (e.g. Rose 1953, Wheeler 1983). Downslope, on the deep, wet peats of the bog itself, it gives way to the *Narthecio-Sphagnetum* with the *Sphagnum auriculatum* bog-pool community in deeper hollows, many of the species of the *Ericetum* running on into the lawn and hummock component of the mire surface, but *Sphagnum compactum* and, to a lesser extent, *S. tenellum*, being replaced by *S. papillosum* and *S. magellanicum*. Beyond this, towards the main axis of the mire, there is often a zone of Littorelletea vegetation or Caricion nigrae poor fen and then a central strip of *Salix-Galium* woodland.

In mires with a little more enrichment, the last community may be replaced by the *Alnus-Carex* woodland from which patchy *Caricetum paniculatae* swamp may run out a little way. Upslope from the *Ericetum*, with the passage to drier acidic soils, there is a transition to Calluno-Ulicetalia heath, though the particular community involved varies according to the region in which the site occurs. In the more continental climate of East Anglia, a sharp switch to the *Calluna-Festuca* dry heath is characteristic but, with the shift into the more oceanic south-west, so-called 'humid heaths' are interposed in the sequence, first the *Ulex minor-Agrostis curtisii* heath in sites around the Hampshire basin, and then from there westwards the *Ulex gallii-Agrostis* heath. Increasingly, in such zonations, both *E. tetralix* and *Molinia*, as well as *Calluna*, maintain their representation through these intermediate vegetation types, so it is generally the downslope limit of *E. cinerea*, the *Ulex* spp. and *A. curtisii* that marks the transition to the Ericion wet heath. Above these zonations, pockets of deeper, well-drained soil can have stands of the *Pteridium-Galium* community or, where there has been some enrichment of the ground, the *Rubus-Pteridium* underscrub or *Ulex-Rubus* scrub, the latter especially characteristic of disturbed places, as along trackways.

Complete zonations of this kind can still be seen in and around the series of New Forest valley mires, such as the Denny/Shatterford system and Wilverley, Holmsley and Matley bogs (Rose 1953, Tubbs 1968, Ratcliffe 1977, NCC New Forest Bogs Report 1984) and on some of the south Dorset heaths, like Morden, though there the *Ericetum* is reduced to a narrow zone around a rather flat-bottomed valley (Ratcliffe 1977). Elsewhere, the sequence of communities is incomplete below. At Thursley Common in Surrey, for example, the *Narthecio-Sphagnetum* terminates the zonation below (Rose 1953, Ratcliffe 1977) and, quite often, it is the *Ericetum* itself which occupies the wettest ground, clothing the bottom of shallow valleys and hollows. This is the case on some of the remaining Sussex commons like Ashdown Forest, over the northern New Forest terraces (Fisher 1975a, b) and on many of the fragments of south Dorset heathland, where there are the additional features of the presence of *Erica ciliaris* in a number of the communities and the close juxtaposition with open-water and maritime vegetation (Chapman 1975, Ratcliffe 1977).

In other sites, the basic sequence of communities is complicated by the interpolation of other vegetation types related to base- or nutrient-richness in the substrates or waters. In some of the New Forest mires, like Cranesmoor for example (Newbould 1960), on Hartland Moor in Dorset (Ratcliffe 1977) and at Aylesbeare Common in Devon (Ivimey-Cook *et al.* 1975), the *Schoenus-Narthecium* community figures prominently along the axes of soligenous water movement and, in the last site, there are also extensive stands of Junco-Molinion vegetation, continuing the floristic trend seen within the *Ericetum* itself in the *Succisa-Carex* subcommunity. Then, in some of the 'mixed mires' developed on north Norfolk commons like Roydon, there can be a much more striking switch to a more calcicolous series of vegetation types, including the *Cirsio-Molinietum*, the *Schoenetum*, tall-herb Phragmitetalia fen and carr (Rose 1953, Daniels & Pearson 1974, Wheeler 1975).

Further north in Britain, these kinds of valley mire zonations become rare and the *Ericetum* often occurs in what is essentially a reverse sequence of communities, giving way above to ombrogenous mire on deeper, wetter blanket peat, generally within the range of the *Ericetum*, the *Calluna-Eriophorum* mire; and passing below to a variety of northern dry heaths, throughout which burning has often resulted in a fairly uniform predominance of *Calluna*, or which have run down to Nardo-Galion grasslands with much *Juncus squarrosus* and *Nardus stricta*.

Where relaxation of burning and grazing allows invasion of shrubs and trees, the *Ericetum* is probably rapidly converted to woodland. Although this process has not been followed in detail, the most natural successor to the community is likely to be the *Betula-Molinia* woodland, *Betula pubescens* and *Salix cinerea* leading the colonisation and *Pinus sylvestris* sometimes making a prominent local contribution where it can readily seed in from nearby plantations. *Molinia* continues to play a major role in the field layer of this woodland, though the ericoids persist only patchily in areas with a more open canopy. Draining can probably speed up this succession, though it may allow a phase of strong *Molinia*-dominance to supervene between the heath and the woodland. In time, the *Betula-Molinia* woodland may give way to the *Quercus-Betula-Deschampsia* woodland.

Combinations of frequent burning, draining and damage due to other operations like military manoeuvres and mineral extraction have led to an irretrievable loss of the *Ericetum* in many areas (Rose 1953, Ratcliffe 1977). Even where stands survive intact, the characteristic zonations in which the vegetation occurs have been truncated and the community can now often be found closely hemmed in by coniferous plantations or intensive agricultural land (e.g. Moore 1962).

Distribution

The *Ericetum* is largely confined to the south and east of Britain though, particularly in lowland England, its distribution has been much fragmented by heathland reclamation. The Typical sub-community is found throughout the south of the country, being especially well represented on the heaths of south-east and south-west England and, increasingly to the north, occurring as impoverished stands. The *Rhynchospora-Drosera* sub-community is much more local, being concentrated in the New Forest and around Poole Harbour, the *Succisa-Carex* sub-community a little less so, occurring through south-west England and perhaps elsewhere. In northern Britain, the *Ericetum* is represented by the *Juncus-Dicranum* sub-community which can be seen as a transition to the *Scirpus-Erica* wet heath, which replaces the *Ericetum* in the sub-montane north and west.

Affinities

The *Ericetum tetralicis* has long been familiar from descriptive accounts as various kinds of 'wet heath' (Rose 1953, Rutter 1955, Williams & Lambert 1959, Newbould 1960, Gimingham 1964b, Bannister 1966, Ivimey-Cook & Proctor 1967). As defined here, the community corresponds well, in general terms, with these vegetation types, and with their phytosociological definition in Ivimey-Cook *et al.* (1975), but it also takes in part of certain *Molinieta* (Rankin 1911b, Summerhayes & Williams 1926, Tansley 1939) and some wet moorland vegetation described from Scotland (Muir & Fraser 1940, Birse & Robertson 1976, Birse 1980).

The community has close relationships with the base-poor *Sphagnum* mires of the Sphagnetalia, particularly the lowland valley bog, the *Narthecio-Sphagnetum*, and is best included with them in the Oxycocco-Sphagnetea. Wet heath of this kind has traditionally been placed in a distinct order, the Sphagno-Ericetalia Br.-Bl. 1948

(= Ericetalia Moore (1964) 1968) with a single alliance, the Ericion tetralicis Schwickerath 1933, within which species such as *Erica tetralix*, *Gentiana pneumonanthe*, *Polygala serpylifolia*, *Sphagnum compactum*, *S. tenellum*, *S. molle* and *Zygogonium ericetorum* show a peak of occurrence. Very similar vegetation has been characterised from neighbouring parts of the Continent as the *Ericetum tetralicis* Schwickerath 1933, some of the sub-associations of which closely parallel types recognised here (e.g. Tüxen 1937, LeBrun *et al.* 1949, Westhoff & den Held 1969). The distinctive vegetation with *Rhynchospora alba*, *Drosera intermedia* and *Lepidotis inundata* has sometimes been separated off from the *Ericetum*, as in the *Lycopodio-Rhynchosporetum albo-fuscae* (Paul 1910) Allorge & Gaume 1925 (Westhoff & den Held 1969). In the north, the community grades, through the *Juncus-Dicranum* sub-community, to its sub-montane counterpart, the *Scirpus-Erica* wet-heath: Birse (1980) united some stands of both these vegetation types in his amended version of the *Narthecio-Ericetum tetralicis* Moore (1964) 1968.

Particularly in the strongly oceanic climate of south-west England, the *Ericetum* shows close affinities with various Calluno-Ulicetalia heaths, into which *E. tetralix* and *Molinia* penetrate far. Through the *Succisa* sub-community, it also grades into the vegetation of the Junco-Molinion. Under certain treatments, these communities and the *Ericetum* can converge into very species-poor *Molinia* vegetation, the phytosociological placement of which presents something of a challenge.

Floristic table M16

	a	b	c	d	16
Erica tetralix	V (3–9)	V (1–7)	V (3–7)	V (1–7)	V (1–9)
Calluna vulgaris	V (1–9)	V (3–7)	IV (2–6)	V (1–9)	V (1–9)
Molinia caerulea	V (3–9)	V (2–8)	V (4–6)	III (2–8)	IV (2–9)
Sphagnum compactum	V (2–8)	II (1–6)	V (2–8)	V (1–8)	IV (1–8)
Potentilla erecta	II (1–3)	V (1–4)		II (1–4)	II (1–4)
Succisa pratensis	I (3–4)	V (1–5)		I (2–3)	I (1–5)
Polygala serpyllifolia	I (1–3)	III (1–4)	II (1–2)	II (1–3)	II (1–4)
Carex panicea	I (1–3)	III (2–5)	II (1–2)	II (1–3)	II (1–5)
Sphagnum auriculatum	I (1–4)	III (4–7)	I (2–4)		I (1–7)
Salix repens	I (1–4)	III (1–5)	I (2)		I (1–5)
Danthonia decumbens		III (2–6)			I (2–6)
Sphagnum papillosum	I (1–5)	II (2–5)	I (1)	I (3)	I (1–5)
Juncus acutiflorus	I (1–4)	II (2–3)			I (1–4)
Myrica gale	I (3–7)	II (5–7)	I (3)	I (6)	I (3–7)
Cirsium dissectum	I (3)	II (3–5)			I (3–5)
Ulex gallii	I (3–7)	II (4–7)			I (3–7)
Serratula tinctoria		II (3–5)			I (3–5)
Ulex europaeus		II (2–4)			I (2–4)
Agrostis curtisii		II (1–7)			I (1–7)
Festuca rubra		II (3–4)			I (3–4)
Juncus effusus	I (3)	II (1–3)			I (1–3)
Luzula multiflora	I (2)	II (2–3)			I (2–3)
Kurzia pauciflora	II (1–4)		IV (2–3)	I (1–2)	I (1–4)
Drosera intermedia	I (3)		IV (1–3)		I (1–3)
Rhynchospora alba			IV (2–4)		I (2–4)
Drosera rotundifolia	II (1–3)		III (1–4)	I (1)	I (1–4)
Campylopus brevipilus	I (2–7)		II (2–3)		I (2–7)
Lepidotis inundata			II (3)		I (3)
Eleocharis multicaulis			II (1–2)		I (1–2)
Cephalozia macrostachya			II (2–3)		I (2–3)
Pinguicula lusitanica			I (3)		I (3)

Hypnum jutlandicum	II (1–4)	II (3–6)	II (1–5)	IV (1–9)	II (1–9)
Cladonia impexa	II (1–7)		III (1–5)	IV (1–9)	II (1–9)
Juncus squarrosus	II (1–5)		I (1–2)	IV (1–6)	II (1–6)
Dicranum scoparium	I (1–4)	I (1)	I (3)	III (1–5)	II (1–5)
Cladonia uncialis	I (1–5)		I (1)	III (1–4)	II (1–5)
Racomitrium lanuginosum			I (1)	III (1–8)	II (1–8)
Diplophyllum albicans				III (1–4)	I (1–4)
Pohlia nutans	I (1–2)	I (1–3)		II (1–6)	I (1–6)
Pleurozium schreberi	I (1–5)	I (2)		II (1–6)	I (1–6)
Campylopus paradoxus	I (1–3)			II (1–5)	I (1–5)
Cladonia arbuscula			I (5)	II (1–9)	I (1–9)
Sphagnum capillifolium			I (3)	II (1–8)	I (1–8)
Cladonia floerkeana			I (1)	II (1–3)	I (1–3)
Cladonia squamosa			I (1–3)	II (1–3)	I (1–3)
Mylia taylori				II (1–7)	I (1–7)
Pleurozia purpurea				II (1–6)	I (1–6)
Lophozia ventricosa				II (1–4)	I (1–4)
Kurzia trichoclados				I (1–4)	I (1–4)
Sphenolobus minutus				I (1–4)	I (1–4)
Huperzia selago				I (1–2)	I (1–2)
Sphagnum tenellum	IV (1–10)	II (3–5)	V (3–7)	III (1–7)	III (1–10)
Scirpus cespitosus	III (1–5)	I (3–4)	III (3–4)	V (1–8)	III (1–8)
Narthecium ossifragum	II (1–5)	I (3)	III (1–6)	III (1–6)	II (1–6)
Eriophorum angustifolium	II (1–8)		III (1–3)	II (1–5)	II (1–8)
Odontoschisma sphagni	I (1–5)		II (2–3)	I (1–3)	I (1–5)
Pinus sylvestris seedling	I (1–3)	I (3)	II (1–2)	I (1–5)	I (1–5)
Cephalozia connivens	I (1–3)	I (1)	I (1)	I (1–2)	I (1–3)
Cephalozia bicuspidata	I (2–3)	I (1–4)	I (1)	I (1–3)	I (1–4)
Ulex minor	I (2–5)	I (9)	I (1)		I (1–9)
Carex echinata	I (2–6)	I (4)		I (1–3)	I (1–6)
Pedicularis sylvatica	I (1)	I (2)		I (1–2)	I (1–2)
Dactylorhiza maculata	I (1)	I (1)		I (1–3)	I (1–3)
Leucobryum glaucum	I (1–6)	I (1)		I (1–7)	I (1–7)
Sphagnum subnitens	I (1–6)	I (3–8)		I (1–5)	I (1–8)
Nardus stricta	I (4)	I (1–2)		I (1–4)	I (1–4)

Floristic table M16 *(cont.)*

	a	b	c	d	16
Gymnocolea inflata	I (2–3)		I (3)	I (1–7)	I (1–7)
Cladonia furcata	I (1–3)		I (1–2)	I (2)	I (1–3)
Sphagnum cuspidatum	I (2–6)		I (2)	I (1)	I (1–6)
Betula pubescens seedling	I (1–2)	I (3–4)			I (1–4)
Cladonia verticillata	I (1–3)		I (3)		I (1–3)
Sphagnum molle	I (1–5)		I (1–2)		I (1–5)
Eriophorum vaginatum	I (2–7)			I (1–4)	I (1–7)
Odontoschisma denudatum	I (1–3)			I (1–3)	I (1–3)
Campylopus pyriformis	I (1)			I (1)	I (1)
Aulacomnium palustre	I (1–6)			I (1–3)	I (1–6)
Number of samples	45	14	10	41	110
Number of species/sample	13 (8–24)	17 (8–27)	17 (10–28)	19 (13–24)	16 (8–28)

a   Typical sub-community
b   *Succisa pratensis-Carex panicea* sub-community
c   *Rhynchospora alba-Drosera intermedia* sub-community
d   *Juncus squarrosus-Dicranum scoparium* sub-community
16 *Ericetum tetralicis* (total)

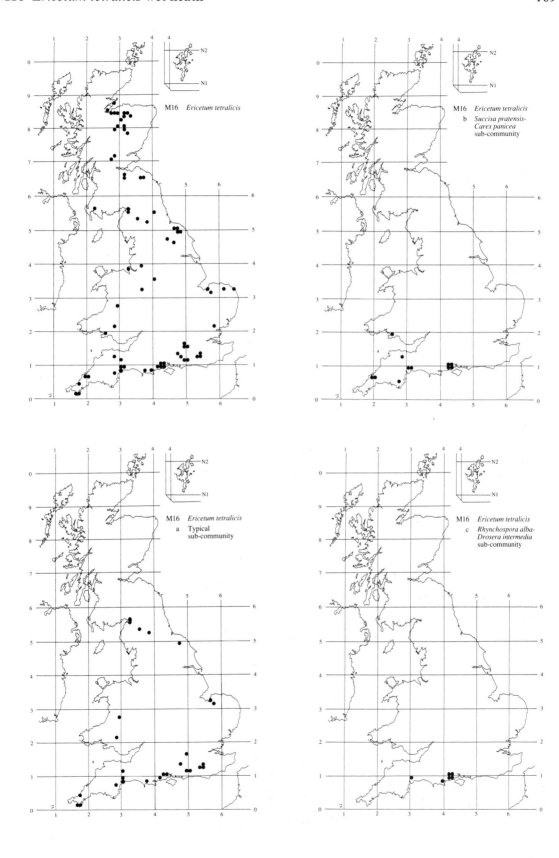

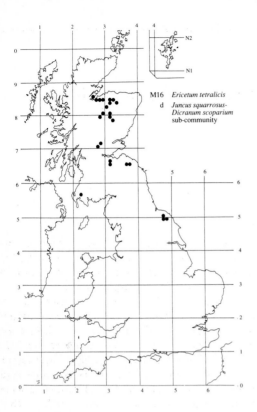

M16 *Ericetum tetralicis*
d   *Juncus squarrosus-*
     Dicranum scoparium
     sub-community

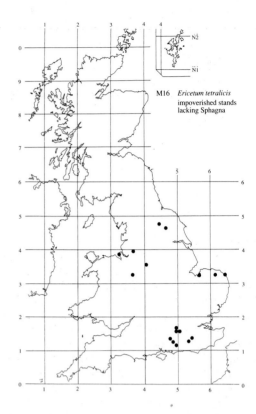

M16 *Ericetum tetralicis*
     impoverished stands
     lacking Sphagna

M17
Scirpus cespitosus-Eriophorum vaginatum blanket mire

Synonymy

Scirpetum cespitosi Watson 1932, Fraser 1933, Tansley 1939 *p.p.*; *Trichophoreto-Eriophoretum typicum* McVean & Ratcliffe 1962, Birks 1973, Evans *et al.* 1977; *Juncus squarrosus* bog McVean & Ratcliffe 1962 *p.p.*; *Pleurozia purpurea-Erica tetralix* Association Br.-Bl. & Tx. 1952 *sensu* Moore 1968 *p.p.*; *Eriophorum vaginatum* bog, low-level facies Edgell 1969; Blanket bog Ward *et al.* 1972; Mire nodum 12 Daniels 1978; *Calluno-Molinietum*: Hill & Evans 1978 *p.p.*; *Vaccinio-Eriophoretum* Hill & Evans 1978 *p.p.*; *Erico-Sphagnetum papillosi* Moore (1964) 1968 emend. Birse 1980 *p.p.*; *Erica tetralix-Sphagnum papillosum* mire Ratcliffe & Hattey 1982 *p.p.*; *Pleurozio-Ericetum tetralicis* Dierssen 1982 *p.p.*; Virgin peatland 1.ii & 1.iii Hulme & Blyth 1984; Disturbed peatland 3.iv Hulme & Blyth 1984.

Constant species

Calluna vulgaris, Erica tetralix, Eriophorum angustifolium, E. vaginatum, Molinia caerulea, Narthecium ossifragum, Potentilla erecta, Scirpus cespitosus, Sphagnum capillifolium, S. papillosum.

Rare species

Campylopus atrovirens var. *falcatus, C. setifolius, C. shawii, Sphagnum imbricatum, S. strictum.*

Physiognomy

The *Scirpus cespitosus-Eriophorum vaginatum* mire is a blanket bog community dominated by mixtures of monocotyledons, ericoid sub-shrubs and Sphagna, the two former groups of plants usually giving the vegetation its distinctive character when it is seen from a distance, but the last often occupying more of the ground, at least in wetter stands. The community can occur as extensive, fairly uniform tracts in which there is a fine-grained alternation of dominance among the species in these different elements from place to place; or, on mires with strong surface undulations, it can comprise the hummock component, with the plants showing a more obvious zonation in relation to the height of the water-table and the vegetation giving way in the hollows to Rhynchosporion pools.

Among the bulkier vascular species, the most common are *Scirpus cespitosus, Eriophorum vaginatum, Molinia caerulea, Calluna vulgaris* and *Erica tetralix*, mixtures of which form an uneven-topped tier, 2–3 dm tall and often rather open. The high frequency of *E. vaginatum* is one of the features which helps distinguish this vegetation from the most closely-related kind of wet heath, the *Scirpus-Erica* community, with which it is often associated; though, in fact, *E. vaginatum* is rarely very abundant here and never a consistent co-dominant, as it is in most types of *Calluna-Eriophorum* mire. *Molinia*, on the other hand, and, to a lesser extent, *Scirpus*, are more frequent than they are in either the *Calluna-Eriophorum* or the *Erica-Sphagnum papillosum* mire, and usually more abundant: together with *Calluna*, they often contribute most of the vascular cover in this community. *E. tetralix* is generally less extensive and, on drier ground, it becomes distinctly patchy but, with *Molinia*, it can extend into somewhat wetter situations than the other dominants and typically replaces *Calluna* as the leading ericoid in transitions to Rhynchosporion hollows. Variation in the proportions of these species is also affected by treatments and different patterns of dominance have sometimes been recognised in the names given to this kind of vegetation, as in the *Scirpetum* of Tansley (1939), the *Calluno-Molinietum* of Hill & Evans (1978) and the *Calluna* mire of Hulme & Blyth (1984).

One other species represented with occasional local abundance in this stratum of the vegetation and, among Sphagnetalia mires, preferential to this community and the *Narthecio-Sphagnetum*, is *Myrica gale*, but it is largely confined to the wetter *Drosera-Sphagnum* sub-community and, even there, is rather patchy. Where *Myrica* is prominent, it can run on from stands of this community into neighbouring areas of *Scirpus-Erica* wet heath, tending to mask floristic distinctions among smaller associates. *Erica cinerea*, *Vaccinium myrtillus* and *Empetrum nigrum* ssp. *nigrum* can also occur, but they are preferential to the drier *Cladonia* and *Juncus-Rhytidiadelphus* sub-communities and typically of low cover. The limited role of the last two of these sub-shrubs is a further contrast with the *Calluna-Eriophorum* mire where, with the Arctic-Alpine *E. nigrum* ssp. *hermaphroditum* and *V. vitis-idaea*, they are of much importance. *Rubus chamaemorus*, an Arctic-Subarctic plant also very characteristic of that community, is likewise very scarce here.

Among the vascular associates, *Eriophorum angustifolium* and *Narthecium ossifragum* maintain high frequencies throughout and each can be found with moderate local abundance, tending to become more noticeable on wetter ground, as around hollows, where the cover of some of the vascular dominants thins. *Drosera rotundifolia* also becomes very common in wetter areas but *Vaccinium oxycoccos* and *Andromeda polifolia*, which often accompany it in the *Erica-Sphagnum papillosum* mire, are characteristically very scarce. On the positive side, the *Scirpus-Eriophorum* mire can be distinguished from most other kinds of Sphagnetalia mires by the constancy of *Potentilla erecta* and the occasional occurrence of *Polygala serpyllifolia*. Other species found at low frequencies throughout are *Pedicularis sylvatica* (including ssp. *hibernica*: Webb 1956), *Huperzia selago*, *Juncus acutiflorus*, *Festuca ovina* and *Carex echinata*, and the community provides an important locus for *C. pauciflora*, though it is not very common. Transitions to Rhynchosporion hollows are often marked by a marginal band of *Rhynchospora alba* and, particularly where there is soligenous influence, *Carex panicea* and *Pinguicula vulgaris* can occur. *Schoenus nigricans* can very occasionally be found in such places, though most occurrences of this plant on our blanket mires are in the *Carex* soakways of *Scirpus-Erica* wet heath, a striking contrast to the situation in western Ireland, where this plant is a constant in the counterpart of the *Scirpus-Eriophorum* mire (Doyle & Moore 1980, Doyle 1982). Nardetalia herbs are generally of restricted occurrence here, but *Juncus squarrosus* and *Nardus stricta* are found in drier situations and become especially frequent, with a number of other preferentials, in the *Juncus-Rhytidiadelphus* sub-community, which represents a floristic transition to the *Calluna-Eriophorum* mire and wet heath.

In the ground layer of the *Scirpus-Eriophorum* mire, Sphagna are of supreme importance though, even in the wettest stands, they are generally not so extensive as in the *Erica-Sphagnum papillosum* mire. Nonetheless, *Sphagnum capillifolium* and *S. papillosum* figure among the community constants and, particularly in the *Drosera-Sphagnum* sub-community, may be accompanied by *S. tenellum*, *S. subnitens* and, rather less frequently, by *S. compactum*, *S. palustre* and *S. auriculatum* in luxuriant carpets covering more than half of the ground. *S. cuspidatum* occurs occasionally throughout and wetter stands provide a locus for species of more restricted distribution like *S. strictum*, *S. fuscum* and *S. imbricatum*, though the last two are rather more characteristic of the *Calluna-Eriophorum* and *Erica-Sphagnum* mires respectively. *S. magellanicum*, a major peat-builder in the *Erica-Sphagnum* mire, is typically scarce here, though it tends to increase in blanket mire vegetation which is transitional to that community (on the Silver Flowe mires, for example: Ratcliffe & Walker 1958). Over more even stretches of wet ground, carpets and tussocks of the Sphagna form an irregular patchwork, variously tinted in shades of green, yellow, pink and ochre, but, with more marked differentiation of hummocks they show a clear zonation in relation to the height of the water-table (see below).

Very characteristically, the *Sphagnum* carpet provides a congenial habitat for a variety of leafy hepatics which occur as scattered shoots or in sometimes quite extensive mats. The community shares with other Sphagnetalia mires *Odontoschisma sphagni* and *Mylia anomala* (common also in the *Erica-Sphagnum papillosum* mire) and *Mylia taylori* (perhaps more typical of eroding *Sphagnum* and bare peat and frequent in some types of *Calluna-Eriophorum* mire). More distinctly preferential to the *Scirpus-Eriophorum* mire (and used to name its Irish equivalent) is *Pleurozia purpurea*, very obvious with its succulent purple or dark-orange shoots. Other hepatics recorded occasionally include *Kurzia pauciflora*, *Diplophyllum albicans*, *Scapania gracilis*, *Calypogeia fissa*, *C. sphagnicola*, *Cephalozia connivens*, *C. media*, *C. macrostachya* and *C. loitlesbergeri* (e.g. Birks 1973).

There is a complete transition within the community from stands of this kind, in which Sphagna and the associated hepatics are especially numerous and abundant, and those where only *S. papillosum* and/or *S. capillifolium* retain any frequency or prominence. In these latter situations, other species become important. *Racomitrium lanuginosum*, for example, is fairly common throughout but it increases in frequency and particularly in abundance on hummock tops and in degraded mires, helping to define the vegetation included in the *Cladonia* sub-community. Then, in the *Juncus-Rhytidiadelphus* sub-community, pleurocarpous mosses, such as *Hypnum jutlandicum*, *Rhytidiadelphus loreus*, *Pleurozium schreberi* and *Plagiothecium undulatum*, together with *Dicranum scoparium* and *Polytrichum commune*, all of them species of low frequency in wetter stands, become a common feature among the Nardetalia herbs. Other species which can be found in two or more of the sub-communities include *Campylo-*

pus atrovirens, C. shawii, C. setifolius, C. paradoxus, Hylocomium splendens and *Breutelia chrysocoma.*

Finally, lichens, especially larger *Cladonia* spp., can be a prominent element in the ground carpet. In general, they tend to follow *Racomitrium lanuginosum* in their frequency and abundance, becoming particularly important on the sides and tops of taller hummocks and over degraded surfaces. *C. impexa, C. uncialis* and *C. arbuscula* are the commonest species overall but individual stands of the *Cladonia* sub-community can show considerable enrichment. *Cornicularia aculeata* also occurs occasionally thoughout.

Sub-communities

***Drosera rotundifolia-Sphagnum* 'spp. sub-community:** *Trichophoreto-Eriophoretum typicum* McVean & Ratcliffe 1962 *p.p.; Pleurozia purpurea-Erica tetralix* Association Br.-Bl. & Tx. 1952 *sensu* Moore 1968 *p.p.; Trichophoreto-Eriophoretum* Birks 1973 *p.p.; Calluno-Molinietum,* wetter facies Hill & Evans 1978; *Erico-Sphagnetum papillosi,* Typical subassociation Moore (1964) 1968 *emend.* Birse 1980 *p.p.; Erica tetralix-Sphagnum papillosum* mire Ratcliffe & Hattey 1982 *p.p.;* Typical virgin mire 1.ii Hulme & Blyth 1984. This sub-community has generally formed the core of previous definitions of this kind of mire and it has the most consistent representation of all the community constants and the richest and most extensive carpets of Sphagna. Among the vascular dominants, mixtures of *Calluna* and *Scirpus* or *Calluna* and *Molinia* usually make up the bulk of the cover overall, with *Eriophorum vaginatum* sometimes showing local abundance on more elevated areas, *Erica tetralix* tending to increase in the wetter, such zonations being especially obvious where hummocks and hollows are strongly differentiated. Occasionally, *Molinia* can be much more obviously dominant over the other species, a situation that seems to be particularly associated with a reduction in the cover of *Scirpus* and, more noticeably, of *E. vaginatum. Myrica* is preferentially frequent in this sub-community and locally abundant, but it is rather irregular in its representation and can be totally absent from some areas (as on Rhum, for example: McVean & Ratcliffe 1962). *Vaccinium myrtillus* and *Empetrum nigrum* ssp. *nigrum* hardly ever occur but *Erica cinerea* can be found occasionally in drier places, where the vegetation grades to the *Cladonia* sub-community.

Among the vascular associates, *Drosera rotundifolia* is strongly preferential here and particularly frequent around wetter hollows, where it can occasionally be accompanied by *D. anglica* or *D. intermedia.* In such situations, too, there can be a little *Carex echinata, C. pauciflora* or *C. limosa* and, in more obvious transitions to Rhynchosporion hollows, the *Drosera-Sphagnum*

sub-community terminates below in a fringe of *Rhynchospora alba,* among which there may be scattered plants of *Juncus bulbosus/kochii, Menyanthes trifoliata* and *Utricularia* spp. Areas with some soligenous influence can show a local abundance of *Carex panicea* and *Pinguicula vulgaris,* with scattered tussocks of *Schoenus nigricans.*

Sphagna are especially varied and extensive in this sub-community and, over undulating ground, they show an obvious zonation over the hummocks and transitions to hollows. The most abundant species are generally *S. capillifolium,* which is concentrated over the hummock sides and tops, and *S. papillosum,* which assumes dominance around the lower fringes of the hummocks and sometimes forms tussocky lawns in flatter wet areas, at around the level of the water-table. *S. compactum* occurs occasionally, mostly among the *S. capillifolium,* with *S. tenellum* and *S. subnitens* frequent among the *S. papillosum. S. auriculatum* and *S. cuspidatum* can also be found in the wetter zone, and it is these which become dominant with the switch to Rhynchosporion pool vegetation.

Leafy hepatics are common among the *Sphagnum* carpet, with *Pleurozia purpurea* and *Odontoschisma sphagni* strongly preferential to this sub-community, *Mylia anomala* and *M. taylori* also occurring occasionally. *Campylopus atrovirens* is also somewhat better represented here than elsewhere in the community and there are records, too, for *C. shawii* and *C. setifolius. Racomitrium lanuginosum* is also frequent, occurring generally as scattered shoots among the Sphagna right down to water-level, but, with the larger *Cladonia* spp., it does not make the prominent contribution to the ground cover typical of the next sub-community.

***Cladonia* spp. sub-community:** *Trichophoreto-Eriophoretum typicum p.p.* & *Rhacomitrium*-rich type McVean & Ratcliffe 1962; *Pleurozia purpurea-Erica tetralix* Association Br.-Bl. & Tx. 1952 *sensu* Moore 1968 *p.p.; Trichophoreto-Callunetum* Birks 1973 *p.p.; Calluno-Molinietum,* drier facies Hill & Evans 1978; Mire nodum 12 Daniels 1978; *Erico-Sphagnetum papillosi,* Typical subassociation Moore (1964) 1968 *emend.* Birse 1980 *p.p.;* Dry virgin mire 1.iii Hulme & Blyth 1984. *Calluna* and *Scirpus* are fairly consistent co-dominants in this sub-community, with *Molinia* and *Erica tetralix* generally playing a subordinate role and *Eriophorum vaginatum* distinctly patchy and usually of low cover. *Myrica* is scarce and never abundant but *Erica cinerea* becomes quite frequent and can be locally prominent. *Drosera rotundifolia* and species associated with transitions to Rhynchosporion hollows are uncommon but there are occasional records for Nardetalia herbs such as *Nardus stricta* and *Juncus squarrosus.*

The ground layer, too, shows distinctive features.

Most obviously, the *Sphagnum* carpet is much impoverished compared with the previous sub-community, with even *S. papillosum* much reduced in frequency and *S. tenellum*, *S. subnitens* and *S. auriculatum* very scarce, leaving *S. capillifolium* as the leading species and this often rather patchy. The associated leafy hepatics of the *Drosera-Sphagnum* sub-community are likewise uncommon here, though *Mylia taylori* and *Diplophyllum albicans* are preferential at low frequencies and locally prominent amongst decaying *Sphagnum* tussocks and on patches of exposed bare peat. Much more obvious, however, is the increased frequency and abundance of *Racomitrium lanuginosum* and *Cladonia* spp., with *C. impexa*, *C. uncialis* and *C. arbuscula* sometimes exceeding the Sphagna in their cover, and *C. coccifera* and *C. pyxidata* occurring less commonly as a crust on bare peat. *Hypnum cupressiforme/jutlandicum* is frequent and *Dicranum scoparium* and *Campylopus paradoxus* are occasionally found.

Juncus squarrosus-Rhytidiadelphus loreus sub-community: *Juncus squarrosus* bog   McVean & Ratcliffe 1962 *p.p.*; *Eriophorum vaginatum* bog, low-level facies   Edgell 1969; *Vaccinio-Eriophoretum*   Hill & Evans 1978 *p.p.*; *Calluna* mire 3.iv   Hulme & Blyth 1984 *p.p. Calluna* and *Scirpus* are again the usual vascular dominants in this kind of *Scirpus-Eriophorum* mire, with *Erica tetralix* and especially *Molinia* reduced in frequency and abundance. *Eriophorum vaginatum*, however, is more common than in the last sub-community and locally of quite high cover. *Myrica* is absent and *Erica cinerea* very scarce but the sub-shrub component is frequently enriched by small amounts of *Vaccinium myrtillus* and, rather less commonly, by *Empetrum nigrum* ssp. *nigrum* and it is in this sub-community that the very few records for *V. vitis-idaea*, *E. nigrum* ssp. *hermaphroditum* and *Rubus chamaemorus* in the *Scirpus-Eriophorum* mire generally occur.

Associated with these features is a marked increase in the frequency of *Juncus squarrosus*, *Nardus stricta*, *Deschampsia flexuosa* and *Carex nigra* with, somewhat less common but still preferential, *Agrostis canina* ssp. *canina*, *Anthoxanthum odoratum* and *Luzula multiflora*. The first four of these species can be found in some moderate abundance but typically they all occur as scattered tufts.

As in the *Cladonia* sub-community, the *Sphagnum* cover consists of but few species with only *S. capillifolium* and *S. papillosum* being frequent and *S. subnitens* occurring occasionally. Here, however, *S. papillosum* is usually the most abundant member of the suite, being best represented in stands where *Erica tetralix* and *Eriophorum vaginatum* also show some prominence and where Nardetalia herbs are more sparse. *Racomitrium* is quite frequent but *Cladonia* spp. are uncommon and the

most distinctive feature of the ground layer is the strong contingent of pleurocarpous and some acrocarpous mosses. *Hypnum cupressiforme/jutlandicum*, *Rhytidiadelphus loreus*, *Pleurozium schreberi* and *Dicranum scoparium* are all very frequent, *Polytrichum commune*, *P. alpestre*, *Plagiothecium undulatum*, *Aulacomnium palustre*, *Ptilidium ciliare*, *Pohlia nutans* and *Campylopus paradoxus* more occasional but still preferential. *Lophocolea bidentata s.l.* and *Lophozia ventricosa* can also sometimes be found. As with most of the preferential herbs, these plants are generally not abundant but the total effect they create can give the vegetation a quite different look from other kinds of *Scirpus-Eriophorum* mire.

Habitat
The *Scirpus-Eriophorum* mire is the characteristic blanket bog vegetation of the more oceanic parts of Britain, occurring extensively on waterlogged ombrogenous peat that has accumulated in the consistently humid climate of the far-west. It is essentially a community of lower altitudes and the composition of the vegetation reflects the relative mildness of the climate, but floristics and structure have also been widely affected by a variety of treatments, including burning, grazing, draining and peat-cutting, and these have often contributed, perhaps with climatic change, to the deterioration and loss of the community.

Blanket peat in Britain is confined to those parts of the country with a consistently wet climate, generally where there are more than 1200 mm of precipitation annually (*Climatological Atlas* 1952) or, more precisely, over 160 wet days yr^{-1} (Ratcliffe 1968, 1977), and where cool and cloudy conditions help maintain high humidity throughout the year, restricting even summer potential water deficit to near-zero. Within this zone, which corresponds by and large with the western and northern uplands, the *Scirpus-Eriophorum* mire is characteristic of lower altitudes where extreme humidity is combined with relative mildness of winter climate. It is most extensive in areas with more than 200 wet days yr^{-1} and over 2000 mm of rain but largely restricted to sites below 500 m, where the annual temperature range is comparatively small. Its range is thus centred in the lower hills of the western Highlands of Scotland and in the Isles, where accumulation of blanket peat has been especially widespread, extending down almost to sea-level, on to moderately steep slopes and over pervious and non-acidic substrates. In areas with a drier climate, such topographic and geological factors increasingly inhibit the development of a peat mantle so, to the south and east, the community tends to be of more restricted occurrence and to penetrate to somewhat higher altitudes where moderately heavy rainfall is maintained. It can be found in south-west Scotland, the Lake District and Wales and on Dartmoor and Bodmin Moor where there are 180–200 wet days yr^{-1} and it occurs locally in the eastern Highlands. But the effect of decreasing oceanicity of climate is especially marked in traversing Scotland and, where blanket peat has accumulated in the wet but harsh climate of high plateaus, both there and all down the Pennines, the *Scirpus-Eriophorum* mire is replaced by the *Calluna-Eriophorum* mire. Altitudinal separation of the two communities is fairly well maintained throughout their ranges: the mean level of the former is around 300 m, that of the latter over 550 m.

Over the blanket peats of the oceanic zone, the *Scirpus-Eriophorum* mire is typically found on deposits that are maintained in a permanently waterlogged state by a high and generally stagnant water-table. It thus usually occurs on deeper peats (2–4 m or so) over flat or gently-sloping ground (mean 4°, range 0–25°), on broader valley bottoms and their immediate surrounds and on low-level plateaus and watersheds. In more rugged country, as in the west-central Highlands and the Lake District, it therefore tends to be of rather restricted occurrence; whereas, on extensive plains, like those of the Sutherland flow country, stands of the community can stretch virtually uninterrupted for many square kilometres, making a major contribution to a bleak landscape that is almost unrivalled in scale through the whole of western Europe (McVean & Ratcliffe 1962, Ratcliffe 1977).

The peats show varying degrees of humification but are typically highly acidic, with a surface pH usually not much above 4 and often less, and very impoverished. The difference which permanent waterlogging of such substrates makes to the character of the vegetation is best seen by comparing the *Scirpus-Eriophorum* mire with the *Scirpus-Erica* wet heath, the Ericetalia community which has virtually the same oceanic distribution, but which is characteristic of better-drained and usually shallower peats (often less than 2 m) on steeper slopes (mean 8°, range 0–42°). The two vegetation types have many species in common, including potential Oxycocco-Sphagnetea dominants like *Scirpus*, *Molinia* and *E. tetralix*, and also *Calluna*, *Potentilla erecta*, *Polygala serpyllifolia* and *Pedicularis sylvatica*. But the switch to Sphagnetalia vegetation in the very wet conditions here is marked by the great increase in the importance of Sphagna to the composition of the vegetation and the maintenance of a thick organic substrate. In the *Scirpus-Erica* wet heath, the *Sphagnum* carpet is rather patchy and *S. capillifolium* is generally the leading species, with increase in abundance and diversity in this element of the vegetation usually being associated with local soligenous influence. In the *Scirpus-Eriophorum* mire, by contrast, fairly luxuriant *Sphagnum* cover is the rule with *S. papillosum*, a species of restricted occurrence in the *Scirpus-Erica* wet heath, becoming of major

significance as a peat-builder on the mire plane, tending to dominate the carpet around the level of the water-table, with other species disposed among it or above and below it where there is differentiation of surface relief. The occurrence of hummocks and hollows here, and of full transitions to Rhynchosporion vegetation in pools, is in fact very variable: it tends to be more pronounced on bogs in the extreme oceanic zone of which the *Scirpus-Eriophorum* mire is most typical, but even there can be quite local. How it develops and what part it plays in bog growth are subjects of considerable discussion: deep peat-cuttings sometimes reveal stratigraphical patterns which suggest a cyclical alternation of hummock and hollow at particular points (e.g. Osvald 1949), but this is not always seen. What does seem certain is that Sphagna have generally made the major contribution to autogenic peat accumulation here, at least on deeper, level bogs (McVean & Ratcliffe 1962).

One other important floristic distinction between the *Scirpus-Eriophorum* mire and the *Scirpus-Erica* wet heath, and likewise signalling the shift from Ericetalia to Sphagnetalia vegetation with more consistent waterlogging, is the constancy of *Eriophorum vaginatum*. However, although remains of this species, together with *E. angustifolium*, *Scirpus* and the ericoids, can be seen in shallower peats where the community extends on to sloping ground (McVean & Ratcliffe 1962), *E. vaginatum* itself is not usually very abundant: it is locally prominent on hummocks but, as noted above, is nothing like so consistently important as in the *Calluna-Eriophorum* mire. This is probably a response to climate: *E. vaginatum* is a circumpolar plant and, in the very extreme oceanic climate of western Ireland, it becomes an even more insignificant component of blanket mire vegetation (Tansley 1939, Moore 1968, Doyle & Moore 1980, Doyle 1982).

The influence of oceanicity on the floristics of the *Scirpus-Eriophorum* mire can be seen in a number of other features of the community which set it apart from the *Calluna-Eriophorum* mire. First, there is the almost total exclusion at these lower altitudes of the Arctic-Subarctic *Rubus chamaemorus* and the Arctic-Alpine *Vaccinium vitis-idaea*, *V. uliginosum* and *Empetrum nigrum* ssp. *hermaphroditum*. Second, on the positive side, there is the constancy and abundance of *Molinia* and the frequent occurrence of *Myrica*, essentially a lowland species, in the *Scirpus-Eriophorum* mire. *E. tetralix* is also better represented here and *E. cinerea* figures on drier peats in the community: both these sub-shrubs are physiologically active in winter and somewhat oceanic in their British distribution (Gimingham 1972). Then, there are differences among the bryophytes, not so much among the Sphagna, where only *S. auriculatum* and the low-frequency *S. strictum* and *S. imbricatum* show preferences for a western climate (Ratcliffe 1968), but in species like *Odontoschisma sphagni*, *Pleurozia purpurea*, *Campylopus atrovirens* and *C. setifolius*. The last three of these tend overall to be more characteristic of wet rocks and banks but, in the extremely wet climate of western Britain, they move in to the blanket mire habitat (Ratcliffe 1968). Various of these floristic trends continue into the blanket mires of western Ireland where, in the *Pleurozio-Ericetum*, *Molinia* is often very prominent and *C. atrovirens* and the leafy hepatics of high frequency (Braun-Blanquet & Tüxen 1952, Moore 1968, Doyle & Moore 1980, Doyle 1982). One important difference between this Irish vegetation and the *Scirpus-Eriophorum* mire is that, in the very oceanic climate typical of the former, *Schoenus nigricans* becomes very common and often abundant on the mire plane, more or less as a physiognomic replacement for *E. vaginatum*. Within the range of the *Scirpus-Eriophorum* mire, by contrast, *Schoenus* is usually confined to soligenous areas where there is some local amelioration of the high concentrations of aluminium ions, to which it is very sensitive (Sparling 1962*b*, 1967*a*, *b*, 1968). It occurs very occasionally in the *Drosera-Sphagnum* sub-community, often with species like *Carex panicea* and *Pinguicula vulgaris*, indicative of modest base-enrichment (pH rising to above 5) but, for the most part, *Schoenus* flushes within tracts of the *Scirpus-Eriophorum* mire are best seen as a particular kind of soligenous *Scirpus-Erica* wet heath.

Much of the floristic character of the *Scirpus-Eriophorum* mire can thus be understood in relation to a gradient of oceanicity, which shows a fairly sudden rise with the altitudinal shift from cold, wet higher plateaus to the more equable western sea-board, and then a further, but gradual increase with the geographical move to western Ireland. Other features of the community seem to be related to the particular hydrological conditions that pertain within blanket peat as opposed to the ombrogenous deposits in raised bogs because, in the latter, the *Scirpus-Eriophorum* mire tends to be replaced by the *Erica-Sphagnum papillosum* mire. This community is concentrated in a somewhat less oceanic zone than the *Scirpus-Eriophorum* mire where massive peat accumulation is partly dependent upon the existence of a topogenous base developed in waterlogged hollows, but there is a considerable geographical and altitudinal overlap between the two communities and their habitats are not sharply separated: within stretches of blanket mire, local areas can have something of the character of raised bog, as on watershed cols or where flows occur over deeper, drift-lined basins, and some raised mires are so extensive as to be locally like blanket bogs (Ratcliffe & Walker 1958, McVean & Ratcliffe 1962, Ratcliffe 1977). In such situations, the communities grade one into another, but one character of fairly general significance for separating them seems to be the preference of *Sphagnum magellanicum* for the *Erica-*

Sphagnum mire: Typically, this is not a major peat-builder in the *Scirpus-Eriophorum* mire. Also, two Continental Northern plants, *Vaccinium oxycoccos* and *Andromeda polifolia*, have their distributions very much centred on lowland, raised bogs of the *Erica-Sphagnum* type, and their conspicuous absence from the *Scirpus-Eriophorum* mire has often been remarked on (e.g. Birks 1973, Moore 1968, Doyle & Moore 1980, Doyle 1982).

Floristic variation within the *Scirpus-Eriophorum* mire can be related in part to differences in the factors already outlined. The *Drosera-Sphagnum* sub-community constitutes the core of the community, occurring throughout the range on the wettest peats, but being especially well developed in areas of highest rainfall and mildest climate. It comprises the bulk of the cover of oceanic blanket bog where this occurs as flat or gently undulating lawns, with *Sphagnum papillosum* and species like *Narthecium ossifragum* and *Drosera rotundifolia* becoming especially conspicuous, and, on mires with more pronounced surface relief, it includes most of the vegetation between the Rhynchosporion pools and the tops of the taller and drier hummocks, with an extensive and rich *Sphagnum* cover disposed in relation to the height of the water-table. It can also include bog runnels in which there is some slight soligenous influence and a modest representation of Caricion nigrae species.

The greater the proportion of drier peats within stretches of the *Scirpus-Eriophorum* mire, the more important is the contribution of the *Cladonia* sub-community. On virgin bogs within the oceanic zone, such a habitat is provided by the tops of the taller hummocks which are far removed from the direct influence of the water-table, so here there is a shift towards *Calluna* and *Scirpus* among the vascular dominants and away from the massive and diverse *Sphagnum* cover typical of wetter situations, its place in the ground layer being taken by *Racomitrium lanuginosum* and larger *Cladonia* spp., with sporadic occurrence of Nardetalia plants. *Erica cinerea* is also able to colonise, being perhaps less restricted on these ombrogenous peats than it is around Sphagnetalia valley bogs in the English lowlands by lower concentrations of toxic ferrous ions (Jones 1971*a*, *b*). Greater surface dryness is also a feature of virgin *Scirpus-Eriophorum* mire developed in areas of somewhat drier climate and many stands of the community in the zone of 180–200 wet days yr^{-1}, notably in south-west Scotland, are predominantly of the *Cladonia* type.

Some natural climatic change to drier atmospheric conditions may also have contributed to the development of the *Cladonia* sub-community on deeper peats but, very often, surface drying of these blanket peats has been accentuated (perhaps sometimes initiated) by treatment and this kind of *Scirpus-Eriophorum* mire has become very extensive, even within areas that still experience an extremely humid climate, because of burning, peat-cutting and draining (e.g. McVean & Ratcliffe 1962, Hulme & Blyth 1984). Burning stretches of the community has a particularly drastic effect on the *Sphagnum* cover, even very wet carpets becoming susceptible to fire-damage in periods of drier weather in spring and summer, and it produces just the kind of dominance by *Scirpus* tussocks and *Racomitrium* hummocks so characteristic of some tracts of the *Cladonia* sub-community (McVean & Ratcliffe 1962). Marginal wastage of relatively undisturbed mantles of this oceanic blanket peat is not so pronounced as in the *Calluna-Eriophorum* mire but, where peat-cutting has occurred, or where there has been some attempt at marginal reclamation for grazing, the *Cladonia* sub-community can spread over drying baulks or the fretted margins of the bogs (e.g. Hulme & Blyth 1984).

Burning, and perhaps grazing, may also have contributed to the distinctive character of the *Juncus-Rhytidiadelphus* sub-community, but this, too, shows some relationship to natural differences in climate. Thus, though many of its preferentials are plants which become common in poor-quality Nardetalia hill-grazings, notably *Juncus squarrosus* and *Nardus stricta*, others are also very frequent plants in the *Calluna-Eriophorum* mire, like *Vaccinium myrtillus*, *Deschampsia flexuosa* and the hypnaceous mosses, and this sub-community includes some stretches of fairly undisturbed blanket bog occurring in environments intermediate between those of the two communities. Most of its occurrences are outside the very oceanic parts of western Britain, in eastern and south-west Scotland, and are at altitudes which have a mean some 250 m above those of the *Drosera-Sphagnum* sub-community. Here, *Eriophorum vaginatum* can assume a greater importance in the vegetation cover and there are very occasional records for such characteristic *Calluna-Eriophorum* associates as *Rubus chamaemorus*, *Vaccinium vitis-idaea* and *V. uliginosum*.

Zonation and succession

Zonations between the different kinds of *Scirpus-Eriophorum* mire and to other vegetation types are most often related to the height of the water-table and the degree of soligenous influence within stretches of blanket peat. Effects of treatment can overlie such transitions and they may precipitate a run-down of the vegetation through wet heath to dry heath and grassland. Without disturbance or any natural shift in the extreme oceanic conditions, the community subsists as a climatic climax.

Internal vegetational patterning on virgin tracts of oceanic blanket bog is most commonly related to the differentiation of surface microrelief. Then, the *Drosera-Sphagnum* sub-community occupies the bulk of the wetter ground, with the *Cladonia* sub-community picking out the tops of the drier hummocks, and Rhynchos-

porion vegetation occurring in the pools. Typically, within the range of the *Scirpus-Eriophorum* mire, this latter is represented by the *Sphagnum auriculatum* community: species like *Eriophorum angustifolium*, *Molinia*, *Narthecium* and *Drosera rotundifolia* may run some way into this vegetation, but there is a pronounced shift in the *Sphagnum* carpet to dominance of *S. auriculatum* and *S. cuspidatum* and stands are commonly marked by a fringe of *Rhynchospora alba*. It is in this kind of situation that species such as *Hammarbya paludosa* and *Scheuchzeria palustris* are recorded within tracts of *Scirpus-Eriophorum* mire. The clarity of differentiation of these pools, and the proportion of the mire surface occupied by their Rhynchosporion vegetation, vary considerably: they are best developed and most extensive in the zone with more than 200 wet days yr^{-1}, becoming less important in the southern part of the range of the community, as in Wales and on Dartmoor (Lindsay *et al.* 1984). But, even in the far north-west, their occurrence is quite variable and seems to be related to local accumulation of waters over depressions in the underlying ground or where there is channelling from the mire surrounds. Quite commonly, it is in just such situations that the *Scirpus-Eriophorum* mire approaches most closely to the *Erica-Sphagnum papillosum* mire, and is sometimes replaced by it (as in certain of the Silver Flowe mires: Ratcliffe & Walker 1958). The relationships between the two communities in these habitats, which are transitional between blanket and raised bogs, need further investigation.

Where there is a thinning of the cover of blanket peat and better drainage, as on steeper ground around valley bottoms or where plateaus give way to fringing hills above or slopes below, the *Scirpus-Eriophorum* mire is typically replaced throughout its range by the *Scirpus-Erica* wet heath. Some important vascular species, like *Scirpus*, *Molinia*, *Calluna* and *E. tetralix*, run on into this vegetation and their dominance throughout may mask other floristic changes, but *Eriophorum vaginatum* declines greatly in frequency, the *Sphagnum* carpet loses its variety and luxuriance, and any differentiation of surface relief is lost as the peat cover becomes drier and thinner. Such zonations can be quite abrupt where there is a fairly marked change of slope, but often they are gradual and the *Sphagnum* sub-community of the *Scirpus-Erica* wet heath, with its modest frequencies of *S. papillosum* and *E. vaginatum*, may then form a transitional zone. And the relative proportions of the two communities vary with regional climate and local topographical modification of it: in drier areas or on south-facing slopes, the whole sequence tends to move downslope, the mire becoming more confined to the flattest ground, the wet heath more extensive.

Such general zonations are complicated by soligenous influence which tends to cut across the transitions down the lines of steeper slope. Most often, in these oceanic blanket mires, areas of more pronounced seepage are marked by the *Carex* sub-community of the *Scirpus-Erica* wet heath, which can form quite extensive stands over slopes with some through-put or narrow, sinuous strips along obvious soakways. These can run through both the wet heath and the *Scirpus-Eriophorum* mire and then out of the bog along the silty margins of streams. They are frequently marked by a local dominance of *Molinia* and *Myrica* but more open stands can provide the most usual locus for *Schoenus nigricans* within British blanket mire.

Altitudinal zonations from the *Scirpus-Eriophorum* mire to the *Calluna-Eriophorum* mire are not very frequent because low- and high-level stretches of flatter ground are often separated by intervening slopes: in the north-west Highlands, for example, the two communities are separated by a zone of the *Scirpus-Erica* wet heath on better-drained blanket peat. But, in some places, generally between 300 and 450 m, the two can grade imperceptibly one into the other and, in eastern Scotland particularly, the *Juncus-Rhytidiadelphus* sub-community represents an intermediate kind of blanket mire.

Treatments, among which burning has probably been of special importance, can modify all these kinds of zonations and induce successional changes in the *Scirpus-Eriophorum* mire. In some cases, where differences of surface-drainage are very marked, burning may actually sharpen up the vegetation boundaries by allowing different species to gain ascendancy on wetter or drier ground: *Eriophorum vaginatum*, for example, may become locally dominant after fire in the *Scirpus-Eriophorum* mire but not in the adjoining heath. In other cases, burning may impose a fairly uniform dominance of *Scirpus* or *Molinia* throughout the sequence, blurring zonations among the associates, and some tracts of such fire-climax vegetation may have been partly or wholly derived from the community.

But, apart from such modifications of dominance, burning has probably played a major part, along with marginal peat-cutting and draining, in the surface-drying of the peats that precipitates more dramatic changes in the vegetation. The *Cladonia* sub-community can represent the first stage in such a development which, with the final elimination of *E. vaginatum* and further impoverishment of the *Sphagnum* carpet, perhaps moves to the *Cladonia* sub-community of the *Scirpus-Erica* wet heath. *Erica cinerea* is the potential vascular dominant that seems to gain ascendancy on such drying peats in more oceanic regions (e.g. Goode & Lindsay 1979, Hulme & Blyth 1984) and *Calluna-Erica* heath may represent an end point in such a run-down. At higher altitudes, in areas with a somewhat drier climate, an analogous trend may involve the conversion of the *Juncus-Rhytidiadelphus* sub-community of the

Scirpus-Eriophorum mire to the *Vaccinium* sub-community of the *Scirpus-Erica* wet heath, where *Vaccinium myrtillus* can become an important sub-shrub perhaps presaging a switch to *Calluna-Vaccinium* heath. Certainly, complex mosaics of intermediate stages in such processes, approximating to various kinds of *Scirpus-Erica* heath, with fragments of *Scirpus-Eriophorum* mire, are of widespread occurrence on deeper peats than would naturally be expected to be clothed with extensive tracts of the latter. Grazing and improvement may take the process further beyond the dry-heath phase to grasslands of various kinds: on Lewis, for example, the crofting townships are fringed by Junco-Molinion swards and *Lolio-Cynosuretum* that have been derived from blanket mire by top-sowing and the addition of shell-sand and ratio fertilisers (Hulme & Blyth 1984).

It is possible that natural climatic change has played some part in the degeneration of blanket peats occupied by the *Scirpus-Eriophorum* mire. This kind of ombrogenous bog appears to have been initiated locally at the Boreal/Atlantic transition about 7000 years ago and to have resumed rapid growth following climatic deterioration between 600 BC to 500 AD, often spreading to replace forest, tree stumps of which are frequently preserved beneath the peat. However, despite some subsequent amelioration of the climate, it probably remains a climax vegetation type in more oceanic parts of the country.

Distribution

The *Scirpus-Eriophorum* mire is largely confined to western Britain, being especially widespread in the western Highlands of Scotland and the western Isles and running down through south-west Scotland, the Lake District, Wales and south-west England. The *Drosera-Sphagnum* sub-community occurs throughout the range, but is particularly extensive in north-west Scotland. The *Cladonia* and *Juncus-Rhytidiadelphus* sub-communities also occur in the west, but they extend the range of the community on to drier peats, in areas with lower rainfall and at higher altitudes, most notably in south-west and eastern Scotland.

Affinities

As defined here, the *Scirpus-Eriophorum* mire represents an expanded version of the vegetation type first described as *Trichophoreto-Eriophoretum typicum* by McVean & Ratcliffe (1962) and later by Birks (1973) and Evans *et al.* (1977), though not by Eddy *et al.* (1969) whose community of that name can be largely subsumed within the *Calluna-Eriophorum* mire. In these schemes *Trichophoreto-Eriophoretum* consists largely of the vegetation included here in the *Drosera-Sphagnum* sub-community, though some samples approach the *Cladonia* sub-community in their composition. The *Calluno-*

Molinietum of Hill & Evans (1978), although named by the pattern of dominance, is essentially *Scirpus-Erio-phorum* mire and it very obviously takes in both these sub-communities, which these authors distinguished as wetter and drier facies. And their *Vaccinio-Eriophore-um* corresponds closely with the *Juncus-Rhytidiadel-phus* sub-community, vegetation previously given only scant recognition in, for example, part of the *Juncus squarrosus* bog of McVean & Ratcliffe (1962). The study of Hebrides mires by Hulme & Blyth (1984) provides a local definition of all three sub-communities, together with transitions to degraded and improved peatland vegetation.

Many accounts of British ombrogenous bogs follow McVean & Ratcliffe (1962) in recognising just two major communities, diagnosed here as the *Scirpus-Eriophorum* mire and the *Calluna-Eriophorum* mire (an expanded version of their *Calluneto-Eriophoretum*). These grade one into the other, through the *Juncus-Rhytidiadelphus* sub-community of the former, but are generally well defined, both floristically and environmentally, the former as our major low-altitude and more oceanic blanket bog, the latter the predominant type of higher altitudes and less oceanic climates. The situation has been complicated by the recognition of a third type of ombrogenous mire, best represented in low-altitude raised bogs in moderately oceanic areas, but showing some geographical overlap with and floristic transition to both the communities. This is the vegetation type which Moore (1968) termed the *Erico-Sphagnetum magellanici*, and which is here called the *Erica-Sphag-num papillosum* mire. Some early accounts of British mires include ombrogenous vegetation which is essen-tially of this kind (e.g. Godwin & Conway 1939, Pearsall 1941, Ratcliffe & Walker 1958), though they did not distinguish it explicitly from the *Scirpus-Eriophorum* type. In fact, the separation of the two communities can be difficult, particularly if individual samples are exa-mined in isolation but, in their typical forms, they are quite distinct: the contrast between them has sometimes been recognised in the conspicuous absence from the

Scirpus-Eriophorum mire of *Vaccinium oxycoccos* and *Andromeda polifolia* (e.g. Birks 1973). Some recent studies have, however, confounded distinctions between the communities: the *Erica-Sphagnum magellanicum* mire of Ratcliffe & Hattey (1982), for example, includes samples better seen as *Scirpus-Eriophorum* mire (as well as much *Scirpus-Erica* wet heath) and the *Erico-Sphag-netum papillosi* of Birse (1980) takes in parts of the *Scirpus-Eriophorum*, *Erica-Sphagnum papillosum* and *Calluna-Eriophorum* mires.

The general similarity between the *Scirpus-Erio-phorum* mire and the blanket bog vegetation of western Ireland has long been recognised (e.g. Tansley 1911, 1939, Osvald 1949, McVean & Ratcliffe 1962) and the two were grouped together by Moore (1968: see also Doyle & Moore 1980) in a single association, the *Pleurozio-Ericetum tetralicis*, first defined from Eire by Braun-Blanquet & Tüxen (1952). Subsequent detailed description of Irish stands of this vegetation type (Doyle 1982) has emphasised the very close relationships, but we have preferred here to maintain a distinction on the basis of the poorer representation in the *Scirpus-Erio-phorum* mire of *Schoenus nigricans*, *Rynchospora alba*, *Drosera anglica*, *Campylopus atrovirens* and the numer-ous algae grouped as *Zygogonium ericetorum*, and its higher frequency of *Eriophorum vaginatum*.

Nonetheless, the *Scirpus-Eriophorum* mire and its Irish counterpart together clearly represent an oceanic extreme within the western European mires of the Sphagnetalia which Moore (1968), following Schwick-erath (1940) and Duvigneaud (1949), grouped into a single alliance, which he termed the Erico-Sphagnion, distinguished from the Sphagnion fusci peatland com-munities of the central European uplands and Scandina-via by the absence of Arctic-Alpine and Boreal plants. In the *Scirpus-Eriophorum* mire, the floristic boundary between the Sphagnetalia and the wet-heath vegetation of the Ericetalia is fairly well defined, though the community is often found in contact with the *Scirpus-Erica* wet heath and can be converted to it with drying of the peats.

Floristic table M17

	a	b	c	17
Scirpus cespitosus	V (1–8)	V (1–8)	V (1–9)	V (1–9)
Calluna vulgaris	V (1–7)	V (1–8)	V (1–9)	V (1–9)
Erica tetralix	V (1–5)	V (1–6)	III (1–6)	V (1–6)
Narthecium ossifragum	V (1–5)	V (1–6)	IV (1–9)	V (1–9)
Eriophorum angustifolium	IV (1–6)	IV (1–5)	IV (1–9)	IV (1–9)
Eriophorum vaginatum	IV (1–8)	III (1–6)	IV (1–5)	IV (1–8)
Potentilla erecta	IV (1–4)	III (1–4)	IV (1–4)	IV (1–4)

Floristic table M17 *(cont.)*

	a	b	c	17
Sphagnum capillifolium	V (1–8)	IV (1–6)	III (1–4)	IV (1–8)
Sphagnum papillosum	IV (1–9)	II (1–8)	IV (1–6)	IV (1–9)
Molinia caerulea	V (1–8)	IV (1–7)	II (1–6)	IV (1–8)
Drosera rotundifolia	IV (1–4)	II (1–4)	I (1–3)	III (1–4)
Sphagnum subnitens	III (1–4)	I (1–4)	II (1–6)	III (1–6)
Pleurozia purpurea	III (1–4)	I (1–3)	I (1–3)	II (1–4)
Sphagnum tenellum	III (1–4)	I (1–4)	I (1–4)	II (1–4)
Odontoschisma sphagni	III (1–3)	I (1–4)	I (1–4)	II (1–4)
Myrica gale	III (1–6)	I (1–8)		II (1–8)
Carex echinata	II (1–4)	I (1–3)	I (1–3)	I (1–4)
Sphagnum auriculatum	II (1–4)	I (1–4)	I (1–3)	I (1–4)
Sphagnum palustre	II (1–5)	I (1–6)	I (1–4)	I (1–6)
Sphagnum compactum	II (1–5)	I (1–2)		I (1–5)
Drosera anglica	I (1–4)			I (1–4)
Schoenus nigricans	I (1–4)			I (1–4)
Racomitrium lanuginosum	III (1–6)	IV (1–10)	II (1–5)	III (1–10)
Cladonia uncialis	II (1–5)	IV (1–4)	I (1–3)	II (1–5)
Cladonia impexa	II (1–5)	IV (1–10)	I (1–3)	II (1–10)
Hypnum cupressiforme	I (1–4)	III (1–7)	II (1–4)	II (1–7)
Cladonia arbuscula	I (1–3)	II (1–9)	I (1–3)	I (1–9)
Erica cinerea	I (1–4)	II (1–6)	I (1–3)	I (1–6)
Mylia taylori	I (1–3)	II (1–4)	I (1)	I (1–4)
Diplophyllum albicans	I (1)	II (1–4)		I (1–4)
Hylocomium splendens	I (1–2)	II (1–4)		I (1–4)
Luzula multiflora		II (1–4)		I (1–4)
Juncus squarrosus	I (1–4)	II (1–4)	IV (1–4)	II (1–4)
Hypnum jutlandicum	I (1–3)	II (1–4)	III (1–4)	II (1–4)
Nardus stricta	I (1–2)	II (1–4)	III (1–4)	II (1–4)
Rhytidiadelphus loreus	I (1–4)	I (1–3)	III (1–4)	II (1–4)
Deschampsia flexuosa	I (1–3)	I (1)	III (1–4)	II (1–4)
Vaccinium myrtillus		I (1–3)	III (1–4)	I (1–4)
Pleurozium schreberi		I (1)	III (1–4)	I (1–4)
Dicranum scoparium		I (1–2)	III (1–3)	I (1–3)
Empetrum nigrum nigrum	I (1–3)	I (1–2)	II (1–4)	I (1–4)
Polytrichum commune	I (1–4)	I (1–4)	II (1–6)	I (1–6)
Carex nigra	I (1)	I (1)	II (1–7)	I (1–7)
Agrostis canina canina		I (1)	II (1–4)	I (1–4)
Plagiothecium undulatum		I (1)	II (1–4)	I (1–4)
Aulacomnium palustre			II (1–3)	I (1–3)
Polytrichum alpestre			II (1–4)	I (1–4)
Vaccinium vitis-idaea			I (1–3)	I (1–3)
Anthoxanthum odoratum			I (1–3)	I (1–3)
Galium saxatile			I (1–2)	I (1–2)
Rhytidiadelphus squarrosus			I (1–4)	I (1–4)

Lophozia ventricosa			I (1–3)	I (1–3)
Pohlia nutans			I (1–2)	I (1–2)
Ptilidium ciliare			I (1–3)	I (1–3)
Polygala serpyllifolia	II (1–4)	II (1–3)	I (1–3)	II (1–4)
Sphagnum cuspidatum	I (1–4)	I (1–7)	I (1–4)	I (1–7)
Campylopus atrovirens	I (1–2)	I (1–5)	I (4)	I (1–5)
Huperzia selago	I (1–3)	I (1–4)	I (1)	I (1–4)
Cornicularia aculeata	I (1–2)	I (1–3)	I (1–3)	I (1–3)
Pedicularis sylvatica	I (1–2)	I (1–2)	I (1–2)	I (1–2)
Mylia anomala	I (1–2)	I (1–3)	I (1–3)	I (1–3)
Breutelia chrysocoma	I (1)	I (1–3)	I (1–3)	I (1–3)
Pinguicula vulgaris	I (1–4)	I (1–5)	I (1)	I (1–5)
Carex panicea	I (1–4)	I (1–3)	I (1–3)	I (1–4)
Festuca ovina	I (1–3)	I (1–3)	I (1–4)	I (1–4)
Campylopus paradoxus	I (1–2)	I (1–3)	I (1–3)	I (1–3)
Calypogeia fissa	I (1–4)	I (1–2)	I (1–2)	I (1–4)
Cladonia coccifera	I (1–2)	I (1–3)	I (1–3)	I (1–3)
Scapania gracilis	I (1–3)	I (1–3)	I (1–3)	I (1–3)
Juncus acutiflorus	I (1–4)	I (1–3)	I (1–3)	I (1–4)
Carex demissa	I (1–3)	I (1–3)	I (1–3)	I (1–3)
Sphagnum magellanicum	I (1–4)	I (1–4)		I (1–4)
Campylopus setifolius	I (1–4)	I (1)		I (1–4)
Campylopus shawii	I (1–3)	I (1–3)		I (1–3)
Sphagnum imbricatum	I (1–3)	I (1–3)		I (1–3)
Kurzia pauciflora	I (1–3)	I (1–3)		I (1–3)
Rhynchospora alba	I (1–2)	I (1–3)		I (1–3)
Sphagnum strictum	I (1–3)	I (1–3)		I (1–3)
Leucobryum glaucum	I (1–3)	I (1–3)		I (1–3)
Number of samples	97	53	44	194
Number of species/sample	21 (10–37)	17 (8–31)	20 (11–38)	20 (8–38)
Herb height (cm)	21 (12–33)	17 (4–25)	20 (6–30)	19 (4–33)
Herb cover (%)	96 (65–100)	92 (60–100)	94 (70–100)	95 (60–100)
Bryophyte height (mm)	42 (20–100)	35 (5–70)	33 (10–50)	37 (5–100)
Bryophyte cover (%)	56 (30–90)	46 (5–90)	41 (2–80)	49 (2–90)
Altitude (m)	216 (8–524)	328 (15–686)	470 (150–880)	304 (8–880)
Slope (°)	4 (0–25)	4 (0–18)	5 (0–24)	4 (0–25)
Soil pH	4.5 (3.6–6.0)	4.4 (3.5–6.4)	4.4 (3.2–6.7)	4.4 (3.2–6.7)

a   *Drosera rotundifolia-Sphagnum* sub-community

b   *Cladonia* sub-community

c   *Juncus squarrosus-Rhytidiadelphus loreus* sub-community

17   *Scirpus cespitosus-Eriophorum vaginatum* blanket mire (total)

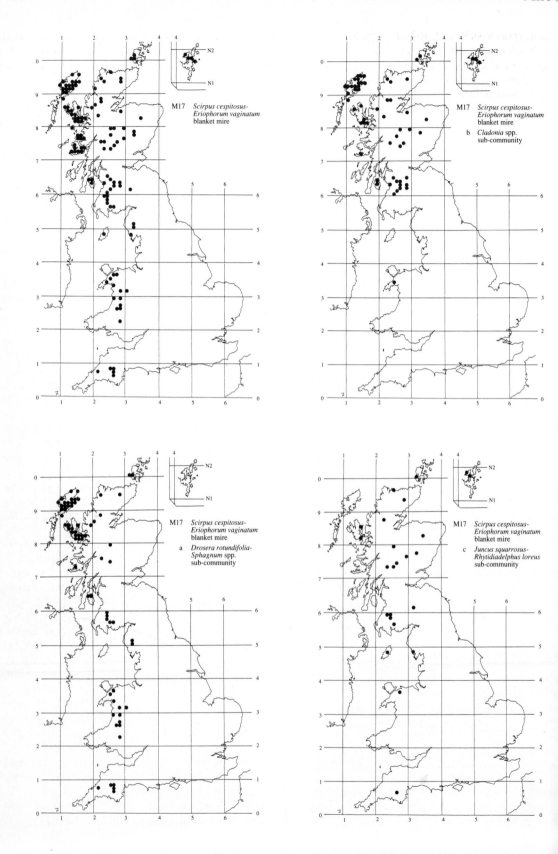

M17 *Scirpus cespitosus-*
Eriophorum vaginatum
blanket mire

M17 *Scirpus cespitosus-*
Eriophorum vaginatum
blanket mire
b   *Cladonia* spp.
sub-community

M17 *Scirpus cespitosus-*
Eriophorum vaginatum
blanket mire
a   *Drosera rotundifolia-*
Sphagnum spp.
sub-community

M17 *Scirpus cespitosus-*
Eriophorum vaginatum
blanket mire
c   *Juncus squarrosus-*
Rhytidiadelphus loreus
sub-community

M18
Erica tetralix-Sphagnum papillosum raised and blanket mire

Synonymy

Eriophoretum vaginati Rankin 1911*a*; *Sphagnetum* Regeneration Complex Tansley 1939, Godwin & Conway 1939 *p.p.*; Marginal *Sphagnetum* Godwin & Conway 1939; *Sphagnum* community, Type A 'moss' Pearsall 1941; Dissected moss Pearsall 1941; *Pleurozia purpurea-Erica tetralix* Assoziation, Sub-assoziation von *Andromeda polifolia* Braun-Blanquet & Tüxen 1952; Flat & hummock communities Ratcliffe & Walker 1958; *Erico-Sphagnetum magellanici*, Sub-Atlantic race Moore 1968 *p.p.*; *Trichophoreto-Eriophoretum*, Typical facies Eddy *et al.* 1969; *Sphagnum papillosum-Erica tetralix* & *Calluna vulgaris-Eriophorum vaginatum* noda, Normal Series Tallis 1973; *Erico-Sphagnetum papillosi* Moore (1964) 1968 emend. Birse 1980 *p.p.*; *Calluna-Eriophorum Sphagnetum* community Bignal & Curtis 1981; *Trichophorum cespitosum-Eriophorum vaginatum* community Bignal & Curtis 1981; *Erica-Sphagnum magellanicum* nodum Ratcliffe & Hattey 1982 *p.p.*; *Erico-Sphagnetum magellanici* Dierssen 1982.

Constant species

Calluna vulgaris, Erica tetralix, Eriophorum angustifolium, E. vaginatum, Sphagnum capillifolium, S. papillosum, S. tenellum, Odontoschisma sphagni.

Rare species

Andromeda polifolia, Sphagnum imbricatum.

Physiognomy

The *Erica tetralix-Sphagnum papillosum* mire is a raised and blanket bog community generally dominated by Sphagna, with ericoid sub-shrubs and monocotyledons often playing a subordinate role, though becoming more important on drier ground and, some of them at least, increasing in prominence with particular kinds of treatment. The vegetation can be found as extensive, undulating carpets comprising irregular mosaics of the different structural elements or, on mires with strongly-differentiated surface microrelief, it can comprise the lawn and hummock components, with the plants more clearly zoned in relation to the height of the water-table and with Rhynchosporion assemblages occupying the wettest hollows.

The bulkier vascular plants typically form a low and patchy canopy, 2 dm or so tall, with *Calluna vulgaris, Erica tetralix* and *Eriophorum vaginatum* being the commonest species, *Scirpus cespitosus* a little less frequent. Often, none of these is truly dominant, small bushes or tussocks occurring scattered through the *Sphagnum* carpet, but, over lawns and hummocks, they can show a stronger pattern of local prominence, *E. tetralix* tending to predominate on wetter ground, where shoots of *Eriophorum angustifolium* can also be abundant, *Calluna, Scirpus* and *E. vaginatum* being concentrated on the drier areas. In such situations, the last two can grow more tussocky, though *Calluna* usually has the higher covers and *E. vaginatum* never really attains the important role it has in the *Calluna-Eriophorum* mire. *Molinia caerulea*, more so than *Scirpus*, is relatively scarce here compared with its prominence in the *Scirpus-Eriophorum* mire, but it can be occasionally abundant, extending down on to wetter flats and becoming more tussocky on hummocks, and it can increase greatly in frequency and cover where better-aerated peats occur within or around stands of the community, as on the drying centres of raised mires, on well-drained rands and where there is some soligenous influence in surface soakways or marginal lagg streams. This is well seen in maps and transects in the studies by Godwin & Conway (1939) of Cors Goch glan Teifi in Dyfed and by Ratcliffe & Walker (1958) and Boatman *et al.* (1981) of the Silver Flowe mires in Galloway. Some stands transitional to such *Molinia*-dominated vegetation can therefore often be included within the community and drying raised mires, which are common, often show a shift in this direction (see below).

Apart from *Calluna* and *E. tetralix*, no other sub-shrub occurs throughout the community. *Myrica gale* is occasionally found in the wetter *Sphagnum-Andromeda* sub-community and, with *Molinia*, it can increase in prominence in soligenous zones, but it is less characteristic here than in the *Scirpus-Eriophorum* mire and rather patchy in its distribution: it becomes commoner in south-west Scotland and parts of Wales, for example, but it is decidedly scarce in Strathclyde (Bignal & Curtis 1981). Drier areas within stands of the community, most notably on the tops of taller and decaying hummocks such as figure frequently in the *Empetrum-Cladonia* sub-community, often have some *Empetrum nigrum* ssp. *nigrum*, but the Arctic-Alpine *E. nigrum* ssp. *hermaphroditum*, *Vaccinium vitis-idaea* and *V. uliginosum* which, together with the Arctic-Subarctic *Rubus chamaemorus*, become so important in the *Calluna-Eriophorum* mire, are all very scarce here: like the *Scirpus-Eriophorum* mire, the *Erica-Sphagnum* mire is largely a low-altitude community. *Vaccinium myrtillus* is similarly very infrequent, even on drying mires.

Of the remaining vascular associates, few are common, and all of these are preferential to the *Sphagnum-Andromeda* sub-community, characteristic of wetter areas on undisturbed mires. In such situations, *Drosera rotundifolia* and *Narthecium ossifragum* occur frequently but both plants are found in other types of ombrogenous bog, being especially common in the *Scirpus-Eriophorum* mire and lower-altitude stands of the *Calluna-Eriophorum* mire, and more distinctive of the *Erica-Sphagnum* mire are two Continental Northern species, *Vaccinium oxycoccos* and *Andromeda polifolia*, whose British distributions are largely centred in the regions where the community is found. *Andromeda*, though somewhat patchy even in the *Sphagnum-Andromeda* sub-community, is particularly diagnostic, occurring in similar vegetation in Ireland (Braun-Blanquet & Tüxen 1952) and in northern Germany (e.g. Moore 1968) though, in mainland Europe, also being found in montane and boreal Sphagnion fusci mires.

Quite often of greater prominence than all these plants, however, are the Sphagna. Their abundance here frequently led early authors to describe the community as a *Sphagnetum* (e.g. Tansley 1939, Godwin & Conway 1939, Pearsall 1941) and the prevailing red-brown tinge of the most important species contributes greatly to the characteristically sombre hue of this vegetation. As in the *Scirpus-Eriophorum* mire, both *Sphagnum papillosum* and *S. capillifolium* are very common throughout, but here *S. tenellum* also attains constancy (though usually in less abundance than these two) and the total cover and luxuriance of the carpet are typically greater. A more obvious preferential feature is the frequent occurrence, particularly in the wetter vegetation of the *Sphagnum-Andromeda* sub-community, of *S. magellanicum*, a species rather unevenly represented in other ombrogenous mires, but here a major peat-builder. *S.*

subnitens, *S. palustre*, *S. cuspidatum*, *S. auriculatum* and, particularly where there is some eutrophication, *S. recurvum*, can also be found occasionally. Among Sphagna of more restricted occurrence in Britain, this community provides an occasional locus for *S. pulchrum* and, sometimes forming pronounced domes, *S. fuscum* and *S. imbricatum*. *S. fuscum* is rather more characteristic in Britain of certain kinds of high-altitude *Calluna-Eriophorum* mire (where it provides a link with the Sphagnion fusci bogs) but *S. imbricatum* seems especially distinctive of this community: it is rather local in its occurrence now (a little less so in equivalent Irish vegetation), but its readily-identifiable sub-fossil remains often figure very prominently in the peats of raised bogs on which the *Erica-Sphagnum* mire is still found (e.g. Godwin & Conway 1939, Pigott & Pigott 1963, Green 1968, Moore 1968).

Over more gently undulating surfaces, the Sphagna are distributed in a rather irregular patchwork but, with increasing differentiation of hummocks and hollows, they show a vertical stratification. *S. papillosum* is primarily a species of the surrounds to wetter depressions and of flats and it often predominates there, with *S. magellanicum* and a little *S. tenellum*. *S. magellanicum* extends a little higher above the water-table than *S. papillosum* but, on hummock sides and tops, both typically give way to an abundance of *S. capillifolium*, and the balance between these species helps define the two sub-communities. In the wettest hollows, semi-aquatic Sphagna become more important in the transition to Rhynchosporion pool vegetation.

Other bryophytes characteristically play a subordinate part in the ground cover but some are frequent and can be locally abundant. As in the *Scirpus-Eriophorum* mire, these can include some leafy hepatics: *Odontoschisma sphagni* and *Mylia anomala* are both very common here and can occur as conspicuous patches among the Sphagna. *Pleurozia purpurea*, on the other hand, strongly preferential among our ombrogenous mires to the most oceanic situations, is generally absent from this vegetation, except in transitional stands (like some of those on the Silver Flowe mires: Ratcliffe & Walker 1958). *Aulacomnium palustre* and *Hypnum cupressiforme/jutlandicum* also occur frequently and there are occasional records for *Pohlia nutans*, *Polytrichum alpestre*, *Campylopus paradoxus* and *Plagiothecium undulatum*. Hypnaceous mosses, notably *Pleurozium schreberi* and *Rhytidiadelphus loreus*, together with *Racomitrium lanuginosum*, are fairly uncommon in wetter situations but they become much more frequent in the *Empetrum-Cladonia* sub-community. It is there, too, that lichens make their strongest contribution to this kind of mire vegetation, larger species like *C. impexa*, *C. uncialis* and *C. arbuscula* being especially good preferentials and together often of moderately high cover.

Sub-communities

Sphagnum magellanicum-Andromeda polifolia sub-community: *Eriophorum vaginati* Rankin 1911a; *Sphagnetum* Regeneration Complex, Stages 3 & 4 Tansley 1939; *Sphagnetum* Regeneration Complex, middle stages Godwin & Conway 1939; *Sphagnum* community, Type A 'moss' Pearsall 1941; Flat communities Ratcliffe & Walker 1958; *Erico-Sphagnetum magellanici*, Sub-Atlantic race Moore 1968 *p.p.*; *Sphagnum papillosum-Erica tetralix* & *Calluna vulgaris-Eriophorum vaginatum* noda, Normal Series Tallis 1973; *Erico-Sphagnetum papillosi*, Typical subassociation Moore (1964) 1968 *emend.* Birse 1980 *p.p.*; *Erica-Sphagnum magellanicum* nodum, *Andromeda polifolia-Rhynchospora alba* complex Ratcliffe & Hattey 1982 *p.p.* All the vascular constants of the community are of high frequency here but very often none of them is dominant and, in the wetter conditions that are characteristic in this kind of *Erica-Sphagnum* mire, *Calluna* is often of rather poor growth and *Eriophorum vaginatum* and *Scirpus* do not show a markedly tussocky habit, though all three of these can increase a little in stature and cover where the vegetation runs up the lower surrounds of well-differentiated hummocks.

Typically, however, it is the Sphagna that make the most obvious immediate impression, forming an extensive and luxuriant carpet in which *S. papillosum*, often here with abundant *S. magellanicum*, predominates around the hollows and over extensive undulant lawns. *S. tenellum* and *S. capillifolium* are also both constant and usually of low cover in wetter situations but the latter can become firmly established on more substantial carpets raised just above the water level and begin to form the hummocks that mark the transition to the drier vegetation of the *Empetrum-Cladonia* sub-community. Much more locally, *S. fuscum* or *S. imbricatum* can be found as prominent hummocks. Then, there is occasionally a little *S. subnitens* or *S. palustre* in the lawn and, in wetter hollows, *S. cuspidatum* or *S. auriculatum*, species which thicken up considerably, replacing *S. papillosum*, in the transition to Rhynchosporion pools. The horizontal and vertical patterning typically found among the Sphagna in this kind of vegetation is very well shown in the quadrats and sections illustrated in Godwin & Conway (1939) and Ratcliffe & Walker (1958). It is probably also best to include in this sub-community the hummock vegetation developed within the extensive *Sphagnum recurvum* lawns of some Cheshire basin mires (e.g. Sinker 1962, Green & Pearson 1968, Tallis 1973a).

Scattered through this ground, in addition to the vascular constants, are frequent individuals of *Drosera rotundifolia*, *Narthecium ossifragum* and, a little less commonly but still strongly preferential, *Vaccinium oxycoccos* and *Andromeda polifolia*. Then, around the

Rhynchosporion pools, *Rhynchospora alba* and *Drosera anglica* can occasionally be found: these, particularly the former, figure very frequently in the stands of Ratcliffe & Hattey (1982) because of the strictly-maintained use of standard sized quadrats which obviously transgressed vegetation boundaries. In the other direction, towards well-differentiated hummocks, *Empetrum nigrum* ssp. *nigrum* is sometimes seen among the thickening *Calluna* and, in the ground layer, the shift towards *S. capillifolium* can be accompanied by the more frequent occurrence of *Cladonia* spp. and *Pleurozium schreberi*. Typically, however, all these species are of low frequency in this sub-community.

Empetrum nigrum ssp. *nigrum-Cladonia* spp. sub-community: *Sphagnetum* Regeneration Complex, Stage 5 Tansley 1939; *Sphagnetum* Regeneration Complex, later stages Godwin & Conway 1939; Marginal *Sphagnetum* Godwin & Conway 1939; Dissected moss Pearsall 1941; Medium & tall hummocks Ratcliffe & Walker 1958; *Erico-Sphagnetum magellanici*, Sub-Atlantic race Moore 1968 *p.p.*; *Trichophoreto-Eriophoretum*, Typical facies Eddy *et al.* 1969 *p.p.*; *Erico-Sphagnetum papillosi*, Typical & *Rhytidiadelphus* subassociations Moore (1964) 1968 *emend.* Birse 1980 *p.p.*; *Erica-Sphagnum magellanicum* nodum, *Sphagnum papillosum-Erica tetralix* type Ratcliffe & Hattey 1982 *p.p.* Where this sub-community forms the medium and taller hummock element of mire vegetation within stretches of the last sub-community, the drawing of a boundary between the two can be quite difficult but, in their extreme forms, they are quite distinct. First, in comparison with the *Sphagnum-Andromeda* sub-community, *Calluna*, *Scirpus* and *E. vaginatum* tend to have higher covers here, the first in particular becoming more vigorous and abundant but the last two also growing markedly more tussocky. Then, there is frequently some *Empetrum* among them, this becoming especially prominent on the tops of decaying hummocks. By contrast, the preferential herbs of the *Sphagnum-Andromeda* sub-community all decline in frequency here, particularly *Drosera rotundifolia* and *Andromeda*, a little less strikingly *Narthecium* and *Vaccinium oxycoccos*.

Second, there are obvious differences in the composition and structure of the ground layer. Among the Sphagna, the balance shifts towards a strong dominance of *S. capillifolium*, with *S. papillosum* still very frequent but much less abundant, *S. tenellum* rather patchy, though locally prominent and *S. magellanicum* only occasional and of low cover. And other mosses now become frequent among the thickening vascular plants, with *Pleurozium schreberi* and *Rhytidiadelphus loreus* especially good preferentials, *Hypnum cupressiforme/jutlandicum* also increasing somewhat and *Racomitrium lanuginosum*, *Dicranum scoparium* and *Polytrichum commune* becoming occasional. *Odontoschisma sphagni* and *Mylia anomala* both remain common but other hepatics, notably *Cephalozia connivens*, *C. bicuspidata*, *Calypogeia muellerana*, *Cladopodiella fluitans* and *Kurzia pauciflora*, now appear with some frequency. The other noticeable difference is the increase in larger *Cladonia* spp., notably *C. impexa*, *C. uncialis* and *C. arbuscula*, each of which can be locally abundant. On especially dry surfaces, as on the tops of the tallest hummocks, even *Sphagnum capillifolium* is reduced, with its shoots decaying and, where *Calluna*, *Scirpus* or *E. vaginatum* do not have a vigorous hold, there is often just a crown of *Empetrum* and *Racomitrium* or encrusting lichens, like *Cladonia gracilis*, *C. coccifera* and *C. pyxidata*, on exposed peat.

Habitat

The *Erica tetralix-Sphagnum papillosum* mire is characteristic of waterlogged ombrogenous peats, typically at low altitudes in the moderately oceanic parts of Britain. It is pre-eminently a community of raised bogs, comprising the bulk of the cover on their active plane, but it can also be found within stretches of blanket mire, and it occurs, too, on acidic topogenous peats in some basin mires. The typical habitat has been very widely affected by various treatments, notably peat-cutting, burning and draining, and these have often modified the vegetation or reduced it to fragmentary stands.

Raised bogs have developed where local accumulation of topogenous deposits has elevated the mire surface above the immediate controlling influence of the ground water-table, leading to the establishment of ombrogenous conditions on the plane. They are found primarily in the lowlands and are particularly associated with the flood-plains of mature river valleys, where flat tracts of alluvium have provided a suitable base for deep and often extensive peat accumulation. Most occur adjacent to large estuaries, as on the Somerset Levels, at Cors Fochno in Dyfed, around Morecambe Bay and at Bowness, Wedholme, Glasson and Kirkconnell on the Solway, where the *Erica-Sphagnum* mire can be found (or once did occur) close to sea-level; but broad valleys far inland, and sometimes at altitudes over 200 m, have also provided locations, as at Cors Goch glan Teifi in Dyfed, Rhos Goch in Powys and alongside the Forth in the Flanders Moss complex (Ratcliffe 1977). Such sites are concentrated in a broad belt running south-west to north-east between the Severn–Humber and Clyde–Moray lines and, though the initiation of raised mire development seems often to have coincided with climatic deterioration at the Boreal/Atlantic transition about 7000 years ago, these bogs subsist today under relatively low rainfall. For the most part, they receive between 800 and 1200 mm annual precipitation (*Climatological Atlas* 1952) with 140–180 wet days yr^{-1} (Ratcliffe 1968), considerably less than either of our two other ombrogenous bogs, the *Scirpus-Eriophorum* and *Calluna-Eriophorum* mires. Characteristically, raised bogs have a domed profile, but the slope of the ground under the *Erica-Sphagnum* mire is usually very near to zero, so run-off is very slight and the active growth of the bog is not so strongly dependent on a high precipitation/ evaporation ratio. Indeed, remnants of raised mire survive in what are now very dry areas, as on Thorne Waste in West Yorkshire/Lincolnshire and at Woodwalton in Cambridgeshire, where annual rainfall is less than 600 mm, and on the former site healthy fragments of the *Erica-Sphagnum* mire can be found.

Increasingly, to the north-west of Britain, individual raised mires lose their integrity within the smothering mantle of blanket peat that has developed in the wetter climate but certain stretches of mire seem to retain something of the character of raised bog, particularly over local depressions or on cols. Such transitional situations provide a suitable location for the *Erica-Sphagnum* mire at unusually high altitudes, up to 500 m or more. They occur occasionally in Wales, at Cors Goch in Powys, for example which provides the most southerly station for *Andromeda*, and on the Pennines on Stainmore (Pearsall 1941) and at Moor House (Eddy et al. 1969), but are especially plentiful in the Border and south-west Scotland, at sites like the Irthinghead mires, on the Silver Flowe (Ratcliffe & Walker 1958) and in Strathclyde (Bignal & Curtis 1981; see also Ratcliffe 1977). At the other extreme, the *Erica-Sphagnum* mire also extends on to some basin bogs, where more strictly topogenous peats have acquired a character verging on the ombrogenous towards their centre. Such sites are quite plentiful over the drift-smeared terrain within the limit of the Final Glaciation and the community is represented on some of them in the north-west Midlands, at Chartley and Wybunbury Mosses, for example, which have a *schwingmoor* character (Sinker 1962, Green & Pearson 1968, Tallis 1973a), and around the Eden valley in Cumbria, at Tarn, Moorthwaite, Cumwhitton and Cliburn Mosses (Ratcliffe 1977).

In this range of site types, the *Erica-Sphagnum* mire is characteristic of virgin surfaces where there is consistent waterlogging, at least on the flats and around the hummock bases, by a high and stagnant water-table. The peats are often deep, typically deeper than under the *Scirpus-Eriophorum* and *Calluna-Eriophorum* blanket mires and sometimes up to 10 m or more. However, although at the base of raised and basin mires they are derived from herbaceous fen vegetation with interposed brushwood, they are uniformly acidic at the surface with a pH of around 4, and oligotrophic. Under such conditions, the Sphagna show very luxuriant growth on the active mire plane and make the major contribution to the accumulating peats, with *Eriophorum vaginatum* constant and confirming the Sphagnetalia character of the vegetation, but generally making the minor contribution to accumulation. The balance of the major structural elements in the vegetation, the Sphagna and the bulky vascular plants, is thus more like that in the *Scirpus-Eriophorum* mire than in the *Calluna-Eriophorum* mire but, if anything, the Sphagna are of greater importance than in the *Scirpus-Eriophorum* mire. For reasons which are unclear, *Sphagnum magellanicum* shows a quite marked preference for this community, frequently joining *S. papillosum* as an abundant component of the undulant flats developed at or just above the level of the water-table.

These flats tend to be more extensive here than in the *Scirpus-Eriophorum* mire and surface relief is often not so strongly differentiated. On many raised mires, Cors Fochno and Glasson Moss being good, more intact examples, the patterning is on rather a large scale and of low amplitude, though the size of the bog seems to have some effect on the relief, smaller raised mires in similar climatic conditions often showing more pronounced undulations, as at Rhos Goch and at Wem Moss in Shropshire, a feature which becomes more exaggerated where rainfall is lower and sporadic, as on Penmanshiel Moor in Berkwickshire (Ratcliffe 1977). And, where the *Erica-Sphagnum* mire extends into more oceanic areas, the pool element can become as extensive and ordered as on some blanket mires, a situation well seen in the Silver Flowe complex (Ratcliffe & Walker 1958, Lindsay *et al.* 1984). Where hummocks and hollows are well differentiated, they show the characteristic zonation of Sphagna and vascular plants in relation to the height of the water-table (see below), but whether bog growth always involves a cyclical alternation of hummock and hollow at particular points is debatable. Although it was from classic raised bogs with the *Erica-Sphagnum* mire that the theory of the 'regeneration complex' was first described in Britain (Osvald 1923, 1939, Tansley 1939, Godwin & Conway 1939), stratigraphical investigation does not always reveal the tell-tale lenses of the different Sphagna and, as in the *Scirpus-Eriophorum* mire, some surface-patterning may be a quasi-permanent feature.

Other distinctive floristic features of the *Erica-Sphagnum* mire reflect the moderately oceanic character of the climate over the region where the community is centred. Compared with the *Scirpus-Eriophorum* mire, for example, *Molinia* is less important here within the bog vegetation itself, though it is often very abundant on drier or better-drained areas in close proximity to the community, and *Sphagnum auriculatum* does not figure so prominently in the transitions to the Rhynchosporion pools. *Potentilla erecta*, *Polygala serpyllifolia* and *Pedicularis sylvatica* are also much reduced on ombrogenous peats with the move to less oceanic conditions, and

Pleurozia purpurea and *Campylopus atrovirens*, though recorded in transitional stands, as on the Silver Flowe (Ratcliffe & Walker 1958) and in similar vegetation in Ireland (Moore 1968), are generally absent. In the other direction, the *Erica-Sphagnum* mire only very occasionally penetrates into the cooler, wet climate characteristic of the *Calluna-Eriophorum* mire, which accounts for the scarcity here of plants like *Rubus chamaemorus*, *Vaccinium vitis-idaea*, *V. uliginosum* or *Empetrum nigrum* spp. *hermaphroditum*, the contrast between the communities being well seen on Stainmore (Pearsall 1941) and at Moor House (Eddy *et al.* 1969). Positive floristic responses to climate are few, but the Continental Northern *Vaccinium oxycoccos* and particularly *Andromeda* are very good preferentials.

Floristic and structural differences between the sub-communities are strongly related to variations in ground moisture and thus partly a reflection of the differentiation of surface relief. The *Sphagnum-Andromeda* sub-community is the more distinct of the two and is consistently associated with wetter conditions, comprising the bulk of the cover where flats and low hummocks predominate. Under such conditions, there is an especially luxuriant carpet of Sphagna, frequent occurrences of lawn plants like *Drosera rotundifolia*, *Narthecium*, *Vaccinium oxycoccos* and, particularly over slightly raised areas, *Andromeda*, poorly-developed tussockiness in *Eriophorum vaginatum* and *Scirpus*, and often puny *Calluna*. Such vegetation parallels that in the *Drosera-Sphagnum* sub-community of the *Scirpus-Eriophorum* mire, containing occasional records for Rhynchosporion plants where there are transitions to bog pools, though generally lacking any indicators of soligenous influence, which is rare on the surface of raised mires. The *Sphagnum-Andromeda* sub-community can also come very close in its floristics to the *Narthecio-Sphagnetum*, the typical Sphagnetalia mire of lowland valley bogs: this is especially true on some of the Cheshire mosses where *Scirpus* and *Eriophorum vaginatum*, two good distinguishing species between the communities, become scarcer.

On somewhat drier peats, the *Sphagnum-Andromeda* sub-community is replaced by the *Empetrum-Cladonia* type, which can therefore be seen on virgin mires on the tops of taller hummocks, as in the later stages of the 'regeneration complex' (Tansley 1939, Godwin & Conway 1939) and the 'medium and tall hummocks' of Ratcliffe & Walker (1958); over more extensive surfaces which are maintained in a drier state, as in the 'marginal *Sphagnetum*' on Cors Goch glan Teifi (Godwin & Conway 1939); and on peats at higher altitudes which have become dissected by erosion channels (Pearsall 1941). In this less strongly-waterlogged environment, dominance among the Sphagna shifts to *S. capillifolium*, *Cladonia* spp. and hypnaceous mosses become prominent among

the increasingly tussocky *Eriophorum vaginatum* and
Scirpus and vigorous *Calluna*, and *Empetrum nigrum*
ssp. *nigrum* spreads on the eroding surfaces. In the
process, the *Erica-Sphagnum* mire loses some of its
distinctive character, *Vaccinium oxycoccos* and particularly *Andromeda* becoming scarce, and the vegetation
closely approaches that of drier forms of the *Scirpus-
Eriophorum* mire. Indeed, drier areas of peat on mires
towards the more oceanic limit of the community, where
Erica cinerea becomes conspicuous as a coloniser, develop vegetation indistinguishable from the *Cladonia* sub-
community of the *Scirpus-Eriophorum* mire, a situation
well seen in Strathclyde (e.g. Bignal & Curtis 1981).

The surface-drying of whole raised mires may be, in
part, a natural phenomenon, recurrence surfaces, often
here with abundant remains of *Sphagnum imbricatum*,
testifying to past episodes of wetter conditions (Godwin
& Conway 1939, Tallis 1961, Walker & Walker 1961,
Green 1968). But there is no doubt that, in very many
cases, the process has been hastened by treatment, more
especially by combinations of peat-cutting, burning and
draining, sometimes with grazing. Very few of our raised
bogs or basin mires, lying as they do in relatively
accessible lowland landscapes, have escaped these kinds
of interference. Much early activity was largely confined
to their margins and often perfunctory in character but,
in some sites, and, in recent years, more universally, the
effects have been more drastic, with extensive drainage
and abandonment, reclamation for agriculture or forestry, or the wholesale stripping of peat, originally for
local burning, now for horticultural use. In extreme
cases, the entire original surface of the mire, or very
nearly all of it, has been removed by peat-cutting, such
that the raised bog vegetation of the *Erica-Sphagnum*
type has been totally lost, as on Holme Fen and at
Woodwalton in Cambridgeshire (Poore 1956*b*), or
severely reduced to fragments, as on Shapwick Heath in
Somerset and at Thorne Waste, or restricted to sections
of the bog, as at Bowness and Wedholme. Elsewhere
cutting has been more confined but attempts at drainage
have lowered the water-table, a situation seen at Kirkconnell Flow and on Flanders Moss, and where this has
been combined with burning and grazing, the effects
have been pronounced. Accidental fires, like those at
Glasson Moss, have sometimes damaged more intact
sites. In some basin mires carrying the *Erica-Sphagnum*
vegetation, there has been a marked eutrophication of
incoming waters by fertiliser run-off which has curtailed
any tendency to the development of ombrogenous nuclei, a phenomenon clearly visible at sites in Cheshire and
Shropshire (Poore & Walker 1959, Sinker 1962, Green
& Pearson 1968, Tallis 1973).

In such situations as these, the *Sphagnum-Andromeda*
sub-community tends to suffer first and most, its extensive *Sphagnum* carpet being very susceptible to surface
drying or burning, recovering only slowly from episodes
of damage or not at all where the effects of interference
are sustained. Often, then, its cover is reduced to wetter
or unburned remnants of the mire surface, though it
does seem to be able to regenerate on new wet areas,
such as abandoned shallow peat-cuttings. At Moorthwaite and Cumwhitton Mosses, for example, two
basin mires in Cumbria, there is vigorous growth over
old cut surfaces and sporadic regeneration can even be
seen in the very dry climate at Thorne Waste (Ratcliffe
1977). And *Andromeda* is one of the distinctive plants of
this kind of *Erica-Sphagnum* mire which can positively
thrive after burning, even though the richness of the
accompanying flora has been lost, at least temporarily
(e.g. Sinker 1962). The often quite marked variation in
the representation of different associates in this sub-
community at particular sites may be attributable to
their differential rates of recovery from frequently complex patterns of interference.

The *Empetrum-Cladonia* sub-community, characteristic as it is of drier peat, may actually be favoured by
certain kinds of treatment, particularly by surface or
marginal drainage which lowers the water-table over
part or all of the mire, allowing its extension from drier
hummocks over what were originally wet lawns and
hollows. In other cases, it can become prominent on
remaining baulks of peat on cut surfaces with the
intervening areas occupied by the regenerating *Sphagnum-Andromeda* sub-community or modified vegetation. Such prominence may, however, be a temporary
interlude prior to extensive invasion by woody plants no
longer held in check by waterlogging (see below). Even
on relatively intact mires, with only slight shifts in the
water regime or infrequent episodes of burning or
grazing, there can be changes in the abundance of
particular species within tracts of reasonably well-pre-
served *Erica-Sphagnum* mire. The local prominence of
Scirpus and *Molinia*, for example, on Cors Goch glan
Teifi may be partly related to past treatment (Godwin &
Conway 1939, Ratcliffe 1977).

Zonation and succession

Zonations between the sub-communities of the *Erica-
Sphagnum* mire are related to the height of the water-
table and transitions to other vegetation types to the
degree of stagnation and to the extent of soligenous
influence. Treatments can modify these patterns greatly
and permit the replacement of the community by other
herbaceous vegetation or by woodland.

Internal patterning within virgin stands of the *Erica-
Sphagnum* mire characteristically reflects the differentiation of surface microrelief. As noted above, this is
generally less pronounced than in the more oceanic
Scirpus-Eriophorum mire, though it shows some variation with climate and also with the size of the bog. Often

the *Sphagnum-Andromeda* sub-community predominates over extensive wet lawns, with the *Empetrum-Cladonia* sub-community marking out raised areas, frequently slight and of irregular disposition, in other cases more prominent and ordered, with a clearer zonation among the plants in relation to the height of the water-table. Where pools occur in the hollows, there are transitions from the *Sphagnum-Andromeda* sub-community to Rhynchosporion vegetation, with a shift in the *Sphagnum* carpet to semi-aquatic species and a reduction in many of the vascular plants characteristic of the mire plane, notably *Calluna*, *Scirpus* and *E. vaginatum*. As on the *Scirpus-Eriophorum* mire, a fringe of *Rhynchospora alba* often marks this switch, but here the Rhynchosporion vegetation is typically represented by the *Sphagnum cuspidatum/recurvum* community, the *Sphagnum auriculatum* type of bog pool generally being confined to sites in areas with a somewhat more oceanic climate, where the mire vegetation itself is of a transitional character (as on the Silver Flowe: Ratcliffe & Walker 1958). Usually, *S. cuspidatum* is the predominant pool species but, where mire surfaces have become eutrophicated, *S. recurvum* often becomes abundant (see below).

Natural zonations from the *Erica-Sphagnum* mire to other vegetation types are often related to the degree of stagnation in the peats and, on raised bogs, this characteristically lessens towards the margins where there is a more steeply-sloping surround or rand of sometimes considerable width. Over this, the freely-drained conditions exclude the *Erica-Sphagnum* mire and it is typically replaced by vegetation in which *Molinia* is the leading species, a pattern well seen in the maps and transects of Cors Goch glan Teifi (Godwin & Conway 1939) and some of the Silver Flowe mires (Ratcliffe & Walker 1958, Boatman *et al.* 1981). The overwhelming dominance of *Molinia*, and the consequent exclusion of associates from this zone, sometimes make it difficult to determine the exact character of this vegetation but it can usually be referred to the Junco-Molinion. Gradations between this and the *Erica-Sphagnum* mire of the bog dome are rarely abrupt and transitional vegetation can sometimes be clearly identified as *Scirpus-Erica* wet heath, often the *Sphagnum* sub-community which has some representation of Sphagnetalia species (as in the 'intermediate bog' of Ratcliffe & Walker 1958). Elsewhere, such gradual transitions can be incorporated within the *Erica-Sphagnum* mire itself, the change being essentially confined to gentle shifts of dominance as the rand is approached (Figure 16).

Rands are sometimes dissected by erosion channels in which there is some clear soligenous influence but, in normal circumstances, these do not eat back on to the bog dome itself and direct transitions from the *Erica-Sphagnum* mire to soligenous mires are rare on raised bogs. Nonetheless, the distinctive vegetation of the lagg zone beyond the rand is an integral part of the large-scale pattern in this kind of habitat and it is typically some kind of Caricion nigrae poor fen, often the *Carex echinata-Sphagnum* mire, with some invasion by woody

Figure 16. Changes in raised mire zonations from more natural sequence of communities (foreground) to that on drying bog surfaces (background).
M1 *Sphagnum auriculatum* bog pool community
M6 *Carex echinata-Sphagnum* mire
M15b *Scirpus-Erica* wet heath, Typical sub-community

M18a *Erica-Sphagnum* bog, *Sphagnum-Andromeda* sub-community
M18b *Erica-Sphagnum* bog, *Empetrum-Cladonia* sub-community
M25a *Molinia-Potentilla* mire, *Erica* sub-community
W4 *Betula-Molinia* woodland

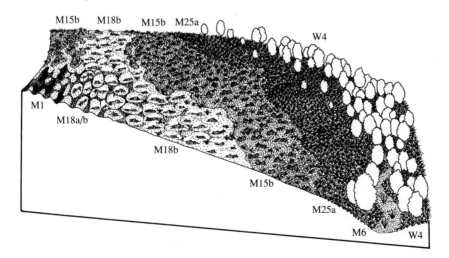

plants producing *Salix-Galium* or *Betula-Molinia* woodland or, where there is some greater degree of base
enrichment, the *Alnus-Carex* woodland. This type of
pattern can be seen at Cors Fochno, Rhos Goch and
Glasson Moss.

Where the *Erica-Sphagnum* mire is represented on
basin mires, the community can be found in closer
juxtaposition with both soligenous and topogenous
vegetation which, with some base- and/or nutrient-
enrichment from the incoming waters, often shows a
slight shift towards a more calcicolous or eutrophic
character. At Llyn in Powys, for example, Cors Graia-
nog in Gwynedd and Cumwhitton, the *Erica-Sphagnum*
mire occurs with the *Potentillo-Caricetum* and there are
sometimes close associations with the *Carex-Sphagnum
warnstorfii* or *Carex-Sphagnum squarrosum* mires or
even with vegetation approaching the *Carex-Calliergon*
mire, again with varying colonisation by *Salix* spp. and
Betula pubescens.

Where the *Erica-Sphagnum* mire occurs in conditions
approaching those of blanket bog, the vegetation pattern may preserve some features of the raised mire
system, with a rand on most of the circumference, or just
on one side (as on Snibe and Brishie bogs, respectively,
in the Silver Flowe complex: Ratcliffe & Walker 1958),
passing gradually over its other edges to blanket mire
proper or, where there is a transition to steeper valley-
side slopes, to *Scirpus-Erica* wet heath on thinner, drier
peats. In other cases, ill-defined patches of the *Erica-
Sphagnum* mire occur embedded within tracts of blanket
bog where the peat mantle deepens over cols or sub-
surface hollows. Usually, at the lower altitudes of which
the community is characteristic, it is the *Scirpus-Eriophorum* mire which provides the context, as in southern
Scottish sites like Mochrum Lochs and Kilquhockadale
Flow (Ratcliffe 1977, Bignal & Curtis 1981), though
towards the Pennines and particularly at higher altitudes, this is replaced by the *Calluna-Eriophorum* mire,
as around Irthinghead on the Cumbria/Northumber-
land border, on Stainmore (Pearsall 1941) and at Moor
House (Eddy *et al.* 1969). At lower levels soligenous
soaks can sometimes be found in close association with
the *Erica-Sphagnum* mire, with a cover of the *Carex* sub-
community of the *Scirpus-Erica* wet heath or some type
of *Carex echinata-Sphagnum* mire. At higher altitudes,
surface erosion features are sometimes found with
systems of gullies fretting the surface and the *Eriophorum angustifolium* community in pools.

Such zonations as these have been very widely affected by treatments, particularly at lower altitudes where
there has been a long history of local, but sometimes
intensive, exploitation of peatlands and where many
raised and basin mires now remain, much modified and
safeguarded only by statutory protection, within intensive agricultural landscapes. The influences of treat-

ments have been very diverse, partly because histories of
interference have been complex, and partly because
different communities in the vegetation patterns are
affected in different ways by peat-cutting, burning and
draining. But some general effects can be discerned.
First, lowering of the water-table, and surface-damage
by burning on raised and basin mires, tend to shift the
balance within the *Erica-Sphagnum* mire towards the
Empetrum-Cladonia sub-community and sometimes
towards more overwhelming dominance by *Calluna*, *E.
vaginatum*, *Scirpus* or, increasing beyond its usual low
cover here, *Molinia*. Such changes can occur over entire
mire surfaces but they often show first around the
margins, where the rand vegetation extends its cover
inwards, and at the centre, such that the *Erica-Sphagnum* mire is reduced to a narrowing ring. Increasing
climatic dryness makes raised mires particularly susceptible to such modifications (Godwin & Conway
1939).

Continuations of such a trend would be expected to
convert the *Erica-Sphagnum* mire to the *Scirpus-Erica*
wet heath with the loss of *Eriophorum vaginatum* and the
luxuriant *Sphagnum* lawns, but particular sites often
show a complex patchwork of vegetation intermediate
between the two, with local dominance of ericoids,
Scirpus and *Molinia*, the balance between which is also
affected by burning and grazing. Establishment of dry
Calluno-Ulicetalia heath may follow but, very commonly, once the surface has ceased to be waterlogged,
invasion by woody plants supervenes. *Betula pubescens*
(with *B. pendula* on drier peats) and *Pinus sylvestris* are
the major colonisers. The former can come to dominate
in *Betula-Molinia* woodland or, where drying out has
proceeded further, in birch-dominated stands of the
Quercus-Betula-Dicranum woodland. The pine, though
usually seeding in from plantation stock, can form a
canopy to vegetation very similar to the natural *Pinus-
Hylocomium* woodland in which *E. tetralix* or *Empetrum* can persist under more open covers and where such
characteristic species as *Pyrola minor* or *Listera cordata*
sometimes appear. Pine woodland of this kind can be
seen colonising the community at Moorthwaite Moss
and Kirkconnell Flow.

Second, peat-cutting often contributes to the surface
drying of raised and basin mires but it also results in
patterns of its own which fret the bog surface with
systems of hollows and baulks. Where the hydrological
regime has not been too grossly disrupted, such uneven
ground may preserve something of the natural pattern
of virgin mires, with the two sub-communities disposed
according to the wetness of the ground, but with an
artificial regularity over the surface. Often, however, the
impact has been more drastic, so that complex mosaics
of secondarily-developed communities remain, with
poor fens,wet and dry heath, bracken stands and *Salix*

and birch woodland all jumbled together: this is well shown over parts of the Somerset Levels and on Thorne Waste.

Eutrophication of the mire surface or of the ground waters around is often an attendant feature of such gross disturbance and even sites which have remained reasonably intact can show marked effects of enrichment from, for example, fertiliser run-off or drift. This is especially prevalent in some of the Cheshire and Shropshire mires, where it has resulted in a spread of *Sphagnum recurvum*, from its usual confines in pools over extensive wet lawns in which the *Sphagnum-Andromeda* sub-community of the *Erica-Sphagnum* mire is reduced to scattered (though sometimes apparently actively-growing) patches on slightly raised areas (Sinker 1962, Green & Pearson 1968, Tallis 1973*a*).

Stratigraphical studies clearly show that, in raised and basin mires, vegetation of the *Erica-Sphagnum* type has developed as the climax of a hydroseral succession, the commonest sequence of which has been open water, swamp, fen, woodland and bog, the stages to fen often taking less than 1000 years, the later phases sometimes much longer (e.g. Walker 1970). Although there is evidence in the peats of periods of drier climate since the initiation of ombrogenous accumulation, the community seems to persist on active surfaces at the present time, provided treatments are not too disruptive of the hydrological regime.

Distribution

The *Erica-Sphagnum* mire is of widespread but local occurrence through the lowlands of Wales and north-west Britain up to the Clyde–Moray line, with a few outlying sites in lowland southern England. In many areas, its extent has been much reduced and its vegetation modified by exploitation of peatlands, but good stands remain in parts of Wales, around the Solway, in the Borders and in south-west Scotland. Both sub-communities occur throughout the range.

Affinities

Although vegetation of this kind was early recognised as distinctive among our range of bog types (e.g. Rankin 1911*a*, Tansley 1939, Godwin & Conway 1939, Pearsall 1941, Osvald 1949), it is generally subsumed within a *Sphagnetum* which included parts of other communities, and the major interest of which was often taken to be its supposed pattern of regeneration. And, being essentially a vegetation type of lowland areas outside north-west

Scotland, it did not figure in McVean & Ratcliffe (1962), whose two major ombrogenous communities, the *Scirpus-Eriophorum* and *Calluna-Eriophorum* mires, have tended to provide the basis for subsequent descriptive accounts.

Although the *Erica-Sphagnum* mire shows a virtually continuous gradation into the *Scirpus-Eriophorum* type, a feature recognised by Braun-Blanquet & Tüxen (1952) and Ratcliffe & Walker (1958), and extends into habitats transitional to the blanket bog context typical of the latter, it shows sufficient peculiar characteristics to be worth recognising as a separate kind of ombrogenous vegetation best developed in the raised mire habitat. Moore (1968) was the first to accord British vegetation of this kind this status, placing it in an *Erico-Sphagnetum magellanici*, a broadly-defined Erico-Sphagnion association centred on the *eigentliches Hochmoor* of Osvald (1923), corresponding roughly with the range of *Sphagneta* diagnosed by Kastner & Flössner (1933) and Schwickerath (1940) and characteristic of the lowlands of north-west Europe. Moore (1968) recognised three geographical races within the community, extending from the extreme Atlantic conditions of central Ireland (where the vegetation had first been described by Braun-Blanquet & Tüxen (1952)), through a sub-Atlantic zone across Britain, Belgium and Germany (e.g. Jonas 1933, Schwickerath 1940, Vanden Berghen 1948), into the sub-continental parts of Europe (e.g. Vanden Berghen 1948, Tüxen & Soyrinki 1958, Jahns 1962). As defined here, using our own data and samples from Birse (1980) and Bignal & Curtis (1981), the *Erica-Sphagnum* mire falls firmly in the central type, distinguished from the first by the low frequencies of *Molinia*, *Pleurozia purpurea* and *Campylopus atrovirens*, and from the third by the high frequencies of *Scirpus* and *Sphagnum papillosum*. We have followed Birse (1980) in using *S. papillosum*, rather than *S. magellanicum*, which is preferential but not constant, to name the community, though his *Erico-Sphagnetum papillosi* also contains vegetation which would here be placed in the *Scirpus-Eriophorum* and *Calluna-Eriophorum* mires. In addition to *S. magellanicum*, the community is separated from these two other ombrogenous Erico-Sphagnion mires by the preferential occurrence of *Andromeda*, *Vaccinium oxycoccos* and the now very local *S. imbricatum*. It can be distinguished from the *Narthecio-Sphagnetum*, the Erico-Sphagnion community which replaces it in valley mires in lowland, southern Britain, by *Andromeda*, *Scirpus* and *Eriophorum vaginatum*.

Floristic table M18

	a	b	18
Calluna vulgaris	V (1–8)	V (1–9)	V (1–9)
Erica tetralix	V (1–7)	V (1–4)	V (1–7)
Eriophorum angustifolium	V (1–6)	V (1–8)	V (1–8)
Sphagnum papillosum	IV (1–8)	V (1–9)	V (1–9)
Eriophorum vaginatum	IV (1–7)	V (1–8)	IV (1–8)
Sphagnum capillifolium	IV (1–6)	V (1–8)	IV (1–8)
Sphagnum tenellum	IV (1–3)	IV (1–5)	IV (1–5)
Odontoschisma sphagni	IV (1–4)	IV (1–3)	IV (1–4)
Sphagnum magellanicum	IV (1–8)	II (1–4)	III (1–8)
Narthecium ossifragum	IV (1–7)	II (1–3)	III (1–7)
Drosera rotundifolia	IV (1–3)	I (1–3)	II (1–3)
Vaccinium oxycoccos	III (1–4)	II (1–3)	II (1–4)
Andromeda polifolia	III (1–3)		I (1–3)
Rhynchospora alba	I (1–4)		I (1–4)
Myrica gale	I (1–4)		I (1–4)
Drosera anglica	I (1–2)		I (1–2)
Empetrum nigrum nigrum	II (1–6)	IV (1–6)	III (1–6)
Pleurozium schreberi	II (1–4)	IV (1–5)	III (1–5)
Cladonia impexa	II (1–3)	IV (1–4)	III (1–4)
Rhytidiadelphus loreus	I (2)	IV (1–4)	II (1–4)
Cladonia uncialis	I (1–4)	III (1–6)	II (1–6)
Cladonia arbuscula	I (1–4)	III (1–3)	II (1–4)
Cephalozia connivens		III (1–3)	II (1–3)
Calypogeia muellerana	I (1)	II (1–3)	I (1–3)
Hypogymnia physodes	I (1)	II (1–3)	I (1–3)
Polytrichum commune	I (1–2)	II (1–3)	I (1–3)
Dicranum scoparium	I (1–3)	II (1–3)	I (1–3)
Cephalozia bicuspidata	I (1–3)	II (1–3)	I (1–3)
Kurzia pauciflora	I (1–4)	II (1–4)	I (1–4)
Cladopodiella fluitans		II (1–3)	I (1–3)
Racomitrium lanuginosum	I (2)	II (1–4)	I (1–4)
Mylia anomala	III (1–6)	III (1–4)	III (1–6)
Aulacomnium palustre	III (1–5)	III (1–3)	III (1–5)
Scirpus cespitosus	III (1–6)	III (1–6)	III (1–6)
Sphagnum cuspidatum	II (1–6)	II (1–2)	II (1–6)
Hypnum jutlandicum	II (1–5)	II (1–3)	II (1–5)
Molinia caerulea	II (1–4)	II (1–4)	II (1–4)
Sphagnum recurvum	I (1–5)	I (1–3)	I (1–5)
Hypnum cupressiforme	I (1–3)	I (1–3)	I (1–3)
Sphagnum palustre	I (2–3)	I (1–2)	I (1–3)
Polytrichum alpestre	I (1–2)	I (1–3)	I (1–3)
Pohlia nutans	I (1–4)	I (1–3)	I (1–4)
Sphagnum subnitens	I (3–4)	I (1–3)	I (1–4)
Rubus chamaemorus	I (1–4)	I (1–2)	I (1–4)

Plagiothecium undulatum	I (1)	I (1–3)	I (1–3)
Vaccinium myrtillus	I (1–3)	I (1–3)	I (1–3)
Vaccinium vitis-idaea	I (1)	I (1–3)	I (1–3)
Sphagnum pulchrum	I (1–4)	I (1–2)	I (1–3)
Deschampsia flexuosa	I (1–3)	I (1)	I (1–4)
Juncus squarrosus	I (1–2)	I (1–3)	I (1–3)
Campylopus paradoxus	I (1–3)	I (1)	I (1–3)
Lophozia ventricosa	I (1–3)	I (1–2)	I (1–3)
Cladonia gracilis	I (1–3)	I (1–3)	I (1–3)
Sphagnum fuscum	I (1–3)	I (1)	I (1–3)
Number of samples	118	77	195
Number of species/sample	15 (8–21)	24 (16–30)	17 (8–30)
Herb height (cm)	25 (15–40)	18 (13–30)	22 (13–40)
Herb cover (%)	71 (10–100)	57 (35–75)	67 (10–100)
Bryophyte height (mm)	56 (20–150)		
Bryophyte cover (%)	63 (2–100)	91 (75–95)	71 (2–100)
Altitude (m)	206 (10–530)	250 (15–550)	209 (10–550)
Slope (°)	0 (0–2)	0 (0–2)	0 (0–2)
Soil pH	4.0 (3.3–4.7)		

a   *Sphagnum magellanicum-Andromeda polifolia* sub-community
b   *Empetrum nigrum nigrum-Cladonia* sub-community
18   *Erica tetralix-Sphagnum papillosum* raised mire (total)

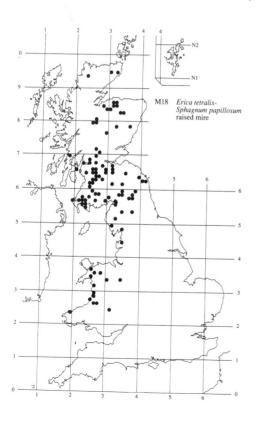

M18 *Erica tetralix-Sphagnum papillosum* raised mire

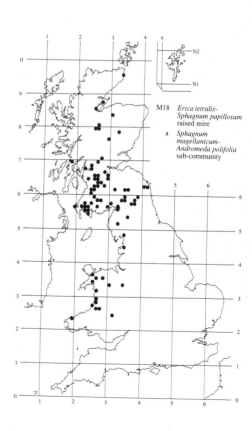

M18 *Erica tetralix-Sphagnum papillosum* raised mire

a   *Sphagnum magellanicum-Andromeda polifolia* sub-community

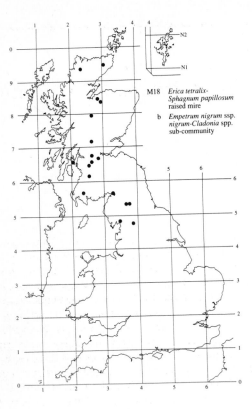

M18   *Erica tetralix-*
      Sphagnum papillosum
      raised mire

   b *Empetrum nigrum* ssp.
      nigrum-Cladonia spp.
      sub-community

M19
Calluna vulgaris-Eriophorum vaginatum blanket mire

Synonymy

Eriophoretum vaginati Smith & Moss 1903, Lewis 1904, Lewis & Moss 1911, Moss 1913, Adamson 1918, Watson 1932, Tansley 1939, all *p.p.*; Mixed moor Pearsall 1938; *Sphagnum* community, Type B 'Moss' Pearsall 1941; *Calluna & Calluna-Eriophorum* Moss Pearsall 1941; *Betula nana* bogs Poore & McVean 1957; Blanket bogs Ratcliffe 1959; *Calluneto-Eriophoretum* McVean & Ratcliffe 1962, Eddy *et al.* 1969, Birks 1973, Meek 1976, Evans *et al.* 1977; *Empetreto-Eriophoretum* McVean & Ratcliffe 1962; *Vaccinio-Ericetum tetralicis* Moore 1962, Birse & Robertson 1976, Dierssen 1982; *Trichophoreto-Eriophoretum* Eddy *et al.* 1969 *p.p.*; *Eriophorum vaginatum* bog Edgell 1969 *p.p.*; *Erica tetralix-Vaccinium oxycoccos & Juncus squarrosus-Deschampsia flexuosa* Series Tallis 1969 *p.p.*; *Rhytidiadelphus loreus-Sphagnum fuscum* Community Birse & Robertson 1976; *Vaccinio-Eriophoretum* Hill & Evans 1978 *p.p.*; Mire noda 10 & 11 Daniels 1978; *Rhytidiadelpho-Sphagnetum fusci* Birse 1980; *Erico-Sphagnetum papillosi* Moore (1964) 1968 *emend.* Birse 1980 *p.p.*; *Calluna-Pleurozium & Calluna-Cladonia* noda Bignal & Curtis 1981.

Constant species

Calluna vulgaris, Eriophorum angustifolium, E. vaginatum, Rubus chamaemorus, Pleurozium schreberi, Sphagnum capillifolium.

Rare species

Arctostaphylos alpinus, Betula nana, Vaccinium microcarpum, Kiaeria starkei.

Physiognomy

The *Calluna vulgaris-Eriophorum vaginatum* mire comprises blanket bog vegetation that is generally dominated by mixtures of *Eriophorum vaginatum* and ericoid sub-shrubs. Sphagna can be prominent over wetter ground but typically this element is not so rich or luxuriant as in the *Scirpus-Eriophorum* or *Erica-Sphag-*

num mires. And the true hummock/hollow relief that can commonly be found in those communities is only rarely developed here, though the surface of the ground is often uneven because of the marked tussockiness of *E. vaginatum*, a structural feature of some importance for the strong contingent of hypnoid mosses, which constitute a further distinctive component over drier surfaces.

E. vaginatum is consistently more important here than in the *Scirpus-Eriophorum* and *Erica-Sphagnum* mires: among our Sphagnetalia communities, it shows a peak of abundance in this kind of vegetation and in the *Eriophorum* mire, which seems often to have been derived from it by impoverishment of the associated flora and with which it was often grouped in early schemes under the heading of *Eriophoretum vaginati* (e.g. Smith & Moss 1903, Lewis 1904*a, b*, Lewis & Moss 1911, Moss 1913, Adamson 1918, Watson 1932, Tansley 1939). *E. vaginatum* is usually at least a co-dominant here, contributing a dull-green colour to the vegetation through much of the year and giving it a particularly striking aspect in June with its cottony fruits. The tussocks, which grow by intravaginal tillering, acquire a robust hemispherical shape and can come to stand proud of the bog surface by 2–3 dm, crowned by the densely-packed living shoots and the remains of the foliage, which dies back annually by February but persists intact for many years (Goodman & Perkins 1968, Polozova 1970, Wein 1973, Chapin *et al.* 1979). The tussocks thus provide a relatively dry microhabitat, though one which is heavily shaded until senescence sets in, which, under normal circumstances, may not be for many decades (Polozova 1970).

The proportion of *E. vaginatum* to the ericoid sub-shrubs is very variable and, though it shows some consistency in the different sub-communities, it is very much affected by certain kinds of treatment, particularly by burning and grazing which can help convert the vegetation into the *Eriophorum* mire, where the cotton-grass is overwhelmingly dominant and the sub-shrubs of much more patchy occurrence. Variation between the two communities is continuous and replacement of the richer by the more impoverished can be temporary, so transitions are common. In general, however, the two can be separated by the very frequent occurrence here of *Calluna*, *Vaccinium myrtillus* and *Empetrum nigrum* ssp. *nigrum* and, at higher altitudes, of the Arctic-Alpine *V. vitis-idaea*, *V. uliginosum* and *E. nigrum* ssp. *hermaphroditum*. Overall, *Calluna* is the most common co-dominant of *E. vaginatum*, but diverse mixtures of these species occur very frequently and typically the sub-shrubs are so abundant and so vigorous, especially over the drier ground, as to give the vegetation the appearance of a heathy moorland, an aspect recognised in Pearsall's (1938, 1968) epithet of 'mixed moor' and in the names *Calluneto-Eriophoretum* and *Empetreto-Eriophoretum* coined by McVean & Ratcliffe (1962). Two other woody Arctic-Alpines, *Arctostaphylos alpinus* and *Betula nana*, both national rarities, can also be found here, preferentially among our Sphagnetalia mires, though they are very local and generally both rather inconspicuous, *Arctostaphylos* being a prostrate plant and *Betula* usually growing low and gnarled.

The distinctly montane character of this element in the vegetation provides a sharp contrast with both the *Scirpus-Eriophorum* and *Erica-Sphagnum* mires, something reinforced by the absence here of *Myrica gale*, essentially a lowland plant. One further difference, which perhaps reflects the generally drier nature of the peats in this community, is the rather restricted occurrence of *Erica tetralix*: this is commoner in stands to the west and at usually lower altitudes, as in parts of Wales (Ratcliffe 1959a, Tallis 1969, Edgell 1969), south-west Scotland (Birse & Robertson 1976, Birse 1980) and through Strathclyde (Bignal & Curtis 1981), where its frequency helps define the *Erica* sub-community, but it is distinctly local through much *Calluna-Eriophorum* mire.

Further contrasts with these other kinds of ombrogenous mire can be seen among the vascular associates. Compared with the strongly oceanic blanket bog vegetation of the *Scirpus-Eriophorum* mire, *Scirpus cespitosus* often plays a subordinate role here, *Molinia caerulea* is strikingly scarce and herbs such as *Potentilla erecta*, *Polygala serpyllifolia* and *Pedicularis sylvatica* are very infrequent. The first two tend to follow *E. tetralix* in their pattern of occurrence, thus helping to give the *Erica* sub-community a transitional floristic character. It is there, too, that *Narthecium ossifragum* and *Drosera rotundifolia*, which can be so prominent in the wetter lawns of both the *Scirpus-Eriophorum* and *Erica-Sphagnum* mires, have most of their relatively sparse occurrences in this community. *Andromeda polifolia* and *Vaccinium oxycoccos*, both good preferentials for the *Erica-Sphagnum* mire, are very uncommon here, although high-altitude *Calluna-Eriophorum* mire provides the

British locus for the rare small cranberry *V. microcarpum*.

On the positive side, the commonest vascular associates in the community are *Eriophorum angustifolium*, a very frequent plant throughout British Oxycocco-Sphagnetea vegetation, and, much more distinctive here, *Rubus chamaemorus*. This Arctic-Subarctic herb is strongly preferential to the *Calluna-Eriophorum* mire, though it is not universally present: it is decidedly scarce in the lower-altitude *Erica* sub-community and sometimes of patchy occurrence elsewhere, being affected by particular treatments. Early or late sampling may also under-estimate its abundance, because it is not ever-green: it passes the winter as dormant buds brought close to the peat surface by rhizome extension, does not put up its shoots until May and shows early senescence of its aerial parts, with their quick death in the first frosts of the autumn. When in bloom, in late May and early June, the large white flowers provide a delightful relief to what is for the most part a rather dreary scene, but flowering is somewhat sporadic and local and, since the flowering period coincides with lactation in deer and hill-sheep, the flowers often get eaten off providing a nutritious bite (Taylor 1971).

Few other features of the vascular flora are distinctive, though *Deschampsia flexuosa* and *Juncus squarrosus* occur occasionally throughout and each can be locally abundant and, at higher altitudes, *Carex bigelowii* becomes frequent. *Nardus stricta*, *Carex nigra*, *Melampyrum pratense* and *Galium saxatile* can be found at low frequencies and the community also provides an occasional locus for the Arctic-Subarctic *Cornus suecica* and the Northern Montane *Listera cordata*. By and large, however, herbs here are few in number and occur as scattered individuals.

The bryophyte flora, by contrast, is rich and often extensive, frequently covering more than 50% of the ground, a further difference between this community and the *Eriophorum* mire. As noted above, though, varied and luxuriant carpets of Sphagna are not the rule in this community. *Sphagnum capillifolium* is sufficiently frequent throughout to qualify as a constant and it can be locally abundant, but even this species is somewhat patchy and others, like *S. papillosum*, *S. subnitens* and *S. tenellum*, prominent in the wet lawns of other ombrogenous mires, and the semi-aquatics *S. cuspidatum* and *S. auriculatum*, which can figure elsewhere in transitions to bog pools, are generally uncommon and rather uneven in their occurrence. Some of these species are better represented in the transitional vegetation of the *Erica* sub-community, where there can be a semblance of hummock/hollow relief; and some higher-altitude *Calluna-Eriophorum* mires also show locally extensive Sphagnum carpets, in which the montane *S. fuscum* has its best representation among British Sphagnetalia communities. In some places, too, generally in the *Erica* sub-community, though also in the *Empetrum* sub-community, *S. recurvum* can be prominent, perhaps in response to some soligenous enrichment.

Typically, however, it is hypnaceous mosses which provide the consistency to the bryophyte element in the *Calluna-Eriophorum* mire. Among these, *Pleurozium schreberi*, *Rhytidiadelphus loreus*, *Hypnum cupressiforme/jutlandicum* and *Plagiothecium undulatum* are all very frequent throughout, with *Hylocomium splendens* becoming common at higher altitudes, and mixtures of these can form extensive mats over drier areas of the ground, particularly over the tops of old *E. vaginatum* hummocks and among the stools of sub-shrubs where the bushes have opened up somewhat. Other mosses recorded frequently are *Dicranum scoparium*, *Polytrichum commune* and *Aulacomnium palustre*, with *Racomitrium lanuginosum*, *Polytrichum alpestre* and *Pohlia nutans* attaining occasional levels in some of the sub-communities and *Campylopus paradoxus* and *Rhytidiadelphus squarrosus* being found sparsely throughout. High-altitude stands sometimes have the rare montane moss *Kiaeria starkei*.

Leafy hepatics are typically not so conspicuous as they are in more oceanic ombrogenous mires and some of the most characteristic species there are either absent from this community (*Pleurozia purpurea*) or much less common (*Mylia anomala*, *Odontoschisma sphagni*). But others occur fairly frequently, such as *Lophozia ventricosa*, *Barbilophozia floerkii*, *Diplophyllum albicans*, *Calypogeia fissa*, *C. muellerana*, *C. trichomanis*, *Cephalozia bicuspidata*, *Mylia taylori* and, particularly at higher altitudes, *Ptilidium ciliare*, and some of these can attain a measure of local abundance.

Finally, among the general features of the community, there is the frequent occurrence of lichens, especially in the higher-altitude *Vaccinium-Hylocomium* sub-community. Larger *Cladonia* spp., such as *C. impexa*, *C. uncialis* and *C. arbuscula*, can be particularly abundant crowning the tops of old *Eriophorum* hummocks and among senescent *Calluna*, but peat-encrusting species, such as *C. squamosa*, *C. bellidiflora*, *C. floerkeana* and *C. gracilis*, also occur and there can be records for *Cetraria islandica*, *Hypogymnia physodes* and *Cornicularia aculeata*.

Sub-communities

Erica tetralix **sub-community:** Blanket bogs Ratcliffe 1959 *p.p.*; *Vaccinio-Ericetum tetralicis* Moore 1962 *p.p.*, Birse & Robertson 1976 *p.p.*; *Eriophorum vaginatum* bog, low-level facies Edgell 1969 *p.p.*; *Erica tetralix-Vaccinium oxycoccos* series, *Plagiothecium-Hylocomium* & *Racomitrium-Cladina* noda Tallis 1969; *Trichophoreto-Eriophoretum* Eddy *et al.* 1969 *p.p.*; *Calluneto-Eriophoretum* Meek 1976; *Calluneto-Erio-*

phoretum typicum Evans *et al.* 1977; *Erico-Sphagnetum papillosi*, *Rhytidiadelphus loreus* subassociation Moore (1964) 1968 *emend.* Birse 1980 *p.p.*; *Calluna-Pleurozium* & *Calluna-Cladonia* noda Bignal & Curtis 1981. This kind of *Calluna-Eriophorum* mire shows a number of floristic features transitional to more oceanic ombrogenous bogs. Among the vascular plants, for example, *E. vaginatum* can be co-dominant with the sub-shrubs, though it is generally less abundant and less prominent structurally than it is in the other sub-communities. *Calluna* often predominates among the woody species, frequently with some *E. nigrum* ssp. *nigrum* and a little *V. myrtillus*, but *E. tetralix* is strongly preferential and is sometimes of moderately high cover. Then, there is frequently some *Scirpus* and this, too, can be locally abundant with prominent tussocks. *Molinia* also occurs occasionally and there can be scattered plants of *Narthecium* and, more sparsely, of *Drosera rotundifolia* or *Vaccinium oxycoccos*. The great infrequency of *Rubus chamaemorus*, which is very local here, is a further distinctive feature.

The bryophyte element is less strikingly peculiar but Sphagna tend to be more consistently abundant than in many tracts of *Calluna-Eriophorum* mire, certainly than in much of the *Empetrum* sub-community, with *S. capillifolium* being quite commonly accompanied by *S. papillosum* and sometimes by *S. tenellum*, *S. subnitens* or *S. compactum*. And there can be a slight suggestion of the development of hummocks and hollows with these species disposed over stretches of wetter or drier ground. Other stands can have an abundance of *S. recurvum*, sometimes with *S. palustre*, in fairly extensive lawns.

The difference between this kind of vegetation and drier types of *Scirpus-Eriophorum* and *Erica-Sphagnum* mire, where hypnaceous mosses and herbs such as *Juncus squarrosus* and *Deschampsia flexuosa* (all quite common in this sub-community) show some increase, is thus a fine one, and especially hard to discern if individual samples are examined in isolation. The context of the samples, that is the overall character of the mire from which they are taken, sometimes helps but the similarity of the vegetation types is a real reflection of the convergence of ombrogenous mire floristics in certain environmental conditions (see below).

Empetrum nigrum **ssp.** *nigrum* **sub-community:** *Eriophoretum vaginati* Smith & Moss 1903, Lewis 1904, Lewis & Moss 1911, Moss 1913, Adamson 1918, Watson 1932, Tansley 1939, all *p.p.*; Mixed moor Pearsall 1938; *Sphagnum* community, Type B 'Moss' Pearsall 1941; *Calluna* & *Calluna-Eriophorum* Moss Pearsall 1941; Blanket bogs Ratcliffe 1959 *p.p.*; *Eriophorum vaginatum* bog, high-level facies Edgell 1969; *Calluneto-Eriophoretum* Eddy *et al.* 1969 *p.p.*, Birks 1973; *Calluneto-Eriophoretum deschampsietosum* & *myrtillosum* Evans *et al.* 1977; *Vaccinio-Eriophoretum* Hill & Evans 1978 *p.p.*; Mire noda 10 & 11 Daniels 1978. This sub-community, long familiar as the richer kind of 'Pennine blanket bog', generally preserves all the typical floristic features of the *Calluna-Eriophorum* mire, though it is very variable in the proportions of the major structural elements and thus in its gross appearance, a feature very well described from Moor House in Cumbria (Welch & Rawes 1966, Eddy *et al.* 1969, Rawes & Hobbs 1979). *E. vaginatum* is usually abundant and structurally prominent and, indeed, on uneroded mires, it can be dominant for some time after episodes of burning, and become more permanently ascendant under intensive grazing. Generally, however, sub-shrubs play a fairly consistent part in the vegetation. *Calluna* is usually the leading species among these and, on virgin mires, this forms the bulk of an open canopy, 2–3 dm tall, the bushes alternating with healthy *Eriophorum* tussocks or growing on moribund ones, and the *Calluna* itself sometimes showing its familiar stages of growth from pioneer through to degenerate where burning or grazing are absent (Watt 1955, Gimingham 1972). There is also often some *V. myrtillus* and/or *E. nigrum* ssp. *nigrum* and both of these, particularly the latter, can increase to very great abundance for some time after burning or where grazing has been withdrawn, or more permanently where drier peat surfaces have been bared by erosion: under these last conditions both *E. vaginatum* and *Calluna* can show much less vigorous growth than usual. At the high altitudes characteristic of this vegetation, such erosion, often aggravated by generations of burning and grazing and, more recently, by drainage schemes, is widespread, the peats often deeply scored by systems of haggs, which introduce a characteristic patterning into the vegetation (e.g. Tansley 1939, Tallis 1964*b*, 1965, 1973*b*, 1985*b*). This sub-community also provides some of the more southerly stations for the Arctic-Alpine sub-shrubs *V. vitis-idaea* and *V. uliginosum*, though these become very much more frequent in the *Vaccinium-Hylocomium* sub-community. In sharp contrast to the *Erica* sub-community, *E. tetralix* is very scarce here, and any increase is often an indication of the local occurrence of a tract of that kind of *Calluna-Eriophorum* mire within the *Empetrum* sub-community (as at Moor House: Eddy *et al.* 1969).

Scirpus and *Molinia* both decline here to their usual low levels of frequency in the community and *Rubus chamaemorus* increases greatly, though it is absent from some far-flung localities for this sub-community, as in North Wales (Ratcliffe 1959*a*, Edgell 1969) and on Skye (Birks 1973) and can be decidedly patchy elsewhere, as on Stainmore (Pearsall 1941) and parts of Moor House (Eddy *et al.* 1969), this local variation sometimes being obviously related to treatment. *Eriophorum angustifolium* occurs very commonly, usually at low covers, and

there is occasionally some *Deschampsia flexuosa*, but other herbs are few and there are no preferentials among this element.

Bryophytes, too, show little that is distinctive, though there is a typically rather impoverished *Sphagnum* flora. *S. capillifolium* remains frequent and it can be quite abundant on the wetter ground around the *E. vaginatum* and beneath the sub-shrubs, though burning and erosion can have a drastic effect on its extent. *S. papillosum* is especially scarce and *S. subnitens* occurs only very occasionally though, as in the *Erica* sub-community, some stands show a local abundance of *S. recurvum* (as in a facies distinguished by Eddy *et al.* 1969).

Over somewhat drier ground, the hypnaceous mosses are particularly abundant, forming extensive mats over the decumbent branches and litter among older *Calluna* bushes and over the crowns of ageing *Eriophorum* tussocks. Where bare peat is exposed, acrocarpous species such as *Dicranum scoparium*, *Campylopus paradoxus* and *Pohlia nutans* become frequent and these are commonly accompanied by such hepatics as *Mylia taylori*, *Kurzia pauciflora*, *Ptilidium ciliare*, *Cephalozia* spp. and *Calypogeia* spp., together with lichens like *Cladonia impexa*, *C. squamosa*, *C. bellidiflora*, *C. floerkeana*, *Cornicularia aculeata* and *Hypogymnia physodes*.

Vaccinium vitis-idaea-Hylocomium splendens sub-community: *Betula nana* bogs Poore & McVean 1957; *Calluneto-Eriophoretum* McVean & Ratcliffe 1962; *Empetreto-Eriophoretum* McVean & Ratcliffe 1962, Dierssen 1982; *Juncus-Deschampsia* Series Tallis 1969 *p.p.*; *Calluneto-Eriophoretum* Eddy *et al.* 1969 *p.p.* This sub-community embraces all of the high-montane blanket mire first comprehensively described by McVean & Ratcliffe (1962). It preserves all the general floristic features of the community but is especially distinctive in the consistent presence of the Arctic-Alpine sub-shrubs *V. vitis-idaea* and *E. nigrum* ssp. *hermaphroditum*, which are frequent throughout, and *V. uliginosum*, which is of somewhat more restricted occurrence. Various mixtures of these species, usually with abundant *Calluna* and generally smaller amounts of *V. myrtillus*, and sometimes with some *E. nigrum* ssp. *nigrum*, are typically co-dominant with *Eriophorum vaginatum* in the familiar kind of mixed canopy, though one which is often less tall than at lower altitudes. The montane character of the vegetation is reinforced by the occurrence of the rarities *Arctostaphylos alpinus*, *Betula nana* and *Vaccinium microcarpum* in some stands.

Among the vascular associates, *Eriophorum angustifolium* is rather less common here than usual but *Rubus chamaemorus* remains very frequent, *Juncus squarrosus* occurs quite often, usually at low covers, and in many stands there is some *Carex bigelowii*. *Scirpus cespitosus* also shows something of a resurgence in frequency here.

The remaining distinctive features of this sub-community are found among the ground layer. First, Sphagna can be quite prominent, with *S. capillifolium* being accompanied in some stands by *S. papillosum* and *S. subnitens*, in others by *S. fuscum*; *S. russowii* is also found very occasionally. Then, the hypnaceous mosses are consistently joined by *Hylocomium splendens* and there is quite commonly some *Racomitrium lanuginosum* and *Polytrichum alpestre*; and the rare *Kiaeria starkei* can sometimes be found. Among the hepatics, *Ptilidium ciliare* is strongly preferential, and *Mylia anomala* a little more common than usual, with *Anastrepta orcadensis* figuring occasionally. The lichen flora, too, is usually well developed, *Cladonia arbuscula* and *C. uncialis* being especially frequent, *C. impexa*, *C. rangiferina* and *Cetraria islandica* occurring somewhat less commonly. Among the available stands, three variants have been recognised.

Betula nana variant: *Betula nana* bogs Poore & McVean 1957; *Calluneto-Eriophoretum*, shrub-rich facies McVean & Ratcliffe 1962. Superficially, this kind of *Calluna-Eriophorum* mire is virtually identical with the typical *Vaccinium-Hylocomium* form, mixtures of *E. vaginatum* and *Calluna* generally being co-dominant, with smaller amounts of *V. myrtillus*, *V. vitis-idaea*, *E. nigrum* ssp. *nigrum* and/or *E. nigrum* ssp. *hermaphroditum*, and sometimes a little *Erica tetralix*, forming a cover 2 dm or so high. Closer inspection, however, reveals the presence of *Betula nana* generally growing low among or beneath the other woody species, sometimes showing more conspicuously where their cover is a little thinner. Some stands, from north of the Great Glen, can also have small amounts of *Arctostaphylos alpinus* growing low over the ground and others, in the north-west Highlands, contain *Vaccinium microcarpum*. *Rubus chamaemorus* is generally of low cover in this variant but, among the other vascular associates, *Juncus squarrosus* can be locally abundant. *Carex bigelowii*, in contrast to the other variants, is absent here.

Bryophytes have extensive cover but, though *Sphagnum* growth seems vigorous, the carpet is made up almost entirely of *S. capillifolium*, very occasionally with some *S. fuscum*. Hypnaceous mosses are more varied, with *Pleurozium schreberi*, *Hylocomium splendens*, *Plagiothecium undulatum* and *Hypnum cupressiforme/jutlandicum* all well represented. *Ptilidium ciliare* is the commonest hepatic, though *Mylia anomala* and *Lophozia ventricosa* occur occasionally. Lichens tend to be less conspicuous than in the other variants but *C. arbuscula* and *C. impexa* can be locally abundant and *C. rangiferina* is found in some stands.

Typical variant: *Calluneto-Eriophoretum, Sphagnum* type and lichen-rich facies McVean & Ratcliffe 1962; *Juncus-Deschampsia* series Tallis 1969 *p.p.*; *Calluneto-Eriophoretum, Empetrum* facies Eddy *et al.* 1969 *p.p.*

Calluna and *E. vaginatum* are generally co-dominant in this variant, with some *V. myrtillus* and *V. vitis-idaea* and, rarely, a little *V. uliginosum*, and *E. nigrum* ssp. *hermaphroditum* usually better represented than *E. nigrum* ssp. *nigrum*. *Rubus chamaemorus* and *Juncus squarrosus* can both be quite abundant and *Carex bigelowii* becomes frequent, though generally at low covers.

Although the contingent of hypnaceous mosses remains an important element in the ground cover, Sphagna can be quite abundant in wetter places, with *S. capillifolium* the best represented species (*S. quinquefarium* sometimes being distinguished) but *S. subnitens* and *S. fuscum* also figuring with local prominence. *Cladonia* spp., too, are common with *C. arbuscula*, *C. uncialis* and *C. rangiferina* occurring throughout and sometimes attaining high cover.

Vaccinium uliginosum-Polytrichum alpestre variant: *Empetreto-Eriophoretum* McVean & Ratcliffe 1962. This is the most distinctive kind of high-montane *Calluna-Eriophorum* mire in which *E. nigrum* ssp. *hermaphroditum* almost totally replaces ssp. *nigrum*, becoming the usual co-dominant of *Eriophorum vaginatum* in place of *Calluna*, which is only very occasional and of low cover. *Vaccinia* are also prominent, with *V. uliginosum* frequently joining *V. myrtillus* and *V. vitis-idaea*. Mixtures of the sub-shrubs and cotton-grass form a very low canopy, often scarcely 1 dm high. Scattered through this are frequent plants of *Rubus chamaemorus* and *Juncus squarrosus* and, now becoming very common, *Carex bigelowii* and *Scirpus cespitosus*. *Vaccinium microcarpum* has also been recorded very occasionally.

In the ground layer, *S. capillifolium* and *S. papillosum* can both have high cover and, less frequently, there can be some *S. fuscum*, *S. subnitens*, *S. tenellum* or *S. russowii*. Hypnaceous mosses are a little less conspicuous, though *Pleurozium schreberi*, *Rhytidiadelphus loreus* and *Hylocomium splendens* all occur quite commonly in small amounts. *Dicranum scoparium* is also frequent, but more striking here is the constancy of *Racomitrium lanuginosum* and *Polytrichum alpestre* and the occasional occurrence of *Kiaeria starkei*. *Ptilidium ciliare* and *Lophozia ventricosa* are the commonest hepatics. Lichens also remain a prominent feature, with frequent records for *Cladonia arbuscula* and *C. uncialis*, and occasional *C. impexa* and *Cetraria islandica*.

Habitat

The *Calluna-Eriophorum* mire is the typical blanket bog vegetation of high-altitude ombrogenous peats that have accumulated in the wet and cold climate of the uplands of northern Britain. The harsh nature of the montane environment is reflected in the floristics of the community, though the vegetation often takes some of its character from the effects of treatments, notably burning, grazing and draining; and these have contributed, perhaps with climatic change, to its modification and, in some places, to the erosion of the peats.

On the mantle of blanket peat that has developed over gentler slopes in those parts of Britain with more than 1200 mm annual precipitation (*Climatological Atlas* 1952) or in excess of 160 wet days yr^{-1} (Ratcliffe 1968), the *Calluna-Eriophorum* mire is characteristic of higher altitudes where high precipitation/evaporation ratios coincide with low temperatures throughout the year. In fact, the climate is not quite so consistently humid as within the range of the *Scirpus-Eriophorum* mire, which is concentrated within the 2000 mm isohyet, with more than 200 wet days yr^{-1}. The *Calluna-Eriophorum* mire does extend into this zone, in the higher reaches of the north-west Highlands, but, for the most part, annual rainfall is between 1200 and 2000 mm, with 160–200 wet days yr^{-1}. But the greater difference between the distributions of the two communities is in relation to temperature. The *Scirpus-Eriophorum* mire is essentially an oceanic blanket bog community, largely confined to lower altitudes with relatively mild winters and a comparatively small annual temperature range. The *Calluna-Eriophorum* mire, by contrast, is characteristic of areas with harsh winters, with February minima for the most part below freezing, and cool summers. Overall, its range coincides closely with those parts of Britain which have annual accumulated temperatures of less than 830 day-degrees C yr^{-1} (Page 1982), an area which takes in most of the higher ground through the Scottish Highlands, in the Southern Uplands, in the Lake District and down the Pennines, but which excludes most of Wales and also Dartmoor, where the climate is wet enough to support blanket mire development, but insufficiently cool for this community to be more than local. The altitudinal separation between the *Scirpus-Eriophorum* and *Calluna-Eriophorum* mires is fairly well maintained throughout their ranges, though its absolute level depends on the degree of oceanicity. In general, the former is restricted to sites below 500 m, with a mean altitude in the available samples of about 300 m, while the latter is generally found above 300 m, with a mean of 550 m, but in the wetter west of Scotland, the dividing line tends to be set rather higher, whereas, in the more continental east, the *Calluna-Eriophorum* mire extends down virtually to sea-level and the *Scirpus-Eriophorum* mire is almost totally excluded.

At these higher altitudes in northern Britain, the *Calluna-Eriophorum* mire is confined to deeper peats, usually more than 2 m thick (at least when uneroded) and sometimes considerably more, on flat or gently-sloping ground, generally 0–10° (mean 4°), though extending on to somewhat steeper slopes in places. It is thus most extensively developed on high-level plateaus and broad watersheds, such as dominate much of the landscape of the eastern Highlands, in the Monadhliat the Forest of Atholl and the Angus Hills (McVean Ratcliffe 1962), parts of the Southern Uplands (Birse Robertson 1976, Birse 1980) and the high Cheviot, a the Pennine summits right down to north Derbyshi (e.g. Moss 1913, Tansley 1939, Pearsall 1941, Eddy *et* 1969), with a few outlying stands in Wales (Ratcli 1959*a*, Edgell 1969, Tallis 1969). In more rugg country, where the climate is suitable but the terra often too steep for the development of this kind vegetation, its occurrence is more restricted: althoug the community is the major kind of blanket bog higher altitudes in the western Highlands, it is th rather patchy there, and it is decidedly scarce in the La District.

The peats are usually well-humified, at least abov highly acidic (with a surface pH often less than 4) a very oligotrophic. But, typically, they are not so tho oughly or consistently waterlogged as in the *Scirpu Eriophorum* mire and, indeed, can become surface-d and oxidised in the summer. This is probably partly d to the drier climate characteristic of the *Calluna-Eri phorum* mire over most of its range, but it is also perha accentuated by the fact that the community often occu on broadly convex summits and slopes which shed wat quite readily. Very frequently, too, such drainage h been sharpened by various kinds of treatment (s below). Thus, although the *Calluna-Eriophorum* mire clearly a Sphagnetalia community, floristically distin from the Ericetalia wet heaths of thinner, drier peat Sphagna are less varied and luxuriant than in th *Scirpus-Eriophorum* or *Erica-Sphagnum* mires, th major peat-builders *S. papillosum* and *S. magellanicu* especially being of restricted occurrence. And the *C luna-Eriophorum* mire only rarely shows the develo ment of the hummock/hollow structure associated wi active differentiation of surface relief within wet *Spha num*-dominated carpets. The nearest the communi comes to this kind of composition and physiognomy in the *Erica* sub-community, where lawn species like *tetralix*, *Narthecium*, *Drosera rotundifolia* and *Vacc nium oxycoccos* are preferential, and this kind of *C luna-Eriophorum* mire is characteristic of habitats th are climatically or topographically transitional to tho typical of our other ombrogenous mires. It is conce trated towards the more oceanic west, often at low altitudes (mean about 400 m), including most of th stands of the community in Wales (Ratcliffe 1959 Edgell 1969, Tallis 1969) and in Strathclyde (Bignal Curtis 1981) and it occurs at higher levels where blank mire of this general kind extends over flat or conca areas of relief (Tallis 1969, Eddy *et al.* 1969) where higher water-table can be maintained. The more consi tent occurrence of *Scirpus* in the *Erica* sub-communi may also reflect the frequency of surface-waterloggi here: it occurs patchily in some other kinds of *Calluna-Eriophorum* mire, but in stands on more surface-dry peats is often only abundant along tracks, where aeration is reduced by trampling.

In the *Erica* sub-community, *Eriophorum vaginatum* often shows some reduction in abundance and vigour out, for the most part, it is the prominence of this species, rather than the luxuriance of Sphagna, which gives the *Calluna-Eriophorum* mire its Sphagnetalian character. The combination of a generally cool climate with a firm, acidic peat substrate, kept generally moist below but not surface-waterlogged, seems to provide this plant with very favourable conditions for its ascendancy (Tansley 1939, Godwin 1975). It can therefore be a major contributor to the accumulating peats under this vegetation, though it has not always been pre-eminent in their development. Its fibrous tussock bases are very resistant to decay and tend to survive in disproportionate abundance as fibrous clods, but earlier views that the deposits here are mainly *E. vaginatum* peat (e.g. Woodhead & Erdtman 1926) have not been borne out by subsequent studies: very often profiles show a transition to peat with abundant *Sphagnum* remains below and layers rich in *E. vaginatum* frequently occur in situations suggestive of climatic or edaphic dryness, so the cotton-grass has probably waxed and waned in abundance since the inception of the development of this kind of blanket mire (e.g. Conway 1954, Tallis 1965, Godwin 1975).

The present climatic and soil conditions experienced by the *Calluna-Eriophorum* mire are reflected in other major features of the vegetation. First, the very poor representation here of *Molinia*, which is virtually confined to the transitional *Erica* sub-community and, even here, not very prominent, and the absence or scarcity of *Myrica*, *Potentilla erecta*, *Polygala serpyllifolia* and *Pedicularis sylvatica*, betoken the shift away from the lowland oceanic environment of the *Scirpus-Eriophorum* mire, where they are all well represented, to a harsher montane habitat. Second, there is the complementary increase in associates whose European distribution is either Arctic-Subarctic, like *Rubus chamaemorus*, or Arctic-Alpine, as with *Vaccinium vitis-idaea*, *V. uliginosum*, *Empetrum nigrum* ssp. *hermaphroditum* and *Carex bigelowii*, all of which occur frequently, and *Betula nana* and *Arctostaphylos alpinus*, which are less common but preferential to this community among our Sphagnetalia mires. The more structurally important of these species are, incidentally, all plants which thrive on acidic but well-drained substrates (e.g. Ritchie 1955a, Taylor 1971, Gimingham 1972) and their prominence here, together with *Calluna*, *Vaccinium myrtillus* and *Empetrum nigrum* ssp. *nigrum*, which have a wider representation on ombrogenous mires, is further testimony to the freedom from surface-waterlogging in the peats.

Much of the floristic variation between the sub-communities and variants of the *Calluna-Eriophorum* mire can be related to the different sensitivities of these species to the temperature regimes within the range of altitudes occupied by the community as a whole. At lower levels, this kind of blanket mire extends some way into areas where the climate is a little more equable, with February minima sometimes above freezing and mean annual maxima over 25 °C. In the *Erica* sub-community, therefore, the montane contingent is very poorly represented and the sub-shrub canopy consists largely of more broadly temperature-tolerant species such as *Calluna*, the cool oceanic *E. nigrum* ssp. *nigrum* and the Continental Northern *V. myrtillus*, though these show somewhat depressed vigour on the wetter peats, with *E. tetralix* being correspondingly more frequent. The woody element thus approaches that found in more oceanic ombrogenous mires, reinforcing the transitional character of the *Erica* sub-community with its more luxuriant *Sphagnum* component, frequent *Scirpus* and occasional *Molinia*.

With the shift to higher altitudes and a cooler climate, the *Empetrum* sub-community becomes the usual form of *Calluna-Eriophorum* mire, sporadically in Wales, but very extensively all up the Pennines, over the Cheviot and in south-east Scotland and into the eastern Highlands. The mean altitude of this sub-community is around 600 m and, at these levels, the winters are harsher and the mean annual maximum temperature generally between 21 and 25 °C. The sub-shrub component remains much as before, except that *E. tetralix*, which retains some measure of physiological activity in the winter (Gimingham 1972), is now much reduced in occurrence, with balance shifting to the other species, all of which perform well on the somewhat drier peats. More strikingly, *Rubus chamaemorus* appears as a constant: the British range of this plant coincides closely with the distribution of the more montane kinds of *Calluna-Eriophorum* mire and is closely confined within the 25 °C mean annual maximum isotherm (Taylor 1971).

V. vitis-idaea has a very similar range in Britain to *R. chamaemorus* (Ritchie 1955a) and it, too, shows a modest rise in frequency in the *Empetrum* sub-community, but it becomes really common with a further rise in altitude and drop in temperatures in the *Vaccinium-Hylocomium* sub-community. This kind of *Calluna-Eriophorum* mire occurs almost exclusively in Scotland, being especially extensive in the Central Highlands with more patchy occurrences in the more rugged country of the north-west Highlands, but it has outliers on the high Cheviot and in the Pennines and transitional stands can be found on the Berwyns. The mean altitude in this sub-community is almost 700 m and virtually everywhere the mean annual maximum temperature is less than 21 °C.

The area bounded by this isotherm coincides closely with the distribution of *E. nigrum* ssp. *hermaphroditum*, which becomes frequent here and gradually replaces ssp. *nigrum* with increasing altitude: the exact upper limit of the latter is hard to fix because of confusion between the two when in the vegetative state, but it is probably around 750 m (Bell & Tallis 1973). Certainly, in the *Vaccinium-Polytrichum* variant, which attains the highest altitudes reached by the *Calluna-Eriophorum* mire, occurring on summits in the Central Highlands with a mean height of over 850 m, it is very much the predominant sub-species and quite often the leading sub-shrub, being frequently co-dominant with *E. vaginatum*, itself a circumpolar plant that retains its abundance and vigour fairly well throughout the whole altitudinal range. Here, too, *V. uliginosum*, another species generally confined within the 21 °C isotherm, becomes constant with the other Vaccinia, and *Calluna* is almost extinguished, reaching its altitudinal limit as a vigorous plant at around 850 m (Gimingham 1960).

In the very harsh climate experienced by the *Vaccinium-Hylocomium* sub-community, the woody cover can be reduced to a very low canopy and the vegetation provides a locus for some other high-montane species. *Carex bigelowii* has a geographical distribution which closely matches those of *E. nigrum* ssp. *hermaphroditum* and *V. uliginosum*, and it follows them here in becoming gradually more prominent in moving to the higher altitudes of the *Vaccinium-Polytrichum* variant. *Vaccinium microcarpum* and *Sphagnum fuscum* occur occasionally in both this and other variants, and with *Betula nana* bring the composition of the vegetation close to that of the North European Sphagnion fusci mires. *B. nana*, however, and *Arctostaphylos alpinus*, seem both to be better represented at not quite the highest altitudinal extreme of the *Vaccinium-Hylocomium* sub-community: they help characterise a lower-level *Betula* variant which is found patchily through the central and north-west Highlands, and where *E. nigrum* ssp. *nigrum* and *Calluna* still retain a frequent presence. The floristic differences between this *Betula* variant and the typical variant may be due partly to treatment (Poore & McVean 1957): the possibility of soligenous influence in the *Betula* variant (Poore & McVean 1957) was discounted by McVean & Ratcliffe (1962).

Treatments, particularly burning and grazing, are of considerable importance in influencing the composition and structure of the *Calluna-Eriophorum* mire throughout its range and particularly where stands form part of unenclosed hill-grazings or grouse-moors. Thus, although the qualitative differences between the sub-communities are set largely by climatic variation, the actual appearance of the vegetation within each altitudinal zone is often the result of human activity. Since all the major structural components of the community, *E.*

vaginatum, the sub-shrubs, *R. chamaemorus*, the bryophytes and the lichens, can be affected by such treatments, and often affected in rather different ways, the range of quantitative variation is considerable. And because treatments have often been applied in complex combinations within a small compass, spatial diversity in individual stands is frequently high. This is shown very clearly in the accounts of the *Calluna-Eriophorum* mire from Moor House, where apparently uniform tracts of the community are revealed on closer inspection to comprise much smaller and quite varied blocks (e.g. Eddy *et al.* 1969, Rawes & Heal 1978, Rawes & Hobbs 1979). This kind of pattern is typical of many stretches of this type of blanket mire.

It is from Moor House that we have the most detailed studies of the impact of burning and grazing on the *Calluna-Eriophorum* mire and, although work there has been largely on the *Empetrum* sub-community (which with the impoverished *Eriophorum* mire, makes up the bulk of the blanket bog on the reserve), the general results of the investigations are of some significance for the community as a whole. With burning, the immediate effects are to destroy a proportion of the above-ground parts of the vegetation, produce fertilising ash and increase light penetration (Rawes & Hobbs 1979). Intense fires may consume all standing material, living or dead, but, if burning occurs in the winter, plants with some measure of dormancy, like *E. vaginatum* and *R. chamaemorus*, may escape the worst effects and the ground layer be better protected against severe damage by the increased wetness of the peat surface. And indeed, it is *E. vaginatum* that typically shows the first response when the sub-shrubs have been burned off, coming to dominate for up to two decades after a fire (Eddy *et al.* 1969, Rawes & Hobbs 1979). *R. chamaemorus*, too, can become very abundant in these early stages, increasing its standing crop and fruit production (Taylor 1971, Taylor & Marks 1971). Over this period, however, the sub-shrubs gradually recover (provided there is no grazing): at Moor House, the mean age of stems of *Calluna*, the usual woody dominant in the blanket bog there, increased through the first decade after a burn and then levelled out. New growth from stools and seed diversified the age-structure of the population with time, though the degenerate phase of growth did not develop because of a smothering of the older stems by the *Sphagnum* carpet (Rawes & Hobbs 1979). *E. nigrum* ssp. *nigrum* and *V. myrtillus*, typical sub-shrubs of the *Empetrum* sub-community, did not figure prominently in post-burn seres at Moor House, but both they and, at higher altitudes, *V. vitis-idaea*, can attain great abundance after fires, provided burning has not been so intense as to destroy their rhizomes: each of these species can thus dominate for some time before being overtopped by *Calluna* (Ritchie 1955a, Ratcliffe

1959a, Gimingham 1964b, 1972, Bell & Tallis 1973). Recovery of *Calluna*, both vegetatively and from seed, proved better at Moor House under short-rotation (10-year) burning than long-rotation (20-year). Indeed, Rawes & Hobbs (1979) reported that most bog species seemed to benefit from the former kind of regime, which appeared to maintain some sort of steady state. Very frequent burning, however, leads to degeneration of the vegetation (see below).

Moderate levels of grazing, too, can maintain a stable diversity in the *Calluna-Eriophorum* mire. The major herbivores on this kind of blanket bog are sheep, with some contribution from deer (mostly red) and smaller mammals (rabbits, hares and voles) and, of particular importance for *Calluna*, grouse and, at higher altitudes, ptarmigan. At Moor House, Rawes & Hobbs (1979) considered that the community could support about 1 sheep per 2.5 ha, without any burning, and continue to produce sufficient food for both stock and grouse.

The most obvious effect of grazing is on the balance between *E. vaginatum* and the palatable ericoids *Calluna* and *V. myrtillus*. *E. vaginatum* is eaten, but its growing points are well protected within the tussocks (Wein 1973), whereas the sub-shrubs can be grazed to extinction under sustained heavy stocking levels (Eddy *et al.* 1969, Rawes & Hobbs 1979) or, conversely, encouraged by enclosure (Rawes 1981, 1983). *R. chamaemorus* also benefits greatly from protection from grazing provided the surrounding vegetation does not become too dense (Taylor 1971, Rawes & Hobbs 1979, Rawes 1983). Less palatable sub-shrubs, on the other hand, like *E. nigrum* ssp. *nigrum* (Bell & Tallis 1973) and *V. vitis-idaea* (Ritchie 1955a), can continue to make a more persistent contribution to the cover on grazed *Calluna-Eriophorum* mire. On shallower peats grazing may also increase the proportion of *Juncus squarrosus* and, where trampling decreases aeration along pathways, *Scirpus cespitosus* may spread. Trampling can also disrupt the *Sphagnum* cover, destroy larger *Cladonia* spp. and favour an increase in encrusting lichens, acrocarpous mosses like *Campylopus flexuosus* and some leafy hepatics on exposed peat surfaces. Such developments produce considerable fine variation between stands of the community and, where grazing is particularly heavy and long-sustained, presage a more substantial change in the character of the vegetation (see below).

It is also possible for episodes of burning and grazing to contribute to the development of erosion within stretches of *Calluna-Eriophorum* mire, a process which can ultimately destroy the vegetation and its substrate, but which, in its less severe manifestations, introduces a further measure of patterning into the community. In fact, erosion of the kind of blanket peat of which the *Calluna-Eriophorum* mire is characteristic is a complex of processes, probably with varied causes (e.g. Johnson

1957, Radley 1962, Tallis 1964b, c, 1965, 1985b). At the higher altitudes to which the community penetrates, wind, rain and frost contribute to an element of natural climatic erosion on exposed summits and ridges, producing networks of bare areas, sometimes showing colonisation by encrusting lichens and bryophytes (Radley 1962, Tallis 1965). More dramatic is the marginal fretting of these bogs, in which extensive systems of gullies run back into the peat, separating upstanding remnant haggs, and sheet-erosion with extensive wastage of the upper layers (e.g. Lewis & Moss 1911, Tansley 1939, McVean & Ratcliffe 1962, Tallis 1965, Eddy *et al.* 1969). Severe erosion of this kind can be seen throughout the range of the *Calluna-Eriophorum* mire, being much more generally widespread than in the *Scirpus-Eriophorum* mire, where the bog margins are not so exposed, but it is especially pronounced through the southern Pennines, affecting about 75% of the blanket peat there (Anderson & Tallis 1981; see also Bower 1960, 1961).

This kind of erosion, too, may have a natural and long-standing component. In a detailed study of a small, but probably quite typical, area, Tallis (1985b) developed an original suggestion of Conway (1954) that mass movement might occur around the bog margins, restoring stability where accumulation had reached some kind of critical limit but exposing bare peat for removal with enhanced local drainage. With the early initiation of peat development in the southern Pennines, such a stage might have been reached a considerable time ago (Conway 1954, Tallis 1985a, b). More recently, erosion seems to have been markedly exacerbated by a variety of biotic influences responsible for the exposure of areas of bare peat, over which surface run-off can be channelled, with increased erosive power. Burning, particularly deep, catastrophic fires, and heavy grazing can both contribute to such developments (Radley 1962, Shimwell 1974, Tallis 1981, Tallis & Yalden 1984) and sheep-tracks (Tallis 1973b) and footpaths (Shimwell 1981) may also disrupt the cover of surface vegetation. In the southern Pennines, too, atmospheric pollution from the great industrial conurbations of Lancashire and Yorkshire (Conway 1949, Tallis 1964b, c, Ferguson & Lee 1983) has undoubtedly played a major part, together with enhanced drying of the peats as a result of water extraction, in the loss of *Sphagnum* from blanket mire vegetation there over the past two centuries, and this must also have exposed extensive areas of bare peat (Tallis 1985b).

Quite often, the bog vegetation in such severely-eroded areas has become itself so impoverished as to fall within the *Eriophorum* mire, but eroded tracts can retain sufficient floristic richness to still qualify as *Calluna-Eriophorum* mire. Typically, however, they exhibit a distinctive mosaic with *R. chamaemorus* often especially abundant around the gully heads, and *E. nigrum* ssp. *nigrum* and *V. myrtillus* becoming more prominent along the sides of the gullies, over the deeply-fretted haggs and on tumbling masses of peat around the bog margins. More gently-sloping gullies with a measure of stagnation may show a regeneration of the mire vegetation but commonly other communities appear here and over thin redistributed peat and any exposed underlying drift (see below).

Zonation and succession

The *Calluna-Eriophorum* mire is characteristically found in zonations and mosaics with wet and dry heaths and grasslands over sequences of increasingly better-drained soils, with soligenous mires often interrupting the patterns where there is local flushing and sometimes relieving the prevailingly calcifuge nature of the vegetation cover. Almost universally, such transitions have been affected by treatments, which can blur the boundaries between the vegetation types, encourage some successional changes among them and introduce new elements of variation, particularly striking where these factors contribute to gross erosion of the peat.

The basic edaphically-related pattern is best seen where there is a gradual thinning of the ombrogenous peat cover to a humic top over gleyed podzols and then a thin layer of mor over freely-draining podzols, where the *Calluna-Eriophorum* mire gives way first to Ericetalia heath, then to Calluno-Ulicetalia heath and/or Nardetalia grasslands. Typically, such sequences are disposed over ground of progressively increasing slope, running downhill where the blanket bog occupies the summits of hills (a frequent occurrence here), running uphill where it clothes valley bottoms and plateaus which give way above to a fringe of steeper ground. The transitions may be on a grand scale, where more or less flat land is a prevailing element in the scenery, as is the case in parts of the Central Highlands, over the rounded summits of the Cheviot and on the high Pennines, where the communities can show a crude altitudinal pattern of replacement one by the other. But, in more rugged terrain, the zonations are on a finer scale, with repeated sequences over the stepped landscapes of grits and shales, for example, or complex mosaics disposed over craggy ground with small boulders. And, as in similar transitions with the *Scirpus-Eriophorum* mire, regional and local climate can influence the proportions of the communities that are represented, with the wetter elements of the sequence extended on to somewhat steeper slopes in areas with higher rainfall or on northern aspects, but contracted in drier sites.

The particular kinds of *Calluna-Eriophorum* mire found in these zonations, and the types of heath and grassland, vary with altitude and, where the bog gives way both above and below to such sequences, different heaths and grasslands may be represented in each, though the general physiognomic character of the transitions is very similar. The shift to Ericetalia wet heath is thus usually marked by a fairly rapid loss of *E. vaginatum* and a further reduction in the often already low cover of Sphagna such as *S. papillosum* in the ground carpet. Where the *Calluna-Eriophorum* mire extends to lower altitudes in the *Erica* sub-community, such transitions can be fairly gradual and very like those found around the *Scirpus-Eriophorum* mire, with the *Sphagnum* sub-community of the *Scirpus-Erica* wet heath forming a marginal zone. But, at higher altitudes, where the *Calluna-Eriophorum* mire is represented by other sub-communities, it is usually drier forms of Ericetalia heath that replace the community as the peat cover thins: often, in such situations, the *Vaccinium* sub-community of the *Scirpus-Erica* wet heath can be found, but at lower altitudes in eastern Scotland, the *Juncus-Dicranum* sub-community of the *Ericetum tetralicis* can mark the transition. In these vegetation types, the most obvious floristic changes, apart from the reduction in *E. vaginatum* and the Sphagna, are increased vigour among the sub-shrubs, an abundance of *Juncus squarrosus* and/ or *Nardus stricta* and the frequent occurrence of acrocarpous mosses such as *Dicranum scoparium* and *Campylopus paradoxus*, with *Racomitrium lanuginosum* and lichens often prominent on bare peat.

Sometimes, this wet-heath zone is of considerable extent but quite frequently, where the margin of the bog is steep, it is reduced in width or totally absent, with an abrupt switch to dry heath, with overwhelming dominance of *Calluna*, or one of the Empetra or Vaccinia Usually, such vegetation can be included within the *Calluna-Vaccinium* or *Vaccinium-Deschampsia* heaths but, at higher altitudes, the *Calluna-Cladonia* or *Calluna-Arctostaphylos* heaths can replace the *Calluna-Eriophorum* mire on steeper, drier slopes, and these in turn may give way to lichen- or moss-dominated vegetation on very exposed summits. At lower altitudes, the zonation often continues beyond the heaths into some kind of Nardetalia grassland and, quite often, there are direct transitions from the community to these vegetation types with the increasing influence of pastoral agriculture (see below).

General patterns of the kinds described above are frequently complicated by the local occurrence of soligenous mires, and sometimes small topogenous basins associated with them. Again, the particular character of this vegetation varies with altitude. At lower levels through the range of the *Erica* and *Empetrum* sub-communities, various kinds of *Carex echinata-Sphagnum* mire, dominated by small sedges or Junci, can be found around flushes and along water-tracks through the bog and into the zones of fringing vegetation. At higher altitudes, this community is replaced in analgous situations by the *Carex-Sphagnum russowii* mire. Where the flushing waters are somewhat less base-poor, the *Carex-Sphagnum warnstorfii* mire occurs, or where there is a marked enrichment, as where the blanket bog has encroached over calcareous substrates, the *Pinguiculo-Caricetum*, *Carici-Saxifragetum* or *Caricetum saxatilis* can mark out sites with soligenous influence, with the *Carex-Calliergon* mire in small waterlogged hollows.

Such variegation often provides the floristically most interesting element of internal patterning on the *Calluna-Eriophorum* mire, because true hummock/hollow relief, with its distinctive zoning of plants and transitions to Rhynchosporion pools, is only rarely encountered. It is sometimes developed in the *Erica* sub-community, which can occur within tracts of the *Empetrum* type where a higher water-table is locally maintained, and then the Rhynchosporion component is generally represented by the *Sphagnum cuspidatum/recurvum* community. In other cases, the *Carex rostrata-Sphagnum recurvum* or *Eriophorum angustifolium* communities can mark out pools.

Much more commonly, internal variation within tracts of the *Calluna-Eriophorum* mire comprises quantitative differences in the major structural elements developed in response to burning and grazing. If such treatments are drastic and frequent or long-maintained, they can induce more substantial qualitative changes in the vegetation: in some cases, such changes can be reversed but they may initiate a run-down of the community and contribute to the destruction of the underlying peats.

Most obviously, frequent burning and heavy grazing contribute to the conversion of the *Calluna-Eriophorum* mire into the *Eriophorum* mire, where ericoids, Sphagna and hypnaceous mosses are of very patchy occurrence and *E. vaginatum* overwhelmingly dominant. Locally-extensive tracts of this impoverished vegetation can be found through much of the range of the *Calluna-Eriophorum* mire, though reversion to the richer blanket bog has been demonstrated after only 15 years of enclosure and freedom from burning at Moor House (Rawes 1983). In the southern Pennines, however, where injudicious treatments have been combined with draining and aerial pollution over the past two centuries, the degeneration of the *Calluna-Eriophorum* mire is particularly widespread and perhaps irremediable. Here, too, erosion of the peats is especially severe, with much bare peat even among the remnants of the richer blanket bog, patchy *Eriophorum angustifolium* pools occurring where water collects and soligenous vegetation approximating to the *Carex echinata-Sphagnum* mire marking out gullies with some moderate water movement. Similar mosaics are very well seen in the map of Moor House in Eddy *et al.* (1969).

In other cases, treatments can convert the *Calluna-*

Eriophorum mire into heath or grassland. The peats here are already fairly dry, so it is often a small step, with artificial drainage, to encourage a succession to wet heath, a change marked most obviously by a reduction in the vigour of *E. vaginatum*, and then to dry heath, with its overwhelming dominance of ericoids. Combined with grazing, such treatment can produce a Nardetalia sward or, with liming and fertiliser application, even something approximating to a Cynosurion pasture. With less assiduous improvement, *Juncus squarrosus* often gains an opportunity to spread in such situations, though enclosure of a *Juncus*-dominated pasture at Moor House showed that such a trend could be reversed, the vegetation beginning to develop into *Calluna-Eriophorum* mire after 25 years (Rawes 1981). The recent subsidising of improvement of hill-grazings has greatly speeded the reclamation of this kind of blanket bog for agriculture and rendered such changes more permanent. In other places, draining has been a prelude to afforestation.

Distribution

The *Calluna-Eriophorum* mire is centred on the higher ground in the Pennines and the Central Highlands of Scotland. The *Empetrum* sub-community is especially extensive in the former area, extending northwards through Cheviot and the Borders into eastern Scotland, but most stands in the Highlands are of the *Vaccinium-Hylocomium* sub-community, which extends to the altitudinal limit of this kind of blanket bog in central Scotland as the *Vaccinium-Polytrichum* variant, with the *Betula* and Typical variants represented at somewhat lower levels and penetrating patchily into the north-west Highlands. At lower altitudes, with a rather more oceanic climate, as through Wales and Strathclyde, the *Erica* sub-community is the usual form.

Affinities

The *Calluna-Eriophorum* mire in this scheme unites the richer kinds of *Eriophoretum vaginati* or 'Pennine blanket bog' (e.g. Smith & Moss 1903, Lewis 1904a, b, Lewis & Moss 1911, Moss 1913, Adamson 1918, Watson 1932, Tansley 1939) with the *Calluneto-Eriophoretum* common in Scotland (Poore & McVean 1957, McVean & Ratcliffe 1962, Birks 1973, Evans *et al.* 1977), also taking in related forms of blanket mire described from Wales (Ratcliffe 1959a, Edgell 1969, Tallis 1969) and south-west Scotland (Meek 1976, Birse & Robertson 1976, Birse 1980, Bignal & Curtis 1981) under a variety of names. The essential differences between these types are preserved here in the *Empetrum*, *Vaccinium-Hylocomium* and *Erica* sub-communities, though the first includes what Eddy *et al.* (1969) described as *Calluneto-Eriophoretum* (and some of their *Trichophoro-Eriophoretum*). The *Vaccinium-Hylocomium* sub-community also subsumes the *Empetreto-Eriophoretum* of McVean & Ratcliffe (1962) and Dierssen (1982) as a distinctive variant.

In the scheme proposed by Moore (1968) for the classification of European mires, this kind of vegetation was termed *Vaccinio-Ericetum tetralicis* (see also Moore 1962) and Birse & Robertson (1976) adopted this name for their lower altitude blanket bog of this kind. But Moore's account was very much based on Irish stands, with some of Pearsall's (1941) data from Stainmore, and the character of this vegetation was prevailingly of the type included here in the *Erica* sub-community. In Birse's (1980) account of Scottish vegetation, his *Vaccinio-Ericetum* became subsumed in the *Erico-Sphagnetum*, though he retained a separate *Rhytidiadelpho-Sphagnetum fusci* for the high-montane vegetation contained in this scheme in the *Vaccinium-Hylocomium* sub-community.

Despite these varied treatments of the different types of this kind of blanket bog, there is sufficient in common among them for them to be retained within a single community. It shows obvious similarities to both the *Scirpus-Eriophorum* and *Erica-Sphagnum* mires and is probably best grouped with them in the Erico-Sphagnion (as Moore (1968) proposed). However, among our ombrogenous communities, it comes closest to the Sphagnion fusci bogs of boreal peatlands, particularly in the *Vaccinium-Hylocomium* sub-community, where *Rubus chamaemorus* and *Empetrum nigrum* ssp. *nigrum* are joined by *E. nigrum* ssp. *hermaphroditum*, *Vaccinium vitis-idaea*, *V. uliginosum*, *Betula nana*, *V. microcarpum* and *Sphagnum fuscum*, all of them species which become prominent to varying degrees through Poland, northeast USSR, Finland and Sweden.

Floristic table M19

	a	b	c	19
Eriophorum vaginatum	V (1–9)	V (1–9)	V (1–8)	V (1–9)
Calluna vulgaris	V (1–9)	V (1–9)	V (1–9)	V (1–9)
Pleurozium schreberi	IV (1–6)	IV (1–6)	V (1–5)	IV (1–6)
Sphagnum capillifolium	IV (1–8)	IV (1–7)	IV (1–9)	IV (1–9)
Eriophorum angustifolium	IV (1–8)	V (1–7)	III (1–4)	IV (1–8)
Rubus chamaemorus	I (1–6)	V (1–9)	V (1–4)	IV (1–9)
Erica tetralix	V (1–4)	I (1–4)	I (1–2)	II (1–4)
Scirpus cespitosus	IV (1–6)	II (1–4)	II (1–5)	II (1–6)
Hypnum jutlandicum	IV (1–8)	I (1–4)		II (1–8)
Narthecium ossifragum	II (1–5)	I (1–2)	I (1–2)	I (1–5)
Molinia caerulea	II (1–7)	I (1–4)		I (1–7)
Drosera rotundifolia	I (1–3)			I (1–3)
Sphagnum compactum	I (1–4)			I (1–4)
Vaccinium oxycoccos	I (1–2)			I (1–2)
Empetrum nigrum nigrum	III (1–5)	V (1–6)	II (1–6)	III (1–6)
Mylia taylori	I (1–3)	II (1–3)	I (1–3)	I (1–3)
Pohlia nutans	I (1)	II (1–3)	I (1–2)	I (1–3)
Cladonia impexa	I (1–4)	II (1–6)		I (1–6)
Cephalozia bicuspidata	I (1–3)	II (1–3)		I (1–3)
Calypogeia trichomanis		II (1–3)	I (1–3)	I (1–3)
Hylocomium splendens	I (1–4)	I (1–2)	V (1–6)	II (1–6)
Cladonia arbuscula	I (4)	II (1–3)	IV (1–8)	II (1–8)
Vaccinium vitis-idaea	I (1–4)	II (1–4)	IV (1–4)	II (1–4)
Juncus squarrosus	II (1–4)	I (1–7)	III (1–5)	II (1–7)
Ptilidium ciliare	I (1–4)	II (1–3)	III (1–5)	II (1–5)
Cladonia uncialis	I (1–4)	I (1)	III (1–4)	II (1–4)
Empetrum nigrum hermaphroditum	I (1)		III (1–6)	II (1–6)
Carex bigelowii		I (1–4)	III (1–5)	II (1–5)

Species				
Vaccinium uliginosum	I (1)	I (1–2)	II (1–4)	I (1–4)
Racomitrium lanuginosum	I (1–5)	I (1–3)	II (1–6)	I (1–6)
Polytrichum alpestre	I (1)	I (1)	II (1–4)	I (1–4)
Sphagnum papillosum	I (1–4)	I (1–6)	II (1–4)	I (1–4)
Sphagnum subnitens	I (1–6)	I (4–5)	II (1–7)	I (1–7)
Cetraria islandica		I (1–3)	II (1–8)	I (1–8)
Sphagnum fuscum		I (1–3)	II (1–3)	I (1–3)
Betula nana		I (1)	II (1–4)	I (1–4)
Polytrichum alpinum			II (1–4)	I (1–4)
Arctostaphylos alpinus			I (1–4)	I (1–4)
Vaccinium microcarpum			I (1–3)	I (1–3)
Vaccinium myrtillus	III (1–8)	III (1–9)	IV (1–5)	III (1–9)
Dicranum scoparium	III (1–4)	III (1–6)	III (1–5)	III (1–6)
Rhytidiadelphus loreus	II (1–4)	III (1–4)	III (1–8)	III (1–8)
Plagiothecium undulatum	III (1–6)	III (1–5)	II (1–2)	III (1–8)
Hypnum cupressiforme	II (1–6)	III (1–4)	II (1–7)	III (1–6)
Deschampsia flexuosa	II (1–8)	II (1–8)	II (1–3)	II (1–7)
Polytrichum commune	II (1–4)	II (1–4)	I (1–4)	II (1–8)
Aulacomnium palustre	II (1–5)	II (1–4)	I (1)	II (1–4)
Lophozia ventricosa	I (1–2)	II (1–3)	II (1–3)	II (1–5)
Sphagnum recurvum	I (1–6)	I (1–8)	I (4)	II (1–3)
Sphagnum palustre	I (1–5)	I (1–6)	I (1–4)	I (1–8)
Potentilla erecta	I (1–2)	I (1–4)	I (1–4)	I (1–6)
Nardus stricta	I (1–8)	I (1–4)	I (1–4)	I (1–4)
Rhytidiadelphus squarrosus	I (1–3)	I (1–4)	I (1–6)	I (1–8)
Barbilophozia floerkii	I (1–2)	I (1–4)	I (1)	I (1–4)
Diplophyllum albicans	I (1–3)	I (1–3)	I (1–2)	I (1–4)
Listera cordata	I (1–2)	I (1–4)	I (1–3)	I (1–3)
Campylopus paradoxus	I (1–4)	I (1–3)	I (1–2)	I (1–4)
Odontoschisma sphagni	I (1–4)	I (1–3)	I (1–2)	I (1–4)
Carex nigra	I (1–4)	I (1–3)	I (1–3)	I (1–4)
Melampyrum pratense	I (1–2)	I (1–3)	I (1–3)	I (1–3)
Cladonia squamosa	I (1–2)	I (1–3)	I (1–2)	I (1–4)
Sphagnum tenellum	I (1–2)	I (1–2)	I (1–4)	I (1–2)
Cladonia bellidiflora	I (1–3)	I (1–3)	I (1)	I (1–3)
Galium saxatile	I (1–3)	I (1–3)	I (1–3)	I (1–3)
Calypogeia fissa	I (1–3)	I (1–3)	I (1–3)	I (1–3)
Pinguicula vulgaris	I (1–2)		I (1–3)	I (1–2)
Calypogeia muellerana	I (1–4)	I (1–3)	I (1–2)	I (1–4)
Cladonia coccifera	I (1–2)	I (1–3)		I (1–3)
Hypogymnia physodes	I (1–3)	I (1–3)		I (1–3)
Cladonia floerkeana	I (1–3)	I (1–3)		I (1–3)
Kurzia pauciflora		I (1–3)		I (1–3)
Cladonia tenuis		I (4)	I (1–3)	I (1–3)
Mylia anomala		I (1–3)	I (1–3)	I (1–4)
Cornus suecica		I (1–3)	I (1–3)	I (1–3)
Festuca vivipara		I (1–6)	I (1–3)	I (1–3)
Huperzia selago		I (1–3)	I (1–2)	I (1–6)
Cornicularia aculeata		I (1–3)	I (1–3)	I (1–3)
Dicranum fuscescens		I (1–3)	I (1–3)	I (1–3)

Floristic table M19 *(cont.)*

	a	b	c	19
Sphagnum robustum		I (1–2)	I (1–4)	I (1–4)
Cladonia gracilis		I (1–3)	I (1–3)	I (1–3)
Blechnum spicant		I (1–2)	I (1–2)	I (1–2)
Dicranum majus		I (4)	I (1–2)	I (1–4)
Carex binervis		I (1–4)	I (1–2)	I (1–4)
Luzula sylvatica		I (1–4)	I (1–3)	I (1–4)
Number of samples	61	53	56	181
Number of species/sample	17 (11–33)	19 (7–33)	20 (10–33)	19 (7–33)
Herb height (cm)	21 (8–40)	23 (11–60)	17 (5–45)	21 (5–60)
Herb cover (%)	83 (6–100)	89 (65–100)	96 (70–100)	89 (6–100)
Bryophyte height (mm)	38 (5–80)	40 (10–100)	33 (10–60)	37 (5–100)
Bryophyte cover (%)	56 (20–100)	39 (2–80)	75 (60–100)	53 (2–100)
Altitude (m)	409 (28–820)	601 (355–978)	691 (457–923)	568 (28–978)
Slope (°)	4 (0–12)	4 (0–18)	6 (0–25)	5 (0–25)
Soil pH	4.3 (3.2–6.2)	3.8 (3.1–4.7)	4.6 (4.1–5.5)	4.2 (3.1–6.2)

a   *Erica tetralix* sub-community
b   *Empetrum nigrum nigrum* sub-community
c   *Vaccinium vitis-idaea-Hylocomium splendens* sub-community
19   *Calluna vulgaris-Eriophorum vaginatum* blanket mire (total)

Floristic table M19c, variants

	ci	cii	ciii
Eriophorum vaginatum	V (1–5)	V (1–8)	V (1–8)
Calluna vulgaris	V (6–9)	V (2–9)	II (1–4)
Pleurozium schreberi	V (1–4)	V (1–5)	IV (1–4)
Sphagnum capillifolium	IV (1–9)	IV (1–6)	IV (1–4)
Eriophorum angustifolium	II (1–4)	II (1–3)	V (1–4)
Rubus chamaemorus	V (1–3)	V (1–4)	V (1–4)
Vaccinium myrtillus	V (1–2)	V (1–5)	V (1–4)
Hylocomium splendens	V (1–6)	V (1–6)	III (1–3)
Vaccinium vitis-idaea	III (1–3)	V (1–3)	IV (1–4)
Cladonia arbuscula	III (1–5)	V (1–8)	IV (1–4)
Betula nana	V (1–4)	I (1)	
Empetrum nigrum nigrum	III (1–4)	II (1–5)	I (1–2)
Hypnum cupressiforme	III (1–4)	II (1–8)	
Arctostaphylos alpinus	III (1–4)	I (4)	
Erica tetralix	III (1–3)		
Cladonia rangiferina	II (1–3)	I (1–3)	
Mylia anomala	II (1–3)	I (1–3)	I (1–3)

Vaccinium microcarpum	II (1–3)		
Vaccinium uliginosum	I (1–3)	I (2–4)	V (1–4)
Carex bigelowii		III (1–6)	V (1–6)
Racomitrium lanuginosum		II (1–4)	V (1–5)
Polytrichum alpestre		I (1–4)	V (1–4)
Scirpus cespitosus	II (1–2)	II (1–5)	IV (1–4)
Sphagnum papillosum		I (1–4)	III (4–6)
Kiaeria starkei			I (1–2)
Empetrum nigrum hermaphroditum	III (1–3)	III (1–4)	IV (4–6)
Juncus squarrosus	IV (1–5)	II (1–5)	III (1–3)
Rhytidiadelphus loreus	III (1–4)	IV (1–8)	III (1–3)
Ptilidium ciliare	III (1–6)	III (1–2)	III (1–3)
Dicranum scoparium	II (1–3)	III (1–6)	III (1–5)
Cladonia uncialis	I (1–3)	IV (1–4)	IV (1–4)
Cladonia impexa	II (1–2)	II (1–4)	II (1–3)
Lophozia ventricosa	II (1–3)	I (1–2)	II (1–3)
Plagiothecium undulatum	II (1–3)	II (1–3)	
Deschampsia flexuosa		II (1–3)	II (1–3)
Cetraria islandica		II (1–3)	II (1–3)
Sphagnum subnitens		II (1–8)	II (1–4)
Sphagnum fuscum	I (1–2)	II (1–4)	I (4–5)
Anastrepta orcadensis	I (1–3)	II (1–2)	I (1)
Melampyrum pratense	I (1–2)	II (1)	
Rhytidiadelphus squarrosus	I (1–3)		II (1–3)
Polytrichum commune	I (1–2)	I (1–4)	I (1–3)
Diplophyllum albicans	I (1–3)	I (1–3)	I (1–3)
Campylopus paradoxus	I (1–2)	I (1–3)	I (1–2)
Aulacomnium palustre	I (1–2)	I (1–2)	I (1–3)
Pinguicula vulgaris	I (1–2)	I (2)	I (2)
Barbilophozia floerkii	I (1–3)	I (1–3)	I (1)
Cornus suecica	I (1–3)	I (1–3)	I (1–3)
Potentilla erecta	I (1–2)	I (1–4)	I (1)
Erica cinerea	I (1–3)	I (1–3)	I (1–3)
Sphagnum robustum	I (1–4)	I (1–2)	I (1–4)
Cladonia gracilis	I (1–3)	I (1–3)	I (1–3)
Cladonia squamosa	I (1–3)	I (1–4)	I (1–3)
Mylia taylori	I (1–3)	I (1)	
Kurzia pauciflora	I (1–3)	I (1–3)	
Cladonia tenuis	I (1–2)	I (2)	
Listera cordata	I (1–3)	I (1–3)	
Calypogeia fissa	I (1–3)	I (1–3)	
Solidago virgaurea	I (1)	I (1–2)	
Carex binervis	I (1–2)	I (1–3)	
Sphagnum palustre	I (1–3)	I (4)	
Trientalis europaea		I (1)	I (1)
Cladonia bellidiflora		I (1–2)	I (1–3)
Sphagnum tenellum		I (1–3)	I (2)
Cornicularia aculeata		I (1–3)	I (1–3)
Festuca vivipara		I (1–3)	I (1–3)
Galium saxatile		I (1–2)	I (2)

Floristic table M19c, variants *(cont.)*

	ci	cii	ciii
Dicranum fuscescens		I (1–3)	I (1–3)
Polytrichum alpinum		I (1–4)	I (1–3)
Number of samples	20	22	12
Number of species/sample	20 (16–27)	21 (12–33)	23 (18–33)
Herb height (cm)	24 (12–45)	15 (5–30)	9 (5–11)
Herb cover (%)	92 (70–100)	98 (85–100)	100
Bryophyte height (mm)		29 (20–40)	40 (10–60)
Bryophyte cover (%)		75	
Altitude (m)	569 (457–766)	705 (495–853)	856 (762–923)
Slope (°)	4 (0–15)	8 (2–25)	5 (2–15)
Soil pH		4.7 (4.1–5.5)	4.3 (4.2–4.3)

ci   *Betula nana* variant
cii   Typical variant
ciii *Vaccinium uliginosum-Polytrichum alpestre* variant

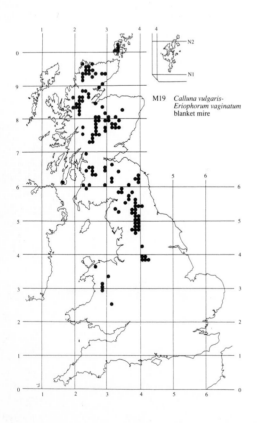

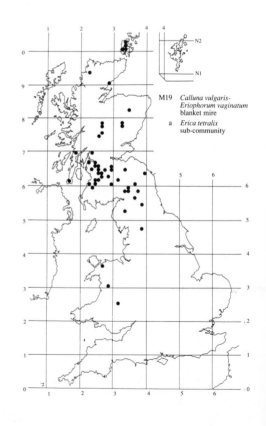

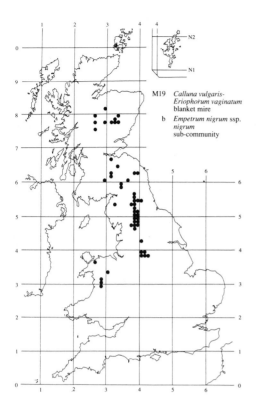

M19   *Calluna vulgaris-*
      Eriophorum vaginatum
      blanket mire
   b *Empetrum nigrum* ssp.
      nigrum
      sub-community

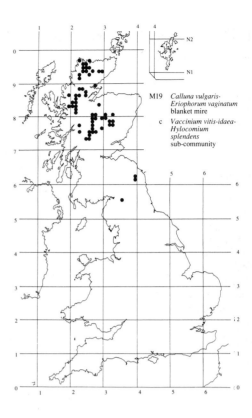

M19   *Calluna vulgaris-*
      Eriophorum vaginatum
      blanket mire
   c *Vaccinium vitis-idaea-*
      Hylocomium
      splendens
      sub-community

M20
Eriophorum vaginatum blanket and raised mire

Synonymy
Eriophoretum vaginati Smith & Moss 1903, Lewis 1904, Lewis & Moss 1911, Moss 1913, Adamson 1918, Watson 1932, Tansley 1939, all *p.p.*; *Eriophorum vaginatum* moss Pearsall 1941; *Eriophorum vaginatum* bog Conway 1949; Raised bog Sinker 1960 *p.p.*, Proctor 1974 *p.p.*; *Eriophoretum* Eddy *et al.* 1969, Fidler *et al.* 1970; *Eriophorum angustifolium-Eriophorum vaginatum* Gesellschaft Dierssen 1982; *Empetrum nigrum* Gesellschaft Dierssen 1982; *Calluna vulgaris* Gesellschaft Dierssen 1982.

Constant species
Eriophorum angustifolium, E. vaginatum.

Physiognomy
The *Eriophorum vaginatum* mire comprises species-poor ombrogenous bog vegetation dominated by *E. vaginatum*, the tussocks of which form an open or closed canopy, 1–3 dm high. Other frequent vascular associates are very few and typically of low cover: the only other constant plant is *E. angustifolium* and this is usually found as scattered shoots, sometimes a little more densely distributed in wetter runnels. Ericoid sub-shrubs in particular are noticeably patchy in their occurrence compared with the *Calluna-Eriophorum* mire. *Calluna vulgaris, Empetrum nigrum* ssp. *nigrum* and *Vaccinium myrtillus* can each be found quite frequently and the last two may occur with some measure of local abundance but they tend not to form a consistently vigorous and mixed component of the vegetation: one or more, quite often all, of them may be lacking or reduced to sparse and puny shoots, with *Calluna* and *V. myrtillus* frequently showing signs of having been nibbled. *Rubus chamaemorus*, another good preferential of the *Calluna-Eriophorum* mire, is likewise unevenly represented here, being only occasional overall and of low cover. *Deschampsia flexuosa* occurs fairly commonly and it can be locally abundant though, where closely-grazed, with its inflorescences eaten off, it can readily be overlooked.

Festuca ovina, Juncus squarrosus, Scirpus cespitosus and *Carex bigelowii* all occur infrequently.

The ground cover is variable in its extent, though typically sparse and patchy and never showing the richness and luxuriance characteristic of virgin Erico-Sphagnion mires. Sphagna, in particular, are scarce with *Sphagnum capillifolium* and *S. papillosum* the most usual species but even these very infrequent and of small cover. Equally obvious is the poor representation of hypnaceous mosses with species such as *Pleurozium schreberi, Rhytidiadelphus loreus, Plagiothecium undulatum* and *Hypnum cupressiforme/jutlandicum* occurring only very uncommonly.

Positive features among the bryophytes are few but *Campylopus paradoxus* is fairly frequent throughout and it can be accompanied by *Dicranum scoparium* in stands with a richer sub-shrub cover. More usual associates are occasional scattered shoots of *Orthodontium lineare, Pohlia nutans, Drepanocladus fluitans* and a variety of leafy hepatics including *Ptilidium ciliare, Calypogeia trichomanis, C. muellerana, Barbilophozia floerkii, Lophozia ventricosa, Gymnocolea inflata, Cephalozia bicuspidata, Cephaloziella hampeana, Diplophyllum albicans* and *Mylia taylori*.

Lichens are typically few in number and of low total cover. Bulkier species, like *Cladonia arbuscula, C. uncialis* and *C. impexa*, can sometimes be found but often there is just a very patchy cover of peat-encrusters, such as *C. chlorophaea, C. floerkeana, C. squamosa* and *C. coccifera*, on the less shaded areas of bare ground.

Sub-communities

Species-poor sub-community. The more impoverished stands included here present one of the gloomiest spectacles among British vegetation. *E. vaginatum* forms the bulk of the vascular cover, with scattered *E. angustifolium*, but only very occasional sub-shrubs and sometimes a little *Deschampsia flexuosa*. The ground between the tussocks is sometimes extensive, often densely-shaded and predominantly bare apart from scattered tufts of *Campylopus paradoxus* and *Orthodontium lineare* (which can also grow on the tussock sides) and hepatics such as *Calypogeia* spp., *Gymnocolea inflata*, *Lophozia ventricosa*, *Cephalozia bicuspidata* and *Cephaloziella hampeana* on the bare peat or cotton-grass litter. *Drepanocladus fluitans* also occurs occasionally with local abundance. Lichens are typically very sparse.

Calluna vulgaris-Cladonia spp. sub-community. The vegetation here is somewhat richer than above and transitional to the *Calluna-Eriophorum* mire in its composition. *E. vaginatum* is not quite so overwhelmingly dominant, though, with *E. angustifolium*, it still remains the most frequent element. But sub-shrubs occur a little more commonly, with scattered bushes of *Calluna*, *V. myrtillus* and, especially likely to show local dominance, *E. nigrum* ssp. *nigrum*. Among this more variegated and less densely-shaded cover, mosses and lichens are rather better represented. *Campylopus paradoxus* and *Dicranum scoparium* occur frequently, though generally in small amounts, and *Pohlia nutans* and *Pleurozium schreberi* are occasional. Larger Cladonias can be found among shrub stools and on moribund cotton-grass tussocks and, on bare ground, there can be a variety of peat-encrusting species. *Cornicularia aculeata* and *Cetraria islandica* also occur sparsely. Although *Mylia taylori*, *Kurzia pauciflora*, *Ptilidium ciliare* and *Calypogeia trichomanis* are occasionally recorded, the hepatics characteristic of damp or shaded runnels in the species-poor sub-community are uncommon here.

Habitat

The *Eriophorum* mire is characteristic of ombrogenous peats on bogs where certain kinds of treatment have become of overriding importance in determining the nature of the vegetation: long-continued and heavy grazing, together with burning, have been of especial significance in its development but draining and aerial pollution have also played a part. The community is commonest on blanket mires, where these factors have often contributed, not only to floristic impoverishment, but also to gross erosion of the underlying peats, but it can also be found more locally on run-down raised bogs.

By and large, the *Eriophorum* mire is a community of higher elevations where the climate is cold and wet. Most stands lie between 500 and 700 m and experience 1200–1600 mm annual precipitation (*Climatological Atlas* 1952), with 160–200 wet days yr^{-1} (Ratcliffe 1968) and mean annual maximum temperatures of 21–25°C (Conolly & Dahl 1970). On gentler slopes, usually from 0 to 10°, within such a zone, blanket peat is widespread and, under natural conditions, the deeper deposits generally support some kind of *Calluna-Eriophorum* mire, typically, at these more modest altitudes, of the

Empetrum sub-community, with its mixed canopy of *E. vaginatum* and sub-shrubs, frequent records for the Arctic-Subarctic *Rubus chamaemorus*, and fairly rich ground cover of Sphagna and hypnaceous mosses. In most cases, the *Eriophorum* mire seems to be a biotically-derived replacement for this kind of blanket bog, showing varying degrees of floristic impoverishment according to the intensity and duration of the treatments to which particular tracts have been subjected. Such replacement can be seen throughout the range of the *Empetrum* sub-community of the *Calluna-Eriophorum* mire, but it is especially extensive in the southern Pennines, where the factors seem to have operated with concerted ferocity: in this region, the *Eriophorum* mire is the prevailing kind of blanket bog over many square kilometres of the Carboniferous grit and shale uplands. Less commonly, the community occurs on ombrogenous peat in raised bogs where, under natural conditions, one would expect to find the *Erica-Sphagnum* mire. The floristic gradation between these two richer Erico-Sphagnion communities and the *Eriophorum* mire is continuous: the *Calluna-Cladonia* sub-community of the latter includes transitional stands which preserve an element of their variety, the species-poor type represents the dismal peak of degradation.

Various treatments can contribute to this impoverishment. Of particular importance for the reduction in the cover of some of the major sub-shrubs are burning and, more significant in the long term, grazing. Burning can result in the total destruction of the above-ground parts and sometimes the stools and rhizomes of the ericoids (Gimingham 1960, Bell & Tallis 1973), stimulating an expansion in the abundance of *E. vaginatum* (Wein 1973, Rawes & Hobbs 1979) and, in the *Calluna-Eriophorum* mire, of *R. chamaemorus* (Taylor 1971, Taylor & Marks 1971). In the absence of grazing, however, the sub-shrubs show a gradual recovery, sometimes first the rhizomatous *V. myrtillus* or *E. nigrum* ssp. *nigrum*, but usually eventually the taller *Calluna* (Eddy *et al.* 1969, Rawes & Hobbs 1979), such that, over a decade or so, something like the original balance of structural components is restored. It is possible that some stretches of blanket bog, like the *Eriophorum* mire, represent temporary post-burn vegetation; and repeated fires may maintain such an impoverished cover by continually setting back sub-shrub regeneration. Judicious burning at regular intervals, however, such as is practised on blanket bogs that contribute to grouse-moors, can stabilise the cover to yield a constant supply of building-phase *Calluna* (e.g. Gimingham 1972).

The more persistent and uncompromising dominance of *E. vaginatum* seen here is more likely to develop where long and intensive grazing has been practised (either alone or with burning). Blanket mire can make up a major part of unenclosed upland and it has long been traditional to turn out sheep on to such ground for summer pasturing. At probably quite modest densities (perhaps about 1 sheep per 2 ha: e.g. Rawes & Hobbs 1979), such grazing can extinguish the palatable ericoids, *Calluna* and *V. myrtillus* and, in the *Calluna-Eriophorum* mire, reduce the cover of *R. chamaemorus* (Eddy *et al.* 1969, Taylor 1971, Rawes & Hobbs 1979) and shift the balance decisively towards *E. vaginatum* which, though it is eaten, has its growing points well protected in the tussocks (Wein 1973). Somewhat lighter or patchy grazing may allow sporadic persistence of sub-shrubs, particularly the less palatable sub-shrub *E. nigrum* ssp. *nigrum* (Bell & Tallis 1973): this is the woody plant which most frequently shows local dominance in the *Eriophorum* mire and, in some places, as on Ilkley Moor in West Yorkshire, it has become abundant over extensive areas (Dalby 1961, Fidler *et al.* 1970, Dalby *et al.* 1971).

Grazing stock can also impoverish ombrogenous mire vegetation by trampling which can disrupt the carpet of Sphagna and damage larger lichens (e.g. Rawes & Hobbs 1979). But, in many stands, dense shading by *E. vaginatum* may be of more importance in reducing the ground cover. In the species-poor sub-community, this is very sparse and patchy, being largely limited to diminutive acrocarps, which can get a hold on bare peat or decaying tussocks, and shade-tolerant hepatics in the runnels. And the modest enrichment seen in the *Calluna-Cladonia* sub-community may be largely due to the sporadic relief from *E. vaginatum*-dominance provided by the occasional sub-shrubs, among the shoots of which there is better light penetration. Even here, however, the rich carpets of hypnaceous mosses so characteristic of the *Calluna-Eriophorum* mire are not seen.

Two other factors may contribute to this general poverty in the ground layer of the *Eriophorum* mire. First, and of especial importance for the marked scarcity of Sphagna, is the dryness of the peats. In the *Erica-Sphagnum* mire, these are maintained in a state of fairly consistent waterlogging, with species such as *S. papillosum* and *S. magellanicum* abundant as active peat-builders. In the *Calluna-Eriophorum* mire, the peat is often surface-dry in summer, so these species are of patchy occurrence, though *S. capillifolium* remains frequent throughout. In the *Eriophorum* mire, they are drier still, often showing surface oxidation and marked acidity, with a pH frequently as low as 3 or so. Such dryness may be partly natural, with climatic change producing a general lowering of the water-table, or erosion resulting in enhanced local run-off, but drainage operations have probably been an important contributory factor, either on a local scale for agricultural improvement or afforestation, or more extensively, as in water-gathering schemes to supply urban or industrial developments. The second factor, resulting from these last activities, is aerial pollution, particularly with oxides of sulphur and nitrogen which have a very adverse effect on most bryophytes and lichens. The great reduction in Sphagna on some blanket mires over the past 200 years is probably largely due to this pollution (Conway 1949, Tallis 1965, Ferguson 1979, Lee 1981, Ferguson & Lee 1983) and it is very striking how the *Eriophorum* mire is concentrated in that part of Britain where blanket bogs are subject to the highest levels of SO_2 (see map in Page 1982). There, in the south Pennines, this community presents a depressing spectacle over the summits between the great industrial conurbations of Lancashire and Yorkshire: fringed by numerous reservoirs and cut through by arterial roads, vegetation and peat are often coated with grime and even the sheep look grey.

It is in this region, too, that the erosion of blanket peat is most severe and widespread in Britain, with up to 75% of the mantle being affected (Anderson & Tallis 1981). Evidence suggests that such erosion has a natural component, perhaps due to marginal instability in peats that have been accumulating longer than elsewhere in the country (Conway 1954, Tallis 1985a, b), but the various biotic factors outlined above have probably been of local significance in exacerbating the processes involved and have perhaps contributed in concert to a general spiral of degradation in the southern Pennines (e.g. Radley 1962, Tallis 1973b, Shimwell 1974, 1981, Tallis & Yalden 1984). Tracts of *Eriophorum* mire in this region and in some other areas of blanket bog (as at Moor House: Eddy *et al.* 1969), show varying degrees of marginal fretting and sometimes the two sub-communities are disposed over this zone in a mosaic: the species-poor type can occupy the core of more extensive upstanding haggs in such situations, with the *Calluna-Cladonia* sub-community running around their margins as a fringe to the drainage channels.

Zonation and succession

The *Eriophorum* mire is typically found as a replacement for the *Calluna-Eriophorum* mire (less often for the *Erica-Sphagnum* mire) in zonations and mosaics with heaths and grasslands over sequences of increasingly better-drained soils, with soligenous mires reflecting the occurrence of local flushing. The extent of replacement and the clarity of the boundaries between the community and remaining tracts of the original mire vegetation depend largely on the intensity of the various factors mediating their interconversion. Such treatments can also affect transitions to neighbouring vegetation types and they have contributed widely to the degradation of entire landscapes in which the community is found.

Around the margins of blanket bogs, on which the

Eriophorum mire most commonly occurs, the thinning of the peat cover is generally marked by a transition to Ericetalia wet heath over gleyed podzols, or Calluno-Ulicetalia heath or Nardetalia grasslands over more free-draining podzols or rankers. Typically, such zonations occur over progressively steepening slopes, usually running downhill from summit plateaus, or sometimes uphill, where flat, peat-covered ground gives way above to rougher terrain. In the south Pennines, these sequences are seen on a grand scale, forming a crude altitudinal pattern, with local reversals where resistant grits are exposed in lines of crags (e.g. Smith & Moss 1903, Moss 1913).

The kind of Ericetalia community most often found in association with the *Eriophorum* mire on blanket bog fringes seems to be the *Juncus-Dicranum* sub-community of the *Ericetum tetralicis* wet heath. This may be of considerable extent where the peat thins gradually or has been eroded into substantial amounts of downwash, but around abrupt mire margins this zone is often curtailed. Then, there may be a more or less direct switch to Calluno-Ulicetalia vegetation, usually some form of *Calluno-Vaccinium* or *Vaccinium-Deschampsia* heath. Very commonly, however, the grazing which plays a major part in the development of the *Eriophorum* mire itself, has converted some or all of this heath vegetation to grassland, leaving strips of sub-shrub vegetation confined to more inaccessible crags and tumbles of boulders below them: this is well seen in the southern Pennines (e.g. Lewis & Moss 1911, Moss 1913). In the poor-quality swards that develop on the grazed ground around the mire, *Juncus squarrosus* often becomes very prominent over the more ill-drained thin peats and peaty podzols, with *Nardus stricta* a leading species on the drier podzols and rankers: the map of Moor House in Eddy *et al.* (1969) shows this kind of pattern very clearly. In other cases, as on Ilkley Moor, the unpalatable *E. nigrum* ssp. *nigrum* has spread in abundance through most of the zonations around the grazed mire fringes, masking the pattern of communities (Fidler *et al.* 1970, Dalby *et al.* 1971).

Strips of *Juncus-Festuca* or *Nardus-Galium* grassland can also be found running through tracts of the *Eriophorum* mire where there is very slight flushing of thin peats or exposed mineral soils; and, if the waters show some amelioration of base-deficiency, small stands of *Festuca-Agrostis-Galium* grassland can occur. Stronger soligenous influence, which maintains the soils in a wetter state, is typically characterised by tracks of Caricion nigrae vegetation, often, at these modest altitudes, the *Carex echinata-Sphagnum* mire, dominated by small sedges or, very frequently, by *Juncus effusus*. These flushes may have Cardamino-Montion springs at their source, and they can unite as they flow through the

mire and the grasslands around to form distinct streams. More stagnant wet areas within stands of the *Eriophorum* mire are generally scarce but occasional pools may have species-poor Rhynchosporion vegetation like that of the *Eriophorum angustifolium, Sphagnum cuspidatum/recurvum* or *Carex rostrata-Sphagnum recurvum* communities.

Two other kinds of surface patterning can be found where the *Eriophorum* mire occurs on blanket bogs. Quite commonly, the community is seen in mosaics with the *Calluna-Eriophorum* mire, the disposition of the two vegetation types and the sharpness of the boundaries between them reflecting the pattern of treatments. Sometimes, the separation of the two along a fence-line provides striking testimony to the importance of grazing in mediating change between them, but often the transitions are less well defined, reflecting gradual reductions in grazing intensity, for example, in moving away from adjacent stretches of more palatable grasslands (e.g. Pearsall 1941, Eddy *et al.* 1969, Rawes 1983). In certain cases, zonations between the two communities may be temporary, their boundaries shifting with variations in grazing or where there is recovery after burning. And, even where the existence of *Eriophorum* mire seems to be well established, it has been shown that the process of impoverishment can be reversed: at Moor House, enclosure and freedom from burning has allowed a convincing progression back to *Calluna-Eriophorum* mire within the space of 25 years (Rawes 1983). Enclosed *Juncus squarrosus* swards there, perhaps derived by very heavy grazing of blanket mire on thinner peats, have also begun to show the same development (Rawes 1981).

It is very doubtful whether the *Eriophorum* mire so common throughout the southern Pennines could show the same kind of recovery, for here, quite apart from the very thorough floristic impoverishment, there is particularly severe erosion of the underlying peats. This introduces a further element of patterning into the vegetation cover, fragmenting the *Eriophorum* mire itself and leading to the development of mosaics of the two sub-communities over the more intact haggs and their wasting margins. It can also precipitate a progression from the *Calluna-Cladonia* sub-community to dry heath along the freely-draining tops of the drainage channels and tumbling masses of dry peat over their sides and, where there is extensive marginal wasting, lead to a replacement of mire vegetation by *Juncus-Festuca* grassland over the redistributed materials. Sometimes, there is very patchy regeneration of the mire within more gently-sloping channels, or where their blockage has induced some stagnation, but usually, where these are not entirely bare, they contain fragments of Caricion nigrae vegetation like the *Carex echinata-Sphagnum* mire, extending back, as erosion progresses, into the heart of the bog.

Where the *Eriophorum* mire occurs more locally on raised bogs, it generally appears to replace the *Erica-Sphagnum* mire and may form a mosaic with surviving fragments of this where treatment has not been so thoroughgoing. Around the mire margins, there is typically a transition to Junco-Molinion vegetation on the rand, with overwhelming dominance of *E. vaginatum* passing to equally uncompromising dominance of *Molinia caerulea*. In much-disturbed situations, the lagg vegetation beyond this may become markedly eutrophicated but, in some sites, impoverished *Eriophorum* mire occurs in close contact with a rich assemblage of soligenous vegetation, as at Malham Tarn (Sinker 1960, Proctor 1974, Adam *et al.* 1975). Drying of the peats under the community in raised bogs, often accentuated by peat-cutting and draining, can favour a spread of Junco-Molinion vegetation which, with removal of grazing, can progress to *Betula-Molinia* woodland.

Distribution
The community can be found locally throughout northern Britain, mainly within tracts of upland blanket bog but more locally on raised bogs. It is especially extensive in the southern Pennines.

Affinities
Impoverishment of Erico-Sphagnion bogs to produce the kind of vegetation included here is a continuous phenomenon so that it is hard to draw the line between the *Eriophorum* mire and its richer progenitors: in early schemes, the community was generally included with the *Calluna-Eriophorum* mire (or, at least with its *Empetrum* sub-community), from which it most usually develops, in an *Eriophoretum* of one kind or another (Smith & Moss 1903, Lewis 1904a, b, Lewis & Moss 1911, Moss 1913, Adamson 1918, Watson 1932, Tansley 1939). But, in its extreme form, this kind of vegetation should not go unrecognised as a distinct unit and the most convenient criteria for separating the communities are the relative prominence of *E. vaginatum* and the palatable ericoids, the frequency of *R. chamaemorus* (within its general geographic range) and the prominence of Sphagna and hypnaceous mosses, as in the treatment of Eddy *et al.* (1969).

Floristic table M20

	a	b	20
Eriophorum vaginatum	V (4–9)	V (4–8)	V (4–9)
Eriophorum angustifolium	V (1–3)	V (1–4)	V (1–4)
Orthodontium lineare	II (1–3)	I (4)	I (1–4)
Calypogeia muellerana	II (1–3)	I (1)	I (1–3)
Lophozia ventricosa	II (1–3)	I (2)	I (1–3)
Cephalozia bicuspidata	II (1–3)	I (1)	I (1–3)
Gymnocolea inflata	II (1–4)		I (1–4)
Cephaloziella hampeana	II (1–3)		I (1–3)
Drepanocladus fluitans	II (1–4)		I (1–4)
Campylopus paradoxus	III (1–3)	IV (1–4)	III (1–4)
Calluna vulgaris	II (1–3)	IV (1–4)	III (1–4)
Empetrum nigrum nigrum	II (2–9)	IV (1–8)	III (1–9)
Vaccinium myrtillus	I (1–3)	III (1–5)	II (1–5)
Dicranum scoparium	I (1)	III (1–4)	II (1–4)
Cladonia arbuscula		III (1–6)	II (1–6)
Mylia taylori		III (1–4)	II (1–4)
Cladonia chlorophaea		III (1–3)	II (1–3)
Cladonia impexa	I (1)	II (1–3)	II (1–3)
Cornicularia aculeata	I (1)	II (1–5)	II (1–5)
Cladonia uncialis	I (1)	II (1–2)	I (1–2)
Cladonia floerkeana	I (1)	II (1–3)	I (1–3)
Rubus chamaemorus	I (1)	II (1–3)	I (1–3)

	a	b	20
Cladonia squamosa	I (1)	II (1–3)	I (1–3)
Pohlia nutans	I (1–3)	II (1–3)	I (1–3)
Diplophyllum albicans	I (1–4)	II (1–3)	I (1–4)
Cetraria islandica		II (1–2)	I (1–2)
Cladonia coccifera		II (1–3)	I (1–3)
Kurzia pauciflora		II (1–4)	I (1–4)
Cladonia impexa		II (1–3)	I (1–3)
Pleurozium schreberi		II (1–2)	I (1–2)
Carex bigelowii		I (1–5)	I (1–5)
Scirpus cespitosus		I (1–4)	I (1–4)
Sphagnum fuscum		I (5)	I (5)
Hypnum jutlandicum		I (1–4)	I (1–4)
Ptilidium ciliare	II (1–3)	II (1–4)	II (1–4)
Calypogeia trichomanis	II (1–3)	II (1–4)	II (1–4)
Deschampsia flexuosa	II (1–4)	II (1–8)	II (1–8)
Barbilophozia floerkei	I (1–3)	I (1–3)	I (1–3)
Festuca ovina	I (1–3)	I (1–5)	I (1–5)
Sphagnum capillifolium	I (1)	I (1–4)	I (1–4)
Juncus squarrosus	I (1–4)	I (1–3)	I (1–4)
Sphagnum papillosum	I (1–4)	I (1)	I (1–4)
Plagiothecium undulatum	I (1)	I (1–3)	I (1–3)
Polytrichum commune	I (1)	I (1–3)	I (1–3)
Number of samples	16	24	40
Number of species/sample	9 (5–13)	12 (9–20)	11 (5–20)
Herb height (cm)	24 (15–30)	16 (10–25)	21 (10–30)
Herb cover (%)	92 (60–100)	90 (55–100)	91 (55–100)
Bryophyte height (mm)	14 (10–30)	30 (10–50)	18 (10–50)
Bryophyte cover (%)	12 (0–40)	26 (1–50)	15 (0–50)
Altitude (m)	522 (370–660)	648 (485–942)	596 (370–942)
Slope (°)	3 (0–15)	3 (0–10)	3 (0–15)
Soil pH	3.1 (3.0–3.4)	3.5 (3.1–3.8)	3.2 (3.0–3.8)

a   Species-poor sub-community

b   *Calluna vulgaris-Cladonia* sub-community

20   *Eriophorum vaginatum* blanket mire (total)

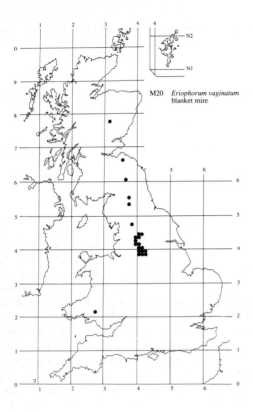

M20 *Eriophorum vaginatum*
blanket mire

M21

Narthecium ossifragum-Sphagnum papillosum valley mire
Narthecio-Sphagnetum euatlanticum Duvigneaud 1949

Synonymy

Sphagnetum Rankin 1911*b p.p.*; *Sphagnum* hummock complex, *Sphagnum papillosum* & *S. rubellum* phases Rose 1953; General bog communities Newbould 1960 *p.p.*; Valley bog Ward *et al.* 1972*a p.p.*; *Sphagnum* lawn bog hummocks NCC New Forest Bogs Report 1984.

Constant species

Calluna vulgaris, Drosera rotundifolia, Erica tetralix, Eriophorum angustifolium, Molinia caerulea, Narthecium ossifragum, Sphagnum papillosum.

Rare species

Agrostis curtisii, Erica ciliaris, Hammarbya paludosa.

Physiognomy

The *Narthecio-Sphagnetum* comprises mire vegetation dominated by carpets of Sphagna with scattered herbs and sub-shrubs, forming extensive lawns or the drier areas within low-amplitude hummock/hollow systems. The dominant *Sphagnum* is generally *S. papillosum*, the cover of which can be very extensive and luxuriant, but there is quite frequently some *S. auriculatum* or *S.recurvum* (very occasionally both) and sometimes *S. cuspidatum* occurs too, all these species tending to increase in prominence in wetter areas, where they may mark a transition to Rhynchosporion pool vegetation. Much more locally, the community provides important stations for *S. magellanicum* and *S. pulchrum*, both scarce species in south-eastern Britain, but sometimes growing in abundance here: the former is plentiful on some New Forest bogs (as on Cranesmoor: Newbould 1960, NCC New Forest Bogs Report 1984), the latter especially striking in some stands around Poole Harbour in Dorset (Ratcliffe 1977). Then, there are occasional records for *S. subnitens* and *S. tenellum*, with *S. capillifolium* occurring infrequently and usually on the tops of drier hummocks. In contrast to the *Ericetum tetralicis*, to which this vegetation often grades on drier ground, *S. compactum* is hardly ever found.

Other mosses are few in number and generally of low cover (*Hypnum jutlandicum* and *Aulacomnium palustre* are sometimes present) but the moist *Sphagnum* carpet provides a very congenial surface for a variety of hepatics, with patches of leafy liverworts often adding to the mosaic of colour over the ground. *Odontoschisma sphagni* and *Kurzia pauciflora* are the commonest species overall, but *Cladopodiella fluitans, Cephalozia macrostachya, C. connivens, C. bicuspidata* and *Calypogeia fissa* have also been recorded. In contrast to the stands described by Rose (1953), however, *Mylia anomala* has not been found very frequently. There can also be scattered thalli of *Aneura pinguis* and various *Riccardia* ssp. including *R. chamedryfolia*. Lichens are typically sparse, though the decaying tops of taller hummocks can provide a habitat for *Cladonia impexa, C. arbuscula* and *C. uncialis*.

The general appearance of the ground layer is thus very similar to that found in the *Erica-Sphagnum* and *Scirpus-Eriophorum* mires, though the surface relief here is often less pronounced than in those communities. The vascular component, too, provides some important floristic differences because, in contrast to other *Erico-Sphagnion* bogs, both *Eriophorum vaginatum* and *Scirpus cespitosus* are rare plants here: within the range of the *Narthecio-Sphagnetum*, both these species are becoming scarce and they are generally confined to the firmer substrates under the *Ericetum tetralicis* wet heath. But other monocotyledons of wet peaty soils remain very frequent and provide a measure of continuity between these different kinds of mires. *Eriophorum angustifolium* is a constant and it can be abundant, the masses of its cottony fruiting heads often enabling stands to be picked out from a distance in June. At closer quarters, it is often the profusion of *Narthecium ossifragum* that strikes the eye, both with its golden flowers in July and again when fruiting in early autumn. Then, there is typically some *Molinia caerulea*, looking rather weak and not tussock-forming in the more stagnant areas, but more vigorous and abundant in better-aerated situations. *Rhynchospora alba* is also very characteristic of one particular kind of *Narthecio-Sphagnetum*, being especially prominent in wetter hollows and around Rhynchosporion pools.

The other consistent structural element in the vegetation comprises woody plants. *Erica tetralix* and *Calluna vulgaris* are both very frequent, though the latter tends to be somewhat patchy in its occurrence, attaining vigour only on the tops of drier hummocks. Typically, these sub-shrubs form a very open canopy here, 1–3 dm tall, and a thickening of their cover often marks the transition to adjacent *Ericetum*. As in that kind of wet heath, though less frequently, *E. tetralix* can be accompanied or replaced by *E. ciliaris* in stands around Poole Harbour (e.g. Chapman 1975). More restricted in its occurrence than the two ericoids, but very noticeable when present in abundance, is *Myrica gale*, which can maintain its contribution to the cover into stands of neighbouring communities, most notably the Junco-Molinion grasslands, thus blurring the transitions.

Other frequent plants are few in number and generally present as scattered individuals. *Drosera rotundifolia* is very common and, at some sites, is accompanied by the more scarce *D. intermedia* or *D. anglica*. *Vaccinium oxycoccos*, by contrast, such a characteristic plant of wetter areas in the *Erica-Sphagnum* mire, is curiously uneven in its occurrence here, being confined to one of the sub-communities. And *Andromeda polifolia*, which often accompanies it on raised mires, is totally absent. There is sometimes a little *Potentilla erecta* and, very occasionally, some *Polygala serpyllifolia*, but the relative scarcity of these species provides a further distinction between the community and the *Scirpus-Eriophorum* mire. Small sedges and rushes are generally infrequent, though *Carex panicea* and *C. echinata* can sometimes be found and there are occasionally tussocks of *Juncus acutiflorus* or *J. effusus*. *C. rostrata* occurs with low frequency in wetter areas and there, too, particularly where there is some soligenous influence, *Potamogeton polygonifolius* and *Eleocharis multicaulis* are sometimes recorded. In some stands, *Phragmites australis* occurs as sparse shoots.

Among species of more restricted distribution in Britain, the community provides an occasional locus for *Pinguicula lusitanica* and, in Devon, introduced *P. grandiflora* survives in this kind of vegetation (Ivimey-Cook 1984). The rare and easily overlooked orchid, *Hammarbya paludosa*, occurs at some sites, typically in the transition zone to Rhynchosporion pool vegetation. *Agrostis curtisii* is found very infrequently on the drier tops of hummocks.

Sub-communities

Rhynchospora alba-Sphagnum auriculatum sub-community: *Sphagnetum* Rankin 1911*b* *p.p.*; *Sphag-* *num* hummock complex, *Sphagnum papillosum* & *S. rubellum* phases Rose 1953 *p.p.*; General bog communities Newbould 1960 *p.p.*; *Sphagnum* lawn bog hummocks NCC New Forest Bogs Report 1984. In this, the most frequently described kind of *Narthecio-Sphagnetum*, the *Sphagnum* carpet is generally dominated by mixtures of *S. papillosum* (with, more locally, *S. magellanicum* or *S. pulchrum*) and *S. auriculatum*. *S. cuspidatum*, *S. subnitens* and *S. tenellum* occur occasionally, but *S. recurvum* is scarce. Hepatics are varied and often abundant with *Odontoschisma sphagni*, *Kurzia pauciflora*, *Cephalozia* spp. and *Aneura pinguis* preferentially frequent.

All the vascular constants of the community retain high frequency in open and mixed mosaics over the ground and, in addition, there is very frequently some *Rhynchospora alba*. *Myrica gale* is also slightly preferential to this sub-community and quite frequently shows local abundance.

Vaccinium oxycoccos-Sphagnum recurvum sub-community: *Sphagnum* hummock complex, *Sphagnum papillosum* & *S. rubellum* phases Rose 1953 *p.p.* In this sub-community, *S. papillosum*, though often abundant, is a little patchier in its dominance of the ground carpet and is quite frequently rivalled in cover by *S. recurvum*, which is strongly preferential here. *S. auriculatum*, by contrast, is much reduced in frequency, though *S. subnitens* and *S. cuspidatum* still make an occasional contribution and there is sometimes a little *S. palustre*. Among the hepatics, *Odontoschisma sphagni* occurs in some stands, but the richness of this element is not so great as in the first sub-community, most of the species occurring very infrequently.

In the vascular element, the most obvious features are the scarcity of *Rhynchospora alba* and the frequency of *Vaccinium oxycoccos*, though even here the latter is of patchy occurrence, being much more common in East Anglian stands than elsewhere. *Potentilla erecta* is somewhat more common in this sub community, too, and there are occasional records for *Carex echinata*, *C. panicea* and *Succisa pratensis*, which give some stands a floristic character transitional to Caricion nigrae mires.

Habitat

The *Narthecio-Sphagnetum* is a community of permanently-waterlogged, acid and oligotrophic peats in the relatively warm and dry, southern lowlands of Britain, where it is especially characteristic of valley mires maintained by a locally high ground water-table. The wetness of the substrate gives the vegetation some protection against the burning and grazing that are (or have been) important features in the heathland that usually surrounds the community, but draining is very deleterious and has severely affected some stands. Modest variation in the trophic state of the ground waters may have some control over floristic differences in the community and this can be accentuated by artificial eutrophication of the habitat.

The consistent saturation of the ground with more base- and nutrient-poor waters gives the vegetation its distinctive Sphagnetalian character, a feature best appreciated by comparing the community with the *Ericetum tetralicis*, the wet heath which replaces it on shallow peats and mineral soils that show seasonal waterlogging. The superficial acidity here is generally similar to that under the wet heath, mostly from 3.5 to 4.5, but the water-table is maintained at or very close to the ground surface throughout the year, favouring the luxuriant growth of peat-building Sphagna. The peat under the community is, in fact, not usually very deep, mostly from 20 to 150 cm, the more extensive accumulations towards the centre of valley mires carrying vegetation influenced by directional through-put. But it is species such as *S. papillosum* with, more locally, *S. magellanicum* and *S. pulchrum*, that comprise the bulk of the cover on the active plane, with semi-aquatic Sphagna becoming abundant in wetter hollows. By and large, these species are all scarce in the *Ericetum*, where extensive cover of more hydrophilous Sphagna is confined to areas of local soligenous influence. Conversely, it is the shift to wetter conditions that accounts for the rarity in the *Narthecio-Sphagnetum* of *S. compactum* and the general restriction of *S. capillifolium* to the drier hummocks. This change in the luxuriance and composition of the *Sphagnum* carpet is matched by the increase in cover and diversity of delicate hepatics, plants intolerant of the periodic surface-drying typical of much *Ericetum*.

Concomitant differences can be seen in the vascular component of the vegetation. Among the herbs, species like *Eriophorum angustifolium*, *Narthecium ossifragum* and *Drosera rotundifolia*, often confined to the areas of wet, bare peat in the *Ericetum*, now become constant; *Rhynchospora alba*, though not frequent throughout here, behaves in the same way. And then there is the shift in the balance of the two common ericoids, *Calluna* and *E. tetralix*, and *Molinia*. All of these are frequent, too, in the wet heath but, in the *Narthecio-Sphagnetum*, *Calluna*, already at something of a disadvantage against *E. tetralix* on the more severely waterlogged soils under the *Ericetum* (Rutter 1955, Bannister 1964*b*, Smart in Gimingham 1972), is usually very obviously subordinate and sometimes distinctly sickly in appearance (Bannister 1964*c*, *d*). And conditions quite often seem to be such that *E. tetralix*, despite its shallow root system (Sheikh 1970), is not able to take full advantage of the consequent lack of competition (Bannister 1966, Gimingham 1972). Poor aeration probably also hinders the growth of *Molinia*, though the general nutrient-poverty of the peats is likely to be of more importance in restricting its abundance and vigour here compared with often

closely-juxtaposed Junco-Molinion vegetation nearer to the lines of soligenous influence in valley mires (e.g. Loach 1966, 1968*a*, *b*, Sheikh & Rutter 1969, Sheikh 1969*a*, *b*, 1970).

Although the quantitative contribution of *Molinia* is under some edaphic constraints, its high frequency through the *Narthecio-Sphagnetum* is one feature which reflects the mild, often distinctly oceanic character of the climate that prevails over the range of the community. This is a vegetation type of southern Britain, found predominantly below 200 m, though reaching higher altitudes towards the south and west, on the fringes of Dartmoor, Exmoor and in Wales, and towards its north-eastern limit, in the North York Moors. Over such a zone, the mean annual maximum temperature is generally in excess of 26 °C (Conolly & Dahl 1970) and, more particularly, the annual accumulated temperature is mostly more than 1400 day-degrees C (Page 1982), with February minima often a degree or more above freezing (*Climatological Atlas* 1952). The montane floristic element typical of northern upland Sphagnetalia bogs, best seen in the *Calluna-Eriophorum* mire, is thus absent here, whereas similarities with north-western oceanic ombrogenous bogs, especially the *Scirpus-Eriophorum* mire, are quite pronounced. Apart from the constancy of *Molinia*, there is the fairly common occurrence of *Myrica* and the occasional representation of plants such as *Drosera anglica*, *D. intermedia*, *Pinguicula lusitanica* and *Hammarbya paludosa*; and, among the Sphagna, the frequent presence of *S. auriculatum* and local occurrence of *S. pulchrum*. With the somewhat less oceanic *Erica-Sphagnum* mire, the *Narthecio-Sphagnetum* shares records for *Vaccinium oxycoccos* and *Sphagnum magellanicum*.

In quite marked contrast to both these communities, however, is the great scarcity of *Scirpus cespitosus* and *Eriophorum vaginatum*: both of these, and especially the latter, are uncommon in southern lowland Britain and, where they do occur, it tends to be on the firmer substrates of the *Ericetum*. One other difference between the communities, which probably has some climatic basis, is that surface relief is less pronounced on the bog plane of the *Narthecio-Sphagnetum* than in the *Scirpus-Eriophorum* or *Erica-Sphagnum* mires. Distinct hummock/hollow systems can be seen, with some patterning among the Sphagna and vascular plants (e.g. Rose 1953, Newbould 1960), and transitions to Rhynchosporion vegetation in the pools (see below), but the relief is typically of rather low amplitude and patterning ill-defined, at least when compared with the mires of the wetter north-west of Britain (Lindsay *et al.* 1984).

The lighter rainfall over the lowland south of the country also has an important control on the kind of habitat favourable to the development of the *Narthecio-Sphagnetum*. In this part of Britain, precipitation falls below the 1200 mm yr^{-1} or 160 wet days yr^{-1} threshold necessary for the accumulation of a mantle of blanket peat (Ratcliffe 1977), so Sphagnetalia vegetation can develop only where topography maintains a locally high water-table. On raised mires, where waterlogged acid peat forms ombrogenously on a topogenous base, and in base-poor basin mires, there is a strong tendency for the bog vegetation to be of the *Erica-Sphagnum* type. The *Narthecio-Sphagnetum*, on the other hand, is very much a community of valley mires developed in catchments of prevailingly acidic substrates where, over sometimes quite gently-undulating topography, base-poor ground waters emerge at impervious bedrocks or superficials and maintain waterlogged conditions in elongated depressions. The hydrological regime in such mires is often quite complex, with percolating waters from the higher ground frequently being channelled along a central soligenous soakway or stream; and there is commonly an artesian element helping to maintain a general high water-table throughout and inducing back-gleying in the surrounds. Within the very distinctive zonation of vegetation types associated with this kind of site, the community characteristically occurs in a belt on shallower marginal peats which are kept consistently wet but where soligenous influence, and therefore any increase in enrichment and aeration, is small.

Valley mires in which the *Narthecio-Sphagnetum* is represented are concentrated on the more acidic elements within the sequences of younger rocks and superficials that underlie the low relief of central-southern and south-eastern England: for example, the Eocene clays, sands and gravels of southern Dorset, Hampshire and Surrey, the sands and clays of the Hastings Beds and Lower Greensand in The Weald and, locally, in west Norfolk, and Triassic sandstones and pebble-beds in Devon and Cornwall. In the New Forest, Pleistocene gravels underlie some valley mires and, further towards the north and west, where the community is found over a variety of bedrocks, glacial drift can play some part in inducing the local drainage-impedence necessary for the development of the community. Increasingly, however, with the move to the wetter climate of the upland fringes, there is a tendency for valley mires to lose their discrete character within the mantle of blanket peat and, in such situations, the *Narthecio-Sphagnetum* is replaced by the *Scirpus-Eriophorum* mire.

The distribution of the two sub-communities shows a clear pattern over this total geographical range, with the *Rhynchospora-S. auriculatum* type being rather strongly confined to the south, where it runs from Devon to Surrey, the *Vaccinium-S. recurvum* type including all the outlying stands sampled to the north and west of this area. This partitioning may have some climatic basis but it is possible that differences between the two kinds of *Narthecio-Sphagnetum* also reflect variations in the trophic state of the peats and waters: in rather similar

vegetation in the *Erica-Sphagnum* mire, large amounts of *S. recurvum* have been related to nutrient enrichment, and the presence of other preferentials like *Carex echinata*, *C. panicea* and *Aulacomnium palustre* is perhaps indicative of some slight amelioration of the generally impoverished conditions. As in the *Erica-Sphagnum* mire (Sinker 1962, Green & Pearson 1968, Tallis 1973a), such eutrophication may be related to human activity, originating from fertiliser run-off or drift, for example, and perhaps such influences have been greater towards the limit of the range of the community.

Towards its heartland, the characteristic occurrence of the community within stretches of unimproved heath has afforded this vegetation some measure of protection against gross disturbance. And the general wetness of the ground provides some insurance against the effects of the burning and grazing that have traditionally been practised in the surrounding landscape. Neither of these is of importance in maintaining the community: indeed, where they do occur, they can induce severe damage, particularly where they have been combined with draining (see below). Valley bogs sometimes show signs of peat-digging, though the deposits are often rather sloppy and little-humified: cut-over areas can diversify the surface patterning within the community and produce new areas for regeneration in mires that are becoming dry.

Zonation and succession

The *Narthecio-Sphagnetum* occurs in mosaics and zonations with other vegetation types in relation to the height of the water-table and the degree of soligenous influence within valley mires, the overall configuration of which is strongly dependent on local topography. With continued autogenic accumulation of peats, the community probably progresses naturally to some kind of woodland and draining may speed this succession. Often, however, together with burning and grazing, it has led to degradation of the vegetation.

The fully-developed sequence of communities in which the *Narthecio-Sphagnetum* occurs is very distinctive (e.g. Rose 1953, Newbould 1960, Ivimey-Cook *et al.* 1975, Wheeler 1983). First, within tracts of the community itself, there can be some measure of surface-relief, with a patterned distribution of the Sphagna and vascular plants over drier and wetter ground and transitions to Rhynchosporion vegetation in pools proper. Characteristically, this latter is of the *Sphagnum auriculatum* type, the pools often surrounded by a zone of *Rhynchospora alba* (sometimes with *R. fusca*) and showing a switch to dominance of *S. auriculatum* and *S. cuspidatum* in the ground carpet, with an appearance of plants like *Menyanthes trifoliata*, *Potamogeton polygonifolius* and *Utricularia* spp. It is in the transition zone to such pools that *Hammarbya* typically occurs.

Where such wetter areas show some measure of soligenous influence, Littorelletea vegetation is often found, with *P. polygonifolius*, *Eleocharis multicaulis*, *Juncus bulbosus/fluitans* and *Hypericum elodes*. These can mark out small seepage areas around the edge of the *Narthecio-Sphagnetum* or water-tracks within stretches of the mire, but quite often the Littorelletea vegetation forms a distinct zone inward of the *Narthecio-Sphagnetum* towards the central mire axis. Continuing this zonation towards the zone of maximal soligenous influence, there can be a strip of Caricion nigrae poor fen and then a central line of *Betula-Molinia* woodland; in valley mires with somewhat more base- and nutrient-enrichment, the last two may be replaced by *Caricetum paniculatae* swamp and *Alnus-Carex* woodland (Figure 17).

In the opposite direction, towards the fringes of the mire, the *Narthecio-Sphagnetum* typically gives way to the *Ericetum tetralicis*. With the move to the periodically-waterlogged shallow peats and peaty gleys that cover the surrounding slopes, the carpet of luxuriant *Sphagnum papillosum* is replaced by patchy *S. compactum* and *S. tenellum*, species such as *Eriophorum angustifolium*, *Narthecium* and *Drosera rotundifolia* thin out and the canopy of ericoids becomes denser. Small patches of wet sandy peat in the transition zone are sometimes picked out by the striking *Rhynchospora-Drosera* sub-community of the *Ericetum*, with its occasional *Lepidotis inundata* and crusts of the purple alga

Zygogonium ericetorum s.l. Beyond the wet heath, the basic sequence continues into dry Calluno-Ulicetalia heath of some kind.

Complete zonations of this type are best seen in some of the New Forest valley mires, as at Wilverley, Holmsley and Matley bogs and in the Denny/Shatterford system (Rose 1953, Tubbs 1968, Ratcliffe 1977, NCC New Forest Bogs Report 1984), and on certain south Dorset heaths (Ratcliffe 1977). In other sites, the sequence of communities is incomplete because of differences in topography and hydrology: at Thursley Common, in Surrey, for example, the *Narthecio-Sphagnetum* terminates the zonation below, forming a rather extensive stretch of bog like a raised mire (Rose 1953, Ratcliffe 1977) and on Aylesbeare Common, in Devon, stretches of the community mark out more strongly waterlogged ground over gently-undulating topography with a complex mosaic of wet and dry heaths (Ivimey-Cook et al. 1975).

Quite frequently, too, the general zonation is complicated by the occurrence of other vegetation types which reflect differences in base- or nutrient-richness of the substrates and waters. Tracts of Junco-Molinion grasslands, sometimes dominated by *Myrica*, can be found at some sites below the *Narthecio-Sphagnetum*, where soligenous influence increases, or cutting through it, where there is some channelling of flow down the slopes (Ivimey-Cook et al. 1975). Then, in certain New Forest mires, notably at Cranesmoor (Newbould 1960), on

Figure 17. Zonation around a typical lowland valley bog in southern England.
M6   *Carex echinata-Sphagnum* mire
M16a   *Ericetum tetralicis* wet heath, Typical sub-community
M21a   *Narthecio-Sphagnetum* valley bog, *Sphagnum-*

Rhynchospora sub-community
M29   *Hyperico-Potametum* soakway
W4   *Betula-Molinia* woodland
H2   *Calluna-Ulex minor* heath
H3   *Ulex minor-Agrostis* heath

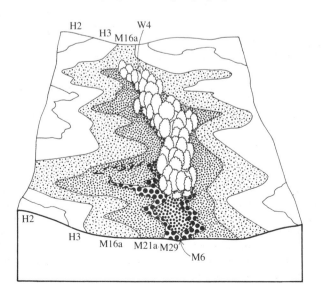

Hartland Moor in Dorset (Ratcliffe 1977) and at Aylesbeare (Ivimey-Cook *et al.* 1975), the *Narthecio-Sphagnetum* gives way to the *Schoenus-Narthecium* community
along the central mire axis. In the 'mixed mires' found
on some north Norfolk commons, like Roydon, where
valleys have cut down through sands and gravels to
Chalky Boulder Clay, there is a more marked shift from
the community to calcicolous fen-meadow and fen, with
vegetation types like the *Cirsio-Molinietum* and
Schoenetum bordering the mire axis (Rose 1953, Daniels
& Pearson 1974, Wheeler 1975).

Towards the north and west, there is an increasing
tendency for the *Narthecio-Sphagnetum* to be less well
characterised floristically and for stands to survive in
fragmentary form within much-altered landscapes.
Even in the south-east, treatments have had some drastic effects and losses of the community, with heathland
reclamation, have probably been extensive. Severe fires
sometimes cause local damage to the *Sphagnum* carpet
with a run-down of the vegetation to the *Ericetum
tetralicis* and, where stock are turned on to the community in large numbers, trampling and manuring may
precipitate a succession to poor fen (Rose 1953). But
draining has been particularly destructive and, even
where this has not been followed by improvement for
agriculture or forestry, it has induced substantial
changes. Essentially, it speeds up what is probably
naturally a rather slow process of surface-drying with
autogenic peat accumulation: lowering the water-table
encourages the invasion of woody plants, among which
Betula spp. and *Pinus sylvestris* are generally prominent,
and makes the vegetation more susceptible to the effects
of burning. Under such conditions, what was once
Narthecio-Sphagnetum often presents a sorry spectacle
of scrubby woodland and impoverished *Ericetum* and
dry heaths.

Distribution
This is a local community of the southern lowlands of
Britain, being best represented by the *Rhynchospora-S.
auriculatum* sub-community in central southern England with the *Vaccinium-S. recurvum* sub-community
widely scattered to the north and west of this area.

Affinities
Although the occurrence of *Sphagnum*-rich bog vegetation in the distinctive zonations of southern British
valley mires has long been recognised (e.g. Rankin
1911*b*) and some descriptive accounts provided (e.g.
Rose 1953, Newbould 1960), the floristic relationships
of the community have not been much explored. Ivimey-
Cook *et al.* (1975) provided the first phytosociological
diagnosis and pointed out the essential similarity of the
vegetation to the *Narthecio-Sphagnetum acutifolii euatlanticum*, described from Brittany by Duvigneaud
(1949). In Moore's (1968) scheme, this was placed
among the wet heaths of the Ericion tetralicis, but,
although the community is typically found in close
association with the *Ericetum tetralicis*, it is better
grouped among the Erico-Sphagnion mires with their
characteristic abundance of peat-building Sphagna.
Among these, it is most closely allied to the *Scirpus-
Eriophorum* mire of the very oceanic far-west of Britain
and can be seen as its soligenous equivalent. Abundance
of *Narthecium* and *Molinia* is also characteristic of the
Norwegian *Narthecio-Sphagnetum* of Dierssen (1982)
which occurs, like the present community, in more or
less minerotrophic situations. However, that vegetation
is found largely in the montane and sub-alpine zones of
western Norway, and it is of a much more northern
character than our community.

Floristic table M21

	a	b	21
Erica tetralix	V (2–7)	V (3–7)	V (2–7)
Molinia caerulea	V (2–9)	V (2–7)	V (2–9)
Eriophorum angustifolium	V (2–6)	V (2–7)	V (2–7)
Narthecium ossifragum	V (1–8)	V (2–8)	V (1–8)
Drosera rotundifolia	V (1–4)	IV (1–4)	V (1–4)
Sphagnum papillosum	V (2–9)	IV (3–7)	V (2–9)
Calluna vulgaris	IV (1–5)	IV (1–7)	IV (1–7)
Sphagnum auriculatum	IV (1–8)	II (1–8)	III (1–8)
Rhynchospora alba	IV (1–5)	I (6)	II (1–6)
Myrica gale	III (1–7)	II (3–6)	II (1–7)
Odontoschisma sphagni	III (1–4)	II (1–3)	II (1–4)
Kurzia pauciflora	III (1–3)	I (1–2)	II (1–3)

	a	b	21
Sphagnum tenellum	II (2–6)	I (2–5)	II (2–6)
Sphagnum subnitens	II (1–4)	I (2–9)	II (1–9)
Phragmites australis	II (1–4)	I (4–6)	I (1–6)
Cephalozia macrostachya	II (1–2)	I (2–3)	I (1–3)
Cephalozia connivens	II (1–4)	I (2)	I (1–4)
Cladopodiella fluitans	II (1–3)		I (1–3)
Drosera intermedia	II (1–3)		I (1–3)
Sphagnum magellanicum	II (1–6)		I (1–6)
Cirsium dissectum	II (1–3)		I (1–3)
Cephalozia lunulifolia	I (1)		I (1)
Aneura pinguis	I (1–2)		I (1–2)
Eleocharis multicaulis	I (2–6)		I (2–6)
Pinguicula lusitanica	I (1–3)		I (1–3)
Sphagnum pulchrum	I (4–7)		I (4–7)
Erica ciliaris	I (5)		I (5)
Hammarbya paludosa	I (1)		I (1)
Sphagnum recurvum	I (5–9)	IV (2–8)	III (2–9)
Potentilla erecta	I (1–3)	III (1–4)	II (1–4)
Vaccinium oxycoccos		III (1–4)	II (1–4)
Carex echinata	I (1–3)	II (3–4)	I (1–4)
Carex panicea	I (1)	II (1–3)	I (1–3)
Aulacomnium palustre	I (2–4)	II (1–4)	I (1–4)
Sphagnum palustre		I (3–4)	I (3–4)
Succisa pratensis		I (2–3)	I (2–3)
Juncus effusus		I (3–5)	I (3–5)
Polytrichum commune		I (3)	I (3)
Nardus stricta		I (4–5)	I (4–5)
Sphagnum cuspidatum	II (1–7)	II (1–8)	II (1–8)
Juncus acutiflorus	II (2–5)	I (3)	I (2–5)
Sphagnum capillifolium	I (2–4)	I (4–6)	I (2–6)
Hypnum jutlandicum	I (2–3)	I (2–4)	I (2–4)
Scirpus cespitosus	I (1–4)	I (4–5)	I (1–5)
Carex rostrata	I (1–3)	I (2–4)	I (1–4)
Cladonia impexa	I (3–5)	I (1–4)	I (1–5)
Polygala serpyllifolia	I (1–2)	I (1)	I (1–2)
Pinus sylvestris sapling	I (1–2)	I (1–2)	I (1–2)
Cephaloziella sp.	I (1)	I (1)	I (1)
Calypogeia fissa	I (2–3)	I (3)	I (2–3)
Cephalozia bicuspidata	I (1–3)	I (1–3)	I (1–3)
Riccardia chamedryfolia	I (2)	I (3)	I (2–3)
Campylopus paradoxus	I (1)	I (3)	I (1–3)
Potamogeton polygonifolius	I (1)	I (4)	I (1–4)
Agrostis curtisii	I (2)	I (1)	I (1–2)
Juncus squarrosus	I (4)	I (1)	I (1–4)
Number of samples	31	24	55
Number of species/sample	15 (8–24)	13 (8–18)	14 (8–24)

a   *Rhynchospora alba-Sphagnum auriculatum* sub-community
b   *Vaccinium oxycoccos-Sphagnum recurvum* sub-community
21 *Narthecio-Sphagnetum euatlanticum* (total)

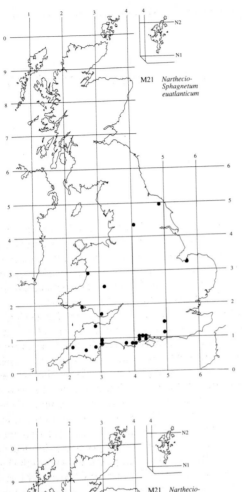

M21 Narthecio-
Sphagnetum
euatlanticum

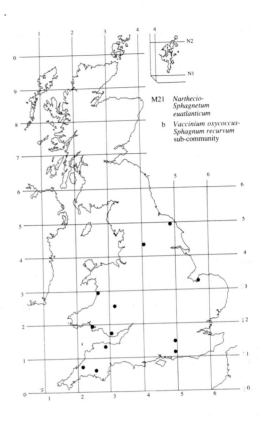

M21 Narthecio-
Sphagnetum
euatlanticum

b   Vaccinium oxycoccus-
Sphagnum recurvum
sub-community

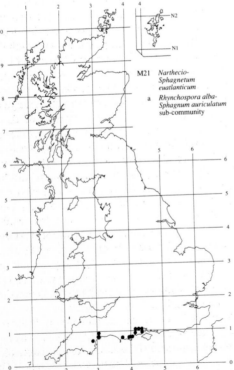

M21 Narthecio-
Sphagnetum
euatlanticum

a   Rhynchospora alba-
Sphagnum auriculatum
sub-community

M22
Juncus subnodulosus-Cirsium palustre fen-meadow

Synonymy
Grass fen   Pallis 1911 *p.p.*; Late *Juncetum*   Clapham
1940 *p.p.*; *Juncetum subnodulosi*   Lambert 1946, 1948;
Fen meadow   Rose 1950, Willis & Jefferies 1959 *p.p.*,
Crompton 1972; Rich-fen meadows   Wheeler 1980*c*
p.p.

Constant species
*Cirsium palustre, Equisetum palustre, Filipendula
ulmaria, Holcus lanatus, Juncus subnodulosus, Lotus
uliginosus, Mentha aquatica, Calliergon cuspidatum.*

Rare species
Peucedanum palustre, Homalothecium nitens.

Physiognomy
The *Juncus subnodulosus-Cirsium palustre* fen-meadow
comprises vegetation which has a readily-recognisable
general stamp but which shows considerable variation
in its floristic composition. The community has a fairly
substantial group of constants and numerous occasional
to frequent species can make the vegetation very rich,
but this associated flora shows considerable variation
from stand to stand, with some plants that are of only
low to moderate frequency throughout showing local
prominence, and widely-represented species sometimes
being locally absent. Often here, such differences reflect
unique and complex histories of mowing and grazing,
treatments of major importance in maintaining this kind
of vegetation.

The most prominent structural element in the
community typically comprises rushes and sedges of
moderate stature, the abundance of which gives the
vegetation the appearance of a rank sward, at least when
viewed from a distance. In ungrazed stands, the herbage
commonly attains 50–80 cm in height, exceptionally up
to 1 m, and it can be dense and rather uniform; where
stock have access, the dominants may be sparser or
distinctly patchy, clumps of the more resistant plants
scattered through the stretches of closer-cropped sward.

Prime among the bulkier species is *Juncus subnodulosus*,
a plant of broad occurrence in a wide variety of fen types
and some swamps, and sometimes only sparsely-repre-
sented here, but able to attain its peak of abundance in
this community and the most usual dominant. It is a far-
creeping rhizomatous perennial, producing an annual
crop of sterile shoots in the autumn, borne singly and
bearing but a single leaf, but arising so closely on the
rhizomes, at 1–3 cm apart, as to be able to produce a very
thick and even cover. These shoots elongate to their full
height in the spring, when the fertile shoots are also
produced, flowering in July and August (a little later to
the north), but then often lodging in autumn winds and
rain to form a thick and heavy mattress. This can
depress the associated flora, though the next crop of
sterile autumn shoots readily penetrates it, their fresh-
green colour making a sharp contrast against the brown
of the decaying mat (Richards & Clapham 1941*d*).

Although *J. subnodulosus* can be locally comple-
mented or replaced by other dominants and, towards
the limits of its rather restricted range, become generally
scarce in or be absent from vegetation of this same
general character, no other rush or sedge attains such a
high overall frequency in the community. Of the additional rushes, the commonest are *J. inflexus*, which is
distinctly tussocky, has grey-green foliage and flowers
earlier than *J. subnodulosus*, and *J. articulatus*, which
often has a decumbent habit here and flowers later; but
though these can be locally abundant, they tend to be
preferential for particular kinds of *Juncus-Cirsium* fen-
meadow. *J. effusus* occurs only occasionally, *J. acutiflorus* rarely, and both are typically of low cover,
although all four of these rushes can become more
consistently common in closely-related types of Molinietalia vegetation.

More striking in this community is the frequent
occurrence and occasional dominance of *Carex acutiformis* or *C. disticha*. Like *J. subnodulosus*, both of these
sedges have extensive rhizome systems and neither is
densely tussocky, though in *C. acutiformis* the shoots are

tufted and robust, with long leaves that are at first distinctly glaucous, so this species can be conspicuous when present in abundance. *C. disticha* is more readily overlooked, having smaller shoots, borne singly or in pairs, but close inspection often reveals some of its bright green foliage and it, too, can locally replace *J. subnodulosus* as the dominant. Much more occasionally, *C. elata* or *C. paniculata* occur here and their bulky tussocks can exert a stronger structural effect on the vegetation. *C. nigra*, though of rather restricted occurrence and usually part of the lower herbage, is sometimes found as the tussocky var. *subcaespitosa*.

Transitions between this kind of fen-meadow and tall-herb Phragmitetea fen are quite common: the two types of vegetation are fairly often found juxtaposed and it is clear that certain treatments can mediate conversion of the latter into the former. One striking difference between them, however, is that the tall helophyte dominants so characteristic of Phragmitetea communities are here reduced to very patchy occurrence. *Phragmites australis* is the commonest and even this is confined to ungrazed stands which are rarely mown or not at all, and then it usually occurs as sparse shoots forming an open upper tier to the vegetation, a metre or more high. *Calamagrostis canescens* and *Phalaris arundinacea* can also be found occasionally at low covers and in some stands a little *Cladium mariscus* may persist.

Generally, however, the strongest link between the vegetation types is provided by the associated herbaceous flora. In summer, for example, the rush and sedge layer can be overtopped by the tall flowering shoots of a variety of dicotyledons. The most frequent of these throughout the community are *Cirsium palustre*, *Filipendula ulmaria*, *Angelica sylvestris*, *Succisa pratensis*, *Eupatorium cannabinum* and *Scrophularia aquatica*, with more local enrichment by *Lythrum salicaria*, *Lysimachia vulgaris*, *Valeriana officinalis*, *Thalictrum flavum*, *Symphytum officinale* and, in Broadland, the nationally rare *Peucedanum palustre*. Quite commonly, though, grazing keeps this component of the vegetation severely in check, too, eliminating some species and reducing the more resistant to closely-nibbled rosettes.

Much more numerous through the community as a whole, and of greater structural significance, are smaller herbs which form an understorey, up to 50 cm or so high, amongst the dominants, reduced in cover and diversity where the rushes and sedges are dense, correspondingly luxuriant and varied in more open herbage, and again showing physiognomic differences according to the intensity of grazing. Among these plants, grasses are of considerable importance, with species such as *Holcus lanatus*, *Festuca rubra* and, less commonly, *Arrhenatherum elatius*, which on drier, ungrazed ground can grow rank and tussocky, and others like *Poa trivialis*, *Agrostis stolonifera*, *Anthoxanthum odoratum* and *Briza media*,

which form mats or smaller tussocks among them or, in grazed swards, make up a good deal of the short turf. *Molinia caerulea* can also be found here but it is rather patchy in its occurrence and very often of low cover, something which helps set the boundary between this community and related vegetation in the *Cirsio-Molinietum*. Other low-frequency grasses which occasionally grow robust and prominent are *Deschampsia cespitosa* and *Festuca arundinacea*. *Lolium perenne*, *Cynosurus cristatus* and *Phleum pratense* also sometimes figure in transitions to improved swards, as for example where stretches of fen-meadow have been subject to draining and improvement, a widespread event.

Among these grasses, the associates comprise species with a wide representation in wetter vegetation, together with plants more especially characteristic of moist grasslands. The commonest are *Mentha aquatica*, a late-flowerer able to grow through even quite thick mats of lodged rush shoots, *Caltha palustris*, *Equisetum palustre*, *Carex panicea*, *Valeriana dioica*, *Hypericum tetrapterum*, *Hydrocotyle vulgaris*, *Lotus uliginosus*, *Lychnis flos-cuculi*, *Cardamine pratensis*, *Ranunculus acris*, *Potentilla erecta*, *Cerastium fontanum* and *Rumex acetosa*, together with scrambling or sprawling *Galium uliginosum*, *G. palustre*, *Vicia cracca* and *Lathyrus pratensis*. In some stands, additional richness is provided by assemblages characteristic of wet fen or swamps, which can occupy hollows, or by herbs of dry pastures or meadows, and quite specialised floras can sometimes be found in damp ruts or on cattle-poached ground.

In what is typically rather dense herbage, bryophytes play a reduced role with only *Calliergon cuspidatum* and *Brachythecium rutabulum* occurring at all commonly throughout. *Plagiomnium undulatum*, *P. rostratum*, *Rhizomnium punctatum*, *Pseudoscleropodium purum* and *Campylium stellatum* can also be found occasionally, but the cover in the ground layer is rarely extensive, with usually just scattered wefts over the stools and litter.

Sub-communities

Typical sub-community: Rich-fen meadows, *Juncus-Centaurea nigra* nodum, *Juncus-Carex hirta-Deschampsia cespitosa* nodum, *Juncus subnodulosus* nodum Wheeler 1980c, all *p.p.* In this most common and widespread kind of *Juncus-Cirsium* fen-meadow there are no preferential floristic features over and above those characteristic of the community as a whole and the general impression when this vegetation is compared with that of other sub-communities is one of rank structure and impoverishment. *J. subnodulosus* is frequent throughout and is the commonest dominant, its general abundance being reflected particularly in the sparse representation of smaller herbs. The commonest associates are thus bulkier grasses, such as *Holcus* *lanatus, Festuca rubra* or, in some localities, *Molinia*, mat-formers like *Agrostis stolonifera* and *Poa trivialis*, taller dicotyledons like *Cirsium palustre, Filipendula ulmaria, Angelica, Succisa* and *Eupatorium*, sprawling plants, and those few shorter species able to tolerate the rank herbage and thick litter, notably *Mentha aquatica* and *Equisetum palustre*. Within this general framework, however, there can be considerable local variation. In some stands, for example, *J. subnodulosus* is supplemented or replaced by other dominants like *J. inflexus, J. articulatus, Carex acutiformis* or *C. disticha*, sometimes with little effect on the associated flora, in other cases with accompanying differences in composition: in certain localities, dry fen-meadow of this general type can have mixtures of these species, together with *C. hirta* and *Deschampsia cespitosa*, a distinctive assemblage which recurs in the *Briza-Trifolium* sub-community (and which Wheeler (1980c) placed together in a *Juncus-Carex-Deschampsia* nodum). Much more locally, where there are very wet areas within Typical *Juncus-Cirsium* fen-meadow, *C. paniculata* can be locally prominent, dominating the structure of the vegetation with its large tussocks. Ungrazed stands can have patchy *Phragmites*, never very dense but catching the eye with its taller shoots, and some areas show invasion by *Epilobium hirsutum* and *Arrhenatherum elatius*. Yet other tracts show floristic transitions to the richer herbage of the next sub-community (vegetation which Wheeler (1980c) united in a *Juncus-Centaurea* nodum) or to related forms of the *Cirsio-Molinietum*.

***Briza media-Trifolium* spp. sub-community:** Rich-fen meadows, *Juncus-Centaurea nigra* nodum *p.p., Juncus-Carex hirta-Deschampsia cespitosa* nodum *p.p., Juncus subnodulosus* nodum *p.p., Juncus-Carex disticha* nodum Wheeler 1980c. *J. subnodulosus* remains the most frequent of the bulkier plants in this sub-community, although *J. articulatus* and *J. inflexus* are both common, and the latter sometimes quite abundant, and *C. disticha* can be locally co-dominant. In general, however, compared with Typical *Juncus-Cirsium* fen-meadow, the rush/sedge tier as a whole here is not so overwhelming in its cover or density.

Correspondingly, the associated flora is that much richer. In the first place, most of the taller dicotyledons of the community, like *Cirsium palustre, Filipendula, Angelica, Eupatorium* and *Succisa*, show an increase in frequency here and are often accompanied by *Centaurea nigra* and *Rumex acetosa*, though, with grazing, which is common in this vegetation, these species are often reduced to rosettes or stumpy stocks and rarely able to put up their flowering shoots. Much more characteristic is a rich suite of plants of small to moderate stature which comprise a dense and varied layer amongst the rush and sedge clumps, up to several dm tall where

grazing is light, forming a much shorter turf where stock are numerous. Grasses are often important in this tier, not only species such as *Holcus lanatus, Festuca rubra* and *Molinia*, which can be nibbled down here to small tufts, and *Agrostis stolonifera* and *Poa trivialis*, well adapted to ramifying close swards, but also less bulky species such as *Briza media* and *Anthoxanthum odoratum*, which are able to survive in the shorter herbage. *Juncus articulatus*, too, though able to grow in ascending clumps, is often found in a semi-prostrate form in this vegetation and, quite commonly, there is some *Carex panicea* in the sward, occasionally also some *C. nigra* and *C. flacca* or, usually with *Deschampsia cespitosa, C. hirta. Cynosurus cristatus* is sometimes found where stands occur among tracts of drier pasture, and *Lolium perenne* and *Phleum pratense* can figure in such situations too.

Intermixed with the grasses is a variety of other herbs. Most frequent among these are *Lotus uliginosus, Lychnis flos-cuculi, Caltha palustris, Ranunculus acris, Valeriana dioica, Potentilla erecta, P. anserina* and *Hypericum tetrapterum*, characteristic of the community as a whole but especially well represented here, and also *Cardamine pratensis, Cerastium fontanum, Trifolium repens, T. pratense, Epilobium parviflorum, Plantago lanceolata, Prunella vulgaris, Ranunculus repens, Triglochin palustre, Rhinanthus minor* and *Dactylorhiza fuchsii*, occurring occasionally elsewhere in the community but strongly preferential to this particular kind of *Juncus-Cirsium* fen-meadow. Other orchids sometimes found here include *D. majalis* ssp. *praetermissa, D. incarnata* and *Epipactis palustris*. Then, sprawling among the herbage, there are frequent plants of *Galium uliginosum, G. palustre, Vicia cracca* and *Lathyrus pratensis*. With the high incidence of grazing, *Phragmites* is scarce here, but wetter patches may show some floristic similarities to the *Iris* sub-community with occasional *Iris pseudacorus, Ranunculus flammula, Hydrocotyle vulgaris* and *Thalictrum flavum*.

***Carex elata* sub-community:** *Juncus subnodulosus-Carex elata* nodum Wheeler 1980c. *Carex elata* is by and large an uncommon plant in fen-meadow vegetation but locally it can occur in some abundance in this kind of *Juncus-Cirsium* community, very occasionally becoming co-dominant with *J. subnodulosus. Phragmites* is also often present, forming an open upper tier and there is usually some *Hydrocotyle vulgaris*, but more specifically preferential are *Potentilla palustris, Equisetum fluviatile, Epilobium palustre, Lythrum salicaria, Dactylorhiza incarnata* and *Thelypteris palustris*. Some stands have *Pedicularis palustris, Menyanthes trifoliata, Ranunculus flammula, R. lingua* and *Berula erecta* (Wheeler 1975). Grasses and the smaller herbs of the *Briza-Trifolium* sub-community are particularly sparsely represented.

Saplings of *Salix cinerea* sometimes gain a hold on the sedge tussocks.

***Iris pseudacorus* sub-community:** *Juncus subnodulosus-Iris pseudacorus* nodum Wheeler 1980*c*. This sub-community shares with the last frequent records for *Phragmites*, again sparsely represented as an open upper tier, and for *Hydrocotyle*, with occasional *Menyanthes*, *Potentilla palustris* and *Equisetum fluviatile*, and once more the smaller herbs of the *Briza-Trifolium* type of fen-meadow are noticeably sparse. *Carex elata* is occasionally found and tussocks of *C. paniculata* are sometimes prominent, and both *C. disticha* and, particularly, *C. acutiformis* can be locally abundant. Generally, though, it is *J. subnodulosus* which remains dominant.

By mid-summer in this vegetation, however, the continuing prominence of the rush/sedge layer is often masked by the flowering shoots of tall dicotyledons which, together with the scattered reed, give a superficial impression of Phragmitetalia fen. Of the general community species, *Cirsium palustre*, *Filipendula*, *Angelica* and *Succisa* all occur commonly and the most frequent of the preferentials is *Iris pseudacorus*. Also distinctive, but rather patchily-represented, are *Lythrum salicaria*, *Lysimachia vulgaris*, *Thalictrum flavum*, *Valeriana officinalis* and, in Broadland, *Peucedanum palustre*, species which, when present in numbers, bring the vegetation particularly close to tall-herb fen. *Calamagrostis canescens*, *Phalaris arundinacea* and *Epilobium hirsutum* can also occur in small amounts and some stands have *Symphytum officinale* and *Calystegia sepium*.

Bryophytes are again rather poorly represented but the usual *Calliergon cuspidatum* and *Brachythecium rutabulum* are sometimes accompanied by *Campylium stellatum* or *Plagiomnium rostratum*.

Habitat

The *Juncus-Cirsium* fen-meadow brings together secondary herbaceous vegetation developed over a variety of moist, base-rich and moderately mesotrophic peats and mineral soils in southern lowland Britain. It can be found in both soligenous and topogenous habitats, either in and around well-developed springs, flushes and mires, or marking out more ill-defined areas of influence of surface or ground waters. Ultimately, however, the community is always dependent on the maintenance of particular kinds of treatment, being derived from a diversity of other sorts of wetland vegetation by mowing and/or grazing, and it owes much in its floristics and physiognomy to the influence of these factors. Its overall distribution and the extent of stands have been much reduced by the abandonment of more traditional agricultural practices and widespread and intensive land improvement.

In the sequence of wet meadows and pastures in which rushes play a prominent part, the *Juncus-Cirsium* fen-meadow occupies one climatic and edaphic extreme. Its geographical range coincides closely with that of its most usual dominant, *J. subnodulosus* (Richards & Clapham 1941*b*, Perring & Walters 1962): although some stands, particularly towards the limit of the range of this species, lack the rush while remaining otherwise quite typical, environmental conditions here seem optimal for *J. subnodulosus* and many records for it in local floras probably refer to occurrences in this kind of vegetation (Wheeler 1975: see, for example, Dony 1967, Newton 1971, Messenger 1971). It is a species of wide distribution through western Europe, though it is essentially a plant of lower altitudes, probably confined in Britain to sites below 150 m (Richards & Clapham 1941*d*), and the prominence of some other members of the community, like *J. inflexus*, *Carex acutiformis*, *C. disticha* and *C. hirta*, reinforces this broad phytogeographic association with the lowland south of Britain. Within these fairly loose limits, however, the range of the rush and of the community are strongly restricted by edaphic factors: thus, it is the distribution of suitably moist, base-rich and not excessively oligotrophic soils that accounts for the general concentration of this kind of vegetation in central and eastern England, rather than any direct influence of a more continental climate. Coincidentally, most stands occur within the 29 °C mean annual maximum isotherm (Conolly & Dahl 1970), but the floristic response in the community to this particular climatic regime is negligible. The only obvious Continental plants occurring here are *Epipactis palustris* and *Oenanthe lachenalii*, and even these are quite scarce; and *Fritillaria meleagris* and *Bromus racemosus*, which are regarded as characteristic of this kind of vegetation on the European mainland, are rare in Britain and usually found in other communities. On the negative side, one might perhaps expect to see *Anagallis tenella* represented more frequently here if suitable soils were common in parts of the country with a more equable climate. But they are not and, in the cool, damper conditions in the western lowlands, where moist acidic soils predominate, the community is largely replaced by the *Juncus-Galium* rush-pasture, a vegetation type that is likewise strongly influenced by treatment, but which has distinctly calcifuge and oceanic features.

In the southern lowlands of Britain, then, the *Juncus-Cirsium* fen-meadow serves as an effective marker of soils which are kept reasonably moist for most of the year and which have moderate to high base-status, with a superficial pH that is almost always between 6 and 8, and usually in the range 6.5–7.5. But the particular situations in which these conditions are met are quite varied and the community can be found over a wide range of profile types, both organic and of predominantly mineral origin, occurring on fen peats, base-rich alluvial soils and gleys, with or without a humic top, and calcareous pelosols (Avery 1980). In lowland Britain, such soils are concentrated in areas where the soil mantle has been directly derived from lime-rich argillaceous bedrocks or heavy-textured superficials, or where there is local impedence of waters draining from calcareous substrates. There is thus a dense clustering of stands in Norfolk and north-west Suffolk, where the Chalk and the overlying Great Chalky Boulder Clay have a strong infuence on pedogenesis and hydrology in and around small mire basins and within the flood-plain mires of Broadland (Pallis 1911, Lambert 1946, 1948). Then, in central England, the ill-drained undulating landscape of north Buckinghamshire, derived from mixtures of boulder clay and glacio-fluvial deposits, is particularly rich in this kind of fen-meadow vegetation (Wheeler 1975, 1983b), while, to the north-west, in the Cheshire–Shropshire Plain, peat-filled hollows in a similar drift-derived scenery provide a further centre. Strings of sites can also be found along the foot of the limestone scarps of central England, notably along the Chalk of the Chilterns, where waters emerge at the junction with the Gault Clay, and, less obviously, along the Oolite up through Bedfordshire and Northamptonshire and into the North York Moors, and, at far flung sites in north Wales, notably in Anglesey, below exposures of Carboniferous Limestone (Wheeler 1975). Between these more important remaining localities are numerous scattered sites which probably represent but a small remnant of a once much denser distribution: further sampling through Northamptonshire, Lincolnshire and South Humberside, where there is the most obvious discrepancy between the occurrence of *J. subnodulosus* and the known range of the community, would be of great assistance in assessing the full extent of this survival.

Despite the close correspondence between the range of the community and the distribution of more base-rich moist soils, the calcicolous aspect of the vegetation is actually rather poorly developed. *J. subnodulosus* itself is the best marker of these conditions (Richards & Clapham 1941d) and, among our rushy meadows and pastures, this plant is strongly preferential here. *J. inflexus* (Richards & Clapham 1941a) and *Carex disticha* (Jermy et al. 1982) are somewhat less confined, and, apart from *C. panicea* and *Briza media*, more diminutive calcicoles are rarely prominent. As a group, these plants are usually sufficient to help separate the *Juncus-Cirsium* fen-meadow from the *Holco-Juncetum*, the closely-related vegetation type that is characteristic of generally less base-rich gleys through the lowlands of Britain, although the communities do intergrade in transitional habitats under the dominance of *J. inflexus*. However, there is, in the *Juncus-Cirsium* fen-meadow, nothing like

the richness and diversity of smaller calcicoles so typical of the base-rich, small-sedge and *Schoenus* fens, despite the fact that the pH of the irrigating waters under such vegetation types is very much the same as here. The difference between the habitats probably lies partly in the nutrient-status of the soils and waters (Wheeler 1975, 1980c). In the Caricion davallianae mires, the concentrations of major nutrients are strikingly low and competition in the consequently rather open swards not intense. Here, by contrast, the growth of smaller herbs is often overwhelmed, not simply by the dominants, but also by the potentially vigorous performance of plants which are mostly mesophytes characteristic of a wide range of ranker swards, such as *Holcus lanatus, Festuca rubra, Poa trivialis, Ranunculus acris, Vicia cracca* and *Lathyrus pratensis* (which are all good Molinio-Arrhenatheretea species), or of a variety of moister vegetation types, like *Cirsium palustre, Equisetum palustre, Filipendula ulmaria, Lotus uliginosus, Lychnis flos-cuculi, Valeriana dioica* and *Angelica sylvestris* (typical of the Molinietalia as a whole). Most of the very frequent, and often the more prominent, plants of the community are thus species tolerant of, rather than demanding of, the high base-status of the soils here. As with the other rush-pastures and meadows, then, the general character of the vegetation is often more obvious than the particular.

Compared with the habitats of the *Holco-Juncetum* and *Juncus-Galium* rush-pasture, however, those in which the *Juncus-Cirsium* fen-meadow are found include a greater proportion of wetter situations. These other communities are essentially vegetation types of gleyed mineral soils, sometimes with a humic top, but not typically waterlogged to the surface, at least not in summer. The *Juncus-Cirsium* fen-meadow does extend on to base-rich soils which are just seasonally water-logged, as around the margins of spring-fens which are back-gleyed by the flushing-waters, below intermittently active seepage lines, in ill-draining hollows where precipitation induces winter surface-water gleying and in some topogenous fens with a fluctuating ground water-table. But many stands are kept much wetter than this, by constant soligenous through-put, by gross impedence in hollows or by more consistent flooding.

Much of the floristic variation in the community can be attributed to differences in the kind and degree of irrigation, and Wheeler (1975) identified two extreme situations with frequent transitions between. The typical spring fen-meadow occurs on sloping ground, sometimes but gently-inclined, in other cases quite steep, where ground waters percolate down to impervious substrates and emerge in clearly-defined, often vigorous, springs or much more diffuse seepage lines. They then run down-slope, sometimes in definite channels or soak-ways which can feed a stream or basin, sometimes in more indeterminate flushes, the waters soaking away again where they cross on to pervious materials. Such fens are found at the foot of scarps and on valley sides, sometimes where there is a sharp break in slope, as around the Gordano valley in Avon (Willis & Jefferies 1959), or in more gentle scenery, as in north Buckinghamshire (Wheeler 1975, 1983b) and in some of the Breckland valleys in East Anglia (Bellamy & Rose 1961, Haslam 1965); and on the slopes around hollows and basins, as in the Anglesey fens and in mires in the Cheshire–Shropshire Plain and Norfolk (Wheeler 1975). They are generally of small extent and can show a quite marked difference between the permanently-waterlogged central area, where there is often an ill-structured, sloppy mineral soil, rich in organic matter above, and the surrounds, where better-defined profiles become less strongly gleyed as one moves away.

Usually, however, such fens are occupied by the Typical or Briza-Trifolium sub-communities in which Molinio-Arrhenatheretea or Molinietalia species account for the bulk of the vegetation cover, sometimes with a Cynosurion element reflecting the influence of grazing stock within the fen itself or in the pastured surrounds: with their large circumference to area ratio, spring fen-meadows are often characterised by a strong floristic overlap with the vegetation around (Wheeler 1975). The difference between these two kinds of Juncus-Cirsium fen-meadow probably lies in their treatment history (see below), but both exhibit a similar variation in their composition according to the degree of waterlogging. For example, such local dominants as Carex disticha (especially prominent in the Briza-Trifolium type), C. acutiformis and C. paniculata (particularly associated with the Typical sub-community), tend to become most abundant in the very wettest springs, on soft muds and peats where there is a more or less continuous through-put. At the opposite extreme is the kind of vegetation represented in Wheeler's (1980c) Juncus-Carex hirta-Deschampsia nodum, especially characteristic of fen edges with transitions to surrounding swards; and his Juncus-Centaurea nodum of summer-dry soils.

These latter forms of the Typical and Briza-Trifolium sub-communities can also be found where the Juncus-Cirsium fen-meadow runs down on to flat ground in badly-drained hollows or valley bottoms with stagnogley or ground-water gley soils. In such situations, described from the Ham area in Kent (Rose 1950), Gordano (Willis & Jefferies 1959) and sites scattered through the Midlands (Wheeler 1975), the stands can be more extensive, forming patches within agricultural fields or occurring as undrained enclaves. Hydrologically, such situations represent a transition to the other distinctive kind of habitat in which this community occurs, where the ground water regime is under topogenous influence. In extreme form, this is well seen around

mire basins, where the humic top to the surrounding gley thickens into a layer of peat, developed under the control of a ground water-table indirectly fed by the springs and seepage lines of the slopes and often showing some seasonal fluctuation, with winter flooding and superficial summer drying. Such conditions are particularly characteristic of small basins in Norfolk, where the Carex elata sub-community forms a fringing zone of wet fen-meadow.

Rather similar are those situations where the Juncus-Cirsium fen-meadow occurs on sometimes deep peats in extensive flood-plain mires with fluctuating waters of a base-rich character, the movement of which may deposit allochthonous mineral matter. Once much more extensive (e.g. Pallis 1911, Lambert 1946, 1948), such stands can still be seen in parts of Broadland, and more fragmentarily over peaty alluvium on terraces in some lime-rich river valleys elsewhere, and typically they are of the Iris sub-community. Both this kind of Juncus-Cirsium fen-meadow and, to a lesser extent, the Carex elata sub-community, are characterised by a shift away from a prominent Molinio-Arrhenatheretea element of grasses and climbers and a supplementing of the Molinietalia tall herbs by a variety of plants usually found in Phragmitetalia fens, Phragmites itself, Carex elata, Calamagrostis canescens, Iris pseudacorus, Valeriana officinalis, Lythrum salicaria and Lysimachia vulgaris.

The striking differences in composition and structure that can be seen among these vegetation types, between, for example, the Iris sub-community on the one hand and the Briza-Trifolium sub-community on the other, are also a function of variations in treatment. The Juncus-Cirsium fen-meadow is a secondary community derived and maintained by mowing and/or grazing, and strictly therefore not always a meadow but often, particularly now, with the demise of large-scale mowing of this kind of vegetation, a pasture. The detailed effects of these treatments here are little understood, and comments in such studies as those of Lambert (1946, 1948) and Willis & Jefferies (1959) and experiments described in Wheeler (1983b) and Wheeler et al. (1985) suggest that they can be subtle and diverse. Certainly, the degree of fine variation in the floristics and physiognomy of the vegetation included here suggests a complex pattern of interactions between treatments and soil factors, among which it is difficult to discern more than very general trends. It is highly likely, too, that many sites have a unique history of management, the details of which are now, of course, often irrecoverable.

However, the general influence of such treatments seems to be as follows. First, they have, both of them, produced a strong measure of floristic convergence in a variety of precursors, such that the resulting vegetation types are now characterised, to a greater or lesser degree, by species tolerant of cropping under low-intensity agricultural regimes, but otherwise of fairly broad ecological amplitude. The substantial group of meadow and pasture mesophytes figures prominently among the most frequent plants of the community and often veils the peculiarities of the vegetation. And the fact that much the same assemblage rises to prominence, under the influence of similar treatments, over less base-rich soils in other parts of the British lowlands, is the main reason why it is difficult to draw hard and fast boundaries between this community and the other Molinietalia pastures and meadows. Sometimes, over soils of a transitional character, a little drier and less base-rich than here, it is the identity of the dominant rush that is the deciding factor and, throughout this community, *J. subnodulosus* provides the strongest differential feature. It is well adapted to survive annual summer mowing, the usual cutting regime that was followed here, by virtue of its phenology (Richards & Clapham 1941*d*), and, though reputedly eaten by cattle (Willis & Jefferies 1959), it is generally resistant to grazing (Richards & Clapham 1941*d*, Lambert 1946, 1948).

These features also mean that *J. subnodulosus* is often one of the strongest threads of continuity between this community and the vegetation from which it has been derived: the second treatment-related characteristic of this fen-meadow is the extent to which its floristics are inherited and the ways in which mowing and grazing have mediated the survival or extinction of the species in its various precursors. Usually, there is no firm evidence of what vegetation preceded the development of this kind of fen-meadow, but it is in the *Iris* sub-community (and to a lesser extent, the *Carex elata* sub-community), that the presumed lines of inheritance are clearest. Essentially, this vegetation represents the middle and lower tiers of drier Phragmitetalia fen, like the *Peucedano-Phragmitetum* or the *Phragmites-Eupatorium* fen, communities which, like the *Iris* fen-meadow, occur on topogenous peats and mineral soils, but which are either primary fen or secondary mowing-marsh, winter-cut for reed or occasionally summer-cut for sedge. Such mowing regimes leave one or other of the potential dominants able to renew in abundance and exert some control over the associated flora by altering the light penetration. Any switch to more regular summer mowing, which was widely practised in areas like Broadland for the harvesting of 'litter' (e.g. Lambert 1946, 1965), is likely to break the cycle of regeneration of secondary Phragmitetalia fen by weakening the dominance of important helophytes like *Phragmites* and allowing resistant plants like *J. subnodulosus* to survive in abundance, particularly on drier ground where it might have some natural edaphic advantage. *J. subnodulosus* is not a formidable shade-caster so taller dicotyledons can persist in some quantity under such conditions, provided they can accumulate sufficient capital via

abundant stem leaves before cutting or through the persistence of rosettes afterwards. And with the regular removal of herbage and the prevention of litter accumulation, smaller fen herbs too are able to thrive. This kind of treatment was undoubtedly responsible for the creation and maintenance of some extensive stands of the community in our larger topogenous fen systems (Lambert 1946), though it must be remembered that annual summer mowing can deflect Phragmitetalia fen into other kinds of vegetation than the *Juncus-Cirsium* fen-meadow, notably the *Cirsio-Molinietum* (Godwin 1929): what controls the particular direction of development is not really known, although the nutrient status of the soil probably plays an important part. Except within managed reserves, like Wicken Fen in Cambridgeshire (Wheeler 1975), such marsh mowing is now very rarely practised, though patches of the *Juncus-Cirsium* fen-meadow in wetter hollows within agricultural enclosures may be included in an annual hay-crop.

Light grazing of Phragmitetalia fen can have a similar effect to mowing for litter and cattle-pasturing of topogenous mires has produced vegetation akin to the *Iris* sub-community on the kind of grazing-levels, embanked and pumped, which Lambert (1948) described at Rockland and Claxton on the Yare in Norfolk. Grazed topogenous fens of such extensive size are now rare, although some stands of the *Carex elata* sub-community show a rich low-growing herbage around mire basins where cattle have access. In general, however, heavy grazing and the attendant trampling favour the elimination of not only *Phragmites*, but also many of the fen associates tolerant of even quite frequent mowing. In grazed *Juncus-Cirsium* fen-meadow the floristic emphasis is thus of a third kind, with but a small inherited element, but a strong appearance of pasture preferentials. This is best seen in the *Briza-Trifolium* sub-community, most frequently developed around spring-fens and in wet field hollows that are grazed throughout the year or at least in summer. The particular character of the preferential element in the vegetation is rather varied but generally it has affinities with Cynosurion swards of varying degrees of base-enrichment, taking in, towards its extremes, more mesophytic calcicoles and some moderate calcifuges. Fairly rich and diverse stretches of such herbage, cropped to a fairly short, rough turf, and kept open, too, by trampling (always likely to produce poaching on these soils), thus form the matrix between the patches of resistant rushes and sedges in this vegetation, the prominence of the elements varying with the initial composition of the sward and the pattern of grazing.

Grazing is nowadays of greater importance in maintaining the *Juncus-Cirsium* fen-meadow than mowing, but many stands bear signs of a fourth kind of treatment, and that is simply neglect, where previous management has fallen into disuetude, as on many mowing-marshes and some grazing-levels, or where it has long been of a rather casual nature, as in wet field bottoms and corners which escape cutting for hay and where stock gain only occasional access. The Typical sub-community seems especially characteristic of such situations and is now the most widely distributed and common kind of *Juncus-Cirsium* fen-meadow, its impoverished floristic composition bearing testimony to the elimination of both Phragmitetalia and Cynosurion associates with the unchecked growth of the rush and sedge dominants.

Zonation and succession

The *Juncus-Cirsium* fen-meadow is found with a wide variety of swamps, tall-herb fens, other mires and grasslands and woodlands in zonations and mosaics which reflect interactions between edaphic variation and treatments. Many such sequences now bear signs of neglect or disturbance and are often truncated above by improvement of the soils, occurring as isolated fragments within intensive agricultural landscapes. The abandonment of traditional treatments has also allowed many tracts to progress to scrub or woodland in renewed successions.

The most extensive sequences in which the *Juncus-Cirsium* fen-meadow is found occur in large topogenous mires, like the flood-plain fens of Broadland, where the *Iris* sub-community once formed an important part of the patchwork of mowing-marsh. Concentrated on the drier fen margins and maintained by annual summer-cutting for litter, such vegetation typically gave way over the more extensively-flooded peats to the *Peucedano-Phragmitetum* or the more species-poor *Phragmites-Eupatorium* fen, different compartments variously treated for crops of *Phragmites*, *Cladium* or *Glyceria maxima*, according to the timing of mowing and the trophic state of the substrates. And, beyond such communities, around the open waters, and penetrating between the compartments along the actively-maintained dykes, were stands of swamps, such as the *Phragmitetum*, the *Glycerietum maximae* or the *Cladietum* (Pallis 1911, Lambert 1946, 1951, 1965). Such vegetation patterns are now in a state of almost universal neglect, with the various dominants reasserting themselves strongly among the secondary fen and *Phragmites* re-invading the fen-meadows (Wheeler 1975). The boundaries between the fen compartments have been further obscured by the choking of dykes and by the invasion throughout of shrubs and trees. Most of the developing woodland approximates to *Salix-Betula-Phragmites* fen carr, the kind of wet forest which develops in most primary hydrarch successions over more base-rich peats in lowland Britain, and towards which secondary successions tend to converge when restraining treatments are withdrawn (Figure 18).

In some stretches of Broadland, around Hickling, Sutton and Wheatfen broads, for example (Wheeler 1975), such mosaics as these can still be seen in less obscured form and at Wicken Fen some of the mown droves have thin strips of this kind of *Juncus-Cirsium* fen-meadow running between the compartments. Elsewhere, smaller fen basins, like those in Norfolk and Anglesey, and alluvial terraces along chalkland streams, have narrow sequences of fen-meadow, fen and swamp of this basic type. Spring-fed hollows in East Anglia, too, can show compressed zonations from the *Caricetum elatae* swamp to the *Carex elata* sub-community. In such cases, grazing often maintains a short sward, although larger sedge tussocks may provide a niche for the scattered invasion of woody plants.

Although grazed *Juncus-Cirsium* fen-meadow was once an extensive component of the vegetation cover on some flood-plain mires, forming a striking contrast to the mowing-marsh around (Lambert 1948), it is now generally found as small stands marking out water-logged hollows, seepage lines and springs within enclosed pastures. Where stock penetrate far and regularly on to the wetter ground, the vegetation is usually of the *Briza-Trifolium* type, although there can be considerable floristic and structural differences both within and between stands according to the pattern of flushing and the intensity of grazing. The sharpness of the boundary between the fen-meadow and surrounding sward is also variable, depending on the gradient of soil moisture and whether there have been attempts to improve the pasture around. Often, the context is a Cynosurion grassland, like the ubiquitous *Lolio-Cynosuretum* or, on less-improved ground, the *Centaureo-Cynosuretum*, both of them agricultural swards of brown soils with free to moderately impeded drainage such as frequently surround flushes and waterlogged hollows with fen-meadow developed over shale and clay landscapes. With such grasslands, the *Briza-Trifolium* sub-community shows a strong floristic continuity through mesophytic herbs such as *Holcus lanatus*, *Festuca rubra*, *Ranunculus acris*, *Trifolium repens*, *Cerastium fontanum*, *Prunella vulgaris*, *Centaurea nigra* and *Cynosurus cristatus*, so the general effect in moving out of the flush is of a loss of rush and sedge dominants and Molinietalia herbs and a supplementing of the general Molinio-Arrhenatheretea element with a prominent Cynosurion component. In other cases, where springs or waterlogged hollows punctuate a landscape dominated by pervious limestones with rendzina soils, the floristic switch around the fen-meadow is to a Mesobromion sward, usually of the *Festuca-Avenula* type, with species such as *Briza media*, *Carex flacca*, *Centaurea nigra* and *Trifolium pratense* providing a link. Sometimes, where base-richness of the soils is pronounced, such grassland may provide a context for a wider range of wetter calcicolous communities around areas of soligenous influence, not only fen-meadow of the *Briza-Trifolium* type, but also the *Schoenetum* or, towards northern Britain, the *Pinguiculo-Caricetum*. In some sites of this type, too, the *Cirsio-Molinietum* fen-meadow can figure in such mosaics, characteristic of somewhat drier soils than these springs and extending on to less base-rich ones. Its floristic similarity to the *Juncus-Cirsium* fen-meadow is considerable and in some cases it may have been derived from it by selective nutrient depletion with long-continued treatment.

Reduction of grazing in the *Juncus-Cirsium* fen-meadow results in an expansion of the bulky dominants and ranker grasses and an overwhelming of the smaller herbs, producing vegetation like that of the Typical sub-community. Patches of this can often be found within stands of the *Briza-Trifolium* sub-community, marking out less-frequented areas of ground, which then acquire some resistance to the reduction of the rough herbage which develops. In other cases, whole springs and hollows, more or less totally free of grazing, in field corners or alongside streams, can be occupied by this kind of fen meadow, sharply delineated from the heavily-cropped pasture around or, where neglect has extended on to drier ground also, grading to some kind of *Arrhenatheretum*, with plants such as *Festuca rubra*, *Holcus lanatus*, *Lotus uliginosus*, *Vicia cracca* and *Lathy-*

rus pratensis providing some floristic continuity, but dominance passing to *Arrhenatherum elatius* and Molinietalia tall herbs being replaced by Arrhenatherion umbellifers such as *Heracleum sphondylium* and *Anthriscus sylvestris*. Quite often, too, in such situations as this, the soils are sufficiently eutrophic to allow plants such as *Epilobium hirsutum* and *Urtica dioica* to gain a hold once grazing is relaxed and dense patches of these may be found among mosaics of these two communities. Indeed, a neat experiment described by Wheeler (1983*b*; see also Wheeler *et al.* 1985) has shown how an enclosed *Juncus-Cirsium* fen-meadow was fairly rapidly replaced by dense *E. hirsutum*, except where pronounced waterlogging of the ground maintained ferrous ions at a level toxic to this tall herb, but still harmless to *J. subnodulosus* and its associates. Enrichment of ground waters by the washing out of fertilisers, now a very common feature of the lowland agricultural landscape in Britain, is likely to increase this kind of development in ungrazed stands.

The Typical sub-community of the *Juncus-Cirsium* fen-meadow, with its rank herbage, is perhaps rather resistant to rapid invasion by woody plants and successional developments subsequent to withdrawal of grazing have not been followed. Wetter spring-fens are likely to be invaded by much the same mixture of shrubs and trees as are characteristic colonists in topogenous fen-

Figure 18. Changes in fen and meadow zonation with improvement and neglect.
In the foreground is a traditional sequence in an open-water transition mire from *Phragmitetum* swamp (S4), through mown *Peucedano-Phragmitetum* (S24) and *Juncus-Cirsium* fen-meadow, *Iris* sub-community (M22d) to grazed *Juncus-Cirsium* fen-meadow, *Briza-*

Trifolium sub-community (M22b) and *Centaureo-Cynosuretum* (MG5). With enclosure and improvement of the drier ground, the last is converted to *Lolio-Cynosuretum* (MG6), while neglect of the mowing marsh leads to spread of rank Typical *Juncus-Cirsium* fen-meadow (M22a) and secondary *Salix-Betula-Phragmites* woodland (W2).

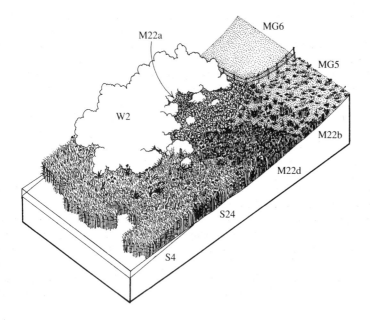

meadow, producing *Salix-Betula-Phragmites* woodland, but even in such cases, many stands are so small as
to be probably marked out in subsequent woodland
cover by just small patches of willow and birch within
predominantly drier vegetation. Stands on ground
which is not quite so thoroughly waterlogged would
perhaps progress eventually to *Fraxinus-Acer-Mercur-
ialis* woodland, whose *Deschampsia cespitosa* sub-
community, when coppiced, produces a field layer which
has some floristic similarity to the *Juncus-Cirsium* fen
meadow. Extensive abandonment of ground, where
such seral processes might be studied, has in fact been
very unusual in the recent decades of vigorous land
improvement and very often now, neglected *Juncus-
Cirsium* fen-meadows remain isolated fragments closely
hemmed in by much-altered pastures or arable land.
Themselves originally of agricultural origin and rarely
of spectacular richness, even when less overgrown, they
nonetheless frequently provide an important remnant
habitat for wetland species which were formerly much
more common.

Distribution
The community is of wide distribution on suitably wet
and base-rich soils through the southern British lowlands with particular concentrations of stands in East
Anglia, north Buckinghamshire and Anglesey. The *Iris*
and *Carex* sub-communities are especially well represented on topogenous mires in the first area but are
much scarcer elsewhere. The *Briza-Trifolium* sub-
community occurs throughout the range but, very frequently, spring-fen stands are of the Typical sub-
community and, with neglect and land improvement,
distribution of the community is becoming increasingly
sparse.

Affinities
It was Wheeler (1975, 1980c) who, in his study of British
rich fens, first gave some coherent shape to the diversity
of vegetation types included in this community. Previously, assemblages of this kind, usually described
under the general heading of 'fen-meadow', had been
noted as components of wetland sequences (Pallis 1911,
Clapham 1940, Rose 1950, Willis & Jefferies 1959), but
rarely described in any detail; or, where subject to closer
scrutiny, had been of a rather particular kind, as in the
mowing-marsh and grazing-level stands of Lambert
(1946, 1948). The treatment here is based heavily on the
data and proposals of Wheeler, although many more
impoverished stands (the commonest type) have been
added and we follow his earlier scheme (Wheeler 1975)
in recognising, as well as the *Carex* and *Iris* types, a
major split among the remainder based on the presence
or absence of the *Briza-Trifolium* preferentials. Many of
the divisions outlined in his later proposals (Wheeler

1980c) are thus relegated to a minor level of variation,
though one that is often quite striking to the eye and
which, as Wheeler himself remarks, is perhaps just a
fraction of the possible complex of responses to edaphic
and treatment differences in this kind of vegetation.

Such differences, which reflect the diversity of vegetation types that have been converted into this fen-
meadow and the peculiarities of site histories, inevitably
make the community very cumbersome, with the individual character of particular stands often being more
noticeable than their floristic similarities. And what they
do have in common is often of a very general nature,
with all the constants of the community and many of the
most frequent associates being plants of fairly wide
ecological tolerance and occurring commonly in other
communities, combined frequently here because of a
floristic convergence produced by the imposition of
mowing and grazing. This characteristic places the
Juncus-Cirsium fen-meadow unequivocally within the
Molinietalia of the Molinio-Arrhenatheretea, but it
makes the boundaries of the communities against other
vegetation types in this order very ill-defined.

Two particular problems should be noted. First, this
community forms a national series with the *Holco-
Juncetum* and the *Juncus-Galium* rush-pasture, which
can be understood in relation to both climatic and
edaphic variation in moving from east to west across the
lowland south of Britain. Floristic shifts through this
series can be seen both among the dominants, where
there is a move from *J. subnodulosus* through *J. inflexus*
to *J. effusus* and *J. acutiflorus*, and among the associates,
where there is a move from more calcicolous to more
calcifuge species and an appearance of some oceanic
herbs, though the rank growth of the dominants and
more bulky companions often masks these peculiarities.
In Britain, too, although the two poles of this sequence
of communities are very different, certain plants which,
on the European mainland would be regarded as diagnostic of the extremes, are much less strictly confined in
their occurrence. Thus, though the *Juncus-Cirsium* fen-
meadow has frequent records for species like *Caltha
palustris*, *Carex disticha*, *Lotus uliginosus* and *Lychnis
flos-cuculi*, which would locate it among the Continental
Calthion communities (Westhoff & den Held 1969),
these all occur elsewhere in the Molinietalia in this
country and some of them even more widely. And other
diagnostic plants of this alliance, such as *Bromus racemosus* and *Fritillaria meleagris*, are rare in Britain and
usually found in other communities. At best, then, the
Juncus-Cirsium fen-meadow can be accommodated into
the European framework with some difficulty, and its
increasing rarity means that, in many places, it is the
Holco-Juncetum that is the more familiar British
approximation to the Calthion. The relationship of the
community to the *Caltha-Cynosurus* flood-pasture, a

third vegetation type which could be referred to this alliance must await further sampling: it has sufficient in common with the *Juncus-Cirsium* fen-meadow to be, perhaps, subsumed within it.

The second diagnostic difficulty is in separating this community from the *Cirsio-Molinietum*, a vegetation type likewise derived by mowing and grazing, with strong floristic links through Molinio-Arrhenatheretea herbs like *Holcus lanatus*, *Festuca rubra*, *Vicia cracca*, and Molinietalia plants such as *Cirsium palustre*, *Filipendula ulmaria*, *Lotus uliginosus* and *Angelica sylvestris*, and with some parallel suites of preferentials in its sub-communities. *J. subnodulosus* can occur quite frequently in the *Cirsio-Molinietum* and *Molinia* is sometimes found in the *Juncus-Cirsium* fen-meadow but, generally speaking, there is a switch in dominance from the one to the other species in these two communities and a preferential occurrence in the *Cirsio-Molinietum*

of plants like *Cirsium dissectum*, *Carex panicea*, *C. hostiana*, *C. pulicaris* and *Succisa pratensis*, in assemblages similar to Continental Junco-Molinion vegetation (Westhoff & den Held 1969).

Among these suites of vegetation types, *J. subnodulosus* shows a definite peak of occurrence in the *Juncus-Cirsium* fen-meadow but this community contains only a fraction of the assemblages that have sometimes been united, in Continental classifications, in a compendious *Juncetum subnodulosi* (after Koch 1926), characterised by dominance of the rush. Among the vegetation types included there, the *Juncus-Cirsium* fen-meadow compares with the lowland stands separated off from this association into a *Crepido-Juncetum subnodulosi* (Libbert 1932, Tüxen 1937, Oberdorfer 1957) and with a variety of Calthion *Feuchtwiesen* described from Germany (Korneck 1963, Krausch 1964, Passarge 1964, Kloss 1965).

Floristic table M22

	a	b	c	d	22
Juncus subnodulosus	V (1–8)	IV (1–8)	V (1–6)	V (1–8)	V (1–8)
Calliergon cuspidatum	IV (1–5)	V (1–3)	V (1–5)	V (1–3)	V (1–5)
Mentha aquatica	IV (1–7)	V (1–3)	III (1–3)	V (1–3)	IV (1–7)
Holcus lanatus	IV (1–6)	V (1–3)	III (1–3)	IV (1–3)	IV (1–6)
Cirsium palustre	IV (1–4)	V (1–3)	IV (1–3)	III (1–3)	IV (1–4)
Equisetum palustre	IV (2–6)	IV (1–3)	IV (1–3)	II (1–3)	IV (1–6)
Filipendula ulmaria	III (1–7)	IV (1–3)	III (1–3)	V (1–3)	IV (1–3)
Lotus uliginosus	III (1–4)	IV (1–3)	III (1–3)	III (1–3)	IV (1–4)
Cardamine pratensis	II (2–3)	IV (1–3)	II (1–3)	II (1–3)	III (1–3)
Juncus articulatus	II (1–8)	IV (1–6)	II (1–3)	II (1–3)	II (1–8)
Cerastium fontanum	II (1–3)	IV (1–3)		II (1–3)	II (1–3)
Briza media	I (1–6)	IV (1–3)		I (1)	II (1–6)
Trifolium repens	I (1–5)	IV (1–3)		I (1–3)	II (1–3)
Trifolium pratense	I (2–4)	IV (1–4)	II (1–3)		II (1–3)
Anthoxanthum odoratum	I (1–5)	III (1–4)	II (1–3)	I (1–3)	II (1–7)
Molinia caerulea	II (1–7)	III (1–4)	II (1–3)	I (1)	II (1–4)
Rumex acetosa	II (1–4)	III (1–3)	II (1–3)	I (1–2)	II (1–6)
Juncus inflexus	II (1–5)	III (1–6)	I (1)	I (1)	II (1–3)
Epilobium parviflorum	II (1–3)	III (1–3)	I (1–2)	I (1–3)	II (1–4)
Dactylorhiza fuchsii	I (1–4)	III (1–3)	I (1–3)	I (1–2)	II (1–4)
Plantago lanceolata	I (1–4)	III (1–3)		I (1–3)	I (1–4)
Prunella vulgaris	I (1–4)	III (1–3)		I (1–3)	I (1–4)
Centaurea nigra	I (2–4)	III (1–2)		I (1–3)	I (1–4)
Ranunculus repens	I (1–4)	III (1–3)			I (1–2)
Triglochin palustre	I (1–2)	III (1–2)	I (3)		I (1–4)
Rhinanthus minor	I (1–4)	II (1–3)		I (1–3)	I (1–3)
Carex flacca	I (1–3)	II (1–3)			I (1–3)
Carex hirta	I (1–3)	II (1–3)			I (1–4)
Juncus effusus	I (1–4)	II (1–3)			I (1–3)
Cynosurus cristatus		II (1–3)			
Phragmites australis	II (1–4)	I (1–5)	IV (1–3)	IV (1–4)	II (1–5)
Hydrocotyle vulgaris	II (1–6)	II (1–3)	V (1–3)	IV (1–3)	III (1–6)

Species	1	2	3	4	5
Lythrum salicaria	I (2-3)	I (1-2)	III (1-3)	III (1-3)	I (1-3)
Menyanthes trifoliata	I (1-2)	I (1)	II (1-4)	II (1-4)	I (1-4)
Myosotis scorpioides	I (1-3)	I (1)	II (1-3)	II (1-2)	I (1-3)
Campylium stellatum	I (1-3)	I (1-2)	II (1-3)	II (1-3)	I (1-3)
Plagiomnium rostratum	I (1)	I (1)	II (1)	II (1-2)	I (1-2)
Carex elata			V (1-4)	I (1-2)	I (1-4)
Potentilla palustris			IV (1-3)	II (1-2)	I (1-3)
Galium palustre	II (3-5)	III (1-3)	IV (1-3)	III (1-3)	II (1-5)
Epilobium palustre	II (1-3)	I (1)	III (1-2)	II (1-3)	II (1-3)
Equisetum fluviatile	I (1-3)	I (1-3)	III (1-3)	II (1-2)	I (1-3)
Dactylorhiza incarnata		II (1-3)	III (1-3)	II (1-2)	II (1-3)
Berula erecta	I (1-3)		II (1-3)	I (1-3)	I (1-3)
Pedicularis palustris	I (1-4)		II (1-3)	I (1)	I (1-4)
Salix cinerea sapling	I (1-3)	I (1-3)	II (1-2)		I (1-3)
Dactylorhiza majalis praetermissa		I (1-3)	II (1-3)	I (1)	I (1-3)
Thelypteris palustris			II (1-3)		I (1-3)
Ophioglossum vulgatum			I (1-2)		I (1-2)
Carex acutiformis	II (1-4)	II (1-4)	I (1)	IV (1-6)	II (1-6)
Iris pseudacorus	I (1-2)	I (1-3)	II (1-3)	V (1-3)	I (1-3)
Ranunculus flammula		II (1-2)	II (1-3)	III (1-3)	I (1-3)
Valeriana officinalis	I (1-4)	I (1)	I (1-3)	III (1-3)	I (1-4)
Lysimachia vulgaris			I (1-3)	III (1-3)	I (1-3)
Thalictrum flavum				III (1-2)	I (1-2)
Epilobium hirsutum	I (1-3)		I (1-3)	II (1-3)	I (1-3)
Calystegia sepium	I (1-3)		I (1-2)	II (1-3)	I (1-3)
Phalaris arundinacea	I (1-4)			II (1-3)	I (1-4)
Carex lepidocarpa		I (1-2)		II (1-3)	I (1-3)
Peucedanum palustre				II (1-3)	I (1-3)
Calamagrostis canescens				II (1-4)	I (1-4)
Symphytum officinale				II (1-3)	I (1-3)
Cladium mariscus				I (1-2)	I (1-2)
Galium uliginosum	IV (1-5)	IV (1-3)	III (1-3)	III (1-3)	III (1-5)
Lychnis flos-cuculi	III (1-3)	IV (1-3)	III (1-3)	IV (1-3)	III (1-3)
Poa trivialis	II (1-4)	IV (1-3)	II (1-3)	IV (1-3)	III (1-4)
Caltha palustris	II (1-4)	III (1-3)	IV (1-3)	IV (1-3)	III (1-4)

Floristic table M22 *(cont.)*

	a	b	c	d	22
Carex panicea	II (1–7)	V (1–3)	IV (1–3)	III (1–3)	III (1–7)
Festuca rubra	III (1–5)	V (1–4)	II (1–3)	III (1–2)	III (1–5)
Agrostis stolonifera	III (2–5)	V (1–5)	II (1–2)	III (1–3)	III (1–5)
Ranunculus acris	II (1–4)	V (1–3)	III (1–3)	III (1–3)	III (1–4)
Valeriana dioica	III (1–4)	IV (1–5)	III (1–3)	III (1–3)	III (1–4)
Vicia cracca	III (1–5)	IV (1–3)	II (1–3)	III (1–3)	III (1–5)
Lathyrus pratensis	III (1–5)	IV (1–3)	III (1–3)	I (1–2)	III (1–5)
Brachythecium rutabulum	II (1–4)	III (1–2)	II (1–2)	IV (1–3)	III (1–4)
Angelica sylvestris	III (1–6)	III (1–3)	II (1–3)	III (1–2)	III (1–6)
Succisa pratensis	II (1–5)	III (1–3)	II (1–3)	III (1–3)	III (1–5)
Carex disticha	II (1–7)	III (1–6)		III (1–6)	III (1–7)
Potentilla erecta	II (1–4)	III (1–3)		IV (1–3)	II (1–4)
Eupatorium cannabinum	II (1–3)	II (1–3)	I (1)	II (1–3)	II (1–3)
Hypericum tetrapterum	II (1–2)	II (1–3)	II (1–3)	I (1–3)	II (1–3)
Arrhenatherum elatius	II (1–4)	II (1–3)	II (1–3)	I (1–3)	II (1–4)
Scrophularia auriculata	II (1–4)	II (1–3)	II (1–3)	I (1)	II (1–4)
Potentilla anserina	I (1–3)	II (1–3)	I (1)	I (1)	I (1–3)
Deschampsia cespitosa	I (1–3)	II (1–3)	I (1)	I (1)	I (1–3)
Carex nigra	I (1–6)	II (1–4)	I (1)		I (1–6)
Pulicaria dysenterica	I (1–6)	II (1–3)		I (1)	I (1–6)
Rumex conglomeratus	I (1–3)	I (1–3)	I (1–3)	I (1–3)	I (1–3)
Myosotis laxa caespitosa	I (1–2)	I (1–3)	I (1–3)	I (1–3)	I (1–3)
Plagiomnium undulatum	I (1–3)	I (1–3)	I (1–3)		I (1–3)
Rhizomnium punctatum	I (1–3)	I (1–3)	I (1–3)		I (1–3)
Oenanthe lachenalii	I (1–3)	I (1–3)	I (1–3)		I (1–3)
Potentilla reptans	I (1–4)	I (1–3)		I (1–3)	I (1–4)
Epipactis palustris	I (1–5)	I (1–3)		I (1–3)	I (1–5)
Lophocolea bidentata s.l.	I (1–2)		I (1–3)	I (1–3)	I (1–3)
Lycopus europaeus	I (1–3)		I (1–3)	I (1–3)	I (1–3)
Scutellaria galericulata	I (1–3)		I (1–3)	I (1–3)	I (1–3)
Eriophorum angustifolium		I (1–6)	I (1–3)	I (1–3)	I (1–6)
Luzula multiflora		I (1–3)	I (1–3)	I (1–3)	I (1–3)

	a	b	c	d	22
Phleum pratense	I (1–3)	I (1–3)			I (1–3)
Veronica beccabunga	I (1–3)	I (1–3)			I (1–3)
Ajuga reptans	I (1–4)	I (1–3)			I (1–4)
Rumex sanguineus	I (1–3)	I (1–3)			I (1–3)
Cirsium arvense	I (1–3)	I (1–3)			I (1–3)
Pseudoscleropodium purum	I (1–3)	I (1–3)			I (1–3)
Achillea ptarmica	I (1–4)	I (1–3)			I (1–4)
Festuca arundinacea	I (1–3)	I (1–3)			I (1–3)
Cratoneuron filicinum	I (1–3)			I (1–3)	I (1–3)
Solanum dulcamara	I (1–3)			I (1–3)	I (1–3)
Lemna minor	I (1–3)		I (1–3)		I (1–3)
Plagiomnium affine	I (1–3)	I (1–3)	I (1–3)		I (1–3)
Cirsium dissectum	I (1)	I (1)		I (1–3)	I (1–3)
Eleocharis palustris	I (1–3)	I (1–3)		I (1–3)	I (1–3)
Carex rostrata			I (1–3)	I (1–3)	I (1–3)
Polygonum amphibium			I (1–3)	I (1–3)	I (1–3)
Stellaria palustris			I (1)	I (1)	I (1)
Eleocharis uniglumis			I (1–3)	I (1–3)	I (1–3)
Number of samples	94	51	10	21	176

a   Typical sub-community

b   *Briza media-Trifolium* spp. sub-community

c   *Carex elata* sub-community

d   *Iris pseudacorus* sub-community

22   *Juncus subnodulosus-Cirsium palustre* fen-meadow (total)

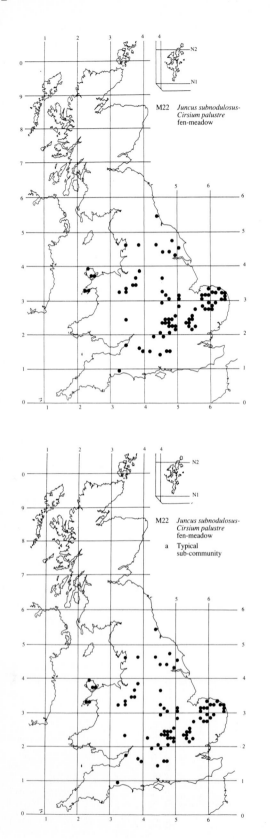

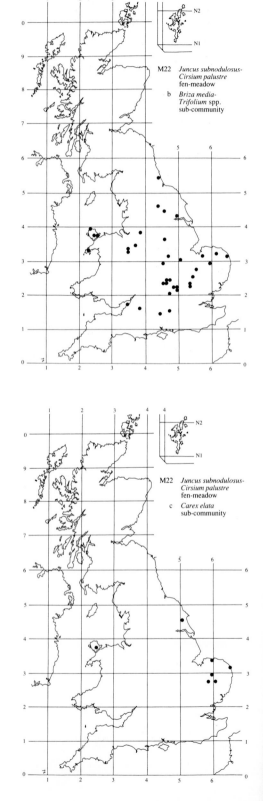

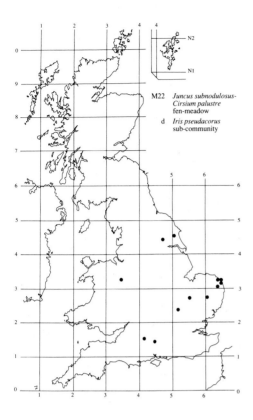

M22   *Juncus subnodulosus-*
       Cirsium palustre
       fen-meadow

d   *Iris pseudacorus*
    sub-community

M23
Juncus effusus/acutiflorus-Galium palustre rush-pasture

Synonymy

Juncus acutiflorus-Acrocladium cuspidatum nodum McVean & Ratcliffe 1962 *p.p.*, Eddy *et al.* 1969; *Juncus acutiflorus* pasture Birse & Robertson 1967, 1973; *Juncus acutiflorus-Acrocladium cuspidatum* sociation Edgell 1969; Species-rich *Juncetum effusi* Eddy *et al.* 1969; *Juncus acutiflorus-Filipendula ulmaria* Association Birks 1973; *Juncus articulatus-Acrocladium cuspidatum* nodum Prentice & Prentice 1975; *Potentillo-Juncetum acutiflori* Birse & Robertson 1976, Birse 1980, 1984; *Juncus acutiflorus-Sphagnum palustre* nodum Adam *et al.* 1977 *p.p.*; *Galium palustre* nodum Daniels 1978 *p.p.*; Fen communities 27 & 28 Meade 1981; *Junco acutiflori-Molinietum* (O'Sullivan 1968) Ratcliffe & Hattey 1982 *p.p.*; Fen meadow Ratcliffe & Hattey 1982 *p.p.*

Constant species

Galium palustre, Holcus lanatus, Juncus effusus/acutiflorus, Lotus uliginosus.

Physiognomy

The *Juncus effusus/acutiflorus-Galium palustre* rush-pasture is a rather ill-defined assemblage of vegetation characterised by the abundance of either *Juncus effusus* or *J. acutiflorus*, sometimes both, in a ground of mesophytic herbs of wide occurrence in moister agricultural grasslands. Diversity among the dominants is not very great (nothing like so locally varied as in the *Juncus-Cirsium* fen-meadow, for example), but the structure of the vegetation can show considerable differences according to the treatment history. And, though the flora is rarely strikingly rich, the associates are quite diverse, both across the whole range of the community and, on a local scale, from site to site and even within large stands. Such variation makes the bounds of this vegetation type hard to fix, both against other rush-dominated communities, a problem that is greater towards the east of its distribution, and, increasingly to the west, against grasslands in which *Molinia caerulea* plays a prominent role.

Of the two rushes which usually dominate here, *J. effusus* is the commoner throughout, extending across the whole range of the community, though being distinctly more frequent and abundant in the more easterly sub-community through which this vegetation grades imperceptibly to the *Holco-Juncetum*. *J. acutiflorus* is much more distinctly western in its occurrence, rivalling or exceeding *J. effusus* in the other of the sub-communities, which is much more distinctive in its floristics and the usual type of this vegetation which has figured in previous accounts (e.g. McVean & Ratcliffe 1962, Birse & Robertson 1967, 1973, 1976, Edgell 1969, Birks 1973, Adam *et al.* 1977, Ratcliffe & Hattey 1982). Among other rushes, none occurs commonly, though some can be locally frequent and abundant. Sometimes, for example, *J. articulatus* supplements or replaces either of the two usual dominants, although there is no indication in the available data that this is a particularly western or coastal phenomenon, as hinted by McVean & Ratcliffe (1962). This species can hybridise with *J. acutiflorus* and intermediates are occasionally found here (McVean & Ratcliffe 1962, Prentice & Prentice 1975), but difficulties of identification among these taxa have probably led to *J. articulatus* being over-recorded in this kind of vegetation in some surveys (as suggested in Meade 1981, for example) and overall it is an uncommon plant in the community. *J. conglomeratus* is likewise rather infrequent, though it, too, can be mistakenly recorded for *J. effusus* var. *compactus*: it has been found at quite high covers at some sites but, generally speaking, is much more characteristic of heathy vegetation where rushes occur with much *Molinia caerulea*. *J. inflexus* and *J. subnodulosus*, species which are best represented on more base-rich soils to the east of Britain, are hardly ever found and their absence provides a good separation of this community from more calcicolous kinds of rush-dominated vegetation in the *Holco-Juncetum* and, more particularly, the *Juncus-Cirsium* fen-meadow.

Although rushes generally dominate, their abundance and stature and the representation of other elements in the vegetation are quite variable. Fairly often, the rushes have very high cover and they commonly attain half a metre in height, sometimes up to twice that, so their shading effect can be quite considerable, especially where the more densely-tussocky *J. effusus* is growing vigorously. In other cases, the rush cover is sparser or more discontinuous, when the associated flora amongst its shoots and between the clumps is denser. In ungrazed and unmown stands, this herbage can grow tall, often showing patchy local abundance among the subordinate species, but, where stock have access, a common occurrence here, the rushes may be closely grazed around and the stretches of intervening sward kept shortly cropped.

Most frequent among the associates is *Holcus lanatus* which is quite often itself so abundant as to give the lower tier of vegetation a rank grassy appearance where growth is less closely checked. But other grasses can be quite common, too, with *Agrostis canina* ssp. *canina*, *A. stolonifera*, *Anthoxanthum odoratum* and *Poa trivialis* occurring throughout with varying degrees of frequency. The first two in particular can show locally high cover and in more open damper places form quite extensive mats, sometimes with a little *Glyceria fluitans*. In drier stands, *Festuca rubra* and *Agrostis capillaris* sometimes become frequent, forming with *Anthoxanthum* the bulk of a sward which can extend out from amongst the rush clumps to form extensive stretches of Nardo-Galion grassland. Then, where there has been some improvement of the surrounding pasture, *Cynosurus cristatus* and *Lolium perenne* can seed in and become well established. Finally, among the grasses, there is *Molinia caerulea*, a species of no more than very local significance in the eastern *J. effusus* sub-community but, towards the west, becoming increasingly common and abundant and making the vegetation of the *J. acutiflorus* sub-community virtually continuous with that of the *Molinia-Potentilla* grassland, a community with which it is often associated around the margins of damper heaths. The similarities and differences between these vegetation types are well seen in the floristic tables produced from Pembrokeshire heaths (Meade 1981) and Welsh wetlands (Ratcliffe & Hattey 1982): typical stands of each can look different enough but, in some cases, the diagnosis of intermediates has to depend on the relative proportions of Junci and *Molinia*.

Intermixed with the grasses is a variety of other herbs which, when present in numbers, can give this vegetation quite a colourful appearance. Some of the commoner species among them are able to grow tall where grazing is not intense and by mid-summer can put up their flowering stems above the level of the rush canopy. *Cirsium palustre* is the commonest of these but *Rumex acetosa*, *Angelica sylvestris* and *Epilobium palustre* also occur quite frequently and, sprawling among them, there is often some *Galium palustre* and *Lotus uliginosus*. Then, among the smaller associates, frequent species are *Mentha aquatica*, *Ranunculus flammula*, *R. repens*, *R.*

acris (these buttercups often disposed in relation to increasing dryness of the ground: Harper & Sagar 1953), *Cardamine pratensis*, *Hydrocotyle vulgaris*, *Viola palustris* and *Stellaria alsine*. *Potentilla erecta* also occurs occasionally, though it is quite strongly preferential for the *J. acutiflorus* sub-community.

Quite frequently, too, there are some sedges in the sward, though the larger species, such as *Carex acutiformis* and *C. disticha*, which are commonly co-dominant in the *Juncus-Cirsium* fen-meadow, and *C. elata* and *C. paniculata*, which occur there locally, are typically absent here. But such smaller sedges as *C. nigra*, *C. echinata* and *C. demissa* make an occasional appearance and can emphasise floristic similarities with the Caricion nigrae poor-fens, among which *J. effusus* and *J. acutiflorus* can also show dominance. In other stands, *C. panicea* is frequent, giving a somewhat basiphilous look to the vegetation, at least in comparison with the calcifuge grasslands and heaths which often surround the community. Where base-enrichment is more pronounced, as in some of the stands described from Moor House (Eddy *et al.* 1969), *C. dioica* and *C. pulicaris* can also be found, though these are never of more than local significance.

Other plants of wide national distribution which can occur in the community include *Trifolium repens*, *Cerastium fontanum*, *Plantago lanceolata* and *Prunella vulgaris*, species which tend to increase in transitions to improved pastures around, or, where grazing is absent, *Filipendula ulmaria*, the local abundance of which can bring the vegetation close to Filipendulion fen in its appearance. Then, in damper hollows, there can be some *Potentilla palustris*, *Equisetum fluviatile* and *Caltha palustris*, while poached areas can have a specialised flora of their own in which *Juncus bulbosus* can be locally prominent.

One further striking occasional in this kind of vegetation and a plant of some phytogeographical significance, is *Carum verticillatum*, which is sometimes found in the *J. effusus* sub-community, but more often in the *J. acutiflorus* type, the range of which takes in the distribution of this local Oceanic West European umbellifer. It is, however, not confined here, occurring also in the *Molinia-Potentilla* grassland and in vegetation which is probably best referred to the *Cirsio-Molinietum*, though very much on its far-western fringes (B.D. Wheeler pers. comm.). Other species which, with *Carum*, were regarded by Braun-Blanquet as characteristic of this kind of Atlantic rush-dominated vegetation, such as *Wahlenbergia hederacea* and *Scutellaria minor*, occur occasionally but are likewise not restricted to the community. *Senecio aquaticus*, a plant regarded as characteristic of Irish vegetation of this same general type (Braun-Blanquet & Tüxen 1952, Ivimey-Cook & Proctor 1966*b*, White & Doyle 1982) is no more than scarce.

Bryophytes are variable in their cover and very sparse among denser herbage but, where the vegetation is a little more open, they can be abundant over the moist surface of the soil, on litter and over the stools of the vascular plants. *Calliergon cuspidatum* is the most frequent species throughout but *Brachythecium rutabulum* and *Rhytidiadelphus squarrosus* can also occur and, less commonly, *Brachythecium rivulare, Pseudoscleropodium purum, Plagiomnium undulatum, Lophocolea bidentata s.l.* and *Pellia epiphylla*. Small patches of *Polytrichum commune* and some of the less demanding Sphagna can sometimes be found but the absence of extensive carpets of these species provides a good diagnostic criterion against the Caricion nigrae poor fens.

Sub-communities

Juncus acutiflorus sub-community: *Juncus acutiflorus-Acrocladium cuspidatum* nodum   McVean & Ratcliffe 1962 *p.p.*, Eddy *et al.* 1969; *Juncus acutiflorus* pasture Birse & Robertson 1967, 1973; *Juncus acutiflorus-Acrocladium cuspidatum* sociation   Edgell 1969; *Juncus acutiflorus-Filipendula ulmaria* Association Birks 1973; *Juncus articulatus-Acrocladium cuspidatum* nodum   Prentice & Prentice 1975; *Potentillo-Juncetum acutiflori* Birse & Robertson 1976, Birse 1980, 1984; *Juncus acutiflorus-Sphagnum palustre* nodum   Adam *et al.* 1977 *p.p.*; *Galium palustre* nodum Daniels 1978 *p.p.*; Fen communities 27 & 28   Meade 1981; *Junco acutiflori-Molinietum* (O'Sullivan 1968) Ratcliffe & Hattey 1982 *p.p.*; Fen meadow   Ratcliffe & Hattey 1982 *p.p.* In this more sharply-defined of the two sub-communities, *J. effusus* remains very common and its dense tussocks can be quite abundant, but it is typically accompanied and usually exceeded by *J. acutiflorus*, not a rush of strongly tufted growth but often forming extensive and thick clumps of shoots from its far-creeping rhizomes. *J. articulatus* (or putative hybrids with *J. acutiflorus*) can be locally prominent and *J. conglomeratus* is occasionally found. Growing among the rushes, the commonest grasses here are *Holcus lanatus* and, preferentially frequent, *Molinia*, and either or both of these can bulk fairly large in a rank second tier to the vegetation in ungrazed stands. But it is the rushes which characteristically retain dominance and a great abundance of *Molinia*, with the kinds of associates generally typical of this community, marks a shift to the *Molinia-Potentilla* grassland. Other grasses to be found here are *Anthoxanthum* and, rather less commonly, *Agrostis canina* ssp. *canina, A. stolonifera, Festuca rubra* and *Poa trivialis*. Then, such community herbs as *Cirsium palustre, Rumex acetosa* and *Angelica* remain frequent and, trailing among them, there is often some *Galium palustre* and *Lotus uliginosus*. *Filipendula ulmaria* is rather more common here, too, than in the *J.*

effusus sub-community and, where there is an abundance of these herbs, together with occasional *Lythrum salicaria* and *Iris pseudacorus*, the appearance of the vegetation is very different from the pastured stands included here. Such transitions to Filipendulion fen can be seen throughout the range of the community, particularly where freedom from grazing is combined with some nutrient enrichment, but this kind of physiognomy is particularly evident in stands described from Skye by Birks (1973, as a *Juncus-Filipendula* Association), where the additional occurrence of plants such as *Crepis paludosa, Cirsium helenioides* and *Trollius europeaus* gives a distinctly northern feel.

More usually, the lower tier of the vegetation has frequent records for *Mentha aquatica, Cardamine pratensis, Ranunculus flammula* and *R. repens* with, preferentially common here, *R. acris, Potentilla erecta, Achillea ptarmica, Equisetum palustre* and *Carex panicea*. In ranker stands, such plants may be reduced to sparse individuals but, where there is some grazing, they can make a more obvious contribution to the shorter sward between the rushes and be locally enriched by Nardo-Galion or Cynosurion herbs according to soil conditions and treatment. In other cases, puddled hollows can have some *Juncus bulbosus, Myosotis laxa* ssp. *caespitosa* and *Potentilla anserina*, or the presence of species such as *Carex nigra, C. echinata, Narthecium ossifragum* and sparse tufts of Sphagna may create the impression of a transition to poor fen. Throughout this diverse range of vegetation included here, *Carum verticillatum* can occasionally be found within its local centres of distribution in Devon, west Wales and western Scotland.

Juncus effusus sub-community: Species-rich *Juncetum effusi* Eddy *et al.* 1969; *Junco acutiflori-Molinietum* (O'Sullivan 1968) Ratcliffe & Hattey 1982 *p.p.*; Fen meadow   Ratcliffe & Hattey 1982 *p.p.* This sub-community is less well-defined than the above and essentially represents a transition between the *J. acutiflorus* sub-community and the *Holco-Juncetum*, the rush-pasture of circumneutral gley soils throughout lowland Britain. Thus, *J. acutiflorus* is generally uncommon here and *J. effusus* the usual dominant, sometimes forming a dense mass of huge spreading tussocks which crowd out many of the associates but quite often occurring as discrete clumps between which are stretches of herbage that, again, can vary between very rank to shortly-cropped. *J. articulatus* can sometimes be found (indeed it is perhaps a little more frequent here than in the other sub-community) and *J. conglomeratus* also occurs in some stands, but *J. inflexus*, a common alternative dominant in the *Holco-Juncetum*, is typically absent. Although *Molinia* is characteristically infrequent, other grasses are quite important in the sward with *Holcus* *lanatus* often being accompanied by *Agrostis canina* ssp. *canina*, *A. stolonifera*, *Poa trivialis* and, weakly preferential, *Deschampsia cespitosa*. *Anthoxanthum* and *Festuca rubra* are a little less common but, where improved pasture forms a surround to stands, a common occurrence, *Cynosurus* and *Lolium perenne* can figure at low covers.

Such plants, together with *Ranunculus repens, R. acris, Cardamine pratensis* and *Lotus uliginosus*, emphasise the close relationship of this vegetation to the *Holco-Juncetum*, but good distinguishing features here are the high frequencies of *Galium palustre, Cirsium palustre, Ranunculus flammula, Mentha aquatica* and·the occasional presence of *Stellaria alsine, Epilobium palustre, Angelica sylvestris, Hydrocotyle vulgaris, Viola palustris* and *Carex nigra*. Also, plants such as *Rumex crispus, R. obtusifolius, Cirsium arvense* and *C. vulgare*, which often get a hold in run-down stands of the *Holco-Juncetum*, and emphasise its closer relationship with the Elymo-Rumicion, are not usually found here.

Habitat

The *Juncus-Galium* rush-pasture occurs over a variety of moist, moderately acid to neutral, peaty and mineral soils in the cool and rainy lowlands of western Britain. It is a community of gently-sloping ground, found around the margins of soligenous flushes and water-tracks and as a zone around topogenous mires and among wet heaths, but it is especially characteristic of and widespread in stretches of ill-drained and relatively unimproved or reverted pasture. And, throughout, it is grazing (occasionally mowing) which ultimately maintains this vegetation against progression to woodland, and which controls much of its floristic and structural character. Draining and other kinds of soil improvement have reduced the extent of the community, particularly at lower altitudes where intensive pastoral agriculture has become prevalent.

Among rush-dominated vegetation of pastures and meadows, this community is found at the opposite climatic and edaphic extreme from the *Juncus-Cirsium* fen-meadow. It has an essentially oceanic distribution, being largely confined to those parts of Britain with cool summers and mild winters. Almost all samples lie beyond the 26 °C mean annual maximum isotherm, which makes a crude separation between the western and northern parts of the country and the south-east (Conolly & Dahl 1970), but few penetrate into areas with a February minimum more than half a degree C or so below freezing (*Climatological Atlas* 1952), which limits the occurrence of the community over much of the Highlands and north-east Scotland, in the Southern Uplands and the Pennines and higher ground in Wales. In more southerly parts of Britain, this kind of vegetation can be found at altitudes up to 400 m or more, but

in general, and increasingly to the north, it is a lowland community with most samples occurring below 200 m. Throughout this zone, the climate is humid, with most sites experiencing over 1200 mm rain annually (*Climatological Atlas* 1952) and more than 160 wet days yr^{-1} (Ratcliffe 1968), and often cloudy skies: such conditions reach a peak in places like Skye, where there can be over 3000 mm rain with more than 220 wet days yr^{-1} (Birks 1973).

The floristic response to such conditions in this vegetation is fairly modest and imprecise. Certainly, within Europe as a whole, species such as *J. acutiflorus, Carum verticillatum, Wahlenbergia hederacea* and *Scutellaria minor* all show Oceanic West European distributions (Matthews 1955) and characterise a distinct trend in the composition of Molinietalia communities. But *J. acutiflorus*, though it shows a clear preference for more westerly vegetation types and, among rush-dominated pastures and meadows is strongly diagnostic of this type, has a wide geographical range in Britain and is of fairly catholic occurrence in relation to our climate. *Wahlenbergia* and *Scutellaria* are much more confined but both, particularly the latter, extend outside the distribution of the *Juncus-Cirsium* rush-pasture, most notably in the warmer and drier parts of central southern and south-east England, and neither is well represented in central Scotland, where the community is widespread. Even within the area of overlap, these species are not common in the community, nor are they confined to it. Geographically, it is *Carum* which shows the best coincidence of occurrence, though this plant is strikingly confined to particular stretches of western Britain (Perring & Walters 1962) and, again, can be found there in other vegetation types than this (B.D. Wheeler, pers. comm.). All these features emphasise the difficulty of characterising a substantial phytogeographic boundary between this kind of vegetation and the rush-pastures and meadows further to the east. The contrast between the *Juncus-Cirsium* fen-meadow and the *J. acutiflorus* sub-community here is clear enough, and this latter vegetation type is the more oceanic of the western kinds of rush-pasture being concentrated at lower altitudes (mean 137 m) down the Atlantic seaboard of Wales and Scotland. But separating the *Juncus-Galium* rush-pasture from the *Holco-Juncetum* is much more difficult. Essentially, the *J. effusus* sub-community of the *Juncus-Cirsium* rush-pasture is a transition to this vegetation, being concentrated in an ill-defined zone to the east of the *J. acutiflorus* type, running through Devon and Cornwall, Wales and into northern Britain and can be found at substantially greater altitudes (mean 195 m). In this region, the prominence of the Oceanic West European element in the flora is weakening somewhat, though change is often visible mainly in terms of a shift in dominance between the two rushes, and this and some other floristic contrasts can be affected by other factors.

A further complication in the definition of the *Juncus-Galium* rush-pasture is that, with the increasing oceanicity that helps sharpen up the differences with the Calthion communities, the boundary with the *Molinia*-dominated vegetation of the Junco-Molinion type breaks down. In western Britain, *J. acutiflorus* and *Molinia* each transgress far into vegetation dominated by the other, and are found associated with a similar range of species, among them *Carum*, *Wahlenbergia* and *Scutellaria minor*, as well as many more general Molinietalia and Molinio-Arrhenatheretea herbs. Such vexatious problems of separating the *J. acutiflorus* subcommunity from various kinds of *Molinia-Potentilla* grassland are a real reflection of a floristic convergence influenced by the oceanic climate of the western part of the country.

Edaphic conditions provide some help in understanding the particular environmental preferences of the *Juncus-Galium* rush-pasture among this range of communities. Within the oceanic lowlands, it can be found on a variety of moderately acid to neutral soils that are kept moist to wet for most of the year, but its distribution is strongly centred on profiles with a measure of drainage impedence, such as have developed on very gentle slopes over argillaceous bedrocks and superficials. In central and north Wales and south-west Scotland, for example, Silurian and Ordovician shales provide important substrates, while Carboniferous and Devonian shales support stands of the community in central and south-east Scotland, south Wales and the South-West Peninsula. In the last area, too, and also in Dumfries & Galloway, this kind of vegetation can be found on profiles derived from granites and, throughout the glaciated areas of western Britain, deposits of heavier-textured drift have often weathered to suitable soils.

The particular kinds of profiles developed in such situations are usually stagnogleys, where high rainfall or water shed from surrounding slopes induces surface-water gleying over impermeable substrates on gentle slopes; or ground-water gleys where, in impeded hollows, there is some shallow fluctuation of the water-table. Over stretches of more subdued scenery, such as have been derived from extensive outcrops of readily-weathering shales or widely mantled by till, such profiles can be the predominant element in the soil cover; in other cases, they mark out areas of more local drainage impedence in rougher terrain. Both kinds of gley can accumulate a humose topsoil and grade to true peats, on to which the community can extend a little way. Thus, it is common over stagnohumic gleys, widespread in areas of higher rainfall, though not usually where these are strongly flushed or at altitudes where the climate is too

inequable. And it occurs over humic gley soils and on shallow peats such as develop in waterlogged hollows and in transitions to streamside alluvium.

The generally moist conditions here are reflected in the associated flora of the community by the high frequencies of species which are of wide occurrence in wetter habitats, such as *Galium palustre*, *Ranunculus flammula*, *Mentha aquatica*, *Agrostis stolonifera*, *Cardamine pratensis* and *Ranunculus repens*, as well as plants that are rather more strongly preferential to Molinietalia communities, like *Cirsium palustre*, *Lotus uliginosus*, *Angelica sylvestris* and, of course, the two rushes. Quite a number of these can be found with some regularity in certain kinds of *Molinia-Potentilla* mire, but the switch from rush- to *Molinia*-dominance in that vegetation seems to mark a tendency to better aeration of the soils and, perhaps, a brisker nutrient turnover. Thus, although the *Juncus-Galium* rush-pasture and *Molinia-Potentilla* mire show some overlap in the kinds of profiles on which they can occur, often being found contiguously on shallow peats, for example, the latter vegetation type seems to be centred on more free-draining, moist acid soils, such as gently-flushed peats and stagnohumic gleys. Under the *Juncus-Galium* rush-pasture, though soils may dry out somewhat above in summer, seasonal stagnation and only moderately mesotrophic conditions seem to be the rule.

One other difference between the edaphic environments of the communities is that the soils here are not quite so acid as under the *Molinia-Potentilla* grassland. Again, there is some overlap in base-richness but, by and large, the superficial reaction under the *Juncus-Galium* rush-pasture is in the range pH 4–6. Certainly, the profiles are sufficiently acid and calcium-poor as to inhibit the occurrence of *J. subnodulosus* and part of the floristic and geographical contrast between this community and the *Juncus-Cirsium* fen-meadow is to do with soil reaction: where the two vegetation types occur in the same region, as on Anglesey, they clearly mark out differing profiles and parent materials. But more calcifuge plants are not numerous or frequent here. *Potentilla erecta* becomes common in the *J. acutiflorus* subcommunity and is very much a marker of this kind of vegetation through lowland Scotland (Birse & Robertson 1967, 1973, 1976, Birse 1980, 1984); and, with plants like *Anthoxanthum odoratum* and *Festuca rubra*, it can provide a strong measure of continuity with the Nardo-Galion swards which often adjoin the *Juncus-Galium* rush-pasture in western Britain (e.g. King 1962, King & Nicholson 1964). *Carum*, too, slightly preferential to the same sub-community, is a marker of less base-rich Molinietalia vegetation (Oberdorfer 1979). Then, in damper stands, species such as *Agrostis canina* ssp *canina*, *Viola palustris*, *Carex nigra*, *C. echinata* and *C demissa* can give the impression of a floristic transition to small-sedge poor-fen vegetation of the Caricion nigrae. Despite the more frequent occurrence of *P. erecta* and *Carum* in the *J. acutiflorus* sub-community the soil reaction there does not appear to be any more acidic than in the *J. effusus* sub-community: indeed, Birks (1973) suggested that, of the two dominants, it was *J. effusus* that indicated a tendency to more pronounced surface acidity that comes with leaching of drier soils under this vegetation. Such puzzles await resolution, but what can be said for the moment is that, even where such calcifuge tendencies are better developed, they are not very strongly expressed and an assessment of the character of the community often depends on comparison with its vegetational context. Among suites of blanket mires, base-poor flushes and acid grasslands, for example, such as provide a background at Moor House (Eddy *et al.* 1969) and on Hoy (Prentice & Prentice 1975), the *Juncus-Galium* rush-pasture can seem mildly basiphilous and sometimes indicate the influence of slightly more calcareous ground waters in such situations as these.

The final very important environmental factor which has an influence on the community is the treatment it receives, which is ultimately responsible for the maintenance of the vegetation, and which exerts a powerful control on its composition and structure, sometimes confusing the effects of climate and soil. For the most part, the community is treated as pasture or, rather, it occurs within grazing land around the upland fringes, constituting a varying proportion of the available herbage, according to natural relief and drainage and the extent of land improvement. Many lowland stands have been subject to draining, fertiliser application and re-seeding, the soils being able to yield potentially heavy grass crops in the wet climate, but towards and beyond the limit of enclosure in western Britain, and, more locally, in areas of traditional, less intensive agriculture, or where common rights have prevailed, such improvement has been uneconomic or hindered and it is here that the most extensive stands are to be found. Often, however, smaller tracts of the community can be found in more intractable areas within generally improved landscapes, as around stretches of the open water or unreclaimed mires and along stream and river banks and, where there has been a history of agricultural decline, such stands have expanded with the reversion of the surrounding pastures.

Both cattle and sheep are turned on to this vegetation and, though each kind of stock produces rather different effects, the general influence of judicious grazing is to keep the rushes in check and reduce the intervening vegetation to a more close-cropped sward. In fact, where the community forms only a proportion of the available herbage, other more productive communities will be preferentially grazed, leaving the *Juncus-Galium* rush-pasture with a fairly rank understorey of herbs, but many of its characteristic dicotyledons are resistant to grazing and will survive heavy cropping as non-flowering rosettes or nibbled shoots. On moister soils, highly susceptible to poaching, especially where cattle are grazed, it may be much more difficult to hinder the dominance of the rushes, since the exposure of moist, bare ground creates ideal conditions for the germination of their seeds, huge quantities of which can remain dormant for long periods in the soil (Milton 1936, 1948, Moore & Burr 1948, Salisbury 1964). Indeed, it is possible that many stands of the community have originated or spread because of this kind of pastoral misman-agement (McVean & Ratcliffe 1962). In drier situations, however, the rushes may be reduced to smaller discrete patches and the intervening stretches of sward take on some of the character of a Nardo-Galion grassland or, where there is an element of eutrophication of the soils by dunging, or by washing or drifting in of fertiliser, of a Cynosurion sward.

Grazing usually also holds in check any marked contribution from taller herbs like *Filipendula* though, in less accessible sites, and particularly where there is some silting, as on river banks, such plants may become more obvious as in the Skye stands described by Birks (1973), and they can maintain their abundance under a mowing regime, provided it is not combined with grazing. In fact, though stands may be included within a hay crop from agricultural enclosures, mowing alone is not a very common treatment of this community, except where it is found on road verges cut to maintain visibility.

Zonation and succession

The *Juncus-Cirsium* rush-pasture occurs in zonations and mosaics with a variety of grasslands, mires and heaths, the disposition of the vegetation types usually reflecting variations in soil conditions and treatments. The community can subsist in non-intensive agricultural landscapes but improvement has destroyed many stands, or altered or truncated zonations. With neglect, improved swards can revert to this kind of vegetation and the community, in turn, progress to woodland.

Around the upland fringes of western Britain, the *Juncus-Galium* rush-pasture is still an important and extensive component of less-improved grazing land, both within and beyond the limits of enclosure. Its contribution to the vegetation cover in such predominantly pastoral landscapes is dependent on the drainage pattern, controlled by such natural factors as geology and relief, so where there has been no improvement of drainage the disposition and size of stands remain effective markers of the extent of impedence over gentle slopes with impervious substrates. And the commonest type of zonation is then to calcifuge grasslands of one sort or another, with increasing freedom of water movement in the profile, a change often indicated by increase in slope, in other cases by a switch in parent materials, or both. Alternations of shales with pervious sandstones present one kind of underlying geological pattern, fairly common in the palaeozoic deposits of western Britain, but frequently it is the distribution of ill-draining drift, lodged on plateaus, moderate slopes and choking small stream-beds, that controls the occurrence of the community.

In edaphic terms, these zonations are usually marked by a change from some sort of gleyed profile under the *Juncus-Galium* rush-pasture, to a brown podzolic soil or podzol, carrying a Nardo-Galion grassland. With increasingly free drainage, and greater tendency to surface eluviation, the rushes and Molinietalia associates drop out, leaving herbs such as *Anthoxanthum*, *Festuca rubra*, *Agrostis capillaris* and *Potentilla erecta*, as mainstays of some kind of *Festuca-Agrostis-Galium* grassland. Junctions can be sharp, where underlying edaphic conditions change suddenly, but often they are fairly gentle, as where drift influence alters gradually or where the penetration of stock creates a transitional mosaic zone between clumps of rushes. The surveys of King (1962; see also King & Nicholson 1964) include some of these intermediate swards from southern Scotland and, in the data from Moor House (Eddy *et al.* 1969), transitions to grasslands with much *Nardus* can be seen, a quite widespread feature at such higher altitudes. At this site, too, and also around Cader Idris (Edgell 1969), some of the surrounding swards are flushed with the more base-rich water which gleys the soils under the *Juncus-Galium* rush-pasture, and here the zonation can be to *Festuca-Agrostis-Thymus* grassland, into which Nardo-Galion herbs and such plants as *Carex panicea*, *Cirsium palustre* and *Agrostis canina* ssp. *canina* can extend.

Although the *Juncus-Cirsium* rush-pasture is only slightly mesophytic, transitions such as these probably involve a switch to soils which are somewhat more oligotrophic than the gleys. Within enclosed agricultural land, however, the surrounding swards have often received some fertilising, even if only from the stock which, in this kind of mosaic, give most of their attention to the drier grasslands, or from artificials such as basic slag. Then, a Cynosurion element may become prominent in the transition, with, say the *Holcus-Trifolium* sub-community of the *Festuca-Agrostis-Galium* grassland surrounding the rush-pasture, or, where there has been more assiduous improvement with some top-sowing, the *Anthoxanthum* sub-community of the *Lolio-Cynosuretum*. Cynosurion species may drift into the rush-pasture, blurring the boundaries, and the next stage of improvement is for draining of the gleys to facilitate the extinction of the rushes and their Molinietalia associates and the reclamation of the entire sward. This has been a very widespread practice, particularly

over less intractable ground at lower altitudes, and though blocking of drains can allow relatively rapid reversion to the *Juncus-Galium* rush-pasture, many stands have been permanently destroyed.

Beyond the limits of enclosure, where the community can be found at moderate altitudes, the stagnohumic gleys on which it characteristically occurs often form intergrade soils between brown podzolic profiles and peats proper (Avery 1980). In some situations, the *Juncus-Galium* rush-pasture can thus be seen as part of a more extensive transition right through from Nardo-Galion swards to blanket mire with the thickening of the organic horizon from mor humus, through humose topsoil to ombrogenous peat. In such places, the community may be preferentially grazed along with stretches of *Festuca-Agrostis-Galium* or *Nardus-Galium* grasslands as the most productive and palatable vegetation types, so the switch to the peat vegetation is often marked by the appearance of ericoids, first in the *Scirpus-Erica* wet heath, then in Erico-Sphagnion mire. In these kinds of zonations, as at Moor House (Eddy *et al.* 1969) and on Cader Idris (Edgell 1969) the *Juncus-Galium* rush-pasture often occurs in quite well-defined stream-side stands running through the tracts of peat. Then, it can be seen as a mesotrophic equivalent to rush-dominated types of *Carex echinata-Sphagnum* mire. Where such streams are surrounded by gently-flushed stretches of peat, and particularly where the vegetation has been burned over and grazed, *Molinia-Potentilla* grassland can form an extensive fringe to the community.

In the lowland landscape, too, the *Juncus-Galium* rush-pasture can be found with wet heath and mires on peats, where soligenous conditions develop in ill-draining valley bottoms with gleying of their surrounds. Indeed, within stretches of more improved agricultural landscapes, such areas, sometimes preserved as common-lands, as on Gower and in Pembrokeshire (Meade 1981), can include most of the transitions of this kind. In these localities, the wet heath can be represented by the *Ericetum tetralicis*, with the *Succisa-Carex* community of which the *Juncus-Galium* rush-pasture has much in common, and this can grade in waterlogged hollows to the *Narthecio-Sphagnetum* valley bog with *Hyperico-Potametum* soakways. Again, *Molinia*-dominated vegetation can supervene in such transitions where gentle slopes have some modest flushing or where burning and grazing have favoured its spread. In the south-west of Britain and south Wales, this may be far-flung stands of the *Cirsio-Molinietum* where there is some base-enrichment in the ground waters but, usually, it is some kind of *Molinia-Potentilla* mire. Where this adjoins the *Juncus-Galium* rush-pasture, transitions between the two can be very gentle or form complex mosaics of dominance by rushes and *Molinia* (e.g. Meade 1981, Ratcliffe & Hattey 1982: site type V).

Topogenous hollows in the lowlands can also show

zonations from the community to vegetation on peat, where the latter accumulates under the influence of a more or less permanently high water-table. In such situations, the *Juncus-Galium* rush-pasture can form a fringe around Phragmitetalia fen, into which the Junci can extend some way, but where continuity is seen mainly through tall dicotyledons like *Cirsium palustre*, *Angelica sylvestris* and *Filipendula ulmaria*, and other herbs such as *Galium palustre, Ranunculus flammula, Mentha aquatica, Cardamine pratensis, Hydrocotyle vulgaris* and *Epilobium palustre*, with a switch in dominance to helophytes, notably *Phragmites australis* or, quite commonly around open-water transitions in the western parts of Britain, *Carex rostrata*. Reed-dominated vegetation is often of the *Phragmites-Eupatorium* fen, the outward spread of *Phragmites* often being curtailed by grazing ahead of its edaphic limit (site types VII & VIII, Ratcliffe & Hattey 1982); while an increase in *C. rostrata*, together with *Potentilla palustris, Menyanthes trifoliata* and *Equisetum fluviatile*, all of them occasional in the *Juncus-Galium* rush-pasture, usually marks a transition to the *Potentillo-Caricetum* (site type IV, Ratcliffe & Hattey 1982). Where stock, usually cattle in these situations, come down to water at such fen stands within enclosed pastures, the *Juncus-Galium* fringe is typically a much-puddled zone.

Transitions from the community to *Phragmites-Eupatorium* fen can also be seen on the unimproved margins of sluggish streams and rivers, where the *Juncus-Galium* rush-pasture can persist on alluvial gleys over the terraces. Here, too, the *Filipendula-Angelica* tall-herb fen is often to be found on soils which are too dry to support dense *Phragmites* but periodically enriched by allochthonous mineral matter deposited in floods. Grazing, again, may have some influence on the transition from the *Juncus-Galium* rush-pasture to this Filipendulion community. Fragments of these kinds of zonations can often be seen alongside field-margin ditches and along damp hedge-banks.

In all these situations, it is usually grazing, much less often mowing, which maintains the *Juncus-Cirsium* rush-pasture; and imprudent pasturing, with its frequent poaching, probably enhances the hold of the community on these ill-draining soils. Where grazing is withheld, however, and there is no cutting or burning of the vegetation, seral progression can be rapid with direct invasion of woody plants, particularly of *Salix cinerea* and *Betula pubescens*, whose light fruits can be windborne for considerable distances, and sometimes also of *Alnus*, if seed-parents are close. Where a persistent phase of *Filipendula*- or *Molinia*-dominance does not supervene, such colonists can quickly come to prevail in stands of *Salix-Galium* woodland, a particularly widespread and common community of gleyed mineral soils in the western lowlands of Britain, or of *Betula-Molinia* woodland. With these two vegetation types, the *Juncus-*

Galium fen-meadow can show considerable floristic overlap, sometimes surviving little changed under more open canopies. It is possible that at lower altitudes such woodlands eventually progress further by the invasion of *Quercus robur*.

Distribution

The community is widespread through the west of Britain from Devon and Cornwall to Skye and Caithness. In Scotland, where the *J. acutiflorus* sub-community prevails, this kind of vegetation is exceedingly common at low to moderate altitudes and much more frequent in the west-central lowlands than our records indicate. It remains common in Wales, though increasingly here there is a switch to the *J. effusus* sub-community, the usual type of *Juncus-Galium* rush-pasture in the South-West Peninsula. In that region, it becomes more local, though it remains abundant in some areas, like the fringes of the granite moors and more intractable parts of the Devonshire Culm Measures.

Affinities

Almost without exception, the rush-pasture vegetation previously described from western Britain has been of the type included here as the *J. acutiflorus* sub-community (McVean & Ratcliffe 1962, Edgell 1969, Eddy *et al.* 1969, Birks 1973, Birse & Robertson 1976, Adam *et al.* 1977, Birse 1980, 1984). Certainly, this is the more sharply characterised of the two kinds, representing an opposite floristic extreme from the *Juncus-Cirsium* fen-meadow but, particularly towards the south of its range, this general type of vegetation shows an ill-defined transition to the *Holco-Juncetum*, through a switch in dominants and a waning of Nardo-Galion elements. It is these intermediate stands which comprise the *J. effusus* sub-community: an alternative treatment, less satisfactory in our view, would be to include such vegetation in the *Holco-Juncetum* as an additional sub-community, leaving the *J. acutiflorus* sub-community to comprise the core of the *Juncus-Cirsium* rush-pasture.

Such difficulties of definition hint at the need for a reassessment of the relationship among Molinietalia vegetation through the whole of Europe now that we have a clearer picture of the range of these kinds of communities in Britain. Much effort has already been expended in debating the phytosociological status of more Atlantic *J. acutiflorus* vegetation such as that included here. One view (Birse & Robertson 1976, Oberdorfer 1977, Birse 1980, 1984) would incorporate it into a Juncion acutiflori alliance, along with equivalent Irish rush-pasture, the *Senecioni-Juncetum*, emphasising the boundary with the Calthion on the basis of such oceanic plants as *J. acutiflorus* itself, *Carum, Wahlenbergia* and *Scutellaria minor*. Against the flora of mainland European Calthion communities, which itself includes some Continental species of fairly strong fidelity, such plants present a sharp phytogeographical contrast. However, when the intervening range of British vegetation of this general type is included, it becomes much more difficult to draw a boundary between these alliances: the extremes are still distinct enough, the intermediates grade with a frustrating continuity. One solution to this problem is to expand the Calthion to include this western rush-dominated vegetation, as O'Sullivan (1976, 1982) and White & Doyle (1982) did. Another is to retain the two alliances, while recognising that, towards the Atlantic seaboard of Europe, the sharp affinities of their species groupings become less precise, and are often heavily overlain by a generalised Molinietalia element encouraged throughout by broadly similar agricultural treatments.

A second problem of definition is that the clearer the differences between this western rush-pasture and the Calthion communities become, the harder it is to separate the former from *Molinia*-dominated vegetation,

which is likewise very widespread in the more oceanic parts of Britain and which tends to show a floristic convergence under the influence of the climate. Thus, although it has been customary to separate the *Molinia-Potentilla* grassland from the *Juncus-Galium* rush-pasture by placing the former in the Junco conglomerati-Molinion alliance, the communities do come very close, with some overlap between *J. acutiflorus* and *Molinia*, and species such as *Carum*, *Wahlenbergia* and *Scutellaria minor* occurring in both. This could be recognised by subsuming both types within a single Juncion acutiflori, and setting the major disjunction among western Molinietalia communities at the Pyrenees, beyond which this alliance gives way to an Anagallido-Juncion acutiflori. But the fact that this alliance is characterised by *Carum*, *Wahlenbergia*, *S. minor*, *Anagallis tenella* and *Lobelia urens*, all of them species whose British ranges fall roughly within our Juncion acutiflori vegetation, brings a new confusion. Clearly, the time is ripe for a further thorough revision of these communities.

Floristic table M23

	a	b	23
Juncus effusus	IV (1–6)	V (2–10)	V (1–10)
Holcus lanatus	IV (1–8)	IV (1–7)	IV (1–8)
Galium palustre	IV (1–4)	IV (1–5)	IV (1–5)
Lotus uliginosus	IV (1–6)	IV (1–5)	IV (1–6)
Juncus acutiflorus	V (2–9)	I (1–5)	II (1–9)
Molinia caerulea	III (1–7)	II (1–5)	II (1–7)
Ranunculus acris	III (1–6)	II (1–5)	II (1–6)
Potentilla erecta	III (1–5)	I (1–4)	II (1–5)
Filipendula ulmaria	II (2–7)	I (2–6)	II (2–7)
Achillea ptarmica	II (1–6)	I (1–5)	I (1–6)
Carex panicea	II (1–4)	I (1–4)	I (1–4)
Equisetum palustre	II (1–4)	I (1–8)	I (1–8)
Carum verticillatum	II (1–4)	I (1–5)	I (1–5)
Aulacomnium palustre	I (1–6)		I (1–6)
Lythrum salicaria	I (1–5)		I (1–5)
Luzula multiflora	I (1–4)		I (1–4)
Taraxacum officinale agg.	I (1)		I (1)
Iris pseudacorus	I (8)		I (8)
Juncus bulbosus	I (2–4)		I (2–4)
Dactylorhiza maculata	I (1–4)		I (1–4)
Narthecium ossifragum	I (2–4)		I (2–4)
Calypogeia fissa	I (2–3)		I (2–3)
Potentilla anserina	I (2–6)		I (2–6)
Lathyrus pratensis	I (3–5)		I (3–5)
Stellaria graminea	I (1–2)		I (1–2)
Myosotis laxa caespitosa	I (1–3)		I (1–3)

Stellaria alsine	I (1–3)	II (1–5)	II (1–5)
Juncus articulatus	I (5–7)	II (1–5)	II (1–7)
Deschampsia cespitosa	I (1–4)	II (1–6)	I (1–6)
Epilobium obscurum		I (1–3)	I (1–3)
Myosotis secunda		I (1–6)	I (1–6)
Eurhynchium praelongum		I (2–6)	I (2–6)
Athyrium filix-femina		I (1–3)	I (1–3)
Luzula campestris		I (2–3)	I (2–3)
Calliergon cordifolium		I (2–8)	I (2–8)
Cirsium palustre	III (1–5)	III (1–5)	III (1–5)
Ranunculus flammula	III (1–5)	II (1–6)	III (1–6)
Agrostis canina canina	II (1–6)	III (2–7)	III (1–7)
Anthoxanthum odoratum	III (1–6)	II (1–6)	II (1–6)
Mentha aquatica	III (1–6)	II (1–7)	II (1–7)
Rumex acetosa	II (1–5)	III (1–5)	II (1–5)
Agrostis stolonifera	II (1–7)	III (1–9)	II (1–9)
Cardamine pratensis	II (1–3)	III (1–7)	II (1–7)
Ranunculus repens	II (2–5)	II (1–5)	II (2–5)
Poa trivialis	II (2–5)	II (1–7)	II (1–7)
Calliergon cuspidatum	II (1–5)	II (1–7)	II (1–7)
Angelica sylvestris	II (1–5)	II (1–6)	II (1–6)
Epilobium palustre	II (1–3)	II (1–3)	II (1–3)
Hydrocotyle vulgaris	II (1–5)	II (1–6)	II (1–6)
Viola palustris	II (1–6)	II (1–5)	II (1–6)
Festuca rubra	II (2–7)	I (1–5)	I (1–7)
Carex nigra	I (2–5)	II (1–5)	I (1–5)
Scutellaria minor	I (1–4)	I (1–5)	I (1–5)
Agrostis capillaris	I (2–6)	I (1–4)	I (1–6)
Rhytidiadelphus squarrosus	I (1–4)	I (1–4)	I (1–4)
Juncus conglomeratus	I (1–4)	I (3–5)	I (3–5)
Brachythecium rutabulum	I (1–6)	I (2–7)	I (1–7)
Lychnis flos-cuculi	I (1–5)	I (2–5)	I (1–5)
Glyceria fluitans	I (1–4)	I (1–6)	I (1–6)
Trifolium repens	I (1–4)	I (1–5)	I (1–5)
Cerastium fontanum	I (1–2)	I (1–4)	I (1–4)
Potentilla palustris	I (2–5)	I (2–5)	I (2–5)
Brachythecium rivulare	I (1–4)	I (1–4)	I (1–4)
Equisetum fluviatile	I (1–3)	I (1–4)	I (1–4)
Plantago lanceolata	I (1–4)	I (1–4)	I (1–4)
Succisa pratensis	I (1–5)	I (1–4)	I (1–5)
Cynosurus cristatus	I (1–4)	I (2–6)	I (1–6)
Prunella vulgaris	I (2–3)	I (1–4)	I (1–4)
Caltha palustris	I (2–5)	I (2–5)	I (2–5)
Senecio aquaticus	I (1–3)	I (1–6)	I (1–6)
Polytrichum commune	I (1–2)	I (1–8)	I (1–8)
Lophocolea bidentata s.l.	I (1–3)	I (1–4)	I (1–4)
Valeriana officinalis	I (3–4)	I (2–4)	I (2–4)
Poa pratensis	I (1–2)	I (2–6)	I (1–6)
Cardamine flexuosa	I (1)	I (1–3)	I (1–3)
Carex ovalis	I (2–4)	I (1–4)	I (1–4)

Floristic table M23 *(cont.)*

	a	b	23
Carex echinata	I (1)	I (2–4)	I (1–4)
Carex demissa	I (1–4)	I (1–3)	I (1–4)
Veronica scutellata	I (1–3)	I (1–3)	I (1–3)
Pellia epiphylla	I (3–4)	I (2–4)	I (2–4)
Polygonum hydropiper	I (1–4)	I (1–6)	I (1–6)
Number of samples	38	62	100
Number of species/sample	21 (6–39)	17 (8–28)	19 (6–39)
Vegetation height (cm)	47 (30–100)	56 (10–130)	52 (10–130)
Herb cover (%)	96 (60–100)	98 (80–100)	98 (60–100)
Bryophyte cover (%)	15 (0–80)	26 (0–80)	21 (0–80)
Altitude (m)	137 (20–390)	195 (10–430)	163 (10–430)
Slope (°)	1 (0–10)	1 (0–30)	1 (0–30)
Soil pH	4.9 (3.7–5.6)	4.9 (3.5–6.2)	4.9 (3.5–6.2)

a   *Juncus acutiflorus* sub-community
b   *Juncus effusus* sub-community
23   *Juncus effusus/acutiflorus-Galium palustre* rush-pasture (total)

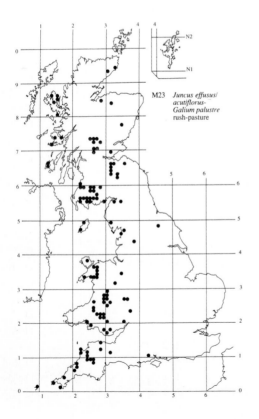

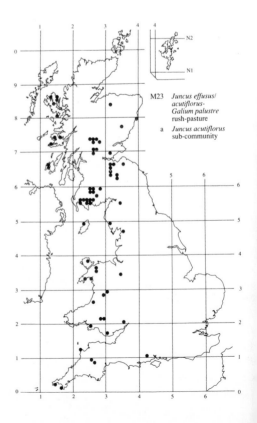

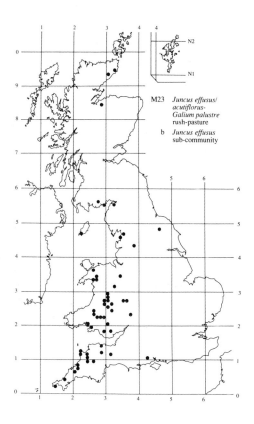

M23 *Juncus effusus/*
      acutiflorus-
      Galium palustre
      rush-pasture
b   *Juncus effusus*
      sub-community

M24
Molinia caerulea-Cirsium dissectum fen-meadow
Cirsio-Molinietum caeruleae   Sissingh & De Vries
1942 *emend.*

Synonymy
Litter Godwin & Tansley 1929; *Molinietum* Godwin 1941, Willis & Jefferies 1959, Ivimey-Cook *et al.* 1975 *p.p.*; *Molinia*-edge Haslam 1965; *Cirsio-Molinietum peucedanetosum* (Sissingh & De Vries 1942) Wheeler 1978; *Cirsio-Molinietum* (Sissingh & De Vries 1942) Wheeler 1980*c p.p.*; Grassy Heath NCC Devon Heathland Report 1980.

Constant species
Carex panicea, Cirsium dissectum, Lotus uliginosus, Molinia caerulea, Potentilla erecta.

Rare species
Hypericum undulatum, Peucedanum palustre, Selinum carvifolia.

Physiognomy
The *Cirsio-Molinietum caeruleae* includes the bulk of the *Molinia caerulea* vegetation in the lowland south-east of Britain. *Molinia* is almost always the dominant plant in the community and it can be very abundant, forming the basis of a rough sward or occurring as a more strongly-tussocky cover, a kind of structure well shown in the classic early account of this vegetation from Wicken Fen in Cambridgeshire (Godwin & Tansley 1929). And there are stands in which the abundance of *Molinia* is so overwhelming that its dense herbage and thick litter reduce the associated flora to scattered individuals of a very few species. Often, however, the number of companions is considerable: although many of the associates are plants of fairly wide distribution among damper meadows and pastures, the most frequent of them comprise a quite distinctive assemblage, and some of the preferentials attain such local prominence as to mask the major contribution which *Molinia* and the other constants make to the community. Much of this character reflects the origin of this kind of vegetation from rather diverse precursors by the application of particular treatments.

In structural terms, the most important associates of *Molinia* are other monocotyledons of medium stature, particularly rushes, which can make quite a substantial contribution to the dense layer of herbage characteristic of the community. Typically, this is 20–60 cm tall, though its physiognomy is strongly dependent on the style and incidence of treatment: in grazed stands, for example, the cover can be shorter than this, with stretches of close-cropped sward running among clumps of more resistant plants; where there is no grazing, and where the vegetation has not been mown for some time, the herbage can be taller and ranker. The trophic state of the substrate also influences the structure of the vegetation, with shorter and more open swards developing over more calcareous and impoverished soils.

The particular rushes represented show quite a strong pattern of regional distribution, because the *Cirsio-Molinietum* extends across a considerable part of southern Britain and has developed in a variety of climatic and edaphic contexts. Thus, through most of the central and eastern parts of its range, where the community is usually found in association with fens, *Juncus subnodulosus* is the characteristic rush, a non-tussocky species and often represented just as a rather sparse cover of shoots, but able to form patches which are locally dense. *J. articulatus* and, less commonly, *J. inflexus* can also be found in these regions, though generally as minor components and often in swards that are shorter than usual and richer in small herbs. To the south and west, by contrast, where the *Cirsio-Molinietum* often develops among heath vegetation, *J. acutiflorus* and *J. conglomeratus*, which are rather local elsewhere in the community, become frequent, the former another far-creeping rush, though sometimes forming extensive clumps, the latter occurring as scattered, spreading tussocks. *J. effusus* can be found occasionally, too, and, in its var. *compactus*, can be confused with *J. conglomeratus*: typically, the stems of the latter are more obviously striate and tend to be markedly splayed out by the inflorescence. Among the rushes, *J. conglomeratus* is quite a good marker of this kind of vegetation but the prominence of all the other species here can make diagnosis of the *Cirsio-Molinietum* difficult, with separation from other Molinietalia communities, like the *Juncus-Cirsium* fen-meadow and the *Juncus-Galium* rush-pasture, being especially problematical. In their typical manifestations, strongly dominated by their respective rush species, these are readily distinguishable from the *Cirsio-Molinietum*, but *Molinia* can occur in both of them and, in transitional stands, separation may be a matter of the relative dominance of the various monocotyledons.

Similar difficulties can arise because of the local prominence of *Schoenus nigricans* in the *Cirsio-Molinietum*. This is only an occasional in the community and generally subordinate to *Molinia* when it does occur, but it can attain moderate abundance and it tends to strike the eye where its growth has become markedly tussocky. Then, the vegetation can take on some of the character of the *Schoenetum*, a community whose distribution overlaps that of the *Cirsio-Molinietum*, and which is often found at the same sites (e.g. Bellamy & Rose 1961, Haslam 1965). A more uncommon physiognomic element, but one that can be quite conspicuous, consists of *Phragmites australis* and/or *Cladium mariscus*, even sparse plants of which can protrude above the level of the *Molinia* and rushes, and which can sometimes be found in such local abundance as to give the impression of tall-herb fen vegetation. Indeed, as the renowned experiments of Godwin (1941) showed, every gradation between Phragmitetalia fen and the *Cirsio-Molinietum* can be produced by different frequencies of mowing: as defined here, the community takes in the *Molinia*-dominated half of Godwin's transitions between what is called 'mixed sedge' and 'litter' in which there was a declining, though sometimes substantial, remnant of reed or sedge.

What helps distinguish the community is the prominence among the associated flora of dicotyledons able to maintain themselves with varying degrees of success, under the frequent summer-mowing or grazing which characterise this vegetation. Some of these plants, such as *Cirsium palustre* and *Angelica sylvestris*, both of which are very frequent throughout the community, can persist in this vegetation as non-flowering rosettes, even in quite close-cropped herbage, but they are able to grow very tall and are well supplied with stem leaves, so they find wide representation, too, far into the Phragmitetalia fens. *Filipendula ulmaria* is of similar growth form and wide occurrence among other kinds of fen vegetation and it likewise can be common here, but it is patchier than *C. palustre* or *Angelica*, being very sensitive to grazing and also of restricted occurrence on the more impoverished substrates on which the *Cirsio-Molinietum* is represented. *Centaurea nigra*, another hemicryptophyte, is likewise confined to more mesotro-

phic situations, though its frequency here is in sharp contrast to tall-herb fens.

Much more strictly limited to the kind of short to moderately tall herbage typical of this community are *Valeriana dioica*, *Succisa pratensis* and *Cirsium dissectum*, three hemicryptophytes the bulk of whose foliage is in basal rosettes. *Succisa*, of course, is of wide occurrence in many vegetation types and *V. dioica*, too, can be found as an important component in other small-sedge fens and fen-meadows, and very occasionally this trio can occur in communities other than this. But *C. dissectum* is strongly preferential here and, although, towards the limits of its British range, stands similar to the *Cirsio-Molinietum* occur without the thistle (B.D. Wheeler, pers. comm.), the distribution of this species and of the community are largely coincidental. In phytogeographic terms, it is an Oceanic West European plant (Matthews 1955) and the community can provide an occasional locus for some other species with this kind of European range, *Scutellaria minor*, *Dactylorhiza maculata* ssp. *praetermissa* and the nationally rare *Hypericum undulatum*.

In stands which have not been heavily grazed in spring or which are not mown early, it is the frequent hemicryptophytes of the community which often provide its most attractive element, putting up their flowering shoots among and above the *Molinia* and rushes. For much of the time, however, they tend to be a rather inconspicuous component of the herbage, growing among a variety of other herbs of small to moderate stature which bulk up the lower tier of the vegetation. In contrast to much tall-herb fen vegetation and the wetter kinds of *Schoenetum*, coarser grasses are often prominent here, with *Holcus lanatus* and *Anthoxanthum odoratum* the most frequent, *Festuca rubra*, *Deschampsia cespitosa* and *Agrostis stolonifera* less common, though sometimes abundant. Then, in transitions to wet heath, there can be some *A. canina* ssp. *canina* or, in grassy base-rich swards, *Briza media*. Where this latter kind of vegetation grades to calcicolous grassland, a quite common feature around some valley and basin mires, *Brachypodium pinnatum* and *Festuca ovina* can transgress some way into the community.

Then, there can be some sedges, though the bulkier species represented in tall-herb fens and/or in the *Juncus-Cirsium* fen-meadow, such as *Carex acutiformis*, *C. disticha*, *C. elata* and *C. paniculata*, are very rare here. Rather, it is smaller species such as *C. panicea* and, somewhat less frequently, *C. hostiana* and *C. pulicaris*, that are most common. Generally, they are of low cover, growing intermixed with the other plants: in contrast with many stands of the *Schoenetum*, they do not typically form the basis of the turf running among the dominants, and the distinctive sedge-lined runnels of that community, in which *C. lepidocarpa* and *C. flacca* are also very common, together with a variety of other calcicoles, are not found here.

Patches within stands of the *Cirsio-Molinietum* can show a structural similarity to such more open and richer swards but, by and large, the character of much of the remaining flora is mesophytic and its texture rather coarse. Thus, although Caricion davallianae plants such as *Epipactis palustris* and *Dactylorhiza incarnata*, which have their best representation in lowland Britain within the *Schoenetum*, are occasionally to be found, they are less frequent overall than species such as *Lotus uliginosus*, *Mentha aquatica*, *Prunella vulgaris*, *Ranunculus acris*, *Hydrocotyle vulgaris* and the scramblers *Vicia cracca* and *Lathyrus pratensis*, plants which are of wide occurrence through damper Molinio-Arrhenatheretea vegetation in Britain.

Other stands have a distinctly heathy aspect. Indeed, *Potentilla erecta* is a constant of the community and, in contrast to the *Schoenetum*, where it is also very frequent but typically epiphytic, it occurs here throughout the sward, sometimes growing on the *Molinia* tussocks, but often rooted in the ground between. And, towards the south and west of Britain, where such sub-shrubs as *Erica tetralix*, *Calluna vulgaris* and *Ulex gallii* also become quite common in the *Cirsio-Molinietum*, the vegetation can approach the *Ericetum tetralicis* in its floristics and structure. Typically, however, grazing and sometimes, in this kind of landscape, burning, keep the cover of these plants in check. The same is true of those shrubs and trees which occasionally get a hold in the community: small saplings of *Salix cinerea* and *Alnus glutinosa* are sometimes to be found but, in normal circumstances, they do not become abundant.

The bryophyte flora of the *Cirsio-Molinietum* is generally poor and of low cover, the dense herbage and thick litter inhibiting its development. The shade-tolerant *Calliergon cuspidatum* is the commonest species throughout, with *Brachythecium rutabulum* also quite frequent and *Campylium stellatum*, *Hypnum cupressiforme*, *Pseudoscleropodium purum*, *Eurhynchium praelongum* and *Thuidium tamariscinum* occasional in various of the sub-communities. In stands which come close to wet heath, *Aulacomnium palustre* and small patches of Sphagna can be found occasionally.

Sub-communities

Eupatorium cannabinum sub-community: *Cladio-Molinietum/Molinietum* transitions Godwin 1941; *Cirsio-Molinietum peucedanetosum* (Sissingh & De Vries 1942) Wheeler 1978; *Cirsio-Molinietum eupatoretosum* (Sissingh & De Vries 1942) Wheeler 1980c. This kind of *Cirsio-Molinietum*, renowned from the mowing experiments performed at Wicken Fen (Godwin 1941), has much in common with the typical form but it shows

some distinct floristic and structural differences which, in extreme stands, bring it very close to Phragmitetalia fen. *Molinia* remains generally dominant overall and it is often accompanied by some *J. subnodulosus* and, in certain localities, by small amounts of *Schoenus*. But, often more conspicuous than these at first sight, are *Phragmites*, which, even when sparse, can have overtopping shoots, and *Cladium*, usually more stunted, but sometimes still abundant in bulky clumps. Then, among the associates, it is often species capable of taller growth that predominate. Providing obvious continuity with tall-herb fen are *Eupatorium cannabinum*, the best single preferential for this kind of *Cirsio-Molinietum*, and less commonly *Lythrum salicaria*, sometimes also *Lysimachia vulgaris* and, at certain sites, the nationally rare *Peucedanum palustre* or even more local *Selinum carvifolia* (Walters 1965, Wheeler 1975).

Cirsium palustre, *Angelica* and *Filipendula* are also common throughout but, as the reed or sedge cover becomes progressively thinner, typical *Cirsio-Molinietum* plants such as *Succisa*, *C. dissectum*, *Centaurea nigra* and *Equisetum palustre* increase in frequency, then smaller species like *Valeriana dioica*, *Carex panicea*, *C. hostiana* and *Potentilla erecta*. The more open conditions also allow *Mentha aquatica* and *Hydrocotyle vulgaris* to spread and *Agrostis stolonifera* becomes frequent, though other grasses are rather scarce. *Lotus uliginosus* and *Vicia cracca* are much more common than in most tall-herb fen and *Galium uliginosum* seems to be more frequent than *G. palustre*. Some stands have *Dactylorhiza incarnata* or *Gymnadenia conopsea*.

Bryophytes can be somewhat more conspicuous here than elsewhere in the community but the species are still few, with only *Campylium stellatum* joining *Calliergon* and *Brachythecium* as a distinctive preferential.

Typical sub-community: Litter Godwin & Tansley 1929; *Molinietum* Godwin 1941, Willis & Jefferies 1959; *Cirsio-Molinietum typicum* (Sissingh & De Vries 1942) Wheeler 1980c. In this, the typical form of the *Cirsio-Molinietum* in all but the south-west of Britain, *Molinia* is generally strongly dominant, although it is often accompanied by smaller amounts of *J. subnodulosus* or *J. articulatus* (rarely both, it seems), less commonly by some *J. conglomeratus*, *J. inflexus*, *J. acutiflorus* or *J. effusus*. *Schoenus* can be found in some stands and, though it is not usually abundant, when occurring along with *J. subnodulosus*, a quite common event here, it can bring the vegetation close in structure to the *Schoenetum*. *Cladium* is very rare but sparse *Phragmites* sometimes overtops the *Molinia* and rush layer.

Other, smaller grasses are generally well represented with *Holcus lanatus* and *Anthoxanthum* frequent, the former sometimes locally abundant, *Festuca rubra*, *Deschampsia cespitosa* and *Agrostis stolonifera* occasional.

Briza media is also strongly preferential to this sub-community and, in transitions to calcicolous swards, *Brachypodium pinnatum* and *Festuca ovina* can sometimes be found. Sedges are also common, particularly in shorter swards of this general type. *C. panicea* and *C. hostiana* both show a peak of occurrence in this kind of *Cirsio-Molinietum*, but *C. pulicaris* is especially frequent and *C. nigra* is preferential too, sometimes occurring in abundance. When numbers of these species occur with semi-prostrate *J. articulatus*, a quite common coincidence, the sward takes on some of the character of a fine-grained Caricion davallianae fen.

Often, however, the herbage is coarser than this and, in less heavily-grazed stands, decidedly rank. The community hemicryptophytes *Succisa*, *Cirsium dissectum*, *C. palustre* and *Angelica* are all very common, with *Valeriana dioica*, *Centaurea nigra* and *Filipendula ulmaria* also frequent. Then, there is often some *Lotus uliginosus* and *Equisetum palustre* and, scrambling among the sward, *Vicia cracca*, *Lathyrus pratensis* and *Galium uliginosum*. *Rumex acetosa* and *Hypericum tetrapterum* are preferential at low frequencies and occasionally some orchids can be found: *Epipactis palustris* and *Gymnadenia conopsea* are the most widespread of these but *Dactylorhiza fuchsii*, *D. incarnata* and *D. majalis* ssp. *praetermissa* also occur. When these taller plants have an opportunity to flower, they can give this vegetation a very colourful aspect, though, where growth of the bulkier monocotyledons proceeds unchecked, certain of them are easily overwhelmed.

Forming a somewhat lower tier or, in close-grazed or recently-mown stands, bulking up an altogether shorter sward, there are frequent plants of *Potentilla erecta*, *Mentha aquatica*, *Hydrocotyle vulgaris* and *Luzula multiflora*, with occasional *Ranunculus acris*, *Cardamine pratensis*, *Prunella vulgaris*, *Plantago lanceolata*, *Trifolium pratense*, *Linum catharticum*, *Leontodon taraxacoides*, *Cerastium fontanum*, *Potentilla reptans*, *P. anserina* and *Triglochin palustris*.

In contrast with the *Juncus-Erica* sub-community, with which this kind of *Cirsio-Molinietum* shares a general abundance of rushes and grasses, sub-shrubs are typically sparse, though at some localities, transitional vegetation can be found with occasional *E. tetralix* and *Calluna*, and it is such stands which often provide an eastern English locus for plants like *D. maculata* ssp. *maculata* and *Platanthera bifolia* (Wheeler 1975).

Bryophytes are usually poorly represented in the coarse herbage with just *Calliergon cuspidatum*, *Brachythecium rutabulum* and *Pseudoscleropodium purum* often being the only species.

Juncus acutiflorus-Erica tetralix sub-community: *Molinietum* Ivimey-Cook *et al.* 1975 *p.p.*; Grassy Heath NCC Devon Heathland Report 1980. In structural terms, this vegetation, which is the usual kind of *Cirsio-Molinietum* in south-western Britain, does not differ greatly from the Typical sub-community, but it does have some distinctive floristic elements. *Molinia* is again the normal dominant and, even from a distance, it generally determines the overall appearance of the vegetation. Rushes are a common feature of the herbage and a locally abundant one, but this sub-community lies, for the most part, beyond the range of *J. subnodulosus* and it is *J. acutiflorus* and *J. conglomeratus*, less often *J. effusus*, that are represented here. *Schoenus* is rarely found and *Phragmites* and *Cladium* are typically absent, but intermixed with the monocotyledons there is frequently some *Erica tetralix*, less often some *Calluna* and *Ulex gallii*, and where these become a little more abundant in relation to the *Molinia* the vegetation can take on a rather heathy aspect.

Indeed, it is mosaics of heaths which provide the usual context for this sub-community in the south-west, and there can be some difficulty in separating this vegetation from soligenous tracts of the *Ericetum tetralicis* (a problem well seen in the NCC Devon Heathland Report 1980). Generally, though, the continuing abundance of the grasses, sedges and dicotyledons of the community provides a distinction. As in the Typical form, *Holcus lanatus*, *Anthoxanthum*, less often *Festuca rubra* and *Agrostis stolonifera*, occur intermixed with the *Molinia*, and there is frequent *Carex panicea* and occasional *C. pulicaris* and *C. hostiana*. *Lotus uliginosus*, *Mentha aquatica*, *Ranunculus acris*, *Luzula multiflora*, *Cardamine pratensis* and *Hydrocotyle vulgaris* all remain occasional to frequent and, by mid-summer, the flowering shoots of *Succisa*, *Cirsium dissectum*, *C. palustre* and *Angelica* provide splashes of colour. Typically, however, *Valeriana dioica*, *Centaurea nigra* and *Filipendula ulmaria* and the tall-fen herbs of the *Eupatorium* sub-community, are not represented here.

Among the preferential associates, there are occasional records for *Dactylorhiza maculata* ssp. *maculata*, *Serratula tinctoria*, *Ranunculus flammula*, *Achillea ptarmica*, *Pedicularis sylvatica*, *Senecio aquaticus* and the Oceanic West European *Scutellaria minor* and the rare *Hypericum undulatum*. This sub-community also probably provides the commonest locus in south-west Britain for *Platanthera bifolia*. Small amounts of *Galium palustre* can often be found and species such as *Viola palustris*, *Agrostis canina* ssp. *canina* and *Narthecium ossifragum* occurring in wetter patches provide a link with poor-fen and bog vegetation. Some stands also show transitions to soakways with *Hypericum elodes*, *Juncus bulbosus* and *Potamogeton polygonifolius*.

Again, bryophytes are not a very conspicuous element of the vegetation but *Calliergon cuspidatum*, *Brachythecium rutabulum*, *Pseudoscleropodium purum*, *Eurhynthium praelongum*, *Thuidium tamariscinum* and *Hypnum cupressiforme* have all been recorded with more distinctly occasional *Aulacomnium palustre*, *Sphagnum subnitens* and *S. auriculatum*.

Habitat

The *Cirsio-Molinietum* is a community of moist to fairly dry peats and peaty mineral soils, circumneutral but only moderately mesotrophic, in the warmer lowlands of southern Britain. It can be found in association with both topogenous and soligenous mires, typically marking out the better-drained fringes of fens and bogs proper or the margins of wet hollows and flushes. Climate and soil together both influence the floristics of the community but this is essentially a secondary vegetation type, derived from a variety of precursors and ultimately maintained by mowing or grazing, and these treatments have marked effects on its composition and structure. Increasingly assiduous reclamation of mire fringes has reduced and fragmented the distribution of the community and other stands have become rank and scrubby with neglect.

The *Cirsio-Molinietum* is a vegetation type of the warmer parts of Britain, almost all the known stands falling within the area where annual accumulated temperatures exceed 1200 day-degrees C (*Climatological Atlas* 1952, Page 1982). Such a zone takes in the lowlands of central and southern England, roughly bounded by a Severn–Humber line, and the coastal fringe of Wales, and it coincides closely with the mainland British distribution of *Cirsium dissectum*, the best single preferential of this kind of vegetation, not only in this country, but also in those parts of Ireland (Brock *et al.* 1978, White & Doyle 1982) and of the Continent (e.g. Westhoff & den Held 1969) where similar conditions prevail. Apart from this species, however, there are no other phytogeographic indicators which are so constant in or faithful to the *Cirsio-Molinietum* with us. *Dactylorhiza majalis* ssp. *praetermissa*, another Oceanic West European plant with a similar distribution in Britain, occurs as an occasional but it is only in south-western parts of the country that this kind of fen-meadow acquires a really Atlantic feel. Here, in the *Juncus-Erica* sub-community, the *Cirsio-Molinietum* extends into the more obviously winter-mild zone of southern Britain, running from the New Forest across to north Devon and taking in south Wales, and provides a locus for such oceanic plants as *Scutellaria minor*, *Hypericum undulatum*, *Platanthera bifolia* and also more widespread and common species such as *J. acutiflorus*, *E. tetralix* and *Ulex gallii*.

Overall, however, there are few plants in the community which can be said to be even of a generally lowland and southern character in their British distribution: many of the most frequent are virtually ubiquitous in this country, except at the highest altitudes and, indeed,

the *Cirsio-Molinietum* provides some southern localities for species which have a mainly northern distribution, such as *Carex hostiana*, *C. pulicaris* and *Angelica*. In fact, the scarcity of this kind of vegetation beyond the limits of the range of *C. dissectum* is not really due to any reliance by a large number of its species on warmer temperatures. It is mainly an edaphic effect, because soils which are suitable for this particular assemblage become increasingly scarce beyond the range of the thistle. Some stands just outside its distribution remain so generally similar to the *Cirsio-Molinietum* that they can properly be included within it (B.D. Wheeler, pers. comm.). Others are quite close in their composition, but are better placed in the north-western equivalent of this community, the *Molinia-Potentilla* mire, to which the *Juncus-Erica* sub-community can be seen as a transition. Still others share many species with the *Cirsio-Molinietum* but, occurring in distinctly cooler and wetter conditions further north, have particular climatically-related floristic elements of their own: this would be true of the *Molinia-Crepis* mire, for example. The northern and western limits of this fen-meadow are thus neither so precise nor so readily explained as might appear at first sight. And further sampling along its general boundary would be very helpful in fixing the position and character of floristic transitions among these vegetation types: of especial value would be an examination of localities around the Humber, sites which have *C. dissectum* (Perring & Walters 1962), but from where the *Cirsio-Molinietum* has not been recorded.

The kinds of soils which support this particular type of fen-meadow are widespread through the warmer, southern lowlands of Britain, though quite local there, even without the extensive losses of suitable habitats that have occurred with land improvement. Most often, the *Cirsio-Molinietum* is found over organic or strongly humic profiles that are of a generally intermediate character in terms of their moisture regime, base-status and nutrient content. These conditions set some broad but important limits on the composition of the community and generally confine its occurrence to situations that are transitional between mires on the one hand and grasslands and dry heaths on the other, though the particular character of these habitats is actually quite varied.

As far as soil moisture is concerned, the profiles here range from fairly moist to quite dry, and though the community can develop around areas of either topogenous or soligenous influence, there is normally no marked seasonal fluctuation in water-level or throughput. The soils are seldom flooded to the surface, even in the wettest parts of the winter, and they can dry out appreciably above in the summer months. Though protected against drought, even in the drier eastern parts of the country, where the Typical or *Eupatorium* sub-communities can experience less than 120 wet days yr^{-1} (Ratcliffe 1968), the profiles are thus consistently better-aerated than the permanently waterlogged or winter-flooded peats which support Phragmitetalia swamps and fens or Sphagnetalia bogs, or many of the strongly-gleyed humic soils carrying transitional communities on their surrounds or in flushes. The *Cirsio-Molinietum* can be found in close association with these vegetation types, but the shift on to the better-drained peats or less strongly gleyed mineral soils which typically support it sees a rise to prominence of those Molinietalia species which provide so much of its character, *Succisa*, *Cirsium dissectum*, *C. palustre*, *Angelica*, *Lotus uliginosus*, *Valeriana dioica*, *Equisetum palustre*, *Filipendula ulmaria*, the various Junci and, of course, *Molinia* itself, and of other plants which it shares with a wide range of moister mesotrophic swards, such as *Holcus lanatus*, *Anthoxanthum odoratum*, *Centaurea nigra* and *Vicia cracca*. This kind of assemblage, distinguishing the Typical sub-community, is the most widely distributed form of *Cirsio-Molinietum* around the fringes of more calcareous mire systems through the central and eastern lowlands of England, more especially on the fringes of topogenous hollows, like those seen in the East Anglian superficials, and of flood-plain mires, whether small alluvial terraces along Chalkland rivers (e.g. Haslam 1965), or the very extensive fens of Broadland (Wheeler 1975, 1978, 1980c) and the Fenland remnants (Godwin & Tansley 1929, Godwin 1929, 1941).

A number of the important Molinietalia species of the community provide a strong continuity with the tall-herb vegetation of topogenous fens and, in the *Eupatorium* sub-community, with its remnants of reed or sedge canopy, the floristic overlap with the Phragmitetalia is very evident. As Godwin (1929, 1941) showed, such vegetation is sometimes the ultimate forebear of the *Cirsio-Molinietum*, though there is an edaphic limit on how far treatments can convert such precursors into this community. Experimental evidence is lacking but Phragmitetalia fen on wetter ground is probably transformed by regular summer-mowing or grazing into the *Juncus-Cirsium* fen-meadow, rather than the *Cirsio-Molinietum*. These two vegetation types sometimes occur together in zonations around mire fringes and topogenous hollows in eastern England and they have a great deal in common, similar treatments encouraging the prominence of mowing- or grazing-tolerant Molinietalia species throughout. But one major difference between the communities, the switch from dominance by rushes and large sedges in the former to dominance by *Molinia* in the latter, is perhaps a further indicator of the move to soils which are better aerated and more resistant to poaching by stock.

Over suites of somewhat more base-poor soils, on to which the *Cirsio-Molinietum* extends locally in the Typi-cal sub-community and more consistently in the *Juncus-Erica* sub-community (see below), its bounds in relation to soil moisture are a little more tightly drawn and perhaps not quite so susceptible to manipulation by treatments. Some important plants of the community, such as *Succisa*, *Potentilla erecta*, *Carex panicea*, *J. acutiflorus* and even *C. dissectum*, can also be found in Ericetalia wet heath, alongside which this *Juncus-Erica* sub-community often occurs in south-west Britain, and *Molinia*, of course, is a common co-dominant there. Some Ericetalia species, too, such as *E. tetralix*, *Narthecium*, *Aulacomnium palustre* and certain *Sphagna*, transgress a little way in the opposite direction, helping to characterise this kind of *Cirsio-Molinietum*. But the area of floristic overlap is relatively small and, within heath and valley bog landscapes, the community is rather strictly confined to those situations which combine accumulation of at least a shallow layer of organic material with enhanced drainage and aeration. Typically, then, it marks out the gently-sloping surrounds to strongly-gleyed and waterlogged hollows over deposits like the Carboniferous shales of Gower and north-west Devon (NCC Devon Heathland Report 1980), the Triassic deposits of the Devonshire Pebble-Bed Commons (e.g. Ivimey-Cook *et al.* 1975) and the Eocene and Oligocene sands and gravels of the New Forest.

In this wetter south-western part of the range of the *Cirsio-Molinietum*, where there is generally in excess of 1000 mm of rain annually and over 140 wet days yr^{-1} (*Climatological Atlas* 1952, Ratcliffe 1968), suitably moist and humic soils can extend the cover of the community a little way on to steeper slopes underlain by more pervious bedrocks, and *Molinia*, in particular, can provide a strong floristic continuity between the fen-meadow and the heaths and grasslands which eventually replace it on drier ground. But any strong tendency to sharpness of drainage is inimical to many of the other most characteristic species of the community. In the drier east, the upper limit of the *Cirsio-Molinietum* tends to be even more sharply defined, because the occurrence of suitable soils is so much more dependent on local drainage conditions.

The regional contrasts in the interaction of climate and soils across the range of the *Cirsio-Molinietum*, which help to define the south-western *Juncus-Erica* type against the other central and eastern sub-communities, are reinforced by variation in the base-status of the soil parent materials and ground waters and in the tendency of the profiles to show surface-leaching. By and large, this is a community of circumneutral soils, with superficial pH generally within the range 5–6.5, very much the optimum for many of our Molinietalia herbs, but also including the lower and upper limits for many more obviously calcicole or calcifuge plants, such that mixtures of species like *Briza media*, *Carex hostiana, C. pulicaris, C. panicea, Potentilla erecta* and *Luzula multiflora* often provide part of the distinctive character of the vegetation. Over central and eastern England, however, the profiles tend towards the more base-rich and calcareous and here the *Cirsio-Molinietum* is essentially a community of the margins of fens fed by lime-rich ground waters. Typically, it occupies the middle range of a sequence of soils which runs from deep, waterlogged fen peats on the one hand to excessively-draining rendzinas on the other, and which can be seen, in varying degrees of completeness where ill-draining hollows occur within landscapes of limestones, especially the Chalk, and calcareous superficials. In the Typical sub-community, which is very characteristic of the fringes of base-rich mires, particularly in East Anglia, such conditions are reflected in the common occurrence of *J. subnodulosus* and *Briza media* with, more locally, *Schoenus, Epipactis palustris* or *Dactylorhiza incarnata* on wetter ground or, in transitions to drier surrounds, *Festuca ovina, Brachypodium pinnatum* or even *Thymus praecox* and *Cirsium acaulon* (Wheeler 1975, 1980c). Some of the former species also find an occasional place in the *Eupatorium* sub-community, a more local kind of *Cirsio-Molinietum* more strictly associated with the margins of flood-plain mires, particularly in Broadland and the Fens, but, although this vegetation type can be found on equally base-rich substrates as the Typical sub-community, the tall growth of the herbage often precludes a strong calcicolous expression among smaller plants. In patches, then, the *Cirsio-Molinietum* can approach Caricion davallianae vegetation in its floristics, although the base-richness of the habitat is probably less than under communities like the *Schoenetum*, which replaces this kind of fen-meadow in spring-fens.

Locally in eastern England, the soils under the *Cirsio-Molinietum* show some measure of surface-leaching where they are further removed from any influence of calcareous ground waters, as around gravel islands within fens, for example, or over remnant peat baulks left upstanding by cutting (Wheeler 1975). And it is then that, in this part of Britain, such plants as *E. tetralix, Calluna* and *Dactylorhiza maculata* ssp. *maculata* make an occasional appearance. In the main, however, these species are confined to the *Juncus-Erica* sub-community whose soils are throughout of somewhat lower pH, usually 4.5–6, something which reflects both the argillaceous bedrocks which typically underlie the *Cirsio-Molinietum* in the south-west of Britain and the wetter climate there. Compared with the surrounding heaths and grasslands, of course, the fen-meadow can appear distinctly basiphilous with its Molinietalia herbs and mildly calcicolous sedges, and it often marks out the influence of bedrocks or seeping waters which are a little enriched in calcium and thus able to counteract the prevailing tendency to eluviation.

The soils beneath the *Cirsio-Molinietum* are probably less impoverished, too, than those which underlie communities like the *Schoenetum* or the *Ericetum tetralicis* and, where grazing and trampling are not too heavy, the fairly luxuriant character of the sward often impresses itself. Compared with the *Juncus-Cirsium* fen-meadow, however, the profiles are probably poorer in major nutrients, or at least selectively depleted, perhaps as a result of long-continued mowing or grazing, with little return to the sward, apart from dunging by stock: this may be another reason for the dominance of *Molinia* as opposed to rushes in this kind of vegetation. But the effects of treatments on the *Cirsio-Molinietum* are much more thorough going than this, because the community owes its maintenance and essential aspects of its floristics and structure to particular patterns of mowing and/ or grazing. In its usual context of mire surrounds in central and eastern England, there is little doubt that many stands of this fen-meadow have been derived from Phragmitetalia fen or less markedly calcicolous tracts of Caricion davallianae fen by annual cutting in the summer months or by grazing. On the most general level, such treatments repeatedly set back progression to scrub and woodland while allowing at least limited further peat accumulation, but more particularly they deflect succession away from vegetation dominated by the typical tall helophytes or bulky cyperaceous plants of mire surrounds. As Godwin (1941) showed in his classic crop-taking experiments, as far as mowing is concerned, it is the frequency and the timing of the treatment that are of crucial importance in determining the extent to which the sere is deflected towards vegetation like the *Cirsio-Molinietum*, because under some cutting regimes, important fen-dominants like *Phragmites* and *Cladium* can maintain themselves: indeed, annual winter-mowing of Phragmitetalia vegetation or three- or four-yearly summer-cutting were the standard treatments for the harvesting of reed or sedge respectively in places like the Norfolk Broads (e.g. Lambert 1951, 1965). The *Cirsio-Molinietum*, on the other hand, formed a major part of the crop known as 'litter', herbaceous vegetation lacking large amounts of the bulkier fen-dominants but with much *Molinia* and *J. subnodulosus*, cut every year between July and October, for fodder and bedding (Godwin & Tansley 1929, Godwin 1929, 1941, 1978, Lambert 1965). And Godwin (1941) demonstrated at Wicken how, over a period of twelve years, treatments which came increasingly close to annual summer mowing in their frequency, were ever more effective in actually converting stands of tall-herb fen, in this particular case *Cladium*-dominated vegetation, into the *Cirsio-Molinietum*; and, further, that some of these floristic and structural changes involved in this transformation could be reversed somewhat when tracts of 'litter' were mown at less frequent intervals or intermediate vegetation simply left uncut. As defined here, the Typical sub-community approximates to what the early workers knew as 'litter' or *Molinietum*, the more or less stable end-point of this deflected succession, while the *Eupatorium* sub-community includes much of the transitional vegetation, in the process of conversion or at some intermediate point of balance under regimes of summer cutting every two or three years. And the gradual floristic change that can be seen running through the sequence from tall-herb fen, through the *Eupatorium* sub-community to the Typical kind of *Cirsio-Molinietum*, largely reflects the extent to which smaller hemicryptophytes are able to capitalise on the absence of shade from plants like *Phragmites* and *Cladium* and of the smothering effects of their bulky litter.

In fact, nowadays, traditional mowing treatments of this kind are almost wholly confined to fen reserves like Wicken, although fragments of the community persisting in field corners may sometimes be taken in with a hay crop. Most often, it is grazing which maintains the community and, though no parallel experiments to those described above have ever been carried out, there seems little doubt that, around the drier fringes of Phragmitetalia vegetation, this treatment has effected changes similar to those seen at Wicken, particularly where palatable *Phragmites* is the dominant fen monocotyledon, as it usually is. Constant removal of herbage and the effects of trampling by the stock, generally cattle in central and eastern England, would both favour the abundance of hemicryptophytes, particularly rosette species, among the *Molinia* tussocks and, though no data are available, variations in the intensity of grazing probably underlie the floristic differences between sampled stands of the *Eupatorium* and Typical sub-communities.

Grazing is of importance, too, in the maintenance of the *Juncus-Erica* sub-community which occurs most often within tracts of heaths and grassland on commons and around the upland fringes of the south and west of Britain, where there is often free access to stock, frequently mixtures of cattle, sheep and ponies, throughout the year. In this type of *Cirsio-Molinietum* there is, as described above, some edaphic limit on the representation of ericoid sub-shrubs but grazing probably helps keep these in check in relation to the *Molinia* and, particularly when combined with burning, such treatment may lead to an extension of the *Cirsio-Molinietum* on to ground that would typically support wet heath.

Zonation and succession

The *Cirsio-Molinietum* is typically found as part of transitions between tall-herb fens or other kinds of mire vegetation on the one hand and grasslands or dry heaths on the other. Essentially, such sequences reflect variations in the soil water-regime but the particular communities involved differ according to the base-status of the mire waters and their surrounding soils and with the

regional climate. And the proportions of the different vegetation types within the zonations are very much influenced by the degree to which treatments have modified the natural patterns. Land improvement has increasingly modified the drier portions of the zonations and often destroyed the community altogether by extending drainage and reclamation right on to areas of mire. In other cases, stands remain intact but have become fragmented and isolated within intensive agricultural landscapes. Even where larger tracts have survived longer, the abandonment of traditional treatments has often left them to revert to rank secondary fen and scrubby woodland.

Within central and eastern England, where the usual context for the community is the surrounds of calcareous mire systems, extensive zonations are now scarce. However, it is still possible to see, at various localities in the flood-plain mires of Broadland (Wheeler 1978, 1980c) and the Fens (Godwin 1978) and along the margins of chalkland rivers in East Anglia (Haslam 1965) and elsewhere (Wheeler 1975, Ratcliffe 1977), transitions from the Typical or *Eupatorium* sub-communities to Phragmitetalia fen. Around the Norfolk Broads, the latter is usually some form of *Peucedano-Phragmitetum*, elsewhere generally the *Phragmites-Eupatorium* fen, and the extent and clarity of the junctions between such vegetation types and the *Cirsio-Molinietum* depend in large measure on how treatments have been imposed and withdrawn over the sequence of progressively drier soils through the mire. In some places, the boundaries between old mowing-marsh compartments can still be marked by fairly abrupt switches from what was fen-meadow mown for litter to tall-herb fen cut for reed or sedge. In parts of Wicken and in other reserves where mowing is used as part of the management of the vegetation, such transitions remain crisp and provide not only a glimpse of the once-widespread patchwork of secondary fen, but also an active indication of how one community can be transformed into another. Often, however, such patterns are now blurred by long neglect, with *Phragmites* re-establishing itself through the *Cirsio-Molinietum* and apparently capricious mixtures of fen and fen-meadow plants occurring in rank jumbles, frequently invaded by scrub. Commonly, then, transitions from the *Cirsio-Molinietum* to Phragmitetalia vegetation have to be seen in the much more compressed sequences found in ill-drained field corners and along marginal ditches, where it is grazing that usually mediates the transition.

Another, still fairly widespread, though increasingly local, zonation to be seen in this part of Britain occurs around the soligenous fen vegetation associated with springs and seepage lines fed by calcareous waters. In such situations, the Typical sub-community often forms a marginal zone around the *Schoenetum* which occupies the central area where base-richness and through-put are at a maximum. Grazing often occurs throughout such zonations and floristic continuity is therefore maintained through resistant small herbs whose edaphic tolerances are fairly broad. What marks the switch in vegetation types is the increasing representation of calcicoles in moving to the heart of the spring-fen, both among the vascular plants and in the bryophytes which are often well represented in the rather open areas of sward between the dominants, and the switch from a preponderance of *Molinia* to *Schoenus*. Withdrawal of grazing in such vegetation results in the development of the ranker and impoverished *Festuca-Juncus* sub-community of the *Schoenetum*, which is very close in its composition to the Typical *Cirsio-Molinietum* and perhaps represents a successional extension of the latter into more calcareous situations with neglect.

Both these types of zonations can be complicated by the occurrence of the other kind of lowland southern fen-meadow, the *Juncus-Cirsium* community alongside the *Cirsio-Molinietum*. As explained above, a combination of edaphic and treatment factors may differentiate these vegetation types, the latter apparently favouring drier ground than the former, though perhaps also being derived from it where mowing or grazing have been applied over long periods of time. Whatever the particular environmental influences on each, the communities are certainly very close and sometimes have to be distinguished simply by the proportions of their usual dominants, *J. subnodulosus* and *Molinia*. Fen helophytes, large sedges and dicotyledons can also occur in the *Juncus-Cirsium* fen-meadow, so very complex patchworks of dominance, superimposed over lower tiers of herbage which share many species in common, have to be expected in such situations: they are a reflection of floristic convergence where agricultural activities have been brought to bear around the more accessible fringes of these mire systems.

Under traditional treatments, and particularly grazing, there was often a strong measure of continuity between such fen-meadows and the vegetation of the drier ground beyond the influence of a fluctuating water-table in topogenous mires or above springs and seepage lines. This is less so now that agricultural improvement has pressed very closely around ill-drained remnants of the landscape, but in some localities good uninterrupted zonations can still be seen. In eastern England, for example, topogenous hollows in drift or springs in the Chalk sometimes pass above to rendzina soils developed from the bedrock, when the Typical sub-community can grade to some sort of Mesobromion sward, usually in this part of Britain, the *Festuca-Avenula* grassland or, towards central England, its derivatives dominated by *Bromus erectus* or *Brachypodium pinnatum*. Species such as *Briza media*, *Agrostis stolonifera*, *Festuca rubra* and some mesophytic dicotyledons like *Prunella vulgaris*,

Ranunculus acris and *Plantago lanceolata*, run on into the calcicolous sward and there may be some transgression of herbs in the opposite direction, but generally speaking the boundary is quite a sharp one. Over less pervious and calcareous bedrocks, but where there has still not been very much improvement of the surrounding grassland, the Typical *Cirsio-Molinietum* can pass to the *Centaureo-Cynosuretum*, with which it is linked through plants like *Holcus lanatus*, *Centaurea nigra*, *Festuca rubra*, *Anthoxanthum odoratum*, *Briza media*, *Potentilla erecta* and *Succisa pratensis*. And, then, where artificial fertilisers have been applied to such swards and the land drained and perhaps top-sown or ploughed and re-seeded, a Cynosurion sward like the *Lolio-Cynosuretum* may surround what are simply fragments of the *Cirsio-Molinietum* which have not yet succumbed to improvement. Outside major areas of calcareous mire vegetation like Broadland, it is often these drier ends of transitions in which the community can be found, marking out remnant topogenous hollows within tracts of pasture.

Towards south-western Britain, these latter kinds of zonations also occur but here there is a shift in the constituent vegetation types towards the more calcifuge. Thus, it is the *Juncus-Erica* sub-community that is the usual type of *Cirsio-Molinietum* that is found in moderately base-rich damper hollows in pastures around the upland fringes, and this can grade on the drier grazed surround to acidophilous sub-communities of the *Lolio-Cynosuretum* or *Centaurea-Cynosuretum* or to mesophytic types of *Festuca-Agrostis-Galium* grassland (NCC Devon Heathland Report 1980), according to the degree of improvement.

More distinctive, however, is the widespread occurrence of the *Juncus-Erica* sub-community within heathland complexes in south-west Britain, most notably on the common lands of the Devon Pebble Beds, on the Culm in the north-west of that county and more locally on Gower. Here, in the kind of sequence well seen in the study of Aylesbeare Common by Ivimey-Cook *et al.* (1975), the *Cirsio-Molinietum* occupies sloping ground around tracts of the *Ericetum tetralicis*, to which it may grade through the *Succisa-Carex* sub-community of the wet heath, and, more locally, of the *Schoenus-Narthecium* mire, both of which in turn can pass over permanently waterlogged ground to the *Narthecio-Sphagnetum*. On drier ground, the *Juncus-Erica* sub-community often grades to some kind of *Ulex gallii-Agrostis* heath, into which *Molinia* runs but where *E. tetralix*, *Calluna* and *U. gallii* increase their cover to become major elements of the vegetation, and from which Molinietalia herbs are mostly excluded.

In such zonations, expansion of the cover of *Molinia* can make the patterns very difficult to interpret because this tends to swamp the representation of any associates which might give some clue as to the original character of the vegetation. Burning followed by heavy grazing is especially likely to favour the spread of *Molinia* on the moister ground in these sequences, and it is quite common for the *Cirsio-Molinietum* to occur with more impoverished vegetation with *Molinia, J. acutiflorus, Potentilla erecta* and perhaps occasional *Cirsium palustre* and *Angelica*, which would probably be grouped within the *Molinia-Potentilla* mire. This is largely a western and northern equivalent to the *Cirsio-Molinietum* but the ranges of the two vegetation types overlap around the upland fringes of the south-west and, in this area of transitional climate, treatments are perhaps quite readily able to convert the one into the other.

In the absence of any kind of treatment, all stands of the community are probably able in theory to progress to scrub or woodland, although reversion to tall-herb fen around topogenous mires or the development of a very dense *Molinia* cover may greatly hinder invasion of woody plants. The ultimate character of any woodland vegetation probably also differs according to the edaphic conditions. In the flood-plain mires of central and eastern England, the natural successor to Phragmitetalia fen is the *Salix-Betula-Phragmites* woodland and it is possible that abandoned fen-meadow may revert to the final stages of this seral line, developing into the *Alnus-Filipendula* sub-community, where the influence of the base-rich ground waters is not too remote. However, long-continued mowing or grazing on somewhat drier ground may be accompanied by the surface-leaching and nutrient-depletion that favour the ultimate establishment of the *Betula-Molinia* woodland over neglected fen-meadow: the *Juncus* sub-community of that forest type can show considerable floristic similarity to the Typical *Cirsio-Molinietum* and birch invasion may effect a ready conversion of the one to the other. To the south-west, too, *Betula pubescens* is the most likely coloniser of ungrazed heaths among which the *Juncus-Erica* sub-community occurs, and there perhaps the *Juncus* or *Sphagnum* sub-communities of the woodland may succeed the fen-meadow. Stands of the *Cirsio-Molinietum* around more base-rich soligenous mires, very local to the west but widespread further east, could perhaps develop into Alno-Ulmion forest, though the *Alnus-Fraxinus-Lysimachia* woodland is essentially an oceanic community rare in the drier parts of Britain.

Distribution

The community is widespread through the lowland south of the country but has become increasingly local with changes in agricultural practice. The Typical sub-community is the commonest kind of *Cirsio-Molinietum* in central and eastern England with the *Eupatorium* sub-community much more confined to East Anglia. In the

south-west, the *Juncus-Erica* sub-community replaces the Typical form and the landscape context of the *Cirsio-Molinietum* shows a shift from more base-rich topogenous mires and spring-fens to heaths and acid grasslands.

Affinities

The classification of British *Molinia*-dominated vegetation raises particular difficulties because *Molinia* becomes increasingly catholic in its floristic associations towards the Atlantic seaboard of Europe, occurring abundantly with us in assemblages which bear little relationship to the various kinds of Molinieta characterised from the Continent (e.g. Passarge 1964), and, by virtue of its often uncompromising dominance, frequently overwhelming such associates as might give a clear clue to the affinities of the herbage. Such species-poverty can be a problem here, and the general prominence of *Molinia* has sometimes led to this vegetation being termed simply a *Molinietum* (e.g. Godwin 1941, Willis & Jefferies 1959, Ivimey-Cook *et al.* 1975) or grouped with other rather different swards for which such a general tag would be equally suitable (e.g. Wheeler 1975). In each of these studies, however, at least some of the stands included are very distinct from the kinds of *Molinia* grassland so extensive in western Britain and from any other mires and heaths in which this grass can be prominent. The more calcicolous and eastern of these, Wheeler (1980c) saw as forming the core of the community which he referred to the *Cirsio-Molinietum* Sissingh & de Vries 1942 (Vanden Berghen 1951, Westhoff & den Held 1969), a Dutch version of the more general *Molinietum caeruleae atlanticum* described from western Europe (Duvigneaud & Vanden Berghen 1945, Duvigneaud 1949, Le Brun *et al.* 1949). As defined here, the *Cirsio-Molinietum* takes in Wheeler's *typicum* and *eupatoretosum* (earlier termed *peucedanetosum*: Wheeler 1978) but excludes most of what he termed a *nardetosum*. Some less basiphilous vegetation of this kind can be accommodated in the Typical sub-community here, but the more obvious trend in this direction is seen in the *Juncus-Erica* type, the south-western *Cirsio-Molinietum*, unsurveyed by Wheeler, but well seen as including the most species-rich *Molinia* swards described from a Devon heathland by Ivimey-Cook *et al.* (1975).

What unites the British vegetation of this kind and brings it within the *Cirsio-Molinietum* as originally defined, is the occurrence of *Molinia* with such species as *Cirsium dissectum, Succisa pratensis, Carex panicea, C. hostiana, C. pulicaris* and *Potentilla erecta* (Westhoff & den Held 1969), an assemblage quite unique among our *Molinia* swards. And the presence, in the south-western type, of *J. conglomeratus* would confirm Continental phytosociologists in their allocation of this community to the Junco conglomerati-Molinion alliance within the Molinietalia (Westhoff & den Held 1969).

Even at the alliance level, however, the affinities of this vegetation are diverse. First, through plants such as the more basiphilous sedges, *Schoenus* and certain orchids, there are clear relationships, particularly in the Typical sub-community, with the Caricion davallianae, and the *Cirsio-Molinietum* is commonly found in the same sites as the *Schoenetum* (e.g. Wheeler 1980c). Second, there is the obvious similarity between the community and Phragmition and Magnocaricion tall-herb fens included in the Phragmitetalia, with the *Eupatorium* sub-community containing vegetation whose transitional character is known to reflect seral interchange between the two (Godwin 1941). Third, there is the close relationship between the *Cirsio-Molinietum* and the other southern lowland fen-meadow, the *Juncus-Cirsium* community, whose general affinities are with the Calthion, but which often intergrades with this Junco-Molinion type through an abundance of general Molinietalia plants encouraged by similar treatments. Inevitably, the direct effects of mowing and grazing, and of neglect, and their influence on the vegetation through soil changes, result in a frustrating degree of floristic convergence among many of the smaller elements of these different kinds of vegetation and a confusing medley of apparently interchangeable dominants through contiguous stands.

Towards the west, there are different problems. Here, the *Cirsio-Molinietum* comes close to Ericion tetralicis vegetation and the kinds of transitions to that alliance seen in the *Schoenus-Narthecium* mire. Usually, however, such gradations are more readily perceived than those between the *Juncus-Erica* sub-community of the fen-meadow and its northern and western equivalent, the *Molinia-Potentilla* grassland. This, too, can probably be placed in the Junco-Molinion alliance, and, though it is often considerably more impoverished than the *Cirsio-Molinietum*, transitions are quite common, particularly in south-western Britain.

Floristic table M24

	a	b	c	24
Molinia caerulea	V (3–6)	V (1–8)	V (4–8)	V (1–8)
Potentilla erecta	V (1–3)	V (1–4)	V (2–4)	V (1–4)
Succisa pratensis	IV (2–6)	V (1–5)	IV (2–5)	V (1–6)
Cirsium dissectum	IV (1–3)	IV (1–4)	IV (1–6)	IV (1–6)
Lotus uliginosus	IV (1–3)	IV (1–3)	IV (2–4)	IV (1–4)
Carex panicea	III (1–3)	IV (2–5)	III (1–4)	IV (1–5)
Galium uliginosum	IV (1–5)	IV (1–4)		III (1–5)
Valeriana dioica	IV (1–3)	IV (1–3)		III (1–3)
Centaurea nigra	III (1–3)	IV (1–4)	I (1–3)	III (1–4)
Juncus subnodulosus	IV (1–5)	III (1–3)		III (1–5)
Equisetum palustre	III (1–3)	III (1–3)		II (1–3)
Vicia cracca	III (1–3)	III (1–4)		II (1–4)
Filipendula ulmaria	III (1–3)	III (1–5)		II (1–5)
Linum catharticum	II (1–3)	II (1–3)		II (1–3)
Gymnadenia conopsea	II (1–3)	II (1–3)		II (1–3)
Eupatorium cannabinum	V (1–3)	II (1–2)		II (1–3)
Phragmites australis	V (1–5)	II (1–8)		II (1–8)
Campylium stellatum	III (1–2)	I (1)	I (2)	I (1–2)
Cladium mariscus	III (1–6)			I (1–6)
Lythrum salicaria	II (1–3)	I (1–2)		I (1–3)
Dactylorhiza incarnata	II (1–2)	I (1–3)		I (1–3)
Holcus lanatus	I (2)	IV (1–4)	III (2–5)	III (1–5)
Anthoxanthum odoratum	I (1)	III (1–4)	III (3–6)	III (1–6)
Hydrocotyle vulgaris	I (1–3)	III (1–3)	II (1–2)	II (1–3)

Species				
Carex pulicaris		III (1–3)	II (1–3)	II (1–3)
Luzula multiflora		III (1–3)	II (1–3)	II (1–3)
Ranunculus acris		II (1–2)	III (2–4)	II (1–4)
Pseudoscleropodium purum		II (1–3)	II (2–6)	II (1–6)
Festuca rubra		II (1–2)	II (4–7)	II (1–7)
Cardamine pratensis		II (1–3)	II (1–2)	II (1–3)
Briza media	I (1)	III (1–4)		II (1–4)
Rumex acetosa	I (1)	II (1)	I (3)	II (1–3)
Epipactis palustris	I (1)	II (1–4)		I (1–4)
Potentilla reptans	I (1)	II (1–3)		I (1–3)
Hypericum tetrapterum	I (1)	II (1–3)		I (1–3)
Cirsium arvense	I (1)	II (1–4)		I (1–4)
Carex nigra		II (1–4)	I (1–4)	I (1–4)
Trifolium pratense		II (1–3)	I (1)	I (1–3)
Polygala vulgaris		II (1–4)		I (1–4)
Leontodon taraxacoides		II (1–3)		I (1–3)
Juncus articulatus		II (1–4)		I (1–4)
Eriophorum angustifolium		II (1–2)		I (1–2)
Dactylorhiza fuchsii		II (1)		I (1)
Galium verum		II (1–3)		I (1–3)
Cerastium fontanum		II (1–3)		I (1–3)
Potentilla anserina		II (1–3)		I (1–3)
Juncus acutiflorus	I (7)	I (1–6)	IV (2–7)	II (1–7)
Juncus conglomeratus		II (1–4)	III (1–7)	II (1–7)
Erica tetralix		II (1–4)	III (2–4)	II (1–4)
Galium palustre		I (1–2)	III (1–3)	II (1–3)
Dactylorhiza maculata		I (1–4)	III (2–3)	II (1–4)
Calluna vulgaris		I (4–5)	II (1–5)	I (1–5)
Juncus effusus		I (1–3)	II (3–5)	I (1–5)
Serratula tinctoria		I (1)	II (1–3)	I (1–3)
Narthecium ossifragum		I (1)	II (2–4)	I (1–4)
Aulacomnium palustre		I (1–3)	II (3–4)	I (1–4)
Ranunculus flammula		I (1–2)	II (1–3)	I (1–3)
Hypnum cupressiforme		I (1)	II (3–4)	I (1–4)
Achillea ptarmica		I (1)	II (2–4)	I (1–4)
Agrostis canina canina			II (3–6)	I (3–6)
Scutellaria minor			II (1–2)	I (1–2)
Viola palustris			II (3–4)	I (3–4)
Sphagnum auriculatum			II (2–7)	I (2–7)
Hypericum undulatum			II (2–3)	I (2–3)
Salix repens			II (1–4)	I (1–4)
Pedicularis sylvatica			II (1–3)	I (1–3)
Senecio aquaticus			II (1–2)	I (1–2)
Thuidium tamariscinum			II (3–5)	I (3–5)
Eurhynchium praelongum			II (1–4)	I (1–4)
Cirsium palustre	III (1–4)	IV (1–5)	III (1–3)	III (1–5)
Angelica sylvestris	III (1–2)	III (1–3)	III (1–5)	III (1–5)
Calliergon cuspidatum	III (1–2)	III (1–3)	III (1–3)	III (1–3)
Mentha aquatica	III (1–4)	III (1–4)	III (2–4)	III (1–4)

Floristic table M24 *(cont.)*

	a	b	c	24
Carex hostiana	III (1–3)	III (1–3)	II (1–6)	III (1–6)
Agrostis stolonifera	III (1–3)	II (1–4)	II (1–4)	II (1–4)
Brachythecium rutabulum	III (1–3)	II (1–3)	II (1–6)	II (1–6)
Prunella vulgaris	II (1–3)	II (1–2)	II (2–3)	II (1–3)
Deschampsia cespitosa	II (1–4)	II (1–3)	I (2)	II (1–4)
Schoenus nigricans	II (1–5)	II (1–4)	I (8)	II (1–8)
Plantago lanceolata	I (1)	II (1–4)	II (1–4)	II (1–4)
Lathyrus pratensis	I (1–3)	II (1–3)	II (1–2)	II (1–2)
Pulicaria dysenterica	I (1)	II (1–3)	II (3–4)	II (1–4)
Rubus fruticosus agg.	I (1)	I (1–3)	I (1–2)	I (1–3)
Salix cinerea sapling	I (1–3)	I (1–3)	I (2)	I (1–3)
Danthonia decumbens		I (1–3)	I (1–5)	I (1–5)
Anagallis tenella		I (1–6)	I (1–3)	I (1–6)
Sphagnum subnitens		I (1–2)	I (4–7)	I (1–7)
Lychnis flos-cuculi		I (1–4)	I (3)	I (1–4)
Drosera rotundifolia		I (1–3)	I (3)	I (1–3)
Alnus glutinosa sapling	I (1)	I (1–2)		I (1–2)
Number of samples	19	33	31	83
Number of species/sample	21 (15–32)	29 (14–52)	19 (9–29)	26 (9–52)

a   *Eupatorium cannabinum* sub-community

b   Typical sub-community

c   *Juncus acutiflorus-Erica tetralix* sub-community

24   *Cirsio-Molinietum caeruleae* (total)

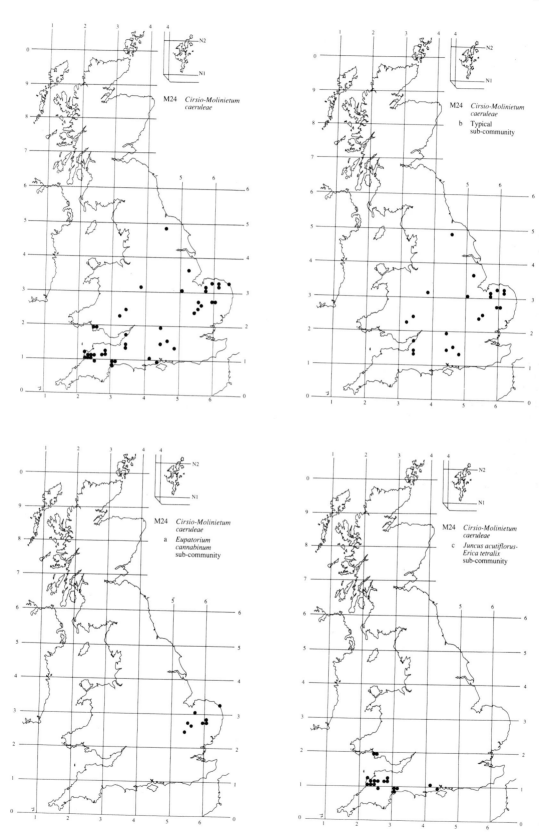

M24 *Cirsio-Molinietum caeruleae*

M24 *Cirsio-Molinietum caeruleae*

b Typical sub-community

M24 *Cirsio-Molinietum caeruleae*

a *Eupatorium cannabinum* sub-community

M24 *Cirsio-Molinietum caeruleae*

c *Juncus acutiflorus-Erica tetralix* sub-community

M25
Molinia caerulea-Potentilla erecta mire

Synonymy
Molinietum caeruleae Moss 1911, Rankin 1911*a, b,* Smith 1911, Crampton 1911, Stapledon 1914, Jefferies 1915, Fraser 1933, Tansley 1939, all *p.p.*; *Molinia-Myrica* nodum McVean & Ratcliffe 1962, Adam *et al.* 1977; *Molinia caerulea* grassland McVean & Ratcliffe 1962, Edgell 1969, Meek 1976, Meade 1981 *p.p.*; *Junco acutiflori-Molinietum* O'Sullivan 1968, Hill & Evans 1978, Ratcliffe & Hattey 1982 *p.p.*; *Festuco-Molinietum anthoxanthosum* Evans *et al.* 1977; *Molinia* flushes Ferreira 1978; Wet Grassy *Juncus* Heath NCC Devon Heathland Report 1980; *Myricetum galis* Birse 1980; *Molinia caerulea-Myrica gale* community Wheeler 1980c *p.p.*; *Molinia-Potentilla erecta* nodum Bignal & Curtis 1981; *Erica vagans-Schoenus nigricans* heath Hopkins 1983 *p.p.*

Constant species
Molinia caerulea, Potentilla erecta.

Rare species
Agrostis curtisii, Erica vagans, Lobelia urens.

Physiognomy
The *Molinia caerulea-Potentilla erecta* mire encompasses vegetation of quite widely differing floristics and physiognomy, but characterised throughout by the overwhelming abundance of *Molinia*. This is a feature which helps distinguish the community from rather similar vegetation types of the Juncion acutiflori and Caricion nigrae, where *Molinia* can occur occasionally but where dominance is usually held by rushes or small sedges, themselves quite frequent here, though typically of subsidiary cover. *Molinia*-dominance is not, however, confined to this community in western Britain: throughout the range of the *Molinia-Potentilla* mire, this grass is a very common element in lowland Erico-Sphagnion blanket mires and Ericion tetralicis wet heaths and it shows a distressing tendency to increase its abundance in these vegetation types in certain circumstances, bring-

ing them very close in composition and structure to this community. Indeed, many references to a *Molinietum* or to '*Molinia* grassland' in the British literature (e.g. Moss 1911, Rankin 1911*a, b,* Smith 1911*a,* Crampton 1911, Stapledon 1914, Jefferies 1915, Fraser 1933, Tansley 1939, McVean & Ratcliffe 1962, Meade 1981) and to other more precisely-defined units, like the *Festuco-Molinietum* of Evans *et al.* (1977) and the *Molinia-Potentilla* nodum of Bignal & Curtis (1981), include vegetation of this kind together with what is here termed *Molinia-Potentilla* mire, and bear ample testimony to the problems of separation of these different communities in the more oceanic parts of the country.

Although sharing a general abundance of *Molinia* with these modified vegetation types, the *Molinia-Potentilla* mire does seem to offer particularly favourable conditions for this grass, and the community includes many of the more extensive and vigorous stands of *Molinia* vegetation in western Britain. Its cover here is often virtually complete, though its growth form can vary from an almost even or gently-undulating sward to a formidably tussocky cover in which individual stools can attain diameters of half a metre or more and even greater heights, with systems of deep litter-lined runnels between. With the emergence of the annual crop of leaves in spring and early summer, the vegetation acquires a fresh-green colour which can mark out stands from a great distance and, in winter, with the dead brown foliage often laid in one direction by the winds, the tussocks give a distinctive wave-like appearance, features well caught in Jefferies' (1915) classic study.

Amongst this dense cover, the associated flora is noticeably poor in species and generally of low total abundance, though it is quite variable. Most common throughout are rushes and a few dicotyledons which help confirm the affinities of the vegetation with the Molinietalia. Among the former, *Juncus acutiflorus* is the most frequent, with *J. effusus* quite well represented too, except in the more acidic situations into which the community penetrates. Each of these can be patchily prominent among the *Molinia*, though typically they do not exceed it in cover and are often represented by just a few scattered shoots, one feature which helps separate the *Molinia-Potentilla* mire from rush-dominated stands of the *Carex echinata-Sphagnum* mire and from the *Juncus-Galium* rush-pasture, in which some of O'Sullivan's (1968a) *Junco acutiflori-Molinietum* is probably best placed. *J. articulatus* also occurs very occasionally but *J. conglomeratus* is very scarce, being more characteristic of heathy Molinietalia vegetation. In the very few eastern English localities where the *Molinia-Potentilla* mire occurs, *J. subnodulosus* is sometimes to be found, as at Roydon Common in Norfolk.

Among the few common dicotyledons of the community, only *Potentilla erecta* is frequent enough to qualify as a constant, and it is the only small herb that occurs more than occasionally throughout this vegetation, being able to survive growing among the dense herbage of the dry hummock tops. More sparsely represented, particularly in wetter, more base-poor situations, where there is heavy grazing or where the dominance of *Molinia* is very uncompromising, are *Lotus uliginosus*, *Succisa pratensis*, *Cirsium palustre* and *Angelica sylvestris*. Rooted between the tussocks, the taller among these hemicryptophytes have flowering shoots which, by mid-summer, can project above the *Molinia* and, where numbers of these plants occur more frequently, along with occasional *Eupatorium cannabinum* or *Filipendula ulmaria*, the community becomes more colourful and structurally varied, sometimes approaching poorer Filipendulion vegetation in its general appearance. In south-western England and south Wales, this kind of *Molinia-Potentilla* mire can also come close to the *Cirsio-Molinietum*, although *Cirsium dissectum* is only very rarely to be found among the taller herbs here, and these two vegetation types are, by and large, geographical replacements of one another.

A further structural element which gains occasional prominence in the *Molinia-Potentilla* mire comprises shrubs and sub-shrubs. *Calluna vulgaris* and, particularly on wetter ground, *Erica tetralix*, can both be quite common, though they are seldom very abundant and usually not very conspicuous, being rather attenuated in growth. Nonetheless, in south-western England in particular, it can sometimes be difficult to separate such vegetation from the *Ericetum tetralicis* wet heath from which expansion of *Molinia* has all but eliminated any distinctive associates. Elsewhere, throughout western Britain, the presence of these ericoids provides a link with the *Scirpus-Erica* wet heath in which *Molinia* can likewise increase its cover, but *Scirpus cespitosus* is only rarely recorded in the *Molinia-Potentilla* mire and such sub-shrubs as *Erica cinerea* and *Vaccinium myrtillus*, which can be commonly found in small amounts in the *Scirpus-Erica* wet heath, hardly ever occur here.

Sometimes, too, in Wales and south-west England, *Ulex gallii* is an occasional in the *Molinia-Potentilla* mire, the bushes scattered but very noticeable in summer when the plant is flowering. *U. europaeus* also occurs in some stands, and it can be locally prominent in this kind of vegetation, as on the Lizard in Cornwall, where both gorses occur, together with small quantities of *Erica vagans*, in a very striking kind of *Molinia-Potentilla* mire transitional to *Erica vagans-Schoenus* heath (and included in that community by Hopkins 1983).

More frequently noted in previous descriptions is the local, but very conspicuous, occurrence here of *Myrica gale*, forming a patchy or sometimes quite extensive and dense over-canopy. The combination of *Molinia* and *Myrica* is very eye-catching and several authors have characterised a vegetation type defined by the constancy of these species. But both have rather wide ecological amplitudes and can be found together with quite diverse suites of associates. In this scheme, therefore, assemblages with abundant *Molinia* and *Myrica* are treated as components of a number of communities: the *Molinia-Potentilla* mire can incorporate most of the *Molinia-Myrica* vegetation of McVean & Ratcliffe (1962) and Wheeler (1980c) and of Birse's (1980) *Myricetum galis*, but it excludes more base-rich vegetation like the *Molinia-Myrica* Association of Birks (1973) and much of that encompassed by Wheeler's (1980c) *Myricetum gale*.

The general absence of more calcicolous herbs among the often impoverished associated flora of the *Molinia-Potentilla* mire provides a further distinction from the *Cirsio-Molinietum* where such plants are few but quite frequent and from grassier stands of the *Schoenetum*, where they are more numerous. *Schoenus nigricans* itself, in fact, is sometimes found here and its local prominence can create a structural resemblance between the community and the *Schoenetum*, but other, more basiphilous cyperaceous plants are poorly represented. Indeed, among denser growths of *Molinia*, smaller sedges of any kind are of only sparse occurrence in the litter-choked runnels and through the community as a whole *Carex panicea*, *C. echinata* and *C. nigra* attain but occasional frequency and usually low cover, with such species as *C. hostiana*, *C. pulicaris*, *C. dioica* and *C. flacca* all rare.

Grasses too, other than *Molinia*, are of limited importance. *Agrostis canina* (presumably mostly ssp. *canina*) and *A. stolonifera* can be found at low frequency throughout and, in more open runnels, can form quite extensive mats. *Holcus lanatus* is also fairly common but it is only in the *Anthoxanthum* sub-community that other species figure prominently: there, *Anthoxanthum odoratum*, *Agrostis capillaris*, *Festuca rubra* and *Danthonia decumbens* all become preferentially common, though rarely very abundant, in swards transitional to Nardo-Galion vegetation. Among this kind of *Molinia-*

Potentilla vegetation, too, can be included some of the grassy herbage in which *Agrostis curtisii* and *Molinia* become frequent on heaths in south-western Britain.

Smaller dicotyledons only exceptionally make more than a scattered appearance in the community. *Viola palustris*, *Hydrocotyle vulgaris*, *Dactylorhiza maculata*, *Ranunculus acris* and *Achillea ptarmica* all occur occasionally and *Mentha aquatica*, *Cardamine pratensis*, *Epilobium palustre*, *Pulicaria dysenterica* and *Narthecium ossifragum* are preferential for particular sub-communities. The *Molinia-Potentilla* mire also provides an occasional locus for such Oceanic West European plants as *Scutellaria minor* and the more local *Carum verticillatum* and is an important context for the nationally rare *Lobelia urens* (Brightmore 1968).

Among the dense herbage, bryophytes are also often sparse, mainly confined to more open areas in the runnels or finding a place on decaying tussocks. *Rhytidiadelphus squarrosus*, *Pleurozium schreberi*, *Pohlia nutans* and *Calliergon cuspidatum* are found occasionally throughout but only in the *Erica* sub-community, where the ground between the tussocks remains wet for much of the year, is this cover enriched. There, a variety of other species, including some Sphagna, become preferentially frequent and abundant.

Sub-communities

***Erica tetralix* sub-community:** *Molinietum caeruleae* Moss 1911, Rankin 1911*a*, *b*, Smith 1911, Crampton 1911, Stapledon 1914, Jefferies 1915, Fraser 1933, Tansley 1939; *Molinia caerulea* grassland McVean & Ratcliffe 1962 *p.p.*, Edgell 1969, Meade 1981 *p.p.*; *Molinia-Myrica* nodum McVean & Ratcliffe 1962, Adam *et al.* 1977, Ratcliffe & Hattey 1982 *p.p.*; *Junco acutiflori-Molinietum*, heathy facies Hill & Evans 1978; *Myricetum galis* Birse 1980; *Molinia caerulea-Myrica gale* community Wheeler 1980*c p.p.*; *Myricetum gale ericetosum* Fischer 1967 *sensu* Wheeler 1980*c*; *Molinia-Potentilla erecta* nodum Bignal & Curtis 1981 *p.p.*; *Molinia caerulea-Juncus squarrosus* nodum Ratcliffe & Hattey 1982 *p.p.* In this, the most widely distributed and commonly described kind of *Molinia-Potentilla* mire, *Molinia* is still very much the dominant but there is a distinct shift in the associated flora towards Ericion tetralicis wet heath. In the first place, *Erica tetralix* is strongly preferential and, though rarely abundant, with frequent *Calluna* it brings a modest variegation to the vascular canopy. Then, although *J. acutiflorus* (only rarely here *J. effusus*) remains common and can be locally abundant, it is often rivalled in frequency by *Eriophorum angustifolium*. Among the sedges, *C. panicea*, *C. echinata* and *C. nigra* all occur occasionally and they sometimes thicken up their cover a little in more open runnels. But, apart from sparse *Holcus lanatus*,

Anthoxanthum, *Festuca rubra* and *Agrostis canina*, grasses are noticeably thin and, though there can be scattered plants of *Succisa*, *Cirsium palustre* and *Lotus uliginosus*, such taller herbs are poorly represented. Among smaller plants, *Viola palustris* and *Hydrocotyle* are sometimes found but more striking here is the occasional presence of such species as *Narthecium*, *Drosera rotundifolia* and *Vaccinium oxycoccos*.

The bryophyte cover, too, shows some distinctive features. *Aulacomnium palustre*, *Polytrichum commune*, *Hypnum jutlandicum* and *Calypogeia fissa* are all preferential at low frequencies but more noticeable is the variety and patchy abundance of Sphagna. *Sphagnum recurvum* and *S. auriculatum* are the commonest species but *S. papillosum*, *S. palustre*, *S. capillifolium* and *S. subnitens* also occur occasionally.

Superimposed on this basic pattern is a variety of local or regional differences. On Roydon Common in Norfolk, for example, one of the few eastern localities for the *Molinia-Potentilla* mire, *J. subnodulosus* replaces *J. acutiflorus* as the charateristic rush in this vegetation. Then, in south-western Britain, *Ulex gallii* extends out of the drier heaths to find an occasional place in this sub-community. More widely distributed are those stands where *Myrica* has attained local abundance over a field layer which is essentially an impoverished version of this kind of *Molinia-Potentilla* mire.

***Anthoxanthum odoratum* sub-community:** *Molinia caerulea* grassland Meek 1976 *p.p.*, Meade 1981 *p.p.*; *Festuco-Molinietum anthoxanthosum* Evans *et al.* 1977; *Molinia* flushes Ferreira 1978 *p.p.*; *Junco-Molinietum caeruleae*, non-heathy facies Hill & Evans 1978; *Junco acutiflori-Molinietum*, *Molinia caerulea* mire variant Ratcliffe & Hattey 1982 *p.p.* Although *Molinia* is still very much the dominant in this sub-community, the sward here tends to be shorter than in other kinds of *Molinia-Potentilla* mire and more varied in its major components. *J. acutiflorus* remains frequent and *J. effusus* reaches occasional frequency but it is grasses which contribute the most distinctive element. *Anthoxanthum* is the best preferential but *Holcus lanatus*, *Agrostis capillaris*, *Danthonia decumbens* and *Festuca rubra* all occur more commonly than usual and *F. ovina*, *Agrostis canina*, *A. stolonifera*, *Nardus stricta*, *Deschampsia cespitosa* and, in the south-west, *A. curtisii* can also be found occasionally. *Luzula campestris* and *L. multiflora* are somewhat commoner than usual, too, and *Carex panicea* and *C. echinata* occur at low frequency.

In this generally grassy ground, other associates can be quite sparse. *Calluna* and *U. gallii* occur occasionally, and *U. europaeus* and *Rubus fruticosus* agg. can also figure, particularly where there has been some disturbance, but *E. tetralix* and *Myrica* are both very uncommon. Then, though *Succisa*, *Lotus uliginosus* and *Cir-*

sium palustre are all more frequent than in the *Erica* sub-community and *Serratula tinctoria* and *Rumex acetosa* weakly preferential, such taller herbs are rarely very abundant and, in grazed stands, the more palatable are often nibbled down to non-flowering rosettes. At scattered localities towards the southern coast of England, *Lobelia urens* is sometimes to be found among this element of the vegetation and in more open but ungrazed stands, where it has a chance to flower, it can provide a striking splash of purplish-blue right through the summer (Brightmore 1968).

Among the dense grassy cover, smaller herbs are poorly represented and, apart from *Potentilla erecta*, only *Viola palustris* and *Hydrocotyle* approach occasional frequency. Bryophytes, too, are sparse with *Pseudoscleropodium purum* and *Rhytidiadelphus squarrosus* the commonest species but even those typically occurring only as scattered small patches. Sphagna are rarely found.

Angelica sylvestris **sub-community:** *Molinia* flushes Ferreira 1978 *p.p.*; Wet Grassy *Juncus* Heath NCC Devon Heathland Report 1980; *Molinia* grasslands Group 10 Meade 1981 *p.p.*; *Molinia-Eupatorium* community Ratcliffe & Hattey 1982; *Erica vagans-Schoenus nigricans* heath, *Ulex europaeus* variant Hopkins 1983. This is the most local but striking kind of *Molinia-Potentilla* mire because, though *Molinia* often grows very vigorously and frequently occurs as substantial tussocks, the cover is often variegated by small clumps of *J. acutiflorus* and *J. effusus* and, in mid-summer, is generally punctuated by the tall flowering shoots of numbers of dicotyledons, notably *Succisa* and *Lotus uliginosus* and then, more preferential to this sub-community, *Angelica* and *Cirsium palustre* with, more occasionally, *Epilobium palustre*, *Filipendula ulmaria*, *Pulicaria dysenterica*, *Valeriana officinalis* and *Centaurea nigra*. Somewhat shorter are *Mentha aquatica*, *Scutellaria minor*, *Cardamine pratensis* and *Equisetum palustre* and then, sprawling among the herbage, there is frequently some *Galium palustre*. *Agrostis stolonifera* and *A. canina* occasionally form spreading mats and scattered tufts of *Holcus lanatus* can also be found but other grasses, and sedges and ground-growing dicotyledons, are scarce.

Some notable local components of this kind of *Molinia-Potentilla* mire are *Carum verticillatum*, which in Britain occurs largely in this vegetation and in the closely-related *Juncus-Galium* rush-pasture, and that more obvious oceanic plant, *Osmunda regalis*, which can reach a substantial size here. *Schoenus nigricans* can also be found and its tussocks are numerous on occasion, though they are not accompanied by the kind of rich basiphilous flora characteristic of the *Schoenetum*. On the Lizard, however, there is some rather different

vegetation with *Molinia* and *Schoenus* that is probably best included here. Hopkins (1983) saw it as part of his *Erica-Schoenus* heath and, along with the two bulky monocotyledons and such species as *Potentilla erecta*, *Angelica*, *Eupatorium* and *Succisa*, it also had bushes of *Ulex europaeus*, *U. gallii* and that distinctive heath of the headland, *Erica vagans*, and frequent *Sanguisorba officinalis*.

Throughout the *Angelica* sub-community bryophytes are again sparse but *Calliergon cuspidatum* and *C. giganteum* can be found as scattered patches.

Habitat

The *Molinia-Potentilla* mire is a community of moist, but well-aerated, acid to neutral, peats and peaty mineral soils in the wet and cool western lowlands of Britain. It occurs over gently-sloping ground, marking out seepage zones and the flushed margins of sluggish streams, water-tracks and topogenous mires, but also extending on to the fringes of ombrogenous bogs. Although both climate and soils influence the composition of the vegetation, treatments, particularly burning and grazing, have probably affected many stands and contributed to the general abundance of *Molinia* in the community, especially where it extends on to less typical habitats. In the upland fringes of the north and west, this kind of vegetation often marks out situations which, with draining, are much prized for coniferous forestry, so much has been lost to this land use.

Geographically, the *Molinia-Potentilla* mire can be seen as a north-western replacement of the *Cirsio-Molinietum*, and some of the floristic differences between the communities can be related to the cooler but relatively mild climate of this part of Britain. Almost all the stands of the *Molinia-Potentilla* mire experience mean annual maximum temperatures of less than 26 °C (Conolly & Dahl 1970), but within this zone, the community is largely confined to the milder lowlands, occurring mostly below 200 m where February minima are generally a degree or so above freezing (*Climatological Atlas* 1952) and thus becoming increasingly local in the less equable reaches of the mountainous west of Scotland. In the warmer, oceanic climate of south-west Britain, there is, in fact, some geographical overlap, and, on circumneutral soils, a close floristic similarity, between the two communities. But much of the *Molinia-Potentilla* mire lies beyond the range of *Cirsium dissectum*, the major preferential of the *Cirsio-Molinietum*, and the more generally oceanic conditions are further reflected in the occurrence, along with abundant and vigorous *Molinia*, of *J. acutiflorus* and, in certain circumstances, *E. tetralix*, *Myrica* and *U. gallii*, together with the more local *Carum verticillatum*, *Scutellaria minor* and *Lobelia urens*, all Oceanic West European plants. Sadly, the easier it becomes to make this separation between our two Junco-Molinion communities, the harder it is to distinguish the *Molinia-Potentilla* mire from the Juncion acutiflori vegetation of north-west Britain, the *Juncus-Galium* rush-pasture.

However, much more obvious and specific than the direct influence of temperature on the *Molinia-Potentilla* mire are the indirect effects of rainfall felt through the soils, which help distinguish the community from these closely-related vegetation types, both in its floristics and in the particular kinds of habitats it characterises. Like both the *Cirsio-Molinietum* and the *Juncus-Galium* rush-pasture, this is a community of generally moist soils, intermediate between the permanently waterlogged on the one hand and the excessively draining on the other. In all three vegetation types, therefore, much of the floristic character is supplied by Molinietalia species such as *Succisa*, *Cirsium palustre*, *Angelica* and *Lotus uliginosus*, and grasses of damp mesotrophic swards like *Holcus lanatus* and *Anthoxanthum*. However, although the soils beneath the *Molinia-Potentilla* mire are frequently just as saturated as those under the *Juncus-Galium* rush-pasture (indeed, quite often they are wetter), they are typically well aerated, with relatively free movement of water through the upper horizons, whereas the latter vegetation is very much associated with a strong measure of impedence in the profile. As far as the floristics of the two communities are concerned, this difference is probably of major importance in affecting the balance between the abundance of *Molinia* and the Junci. Their proportions vary in a virtually continuous gradation from one vegetation type to the other, and the prominence of each can be influenced by factors other than drainage conditions, but *Molinia* is strongly favoured by a wet but not waterlogged environment (Jefferies 1915, Rutter 1955, Loach 1966), while the Junci thrive on markedly gleyed soils. In fact, judging by the frequently very vigorous growth of *Molinia* in this community, the soil moisture regime seems to be well-nigh ideal for this grass, and its resulting abundance has a marked effect on the frequency and disposition of equally well-suited but less bulky and competitive associates.

In the dry climate of south-east England, such strong representation of *Molinia* in the *Cirsio-Molinietum* is of rather local occurrence, marking out the sloping fringes of topogenous hollows and flood-plain mires, and the margins of springs and flushes, where thin peats and peaty mineral soils are protected from summer drought but kept free from continuous waterlogging. In the wetter west of Britain, however, beyond the 1000 mm annual isohyet (*Climatological Atlas* 1952), where there are at least 140 wet days yr^{-1} (Ratcliffe 1968), such edaphic conditions are met much more widely, so *Molinia*-dominance can be maintained there over a broader range of physiographies and soils. The beginning of this

extension can be seen in south-western stands of the *Cirsio-Molinietum*, which occur a little beyond the usual strict confines of that community, but it is in the *Molinia-Potentilla* mire that this development reaches its full expression.

Throughout western Britain, the community can thus be found over moist slopes in a variety of landscape settings. Many of the more clearly-defined stands continue to mark out tracts of obvious soligenous influence or quickening drainage due to variations in relief or in the permeability of underlying bedrocks or drift, around the heads and sides of moorland streams for example (Jefferies 1915, NCC Devon Heathland Report 1980), or around the margins of valley or basin mires (Ratcliffe & Hattey 1982) or on the periodically-flooded margins of lake sides or small rivers (McVean & Ratcliffe 1962). In such situations, it is usually shallow, well-humified peats that underlie the community, very much as beneath the *Cirsio-Molinietum*, but becoming much more common in the gently-undulating scenery of the Silurian, Ordovician and Carboniferous shales and grits, and their drift mantle, through the upland fringes of south-west England, Wales and southern Scotland. But, with the higher rainfall the *Molinia-Potentilla* mire is often found over open slopes which channel seepage waters in a rather ill-defined fashion and it frequently extends on to somewhat steeper ground, cut into more pervious substrates and carrying brown podzolic soils or even podzols proper, profiles which are free-draining but which have a more or less permanently moist humic top. And, where precipitation has been such as to favour the accumulation of ombrogenous peats, now independent of a ground water-table, the community can extend on to the margins of blanket and raised mires, providing the slope of the ground or flushing ameliorate the stagnant conditions.

However, even in the very wettest situations in which it is found, where the ground between the *Molinia* can be under water for much of the winter and spring, continuing good aeration helps mark off the community from the vegetation of the Caricion nigrae or Oxycocco-Sphagnetea. It is in the *Erica* sub-community that the *Molinia-Potentilla* mire makes its nearest approach to such assemblages, with *Eriophorum angustifolium* *Narthecium*, various Sphagna and some other poor-fen bryophytes marking the floristic transition, but such associates are generally scattered and patchy in their occurrence, even where the shade from the *Molinia* is not too dense, and such species as *Scirpus cespitosus* and *Eriophorum vaginatum* and extensive carpets of peat-building Sphagna are not typical here. At the opposite extreme, where the *Molinia-Potentilla* mire extends on to moist peaty mineral soils, it is usually represented by the *Anthoxanthum* sub-community which lacks the above preferentials and, with its distinctive suite of grasses, come close to damp Nardo-Galion grasslands.

One other important diagnostic feature of the habitat of the western *Molinia* vegetation included here is that the soils are somewhat more base-poor than those of the *Cirsio-Molinietum*. Both communities can be found over circumneutral profiles, which is part of the reason why Molinietalia herbs flourish so well in each, but, in eastern Britain at least, the *Cirsio-Molinietum* is very much a vegetation type of more base-rich and calcareous fens. To the south-west, where the climate is still warm, but where rainfall is higher, it extends in the *Juncus-Erica* sub-community on to argillaceous bedrocks of somewhat lower pH but, for the most part, through western Britain, where siliceous substrates prevail and the tendency to surface-leaching is high, *Molinia*-dominance is seen in the *Molinia-Potentilla* mire over soils of pH 4–5.5. Such plants as *Carex hostiana*, *C. pulicaris* and *Briza media* which can make a frequent contribution to the *Cirsio-Molinietum* are therefore of very limited occurrence here: even in the *Anthoxanthum* and *Angelica* sub-communities which can mark out soils where the base-richness approaches or exceeds pH 6, yet a mildly calcicolous aspect to the vegetation is exceptional. At the opposite extreme, where the *Anthoxanthum* sub-community extends a little way on to drier podzols or where, more commonly, the *Erica* sub-community is found in flushes with more markedly base-poor waters, of pH well below 4, the general appearance of the associated flora is quite strongly calcifuge. Even where the substrate is nearer to neutral in its reaction, the acidophile character of the vegetation can be pronounced if plants are able to get a hold in the leached tops of the *Molinia* tussocks.

The irrigation which often marks off the habitat of the *Molinia-Potentilla* mire from the grassland, heath or bog which surrounds it probably also brings modest nutrient enrichment and, compared with the Caricion nigrae vegetation, too, which can occupy similar flushes and water-tracks, the waters and peats here are less impoverished. Among wet heaths in Hampshire, for example, Loach (1966, 1968*a*, *b*) found that, compared with *Calluna* and *Erica tetralix*, *Molinia* occupied the soils with the greatest nutrient supply and the lowest carbon:nitrogen ratio. Nutrients were concentrated in the surface horizons, perhaps because of the activity of the deep-rooted grass itself (e.g. Jefferies 1915) and seasonal turnover was rapid. On the Lizard, Hopkins (1983) also noted that increase of *Molinia* among *Erica-Schoenus* heath was related to larger amounts of potassium and phosphorus than usual.

In this respect, the quite common coupling of *Molinia* with *Myrica* here is noteworthy for this shrub has symbiotic root-nodule bacteria which can annually fix nitrogen equivalent to roughly half that in the current year's shoots (Sprent *et al.* 1978). The additional shade

of the *Myrica*, together with the luxuriant cover of the *Molinia* encouraged by enrichment, can make such vegetation particularly species-poor.

Although edaphic conditions in the community seem generally very favourable for the vigour of *Molinia* and though, once established in abundance, the grass may be able to maintain itself as a well-entrenched dominant, treatments probably play some part in its widespread ascendancy and persistence over suitable soils in western Britain, and are perhaps largely responsible for the development of the *Molinia-Potentilla* mire over ground that would naturally carry some other kind of mire or wet-heath vegetation, quite a common occurrence. Burning and grazing have probably been of especial importance since these can both leave *Molinia* largely unaffected while destroying or keeping in check less robust perennial associates or more vulnerable palatable plants, such as the tall hemicryptophyte dicotyledons and the sub-shrubs. Such effects reinforce the impoverished character of the community in many cases and help maintain it against colonisation by ericoids, gorse and probably also, in suitable situations, by *Myrica* which, though not quickly eliminated by such treatments, grows with reduced stature and is perhaps eventually lost (McVean & Ratcliffe 1962). And ultimately, burning and grazing probably hold many stands as plagioclimax vegetation when they would otherwise be directly invaded by trees and progress to woodland (see below). Draining, too, may be important in the establishment of the community from mire vegetation of previously waterlogged peats, either on the fringes of blanket mire or over the rand of raised mires, where it can augment the effects of a drying climate on the lowering of the water-table (Godwin & Conway 1939). In such circumstances, the survival of the *Molinia-Potentilla* mire may be more precarious, further drainage precipitating the degeneration of the *Molinia* and colonisation by trees. The potentially favourable soil conditions produced by surface drying and peat oxidation means that the ground under the community has often attracted the attention of foresters and many stands have been drained and ploughed for the establishment of conifers.

Within the community, burning and, particularly, grazing are of some importance in maintaining the floristic differences seen in the *Anthoxanthum* and *Angelica* sub-communities. The former, with its more mixed grassy herbage is typically found where *Molinia-Potentilla* mire on drier ground has been open to grazing. Continuing predation by stock can keep such swards fairly short and diverse, though where grazing is relaxed the more robust grasses can grow up with the *Molinia* to produce a rank, more species-poor sward retaining this general floristic character. Freedom from grazing on somewhat moister ground favours the development of the *Angelica* sub-community with its more prominent contingent of tall dicotyledons among the *Molinia* tussocks, though some of these will persist as non-flowering rosettes with the introduction of stock.

Zonation and succession

The *Molinia-Potentilla* mire is a common and widespread element in the heath and moorland scenery of western Britain, occurring with other mire communities, heaths and grasslands in zonations and mosaics under the primary influence of edaphic conditions, though much affected by treatments. At lower altitudes, improvement has often grossly altered or truncated such sequences, leaving them isolated within intensive agricultural landscapes. On somewhat higher ground, afforestation has often been centred on slopes previously occupied by the community, though many tracts remain in more intact zonations and have extended their cover on to much-burned and grazed ground. Such treatments play an important part in preventing seral progression to scrub woodland.

The simplest patterns are to be seen where the *Molinia-Potentilla* mire marks out tracts of moist but well-aerated ground within sequences from grassland or dry heath to bog over ground which becomes increasingly ill-drained and eventually permanently waterlogged. In the less humid parts of its range, in south Wales and through south-west England, where such zonations are strongly dependent on topography, it is thus commonly found on the slopes around valley bogs, its exact position in the sequence of vegetation types depending on the conformation of the ground and the way in which seepage is affected by the permeability of the underlying deposits. Generally speaking, it occurs in such zonations at or around the level of the wet heath, interposed between the drier heaths on the podzols of the interfluves and the bog vegetation of the waterlogged peats in the valley bottoms. Where the ground drops away fairly quickly from among the former, gathering substantial amounts of seepage and channelling it down the slope, then there can be a direct switch to the *Molinia-Potentilla* mire over the flushed zones. In south-west Britain, *Molinia* finds strong representation among communities like the *Ulex minor-Agrostis* and *Ulex gallii-Agrostis* heaths, but transitions from these vegetation types are usually fairly well marked in a decisive shift in dominance to the grass, with a much reduced cover of subshrubs. Quite often, however, where there are extensive tracts of flatter and ill-drained ground around the drier heaths, the *Ericetum tetralicis* wet heath supervenes between them and the *Molinia-Potentilla* mire, and then boundaries can be much less well defined, particularly if there is some extension of the soligenous influence back into the wet heath. Then, the *Angelica* sub-community

of the *Molinia-Potentilla* mire can grade imperceptibly to the *Succisa-Carex* sub-community of the *Ericetum*, taller herbs, clumps of rushes and an over-canopy of *Myrica* sometimes providing a strong visual continuity throughout and masking the gradual move in dominance from the grass to a more mixed cover of *Molinia* and ericoids.

Downslope, where the levelling of the ground in such terrain leads to permanent waterlogging and the accumulation of deeper acid peats, the *Molinia-Potentilla* mire is typically replaced by the *Narthecio-Sphagnetum* valley bog. To such vegetation, the *Erica* sub-community can form a gradual transition and *Molinia* itself persists with high frequency into the bog, but it becomes non-tussocky and grows ever more weakly in the stagnant conditions and the *Sphagnum* cover, dotted with abundant *Narthecium* and *Drosera rotundifolia*, extends to form a luxuriant carpet. In many larger valley mires of this kind, there is also a longitudinal zonation between these two communities, the *Narthecio-Sphagnetum* being concentrated towards the stagnant head of the valley and giving way downstream, with increasing channelling of the waters, to the *Molinia-Potentilla* mire, though sometimes persisting over a marginal zone of stagnation between it and the surrounding *Ericetum tetralicis*.

Further complexities arise in such patterns where there is some variation in the base-richness of the seepage waters because of geological heterogeneity. A slight rise in pH within such flushed ground in south-west Britain is generally marked by the occurrence of the *Cirsio-Molinietum* which, in its distinctive *Juncus-Erica* sub-community, comes very close in its composition to the *Molinia-Potentilla* mire and which can occur adjacent to it or replace it in neighbouring, but more base-rich, seepage zones. *Schoenus* can figure patchily in such situations and sometimes thickens its cover in strongly irrigated mires as the dominant in the distinctive *Schoenus-Narthecium* community or, where the waters are more base-rich, in vegetation resembling the *Schoenetum* though usually without *J. subnodulosus*.

Sequences of this kind, occasionally complete, often in part and sometimes disposed in complex mosaics over gently undulating terrain with complex hydrology, are a very characteristic feature of the heathlands of south-west Britain. The *Molinia-Potentilla* mire can be seen, represented to a greater or lesser degree, where such patterns have developed over the Eocene clays, sands and gravels of the New Forest and Dorset (Ratcliffe 1977), the Triassic deposits of the Devon Pebble-Bed Commons (well shown in Ivimey-Cook *et al.* 1975), the Carboniferous rocks of north-west Devon (NCC Devon Heathland Report 1980) and Gower, and on Silurian and Ordovician rocks, variously smeared with drift, through west Wales (Meade 1981). And analogous, though rather striking, zonations are to be found over the serpentine and gabbro of the Lizard (Hopkins 1983), where the community occurs among dry and wet heaths in both of which *Erica vagans* figures prominently.

Northwards in this western part of Britain, the basic pattern of vegetation types in such sequences is pre-served but, with the increasing rainfall, there is an extension of the wetter components of the zonation out of the narrow topographic confines characteristic of the south-west, and a switch to the sub-montane analogues of many of the communities. Even on Gower, much of the wet heath begins to resemble the *Scirpus-Erica* community, which replaces the *Ericetum tetralicis* in north-west Britain and which mantles many of the thinner peat soils through the upland fringes, forming an intergrade between a variety of sub-montane heaths and grasslands on podzolic profiles to blanket bog, usually the *Scirpus-Eriophorum* mire, on deep ombroge-nous peats. Typically, in this kind of landscape, the *Molinia-Potentilla* mire marks out water-tracks running through this zonation, channelling seepage from the bog or wet heath and carrying it downslope through the drier heaths and grasslands, or gathering it from shedding slopes above the bog and debouching it on to its surface (e.g. McVean & Ratcliffe 1962, Edgell 1969). Again, though *Molinia* occurs frequently in the bog and wet-heath vegetation in such zonations, its abundance usually increases very markedly over the soligenous zones, so the boundary of the community is often clear. However, as before, where seepage extends back into the wet heath, transitions can be less well defined, with the *Molinia-Potentilla* mire grading to the *Carex* sub-community of the *Scirpus-Erica* wet heath: as in south-ern valley bogs, *Myrica* can form a canopy over both these vegetation types. And, there can be analogous complications along the length of seepage tracts, with the *Carex echinata-Sphagnum* mire and the *Juncus-Galium* rush-pasture replacing the *Molinia-Potentilla* mire according to the base-poverty of the soils and flushing waters and the extent of waterlogging as the ground levels and falls away over the hillsides.

Similar kinds of vegetation, though disposed in rather different fashion, can be seen where the *Molinia-Poten-tilla* mire occurs on raised bogs in western Britain. Here, it again marks out the better-drained tracts of the mire surface, typically dominating the rand which surrounds the active plane, itself generally carrying some form of the *Erica-Sphagnum papillosum* mire, sometimes fringed by a zone of *Scirpus-Erica* wet heath, a pattern seen in classic form at Cors Goch glan Teifi (Godwin & Conway 1939) and on a smaller scale at Malham (Proctor 1974, Adam *et al.* 1975). Where well-aerated conditions are maintained in the marginal lagg of such bogs, the

Molinia-Potentilla mire can extend its cover into this, but more consistent waterlogging there is often marked by the occurrence of some kind of *Carex echinata-Sphagnum* mire or rush-pasture.

Patterns of this kind are often further complicated by treatments, particularly by burning and grazing, which have been a traditional element in the management of both the lowland heath and the upland fringes in which this community is found. Practised with care, these treatments, either alone or in concert, do little damage to the *Molinia-Potentilla* mire: indeed, they probably often ultimately prevent seral progression and, on moister soils, can reinforce the uncompromising dominance of *Molinia* by helping eliminate its associates or competi-tors. And it is this latter effect that is seen most fre-quently in the kinds of zonations discussed above because, where the surrounding wet heaths are subject to burning and then grazed, they show a strong tendency to develop into the *Molinia-Potentilla* mire, extending its cover outside what would be its normal edaphic bounds. Such a succession can extend on to the deeper peats of the bog themselves and is especially likely where artificial drainage has sharpened up the loss of water from the mire plane. This was already visible at Cors Goch glan Teifi at the time of Godwin & Conway's (1939) study, markedly enhancing what was perhaps a natural tendency with the drying climate, and it is now very evident on, for example, the raised and blanket mires surveyed through Strathclyde by Bignal & Curtis (1981). It can be seen, too, where longitudinal drains have been put through some of the valley mires in the New Forest.

Over drier ground, where the *Molinia* tends to be not quite so abundant and not so strongly hummocky, grazing can perhaps effect more obvious changes in the community itself. The *Anthoxanthum* sub-community, for example, characteristic of grazed sites, has much in common with moist Nardo-Galion swards and conti-nued pasturing could perhaps effect a transition to that kind of grassland and, combined with draining and the application of fertiliser, convert the *Molinia-Potentilla* mire into a Cynosurion sward. However, where there is a tendency towards poaching of the ground here, particu-larly likely where moist mineral soils are heavily grazed with cattle, there is a strong likelihood of rushes spread-ing with the eventual development of the *Juncus-Galium* rush-pasture. Mosaics of these two vegetation types are very common on some of the heavily-grazed Gower commons on soils derived from Carboniferous shales.

The common occurrence of scattered shrubs and trees in association with the *Molinia-Potentilla* mire, particu-larly on lowland commons where seed-parents are often close at hand, suggests that, where treatments are relaxed and where the herbage has not grown too rank, succession to scrub and woodland could be quite rapid. *Salix cinerea* and *Betula pubescens* are the commonest woody species found in such situations though, further north, *S. aurita* becomes more frequent (McVean & Ratcliffe 1962, Prentice & Prentice 1975). The likely development in such seres would be the *Betula-Molinia* woodland, in the more open flushed stands of which many of the characteristic herbs here survive.

The potential suitability for forestry of the soils under the *Molinia-Potentilla* mire, often better-drained and somewhat richer than the surrounding ground, means that tracts of the community have often been replaced, in whole or part, by coniferous plantations. This is particularly widespread in the upland fringes of the north-west where such forests have been established, often with little care for the natural configuration of the ground, cutting across sequences of wet heath and mire, but it can also be seen on certain lowland heaths, notably over some New Forest zonations. Elsewhere in the lowlands, other tracts of the community, together with neighbouring vegetation, have been lost to agricultural improvement or left as isolated fragments in intensively pastoral landscapes.

Distribution

The community occurs throughout western Britain, and is especially frequent in south-west England, Wales and southern Scotland. The *Erica* sub-community has been recorded most widely and its stands can be extensive; the facies with *Myrica* is local, though it can be very common, as in the New Forest and in parts of Wales and the west Highlands. The *Anthoxanthum* sub-community occurs scattered throughout the range but is particularly well represented in Wales. The *Angelica* sub-community is more concentrated in south-west England and south and west Wales.

Affinities

Defining the *Molinia-Potentilla* mire from among the variety of British vegetation types in which *Molinia* figures prominently is a troublesome job. Separation from the *Cirsio-Molinietum* is the first problem, though an easier one because, although the two communities come very close in their floristics and habitat in the south-west, and have both been described as Molinieta (Ivimey-Cook *et al.* 1975), some fairly consistent differences can be seen throughout. In moving from the warmer and drier south-east of Britain, where suitably moist but aerated soils are of local occurrence and often tending towards base-richness, to the cooler and wetter west, where they are more widespread but often rather acidic, the shift from the one community to the other is marked by the loss of species such as *Cirsium dissectum*, *Carex hostiana*, *C. pulicaris* and *Briza media*, together with Caricion davallianae and Mesobromion occasion-

als, and the more frequent occurrence of plants of Ericion wet heaths, Caricion nigrae poor fens and Nardo-Galion grasslands in what are often larger stands within a broader landscape context.

The change in climatic and soil conditions between these two kinds of *Molinia* vegetation is also accompanied by a switch from *Juncus subnodulosus* to *J. acutiflorus* as the commonest associated rush. And just as there can be problems in the east of Britain separating the *Cirsio-Molinietum* from Calthion fen-meadow, so there is a continuous gradation in the west between the *Molinia-Potentilla* mire and the *Juncus-Galium* rush-pasture. The two vegetation types have many associates in common and are often found in close proximity on heaths and around the upland fringes but, although *Molinia* and *J. acutiflorus* each transgress far into swards dominated by the other, the balance between them is a reasonably good guide to the separation of the communities, strongly reflecting the drainage conditions favoured by each. The frequent association of *Molinia* with *J. acutiflorus* in the *Molinia-Potentilla* mire has sometimes led to the community being termed a *Junco-Molinietum* (Hill & Evans 1978, Ratcliffe & Hattey 1982), though, as applied by O'Sullivan (1982), who originally coined the epithet (Tüxen & O'Sullivan 1964 in O'Sullivan 1968*a*), this vegetation type includes mainly rush-dominated swards.

Reliance on the dominance of *Molinia* to help delimit the *Molinia-Potentilla* mire brings problems of its own because, in the more oceanic parts of Britain, this grass becomes increasingly frequent and abundant in a wide variety of vegetation types which thus show floristic convergence into impoverished kinds of herbage whose most obvious shared characteristic is simply *Molinia*-dominance. Although the *Molinia-Potentilla* mire does take in much of the species-poor *Molinia* vegetation of western Britain, including some which, under the combined influence of climate and various treatments, has been derived from other communities, its bounds are drawn more tightly and precisely than those of most of the Molinieta and '*Molinia* grasslands' erected to include such a compendious assemblage. Such communities, gathered together in Tansley (1911, 1939), often took in almost any kind of vegetation in which *Molinia* played a leading role, irrespective of the associates represented with it and, in some later studies too, fairly diverse vegetation types are gathered together under such general headings (Edgell 1969, Meek 1976, Meade 1981, Ratcliffe & Hattey 1982). In this scheme, as on a smaller scale in McVean & Ratcliffe (1962), some *Molinia*-dominated vegetation is retained within such communities as the *Scirpus-Erica* wet heath and, towards the east, the *Ericetum tetralicis*, with the *Molinia-Potentilla* mire being centred on assemblages in which such plants as *Succisa*, *Lotus uliginosus*, *Cirsium palustre* and *Ange-*

lica play a consistent if sometimes sparse role. Despite the still fairly broad character of the community, with its clear floristic links with Ericion wet heath, Caricion nigrae poor fens and Nardo-Galion grasslands, this group of plants, and the absence of more than locally important *Sphagnum* carpets, ericoid canopies and small-sedge or short calcifugous swards, provides a reasonably good definition and can even help integrate such superficially distinct assemblages as some *Myrica* vegetation (McVean & Ratcliffe 1962, Adam *et al.* 1977, Birse 1980, Wheeler 1980c).

At a higher phytosociological level, where precise distinctions have been the subject of long debate (e.g.

Duvigneaud 1949, Westhoff & den Held 1969), the status of this community is more difficult to assess. Although it is somewhat paradoxical that the vegetation in which *Molinia* is most generally abundant in Britain is the kind which often least clearly belongs to the Molinietalia, the preferential occurrence of the assemblage noted above helps locate this impoverished community in that order. And, within it, it is probably best placed, with the *Cirsio-Molinietum*, in the Junco-Molinion. But, certainly, looking towards Europe from western Britain, the perspective on these vegetation types is very different from that which has grown up on the Continent.

Floristic table M25

	a	b	c	25
Molinia caerulea	V (4–10)	V (2–9)	V (6–10)	V (2–10)
Potentilla erecta	IV (1–4)	V (2–6)	V (1–5)	V (1–6)
Erica tetralix	IV (1–7)	I (2–5)	I (4)	II (1–7)
Eriophorum angustifolium	III (1–5)		I (3)	I (1–5)
Myrica gale	II (2–9)	I (8)	I (8)	II (2–9)
Sphagnum recurvum	II (1–9)		I (4)	I (1–9)
Narthecium ossifragum	II (1–7)		I (5)	I (1–7)
Sphagnum auriculatum	II (1–6)	I (2–4)		I (1–6)
Calypogeia fissa	II (1–3)	I (1)	I (4)	I (1–3)
Polytrichum commune	II (1–7)			I (1–7)
Aulacomnium palustre	II (1–5)			I (1–5)
Hypnum jutlandicum	II (1–9)			I (1–9)
Sphagnum papillosum	II (2–6)			I (2–6)
Sphagnum palustre	I (1–6)			I (1–6)
Sphagnum capillifolium	I (1–6)			I (1–6)
Vaccinium oxycoccos	I (1–4)			I (1–4)
Juncus subnodulosus	I (2–5)			I (2–5)
Anthoxanthum odoratum	I (2–4)	IV (1–5)	I (1–3)	II (1–5)
Holcus lanatus	I (1–6)	III (1–6)	II (1–4)	II (1–6)
Agrostis capillaris		III (1–5)		II (1–5)
Festuca rubra	I (2–4)	II (2–5)	I (1–4)	I (1–5)
Luzula multiflora	I (1–3)	II (1–3)	I (2)	I (1–3)
Viola palustris	I (2–4)	II (1–5)	I (1–4)	I (1–5)
Rumex acetosa		II (3–4)	I (3–4)	I (3–4)
Danthonia decumbens		II (1–3)	I (1–3)	I (1–3)
Pseudoscleropodium purum		II (1–4)		I (1–4)
Serratula tinctoria		II (1–2)		I (1–2)
Luzula campestris		II (1–4)		I (1–4)
Carex flacca		I (2–5)		I (2–5)
Agrostis curtisii		I (1–6)		I (1–6)
Lobelia urens		I (1–3)		I (1–3)
Angelica sylvestris		I (1–2)	V (1–6)	II (1–6)

Floristic table M25 *(cont.)*

	a	b	c	25
Cirsium palustre	I (1–4)	II (1–4)	IV (1–4)	II (1–4)
Juncus effusus	I (1–3)	II (1–4)	III (3–5)	II (1–5)
Galium palustre			III (1–4)	II (1–4)
Epilobium palustre			II (1–3)	I (1–3)
Mentha aquatica			II (1–4)	I (1–4)
Eupatorium cannabinum			II (1–5)	I (1–5)
Filipendula ulmaria			II (3–5)	I (3–5)
Scutellaria minor			II (1–3)	I (1–3)
Pulicaria dysenterica			I (4)	I (4)
Cardamine pratensis			I (1–3)	I (1–3)
Schoenus nigricans			I (5–7)	I (5–7)
Calliergon giganteum			I (3–4)	I (3–4)
Equisetum palustre			I (1–3)	I (1–3)
Osmunda regalis			I (4–6)	I (4–6)
Carum verticillatum			I (1–3)	I (1–3)
Dryopteris carthusiana			I (3–6)	I (3–6)
Juncus acutiflorus	III (1–6)	III (3–7)	III (1–5)	III (1–6)
Succisa pratensis	II (1–4)	III (1–6)	III (1–5)	III (1–6)
Lotus uliginosus	I (2–4)	III (1–4)	III (1–4)	III (1–4)
Ulex gallii	II (1–4)	II (4–5)	I (1–4)	II (1–5)
Carex echinata	II (1–4)	II (2–3)	I (3)	II (1–4)
Carex panicea	II (1–6)	II (3)	I (3)	II (1–6)
Calluna vulgaris	II (1–7)	II (2–6)		II (1–7)
Agrostis canina	I (1–7)	II (2–5)	II (1–6)	II (1–7)
Agrostis stolonifera		II (1–4)	II (3–5)	II (1–5)
Carex nigra	I (1–5)	I (1–3)	I (2)	I (1–5)
Rhytidiadelphus squarrosus	I (1–4)	I (2–4)	I (3)	I (1–4)
Hydrocotyle vulgaris	I (3–6)	I (3–7)	I (3)	I (3–7)
Dactylorhiza maculata	I (1–3)	I (1–2)	I (2)	I (1–3)
Festuca ovina	I (1–3)	I (1–4)	I (3)	I (1–4)
Pleurozium schreberi	I (1–4)	I (3)		I (1–4)
Juncus articulatus	I (2–3)	I (4–5)		I (2–5)
Nardus stricta	I (1–4)	I (2)		I (1–4)
Pohlia nutans	I (1–4)	I (3)		I (1–4)
Ranunculus acris	I (2–3)	I (3)		I (2–3)
Sphagnum subnitens	I (1–6)		I (3)	I (1–6)
Cephalozia bicuspidata	I (1–3)		I (3)	I (1–3)
Phragmites australis	I (2–5)		I (6)	I (2–6)
Menyanthes trifoliata	I (1–4)		I (4)	I (1–4)
Lophozia ventricosa	I (1–4)		I (1)	I (1–4)
Ranunculus flammula		I (3)	I (3)	I (3)
Calliergon cuspidatum		I (1–7)	I (4–5)	I (1–7)
Ulex europaeus		I (2–3)	I (1–6)	I (1–6)
Rubus fruticosus agg.		I (2–4)	I (1–4)	I (1–4)
Eurhynchium praelongum		I (2)	I (3)	I (2–3)

	a	b	c	25
Centaurea nigra		I (2–3)	I (1–5)	I (1–5)
Valeriana officinalis		I (2–4)	I (3–5)	I (2–5)
Hypericum pulchrum		I (3)	I (1–3)	I (1–3)
Deschampsia cespitosa		I (2–4)	I (5)	I (2–5)
Salix cinerea		I (3)	I (2–6)	I (2–6)
Achillea ptarmica		I (4)	I (4)	I (4)
Hypericum tetrapterum		I (2)	I (1–3)	I (1–3)
Lophocolea bidentata s.l.		I (3)	I (2–3)	I (2–3)
Number of samples	59	159	69	287

a   *Erica tetralix* sub-community
b   *Anthoxanthum odoratum* sub-community
c   *Angelica sylvestris* sub-community
25 *Molinia caerulea-Potentilla erecta* mire (total)

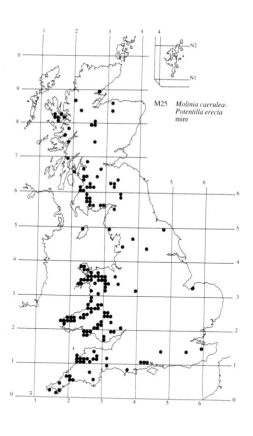

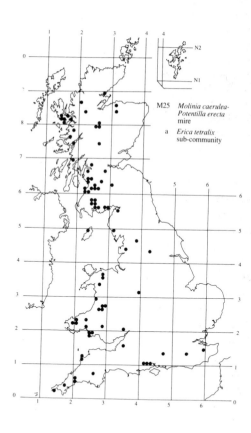

282

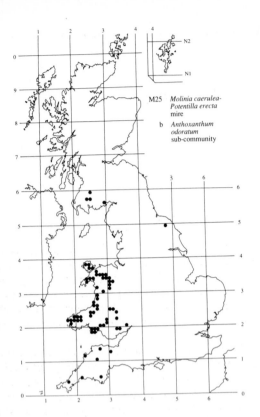

M25   Molinia caerulea-
      Potentilla erecta
      mire

b   Anthoxanthum
      odoratum
      sub-community

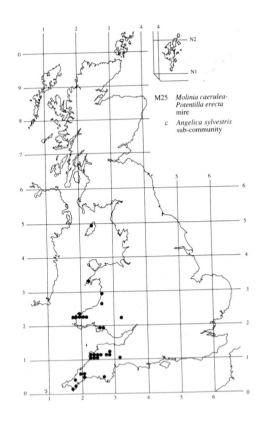

M25   Molinia caerulea-
      Potentilla erecta
      mire

c   Angelica sylvestris
      sub-community

M26
Molinia caerulea-Crepis paludosa mire

Synonymy

Mixed fen Holdgate 1955*b* *p.p.*; *Carex nigra-Sanguisorba officinalis* nodum Proctor 1974 *p.p.*; *Carex nigra-Sanguisorba officinalis* community Wheeler 1980*c*; *Crepido-Juncetum* Jones 1984.

Constant species

Carex nigra, C. panicea, Crepis paludosa, Equisetum palustre, Filipendula ulmaria, Molinia caerulea, Potentilla erecta, Ranunculus acris, Succisa pratensis, Valeriana dioica, Calliergon cuspidatum.

Rare species

Primula farinosa.

Physiognomy

The *Molinia caerulea-Crepis paludosa* mire is a striking vegetation type, well-defined by a substantial block of constants and frequent companions, but showing quite considerable variation in its associated flora and embracing stands which, at one extreme, resemble swamp and, towards the other, are of a rank grassy character. *Molinia caerulea* is almost always present and it is often abundant, being the commonest dominant overall and frequently growing in a strongly-tussocky form, creating an irregular canopy up to 50 cm or so tall, and exerting a marked influence on the character and distribution of the associates. Among these are some other plants which can make a substantial contribution to the herbage. *Carex nigra*, for example, is a constant and it quite frequently occurs as prominent tufts or tussocks (var. *subcaespitosa*), sometimes rivalling or exceeding *Molinia* in its cover, particularly in the *Sanguisorba* sub-community (Proctor 1974). *C. panicea* can also be fairly abundant and *C. pulicaris* occurs commonly, too, though not usually as dense tufts. Then, in stands which are transitional to swamp, there can be some *C. appropinquata*, as at Malham Tarn in North Yorkshire (Proctor 1974), or *C. rostrata*, as at Sunbiggin Tarn in Cumbria (Holdgate 1955*b*). In the *Festuca* sub-

community, by contrast, it is taller rushes and grasses which, together with *Molinia*, provide the bulk of the cover, *Juncus acutiflorus* forming locally dense patches, *J. conglomeratus* occurring as scattered tussocks. *J. articulatus* is also occasional throughout, though it often grows as a semi-decumbent plant and is rarely abundant.

The other important structural element typical of the community as a whole comprises hemicryptophyte dicotyledons which are capable of moderate to tall growth among the *Molinia*, sedges and rushes, but many of which can also subsist as rosettes in grazed stands. Most frequent among this group are *Succisa pratensis, Filipendula ulmaria, Valeriana dioica, Cirsium palustre, Caltha palustris* and, adding a particularly distinct regional look, the Continental Northern *Crepis paludosa* and Northern Montane *Trollius europaeus*. Also common, though preferential to different sub-communities, are *Sanguisorba officinalis* and *Angelica sylvestris, Centaurea nigra, Leontodon hispidus, Geum rivale* and *Lychnis flos-cuculi*. When mixtures of these species are present in some numbers, the vegetation can present a splendid sight by mid-summer, with the inflorescences raised above the herbage and set off against the dull-green and russet background.

Less conspicuous, but also very frequent in the community, is *Equisetum palustre*, and some smaller herbs are quite common too, but beneath the thick herbage of the dominants and tall herbs, these are rather few in number and typically of low cover. The most frequent are *Potentilla erecta, Ranunculus acris* and *Anemone nemorosa*, with *Ranunculus auricomus* occasional, and a variety of Cynosurion species more common in grazed stands. The *Molinia-Crepis* mire also provides a locus for *Primula farinosa*, another Northern Montane species. Even in its richest expression, however, this vegetation does not have the kinds of varied calcicolous turf included in communities like the *Pinguiculo-Caricetum*.

In marked contrast to those swards, too, bryophytes

are only exceptionally prominent here. The shade-toler-ant *Calliergon cuspidatum* is the most frequent, with *Lophocolea bidentata s.l.*, *Thuidium tamariscinum*, *Campylium stellatum* and *Pseudoscleropodium purum* occasional to common, *Ctenidium molluscum*, *Plagiochila asplenoides*, *Campylium elodes*, *Climacium dendroides* and *Plagiomnium undulatum* variously represented in the two sub-communities.

Sub-communities

Sanguisorba officinalis **sub-community:** Mixed fen Holdgate 1955*b p.p.*; *Carex nigra-Sanguisorba officinalis* nodum   Proctor 1975 *p.p.*; *Carex nigra-Sanguisorba officinalis* community   Wheeler 1980*c*. Although this kind of *Molinia-Crepis* mire is generally less species-rich than the *Festuca* sub-community, it is often much more distinctive to the eye. *Molinia* and *C. nigra* are usually the most abundant plants, one or both of them dominating in what is typically a dense and quite tall cover, though, at some sites, *C. rostrata* or *C. appropinquata* have been recorded in some abundance in this kind of vegetation. At Sunbiggin Tarn, for example (Holdgate 1955*b*), the prominence of the former, together with frequent *Potentilla palustris* and *Menyanthes trifoliata*, plants which are normally rather scarce in the community, gives the vegetation the appearance of a swampy poor fen; while, at Malham (Proctor 1974), the robust tussocks of the latter are locally dominant in transitions to Phragmitetalia fen. *Phragmites australis* itself is very occasionally present, though typically as sparse shoots.

Smaller sedges can also occur, with *C. panicea* and *C. pulicaris* common, *C. flacca* occasional and, in some stands, *C. disticha* and *C. hostiana* have been recorded. *Juncus articulatus* is sometimes found, too, though not at high cover and other rushes are rare. Apart from sparse *Agrostis stolonifera*, grasses are likewise poorly represented and, among the smaller dicotyledons, only *Potentilla erecta*, *Ranunculus acris* and *Anemone nemorosa* occur with any frequency.

Much more prominent are the taller herbs among which, as well as the community species, there is very commonly some *Sanguisorba officinalis* and *Angelica sylvestris* and also, coming close in this vegetation to its northern limit in Britain, the Continental *Serratula tinctoria*. *Galium palustre* and *G. uliginosum* can also occur trailing among the herbage.

Bryophytes show a rather patchy development, but are generally rather more varied and abundant than in the *Festuca* sub-community. *Calliergon cuspidatum*, *Lophocolea bidentata s.l.*, *Thuidium tamariscinum* and *Campylium stellatum* are frequent with, preferential here, *Ctenidium molluscum*, *Plagiochila asplenoides*, *Campylium elodes* and *Aulacomnium palustre*, usually occurring as small pads or tufts over damp litter and soil.

Festuca rubra **sub-community:** *Crepido-Juncetum* Jones 1984. The general appearance of this kind of *Molinia-Crepis* mire is altogether more grassy. *Molinia* is the usual dominant, forming the bulk of a rough sward, sometimes quite tall and rank, in other cases more shortly-cropped, but along with it there is usually some *Festuca rubra*, *F. ovina*, *Briza media*, *Anthoxanthum odoratum*, *Holcus lanatus* and *Deschampsia cespitosa*, none of them consistently abundant, but each able to attain some measure of local prominence. Sedges are common, too, though these are exclusively the smaller species and generally not so strikingly abundant as in the *Sanguisorba* sub-community. *C. nigra* in particular often plays a subordinate role here, though, along with *C. panicea* it can show moderately high cover. *C. flacca* and *C. pulicaris* also occur quite frequently, generally as scattered tufts. Then, there are quite commonly some rushes: as before, *J. articulatus* can be found here, though strongly preferential are the more conspicuous *J. acutiflorus* and *J. conglomeratus*, the former especially showing local abundance and sometimes being co-dominant with *Molinia*.

Taller dicotyledons remain frequent in this sub-community and though *Sanguisorba*, *Angelica* and *Serratula* are all scarce, *Geum rivale*, *Centaurea nigra* and *Leontodon hispidus* are rather more common than in the *Sanguisorba* type, and, in shorter swards, these and many of the community dicotyledons occur as non-flowering rosettes. In such stands, they are often accompanied by *Prunella vulgaris*, *Plantago lanceolata* and *Trifolium repens*. *Lathyrus pratensis* can be conspicuous in somewhat ranker herbage and there is occasionally some *Rhinanthus minor* and *Lychnis flos-cuculi*.

Bryophytes are often poorly represented in the dense turf here, but *Calliergon cuspidatum*, *Pseudoscleropodium purum* and *Lophocolea bidentata s.l.* remain frequent and are sometimes accompanied by *Plagiomnium undulatum*, *Hylocomium splendens* and *Climacium dendroides*.

Habitat

The *Molinia-Crepis* mire is a very local community of moist, moderately base-rich and calcareous peats and peaty mineral soils in the sub-montane northern Pennines. It can be found as an apparently stable component of topogenous sequences around open waters and mires, but it also occurs in soligenous situations on flushed slopes, where it is more often subject to some grazing and able to persist in traditionally-treated pasture. Draining and sward improvement have probably destroyed many smaller stands.

This community represents a northern and altitudinal extreme in Britain of the richer kind of *Molinia*-tall herb vegetation. Beyond the warmer, southern lowlands of England and Wales, the *Cirsio-Molinietum* begins to lose its floristic integrity, with the disappearance of cold-sensitive species such as *Cirsium dissectum*, *Dactylorhiza majalis* ssp. *praetermissa* and other more exclusively south-western oceanic plants. And, even though many of the frequent associates of that fen-meadow remain common well into Scotland, the edaphic conditions and treatments necessary for their occurrence together in this general kind of vegetation coincide only very locally. Suitably moist, circumneutral to moderately base-rich, organic soils become decidedly scarce in the northern lowlands of Britain and such unimproved fragments as there are rarely support herbaceous vegetation, more often now neglected scrub or woodland. Where congenial situations do occur in the north of the country, most obviously on the drift-smeared surfaces of the Carboniferous Limestone, they are at sufficiently high altitudes for the climate to produce a distinct floristic response of its own.

Stands of the *Molinia-Crepis* mire are few and far between, but they all occur around the fringes of the north Pennine uplands, between about 250 and 450 m, experiencing a climate which is generally cold, wet and cloudy (Manley 1936, 1942). Annual precipitation is, in fact, not much greater than for south-western stations of the *Cirsio-Molinietum*, being around 1600 mm (*Climatological Atlas* 1952), with 160 wet days yr^{-1} or so (Ratcliffe 1968). But annual accumulated temperatures are sometimes only a half of those in southern Britain, generally less than 800 day-degrees C (Page 1982), with long bitter winters and late springs, brief overcast summers and a stormy end to what is one of the shortest growing seasons in England and Wales (Manley 1940, Pigott 1956a, Smith 1976). Floristically, then, although such general lowland plants as *Molinia*, *Carex panicea* and the Junci can still grow vigorously here, and though the community even provides some northern stations for the Continental *Serratula tinctoria*, it is the frequency of such plants as the Continental Northern *Crepis paludosa* and the Northern Montane *Trollius europaeus* which provides the sharpest phytogeographical distinction from the *Cirsio-Molinietum*. *Sanguisorba officinalis*, too, though not universally common here, and of rather peculiar overall distribution in Britain, is much more characteristic of northern vegetation of this general type than of the southern.

For the rest, however, many of the frequent plants of the *Molinia-Crepis* mire are shared with the *Cirsio-Molinietum* and the prominence among them of Molinietalia species is partly a reflection of the edaphic conditions which both communities favour. As in southern Britain, this kind of vegetation is typical of situations intermediate between swamps and wet mires on the one hand and dry grasslands on the other, where soil moisture is such as to favour the accumulation or maintenance of fen peat, or at least a humic top over a mineral base, but not such as to provide a continually waterlogged edaphic environment. Conditions are thus very favourable for the vigour of plants like *Molinia*, *Succisa*, *Filipendula*, *Equisetum palustre*, *Valeriana dioica*, *Angelica*, *Cirsium palustre*, *Serratula tinctoria* and the Junci as well as the *Molinia-Crepis* preferentials, *Trollius* and *Sanguisorba*, whose general affiliations are also with the Molinietalia, and some other distinctive herbs of the community, such as *Anemone nemorosa* and *Ranunculus auricomus*.

In fact, although the *Molinia-Crepis* mire does stop short of soils which are badly aerated for the whole of the year, it does extend a little further into wet situations than the *Cirsio-Molinietum*. The substrate here is always rather firm but is normally kept at least damp to the surface and, in winter, can be decidedly wet above. Such conditions are best seen in the *Sanguisorba* sub-community which is typically a vegetation type of fen peats in topogenous hollows, being found around the margins of open-water transition mires and waterlogged depressions. Here, there can even be some standing water among the *Molinia* and *Carex* in the winter, and fluctuations of the water-table in such situations probably encourage the development of a strongly tussocky structure in these plants, but these movements are not very substantial and the vegetation is probably never deeply flooded. However, it is in this sub-community that the *Molinia-Crepis* mire approaches most closely the kinds of vegetation seen on sloppy peats or as floating rafts over shallow open water, with plants such as *Carex appropinquata*, *C. rostrata*, *Phragmites*, *Potentilla palustris*, *Menyanthes trifoliata* and *Equisetum fluviatile* weakly preferential, together with the fen bedstraws *Galium palustre* and *G. uliginosum*. More open damp surfaces here also offer an opportunity for the development of bryophytes, though this is still very patchy under the dense and tall herbage of the vascular cover.

The *Molinia-Crepis* mire can also be found on quite wet soils where there is flushing over slopes cut into deposits of varying permeability. Typically, however, it occurs on the gently-irrigated fringes of such soligenous areas, where the profiles are some sort of stagnohumic gley, often with a substantial peaty topsoil, but not waterlogged to the surface all the year round. In such situations, the *Festuca* sub-community is the usual form, and this kind of *Molinia-Crepis* mire can extend on to ground that is substantially drier than under the *Sanguisorba* sub-community. Phragmitetalia associates are thus lacking and the vegetation often resembles a rough mesotrophic sward with such plants of damp Molinio-Arrhenatheretea grasslands as *Holcus lanatus*, *Festuca rubra*, *Lathyrus pratensis*, *Deschampsia cespitosa* and *Ranunculus acris* becoming very frequent.

In marginal habitats of this kind, further out of the reach of strong influence by base-rich ground waters, and particularly where there is some more free-draining component to the soil parent material such as coarse-textured drift, the high rainfall can induce some slight surface leaching in the profiles, such that pH under the *Festuca* sub-community sometimes falls below 6. Indeed, although values approaching pH 7 have been recorded under this vegetation, and, beneath the *Sanguisorba* sub-community, are generally above pH 6.5, extreme base-richness and very high calcium contents of substrates and waters are not the rule here. The floristic contrasts with vegetation types of more calcareous fen peats or stagnohumic gleys are well seen at Sunbiggin Tarn (Holdgate 1955*b*), Malham Tarn (Proctor 1974) and in the range of flushed sites described from Teesdale by Jones (1984), in all of which there is a shift from Molinietalia vegetation of the *Molinia-Crepis* type to Caricion davallianae fens with the increase in base-richness. The communities can have much in common in their background floristics and certain more catholic Molinietalia plants are frequent in the Caricion fens, but the general absence of calcicoles, whether vascular plants or bryophytes, within the *Molinia-Crepis* mire is very striking.

One other important edaphic difference is that the soils beneath the *Molinia-Crepis* mire are somewhat more productive than those under Caricion davallianae small-sedge fens, although they are probably not more than moderately rich in major nutrients and any marked increase in trophic state, whether from silt deposition along soligenous tracks or from the application of chemical fertilisers to agricultural swards, results in floristic changes in the vegetation. But potentially quite large amounts of herbage can be produced by the community and, being characteristic of the more accessible parts of mires, it has often been subject to some form of treatment, particularly where it occurs as the *Festuca* sub-community on open slopes. In some places this kind of *Molinia-Crepis* mire developed around flushes within agricultural enclosures is included in a hay-crop harvested under a traditional regime of meadow management, a scarce practice now but one that is still extant in some valleys around the fringes of the north Pennines. Generally, however, and in contrast to the case with the *Cirsio-Molinietum*, at least as it was treated in the past, the *Molinia-Crepis* mire is not now mown (Jones 1984). It is, however, quite often open to some grazing and, being concentrated on drier ground, such treatment accentuates the floristic and structural differences between the two sub-communities. Thus, although potentially tall herbs are still very frequent in the *Festuca* type, some, such as *Sanguisorba*, *Angelica* and *Serratula* are largely extinguished and others, notably *Filipendula*, much reduced in cover. And those which remain constant, or which, like *Geum rivale*, *Leontodon hispidus* and *Centaurea nigra*, show an increase, are plants which can subsist as non-flowering rosettes even in very short turf. Close nibbling also increases the richness of the sward by favouring smaller grasses and dicotyledons which are unable to compete in the taller herbage.

Two other ancillary effects of grazing influence this vegetation. First, trampling, like the close-cropping of the herbage, can help to keep the sward among the tussocks more open and varied, although on heavier wet soils, as around flushes, there is always the risk of poaching and an increase in drainage-impedence, features which favour the kind of patchy spread of rushes seen in the *Festuca* sub-community, and which may precipitate a run-down to Juncion acutiflori vegetation, particularly if cattle are turned out in wet spring weather (see below). Second, there is the effect of dunging, again especially marked where cattle, rather than sheep, are grazed. The enrichment this produces may play some part in the more mesophytic character of the vegetation of the *Festuca* sub-community and might also encourage floristic changes were nutrients to accumulate over many years.

Where the *Molinia-Crepis* mire occurs around open-water transitions, it is less often accessible to grazing and the taller, more luxuriant herbage of the *Sanguisorba* sub-community probably experiences no more than light cropping. The development of a very rank and hummocky cover in such situations probably hinders the invasion of woody plants (Proctor 1974), such that the community acquires a certain stability. On drier ground, however, sudden withdrawal of grazing would probably allow rapid colonisation by woody plants of the shorter, more open sward, such that the vegetation can here be considered a plagioclimax.

Zonation and succession

The *Molinia-Crepis* mire can be found in zonations and mosaics with other kinds of fen vegetation, with grasslands and woodland, disposed according to variations in the height of the water-table, in the base-richness, calcium content and nutrient status of the waters and soils, and in the pattern of treatment. Around open waters, it may be a natural part of primary hydrarch successions with some stability but, on drier ground, grazing probably plays some part in its maintenance. Pastoral mismanagement or land improvement can both destroy this vegetation, and have probably contributed to its present very local distribution.

The simplest kind of sequence in which the community can be seen relates to differences in the water-regime of underlying soils. At both Sunbiggin (Holdgate 1955*b*) and Malham (Proctor 1974), for example, the *Molinia-Crepis* mire can be seen in complex mosaics, but part of the floristic variation is controlled by the degree of waterlogging in the peats. On wetter ground in these sites, the *Sanguisorba* sub-community, already with a modest representation of Phragmitetalia plants, grades to tall-herb fen, typically in these northern basins, of the *Potentillo-Caricetum rostratae*. There, dominance switches to large sedges, notably *C. rostrata* but sometimes also *C. appropinquata* or a tussocky form of *C. nigra*, occasionally with some rushes or an overtopping canopy of *Phragmites*. *Molinia* extends hardly at all into this vegetation and the most common associates are plants like *Potentilla palustris*, *Menyanthes trifoliata* and *Equisetum fluviatile*, but some continuity is maintained through such associates as *Angelica sylvestris*, *Cirsium palustre*, *Caltha palustris* and *Galium palustre*.

In such situations as these, however, where the mire waters are draining from catchments in which calcareous rocks predominate, here Carboniferous Limestone, wetter soils are also often more base-rich and more oligotrophic so that Caricion davallianae vegetation is a characteristic element of the sequences, sometimes replacing the Phragmitetalia fen, or occurring alongside it, where there are fine-scale variations in the quality of the incoming waters. Typically, in sites like Sunbiggin and Malham, and particularly on rather sloppy peats, the community is the *Carex-Calliergon* mire, many of whose plants are shared with the *Potentillo-Caricetum*, but which has a less prominent mesophytic element and a stronger contingent of calcicoles. Continuity with the *Molinia-Crepis* mire can thus be quite substantial, through plants tolerant of the middle range of pH and nutrient status, such as *Molinia* itself, *Carex panicea*, *C. nigra*, *Succisa*, *Angelica*, *Cirsium palustre*, *Valeriana dioica*, *Caltha*, *Equisetum palustre* and *Filipendula*. But with the move on to more base-rich and wetter soils, dominance passes to such sedges as *Carex diandra* with *C. lepidocarpa*, *C. demissa* and *C. dioica* appearing among the associates, and the patchy bryophyte cover of the *Molinia-Crepis* mire is replaced by lush carpets of larger *Calliergon* spp., with *Campylium stellatum*, *Scorpidium scorpioides*, *Drepanocladus revolvens* and *Bryum pseudotriquetrum*.

On firmer, but still very wet and base-rich, peat and peaty mineral soils, the *Molinia-Crepis* mire grades to a different kind of Caricion davallianae fen. In mires like those at Sunbiggin and Malham, such situations are usually found around the margins, where there is flushing from the surrounding slopes through a stagnohumic gley and down on to the fen peat itself and then the *Sanguisorba* sub-community may extend up from the mire surface to form the flush surrounds. Over more open slopes, such a marginal position around this kind of Caricion davallianae flush is very much the characteristic location of the *Festuca* sub-community. Almost always, the Caricion community is the *Pinguiculo-Caricetum dioicae*, a small-sedge fen which, like flush stands of the *Molinia-Crepis* mire, is often kept short by grazing. Structural and floristic similarities between these vegetation types are thus often very considerable: tall herbs are rarely a feature of the *Pinguiculo-Caricetum*, but such plants as *Molinia*, *C. panicea*, *C. nigra*, *Succisa*, *Potentilla erecta*, *Equisetum palustre*, *Prunella vulgaris*, together with *Juncus articulatus* and the grasses *Briza media*, *Festuca ovina* and *Anthoxanthum*, occur commonly in both communities. The calcicole contingent in the *Pinguiculo-Caricetum* is, however, generally more striking than in the *Carex-Calliergon* mire, so close inspection will usually reveal that transitions from the *Molinia-Crepis* mire are marked by a general switch to dominance by mixtures of *C. panicea* with *C. dioica*, *C. lepidocarpa*, *C. hostiana*, *C. flacca* and *C. pulicaris*, a variety of smaller herbs such as *Pinguicula vulgaris*, *Selaginella selaginoides*, *Linum catharticum*, *Parnassia palustris*, and extensive patches of bryophytes like *Bryum pseudotriquetrum*, *Drepanocladus revolvens*, *Campylium stellatum*, *Aneura pinguis* and *Ctenidium molluscum*.

In the opposite direction in such sequences as these, increasing dryness of the ground is often marked by a shift from the *Molinia-Crepis* mire to some kind of Nardo-Galion sward over gently-flushed brown soils developed from the drift cover that typically plasters the surrounding slopes. Through the north Pennines, such vegetation is usually the *Festuca-Agrostis-Thymus* grassland, into which grasses like *Festuca rubra*, *F. ovina*, *Anthoxanthum* and *Briza*, together with *Carex panicea*, *Potentilla erecta*, *Ranunculus acris* and *Prunella vulgaris*, run on with high frequency. Where the drift-cover is thinner and the sward more strongly influenced by the free-draining calcareous bedrock, this is replaced by the *Sesleria-Galium* grassland, with a more marked calcicolous element.

Quite often, transitions of these kinds are further complicated by greater natural variations in the trophic state of the soils. This is clearly seen around the margins of fen complexes like those at Malham where incoming streams winding through the *Molinia-Crepis* mire slacken their flow and drop quantities of rich silt, but it sometimes occurs, too, at the foot of slopes where flushing waters debouch on to the flatter ground below. Then, along the levees or deltas of the deeper mineral material, the community can grade to the *Filipendula-Angelica* mire in which mixtures of tall mesophytic dicotyledons predominate.

Such complexities as these and the way in which even simple transitions can be fragmented into mosaics with variations in water-table and base-status over mire surfaces, can obscure the fact that the *Molinia-Crepis* community may play a part in primary successions around open waters. At Malham, for example, stratigraphy reveals that this kind of vegetation can succeed more calcicolous fen with progressive terrestrialisation (Proctor 1974). Subsequent stages in succession have not been followed, though where the herbage does not become too rank, it is likely that woody plants, notably *Salix* spp. and *Betula pubescens*, could develop fairly quickly into a scrubby canopy. Towards the limit of the spread of the *Molinia-Crepis* mire on to wetter substrates, such a cover would probably be of the *Salix-Carex rostrata* type, the usual woodland of wetter basin peats in northern Britain, where mesotrophic and moderately base-rich conditions prevail. Over drier ground, which might develop eventually within tracts of the above woodland from ombrotrophic nuclei, the *Betula-Molinia* woodland might succeed this. This vegetation type and, over somewhat more base-rich ground, the *Alnus-Fraxinus-Lysimachia* woodland, can also be found over slope flushes around the Pennine fringes, where otherwise the *Molinia-Crepis* mire might be expected, and these too may represent natural successors to it, showing considerable floristic similarities.

In drier situations, grazing undoubtedly plays some part in arresting these seral progressions and some stands may represent the more or less direct result of past woodland clearance and pasturing. In this respect, the presence of plants like *Anemone nemorosa* and *Ranunculus auricomus* in the community is interesting, and the general similarity to the *Anthoxanthum-Geranium* grassland, essentially a mown derivative of the flush woodlands mentioned above, noteworthy. Under moderate and judicious grazing, the *Molinia-Crepis* mire is probably stable but careless sward management or improvement can produce marked changes. With heavy grazing over impeded soils, there is a strong tendency for rushes to become abundant and dunging may help effect a shift towards the *Juncus-Galium* rush-pasture, the widespread Juncion acutiflori vegetation of mesotrophic gleyed soils in north-west Britain. In drier situations, and particularly where there has been some enrichment with manure or chemical fertilisers, the sward may begin to approach a mesotrophic Nardo-Galion grassland or a Cynosurion community and top-sowing or ploughing and re-seeding can fully effect such a transformation over less intractable ground. In some places, rougher terrain has given the *Molinia-Crepis* mire enough protection to ensure its isolated survival but destruction with agricultural improvement has probably been widespread.

Distribution

The community is very local in the north Pennines and Lake District, the *Sanguisorba* sub-community surviving most extensively at Sunbiggin and Malham Tarns, the *Festuca* sub-community occurring in a more scattered distribution through the dales along the upland fringes. Similar vegetation has been reported from some other localities, such as certain of the Whitlaw Mosses in the Southern Uplands (Wheeler 1975), and assiduous searching should produce more stations.

Affinities

Although the *Molinia-Crepis* mire has obvious floristic links with various vegetation types, such as Phragmitetalia tall-herb fens, Caricion davallianae communities, Nardo-Galion grasslands and traditional hay-meadows, its general affinities are clearly with the Molinietalia. This is still true even when the unpublished data of Jones (1984), which forms the basis of the grassier *Festuca* sub-community, are included with that previously described by Holdgate (1955b) and first given systematic phytosociological treatment by Proctor (1974), and which here constitutes the more distinctive *Sanguisorba* sub-community. Compared with other British vegetation of this same general kind, however, the particular relationships of the community are less clear. It is obviously rather different from the *Cirsio-Molinietum* and its sub-montane character makes it difficult to incorporate the community in the Junco-Molinion, essentially an alliance of more oceanic fen-meadows and rush-pastures. If anything, it is more like some of the central European vegetation placed in the Calthion or the Molinion (e.g. Oberdorfer 1957, 1983) and some of the calcareous pastures described from southern Sweden by Regnéll (1980).

Floristic table M26

	a	b	26
Molinia caerulea	V (4–8)	V (1–8)	V (1–8)
Crepis paludosa	V (1–3)	V (1–5)	V (1–5)
Carex nigra	V (2–8)	V (1–6)	V (1–8)
Carex panicea	V (2–6)	V (1–5)	V (1–6)
Valeriana dioica	V (2–5)	V (1–5)	V (1–5)
Succisa pratensis	V (3–4)	IV (1–5)	IV (1–5)
Equisetum palustre	V (1–4)	IV (1–5)	IV (1–5)

Potentilla erecta	IV (1–3)	IV (1–3)	IV (1–3)
Filipendula ulmaria	IV (2–4)	IV (1–3)	IV (1–4)
Calliergon cuspidatum	IV (1–3)	IV (1–5)	IV (1–5)
Ranunculus acris	III (1–2)	V (1–4)	IV (1–4)
Sanguisorba officinalis	V (3–6)	II (1–3)	III (1–6)
Angelica sylvestris	V (1–4)	II (1–4)	III (1–4)
Serratula tinctoria	III (2–4)	I (1–2)	II (1–4)
Galium palustre	III (1–3)		II (1–3)
Ctenidium molluscum	III (1–3)		II (1–3)
Plagiochila asplenoides	III (2–4)		II (2–4)
Campylium elodes	III (1–3)		II (1–3)
Aulacomnium palustre	III (1–3)		I (1–3)
Carex disticha	II (1)		I (1)
Agrostis stolonifera	II (2–3)		I (2–3)
Equisetum fluviatile	II (1–2)		I (1–2)
Galium uliginosum	II (1–2)		I (1–2)
Menyanthes trifoliata	II (2–4)		I (2–4)
Phragmites australis	II (1–2)		I (1–2)
Carex hostiana	I (5)		I (5)
Carex appropinquata	I (8)		I (8)
Potentilla palustris	I (1)		I (1)
Festuca rubra	I (2)	V (1–5)	III (1–5)
Briza media		V (1–4)	III (1–4)
Holcus lanatus		V (1–4)	III (1–4)
Lathyrus pratensis		IV (1–4)	III (1–4)
Deschampsia cespitosa		IV (1–2)	III (1–2)
Geum rivale	I (2)	IV (1–4)	III (1–4)
Anthoxanthum odoratum	I (2)	IV (1–3)	III (1–3)
Juncus acutiflorus		IV (1–5)	III (1–5)
Festuca ovina	I (1)	III (1–4)	II (1–4)
Climacium dendroides	I (4)	III (1–4)	II (1–4)
Juncus conglomeratus	I (1)	III (1–2)	II (1–2)
Carex echinata	I (1)	III (1)	II (1)
Prunella vulgaris		III (1–4)	II (1–4)
Plagiomnium undulatum		III (1–3)	II (1–3)
Leontodon hispidus		II (1–3)	I (1–3)
Plantago lanceolata		II (1–2)	I (1–2)
Rhinanthus minor		II (1–2)	I (1–2)
Trifolium repens		II (1–2)	I (1–2)
Lychnis flos-cuculi		II (1)	I (1)
Hylocomium splendens	I (1)	II (1–4)	I (1–4)
Cirsium palustre	III (1)	III (1–2)	III (1–2)
Carex pulicaris	III (2–3)	III (1–2)	III (1–3)
Caltha palustris	III (1–2)	III (1–3)	III (1–3)
Trollius europaeus	III (2–3)	III (1–2)	III (1–3)
Lophocolea bidentata s.l.	III (1–3)	III (1–4)	III (1–4)
Anemone nemorosa	III (2–3)	III (1–2)	III (1–3)
Thuidium tamariscinum	III (1–6)	II (1–4)	III (1–6)
Juncus articulatus	III (1–2)	III (1–4)	III (1–4)

Floristic table M26 *(cont.)*

	a	b	26
Campylium stellatum	III (3–5)	II (1–6)	III (1–6)
Carex flacca	II (1–2)	III (1–5)	III (1–5)
Rhytidiadelphus squarrosus	II (1–2)	III (1–4)	III (1–4)
Pseudoscleropodium purum	II (2–4)	III (1–2)	III (1–4)
Centaurea nigra	II (1–2)	III (1–5)	III (1–5)
Ranunculus auricomus	II (1–2)	II (1–2)	II (1–2)
Rumex acetosa	II (1)	II (1)	II (1)
Primula farinosa	I (1)	I (1–5)	I (1–5)
Number of samples	7	13	20
Number of species/sample	26 (20–32)	34 (17–52)	31 (17–52)
Herb cover (%)	90 (80–100)		
Bryophyte cover (%)	13 (10–30)		
Altitude (m)	370	344 (260–450)	353 (260–450)
Soil pH	6.5	6.1 (5.6–6.8)	

a   *Sanguisorba officinalis* sub-community
b   *Festuca rubra* sub-community
26 *Molinia caerulea-Crepis paludosa* fen (total)

M27
Filipendula ulmaria-Angelica sylvestris mire

Synonymy

Ulmaria society Pearsall 1918; *Filipendula ulmaria* consocies Tansley 1939; *Filipendula* communities Proctor 1974; Tall herb fens Adam *et al.* 1975 *p.p.*; *Filipendulo-Iridetum pseudacori* Adam 1976 *p.p.*, Adam *et al.* 1977 *p.p.*; *Epilobium hirsutum-Filipendula ulmaria* community Wheeler 1980c *p.p.*; *Valeriano-Filipenduletum* (Sissingh 1946) Birse 1980; Fen meadow Ratcliffe & Hattey 1982 *p.p.*; *Junco-Filipenduletum* Jones 1984 *p.p.*

Constant species

Filipendula ulmaria.

Physiognomy

Filipendula ulmaria is a frequent and locally abundant member of a variety of herbaceous vegetation types, notably Phragmitetalia fens, damp Arrhenatheretalia swards and the rush-pastures and fen-meadows of the Molinietalia. But it is also widely encountered as a more overwhelming component of vegetation in which the dominants of the above communities, tall helophytes, bulky sedges, rushes and rank grasses, are, if present at all, relegated to a subordinate role. Most of this kind of vegetation is included in the *Filipendula ulmaria-Angelica sylvestris* mire.

   F. ulmaria is a shortly rhizomatous perennial which puts up an annual crop of leafy shoots and, under the favourable conditions characteristic here, these grow tall and luxurious, often reaching a metre or more in height and becoming densely dominant over extensive areas where clonal patches coalesce. In the deep shade cast by this herbage, the associated flora, though quite varied in composition from stand to stand, is frequently poor in species: *F. ulmaria* is the sole constant of the community, indeed the only plant to attain more than occasional frequency throughout, and even the commoner companions are often found only as scattered individuals or in dispersed clumps.

   Among these, the most frequent species are other tall herbs which are able to grow up among the meadowsweet and which, by mid-summer, can present a colourful spectacle with their flowering shoots projecting above the canopy of foliage. The commonest plants among this group are *Angelica sylvestris*, *Valeriana officinalis* and *Rumex acetosa* and, in the more striking vegetation of the *Valeriana-Rumex* sub-community, these become preferentially frequent and are often accompanied by *Lychnis flos-cuculi*, *Succisa pratensis*, *Geum rivale* and sprawls of *Galium palustre*. In other stands, such species are more scarce but *Urtica dioica* is very common and, with *Cirsium arvense*, *Epilobium hirsutum*, *Eupatorium cannabinum* and *Vicia cracca*, helps to characterise an *Urtica-Vicia* sub-community. Then, at low frequency throughout the *Filipendula-Angelica* mire, there can be scattered *Lythrum salicaria*, *Rumex crispus*, *R. sanguineus*, *Epilobium palustre*, *Equisetum palustre*, *E. arvense* and *E. fluviatile*. Particularly towards the west, *Iris pseudacorus* and *Oenanthe crocata* can also figure with some local prominence, though these species are not nearly so consistent or abundant a feature here as in the *Filipendulo-Iridetum*, the closely-related tall-herb vegetation of our Atlantic seaboard.

   As for the bulky monocotyledons that can be found as dominants over *Filipendula* in other fen vegetation, only *Phragmites australis* occurs with any frequency here and it is strongly associated with the more eutrophic lowland stands included in the *Urtica-Vicia* sub-community. Even there, it is always less abundant overall than the meadowsweet, though, where locally dense, it gives the vegetation a structure transitional to the *Phragmites-Urtica* fen, in close association with which this kind of *Filipendula-Angelica* mire is often found. Other swamp and fen dominants occur so sparsely as to rarely give rise to any problems of diagnosis but *Phalaris arundinacea* is found occasionally throughout and locally there can be records for big sedges like *Carex acutiformis*, *C. rostrata*, *C. vesicaria*, *C. paniculata* and *C. appropinquata*.

   Rushes, too, are likewise few in number and uneven in their occurrence through the community. *Juncus effusus* is the most common species overall, though it is markedly preferential for more acidic soils in western Britain where it helps characterise the *Juncus-Holcus* sub-community, and even there it is rarely of high cover. *J. acutiflorus*, *J. articulatus* and *J. conglomeratus* are all found much less frequently and *J. inflexus* and *J. subnodulosus* are very scarce. Among the bulkier grasses, the occurrence of occasional tussocks of *Molinia caerulea* along with clumps of rushes can give the appearance of Junco-Molinion vegetation, but the balance of dominance still lies with *F. ulmaria*. Other rank grasses can play a subordinate role, too, although they usually show a preference for particular kinds of *Filipendula-Angelica* mire: *Holcus lanatus*, for example, is more frequent in the *Juncus-Holcus* sub-community, whilst *Arrhenatherum elatius* often helps define the *Urtica-Vicia* type. *Dactylis glomerata* and *Festuca rubra* also occur very occasionally and, where grasses as a group bulk a little larger than usual, the vegetation can approach a damp Arrhenatherion sward in its composition. Typically, however, these species are all found as scattered individuals, often attenuated beneath the tall-herb canopy, or growing more vigorously where the cover thins out a little and at the margins of stands.

Other grasses of smaller stature can occasionally contribute to a lower tier of vegetation though, with the dense shade, smaller herbs are usually few and far between, often noticeable only if they make some growth before the full emergence of the meadowsweet shoots, or where they spread on to stretches of newly deposited silt where stands are flooded. *Poa trivialis* is the most frequent member of this group, with *Agrostis stolonifera*, *A. canina* (presumably ssp. *canina*) and *Anthoxanthum odoratum* occurring less commonly. Then, among smaller dicotyledons, there can be some *Ranunculus repens*, *Mentha aquatica*, *Lotus uliginosus* and *Caltha palustris*, with more occasional *Ranunculus acris*, *Cardamine pratensis*, *C. flexuosa*, *Potentilla anserina* and *Polygonum hydropiper*.

Bryophytes, too, are often few in number and of low cover, although they can catch the eye in winter and early spring when their fresh-green patches are revealed growing over the exposed bare ground and herb stools. *Brachythecium rutabulum* (probably including some *B. rivulare*), *Calliergon cuspidatum*, *Eurhynchium praelongum* and *Lophocolea bidentata s.l.* are all found occasionally, and with somewhat greater regularity and luxuriance in the *Valeriana-Rumex* sub-community, where a variety of other low-frequency preferentials enriches this element of the vegetation.

Woody plants seem to get a hold only with difficulty in denser stands of the *Filipendula-Angelica* mire, but *Salix cinerea* seedlings can sometimes be found in the ground layer and these occasionally get away where the cover is more open. In drier stands, there can also be some patchy *Rubus fruticosus* agg.

Sub-communities

***Valeriana officinalis-Rumex acetosa* sub-community:** *Filipendula* communities Proctor 1974; Tall herb fens Adam *et al.* 1975 *p.p.*; *Valeriano-Filipenduletum* (Sissingh 1946) Birse 1980; *Junco-Filipenduletum* Jones 1984 *p.p.* *F. ulmaria* is still very much the most abundant plant here and it can be so overwhelmingly dominant as to make this vegetation as poor in species as any kind of *Filipendula-Angelica* mire. But overall it is characterised by the preferential occurrence of quite a variety of associates and when present in numbers, even if only as scattered individuals, these can produce a distinctive effect. Some are common throughout the sub-community, almost rivalling *F. ulmaria* in frequency and with tall flowering shoots. Foremost among these are *Angelica* and *Valeriana officinalis*, both able to make luxuriant growth here and attain a measure of local abundance, together with *Rumex acetosa* and, rather less common, *Lychnis flos-cuculi*. *Succisa* and *Geum rivale* are also preferential at lower frequencies and then there can be occasional records for taller community plants like *Cirsium palustre*, *Lythrum salicaria*, *Rumex crispus*, *Potentilla palustris* and *Equisetum fluviatile*. Towards its southern limit, this kind of *Filipendula-Angelica* mire can also have some *Sanguisorba officinalis* and, at higher altitudes, as at Malham Tarn in North Yorkshire (Proctor 1974, Adam *et al.* 1975) and in Teesdale (Jones 1984), it provides a locus for such Continental Northern herbs as *Crepis paludosa*, *Alchemilla glabra* and *Cirsium helenioides*, and for the Northern Montane *Trollius europaeus*.

Providing occasional enrichment a little below the level of the tall herbs, there can be some *Caltha palustris*, *Ranunculus flammula*, *R. repens*, *R. acris*, *Cardamine flexuosa*, *C. pratensis*, *Mentha aquatica*, *Stellaria alsine* and *Ajuga reptans*, thickening up a little where the meadowsweet is not too dense. *Galium palustre* is very common trailing among the herbage, *G. uliginosum* much less frequent, and occasionally *Lathyrus pratensis* can be found.

Apart from *Poa trivialis*, which is preferential to this sub-community, though generally of low cover, grasses and rushes are infrequent. *Phalaris* and *Phragmites* sometimes put up sparse tall shoots or, very occasionally, thicken up into small clumps, and scattered tussocks of *Juncus effusus* are sometimes to be seen. Sedges are a little more frequent, though even among these there is little consistency of occurrence and they rarely rival *F. ulmaria* in abundance. However, *Carex rostrata* is quite common and, when present with *Menyanthes* and *Potentilla palustris*, can bring the vegetation close to the *Potentillo-Caricetum* in its composition. *C. nigra* also occurs occasionally and, where there are marked fluctuations of water-level, it can grow in its strongly-tussocky form (var. *subcaespitosa*), catching the eye,

though rarely of very high total cover. *C. vesicaria* is
another sedge which can attain some local prominence,
particularly in Scottish localities, and, in other stands,
C. acutiformis and the Continental Northern *C. disticha*
or *C. appropinquata* have been found.

Bryophytes are somewhat better developed in this
sub-community than in the others, though their cover is
still rather patchy. *Brachythecium rutabulum* (with some
B. rivulare) is the most frequent, but the other community species can also be found, together with occasional *Rhizomnium punctatum*, *Plagiomnium elatum*,
P. undulatum, *Chiloscyphus pallescens* and *Thuidium
tamariscinum*.

Urtica dioica-Vicia cracca sub-community: *Epilobium
hirsutum-Filipendula ulmaria* community Wheeler
1980c *p.p.* *F. ulmaria* and a variety of other tall herbs
again compose the major structural element of the
vegetation here and, if anything, the cover of these
plants is denser than in the above, further reducing the
contribution from smaller species. But, although *Angelica*, *Cirsium palustre* and *Lythrum salicaria* remain
occasional, *Valeriana officinalis*, *Rumex acetosa* and
many of the preferentials of the first sub-community are
either very scarce or totally absent. *Urtica dioica*, by
contrast, becomes very common and, together with
occasional *Eupatorium cannabinum* and *Epilobium hirsutum*, can form small clumps among the meadowsweet,
producing a characteristically patchy variegation to the
canopy. Then, scattered throughout, there can be plants
of *Cirsium arvense* and *Centaurea nigra* and, replacing
Galium palustre as the typical sprawler, there is often
some *Galium aparine* and *Vicia cracca*, with occasional
Calystegia sepium.

Phragmites is also quite commonly found here and,
though typically much less abundant than *F. ulmaria*, it
can thicken up locally and, where it is accompanied by
scattered individuals of such species as *Lysimachia
vulgaris*, *Lycopus europaeus* or *Thelypteris palustris*, the
floristic resemblance to Phragmitetalia fen is close. In
other stands, the vegetation takes on a rank grassy look,
with tussocks of *Arrhenatherum elatius* scattered
beneath the tall dicotyledons, some *Holcus lanatus* and
occasional *Heracleum sphondylium* and *Lotus uliginosus*. In drier stands of this kind, *Rubus fruticosus* agg. can
get a substantial hold. In yet other situations, scattered
clumps of rushes can be found, with *J. effusus*, *J.
acutiflorus*, *J. articulatus*, *J. conglomeratus*, *J. inflexus*
and *J. subnodulosus* all recorded sparsely.

Small herbs are few and of low cover, with just occasional *Mentha aquatica*, *Prunella vulgaris*, *Potentilla
reptans* and *P. anserina*. And, though litter turnover can
be brisk here, with quite extensive patches of bare ground
exposed among the stools by late winter, bryophytes are
typically very sparse with only *Brachythecium rutabulum*
and *Lophocolea bidentata s.l.* figuring occasionally.

Juncus effusus-Holcus lanatus sub-community: *Filipendulo-Iridetum pseudacori* Adam 1976 *p.p.*, Adam *et al.*
1977 *p.p.*; *Valeriano-Filipenduletum* (Sissingh 1946)
Birse 1980 *p.p.*; Fen meadow Ratcliffe & Hattey 1982
p.p. *F. ulmaria* is still the most abundant species in this
sub-community, though its dominance is not quite so
overwhelming as in the other kinds of *Filipendula-
Angelica* mire. Other tall herbs such as *Angelica*, *Valeriana officinalis*, *Cirsium palustre* and *Rumex acetosa*
occur occasionally, but of greater structural importance
is the presence of rushes and grasses among the meadowsweet, sometimes with moderate abundance. *Juncus
effusus* and *Holcus lanatus* are both constant, *J. acutiflorus* and *Molinia* occasional and other grasses recorded
less frequently include *Anthoxanthum odoratum*, *Agrostis stolonifera*, *A. canina* and *Poa trivialis*. Such plants,
along with frequent *Mentha aquatica* and *Lotus uliginosus* and occasional *Achillea ptarmica*, bring the vegetation close to a rank Junco-Molinion sward. In other
stands, *Iris pseudacorus* and/or *Oenanthe crocata* can be
prominent and, around the upper limit of salt-marshes
in western Britain, this kind of *Filipendula-Angelica* mire
can grade to the *Filipendulo-Iridetum* (Adam 1976,
Adam *et al.* 1977).

Habitat

The *Filipendula-Angelica* mire is typically found where
moist, reasonably rich, circumneutral soils occur in
situations protected from grazing. It is to be seen
throughout the lowlands of Britain, in both soligenous
and topogenous mires, being especially typical of the
silting margins of slow-moving streams and soakways,
the edges of flushes and damp hollows, but it also occurs
widely in artificial habitats, as along dykes and roadside
ditches, and around ponds and pools with richer waters.
Natural climatic and edaphic differences across its
extensive range have some influence on floristic variation in the community, but this is also affected by
agricultural practices and, in many places, draining and
grazing have reduced this kind of vegetation to small
remnants.

The *Filipendula-Angelica* mire is, first, a community
of moist ground, occurring on a variety of mineral and
organic soils which are kept damp for much or all of the
year. It can be found, along stream edges, right down to
the water side, and it often experiences substantial
seasonal and short-term fluctuations in the level of the
water-table (Spence 1964, Proctor 1974) but, generally
speaking, it is typical of situations between permanent
standing waters and deeply winter-flooded ground on
the one hand and, on the other, land which is more than
superficially dry for long periods of time. Much of the
overall floristic character of the vegetation is due to the
ability of *F. ulmaria* to become dominant on such
intermediate ground over species which have their peak
of abundance at the extremes of soil moisture con-

ditions. Towards the wetter limit of the *Filipendula-Angelica* mire, for example, certain of the tall helophytes and bulky sedges which characterise the Phragmitetalia swamps and fens of open waters and their extensively-inundated surrounds can maintain moderate frequency, though they rarely occur with any vigour. This sparse cover is generally the maximum extent of the contribution here of such species as *Phragmites* (in the *Urtica-Vicia* sub-community) and *Carex rostrata* (in the *Valeriana-Rumex* type), though they often thicken up, on adjacent wetter ground, to form an overtopping canopy in floristically-similar fen vegetation (e.g. Proctor 1974, Adam *et al.* 1975). Towards the other extreme, where the *Filipendula-Angelica* mire runs on to drier ground, it is the increase in frequency of bulky grasses of the Molinio-Arrhenatheretea, such as *Arrhenatherum*, *Molinia* and *Holcus lanatus*, or of tall herbs such as *Urtica*, well seen in both the *Urtica-Vicia* and *Juncus-Holcus* sub-communities, that marks the environmental limits of the community.

Over the middle ground, a variety of other plants apart from *F. ulmaria* is able to thrive under the moist conditions but the uncompromising dominance of the meadowsweet ensures that these attain only limited frequency and abundance. Other tall hemicryptophytes fare best, being able to keep pace with the expansion of the leafy canopy, and such associates as *Angelica*, *Rumex acetosa*, *Cirsium palustre*, *Lychnis flos-cuculi*, *Succisa pratensis* and *Lythrum salicaria*, even though they are often reduced to scattered individuals among the dense cover, help confirm the general Molinietalia character of the vegetation. Some other broadly-distributed tall wetland plants, like *Valeriana officinalis* and, to the west, *Iris pseudacorus*, can also be well represented and plants like *Galium palustre* and *G. aparine* are able to reach the light by sprawling over the other herbage. Smaller herbs, though, such as *Mentha aquatica* and *Poa trivialis*, together with some of the more diminutive sedges, for which conditions here are probably otherwise very congenial, are much reduced or simply overwhelmed.

Two other habitat features favour this luxuriant growth of tall herbs in general and *F. ulmaria* in particular. The first is the at least moderate nutrient-richness of the substrates, for this is a community of soils which are either naturally or, as a result of treatments, maintained in a mesotrophic or eutrophic state. To a great extent, it is this edaphic requirement, rather than any direct influence of climate, which limits the *Filipendula-Angelica* mire largely to the lowland parts of the country. It is a community that occurs throughout Britain, and it can be found almost down to sea-level in the west, but only very locally does it extend above 200 m. It is true that some of its associates show reduced growth in the cooler temperatures at greater altitudes, but certain of its most characteristic plants can be found at much higher levels in, for example, montane ledge vegetation. However, although soils which are suitably moist for the development of the *Filipendula-Angelica* mire are widespread in the upland zone of the country, this assemblage is largely excluded from them by their impoverished character, derived as they often are from siliceous rocks and rarely showing any accumulation of fine mineral detritus in the harsh eroding landscapes. Where such processes do occur at higher altitudes, as along the more sluggish stretches of streams or around the sheltered shores of lakes, the *Filipendula-Angelica* mire can form a quite striking addition to suites of poor fens and calcicolous mires on the poorer substrates. This is well seen at Malham Tarn, where the community occupies small levees alongside the input streams and where Proctor (1974) has shown that the limits of this kind of vegetation are probably set by the greater phosphate requirements of its predominant species.

In the more subdued landscapes at lower altitudes in northern and western Britain, where depositional habitats become more common, the *Filipendula-Angelica* mire consequently increases in frequency, occurring widely over mesotrophic peaty gleys and peaty alluvial soils alongside streams, around pools and in flushes. In northern England and through southern and eastern Scotland, the usual form of the community is the *Valeriana-Rumex* type, vegetation which is largely confined to the cooler parts of Britain, where mean annual maximum temperatures are generally less than 24 °C (Conolly & Dahl 1970) and February minima often below freezing (*Climatological Atlas* 1952). At the highest altitudes to which this sub-community penetrates, the appearance of such Continental Northern and Northern Montane plants as *Crepis paludosa*, *Alchemilla glabra*, *Cirsium helenioides* and *Trollius europaeus* indicates a floristic transition to Cicerbition alpini ledge vegetation.

Down the western fringe of the country, along the Atlantic seaboard of Scotland, through Wales and into the South-West Peninsula, suitably moist and mesotrophic soils continue to be widely present through the lowlands, but the climate is somewhat wetter and distinctly milder. Here, annual rainfall often exceeds 1600 mm (*Climatological Atlas* 1952), with more than 160 wet days yr^{-1} (Ratcliffe 1968), so the tendency to leaching is more pronounced, soils under the community frequently having a pH that is nearer 5 than 6, the mean of the community as a whole. Moreover, winter temperatures are high, with February minima usually above freezing, quite markedly so in the south (*Climatological Atlas* 1952). Such shifts are reflected in the *Filipendula-Angelica* mire by the occurrence of the *Juncus-Holcus* sub-community throughout this region, species such as *J. effusus*, *Holcus lanatus*, *Molinia* and *Lotus uliginosus* representing a clear transition to the Junco-Molinion. And very close to the coast, where the community can penetrate on to the flushed upper margins of salt-marshes, well out of reach of tidal influence but under the strong influence of the oceanic climate, *Iris* and *Oenanthe crocata*, Oceanic West European plants, can make a distinctive contribution.

It is possible that the southern and eastern margins of these two kinds of *Filipendula-Angelica* mire might be drawn a little more generously were suitable edaphic conditions to extend further into central and eastern England. Moist profiles are certainly widespread there, over the alluvium deposited along mature rivers and around lakes and pools, on the fen peats of flushes and topogenous mires, which often also receive some alloch-thonous mineral material, and in damp hollows with stagnogleys and pelosols developed from the argilla-ceous bedrocks and superficials that underlie the exten-sive undulating landscapes of this part of Britain. Here, too, artificial habitats are particularly widespread with numerous ditches and ponds. But, with the intensive agricultural activity that has long been characteristic of the Midlands and south-east England, even the more natural of these soils are frequently disturbed and artificially enriched by fertiliser run-off or drift, by contamination with other kinds of eutrophic effluent, by dumping of dredgings or by surface oxidation of drying organic matter. In such habitats as these, it is the *Urtica-Vicia* sub-community, with its characteristically patchy cover of eutrophic tall herbs, that is the usual kind of *Filipendula-Angelica* mire.

One further environmental feature essential for the development of any of these vegetation types is freedom from any grazing more than the very light or sporadic. *F. ulmaria* is highly palatable and the predations of large herbivores can quickly reduce it to nibbled stumpy stocks with a few tufts of diminutive leaves, and then eliminate it altogether. And, although some of the other taller hemicryptophytes that give this vegetation its distinctive character can survive in grazed stands as non-flowering rosettes, the real beneficiaries of such treat-ments are the unpalatable bulky plants, notably rushes, which then come to dominate over those smaller dicoty-ledons and grasses able to persist in close-cropped swards. Such assemblages would not qualify for inclu-sion here and can often be found replacing the *Filipen-dula-Angelica* mire where moist, mesotrophic soils have been enclosed for pasture (see below). The natural wetness of the ground under this community is rarely such as to itself prevent access by stock, so stands often persist in intensive pastoral landscapes only outside such enclosures, around unreclaimed mires and flushes, for example, in wet field bottoms and edges which have been fenced off, and alongside streams and ditches between pasture and boundaries. Where the community

occurs in roadside ditches, it may be subject to occasio-nal mowing, though regular inclusion in a hay-crop probably favours a more varied pattern of dominance than is characteristic here.

Zonation and succession

The community can be found in a wide variety of zonations and mosaics with swamps, other kinds of mire, tall-herb vegetation and grasslands, among which the major lines of floristic variation are governed by interactions between soil moisture and trophic state and by differences in treatment. Scrub and woodland also often occur in close association with the *Filipendula-Angelica* mire and can undoubtedly develop from it, though, in many cases, this kind of vegetation seems somewhat resistant to the invasion of woody plants and able to persist in a fairly stable state, provided edaphic conditions and treatment remain unchanged.

In the more natural vegetation sequences in which the *Filipendula-Angelica* mire is to be seen, it typically occupies a zone between standing or sluggish open waters on the one hand, and tall-herb communities on the other. Alongside small streams, as seen at Black Beck in the Esthwaite fens (Pearsall 1918, Tansley 1939) and at Malham (Proctor 1974), it can actually run right down to the water's edge, forming a strip over the intermittently flooded ground and often helping to accumulate small levees. In other cases, it can give way below to emergent vegetation on the permanently sub-merged ground or that subject to extensive winter flooding. To the north of Britain, where the *Valeriana-Rumex* sub-community is a common component of open-water transitions around lakes and along streams, it is the *Caricetum rostratae* which often replaces it on wetter ground or, more locally, the *Caricetum vesicariae*, and, with these vegetation types, there can be a measure of floristic continuity, some of the associates of the *Filipendula-Angelica* mire extending out under the dominant sedge (Spence 1964, Proctor 1974, Adam *et al.* 1975). Such swamps also extend into western Britain and can be found abutting on to the *Juncus-Holcus* sub-community where this occurs around open waters. To the south and east, by contrast, where the *Urtica-Vicia* sub-community is typical, it is the *Phragmitetum austra-lis* that most often replaces it on wetter ground or, more occasionally, the *Caricetum paniculatae* or communities like the *Sparganietum erecti* or *Typhetum latifoliae*; and, again, there can be some gradations between the vege-tation types.

Along stream sides and drainage ditches and around small pools and ponds which are maintained in an open condition, particularly common habitats for the *Filipen-dula-Angelica* mire to the south of Britain, zonations such as these can be very compressed, with little more than fragments of the various communities occurring in a belt of very small overall width. In more extensive open-water transitions, on the other hand, such as those found around some lakes in northern Britain and in Broadland, the ground between the swamps and the *Filipendula-Angelica* mire can be occupied by a wide zone of what is essentially Phragmitetalia fen, in which the tall helophytes and bulky sedges of the swamps retain their dominance but are accompanied by a richer suite of associates which shows considerable continuity with the meadowsweet vegetation. To the north of the country, such fen is typically of the *Potentillo-Caricetum rostratae* type; to the south and east, it can generally be grouped in the *Phragmites-Eupatorium* fen or, in Broadland the *Peucedano-Phragmitetum*, or very often now, around disturbed and eutrophicated water-margins, the *Phragmites-Urtica* fen.

However, only where the substrates of the fen hinterland are maintained in a reasonably rich state, either by fresh deposition of silt in floods, or by input of nutrients from the landward edge, will they be able to sustain the *Filipendula-Angelica* mire in such extended zonations as these and, at some sites with large tracts of fen, the community is very much restricted to the edges of streams, soakways or dykes, running through, or cut through, the flood-plain deposits. Both kinds of pattern can be seen at Malham, where complex variations in water-depth and nutrient- and base-status across the fens result in particularly varied mosaics of rich and poor fens disposed over the peats between the channels and grading, with varying degrees of abruptness, to the *Filipendula-Angelica* mire on the richer, circumneutral deposits (Proctor 1974).

Even in more extensive and natural zonations such as these, the sequence of vegetation types rarely continues on to drier ground without some suspicion of modification as a result of man's activities. Very locally in northern Britain, it is possible to see transitions from the *Valeriana-Rumex* sub-community to the *Molinia-Crepis* mire on ground that is subject to less marked fluctuations in water level, as at Malham (Proctor 1974), and, around flushes on the lower valley sides of some Pennine dales, such zonations can continue into stretches of *Anthoxanthum-Geranium* grassland, a community which is often cut for hay, but which is otherwise little improved (Jones 1984). Very commonly through northern and western Britain, however, even where the surrounds of mires and flushes with the *Filipendula-Angelica* mire have escaped marked improvement, they are heavily grazed, so many stands of the *Valeriana-Rumex* and *Juncus-Holcus* sub-communities survive sharply marked off from surrounding Nardo-Galion swards by artificial boundaries, remaining largely within fenced-off damp hollows or field corners, and persisting otherwise only in very fragmentary fashion along stream sides running through pasture.

The situation within southern and eastern England is very similar, except that there the agricultural landscape is more intensive and largely given over in certain areas to arable cropping with its attendant eutrophication of the drainage waters by fertiliser run-off. More natural zonations probably run from the *Urtica-Vicia* sub-community to some kind of *Arrhenatheretum*, and there can be considerable overlap between the two vegetation types. Paradoxically, such transitions are now most easily seen within the artificial roadside verge habitat, where damp ditches and drier banks often carry these two closely juxtaposed. In many places, however, the *Urtica-Vicia* sub-community gives way, not to such grasslands, but to eutrophic tall-herb vegetation dominated by such plants as *Urtica* and *Epilobium hirsutum*, which forms a fringe on the disturbed and enriched edges of mires and drainage ditches, set within intensive pastoral or arable landscapes.

Where more traditional regimes of mowing and grazing have been extended down, around mires and flushes, on to the moister soils normally occupied by the *Filipendula-Angelica* mire, further variations on these patterns can be seen. Detailed studies of the effects of such treatments on the community have never been undertaken, but it seems likely that grazing, and perhaps also regular summer-mowing, convert it to the kinds of vegetation, generally dominated by rushes, included in the Calthion to the east of Britain, and in the Juncion acutiflori to the west and north. Where the *Urtica-Vicia* sub-community occurs in spring-fens, for example, it can often be seen alongside the *Juncus-Cirsium* rush-pasture in those areas where stock have access or which, in the past, were mown for litter. Similarly, to the west, a very common context for the *Juncus-Holcus* sub-community is stretches of the *Juncus-Galium* rush-pasture, which form a transitional zone between the *Filipendula-Angelica* mire and the surrounding Nardo-Galion sward where stock have had access to the damper flush surrounds. With further cropping and selective nutrient depletion, such developments could proceed further with the expansion of *Molinia* in Junco-Molinion swards.

Even where such treatments have been withheld for long periods of time, progression of the *Filipendula-Angelica* mire to woodland seems slow, perhaps because of the overwhelming dominance of the meadowsweet, and, quite commonly, stands of the community can be found closely hemmed in by a cover of trees or shrubs. However, observation suggests that invasion of woody plants can take place and it seems likely to result in the development of either drier forms of the *Salix-Betula-Phragmites* woodland, particularly in the south and east, or, to the west and north, the *Salix-Galium* or *Alnus-Fraxinus-Lysimachia* woodlands. Drying and disturbance of the substrate probably favours progression to the *Alnus-Urtica* woodland.

Distribution

The *Filipendula-Angelica* mire occurs throughout low-land Britain with the sub-communities showing clear regional associations: the *Valeriana-Rumex* type is the usual form of the community in northern England and southern and eastern Scotland, the *Juncus-Holcus* type down the western seaboard of the country and the *Urtica-Vicia* type in central, southern and eastern Britain.

Affinities

Except in the study of particular sites (e.g. Pearsall 1918, Tansley 1939, Proctor 1974, Adam *et al.* 1975), the kinds of vegetation included here have figured very little in accounts of British plant communities and have usually been defined in terms of the dominance of meadow-sweet. This scheme brings together data from a variety of existing sources, as well as including new samples, and it confirms the floristic integrity of the richer kind of *F. ulmaria* vegetation characterised here as the *Valeriana-Rumex* sub-community, first described in detail from this country by Proctor (1974), and considered by him and by Birse (1980) to be essentially similar to the *Valeriano-Filipenduletum*, an association described by Sissingh (1946) from Holland and north-west Germany with closely-related assemblages in Eire (Braun-Blan-quet & Tüxen 1952), Belgium (Duvigneaud 1958) and northern France (Géhu 1961). But it also seems sensible to retain, within the same general unit, the transitional kinds of vegetation placed here in the *Urtica-Vicia* and *Juncus-Holcus* sub-communities. The community as a whole cannot then be uniquely defined in terms of anything other than the overwhelming abundance of *F. ulmaria*, but in qualitative terms there is nothing pecu-liar even about the richer assemblage of the *Valeriana-Rumex* type: what is special about these vegetation types is the persistence of the associates, albeit often in an attenuated form, among a dense canopy of meadow-sweet, the luxuriance of which is a real reflection throughout of a particular combination of ecological conditions.

The general affinities of the community, indicated by such species as *F. ulmaria* itself, *Angelica, Cirsium palustre, Equisetum palustre, Lotus uliginosus* and *Jun-cus effusus*, are with the damp herbaceous communities of the Molinietalia and, in a broad context, the vege-tation shows links with both the Phragmitetalia swamps and fens and the grassy swards of the Arrhenatheretalia, transitions to which are mediated essentially by edaphic differences. Species such as *Carex rostrata* and *Phrag-mites* (in wetter stands of the *Valeriana-Rumex* and *Urtica-Vicia* sub-communities) thus provide a connec-tion with the former, plants like *Arrhenatherum elatius* and *Holcus lanatus* (in drier tracts of the *Urtica-Vicia* and *Juncus-Holcus* types) with the latter, and many of the associated herbs of the *Filipendula-Angelica* mire extend far in both these directions, as an understorey in both fens and wet grasslands.

More narrowly, the community is particularly close to the vegetation included in the Calthion and the Juncion acutiflori, other Molinietalia alliances in which the dominance of coarse monocotyledons, notably rushes, is under the strong control of treatments charac-teristically absent in sites with the *Filipendula-Angelica* mire. Phytogeographical trends in the community to some extent reflect those among these other kinds of Molinietalia vegetation, with the *Juncus-Holcus* sub-community representing a clear transition to the oceanic *Juncus-Galium* rush-pasture towards western Britain, the other types coming closer to the more continental *Juncus-Cirsium* rush-pasture in the eastern part of the country.

Two further features confuse these relationships. First, where the *Valeriana-Rumex* sub-community at-tains higher altitudes in northern Britain, it begins to show a floristic affinity to the tall-herb vegetation of the Cicerbition alpini in which Northern Montane and Continental Northern plants become important. More widely, in the intensive agricultural landscape of south-ern and eastern Britain, the effect of disturbance and enrichment of the habitats of the *Filipendula-Angelica* mire is to bring it very close, in the *Urtica-Vicia* sub-community, to the weedy eutrophic tall-herb vegetation of the Artemisietea.

The striking influence of the dominance of *F. ulmaria* has led phytosociologists to group the kind of vegetation included here in a separate Molinietalia alliance, the Filipendulion. No British plants apart from meadow-sweet show a qualitative or quantitative preference for this group and only one other vegetation type, the *Filipendulo-Iridetum*, could properly be regarded as belonging to it.

Floristic table M27

	a	b	c	27
Filipendula ulmaria	V (4–10)	V (5–10)	V (4–8)	V (4–10)
Angelica sylvestris	IV (1–4)	II (1–5)	II (1–7)	III (1–7)
Valeriana officinalis	IV (2–7)	I (1–5)	II (2–6)	II (1–7)
Rumex acetosa	IV (1–4)	I (1–3)	II (1–4)	II (1–4)
Galium palustre	IV (1–4)	I (2–4)	II (2–4)	II (1–4)
Brachythecium rutabulum	IV (3–6)	I (3–4)	I (3–7)	II (3–7)
Ranunculus repens	III (1–7)	I (3–4)	II (2–4)	II (1–7)
Poa trivialis	III (1–4)	I (1–5)	II (3–5)	II (1–5)
Lychnis flos-cuculi	III (1–5)	I (3–4)	I (1–3)	II (1–5)
Caltha palustris	III (1–6)	I (3–4)	I (3–4)	II (1–6)
Lophocolea bidentata s.l.	III (1–5)		I (3)	I (1–5)
Carex rostrata	III (1–5)		I (1)	I (1–5)
Calliergon cuspidatum	II (1–8)		II (1–4)	I (1–8)
Eurhynchium praelongum	II (1–6)		II (3–5)	I (1–6)
Equisetum palustre	II (1–2)	I (1–5)	I (4)	I (1–5)
Lathyrus pratensis	II (1–3)	I (2–3)	I (1–4)	I (1–4)
Ranunculus acris	II (1–5)	I (1–3)	I (1–4)	I (1–5)
Cardamine flexuosa	II (1–3)		I (2–3)	I (1–3)
Cardamine pratensis	II (1–3)		I (3)	I (1–3)
Carex nigra	II (1–5)		I (1)	I (1–5)
Geum rivale	II (1–7)			I (1–7)
Stellaria alsine	II (1–4)		I (2)	I (1–4)
Rhizomnium punctatum	II (2–4)			I (2–4)
Sanguisorba officinalis	II (1–3)			I (1–3)
Succisa pratensis	II (1–4)			I (1–4)
Crepis paludosa	II (1–6)			I (1–6)
Plagiomnium elatum	I (1–4)			I (1–4)
Carex disticha	I (1–3)			I (1–3)
Alchemilla glabra	I (2–5)			I (2–5)
Myosotis scorpiodes	I (1–2)			I (1–2)
Cirsium helenioides	I (3–4)			I (3–4)
Trollius europaeus	I (2–6)			I (2–6)
Ranunculus flammula	I (2–3)			I (2–3)
Carex panicea	I (1–3)			I (1–3)
Carex vesicaria	I (5–6)			I (5–6)
Calliergon cordifolium	I (1–5)			I (1–5)
Plagiomnium undulatum	I (3–4)			I (3–4)
Rhytidiadelphus squarrosus	I (1–4)			I (1–4)
Cochlearia pyrenaica	I (1–4)			I (1–4)
Ajuga reptans	I (1–3)			I (1–3)
Chiloscyphus pallescens	I (1–3)			I (1–3)
Thuidium tamariscinum	I (1–5)			I (1–5)
Carex appropinquata	I (1–3)			I (1–3)
Urtica dioica	I (1–6)	III (2–5)	I (3)	II (1–6)
Vicia cracca	I (1–3)	III (1–6)	I (2)	II (1–6)

Phragmites australis	I (3–5)	III (3–8)	I (3–4)	II (3–8)
Arrhenatherum elatius	I (3–4)	III (1–5)	I (3)	II (1–5)
Cirsium arvense	I (1–3)	II (1–5)	I (3)	II (1–5)
Galium aparine	I (1–3)	II (3–5)		I (1–5)
Centaurea nigra	I (1–4)	II (1–4)		I (1–4)
Eupatorium cannabinum		II (1–7)	I (2)	I (1–7)
Epilobium hirsutum		II (1–7)	I (3)	I (1–7)
Calystegia sepium		I (4–5)		I (4–5)
Elymus repens		I (3)		I (3)
Thalictrum flavum		I (3–7)		I (3–7)
Prunella vulgaris		I (3–5)		I (3–5)
Juncus inflexus		I (2–5)		I (2–5)
Potentilla reptans		I (2–4)		I (2–4)
Polygonum amphibium		I (2–4)		I (2–4)
Juncus subnodulosus		I (2–7)		I (2–7)
Juncus effusus	II (1–5)	I (5)	V (2–6)	III (1–6)
Holcus lanatus	I (1–4)	I (1–6)	IV (1–6)	II (1–6)
Mentha aquatica	II (1–4)	II (1–5)	III (2–6)	II (1–6)
Lotus uliginosus		II (1–5)	III (2–5)	II (1–5)
Iris pseudacorus	I (4–5)	I (2–4)	II (5–7)	I (2–7)
Achillea ptarmica	I (1–2)		II (1–4)	I (1–4)
Anthoxanthum odoratum	I (1–4)		II (3–4)	I (1–4)
Agrostis stolonifera			II (3–4)	I (3–4)
Molinia caerulea			II (3–6)	I (3–6)
Stellaria graminea			I (1–3)	I (1–3)
Carex paniculata			I (4–5)	I (4–5)
Pteridium aquilinum			I (4–5)	I (4–5)
Agrostis canina			I (2–4)	I (2–4)
Senecio aquaticus			I (2–3)	I (2–3)
Epilobium obscurum			I (1)	I (1)
Hydrocotyle vulgaris			I (3–5)	I (3–5)
Sparganium erectum			I (2–3)	I (2–3)
Cirsium palustre	II (1–3)	II (1–5)	II (1–3)	II (1–5)
Galium uliginosum	I (1–3)	I (3–5)	I (1–4)	I (1–5)
Phalaris arundinacea	I (4–7)	I (3–5)	I (4)	I (3–7)
Lythrum salicaria	I (1–5)	I (1–5)	I (1–3)	I (1–5)
Rumex crispus	I (1)	I (1–4)	I (1–4)	I (1–4)
Juncus acutiflorus	I (1–3)	I (1–4)	I (2–7)	I (1–7)
Dactylis glomerata	I (1–6)	I (2–4)		I (1–6)
Equisetum arvense	I (1–5)	I (2–4)		I (1–5)
Carex acutiformis	I (3–6)	I (3–4)		I (3–6)
Epilobium palustre	I (1–2)		I (2)	I (1–2)
Potentilla palustris	I (1–3)		I (4)	I (1–4)
Festuca rubra	I (2–4)		I (1–4)	I (1–4)
Poa pratensis	I (1–5)		I (1)	I (1–5)
Viola palustris	I (1–2)		I (1)	I (1–2)
Equisetum fluviatile	I (1–2)		I (1)	I (1–2)
Rubus fruticosus agg.		I (2–5)	I (3)	I (2–5)
Heracleum sphondylium		I (1–4)	I (1)	I (1–4)

Floristic table M27 *(cont.)*

	a	b	c	27
Potentilla anserina		I (1–3)	I (2–4)	I (1–4)
Oenanthe crocata		I (2–3)	I (5–6)	I (2–6)
Juncus articulatus		I (2–5)	I (5)	I (2–5)
Lysimachia vulgaris		I (2–6)	I (1)	I (1–6)
Glyceria maxima		I (3–5)	I (2)	I (2–5)
Rumex sanguineus		I (3–4)	I (2)	I (2–4)
Polygonum hydropiper		I (3–5)	I (3)	I (3–5)
Phleum pratense		I (2–4)	I (3)	I (2–4)
Juncus conglomeratus		I (3–6)	I (4)	I (3–6)
Pulicaria dysenterica		I (3)	I (2)	I (2–3)
Number of samples	29	41	18	88
Number of species/sample	17 (8–28)	14 (6–33)	15 (9–22)	15 (6–33)
Vegetation height (cm)	85 (48–130)	109 (20–200)	92 (20–150)	102 (20–200)
Herb cover (%)	96 (80–100)	100	99 (85–100)	99 (80–100)
Ground cover (%)	25 (0–70)	1 (0–10)	7 (0–50)	8 (0–70)
Altitude (m)	259 (2–378)	45 (4–246)	119 (4–320)	145 (2–378)
Soil pH	5.8 (5.7–6.0)	6.5 (5.4–7.5)	5.3 (4.4–6.1)	6.0 (4.4–7.5)

a   *Valeriana officinalis-Rumex acetosa* sub-community
b   *Urtica dioica-Vicia cracca* sub-community
c   *Juncus effusus-Holcus lanatus* sub-community
27 *Filipendula ulmaria-Angelica sylvestris* mire (total)

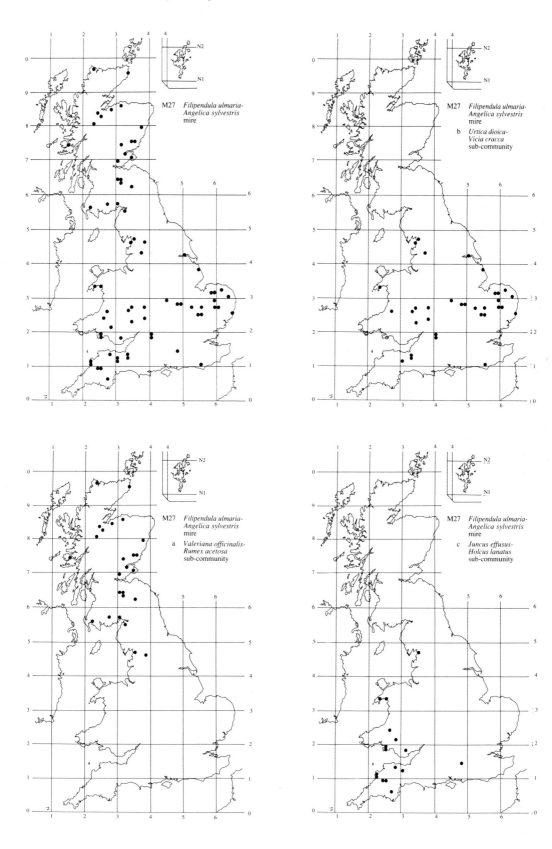

M27   *Filipendula ulmaria-Angelica sylvestris* mire

M27   *Filipendula ulmaria-Angelica sylvestris* mire

b   *Urtica dioica-Vicia cracca* sub-community

M27   *Filipendula ulmaria-Angelica sylvestris* mire

a   *Valeriana officinalis-Rumex acetosa* sub-community

M27   *Filipendula ulmaria-Angelica sylvestris* mire

c   *Juncus effusus-Holcus lanatus* sub-community

M28

Iris pseudacorus-Filipendula ulmaria mire
Filipendulo-Iridetum pseudacori   Adam 1976 emend.

Synonymy
Iridetum Gilham 1957*b*; Salt-marsh/swamp transitions Gimingham 1964*a p.p.*; *Iris pseudacorus* Community Birse 1980.

Constant species
Filipendula ulmaria, Iris pseudacorus, Oenanthe crocata, Poa trivialis.

Physiognomy
Iris pseudacorus and *Oenanthe crocata* can be found as occasionals in western stands of the *Filipendula-Angelica* mire, but in the *Filipendulo-Iridetum pseudacori* they are both constant except in the far north of the range of the community: *O. crocata* does not extend much into Sutherland and it is absent from Orkney and Shetland (Perring & Walters 1962), although essentially similar vegetation to this does occur there (Birse 1980). In its characteristic form it is an often luxuriant and species-rich community in which *I. pseudacorus* is typically much more abundant than *F. ulmaria*, often a clear physiognomic dominant and particularly striking when it puts up its yellow flowers in late spring and summer. The flag is often almost a metre or so high and among this the *F. ulmaria* and *O. crocata* occur as scattered plants, though sometimes showing local abundance, a feature which is especially noticeable with the water-dropwort, the leafy shoots of which can thicken up to form almost pure patches with their stout flowering stems sticking up above the general level of the canopy.

There are nearly always some other tall herbs, though only *Lycopus europaeus, Rumex crispus* and *Scutellaria galericulata* occur with any frequency throughout the community, other species such as *Rumex acetosa, Lychnis flos-cuculi, Angelica sylvestris, Valeriana officinalis, Cirsium palustre, C. arvense,* and *Urtica dioica* often becoming common and conspicuous, though preferentially in particular kinds of *Filipendulo-Iridetum*. Other tall herbs, such as *Lythrum salicaria* and *Stachys palustris*, are never more than occasional, though this community can provide an important locus for them in northern Britain as they become more confined to the coast where suitable habitats are more common than inland. Then, sprawling among this herbage, there is often some *Galium palustre* or *G. aparine*.

Rushes and grasses frequently make an important contribution to lower tiers of the vegetation. *Juncus effusus* and *J. acutiflorus* are both quite common throughout, though especially characteristic of one of the sub-communities, where their bulky tussocks can become quite abundant. Among the grasses, *Poa trivialis* and *Agrostis stolonifera* are the most frequent species overall, and fairly often they form spreading mats or extensive carpets over the ground. Less common, and usually growing as scattered tussocks or clumps, are *Festuca rubra, Holcus lanatus, Elymus repens, Anthoxanthum odoratum, Arrhenatherum elatius* and *Dactylis glomerata*. Much more local, though sometimes growing lush and abundant where the ground is flushed with fresh water, is *Catabrosa aquatica*.

Among this cover can be found a variety of dicotyledons of small to moderate stature. Some typically occur as scattered plants, like *Ranunculus acris, Caltha palustris, Stellaria alsine, Mentha aquatica* and *Hydrocotyle vulgaris*, all of which are frequent through the community, while others, such as *Ranunculus repens* and *Potentilla anserina*, can form far-spreading stoloniferous mats with quite high local cover. Then, on patches of wet and more open ground, there can be locally prolific seeding of annuals such as *Polygonum hydropiper, Montia fontana, Stellaria media* and, particularly on cattle-poached mud, *Ranunculus sceleratus*. On transitions to salt-marsh, *Atriplex prostrata* and *Matricaria maritima* can also become frequent, with *Samolous valerandi, Oenanthe lachenalii* and halophytic herbs.

Bryophytes are usually few in number but they can form an extensive mat on occasion. *Eurhynchium praelongum* is the commonest species throughout, but *Calliergon cuspidatum* and *Rhytidiadelphus squarrosus* can also be found.

Sub-communities

***Juncus effusus-J. acutiflorus* sub-community.** This is the richest kind of *Filipendulo-Iridetum* in which other dicotyledons, rushes and grasses form a fairly consistent associated flora among the constants. *I. pseudacorus* is generally a clear dominant, although both *O. crocata* and *F. ulmaria* can be patchily abundant and *J. effusus* and/or *J. acutiflorus*, which are more common here than in the other sub-communities, can occur as prominent tussocks. Among other tall herbs, too, there are some strong preferentials: in addition to community species such as *Lycopus*, *Rumex crispus* and *Scutellaria galericulata*, *Rumex acetosa*, *Cirsium palustre*, *Epilobium palustre*, *Lychnis flos-cuculi*, *Angelica sylvestris* and *Valeriana officinalis* all occur occasionally to frequently and, when present in numbers, they give the vegetation a rich and colourful texture.

Somewhat smaller are frequent plants of *Ranunculus acris*, *Caltha palustris*, *Lotus uliginosus*, *Myosotis laxa* ssp. *caespitosa*, *Cardamine pratensis*, *Stellaria alsine*, *Mentha aquatica* with scattered tussocks of *Festuca rubra*, *Holcus lanatus*, *Anthoxanthum odoratum*, *Poa pratensis*, *Elymus repens* and *Carex otrubae*, and trails of *Galium palustre* spreading through the herbage. Then, beneath, there can be a patchy ground carpet of *Poa trivialis*, *Agrostis stolonifera*, *Ranunculus repens* and *Potentilla anserina* with, in more open, damp patches, annual herbs.

***Urtica dioica-Galium aparine* sub-community.** The vegetation here, though usually as tall and luxuriant as that of the *Juncus* sub-community, is noticeably less species-rich. *I. pseudacorus* is still very much the dominant and both *O. crocata* and *F. ulmaria* remain frequent and locally prominent but, apart from the community plants *Lycopus* and *Scutellaria galericulata*, the only really common taller dicotyledons are *Urtica dioica* and *Cirsium arvense* and, when these are present in abundance, the former as sometimes extensive patches, the latter as often numerous scattered individuals, they can give the herbage a distinctly weedy look. *Rumex crispus*, *R. acetosa* and *Cirsium palustre*, and both *J. effusus* and *J. acutiflorus*, can all be found occasionally, but species like *Angelica*, *Valeriana officinalis* and *Lychnis flos-cuculi* are scarce, and *Galium palustre* is almost always replaced by *G. aparine*. Sometimes, too, there is a little *Stachys palustris* or *Geranium robertianum*.

Grasses are often quite a conspicuous element in the vegetation, both *Poa trivialis* and *Agrostis stolonifera* very common and occurring as patchy carpets. *Elymus repens* is also preferentially frequent here, forming scattered tufts or sometimes extensive mats, and tussocks of *Arrhenatherum* and less commonly, *Dactylis glomerata* can sometimes be found beyond the reach of tidal flooding which occasionally inundates the lower reaches of stands of this vegetation. On tidal litter, mainly decaying brown algae, deposited among the stools, annuals can show a flush of growth, with *Stellaria media* being particularly frequent. *Ranunculus repens* and *Potentilla anserina* can spread over the ground and scattered plants of *Stellaria alsine*, *Mentha aquatica*, and *Hydrocotyle* are sometimes found.

***Atriplex prostrata-Samolus valerandi* sub-community.** Although *I. pseudacorus* and *O. crocata* can both be abundant in this kind of *Filipendulo-Iridetum*, the taller associates prominent elsewhere in the community are noticeably lacking: even *F. ulmaria* is scarce here and occasional scattered plants of *Lycopus* and *Rumex crispus* often represent the limit of the development of this particular element. The cover of the herbage can be a little more open than usual, although among the smaller plants, too, there are some distinctive features. The commonest grasses, for example, are *Agrostis stolonifera* and *Festuca rubra*, the former in particular often showing quite extensive ground cover, with *Poa trivialis*, usually a constant of the community, very scarce. Then, though *Stellaria alsine*, *Mentha aquatica* and *Hydrocotyle* can be found occasionally, more common is a group of preferentials tolerant of saline habitats. *Atriplex prostrata* is the most frequent of these and it can be quite numerous over litter and patches of exposed ground, but *Samolus valerandi*, *Oenanthe lachenalii*, *Matricaria maritima*, *Triglochin maritima* and *Glaux maritima* have also been recorded in some stands. Bryophytes are typically more sparse than elsewhere in the community, with occasional *Eurhynchium praelongum* often being the sole representative.

Habitat

The *Filipendulo-Iridetum* is confined to moist, more nutrient-rich soils along the oceanic seaboard of Britain. It is especially characteristic of the fresh-water seepage zone along the upper edge of salt-marshes in the sheltered sea-lochs of western Scotland, but it can be found in a variety of other habitats which combine congenial edaphic conditions with a mild climate. It thus occurs down the west coast over moist stabilised shingle and in wetter hollows and flushes on raised beach platforms and over gentle cliff slopes, the peculiarities of these situations being reflected in the floristics of the different sub-communities. It does not seem to be heavily grazed as a rule, but looks nonetheless to be a relatively stable vegetation type, perhaps persisting long before any progression to scrub or woodland.

This community can be seen as the oceanic counterpart of the *Filipendula-Angelica* mire. That vegetation type occurs throughout lowland Britain, its range being controlled largely by the distribution of suitably moist and rich soils. The *Filipendulo-Iridetum* is found over similar profiles but only in the lowlands of the far-west of the country where the climate is at its most equable. There, in a zone running down from Orkney and Shetland, skirting the north and west coasts of Scotland and taking in the extremities of Anglesey and Lleyn, west Dyfed and Cornwall, February minima are above freezing (*Climatological Atlas* 1952) and the annual range of temperatures is the smallest of any part of Britain (Page 1982). Among the more common plants of the community, it is *I. pseudacorus*, a generally Atlantic species in Europe, and *O. crocata*, more strictly Oceanic West European, which provide the clearest floristic response to such climatic conditions and the best single separation from the *Filipendula-Angelica* mire. Less pronounced, there is the frequency through much of the *Filipendulo-Iridetum* of *Juncus effusus* and *J. acutiflorus*, species which, by and large, only become really prominent in meadowsweet vegetation in the more westerly parts of the country.

Apart from the two oceanic constants, the community shares many species with the *Filipendula-Angelica* mire, though the less deep shade cast by the flag canopy here means that the associated flora is rather more consistent and abundant than beneath dense meadowsweet. Much of this common character comes from the nature of the soils that both vegetation types favour, with species of Molinio-Arrhenatheretea damp grasslands and mires predominating. Thus the frequency of plants such as *Festuca rubra*, *Holcus lanatus*, *Poa trivialis*, *Ranunculus acris*, *Rumex acetosa*, *Angelica sylvestris*, *Cirsium palustre*, *Equisetum palustre*, *Filipendula ulmaria*, *Juncus effusus*, *J. acutiflorus*, *Lotus uliginosus* and *Lychnis floscuculi*, which provide so much of the general character of this vegetation, and which become particularly important in the most distinctive of the sub-communities, the *Juncus* type, all reflect the moist, and at least moderately nutrient-rich, character of the profiles.

However, though this is a community of soils which are kept damp or wet through much or all of the year, and which are maintained in a mesotrophic or eutrophic state, the particular situations in which these general conditions are met are quite varied. Sometimes, the *Filipendulo-Iridetum* can be found around the kind of waterlogged hollows and flushes that, further inland, would carry the *Filipendula-Angelica* mire and, in sites set a little way back from the coast, the two vegetation types can come very close in their floristics: the *Filipendulo-Iridetum* can be found in this kind of setting, marking out wetter stretches of ground in marginal pastoral land in Anglesey and over the raised beach platforms of western Scotland. With the move northwards, however, the zone that is congenial for the development of the community becomes progressively narrower, as either unsuitably impoverished soils

stretch almost to the edge of flatter coasts, or high mountains rising close to the sea bring a quick shift to an inhospitable climate. In this part of Britain, therefore, the *Filipendulo-Iridetum* is effectively squeezed into maritime habitats, forming a fringe to salt-marshes and shingle beaches, though there are no species running through the community as a whole which indicate a maritime influence in the strict sense, that is, a pronounced input of sea-salts from spray or inundating waters.

In such situations as these, the community can be found over a variety of soil types, ranging from virtually pure shingle through thin layers of organic matter resting unconformably on gravel or sand to orthodox gleys. In many cases, however, there is obvious seepage of fresh-water through the upper horizons of the soil and over the surface, either from springs or flushes which can debouch right down to high water mark, or from drainage waters running in a more ill-defined fashion from slopes above. Such wetter conditions, frequently juxtaposed with saline habitats, are what encourage the occurrence here of species such as *Lycopus* and *Scutellaria galericulata*, common enough in swamps and fens further south in Britain, but increasingly confined to the west in Scotland, as suitable habitats become scarce. And the particular kind of gentle surface-water flushing that often characterises these situations encourages a further floristic element in many stands, with Elymo-Rumicion plants such as *Agrostis stolonifera*, *Potentilla anserina* and *Rumex crispus* quite frequent throughout, and Bidentetea species, such as the annuals *Polygonum hydropiper* and *Ranunculus sceleratus*, and *Catabrosa aquatica*, able to capitalise on the occasional exposure of areas of bare wet ground.

Provided such flushing is fairly constant, the *Filipendulo-Iridetum* can extend down into the zone of the upper shore which is subject to rare flooding by exceptional spring tides, and even the *Juncus* sub-community, which is the usual form of the *Filipendulo-Iridetum* in flushes beyond extreme high water mark, can be found in such situations. Towards the lower limit of the occurrence of the community, however, where periods of dry weather may produce locally high salinities (Gillham 1957b), the *Atriplex-Samolus* type is particularly characteristic, with the salt-tolerant Bidentetea species *Atriplex prostrata* becoming frequent and Asteretea halophytes like *Glaux* and *Triglochin maritima* able to persist.

In stands with vigorous flushing, the through-put of water can bring some enrichment with dissolved salts or by the deposition of silty material washed down from above. Where the *Filipendulo-Iridetum* occurs on the upper edge of the shore, however, it may receive more obvious additions of nutrients in the form of tidal litter, algal remains and other drift material getting trapped among the herbage and slowly rotting. The *Urtica-Vicia* sub-community, with its prominent contingent of eutrophic plants like *Urtica* and *Galium aparine*, and weedy species able to spread on the matted detritus, such as *Cirsium arvense* and *Stellaria media*, is especially typical of such situations and it can even be found over raw shingle, provided this is kept moist and frequently enriched with drift. In more saline conditions, such vegetation can overlap a little in its floristics with the *Atriplex-Samolus* sub-community, whereas on the upper fringes of stands, the appearance of tussocks of *Arrhenatherum* and *Dactylis* create the impression of a wet and weedy mesotrophic grassland.

Although the *Filipendulo-Iridetum* is often found on or adjacent to salt-marshes that are open to stock or within coastal pastures, it seems itself to suffer few effects of grazing. *I. pseudacorus* is sometimes eaten when other herbage is in short supply but it is not particularly palatable and *O. crocata*, of course, is deadly; and it may be that an abundance of these plants affords some protection to more vulnerable ones like *F. ulmaria* and some of the other hemicryptophyte dicotyledons. Around the edges of stands, where stock push in to reach drinking water, trampling may create poached patches which afford a chance for the spread of annuals and dunging can result in locally-enriched conditions, but the herbage seems rarely to be grazed hard back. At the same time, woody plants seem to get a hold only with difficulty so that the vegetation remains essentially stable.

Zonation and succession

The *Filipendulo-Iridetum* can be found around springs and flushes close to the coast in zonations which largely reflect variation in soil moisture, acidity and trophic state and differences in treatment on the surrounding ground, but the most striking vegetation patterns are to be seen where the community occurs on the upper part of the sea-shore where it can pass to salt-marsh and shingle communities with an increase in maritime influence. In such habitats particularly, the *Filipendulo-Iridetum* appears to be a fairly natural vegetation type and one which develops but slowly to woodland.

In situations set back a little from the coast, the community can be found in very much the same kind of zonations as the *Filipendula-Angelica* mire, especially its *Juncus-Holcus* sub-community. In the more oceanic parts of the country, the *Juncus* sub-community of the *Filipendulo-Iridetum* replaces that vegetation type, occupying gleyed mesotrophic soils in flushes and hollows over gently-sloping ground near the coast. Usually, the landscape context in this kind of situation is one of marginal pastoral agriculture, with the community surrounded by relatively unimproved, grazed swards of the Nardo-Galion, like the *Festuca-Agrostis-Galium* grass-

land, or of the Junco-Molinion, like the *Molinia-Potentilla* mire, or, where there is a sharper transition to damp, nutrient-poor peat, by the *Scirpus-Erica* wet heath. Where grazing stock penetrate some way on to the wetter, mesotrophic soils around the flushes, a zone of the *Juncus-Galium* rush-pasture may be interposed between the *Filipendulo-Iridetum* and the drier grasslands around, Molinietalia herbs such as *Filipendula ulmaria*, *Angelica* and *Cirsium palustre* running out far into the rush-dominated vegetation together with ranker grasses like *Holcus* and *Festuca rubra*. Where the land has been improved a little by fertiliser application and re-seeding, then the whole sequence of vegetation types can be shifted somewhat, with the *Filipendulo-Iridetum* passing to the *Holco-Juncetum* and thence to *Lolio-Cynosuretum*. Such patterns can be seen on Anglesey and have been described from Mull and Iona (Gillham 1957b) and from Arran (Adam *et al.* 1977). Over the distinctive machair topography that has developed over the raised beaches in the latter sites and elsewhere in western Scotland, a local variation on these kinds of zonations can be seen where the influence of the consolidated shell-sand overlying much of the ground becomes paramount: then the *Festuca-Galium* dune grassland can become prominent in the sequence, partly or wholly replacing the Nardo-Galion swards around the *Filipendulo-Iridetum*.

Only in a few places is the *Filipendulo-Iridetum* contiguous with both these kinds of communities on the drier and/or less nutrient-rich soils to the landward side and also more maritime vegetation types of salt-marsh and shingle below. Very often now, where the community occurs in these latter kinds of habitats, its upper boundary is an artificial one, either a road or an abrupt transition to improved pasture or cultivated ground. Down-shore, however, the zonations are often very distinctive, particularly around the margins of the western Scottish sea-lochs. Here, Adam (1978, 1981) has described a characteristic regional assemblage of salt-marsh vegetation types, not numerous but, taken together, often species-rich, and frequently developed over rather unusual substrates, highly organic and sometimes amounting to just a thin layer of peat resting on shingle. Such sequences can be seen in whole or part in sites like Loch Ranza on Arran, Loch Gruinart and the Sorn estuary on Islay, Loch Creran in Argyll, Loch Ainort and Loch na h'Airde on Skye, Loch Kishorn, Loch Torridon and the Broom lochs in Wester Ross, on the Sutherland coast and in the Outer Hebrides (Adam *et al.* 1977, Adam 1978), as well as on Mull and Iona (Gillham 1957b) and Shetland (Hilliam 1977, Roper-Lindsay & Say 1986). In the more sheltered of these situations, the *Filipendulo-Iridetum* usually gives way below to one of the richer forms of the *Juncetum gerardi*, often the *Carex* sub-community, less commonly the

Leontodon type and, although *I. pseudacorus*, *O. crocata* and *F. ulmaria* all stop rather abruptly with the move on to more frequently-inundated ground, there is often a strong continuity between the vegetation types in the lower tier of the herbage, with such species as *Agrostis stolonifera* and *Potentilla anserina* running throughout as a patchy carpet and *Festuca rubra* occurring in both. Both the *Juncus* and *Atriplex-Samolus* sub-communities of the *Filipendulo-Iridetum* can be found in such zonations, though the latter type penetrates further into more saline situations and shows a stronger affinity with the salt-marsh swards through the occurrence of Asteretea herbs. At some sites, the junction between the *Juncetum gerardi* and the *Filipendulo-Iridetum* is made more striking by the scattered presence of small patches of the *Blysmetum rufi* in wet shingly hollows and flushes, and then down-shore there is typically a transition to the *Puccinellia*-turf fucoid sub-community of the *Puccinellietum*. On more exposed coasts, the *Juncetum gerardi* may be the only real salt-marsh vegetation present below the *Filipendulo-Iridetum* and even it may be fragmented into small patches of grasses and *Potentilla anserina* with Bidentetea herbs like *Atriplex prostrata* and *Polygonum hydropiper* scattered on the bare substrate (Gillham 1957*b*).

In these Scottish sea-lochs, the salt-marsh often seems to have developed over a shingle base but where such material directly underlies the vegetation, and particularly where considerable amounts of drift-line detritus are deposited, there is a tendency for the *Filipendulo-Iridetum* to be represented by the *Urtica-Galium* sub-community and for it to grade to eutrophic strandlines where the shingle becomes drier. *Galium aparine*, *Rumex crispus* (often obviously var. *littoreus*), *Atriplex prostrata*, *Stellaria media*, *Agrostis stolonifera*, *Festuca rubra* and *Elymus repens* can all provide continuity with the open assemblages of the *Atriplex-Urtica* strandline, a further community which adds regional distinction to these western Scottish zonations.

The accumulation of marsh deposits to heights capable of supporting the non-halophytic vegetation of the *Filipendulo-Iridetum* can be considered part of the continuous process of sediment accumulation, but it is hard to see the community as an end-point of marsh succession when it depends so obviously on continuous irrigation by fresh-water draining from the land. It is, rather, an oceanic mire type which happens to find its most striking expression on the maritime fringe, at the junction between fresh and saline habitats. Whether it can progress further in seral developments to scrub and woodland is a separate issue, though it is true that, on the salt-marsh fringe, exposure to winds and rare episodes of tidal flooding may make invasion of woody plants particularly unlikely. In some coastal situations, the *Filipendulo-Iridetum* can be found closely juxtaposed with alder-dominated woodland, though there is no direct evidence of the one having developed into the other. Further inland, where the community can be seen among stretches of wet scrub and woodland, such successions are perhaps more likely, though even there the dense canopy of the herbs may cast a shade inimical to seedlings and young saplings.

Distribution

The *Filipendulo-Iridetum* is largely confined to the west coast of Britain and is especially well developed in Scotland. Scattered stands occur in south-west England and west Wales but here the community may have been widely destroyed in its salt-marsh habitat by interference with the transitional upper zones. In Scotland, from Arran round to north Sutherland, and into Orkney and Shetland, it is probably virtually ubiquitous in suitable situations, with scattered localities too down the north-east coast.

Affinities

This colourful community received but brief mention in the literature (Gillham 1957*b*, Gimingham 1964*a*, Slack 1970) before the detailed accounts of Adam (1976, 1978, 1981) and Adam *et al.* (1977). The treatment here relies largely on published and unpublished data from Adam, although it also uses material from Hilliam (1977, see also Roper-Lindsay & Say 1986) and Birse (1980). Adam *et al.* (1977) recognised considerable heterogeneity within the *Filipendulo-Iridetum* as then defined, with the possible characterisation of two noda. With the additional data now available, and with a wider perspective on mire vegetation of this general kind throughout Britain, it is certainly sensible to retain a community of this type, though it can be more precisely defined than in the earlier accounts, with some of the samples included there transferred to the *Filipendula-Angelica* mire. It has also become possible to define the three sub-communities but, although they seem clearly influenced by particular environmental conditions, further sampling is needed to understand these relationships fully.

In general terms, the affinities of the *Filipendulo-Iridetum* are clearly with the Molinio-Arrhenatheretea but beyond that its placement is problematical. Adam *et al.* (1977) argued that its distinctively oceanic character favoured a location in the Anagallido-Juncetalia, a new order which Braun-Blanquet (1967) proposed for a range of mire, flush and wet-meadow vegetation types in northern Spain, but which they suggested could be extended around the western seaboard of Europe. Certainly, similar vegetation to the *Filipendulo-Iridetum* has been reported from other oceanic areas, among coastal rocks in western Norway (Nordhagen 1922, Skogen 1971, 1973) and on shingle there (Nordhagen 1940, Dahl & Hadač 1941) and from western Ireland (Praeger

1934), from northern Spain (Tüxen & Oberdorfer 1958, Bellot 1966) and as far afield as north Africa (Dahlgren & Lassen 1972). But such a solution seems unnecessarily drastic, for the community could be accommodated fairly easily in the Filipendulion alliance of the Molinietalia. Certainly, it is distinctively oceanic in comparison with the *Filipendula-Angelica* mire which includes most British Filipendulion vegetation, but this trend can be seen too in the Juncion acutiflori, the Calthion and the Cynosurion, in each of which *I. pseudacorus* increases with the shift westwards.

Floristic table M28

	a	b	c	28
Iris pseudacorus	V (2–9)	V (7–9)	V (6–8)	V (2–9)
Oenanthe crocata	V (1–7)	IV (2–6)	IV (3–8)	IV (1–8)
Filipendula ulmaria	V (2–7)	IV (3–5)	I (6)	IV (2–7)
Poa trivialis	IV (3–7)	V (2–6)	I (4)	IV (2–7)
Juncus effusus	V (2–4)	II (2–4)		III (2–4)
Ranunculus acris	IV (2–4)	II (2–3)	II (2–3)	III (2–4)
Rumex acetosa	IV (2–5)	II (3–4)	I (1)	III (1–5)
Juncus acutiflorus	IV (1–5)	II (3–4)	I (2)	III (1–5)
Galium palustre	IV (2–4)	II (3)		III (2–4)
Cirsium palustre	IV (1–3)	II (1–2)		III (1–3)
Caltha palustris	III (1–4)	I (2)	II (1–3)	II (1–4)
Calliergon cuspidatum	III (3–6)	I (4)	I (6)	II (3–6)
Anthoxanthum odoratum	III (2–4)	II (3–4)		II (2–4)
Lotus uliginosus	III (2–5)	I (3)	I (3)	II (2–5)
Epilobium palustre	III (2–4)	I (2)		II (2–4)
Lychnis flos-cuculi	III (2–3)		I (2–3)	II (2–3)
Cardamine pratensis	II (2–3)	I (2)	I (3)	II (2–3)
Angelica sylvestris	II (2–3)	I (2–3)		I (2–3)
Valeriana officinalis	II (2–4)	I (3)		I (2–4)
Myosotis laxa caespitosa	II (2–3)		I (2)	I (2–3)
Poa pratensis	II (2–4)			I (2–4)
Vicia cracca	I (2–4)			I (2–4)
Festuca arundinacea	I (2–4)			I (2–4)
Lathyrus pratensis	I (3–4)			I (3–4)
Phragmites australis	I (3–4)			I (3–4)
Galeopsis tetrahit	I (2)			I (2)
Juncus inflexus	I (3–4)			I (3–4)
Polygonum hydropiper	I (1–4)			I (1–4)
Lophocolea bidentata s.l.	I (2)			I (2)
Carex ovalis	I (3)			I (3)
Galium aparine	I (2–3)	IV (2–5)	II (3–4)	II (2–5)
Urtica dioica	I (1–2)	IV (3–8)	I (3)	II (1–8)
Cirsium arvense	I (2–3)	IV (2–4)		II (2–4)
Elymus repens	II (2–3)	III (3–4)	I (2)	II (2–4)
Stellaria media	II (1–4)	III (3–4)		II (1–4)
Arrhenatherum elatius	I (2–3)	II (2–5)		II (2–5)
Dactylis glomerata	I (2)	II (2–3)		I (2–3)
Plantago lanceolata	I (2)	II (2)		I (2)
Stachys palustris	I (3)	II (3)		I (3)

Floristic table M28 *(cont.)*

	a	b	c	28
Geranium robertianum		II (2–4)		I (2–4)
Atriplex prostrata		I (2)	IV (1–3)	I (1–3)
Samolus valerandi			III (1–3)	I (1–3)
Matricaria maritima	I (1–2)	I (2–3)	II (2–3)	I (1–3)
Oenanthe lachenalii		I (2)	II (3)	I (2–3)
Triglochin maritima			II (2–3)	I (2–3)
Glaux maritima			II (2)	I (2)
Agrostis stolonifera	III (2–6)	IV (3–5)	IV (3–5)	III (2–6)
Eurhynchium praelongum	III (2–5)	III (3–4)	II (1–3)	III (1–5)
Lycopus europaeus	III (2–7)	III (2–4)	II (1–3)	III (1–7)
Stellaria alsine	III (2–3)	III (2–3)	II (1–2)	III (1–3)
Rumex crispus	III (1–3)	II (2–3)	II (3)	II (1–3)
Festuca rubra	III (3–5)	II (3)	II (2–3)	II (2–5)
Ranunculus repens	III (2–5)	III (3–4)	I (3)	III (2–5)
Potentilla anserina	III (2–6)	III (3–4)	I (5)	III (2–6)
Mentha aquatica	III (2–6)	II (3)	II (2–4)	II (2–6)
Scutellaria galericulata	III (2–5)	III (3–7)		III (2–7)
Holcus lanatus	III (3–6)	III (3–4)		III (3–6)
Hydrocotyle vulgaris	II (3–6)	II (3)	II (3–4)	II (3–6)
Carex otrubae	II (2–4)		II (2)	II (2–4)
Rhytidiadelphus squarrosus	II (2–5)	II (2–3)		II (2–5)
Equisetum palustre	I (2–3)	I (3)	I (3)	I (2–3)
Montia fontana	I (2–4)	I (2)		I (2–4)
Deschampsia cespitosa	I (2)	I (2–5)		I (2–5)
Lythrum salicaria	I (2–3)	I (2–3)		I (2–3)
Galium uliginosum	I (3)	I (3)		I (3)
Rumex conglomeratus	I (2–5)	I (4)		I (2–5)
Eleocharis palustris	I (1–2)		I (2)	I (1–2)
Scirpus maritimus	I (3)		I (3)	I (3)
Number of samples	16	9	5	30
Number of species/sample	30 (19–42)	20 (10–26)	18 (15–21)	25 (10–42)
Vegetation height (cm)	75 (30–175)	88 (20–120)	80 (40–120)	80 (20–175)
Vegetation cover (%)	99 (95–100)	100	88 (75–100)	95 (75–100)

a   *Juncus* spp. sub-community
b   *Urtica dioica-Galium aparine* sub-community
c   *Atriplex prostrata-Samolus valerandi* sub-community
28   *Filipendulo-Iridetum pseudacori* (total)

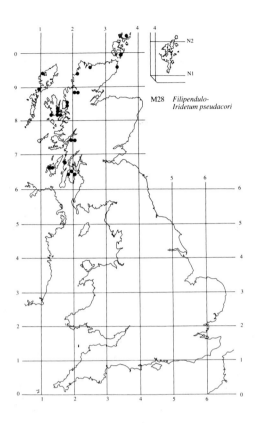

M28   *Filipendulo-
Iridetum pseudacori*

M29
Hypericum elodes-Potamogeton polygonifolius soakway
Hyperico-Potametum polygonifolii (Allorge 1921) Braun-Blanquet & Tüxen 1952

Synonymy
Valley moor plashes Rankin 1911*b p.p.*; *Hyperico-Potamogetonetum sensu* Birse 1980, Ratcliffe & Hattey 1982, Dierssen 1982; Flushing Water Community NCC New Forest Bogs Report 1984 *p.p.*

Constant species
Hypericum elodes, Juncus bulbosus/kochii, Potamogeton polygonifolius, Ranunculus flammula, Sphagnum auriculatum.

Rare species
Galium debile, Pilularia globulifera.

Physiognomy
The *Hyperico-Potametum polygonifolii* has a very distinctive appearance, typically consisting of low creeping or floating mats of *Hypericum elodes* and *Potamogeton polygonifolius*, the former prominent with its soft tomentose shoots, the flowering ones shortly erect, against the olive-green foliage of the latter, lying flat on the surface of the water or the wet soil of the small runnels and pools in which this vegetation characteristically grows. Very often, except where the ground has been badly trampled by grazing animals, a feature of some of the stands included in Ratcliffe & Hattey (1982) and the NCC New Forest Bogs Report (1984), these two constants are set in a carpet of more or less submerged *Sphagnum auriculatum*, sometimes accompanied by *S. cuspidatum, S. palustre* or *S. recurvum*, and very occasionally with small patches of *S. papillosum* growing on the slightly raised areas. But other bryophytes are typically sparse: some stands show a modest local abundance of *Polytrichum commune* or *Aulacomnium palustre*, and *Drepanocladus exannulatus, D. revolvens* and *Calliergon cuspidatum* can also be found, but enrichment of this element of the vegetation is rare.

Neither is the associated vascular flora very extensive or consistent, being frequently confined to scattered plants of a variety of bog and poor-fen herbs, growing among the dominants or projecting above them as a very open canopy. Only *Juncus bulbosus* (very occasionally recorded as *J. kochii*) and *Ranunculus flammula* attain constancy among the companions but, of smaller species, *Hydrocotyle vulgaris, Anagallis tenella, Drosera rotundifolia, Narthecium ossifragum* and *Galium palustre*, all occur with moderate frequency, together with such sedges as *Carex demissa, C. echinata, C. panicea* and *C. nigra*. Then, there can be sparse shoots or small patches of *Molinia caerulea, Agrostis canina* ssp. *canina, Juncus articulatus, J. effusus, J. acutiflorus, Eleocharis multicaulis, Eriophorum angustifolium* and *Rhynchospora alba. Carex rostrata* can also be found in some stands, occasionally with small amounts of *Equisetum fluviatile, Menyanthes trifoliata* and *Potentilla palustris*. Typically, all these plants remain of minor structural importance although many of them are abundant in other vegetation types that can occur in close association with the *Hyperico-Potametum*, so they sometimes figure a little more prominently in samples from stands that are found in fine mosaics or more ill-defined zonations.

Apart from *H. elodes*, the community also provides an occasional locus for other Oceanic West European plants like *Scutellaria minor* and, in south-west Scotland, *Carum verticillatum* (Birse 1980). *Scirpus fluitans*, largely confined in this country to more westerly localities, can also be found here, sometimes with very considerable abundance, co-dominant in the floating carpet. And the nationally-rare *Galium debile* is confined to the *Hyperico-Potametum* and closely-related vegetation in its major British locality in the New Forest in Hampshire, and was formerly recorded in this kind of soakway in Devon: it closely resembles *G. palustre*, but can be distinguished from it by its narrower leaves, particularly on submerged shoots, and by the fact that its fruiting shoots are never reflexed.

Other interesting plants that can be found in association with *Hyperico-Potametum* species are *Baldellia ranunculoides*, looking rather like a slim version of

Alisma plantago-aquatica and, in the same way, growing emergent from shallow pools and soaks, and a more truly aquatic element with *Apium inundatum*, *Sparganium angustifolium*, *Myriophyllum alternifolium* and the rare fern *Pilularia globulifera*. For the present, such distinctive stands have been retained here, but further sampling may well characterise them as a distinct sub-community of the *Hyperico-Potametum* of wetter and muddier situations or as a separate community altogether.

Habitat

The *Hyperico-Potametum* is characteristic of shallow soakways and pools in peats and peaty mineral soils with fluctuating waters, moderately acid to neutral and probably quite oligotrophic. It is a community of southern and western Britain, occurring in seepages and runnels around mires and in heathland pools at moderate altitudes. It is sometimes grazed and trampled.

The situations occupied by this vegetation are very distinctive but little understood. The waters are typically fresh and clear, still or only gently-flowing and with a pH in available samples generally between 4 and 5.5. Calcium concentrations are probably low in most cases and the character of the vegetation suggests that low availability of phosphorus and relatively slow turnover of nitrogen limit growth, but these suppositions would well repay investigation. Certainly, there is a general similarity in the associated flora with the bog-pool communities of the Rhynchosporion and the poorer water-tracks of the Caricion nigrae, with the extensive *Sphagnum* carpet and scattering of bog and poor-fen herbs. But one important difference between the habitats is that here there is a fluctuating water-table, subject to seasonal variation, and perhaps also to more short-term changes through the year. The vegetation is thus inundated to shallow depth for much of the time, but at or above the water-level for most of the summer months, particularly in smaller hollows which can dry out completely then. It is such conditions that can be so well exploited by the distinctive dominants of the community, *H. elodes* and *P. polygonifolius*, and also by *J. bulbosus/kochii* and the less common *Scirpus fluitans*, all rhizomatous or nodal-rooting plants able to form extensive mats, whether submerged or over the surface of damp ground. And differences in the degree of wetness probably have some influence on floristic variation within the community, with such species as *Apium inundatum*, *Sparganium angustifolium*, *Baldellia* and *Pilularia* characteristic of a lower tier in a depth-related zonation.

In general terms, then, this type of vegetation is typical of more base- and nutrient-poor soils where fluctuation of the water-table, often also with a modest through-put, ameliorates the extremes of stagnant waterlogging and superficial drying-out. But it also shows quite a striking geographical confinement, being characteristic of the warm oceanic parts of the country, where February minima are usually at least a degree above freezing (*Climatological Atlas* 1952). Such a zone, which runs along the south coast of England and then takes in the South-West Peninsula, most of the Welsh lowlands, the fringe of Cumbria and the western Scottish seaboard, corresponds more or less with the distribution of *H. elodes*, an Oceanic West European plant which is strongly preferential to this kind of vegetation throughout its range (Braun-Blanquet & Tüxen 1952, Schoof van Pelt 1973, Dierssen 1975). Other oceanic features of the community are the abundance of *Sphagnum auriculatum*, also the typical species in western Rhynchosporion and Caricion nigrae vegetation, and the occasional occurrence of *Scutellaria minor* and *Carum verticillatum*.

Throughout this region, by far the most widespread kind of habitat where the general requirements of the *Hyperico-Potametum* are met, is provided by seepage areas on or around valley mires and base-poor flushes at low to moderate altitudes and along the shelving margins of slow-moving moorland streams, where the soils are typically peats or gleys with a humic top. The community can also be found much more locally over the latter type of profile in shallow seasonal pools on heathlands. These habitats, and particularly the pools, often present small, fragmented and isolated situations, something which probably has an influence on the considerable variation that can be seen in the composition of stands from place to place, with fine local differences in environmental conditions and also chance influencing colonisation.

Trampling by grazing animals, which often have free access across the kind of landscapes in which the community occurs, can also play a part in keeping the vegetation open and varied, though heavy poaching can be deleterious to the *Sphagnum* carpet (Ratcliffe & Hattey 1982, NCC New Forest Bogs Report 1984). Grazing may also set back direct invasion of the ground by woody plants. Conditions are anyway too impoverished to support colonisation by an abundance of the grazing-sensitive tall herbs characteristic of fens, but the moist soil surface can present a congenial environment for the germination of the seeds of trees such as *Salix cinerea* and seedlings could perhaps get away if they were to escape both flooding and being bitten off.

Zonation and succession

The *Hyperico-Potametum* typically occurs as small and sometimes fragmentary stands in close association with bog, wet heath, poor fen and aquatic vegetation in zonations and mosaics which probably reflect the duration and extent of inundation and the base- and nutrient-richness of the ground waters. With continuing impoverished conditions, it is probably essentially stable, though in transitional situations grazing may be important in curtailing any successional developments.

In its most common situation on valley mires in southern and south-western Britain, the community is usually found within or around tracts of the *Narthecio-Sphagnetum*, the characteristic vegetation type of the acid peats of this region accumulating under stagnant waterlogging. Where lines of surface seepage across the peats are clear, the mosaic of communities can be accordingly well-defined, with patches of the dominant *H. elodes* and *P. polygonifolius* picking out the stands, scattered over or snaking across the bog. But, in other cases, where the influence of the trickling and fluctuating waters is diffuse, transitions are much less obvious, with *Sphagnum auriculatum* and such herbs as *Narthecium*, *Drosera rotundifolia*, *Eriophorum angustifolium* and *Rhynchospora alba* providing strong continuity with the wetter lawns of the mire surface and its *Sphagnum auriculatum* bog pools.

Quite often, these kinds of *Hyperico-Potametum* soakways are concentrated in a distinct zone on such valley mires. Sometimes, where waters seeping from the slopes around are channelled along the bog edge, the community marks out a marginal belt between the *Narthecio-Sphagnetum* and the *Ericetum tetralicis* wet heath that characteristically occupies the seasonally gleyed soils of slightly higher ground. In other situations, it can occur along the inner side of the *Narthecio-Sphagnetum*, between it and an axial strip of swamp and woodland: here there can also be some seepage and fluctuation of waters, but little enrichment beyond the immediate fringes of the central line of the mire where flow is concentrated.

In both these types of zonation, variations on which are well shown in the NCC New Forest Bogs Report (1984), the *Hyperico-Potametum* is usually clearly marked off from the wet heath, swamp and woodland communities that bound it on the side away from the *Narthecio-Sphagnetum*. But, where there is gradual enrichment of the waters as they flow from less impoverished ground, or concentrate the nutrients from ever-wider areas, transitions can be more complex. Then, some kind of poor-fen or Junco-Molinion vegetation often occurs in close association with the community. Among the former, the *Carex echinata-Sphagnum* and *Carex rostrata-Sphagnum* mires can both be found grading to the *Hyperico-Potametum*, the *Sphagnum* carpet running throughout, but species such as *C. rostrata*, *C. demissa*, *C. echinata*, *C. nigra*, *C. panicea*, *Juncus effusus* or *J. acutiflorus* thickening up as dominants or co-dominants. *Molinia* can also be locally prominent within such vegetation, but it is much more abundant, often as vigorous tussocks, in the *Molinia-Potentilla* mire and the *Cirsio-Molinietum*, both of which, and particularly the former, can mark out zones of soligenous enrichment on somewhat drier ground adjacent to the *Hyperico-Potametum*, with associates such as *Succisa pratensis*, *Angelica sylvestris* and *Cirsium palustre* becoming frequent.

Similar patterns to these latter can be found throughout the south and west of Britain, on lowland heaths and the moorland fringes, where the community marks out seepage and fluctuation zones around base-poor flushes and basins (e.g. Ratcliffe & Hattey 1982). Locally, the *Hyperico-Potametum* is also seen in seasonally-flooded pools within tracts of wet heath and poor fen and, where these are sufficiently deep, there can be the kind of zonation within stands of the community mentioned earlier. Such layering represents the beginning of a transition to truly aquatic vegetation that occurs more fully where soakaways run out of tracts of poor fen and heath into the margins of permanent streams and pools. In the former, where there is an increase in water flow, the aquatic vegetation is often sparse, but around the shelving shores of pools and lakes, the *Hyperico-Potametum* can pass downwards to other Littorelletea communities with plants like *Littorella uniflora* itself, *Myriophyllum alternifolium* and *Isoetes lacustris* and *I. setacea*.

Even in their more fully-developed form, such sequences probably rarely, if ever, represent a hydroseral succession and in most situations the community appears to be a stable vegetation type marking out zones where accumulation of any nutrient capital, whether by concentration of dissolved salts or by silting, is unlikely to proceed very far. Where there is some modest enrichment, grazing and trampling may help continually set back any tendency to succession.

Distribution
Available samples of the *Hyperico-Potametum* extend in a well-defined zone from west Surrey, through the New Forest to the South-West Peninsula, up through Wales and into Galloway. It seems likely that the full range of the community roughly matches that of *H. elodes*, so occurrences might be expected further north through the Inner and Outer Hebrides and at its scattered localities in eastern England.

Affinities
This is a little-described vegetation type but one which has been readily characterised from among the suites of bog and poor-fen communities with which it is usually found (Ratcliffe & Hattey 1982, NCC New Forest Bogs Report 1984) and which has also been recorded in almost identical form from Ireland (Braun-Blanquet & Tüxen 1952), where it has provided one locus for the introduced *Hypericum canadense* (Webb 1957, Webb & Halliday 1973), and from other parts of western Europe (Schoof van Pelt 1973, Dierssen 1975), where it has been

designated as a modified version of the *Hyperico-Potametum polygonifolii* (= *oblongi*) Allorge (1926).

Such vegetation clearly belongs among the Littorelletea but there are considerable difficulties in characterising it from among what are probably best regarded as impoverished stands lacking one or other of the few constants, and in separating it from stands in which such associates as *Eleocharis multicaulis*, *Baldellia ranuncu-* *loides*, *Pilularia globulifera* and *Scirpus fluitans* are dominant with some of the same occasionals as are recorded here. This kind of vegetation has certainly been under-sampled in the project and further study may well characterise the variety of communities recognised in Ireland by White & Doyle (1982). Meanwhile, related stands are grouped together in the next ill-defined assemblage.

Floristic table M29

Hypericum elodes	V (3–9)	*Sphagnum cuspidatum*	I (3–7)
Potamogeton polygonifolius	IV (2–7)	*Succisa pratensis*	I (2–4)
Ranunculus flammula	IV (1–6)	*Utricularia minor*	I (2–4)
Sphagnum auriculatum	IV (1–10)	*Campylium stellatum*	I (1–4)
Juncus bulbosus	IV (2–5)	*Potentilla erecta*	I (1–3)
Hydrocotyle vulgaris	III (2–5)	*Sphagnum papillosum*	I (1–3)
Molinia caerulea	III (1–5)	*Rhynchospora alba*	I (1–2)
Eleocharis multicaulis	III (1–5)	*Cirsium dissectum*	I (3–4)
Eriophorum angustifolium	III (2–4)	*Nardus stricta*	I (3)
Agrostis canina	III (3–7)	*Juncus acutiflorus*	I (1–2)
Anagallis tenella	II (1–4)	*Myosotis secunda*	I (1–3)
Juncus articulatus	II (2–5)	*Pedicularis palustris*	I (1)
Carex demissa	II (1–4)	*Juncus bufonius*	I (3–4)
Carex echinata	II (1–4)	*Equisetum palustre*	I (3–4)
Carex panicea	II (1–5)	*Drosera intermedia*	I (3–4)
Drosera rotundifolia	II (1–3)	*Aneura pinguis*	I (2)
Juncus effusus	II (1–8)	*Drepanocladus revolvens*	I (2)
Carex nigra	II (1–5)	*Carex hostiana*	I (4–6)
Narthecium ossifragum	II (1–4)	*Salix cinerea* seedling	I (1–2)
Equisetum fluviatile	II (1–3)	*Aulacomnium palustre*	I (5–7)
Galium palustre	II (1–4)	*Calliergon cuspidatum*	I (2–4)
Lotus uliginosus	II (1–4)	*Potentilla palustris*	I (1–2)
Drepanocladus exannulatus	I (2–3)	*Sphagnum palustre*	I (4–5)
Erica tetralix	I (1–3)	*Glyceria fluitans*	I (2–3)
Menyanthes trifoliata	I (2–3)	*Eleocharis palustris*	I (2–4)
Viola palustris	I (1–5)	*Pilularia globulifera*	I (3–5)
Holcus lanatus	I (1–3)	*Sparganium angustifolium*	I (2)
Scutellaria minor	I (2–4)	*Apium inundatum*	I (1–7)
Polytrichum commune	I (2–4)	*Baldellia ranunculoides*	I (2)
Agrostis stolonifera	I (2–4)	*Galium debile*	I (3)
Scirpus fluitans	I (1–8)	Number of samples	29

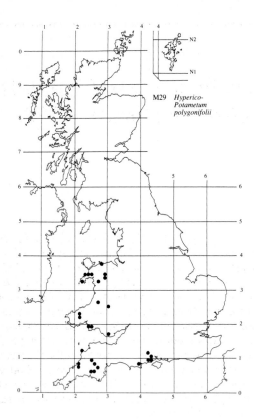

M29   *Hyperico-*
     Potametum
     polygonifolii

M30

Related vegetation of seasonally-inundated habitats Hydrocotylo-Baldellion   Tüxen & Dierssen 1972

Other vegetation of the same general type as the *Hyperico-Potametum* and characteristic of similar, seasonally-inundated habitats with rather base-poor and only moderately enriched waters, undoubtedly occurs in Britain, but it has been very poorly sampled. Some, lacking *Hypericum elodes* but otherwise essentially the same as the *Hyperico-Potametum*, could perhaps be regarded as impoverished stands of that community, though it must be noted that species such as *Potamogeton polygonifolius*, *Eriophorum angustifolium*, *Juncus bulbosus/kochii* and *Sphagnum auriculatum* also occur with some frequency in bog-pool and poor-fen vegetation.

Then, there are stands in which *Eleocharis multicaulis* is strongly dominant with little or no *H. elodes* or *P. polygonifolius*, and some at least of these look very similar to the *Eleocharitetum multicaulis* Tüxen 1937 which has been recorded from Eire (Braun-Blanquet & Tüxen 1952, Brock *et al.* 1978, van Groenendael *et al.* 1979; see also Ivimey-Cook & Proctor 1966*b*) and from other parts of western Europe (Schoof van Pelt 1973, Dierssen 1975, 1982). White & Doyle (1982) list *Deschampsia setacea* as a characteristic species of such vegetation and this national rarity is certainly typical, in Britain, of this general kind of habitat. *Scirpus fluitans* can also be found dominating in swards which lack some of the most typical *Hyperico-Potametum* plants and, in the New Forest and Cornwall, *Baldellia ranunculoides* is a frequent and conspicuous component of low-growing vegetation in seasonally-wet pools. In the Burren, the latter kind of assemblage was designated the *Baldellio-Littorelletum* by Ivimey-Cook & Proctor 1966*b*, although there the habitat was characterised by base-rich and calcareous waters.

In the latest revision of the Littorelletalia by Dierssen (1975), all these vegetation types are grouped together with the *Hyperico-Potametum* in the Hydrocotylo-Baldellion alliance, a group comprising assemblages of mesotrophic to oligotrophic, and periodically-fluctuating waters. Other Littorelletalia communities, mostly falling in the Isoetion lacustris and Lobelion dortmannae, are dealt with in the chapter on aquatic vegetation.

M31
Anthelia julacea-Sphagnum auriculatum spring
Sphagno auriculati-Anthelietum julaceae Shimwell 1972

Synonymy
Anthelia julacea-Deschampsia cespitosa provisional nodum McVean & Ratcliffe 1962; *Anthelia julacea* banks Birks 1973.

Constant species
Deschampsia cespitosa, Anthelia julacea, Marsupella emarginata, Scapania undulata, Sphagnum auriculatum.

Rare species
Anthelia juratzkana, Pohlia ludwigii.

Physiognomy
In the *Sphagno auriculati-Anthelietum julaceae, Anthelia julacea* more than justifies the claim of McVicar (1912) that it is 'the most conspicuous hepatic of our highland mountains'. For though its individual shoots are thread-like, they are here massed into extraordinarily robust and dense tufts, cushions or huge swelling masses, up to a metre or more thick and sometimes covering as much as several square metres in total extent. Dull and glaucous when kept wet, as they typically are, or occasionally drying out to a pale grey colour, the larger mats and banks can scarcely be missed, even from a distance.

In terms of its associated flora, however, this is generally a rather species-poor assemblage, with vascular plants being especially sparse. Among other bryophytes, *Sphagnum auriculatum, Marsupella emarginata* (both var. *emarginata* and var. *aquatica*) and *Scapania undulata* are all constant and, though none rivals *A. julacea* itself in abundance, the first in particular can be found as patches of moderate size, often prominent by virtue of their coppery hue. *Racomitrium lanuginosum* and *Philonotis fontana* are also quite frequent, with *Calliergon sarmentosum, Campylopus atrovirens, Polytrichum commune* and *Racomitrium fasciculare* occurring occasionally. Rare bryophytes recorded here include *Anthelia juratzkana*, which seems totally to replace *A. julacea* in some stands (McVean & Ratcliffe 1962), and *Pohlia ludwigii*.

The commonest vascular plant in the community is *Deschampsia cespitosa* (presumably sometimes ssp. *alpina* at higher altitudes) and this is typically found as small tufts set in the bryophyte cushions. Scattered plants of *Nardus stricta* also occur quite often with occasional *Narthecium ossifragum, Pinguicula vulgaris, Carex demissa* and *Saxifraga stellaris*. Less frequent are *Eriophorum angustifolium, Carex bigelowii, C. nigra, Festuca vivipara, Agrostis canina, A. stolonifera, Juncus bulbosus, Thalictrum alpinum* and *Viola palustris*.

The available data do not indicate any well-marked sub-divisions within the community, though there is a suggestion that *Racomitrium lanuginosum, Nardus* and *Narthecium* might define one sub-group, with *Philonotis fontana, Calliergon sarmentosum, Carex demissa* and *Pinguicula vulgaris* preferential to another.

Habitat
The *Sphagno-Anthelietum* is a montane community of north-west Britain, typical of often skeletal mineral and organic soils kept more or less permanently wet by the trickling of acid and oligotrophic waters, frequently derived, at higher altitudes, from snow-melt.

A. julacea occurs at moderate to high altitudes, from about 400 m to over 1000 m in the available samples, in Wales, north-western England, south-west Scotland and up through most of the Highlands, a region characterised by mean annual maximum temperatures generally less than 25 °C and considerably lower through most of north-west Scotland (Conolly & Dahl 1970). Although it can be found in this zone as an occasional in other kinds of vegetation, it typically dominates here around spring-heads and over constantly-irrigated soils, often of a fragmentary character. Thus, though some of the profiles could be classed as stagnogleys or alluvial gleys (Avery 1980, Birse 1984), many are little more than raw accumulations of mineral and organic detritus held beneath the mass of vegetation. Stands can be found on virtually level ground or gently shelving slopes in hollows and corries cut into granite, gneiss and acid schists, or over somewhat steeper banks and stabilised masses of debris.

The irrigating waters are characteristically acid, of pH 4.5–5.0 in the available samples, and low in carbonates and exchangeable bases. At higher altitudes, melt-water contributes substantially to seepage in the late spring and summer months and snow-lie over this vegetation can be quite substantial, though typically not so long as with the *Pohlietum glacialis* or snow-bed vegetation proper. Even though the ground is kept more or less continually wet, however, and though the community often marks out distinct spring-heads, irrigation is probably not so vigorous as in the *Philonoto-Saxifragetum stellaris*.

Zonation and succession
The community is typically found marking out areas of less vigorous seepage among tracts of montane grasslands and grass-heaths or among snow-bed vegetation, and in such situations it can form intimate mosaics with the *Philonoto-Saxifragetum* where there is variation in the rate of flow. It is an essentially stable community in the harsh environment in which it characteristically occurs.

Distribution
The *Sphagno-Anthelietum* is a local but widespread community through much of Scotland, extending from the Central Highlands west to Skye and South Uist and northwards to the Loch Broom area. It can be found too, often at somewhat lower altitudes, through south-west Scotland and in the Lake District and Snowdonia.

Affinities
This distinctive vegetation type, defined here as an expanded though essentially similar version of the assemblage first characterised by McVean & Ratcliffe (1962), is rather isolated syntaxonomically. Species such as *Scapania undulata* and *Marsupella emarginata* provide a link with submerged bryophyte communities of fast-flowing mountain streams, but closer affinities are with the *Philonoto-Saxifragetum*. However, the absence of *Montia fontana* and many of the typical plants of the Cardamino-Montion is very noticeable and Shimwell (1972) proposed a new alliance, the Anthelion julaceae, within the Montio-Cardaminetalia, to contain the community.

Floristic table M31

Anthelia julacea	V (7–10)	*Polytrichum alpinum*	I (2)
Sphagnum auriculatum	V (1–5)	*Racomitrium heterostichum*	I (1–3)
Marsupella emarginata	V (1–7)	*Juncus bulbosus*	I (1–3)
Scapania undulata	IV (1–5)	*Diplophyllum albicans*	I (1)
Deschampsia cespitosa	IV (1–5)	*Juncus squarrosus*	I (2)
		Drepanocladus exannulatus	I (1–3)
Racomitrium lanuginosum	III (1–7)	*Saussurea alpina*	I (2)
Nardus stricta	III (1–6)	*Cerastium fontanum*	I (1)
Philonotis fontana	II (1–4)	*Carex echinata*	I (3)
Pinguicula vulgaris	II (1–2)	*Plantago maritima*	I (3)
Narthecium ossifragum	II (1–7)	*Ranunculus acris*	I (2)
Carex demissa	II (1–4)	*Ranunculus flammula*	I (1)
Calliergon sarmentosum	II (2–5)	*Taraxacum officinale* agg.	I (2)
Saxifraga stellaris	II (1–3)	*Pohlia ludwigii*	I (3)
Festuca vivipara	I (1–3)	*Nardia scalaris*	I (4)
Campylopus atrovirens	I (1–6)	*Barbilophozia floerkii*	I (3)
Polytrichum commune	I (1)	*Molinia caerulea*	I (2–3)
Eriophorum angustifolium	I (3–4)	*Blindia acuta*	I (1–2)
Racomitrium fasciculare	I (1–3)	*Huperzia selago*	I (1–2)
Agrostis canina	I (3–4)	*Deschampsia flexuosa*	I (1)
Carex bigelowii	I (1–4)		
Carex nigra	I (1–2)	Number of samples	28
Thalictrum alpinum	I (3)	Number of species/sample	13 (9–20)
Viola palustris	I (1–2)		
Agrostis stolonifera	I (1–2)	Vegetation cover	96 (65–100)
Mylia taylori	I (1–4)	Altitude (m)	712 (442–1052)
Aneura pinguis	I (1–2)	Slope (°)	10 (0–30)

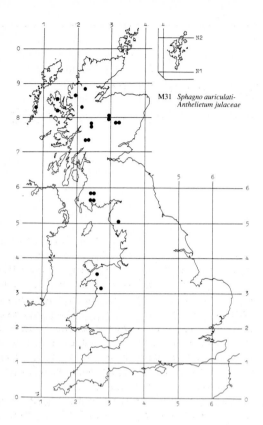

M31 *Sphagno auriculati-*
      Anthelietum julaceae

M32
Philonotis fontana-Saxifraga stellaris spring
Philonoto-Saxifragetum stellaris   Nordhagen 1943

Synonymy

Philonotis association Smith 1911 *p.p.*; *Philonotis-Saxifraga stellaris* bryophyte flush Pearsall 1950; *Philonoto-Saxifragetum stellaris sensu* McVean & Ratcliffe 1962, Eddy *et al.* 1969, Birks 1973; *Poa annua-Montia fontana* nodum Eddy *et al.* 1969; *Philonotis fontana-Saxifraga* nodum Edgell 1969; *Philonotis* flushes Meek 1976; *Philonotis* springs Ferreira 1978; *Philonotis-Chrysosplenium-Poa subcaerula* hanging mats Ferreira 1978; *Chrysosplenium-Montia* springs Ferreira 1978; *Scapania undulata-Philonotis fontana* nodum Huntley 1979; *Montio-Philonotidetum fontanae* (Bük & Tüxen in Bük 1942) Birse 1980; *Nardia compressa* Community Birse 1980 *p.p.*

Constant species

Saxifraga stellaris, Philonotis fontana.

Rare species

Alopecurus alpinus, Cerastium cerastoides, Epilobium alsinifolium, Koenigia islandica, Myosotis stolonifera, Phleum alpinum, Sedum villosum, Bryum schleicheri var. *latifolium, B. weigelii, Oncophorus virens, Pohlia ludwigii, P. wahlenbergii* var. *glacialis, Splachnum vasculosum, Scapania paludosa, Tritomaria polita.*

Physiognomy

The *Philonoto-Saxifragetum stellaris* comprises bryophyte-dominated springs, flushes and rills of very striking appearance. *Philonotis fontana* is the usual dominant, its fresh-green picking out stands of the community from a distance, but often abundant too are *Dicranella palustris*, luxuriant and golden-green, and *Scapania undulata*, forming patches which range from vivid-green to reddish-purple (those at the latter extreme once being classified as *S. dentata*, as in McVean & Ratcliffe's (1962) samples). The cover is sometimes broken by patches of wet soil, but often densely swollen into hummocks or plush mats or forming hanging carpets over steep dripping ground. Less consistent through the community as a whole, but also able to show local prominence, catching the eye as splashes of pink, bronze or deep green, are *Sphagnum auriculatum, Scapania uliginosa, Calliergon sarmentosum, Drepanocladus exannulatus, D. fluitans* and, not very frequent overall but rather characteristic of this vegetation, *Jungermannia exsertifolia*. *Bryum pseudotriquetrum* is also quite common and patchily abundant but it is not so consistent here as in more base-poor springs and *Cratoneuron commutatum* and *C. filicinum*, typical dominants in such situations, are noticeably scarce.

Other bryophytes found occasionally at generally low covers include *Nardia scalaris* and *N. compressa*, the latter especially where the community occurs around bouldery rills, *Polytrichum commune, Calliergon cuspidatum, C. stramineum, Chiloscyphus polyanthos, C. pallescens, Brachythecium rivulare, B. plumosum, Rhizomnium punctatum, Sphagnum squarrosum, Pellia epiphylla* and *Aneura pinguis*. Then, among rarer taxa, the vegetation provides a locus for *Oncophorus virens, Bryum weigelii, B. schleicheri* var. *latifolium, Scapania paludosa* and *Tritomaria polita* (Eddy *et al.* 1969, Birks 1973, Birse 1980). *Splachnum vasculosum*, and perhaps also *Aplodon wormskjoldii*, two species characteristic of wet rotting animal dung, are particularly associated with this kind of spring (Pigott 1956a, McVean & Ratcliffe 1962). *Pohlia wahlenbergii* var. *glacialis* and *P. ludwigii* can also be found in some stands, though they are much more typical of the closely-related *Pohlietum glacialis* snow-bed springs.

The associated vascular flora of the *Philonoto-Saxifragetum* is rather varied in its composition and cover, though typically it is species-poor and of low total abundance. The only constant species is *Saxifraga stellaris* but, though this is by no means confined to this community, its scattered rosettes and delicate white flowers, set off against the moss carpet, are very characteristic and it provides a strong floristic link with similar Arctic-Alpine flushes in other parts of Europe. Its British distribution is generally coincident with that of the community, though, both within its heartland and towards the fringes of its range, stands can be found from which it is lacking but which are otherwise of this same general character (e.g. Meek 1976, Birse 1980).

Other herbs frequent throughout are few but *Deschampsia cespitosa* (with ssp. *alpina* at higher altitudes: McVean & Ratcliffe 1962) is often found in small quantities and there are commonly some scattered plants of *Stellaria alsine*. More occasional overall are *Festuca rubra, Anthoxanthum odoratum, Agrostis stolonifera* and *A. canina*, all usually at low cover, with sparse *Viola palustris, Nardus stricta, Carex bigelowii* and *C. panicea*. However, in stands over substrates which are perhaps less base-poor, there is a distinctive enrichment in this element of the vegetation, with *Montia fontana* and *Chrysosplenium oppositifolium* becoming very frequent and locally abundant and *Caltha palustris* and *Cardamine pratensis* heading a substantial list of associates which help diagnose this less impoverished kind of spring.

Of rarer vascular plants, the *Philonoto-Saxifragetum* provides a locus for a variety of high-montane species. Among Arctic-Alpines, this is an important vegetation type for *Epilobium alsinifolium* and the rather more widespread *E. anagallidifolium*, for *Cerastium cerastoides* (though this is more frequently found in the *Pohlietum glacialis*) and for *Phleum alpinum*. In some sites, as at Caenlochan, this last can be found around these springs with another very rare grass, the Arctic-Subarctic *Alopecurus alpinus*, which seems to prefer the somewhat wetter spots in the middle of the rills (Raven & Walters 1956). *Sedum villosum* and *Myosotis stolonifera* have also been recorded here and, around The Storr on Skye, where it was first found only in 1934, *Koenigia islandica* can sometimes be seen on stony patches within flushes of this kind (Raven & Walters 1956, Birks 1973).

Sub-communities

***Sphagnum auriculatum* sub-community:** *Philonoto-Saxifragetum, Sphagnum* facies McVean & Ratcliffe 1962. In this, very much the more species-poor kind of *Philonoto-Saxifragetum*, the bryophyte mat consists of mixtures of *P. fontana, S. undulata* and *D. palustris* with *Sphagnum auriculatum* strongly preferential and often abundant, *Calliergon sarmentosum* and *Scapania uliginosa* occasional and locally prominent, and *Polytrichum commune* and *Hygrohypnum ochraceum* also occurring at low frequencies. Among the vascular plants, only *S. stellaris* and *D. cespitosa* are constant but other grasses, such as *Agrostis stolonifera, A. capillaris, Anthoxanthum* and *Festuca rubra* are quite frequent as scattered tufts and *Stellaria alsine* and *Viola palustris* are occasional.

***Montia fontana-Chrysosplenium oppositifolium* sub-community:** *Philonoto-Saxifragetum*, species-rich facies McVean & Ratcliffe 1962. Vascular plants in particular are more numerous and varied in this sub-community, although bryophytes still generally have overall dominance. As before, *P. fontana, D. palustris* and *S. undulata* are all very common and each, especially the first, can be abundant in the carpet, though *S. auriculatum* and the other low-frequency bryophytes preferential to the *Sphagnum* sub-community are very scarce. Here, by contrast, *Bryum pseudotriquetrum* becomes frequent and *Jungermannia exsertifolia* occasional and each can have locally high covers. *Drepanocladus exannulatus, D. fluitans* and *D. revolvens* are all occasional and sometimes quite abundant and there can be records too for *Calliergon cuspidatum, C. stramineum, Chiloscyphus polyanthos, C. pallescens, Brachythecium rivulare, Pellia epiphylla* and *Aneura pinguis*.

Striking among the herbs is the high frequency, along with *S. stellaris*, of *Montia fontana* and *Chrysosplenium oppositifolium*, which occur very occasionally in the *Sphagnum* sub-community but which are here constant and sometimes abundant among the bryophytes. Then, along with the community species *Stellaria alsine*, there are very often some plants of a diminutive form of *Caltha palustris* (sometimes elevated to ssp. *minor*: Clapham *et al.* 1962), of *Cardamine pratensis* and occasionally of *Ranunculus flammula, R. acris, Cerastium fontanum* and *Equisetum palustre*. *Epilobium palustre* can sometimes be found but more distinctive is the quite frequent occurrence of *E. alsinifolium* and *E. anagallidifolium*, the small Arctic-Alpine willow-herbs with noticeably drooping flowers.

Grasses and sedges can be quite common, too, with *Deschampsia cespitosa* often joined by *Anthoxanthum* and *Agrostis canina*, less commonly by *Festuca rubra, F. ovina, Poa annua, P. trivialis* and *P. subcaerulea*, and *Carex nigra, C. echinata, C. demissa, C. bigelowii, C. panicea* and *Eriophorum angustifolium*. Typically, these all occur at fairly low covers, though occasionally members of this group can be more abundant: in some stands at Moor House in Cumbria, for example (Eddy *et al.* 1969), *Poa annua* is especially prominent, and small sedges and grasses can thicken up around the edges of springs where there is a transition to flushed grassland.

With the fairly modest base-enrichment that seems to be characteristic of even these richer stands of the *Philonoto-Saxifragetum*, calcicoles are not strongly represented but some springs can be found in which species such as *Cratoneuron* spp., *Pinguicula vulgaris, Selaginella selaginoides, Carex pulicaris* and *Saxifraga hypnoides* bring the community close to Cratoneurion vegetation.

Habitat

The *Philonoto-Saxifragetum* is a community of springs and rills at moderate to high altitudes where there is continuous irrigation with circumneutral and oligotrophic waters. The harsh montane environment has a strong influence on the composition of the community and, though stands can be grazed and trampled, climatic and soil conditions probably play the major part in maintaining the vegetation as an effective climax.

This is one of the most common and widespread types of spring vegetation in the uplands of north-west Britain and it is ultimately dependent on the kind of sustained and fairly vigorous irrigation by ground waters that is commonplace in the rainier parts of the country. Throughout the range of the community, annual precipitation is almost everywhere in excess of 1600 mm (*Climatological Atlas* 1952), with at least 180 wet days yr⁻¹ (Ratcliffe 1968), and, in this zone, it marks out places where such heavy and consistent rainfall feeds permanent springs of a well-defined character, more diffuse flushes and seepage lines, rills and small streams and occasionally steep, dripping ground. In some sites, too, snow-melt is an important source of irrigating waters, though this community is not so consistently associated with such situations as is the closely-related *Pohlietum glacialis*.

Where flushing is more vigorous, the *Philonoto-Saxifragetum* can be found on almost level ground but sloping sites are more usual, either over evenly aggraded hill slopes or valley sides or where there are declivities as at the foot of screes or cliffs. The soils are often of a primitive character, sometimes little more than fragmentary accumulations of silt among stones with decaying organic matter beneath the bryophyte carpet, but the community can also be found on flushed peats and over gleys around springs. The profiles are, however, typically waterlogged to the surface or often submerged for most of the year, and the constancy of irrigation is reflected in the general luxuriance of the mosses and liverworts and in the prevalence among them of species such as *Philonotis fontana*, *Dicranella palustris*, *Scapania undulata*, *Bryum pseudotriquetrum*, *Brachythecium rivulare* and *Calliergon* spp., characteristic of sodden or continually splashed ground in a variety of flush and stream habitats.

The number of other plants, particularly vascular species, which find such conditions congenial is limited, but there are two further factors which constrain the character of the associated flora. The first is temperature, both of the air and of the spring waters, for this is a community of the colder reaches of our uplands, limited in general to sites above 450 m and reaching over 1000 m, in north Wales, the Pennines and the Lake District, southern Scotland and the Highlands. Throughout this range, annual accumulated temperatures are generally less than 800 day-degrees C (Page 1982) with mean annual maxima always below 24 °C, less than 22 °C over much of the heartland of the community in Scotland (Conolly & Dahl 1970). The irrigating waters are likewise consistently cold, though probably not as frigid as

in the *Pohlietum glacialis*. The best single indicator of such conditions here is *Saxifraga stellaris*, but other Arctic-Alpines like *Epilobium alsinifolium*, *E. anagallidifolium* and the rarer montane herbs and bryophytes show a similar response and help give this kind of spring its distinctive character. *Myosotis secunda*, on the other hand, which towards the northern part of its European range is not really a plant of high mountains, and *Ranunculus omiophyllus*, an Oceanic West European species, hardly ever figure here, though they occur commonly enough in similar Montion springs in the less cold upland fringes of south-west Britain.

A few other herbs of this community have a widespread distribution throughout Britain but show tolerance of the lower temperatures here. *Deschampsia cespitosa* (with ssp. *alpina* at higher altitudes) and *Stellaria alsine* are the most common of these, the latter in particular providing floristic continuity with similar vegetation at lower altitudes. Apart from these, however, the herbaceous element of the community is further limited by the chemical character of the irrigating waters, particularly by their base-poverty, perhaps also by their poor trophic state. By and large, the waters here are circumneutral, with pH in available samples ranging from about 4.5 to 6.0, and the vegetation is thus found over a wide variety of rock types, including shales, sandstones, quartzites, schists and granite, but it is generally absent from limestone landscapes except where there is seepage from patches of decalcified drift or weathering residues. This is what strictly limits the contribution of calcicolous herbs and bryophytes here, a potentially large spring and flush flora, even at these altitudes.

Even within the fairly narrow range of base-richness encompassed by the community, soil reaction seems to have some effect on floristic variation within it. Thus, the *Sphagnum* sub-community, with its very impoverished vascular element, occurs mainly on the harder acidic quartzites and sandstones of the north-west Highlands and indeed may be the only type of *Philonoto-Saxifragetum* in that area. The *Montia-Chrysosplenium* sub-community, on the other hand, is associated with a diversity of substrates perhaps only marginally more base-rich but sufficiently so to favour the quite luxuriant growth of *Montia fontana*, *Chrysosplenium oppositifolium* and its other numerous preferentials. The irrigating waters in this kind of spring may also be somewhat less impoverished in major nutrients than those flushing the *Sphagnum* sub-community.

In general, however, conditions here are oligotrophic and this, together with the harsh climatic regime is probably sufficient to prevent any seral succession. Stands are often open to grazing, though, and, particularly at lower altitudes, cropping of the herbage and trampling of the ground may help set back the growth of any invading shrubs or trees.

Zonation and succession

The springs and flushes that characteristically support the *Philonoto-Saxifragetum* can arise in and flow through a wide variety of vegetation types around our upland fringes and through the montane zone. Either on its own, or in association with other kinds of spring, it can pass to mire, heath or grassland or, at higher altitudes, to bryophyte and lichen communities and snow-bed vegetation. It is essentially a permanent community though, at lower altitudes, could perhaps show some successional development in ungrazed situations.

Very often, the *Philonoto-Saxifragetum* itself marks out the head of a spring or core of a flush, frequently occurring as a small stand of just a few square metres' extent, sometimes much less, or extending as a very narrow strip down rills and the sides of small streams. Quite commonly, a series of such stands will occur across a slope or at its foot, marking some geological disjunction, and, where these are grouped close to one another, they can form a single complex mosaic with the local vegetation types. In very vigorous springs, particularly those on steeper ground where the gradient of waterlogging can die away rapidly, boundaries between the *Philonoto-Saxifragetum* and its context can be very sharp with next to no floristic continuity between the vegetation types: such stark patterning is often visible in a marked colour contrast between the fresh-green of the spring and the dull-green or russet of its surrounds. In other situations, flushing is more diffuse, with the stands more extensive but much less well-defined from their context, something which is again often visible to the eye from a distance but evident at close quarters in a gradual zonation between the community and its neighbouring vegetation.

Even in these latter situations, however, the typical bryophytes of the *Philonoto-Saxifragetum* and species such as *Montia fontana* and *Chrysosplenium oppositifolium* rarely extend far out of the spring or flush core and continuity is generally provided by the grasses and sedges that occur in the community. Transitions thus tend to be most gradual where flushes occur within tracts of Nardo-Galion swards such as the *Festuca-Agrostis-Galium* and *Nardus-Galium* grasslands which extend into the higher reaches of the sub-montane zone. More abrupt switches can be seen where the *Philonoto-Saxifragetum* occupies springs emerging in a variety of Nardo-Callunetea dry heaths, Ericion tetralicis wet heath and Erico-Sphagnion blanket mire and among Caricetea curvulae communities of high altitudes.

Complications arise where the *Philonoto-Saxifragetum* passes to other vegetation types of waterlogged ground where the water-flow is less vigorous or where there is some increase in base- or nutrient-richness along the length of a flush or water-track, and such communi-

ties can form a transition zone around the spring core. At lower altitudes, the *Philonoto-Saxifragetum* can occur at the head of *Carex echinata-Sphagnum* mires but in the montane zone it is more often found in association with the *Carex-Sphagnum russowii* mire which can surround a spring or rill and grade from it to *Calluna-Eriophorum* mire. In other cases, high-altitude stands occur with vegetation types of snow-beds or melt-water flushes and streams, such as the *Sphagno-Anthelietum*, in which there is less vigorous irrigation, and the *Pohlietum glacialis*, where there is longer snow-lie and colder waters.

Distribution

The community is common and widespread above 450 m through the Scottish Highlands, the Southern Uplands, the Lake District and north Wales and over the non-calcareous parts of the Pennines. At lower altitudes throughout its range and particularly towards the southern limit of its distribution, fragmentary stands can be found which lack the more specifically montane element but which are otherwise of the same general character.

Affinities

This kind of spring vegetation was early recognised as of a distinctive character (*e.g.* Smith 1911*b*, Pigott 1956*a*, Pearsall 1968) though not systematically described until the account of Scottish Highland stands provided by McVean & Ratcliffe (1962). The description here has used their data, together with samples from other parts of Scotland (Birks 1973, Meek 1976, Ferreira 1978, Huntley 1979, Birse 1980), Wales (Edgell 1969) and northern England (Eddy *et al.* 1969), characterising more sharply what McVean & Ratcliffe recognised as a crude distinction between stands with *Sphagnum auriculatum* and the rest. The former could perhaps be seen as a floristic transition to the spring vegetation of the Anthelion but, in general, the affinities of the community are clearly with the Cardamino-Montion and, more particularly, with the sub-alliance Montion which includes higher-altitude springs and flushes of unshaded situations in which such plants as *Montia fontana*, *Saxifraga stellaris*, *Stellaria alsine*, *Epilobium alsinifolium*, *Philonotis fontana*, *Bryum weigelii* and *B. schleicheri* occur. For us, the *Philonoto-Saxifragetum* constitutes the most widespread and common of our Montio-Cardaminetea communities and is therefore the standard against which related vegetation types can be judged. From this perspective, stands with little more than *P. fontana* and *D. palustris*, which are widespread in the uplands of south Wales and also found in south-west England, look to be very fragmentary forms of the community. *Bryum weigelii* springs, which have been recorded from the Long Mynd in Shropshire, could also perhaps be regarded in this light. Further work on such vegetation would be well worthwhile.

Meanwhile, it is possible to see the *Philonoto-Saxifragetum* as a fairly well-defined and consistent unit with its closest relative in the *Montia-Ranunculus* spring which largely replaces it in the more lowland and oceanic climate of south-west Britain. It is fairly well marked off from the more base-rich springs and flushes of the Cratoneurion by the absence or great scarcity here of *Cratoneuron* spp. themselves, *Philonotis calcarea* and *Saxifraga aizoides*, although some of the more catholic Montio-Cardaminetea herbs maintain their frequency in both communities. As McVean & Ratcliffe (1962) noted, the vegetation included in the community is virtually identical with the *Philonoto-Saxifragetum* as first described by Nordhagen (1943) from Norway and later by Dahl (1956), Persson (1961) and Fransson (1963) from other parts of Scandinavia, from Greenland (Böcher 1954), from Iceland (Hadač 1971) and from central Europe (Braun-Blanquet 1948, Oberdorfer 1957), and incorporating essentially similar vegetation described in an early account from the Faroes (Ostenfeld 1908).

Floristic table M32

	a	b	32
Philonotis fontana	V (1–10)	V (1–8)	V (1–10)
Saxifraga stellaris	IV (1–5)	IV (1–5)	IV (1–5)
Sphagnum auriculatum	IV (1–10)	I (1–5)	II (1–10)
Agrostis stolonifera	III (1–5)	I (1–4)	II (1–5)
Calliergon sarmentosum	II (1–9)	I (1–9)	I (1–9)
Polytrichum commune	II (1–3)	I (1–3)	I (1–3)
Scapania uliginosa	II (1–9)		I (1–9)
Agrostis capillaris	II (1–5)		I (1–5)
Juncus bulbosus	I (1–5)		I (1–5)
Hygrohypnum ochraceum	I (1–3)		I (1–3)
Jungermannia sphaerocarpum	I (1)		I (1)
Sphagnum teres	I (1)		I (1)
Montia fontana	I (1–4)	IV (1–8)	III (1–8)
Chrysosplenium oppositifolium	I (1–5)	IV (1–5)	III (1–5)
Anthoxanthum odoratum	II (1–3)	III (1–5)	II (1–5)
Agrostis canina canina	I (1–4)	III (1–4)	II (1–4)
Caltha palustris	I (1–4)	III (1–4)	II (1–4)
Bryum pseudotriquetrum		III (1–6)	II (1–6)
Cardamine pratensis		III (1–4)	II (1–4)
Carex nigra		III (1–5)	II (1–5)
Epilobium alsinifolium		III (1–2)	II (1–2)
Epilobium palustre	I (1–2)	II (1–3)	II (1–3)
Eriophorum angustifolium	I (1–3)	II (1–3)	II (1–3)
Epilobium anagallidifolium	I (1–3)	II (1–2)	II (1–3)
Cerastium fontanum		II (1–4)	II (1–4)
Jungermannia exsertifolia		II (1–7)	II (1–7)
Calliergon cuspidatum		II (1–4)	II (1–4)
Ranunculus acris	I (1–3)	II (1–4)	I (1–4)
Carex echinata	I (1–4)	II (1–3)	I (1–4)
Carex demissa	I (1–3)	II (1–6)	I (1–6)
Chiloscyphus polyanthos	I (1–3)	II (1–4)	I (1–4)
Ranunculus flammula		II (1–3)	I (1–3)
Poa annua		II (4–8)	I (4–8)
Equisetum palustre		II (1–3)	I (1–3)

Floristic table M32 *(cont.)*

	a	b	32
Pellia epiphylla		II (1–3)	I (1–3)
Festuca ovina		II (1–4)	I (1–4)
Brachythecium rivulare		II (1–4)	I (1–4)
Cratoneuron filicinum		II (1–6)	I (1–6)
Sagina procumbens		I (1–4)	I (1–4)
Poa trivialis		I (1–4)	I (1–4)
Aneura pinguis		I (1–3)	I (1–3)
Alopecurus alpinus		I (1–4)	I (1–4)
Rumex acetosa		I (1–4)	I (1–4)
Calliergon stramineum		I (1–3)	I (1–3)
Drepanocladus revolvens		I (1–4)	I (1–4)
Juncus articulatus		I (1–4)	I (1–4)
Cardamine flexuosa		I (1–4)	I (1–4)
Bryum weigelii		I (1–4)	I (1–4)
Alchemilla glabra		I (1–4)	I (1–4)
Veronica serpyllifolia		I (1–3)	I (1–3)
Drepanocladus fluitans		I (1–7)	I (1–7)
Selaginella selaginoides		I (1–3)	I (1–3)
Pinguicula vulgaris		I (1–3)	I (1–3)
Galium palustre		I (1–3)	I (1–3)
Geum rivale		I (1–3)	I (1–3)
Marchantia polymorpha		I (1–3)	I (1–3)
Chiloscyphus pallescens		I (1–3)	I (1–3)
Saxifraga hypnoides		I (1–4)	I (1–4)
Thalictrum alpinum		I (1–4)	I (1–4)
Polygonum viviparum		I (1–4)	I (1–4)
Plagiomnium ellipticum		I (1–4)	I (1–4)
Juncus effusus		I (1–4)	I (1–4)
Sphagnum subnitens		I (1–5)	I (1–5)
Myosotis stolonifera		I (1–4)	I (1–4)
Carex pulicaris		I (1–4)	I (1–4)
Juncus triglumis		I (1–4)	I (1–4)
Prunella vulgaris		I (1–6)	I (1–6)
Euphrasia frigida		I (1–3)	I (1–3)
Cratoneuron commutatum		I (1–3)	I (1–3)
Brachythecium plumosum		I (1–4)	I (1–4)
Bellis perennis		I (1–3)	I (1–3)
Scapania undulata	IV (1–7)	III (1–4)	III (1–7)
Deschampsia cespitosa	IV (1–4)	III (1–4)	III (1–4)
Dicranella palustris	III (1–9)	IV (1–9)	III (1–9)
Stellaria alsine	II (1–3)	III (1–4)	III (1–4)
Festuca rubra	II (1–4)	II (1–4)	II (1–4)
Drepanocladus exannulatus	II (1–9)	II (1–6)	II (1–9)
Viola palustris	I (1–3)	I (1–3)	I (1–3)
Nardus stricta	I (1–3)	I (1–4)	I (1–4)

Carex bigelowii	I (1–3)	I (1–3)	I (1–3)
Carex panicea	I (1–3)	I (1–4)	I (1–4)
Nardia scalaris	I (1–3)	I (1–4)	I (1–4)
Leontodon autumnalis	I (1–3)	I (1–3)	I (1–3)
Rhizomnium punctatum	I (1–3)	I (1–3)	I (1–3)
Sphagnum squarrosum	I (1–3)	I (1–5)	I (1–5)
Rhytidiadelphus squarrosus	I (1–3)	I (1–8)	I (1–8)
Juncus squarrosus	I (1–3)	I (1–3)	I (1–3)
Number of samples	28	52	80
Number of species/sample	11 (5–24)	22 (6–42)	18 (5–42)
Altitude (m)	729 (210–1129)	686 (138–976)	701 (138–1129)
Slope (°)	7 (0–90)	14 (0–45)	12 (0–90)

a   *Sphagnum auriculatum* sub-community
b   *Montia fontana-Chrysosplenium oppositifolium* sub-community
32 *Philonoto-Saxifragetum stellaris* (total)

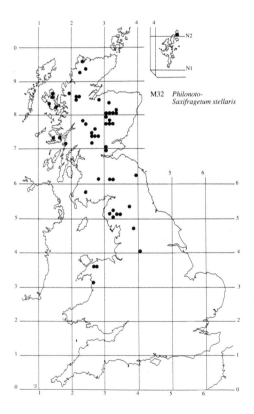

M32 *Philonoto-Saxifragetum stellaris*

M33
Pohlia wahlenbergii var. *glacialis* spring
Pohlietum glacialis   McVean & Ratcliffe 1962

Synonymy
Pohlia 'glacialis' spring   Ratcliffe 1964; *Pohlia albicans* var. *glacialis* spring   Ratcliffe 1977.

Constant species
Deschampsia cespitosa, Saxifraga stellaris, Pohlia ludwigii, P. wahlenbergii var. *glacialis*.

Rare species
Alopecurus alpinus, Cerastium cerastoides, Epilobium alsinifolium, Phleum alpinum, Pohlia ludwigii, P. wahlenbergii var. *glacialis*.

Physiognomy
Pohlia wahlenbergii var. *glacialis* can be found as an infrequent and low-cover associate in a variety of vegetation types of wet ground through the uplands of north-western Britain but, in the *Pohlietum glacialis*, it dominates in spongy carpets, often of quite small extent, but exceptionally covering up to 200 m², of a bright apple-green colour that makes the stands readily recognisable from a distance. Few other bryophytes occur with any frequency and none is consistently abundant. There is commonly a little *P. ludwigii* and *Philonotis fontana* sometimes attains a measure of prominence, though typically it is of nothing like such high cover as in the *Philonoto-Saxifragetum*. Other bryophytes recorded occasionally are *Hygrohypnum luridum, Bryum weigelii, Calliergon stramineum, Scapania undulata, S. uliginosa, Dicranella palustris* and *Marchantia alpestris*.

In this carpet, vascular plants are few in number and typically of low cover. Only *Deschampsia cespitosa* (presumably ssp. *alpina* at the high altitudes characterised by this community) and *Saxifraga stellaris* are constant, but the rare Arctic-Alpine *Cerastium cerastoides* is quite often found and there can also be some *Stellaria alsine, Chrysosplenium oppositifolium, Epilobium anagallidifolium, Veronica serpyllifolia* var. *humifusa* and *Rumex acetosa*. Other rare plants which find an occasional locus here are *Epilobium alsinifolium, Alopecurus alpinus* and *Phleum alpinum*.

Habitat
The *Pohlietum* is strictly confined to spring-heads associated with the late snow-beds of the higher reaches of the Scottish Highlands, where there is vigorous irrigation by cold, oligotrophic waters.

Although *P. wahlenbergii* var. *glacialis* occurs in small amounts over quite a wide range of altitudes through the uplands of north Wales, Cumbria and Scotland, it is found with the kind of dominance characteristic here only within the high-montane zone, at altitudes generally above 850 m, where mean annual maximum temperatures do not exceed 21 °C (Conolly & Dahl 1970). Within this area, which includes the central and north-western Highlands of Scotland, the community is further restricted to situations where snow lies longest. Precipitation is heavy throughout the region, with more than 1600 mm annually (*Climatological Atlas* 1952) and, with the bitter winter temperatures at higher altitudes, much of this falls as snow, persisting long everywhere but especially so over north- and east-facing slopes. The majority of the stands of the *Pohlietum* are from such aspects and the community is especially well developed in association with those extensive late snow-beds found in the great sunless amphitheatres of the corries in the Cairngorms, Ben Alder and Creag Meagaidh in the central Highlands and, further north-west, in the Affric-Cannich hills and on Beinn Dearg.

Typically, in these localities and at other sites where the community occurs less extensively, as on Beinn Laoigh, Bidean nam Bian, around Lochnagar and in the Monar Forest (Ratcliffe 1977), the rocks from which the springs emerge are acidic and calcium-poor, usually schists, granulites and grits of the Moine series, granites or lavas and agglomerates. So the flushing waters, and the often sloppy, ill-structured mixtures of mineral and organic matter held beneath the moss carpet, are base-poor and oligotrophic. And it is these general climatic and edaphic features which determine the overall character of the community, with its cold-tolerant plants such as *Deschampsia cespitosa, Stellaria alsine, Chrysosplenium oppositifolium* and *Philonotis fontana*, and definite montane species like *Saxifraga stellaris*, *Cerastium cerastoides*, *Epilobium anagallidifolium* and *Bryum weigelii*. This much, and the noticeable lack of calcicolous plants, the *Pohlietum* shares with the *Philonoto-Saxifragetum*.

The difference between the two kinds of spring vegetation is best seen among the bryophyte element where there is a switch from dominance by *Philonotis fontana*, *Dicranella palustris* and *Scapania undulata* to *Pohlia wahlenbergii* var. *glacialis* and *P. ludwigii*. All these species are capable of luxuriant growth in vigorous oligotrophic springs, but the latter two become prevalent where the water temperatures are lower: not all cold springs have the *Pohlietum* but, in those which do, the water temperature is consistently below 4 °C (McVean & Ratcliffe 1962). The *Philonoto-Saxifragetum* has a much broader geographical range than the *Pohlietum*, its mean altitude is some 300 m lower and, though it can be found in springs fed by melting snow, its association with that habitat is by no means as exclusive as with the *Pohlietum*.

Zonation and succession

The *Pohlietum* is typically found with a very distinctive suite of vegetation types of late snow-beds, variation among which can be related partly to the wetness of the ground. It is sometimes the only kind of spring in such situations, but other related communities sometimes occur with it where the irrigating waters become less frigid or vigorous.

Pohlietum springs can vary considerably in size, shape and numbers, but they generally occur towards the base of snow-beds, giving way sharply above to vegetation of more freely-draining ground, which may not be fully exposed by the melting snow until early summer. Here, the *Polytrichum-Kiaeria* community is very characteristic, providing a further locus for *Saxifraga stellaris* but otherwise showing little floristic continuity with the *Pohlietum* even in those patches where there is intermittent irrigation by melt-water or rain. Often, too, there are stands of the *Salix-Racomitrium* community on moister soils with much solifluction. Then, around these, on ground with less extensive snow-lie, there can be zones of the *Deschampsia-Galium* community and the *Carex-Racomitrium* and *Carex-Polytrichum* montane heaths.

With increasing distance from the spring-head, where snow-lie is not so long and the irrigating waters a little warmer, though still flowing vigorously, the *Pohlietum* may give way to the *Philonoto-Saxifragetum* along the melt-water rills and on the flushed ground around. Here, continuity among the herbs and some of the associated bryophytes is considerable, though there is the marked shift in dominance in the ground carpet. Where water-flow is considerably reduced, there can also be flushed banks of the *Sphagno-Anthelietum julaceae*.

In patterns such as these, the *Pohlietum* forms an integral part of a complex of vegetation types associated with one of our most extreme habitats. Particularly fine suites of the communities can be seen in the Cairngorms, on the upper slopes of Ben Alder and Aonach Mor, in the Creag Meagaidh corries and on the high slopes of the Affrich hills, all in Inverness and on Beinn Deargh and Am Faochagach in Ross.

Distribution

The community occurs widely but very locally through the central and north-western Scottish Highlands.

Affinities

Although Dixon (1954), in his description of what he knew as *Webera albicans* var. *glacialis*, noted the tendency of this moss to form striking patches around high-montane springs, these stands were not systematically described until the survey of McVean & Ratcliffe (1962). Dahl (1956), however, had already noted the similarity of this kind of Scottish vegetation to cold springs described by him and others (e.g. Samuelsson 1934, Nordhagen 1943, Vigerust 1949) from the Norwegian mountains. There, his *Mniobryo-Epilobietum* was likewise dominated by *P. wahlenbergii* var. *glacialis* (once known also as *Mniobryum wahlenbergii*) and shared some of the same associates, notably *Saxifraga stellaris* and *Cerastium cerastoides*, although in Norway *Epilobium anagallidifolium* is replaced by *E. hornemannii*.

As McVean & Ratcliffe (1962) noted, there is considerable qualitative similarity between the *Pohlietum* and the *Philonoto-Saxifragetum*, and they followed Scandinavian workers in locating both communities in the same Mniobryo-Epilobion alliance. More recent revisions of the Montio-Cardaminetea would place them together in the Montion sub-alliance of the Cardamino-Montion.

Floristic table M33

Pohlia wahlenbergii var. *glacialis*	V (6–10)
Deschampsia cespitosa	V (1–4)
Saxifraga stellaris	IV (2–3)
Pohlia ludwigii	IV (2–4)
Cerastium cerastoides	III (1–3)
Philonotis fontana	III (1–7)
Chrysosplenium oppositifolium	III (2–4)
Stellaria alsine	III (3)
Epilobium anagallidifolium	III (3)
Hygrohypnum luridum	II (2–3)
Bryum weigelii	II (1–3)
Calliergon stramineum	II (6)
Scapania undulata	II (1–2)
Scapania uliginosa	II (1–4)
Marchantia polymorpha	II (2–4)
Veronica serpyllifolia	II (2)
Rumex acetosa	II (1)
Drepanocladus exannulatus	I (2–4)
Poa annua	I (2–3)
Dicranella palustris	I (2–6)
Scapania nemorosa	I (4)
Sphagnum squarrosum	I (3)
Festuca rubra	I (2)
Montia fontana	I (4)
Agrostis capillaris	I (3)
Number of samples	10
Number of species/sample	10 (5–22)
Vegetation cover (%)	96 (75–100)
Altitude (m)	992 (868–1083)
Slope (°)	23 (0–35)

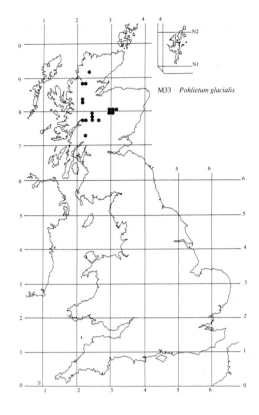

M33   *Pohlietum glacialis*

M34
Carex demissa-Koenigia islandica flush

Synonymy
Koenigia islandica-Carex demissa nodum   Birks 1973.

Constant species
Carex demissa, Deschampsia cespitosa, Juncus triglumis, Koenigia islandica, Saxifraga stellaris, Blindia acuta, Scapania undulata.

Rare species
Juncus biglumis, Koenigia islandica, Sedum villosum.

Physiognomy
Since it was first found in Britain in 1934 and finally accurately determined in 1950 (Burtt 1950, Raven & Walters 1956), *Koenigia islandica* has been observed on and around The Storr in Skye, where it was first collected, on other parts of the Trotternish ridge there (Birks 1973) and on Mull (Birse 1984), in a variety of vegetation types. In certain cases it is best seen as an occasional, sometimes of quite high local cover, in communities which occur widely elsewhere: some stands, for example, can be placed in the *Philonoto-Saxifragetum*, while others are very similar to the *Festuca-Agrostis-Alchemilla* or *Carex-Polytrichum* grass-heaths. Others, however, are more peculiar and it is these which are gathered into this *Carex demissa-Koenigia islandica* community.

It is an open vegetation type, with a bryophyte-dominated carpet broken by areas of wet, silty and stony ground. *Scapania undulata, Calliergon sarmentosum* and *Blindia acuta* are all common and each can be abundant, with occasional patches of *Dicranella palustris, Philonotis fontana, Drepanocladus revolvens, Marsupella aquatica* and *Sphagnum auriculatum*. Scattered through this mat and over the rills themselves are plants of *Carex demissa, Koenigia, Deschampsia cespitosa, Saxifraga stellaris, Juncus triglumis, J. bulbosus* and the rare *J. biglumis* and *Sagina saginoides*. All of these are generally of low cover, though many can show a measure of abundance and *Koenigia* itself, though its individual plants are little more than a few centimetres across, can cover quite an area of ground.

Habitat
In common with many of the other situations in which *Koenigia* is found, both in this country and elsewhere, the *Carex-Koenigia* community occurs on ground which is kept periodically moist by circumneutral and oligotrophic waters. Typically, it is found in open silty or stony flushes fed by vigorous seepage from springs issuing at moderately high altitudes, over 500 m, from the basalt of the Trotternish ridge. In their base-status, with pH values around 6.0, and their low cation content, the waters are of similar character to those which feed the *Philonoto-Saxifragetum*, which community indeed often occupies the spring-heads above the flushes. And the occurrence in the *Carex-Koenigia* flushes of Arctic-Alpines such as *Saxifraga stellaris, Juncus biglumis, J. triglumis* and *Sagina saginoides* testifies to the similarly harsh montane conditions that both vegetation types favour. But, at least as far as *Koenigia* is concerned, the climate, and vigorous flushing, probably have their major effect through helping to maintain the open nature of the habitat: *Koenigia* is found throughout the Arctic-Subarctic zone but only in situations where there is freedom from the competition that easily overwhelms its diminutive growth.

Zonation and succession
Most often, the *Carex-Koenigia* community forms a flush zone over the open rills that spread out below a *Philonoto-Saxifragetum* spring-head, the whole complex grading around to *Carex-Polytrichum* or *Carex-Racomitrium* heaths.

Distribution
The community is confined to Skye, where it occurs scattered along the Trotternish ridge extending several kilometres north of The Storr.

Affinities

This vegetation, the description of which is based entirely on the account of Birks (1973), is very similar in its general character to the *Philonoto-Saxifragetum*. More so than there, the presence of species such as *Carex demissa*, *Juncus triglumis* and *Blindia acuta* suggests some affinity with montane Caricion davallianae mires but, by and large, this community belongs among our Montion springs and flushes where plants like *Saxifraga stellaris*, *Calliergon sarmentosum*, *Scapania undulata* and *Deschampsia cespitosa* are characteristic. Rather similar vegetation types occur in Scandinavia, from where a *Carex rufina-Koenigia-Acrocladium sarmentosum* Association has been described by Gjaerevøll (1956) and Lid (1959), and in Iceland (Sörenson 1942, Steindórsson 1963, Hadač 1971). In Iceland, the distinctive conjunction of *Koenigia* with *Saxifraga stellaris*, *Juncus triglumis* and *Deschampsia cespitosa* ssp. *alpina*, led Sörensen (1942) to suggest the erection of a Koenigio-Microjuncion alliance. The *Sagina nodosa-Koenigia islandica* Association which Birse (1984) described from Mull, may also belong to the Montio-Cardaminetea, but it is rather different from this *Carex-Koenigia* flush and needs further sampling.

Floristic table M34

Carex demissa	4 (3–4)
Koenigia islandica	4 (3–6)
Deschampsia cespitosa	4 (3–5)
Saxifraga stellaris	4 (1–6)
Scapania undulata	4 (2–8)
Blindia acuta	4 (5–8)
Juncus triglumis	4 (3–4)
Juncus biglumis	3 (2–5)
Calliergon sarmentosum	3 (4–5)
Sagina saginoides	2 (2–3)
Juncus bulbosus	2 (3)
Juncus squarrosus	1 (2)
Carex nigra	1 (4)
Cochlearia officinalis	1 (1)
Euphrasia frigida	1 (3)
Pinguicula vulgaris	1 (1)
Ranunculus acris	1 (3)
Dicranella palustris	1 (3)
Drepanocladus revolvens	1 (3)
Philonotis fontana	1 (4)
Sphagnum auriculatum	1 (3)
Marsupella emarginata	1 (4)
Carex dioica	1 (1)
Luzula spicata	1 (3)
Racomitrium heterostichum	1 (3)
Number of samples	4
Number of species/sample	13 (10–15)
Vegetation cover (%)	47 (30–60)
Altitude (m)	625 (510–675)
Slope (°)	4 (0–8)

M35
Ranunculus omiophyllus-Montia fontana rill

Constant species

Montia fontana, Ranunculus flammula, R. omiophyllus, Sphagnum auriculatum.

Physiognomy

Ranunculus omiophyllus-Montia fontana rills typically have a rather crowded, though not always continuous, cover of vascular plants and bryophytes, much of the growth often submerged in the shallow waters, with a floating or shortly-emergent canopy. *Ranunculus omiophyllus* is often abundant, its delicate white summer flowers set off against the dark green of the floating leaves, and there is very frequently some *Montia fontana*. Then, the bronze-coloured floating leaves of *Potamogeton polygonifolius* are commonly prominent and there can be local patches of *Agrostis stolonifera, Glyceria fluitans, Juncus bulbosus, J. articulatus* and *Callitriche stagnalis* and scattered plants of *Ranunculus flammula, Myosotis secunda* and *Stellaria alsine* with, more occasionally, *Ranunculus repens, Equisetum palustre, Hydrocotyle vulgaris, Galium palustre* and *Lotus uliginosus. Juncus bufonius* and *Scirpus setaceus* can also sometimes be seen on open patches of wet mud.

Bryophytes quite commonly make a substantial contribution to the cover, though the frequent species are very few. Often, there are red-brown clumps of *Sphagnum auriculatum* growing semi-submerged and fresh-green patches of *Philonotis fontana* but, apart from occasional *Polytrichum commune*, other species are sparse with just scattered records for such plants as *Calliergon cuspidatum, C. stramineum, Drepanocladus exannulatus, D. vernicosus, Scapania irrigua* and *Rhytidiadelphus squarrosus*.

Habitat

This community is typical of spring-heads and rills at moderate altitudes in south-western Britain, where there is irrigation with circumneutral and probably quite oligotrophic waters.

Phytogeographically, the *Ranunculus-Montia* community can be seen as an oceanic replacement for the *Philonoto-Saxifragetum*. It has been recorded only from south-western England, Wales and from around the Lake District, though further sampling may well reveal that it occurs throughout the British range of *R. omiophyllus*, an Oceanic West European plant which is found in central southern and south-west England, through much of north-west England and south-west Scotland. All known stands of the community fall within that part of the country with mild winters, where February minima are by and large more than 1 °C above freezing (*Climatological Atlas* 1952) though, apart from the presence of *R. omiophyllus* and the generally Atlantic *S. auriculatum* such conditions make themselves felt here mostly in a negative way, with the very obvious exclusion of species characteristic of montane springs, such as *Saxifraga stellaris, Epilobium anagallidifolium, E. alsinifolium, Pohlia ludwigii, P. wahlenbergii* var. *glacialis* and *Bryum weigelii*.

The continuing prominence of plants like *Montia fontana, Juncus bulbosus, Ranunculus flammula, Philonotis fontana* and *Sphagnum auriculatum*, strongly reflects the character of the irrigating waters here which, as in the Montion springs, are typically rather base- and nutrient-poor, with pH values ranging from 4.5 to 6.5. Throughout the south-west, springs and rills of this kind are widespread over the acidic rocks which comprise the bulk of the uplands and here the community is often seen as a component of moorland vegetation, generally between 250 and 450 m, in areas such as Dartmoor and Bodmin Moor, where granite underlies the stands, and around the lower reaches of the Welsh and Cumbrian hills, where there is a wider variety of suitable rocks and drift. Irrigation can be quite vigorous though the community can subsist in gentle trickles of water.

Zonation and succession

The *Ranunculus-Montia* community can be found among a wide variety of vegetation types on the drier peats and acidic mineral soils around its springs and rills. A common context is provided by the *Scirpus-Eriophorum* blanket mire or *Scirpus-Erica* wet heath over thin peats, by drier heaths like the *Ulex gallii-Agrostis* heath in the south-west and the *Calluna-Erica* and *Calluna-Vaccinium* heaths in Wales, and derived Nardo-Galion swards maintained by grazing.

Quite often, the community can pass downstream, in more sluggish and impermanent rills, to the *Hyperico-Potametum*.

Distribution

Commonest in south-western England and Wales, the *Ranunculus-Montia* community may well extend into other parts of the warmer oceanic region of Britain.

Affinities

This kind of vegetation has attracted little attention in the literature, apart from rather informal descriptions of moorland streams (e.g. Tansley 1911), though it is unique in its floristics. The presence of such plants as *P. polygonifolius*, *J. bulbosus* and *R. flammula* brings the community close to the *Hyperico-Potametum*, which has a similar British distribution, and Westhoff & den Held (1969) regard *R. omiophyllus* as a character species of the Potamion graminei in the Potametea. On balance, however, it seems preferable to locate the *Ranunculus-Montia* rill in neither that alliance nor among the Hydrocotylo-Baldellion vegetation but with the other flushes of base-poor, oligotrophic waters, in the Montion. The community can then be seen as an oceanic counterpart of the *Philonoto-Saxifragetum* and *Sphagno-Anthelietum*, from which it can generally be separated by the absence of montane plants. Like those communities, it grades to species-poor stands, in this case with little more than swelling masses of *Sphagnum auriculatum* with a very few scattered herbs, which it can be very difficult to classify.

Floristic table M35

Ranunculus omiophyllus	V (3–9)	*Carex nigra*	I (1–2)
Montia fontana	IV (3–8)	*Poa annua*	I (2)
Sphagnum auriculatum	IV (2–10)	*Trifolium repens*	I (2–3)
Ranunculus flammula	IV (2–5)	*Scirpus setaceus*	I (3)
		Sagina procumbens	I (3)
Agrostis stolonifera	III (3–5)	*Carex panicea*	I (1–2)
Juncus bulbosus	III (2–6)	*Calliergon cuspidatum*	I (1–3)
Juncus articulatus	III (1–4)	*Rhytidiadelphus squarrosus*	I (1–2)
Philonotis fontana	III (1–2)	*Nardus stricta*	I (3)
Myosotis secunda	III (2–5)	*Molinia caerulea*	I (2)
Potamogeton polygonifolius	III (4–9)	*Holcus mollis*	I (3)
Callitriche stagnalis	II (2–7)	*Chamaemelum nobile*	I (6)
Juncus bufonius	II (1–3)	*Alopecurus geniculatus*	I (4)
Glyceria fluitans	II (2–5)	*Carex echinata*	I (2)
Stellaria alsine	II (2–3)	*Cardamine pratensis*	I (4)
Poa trivialis	II (2–4)	*Scapania irrigua*	I (2)
Polytrichum commune	II (2–3)	*Drepanocladus exannulatus*	I (2)
Equisetum palustre	I (3–5)	*Epilobium palustre*	I (3)
Hydrocotyle vulgaris	I (4)	*Brachythecium rivulare*	I (1)
Agrostis canina	I (2–3)	*Calliergon stramineum*	I (1)
Ranunculus repens	I (2–4)	*Drepanocladus vernicosus*	I (1)
Galium palustre	I (2–4)		
Lotus uliginosus	I (2–4)	Number of samples	12
Juncus effusus	I (2)		

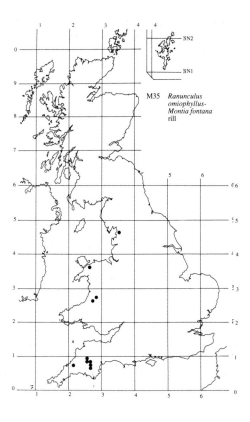

M35 *Ranunculus
omiophyllus-
Montia fontana*
rill

M36
Lowland springs and streambanks of shaded situations
Cardaminion (Maas 1959) Westhoff & den Held 1969

There is a clear contrast, among the Cardamino-Montion springs and flushes of acid to circumneutral habitats, between the upland communities described above, where *Montia fontana*, *Saxifraga stellaris* and *Philonotis fontana* are conspicuous and the vegetation of lowland and often shaded situations, in which *Chrysosplenium oppositifolium* occurs with such bryophytes as *Hookeria lucens*, *Rhizomnium punctatum*, *Trichocolea tomentella*, *Pellia epiphylla* and *Conocephalum conicum*. Such assemblages were used by Westhoff & den Held (1969) to diagnose distinct sub-alliances within the Cardamino-Montion, the sub-montane and montane Montion and the lowland Cardaminion. The latter kind of vegetation has not been separately sampled in this survey but it figures in the field and ground layers of various wet woodlands, notably the *Alnus-Carex*, *Alnus-Urtica* and *Alnus-Fraxinus-Lysimachia* types, where it is distinctive of seepage lines and damp stream banks, quite often with *Cardamine flexuosa*, *C. amara* and *Chrysosplenium alternifolium* (see also Oberdorfer 1977). Similar mixtures of plants can also be found widely through lowland Britain, especially in the wetter west and around the upland fringes, along stream sides and wet banks which were probably once wooded but where shade is now provided by tall herbs or by virtue of the aspect of the sites. These need further sampling to see if the kinds of communities identified on the Continent occur here.

M37
Cratoneuron commutatum-Festuca rubra spring

Synonymy

Tufaceous mounds Holdgate 1955a p.p.; *Cratoneuron commutatum-Saxifraga aizoides* nodum McVean & Ratcliffe 1962, Birks 1973; *Cratoneuron* springs Ferreira 1978; *Cratoneuron* drip-zones Ferreira 1978; *Cratoneuron commutatum* Community Birse 1980; *Cratoneuron filicinum* Community Birse 1980.

Constant species

Festuca rubra, Bryum pseudotriquetrum, Cratoneuron commutatum.

Physiognomy

Cratoneuron commutatum occurs frequently and with a measure of local abundance in a variety of calcareous mires, but in the *Cratoneuron commutatum-Festuca rubra* spring it is consistently dominant in large swelling masses, often forming prominent mounds or banks, of a golden-green colour grading to orange-brown. In some stands, which seem to preserve the same general floristic composition, *C. filicinum* accompanies or totally replaces it: the two species can show intergradations, and indeed both exhibit wide intraspecific variation (Bell & Lodge 1963, Smith 1978) but, by and large, *C. filicinum* is a smaller and somewhat stiffer plant and its stem leaves are plicate.

Other bryophytes can make a contribution to the mat, though typically it is a minor one. However, *Bryum pseudotriquetrum* is very common and occasionals include *Philonotis fontana* and the distinctly calcicolous *P. calcarea, Aneura pinguis, Pellia endiviifolia, Drepanocladus revolvens, Gymnostomum recurvirostrum, G. aeruginosum, Brachythecium rivulare* and *Dicranella palustris*. Very typically, there is some tufa deposition among the bryophyte shoots, which lends the mat a distinctive crunchy texture and allows it to build up into mounds.

The vascular element of the vegetation is typically species-poor and of low total cover with, in many stands, just a few scattered herbs. But there is consider-

able variation in this associated flora from place to place and, particularly where stands are developed over gently-sloping ground, a richer and more extensive herb layer can be found, such that the vegetation comes close to the *Cratoneuron-Carex* spring or to a Caricion davallianae flush. Often, however, the only species present are *Festuca rubra, Cardamine pratensis* and *Saxifraga aizoides*. The last can be quite conspicuous here, particularly when it has its yellow summer flowers, but it is generally present only in small amounts and is totally absent from springs of this kind throughout southern Scotland (e.g. Meek 1976, Ferreira 1978) and in Wales.

Occasional herbs include *Agrostis stolonifera, Deschampsia cespitosa, Equisetum palustre, Chrysosplenium oppositifolium, Poa trivialis, Carex panicea, C. nigra, C. dioica* and the rare *Epilobium alsinifolium* and *Equisetum variegatum*.

Habitat

This is a community of ground kept permanently moist by irrigation with base-rich, calcareous and generally oligotrophic waters. It is widespread but local throughout the cooler and wetter north-western uplands of Britain where springs and seepage lines occur in areas of lime-rich bedrocks. Trampling and grazing can have an adverse effect on the bryophyte carpet but in inaccessible positions the community is essentially permanent.

Like the *Philonoto-Saxifragetum*, this is a vegetation type dependent on the kind of sustained irrigation common in areas of higher rainfall. It is best developed in those parts of the country where there are more than 1600 mm precipitation annually (*Climatological Atlas* 1952), with in excess of 180 wet days yr^{-1} (Ratcliffe 1968), and here it can be found marking out springheads, seepage lines and drip-zones, where waters emerge along bedding-planes or at junctions with impervious substrates. Provided the ground is kept permanently sodden by the trickling or splashing waters, the community can occur even over vertical surfaces and bare rock, able to hang down in curtain-like masses by virtue of the pleurocarpous habit of the *Cratoneuron* spp. and the binding of the dead shoots by tufa. In other cases, the vegetation has more of the character of a flush, with the dominant moss forming a carpet over more gently-sloping ground.

In all cases, however, the habitat is a base-rich and calcareous one, with pH values generally around 7 and, in stands sampled by McVean & Ratcliffe (1962), dissolved calcium levels between 23 and 39 mg l^{-1}. Throughout the uplands, then, the community is confined to lime-rich bedrocks or superficials, occurring over limestones, such as those of the Carboniferous in the Pennines (Holdgate 1955a, Pigott 1956a), and of Cambrian/Ordovician and Jurassic provenance on Skye (Birks 1973), and marking out limey partings in Ordovician/Silurian shales in the Moorfoot cleughs (Ferreira 1978). Stands can also be found on more basic igneous rocks, among the Borrowdale Volcanics, for example, and on calcareous metamorphic rocks, most notably on the Dalradian meta-sediments in the Scottish Highlands, where schists and epidiorites are especially important substrates and, more locally, among the Moine series (McVean & Ratcliffe 1962, Birse 1980). The community spans a wide altitudinal range, though it tends to be absent from the very highest ground where extremely heavy rain has surface-leached the weathering products of even calcareous rocks. The character of the soil mantle is varied: in many cases the profiles are primitive and fragmentary, amounting to little more than mineral detritus and rotting moss remains while, in other stands, there can be a more substantial base of silty, humus-rich mud beneath the mat. Commonly, too, there is the deposition of tufa among the moss shoots.

The calcareous character of this spring environment is reflected in the vegetation in the overwhelming dominance of the *Cratoneuron* spp., and the occasional occurrence of mosses such as *Philonotis calcarea* and *Gymnostomum recurvirostrum*, and the presence among the vascular component of *Saxifraga aizoides* and, much more occasionally, plants like *Carex dioica*, *C. pulicaris*, *Pinguicula vulgaris* and *Equisetum variegatum*. Where numbers of these are present together, they can give something of the richness and diversity of a Caricion davallianae flush, but the overwhelming dominance of the moss carpet typically precludes more than local expression of this. The freedom from grazing that occurrence on steeper slopes brings probably plays a large part in maintaining such dominance: in many base-rich flushes, trampling and nibbling of the herbage are important agents in maintaining an open, varied turf.

There is one further environmental feature that exerts some influence on the character of the vegetation. It is temperature, particularly the coolness of the air and flushing waters during the summer through much of the range of the community, which encourages the precipitation of tufa and imparts an Arctic-Alpine feel to the modest vascular flora. Thus, within the heart of the range of the *Cratoneuron-Festuca* spring, mean annual maximum temperatures rarely exceed 23 °C (Conolly & Dahl 1970) with annual accumulated temperatures usually below 500 day-degrees C (Page 1982). The Arctic-Alpine *S. aizoides* is the best reflection of such conditions here, although its range shows a striking gap in southern Scotland with outlying populations in the colder reaches of the Lake District and the north Pennines (Perring & Walters 1962). Other, less common, plants of this character are *Epilobium alsinifolium* and *Thalictrum alpinum*. Beyond the range of such species, springs dominated by *Cratoneuron* spp. can certainly be found and further sampling may reveal a distinct syntaxon of such vegetation. *C. filicinum* is certainly the more common of the two species at lower altitudes and it seems to be characteristic of somewhat more eutrophic irrigating waters (Oberdorfer 1977, Birse 1980). For the moment, however, it seems best to retain all the sampled variation within the same unit.

Zonation and succession

The *Cratoneuron-Festuca* spring is generally found as an integral part of very characteristic sequences of our more calcicolous upland communities, developed over ground that is progressively drier. Topographic and edaphic variation over exposures of lime-rich substrates can modify and fragment such zonations and grazing and trampling also often play a part in influencing the vegetation pattern.

Very commonly, the *Cratoneuron-Festuca* community marks out springs and seepage lines within tracts of calcicolous grasslands developed over drier rendzina soils or calcareous brown earths. Stands are typically small, frequently less than 1 m², though more extensive banks and curtains can sometimes be found, and where there is a repeated pattern of irrigation at points along a slope, whole series of springs can pick this out. In rockier situations, where seepage occurs directly from bedding planes in exposures, the stands are often well defined, with a sharp disjunction from the surrounds but, with more diffuse irrigation, particularly over gentle slopes that are open to grazing, much more gradual transitions can be seen.

The kinds of sward that surround the *Cratoneuron-Festuca* spring vary with the region and, within regions, with altitude. A common context is provided by the *Festuca-Agrostis-Thymus* grassland, the typical plagio-climax pasture of moist calcareous soils through much of the north-western uplands of Britain and a community which, on the flushed surrounds of springs, can grade to the *Cratoneuron-Festuca* vegetation through its *Carex* or *Saxifraga-Ditrichum* sub-communities. On the limestones of the north Pennines, exactly parallel sequences can be seen between the springs and the *Sesleria-Galium* grassland which largely replaces the *Festuca-Agrostis-Thymus* grassland in this part of Britain, and which also has a *Carex* sub-community of moister soils. On the metamorphosed sugar-limestone of Upper Teesdale, the very distinctive *Carex capillaris-Kobresia* sub-community of the *Sesleria-Galium* grassland can be found in such patterns.

Small-sedge fens of the *Pinguiculo-Caricetum* also often occur on the strongly-flushed ground around or below *Cratoneuron-Festuca* springs and, where this community is interposed between the spring and the grassland, much more gradual transitions can be seen through the sequence of vegetation types with bryophytes such as *Cratoneuron commutatum*, *Bryum pseudotriquetrum* and *Aneura pinguis* running on into the flush as a broken carpet among the now much richer herb layer.

At higher altitudes, in the Scottish Highlands, for example, the *Cratoneuron-Festuca* spring can extend up to almost 1000 m and here it can be found among the more montane calcicolous communities like the *Festuca-Alchemilla-Silene* dwarf herb vegetation, an important context over the mica-schist slopes and banks of the Dalradian deposits. At these higher altitudes, too, there is a tendency for the *Pinguiculo-Caricetum* to be replaced by the *Cariceto-Saxifragetum* over the strongly-irrigated ground of open stony flushes below *Cratoneuron-Festuca* springs.

In more gradual transitions between the community and the surrounds, developed over gentler slopes, trampling by grazing stock or deer often plays an important part in maintaining the characteristically open conditions of the flushed soils. Typically, the *Cratoneuron-Festuca* spring itself is inaccessible to such treatment and heavy trampling is undoubtedly deleterious to the bryophyte mat. Such influences may indeed help convert the vegetation to a Caricion davallianae flush where springs occur on ground that is opened to pasturing, with an increase in the cover and diversity of smaller herbs able to exploit the lack of competition. The much richer *Cratoneuron-Carex* community described largely from Moor House in Cumbria (Eddy *et al.* 1969; see also Huntley 1979) may represent a stage in such a process.

In most circumstances, the *Cratoneuron-Festuca* spring appears to be a permanent community maintained by the edaphic and climatic conditions of the environment and, even where long ungrazed, it is unlikely that progression to any kind of woody vegetation is easy because of the impoverished character of the waters and soils. Where conditions are a little more eutrophic, however, lack of grazing could perhaps permit the spread of taller herbs. Over inaccessible dripping cliffs, the *Cratoneuron-Festuca* spring can sometimes be found with the *Alchemilla-Saxifraga* tall-herb community, in which *C. commutatum* and *B. pseudotriquetrum* form a patchy ground layer beneath tumbling masses of *S. aizoides* and a variety of calcicolous and mesophytic montane herbs.

Distribution

The community can be found over more lime-rich rocks throughout the north-western uplands of Britain, with its more Arctic-Alpine element best developed in the Scottish Highlands with outliers in the Lake District and Upper Teesdale. Springs dominated by *Cratoneuron* spp. also occur widely, though often very locally, in the British lowlands and further sampling of these is needed to provide a complete definition of this kind of vegetation.

Affinities

Although *C. commutatum* found early mention among general discussions of spring vegetation (e.g. Tansley 1911), this kind of community was not formally described until McVean & Ratcliffe's (1962) survey of stands in the Scottish Highlands. These form the core of the *Cratoneuron-Festuca* community as defined here, but such further sampling as has been undertaken shows that there is a gradation between the more Arctic-Alpine stands and *Cratoneuron* vegetation found at lower altitudes. Although more work is needed here, the general affinities of the vegetation are clear enough. They lie between the springs of the Cardamino-Montion, sharing such species as *Philonotis fontana*, *Cardamine pratensis*, *Deschampsia cespitosa* and *Chrysosplenium oppositifolium*, and the calcicolous small-sedge fens of the Caricion davallianae, where such plants as *Cratoneuron commutatum*, *Bryum pseudotriquetrum* and *Aneura pinguis* occur with a variety of basiphile herbs. Such vegetation is traditionally placed in the Cratoneurion to which *Cratoneuron* spp., *Philonotis calcarea* and *Saxifraga aizoides* are regarded as preferential (Ellenberg 1978).

Floristic table M37

Cratoneuron commutatum	V (4–10)	Festuca ovina	I (2–4)
Festuca rubra	IV (1–6)	Cochlearia officinalis	I (3–8)
Bryum pseudotriquetrum	IV (1–5)	Montia fontana	I (3–4)
		Rumex acetosa	I (1–3)
Cardamine pratensis	III (1–5)	Thalictrum alpinum	I (1–4)
Saxifraga aizoides	III (1–7)	Tussilago farfara	I (2–3)
Cratoneuron filicinum	II (2–9)	Dicranella palustris	I (1–3)
Agrostis stolonifera	II (2–5)	Selaginella selaginoides	I (1–3)
Philonotis fontana	II (1–6)	Carex demissa	I (1–4)
Deschampsia cespitosa	II (1–4)	Juncus bulbosus	I (1–3)
Equisetum palustre	II (1–5)	Carex flacca	I (2–3)
Epilobium alsinifolium	II (2–5)	Carex lepidocarpa	I (2–7)
Chrysosplenium oppositifolium	II (2–7)	Alchemilla glabra	I (2)
Aneura pinguis	II (1–3)	Jungermannia exsertifolia	I (2)
Carex panicea	II (2–4)	Equisetum variegatum	I (2–3)
Carex nigra	II (1–6)	Carex pulicaris	I (2)
Poa trivialis	II (2–6)	Caltha palustris	I (2–3)
Carex dioica	II (2–4)	Crepis paludosa	I (2–3)
Juncus articulatus	I (1–3)	Euphrasia officinalis agg.	I (1–3)
Pinguicula vulgaris	I (1–4)	Hypericum pulchrum	I (1–2)
Poa pratensis	I (1–4)	Blindia acuta	I (2–3)
Sagina procumbens	I (1–3)	Breutelia chrysocoma	I (1)
Polygonum viviparum	I (1–5)	Gymnostomum aeruginosum	I (2–5)
Taraxacum officinale agg.	I (1–3)	Conocephalum conicum	I (1)
Drepanocladus revolvens	I (2–3)	Holcus lanatus	I (1–4)
Agrostis capillaris	I (2–4)	Triglochin palustris	I (2)
Agrostis canina	I (2)	Eriophorum angustifolium	I (1–3)
Pellia endiviifolia	I (1–5)	Sagina nodosa	I (1–3)
Ctenidium molluscum	I (2–5)	Holcus mollis	I (2–4)
Gymnostomum recurvirostrum	I (1–3)		
Cerastium fontanum	I (1–2)	Number of samples	33
Brachythecium rivulare	I (1–5)		

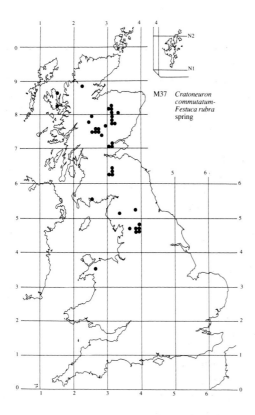

M37 *Cratoneuron commutatum-Festuca rubra* spring

M38
Cratoneuron commutatum-Carex nigra spring

Synonymy

Bryophyte flushes Pigott 1956*a p.p.*.; *Carex panicea-Campylium stellatum* nodum McVean & Ratcliffe 1962 *p.p.*; *Cratoneuron-Carex nigra* nodum Eddy *et al.* 1969; *Carex nigra-Equisetum palustre* nodum Huntley 1979 *p.p.*; *Saxifraga aizoides-Tussilago farfara* nodum Huntley 1979 *p.p.*; *Saxifraga aizoides-Juncus triglumis* nodum Huntley 1979 *p.p.*

Constant species

Agrostis canina, Cardamine pratensis, Carex demissa, C. nigra, C. panicea, Festuca rubra, Leontodon autumnalis, Polygonum viviparum, Selaginella selaginoides, Trifolium repens, Bryum pseudotriquetrum, Cratoneuron commutatum, C. filicinum, Philonotis fontana.

Rare species

Epilobium alsinifolium, E. nerteroides, Saxifraga hirculus, Oncophorus virens.

Physiognomy

While the *Cratoneuron commutatum-Carex nigra* spring preserves the same pattern of generally overwhelming dominance by *Cratoneuron commutatum* (again occasionally supplemented or replaced by *C. filicinum*) as in the *Cratoneuron-Festuca* community, the associated flora here is very much richer. This is partly to be seen among the bryophytes, where a variety of other species finds frequent or occasional representation in the mat, though not generally with very much abundance. *Bryum pseudotriquetrum*, however, and *Philonotis fontana*, which are the commonest among these plants, can have moderately high cover and there are many companions which can occur locally as quite prominent patches: *Aneura pinguis, Plagiomnium elatum, Fissidens adianthoides, Philonotis calcarea, Ctenidium molluscum, Rhizomnium pseudopunctatum, Campylium stellatum, Drepanocladus revolvens* and *Cinclidium stygium*. Among rarer species, the community also provides a locus for *Oncophorus virens* and *Meesia uliginosa*.

But it is among the vascular element that the increased richness of the vegetation is most evident, for here a large number of species make a frequent, even if usually low-cover, contribution. Small sedges are particularly noticeable, with *Carex demissa, C. nigra* and *C. panicea* constant, *C. pulicaris, C. flacca* and *C. dioica* also common. The first three in particular can be quite abundant and, when present in fairly extensive mixtures, these sedges can bring the vegetation close to a Caricion davallianae mire in its structure and composition. Then, there are frequent scattered plants of *Cardamine pratensis, Selaginella selaginoides, Leontodon autumnalis, Polygonum viviparum, Trifolium repens, Cirsium palustre, Ranunculus flammula, Sagina nodosa, Juncus triglumis, J. articulatus, J. bulbosus, Cerastium fontanum, Prunella vulgaris, Caltha palustris* (often in its diminutive montane form), *Galium palustre, Equisetum palustre, Achillea ptarmica, Cochlearia officinalis* (often recorded as ssp. *alpina*), *Triglochin palustris, Ranunculus acris, Anthoxanthum odoratum, Festuca ovina, Epilobium anagallidifolium* and, in northern England, where it has become very much at home on wet stony ground, the New Zealand introduction *E. nerteroides*. Rare herbs occasionally recorded are *Sedum villosum, Minuartia verna, Alopecurus alpinus, Epilobium alsinifolium* and, in parts of Upper Teesdale, though not in its Scottish localities, this community also provides the characteristic locus for the very rare *Saxifraga hirculus*. Looking at first sight like the much commoner *S. aizoides* when in flower, this plant is easily overlooked because its flowering stems are often cropped by sheep and its leafy shoots are rather inconspicuous among the moss carpet (Ratcliffe 1978).

Habitat

This kind of vegetation is confined to montane springs and flushes strongly irrigated by base-rich, calcareous and oligotrophic waters. It is very local in the northern Pennines and central Highlands of Scotland and, though the harsh climatic and edaphic conditions exert a strong influence on the structure and composition of the vegetation, heavy grazing may play a major role in maintaining the distinctive richness of the community.

Like the *Cratoneuron-Festuca* spring, this vegetation marks out sites of sustained irrigation with waters draining from lime-rich bedrocks, notably the Carboniferous Limestone around Upper Teesdale (Pigott 1956a, 1978a, Eddy *et al.* 1969) and certain of the Dalradian meta-sediments in central Scotland (McVean & Ratcliffe 1962, Huntley 1979). Such areas experience more than 1600 mm rain annually (*Climatological Atlas* 1952), with more than 180 wet days yr^{-1} (Ratcliffe 1968) and here the community can be found where excess surface water emerges in well-defined springs or more diffuse seepage lines, producing more or less permanently saturated stagnogley soils, usually with a shallow peaty top-soil (Pigott 1956a, 1978a, Eddy *et al.* 1969). Detailed data on pH and calcium content are sparse but both are probably quite high and tufa-encrustation among the moss mat is often seen. Certainly among both bryophytes and vascular plants, there are obvious calcicolous elements. Apart from the generally overwhelming dominance of *Cratoneuron commutatum* or *C. filicinum*, such species as *Bryum pseudotriquetrum*, *Aneura pinguis*, *Fissidens adianthoides*, *Philonotis calcarea*, *Ctenidium molluscum*, *Cinclidium stygium*, *Drepanocladus revolvens* and *Campylium stellatum* all reinforce this general aspect of the vegetation, together with such herbs as *Carex pulicaris*, *C. flacca*, *C. dioica*, *Briza media*, *Juncus triglumis*, *Triglochin palustris* and *Selaginella selaginoides*.

The other general feature of this kind of vegetation is its montane character. The *Cratoneuron-Carex* spring is a high-altitude community, found mostly above 650 m in areas where mean annual maximum temperatures rarely exceed 23 °C (Conolly & Dahl 1970) with annual accumulated temperatures usually below 500 day-degrees C (Page 1982). Arctic-Alpine plants such as *Polygonum viviparum*, *Juncus triglumis*, *Epilobium anagallidifolium* and *E. alsinifolium* are thus a frequent feature here. *Saxifraga hirculus*, it should be noted, though now surviving at higher altitudes in mainland Britain, is actually pre-alpine through Europe as a whole (Pigott 1956a) and, in those more low-level English localities from which it has been eradicated by drainage (Perring & Farrell 1977), probably occurred in quite a wide variety of vegetation types.

A third characteristic of the environment of the *Cratoneuron-Carex* spring is that it is grazed, and trampling and cropping of the herbage by sheep and deer, which have free access over the gently-sloping unenclosed ground on which these springs are usually found, is responsible for the most obvious floristic differences between the community and the *Cratoneuron-Festuca* spring. Not only is the former more species-rich than the latter, something which reflects the more open character of the sward, but its enrichment is of the kind which

reaches its full expression in such vegetation as the *Pinguiculo-Caricetum*, the characteristic small-sedge mire of grazed, base-rich flushes. The frequency of such plants as *Carex nigra*, *C. panicea*, *C. demissa*, *Leontodon autumnalis*, *Trifolium repens* and *Selaginella*, together with the greater variety among the bryophytes, helps place the *Cratoneuron-Carex* spring as a floristic intermediate between these two vegetation types and hints at its possible seral developments.

Zonation and succession

The *Cratoneuron-Carex* spring typically occurs as a local replacement for the *Cratoneuron-Festuca* spring in sequences of grazed calcicolous communities in upland Britain, where much of the floristic variation can be related to differences in topography and soils.

Commonly, therefore, the community picks out definite springs or flushes within stretches of calcicolous swards developed over drier rendzinas or brown soils. Individual stands are often small, although where vigorous irrigation runs down-slope flushed zones can be very elongated and springs can be very numerous within quite small areas, occurring repeatedly along valley sides. Where they emerge among rockier topography, the stands can be well defined from their surrounds but often here there is quite a gradual transition to the grassland, the extent and richness of the bryophyte mat declining but many of the small herbs retaining high frequency and cover into the sward. In Upper Teesdale, it is generally the *Sesleria-Galium* grassland which provides the context for the *Cratoneuron-Carex* spring, grading to it through the *Carex* or *Carex-Kobresia* sub-communities. Over somewhat less base-rich profiles both there and in Scotland, the *Festuca-Agrostis-Thymus* grassland can be found in similar sequences and this, too, can show transitions to the spring via its intermediate *Carex* sub-community. Often, a zone of the *Pinguiculo-Caricetum* is interposed over the strongly-flushed ground between the spring and the wet grassland, further attenuating the floristic gradation over the space of several metres, with quite gentle shifts among the bryophyte and herb layer in moving from the stagnogley to the rendzinas. At higher altitudes in the central Highlands, such vegetation types can be replaced by the *Cariceto-Saxifragetum* and the *Festuca-Alchemilla-Silene* dwarf-herb community around *Cratoneuron-Carex* springs.

Trampling and grazing in these kinds of zonations play an important part in keeping the vegetation open and varied and may be vital in maintaining the rich calcicolous flora characteristic of the community (Eddy *et al.* 1969). Indeed, such treatment may help mediate a succession from the more species-poor *Cratoneuron-Festuca* spring to this kind of vegetation but the continuing extreme wetness of the ground probably prevents further development to a Caricion davallianae sward. Withdrawal of grazing is unlikely to be accompanied by a pronounced expansion of mesophytic herbs because of the impoverished soil conditions, although it is possible that some limited increase in plants such as *Saxifraga aizoides*, *Alchemilla glabra*, *Filipendula ulmaria* and *Geum rivale* could produce a fragmentary tall-herb vegetation.

Distribution

The *Cratoneuron-Carex* spring occurs locally around Upper Teesdale and in the central Scottish Highlands.

Affinities

Although a variety of *Cratoneuron*-rich stands described by McVean & Ratcliffe (1962) and Huntley (1979) from the Scottish Highlands can be accommodated in this community, the bulk of the available data originates from around Upper Teesdale, particularly Moor House, where this kind of vegetation was recognised as distinct by Eddy *et al.* (1969). It can be seen as intermediate between the *Cratoneuron-Festuca* spring and Caricion davallianae fens like the *Pinguiculo-Caricetum* and accommodated with the former in the Cratoneurion alliance.

Floristic table M38

Cratoneuron commutatum	V (1–10)	Philonotis calcarea	III (1–5)
Carex nigra	V (1–9)	Alchemilla glabra	III (1–4)
Carex panicea	V (1–6)	Ctenidium molluscum	III (1–6)
Cardamine pratensis	V (1–3)	Carex pulicaris	III (1–4)
Bryum pseudotriquetrum	V (1–5)	Juncus bulbosus	III (1–5)
Leontodon autumnalis	V (1–3)	Bellis perennis	II (1–3)
Festuca rubra	IV (1–3)	Valeriana dioica	II (1–3)
Carex demissa	IV (1–6)	Sedum villosum	II (1–3)
Selaginella selaginoides	IV (1–3)	Saxifraga hirculus	II (1–4)
Philonotis fontana	IV (1–4)	Veronica scutellata	II (1–3)
Trifolium repens	IV (1–3)	Epilobium alsinifolium	II (1–3)
Polygonum viviparum	IV (1–3)	Viola riviniana	II (1–3)
Agrostis canina	IV (1–3)	Briza media	II (1–3)
Cratoneuron filicinum	IV (1–9)	Filipendula ulmaria	II (1–3)
		Rhizomnium punctatum	II (1–3)
Cirsium palustre	III (1–3)	Cinclidium stygium	II (1–3)
Aneura pinguis	III (1–3)	Geum rivale	II (1–3)
Ranunculus flammula	III (1–5)	Carex dioica	II (1–4)
Epilobium palustre	III (1–3)	Calliergon cuspidatum	II (1–6)
Juncus triglumis	III (1–3)	Drepanocladus revolvens	II (1–4)
Sagina nodosa	III (1–3)	Poa trivialis	II (1–3)
Cerastium fontanum	III (1–3)	Ranunculus repens	II (1–3)
Prunella vulgaris	III (1–3)	Veronica beccabunga	II (1–3)
Juncus articulatus	III (1–5)	Veronica serpyllifolia	II (1–3)
Caltha palustris	III (1–4)	Potentilla erecta	II (1–3)
Galium palustre	III (1–3)	Viola palustris	II (1–3)
Equisetum palustre	III (1–3)	Leiocolea bantriensis	II (1–3)
Achillea ptarmica	III (1–3)	Oncophorus virens	II (1–3)
Cochlearia officinalis	III (1–3)	Taraxacum officinale agg.	II (1–3)
Epilobium anagallidifolium	III (1–3)	Euphrasia confusa	II (1–3)
Triglochin palustris	III (1–3)	Agrostis stolonifera	II (1–4)
Anthoxanthum odoratum	III (1–3)	Chiloscyphus polyanthos	II (1–3)
Plagiomnium elatum	III (1–3)	Alopecurus geniculatus	II (1–3)
Fissidens adianthoides	III (1–5)	Carex flacca	II (1–3)
Campylium stellatum	III (1–3)	Jungermannia atrovirens	II (1–3)
Festuca ovina	III (1–4)	Deschampsia cespitosa	II (1–5)
Ranunculus acris	III (1–3)		

Brachythecium rivulare	II (1–3)	Sagina procumbens	I (1–3)
Nardus stricta	II (1–3)	Eriophorum angustifolium	I (1–3)
Rhizomnium pseudopunctatum	II (1–3)	Alchemilla filicaulis vestita	I (1–3)
Carex echinata	II (1–5)	Parnassia palustris	I (1–3)
Meesia uliginosa	I (1–3)	Moerckia hibernica	I (1–3)
Thuidium delicatulum	I (1–3)	Plantago lanceolata	I (1–3)
Potentilla palustris	I (1–3)	Achillea millefolium	I (1–3)
Carex lepidocarpa	I (1–3)	Saxifraga stellaris	I (1–3)
Linum catharticum	I (1–3)	Chiloscyphus pallescens	I (1–3)
Epilobium nerteroides	I (1–3)	Gymnostomum aeruginosum	I (1–3)
Minuartia verna	I (1–3)	Fissidens osmundoides	I (1–3)
Leiocolea alpestris	I (1–3)	Juncus acutiflorus	I (1–3)
Poa subcaerulea	I (1–3)		
Climacium dendroides	I (1–3)	Number of samples	26
Thymus praecox	I (1–3)	Number of species/sample	42 (25–57)
Carex rostrata	I (1–3)		
Pellia endiviifolia	I (1–3)	Altitude (m)	693 (503–853)
Hylocomium splendens	I (1–3)	Slope (°)	8 (4–10)

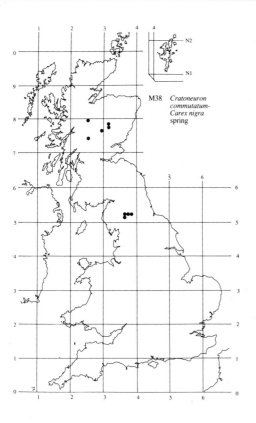

M38 *Cratoneuron commutatum-Carex nigra* spring

HEATHS

INTRODUCTION TO HEATHS

The sampling of heath vegetation

Extensive tracts of heath vegetation can be hard to find, fragmented now among improved lowland landscapes, and other stretches are difficult of access on the tops of sea-cliffs or remote mountain summits. But whether in such situations, or among the great expanses of heath that we have surveyed in the sub-montane zone of the north and west of Britain, we have not found this kind of vegetation awkward to sample. Again, though, in describing the variation encountered, we must distinguish between heath assemblages in the stricter sense and other sorts of plant community that can be found in intimate association with them on what are traditionally known as heaths or heathlands in the broader landscape sense. These were all sampled as part of our survey programme, but only the former are included in this section of the work. Here, then, heaths are taken to be vegetation types in which sub-shrubs play the most important structural role, albeit sometimes in a dwarfed or broken canopy, with such species as *Calluna vulgaris* and other ericoids, *Vaccinium*, *Empetrum* and *Arctostaphylos* spp., *Loiseleuria procumbens*, *Ulex minor* and *U. gallii* the usual dominants, alone or in various combinations. Thus, for the accounts of associated grasslands, moss and lichen vegetation, mires and scrub, in which such plants can be common but generally subordinate in their cover or at least accompanied by other prominent structural elements, readers will have to consult other sections of the work. Most confusing, perhaps, in this respect is the inclusion of much of what would generally be recognised as wet-heath vegetation among the mires. However, there are good floristic grounds for such a decision and, in this case as elsewhere, we have tried to remain sensitive to the close relationships between communities whose descriptions are, for one reason or another, widely separated in these volumes, and to their dependence upon the complex of environmental factors operative in the broader heathland scene where they can be found together. Some readers may also expect to find *Ulex europaeus* and *Cytisus scoparius* vegetation here,

whereas we have thought it more sensible to include it among the scrubs, while *Dryas octopetala* communities figure with calcicolous grasslands. Once more, the contents pages and indices of the volumes should help locate particular vegetation types of interest.

Following our general principle, samples were located among stretches of heath vegetation using floristic and structural homogeneity as the sole criterion. Species-rich assemblages or those with a restricted distribution were not given any undue attention and no special attempts were made to place samples so as to include plants thought indicative of particular floristic trends or environmental conditions. In fact, it was an explicit aim of the survey to try and locate the more peculiar or rare kinds of sub-shrub vegetation within their general context, so many tracts of impoverished and run-down heath, all that there is now in many parts of the country, were visited for sampling. Generally speaking, it was not difficult to discern homogeneity in heath vegetation, although account often had to be taken of some patterning among the plants, with herbs and cryptogams disposed according to the growth form of the dominant sub-shrubs, these themselves sometimes distributed in an ordered fashion in relation to such things as exposure to wind or frost-heave. Even the more dwarfed vegetation included here could also show some tendency to layering of the different elements though, as with the more obviously stratified heaths where sub-shrubs had grown tall over field and ground layers, all the components were always recorded together in a single sample. Clearly, in heaths, such horizontal or vertical structuring often reflected the influence of burning and, though stands could not be revisited within the short time-span of the survey, every attempt was made to include samples of obviously different ages since the last burn.

Usually, samples of 4×4 m were the most appropriate to take account of the structural scale and patterning within heaths, with 2×2 m being sometimes better for dwarfed sub-shrub vegetation, 10×10 m occasionally needed where the cover was more grossly structured.

Where stands were of an unusual shape, as over exposed brows, around small snow-beds or on patterned ground, irregularly-shaped samples of the relevant area were employed. In fine mosaics, where small patches of heath occurred among a matrix of grassland or mire vegetation, for example, it was sometimes necessary to treat these diminutive stands as the sample.

As always, recording of floristic data involved scoring the quantitative contribution of all vascular plants, bryophytes and macrolichens, special care being taken with the identification of smaller hepatics and lichens, an important element in some heaths. However, the occasional occurrences of *Rubus fruticosus* and *Euphrasia officinalis* were recorded to the aggregate, very young birch and oak sometimes just to *Betula* or *Quercus* sp. and *Hypnum cupressiforme*, apart from what is now known as *H. jutlandicum*, to *H. cupressiforme s.l.* Infertile material of *Kurzia* and *Weissia* was referred to the genus and *Lophocolea bidentata s.l.* may include some sterile *L. cuspidata*. Epiphytic cryptogams on the twigs of sub-shrubs were not recorded as a rule.

In addition to these numerical data, notes were often made on the physiognomy of the vegetation including such things as the growth phase of any *Calluna*, the structural arrangement of other sub-shrubs and any tendency to patterning or layering among the associates or indications of phenological rhythms. Spatial relationships with neighbouring vegetation types were detailed, particularly where these were gradual or intricate, together with any suggestions of successional change (Figure 19).

Along with the usual records of altitude, slope, aspect, geology and soil type, it was also extremely helpful for the interpretation of the floristic data to provide some qualitative indication of other aspects of the environment. Thus, notes were often made on such things as topography, particularly as it affected shelter from or exposure to wind and, in coastal sites, salt-spray, the deposition and persistence of snow, or shade from insolation; the presence of inherited patterned-ground features or the active influence of freeze-thaw and solifluction; the pattern of drainage of ground water or precipitation; and the effect of any biotic factors. In some cases, wild herbivorous mammals or birds had an obvious impact on the vegetation but, very commonly, anthropogenic influences were of great importance, especially through burning and grazing treatments, but also in the practice, or usually now the neglect, of traditional activities like the gathering of bracken or gorse and peat-cutting, and the increasing exploitation of heaths for leisure and extraction of underlying mineral resources. As throughout the survey, there was

Figure 19. Two completed sample cards from heaths.

NVC record sheet 10/81

Location	Grid reference	Region	Author
Lakenheath Warren Suffolk	TL (52) 761814	SE	JPH

Site and vegetation description

Heath of this kind is locally extensive at Lakenheath but much heather seems to have grown luxuriant or leggy with decline of grazing. Here the bushes are quite mature with scattered moribund individuals but there are stretches of open turf between where hummocky Festuca ovina & cryptogams are abundant, the lichens in particular forming an extensive and diverse cover. Heathy mosaics of this kind pass locally to areas of lichen-rich Festuca-Agrostis grassland or, where there is some surface disturbance, to more open sward with many diminutive ephemerals. Elsewhere there are transitions to Carex arenaria stands on somewhat looser sand. Mosaics with Pteridium also occur & there are scattered patches of Epilobium angustifolium. Interestingly, Deschampsia flexuosa is spreading in some of this grassy Calluna heath.

Date	Sample no.
13.5.1976	741
Altitude	**Slope**
17 m	0 °
Aspect	**Soil depth**
- °	cm
Stand area	**Sample area**
10 m x 20 m	2 m x 2 m

Layers: mean height: m m 50 cm 40 mm

Layers: cover: % % 90 % 60 %

Geology: Sandy Glaciofluvial drift with aeolian sediments above

Species list

Calluna vulgaris	7
Festuca ovina	6
Galium saxatile	3
Rumex acetosella	3
Dicranum scoparium	6
Hypnum cupressiforme	6
Ptilidium ciliare	3
Cladonia fimbriata	4
Cladonia pyxidata	3
Cladonia gracilis	3
Cornicularia aculeata	3
Cladonia bellidiflora	4
Cladonia squamosa	3
Hypogymnia physodes	3

Soil profile: Brown sand with signs of podzolisation in places

NVC record sheet 10/81

Location	Grid reference	Region	Author
Cauldrns, Yesnaby Orkney	HY(130) 223166	Scotland	AJCM

Site and vegetation description

Dwarfed patchy cover of wind-pruned & close-grazed Calluna with some Empetrum, forming irregular hummocks with much bare & stony ground between on more exposed stretches of cliff-top ablation surface (1). In other places, as here, a little grassier though still species-poor and with just sparse & grazed-down associates, apart from plantains which are locally plentiful. Sub-shrubs thin out quickly in plantain-rich sward to seaward (2), with open Festuca-Armeria turf on eroding cliff-edge (3). Wet areas of flat ground throughout have a sort of salt-marsh vegn. with much A. stolonifera & Glaux. (4).

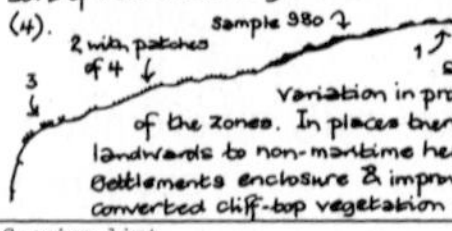

This general sequence continues southwards with some variation in proportions & clarity of the zones. In places there is a transition landwards to non-maritime heath but around the settlements enclosure & improvement have often converted cliff-top vegetation to pasture.

Date	Sample no.
2.7.1976	980
Altitude	**Slope**
24 m	5 °
Aspect	**Soil depth**
230 °	24 cm
Stand area	**Sample area**
50 m x 100 m	2 m x 2 m

Layers: mean height: m m 4 cm mm

Layers: cover: % % 100 % %

Geology: Devonian Old Red Sandstone flags and shales

Species list

Calluna vulgaris	6
Empetrum nigrum nigrum	5
Festuca rubra	5
Plantago lanceolata	4
Plantago maritima	5
Thymus praecox	2
Agrostis stolonifera	2
Lotus corniculatus	5
Scilla verna	2
Cerastium fontanum	1
Carex panicea	4
Euphrasia officinalis	2
Trifolium repens	2

Soil profile: Humic ranker, very sandy below

pH 0-5cm = 5.1

little time to quantify these effects or research site histories, but we were much helped in some situations by detailed investigations in existing studies of heath vegetation or individual species.

Some previous work was especially valuable in providing samples from heaths additional to those collected by our own research team. As so often, large quantities of data from throughout the Scottish Highlands (McVean & Ratcliffe 1962 and unpublished) and from the Scottish lowlands and Southern Uplands (Birse & Robertson 1976, Birse 1980, 1984 and unpublished) were very welcome, along with material from more substantial site surveys like those of Moor House (Eddy *et al.* 1969), the Cornish coast (Malloch 1971), Skye (Birks 1973), the Lizard (Hopkins 1983), the Hebrides (Hulme & Blyth 1984) and Shetland (Roper-Lindsay & Say 1986). Continuing sampling around our coastal cliffs by Malloch has substantially augmented our coverage of maritime heaths, and we have also been able to make use of a large number of smaller published and unpublished studies of particular sites or heath plants. Some internal NCC reports (NCC Devon Heathland Report 1980, NCC South Gower Coast Report 1981, NCC Pembrokeshire Heaths Report 1981) and other investigations (Urquhart & Gimingham 1979), while not providing samples for inclusion in the analyses, offered valuable detail about the distribution and character of heath vegetation in particular sites and regions.

In combining existing data with our own, differing recording conventions sometimes affected the way in which taxa were named. For example, not all previous studies distinguished what is now called *Agrostis vinealis*, so we had to retain the older convention of using *A. canina* ssp. *montana* where this taxon was separately scored from ssp. *canina*, with *A. canina* subsuming records where such a distinction was not made. Similarly, we have used *Empetrum nigrum* to include those records for crowberry which did not separate the subspecies, and *Festuca ovina* probably includes some *F. vivipara*, though we have listed this separately where it was unequivocally identified.

Altogether, about 2300 samples were available for analysis with a good geographical spread (Figure 20) and adequate coverage of most of the kinds of heath represented in Britain.

Data analysis and the description of heath communities

In the usual fashion, only the quantitative floristic records for the samples were used in the data analysis and characterisation of the vegetation types, with no weighting given to rare plants or supposed indicator species of any kind. By and large, the communities that emerged were defined by unique combinations of sub-shrubs, herbs and cryptogams with dominance of individual species rarely adequate for making more than crude distinctions. Indeed, with the high frequency of *Calluna* through many kinds of heath, and especially with its tendency to dominate in a wide variety of situations in the early and middle years after burning, such simple quantitative criteria can be a real distraction from understanding the subtleties of floristic variation among these vegetation types. This kind of biotically-mediated convergence to heather-dominance is very widespread through the British lowlands and sub-montane zone and means that the description of heath vegetation at any particular site ought to be based on samples from all age-ranges of growth among the sub-shrubs. With uncontrolled burning or extensive neglect of these kinds of vegetation, however, large tracts of heath can show a distressingly similar impoverishment and uniformity, and become very hard to distinguish from one another within a broadly-defined *Callunetum*. This clearly has profound implications for the conservation of the full variety of these heaths in Britain.

Altogether, 22 communities were characterised: five types of lowland dry heath with three related transitions to damper sub-shrub vegetation, two kinds of maritime heath of sea-cliffs and sand-dunes, four sub-montane heaths from the upland fringes of the north and west,

Figure 20. Distribution of samples available from heaths.

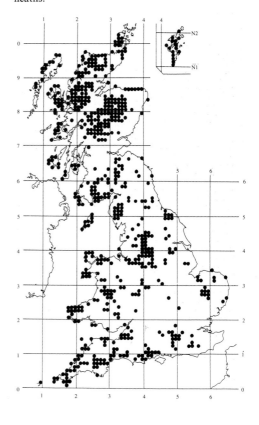

two transitions to more chionophilous vegetation in the sub-alpine zone and six lichen- or *Racomitrium lanuginosum*-rich montane heaths. Almost all the heaths were sufficiently diverse to warrant further division, some sub-communities so distinguished being transitional to other kinds of heath, others grading to closely-related grasslands, mires or various types of montane vegetation. Such floristic boundaries have been carefully checked for overlaps and gaps, and overall an attempt has been made to relate the range of communities to those described in the previous extensive surveys of Bridgewater (1970), Gimingham (1972), and Webb (1986). These and other studies in the literature have also provided a wealth of information to help interpret the floristic variation that has emerged, and to assist in understanding how heaths work. We have incorporated such material as seemed most appropriate, though the general relevance of some observations, as on the comparative physiology of the ericoids, or the responses of the commoner sub-shrubs to burning, makes it difficult to avoid repetitious reference through the text. We have sometimes avoided lengthy duplication of this kind by concentrating consideration of such issues under one or a few of the community accounts, as with the impact of fire on sub-montane sub-shrub vegetation, which is dealt with at greatest length in the descriptions of the *Calluna-Vaccinium* and *Calluna-Arctostaphylos uva-ursi* heaths. As usual, we have tried to relate the vegetation types to their closest equivalents in other parts of Europe, though our results confirm the suggestions of some previous authors that the inclusion of British heaths within a European overview challenges some of the existing phytosociological ideas about the major lines of variation among such vegetation types. The perspective looking towards the Continent from our own generally oceanic standpoint is rather different from that hitherto proclaimed as normative from the opposite direction.

Lowland dry heaths

Much of our lowland heath vegetation is relatively species-poor and the different types can be understood largely as permutations on a fairly limited range of dominants and associates representing biotically-derived replacements of the more calcifuge of our oak-birch and beech forests. Nonetheless, the communities compose with some felicity into a series that can be related broadly to variations in regional climate, soils and treatments (Figure 21). On drier acid sands and podzols through the generally warmer lowlands of Britain, five kinds of heath can be distinguished, with canopies made up of various combinations of *Calluna*, *Erica cinerea*, *Ulex minor* and *U. gallii*, and no more than a local contribution from *E. tetralix* and *Vaccinium myrtillus*. The sequence is best understood by starting in eastern

England with the *Calluna-Festuca ovina* heath (H1), the most continental of our sub-shrub communities in southern Britain and an impoverished relative of the Genisto-Callunion heaths of north Germany and the Low Countries. With us, plants like *Genista anglica* and *G. pilosa* are not especially characteristic and *Calluna* is generally the only sub-shrub, *E. cinerea* and the Ulices being largely excluded from suitable soils in this part of the country where the annual temperature range is large and the rainfall very low. Positive floristic responses to such conditions are scarcely seen in the vegetation, though some more continental ephemerals make a sporadic appearance on more open patches of drought-prone soil among the sub-shrubs and tussocks of *F. ovina*. For the most part, however, it is a few species of moss and lichen which bring variety to this kind of heath and even this modest enrichment has been extinguished in many places with the abandonment of burning and the demise of rabbits with myxomatosis. The community now has a very fragmented distribution but bigger surviving stands in the Breckland still provide something of the traditional heathland scene in this region of Britain, with the added interest at some sites of intricate grass/heath mosaics on patterned ground and transitions to inland dune vegetation.

The shift south and west towards The Weald and the Hampshire Basin, with their less extreme temperature range and higher rainfall, is marked by the appearance of Oceanic West European plants among the heathland flora. Most widespread among these is *E. cinerea*, the importance of which among the remaining drier heaths gives them a general similarity with the Ericion cinereae communities, the eu-oceanic sub-shrub assemblages of north-west Europe. Equally obvious in south-east England, however, at least where grazing is not too severe, is the less oceanic of our two heathland gorses, *U. minor*, and it is generally mixtures of these two, together with *Calluna* and a sparse tussocky cover of *Deschampsia flexuosa*, rather than *F. ovina* in these somewhat more humid conditions, that make up the bulk of the cover in the *Calluna-U. minor* heath (H2). Again, however, heathland reclamation and neglect of traditional treatments through much of the region have made for many fragmentary stands of rank or scrubby vegetation, though some commons and the New Forest preserve larger tracts.

Further west again, moving beyond Poole Harbour, down into the South-West Peninsula and Wales, *U. minor* is replaced in such drier heath vegetation by *U. gallii*, the rather sharp vicarism between the two species of gorse being hard to account for in simple ecological terms, but perhaps reflecting differing susceptibility to winter cold as well as varying responses to edaphic conditions and some element in conditions past. The *Calluna-U. gallii* heath (H8) is thus generally characteristic of the more equable regions of our southern and western lowlands, extending north to the Isle of Man and then curling round from north Wales across the fringes of the southern Pennines on to the mildly oceanic coast of East Anglia. Moving through this zone, however, rainfall is very variable and, where the community extends into the more humid and cooler upland fringes, *V. myrtillus* and hypnoid mosses make a more widespread contribution to this vegetation than their rather local occurrences in the *Calluna-U. minor* heath. A grassy ground of plants with Nardo-Galion affinities, usually mixtures of *F. ovina*, *Agrostis capillaris*, *Potentilla erecta* and *Galium saxatile*, is also often well represented here, especially in unimproved pastures and, where this kind of vegetation extends on to calcareous bedrocks mantled by superficials, it can acquire a further enrichment by calcicoles that often earns it the epithet 'limestone heath'. The distinctive character of these two gorse heaths has favoured their separation from the Ericion communities into either a single Ulicion alliance or separate Ulicion nanae and Ulicion gallii alliances, groups that encompass vegetation extending from northern France down into the Iberian peninsula.

A rather particular local expression of this general kind of vegetation is to be seen within the range of the *Calluna-U. gallii* heath on the Lizard in Cornwall where mixtures of *Calluna*, *E. cinerea* and *U. gallii* are often subordinate to a striking abundance of *Erica vagans* and *Ulex europaeus*. Here, the climate is at a warm oceanic extreme for Britain, *E. vagans* providing a clear link with the Lusitanian flora of more southerly parts of the western seaboard of Europe, though for the most part it is interactions between the very equable, humid conditions, the soil parent materials on the headland and treatments that determine the floristic character of the *E. vagans-U. europaeus* heath (H6). Rather unusually for our lowland heaths, this is a vegetation type of brown earths, probably a major factor in the pre-eminence here of *U. europaeus*, not generally a plant of heaths except where disturbance has brought some amelioration of highly impoverished conditions. Such soils have escaped cultivation on the Lizard on steeper, shedding slopes around valley heads and sides, and usually over the more lime-poor and oligotrophic

Figure 21. The circle of lowland dry heaths running from continental East Anglia through the oceanic south-west and up to the Pennine fringes.

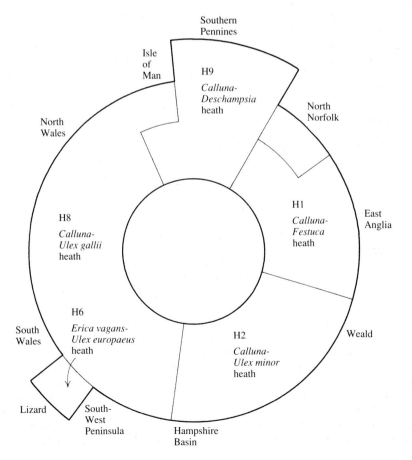

serpentine. The free-draining, fairly base-rich character of the soils thus favours a good representation of more mesophytic associates such as *Viola riviniana*, *Carex flacca*, *Dactylis glomerata*, *Stachys betonica*, *Filipendula vulgaris* and *Lotus corniculatus*, although the incidence of burning and grazing considerably modify the expression of this and other elements in the flora. The very distinctive character of this kind of heath has led some authors to locate it among the Ulici-Ericion ciliaris communities of north-west France and western Ireland.

A fifth community completes a circle of distribution among drier heaths in the lowlands and upland fringes extending across the Midlands and northern England where, in a cooler and consistently wetter climate, *U. gallii* becomes scarce and even *E. cinerea*, a much less narrowly oceanic plant, peters out. *V. myrtillus*, by contrast, now begins to assume the important role it has through our northern upland heaths, with *V. vitis-idaea* and *Empetrum nigrum* also figuring locally. Equally obvious, however, in this *Calluna-Deschampsia flexuosa* heath (H9) is a very poor expression of the cryptogam flora that one might expect, given the climatic conditions of this part of the country. Much more widespread, in fact, than hypnoid mosses are the acrocarps *Pohlia nutans* and *Orthodontium lineare*, species renowned for their tolerance of atmospheric pollution, with certain hepatics, such as *Gymnocolea inflata*, though hardly any encrusting lichens, taking advantage of bare patches of the often rather slimy mor humus. Centred around our

Figure 22. The increasing importance of moister lowland heaths with the climatic shift to the oceanic south-west.

great industrial conurbations, this vegetation is very similar to impoverished heaths found around grossly-polluted areas in the Low Countries and Germany and could be seen as a modified continental assemblage, perhaps in a distinct Pohlio-Callunion alliance. Alternatively, it can be regarded as a polluted vicariant of the *Calluna-Vaccinium* heath (H12) at its southern limit.

Transitions to wet heaths in the lowlands

At the continental extreme of climatic variation across lowland Britain, the distinction between dry and wet heaths is very clear, both in community composition and in the distribution of the two kinds of vegetation on the ground. With the shift to the more oceanic conditions of the south-west, however, the boundary becomes increasingly diffuse because wet-heath plants are able to transgress ever more extensively on to more free-draining acid soils from which, in drier climates, they are excluded by susceptibility to drought (Figure 22). Transitional types of heath vegetation thus become interposed between drier and wetter communities in this part of Britain, with *E. tetralix*, *Molinia caerulea* and *Agrostis curtisii* playing an important role alongside *Calluna*, *E. cinerea* and one or other of the gorses. Our scheme recognises two such humid heaths, as they are sometimes called, parallel to the two dry gorse heaths, with the *U. minor-Agrostis* heath (H3) characteristic of the moderately oceanic area around the Hampshire Basin, the *U. gallii-Agrostis* heath (H4) replacing it in the more equable south-west of England, though extending surprisingly little into south Wales. Typically, at lower altitudes, these kinds of vegetation form a fringe on periodically-gleyed podzols around wet heaths and

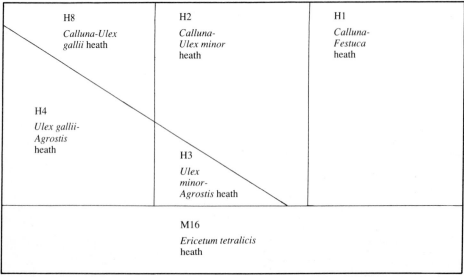

valley bogs but, where the *U. gallii-Agrostis* heath extends on to the higher moorlands of the south-west, with their distinctly cooler and wetter climate, it can be found on more humic intergrades to the ombrogenous peats that mantle the summit plateaus. Throughout the ranges of both communities, however, burning and grazing remain of considerable importance in the local expression of the vegetation, with *A. curtisii* or, on moister ground, *Molinia* often becoming very abundant in the early post-burn years, Nardo-Galion herbs increasing where stands on more free-draining soils are grazed. As with its drier analogue, the *U. gallii-Agrostis* heath can also have a mild calcicolous aspect among its associates where it extends on to more base-rich ground. The general similarity of these communities to the other gorse heaths argues for their inclusion within one or more Ulicion alliances, while their transitional relationship to wetter heaths could be recognised by locating them in a Ulicio-Ericion tetralicis (Figure 23).

Both the *U. minor-Agrostis* and *U. gallii-Agrostis* heaths can provide a locus for the rare *Erica ciliaris* and its hybrids with *E. tetralix* and, more unusually, where

the latter community extends on to the Lizard, it can have small amounts of *E. vagans*. Generally, though, wetter stagnogleys derived from the base-rich but calcium-poor bedrocks of this very oceanic headland carry a distinctive *E. vagans-Schoenus nigricans* heath (H5). Here, *E. vagans* is much more consistently abundant, with some *E. tetralix* and *U. gallii*, but just occasional *E. cinerea* and *Calluna*, often much *Molinia*, and a striking group of associates among which *S. nigricans*, *Serratula tinctoria*, *Succisa pratensis*, *Anagallis tenella*, *Carex pulicaris* and *Sanguisorba officinalis* are the most common. More open runnels between the *Schoenus* and *Molinia* tussocks can have a varied bryophyte flora or, where there is shallow flooding for much of the year, an assemblage with Hydrocotylo-Baldellion affinities. This community is much less readily located in the Ulicio-Ericion ciliaris than the *E. vagans-Ulex* heath, and is perhaps best understood in relation to the wet heaths of the Ericion tetralicis into which, with increasing oceanicity of climate, *S. nigricans* begins to penetrate on moving towards the west of Britain.

Maritime heaths

Both the *Calluna-U. gallii* heath and, within its very restricted range, the *E. vagans-Ulex* heath can extend on

Figure 23. Transition to wet heaths in the more oceanic parts of lowland Britain.

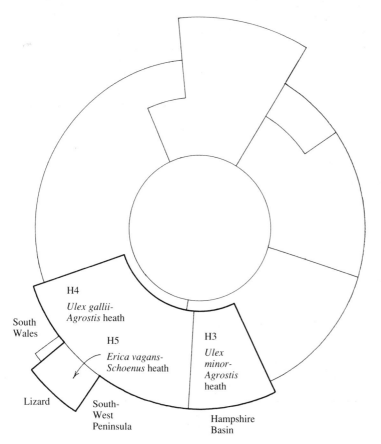

to more sheltered stretches of coastal cliff top where
species such as *Plantago maritima* and *Scilla verna* can
figure occasionally in more grassy stands. In general,
though, where salt-spray deposition rises above even
moderate levels on cliffs around the coasts of Britain,
more base-poor brown earths and rankers support a
more consistently maritime community, the *Calluna-
Scilla verna* heath (H7). Here, the sub-shrubs are generally few in number and often rather inconspicuous,
being reduced to a low, wind-shaped mat that can
become distinctly patchy in more exposed and rocky
situations. However, *Calluna* maintains its constancy
throughout, with *E. cinerea* also very common, though
replaced as the usual co-dominant by *E. tetralix* and
Empetrum nigrum around the cooler and wetter coasts of
Scotland. More locally, the community provides a locus
for the distinctive prostrate ecotypes of *Cytisus scoparius*, *Genista tinctoria* and the rare *G. pilosa* and diminutive forms of *Juniperus communis* ssp. *communis*.
Throughout, plants such as *Festuca ovina*, *Holcus lanatus*, *Lotus corniculatus*, *Hypochoeris radicata*, *Thymus
praecox*, *Plantago lanceolata* and *Potentilla erecta* contribute, along with *Scilla verna* and *Plantago maritima*,
to quite rich grassy swards, with additional suites of
herbs in more spray-drenched situations and on more
base-rich, more base-poor or more mesotrophic soils.

On exposed stretches of sea-cliffs, the *Calluna-Scilla*
heath can probably be seen as a climatic climax,
although the development of many stands has very
likely been influenced by pasturing. Of much more
clearly primary origin, and therefore rather exceptional
among our lowland sub-shrub communities, is the *Calluna-Carex arenaria* heath (H11) found on coastal dunes
made up of stabilised quartzitic sands or more lime-rich
sediments that have become leached with the passage of
time. Even here, though, it seems that episodes of relief
from grazing by rabbits or stock are essential for the
sub-shrubs to become established among the fixed dune
swards that develop as the dominance of *Ammophila
arenaria* or *C. arenaria* wanes. *Calluna* is the only
constant sub-shrub through the community as a whole,
though around our more oceanic coasts it is often
preceded in invasion by *Erica cinerea*, which can remain
abundant where the heather does not become too leggy.
In many stands, too, a variety of encrusting lichens is to
be seen among the grassy areas with acrocarpous mosses
and annuals colonising open patches in the turf. Much
more locally, around the cooler and wetter coasts of
northern and eastern Scotland, *Empetrum nigrum* ssp.
nigrum is an early coloniser, sometimes on locally
mobile sand, with Nardo-Galion herbs and hypnoid
mosses giving a rather different look to the vegetation.
Except where the substrates are very impoverished or
droughty, or remote from seed-parents, such heath
probably progresses fairly readily to scrub or woodland

if grazing ceases. Despite their various peculiarities,
both the *Calluna-Carex* and *Calluna-Scilla* heaths are
probably best accommodated in a broadly-defined Ericion cinereae.

Sub-montane heaths of the north and west
Around the upland fringes of northern and western
Britain, free-draining acidic soils able to support drier
heath vegetation are much more widely distributed than
in the lowlands, occurring extensively on hill slopes cut
into the prevailingly lime-poor bedrocks and over more
permeable superficials. However, at the more moderate
altitudes where heaths continue to provide a biotically-
derived replacement for forest, the prevalence of heavy
pasturing often severely restricts their contribution to
the landscape or gives them a distinctly grassy aspect,
particularly on somewhat better or marginally improved soils. Moreover, the continuing use of periodic
burning to renew sub-shrub growth for either stock or
grouse has favoured a widespread floristic impoverishment and convergence into a generalised heather-dominated moorland.

Nonetheless, it is still possible to detect some broad
climate-related patterns among the communities. The
floristic trends characteristic of the Ericion dry heaths,
for example, continue northwards beyond the geographical limit of *U. gallii* in the *Calluna-Erica cinerea* heath
(H10). This 'Atlantic heather moor', as it has been
called, includes most of the dry, non-maritime sub-
shrub vegetation on rankers, acid brown earths, brown
podzolic soils and podzols through the more equable
lowlands and upland margins of north-west Britain.
Here, the cool oceanic conditions are reflected in the
limited contribution of *Vaccinium* and *Empetrum* spp. to
the canopy and in the frequency among the associates of
Molinia, *Carex binervis*, *C. pilulifera* and *Blechnum
spicant*. Heath of this kind, with mixtures of *Calluna* and
E. cinerea usually dominating, and with groups of
associates reflecting differences in soil conditions and
treatments, does extend into eastern Scotland, but it
characteristically marks out lower ground there, or
south-facing slopes, where there is some local amelioration of harsher climatic conditions. More generally,
though, with the shift to a colder environment on
somewhat higher ground through the hills of the north
and west, and especially away from our more oceanic
uplands (Figure 24), this community is replaced by the
Calluna-Vaccinium myrtillus heath (H12). This can
share many of the herbs and bulky pleurocarps typical
of less bushy *Calluna-Erica* vegetation, but *V. myrtillus*,
V. vitis-idaea and *Empetrum nigrum* ssp. *nigrum* are the
usual dominants along with *Calluna*, and more oceanic
associates are scarce. This is our most widespread and
extensive Myrtillion heath, although its floristics and
structure are often controlled not by climatic or soil

variation across its range, but by response to burning, for it is a major component of sheep ranges and grouse-moors, often occurring in patchworks of stands whose composition reflects particular combinations and timings of treatments.

In broad terms, the *Calluna-Vaccinium* heath can be regarded as the British 'Boreal heather moor', although a much more striking response to cold continental conditions is seen in the *Calluna-Arctostaphylos uva-ursi* heath (H16), a community found at similar altitudes but strongly centred on the east-central Highlands of Scotland, where the climate is much drier than to the north-west and has greater extremes of temperature variation. Similar mixtures of sub-shrubs as in the *Calluna-Vaccinium* heath form the bulk of the canopy and leggier stands on poorer soils can grade imperceptibly to that vegetation, but *A. uva-ursi* is characteristic of more open areas of the patchwork and, particularly where there is

some relief from edaphic impoverishment, a very distinctive suite of herbaceous associates occurs: typical, then are the Northern Montane *Listera cordata*, *Trientalis europaea* and *Antennaria dioica* and the Continental Northern *Pyrola media*, along with *Anemone nemorosa*, *Lathyrus montanus*, *Viola riviniana* and *Hypericum pulchrum*. However, burning is again very common among this vegetation, which can make an important contribution to our more productive grouse-moors, and the survival of this rich assemblage from one burn cycle to the next seems to be a complex and precarious affair. Injudicious burning on poorer soils might also contribute to a more permanent run-down of this vegetation which, with the closely-related pine, juniper and oak-birch woods, makes up a very arresting regional complex of vegetation types and brings our Myrtillion heaths close to Scandinavian sub-shrub vegetation.

Heath of the *Calluna-Vaccinium* type can occur locally in the more oceanic parts of Britain, but a much more distinctive community with Myrtillion affinities is characteristic of low to moderate altitudes through the

Figure 24. Atlantic and boreal trends in the heather moors of the north-west uplands.

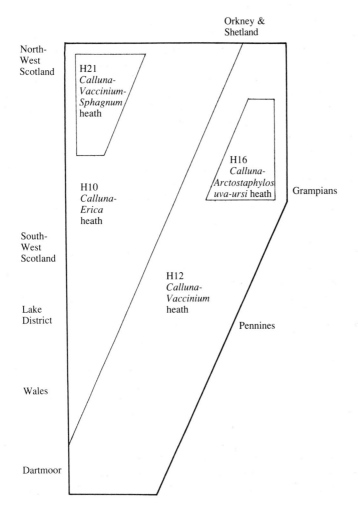

north-west Highlands of Scotland. There, in situations where the cool but equable climate is enhanced by shade and shelter among crags and blocky talus, and where there is relief from burning, the *Calluna-Vaccinium-Sphagnum capillifolium* heath (H21) can be found. Broadly oceanic plants such as *E. cinerea*, *Blechnum spicant* and *Molinia* can again be quite common among the patchworks of *Calluna*, *V. myrtillus* and *Empetrum nigrum*, but the damp peaty soils also support *S. capillifolium* among a carpet of bulky pleurocarps, and the consistently humid atmosphere of the most sheltered stands favours the development of a rich and varied assemblage of oceanic liverworts, the so-called 'mixed northern hepatic mat' which can include many rarities. Other more far-flung localities give this community some representation in southern Scotland and the Lake District, but here there are just fragments of this diversity among the ground flora.

Sub-alpine transitions to snow-bed vegetation

The sort of mixed sub-shrub cover characteristic of less assiduously managed stands of *Calluna-Vaccinium* heath extends on to higher ground through Wales, northern England and Scotland, and particularly in the central and eastern Highlands, in the *Vaccinium-Deschampsia* heath (H18). Towards its lower limits, this vegetation has probably been derived by burning and grazing, but at higher altitudes it is probably a natural form of grassy, moss-rich heath on free-draining, base-poor and circumneutral soils where the cold, wet character of the montane climate is locally enhanced by a sunless aspect and moderate snow-lie on the sheltered slopes. Most stands occur below the altitudinal limit of *Calluna* though, at these levels, heather is better represented on more exposed and drier slopes, and it is generally *V. myrtillus*, *V. vitis-idaea* and *Empetrum nigrum* that compose the bulk of the often rather open sub-shrub cover, with frequent *Alchemilla alpina* and, very locally, the rare *Phyllodoce caerulea*. Nardetalia herbs and hypnoid mosses remain important, the latter being especially abundant in shady and sheltered situations in and around shallow snow-beds.

This chionophilous trend becomes much more obvious around these altitudes in the Grampians in the *Vaccinium-Rubus chamaemorus* heath (H22), characteristic of hollows and lee slopes with locally prolonged snow-lie. Similar mixtures of sub-shrubs as above dominate here, but the Arctic-Subarctic *R. chamaemorus* and *Cornus suecica* are very frequent associates, the sodden raw humic soils or shallow peats also often supporting sparse shoots of *Eriophorum vaginatum* and various Sphagna, particularly *S. capillifolium* and *S. quinquefarium*, among a typically lush mat of hypnoid mosses. Both the *Vaccinium-Deschampsia* and *Vaccinium-Rubus* heaths extend into the north-west Highlands where the

moist, equable climate provides the sheltered humid conditions that these vegetation types require and, in the latter community in particular, enrichment with broadly oceanic hepatics can bring the composition close to that of the *Calluna-Vaccinium-Sphagnum* heath. In general, though, the mildly chionophilous character of these assemblages would argue for their inclusion among the Phyllodoco-Vaccinion myrtilli communities described from Scandinavia.

Montane heaths with lichens and mosses

Freedom from the moist, sheltered conditions favoured by this Phyllodoco-Vaccinion vegetation allows *Calluna* to extend in some abundance to high altitudes and, in Scottish mountains with a cold continental climate, it is the usual vascular dominant in a *Calluna-Cladonia arbuscula* heath (H13) of exposed ridges and summits in the low-alpine zone. Arctic-Alpine sub-shrubs like *V. vitis-idaea*, *E. nigrum* ssp. *hermaphroditum* and *Loiseleuria procumbens* can also occur with some frequency, along with sparse *V. myrtillus* and *C. bigelowii*, but typically it is lichens which impress with their abundance among a very dwarfed woody mat: here, species such as *Cladonia arbuscula*, *C. uncialis*, *C. rangiferina*, *Cornicularia aculeata*, *Cetraria islandica* and *Alectoria nigricans* can all be constant and plentiful. In this harsh environment, too, the impact of wind and frost on the shallow, humic soils and vegetation can result in some striking mosaics over the patterned ground (Figure 25).

On such exposed crests and ridges at similar altitudes in the cold but more humid conditions of north-west Scotland, lichen-rich heath of this general kind is often distinguished by the presence of *Arctostaphylos alpinus*, frequently with *Loiseleuria*, among a stunted cover of *Calluna* and *E. nigrum*, again with sparse *V. myrtillus*. Such *Calluna-A. alpinus* heath (H17) experiences a less extreme climate than the *Calluna-Cladonia* heath and, at lower altitudes where *E. nigrum* ssp. *hermaphroditum* is replaced by ssp. *nigrum*, with *Erica cinerea* and *Scirpus cespitosus* becoming common, there is a broadly oceanic feel to the vegetation.

Distinctive mixtures of such plants among a low canopy of *Juniperus communis* ssp. *nana* are also characteristic of the low-alpine zone of the north-west Highlands on slopes blown clear of snow, but with some shelter from the harshest winds and from insolation. The *Calluna-Juniperus* heath (H15) can have a cryptogam complement essentially the same as the *Calluna-Cladonia* and *Calluna-A. alpinus* heaths, but characteristically there are also some oceanic hepatics, *Pleurozia purpurea* and *Diplophyllum albicans* being especially distinctive with rare taxa sometimes present too. This kind of vegetation is extremely vulnerable to burning and the present patchy distribution is probably a relic of a more widespread representation from Skye to Cape

Wrath with possible fragments elsewhere in the north-west in the past.

Much more extensive than the *Calluna-A. alpinus* and *Calluna-Juniperus* heaths over windswept plateaus and ridges at moderate altitudes in the cool oceanic mountains of north-west Scotland is the *Calluna-Racomitrium* heath (H14). Even in the harsher climate of the eastern Highlands, conditions can be such as to make the ascendancy of lichens a finely balanced matter but, with the shift to the decidedly more humid atmosphere in the north-west, *R. lanuginosum* gains a strong competitive edge, dominating in this vegetation as a thick woolly carpet among open patchworks of *Calluna*, *E. nigrum*, *A. uva-ursi* and *A. alpinus*. *Erica cinerea* again becomes common, along with *Carex pilulifera* and *Scirpus cespitosus* and, though stands on higher ground can grade to *Calluna-Cladonia* heath, the occasional presence of Atlantic bryophytes in more sheltered situations towards lower ground presages a gradation to sub-montane oceanic heaths.

On higher ground throughout the Scottish Highlands and more locally elsewhere, the switch from lichen- to *Racomitrium*-dominance among the cryptogams of montane *Calluna* heaths is matched among vegetation where *V. myrtillus* is the most important sub-shrub. In the more continental mountains of the eastern and central Highlands, for example, a *Vaccinium-Cladonia arbuscula* heath (H19) gradually replaces the *Calluna-Cladonia* heath in passing up through the low-alpine zone, *Calluna* yielding to *V. myrtillus* and to a lesser extent *V. vitis-idaea*, though these themselves are often subordinate to the lichen carpet. *E. nigrum* ssp. *hermaphroditum* is common here, along with the Arctic-Alpine *Carex bigelowii*, but *Erica cinerea* and *Molinia*, which sometimes find a place at lower altitudes in the *Calluna-Cladonia* heath, are absent. Chionophobous plants also tend to be sparse here because the *Vaccinium-Cladonia* heath often marks out fairly sheltered situations where snow settles and sometimes persists for fairly long periods.

Figure 25. Bilberry (*a*) and heather (*b*) heaths with lichens and *Racomitrium* on boreal and atlantic mountains in the Scottish Highlands.

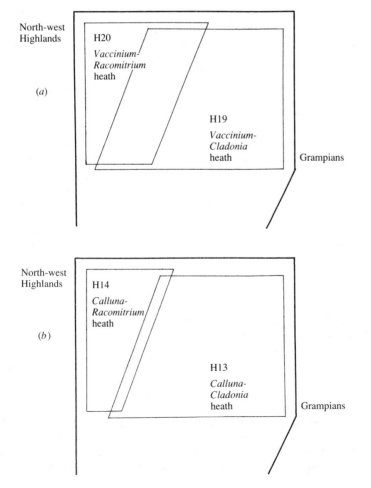

However, the response to more generally humid conditions is clearer where high-altitude bilberry heath extends into the oceanic north-west Highlands. Here, the *Vaccinium-Cladonia* heath persists locally, but is generally replaced by the *Vaccinium-Racomitrium* heath (H20) on higher, cloud-ridden slopes and summits with a fair exposure to wind. Mixtures of *V. myrtillus* and *E. nigrum* ssp. *hermaphroditum* generally provide the dominant vascular element, with some *V. vitis-idaea* and quite often *Alchemilla alpina* and *Diphasium alpinum*. *R. lanuginosum* is often accompanied by *Polytrichum alpinum* but a variety of bulky hypnoid mosses is also characteristic. Moreover, as so often in this part of Britain, there can be associates indicative of the humid and equable character of the climate, shady situations among boulders in the wettest part of the range providing a suitable niche for the appearance of the 'mixed northern hepatic mat'. However, despite the considerable variety among these six heaths, they can all be seen as part of the Caricetea curvulae, the class of lichen- and moss-rich vegetation transitional to fell-field assemblages.

KEY TO HEATHS

With something as complex and variable as vegetation, no key can pretend to offer an infallible short cut to diagnosis. The following should thus be seen simply as a crude guide to identifying the types of heath vegetation in the scheme and must always be used in conjunction with the data tables and community descriptions. It relies on floristic (and, to a lesser extent, physiognomic) features of the vegetation and demands a knowledge of the British vascular flora and, in many cases here, of bryophytes and macrolichens. It does not make primary use of any habitat features, although these may provide a valuable confirmation of a diagnosis.

Because the major distinctions among the vegetation types in the classification are based on inter-stand frequency, the key works best when sufficient samples of similar composition are available to construct a constancy table. It is the frequency values in this (and, in some cases, the ranges of abundance) which are then subject to interrogation with the key. Most of the questions are dichotomous and notes are provided at particularly awkward choices or where confusing vegetation patterns are likely to be found.

Samples should always be taken from homogeneous stands with 4 × 4 m generally sufficing to encompass the scale of the vegetation. Complex mosaics may demand samples of identical size but irregular shape, while very small stands can be sampled in their entirety.

---

1   *Ulex gallii* or *U. minor* constant and often abundant with *Calluna vulgaris* and *Erica cinerea*   2

Calluna and *E. cinerea* can be frequent and abundant but without *U. gallii* or *U. minor*   20

> With grazing, *U. minor* can become very sporadic in heaths which rightly follow 2. Towards its northern limit and on some sheltered sea cliffs, *U. gallii* may also occur very occasionally in heaths which rightly follow 20.

---

2   *U. minor* present without *U. gallii*   3

U. gallii present without *U. minor*   8

> There is a slight overlap in the distributions of the two *Ulex* species around Poole Harbour in Dorset where they are very occasionally found together in heaths. With non-flowering plants, it can also be difficult to distinguish the species.

---

3   *Deschampsia flexuosa* very common with *Molinia caerulea* and *Erica tetralix* occasional where the vegetation extends on to seasonally waterlogged ground but *Agrostis curtisii* very rare

H2 *Calluna vulgaris-Ulex minor* heath   4

D. flexuosa very scarce but *A. curtisii* constant and often abundant with *Molinia* and *E. tetralix* (or locally *E. ciliaris*)

H3 *Ulex minor-Agrostis curtisii* heath   6

4   *Vaccinium myrtillus* constant and locally abundant with frequent *Pteridium aquilinum* and scattered seedlings and saplings of oak, birch and pine

H2 *Calluna vulgaris-Ulex minor* heath
Vaccinium myrtillus sub-community

Pteridium can be quite common but *V. myrtillus* and young trees very scarce   5

5   *Molinia* very common and often almost totally displacing *D. flexuosa* with *E. tetralix* frequent and sometimes exceeding *E. cinerea*

H2 *Calluna vulgaris-Ulex minor* heath
Molinia caerulea sub-community

Molinia occasional at most and usually of low cover
with *D. flexuosa* very common and occasionally accompanied by *Festuca rubra*; *Dicranum scoparium* and *Hypnum ericetorum*, together with *Cladonia fimbriata, C.
coccifera* and *C. chlorophaea*, patchily prominent in
more open areas

H2 *Calluna vulgaris-Ulex minor* heath
Typical sub-community

6   *Calluna* and *E. tetralix* reduced in frequency and
cover and *Molinia* and *U. minor* somewhat patchy in
occurrence but *A. curtisii* very abundant with *U. europaeus* common at generally low covers and *Viola lactea*
often persisting sporadically after its colonisation of
open ground

H3 *Ulex minor-Agrostis curtisii* heath
Agrostis curtisii sub-community

A. curtisii very frequent but not extensive and *U. europaeus* occasional at most among mixed or *Calluna*-
dominated canopies   7

7   *Polygala serpyllifolia* common with a patchy
cover of bryophytes and lichens on more open areas
with *Campylopus brevipilus, Polytrichum juniperinum,
Cladonia impexa, C. floerkeana* and *C. coccifera*

H3 *Ulex minor-Agrostis curtisii* heath
Cladonia spp. sub-community

Above species very sparse among usually dense sub-
shrub canopies

H3 *Ulex minor-Agrostis curtisii* heath
Typical sub-community

8   *Erica vagans* and *U. europaeus* constant and
usually co-dominant with some of *Carex flacca, Viola
riviniana, Filipendula vulgaris* and *Stachys betonica*; only
on Lizard Peninsula in Cornwall

H6 *Erica vagans-Ulex europaeus* heath   9

E. vagans and *U. europaeus* very occasional and not with
the above associates   12

9   *Molinia* and *Potentilla erecta* constant with *Serratula tinctoria* frequent, but *F. vulgaris* rather uncommon
                                                                          10

F. vulgaris very frequent but *Molinia* absent and *P.
erecta* and *S. tinctoria* only occasional   11

10   *A. curtisii* constant and often very abundant,
particularly after burning, with *Calluna* unusually
scarce; *Carex panicea, Hypericum pulchrum, Viola lactea*
and *Polygala serpyllifolia* frequent

H6 *Erica vagans-Ulex europaeus* heath
Agrostis curtisii sub-community

A. curtisii and above associates absent but *E. tetralix*
constant in small amounts and *Sanguisorba officinalis*
and *Schoenus nigricans* frequent

H6 *Erica vagans-Ulex europaeus* heath
Molinia caerulea sub-community

On gentler slopes where deep soils receive some
slight flushing or are somewhat gleyed below,
there may be difficulty in separating this vegetation from the *Erica vagans-Schoenus* heath, but
U. europaeus and *E. cinerea* disappear there while
Molinia and *Schoenus* assume a much greater
structural importance with the appearance of a
distinctive runnel flora.

---

11   Sub-shrub canopy usually low and open, some-
times decidedly patchy, with a rich flora between the
bushes including *Festuca ovina, Koeleria macrantha,
Carex caryophyllea, Thymus praecox, Scilla verna,
Hypochoeris radicata, Lotus corniculatus, Galium verum,
Leontodon taraxacoides* and *Jasione montana*

H6 *Erica vagans-Ulex europaeus* heath
Festuca ovina sub-community

Sub-shrub canopy usually extensive with occasional
Rubus fruticosus agg., *Prunus spinosa* and *Pteridium* and
scattered *Teucrium scorodonia* and *Geranium sanguineum* but above listed associates very sparse in more
recently regenerating stands

H6 *Erica vagans-Ulex europaeus* heath
Typical sub-community

12   *Molinia* and *A. curtisii* constant with *E. tetralix*
generally very frequent

H4 *Ulex gallii-Agrostis curtisii* heath   13

Molinia occasional at low covers but *A. curtisii* and *E.
tetralix* very infrequent

H8 *Calluna vulgaris-Ulex gallii* heath   1

13   *E. tetralix* very infrequent in vegetation usually
dominated by *A. curtisii* or *E. cinerea*

H4 *Ulex gallii-Agrostis curtisii* heath
Agrostis curtisii-Erica cinerea sub-community

E. tetralix a frequent component of the sub-shrub canopy 14

14 *E. cinerea* rather patchy but *V. myrtillus* common as sparse shoots in a grassy heath with frequent *F. ovina*, *Danthonia decumbens*, *Agrostis capillaris*, *Galium saxatile* and occasional *Carex pilulifera* and *C. binervis*

H4 *Ulex gallii-Agrostis curtisii* heath
Festuca ovina sub-community

Newly-burned stands of this kind of vegetation may become overwhelmed by the establishment of *Agrostis curtisii* grassland where the grass is dominant over the sub-shrubs. In other places around the upland margins of south-west Britain, grazing can mediate the development of ill-defined mosaics of the *Festuca* sub-community and *Festuca-Agrostis-Galium* grassland from which the sub-shrubs have been virtually eliminated. Yet other stands on more base-rich substrates can show modest enrichment with calcicoles in what has been termed 'limestone heath'.

E. cinerea and *E. tetralix* both very common but above associates occasional at most in less grassy heath 15

15 *Scirpus cespitosus* constant in generally small amounts in vegetation usually dominated by *Calluna* but with other sub-shrubs and grasses locally abundant; *Dicranum scoparium* and *Leucobryum glaucum* frequent with occasional *Cladonia impexa* and *C. uncialis*

H4 *Ulex gallii-Agrostis curtisii* heath
Scirpus cespitosus sub-community

S. cespitosus absent from vegetation variously dominated by one or more of the sub-shrubs and grasses and with *E. vagans* or *E. ciliaris* locally abundant

H4 *Ulex gallii-Agrostis curtisii* heath
Erica tetralix sub-community

On periodically waterlogged ground among wetter heaths in the south-west, this vegetation can pass gradually to the *Ericetum tetralicis*, but *E. cinerea*, *U. gallii* and *A. curtisii* generally persist there only where soligenous influence maintains good aeration. In similar situations around the thinner margins of drift patches on the Lizard in

Cornwall, comparable hazy zonations to the *Erica-Schoenus* heath can be seen.

16 *Danthonia*, *Anthoxanthum odoratum*, *Festuca rubra* and *Potentilla erecta* frequent 17

P. erecta can be quite common but these other species scarce 18

17 *Sanguisorba minor*, *Carex flacca*, *Helianthemum nummularium* and *Plantago lanceolata* frequent with occasional *Galium verum*, *Lotus corniculatus*, *C. caryophyllea*, *Linum catharticum*, *Brachypodium sylvaticum*, *Hypericum pulchrum* and *Stachys betonica*

H8 *Calluna vulgaris-Ulex gallii* heath
Sanguisorba minor sub-community

Where this kind of vegetation occurs on superficials over limestones, the thinning of the mantle can be marked by complex mosaics with or gradual transitions to the local Mesobromion swards. Generally, the proportions of sub-shrubs and associated calcifuges to basiphilous herbs will effect a diagnosis, but some situations may remain problematic.

Combinations of these species not present

H8 *Calluna vulgaris-Ulex gallii* heath
Danthonia decumbens sub-community

18 *Scilla verna*, *Plantago maritima*, *Hypochoeris radicata* and *Thymus praecox* frequent

H8 *Calluna vulgaris-Ulex gallii* heath
Scilla verna sub-community

Such species usually very scarce 19

19 *E. cinerea* reduced in frequency but *V. myrtillus* constant in small amounts with *D. flexuosa* frequent and *Nardus stricta*, *Pleurozium schreberi* and *Rhytidiadelphus squarrosus* occasional

H8 *Calluna vulgaris-Ulex gallii* heath
Vaccinium myrtillus sub-community

E. cinerea remains common but *V. myrtillus* and other listed species scarce

H8 *Calluna vulgaris-Ulex gallii* heath
Species-poor sub-community

20   *E. vagans* constant and often abundant with frequent *Schoenus*, *Molinia* and *E. tetralix* and, in runnels between, mixtures of *C. pulicaris*, *C. flacca*, *C. panicea*, *C. hostiana*, *Anagallis tenella*, *Succisa pratensis*, *Serratula tinctoria* and *Sanguisorba officinalis*; only on Lizard Peninsula in Cornwall

> **H5** *Erica vagans-Schoenus nigricans* heath   21

E. vagans, *E. tetralix*, *Molinia* and *S. pratensis* occasional but not with the above associates   22

21   Runnels between dominants usually flooded to shallow depth for much of year and with frequent *Eleocharis multicaulis*, *Eriophorum angustifolium*, *Drosera rotundifolia*, *Pinguicula lusitanica* and, in ungrazed stands, scattered *Phragmites australis*

> **H5** *Erica vagans-Schoenus nigricans* heath
> *Eleocharis multicaulis* sub-community

Vegetation variable in composition and structure according to burning and other treatments, but above species very rare

> **H5** *Erica vagans-Schoenus nigricans* heath
> Typical sub-community

22   Sub-shrubs often inconspicuous at first sight with a very low and often discontinuous cover of wind-trimmed *Calluna* and *E. cinerea* (and locally some other species) among which are frequent *Festuca ovina*, *F. rubra*, *Holcus lanatus*, *Scilla verna*, *Plantago maritima*, *P. lanceolata*, *Thymus praecox*, *Hypochoeris radicata* and *Lotus corniculatus*

> **H7** *Calluna vulgaris-Scilla verna* heath   23

Above sub-shrubs and grasses can be present but not with *Scilla verna* and the other listed associates   27

23   *E. cinerea* and *Hypochoeris* reduced in frequency but *E. tetralix* and/or *Empetrum nigrum* common and *P. maritima* and *P. lanceolata* often very conspicuous; *Anthoxanthum* and *Agrostis capillaris* frequent and *Carex panicea* and *C. nigra* occasional   24

A. capillaris and *Anthoxanthum* occasional but *E. cinerea* and *Hypochoeris* remain very frequent and *E. tetralix*, *Empetrum*, *C. panicea* and *C. nigra* rare   25

24   *E. tetralix* constant with occasional *Empetrum* and frequent *Danthonia* and *Succisa*; *Molinia*, *Nardus* and *Salix repens* locally prominent

> **H7** *Calluna vulgaris-Scilla verna* heath
> *Erica tetralix* sub-community

Empetrum constant with *E. tetralix*, *Danthonia* and *Succisa* only very occasional but *Trifolium repens* and *Luzula multiflora* quite common

> **H7** *Calluna vulgaris-Scilla verna* heath
> *Empetrum nigrum* sub-community

25   *Armeria maritima* and *Sedum anglicum* constant and often abundant with frequent *Dactylis glomerata*, *Anthyllis vulneraria* and *Jasione montana*, occasional *Plantago coronopus* and *Silene vulgaris* ssp. *maritima* and ephemerals sporadically represented on patches of bare ground

> **H7** *Calluna vulgaris-Scilla verna* heath
> *Armeria maritima* sub-community

Dactylis and *Anthyllis* occasional but other listed associates rare   26

26   *Viola riviniana*, *Polygala vulgaris*, *C. flacca* and *C. caryophyllea* frequent and *Achillea millefolium*, *Leontodon taraxacoides*, *Galium verum*, *Stachys betonica* and *Serratula tinctoria* occasional with *E. vagans* and *U. europaeus* locally prominent among the sub-shrubs

> **H7** *Calluna vulgaris-Scilla verna* heath
> *Viola riviniana* sub-community

The *Viola* and *Armeria* sub-communities of the *Calluna-Scilla* heath are often found in mosaics on cliff tops where there are patchworks of deeper and shallower soils and, with further thinning of the mantle around rock exposures, the community can be replaced by the *Armeria-Cerastium* vegetation. Transitions can be gradual, but the final disappearance of remnant sub-shrubs, together with more consistent occurrence of therophytes, should help discern boundaries.

V. riviniana occasional but the other listed associates rare among an often rather impoverished *Calluna*-dominated cover

> **H7** *Calluna vulgaris-Scilla verna* heath
> *Calluna vulgaris* sub-community

27   *Carex arenaria* constant and *Ammophila arenaria* very frequent among *Calluna* and *E. cinerea* or *Empetrum nigrum*

> **H11** *Calluna vulgaris-Carex arenaria* heath   28

C. arenaria sometimes present but not with *Ammophila*, *E. cinerea* or *Empetrum*   29

28 *Erica cinerea* constant and *Aira praecox* frequent but *Empetrum nigrum* ssp. *nigrum*, *Agrostis capillaris* and *Galium saxatile* rare; *Dicranum scoparium* common with some of *Cladonia furcata*, *C. floerkeana*, *C. pyxidata*, *C. gracilis* and *C. foliacea* in an often extensive carpet of lichens

 H11 *Calluna vulgaris-Carex arenaria* heath
 Erica cinerea sub-community

E. cinerea and *A. praecox* only occasional at most but *Empetrum nigrum* ssp. *nigrum*, *Agrostis capillaris* and *G. saxatile* very frequent; *Cladonia arbuscula* and *C. impexa* can be patchily prominent but lichen-rich stretches of turf are not characteristic

 H11 *Calluna vulgaris-Carex arenaria* heath
 Empetrum nigrum ssp. *nigrum* sub-community

Above associates all scarce in impoverished mixtures of *Calluna* and *C. arenaria*

 H11 *Calluna vulgaris-Carex arenaria* heath
 Species-poor sub-community

29 *Calluna* generally the only sub-shrub and associated flora often very species-poor with scattered tussocks of *Festuca ovina* and patches of *Hypnum cupressiforme* and *Dicranum scoparium* the only really consistent elements, though lichens and therophytes may give some local variety

 H1 *Calluna vulgaris-Festuca ovina* heath 30

Calluna, F. ovina and the mosses can occur but typically with other sub-shrubs, herbs and bryophytes 33

Where heaths are heavily grazed or have been recently burned, or where the cover of *Calluna* has grown dense and rank, this can be a difficult distinction to make.

---

30 Cover of *Calluna* rather open with some of *Cladonia uncialis*, *C. fimbriata*, *C. pyxidata*, *C. impexa*, *C. squamosa*, *Cornicularia aculeata* and *Hypogymnia physodes* locally abundant on mor and bare ground 31

Lichens poorly represented 32

31 *Hypnum cupressiforme* very common and abundant among collapsed *Calluna* bushes and in grassy patches between

 H1 *Calluna vulgaris-Festuca ovina* heath
 Hypnum cupressiforme sub-community

Bryophytes rather patchy but lichens often extensive with *Cladonia furcata* and *C. macilenta* common

 H1 *Calluna vulgaris-Festuca ovina* heath
 Hypogymnia physodes-Cladonia impexa sub-community

32 *Carex arenaria* constant and sometimes quite abundant

 H1 *Calluna vulgaris-Festuca ovina* heath
 Carex arenaria sub-community

Teucrium scorodonia frequent with *Senecio jacobaea*, *Agrostis capillaris* and *Galium saxatile* occasional

 H1 *Calluna vulgaris-Festuca ovina* heath
 Teucrium scorodonia sub-community

Very impoverished rank canopies of *Calluna*

 H1 *Calluna vulgaris-Festuca ovina* heath
 Species-poor sub-community

33 *Calluna* frequently the only sub-shrub and associated flora often species-poor with scattered tufts of *Deschampsia flexuosa* and patches of *Pohlia nutans* the only really consistent elements

 H9 *Calluna vulgaris-Deschampsia flexuosa* heath
 34

Calluna and *D. flexuosa* can occur but typically with a richer associated flora 38

Where heaths are heavily grazed or have been recently burned, or where the cover of *Calluna* has grown dense and rank, this can be a difficult separation to make.

---

34 *Vaccinium myrtillus* or, more locally, *V. vitisidaea*, *V. × intermedium* or *Empetrum nigrum* common and locally abundant with occasional to frequent *Campylopus paradoxus*, *Gymnocolea inflata*, *Barbilophozia floerkii*, *Cladonia chlorophaea*, *C. floerkeana* and *C. squamosa*

 H9 *Calluna vulgaris-Dechampsia flexuosa* heath
 Vaccinium myrtillus-Cladonia spp. sub-community

In the southern Pennines, it can be difficult to distinguish this kind of vegetation from more species-poor stands of the *Calluna-Vaccinium* heath but bulky hypnoid mosses like *Pleurozium schreberi*, *Hylocomium splendens* and *Rhytidiadelphus loreus* are very common there.

V. myrtillus and *C. paradoxus* at most occasional and other listed species rare   35

35   *Molinia* constant at low cover

   H9 *Calluna vulgaris-Deschampsia flexuosa* heath
   Molinia caerulea sub-community

Molinia absent   36

36   *D. flexuosa* often especially abundant with occasional *Holcus mollis* and *Festuca rubra*; *Galium saxatile* and *Potentilla erecta* frequent with occasional *Rumex acetosella*

   H9 *Calluna vulgaris-Deschampsia flexuosa* heath
   Galium saxatile sub-community

D. flexuosa may be abundant but not with these associates   37

37   *Hypnum cupressiforme* and *Dicranum scoparium* common and sometimes abundant

   H9 *Calluna vulgaris-Deschampsia flexuosa* heath
   Hypnum cupressiforme sub-community

Calluna and *D. flexuosa* often the only plants with occasional *P. nutans*

   H9 *Calluna vulgaris-Deschampsia flexuosa* heath
   Species-poor sub-community

38   *Calluna* and *E. cinerea* constant with *Potentilla erecta*, and with Vaccinia and Empetra usually only occasional and hypnoid mosses of but local significance

   H10 *Calluna vulgaris-Erica cinerea* heath   39

E. cinerea can be frequent but, if so, then Vaccinia, Empetra, *Arctostaphylos* spp. or *Juniperus communis* present and hypnoid mosses prominent   42

This can be a difficult separation to make, especially in southern and eastern Scotland where there is some distributional overlap and floristic gradation between the *Calluna-Erica* and *Calluna-Vaccinium* heaths.

---

39   Sub-shrub canopy often short in a grassy heath with frequent *F. ovina*, *F. rubra*, *A. capillaris*, *Anthoxanthum* and *G. saxatile* and occasional *Campanula rotundifolia*, *Succisa* and *H. pulchrum*; *Dicranum scoparium*, *Pleurozium* and *Hylocomium splendens* patchily represented   40

Combinations of such species rare   41

40   *Danthonia* very common with occasional to frequent *Carex pulicaris*, *Viola riviniana*, *Linum catharticum*, *Prunella vulgaris* and *Primula vulgaris*

   H10 *Calluna vulgaris-Erica cinerea* heath
   Thymus praecox-Carex pulicaris sub-community

Combinations of such species rare

   H10 *Calluna vulgaris-Erica cinerea* heath
   Festuca ovina-Anthoxanthum odoratum sub-community

41   *Empetrum nigrum* ssp. *nigrum* quite common and *Scirpus cespitosus* patchily prominent with frequent *Carex panicea* and *C. pilulifera* and occasional *Huperzia selago*; *Racomitrium lanuginosum* common and often abundant among degenerating bushes with patches of *Cladonia uncialis* and *C. impexa*

   H10 *Calluna vulgaris-Erica cinerea* heath
   Racomitrium lanuginosum sub-community

E. nigrum spp. *nigrum* and *S. cespitosus* occasional at most but *Molinia* very common at low covers with frequent *C. binervis* and occasional *Juncus squarrosus*; *Campylopus paradoxus*, *Sphagnum capillifolium* and *Diplophyllum albicans* occasional to frequent

   H10 *Calluna vulgaris-Erica cinerea* heath
   Typical sub-community

Throughout the more oceanic parts of northern Britain, these two kinds of *Calluna-Erica* heath can be found in complex mosaics with and gradual transitions to the *Scirpus-Erica* wet heath on ground with periodic drainage impedence and the accumulation of a humic top. The latter vegetation can generally be distinguished by a replacement of *E. cinerea* by *E. tetralix*, the increasing prominence of *Molinia* and *S. cespitosus*, an eclipse of smaller grasses and the replacement of the mosses here by at least some Sphagna.

---

42   *Arctostaphylos uva-ursi* constant and moderately abundant in more open areas of *Calluna* cover, and often accompanied by *Vaccinium vitis-idaea*; *Genista anglica* frequent, *Listera cordata* occasional and *Pyrola media* sometimes present

   H16 *Calluna vulgaris-Arctostaphylos uva-ursi* heath   43

A. uva-ursi can occur but not with the above associates   45

43   *A. uva-ursi* often quite extensive in patchworks with *Calluna* and scattered *E. cinerea* and Vaccinia; *F. ovina* common with occasional *A. capillaris* and *Anthoxanthum*; rich associated herb flora includes frequent *P. erecta*, *Pyrola media*, *V. riviniana*, *Lathyrus montanus*, *Hypericum pulchrum*, *Anemone nemorosa*, *Trientalis europaeus*, *G. saxatile* and *L. corniculatus*

> **H16** *Calluna vulgaris-Arctostaphylos uva-ursi* heath
> *Pyrola media-Lathyrus montanus* sub-community

Calluna usually a strong dominant with *A. uva-ursi* subordinate and above associates occasional at most
                                                                                             44

As the *Pyrola-Lathyrus* sub-community recovers from burning, the *Calluna* canopy gradually closes with an eclipse in the cover of *A. uva-ursi* and an attenuation of the associated flora, making this distinction hard to discern.

---

44   *V. myrtillus* commonly accompanies *V. vitis-idaea* in a sparse understorey with *E. nigrum* ssp. *nigrum* intermingled or patchily prominent; hypnoid mosses often extensive with *Cladonia arbuscula* and *C. rangiferina* also occasional

> **H16** *Calluna vulgaris-Arctostaphylos uva-ursi* heath
> *Vaccinium myrtillus-V. vitis-idaea* sub-community

V. myrtillus absent, *V. vitis-idaea* and *E. nigrum* ssp. *nigrum* scarce and hypnoid mosses somewhat patchy; *S. cespitosus* and *C. pilulifera* common and lichens extensive on areas of bare ground with frequent *C. uncialis*, *C. impexa*, *C. floerkeana*, *C. coccifera* and *C. squamosa*

> **H16** *Calluna vulgaris-Arctostaphylos uva-ursi* heath
> *Cladonia* spp. sub-community

It can be difficult to separate this kind of vegetation from certain sorts of *Calluna-Cladonia* heath where *A. uva-ursi* is sometimes found, but *E. cinerea* and hypnoid mosses are scarce there.

---

45   *Juniperus communis* ssp. *nana* constant and usually dominant with frequent *A. uva-ursi*, *A. alpinus* and *S. cespitosus*; *Pleurozia purpurea*, *Frullania tamarisci* and *Diplophyllum albicans* common and sometimes abundant

> **H15** *Calluna vulgaris-Juniperus communis* ssp. *nana* heath

J. communis ssp. *nana* can occur but not usually in any abundance and without the characteristic bryophyte flora
                                                                                             46

Degraded stands of *Calluna-Juniperus* heath, often the product of burning, can have a very patchy cover of *Juniperus* and only fragments of the characteristic bryophyte flora, thus making this separation hard.

---

46   *A. alpinus* constant though generally subordinate to *Calluna* with occasional *A. uva-ursi* and local *Juniperus*.

> **H17** *Calluna vulgaris-Arctostaphylos alpinus* heath   47

A. alpinus only a very local feature   48

47   *E. nigrum* ssp. *hermaphroditum* and *Loiseleuria procumbens* constant with occasional *V. vitis-idaea* and *J. communis* ssp. *nana*, frequent *Carex bigelowii*, *Diphasium alpinum* and *Antennaria dioica*, *Cladonia uncialis*, *C. arbuscula*, *C. gracilis*, *C. pyxidata*, *C. bellidiflora*, *Cetraria glauca*, *C. islandica*, *Alectoria nigricans* and *Sphaerophorus globosus*

> **H17** *Calluna vulgaris-Arctostaphylos alpinus* heath
> *Loiseleuria procumbens-Cetraria glauca* sub-community

Loiseleuria occasional but *E. nigrum* ssp. *hermaphroditum* replaced by ssp. *nigrum* with *Erica cinerea* becoming common; *P. erecta* very frequent with *S. cespitosus*, *C. pilulifera* and *Nardus* quite often found; *C. uncialis* and *C. arbuscula* remain common but lichen flora not so varied or abundant

> **H17** *Calluna vulgaris-Arctostaphylos alpinus* heath
> *Empetrum nigrum* ssp. *nigrum* sub-community

48   Dwarfed mat of sub-shrubs with lichens at least co-dominant and including some of *Cetraria islandica*, *Cornicularia aculeata*, *Cladonia arbuscula*, *C. uncialis*, *C. rangiferina* and *Alectoria nigricans*; *R. lanuginosum* can be very common and locally prominent but is not overwhelmingly dominant
                                                                                             49

Cornicularia aculeata, *Cladonia arbuscula*, *C. uncialis* and sometimes other lichens can be frequent but not generally in any abundance
                                                                                             54

49   *Calluna* the commonest and usually the dominant sub-shrub, with *V. myrtillus* and *V. vitis-idaea* frequent but subordinate in cover; *C. bigelowii* can be quite common but not abundant and *Dicranum fuscescens* rare

> **H13** *Calluna vulgaris-Cladonia arbuscula* heath
>   50

V. myrtillus constant and *V. vitis-idaea* frequent with the former generally the most abundant sub-shrub, *Calluna* being uncommon and always subordinate in cover; *C. bigelowii* constant and often abundant with *D. fuscescens* frequent

> **H19** *Vaccinium   myrtillus-Cladonia   arbuscula* heath
>   52

50   Vaccinia very common and *Loiseleuria* frequent in virtually prostrate mat with *Cladonia crispata, C. coccifera, Ochrolechia frigida* and *Thamnolia vermicularis* often found

> **H13** *Calluna vulgaris-Cladonia arbuscula* heath
> *Cladonia crispata-Loiseleuria procumbens* sub-community

Vaccinia less frequent and *Loiseleuria* and listed lichens occasional at most   51

51   *Deschampsia flexuosa* very common with occasional *Diphasium alpinum* and *Juncus trifidus*; *Cetraria nivalis* frequent and *Cladonia gracilis* and *C. bellidiflora* occasional

> **H13** *Calluna vulgaris-Cladonia arbuscula* heath
> *Empetrum nigrum* ssp. *hermaphroditum-Cetraria nivalis* sub-community

D. flexuosa only occasional and *C. bigelowii* unusually scarce but *E. cinerea* and *E. tetralix* occasional among the sub-shrubs and *S. cespitosus* and *Molinia* sometimes found; *Cetraria nivalis* and *Cladonia gracilis* uncommon but *C. arbuscula* and *C. rangiferina* especially abundant with occasional *C. impexa*

> **H13** *Calluna vulgaris-Cladonia arbuscula* heath
> *Cladonia   arbuscula-C.   rangiferina*   sub-community

Even with a fairly abrupt shift from higher, wind-lashed ground to lower, sheltered slopes, there can be considerable continuity in qualitative composition between this kind of vegetation and the *Calluna-A. uva-ursi* and *Calluna-Vaccinium* heaths where quite varied lichen floras can become locally important. In such zonations, it

may be necessary to rely on the proportions of the dominants to effect a diagnosis.

52   *V. myrtillus* and *C. bigelowii* usually co-dominant among the vascular plants with *F. ovina/vivipara*; *G. saxatile* frequent and *P. erecta* occasional

> **H19** *Vaccinium   myrtillus-Cladonia   arbuscula* heath
> *Festuca ovina-Galium saxatile* sub-community

F. ovina and *G. saxatile* can occur occasionally but not in any abundance   53

53   *R. lanuginosum* sometimes quite abundant among mixed and rather patchy lichen cover with occasional *Kiaeria starkei*; *Alchemilla alpina* frequent in small amounts with occasional *Salix herbacea* and *Juncus trifidus*

> **H19** *Vaccinium   myrtillus-Cladonia   arbuscula* heath
> *Racomitrium lanuginosum* sub-community

Empetrum nigrum ssp. *hermaphroditum* becomes very common among the sub-shrubs with lichens overwhelmingly abundant in the mat, *Cladonia rangiferina, C. gracilis* and *Ochrolechia frigida* being preferentially frequent, *C. bellidiflora, C. pyxidata* and *Cetraria nivalis* more occasional; *R. lanuginosum* frequent but not abundant and often accompanied by small amounts of *Pleurozium schreberi, Dicranum scoparium, Ptilidium ciliare* and, less commonly, *Rhytidiadelphus loreus*

> **H19** *Vaccinium   myrtillus-Cladonia   arbuscula* heath
> *Empetrum nigrum* ssp. *hermaphroditum-Cladonia* spp. sub-community

In more sheltered and humid situations, both these kinds of vegetation, and particularly the *Racomitrium* sub-community, can come close to the more chionophilous *Vaccinium-Deschampsia* heath in composition, and hillsides in the low-alpine zone can show complex patchworks of the different assemblages disposed over hollows and brows.

54   Dwarfed sub-shrubs often patchy in cover with *R. lanuginosum* constant and very abundant in a thick woolly carpet   55

R. lanuginosum can be common but not generally with more than local abundance   61

55   *Calluna* the commonest and usually the most abundant sub-shrub with *V. myrtillus* never more than occasional and always subordinate and *V. vitis-idaea* rare

> **H14** *Calluna vulgaris-Racomitrium lanuginosum* heath   56

V. myrtillus constant and *V. vitis-idaea* frequent though both quite often subordinate to *Empetrum nigrum* ssp. *hermaphroditum*, with *Calluna* typically scarce, except at lower altitudes

> **H20** *Vaccinium myrtillus-Racomitrium lanuginosum* heath   58

56   *A. uva-ursi* constant with frequent *A. alpinus*, *E. nigrum* ssp. *nigrum* and *Erica cinerea*; *S. cespitosus* common with *Molinia* and *C. binervis* occasional; *Dicranum scoparium*, *Diplophyllum albicans* and *Pleurozium* frequent

> **H14** *Calluna vulgaris-Racomitrium lanuginosum* heath
> *Arctostaphylos uva-ursi* sub-community

Empetrum nigrum ssp. *nigrum* and *Erica cinerea* at most occasional and *Arctostaphylos* spp. and listed herbs and bryophytes very scarce   57

57   *E. nigrum* ssp. *hermaphroditum* constant with frequent *Nardus* and *Diphasium alpinum* and, among a sometimes quite extensive lichen cover, *Cetraria islandica*, *Cladonia gracilis* and *Ochrolechia frigida*

> **H14** *Calluna vulgaris-Racomitrium lanuginosum* heath
> *Empetrum nigrum* ssp. *hermaphroditum* sub-community

E. nigrum ssp. *hermaphroditum* and listed herbs and lichens occasional at most but *F. ovina* and *Agrostis canina* very common, sometimes with *Antennaria dioica*, *C. panicea*, *T. praecox* and *Euphrasia micrantha*

> **H14** *Calluna vulgaris-Racomitrium lanuginosum* heath
> *Festuca ovina* sub-community

58   *V. vitis-idaea* very scarce but *Calluna*, *E. cinerea* and *J. communis* ssp. *nana* occasional with *Alchemilla alpina* sometimes co-dominant; *P. erecta*, *T. praecox*, *V. riviniana* and *C. pilulifera* frequent with *S. cespitosus*, *C. binervis* and *Salix herbacea* occasional

> **H20** *Vaccinium myrtillus-Racomitrium lanuginosum* heath
> *Viola riviniana-Thymus praecox* sub-community

Where this vegetation type extends the range of the *Vaccinium-Racomitrium* heath to its lowest altitudes along the north-western seaboard of Scotland, it can be difficult to partition samples between this community and the *Calluna-Racomitrium* heath and, in such situations, it may be necessary to rely on the proportions of the sub-shrubs to effect a diagnosis.

V. vitis-idaea very common and all other listed associates scarce   59

59   Rich and luxuriant patchwork of bryophytes present among the *Racomitrium* carpet with *Dicranum scoparium*, *Plagiothecium undulatum*, *Sphagnum capillifolium*, *Mylia taylori*, *Diplophyllum albicans*, *Pleurozia purpurea*, *Bazzania tricrenata*, *Scapania gracilis*, *S. ornithopodioides*, *Anastrepta orcadensis* and *Anthelia julacea* frequent

> **H20** *Vaccinium myrtillus-Racomitrium lanuginosum* heath
> *Bazzania tricrenata-Mylia taylori* sub-community

D. albicans and *A. orcadensis* sometimes found but combinations of these other bryophytes rare   60

60   *R. lanuginosum* reduced in cover with dominance often passing to mixtures of *Pleurozium*, *Hylocomium splendens* and *Rhytidiadelphus loreus*

> **H20** *Vaccinium myrtillus-Racomitrium lanuginosum* heath
> *Rhytidiadelphus loreus-Hylocomium splendens* sub-community

Bulky pleurocarps can be quite common though not abundant among the dominant *Racomitrium* carpet but the lichen flora is somewhat richer than usual with frequent *Cladonia gracilis*, *Cetraria islandica* and *Cornicularia aculeata* and occasional *Cladonia leucophaea*, *Sphaerophorus globosus* and *Alectoria nigricans*

> **H20** *Vaccinium myrtillus-Racomitrium lanuginosum* heath
> *Cetraria islandica* sub-community

This kind of vegetation can look very much like a sub-shrub facies of *Carex-Racomitrium* summit vegetation and often forms intimate mosaics with it over mountain sides, the former as small patches in hollows or over stretches of blocky detritus among the lower reaches of the latter. Usually the abundance of *V. myrtillus* and *E.*

nigrum ssp. *hermaphroditum* will serve to distinguish the *Vaccinium-Racomitrium* heath.

---

61   *Rubus chamaemorus* and *Cornus suecica* constant as scattered individuals with *Eriophorum vaginatum* a distinctive occasional

> **H22** *Vaccinium myrtillus-Rubus chamaemorus* heath   62

R. chamaemorus, *C. suecica* and *E. vaginatum* all rare   63

62   *Carex bigelowii* frequent in small amounts with a rich and extensive patchwork of cryptogams, among which *R. lanuginosum, Plagiothecium undulatum, Ptilidium ciliare, Anastrepta orcadensis, Barbilophozia floerkii, Cladonia bellidiflora, C. uncialis, C. leucophaea, C. gracilis* and *C. impexa* are very common

> **H22** *Vaccinium myrtillus-Rubus chamaemorus* heath
> *Plagiothecium undulatum-Anastrepta orcadensis* sub-community

C. bigelowii and above listed cryptogams all very scarce but *G. saxatile* and *Blechnum* frequent as scattered individuals and *Polytrichum commune* very common among usually plentiful hypnaceous mosses

> **H22** *Vaccinium myrtillus-Rubus chamaemorus* heath
> *Polytrichum commune-Galium saxatile* sub-community

63   *Blechnum spicant, Solidago virgaurea, Listera cordata* and *Sphagnum capillifolium* frequent among the associates

> **H21** *Calluna vulgaris-Vaccinium myrtillus-Sphagnum capillifolium* heath   64

Combinations of these species rare   65

64   *E. nigrum* ssp. *hermaphroditum* frequent and locally abundant among the sub-shrubs with especially rich and luxuriant cryptogam carpets among which there is frequent *R. lanuginosum, Mylia taylori, Scapania gracilis, Bazzania tricrenata, Pleurozia purpurea, Diplophyllum albicans, Anastrepta orcadensis, Mastigophora woodsii, Herbertus aduncus* ssp. *hutchinsiae, Cladonia uncialis* and *C. arbuscula*

> **H21** *Calluna vulgaris-Vaccinium myrtillus-Sphagnum capillifolium* heath
> *Mastigophora woodsii-Herbertus aduncus* ssp. *hutchinsiae* sub-community

E. nigrum ssp. *hermaphroditum* local and combinations of listed cryptogams rare but *Dicranum scoparium* common with frequent scattered fronds of *Pteridium aquilinum*

> **H21** *Calluna vulgaris-Vaccinium myrtillus-Sphagnum capillifolium* heath
> *Calluna vulgaris-Pteridium aquilinum* sub-community

Both kinds of *Calluna-Vaccinium-Sphagnum* heath can grade to *Vaccinium-Deschampsia* and *Vaccinium-Rubus* heaths in hollows where there is longer snow-lie, boundaries with the latter being especially difficult to discern where there is strong continuity among the hepatic flora.

---

65   *Calluna* constant and the usual dominant with *Erica cinerea* quite common

> **H12** *Calluna vulgaris-Vaccinium myrtillus* heath   66

Calluna occasional at most with scattered plants lacking in vigour and *E. cinerea* very rare

> **H18** *Vaccinium myrtillus-Deschampsia flexuosa* heath   68

66   *V. vitis-idaea* and *Empetrum nigrum* ssp. *nigrum* frequent with occasional *J. squarrosus* and *Blechnum*   67

V. vitis-idaea and *E. nigrum* ssp. *nigrum* both scarce in rather species-poor heath usually overwhelmingly dominated by *Calluna*

> **H12** *Calluna vulgaris-Vaccinium myrtillus* heath
> *Calluna vulgaris* sub-community

67   *Calluna* not nearly so abundant as usual and other sub-shrubs of generally moderate cover in a grassy heath with frequent *F. ovina, A. capillaris* and *Nardus, G. saxatile* and *P. erecta* and occasional *Carex pilulifera, Campanula rotundifolia* and *Polygala serpyllifolia*; bryophytes and lichens not very extensive and often patchy

> **H12** *Calluna vulgaris-Vaccinium myrtillus* heath
> *Galium saxatile-Festuca ovina* sub-community

Grazing-mediated transitions between this vegetation type and *Festuca-Agrostis-Galium* grassland are very common in upland pastures throughout the north and west of Britain and it may be necessary to rely on the abundance of sub-shrubs to effect a diagnosis.

Sub-shrub cover extensive and varied with *V. vitis-idaea* especially frequent and above-listed herbs typically scarce; cryptogam flora quite varied and often abundant with bulky pleurocarps particularly prominent, *Hylocomium* often joining *Pleurozium* and *Hypnum jutlandicum*, *Cladonia impexa*, *C. uncialis* and *C. pyxidata* common

> **H12** *Calluna vulgaris-Vaccinium myrtillus* heath
> *Vaccinium vitis-idaea-Cladonia impexa* sub-community

68 Sub-shrubs often co-dominant with *Alchemilla alpina* and/or grasses among which *F. ovina*, *Agrostis capillaris* and *Anthoxanthum* are especially common with occasional *Nardus*, *Danthonia* and *Deschampsia cespitosa*; *Luzula campestris*, *Carex pilulifera* and *C. binervis* also common with *Potentilla erecta*, and occasional *Campanula rotundifolia*, *Viola riviniana* and *Ranunculus acris*

> **H18** *Vaccinium myrtillus-Deschampsia flexuosa* heath
> *Alchemilla alpina-Carex pilulifera* sub-community

Sub-shrubs generally strongly dominant with the listed herbs occasional at most 69

69 *Calluna* quite common among the sub-shrubs with *Blechnum* frequent; mosses forming an extensive ground carpet with *Hylocomium* and *R. loreus* joining *Pleurozium* and *D. scoparium*, and occasional *Plagiothecium undulatum*, *D. majus*, *H. umbratum* and *Sphagnum capillifolium*

> **H18** *Vaccinium myrtillus-Deschampsia flexuosa* heath
> *Hylocomium splendens-Rhytidiadelphus loreus* sub-community

Calluna sparse and *Blechnum* absent with pleurocarps rather patchy but *R. lanuginosum* common and lichens quite conspicuous. *Cladonia arbuscula* being locally abundant, *C. impexa* and *C. uncialis* occasional

> **H18** *Vaccinium myrtillus-Deschampsia flexuosa* heath
> *Racomitrium lanuginosum-Cladonia* spp. sub-community

COMMUNITY DESCRIPTIONS

H1
Calluna vulgaris-Festuca ovina heath

Synonymy
Callunetum arenosum Tansley 1911 *p.p.*; *Callunetum* Farrow 1915, Watt 1936; *Callunetum arenicolum* Tansley 1939 *p.p.*

Constant species
Calluna vulgaris, Festuca ovina, Dicranum scoparium, Hypnum cupressiforme s.l.

Physiognomy
The *Calluna vulgaris-Festuca ovina* heath is a heather-dominated community, very poor in vascular associates, though sometimes showing a modest diversity among the bryophytes and, more especially, the lichens. *Calluna vulgaris* is usually the only woody species present and invariably the most abundant. More particularly, *Erica cinerea, Ulex minor* and *U. gallii*, which become important constituents of the sub-shrub component of dry heaths further to the south and west, are largely excluded here. They sometimes figure in stands on or near the East Anglian coast, and the first two can occasionally be found in the few stations for the community south of the Thames, but the general scarcity of these species furnishes a good diagnostic criterion, provided that it is remembered that they can be locally eliminated from richer heath types to produce the kind of virtually pure *Calluna* that is usual here. Another general absentee is *Erica tetralix*: this becomes increasingly prominent through the whole range of wet and humid heaths towards south-west Britain but is very scarce here and strictly limited to those rather unusual situations where there is a transition to soils with impeded drainage. *Ulex europaeus* is likewise uncommon, though it can become locally abundant where there has been some disturbance, not necessarily recent, as along trackways or around the margins of plantations.

However, although it is *Calluna* which gives this community its distinctive stamp, the cover and height of the canopy are very variable. This affects not only the gross appearance of the vegetation, which can have a canopy of living heather anything between 10 and 100 cm tall, covering 50–100% of the ground, but it also has a controlling influence on the variety, abundance and distribution of the associated flora, which is for the most part confined to areas between the *Calluna* clumps and to the centre of those bushes which are collapsing with age or showing early regeneration from the stools or seed. It was from this kind of heath that Watt (1955) first described the now classic phases of heather development, 'pioneer', 'building', 'mature' and 'degenerate', and, in the absence of competing sub-shrubs, the influence of the structural changes in the canopy on the waxing and waning of the herbs and cryptogams is often very clear. Where these phases succeed one another in uninterrupted cycles here, the result is an uneven-aged cover of heather with considerable physiognomic heterogeneity and a correspondingly complex variation in the distribution of the associates. In other cases, sudden perturbations to the growth of the heather, due to a variety of external circumstances, can produce a uniformity of structure and floristics within stands, but make one tract of the community look very different from another. And variations in grazing intensity can also modify the shape and extent of the heather canopy, influencing the proportion of the sub-shrub cover to the other components.

By and large, the contribution of the herbaceous element among anything other than pioneer *Calluna* or heather severely checked by grazing is small, both in cover and in floristic diversity; and, where it becomes more extensive, it is often better to regard the vegetation as a mosaic of this kind of heath and one of a number of associated communities. Typically, here, there is no extensive development of a grassy ground among the heather bushes. *Festuca ovina* is very common through the community as a whole, but it is usually present as scattered tussocks, often of less than 30% total cover and completely extinguished from beneath vigorous building or mature *Calluna*. Other grasses are few and similarly confined: *Agrostis capillaris* occurs occasionally, sometimes with local abundance, and it is usually mixtures of these two species which, thickening up, form the matrix of associated grasslands. In some stands, too, there is a little *Deschampsia flexuosa*, a grass which seems to have increased its representation among this kind of heath since myxomatosis (Ratcliffe 1977). *Agrostis curtisii*, by contrast, which, with *Erica cinerea*, *Ulex minor* and *U. gallii*, becomes very common in dry heaths further to the south and west, is never found; and *Molinia caerulea*, which follows *E. tetralix* in transgressing into drier heaths in more oceanic areas, is very scarce. More mesophytic grasses, such as *Holcus lanatus*, occur only very occasionally, and more basiphile species, like *Avenula pratensis* and *Koeleria macrantha*, are found only in those situations where the community is closely juxtaposed with calcicolous swards, an unusual, though rather distinctive, occurrence.

Associated dicotyledons are few and patchy, being again largely confined to areas of more open heather cover, usually among pioneer growth, or within included patches of sward. *Teucrium scorodonia* is particularly distinctive of one sub-community and *Senecio jacobaea* also occurs occasionally. Then, especially where the grasses thicken up a little, they may be accompanied by species such as *Galium saxatile*, *Cerastium fontanum*, *Campanula rotundifolia* or *Luzula campestris*. Or, where patches of bare soil have been exposed by the decay of mor within degenerate *Calluna*, or where herbivores or human activity have scuffed the ground, *Rumex acetosella* and ephemerals such as *Aphanes arvensis*, *Teesdalia nudicaulis*, *Myosotis ramosissima* or the annual grass *Aira praecox* can sometimes be found. In general, though, this kind of enrichment is modest and sporadic.

Finally, among the vascular plants, there are two species which can play a more important local role, their abundance in relation to *Calluna* often reflecting successional shifts in dominance. The first is *Pteridium aquilinum*. This is typically only occasional within the heath and usually restricted to sparse fronds but, where vigorous marginal growth of the bracken enters a zone of heather in a more open and less competitive condition, it can produce a closed overtopping canopy and come to dominate; conversely, the bracken itself may be overwhelmed where its own degeneration coincides with a local rise in *Calluna* vigour. The results of such competitive interactions, again first described by Watt (1955), are complex mosaics of bracken and heather-dominance which are often hard to separate into discrete stands of *Calluna-Festuca* heath and the *Pteridium-Galium* community. The second plant of this kind, *Carex arenaria*, is more local in its occurrence and its abundance more clearly related to particular habitat conditions: its presence in quantity helps define one sub-community that is essentially transitional to dune vegetation.

The bulk of the remaining floristic variation in the *Calluna-Festuca* heath is found among the bryophytes and lichens. Again, these plants are of variable abundance, and often very patchy within individual stands, being generally sparse in denser heather, more extensive and diverse among open covers and in some cases showing fairly precise patterns of occurrence according to the growth phases of the *Calluna*. Few species occur throughout the community though, among the mosses, *Hypnum cupressiforme* (often recorded as *H. jutlandicum*) is very common and can be found occasionally even among building or mature heather. *Dicranum scoparium* is less persistent in such situations and intolerant of loose sandy substrates but it occurs frequently in more open areas and these two species usually form the bulk of the bryophyte cover among the degenerate and pioneer *Calluna*. In such situations, too, species such as *Hylocomium splendens*, *Pleurozium schreberi*, *Ptilidium ciliare* and *Dicranella heteromalla* have most of their occasional occurrences. *Polytrichum juniperinum* and *P. piliferum* can also figure with some abundance on patches of bare mineral soil among regenerating heather or on rabbit-scuffed ground.

Lichens often rival or exceed the mosses in their cover and variety, though they are more strictly confined to the open degenerate and pioneer phases. *Cladonia* spp. are particularly prominent, with encrusting species, such as *C. pyxidata*, *C. squamosa* and *C. fimbriata*, occurring over patches of bare ground, sometimes with *Cornicularia aculeata*, bulkier species like *C. impexa*, *C. furcata* and *C. arbuscula* being especially abundant among the decumbent branches of old collapsed *Calluna*. On these stems, *Hypogymnia physodes* is very characteristic.

Sub-communities

Hypnum cupressiforme **sub-community.** The *Calluna* cover in this sub-community is often less than complete and only moderately tall, usually from 3 to 5 dm high, sometimes showing vigorous young growth, though often with many bushes in the degenerate phase. *Erica cinerea* can occasionally be found, though the most prominent associates are typically *Festuca ovina*, which occurs as scattered tussocks, and the cryptogams, whose total cover often exceeds 50%. Among the mosses, both *Hypnum cupressiforme s.l.* and *Dicranum scoparium* are very frequent, the former especially conspicuous as extensive mats in the more open areas, both among the collapsed heather stems and in any small grassy patches between the *Calluna*. Lichens are also abundant on the mor and bare mineral soil exposed within the degenerating bushes with encrusting species, such as *Cladonia pyxidata*, *C. squamosa*, *C. fimbriata* and *C. gracilis* usually predominating, bulkier species like *C. impexa*, *C. uncialis* and *C. arbuscula* being less frequent.

Hypogymnia physodes occurs occasionally on the collapsed heather stems.

Hypogymnia physodes-Cladonia impexa sub-community.
The heather cover here is very much the same, in total cover and height, to that above, but degenerate bushes predominate and the contribution of *Festuca ovina*, *Hypnum cupressiforme s.l.* and *Dicranum scoparium* is more uneven, the mosses occurring as small patches among the collapsed stems. Lichens, by contrast, are very abundant in these areas, with encrusting species being joined and exceeded in cover by *Cladonia impexa*, *C. furcata* and occasional *C. macilenta*. *Hypogymnia physodes* also has its best representation in this sub-community, frequent small thalli growing on the old, decumbent heather stems.

Teucrium scorodonia sub-community. Degenerate bushes are scarce here and the heather cover is generally vigorous and extensive, though with some more open included areas. Within these, scattered plants of *Teucrium scorodonia* or *Senecio jacobaea* can sometimes be found, or small stretches of turf, with *Festuca ovina*, *Agrostis capillaris*, *Deschampsia flexuosa* and scattered *Rumex acetosella*, *Galium saxatile* and *Cerastium fontanum*. Lichens are very infrequent but *Hypnum cupressiforme s.l.* and *Dicranum scoparium* remain common among the grasses or, in the case of the former, in moderately dense heather.

Carex arenaria sub-community. In this sub-community, the heather is often tall and somewhat open with scattered plants or denser, intervening patches of *Carex arenaria*, which is sometimes so abundant as to be co-dominant. *Festuca ovina* occurs as sparse individuals and there is sometimes a little *Rumex acetosella*, but other herbs are scarce. Among the cryptogams, only *Hypnum cupressiforme s.l.* is frequent.

Species-poor sub-community. Unbroken canopies of dense and often tall heather, up to 50 cm or more high, are characteristic here, among which virtually no associates can survive, apart from occasional sparse plants of *Hypnum cupressiforme s.l.*

Habitat
The *Calluna-Festuca* heath is confined to base-poor and oligotrophic sandy soils in the more continental lowlands of eastern England. The regional climate has an important influence on the general composition of the community by excluding species of more oceanic heaths and, in the absence of these, the controlling effect of the growth pattern of the heather on floristic variation in the community is often very obvious. This dominance is sometimes itself adversely affected by climate or insect

damage and, particularly in the past, has been widely perturbed by burning and grazing. The demise of these traditional treatments has often been a prelude to improvement for agriculture and forestry which has much reduced and fragmented tracts of the community.

This kind of heath is characteristic of those parts of Britain with low rainfall and a wide range of annual temperatures. For the most part, precipitation is less than 650 mm yr^{-1} (*Climatological Atlas* 1952) with fewer than 120 wet days yr^{-1} (Ratcliffe 1968), and a slight tendency for a summer rainfall maximum. As over much of southern Britain, annual accumulated temperatures are generally high, but the contrast between temperature extremes is the most marked of anywhere in the country, with mean annual maxima above 30 °C (Conolly & Dahl 1970), but February minima often below freezing. Such strongly continental conditions prevail over much of East Anglia (lessening a little in severity towards the northern and western coasts), up around the Wash, westwards towards the Chilterns and then south some way beyond the Thames, and they severely restrict the distribution in eastern England of a variety of Oceanic West European species which are of major importance in heaths in parts of Britain with a more equable climate. The *Calluna-Festuca* heath thus takes much of its floristic character from the exclusion of *Erica cinerea*, *Ulex minor*, *U. gallii* and *Agrostis curtisii* from potentially suitable soils in this region, because of sensitivity either to low temperatures in winter, when a measure of physiological activity may be retained (as in *E. cinerea*: Bannister 1965, Gimingham 1972) or to low rainfall (as, perhaps, in *A. curtisii*: Ivimey-Cook 1959); and also from the increasing scarcity in eastern England, and progressive confinement to the kinds of wetter soils quite uncharacteristic here, of *Erica tetralix* and *Molinia caerulea*. A positive response to such continental climatic conditions is scarcely visible in the community, though it can sometimes be seen in the presence of species like *Teesdalia*, which are very distinctive of the open grasslands of this part of Britain, but restricted here by the overwhelming dominance of *Calluna*.

Within this climatic zone, the *Calluna-Festuca* heath is limited to acid and impoverished soils, which have sometimes been derived from arenaceous bedrocks, as in scattered localities in Lincolnshire and around The Weald, but which for the most part have developed from sandy glacio-fluvial drift, sometimes supplemented by aeolian sand and widely affected by periglacial sorting: such heterogeneous parent materials underlie most of the stands in the core of the community's range, in the Breckland, on the north Norfolk commons and along the Suffolk coast. In a few places on the Norfolk coast, the soils have been derived from dune sand.

By and large, the profiles under the community are brown sands, that is non-calcareous brown soils in which sandy or sandy-skeletal materials predominate in the upper horizons (Avery 1980, *Soil Survey* 1983), including series like the Newport soils, which are extensive on the Suffolk coast, and the Worlington, which is of particular importance in Breckland (Corbett 1973, Hodge *et al.* 1984). Often, the parent material is naturally acidic but, even where it is less prevailingly so, decalcification in the free- to excessively-draining profiles has resulted in consistently low surface pH, and there are frequently signs of podzolisation, which process is accentuated by the acidifying properties of *Calluna* and its accumulating mor (Gimingham 1972, Corbett 1973). The community thus shows only very local and limited relief of its generally calcifuge and oligotrophic character, the very sparse representation of mesotrophic or calcicolous species being usually limited to situations where the heath occurs in edaphically-related mosaics, and weedy mesotrophic plants figuring only where there has been some soil disturbance, the occurrence of which can be quite frequent on unenclosed commons and heaths or where rabbits are numerous. More marked instability of the surface of very sandy profiles, with erosion and deposition of loose material by wind, provides a more obvious kind of edaphic variation and is influential in the occurrence of the *Carex arenaria* sub-community. *C. arenaria* is favoured by fresh sand deposition (Noble 1982) and this kind of *Calluna-Festuca* heath can be found in coastal and a now very few inland dune systems (e.g. Wangford Warren and Foxhole Heath in Suffolk and on some heaths around Gainsborough), where there is, or has been, mobile sand (e.g. Watt 1937).

Natural decline of the heather cover with age, or its sudden demise because of environmental change, may play some part in exposing the soil surface to erosion and thus contribute to the rise of *C. arenaria* and the loss or exclusion of other species intolerant of a shifting substrate. In the remainder of the community the soil surface is generally stable, but it is such changes in the character of the heather cover that have the most obvious direct influence on floristic variation by mediating differences in light, temperature and humidity over the surface of the mor and mineral soil (Watt 1955, Gimingham 1960, Barclay-Estrup & Gimingham 1969, Barclay-Estrup 1971). Detailed data on such factors are not available for the samples, but simple observation suggests that the Species-poor sub-community is typical of building and mature heather, the *Hypogymnia-Cladonia* of opening, degenerate bushes and the *Hypnum* of those areas with more advanced recolonisation by *Festuca* and young *Calluna*. The *Teucrium* sub-community may also represent a regeneration phase, perhaps sometimes with persistent inclusions of grassy sward.

Cyclical growth and degeneration of the heather can provide marked internal variation within stands of the community but more even-aged tracts, in which one or other of the phases, with its typical suite of associates, predominates, are widespread. In the past, when more intensive use of lowland heaths was customary, a patchwork of such stretches of heather was often deliberately maintained by burning, to ensure a continuous supply of vigorous growth for grazing stock: such treatment continues at Berner's Heath in the Lakenheath–Elveden complex in Norfolk but it has largely fallen into disuse, although accidental fires do still occur. Even without burning, however, catastrophic death of the *Calluna* can short-circuit the growth cycle, where the bushes succumb wholesale to defoliation by the heather beetle (*Lochmaea saturalis*), to drought which is occasionally severe in a region where crop growth often necessitates irrigation, or to frost, which can occur into late May or early June in this part of Britain (e.g. Farrow 1915, Watt 1971, Ratcliffe 1977). Marrs (1986) suggested that such events might constitute a fairly normal exogenous control on *Calluna* regeneration in susceptible populations and account for the kind of age-structure seen on Cavenham Heath in Breckland, for example, one of the largest intact blocks of this kind of vegetation in East Anglia, but composed almost entirely of extensive more or less even-aged stands, despite the absence of any burning for at least 25 years. As outlined below, the predominance of uncompetitive *Calluna* in such tracts can speed seral progression to other vegetation types, notably bracken or woodland.

Apart from burning, the other major biotic factor which can affect the physiognomy of the *Calluna-Festuca* heath is grazing itself, either by stock, now much less widely or intensively pastured on this kind of heath, or by wild herbivores, occasionally deer (as on the Cavenham–Tuddenham heaths) but much more importantly in the past, rabbits. In some areas, rabbits have again become very common, but in many sites, as in parts of Lakenheath–Eriswell Warrens and on Thetford Heath, the rank growth of the heather in this community since myxomatosis is very evident. Consistent heavy grazing of this vegetation keeps the *Calluna* canopy firmly in check, favouring the establishment of Nardo-Galion herbs in accessible places between the bushes (as seen in some stands of the *Teucrium* sub-community) and ultimately extinguishing the heather in a seral shift to close-cropped calcifuge swards dominated by such vascular species or, in extreme cases, lichen-dominated vegetation or open eroding ground. Grazing by cattle, as opposed to the more usual sheep, which still occurs at Eriswell Warrens, can lead to eutrophication of the soils under this kind of heath and the occurrence of rank weeds on scuffed areas of soil.

Very often, however, abandonment of traditional heathland management has been followed by reclamation and improvement of the land under this community such that, within its possible geographical range, stands now generally survive as much-reduced and fragmented remnants within a landscape given over largely to agriculture and forestry. Although the soils here present considerable problems for cropping, being droughty, lime-deficient, often copper-deficient and susceptible to wind erosion, they have the advantage of being easy to cultivate and much of them has gone to arable, mostly barley and sugar beet (Hodge *et al.* 1984). Yields, though, are low and, as an alternative productive landuse, coniferous afforestation has been widely pursued, mostly with *Pinus sylvestris* or, increasingly, *P. nigra* var. *maritima*. Other losses have been to airfield and military training areas, though the latter use, as around Stanford in Breckland, can provide a measure of protection by its prohibition on other kinds of activity. Even where more substantial tracts of the *Calluna-Festuca* heath survive, however, it should not necessarily be assumed that these are very old. As Sheail (1979) has shown in the Stanford Practical Training Area, changes in agricultural custom and economics can produce a very complex pattern of shifting land use within a relatively small period of time, such that only a very small proportion of the considerable extent of grassland and heath there is more than a century old. More long-established stands of the *Calluna-Festuca* heath are unlikely to be floristically richer than younger ones but they can show much more clearly some of the most interesting aspects of the community, namely its place in a sequence of vegetation and soil types that preserves evidence of post-Glacial climatic change and a kind of land use dating back to prehistoric times.

Zonation and succession

The *Calluna-Festuca* heath is found with a variety of other vegetation types in zonations and mosaics which reflect patterns of edaphic variation and long and diverse treatment histories. Very often now, the neglect of traditional styles of heathland management has blurred the boundaries in such sequences and allowed seral developments previously held in check. More drastic heathland improvement has truncated and fragmented many zonations and stands of the community are frequently abruptly marked off from surrounding arable land or coniferous plantations.

Even the relatively small surviving tracts of heathland in eastern England sometimes stretch over a range of soil parent materials, with either sequences of diverse solid deposits or, very often, a cover of heterogeneous drift, variable in thickness, and thus in the extent to which it masks the influence of the underlying bedrock, and in its own chemical and physical characteristics. The soil mantle is therefore often quite varied, with profiles differing in such features as base-richness, texture and permeability, even within a small compass (e.g. Watt

1936, 1940, Corbett 1973, Hodge *et al.* 1984). Over such sequences of soils, the *Calluna-Festuca* heath has become established on the more acidic and free-draining, accentuating the tendency towards podzolisation and often marked off from neighbouring vegetation types by shifts in one or more of these edaphic factors.

In a few localities, soil moisture is a controlling variable, the community representing the dry-heath element in the kind of pattern so characteristic of the surrounds to southern valley mires (e.g. Rose 1953), where there is a zonation through wet heath to mire vegetation in relation to a locally high ground watertable. On some of the west Norfolk commons, at Sandringham Warren and Roydon Common, for example (Daniels & Pearson 1974, Ratcliffe 1977), the *Calluna-Festuca* heath occupies some of the driest of the sequence of acidic soils, passing on seasonally-waterlogged mineral profiles to the *Ericetum tetralicis*, which then gives way to a variety of mire communities, depending on the base-richness of the ground waters. Although *Calluna* maintains its frequency within the *Ericetum*, its vigour is usually depressed and the boundary between the two kinds of heath is generally well marked by the confinement of *Erica tetralix* and *Molinia* to the wetter ground, a restriction which reflects the continental character of the climate.

Within the range of the *Calluna-Festuca* heath, zonations of this kind are less common than in central southern and south-west Britain and more often the prevailing kind of edaphic variation over the tracts of heathland here is in base-richness among profiles which are more or less uniformly free-draining. Such differences depend upon variations in the amount of calcareous material in the superficials and the proximity to the surface of any limy bedrocks, most importantly through much of East Anglia, the Chalk, and they are particularly influential in the Breckland mosaics of heath and grassland renowned from Watt's (1936, 1940) accounts. Such patterns, still to be seen in whole or part on the remaining fragments of the Stanford–Wretham, Icklingham and Lakenheath–Elveden heaths (Ratcliffe 1977), involve transitions from the *Calluna-Festuca* heath, through calcifuge Nardo-Galion swards, to the *Festuca-Avenula* and *Festuca-Hieracium-Thymus* grasslands, with an edaphic shift from Worlington brown sands with some podzolisation, through Methwold brown calcareous sands, to Newmarket brown rendzinas (Corbett 1973, Hodge *et al.* 1984). Floristic variation within the sequence is more or less continuous and though the limit of the heath is well marked by the bounds of the dominant *Calluna*, fine mosaics of the community with patches of included grassland are common (though they lose much of their clarity if grazing is relaxed: see below). The degree of order within these soil and vegetation mosaics is, in fact, very vari able, dependent as it is on the great heterogeneity of the cover of superficials, but some especially striking patterns have resulted where the materials have undergone periglacial sorting to produce such features as polygons or stripes. These occur quite widely over the gently-undulating landscape of East Anglia and are often very clearly delineated on unimproved land, because the *Calluna-Festuca* heath typically picks out the areas of deeper and more acidic sandy infill, while the polygon centres or areas between the stripes are occupied by the grasslands (Williams 1964, Corbett 1973, Curtis *et al.* 1976). Especially good examples of the striped kind of patterning can be seen at Eriswell High Warren, on Thetford Heath and around Grimes Graves (Ratcliffe 1977).

Two other kinds of floristic variation can confuse these soil-related patterns. These, too, have some kind of edaphic basis but they are also a reflection of competitive interactions between *Calluna* and two other possible dominants, *Pteridium* and *Carex arenaria*, on the dry acidic soils which the community favours, so the spatial patterns that result are often but a temporary indication of shifting successional developments. *Pteridium* is generally confined to deeper soils, free from parching, but toing and froing of dominance between it and *Calluna* is in large measure dependent upon coincidence of a vigorous, expansive phase in the one with a declining, weakly-competitive phase in the other (Watt 1955), a pattern of growth which is endogenous, though strongly influenced by external biotic or climatic factors (Watt 1971, Marrs 1986). This is why mosaics between the *Calluna-Festuca* heath and the *Pteridum-Galium* community are so diverse on the ground and variable through time. Zonations to the *Carex arenaria* or *Carex-Cladonia* dune communities are likewise related to competition between heather and the sedge, though the greater tolerance of the latter to shifting sand often means that transitions from one vegetation type to the other reflect present or past instability of the sandy soil surface (Watt 1937).

Patchworks of grassland, heath, bracken and sand-sedge compose the characteristic vegetation cover of the heathlands of eastern England and were in the past maintained by the traditional burning and grazing treatments, with rabbits, long encouraged in the warrens of this region, providing a very important additional grazing component. The burning renewed the heather at regular intervals, probably helped keep bracken in check and repeatedly set back any invasion of shrubs or trees; the grazing helped maintain the grassland against expansion of the heath, sharpening up the boundaries between them. The abandonment of these treatments and the demise of rabbits in the myxomatosis epidemic have allowed the grasslands and the heather to grow rank in many places, so that transitions between the

communities have been obscured. They have also permitted seral progressions to scrub and woodland, patches of which, sometimes very dense and extensive, have become common on these heathlands. The most important woody invaders on the dry, acidic soils characterised by the *Calluna-Festuca* heath are *Betula pendula* and the pines, of which there are now such abundant seed-parents in the plantations. Oak, often *Quercus robur* over the range of the community, sometimes figures if there are mature trees fairly close at hand and locally *Rhododendron ponticum* has become common, as at Sandringham Warren, for example. Colonisation of vigorous, dense *Calluna* by these species, particularly birch and pine, is very difficult but degenerate heather, experiencing no burning or grazing, offers ample open ground. Once well established, such trees can shade out the heather and most of its heathland associates and come to dominate in *Quercus-Betula-Deschampsia* woodland, the climax forest of base-poor, free-draining soils in lowland Britain. Natural gaps or cleared areas within such woodland can show a temporary resurgence of the heath, though there is a strong tendency for the *Pteridium-Galium* community to become ensconced in such places or for *Ulex-Rubus* scrub to develop in disturbed areas. Gorse scrub of this kind is also sometimes found around coniferous plantations and often runs across the heathland in strips marking out the edges of trackways.

Distribution

The *Calluna-Festuca* heath occurs widely throughout the eastern lowlands of England, though it is now often very local. More extensive stands survive in Breckland where something of the traditional heathland scenery of this part of the country can still be glimpsed.

Affinities

Impoverished heaths dominated exclusively by *Calluna* can be found throughout Britain and, though these pose a classificatory problem when they are considered out of their context, they should certainly be distinguished from the *Calluna-Festuca* heath, the purity of whose sub-shrub canopy is a real reflection of climatic conditions in eastern England rather than local differences in treatment. This is the most continental of our heaths, though its lack of any positive floristic element reflecting this affinity has meant that it has generally been included within a broadly-defined lowland *Callunetum* (e.g. Tansley 1911, 1939) or marked out as special for other reasons, such as the clarity of its internal dynamics or competitive relationships or because of its place within a regional sequence of communities (Watt 1936, 1940, 1955), features of considerable importance to the conservation value of this vegetation.

The community grades floristically to the *Calluna-*

Erica and *Calluna-Ulex minor* heaths but phytosociologically is best placed in the Genisto-Callunion alliance, which includes the Calluno-Ulicetalia heaths of North Germany, Belgium and The Netherlands in which *Genista pilosa*, *G. anglica* and *G. germanica* are characteristic. In general, however, the *Calluna-Festuca* heath is more species-poor than associations like the *Genisto pilosae-Callunetum* Oberdorfer 1938, the *Genisto anglicae-Callunetum* Tüxen 1937 and the *Genisto germanicae-Callunetum* Oberdorfer 1957. *G. pilosa* used to occur on Suffolk heaths and *G. anglica* is very occasionally found in the community though, through much of East Anglia, it is more characteristic of Ericetalia wet heaths.

Floristic table H1

	a	b	c	d	e	1
Calluna vulgaris	V (7–10)	V (7–10)	V (7–10)	V (2–10)	V (8–10)	V (2–10)
Hypnum cupressiforme	V (3–6)	II (2–3)	IV (4–6)	V (4–6)	II (3–6)	IV (2–6)
Festuca ovina	IV (1–6)	II (2–4)	V (2–6)	IV (3–4)		IV (1–6)
Dicranum scoparium	V (3–6)	III (3–6)	IV (3–5)	II (2–4)		IV (2–6)
Cladonia uncialis	III (2–4)	II (1–5)				II (1–5)
Cladonia fimbriata	II (3–4)	II (3)	I (4)			I (3–4)
Erica cinerea	II (3–4)	II (2–5)		I (4)		I (2–5)
Cladonia pyxidata	II (3)	II (2–4)				I (2–4)
Cornicularia aculeata	II (3–4)	II (3–6)				I (3–6)
Cladonia squamosa	II (3–5)	I (2–3)				I (2–5)
Cladonia tenuis	I (4)	I (3)				I (3–4)
Erica tetralix	I (3)	I (2–3)				I (2–3)
Cladonia arbuscula	I (3)	I (3–6)				I (3–6)
Pleurozium schreberi	I (6)	I (7)				I (6–7)
Hypogymnia physodes	III (3–4)	IV (2–4)	I (2)	I (3)		II (2–4)
Cladonia impexa	II (2–6)	IV (3–7)				II (2–7)
Cladonia furcata		III (3–6)				I (3–6)
Dicranella heteromalla	I (4)	II (2–4)			I (2)	I (2–4)
Cladonia macilenta		II (2–4)				I (2–4)
Cladonia coniocraea		I (4)				I (4)
Rumex acetosella			II (2–4)	II (2–4)		II (2–4)
Ulex europaeus			I (6)	I (5)		I (5–6)
Deschampsia flexuosa			I (4–5)	I (5)		I (4–5)
Holcus lanatus			I (4)	I (3)		I (3–4)
Campanula rotundifolia			I (4)	I (3)		I (3–4)
Psuedoscleropodium purum			I (1–4)	I (5)		I (1–5)

Floristic table H1 (*cont.*)

	a	b	c	d	e	1
Teucrium scorodonia	I (3)		III (2–3)			I (2–3)
Galium saxatile			II (2–3)	I (3)		I (2–3)
Agrostis capillaris			II (3–6)	I (4)		I (3–6)
Cerastium fontanum			II (3–4)			I (3–4)
Senecio jacobaea			II (3)			I (3)
Carex arenaria	I (3)	I (3)	I (3)	V (2–10)		II (2–10)
Pteridium aquilinum	I (2)		I (4)	I (4)	I (3)	I (2–4)
Hylocomium splendens	I (4)		I (1)	I (5–6)		I (1–6)
Ptilidium ciliare	I (3)		I (1)			I (1–3)
Polytrichum juniperinum	I (3)		I (3)			I (3)
Number of samples	8	8	10	7	9	42
Number of species/sample	9 (5–14)	9 (5–12)	10 (5–22)	7 (4–11)	3 (1–8)	7 (1–22)
Shrub/herb height (cm)	37 (10–60)	39 (30–50)	38 (10–60)	53 (20–100)	59 (25–100)	44 (10–100)
Shrub/herb cover (%)	89 (50–100)	84 (50–100)	96 (85–100)	100	94 (75–100)	95 (50–100)
Ground cover (%)	54 (30–80)	50 (25–90)	28 (2–70)	21 (0–80)	4 (0–30)	31 (0–80)
Altitude (m)		30 (1–76)				

a   *Hypnum cupressiforme* sub-community
b   *Hypogymnia physodes-Cladonia impexa* sub-community
c   *Teucrium scorodonia* sub-community
d   *Carex arenaria* sub-community
e   Species-poor sub-community
1   *Calluna vulgaris-Festuca ovina* heath (total)

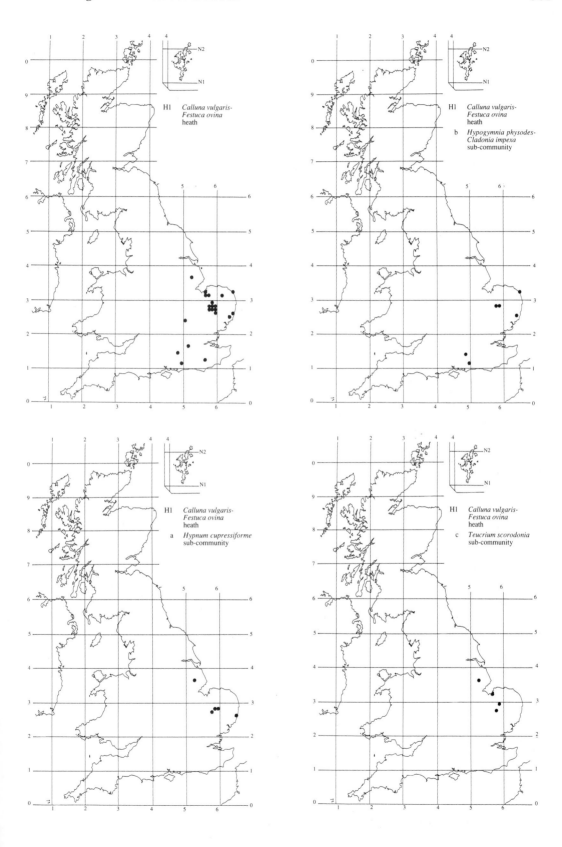

H1   *Calluna vulgaris-*
        Festuca ovina
        heath

H1   *Calluna vulgaris-*
        Festuca ovina
        heath
b   *Hypogymnia physodes-*
        Cladonia impexa
        sub-community

H1   *Calluna vulgaris-*
        Festuca ovina
        heath
a   *Hypnum cupressiforme*
        sub-community

H1   *Calluna vulgaris-*
        Festuca ovina
        heath
c   *Teucrium scorodonia*
        sub-community

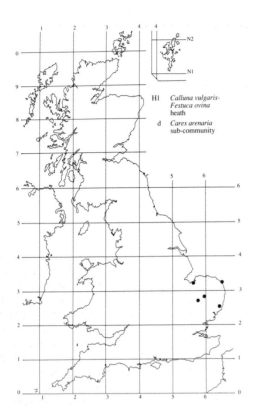

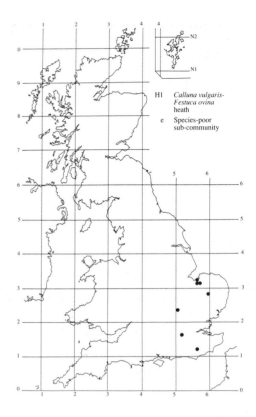

H2
Calluna vulgaris-Ulex minor heath

Synonymy
Callunetum arenosum Tansley 1911 *p.p.*; *Calluna-Ulex-Erica* heath Fritsch & Parker 1913, Fritsch & Salisbury 1915, Haines 1926; *Callunetum arenicolum* Tansley 1939 *p.p.*; Dry heath assemblage Newbould 1960; Dry heath Harrison 1970 *p.p.*; *Calluna-Ulex minor* heaths Gimingham 1972 *p.p.*

Constant species
Calluna vulgaris, Deschampsia flexuosa, Erica cinerea, Ulex minor.

Rare species
Agrostis curtisii, Genista pilosa.

Physiognomy
The *Calluna vulgaris-Ulex minor* heath is generally dominated by *Calluna vulgaris*, though with both *Erica cinerea* and *Ulex minor* playing a very frequent and sometimes prominent role in the sub-shrub canopy and together providing an important floristic distinction from the more continental heath vegetation of much of East Anglia. The canopy is very variable in height, from 1 to 8 dm or more, and, as always where *Calluna* is an important species, its structure can be much affected by the growth phases of the heather plants, and whether the individuals in a particular stand are of even or uneven age. Where burning still occurs, as in parts of the New Forest, tracts of the community can show the characteristic patchwork of swales with the heather in various stages of recovery from pioneer through to building, though generally not beyond if burning is being judiciously practised to maximise grazing value. But, often now, burning has ceased or is accidental and sporadic and grazing not pursued, so many stands have a cover of very leggy *Calluna*.

Recovery of the vegetation after degeneration of the heather, or after an episode of burning, can also see marked changes in the proportions of the two other common sub-shrubs. *E. cinerea* is a prolific seeder and, particularly where patches of mineral soil have been exposed, as after fires, it can outstrip *Calluna* at first, even when most of its original plants have been destroyed (Fritsch & Salisbury 1915, Gimingham 1949, 1972). *Calluna* generally comes to dominate eventually, but *E. cinerea* can retain a co-dominant role, particularly perhaps on drier soils on slopes facing south or south-west (Fritsch & Parker 1913, Fritsch & Salisbury 1915), and it can tolerate some shade from heather (Bannister 1965). In some stands, however, its cover is very low and it can be totally absent, especially where there is a tendency for the soils to experience some seasonal waterlogging, as in the *Molinia* sub-community.

U. minor shows a similar variation in its abundance and, in some areas like the New Forest, where grazing still occurs, it seems to have been widely reduced in dry heaths of this kind so that they have taken on the composition of the much more widespread and largely northern *Calluna-Erica cinerea* heath. In its characteristic form, however, the *Calluna-Ulex minor* heath always has at least some dwarf gorse, being one of only two British heath communities in which this geographically restricted species plays a consistently frequent part. Not that the plants are always particularly dwarfed (e.g. Skipper 1922, Proctor 1965): procumbent individuals in open stretches of heath can form mats barely 5 cm tall and, after fires, there can be prolific production of low shoots from surviving stools, but the species is very plastic and bushes up to a metre or more can be found in more sheltered places, with the species showing local co-dominance with the *Calluna*. Usually, however, it plays a subsidiary role and, being relatively shade-tolerant, may form a patchy understorey beneath the level of the taller heather. In denser shade, it rarely flowers, which deprives the vegetation of a glorious splash of pale yellow, intermixed with the purples of the ericoids, in late summer and early autumn, and makes it difficult to tell this gorse from *U. gallii*. In fact, the latter is very scarce in this community, showing an almost exclusive geographical vicarism with *U. minor*: it is mostly around Poole Harbour in Dorset that the two gorse species occur together in these kinds of heaths (Proctor 1965).

No other sub-shrub occurs frequently throughout the community as a whole. *U. europaeus* can be found occasionally, and sometimes with local abundance, often along disturbed tracks or around old settlements, though in parts of the New Forest a little more widely (Tubbs & Jones 1964). Of greater significance for defining floristic variation within the community are *Erica tetralix* and *Vaccinium myrtillus*, both fairly uncommon overall but each preferentially frequent in particular sub-communities, the former where this kind of heath extends on to seasonally-waterlogged soils, the latter perhaps where rainfall is somewhat higher. *Genista anglica* also occurs very occasionally throughout and in Ashdown Forest in Sussex the community provided the most inland locus for *G. pilosa*, now extinct there.

Other consistent floristic features of the community are also very few. Apart from the sub-shrubs, only *Deschampsia flexuosa* is constant and even that is somewhat patchy in its occurrence, being shade-tolerant, though tending to be depressed by accumulation of litter and thus often largely confined to areas between the woody plants. Where there is some grazing, its abundance in such accessible places can be encouraged and it can also spread extensively after burning (Tansley 1939). On wetter soils here, it tends to be replaced by *Molinia caerulea*, not usually present in much quantity, though prone to expand where there has been very frequent burning followed by heavy grazing on such damper ground. In marked contrast to heath with *U. minor* further to the west, however, *Agrostis curtisii* is only very occasional in this community: some of its Surrey occurrences are in this kind of heath but generally it is not an important species. And, compared with the more continental *Calluna-Festuca* heath of eastern England, *Festuca ovina* and *Agrostis capillaris* are scarce.

Among other herbaceous plants, only *Pteridium aquilinum* occurs with any frequency, being occasional overall and preferential for one particular sub-community. It is usually not very abundant but, when its fronds are fully unfurled by mid-summer, it can look conspicuous and thick patches of bracken, which would be classified in the *Pteridium-Galium* community, occur commonly in close association with the *Calluna-Ulex minor* heath. Scattered individuals of *Potentilla erecta* or *Galium saxatile* can sometimes be found in more open areas and one very distinctive plant that can be abundant after burning is *Cuscuta epithymum*, its very slender reddish stems able to attach parasitically to almost all the species mentioned above.

A further important element of the vegetation in some stands is seedling and sapling trees, with *Quercus* spp., particularly *Q. robur*, *Betula* spp. and *Pinus sylvestris*,

especially able to gain a hold on more open areas of ground and quick to get away if there is no burning or grazing.

Often now, in the absence of such regular burning, the ground layer of the *Calluna-Ulex minor* heath is very patchy and more or less limited to the cores of degenerate heather bushes, where the characteristic sequences of mosses and lichens can be seen, *Dicranum scoparium* and *Hypnum jutlandicum* being the most frequent bryophytes overall, *Polytrichum piliferum* and *P. juniperinum* being less common, though showing local abundance in the early stages of colonisation of mineral ground. Peat-encrusting *Cladonia* spp., sometimes with larger taxa like *C. furcata* and *C. arbuscula*, can also figure in such places, together with *Hypogymnia physodes* on old heather stems (e.g. Watt 1955). On burned ground, a much more extensive sequence of colonisation can be observed (e.g. Fritsch & Salisbury 1915), with sequences of algae first to appear, then mosses such as *Ceratodon purpureus*, *Funaria hygrometrica* and *Polytrichum piliferum*, and *Cladonia* spp. developing as the sub-shrubs regenerate from sprouting stools and seed.

Sub-communities

Typical sub-community. In this kind of *Calluna-Ulex minor* heath, *Calluna* is generally strongly dominant with subsidiary amounts of *U. minor* and *E. cinerea*, the latter especially being very variable in its cover and sometimes absent, though both occasionally attaining co-dominance. Neither *V. myrtillus* nor *E. tetralix* occur among the sub-shrubs and *Molinia* is typically scarce. Indeed, no other plants occur consistently throughout, although *Deschampsia flexuosa* is quite common at usually low covers and sometimes accompanied or replaced here by *Festuca rubra*. *Pteridium* and tree seedlings or saplings are infrequent. Bryophytes and lichens can be locally conspicuous among older heather or after burning, with *Cladonia fimbriata*, *C. coccifera*, *C. chlorophaea* and *C. arbuscula* all showing a slight preference for this sub-community.

***Vaccinium myrtillus* sub-community.** *Calluna* is a little less overwhelmingly dominant here, the sub-shrub canopy usually consisting of mixtures of heather, with sometimes substantial amounts of *U. minor*, *E. cinerea* and, strongly preferential to this sub-community, *V. myrtillus*. *Pteridium* is also rather more frequent than elsewhere in the community and, among somewhat more open areas of this diverse cover, there is generally some *Deschampsia flexuosa* and occasionally a little *Molinia*. *Gaultheria shallon*, a North American shrub planted in Britain for game cover and naturalised, can be found quite widely in this vegetation in Surrey. Young trees are also strongly preferential, with oak and birch seedlings and small saplings frequent and small pines locally prominent. *Hypnum jutlandicum* and *Dicranum scoparium* can occasionally be found but lichens are sparse.

Molinia caerulea sub-community. Again, *Calluna* can be very abundant in this sub-community, usually with smaller amounts of *U. minor* and particularly of *E. cinerea*, which is sometimes joined or replaced by *E. tetralix*. More obviously preferential is *Molinia*, which almost totally displaces *Deschampsia flexuosa*. *Pteridium* is uncommon and young trees very rarely found. The ground layer, too, is particularly sparse in this kind of heath.

Habitat
The *Calluna-Ulex minor* heath is characteristic of impoverished acid soils, predominantly free-draining, in south-east and central southern England. The slight tendency towards an oceanic climate in this region is reflected in the general composition of the community, although in some areas the vegetation still takes much of its structural, and some of its floristic, character from the traditional burning and grazing treatments. Elsewhere, neglect of these activities means that this kind of heath is now in various stages of progression to woodland or found as small remnants, fragmented and isolated by improvement for agriculture or forestry.

With the geographical shift south and west from the range of the *Calluna-Festuca* heath, centred on East Anglia, the climate takes on a distinctly less continental character. Thus, though annual accumulated temperatures are within much the same range as there, mean annual maxima are generally lower, for the most part between 27 and 30 °C (Conolly & Dahl 1970). More importantly, the winters are considerably milder, with February minima often one or more degrees C above freezing (*Climatological Atlas* 1952), so the annual temperature range is considerably reduced. These more equable conditions are marked in the vegetation by the appearance of some of the Oceanic West European species that become a characteristic feature of the series of heaths running around the Atlantic seaboard of Britain. Among those typical of drier soils, *Erica cinerea* and *Ulex minor* show the most consistent eastward penetration south of the Thames: it is their constancy which provides the most obvious floristic distinction between this community and the *Calluna-Festuca* heath and probably their sensitivity to winter cold which sets the geographical boundary between the two (e.g. Bannister 1965, Gimingham 1972). Westwards, it is harder to see any climatic explanation for the limit to the range of the *Calluna-Ulex minor* heath. Beyond Poole, it is replaced on similar soils by the *Calluna-Ulex gallii* heath, a switch that essentially involves the replacement

of one gorse by the other (Proctor 1965). This fairly sharp vicarism is also found in north-west France (des Abbayes & Corillion 1949, Corillion 1950, 1959) and perhaps also in northern Spain (Proctor 1965) and it may reflect greater tolerance of winter cold by *U. minor*, but other present or past climatic conditions, or some edaphic factors, could be involved.

The other distinctive feature of the climate within the range of the community compared with that of East Anglia, indeed, with that over much of the central and eastern lowlands of Britain, is that it is decidedly wetter, with more than 120 wet days yr^{-1} (Ratcliffe 1968) and annual precipitation almost everywhere in excess of 800 mm (*Climatological Atlas* 1952). This isohyet shows a close correspondence with the British distribution of *Vaccinium myrtillus*, another sub-shrub absent from the *Calluna-Festuca* heath but sporadically represented here and in more oceanic lowland heaths provided, of course, there is little or no grazing. Such treatment may play some part in restricting its occurrence within the *Calluna-Ulex minor* heath; certainly, absence of grazing is implicated in the frequent association of bilberry with young trees in the *Vaccinium* sub-community. It should be noted, though, that this kind of vegetation, like other bilberry-rich lowland heaths, does tend to be found at somewhat higher altitudes than usual (mean almost 140 m, compared with 75–100 m for the other sub-communities), and is concentrated in those parts of The Weald where rainfall approaches or exceeds 1000 mm (*Climatological Atlas* 1952).

Over the range of acid soils occupied throughout this climatic zone by heaths in which *Calluna*, *E. cinerea* and *U. minor* play an important role, this community is characteristic of the more free-draining profiles, developed from pervious arenaceous or pebbly parent materials. In the High Weald, the higher reaches of the Cretaceous Hastings Sands and Ashdown Sands form important substrates in Ashdown Forest, and then around the western rim of The Weald, the Folkstone and Sandgate Beds, and the less lime-rich stretches of the Hythe Beds underlie stands on the Lower Greensand dip slope, as on Iping and Ambersham Commons in Sussex and Thursley and Hankley Commons in Surrey. Almost contiguous with these last sites is the extensive expanse of Eocene Bagshot and Bracklesham Beds which support many tracts of the community running north and west into the Thames basin, as on Chobham Common, for example. Eocene sands and gravels, overlain in parts by Plateau Gravels and river terrace drift, are also the characteristic substrate through the New Forest and around Poole Harbour (e.g. Wooldridge & Goldring 1953, Ratcliffe 1977).

Under the *Calluna-Ulex minor* heath, such parent materials have typically given rise to some kind of podzolic profile, either a classic humo-ferric podzol, like those of the Shirrell Heath series, particularly important over the Lower Greensand, or, where there is more material of the finer fractions in the profile, a palaeo-argillic podzol, as of the Southampton series, a common soil type over Eocene and more recent deposits (*Soil Survey* 1983, Jarvis *et al.* 1984). Such profiles are highly acidic, superficial pH beneath the community being between 3.5 and 4.5, and generally very impoverished, features reflected in the thoroughly calcifuge character of the vegetation and the poor representation of meso-phytic plants. In fact, Tubbs (1968) has suggested that, in the New Forest, the soils under this kind of heath are not quite so bereft of nutrients as in, for example, the Poole basin, something of importance to the readiness with which the vegetation can progress to different kinds of woodland (see below).

The other general characteristic of these soils is that they are relatively free-draining; indeed, in the warm dry summers typical of the region, they can be distinctly droughty (Jarvis *et al.* 1984). It is this edaphic feature which provides the major distinction between the habi-tat of the *Calluna-Ulex minor* heath and the *Ulex minor-Agrostis curtisii* heath. The two overlap considerably in their geographical range but, in the latter, the shift to seasonally-waterlogged gley-podzols is marked by the consistent appearance, along with *Calluna*, *E. cinerea* and *U. minor*, of *E. tetralix*, *Molinia* and *A. curtisii*. Characteristically, these species are of restricted occur-rence here, though the boundary between the two vege-tation types is not a hard and fast one, because the development of an argillic B horizon or a B_{Fe} pan in these podzols can impede drainage and result in local surface-water gleying in winter: the *Molinia* sub-community, with its partial replacement of *E. cinerea* and *Deschamp-sia flexuosa* by *E. tetralix* and *Molinia*, is characteristic of such transitional situations.

Whether the development of such generally poor soils was already in train before the extensive establishment of dry heath in central southern and south-east England is still debatable (e.g. Dimbleby 1962, Tubbs & Dimb-leby 1965, Godwin 1975, Haskins 1978), but what seems certain is that the spread of vegetation of this type was encouraged by human activity, perhaps as early as in the Mesolithic Period in some areas, and certainly by the Bronze Age, and that the process of podzolisation was enhanced under the heath canopy. And treatments continue to be of prime importance in maintaining the *Calluna-Ulex minor* heath, rather than the harsh char-acter of the edaphic environment itself: this is essentially a plagioclimax vegetation type which, when released from a regime of grazing and/or burning, progresses to woodland, a development which can already be seen in the *Vaccinium* sub-community with its crop of tree seedlings and preferentially frequent bilberry and bracken (see below).

The decline of traditional styles of heathland exploi-tation means that such stands of the *Calluna-Ulex minor* heath as remain often have a very leggy sub-shrub canopy in which heather is the overwhelming dominant. Burning of this vegetation can help maintain a more variegated canopy in which there is opportunity for a considerable local expansion in the cover of *E. cinerea*, *U. minor* or *D. flexuosa* on drier ground, or *Molinia* on wetter, before the regrowth or re-establishment from seed of the *Calluna*, and a coincident colonisation by a series of algae, bryophytes and lichens on the exposed soil, a process described in detail from Hindhead Common by Fritsch & Salisbury (1915). On most sites in south-eastern England, however, burning is now spora-dic, accidental and uncontrolled so the kind of fire-derived mosaic which they observed in stands of the *Calluna-Ulex minor* heath is rare there. It can, though, still be seen in tracts of the community in the New Forest, where burning is the major factor perpetuating a range of dry and wet heaths: here, the Forestry Commis-sion burns on rotations of 6–12 years to reduce the risk of accidental fires and maintain an irregular patchwork of younger sub-shrub growth and grass for grazing stock (Tubbs 1968).

In fact, the contribution of palatable grasses to the post-burn succession in this community is generally less prominent than in the *Ulex minor-Agrostis* heath, its near relative on somewhat wetter ground. But grazing has undoubtedly contributed to the maintenance of the community in the past by curtailing the invasion of trees, and Webb (1986) has suggested that, in the lowlands, this factor has been of greater long-term importance than burning in preserving the cover of heath vege-tation. As with burning, however, this practice is vir-tually defunct, apart from in the New Forest where the various heath communities provide the bulk of the unenclosed land that is still exploited by the traditional mixture of cattle and 'heath-cropper' ponies, together with some deer (Tubbs 1968). Apart from controlling colonisation by tree seedlings, such grazing affects the proportions of the different sub-shrubs: it is perhaps a major cause of the scarcity of *U. minor* in some stands, the soft young shoots being very palatable, and it tends to favour *Calluna* as against *E. cinerea*. In more humid areas, it is probably the major factor affecting the abundance of *Vaccinium myrtillus* in this community.

Over the potential range of the *Calluna-Ulex minor* heath, the often fragmentary scatter of remaining stands also reflects losses of this kind of vegetation by a more deliberate change of land-use. Some of the Surrey commons, for example, were deregulated by the Minis-try of Defence, leaving only Chobham as an extensive tract of the very characteristic landscape of the Eocene sands and gravels. Other tracts close to London persist only as part of the rough on golf courses. Then, there has been extensive afforestation with conifers in some places: despite the inherent infertility of the soils under the community, tine-ploughing to break up the podzol pans and application of nitrogen and phosphorus has facilitated successful cropping with *Pinus sylvestris*, *P. nigra* var. *maritima* and *Tsuga heterophylla*.

Zonation and succession

Zonations between the *Calluna-Ulex minor* heath and other vegetation types are generally related to edaphic variation or to differences in treatment. Burning and grazing can modify soil-related patterns and also create effects of their own: most obviously they mediate the invasion of this vegetation by trees and very often now, with their decline, the community survives as a decreasing element in patchworks of woodland, heath and bracken. Elsewhere, stands give way abruptly to agricultural land, coniferous forest or encroaching settlements.

In the traditional kind of heath landscape of central southern and south-east England, preserved extensively now only in the New Forest, the imposition of grazing and burning over large areas enables the effect of edaphic variation to be seen in a very distinctive sequence of sub-shrub communities. Within this pattern, the *Calluna-Ulex minor* heath occupies the most free-draining of the series of acid and impoverished soils, typically podzols which show no tendency, or only a slight one, towards surface-water gleying in winter. With the shift to profiles with somewhat more severe or consistent drainage impedence, as where there is a strong pan development or a markedly argillic B horizon, the community typically gives way, through the Dorset and Hampshire area, to the *Ulex minor-Agrostis curtisii* heath. The boundary between the vegetation types can be ill-defined, but, in this moderately oceanic part of Britain, the move on to soils which show even only a slightly greater tendency to gleying is marked by the consistent appearance of *E. tetralix* along with *E. cinerea*, and the replacement of *D. flexuosa* by *Molinia* and *A. curtisii*. The sequence of vegetation types then continues with the *Ericetum tetralicis*, on seasonally surface-waterlogged ground, and the *Narthecio-Sphagnetum* on valley-mire peat accumulated in elongated hollows.

The clarity of this pattern, characteristic of the undulating sand and gravel landscapes of this part of Britain (Rose 1953) and described in detail from Cranesmoor by Newbould (1960), depends on the exact conformation of the ground and its hydrology: over generally less impeded surfaces, the drier elements in the sequence prevail; where wetter hollows predominate, the role of the *Calluna-Ulex minor* heath is much reduced and, in areas of complex topography, fine mosaics of the communities can replace ordered zonations. Geological heterogeneity, too, can introduce additional variation in

soils and vegetation: patches of less sharply-draining sands or clay-sands, for example, can weather to deep acid brown earths with a cover of the *Pteridium-Galium* community, and stretches of somewhat more fertile soils often have stands of the *Ulex-Rubus* scrub, as in the New Forest 'brakes'. Although burning and grazing play a crucial role in preserving the generally open character of this landscape, they can affect the soil-related pattern. Among the heaths, for example, such treatments can favour a general dominance throughout of *Calluna*, the abundance of which may swamp variation in the subordinate preferentials of the different communities; and certain regimes of burning and grazing may allow the replacement of the heaths by Nardo-Galion or Junco-Molinion grasslands or alter the balance between heather and bracken. Disturbance of the soils, as around settlements or plantations or along tracks, can also lead to a spread of *Ulex-Rubus* scrub within the *Calluna-Ulex minor* heath.

Such sequences of heath communities, intermixed with bracken, gorse and acid grasslands, can still be seen, though not so extensively, around Poole Harbour, where the *Calluna-Ulex minor* heath is well represented in the Arne area, and on Iping Common in Sussex, Thursley, Hankley and Chobham Commons in Surrey (Ratcliffe 1977); at Ambersham Common and in Ashdown Forest in Sussex, wetter heaths and mire tend to prevail, though in this part of Britain, the *Ulex minor-Agrostis* heath is absent, the *Calluna-Ulex minor* heath passing directly to the *Ericetum tetralicis*, sometimes with the *Molinia* sub-community as a transition zone. In many of these and other smaller sites in the south-east, too, the soil-related zonation is overlain by patterns resulting from the neglect of burning and grazing, and other traditional heathland treatments like the cutting of bracken for bedding and of gorse for fuel. In the rainier parts of The Weald, the *Vaccinium* sub-community is typical of such neglected dry heaths, with its tree seedlings able to get away quickly in less densely-shaded areas. The commonest woody invaders in such situations are birch, particularly *B. pendula* on these drier soils, and oak, generally *Q. robur*, though with *Q. petraea* locally important. Pines, particularly *P. sylvestris*, can also be very abundant where seed-parents are close and other colonisers include occasional *Sorbus aucuparia* and, particularly striking in ungrazed or unburned stands in the New Forest, *Ilex aquifolium* (Peterken & Tubbs 1965, Tubbs 1968). On some Surrey commons, *Amelanchier lamarckii* has become thoroughly naturalised among birch thickets developing among the *Calluna-Ulex minor* heath, easily overlooked in summer but very obvious in April with its lovely white flowers and again with its crimson autumn foliage.

The immediate product of this kind of invasion is the characteristic 'oak-birch heath' described from south-east England (Tansley 1911, 1939, Wooldridge & Gold-ring 1953) which is actually open, immature *Quercus-Betula-Deschampsia* woodland, locally dominated in its early stages by patchworks of oak, birch, pine or holly, but showing considerable floristic continuity with the heath in its field layer, *Calluna* and *U. minor* persisting in more open areas, *Pteridium*, *Deschampsia flexuosa* and *V. myrtillus* maintaining high cover even under the deepening shade of the closing canopy. Oak is typically the eventual dominant in this kind of woodland which, even within the supposed natural range of beech (in which the *Calluna-Ulex minor* heath falls), often seems to have a stability of its own, perhaps not irrevocably yielding to the *Fagus-Deschampsia* woodland, even though this is often taken to be the natural climax on base-poor soils in this region. In fact, although young beech can sometimes be found in this community, and though this species can attain dominance by such invasion (as in some of the New Forest woodlands: Peterken & Tubbs 1965, Tubbs 1968), colonisation by *Fagus* is a sporadic and localised event, depending on the occurrence of good mast years in mature trees close to stands of the *Calluna-Ulex minor* heath (Watt 1923, 1924, 1925). Such a line of succession may thus often remain a theoretical pathway.

Distribution

The community occurs from the Poole Harbour area in the west through the New Forest, where stands are particularly numerous and extensive, to Surrey and the High Weald in the east, where they occur in more local and fragmented stretches of heathland.

Affinities

In early descriptive accounts, the series of heaths of drier and intermittently gleyed soils in lowland Britain were grouped together in a compendious *Callunetum*, within which local variation among the sub-shrub associates was informally recognised (Tansley 1911, 1939), in this case on the presence in south-east and central southern England of *U. minor*. More systematically, this community can be regarded as the least extreme of the range of heaths that replaces the *Calluna-Festuca* heath in moving into the more oceanic southern and western seaboard of Britain. The presence of *E. cinerea* helps locate this kind of vegetation generally among the eu-oceanic heaths of north-west Europe; the occurrence of *U. minor* places it more precisely with the gorse heath of southern Britain and the Atlantic fringe of France and the Iberian peninsula (Böcher 1943). Phytosociologi-cally, such communities have been grouped in one or more distinct alliances, *U. minor* and *U. gallii* sometimes being grouped together as character species, as in Bridgewater's Ulicion (in Gimingham 1972), sometimes separated, as in the split between the Ulicion nanae Duvigneaud 1944 and the Ulicion gallii des Abbayes & Corillion 1949. Whether the striking vicarism between these two gorse species is recognised at alliance level or not, the *Calluna-Ulex minor* heath is certainly best separated from the *Calluna-Ulex gallii* heath, which replaces it on similar soils beyond south Dorset, into the South-West Peninsula, Wales, the southern Pennines and round into the more oceanic parts of the Norfolk coast. In contrast to their continental equivalents, *U. europaeus* does not figure prominently in these vege-tation types, perhaps because they have been longer established on very impoverished soils, though transi-tions to the Ulici-Cytision *Ulex-Rubus* scrub can be seen on disturbed ground.

Towards the drier parts of its range, the *Calluna-Ulex minor* heath shows quite a well-defined, edaphically-related boundary with wetter Ericetalia heath, based on the almost totally mutual exclusion of *E. cinerea* and *E. tetralix*. Increasingly to the west, this floristic junction becomes less clear, with the *Ulex minor-Agrostis curtisii* heath occupying an intermediate position.

Floristic table H2

	a	b	c	2
Calluna vulgaris	V (7–10)	V (7–10)	V (6–9)	V (6–10)
Ulex minor	V (2–7)	V (2–7)	V (2–5)	V (2–7)
Erica cinerea	IV (2–7)	V (3–8)	IV (3–6)	V (2–8)
Deschampsia flexuosa	IV (2–5)	V (1–5)	II (2–3)	IV (1–5)
Dicranum scoparium	II (1–4)	II (1–3)		II (1–4)
Hypnum jutlandicum	II (2–5)	II (1–3)		II (1–5)
Festuca rubra	II (2–3)	I (2)		I (2–3)
Cladonia fimbriata	II (2–3)			I (2–3)
Cladonia coccifera	I (1–2)			I (1–2)
Campylopus paradoxus	I (1–2)			I (1–2)

Cladonia chlorophaea	I (1–3)		I (1–3)
Cladonia arbuscula	I (1–3)		I (1–3)
Polytrichum juniperinum	I (2–3)		I (2–3)
Pteridium aquilinum	II (1–2)	IV (2–7)	II (2–3)	III (1–7)
Vaccinium myrtillus		V (2–7)		II (2–7)
Quercus spp. seedling	I (2)	III (1–2)		II (1–2)
Betula spp. seedling	I (1–3)	III (1–2)		II (1–3)
Pinus sylvestris seedling		I (1–2)		I (1–2)
Sorbus aucuparia sapling		I (2)		I (2)
Rubus fruticosus agg.		I (2)		I (2)
Molinia caerulea	II (2–7)	III (1–5)	V (3–6)	III (1–7)
Erica tetralix	I (2)		III (3–5)	I (2–5)
Genista anglica		I (2)	I (2–3)	I (2–3)
Agrostis curtisii	I (2–3)	I (3–5)	I (3)	I (2–5)
Ulex europaeus	I (2–5)	I (5)	I (3–6)	I (2–6)
Cuscuta epithymum	I (2)	I (4)	I (4)	I (2–4)
Cladonia furcata	I (2–7)	I (2)	I (2–5)	I (2–7)
Festuca ovina	I (2)	I (1)	I (3)	I (1–3)
Potentilla erecta	I (3)	I (2)	I (2)	I (2–3)
Agrostis canina montana	I (2)	I (2)		I (2)
Cladonia impexa	I (2–6)	I (2)		I (2–6)
Polytrichum piliferum	I (2–3)	I (1)		I (1–3)
Kurzia sp.	I (1)		I (2)	I (1–2)
Ulex gallii		I (2)	I (3)	I (2–3)
Number of samples	22	9	8	38
Number of species/sample	8 (4–13)	11 (8–16)	7 (6–10)	8 (4–16)
Vegetation height (cm)	49 (20–80)	23 (15–30)	53 (15–75)	47 (15–80)
Shrub/herb cover (%)	96 (80–100)	97 (70–100)	98 (90–100)	97 (70–100)
Ground cover (%)	9 (0–60)	8 (0–30)	3 (0–20)	8 (0–60)
Altitude (m)	76 (5–220)	138 (120–215)	99 (30–168)	93 (5–220)
Slope (°)	6 (0–30)	4 (0–25)	0	4 (0–30)

a   Typical sub-community

b   *Vaccinium myrtillus* sub-community

c   *Molinia caerulea* sub-community

2   *Calluna vulgaris-Ulex minor* heath (total)

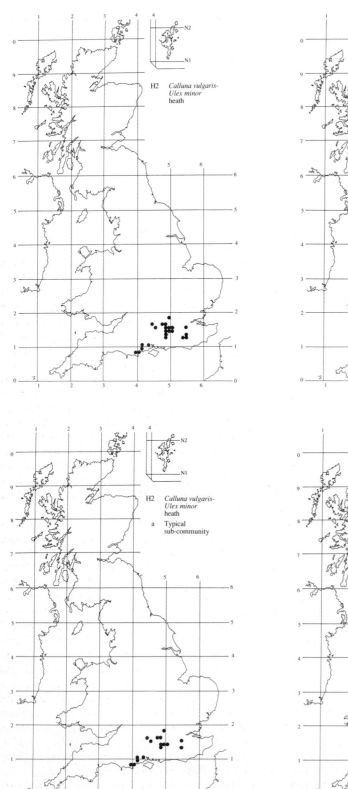

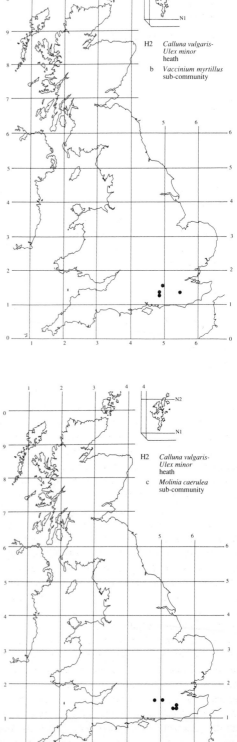

H3
Ulex minor-Agrostis curtisii heath

Synonymy
Callunetum arenosum Tansley 1911 *p.p.*; *Callunetum arenicolum* Tansley 1939; *Agrostis setacea-Ulex minor* heath Ivimey-Cook 1959; *Agrosto setaceae-Ulicetum minoris* Bridgewater 1970; *Calluna-Ulex minor* heaths Gimingham 1972 *p.p.*

Constant species
Agrostis curtisii, Calluna vulgaris, Erica cinerea, E. tetralix, Molinia caerulea, Ulex minor.

Rare species
Agrostis curtisii, Erica ciliaris, Viola lactea.

Physiognomy
The *Ulex minor-Agrostis curtisii* heath contains virtually all the sub-shrub vegetation in which these two species occur together as important components. Throughout the eastern part of its range, and particularly on some of the Surrey commons, *A. curtisii* can figure occasionally in the *Calluna-Ulex minor* heath, in which mixtures of *Calluna, U. minor* and *Erica cinerea* generally predominate but, in the present community, it is a consistent feature and has, as additional constant associates, *Erica tetralix* and *Molinia caerulea*. Even when there are few other companions, therefore, and this is typically not a species-rich kind of vegetation, the cover tends to be a little more diverse than in the *Calluna-Ulex minor* heath, particularly as many stands still experience some burning and grazing, treatments which affect both the floristics and structure of the community.

Often, therefore, the sub-shrub canopy in this kind of heath is fairly low, usually 2–3 dm tall, and though *Calluna* quite frequently dominates, especially in stands which have not been burned for some time, degenerate and leggy heather is not so common as in some other, more neglected, heath communities. With its growth kept in check somewhat, the other sub-shrubs maintain a more or less consistent contribution throughout, though their proportions are very variable from stand to stand. Compared with less oceanic heaths, the most unusual feature of the woody cover is the occurrence together of *Erica cinerea* and *E. tetralix*, species which, in areas of less equable climate, are fairly rigidly partitioned into dry- and wet-heath vegetation respectively. Both can grow vigorously here, although *E. cinerea* is the more abundant species overall and is especially likely to become prominent after burning or on disturbed ground where, provided the soil is reasonably dry, it can temporarily outstrip *Calluna* in its regenerative ability. But *E. tetralix* can also have high cover locally, particularly where the community extends on to somewhat more strongly-gleyed soils, where an increase in its abundance can mark the beginning of a transition to the *Ericetum tetralicis* wet heath. In such slightly wetter situations on some of the south Dorset heaths, in the area around Poole Harbour, the community also provides a locus for the nationally rare *E. ciliaris*, accompanying or sometimes replacing *E. tetralix* and hybridising with it (Chapman 1975).

Into this kind of vegetation, *U. minor* maintains its high frequency throughout, though its abundance is very variable. It can show prolific sprouting from surviving stools after burning, remaining prominent for some time before being overtaken by the ericoids, and even then being able to persist as a second tier to the vegetation by virtue of its moderate shade-tolerance; whereas, with heavy grazing, it can be much reduced or perhaps even eliminated. But its constancy overall provides an important distinction between the community and the *Ulex gallii-Agrostis curtisii* heath which replaces it to the west of Poole Harbour and which differs most obviously in the replacement of one gorse by the other. By definition, then, *U. gallii* is very scarce here and the sharp vicarism between the two species means that they can actually occur together in this type of heath only in a very few localities in south Dorset (Proctor 1965). *U. europaeus*, by contrast, can be quite frequent, though it is often obviously associated with disturbed areas, when it can show some local abundance.

In stands which have not been burned for some time, the two constant grasses of the community, *A. curtisii* and *Molinia*, are generally found as scattered tufts or tussocks among the regenerating sub-shrub canopy, thickening up locally as more extensive persistent clumps. But both species, and especially in this community, *A. curtisii*, can be much more abundant: this grass can seed prolifically on to nearby newly-burned ground or otherwise disturbed areas and rapidly form more or less continuous stretches of dense wiry tussocks among which the sub-shrubs have but a sparse representation, as in the *Agrostis* sub-community (Ivimey-Cook 1959). *Molinia*, too, can become locally prominent after fire by regrowth of surviving tussocks, but overall it tends to be the subordinate species; unlike *A. curtisii*, however, it often extends out from this community on to more consistently waterlogged ground occupied by the *Ericetum tetralicis*. In contrast to the *Calluna-Ulex minor* heath, *Deschampsia flexuosa* is very scarce here and, indeed, no other grasses occur at all frequently, though *Agrostis capillaris*, *Festuca ovina* and *Danthonia decumbens* can be found on occasion.

Other frequent vascular species, too, are very few in number. *Pteridium aquilinum* occurs occasionally but, though it is often found in close association with the community in dense stands of the *Pteridium-Galium* vegetation, it is typically of sparse cover within the heath itself. Other herbs are also usually found as scattered individuals: there is sometimes a little *Potentilla erecta*, *Polygala serpyllifolia* or *Carex pilulifera* and *Cuscuta epithymum* can occasionally be found growing parasitically on a variety of hosts (Tansley 1939). Then, in disturbed situations, where fire or trampling have opened up the cover of vegetation, a very characteristic plant is *Viola lactea*. The frequency of burning and grazing, however, means that seedlings and saplings of trees are scarce: young birch, *Quercus robur* or *Pinus sylvestris* can sometimes be found but they rarely get away.

The pattern of burning also has a major influence on the richness and diversity of the ground flora which, in general, is not very prominent here, though markedly better developed when the cover of sub-shrubs is destroyed by fire or where, with the natural degeneration of the heather in older neglected stands, the canopy opens up somewhat. Among the bryophytes, *Campylopus brevipilus* is perhaps one of the most distinctive species here and it can be accompanied by *C. paradoxus*, *Polytrichum juniperinum*, *Dicranum scoparium*, *Hypnum jutlandicum* and *Leucobryum glaucum*. Lichens, too, can become conspicuous in such situations, particularly *Cladonia impexa* and peat-encrusting species such as *C. floerkeana*, *C. coccifera* and *C. pyxidata*. Old heather stems are often colonised by *Hypogymnia physodes*.

Sub-communities

Typical sub-community. In this kind of *Ulex minor-Agrostis* heath, the sub-shrubs typically form an extensive canopy, often with *Calluna* as an overwhelming dominant, sometimes with a more mixed cover, though with the grasses usually subordinate. *Ulex europaeus* figures occasionally and *Erica ciliaris* can be found in south Dorset vegetation which is otherwise little different apart from a reduction in the vigour of *Erica cinerea*. Other species are few although *Potentilla erecta* is preferential at low frequency and there is sometimes a little *Pteridium*, *Carex pilulifera*, *Polygala serpyllifolia* or *Cuscuta epithymum*. Towards the wettest limit for this kind of vegetation, *Schoenus nigricans* has been recorded. Bryophytes and lichens are noticeably sparse, with just very occasional *Hypnum jutlandicum*, *Leucobryum* and *Cladonia* spp.

Cladonia spp. sub-community. Here, the sub-shrub canopy is a little more open than in the above and, although *Calluna* is often the leading species, dominance is frequently shared between the woody plants and the grasses. *Polygala serpyllifolia* is quite strongly preferential but equally noticeable is the patchy cover of bryophytes and lichens on exposed areas of litter and mor. Among the former, *Campylopus brevipilus* and *Polytrichum juniperinum* are the most frequent species with *Cladonia impexa*, *C. floerkeana*, *C. coccifera* and *C. pyxidata* common among the latter.

Agrostis curtisii sub-community. This sub-community has the most strikingly different kind of vascular cover, with *A. curtisii* very abundant and *E. cinerea* often co-dominant, but with *U. minor* and *Molinia* somewhat patchy and *Calluna* and *E. tetralix* much reduced in both frequency and cover. *U. europaeus* is strongly preferential, though not often abundant, and, although the vegetation cover is often high, the early stages of the development of this sub-community often allow opportunity for colonisation by *Viola lactea*, scattered plants of which persist as the dominants expand.

Habitat

The *Ulex minor-Agrostis* heath is the characteristic sub-shrub community of impoverished acid soils which are protected against parching by a measure of drainage impedence and a moderately oceanic climate. This combination of environmental conditions is reflected in the general floristics of the community though, within its relatively small geographical range, burning and grazing still often exert an important measure of control on its composition and structure.

This kind of heath is more or less confined to the southern parts of Hampshire and Dorset, a region which shows a further shift in climatic conditions, compared with the High Weald and Thames basin, towards the warm oceanic environment characteristic of the far south-west. Annual rainfall, at between 800 and 1000 mm, is very much as over the whole of central southern and south-eastern England, and fairly well distributed throughout the year, though with a distinct winter peak (*Climatological Atlas* 1952, Chandler & Gregory 1976). Annual accumulated temperatures, too, are of much the same order as through the whole of southern Britain but the winters are noticeably mild, with February minima at least 1.5 °C above freezing and, in sites close to the coast, winter accumulated temperatures (December–March) above freezing (Page 1982).

Differences in sensitivity to winter cold may play some part in determining the distribution of the two gorses, *U. minor* and *U. gallii*, over this part of Britain (Proctor 1965): the general boundary between these species, running roughly from Salisbury to Dorchester, forms the western limit to this community, beyond which it is replaced by the *Ulex gallii-Agrostis curtisii* heath. The *Ulex minor-Agrostis* heath is thus one of two sub-shrub communities, the other being the *Calluna-Ulex minor* heath, that lie between the more oceanic *U. gallii* heaths of the south-west and the continental *Calluna-Festuca* heath of East Anglia. But the really distinctive feature here is the occurrence, together with *U. minor* and the more broadly oceanic *E. cinerea*, which are found together in both these heaths, of *E. tetralix* and *Molinia* and the more localised Oceanic West European grass, *A. curtisii*: this particular characteristic is related to climatic and soil conditions working together.

E. tetralix, like *E. cinerea*, retains a measure of physiological activity in winter and shows a broadly similar national distribution, becoming markedly infrequent towards the more continental east of Britain (Bannister 1965, 1966, Gimingham 1972). And, more particularly, the drier the climate, the greater the tendency for these two species to show a sharp edaphic separation, *E. cinerea* being confined to dry heaths of free-draining acidic soils, *E. tetralix* to wet heaths on strongly-impeded profiles, a relationship which probably reflects competitive interactions between the two, and with *Calluna* an associate throughout the edaphic range (Bannister 1964*d*, Gimingham 1972). With increasing oceanicity towards south-western Britain, not only do both species become more frequent but this exclusivity tends to break down somewhat, the wetter climate helping to maintain even free-draining acid soils in a rather moister state.

But edaphic conditions themselves, and topography, contribute to the development of this intermediate kind of heath habitat. In this central southern part of Britain, the community is one of a suite of heaths developed over base-poor and oligotrophic soils, which are extensively represented through the Hampshire Basin and around Poole Harbour on Tertiary sands and gravels and superimposed Plateau Gravels. Over the range of profiles developed from these parent materials, the *Ulex minor-Agrostis* heath occupies the middle range between excessively-draining humo-ferric podzols, typically supporting the *Calluna-Ulex minor* heath, and those mineral soils which are seasonally waterlogged to the surface, which have the *Ericetum tetralicis*. Between these extremes, the soils are often kept moist above, particularly in winter, by a measure of drainage impedence due to the development of an impervious B_{Fe} pan or to the presence of much fine fraction material in an argillic B horizon. Over the typically gently undulating topography of the terraces of the Hampshire/Poole basin, such intergrade stagnogley-podzols, gley-podzols and palaeo-argillic podzols are widespread on areas with a perched water-table and over the gently-sloping surrounds to wetter hollows (*Soil Survey* 1983, Jarvis *et al*. 1984). Here, the soil surface is not so consistently wet and reducing as to prohibit germination and growth of *E. cinerea* (Bannister 1964*a*, *d*, 1965, Jones & Etherington 1970, Jones 1971*a*, *b*), but sufficiently inhibiting of its vigour, and that of *Calluna*, as to allow *E. tetralix* to maintain itself (Rutter 1955, Bannister 1966, Gimingham 1972).

The combination of reasonably high levels of soil moisture with good aeration is also important for the two characteristic grasses of the community. *Molinia* tends to follow *E. tetralix* in its transgression into somewhat drier heaths in south-west Britain, though its vigour here, as in the *Ericetum*, is often hindered by the generally poor trophic state of the soils (Loach 1966, 1968*a*, *b*, Sheikh & Rutter 1969, Sheikh 1969*a*, *b*, 1970). *A. curtisii* is a more restricted species, both geographically and ecologically, than *Molinia*, but its limits of growth seem to be partly set by a balance of soil and climatic moisture that is well met on these kinds of profile in south-west Britain (Ivimey-Cook 1959) and this type of heath represents one of its major loci, particularly towards its eastern British limit. For *Deschampsia flexuosa*, on the other hand, conditions are clearly inimical: over the sequence of soils, the appearance of this grass typically marks the transition to the *Calluna-Ulex minor* heath on free-draining podzolic profiles, and soil moisture may be a critical factor in limiting its growth here.

Climatic and edaphic conditions thus play a major role in determining the general floristics of the community and its demarcation from closely-related heaths, but the appearance of any particular tract is often affected by treatment because a large proportion of the stands occurs within the New Forest where burning and grazing of the heaths are still practised. Burning is carried out here by the Forestry Commission on rotations of

6–12 years, to effect some control on accidental fires close to the woodlands and to regenerate a palatable bite for the cattle and ponies that have access to the unenclosed parts of the Forest (Tubbs 1968). Its general effect is to curtail the mature and degenerate phases of the *Calluna* growth cycle (Watt 1955) and, with grazing, to set back repeatedly any invasion of trees and seral progression to woodland. More particularly, burning helps maintain heterogeneity among the vascular component by providing occasional opportunities for the temporary local dominance of sub-shrubs or grasses before *Calluna* once more exerts its general pre-eminence. Among the former here, both *U. minor*, whose stools often survive fires to sprout vigorously afterwards, and *E. cinerea*, which also regenerates prolifically from seed on somewhat drier soils, can outstrip *Calluna* in the early years following burning (Gimingham 1972). Of the grasses, *Molinia* tussocks often survive less intense fires and show healthy regrowth with the flush of fertilising ash, but *A. curtisii* frequently shows the more dramatic response, with its ability to seed into open ground in great profusion (Ivimey-Cook 1959). Burning also influences the contribution of the ground flora because, though it pre-empts natural collapse of the older heather bushes which exposes mor and decaying branches for colonisation by bryophytes and lichens, it repeatedly creates areas of burned litter, humus and raw mineral material for invasion. Pleurocarpous mosses, like *Hypnum cupressiforme s.l.*, tend to be less conspicuous in this kind of heath, whereas acrocarpous invaders, such as *Campylopus brevipilus* and *Polytrichum juniperinum*, and *Cladonia* spp., particularly peat-encrusters, are frequent and locally prominent, sometimes for considerable periods before regrowth of the vascular plants. Floristic differences of these kinds are what separate the Typical and *Cladonia* sub-communities and, though no heather-ageing was carried out, it is highly likely that the former, with its more consistent dominance of *Calluna* and impoverished cryptogam flora, includes older stands, the latter those more recently burned.

Grazing interacts with the effects of burning, although very little systematic information is available as to how this works in this kind of heath. The most nutritious bite in this vegetation is provided by *Molinia* and, after fires, which are almost always in March where burning is deliberate, the speedy regrowth of this grass can offer valuable herbage. *A. curtisii* sprouts a little later from surviving tussocks but, although new growth or seedlings may be eaten, the tough wiry shoots of older plants may be relatively unpalatable (Ivimey-Cook 1959). Among the sub-shrubs, *U. minor* may be the most sensitive species and its relative scarcity in some areas could be related to heavy grazing. And, of course, herbivores are of major importance in cropping any tree seedlings that get a hold on areas of open ground exposed by fires.

It is likely that the striking abundance of *A. curtisii* and *E. cinerea* in the *Agrostis* sub-community is also sometimes related to the rapid colonisation of burned areas, but this kind of *Ulex minor-Agrostis* heath is also found where there has been physical disturbance of the soils, as along trackways, roadsides and railway embankments, and over the spoil from gravel and clay workings: in some areas, the underlying deposits contain materials of considerable economic value. Disruption of the existing vegetation cover here also provides ideal conditions for the appearance of *Viola lactea*, a species very intolerant of competition for its establishment and now largely confined to disturbed, acid soils within stretches of heathland in south-west England (Moore 1958). Its seeds may be dispersed by ants, an agent of some importance, too, for *U. europaeus* which is also preferential and whose frequency perhaps denotes some modest enrichment of the soils with disturbance. By and large, the profiles under the *Ulex minor-Agrostis* heath are impoverished, though perhaps not so starkly as beneath the *Calluna-Ulex minor* heath, particularly where they are argillic below, and aeration of the surface mor and disruption of the horizons in this sub-community may release some nutrients.

Zonation and succession

The *Ulex minor-Agrostis* heath is typically found as part of soil-related sequences in which the degree of drainage impedence is the most important governing factor. Burning and grazing can modify these sequences and their intensity ultimately controls succession to woodland. Although such treatments continue in many stands, much heath vegetation of this kind has been lost by abandonment of traditional land use and soil improvement for agriculture or forestry so surviving tracts can be much fragmented and sharply delineated from their surrounds.

Over more extensive stretches of heathland vegetation in central southern England, as in the New Forest where burning and grazing still maintain suites of sub-shrub communities over mosaics of different soil types, the *Ulex minor-Agrostis* heath occupies a distinct position on ground that is too dry for the *Ericetum tetralicis* wet heath and too moist for the *Calluna-Ulex minor* dry heath. In the former direction, an increased tendency for seasonal surface waterlogging in mineral soils is marked by the virtual extinction of *U. minor*, *E. cinerea* and *A. curtisii*, except in places where slight soligenous influence ameliorates the lack of aeration. *Calluna*, *Molinia* and *E. tetralix* maintain their frequency with the passage to the *Ericetum*, the last species especially benefiting from reduced competition from other sub-shrubs. On these moister soils, too, species such as *Scirpus cespitosus* and *Narthecium ossifragum*, which are very rare in the *Ulex minor-Agrostis* heath, begin to make a frequent appearance with *Sphagnum compactum* and *S. tenellum* becoming common in the ground layer. Such transitions as this are best seen around the margins of the elongated hollows that have been eroded into the sands and gravels of this part of Britain, the zonation sometimes continuing downslope over permanently-waterlogged ground with accumulating peat into the *Narthecio-Sphagnetum* valley bog. This is the kind of classic sequence described by Rose (1953) and Newbould (1960) and well shown around Cranesmoor, Denny, Wilverley and Holmsley bogs in the New Forest (Ratcliffe 1977).

In the other direction, the move from gleyed podzols of various kinds on to sharply-draining humo-ferric podzols sees a disappearance from the vegetation of *E. tetralix*, *Molinia* and *A. curtisii*, with *Calluna*, *E. cinerea* and *U. minor* continuing and, together with *Deschampsia flexuosa*, making up the bulk of the cover of the *Calluna-U. minor* heath which terminates the sequence developed in relation to soil moisture. Away from the sharply-defined valley-mire hollows, the vegetation pattern may be less well ordered and incomplete or with the drier or wetter elements prevailing according to local topography and hydrology. Over the landscape of the New Forest, for example, dry heath predominates over the well-drained slopes of Tertiary sands in the higher north-west with wetter heath restricted to ill-drained fragments of Plateau Gravel terraces; in the lower southern part of the area, damp heath is more abundant though most of the terraces are here better drained and have dry heath (Lambert & Manners 1964, Fisher 1975a, b). And the pattern may be further complicated by the local pre-eminence of other edaphic factors where parent materials and soils change. Well-drained acid brown earths, for example, weathered from sandy or clay-sandy substrates, often have patches of the *Pteridium-Galium* community, and more fertile or disturbed ground, the *Ulex-Rubus* scrub, to which the *Agrostis* sub-community may grade, in the New Forest gorse 'brakes'.

Although this part of Britain is quite exceptional in the extent to which it still retains large areas of these vegetation types disposed in extensive zonations, essentially the same mixture of communities is to be seen in the other major centre for the *Ulex minor-Agrostis* heath, around Poole Harbour. Here, on the Hartland–Arne and Studland–Godlingston heaths, originally continuous stretches of sub-shrub vegetation have been much fragmented (Moore 1962, Webb 1986), though there are the additional features of the presence of *E. ciliaris* in this and other communities and the striking juxtaposition of the heaths with open fresh-water and maritime vegetation types.

In the long term, burning and grazing, which played a vital part in the development of this kind of open heath landscape, perhaps as early as the Mesolithic and certainly by the Bronze Age (e.g. Dimbleby 1962, Tubbs

1968, Haskins 1978, Webb 1986), are essential for its maintenance against succession to woodland. These treatments can confuse the soil-related boundaries between the heath types, as where burned or grazed areas cut across the major line of variation and favour the temporary expansion of some species common to more than one of the communities, such as *E. cinerea* or *Molinia*, or the more general long-term predominance of *Calluna* throughout. They can also mediate transitions to grasslands dominated by *Molinia* or *A. curtisii*, which can develop a stability of their own (e.g. Tubbs 1968). But their ultimate effect is repeatedly to set back invasion of trees, most characteristically on these soils both species of birch, *Quercus robur* and *Pinus sylvestris*, the last often abundantly represented in plantations established among stands of the community. Succession has not been followed but is likely to result in the development of drier *Betula-Molinia* woodland, usually dominated by poorly-grown birch. Under the rather open canopy typical of this community, *Molinia* persists in abundance with *A. curtisii* less shade-tolerant, and sub-shrubs usually confined to gaps.

Distribution

The *Ulex minor-Agrostis* heath is confined to south Dorset and Hampshire.

Affinities

In general terms, this community falls within the more southerly group of Böcher's (1943) eu-oceanic *E. cinerea* heaths in which one or other of the gorse species, *U. minor* or *U. gallii*, makes an important contribution. On one view, it can be placed with the *Calluna-U. minor* heath as part of the sub-shrub vegetation characteristic of the moderately oceanic area between the Poole basin and the High Weald, as in the treatments of Tansley (1911, 1939), who distinguished this kind of heath within a broadly-defined *Callunetum*, and Gimingham (1972), who recognised a more narrow vegetation type with *Calluna* and *U. minor*. Phytosociologically, such communities can be located in the Ulicion nanae Duvigneaud 1944, marked off from the Ulicion gallii Des Abbayes & Corillion 1949 at the sharply-defined line of geographical separation of the two gorse species, a vicarism maintained throughout western Europe.

In another direction, the *Ulex minor-Agrostis* heath has floristic affinities which bridge this gap. West of Poole, it is replaced by the *Ulex gallii-Agrostis* heath, but the two share the constancy of *E. tetralix* and *Molinia*, now beginning to transgress on to more free-draining soils with the increasing oceanicity of the climate, and thus bringing the vegetation a little closer to the Ericion tetralicis wet heaths, and also of the geographically and ecologically more restricted *A. curtisii* (Ivimey-Cook 1959).

Floristic table H3

	a	b	c	3
Ulex minor	V (3–7)	V (3–6)	V (1–5)	V (1–7)
Agrostis curtisii	V (3–8)	V (3–6)	V (8–9)	V (3–9)
Molinia caerulea	V (2–7)	V (1–7)	IV (1–5)	V (1–7)
Erica cinerea	V (1–8)	V (2–6)	V (5–9)	V (1–9)
Calluna vulgaris	V (2–10)	V (4–8)	II (2–4)	IV (2–10)
Erica tetralix	V (2–8)	V (1–6)	II (1–2)	IV (1–8)
Potentilla erecta	II (2–4)	I (2)	I (1)	I (1–4)
Erica ciliaris	II (5–10)	I (2)		I (2–10)
Betula spp. sapling	II (1–5)			I (1–5)
Cuscuta epithymum	I (3–5)			I (3–5)
Zygogonium ericetorum	I (2–3)			I (2–3)
Danthonia decumbens	I (2–3)			I (2–3)
Hypericum pulchrum	I (2–3)			I (2–3)
Schoenus nigricans	I (3–7)			I (3–7)
Campylopus brevipilus	I (2–3)	III (2–4)		II (2–4)
Cladonia impexa	I (1–7)	III (1–4)		II (1–7)
Cladonia floerkeana	I (2–3)	III (1–3)		II (1–3)
Polygala serpyllifolia	I (2)	III (1–3)	I (2)	II (1–3)
Polytrichum juniperinum	I (2)	II (1–2)		I (1–2)
Kurzia sp.		II (1–3)		I (1–3)
Cladonia coccifera		II (1–3)		I (1–3)
Cladonia pyxidata		II (1–2)		I (1–2)
Campylopus paradoxus		I (2–3)		I (2–3)
Cladonia crispata		I (1–3)		I (1–3)
Ulex europaeus	II (2–10)	II (1–5)	V (1–4)	III (1–10)
Viola lactea	I (3)		III (1–3)	I (1–3)
Pteridium aquilinum	I (2–3)	I (1–5)	I (1–4)	I (1–5)
Carex pilulifera	I (2–3)	I (2)	I (1)	I (1–3)
Hypnum jutlandicum	I (2–8)	I (1–7)		I (1–8)
Leucobryum glaucum	I (2–3)	I (5)		I (2–5)
Hypogymnia physodes	I (1–2)	I (2)		I (1–2)
Cladonia uncialis	I (2)	I (2–4)		I (2–4)
Cladonia arbuscula	I (2)	I (1)		I (1–2)
Quercus robur sapling	I (2)		I (1)	I (1–2)
Number of samples	26	18	9	53
Number of species/sample	9 (6–13)	11 (8–17)	7 (6–9)	9 (6–17)
Vegetation height (cm)	30 (25–40)	25 (20–35)	30	28 (20–40)
Shrub/herb cover (%)	99 (85–100)	92 (60–100)	100	96 (60–100)
Ground cover (%)	8 (0–70)	11 (1–85)	1 (0–1)	8 (0–85)
Altitude (m)	43 (5–180)	57 (25–150)	28 (25–30)	46 (5–180)
Slope (°)	4 (0–20)	6 (0–20)	0	3 (0–20)

a   Typical sub-community
b   *Cladonia* spp. sub-community
c   *Agrostis curtisii* sub-community
3   *Ulex minor-Agrostis curtisii* heath (total)

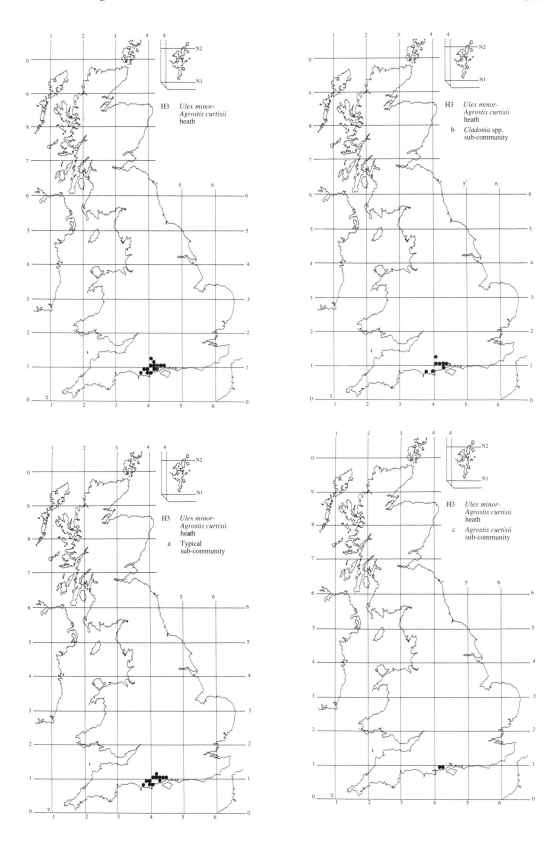

H3 *Ulex minor-Agrostis curtisii* heath

H3 *Ulex minor-Agrostis curtisii* heath
b *Cladonia* spp. sub-community

H3 *Ulex minor-Agrostis curtisii* heath
a Typical sub-community

H3 *Ulex minor-Agrostis curtisii* heath
c *Agrostis curtisii* sub-community

Synonymy

Callunetum arenosum western heaths Tansley 1911 *p.p.*; Somerset upland heath Watson 1932 *p.p.*; *Callunetum* south-western heaths Tansley 1939 *p.p.*; *Agrostis setacea* 'Short Heath' Coombe & Frost 1956a, Marrs & Proctor 1978; *Agrostidetum setaceae cornubiense* Coombe & Frost 1956a, Ivimey-Cook 1959, Ivimey-Cook *et al.* 1975; *Erica vagans-Erica tetralix* 'Wet Heath' Coombe & Frost 1956a; *Agrosto setaceae-Ulicetum gallii* Bridgewater 1970; Heath, Grassland with Gorse & Grass-heath Ward *et al.* 1972a, all *p.p.*; *Calluna-Ulex gallii* heaths Gimingham 1972 *p.p.*; Intermediate Tall/Short Heath Marrs & Proctor 1978; Species-poor *Erica cinerea* Heath NCC Devon Heathlands Report 1980; Species-poor Dry *Agrostis setacea* Heath NCC Devon Heathlands Report 1980; *Ulex europaeus-Molinia caerulea* limestone heath NCC South Gower Coast Report 1981 *p.p.*; *Agrostis curtisii* heath Hopkins 1983.

Constant species

Agrostis curtisii, Calluna vulgaris, Erica cinerea, E. tetralix, Molinia caerulea, Potentilla erecta, Ulex gallii.

Rare species

Agrostis curtisii, Carex montana, Erica ciliaris, E. vagans.

Physiognomy

In its general floristics, the *Ulex gallii-Agrostis curtisii* heath is very similar to the *Ulex minor-Agrostis* heath, the major difference being the replacement of the one gorse by the other. Without close inspection of the petals and calyx, the different lengths of which are diagnostic, separation of these two genetically variable and vegetatively plastic species can be difficult (Proctor 1965, Wigginton & Graham 1981), but they show an almost perfect vicarism in their range: they do not occur together here, the western limit of *U. minor* in east Dorset serving as a boundary between the two communities. Apart from this, however, the heaths share five constants, *Calluna vulgaris, Erica cinerea, E. tetralix, Molinia caerulea* and *Agrostis curtisii* and, together with *U. gallii*, these generally account for the bulk of the vascular cover. But their proportions and structural arrangements are very variable so the gross appearance of the vegetation can differ quite markedly from stand to stand, something reflected in the use of epithets such as 'short heath' (Coombe & Frost 1956a) and 'grass heath' (Ward *et al.* 1972a). Quite often, sub-shrubs and grasses comprise an intimately mixed canopy of continuous cover, compact and springy and quite low, sometimes little more than 1 dm high; in other cases, the elements may be of similar short stature but disposed in a more obvious mosaic with clumps of sub-shrubs separated by small stretches of sward in which grasses predominate. Then, again, some stands have a much taller canopy of woody species, half a metre or so high, with the other vascular associates relegated to a patchy understorey. And there can be quite extensive areas of barer ground breaking up the cover into discrete islands. Some of these differences are fairly consistent and, accompanied by the presence of preferential species, help to distinguish the various sub-communities. Others, particularly those related to burning and grazing, which are quite common here but often not practised very systematically, can be more disordered or ephemeral.

Of the common sub-shrubs, *Calluna* and *U. gallii* are the most consistently represented, and both can be quite abundant. Indeed, when there is a dominant species here, it is most frequently *Calluna*, the vigour of which is especially noticeable several years after a fire, and whose abundance is favoured in the long term by a regime of more regular burning, such that in areas like the Dartmoor fringes with its tradition of swaling, this community can make some contribution to vegetation which has been grouped within a general *Callunetum* (e.g. Tansley 1911, 1939) or heather moor (Harvey & St. Leger-Gordon 1974). *Calluna* can dominate, too, as leggy bushes, in stands that have been long unburned and ungrazed, whereas, in the immediate aftermath of a fire or under heavy herbivore pressure, it can be much reduced. On moister soils occupied by this kind of heath, its contribution is also less important. *U. gallii* is likewise very variable in its cover, though it is only exceptionally a dominant and, in the mixed grassy-heath kind of canopy, where it is often kept in check by grazing, its low shoots may be more perceptible to the ankle than to the eye, at least outside its late summer and autumn flowering season. After burning, however, it can show a considerable increase in cover provided herbivores do not continually nibble off the soft young shoots sprouting from the stools and eat any seedlings. *U. europaeus*, by contrast, is very scarce in this community and almost always in disturbed places.

As in the *Ulex minor-Agrostis* heath, the frequent occurrence together here of *Erica cinerea* and *E. tetralix* provides one distinctive difference between this vegetation and the corresponding dry heath of the region. In fact, neither is as frequent overall as *Calluna* or *U. gallii* and in some sub-communities one or the other can be markedly patchy in its occurrence. There is a reciprocal element in this behaviour, *E. cinerea* being less abundant on moister soils and less sunny aspects (as in the *Erica tetralix* and *Scirpus* sub-communities), *E. tetralix* being less competitive on drier ground, but superimposed on this is the better response of the former to burning, after which it can be dominant for some time before *Calluna* supervenes (a feature best seen here in the *Erica cinerea-Agrostis* sub-community). In fact, quite often, more or less equal amounts of the two species occur intermingled, and both can show a general reduction in relation to grasses in stands which are heavily grazed (as in the *Festuca* sub-community).

Four other sub-shrubs are of more restricted occurrence in this kind of heath, though they are preferential or differential for particular sub-communities and can give the vegetation a distinctive appearance. The first is *Vaccinium myrtillus*, becoming generally more common in this part of Britain with its wetter regional climate, and especially likely to be encountered here at higher altitudes where rainfall is locally greater. Such situations are better represented in the *Festuca* and *Scirpus* sub-communities, though in both these kinds of heath grazing can reduce the cover of this palatable sub-shrub to very sparse scattered shoots. Then, stands of the wetter *Erica tetralix* sub-community can have some *Salix repens*, though it is rarely abundant.

The other two species, the nationally-rare *Erica ciliaris* and *E. vagans*, are very local in occurrence, though often more obvious than either of the above when present. Both tend to favour situations near the limit of soil moisture tolerated by this vegetation, found in the *Erica tetralix* sub-community. For *E. ciliaris*, this kind of heath, together with the *Ericetum tetralicis* which replaces it on more strongly waterlogged ground, provides the locus for its Devon and Cornish occur-

rences and it can occur in abundance here, accompanying or replacing *E. tetralix*. *E. vagans* can show similar prominence on the Lizard, in Cornwall, to which it is strictly confined in mainland Britain, in *Ulex gallii-Agrostis* heath that is transitional, floristically and on the ground, to the *Erica vagans-Schoenus* heath, characteristic of more base-rich gleys and one of the two communities in which this ericoid plays a major role on this headland.

Even in stands where sub-shrubs are more obviously dominant, the two community grasses, *A. curtisii* and *Molinia* almost always make some contribution to the vascular cover and quite often they rival or exceed the woody plants in abundance. *A. curtisii*, in particular, is an important plant here and the *Ulex gallii-Agrostis* heath provides its major locus throughout its range west of Poole Harbour. It can become especially abundant after burning (as in the *Agrostis-Erica* sub-community) and, being relatively unpalatable, will persist even where this treatment is combined with grazing, when it can be a dominant in grassy heath (seen in the *Festuca* sub-community). *Molinia* is rather less consistent throughout and in general not as abundant as *A. curtisii*. It tends to follow *E. tetralix* in its preference for moister soils, though it can increase greatly after fires, but, being palatable, may not be so prominent in grazed stands.

Other grasses, too, and some cyperaceous plants, are among the more common and occasionally abundant associates in the *Ulex gallii-Agrostis* heath, a feature that, in general, is not met with in the *Ulex minor-Agrostis* heath. None is frequent throughout but as preferentials for some of the sub-communities they make an obvious contribution. Among the grassier heaths included here (and predominating in the *Festuca* sub-community), *Festuca ovina* and *Danthonia decumbens* are particularly important, with *Agrostis capillaris* and *A. canina* ssp. *montana* more occasional, *Nardus stricta*, *Festuca rubra*, *Anthoxanthum odoratum* and *Deschampsia flexuosa* sparse and less obviously preferential. Among sedges found in this kind of heath, *Carex binervis* and *C. pilulifera*, though not very frequent, are rather characteristic of these grassier stands and, where the soils are a little more base-rich, *C. flacca* or, on moister ground, *C. panicea* can occur. The rare *C. montana*, a sedge that is rather catholic in its floristic affinities in Britain, has been recorded in this vegetation in South Wales (Ivimey-Cook 1959). Playing a more prominent part than any of these sedges, however, in stands on cooler and moister west-facing slopes, is *Scirpus cespitosus* (the frequency of which helps define a *Scirpus* sub-community).

Apart from *Potentilla erecta*, which occurs as a constant with reasonable consistency in all the sub-communities, though even then only as scattered plants, dicotyledonous herbs are not a frequent component of this vegetation overall. *Polygala serpyllifolia* and *Pedicularis sylvatica* occur occasionally, and sometimes (as among the Lizard stands in Hopkins 1983) with locally high frequency. *Viola lactea* can also become common in disturbed situations and, among grassier heaths, *Galium saxatile* is weakly preferential. Where the soils are drier and more base-rich, species such as *Thymus praecox* and *Helianthemum nummularium* can occur intermixed with the heath calcifuges in vegetation which has been included in 'limestone heath' (e.g. Ivimey-Cook 1959).

Finally, among the vascular plants found here, there is *Pteridium aquilinum*. This is only occasional throughout and often represented just by scattered fronds but, where this kind of heath extends on to deeper soils that are kept free of any waterlogging, it can become locally abundant and it frequently dominates in stretches of the *Pteridium-Galium* community closely juxtaposed with the heath and often spreading through associated grasslands.

The ground cover in the *Ulex gallii-Agrostis* heath is rather variable in its extent and diversity, though rarely very abundant or showing any marked species-richness. No bryophyte or lichen occurs with any frequency throughout and, in ranker grassy swards or among vigorous building sub-shrubs, total cover can be very low. Then, there may be just a little *Hypnum cupressiforme s.l.* or *Dicranum scoparium*, the two mosses which are found most commonly overall, though even these are quite often absent altogether. More open areas may have some *Campylopus paradoxus*, the introduced *C. introflexus* or, sometimes showing marked local frequency (as on Aylesbeare Common in Devon: Ivimey-Cook *et al.* 1975), *C. brevipilus*; and these can be accompanied by *Racomitrium lanuginosum* and hepatics such as *Calypogeia fissa*, *Cephalozia bicuspidata* and *C. connivens*. With peat-encrusting *Cladonia* spp., such as *C. floerkeana*, *C. coccifera*, *C. chlorophaea*, and larger taxa such as *C. impexa*, *C. crispata*, *C. uncialis* and *C. arbuscula*, such a suite can attain prominence on bare ground in the middle years of regrowth after burning, before the sub-shrub or grass cover has become extensive, or, more locally, within the centre of degenerate heather bushes. New fires destroy such a ground flora which shows a temporary replacement by species such as *Polytrichum juniperinum*, *Ceratodon purpureus* and *Funaria hygrometrica*, which can be patchily abundant for some time.

Sub-communities

***Agrostis curtisii-Erica cinerea* sub-community:** Species-poor *Erica cinerea* heath NCC Devon Heathlands Report 1980. Unusually among the vegetation included within the *U. gallii-Agrostis* heath, this sub-community shows an almost total absence of *E. tetralix*.

Otherwise, however, all the constants remain frequent, though there is quite a strong tendency for *A. curtisii* to be dominant. In contrast to the *Festuca* sub-community, however, such a trend is not accompanied by a general increase in the cover of grasses: *Molinia* is rarely abundant and, though *Danthonia* occurs occasionally, *A. curtisii* usually forms a virtually pure and densely-tussocky sward in such stands in which even the sub-shrubs are reduced to sparse shoots. In other cases, *E. cinerea* or, more rarely, *U. gallii* or *Calluna*, have the highest cover and then *Potentilla erecta* and bryophytes and lichens are somewhat better represented, though, by and large, both vascular and cryptogam associates are few in number throughout.

***Festuca ovina* sub-community:** Grass-heath Ward *et al.* 1972 *p.p.*; Species-poor *Erica cinerea* heath NCC Devon Heathlands Report 1980 *p.p.*; Species-poor Dry *Agrostis setacea* Heath NCC Devon Heathlands Report 1980. As in the previous sub-community, *A. curtisii* is often the most abundant species here, though it is not quite so uncompromisingly dominant: most commonly, it is the leading component of a fairly rich mixture of sub-shrubs and herbs for which the term grass-heath is probably the most appropriate. Among the woody plants, both *U. gallii* and *Calluna* retain very high frequencies and the latter can have moderately high cover. *E. cinerea*, however, is much more patchy in its occurrence and quite often totally absent and *E. tetralix*, though considerably more common than in the first sub-community, is still not constant; and usually neither of these plants is abundant. Likewise, *Vaccinium myrtillus*, which is more frequent here than in any other kind of *Ulex gallii-Agrostis* heath, is often present as sparse shoots, and these commonly show signs of having been nibbled.

More strictly preferential, and generally more abundant, is *Festuca ovina* which, with the constant *A. curtisii* and *Molinia caerulea*, frequent *Danthonia decumbens* and occasional *Agrostis capillaris* and *A. canina* ssp. *montana*, occurs in intimate mixtures with the sub-shrubs or as a grassy matrix between the bushes, the height and arrangement of these different elements strongly reflecting the particular grazing regime. Also rather distinctive of this kind of vegetation, though not occurring very commonly, are *Carex pilulifera* and *C. binervis*, usually found as scattered tufts in the sward, though with the former sometimes locally abundant after burning. On moister ground, there can also be some *C. panicea*, and, in south Wales, *C. montana* has been recorded (Ivimey-Cook 1959). Among the other associates, *Potentilla erecta* is very common and frequently accompanied here by *Galium saxatile*. Locally, there can be a patchy over-canopy of *Pteridium*, though this does not reach its full development until July. Then, in some stands, where the soils are somewhat more base-rich than is usual for this kind of vegetation, there is a distinctive enrichment by species such as *Thymus praecox*, *Helianthemum nummularium* and *Carex flacca*, producing what has sometimes been included in a 'limestone heath' community (e.g. Ivimey-Cook 1959).

Among denser covers in this sub-community, where the herbage is matted with grass litter, bryophytes and lichens can be very sparse but, where the sward is nibbled shorter or among gaps in the sub-shrub canopy, *Hypnum cupressiforme s.l.*, *Pleurozium schreberi*, *Pseudoscleropodium purum* and *Dicranum scoparium* can occasionally be found and, out of reach of trampling, *Cladonia impexa*.

Erica tetralix sub-community: *Agrostis setacea* 'Short Heath' Coombe & Frost 1956a, Marrs & Proctor 1978; *Agrostidetum setaceae cornubiense* Coombe & Frost 1956, Ivimey-Cook 1959, Ivimey-Cook *et al.* 1975; *Erica vagans-Erica tetralix* 'Wet Heath' Coombe & Frost 1956a; Intermediate Tall/Short Heath Marrs & Proctor 1978; *Agrostis curtisii* heath Hopkins 1983. This kind of *Ulex gallii-Agrostis* heath shows a strong general uniformity of composition, with some striking local floristic peculiarities and quite a wide range of physiognomic variation. Both of the grasses and sub-shrubs of the community, including now *E. tetralix*, have very high frequency here, and each can be abundant, either in mixed canopies or in covers where one or other of the species is more obviously dominant. Sometimes, too, there can be a little *V. myrtillus*. *Potentilla erecta* occurs frequently and there can be occasional *Danthonia decumbens*, *Polygala serpyllifolia* or *Carex panicea*. Among stands on the Lizard, from where this vegetation was first described as 'Short Heath' or *Agrostidetum setaceae cornubiense* by Coombe & Frost (1956a), the last two species, together with *Pedicularis sylvatica*, *Dactylorhiza maculata* ssp. *ericetorum* and *Salix repens* show increased local frequency.

More striking, however, is the occasional occurrence of *E. ciliaris* or *E. vagans* in this kind of heath. In some of its Devon and Cornish stations, the former can be very abundant here, sometimes totally replacing *E. tetralix*, though not usually being accompanied by any other floristic peculiarities. *E. vagans*, too, can be found in this vegetation on the Lizard as a generally low-cover member of an otherwise unchanged vascular cover. But its typical position when it occurs here is in spatial transitions to the *Erica vagans-Schoenus* heath and it is occasionally accompanied by some of its associates in that vegetation type, notably *Serratula tinctoria* and *Sanguisorba officinalis*. Such more distinctive assemblages (characterised as intermediate vegetation by Coombe & Frost 1956a, Marrs & Proctor 1978 and Hopkins 1983) could be recognised as a variant.

As usual in this community, bryophytes and lichens show a varying representation here, with the most diverse and extensive covers developing on patches of bare ground that have remained unburned for some time but uncolonised by a dense growth of vascular plants. In many stands on the Lizard, peculiar drainage conditions help to maintain such more open areas permanently. Not only is this kind of heath of very short stature in this district, but the cover of grasses and sub-shrubs is broken into discrete clumps, between which are pans, waterlogged in winter but often parched in summer. These support but a few sparse shoots of *Molinia* and *Carex panicea* but can develop a varied cover of mosses and lichens (Hopkins 1983).

Scirpus cespitosus sub-community. The constant occurrence in this sub-community of the strongly preferential *Scirpus cespitosus* with very frequent *Calluna*, *Molinia* and *E. tetralix* gives something of the appearance of a *Scirpus-Erica* wet heath, especially those types which, as here, have fairly common *E. cinerea* and *V. myrtillus*. But the continuing constancy of *U. gallii* and *A. curtisii* provide a good distinction, and the latter especially is sometimes very conspicuous in this vegetation with an extensive cover of dense tussocks. In other stands *Molinia* is abundant, though most often it is *Calluna* which dominates.

Other distinctive features of this sub-community are very few but *Dicranum scoparium* is rather frequent and, more obviously, *Leucobryum glaucum* is preferential, its pale green hummocks sometimes having high cover. Lichens, too, may be conspicuous with occasional records for *Cladonia impexa* and *C. uncialis*.

Habitat

The *Ulex gallii-Agrostis* heath is confined to the warm oceanic parts of south-west Britain where it occurs on a variety of moist, acid soils. Climatic and edaphic conditions combine to influence the general character of this vegetation and interactions between them, in relation to altitude, aspect and surface relief, are partly responsible for the floristic differences seen in the various sub-communities. In most situations, however, burning and grazing are of great importance to the maintenance of the community against succession to woodland and they have a marked effect on the floristics and physiognomy of the vegetation. Other past treatments, like cultivation and abandonment, may also have influenced the composition and distribution of this kind of heath and, in more recent times, intensive improvement for agriculture has reduced and fragmented its extent.

Compared with the climate of central southern England, the conditions experienced by this community show a further shift towards an oceanic extreme, with a more equable temperature regime and increased humidity. Moving west from Poole Harbour, mean annual maximum temperatures begin to fall away a little, to 27 °C or less, from the high values that prevail over much of central and eastern England (Conolly & Dahl 1970), but the winters are markedly less severe: throughout a deep fringe all around the south-western seaboard, February minima are at least 2 °C above freezing (*Climatological Atlas* 1952) and there are fewer than 40 frosts per year (Page 1982). The floristic differences which mark this temperature shift are actually small. Already, in central southern England, a strong Oceanic West European character is becoming visible, with species such as *Erica cinerea*, *E. tetralix* and *A. curtisii* occurring together as constants in the *U. minor-Agrostis* heath, and *E. ciliaris* figuring as a distinctive occasional. That basic pattern continues here, the major general difference between the communities being the switch from *U. minor* to *U. gallii*. The sharp boundary between these two species along the south coast, in east Dorset, marks the geographical division between the heath types and it may reflect a greater need by *U. gallii* for the milder conditions that prevail to the west of the line (Proctor 1965).

More obvious, however, than any simple floristic response to the more oceanic character of the climate through south-west Britain, is the fact that, in the more equable and moist environment, this kind of heath is able to extend to altitudes and on to soils which, in a harsher and drier climate, would be uncongenial and occupied instead, where sub-shrub vegetation has developed, by different types of upland heath or lowland dry heath. As it is, with the relatively mild winter temperatures, the same basic assemblage of Oceanic West European plants occurs in this community, not only on sites close to sea-level like the Lizard and parts of the Gower coast, and on inland commons at low altitudes, as in east Devon, but also up to levels of 500 m or so on the moorland fringes of Dartmoor, Bodmin Moor and Exmoor, something which gives the heath vegetation of much of this part of Britain a strong uniformity. And, within this general framework, the floristic differences to be seen in the various sub-communities can be understood in relation to interactions between soil conditions and the increase in rainfall on moving through this altitudinal range.

Like its eastern counterpart, the *U. minor-Agrostis* heath, this is a vegetation type of acid soils that are too moist for dry heath but not so consistently waterlogged as to be able to sustain wet heath. And, as in that community, this intermediate edaphic character is marked by the distinctive coincidence of *E. cinerea* with *E. tetralix* and *Molinia*, species which, in the more continental parts of Britain, are rather sharply partitioned into dry and wet heaths respectively (Rutter 1955, Bannister 1965, 1966, Gimingham 1972); and, again, by

the constant contribution from *A. curtisii*, a grass typical of moist but not waterlogged soils (Ivimey-Cook 1959). In the drier climate of central southern England, such soil conditions are often maintained by some impedence to drainage in brown earths or podzolic profiles with an argillic B horizon or impervious pan. And this is quite often the case here, particularly at lower altitudes where the annual precipitation is in the order of 1000 mm (*Climatological Atlas* 1952) with around 140 wet days yr^{-1} (Ratcliffe 1968), that is, not very much greater than that in parts of, say, the New Forest. On the Devonshire Pebble-Bed commons, for example, which run northwards to the east of the Exe, the *U. gallii-Agrostis* heath occurs on stagnogleys and gleyed podzols developed over the gently-dissected surface of the Triassic dip slope, a very similar topographic and edaphic context to that of moist heaths further to the east (Ivimey-Cook *et al.* 1975). On the Lizard, too, where the community marks out stretches of base-poor loess and Crousa Gravels deposited over serpentine and gabbro (Coombe & Frost 1956b, Hopkins 1983), sub-surface drainage impedence over the more or less level platform of the headland results in extensive gleying, with surplus winter rain draining away laterally and only slowly (Staines 1984, Findlay *et al.* 1984). In such situations as these, the *Erica tetralix* sub-community is especially characteristic, functioning very much as a western continuation of typical *U. minor-Agrostis* heath, occurring in very similar lowland heath landscapes and again providing an occasional locus on wetter soils for *E. ciliaris*. Additionally here, in some Lizard stands, there is the rather striking occurrence of *E. vagans* and some of its associates from the heaths endemic there, but this has been clearly related (Coombe & Frost 1956a, Hopkins 1983) to a thinning of the drift mantle over the base-rich serpentine or gabbro, with a rise in superficial pH from less than 5 to just above: in its general floristic and edaphic features, the 'Short Heath' of this area thus clearly belongs here.

Some of these lowland gleyed soils show a shallow accumulation of mor humus beneath sub-shrub covers that have not been burned for some time but, with a shift to higher ground, where annual rainfall increases towards 1600 mm (*Climatological Atlas* 1952) with over 160 wet days yr^{-1} (Ratcliffe 1968), there is a strong tendency for profiles with impeded drainage to develop a humose top-soil. Around the fringes of Dartmoor, Bodmin Moor and Exmoor, therefore, over ill-draining stretches of Devonian and Carboniferous shales and mudstones and granite, the *U. gallii-Agrostis* heath is often found on stagnohumic gleys that form an intergrade between mineral stagnogleys and the thick ombrogenous peats mantling the highest and wettest ground. Here, the *Scirpus* sub-community is the characteristic type, with *Scirpus* itself and, to a lesser extent,

Vaccinium myrtillus, both Continental Northern plants, and the general moorland context, giving the vegetation a rather different feel from that of the lowland *Erica* sub-community.

One other consequence of the increased rainfall characteristic of the higher ground of north-west Somerset and the heartlands of Devon and Cornwall is that there the *U. gallii-Agrostis* heath is able to extend on to more free-draining soils that are kept moist as much by high precipitation as by any drainage impedence: indeed, such conditions probably represent the edaphic optimum for *A. curtisii* (Ivimey-Cook 1959). Quite commonly, then, over more pervious Devonian or Carboniferous sandstones and coarse granitic debris or boulders, the community can be found on moist brown podzolic profiles or podzols proper. In such situations, the *Festuca* sub-community is particularly distinctive with *V. myrtillus* again indicating the wetter and somewhat cooler conditions which prevail at higher altitudes, though much of the character of this kind of *U. gallii-Agrostis* heath reflects the treatment which the vegetation has received on this quite well-drained ground over the fringing upland slopes (see below). Indeed, this is the type of grass-heath likely to develop throughout the region wherever grazing has been applied to the community on soils which show some relief from drainage impedence: it is the usual sub-community found, down to quite low altitudes, over loess-smeared Carboniferous Limestone in south Wales, for example, where incomplete masking of the base-rich bedrock sometimes permits the development of stands with a modest admixture of calcicoles (e.g. Ivimey-Cook 1959, NCC South Gower Coast Report 1981).

As with other kinds of lowland heath, human activity has undoubtedly played an important part in the extensive development of this community over stretches of impoverished acid soils and, within the general floristic framework set by climatic conditions, treatments continue to have an important effect on the composition and structure of the vegetation and on the disposition and extent of stands through the region. With a community like the *U. gallii-Agrostis* heath, which spans a fair range of altitudes and makes a contribution to different types of agricultural landscape, the pattern of treatments from place to place has probably been very varied, and there are certainly differences today in intensity and type of use. Even within a particular area, the history of exploitation can be very complex: on the Lizard, for example, Hopkins (1983) described the community from a patchwork of heaths that, in Napoleonic times, was actually less extensive than at present, probably because of wartime reclamation, and which, even in the relatively short time since then, has experienced regimes of burning and grazing, gorse-cutting and turf-paring, with bouts of neglect and more recent intensive improvement for farming or forestry and enclosure for military training and telecommunications purposes.

The most drastic of the traditional treatments is burning, though on the Lizard, this is not now a deliberate practice outside the enclosed and cliff-top heaths and, even there, where it is used to encourage a flush of new herbage and control the spread of *U. europaeus* and *Prunus spinosa* over the stretches of deeper soil, many fires are started outside the close season and allowed to burn uncontrolled. In the unenclosed inland heaths, where the *Erica tetralix* sub-community makes a bigger contribution to the vegetation cover, accidental fires are frequent, perhaps more so now with the accumulation of litter and dead wood following the demise of grazing and fuel-gathering, the overgrowth of turf-pits, which provided breaks, and the increase in visitors (Hopkins 1983). In the less rainy weather of summer, the soil surface, even of the stagnogleys, can become parched and the vegetation become very inflammable, so intense fires often clear the ground completely, leaving much exposed mineral soil. After burning, Hopkins (1983) noted that eutrophic weeds, such as *Epilobium angustifolium* and *Chenopodium rubrum*, could figure briefly but the first real stage in recolonisation was the spread of acrocarpous mosses, which occupied the ground for some years before the reappearance of the characteristic suite of bryophytes and *Cladonia* spp. In fact, Hopkins (1983) considered that even this assemblage might represent an impoverished version of the potential cryptogam flora, its further development curtailed by frequent fires. And, in this area, Coombe & Frost (1956a) thought that burning was an important contributory factor, maintaining the distinctive open pans in which wind erosion could hinder extensive recolonisation by vascular plants.

In less exposed situations, burning tends to favour the eventual resurgence of dominance by *Calluna*, in both the *Erica tetralix* and *Scirpus* sub-communities of this kind of heath and, where practised judiciously, can maintain a continuous supply of vigorous building-phase heather through a mosaic of swales. In fact, as on the Lizard, burning of many stands of the community at lower altitudes is generally now an accidental and uncontrolled affair and often infrequent, so many stands have a rank and leggy *Calluna* canopy. In some areas, however, as around the periphery of Dartmoor, swaling is still a regular practice, if not always as carefully controlled as it might be, maintaining heather as the eventual dominant over many hectares (Ward *et al.* 1972a, Harvey & St. Leger-Gordon 1974).

In the early stages of regeneration, however, burning can favour marked local heterogeneity in the canopy of both the *Erica* and *Scirpus* sub-communities, by allowing opportunity for *A. curtisii* to regrow from surviving stools and to seed in, often in great profusion, to bare areas. *E. cinerea*, too, can show some temporary prominence in the early years of recovery. Some of this variety in dominance can be readily accommodated within these kinds of *U. gallii-Agrostis* heath but the *Agrostis-Erica* sub-community seems to consist mainly of stands of the community where *Calluna* and *E. tetralix* have been more thoroughly eclipsed by their rivals. It can be found throughout the geographical and altitudinal range of the heath and over a variety of soils.

One other important treatment-related development in the *U. gallii-Agrostis* heath is the response to grazing. On the Lizard, grazing is now confined to the cliff-top heaths and the inland enclosures and actually seems to be increasing in the latter with the switch from dairying to beef production, for which, with some time on improved pasture plus supplementary feeding, this kind of vegetation can provide a reasonable bite. On the unenclosed heaths, however, grazing seems to have ceased fifty or so years ago, with the combination of low economic returns, the failure of the commoners to protect their rights and the neglect in wartime (Hopkins 1983). On the upland fringes, by contrast, the community continues to provide extensive pasturage on the slopes between the enclosed and often improved grasslands below and the wet heath and blanket mire above, with grazing by sheep, cattle and ponies (Ward *et al.* 1972a, Harvey & St. Leger-Gordon 1974). Often combined with a burning regime, such consistent herbivore pressure favours the development, particularly on the more free-draining soils, of the *Festuca* sub-community, with its low total cover of sub-shrubs, abundance of *A. curtisii*, *Festuca ovina* and other grasses and somewhat richer total vascular flora.

Zonation and succession

The *U. gallii-Agrostis* heath is typically found in zonations with other heath communities and mires over sequences of soils which differ mainly in their moisture regime, occasionally in their base-richness. The particular sub-communities, and the other vegetation types, involved in these patterns vary with altitude and are throughout subject to modification by treatments. Throughout most of its range, such human interference has maintained this vegetation against progression to woodland, though abandonment of traditional treatments has in recent years often been a prelude to improvement for agriculture or forestry with permanent loss of the community.

In its typical lowland setting, best seen now on the Devonshire Pebble-Bed commons, from the largest of which, at Aylesbeare, the vegetation pattern was described by Ivimey-Cook *et al.* (1975), the *U. gallii-Agrostis* heath is found in very much the same general kind of sequence as is characteristic of stretches of impoverished acid soils further east, around Poole, in the New Forest and on the commons of Surrey and Sussex. Essentially, it replaces the *U. minor-Agrostis* heath as the intermediate component of the zonation from dry to wet heath developed over soils with an increasing tendency to surface-waterlogging, except that, in the wetter climate typical of this more westerly region, the contribution of the dry heath becomes proportionately less prominent. Generally, at these lower altitudes, the *U. gallii-Agrostis* heath is represented by the *Erica tetralix* sub-community and, as in the Aylesbeare transect (Ivimey-Cook *et al.* 1975), this can occupy the bulk of the less waterlogged ground. Where more sharply-draining acidic soils are present, however, it gives way to drier heath of the *Calluna-Ulex gallii* type, a community into which *Calluna*, *E. cinerea* and *U. gallii* run as consistent components but from which *E. tetralix*, *Molinia* and usually also *A. curtisii* are excluded. In the opposite direction, the *U. gallii-Agrostis* heath gives way to some kind of *Ericetum tetralicis* with the move to mineral soils with seasonal surface waterlogging. Here, it is *Calluna*, *E. tetralix* and *Molinia* which provide the major continuity of cover, although *A. curtisii*, *E. cinerea* and *U. gallii* can run some way into the wet heath if soligenous influence, even gentle flushing, maintains good aeration. Such more gradual transitions can be seen to some extent at Aylesbeare (Ivimey-Cook *et al.* 1975) but they are more obvious at some of the sites included in the NCC Devon Heathlands Report (1980), at Hares Down, for example. In the full sequence of vegetation types in this kind of site (Rose 1953), the zonation usually continues over permanently-waterlogged peat into the *Narthecio-Sphagnetum* valley bog and, over gently-undulating topography, the whole pattern may be laid out as elongated, concentric zones. Often, however, the physiography is more complex than this and parent material heterogeneity can introduce further variations in the moisture or mineral regime. Commonly, as at Aylesbeare, there is some soligenous influence down the slopes such that Junco-Molinion vegetation intervenes through the zonation or, along soakways, the *Schoenus-Narthecium* mire can occur. And, as on heathlands in central southern England, deeper brown earth soils can have patches of the *Pteridium-Galium* community and on disturbed areas the *Ulex-Rubus* scrub.

In such systems as these, the measure of base-enrichment is usually modest and dependent upon the concentration of sub-surface drainage waters from any local more calcareous substrates. On the Lizard, the *U. gallii-Agrostis* heath occurs in a rather different and peculiar situation, forming the most calcifuge element in a series of sub-shrub communities whose general character is controlled by the underlying serpentine and gabbro. Only where the dominating influence of these bedrocks is masked by acidic loess or Crousa Gravels are the soils sufficiently base-poor to support the *U. gallii-Agrostis* heath in this area, and from the deeper deposits, *Erica vagans*, the characteristic species through much of the Lizard heathland, is excluded. Around the thinner margins of the drift patches, however, it appears in a distinctive strip, usually 2–10 m wide, of the vegetation transitional to the *E. vagans-Schoenus* heath of the base-rich gleys developed directly over the igneous parent materials (Coombe & Frost 1956*a*, Hopkins 1983). On loess-contaminated brown earths, heath transitional to the other major sub-shrub community of the inland areas, the *E. vagans-Ulex europaeus* heath, can sometimes be found (Hopkins 1983).

With the move to higher altitudes and a wetter climate, the *Scirpus* sub-community tends to replace the *Erica* sub-community and it can sometimes be found in the kind of dry–wet heath sequence typical of the lowlands, around the valley bogs of Dartmoor. On these upland fringes, however, there is a tendency for the whole character of the vegetation pattern to shift towards the sub-montane. In the first place, around the elongated topogenous hollows that drain radially across

the fringes of Dartmoor, the *Scirpus* sub-community often gives way, on wetter ground, to some type of *Scirpus-Erica* wet heath or Junco-Molinion vegetation which then passes, in the water-tracks, to *Carex echinata-Sphagnum* mire, often rush-dominated. And, in the other direction, towards drier, sometimes rocky, ground, it is replaced by the *Calluna-Vaccinium* heath. Second, on a grander scale, the *U. gallii-Agrostis* heath makes a major contribution to the zone of marginal sub-shrub vegetation that runs concentrically all around Dartmoor, grading above, through *Scirpus-Erica* wet heath to the *Scirpus-Eriophorum* blanket mire and passing below to *Festuca-Agrostis* grassland and improved pasture (Figure 26).

Similar gross landscape patterns, though not so intact or clear, can be seen over the other major upland areas of the south-west, Bodmin Moor and Exmoor, and they reflect the tendency, in the more humid climate, for gentle slopes to accumulate a mantle of ombrogenous peat and for wet and dry heath to be pushed on to the steeper margins. But there is also a strong element of

Figure 26. Zonations of heaths and mires on the lowland commons and moorland fringes of south-west England.
H4c   *Ulex gallii-Agrostis* heath, *Erica* sub-community
H4d   *Ulex gallii-Agrostis* heath, *Scirpus* sub-community
H8b   *Calluna-Ulex gallii* heath, *Danthonia* sub-community

M15d   *Scirpus-Erica* wet heath, *Vaccinium* sub-community
M16a   *Ericetum tetralicis* wet heath, Typical sub-community
M17a   *Scirpus-Eriophorum* mire, *Drosera-Sphagnum* sub-community
M21a   *Narthecio-Sphagnetum* mire, *Sphagnum-Rhynchospora* sub-community

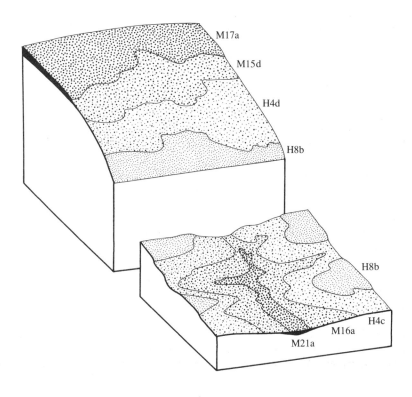

cultural influence in the zonations, with sub-shrub vegetation becoming confined to a narrow belt of relatively unimproved land, edaphically impoverished and sometimes topographically intractable, above the limit of more intensive agriculture. Traditionally around these upland fringes, the heath zone has provided low-grade pasture, periodically renewed by burning, so it is often the *Festuca* sub-community that occurs in these kinds of zonations. And, in such situations, it is very obvious that grazing can mediate a conversion from *U. gallii-Agrostis* heath in which sub-shrubs predominate, through the grass-heath of this sub-community to calcifugous grassland of the *Festuca-Agrostis-Galium* type. The upland margins in the south-west often consist of ill-defined mosaics of these vegetation types (e.g. Ward *et al.* 1972*a*) and enclosures below frequently contain improved pasture which has been derived from the heath and grassland by further treatment, liming, fertilising and top-sowing, which can effect a fairly ready conversion to calcifuge types of *Lolio-Cynosuretum*. Injudicious grazing can, however, create problems along the way, permitting the spread of *Nardus stricta*, for example, or *Pteridium*. On Dartmoor, the pasturing of the heaths by mixed herds of cattle, sheep and ponies may help keep these in check: ponies nibble out the centre of *Nardus* tussocks (Havinden & Wilkinson 1970) and cattle can trample out spreading bracken (Sayer 1969).

Careless burning, too, can precipitate a run-down of the *U. gallii-Agrostis* heath. The spread of *A. curtisii*, a relatively unpalatable grass (Ivimey-Cook 1959, Ward *et al.* 1972*a*), is a common problem in newly-burned heaths, and with frequent fires it may become more or less permanently dominant in vegetation of the *Agrostis-Erica* type. In other cases, *Molinia* may become very abundant in a shift to Junco–Molinion vegetation. And, even where there is regeneration of the sub-shrubs, repeated burning can impoverish the cryptogam flora, and uncontrolled fires destroy the characteristic mosaic of dominance seen in well-managed swales.

Through most of its range, however, burning and grazing have been essential to the maintenance of the vegetation against progression to woodland. Except in situations like the Lizard, where exposure to high and frequent winds is combined with a general scarcity of seed-parents, invasion of shrubs and trees occurs readily in the absence of treatments. Patchy development of the *Ulex-Rubus* scrub is often seen over disturbed ground in the *U. gallii-Agrostis* heath, along trackways, field margins and the edges of plantations, but of greater long-term significance is colonisation by birch which can fairly quickly form an open canopy to developing *Betula-Molinia* woodland. Very commonly, however, the abandonment of traditional treatments in the lowlands has been followed eventually by improvement and conversion of the land to pasture or plantation.

Distribution

The community is confined to south-west Britain, beyond a line from mid-Dorset to the Quantocks, and including parts of the south Wales seaboard. The *Erica* sub-community is especially characteristic of lower altitudes, being well represented on the Devon Pebble Bed commons, the lower fringes of Dartmoor and Bodmin Moor and on the Lizard. At higher levels, particularly on Dartmoor and Exmoor, it is replaced by the *Scirpus* and *Festuca* sub-communities, though the latter type can also be found throughout the region where this kind of heath occurs on free-draining soils with grazing. The *Agrostis-Erica* sub-community also occurs throughout.

Affinities

In the sequence of non-maritime heaths that runs around the southern and western seaboard of Britain, this community represents an oceanic extreme, early recognised as distinctive among lowland heather-dominated vegetation (e.g. Tansley 1911, 1939, Watson 1932) by virtue of the coincident occurrence of *U. gallii* and *A. curtisii*, though first given detailed definition from its rather unusual context on the Lizard (Coombe & Frost 1956*a*, Hopkins 1983). Thereafter, Ivimey-Cook *et al.* (1975) set this kind of heath in its typical lowland framework as part of the sequence of dry and wet heaths around southern valley mires.

In that kind of setting, the floristic character of the *U. gallii-Agrostis* heath as a western replacement for the *U. minor-Agrostis* heath is very clear, despite the difficulty of giving precise ecological meaning to the vicariant switch from the one gorse to the other (Proctor 1965): this is one of the pair of heaths that, with the increasing oceanicity of climate towards the south-west, becomes interposed between the wet and dry types, seeing a transgression of *E. tetralix* and *Molinia* on to ever more free-draining soils where they coexist with *E. cinerea* and the gorse species and providing the major British locus for *A. curtisii*. On this view, the community takes its place within the Atlantic wet heaths of Böcher's (1943) Ulicio-Ericion tetralicis or the Ulicion of Bridgewater (in Gimingham 1972), grading in one direction to drier heaths of an Ericion cinereae, in the other to the wet heaths of the Ericion tetralicis. Alternatively, one could stress the continuity with the drier western sub-shrub vegetation of the *Calluna-U. gallii* heath, as in Gimingham's (1972) grouping, an approach that would reflect the Continental division of north-west European heaths of this type into a less oceanic Ulicion nanae Duvigneaud 1944 and a more oceanic Ulicion gallii des Abbayes & Corillion 1940, the latter containing equivalent vegetation of the western French seaboard, like the *Ulici gallii-Ericetum ciliaris* Géhu 1973.

Floristic table H4

	a	b	c	d	4
Ulex gallii	V (1–8)	V (1–5)	V (1–8)	V (1–6)	V (1–8)
Agrostis curtisii	V (2–9)	V (1–8)	V (1–9)	V (1–8)	V (1–9)
Calluna vulgaris	V (1–8)	V (1–6)	V (2–10)	V (3–9)	V (1–10)
Molinia caerulea	IV (1–8)	IV (1–8)	V (2–7)	V (4–8)	V (1–8)
Erica cinerea	V (1–8)	III (1–4)	V (1–7)	IV (1–5)	IV (1–8)
Potentilla erecta	IV (1–5)	V (1–3)	III (1–5)	IV (1–4)	IV (1–5)
Erica tetralix	I (1–3)	III (1–4)	IV (1–8)	V (1–7)	IV (1–8)
Festuca ovina		V (1–8)	I (2)	I (1)	II (1–8)
Vaccinium myrtillus	II (1–8)	IV (1–4)	II (1–5)	II (1–3)	II (1–8)
Danthonia decumbens	II (1–4)	III (1–5)	I (1–5)	I (2–3)	I (1–5)
Galium saxatile	II (1–4)	III (1–3)	I (2–3)		I (1–4)
Pteridium aquilinum	I (2–3)	II (1–7)	I (3–7)	I (3–5)	I (1–7)
Carex binervis	I (2)	II (1–3)	I (2–3)	I (1–3)	I (1–3)
Carex pilulifera	I (1)	II (1)	I (2–3)		I (1–3)
Agrostis capillaris	I (2–3)	II (1–5)	I (1–2)		I (1–5)
Pleurozium schreberi	I (7)	II (1–3)	I (1)		I (1–7)
Agrostis canina montana	I (1–2)	II (2–3)	I (1)		I (1–3)
Pseudoscleropodium purum	I (2–4)	II (1–3)	I (2)		I (1–4)
Rhytidiadelphus squarrosus		II (1)			I (1)
Carex flacca		I (1–2)			I (1–2)
Helianthemum nummularium		I (1–3)			I (1–3)
Lotus corniculatus		I (1–3)			I (1–3)
Thymus praecox		I (1–3)			I (1–3)
Carex montana		I (2–4)			I (2–4)
Carex panicea	I (1)	I (1–2)	II (1–5)	I (1)	I (1–5)
Erica ciliaris			II (5–10)		I (5–10)
Erica vagans			II (2–6)		I (2–6)
Salix repens			I (2–3)		I (2–3)
Dicranum scoparium	I (1–5)	III (1–6)	II (1–3)	IV (1–4)	II (1–6)
Scirpus cespitosus		I (1)		V (1–5)	I (1–5)
Leucobryum glaucum		I (1–5)	I (1–6)	III (1–7)	I (1–7)

Floristic table H4 *(cont.)*

	a	b	c	d	4
Polygala serpyllifolia	II (1–2)	I (1)	II (1–3)	II (1–3)	II (1–3)
Hypnum cupressiforme s.l.	II (1–5)	II (1–3)	I (1–8)	I (1–8)	I (1–8)
Nardus stricta	I (1–3)	II (1–4)	I (1–4)	II (2–5)	I (1–5)
Cladonia impexa	I (1–10)	II (1–5)	I (1–9)	II (1–8)	I (1–10)
Cladonia floerkeana	I (1)	I (1)	I (2–3)	I (2)	I (1–3)
Festuca rubra	I (1)	I (1–6)	I (2–3)	I (3–4)	I (1–6)
Campylopus paradoxus	I (2–4)	I (1–2)	I (1–4)	I (4)	I (1–4)
Cephalozia bicuspidata	I (2)	I (1)	I (1–3)	I (2)	I (1–3)
Pohlia nutans	I (1–2)	I (1)	I (2–7)	I (2)	I (1–7)
Deschampsia flexuosa	I (1–7)	I (1–2)	I (1)		I (1–7)
Anthoxanthum odoratum	I (1)	I (1–2)	I (3)		I (1–3)
Calypogeia fissa	I (1)	I (1)	I (2)		I (1–2)
Pedicularis sylvatica	I (1)	I (1)	I (3)		I (1–3)
Lophocolea bidentata s.l.	I (2)	I (1)	I (2–3)		I (1–3)
Hypogymnia physodes	I (1)		I (1–5)	I (1–2)	I (1–5)
Cladonia coccifera		I (1)	I (1–3)	I (2)	I (1–3)
Cladonia chlorophaea		I (1–2)	I (1–3)	I (1–2)	I (1–3)
Juncus squarrosus		I (1)	I (4)	I (1)	I (1–4)
Cladonia crispata		I (1)	I (2–3)	I (2)	I (1–3)
Cladonia arbuscula	I (1)	I (1–2)	I (1–2)		I (1–2)
Polytrichum juniperinum	I (4)	I (8)			I (4–8)
Calypogeia muellerana	I (3)		I (2)		I (2–3)
Viola lactea	I (1)		I (2)		I (1–2)
Campylopus introflexus	I (5)		I (2)		I (2–5)
Cuscuta epithymum		I (1)	I (1–2)		I (1–2)
Racomitrium lanuginosum		I (1)	I (1–3)		I (1–3)
Polygala vulgaris		I (1)	I (2–3)		I (1–3)
Diplophyllum albicans		I (1)	I (5–7)		I (1–7)
Luzula multiflora		I (1)	I (1–2)		I (1–2)
Cladonia fimbriata			I (4)	I (3)	I (3–4)
Cladonia uncialis			I (3)	I (1–2)	I (1–3)
Cladonia furcata			I (1–6)	I (2)	I (1–6)

	a	b	c	d	4
Cladonia subcervicornis			I (1–4)	I (2)	I (1–4)
Number of samples	27	36	77	32	172
Number of species/sample	10 (5–17)	14 (6–19)	9 (6–13)	11 (8–19)	11 (5–19)
Shrub/herb height (cm)	27 (5–70)	22 (3–50)	22 (5–45)	23 (6–40)	23 (3–70)
Shrub/herb cover (%)	94 (70–100)	95 (65–100)	95 (75–100)	92 (70–100)	95 (65–100)
Ground layer cover (%)	14 (0–100)	8 (0–90)	11 (0–80)	22 (1–70)	10 (0–100)
Altitude (m)	227 (50–430)	283 (24–500)	206 (46–435)	257 (150–400)	235 (24–500)
Slope (°)	6 (0–60)	5 (0–10)	4 (0–20)	4 (0–5)	5 (0–60)

a *Agrostis curtisii-Erica cinerea* sub-community
b *Festuca ovina* sub-community
c *Erica tetralix* sub-community
d *Scirpus cespitosus* sub-community
4 *Ulex gallii-Agrostis curtisii* heath (total)

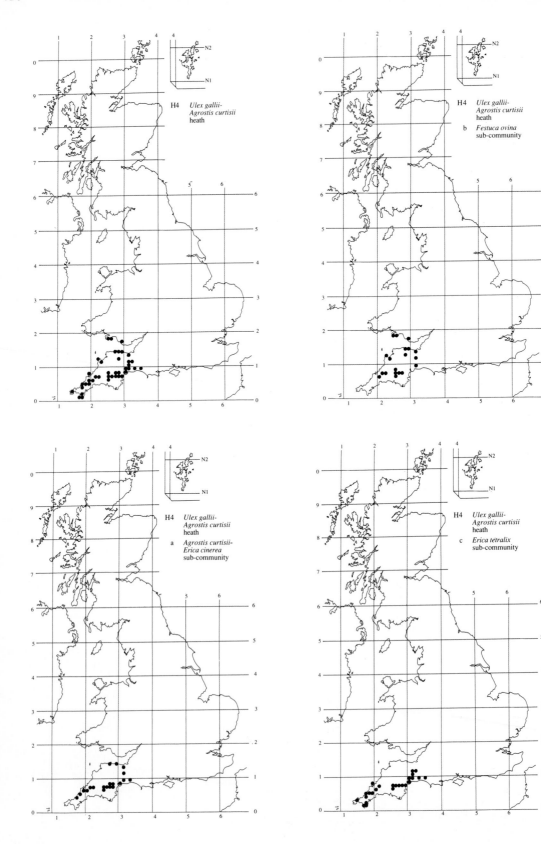

H4   *Ulex gallii-Agrostis curtisii* heath

H4   *Ulex gallii-Agrostis curtisii* heath

b   *Festuca ovina* sub-community

H4   *Ulex gallii-Agrostis curtisii* heath

a   *Agrostis curtisii-Erica cinerea* sub-community

H4   *Ulex gallii-Agrostis curtisii* heath

c   *Erica tetralix* sub-community

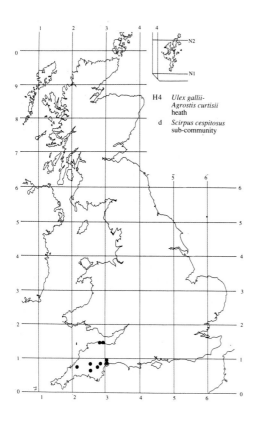

H4   *Ulex gallii-Agrostis curtisii* heath

d   *Scirpus cespitosus* sub-community

H5
Erica vagans-Schoenus nigricans heath

Synonymy

Erica vagans-Schoenus nigricans 'Tall Heath' Coombe & Frost 1956*a*; *Erica vagans* heath Gimingham 1972 *p.p.*; *Ulici gallii-Ericetum vagantis* Bridgewater 1970 *p.p.*; *Ulici maritimi-Ericetum vagantis* (Géhu 1962) Géhu & Géhu 1973 *emend.* Bridgewater 1980; *Erica vagans-Schoenus nigricans* heath Hopkins 1983 *p.p.*

Constant species

Anagallis tenella, Carex pulicaris, Erica tetralix, E. vagans, Festuca ovina, Molinia caerulea, Potentilla erecta, Schoenus nigricans, Serratula tinctoria, Succisa pratensis, Ulex gallii, Campylium stellatum.

Rare species

Agrostis curtisii, Erica vagans, Scilla verna.

Physiognomy

The *Erica vagans-Schoenus nigricans* heath is one of the two sub-shrub communities to which the nationally-rare Oceanic West European *Erica vagans* makes a constant and often prominent contribution, its long spikes of flowers adding a distinctive pale pink to the splash of colour provided by the ericoids in July through to September. Often, however, it is the consistent presence of *Schoenus nigricans* which marks out stands of this vegetation, its tall greyish semi-evergreen foliage standing a little proud of the rest of the cover (hence Coombe & Frost's (1956*a*) epithet 'Tall Heath'). In fact, *Schoenus* can be found as a rare associate in wetter tracts of various lowland heath types, but here it is an invariable member of the community and one of the most important structural components of the vegetation, occurring, often in abundance and generally in a strongly-tussocky form, with dense caespitose clusters of shoots growing on bulky rootstocks (Sparling 1962*a*, 1968, Wheeler 1975). *Molinia caerulea*, too, which is also constant and frequently of high cover, typically occurs as pronounced tussocks and *E. vagans*, together with the next most common sub-shrub, *E. tetralix*, often forms compact

bushes. Thus, although these four species generally dominate in intimate mixtures, forming a canopy 2–4 dm tall, there is often beneath this a well-defined system of runnels between the individual plants. These are flooded to a depth of a few centimetres for long periods during the winter and, more briefly, throughout the year after heavy rain, and so can provide a microhabitat different in character from the sides and tops of the *Schoenus* and *Molinia* tussocks, which are far removed from the influence of the ground waters.

Among other lowland heath sub-shrubs, only *Ulex gallii* occurs with any frequency throughout and, though occasionally co-dominant, it is typically subordinate in cover to the above, but conspicuous in late summer with its contrasting yellow flowers. *Calluna vulgaris*, however, is reduced to occasional and, with *E. cinerea*, which is scarce, is usually found as rather spindly plants rooted in the tussock sides. *Genista anglica* can also occur quite frequently: indeed, among the Lizard heaths, it is strongly preferential to this community, though is absent from apparently suitable sites on gabbro (Proctor 1971, Hopkins 1983). It, too, grows on the *Schoenus* and *Molinia* and its straggling shoots can become particularly abundant in the years after burning. *Salix repens* is occasional at low cover. In contrast to the other major kind of *E. vagans* vegetation, the *E. vagans-Ulex europaeus* heath, *U. europaeus* is uncommon here and usually confined to disturbed situations: among Hopkins's (1983) samples, it made a small contribution on the drier ground of old cultivation areas. His *Ulex europaeus* variant (equivalent to an intermediate 'Tall/ Mixed Heath' in Coombe & Frost 1956*a*) in which this gorse is much more prominent, with much strongly-tussocky *Molinia*, but reduced frequencies of a number of important species in the community, is best regarded as a type of *Molinia-Potentilla* mire in which *E. vagans* has a local role.

In undisturbed stands, not burned for some time and without any grazing, *Schoenus* and *E. vagans* tend to become more overwhelmingly dominant, and the vegetation choked with *Schoenus* and *Molinia* litter. In such situations, even those common associates, like *Potentilla erecta*, which are concentrated on the sides and tops of the tussocks, or which, like *Festuca ovina*, often occur there, can be crowded out. More striking still, the densely-shaded runnels can be virtually bare, such that the vegetation is very impoverished. Where litter and standing material have been destroyed by fire, on the other hand, or their accumulation kept in check by trampling and grazing, the associated flora is more extensive and richer. Among smaller vascular plants in the runnels, sedges are often important, with *C. pulicaris* constant, *C. panicea* and *C. flacca* frequent and, on the gabbro (Proctor 1971), *C. hostiana*, forming the kind of open sward that is also seen in many stands of the *Schoenetum*. And, as there, *Anagallis tenella* occurs commonly, forming quite extensive mats, particularly following fires. Even in the most well colonised and diverse runnels, however, there is not that striking richness among the vascular element typical of well-developed *Schoenetum* and many of the widely-tolerant small herbs frequent there, such as *Mentha aquatica*, *Equisetum palustre* and *Hydrocotyle vulgaris*, as well as more calcicolous plants, are absent here. Among taller herbs, too, there is a restricted flora, though this element is a little better developed and, as in the *Schoenetum*, includes some orchids. Thus, *Serratula tinctoria* and *Succisa pratensis* are constant, *Sanguisorba officinalis* frequent, and *Hypericum pulchrum*, *Stachys betonica*, *Pedicularis sylvatica*, *Dactylorhiza maculata* ssp. *ericetorum*, *D. incarnata* ssp. *incarnata* and *Gymnadenia conopsea* all occasional with, on gabbro again, *Platanthera bifolia*. Other low-frequency associates include *Polygala vulgaris*, *Juncus acutiflorus*, *J. maritimus* and *Galium uliginosum*. On the permanently moist ground characteristic of the *Eleocharis* sub-community, plants such as *E. multicaulis*, *Eriophorum angustifolium*, *Drosera rotundifolia* and *Pinguicula lusitanica* provide a floristic link with the *Schoenus-Narthecium* mire, although *Narthecium ossifragum* itself is rare here. In ungrazed heaths on wetter ground, particularly of this *Eleocharis* type, *Phragmites australis* can figure as scattered shoots or even as a quite dense overtopping canopy, giving a superficial impression of Phragmitetalia fen.

Bryophytes vary considerably in their contribution to the diversity and cover of this vegetation, though in the data of Hopkins (1983) they are a rather more consistent feature than in the stands originally described by Coombe & Frost (1956a), something which may reflect the impact of treatments because, as with the smaller vascular plants, the bryophytes are much reduced by litter accumulation and shade. In more open runnels, however, *Campylium stellatum* is very frequent and sometimes quite abundant with, fairly often, some *Ric-*

cardia multifida, *R. sinuata* and, over gabbro, *Scorpidium scorpioides*. In contrast with the *Schoenetum*, however, runnel species such as *Aneura pinguis*, *Drepanocladus revolvens*, *Cratoneuron commutatum* and *C. filicinum* are not usually found here, and the larger Mniaceae characteristic of the tussock sides in that community are likewise scarce, though *Fissidens adianthoides* can occasionally be found in such situations. On the other hand, the Sphagna typical of the tussock tops in the *Schoenus-Narthecium* community are also absent in this heath.

A distinctive feature of many runnels after wet weather is swollen gelatinous granules of blue-green algae, particularly *Scytonema ocellatum* with some *Stigonema ocellatum*, *Nostoc* sp. and *Zygogonium ericetorum* (Coombe & Frost 1956a).

Sub-communities

Typical sub-community: Typical 'Tall Heath' Coombe & Frost 1956a; Typical *Erica vagans-Schoenus nigricans* heath Hopkins 1983. This vegetation has all the general features of the community described above with no additional preferential species. The tussock/runnel structure is often well-defined, though stands vary considerably in species-richness which is strongly dependent upon time since burning and other treatments. Some floristic peculiarities can be seen where this kind of heath occurs over gabbro, as opposed to serpentine, with *Carex hostiana*, *Platanthera bifolia* and *Scorpidium scorpioides* appearing, *Genista anglica* dropping out (Proctor 1971, Hopkins 1983): such differences could form the basis of recognising two variants (see below).

Eleocharis multicaulis **sub-community:** 'Tall Heath' variants Coombe & Frost 1956a p.p.; *Erica vagans-Schoenus nigricans* heath, *Eleocharis multicaulis* variant Hopkins 1983. *Schoenus* is generally an obvious dominant here, with *Molinia*, *E. vagans* and *E. tetralix* somewhat less abundant, *Calluna* and *E. cinerea* usually absent. Among smaller associates, *Carex panicea* is also typically missing. On the positive side, this vegetation is characterised by the preferential frequency of *Eleocharis multicaulis*, *Eriophorum angustifolium*, *Drosera rotundifolia*, *Pinguicula lusitanica* and *Dactylorhiza incarnata* ssp. *incarnata*, growing in runnels that are usually flooded to shallow depth through most of the year. Where there is no grazing, *Phragmites* is quite common and can be locally abundant.

Habitat

The *E. vagans-Schoenus* heath is confined to wet, base-rich but calcium-poor mineral soils and shallow peats on the Lizard in Cornwall. Interactions between the warm, oceanic climate of the area and its distinctive parent materials have an important influence on the floristics of the community but the composition and physiognomy of particular stands are much affected by burning and sometimes also by grazing. Other past treatments, like the cutting of 'turf' have probably also influenced the appearance and distribution of the vegetation and, though the wet and impoverished character of the soils has previously afforded some protection against cultivation of the ground, the community is vulnerable to modern techniques of land improvement. Much of the remaining extent of this unusual vegetation type now has statutory or voluntary protection.

The climate of the extreme tip of Cornwall is warm, wet and windy (Coombe & Frost 1956a, Malloch 1970, Hopkins 1983). Mean annual maximum temperatures are a little cooler, at 25 °C, than those through most of the South-West Peninsula (Conolly & Dahl 1970), but the winters are the mildest on the British mainland, with February minima usually 4 °C above freezing (*Climatological Atlas* 1952), accumulated winter temperatures (December to March) of more than 55 °C (Page 1982) and, on average, less than 20 frosts annually, with these few being generally confined between mid-December and the beginning of March. The annual range of mean monthly temperatures is about 9 °C and the effective growing season is virtually year-round (Fairburn 1968).

However, despite the frequency of strong westerly winds (*Climatological Atlas* 1952), the rainfall of this part of Cornwall is not especially high, annual precipitation of around 900 mm (Meteorological Office 1977) being similar to levels recorded through much of lowland south-west Britain and very much less than that in more northerly oceanic areas. The annual distribution of rainfall also shows a clear winter peak and a minimum between April and July. Combined with the high summer insolation, this means that there is a potential water deficit in spring and early summer comparable with that in many more easterly and inland areas (Malloch 1970, Hopkins 1983). Along a west–east axis across southern Britain, therefore, the area has a markedly oceanic climate; along a south–north axis up the western seaboard, it exhibits certain Mediterranean tendencies (Coombe & Frost 1956a).

The general floristic character of the community reflects these climatic features. In its frequent combination of *E. tetralix* with *U. gallii*, this kind of vegetation continues the Oceanic West European trend already visible in the *U. gallii-Agrostis* heath, which has a wide distribution through south-west England. But here, there is the additional striking presence of *E. vagans*, a species virtually restricted in Britain to this area and one which adds a more specifically 'Lusitanian' feel to the community, representing a further stage in the progressive confinement of *Erica* spp. to the western seaboard of Europe, which continues in Ireland with *E. mediterranea*

and *E. mackiana* (Matthews 1955). Less typical of the community as a whole, and also occurring widely in cooler oceanic areas further north, but generally 'Lusitanian' in its European range, there is *Pinguicula lusitanica*. And the community provides a locus, too, for the Oceanic Southern *Anagallis tenella* and for *Serratula tinctoria* which, though Continental through Europe as a whole, has a strongly south-western distribution in Britain.

It is, however, the indirect effects of climate, working on the parent materials, that have the more obvious impact on the composition of the vegetation because, within this climatic zone, the *E. vagans-Schoenus* heath is rather strictly confined to soils that are, in the first place, wet and, second, moderately base-rich yet calcium-poor, features which derive in this area from the distinctive lithology of serpentine and gabbro, the former of major importance through the Lizard, the latter occurring as a smaller intrusion towards the east of the headland (Flett 1946). The resistance of these rocks to sub-aerial erosion means that the landscape of the Lizard is still dominated by the wave-cut platform, supposedly cut in Pliocene–Pleistocene times (Balchin 1964), over which gentler slopes predominate and which shows only a gradual rise on moving inland from the encircling cliffs. Much of this surface has a somewhat indistinct drainage pattern, with wide and shallow basins gathering water by percolation and surface run-off and feeding relatively unbranched streams that cut down sharply near the coast in deep 'cove-valleys' (Hopkins 1983). Over much of the ground, therefore, the soils shed excess winter rain but slowly and, where the serpentine and gabbro are free of a mantle of pervious drift, drainage is further impeded by the fact that these rocks weather to produce a large proportion of sticky clays (Butler 1953). It is on the stagnogleys of the Croft Pascoe Association (Staines 1984, Findlay *et al.* 1984), derived from serpentine and gabbro over slopes of less than 5° or so, that the *E. vagans-Schoenus* heath mainly occurs. Here, there is marked seasonal waterlogging with, on level ground, a few centimetres of water standing in the runnels long through the winter, then disappearing in the drier spring. In the *Eleocharis* sub-community, the vegetation extends a little way on to ground that is more or less permanently waterlogged around streams and more ill-defined drainage lines, and in such situations there can be a thicker humic top or a layer of peat in what is approaching a topogenous fen.

The other distinctive feature of these profiles lies in their chemistry. Serpentine, the more extensive parent material, is petrologically heterogeneous but made up largely of ferromagnesian silicates and fairly uniform in its decomposition products, giving rise to soils that have a superficial pH of between 5.5 and 7.5, but in which magnesium predominates over calcium; the profiles are also rich in chromium and nickel, but poor in aluminium, potassium and phosphorus (Flett 1964, Coombe & Frost 1956*b*, Malloch 1970, Hopkins 1983). In fact, among British serpentine soils, those on the Lizard have a little more calcium than usual (Proctor & Woodell 1971), perhaps because of the presence of augite in the rocks (Coombe & Frost 1956*b*), perhaps because of some salt-spray deposition (Hopkins 1983) although this dies off rapidly inland and must be slight over most of the stands (Malloch 1970). It is, however, still insufficient to dominate the cation-exchange system. The gabbro soils, on the other hand, have a lower magnesium:calcium ratio, and also less chromium and nickel (Staines 1984), features which may play a part in the uneven distribution through the community of *Genista anglica* (mainly on serpentine), *Carex hostiana*, *Platanthera bifolia* and *Scorpidium scorpioides* (almost always on gabbro) (Proctor 1971, Hopkins 1983). Generally, however, the profiles are similar in providing an edaphic environment which is moderately base-rich, but not overwhelmingly calcareous, and also oligotrophic.

Some of the floristic characteristics of the community relate more obviously to the wetness of the soils. On the negative side, for example, a number of species which, on climatic grounds, might be expected to figure prominently here, are scarce. In comparison with the *U. gallii-Agrostis* heath, which extends through the South-West Peninsula but which is typical of more free-draining profiles, *Agrostis curtisii*, *Erica cinerea* and even the ubiquitous *Calluna*, are infrequent here and, when they are found, are typically growing epiphytically on the *Schoenus* or *Molinia* tussocks, where they are protected from any adverse effects of prolonged waterlogging (Ivimey-Cook 1959, Bannister 1965, Jones & Etherington 1970, Jones 1971*a*, *b*). On the Lizard, the edaphic preferences of the two heath types are sharply illustrated where the serpentine or gabbro become mantled with more pervious loess or Crousa Gravels, over which the *E. vagans-Schoenus* heath is replaced by the *U. gallii-Agrostis* type. This floristic difference, together with the corresponding vigour of *E. tetralix* and *Molinia* here, particularly in the Typical sub-community where the soils become aerated in spring and summer, confirms the general similarity of this kind of vegetation to the *Ericetum tetralicis* wet heath.

The presence of *Schoenus nigricans* provides one striking difference between the two communities, although it is not as surprising as all that. Certainly, though combinations of *Schoenus* and *Molinia* occur commonly elsewhere, sometimes also with *E. tetralix*, mixtures of these three species with *U. gallii*, even without the additional presence of *E. vagans*, are unusual: this is true not only of Britain but throughout western Europe (Coombe & Frost 1956*a*). But, on edaphic and climatic grounds, there is nothing especially

odd about the vigour of *Schoenus* in the community. In general, it is a lowland plant, restricted to areas with a February minimum above freezing (Sparling 1968), so it can thrive in the relatively frost-free climate of the Lizard. And, though it is characteristic of wet and oligotrophic soils, these can vary greatly in their base-status and calcium carbonate content. Through much of mainland Britain, and particularly in the more continental east, it is characteristic of base-rich and calcareous soligenous mires, occurring pre-eminently in the *Schoenetum*. But, with the move to a more oceanic climate, it extends first into more base-poor soligenous vegetation, like the *Schoenus-Narthecium* mire of south-west Britain and flushed *Scirpus-Erica* wet heath of western Scotland, and then, in Ireland, on to the acidic ombrogenous peats of blanket mires. Within this sequence, the habitat of the *E. vagans-Schoenus* heath occupies an intermediate position between those of the *Schoenetum* and the *Schoenus-Narthecium* mire as far as base-status and calcium content are concerned. The calcium content is very much less than that of the *Schoenetum* soils, though the substitution of magnesium in the exchange complex means that there is some considerable overlap in the pH range of the two communities; in the *Schoenus-Narthecium* mire, on the other hand, pH is almost always lower than here because of the general cation-poverty of the substrates. The increased tolerance in *Schoenus* of more acidic and less calcareous environments in moving further west was attributed by Sparling (1962*b*, 1967*a*, *b*) to the greater oceanicity of the climate, perhaps working through dilution of the aluminium levels in the substrate. Quite apart from the relative dampness of the environment in the Lizard, the serpentine soils are also inherently poor in aluminium (Flett 1946).

Furthermore, the wet and oligotrophic character of these soils is also reflected in the herbaceous and cryptogam associates of the community. And, once again, given the edaphic conditions here, there is nothing especially unusual about the suite of species that marks out this community both from the other kinds of *Schoenus* vegetation in Britain and among the oceanic heaths of the south-west; in particular there is no floristic feature that is uniquely related to the predominance of serpentine among the soil parent materials. The group of species, *Anagallis tenella*, *Serratula tinctoria*, *Succisa pratensis*, *Carex panicea*, *C. pulicaris*, *C. flacca*, *Sanguisorba officinalis* and, occurring less frequently, *Stachys betonica* and *Hypericum pulchrum*, is of fairly diverse floristic affinities but they are all characterised by a tolerance of or, in some cases, a distinct preference for, soils of intermediate base-status and calcium content; and they can also be found together, in various combinations, in, for example, Molinietalia poor fens, surface-leached but moist calcicolous grasslands, and in calcifugous grasslands and heaths in which there is some modest amelioration of base-poor conditions by flushing. Compared with the *Schoenetum*, there is a poor representation of more strictly calcicolous plants: among the sedges, *C. hostiana* occurs on the gabbro but, even among the bryophytes, which are in close contact with the ground waters in the runnels, it is species such as *Campylium stellatum*, *Riccardia multifida* and *Riccardia sinuata* which predominate. On the other hand, although the *Eleocharis* sub-community comes a little closer to the *Schoenus-Narthecium* mire than Typical *E. vagans-Schoenus* heath, even there, base-intolerant herbs and Sphagna are absent, maintaining the negative definition in the other direction. This group of species also provides a good distinguishing feature from other lowland heaths and, occurring together with more calcifugous plants like the ericoids and *Potentilla erecta*, produces a striking impression that is really only closely approached in flushed wet heath of the *Succisa-Carex panicea* sub-community of the *Ericetum tetralicis*, a vegetation type of wet gravels and sands of similar pH.

In the past, the wet and oligotrophic character of the soils at present occupied by the *E. vagans-Schoenus* heath probably afforded some protection against the primitive type of land improvement – paring, burning, scattering of fish-waste and seaweed, followed by a few seasons' cultivation – that seems to have occurred both inside and outside the enclosed fields on the Lizard, though it is possible that the tussocky character of the vegetation masks any remaining surface signs of such activity (Coombe & Frost 1956a, Hopkins 1983). It seems likely, however, that much of the ground was subject to the cutting of 'turf' which, particularly in the fourteenth and fifteenth centuries, was of great importance as a fuel for tin-smelting and which was also used domestically (Hopkins 1983). Exactly what this 'turf' was, whether always a distinct layer of peat or sometimes more of a thin paring of humus from the soil surface, is unknown (Coombe & Frost 1956a), but it looks as if the *E. vagans-Schoenus* heath is largely a secondary vegetation type which has developed on the wetter surfaces between the mounds of loess and Crousa Gravels after perhaps several extensive bouts of such stripping (Hopkins 1983, Staines 1984). In some cases, the development of more extensive runnels or pans within stands of the community could still mark out surface irregularities produced by such activities.

How important burning was in the past is unknown. Certainly, today, it is of major significance to the character of this vegetation though, except within the small proportion of enclosed heathland, such fires are now uncontrolled. With the accumulation of litter and dead sub-shrub branches, often substantial in the absence of grazing and gorse-gathering, and with the demise of 'turf'-cutting which created fire-breaks, and the increase in visitors, burning is, however, frequent. The effect of light, winter burns, especially where these are well controlled within enclosed stands, is slight but, in spring and summer, when the soil surface and litter begin to dry out, fires can be ferocious and, where accidentally started, difficult to bring under control. Not that the effects of such burning need be detrimental: indeed, ultimately, they help maintain the community against the invasion of shrubs and trees (see below), and, by destroying the large quantities of *Molinia* and *Schoenus* litter that can accumulate and by setting back the development of a closed sub-shrub canopy, they create an opportunity for the short-term expansion of certain species, such as *Genista anglica*, *Anagallis tenella* and *Campylium stellatum* and, more generally, increase the species-richness of the vegetation. This is particularly true of stands where *Schoenus* and *Molinia* show a strongly-tussocky structure (something which may be related to fluctuation of the water-table: Hopkins 1983), where burning allows a fuller exploitation of the various microhabitats of the runnels, tussock sides and tussock tops. In such situations, the diversity of the community is strongly related to the interaction between time since the last burn and the degree of contrast between the wet and more base-rich runnels and dry and more base-poor tussocks.

Grazing which, since the 1930s and 1940s, has occurred only in the enclosed heaths of the Lizard, might be expected to produce similar effects to fire by preventing the accumulation of standing herbage and litter, but Hopkins (1983) found it difficult to assess its precise impact because of the fact that it is always accompanied by burning. Although he could detect no significant floristic difference between grazed and ungrazed stands, it is likely that trampling by stock enhances the development of the tussock/runnel structure, especially where the animals have access while the ground is moist.

The seasonal waterlogging of the ground means that, without drainage, which is anyway difficult over such gently-sloping land, the period of risk-free cultivation of the soils under the *E. vagans-Schoenus* heath is short. And, though the mild, moist climate promises potentially large yields of grass, there is often the continuing problem of poaching for much of the year. In addition to this, the inherent oligotrophic character of the soils necessitates the application of fertilisers (Findlay *et al.* 1984). Much of the cultivation on the Lizard has thus been confined to the more naturally eutrophic and better-drained soils beneath the *E. vagans-Ulex europaeus* heath and the loess-derived profiles under the *U. gallii-Agrostis* heath, although there has probably been some loss of the community over the more calcium-rich gabbro. Many remaining tracts fall within nature reserves or notified sites.

Zonation and succession

The *E. vagans-Schoenus* heath forms a major component of the mosaic of sub-shrub communities and other vegetation types developed on the Lizard in relation to variations in soil moisture, base-status and trophic level and the amount of salt-spray deposition. The pattern of past and present treatments often influences the effect of these factors; and burning and, to a lesser extent, grazing still also play a major role in maintaining the community in its typical form.

One of the more gradual kinds of transitions is found where the soils retain the generally wet and base-rich character associated with the *E. vagans-Schoenus* heath, but seem to become somewhat more fertile with, in particular, increased availability of phosphate (Hopkins 1983), a feature which may reflect a switch in bedrock from serpentine to gabbro and somewhat less lengthy winter flooding, but which can also be associated with disturbance alongside walls, ditches and roadways. Then the community is replaced by the *Molinia-Potentilla* mire in which *Schoenus* can retain some representation, but where dominance shifts markedly to *Molinia*, with *E. vagans*, *E. tetralix* and *U. gallii* still present quite frequently but usually exceeded in abundance by *U. europaeus*, its bushes rooted in the often very large *Molinia* tussocks. Among the associates, such distinctive smaller herbs as *Anagallis tenella*, *Carex pulicaris* and *Festuca ovina* are usually overwhelmed by the accumulation of thick *Molinia* litter, and taller plants like *Serratula* and *Succisa* are replaced by *Angelica sylvestris* and *Eupatorium cannabinum*.

In such vegetation on the Lizard (which Hopkins (1983) included within the *E. vagans-Schoenus* heath as a *Ulex* variant), the climate is sufficiently mild for *Cladium* to gain a local hold in sites which are flooded only for very brief periods, as in the Kynance valleys, on Goonhilly Down and at Main Dale. Elsewhere, increased abundance and vigour of *Cladium* and the appearance of *Phragmites* on permanently-flooded soils along stream sides and in hollows marks a continuation of the zonation into sedge-dominated *Phragmites-Eupatorium* fen or *Cladietum marisci* swamp. In other places, the *E. vagans-Schoenus* heath grades more directly to fen vegetation through the *Eleocharis* sub-community, *Phragmites* becoming more abundant and vigorous in ungrazed transitions. Except where *Cladium* is overwhelmingly dominant, the *Molinia-Potentilla* mire and fens of such sequences seem to be more susceptible to shrub invasion than the *E. vagans-Schoenus* heath itself, perhaps because of their more eutrophic soils and, in the absence of burning, they are readily colonised by *Salix cinerea* to form stands of the *Salix-Galium* woodland (Hopkins 1983).

In other situations, the passage to the *Molinia-Potentilla* mire can denote a transition, over the lower flushed slopes of the shallow valleys that wind through the interior of the headland, to the *E. vagans-Ulex* heath, the community centred on the more free-draining base-rich brown earths of shedding ground on the serpentine and gabbro. In the *Molinia* sub-community of that kind of heath, typical of slightly gleyed profiles above the limit of seasonal inundation, *Schoenus* is reduced to low cover and *Molinia*, though abundant, is not strongly tussocky. In addition to *E. vagans*, *E. tetralix* and *U. gallii* remain frequent and *Serratula tinctoria* and *Stachys betonica* occur occasionally. But, again, *U. europaeus* becomes very common and prominent and, on the somewhat drier ground, there is an increase in *E. cinerea* and *Calluna*. Among the herbs, the appearance of *Sanguisorba officinalis* and *Viola riviniana* is the most obvious indication of the switch from one community to another. This vegetation may in turn give way to Typical *E. vagans-Ulex* heath over free-draining brown earths with local patches of *Ulex-Rubus* scrub or *Prunus* scrub where freedom from burning has permitted invasion of other woody plants.

By and large, it is through a zone of the *E. vagans-Ulex* heath that the *E. vagans-Schoenus* heath grades to the maritime *Calluna-Scilla* heath of the coastal cliff-top fringe. Locally, however, in very sheltered situations, the *E. vagans-Schoenus* heath can itself be found in cliff-top flushes, though it should be noted that such vegetation is different in composition from the *Schoenus*-dominated flushes found in exposed situations near the foot of cliffs on the Lizard: these are best placed in the *Festuca-Daucus* grassland.

More frequent than transitions to drier, base-rich soils, many of which, at least on the gabbro, are now cultivated and thus marked by an abrupt switch to improved grassland, are zonations to the base-poor and somewhat better-drained profiles developed from the loess and Crousa Gravels which are superimposed over the ultrabasic bedrocks. Characteristically, such soils are occupied by the *Ulex gallii-Agrostis* heath which thus forms irregular patches on the gentle interfluves of the peninsula, surrounded and broken up by stretches of the *E. vagans-Schoenus* heath. Over the margins of less well-defined areas of superficials, where the cover thins gradually or where loess or gravel has been mixed with the underlying base-rich parent material in bouts of past cultivation, the transition between the two heath types is correspondingly indistinct with *E. vagans*, *Serratula* and *Sanguisorba officinalis* running into the *U. gallii-Agrostis* heath in a marginal zone 2–10 m wide (Hopkins 1983). In other cases, steeper-edged banks of loess or gravel show an abrupt replacement of the one heath by the other, *E. vagans* and *Schoenus* both disappearing, together with most of their herbaceous associates, *E. cinerea*, *Calluna* and *Agrostis curtisii* increasing in frequency and cover to become co-dominant with the *E. tetralix*, *U. gallii* and *Molinia*.

This basic soil-related pattern has been affected by treatments in a variety of ways. Some activities, strongly controlled by the natural drainage quality or fertility of the ground, have had a more ordered impact which has often confirmed the edaphic influence, as in the case of 'turf' stripping, largely confined to the humic stagno-gleys of the serpentine and gabbro, or the primitive shifting cultivation of the past which seems to have been concentrated on the thinner deposits of loess and over the more nutrient-rich brown earths on the gabbro. Enclosure for grazing and the present pattern of burning, on the other hand, cut across the edaphic boundaries, so their effects can confuse the zonations described above. Fires, for example, particularly severe summer burns, can impose a structural uniformity within a tract of heathland, whose soil-related mosaic may take some years to grow out again. Within the communities of wetter soils here, *Molinia* is the plant likely to become temporarily dominant across such boundaries following fires, though on the poorer soils its vigour is reduced. Burning, with subsequent severe grazing, especially on more eutrophic soils or ones enriched with fertilisers, is likely to convert the *E. vagans-Schoenus* heath (and wetter stands of the *E. vagans-Ulex* and *U. gallii-Agrostis* heaths) into *Molinia-Potentilla* mire.

For the most part, however, burning, sometimes combined with grazing in the enclosures, has the effect of maintaining the different sub-shrub communities in a complex patchwork of vegetation in various stages of cyclical regeneration. Except on the most infertile soils, such treatments probably play some part in setting back the invasion of shrubs and trees though, over much of the headland, the scarcity of seed-parents and exposure to severe winds militate against rapid development of scrub and woodland. On the wetter soils, the natural eventual successor to the *E. vagans-Schoenus* heath is probably *Salix-Galium* woodland or *Salix-Betula-Phragmites* woodland; on seasonally-waterlogged profiles, some kind of *Betula-Molinia* woodland seems a likely development.

Distribution
The community is confined to the Lizard in Cornwall, where it makes up a major proportion of the open and enclosed heaths of the hinterland of the peninsula.

Affinities
The very striking character of the *E. vagans-Schoenus* heath, only vaguely hinted at in early studies (Tansley 1911, 1939), was first revealed in detail by Coombe & Frost (1956a) whose descriptive account has recently been expanded in a systematic phytosociological investigation of the vegetation of the Lizard by Hopkins (1983). The diagnosis here relies heavily on Hopkins's data and in essence the community differs only in the transfer of his *Ulex* variant into the *Molinia-Potentilla* mire.

In their discussion, Coombe & Frost (1956a) recognised the distinctively mixed nature of the communities,

though, at the time of their work, accounts of related heaths and mires were insufficiently detailed to permit more than a brief comparison of the vegetation types: now, these relationships are a little clearer, though this kind of heath remains difficult to place in existing phytosociological frameworks, a problem it shares with the other sub-shrub vegetation in which *E. vagans* plays a prominent role in Cornwall and north-west France (e.g. Dupont 1973, Vanden Berghen 1973, Hopkins 1983). In a more generalised scheme, like that of Böcher (1943), there is little difficulty in locating the British *E. vagans* heaths in an intermediate position in the sequence of communities running down the western oceanic seaboard of Europe, with affinities northwards to the *Ulex* heaths of the Ericion cinereae and southwards to the Ericion scopariae heaths. Böcher (1943) himself favoured recognising an '*Erica vagans* group' in the latter alliance, though he had very little data on which to base his diagnosis (Gadeceau 1903, Rübel 1930). Coombe & Frost (1956a) echoed this general view but, with their more detailed knowledge of the Lizard communities, considered that only the *E. vagans-Ulex* heath could be readily incorporated in the Ericion scopariae. They also stressed the floristic relationships that can be seen when British sub-shrub communities are examined on an east–west, rather than a north–south, axis. In such a light, the *E. vagans* heaths clearly belong among the *U. gallii* communities which replace the *U. minor* complex on moving westwards: this kind of perspective would locate them in the Ulicion gallii of des Abbayes & Corillion (1949). Alternatively, much more weight can be given to the presence of *E. vagans* itself as in the scheme of Géhu (1975), where this species and *E. ciliaris* are diagnostic of a Ulici-Ericion ciliaris, an alliance confined to south-west England and north-west France (with small outliers in Ireland: White & Doyle 1982).

At the present time, without a comprehensive examination of variation in heaths throughout western Europe, none of these solutions seems entirely satisfactory. One difficulty is that, within the range of British sub-shrub communities, the presence of *E. vagans* in both this and the *E. vagans-Ulex* heath is not accompanied by any other floristic features that mark out the vegetation types as distinctly southern European. Certainly, the geographical confinement of *E. vagans* to the Lizard is very striking, but the overall floristic affinities of the communities it occurs in are not directly related to any climatic peculiarities of the peninsula itself, rather to the generally oceanic conditions of the south-western part of Britain. Among the dry heaths, therefore, the communities are closest to the *U. gallii-Agrostis* and *Calluna-U. gallii* heaths.

However, the other difficulty in locating the *E. vagans* heaths together is that they are rather different from one another. If the *E. vagans-Schoenus* heath is considered on its own, its closest relationships among other sub-shrub communities are much more obvious: they lie with the wet heaths of the Ericion tetralicis, particularly those types in which base-poverty is relieved by flushing, as in the *Succisa-Carex* sub-community of the *Ericetum tetralicis* and the *Carex* sub-community of the *Scirpus-Erica* wet heath, in the latter of which *Schoenus* is a very occasional local dominant.

Alternatively, as suggested by Coombe & Frost (1956*a*) and favoured by Hopkins (1983), the *E. vagans-Schoenus* heath can be related to other vegetation in which *Schoenus* plays a major role, more particularly to what the former authors knew as the *Schoeneto-Juncetum* of Lemée (1937), but which has been more recently defined in Britain, largely as a result of the work of Wheeler (1975, 1980*c*). In Wheeler's *Schoeno-Juncetum*, sub-shrubs like *E. tetralix* and *Calluna* can make a locally prominent contribution, together with some poor-fen and wet-heath plants, in an *ericetosum*, and Hopkins (1983) proposed incorporating the *E. vagans-Schoenus* heath as a new sub-community intermediate between this type of *Schoeno-Juncetum* and the *serratuletosum* of Wheeler (1980*c*). In our scheme, in fact, Wheeler's (1980*c*) *ericetosum* is largely transferred to the *Schoenus-Narthecium* mire and the floristic affinities of the *E. vagans-Schoenus* heath to both this and our *Carex-Serratula* sub-community of a revised *Schoenetum* have already been remarked upon. But the differences in both directions are also quite strong and the community could not readily be subsumed into either. It is better, then, to recognise the uniqueness of this kind of heath but to acknowledge that its peculiar character derives from various sources, not simply from the oceanic conditions, nor particularly from the predominance of serpentine among the substrates, but from a complex of climatic and edaphic features which overlap and interact in this particular way only on the Lizard.

Floristic table H5

	a	b	5
Erica vagans	V (3–8)	V (1–5)	V (1–8)
Schoenus nigricans	V (4–8)	V (5–8)	V (4–8)
Erica tetralix	V (1–6)	V (2–5)	V (1–6)
Serratula tinctoria	V (1–4)	V (2–5)	V (1–5)
Molinia caerulea	V (2–9)	IV (4–7)	V (2–9)
Anagallis tenella	IV (1–4)	V (1–3)	IV (1–4)
Campylium stellatum	IV (1–5)	V (2–5)	IV (1–5)
Succisa pratensis	IV (1–4)	V (2–4)	IV (1–4)
Carex pulicaris	IV (1–3)	IV (1–3)	IV (1–3)
Festuca ovina	V (1–5)	III (1–3)	IV (1–5)
Potentilla erecta	IV (1–4)	III (2–3)	IV (1–4)
Ulex gallii	IV (1–7)	III (1–6)	IV (1–7)
Genista anglica	III (1–3)	I (3)	II (1–3)
Carex panicea	III (2–4)		II (2–4)
Fissidens adianthoides	II (1–2)	I (1–2)	II (1–2)
Platanthera bifolia	II (1–3)	I (2)	I (1–3)
Calluna vulgaris	II (1–4)		I (1–4)
Dactylorhiza maculata ericetorum	II (1–3)		I (1–3)
Calypogeia fissa	II (1–3)		I (1–3)
Pedicularis sylvatica	II (1–3)		I (1–3)
Salix repens	II (2–4)		I (2–4)
Danthonia decumbens	II (1–3)		I (1–3)
Erica cinerea	I (2–7)		I (2–7)
Agrostis curtisii	I (1–4)		I (1–4)
Plantago maritima	I (1–3)		I (1–3)
Thymus praecox	I (1–2)		I (1–2)
Scutellaria minor	I (2–3)		I (2–3)
Linum catharticum	I (1–3)		I (1–3)

Floristic table H5 *(cont.)*

	a	b	5
Agrostis canina montana	I (1–3)		I (1–3)
Carex flacca	III (2–4)	IV (3–4)	III (2–4)
Eleocharis multicaulis	I (2)	V (1–4)	I (1–4)
Eriophorum angustifolium		V (1–5)	I (1–5)
Pinguicula lusitanica	I (1)	IV (1–3)	I (1–3)
Drosera rotundifolia	I (2)	IV (1–3)	I (1–3)
Phragmites australis	I (1–5)	III (2–3)	I (1–5)
Dactylorhiza incarnata incarnata		III (1–2)	I (1–2)
Riccardia multifida	III (1–3)	III (1–2)	III (1–3)
Riccardia sinuata	III (1–2)	III (1–3)	III (1–3)
Sanguisorba officinalis	III (1–4)	III (1–4)	III (1–4)
Carex hostiana	II (2–4)	III (2–3)	II (2–4)
Juncus acutiflorus	II (1–3)	III (1–2)	II (1–3)
Polygala vulgaris	II (1–3)	III (1–3)	II (1–3)
Scorpidium scorpioides	II (2–5)	II (2–3)	II (2–5)
Gymnadenia conopsea	II (1–2)	II (2–3)	II (1–3)
Hypericum pulchrum	II (1–2)	II (1)	II (1–2)
Stachys officinalis	I (1–2)	II (1)	I (1–2)
Galium uliginosum	I (1–2)	II (1)	I (1–2)
Ulex europaeus	I (4–5)	I (2)	I (2–5)
Angelica sylvestris	I (2)	I (2)	I (2)
Eupatorium cannabinum	I (4)	I (1)	I (1–4)
Eurhynchium praelongum	I (1–3)	I (1)	I (1–3)
Juncus maritimus	I (3–5)	I (4)	I (3–5)
Number of samples	51	9	60
Number of species/sample	19 (10–29)	20 (16–27)	19 (10–29)
Vegetation height (cm)	34 (14–90)	29 (17–45)	33 (14–90)
Vegetation cover (%)	98 (85–100)	91 (65–100)	97 (65–100)
Slope (°)	2 (0–5)	1 (0–4)	2 (0–5)
Soil pH	6.5 (5.4–7.2)	6.7 (6.5–7.1)	6.5 (5.4–7.2)

a   Typical sub-community
b   *Eleocharis multicaulis* sub-community
5   *Erica vagans-Schoenus nigricans* heath (total)

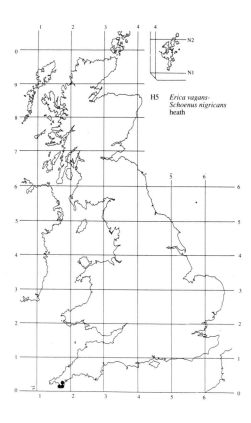

H5 *Erica vagans-
Schoenus nigricans*
heath

H6
Erica vagans-Ulex europaeus heath

Synonymy
Erica vagans-Ulex europaeus 'Mixed Heath' Coombe
& Frost 1956a *p.p.*; *Ulici maritimi-Ericetum vagantis*
(Géhu 1962) Géhu & Géhu 1973 *emend.* Bridgewater
1980; *Erica vagans-Ulex europaeus* heath Hopkins
1983 *p.p.*

Constant species
Agrostis canina ssp. *montana, Carex flacca, Erica
cinerea, E. vagans, Filipendula vulgaris, Ulex europaeus,
U. gallii, Viola riviniana.*

Rare species
*Agrostis curtisii, Allium schoenoprasum, Erica vagans,
Juncus capitatus, Scilla verna, Trifolium bocconei.*

Physiognomy
The *Erica vagans-Ulex europaeus* heath is a distinctive
kind of sub-shrub vegetation, but one that is rather
variable in its floristics and structure, a feature recognised in Coombe & Frost's (1956a) epithet of 'Mixed
Heath'. In fact, as defined here, the community excludes
some of the vegetation which these authors and also
Hopkins (1983) considered as falling within their *Erica-
Ulex* heaths, but it still encompasses stands which differ
quite markedly from one another, variation which can
be related partly to edaphic and local climatic factors
and partly to treatment, especially the frequency and
intensity of burning. The influence of the former factors
can be seen in the ordered differentiation of the sub-
communities; the effects of treatment tend to cross-cut
this variation, resulting in complex mosaics of vegetation in all stages of regeneration.

The most obvious constant feature of the community
is the mixed canopy of sub-shrubs in which *Erica vagans*
and *Ulex europaeus* are the usual co-dominants, a
combination which is not confined to this kind of heath
but which is most often found here. In newly-burned
stands, the cover of these and the other sub-shrubs can
be low, with masses of new shoots sprouting from half-
buried stools. And in extreme environmental conditions, where this heath is found as the *Festuca* sub-
community over summer-parched soils in exposed situations, even old stands can have small, scattered wind-
pruned bushes no more than 1 dm high and with a total
cover of less than 20%. Generally, however, the canopy
grows taller than this, most often 3–6 dm high, even
more in sheltered sites, and is much more extensive,
covering 60–90% of the ground. In the immediate
aftermath of fires, *E. vagans*, which regenerates rapidly,
may be the most prominent species but, with the passage
of time, *U. europaeus* matches it in cover and height and
usually overtops it after 7 or 8 years.

The two other constant sub-shrubs, *U. gallii* and *E.
cinerea*, can also be present in some abundance though,
except where sheltered, they tend to be shorter in stature
than the above and, indeed, in very dense covers, *U.
gallii* can be totally supressed. *Calluna vulgaris* is also
rather poorly represented in such stands and in some
recently-burned tracts, so much so that, overall, it fails
to achieve constancy and, even in the more open kinds of
heath, its cover tends to be low. In those rare stands
which have escaped burning for long periods, the sub-
shrubs can become degenerate and leggy, and are occasionally accompanied by brambles and *Prunus
spinosa*, features which give the vegetation some of the
character of *Ulex-Rubus* or *Prunus* scrub, as in Coombe
& Frost's (1956a) 'old Mixed Heath' and in stands which
Hopkins (1983) described from Kynance. The persistence of *E. vagans*, a long-lived species which can outlast
U. europaeus perhaps five-fold, can help separate such
transitional vegetation.

In contrast to the *E. vagans-Schoenus* heath, *E. tetralix* is only occasional overall here and is very much
confined to the *Molinia* sub-community of wetter soils,
where high frequency of *Molinia caerulea* itself, and of
Schoenus nigricans, Serratula tinctoria and *Sanguisorba
officinalis* also give some indication of a floristic overlap
with this other major kind of *E. vagans* heath. By and
large, however, the two communities are well marked off

from one another in their associated floras: apart from the sub-shrubs which they have in common, only *Carex flacca*, *Potentilla erecta* and *Polygala vulgaris* occur at all frequently in both heaths and the prevailing element among the herbs here comprises plants of well-drained, moderately base-rich soils. Some of these species are also quite well represented in the more mesophytic kinds of *U. gallii-Agrostis* heath, and the *E. vagans-Ulex* heath shows some measure of continuity with that vegetation, particularly where it extends on to loess-contaminated soils in the *Agrostis* sub-community. In general, however, *A. curtisii* is not a common plant here.

Apart from *Carex flacca*, *Polygala vulgaris* and *Potentilla erecta*, the most common and distinctive herbs of the *E. vagans-Ulex* heath as a whole are *Viola riviniana*, *Filipendula vulgaris*, *Stachys betonica*, *Hypochoeris radicata*, *Agrostis canina* ssp. *montana*, *Dactylis glomerata* and, here on the Lizard, extending further into the inland heaths than usual, *Scilla verna*. In stands burned relatively recently, where the sub-shrubs have not yet coalesced into a more or less continuous canopy, most of these species will be found and sometimes they can form a fairly extensive and luxuriant understorey. With the increasing shade and accumulation of litter that comes with age, however, the flora becomes more and more confined to areas between the bushes, and then very attenuated, such that older stands have just a few scattered individuals of the more tolerant species, *Carex flacca*, *Stachys*, *Viola*, *Filipendula* and *Scilla*, plants of the last two often looking particularly etiolated and with much reduced flowering.

On shallower soils, where more inhospitable edaphic conditions keep the sub-shrubs in check, richness among the associated flora can be maintained even where burning is infrequent, especially if grazing also helps keep the herbage short. Then, the above species can be joined by *Festuca ovina*, *Thymus praecox*, *Lotus corniculatus*, *Galium verum*, *Jasione montana*, *Danthonia decumbens* and *Brachypodium sylvaticum* with, in the *Festuca* sub-community, a long list of further preferentials, some common, some just occasional, but together constituting a vegetation which looks very different from the stands dominated by tall sub-shrubs. In such heath, too, which is generally situated close to the cliff tops, some species more characteristic of maritime vegetation can occasionally figure, *Plantago maritima*, for example, together with *Genista pilosa*. In this scheme, however, heath in which such plants occur consistently with *E. vagans*, and often with much reduced frequencies for *U. europaeus* and *U. gallii* (what both Coombe & Frost (1956a) and Hopkins (1983) called the *Genista* variant) is considered as part of the *Calluna-Scilla* heath.

In the seasons immediately after burning, a different and more chaotic kind of diversity is to be seen in the flora that develops among the regenerating sub-shrubs

and the new shoots of the characteristic perennial herbs of the community. In the first autumn and spring, the most obvious feature is often the abundance and variety of ephemerals on the stretches of open ground. Some of these, such as *Aira caryophyllea* and *Centaureum erythraea*, are species which also find a place in bare patches within mature stands; others, such as *Cochlearia danica*, *Trifolium striatum*, *T. bocconei*, *Euphrasia tetraquetra* and various *Cerastium* spp., seed in from the more open heaths and therophyte vegetation around rocky cliff tops; yet others are widespread weed species like *Anagallis arvensis*, *Cardamine hirsuta*, *Myosotis arvensis*, *Spergula arvensis*, *Poa annua*, *Senecio vulgaris*, *Sonchus oleraceus* and *Cirsium vulgare*, or more local ephemerals which find occasional representation in various kinds of open vegetation, such as *Aphanes microcarpa*, *Erodium maritimum* and the introduced *Coronopus didymus*. These may then be succeeded, before the heath regains its characteristic balance of woody and herbaceous plants, by a temporary abundance of sub-community preferentials like *Molinia* or *A. curtisii* or community occasionals such as *Leucanthemum vulgare* or *Holcus lanatus*. Mixtures of various of these species were characteristic in Hopkins (1983) of a 'recently burned heath nodum'.

The continuing frequency of burning and the dense shade and thick litter in older stands both tend to inhibit the development of any extensive cryptogam flora in the *E. vagans-Ulex* heath, either on bare soil or growing corticolously on degenerate woody plants. No bryophytes or lichens occur commonly throughout, though *Hypnum cupressiforme s.l.* can sometimes be found, particularly in the shorter swards characteristic of the *Festuca* sub-community, where *Cladonia* spp. can also be well represented.

Sub-communities

Typical sub-community: *Erica vagans-Ulex europaeus* 'Mixed Heath', Typical, Regenerating and Old forms Coombe & Frost 1956a; *Erica vagans-Ulex europaeus* heath, Typical variant and recently-burned nodum and *Agrostis curtisii* variant Hopkins 1983 *p.p.* Typical *E. vagans-Ulex* heath shows the complete range of floristic and structural variation related to burning of this vegetation but, in stands a few years into the regeneration cycle, there is usually a well-developed sub-shrub canopy, several decimetres tall and with high cover, in which *E. vagans* and *U. europaeus* usually predominate with smaller amounts of *E. cinerea*, *U. gallii* and *Calluna*. *Viola riviniana*, *Carex flacca*, *Filipendula vulgaris*, *Stachys betonica* and *Potentilla erecta* are the commonest herbs and there are frequently some tufts of *Agrostis canina* ssp. *montana* and *Dactylis glomerata* with occasional *Anthoxanthum odoratum* or even locally abundant *A. curtisii*. Richer stands can have occasional records for *Geranium sanguineum* and *Teucrium scorodonia*, the only preferential dicotyledons here, and also for *Festuca ovina*, *Scilla verna*, *Thymus praecox*, *Polygala vulgaris*, *Hypochoeris radicata* and *Lotus corniculatus* but, with increasing age, even the constant herbs of the community begin to thin out, leaving little more than a dense woody cover. Among leggier sub-shrubs, *Rubus fruticosus* agg. and *Prunus spinosa* can occasionally be found, or sparse *Pteridium*.

Festuca ovina sub-community: *Erica vagans-Ulex europaeus* 'Mixed Heath', Typical form   Coombe & Frost 1956a *p.p.*; *Erica vagans-Ulex europaeus* heath, *Festuca ovina* variant   Hopkins 1983. In this sub-community, the abundance and height of the sub-shrub canopy are generally less than in the typical form: their total cover is often below 50% and they commonly attain less than 2 dm with, in more extreme cases, small, discrete bushes, wind-pruned and sometimes with obvious signs of droughting, confined to the deeper soils of crevices in the underlying bedrock. With their vigour thus inhibited, the associated flora tends to be richer here, particularly where it is grazed, when the vegetation can present the appearance of a varied grassy sward forming a mosaic among the bushes. Apart from *Potentilla erecta*, the common herbs of the community are all well represented and, compared with the typical kind of heath, *Festuca ovina*, *Thymus praecox*, *Scilla verna*, *Hypochoeris radicata*, *Polygala vulgaris* and *Lotus corniculatus* show consistently higher frequencies. In addition, *Danthonia decumbens*, *Koeleria macrantha*, *Aira caryophyllea*, *Carex caryophyllea*, *Galium verum* and *Leontodon taraxacoides* are strongly preferential with, occurring less commonly, *Jasione montana*, *Anthyllis vulneraria*, *Daucus carota* and *Holcus lanatus*. Other occasional herbs include *Hypericum pulchrum*, *H. humifusum*, *Pedicularis sylvatica*, *Festuca rubra*, *Plantago maritima* and *Centaureum erythraea*. In the more open conditions, cryptogams are rather better represented than usual, with *Hypnum cupressiforme s.l.* frequent, *Trichostomum brachydontium* and *Weissia* spp. occasional, and various *Cladonia* spp. common, particularly *C. impexa*, *C. fimbriata*, *C. furcata* and *C. subrangiformis*.

Agrostis curtisii sub-community: *Erica vagans-Ulex europaeus* 'Mixed Heath', Typical form   Coombe & Frost 1956a *p.p.*; *Erica vagans-Ulex europaeus* heath, *Agrostis curtisii* variant   Hopkins 1983 *p.p.* Apart from occasional occurrences in otherwise fairly orthodox stands of the Typical sub-community, *A. curtisii* is largely confined to this kind of *E. vagans-Ulex* heath and it can be very abundant here, especially after burning. *Molinia*, too, is frequent, and the two grasses sometimes dominate beneath the usually rather open canopy of *E.*

vagans, *E. cinerea* and the two gorse spp. *Calluna* is unusually scarce in this sub-community but *E. tetralix* can occasionally be found. Among the herbs, too, there are some obvious omissions with *Viola riviniana* totally absent and *Filipendula vulgaris* scarce. *Danthonia decumbens* and *Potentilla erecta*, on the other hand, occur very frequently, *Stachys betonica* and *Serratula tinctoria* are common and, more strikingly preferential, *Hypericum pulchrum*, *Viola lactea* and *Polygala serpyllifolia*.

Molinia caerulea sub-community: *Erica vagans-Ulex europaeus* 'Mixed Heath', *Molinia caerulea* variant   Coombe & Frost 1956a; *Erica vagans-Ulex europaeus* heath, *Molinia caerulea* variant   Hopkins 1983. *E. vagans* and *U. europaeus* retain high frequency and abundance here, with generally smaller amounts of *U. gallii*. Both *E. cinerea* and *Calluna* have somewhat reduced frequencies and covers but more striking in this element of the vegetation is the constant presence of a little *E. tetralix*. In stands not recently burned, the sub-shrub canopy shows the kind of vigorous growth usual in the Typical sub-community but here the most distinctive feature in the associated flora is the occurrence beneath of *Molinia*, usually not strongly tussocky but quite often abundant. The shade of the sub-shrubs and the thick *Molinia* litter depress the rich development of other herbs and generally only *Carex flacca*, *Viola riviniana*, *Potentilla erecta* and *Stachys betonica* occur with any frequency among the community species. Preferentially, however, there can be a little *Serratula tinctoria*, *Sanguisorba officinalis*, *Schoenus nigricans* and *Carex pulicaris*.

Habitat

The *E. vagans-Ulex* heath is confined to the Lizard in Cornwall where it is characteristic of free-draining brown earths, usually quite base-rich but calcium-poor and fairly oligotrophic. Edaphic variation and local differences in the generally warm and sunny oceanic climate have a strong influence on floristic diversity within the community but treatments, especially burning and, to a lesser extent, grazing, also have a marked effect on the composition and physiognomy of the vegetation. Preferential cultivation of the more fertile soils developed over gabbro and schists mean that the community survives most extensively over serpentine and many tracts of this unique vegetation type are now included within statutory or voluntary reserves.

   This community occurs within the same part of Cornwall as the *E. vagans-Schoenus* heath, an area of Britain with a strikingly oceanic climate (Coombe & Frost 1956a, Malloch 1970, Hopkins 1983). Here, on the Lizard, the winters are very mild, with accumulated temperatures between December and March of more than 55 day-degrees C (Page 1982) and usually less than 20 frosts annually (*Climatological Atlas* 1952), resulting in an almost year-round growing season (Fairburn 1968). The summers are warm, with mean annual maximum temperatures around 26 °C (Conolly & Dahl 1970), and insolation is high. This, coupled with the fact that precipitation, which stands at about 900 mm annually, shows a clear minimum in spring and early summer (Meteorological Office 1977), means that there is a marked potential water deficit at this time of the year (Malloch 1970, Hopkins 1983).

The direct influence of these climatic features can be seen in the general composition of the community. Along an east–west axis of floristic variation among our heath types, the *E. vagans-Ulex* heath, like the *E. vagans-Schoenus* heath, continues the Oceanic West European trend that can be seen already in the widespread southwestern *U. gallii-Agrostis* heath where *U. gallii*, together with *E. cinerea* and/or *E. tetralix* maintain a strong representation. And again, here, there is the distinctive high frequency and abundance of *E. vagans*, for which this community provides the second of its two major loci in Britain, continuing the oceanic trend that runs on into Ireland where *E. mediterranea* and *E. mackiana* figure in the heaths, and representing a link with the 'Lusitanian' flora of more southerly parts of the western seaboard of Europe (Matthews 1955).

For the most part, however, it is interactions between climate and the soil parent materials that control the floristic distinctions between the two most important kinds of *E. vagans* heath and which help distinguish this community from other types of sub-shrub vegetation. For, whereas the *E. vagans-Schoenus* heath is essentially a community of wet, base-rich soils, the profiles here, though of roughly similar pH (generally between 5 and 7), are typically free-draining. The parent materials beneath the two heaths are often identical but the *E. vagans-Ulex* heath extends but a short way on to the flat and gently-undulating ground of the Lizard hinterland where seasonally-waterlogged stagnogleys predominate on the drift-free areas (Coombe & Frost 1956*b*, Hopkins 1983, Staines 1984). Rather, it is typically found over steeper, shedding slopes, the kind of ground that is of little extent over much of the plateau of the peninsula where the valleys are wide and shallow, but which becomes much more extensive around the coves where the streams cut down sharply to the sea, and which runs too around much of the cliff tops of the headland. The distribution of this community within the Lizard is thus predominantly coastal, though it is not strictly speaking a maritime heath and is replaced on slopes exposed to salt-spray by the *Calluna-Scilla* heath (see below).

The characteristic soils of such situations are well-drained and loamy, sometimes showing a clear differentiation of a B horizon, when they can be classed as brown earths, but quite often comprising a fairly homogeneous A horizon over rotting bedrock, when they should more strictly be termed rankers (Avery 1980): for the most part, they probably fall within the Kynance and Black Head series mapped within the Lizard Croft Pascoe association (Staines 1984, Findlay *et al.* 1984). Over the schists and gabbro of the headland, however, such profiles are moderately lime-rich and fairly fertile so, around Predannack and Lizard Point, where the former rock is exposed, and north of Coverack, where the latter runs to the sea, many stretches have been taken into cultivation, except where the topography has proved too intractable. Mostly, then, the *E. vagans-Ulex* heath survives over serpentine, where the brown earths and rankers are more oligotrophic, and where the exchange complex is dominated by magnesium rather than calcium, though where, free of drift contamination, base-status remains high.

However, it is not the particular ultrabasic character of the profiles that has the most obvious influence on the vegetation, but rather their generally base-rich and free-draining nature, although, combined with other environmental features on the Lizard, this does produce some rather peculiar effects. First, there is the prominence of *U. europaeus* which, over the headland, is one of the best single indicators of the distribution of these profiles out of reach of heavy salt-spray deposition (Coombe & Frost 1956*a*). In no other kind of heath on the Lizard, indeed throughout Britain, is it so consistent a member: in many areas, of course, more free-draining soils have been even more extensively brought into cultivation, such that *U. europaeus* now most often marks out places where disturbance has brought some amelioration of highly impoverished and/or waterlogged conditions within heath communities where it would not normally find a place. In fact, this is sometimes the case on the Lizard itself and the *E. vagans-Ulex* heath as a whole can sometimes indicate old areas of cultivation, boundaries and tracks on more ill-drained parts of the plateau.

Then, there is the move in the associated flora away from the waterlogging-tolerant assemblage characteristic of the *E. vagans-Schoenus* heath. *E. cinerea*, for example, largely replaces *E. tetralix* here and *Molinia*, *Schoenus*, *Serratula* and *Sanguisorba* are all of restricted occurrence. These species occur together with the typical plants of the *E. vagans-Ulex* heath only in the *Molinia* sub-community which represents one extreme of this kind of heath. It is characteristic of more gentle slopes and deeper soils than usual (Hopkins 1983) and extends the distribution of the community some way on to the flatter cliff tops and valley surrounds, where the brown earth soils are still free-draining but gently flushed or where there is some slight gleying below. Even in this vegetation, however, *Molinia* and *Schoenus* do not have the strong structural influence that they exert in the *E.*

vagans-Schoenus heath where they are markedly tussocky: perhaps this is because here fluctuation of the ground waters never breaches the soil surface (Hopkins 1983).

By and large, however, it is more mesophytic plants, tolerant of oligotrophic soils which are circumneutral or fairly base-rich but not lime-saturated, that characterise the community. Species such as *Carex flacca, Viola riviniana, Filipendula vulgaris, Agrostis canina* ssp. *montana, Stachys betonica, Dactylis glomerata, Polygala vulgaris, Lotus corniculatus, Festuca ovina* and *Thymus praecox* all fall into this category and constitute the core of the associated herbaceous flora. They are of fairly diverse floristic affinities and some are much more catholic than others, though interestingly those other vegetation types where they occur together occupy similar edaphic environments, as in those grasslands over limestones where moderately high rainfall induces some slight surface-leaching of calcium.

In the Typical sub-community, it is these species which provide the characteristic enrichment to be seen in stands of moderate age. In the *Festuca* sub-community, the *E. vagans-Ulex* heath extends towards another edaphic extreme, where the slopes attain the steepest occupied by the community and the soils become most shallow and sharply draining. Indeed, the rankers which predominate here often become parched in the dry and sunny spring and early summer, so the sub-shrubs are often confined to deeper pockets of soil within crevices in the underlying bedrock, have their vigour held in check and are sometimes killed in years of severe drought (Hopkins 1983). It is the consequent lack of shade from the bushes and the limited accumulation of litter that are partly responsible for the increased richness of the associated flora in this kind of *E. vagans-Ulex* heath, because species which would normally succumb early to canopy closure can maintain themselves, even in the absence of burning. Such conditions, together with the generally impoverished character of the soils and the grazing of the herbage by stock, help produce a relatively short sward, in which the species characteristic of the Typical sub-community increase in frequency and where there is a more strongly preferential appearance of a variety of other herbs found in oligotrophic and fairly base-rich pastures: *Galium verum, Carex caryophyllea, Koeleria macrantha, Plantago lanceolata, Leontodon taraxacoides, Anthyllis vulneraria, Danthonia decumbens.* Open patches of dry soil also provide a niche for *Jasione montana, Aira caryophyllea* and *Centaureum erythraea* and among the cropped turf there are more frequent records for *Hypnum cupressiforme s.l.*, for acrocarps of dry soils like *Trichostomum brachydontium* and *Weissia* spp., and for *Cladonia* spp.

It is probably also the reduced competition arising from such factors in this kind of *E. vagans-Ulex* heath

that helps maintain the frequency of *Scilla verna* further away from the immediate vicinity of the cliff tops in the Lizard than happens elsewhere. Certainly, salt-spray deposition has little if any effect here. The strong and frequent westerly winds do extend maritime influence some distance inland on the more exposed coasts of the headland but deposition dies away quickly on moving away from the cliffs (Malloch 1970) and, though the amount of sodium in the soils of the *Festuca* sub-community is often greater than in the Typical kind of heath (Hopkins 1983), it is nothing like the concentrations found beneath the *Calluna-Scilla* heath. However, floristic transitions between the communities can be seen in what both Coombe & Frost (1956a) and Hopkins (1983) termed the *Genista pilosa* variant of *E. vagans-Ulex* heath, but what is here considered as part of the *Calluna-Scilla* heath. *E. vagans* can run some way into this vegetation but *U. europaeus* and, to a lesser extent, *U. gallii* are reduced by their sensitivity to salt-spray (see below).

Although the flora of the *E. vagans-Ulex* heath is never really markedly calcicolous, the base-rich character of the serpentine maintains at least some influence throughout the community and, on the Lizard, a pronounced shift to calcifuge sub-shrub vegetation is associated with the occurrence of a mantle of acid superficials, local Crousa Gravels and the more widely distributed loess, on which the *U. gallii-Agrostis* heath is characteristic. The *Agrostis* sub-community of the *E. vagans-Ulex* heath represents a transition to such vegetation over some of the deepest and most base-poor profiles on to which the community extends. It is concentrated around the heads of the cove valleys and around the rock outcrops on the inland valley sides where, over fairly gentle slopes, there seems to have been some loessial contamination. It can also be found over old 'lazy beds' where intermixing of serpentine and loess may have occurred (Hopkins 1983).

Superimposed upon this fairly well-defined pattern of soil-related variation, there is the impact of treatments, particularly burning. On the Lizard, deliberate burning now occurs only within the enclosed heathy pastures and in the cliff-top grazings but, even there, fires are frequently started out of season and are badly controlled. Over the open inland heaths, fires are accidental, though quite frequent and mostly occurring in summer. Of the different kinds of *E. vagans-Ulex* heath, it is the Typical, *Agrostis* and *Molinia* sub-communities that are most frequently affected, because of their extensive woody canopies and often thick accumulations of litter which, in the pastures, reduce the grazing value of the land and which, even in normal springs and summers, become tinder-dry and very fire-prone. The *Festuca* sub-community, with its discontinuous cover of bushes and little litter, does not need clearing so often and, when ignited accidentally, is not so likely to be completely consumed, despite the often parched character of the ground.

The detailed effects of burning depend very much on the frequency and intensity of the fires and the character of the particular kind of heath that is burned (Coombe & Frost 1956a) but, in general, it repeatedly rejuvenates the vegetation, setting back any tendency to succession (see below) and offers a chance for temporary exploitation of the cleared ground by opportunists among the usual heath flora and a wide variety of adventives. *E. vagans* is often the first of the sub-shrubs to respond to fire, producing new shoots 1–2 dm tall in the same season as a spring burn though flowering a little later than usual from September to November (Coombe & Frost 1956a). The gorse species also sprout readily from surviving stools, though their response is a little slower and the soft young shoots are often consumed by stock. Unlike *E. vagans*, however, both the gorses readily regenerate from seed.

In the first autumn and spring after burning, it is often the abundance and variety of the ephemerals that catch the eye and then there can be a phase in which *E. cinerea* becomes prominent or some of the grasses, notably *A. curtisii* or *Molinia*, both of the latter able to remain dominant for some considerable time. As these wane, but before the canopy of sub-shrubs becomes extensive, the perennial herbs of the community make their most plentiful and varied contribution to the vegetation, thereafter becoming sparser and, in many cases, being totally extinguished.

In the enclosed heaths and on the cliff tops, grazing by stock can also play some part in maintaining the vegetation, though it is difficult to assess its role exactly because it occurs either in conjunction with burning (as in the enclosures) or where edaphic conditions themselves tend to produce similar effects as grazing (as on the cliff tops). But its major effects will be felt on the younger shoots of the gorse spp., and on *Calluna*, an important consideration during the early stages of regeneration, and on the herbaceous element of the community. *E. vagans*, *E. cinerea* and the tougher, older shoots of the gorse spp. are rather unpalatable (indeed, this is probably an important reason why the community subsists so well in the pastures) but, where there are stretches of grassy sward, the community can provide a valuable bite and grazing can perhaps help maintain mosaics of bushes and herbaceous vegetation and keep the latter element diverse.

Zonation and succession

The community occurs in mosaics and zonations in the open and enclosed heaths of the Lizard where transitions can be related to variations in soil moisture, base-status and trophic level and the amount of salt-spray

deposition. The effects of burning and, to some extent, grazing are superimposed on to these patterns and, except in extreme edaphic and climatic conditions, these factors probably play an important part in maintaining the sub-shrub vegetation against successional change.

The floristic contrasts between the two major kinds of *E. vagans* heath, the *E. vagans-Ulex* and the *E. vagans-Schoenus* types, developed in relation to the degree of drainage impedence in base-rich soils, is sometimes expressed as a zonation between the communities, over ground that becomes more gently sloping and more prone to seasonal waterlogging within the shallow drainage basins and over the undulating surface of the plateau itself. In such situations, the *Molinia* sub-community of the *E. vagans-Ulex* heath may form a transition zone between the vegetation types, *E. tetralix* and *Molinia* appearing first, then *Schoenus* increasing in frequency, but *U. europaeus* maintaining its cover quite well. Then, with the move into the *E. vagans-Schoenus* heath, the latter fades, together with many of the attendant herbs, and *Schoenus* and *Molinia* begin to show the strongly-tussocky structure typical of seasonally-inundated ground, *E. cinerea* and *Calluna* becoming confined to the tussocks and the runnels developing their characteristic rich flora. Sometimes, where there is a switch from serpentine to gabbro or in sites where there has been some disturbance, the *Molinia-Potentilla* mire may occur within such sequences. In such vegetation, *Molinia* is strongly dominant over the somewhat more fertile soils, but *Schoenus* can be found occasionally and both *E. vagans* and *U. europaeus* can occur rooted in the tussocks. Separating this kind of vegetation from transitional heath stands can present some difficulties.

The other quite common type of gradual zonation occurs where the base-rich parent materials which give rise to the free-draining brown earths supporting *E. vagans-Ulex* heath are overlain by banks of acid loess. Where contamination with such aeolian material is only slight, the *Agrostis* sub-community is characteristic but this can give way to *U. gallii-Agrostis* heath over deeper deposits. Then *E. vagans* peters out fairly quickly, *U. europaeus* becomes confined to disturbed areas and the fairly rich assemblage of associates is replaced by a restricted suite of more calcifugous plants.

Neither the *U. gallii-Agrostis* nor the *E. vagans-Schoenus* heaths are often found running continuously up to the maritime heath zone of the Lizard because the soils of which they are most characteristic are mainly confined to the inner parts of the peninsula. On more fertile gabbro and schist soils, moreover, the maritime sequence is often truncated above the *Calluna-Scilla* heath by cultivated land bounded by an enclosure wall. On the serpentine, however, where the soils are more oligotrophic, the zonation can run uninterrupted from the maritime heath on to the plateau. In such transitions, which are especially well developed on the more exposed west coast of the Lizard and, more locally, around the cove valleys, the *E. vagans-Ulex* heath typically occupies a position inland of the *Calluna-Scilla* heath. *E. vagans*, which is mildly salt-tolerant, can extend some way in to the *Viola* sub-community of the maritime heath, though the gorse spp., particularly *U. europaeus*, become much patchier with increasing exposure until eventually it is *Calluna* and *E. cinerea* that are providing the bulk of an often stunted and wind-pruned sub-shrub cover. But, apart from the appearance of *Plantago maritima*, there is considerable continuity between the herbaceous element of the *Festuca* sub-community of the *E. vagans-Ulex* heath, the usual cliff-top type, and the *Viola* sub-community of the *Calluna-Scilla* heath, with species like *Festuca ovina*, *Scilla verna*, *Viola riviniana*, *Carex flacca*, *C. caryophyllea*, *Polygala vulgaris*, *Koeleria macrantha*, *Plantago lanceolata*, *Hypochoeris radicata*, *Lotus corniculatus* and *Thymus praecox* all occurring in both. *Genista pilosa* can be quite a good marker of the transition and both Coombe & Frost (1956a) and Hopkins (1983) used this species to name the vegetation of this intermediate zone. Where the zonation continues inland from the *E. vagans-Ulex* heath into the *E. vagans-Schoenus* and *U. gallii-Agrostis* types, the sequences constitute some of the most extensive and valuable tracts of coastal sub-shrub vegetation that we have.

On the rocky cliff tops, where parching of the soils in spring and summer, inherent infertility, exposure to wind and grazing all contribute to the inhospitable character of the environment, the *Festuca* sub-community may represent a natural climax kind of *E. vagans-Ulex* heath (Hopkins 1983). In other situations, however, it is burning and grazing which ultimately maintain the vegetation as a plagioclimax, though the poor quality of the serpentine soils and the shortage of seed-parents over much of the headland would probably make succession to scrub slow. As noted above, older stands of the community do show some transitional features to *Ulex-Rubus* or *Prunus* scrubs, though Coombe & Frost (1956a) considered that progression from heath to the latter was not clearly supported by field evidence.

Distribution
The *E. vagans-Ulex* heath occurs only on the Lizard in Cornwall, where the Typical and *Festuca* sub-communities occur mostly towards the coastal fringe, the *Molinia* sub-community more extensively inland.

Affinities
Like the *E. vagans-Schoenus* heath, this community was first described in detail by Coombe & Frost (1956a) and subject to phytosociological scrutiny by Hopkins

(1983). This account relies heavily on the data of the latter but, although agreeing with these workers in according this heath a central place among the range of sub-shrub vegetation in this part of Britain, it draws the bounds of the community a little tighter. Essentially, the *E. vagans-Ulex* heath of this scheme corresponds with the Typical, *Festuca*, *Molinia* and *Agrostis* variants of Hopkins (1983) but transfers the *Genista* variant to the *Calluna-Scilla* heath, a community which has a wider distribution but which, on the Lizard, provides a further locus for *E. vagans*.

And, as with the *E. vagans-Schoenus* heath, there is some difficulty in locating this second major *E. vagans* community within phytosociological frameworks of heath variation. In a British context, this kind of vegetation is certainly unique in its floristics but there is little to justify considering it as widely separated from other *Erica cinerea* and *Ulex gallii* heaths of the southwest: it grades to both the *U. gallii-Agrostis* and *Calluna-U. gallii* heaths, showing particularly strong continuity with the latter, and could be grouped with them in the Ulicion gallii des Abbayes & Corillion 1949. Such an affiliation would emphasise the phytogeographical position of the community on an east–west axis of variation, in particular stressing the separation of this vegetation from the *U. minor* heaths further east. Were more weight to be given to the presence of *E. vagans*, the community could be placed in a group like Géhu's (1975) Ulici-Ericion ciliaris, an alliance of heaths confined to the far west of England and north-west France, with small outliers in Ireland (White & Doyle 1982).

Other authors have stressed the position of the *E. vagans-Ulex* heath along a north–south axis down the western seaboard of Europe. On such a view, the community is seen by some, not as a most southerly member of the Ericion cinereae heaths, but as the most northerly representative of the Ericion scopariae, the euoceanic southern alliance which, running down through France, Spain and Portugal gives way around the Mediterranean to the maquis (Böcher 1943, Coombe & Frost 1956a, Gimingham 1972). Certainly, in contrast to the *E. vagans-Schoenus* heath, this community does have obvious counterparts on the European mainland, like the Breton vegetation described by Gadeceau (1903) and Géhu & Géhu (1973), though these heaths can lack *U. gallii* and a variety of other *E. vagans-Ulex* species and they also often contain *E. ciliaris*; the Ulici-Ericetum vagantis of Géhu & Géhu (1973) is also distinctly more maritime than most British vegetation with *E. vagans* (see also Bridgewater 1970).

Further south, in the western French Pyrenees, Vanden Berghen (1973) has described a *Pteridio-Ericetum vagantis* which, despite its dominance by bracken, maintains a variety of floristic features in common with the *E. vagans-Ulex* heath (see also Jovet 1941). This was placed in the Ericion umbellatae, though, from Iberian studies by Braun-Blanquet *et al.* (1964), Rodriguez (1966) and Braun-Blanquet (1967), Hopkins (1983) considered the community only peripheral to this alliance. This last author, too, saw the floristic trend running on into the more Continental *E. vagans-E. cinerea-U. minor* heath described by Duvigneaud (1966) from south-east France and suggested that all these communities might be united in a new alliance characterised by *E. vagans*, *E. cinerea*, *Filipendula vulgaris*, *Genista pilosa*, *G. tinctoria*, *Serratula tinctoria*, *Stachys betonica* and *Viola riviniana*.

Floristic table H6

	a	b	c	d	6
Erica vagans	V (5–8)	V (1–8)	V (4–6)	V (5–7)	V (1–8)
Ulex europaeus	V (2–9)	V (1–5)	V (4–8)	V (4–7)	V (1–9)
Erica cinerea	V (2–8)	V (2–5)	V (2–6)	IV (2–5)	V (2–8)
Ulex gallii	IV (1–6)	IV (1–6)	V (1–6)	V (2–5)	IV (1–6)
Carex flacca	IV (1–5)	III (1–5)	IV (1–3)	V (1–3)	IV (1–5)
Viola riviniana	V (1–4)	IV (1–3)		IV (1–3)	IV (1–4)
Agrostis canina montana	IV (1–5)	IV (1–4)	V (1–5)	II (1–3)	IV (1–5)
Filipendula vulgaris	IV (1–5)	IV (1–4)	II (1–2)	II (1–2)	IV (1–5)
Rubus fruticosus agg.	II (1–3)	I (1)	I (1–3)		I (1–3)
Teucrium scorodonia	II (1–5)	I (1–2)	I (1–3)		I (1–5)
Prunus spinosa	II (1–4)	I (1)			I (1–4)
Geranium sanguineum	II (1–5)				I (1–5)
Pteridium aquilinum	II (1–2)	I (1)			I (1–2)
Polygala vulgaris	II (1–3)	V (1–3)	II (1–2)	II (1–3)	III (1–3)
Thymus praecox	II (1–3)	V (2–6)	I (2)	II (1)	III (1–6)
Scilla verna	II (1–2)	V (1–3)	II (1–3)	I (2)	III (1–3)
Hypochoeris radicata	II (1–2)	V (1–4)	I (1)		III (1–4)
Hypnum cupressiforme	II (2–5)	V (1–5)	I (2–3)		III (1–5)
Lotus corniculatus	II (1–3)	IV (1–4)	II (1–2)		III (1–4)
Dactylis glomerata	III (1–4)	IV (1–5)			III (1–5)
Danthonia decumbens	I (1–3)	IV (1–4)	IV (1–3)	I (2)	III (1–4)
Festuca ovina	I (1–2)	V (1–7)	I (1)	I (1)	III (1–7)
Galium verum	I (1–3)	V (1–3)	I (1–2)	I (1)	III (1–3)
Carex caryophyllea	I (1–2)	V (1–4)	I (1–2)		II (1–4)
Plantago lanceolata	I (2–3)	V (1–5)	I (1–2)		II (1–5)
Leontodon taraxacoides	I (1–4)	V (1–4)	I (1)		II (1–4)
Koeleria macrantha	I (1–2)	V (1–5)			II (1–5)
Aira caryophyllea	I (3)	IV (1–3)			II (1–3)
Jasione montana	I (2–3)	III (1–3)			II (1–3)
Anthyllis vulneraria	I (2–3)	III (1–4)			II (1–4)
Daucus carota	I (1–2)	III (1–3)			II (1–3)

Species	Col 1	Col 2	Col 3	Col 4	Col 5
Holcus lanatus	I (1-4)	III (1-3)			II (1-4)
Cladonia impexa	I (1-3)	III (1-5)			II (1-5)
Hypericum humifusum	I (2)	II (1-3)			I (1-3)
Cladonia fimbriata	I (3)	II (1-3)			I (1-3)
Festuca rubra	I (1-3)	II (1-3)			I (1-3)
Trichostomum brachydontium	I (1)	II (1-3)			I (1-3)
Weissia sp.	I (3)	II (1-2)			I (1-2)
Cladonia furcata		II (1-3)			I (1-3)
Dicranum scoparium		II (1-3)			I (1-3)
Cladonia subrangiformis		II (2-4)			I (2-4)
Plantago maritima		II (1-3)			I (1-3)
Centaurium erythraea		II (1-2)			I (1-2)
Potentilla erecta	III (1-4)	I (2-3)	V (2-4)	V (1-3)	III (1-4)
Molinia caerulea	I (2-3)		IV (2-7)	V (3-7)	II (2-7)
Serratula tinctoria	I (1-4)	II (2-5)	III (1-3)	III (2-3)	II (1-5)
Agrostis curtisii	I (1-4)	I (1)	V (3-7)		II (1-7)
Carex panicea	I (2-3)	I (1-3)	IV (1-3)		II (1-3)
Hypericum pulchrum	I (1)	II (1-2)	IV (1-2)		II (1-2)
Viola lactea	I (2)	I (1-2)	III (1-3)		I (1-3)
Polygala serpyllifolia			II (1-3)		I (1-3)
Erica tetralix	I (1-2)		II (4-5)	V (1-5)	I (1-5)
Sanguisorba officinalis	II (1-4)	I (2)		IV (2-4)	I (1-4)
Schoenus nigricans				III (3-4)	I (3-4)
Carex pulicaris	I (1-2)		I (2-3)	II (3)	I (1-3)
Stachys betonica	III (1-3)	IV (1-3)	IV (1-4)	III (1-3)	III (1-4)
Calluna vulgaris	III (1-5)	IV (1-6)	I (2-4)	III (3-4)	III (1-6)
Brachypodium sylvaticum	I (1)	III (1-2)		II (1-2)	II (1-2)
Pedicularis sylvatica	I (2-3)	II (1-3)	II (1-3)		I (1-3)
Anthoxanthum odoratum	II (1-3)	II (1-4)			II (1-4)
Leucanthemum vulgare	II (1-3)	II (1-3)			II (1-3)
Agrostis capillaris	I (1-2)	I (1-4)	I (1)		I (1-4)
Hypnum jutlandicum	I (2)	I (1-3)		I (1)	I (1-3)
Solidago virgaurea	I (1)	I (1)			I (1)

Floristic table H6 *(cont.)*

	a	b	c	d	6
Campylopus paradoxus	I (2)	I (2)			I (2)
Pimpinella saxifraga	I (1–2)	I (1)			I (1–2)
Number of samples	38	30	13	7	88
Number of species/sample	16 (6–27)	34 (26–48)	19 (12–25)	15 (11–20)	24 (6–48)
Vegetation height (cm)	39 (12–75)	18 (6–32)	31 (12–46)	34 (26–43)	29 (6–75)
Vegetation cover (%)	97 (82–100)	83 (45–100)	97 (75–100)	100	92 (45–100)
Slope (°)	15 (0–36)	20 (0–72)	10 (0–66)	3 (0–8)	15 (0–72)
Soil pH	5.8 (4.8–7.1)	6.5 (5.8–7.3)	5.4 (4.8–6.4)	6.0 (5.2–6.4)	6.1 (4.8–7.3)

a Typical sub-community
b *Festuca ovina* sub-community
c *Agrostis curtisii* sub-community
d *Molinia caerulea* sub-community
6 *Erica vagans–Ulex europaeus* heath (total)

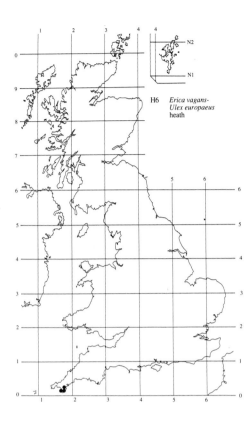

H6 *Erica vagans-*
Ulex europaeus
heath

H7
Calluna vulgaris-Scilla verna heath

Synonymy

Festuca ovina-Calluna 'Rock Heath'　Coombe & Frost 1956a *p.p.*; *Erica vagans-Ulex europaeus* 'Mixed Heath', *Genista pilosa* variant　Coombe & Frost 1956a; *Calluno-Scilletum vernae*　Malloch 1971 *p.p.*; Maritime heath　Urquhart & Gimingham 1979 *p.p.*; *Genisto maritimae-Ericetum cinereae*　Bridgewater 1970; *Anthyllio corbierei-Ericetum cinereae*　Bridgewater 1970; *Festuca ovina-Calluna vulgaris* heath Hopkins 1983; *Erica vagans-Ulex europaeus* heath, *Genista pilosa* variant Hopkins 1983.

Constant species

Calluna vulgaris, Erica cinerea, Festuca ovina, Holcus lanatus, Hypochoeris radicata, Lotus corniculatus, Plantago lanceolata, P. maritima, Potentilla erecta, Scilla verna, Thymus praecox.

Rare species

Allium schoenoprasum, Astragalus danicus, Erica vagans, Euphorbia portlandica, Genista pilosa, Herniaria ciliolata, Isoetes histrix, Minuartia verna, Primula scotica, Scilla autumnalis, S. verna, Spiranthes spiralis, Trifolium bocconei, T. occidentale.

Physiognomy

In the *Calluna vulgaris-Scilla verna* heath, sub-shrubs are a consistent feature of the vegetation, though they are not always very obvious at first sight. For one thing, the canopy is typically of very short stature, rarely over 20 cm and usually less than 10 cm, with the bushes often wind-shaped over the unconformities of the ground. Then, the cover of the woody plants is rarely continuous and quite often rather open, particularly in unsheltered or rocky situations where it can be reduced to scattered bushes with stretches of grassy sward or exposures between. Even where the sub-shrubs are more extensive, their branches are commonly inter-penetrated by the herbs and, in grazed sites, which occur throughout the range of the community but which are especially prevalent to the north, the whole vegetation can be trimmed to a neat, tight mat, just 2 or 3 cm high.

Of the sub-shrubs, *Calluna vulgaris* is the most frequent throughout and the commonest dominant though, on drier soils, it is characteristically accompanied by *Erica cinerea* and this, too, can be abundant. Where the community extends on to wetter ground, as it often does in the rainier north of the country, the latter is much reduced in frequency and *E. tetralix* and/or *Empetrum nigrum* ssp. *nigrum* (generally one or the other) are the more usual associates of *Calluna* and sometimes co-dominant with it.

No other woody species occurs frequently throughout. *Ulex gallii* can occasionally be found but it is generally excluded by the salt-spray characteristic here and the vegetation which Malloch (1971) included in a *U. gallii* sub-association of his *Calluno-Scilletum* is in this scheme transferred to the essentially non-maritime *Calluna-U. gallii* heath. *U. europaeus* is even more sensitive to salt-spray and it occurs only occasionally and then on deeper and better-drained soils. On the Lizard, in Cornwall, however, the nationally rare *E. vagans* can extend a little way into the zone occupied by this community, in vegetation which Coombe & Frost (1956a) and Hopkins (1983) included within their *E. vagans-Ulex* heaths. At this site, too, and at various other places around the south-west coast of Britain, the *Calluna-Scilla* heath provides one locus for prostrate ecotypes of *Cytisus scoparius, Genista tinctoria* and the rare *G. pilosa. Juniperus communis* (probably ssp. *communis*, though often of diminutive habit) also survives very locally in the community in this part of the country and at a few sites in south-west Scotland. On wetter ground, *Salix repens* can be locally prominent.

Among the herbaceous associates growing among and between the sub-shrubs, grasses often bulk fairly large, though the species involved are few and the plants are usually found as small tussocks or as rather indistinguishable components of a closely-nibbled sward. *Festuca ovina* (including some records for *F. vivipara* in

Scotland) is by far the most frequent grass, though *F. rubra* also occurs quite commonly and the two may be difficult to separate in short turf. Very common, too, though not generally of high cover, is *Holcus lanatus*, often accompanied by *Dactylis glomerata* on drier soils, or by *Danthonia decumbens* on moister ground. In the wetter, northern heaths of this community, too, *Agrostis capillaris* and *Anthoxanthum odoratum* become very common and they can be quite abundant. *Molinia caerulea*, however, is rather infrequent, even where the vegetation makes its closest approach to Ericion tetralicis heath in the *Erica* sub-community, and, in south-west Britain, that common companion of *Molinia* in damper heaths, *Agrostis curtisii*, typically stops short of the zone occupied by this community. Scattered patches of bare soil can provide an occasional niche for the annual grasses *Aira praecox* and *A. caryophyllea*.

In this grassy ground is found a variety of other herbs which help make this community one of our most consistently rich lowland heaths. Most distinctive among the constants are *Plantago maritima* and *Scilla verna*: these two species also occur fairly often in more maritime stands of the *Calluna-U. gallii* heath and *S. verna* extends through much of the *E. vagans-Ulex* heath on the Lizard, but nowhere else among British sub-shrub communities are they both so frequent overall. *Scilla verna* is typically present as scattered individuals, striking in spring with its violet-blue flowers, but in leaf only in the wetter months of the year and, even then, easily missed amongst the grasses. *S. autumnalis*, which can also be found here at a few localities in south-west England, shows a similar pattern of foliar development, though its bractless and more purple flowers appear from July to September. *Plantago maritima*, on the other hand, is often very conspicuous, its rosettes being especially numerous in heavily-grazed swards, and other hemicryptophytes occur frequently too, *P. lanceolata* being common throughout, *Hypochoeris radicata* tending to favour drier soils. *Potentilla erecta* is common on all but the most parched and maritime soils, but usually more abundant are mats of *Lotus corniculatus* and the chamaephyte *Thymus praecox* which add prominent splashes of colour to the vegetation in summer. Also noticeable in this respect, though more strongly confined than these species to drier soils, is *Anthyllis vulneraria*, banks of which can be locally dominant over broken ground and very striking. Finally, among this group, are the *Euphrasia* spp., of which *E. tetraquetra* is the most common, these finding a place in small patches of open ground.

Then, there are species which, though occurring occasionally throughout, are rather unevenly represented in the different sub-communities, being to some degree or other preferential for particular environmental conditions. Certain sedges belong among this group, *Carex*

flacca and *C. caryophyllea* favouring the more free-draining and base-rich profiles, *C. panicea* and *C. nigra* extending on to wetter and more acidic soils. *Viola riviniana* and *Succisa pratensis* are also more frequent on moister ground with, less commonly, *Trifolium repens*, *Cerastium fontanum*, *Luzula campestris* and *Leontodon autumnalis*. In drier situations, *Daucus carota* ssp. *gummifer*, *Polygala vulgaris* and *Leontodon taraxacoides* are occasionally found and, on the most parched and exposed ground on cliff tops, some more maritime species and therophytes.

Such trends as these, among both the woody and herbaceous plants, help define the different sub-communities of this quite variable kind of heath. Locally, stands can gain further individuality by the occurrence of rare species which, in this widespread community, show a diversity of phytogeographical affinities. On the Lizard, for example, the community can provide a locus not only for *E. vagans*, but also for *Herniaria ciliolata*, *Isoetes histrix*, *Trifolium occidentale* and *T. bocconei* and for the peculiarly far-flung *Minuartia verna*; extending somewhat further around the Cornish coast is *Scilla autumnalis* and, spreading much further northwards, *Euphorbia portlandica* and *Spiranthes spiralis*. Then, on the north Scottish coast, *Primula scotica* can be found here and, particularly on the coast of north-east Scotland, *Astragalus danicus*.

The general richness of the vascular flora is not matched by the cryptogams, which are very few in number and never of high cover. Among the mosses, only *Hypnum cupressiforme* attains moderate frequency, though rockier stands can have some *Trichostomum brachydontium* and, on base-rich substrates, *Homalothecium sericeum* (as in Hopkins' (1983) samples). *Frullania tamarisci* is the commonest hepatic and, even then, occurs fairly infrequently. *Cladonia rangiformis*, *C. chlorophaea*, *C. impexa* and *C. subrangiformis* can occasionally be found, usually in small amounts.

Sub-communities

***Armeria maritima* sub-community:** *Festuca ovina-Calluna* 'Rock Heath' Coombe & Frost 1956a *p.p.*; *Calluno-Scilletum vernae*, *Armeria maritima* subassociation Malloch 1971; *Festuca ovina-Calluna vulgaris* heath Hopkins 1983. The canopy of sub-shrubs here is generally less extensive than in other sub-communities, usually having a mosaic of more open areas or quite often being reduced to discrete patches of bushes, wind-pruned and almost prostrate, growing in pockets of soil among rock outcrops. Even *Calluna* cannot maintain high cover in this vegetation in very exposed situations and *E. cinerea* shows a general reduction in both frequency and abundance though, with a measure of shelter, it can attain co-dominance. The other common woody species of the community, *E. tetralix*, *Empetrum* and the *Ulex* spp., are all either extremely scarce or absent though, at some sites, scattered bushes of *Juniperus* can be found, and, in south-west England, prostrate ecotypes of *Cytisus*, *Genista tinctoria* and *G. pilosa*.

Growing among and between the low bushes, *Festuca ovina*, sometimes with *F. rubra*, is the usual dominant in the herbaceous component, frequently with a little *Holcus lanatus* and, weakly preferential to this subcommunity, *Dactylis glomerata*. Compared with maritime heath in northern Britain, however, *Agrostis capillaris* and *Anthoxanthum* are rather infrequent and *Danthonia*, too, occurs only rarely. Sedges are also poorly represented. Apart from *Potentilla erecta*, which is unusually scarce, all the other community constants are well represented. *Scilla verna* (accompanied at some sites in the south-west by *S. autumnalis*) is very common at low cover and there are frequent plants of *Hypochoeris radicata*, *Plantago lanceolata* and *P. maritima*, with *P. coronopus* also figuring here occasionally, these rosette hemicryptophytes being particularly abundant where grazing is locally heavy. Often, however, it is the extensive patches of *Lotus corniculatus* and *Thymus praecox* that catch the eye, together with a variety of preferential herbs characteristic of maritime vegetation of rocky outcrops. Important among these are *Armeria maritima* and *Sedum anglicum*, both of which, growing as chamaephytes, can form prominent mats or cushions. *Anthyllis vulneraria*, too, is rather more common here than elsewhere in the community and can be locally abundant and occasionally there are a few plants of *Silene vulgaris* ssp. *maritima*. Then, *Jasione montana* is very frequent and small patches of bare ground, especially common around the rock exposures where the thin mantle of soil becomes easily parched, often have some ephemerals, such as *Aira caryophyllea*, *A. praecox*, *Centaurium erythraea*, *Cerastium diffusum* ssp. *diffusum*, *Bromus hordeacus* ssp. *ferronii* and, on the Lizard, the rare annual clovers *Trifolium bocconei* and *T. occidentale* (Coombe & Frost 1956*a*, Coombe 1961, Hopkins 1983).

Other species occurring occasionally here include *Daucus carota* ssp. *gummifer*, *Rumex acetosa*, *Polygala vulgaris*, *Viola riviniana*, *Leontodon taraxacoides*, *Leucanthemum vulgare* and *Hypericum humifusum*. On the Lizard, the sub-community also provides a locus for *Herniaria ciliolata*, *Minuartia verna* and *Isoetes histrix*. Somewhat more widely distributed is *Allium schoenoprasum* and, at some Scottish sites, *Astragalus danicus* can be found.

On shallow soils among rocks, mosses such as *Trichostomum brachydontium*, *Weissia* spp. and *Homalothecium sericeum* may be quite common here, together with a variety of *Cladonia* spp. (as in Hopkins's 1983 data).

Viola riviniana sub-community: *Erica vagans-Ulex europaeus* 'Mixed Heath', *Genista pilosa* variant Coombe & Frost 1956*a*; *Calluno-Scilletum vernae*, *Viola riviniana* subassociation Malloch 1971; *Erica vagans-Ulex europaeus* heath, *Genista pilosa* variant Hopkins 1983. The sub-shrub canopy in this sub-community is generally more extensive than in the last, though not usually any taller, and both *Calluna* and *E. cinerea* are very frequent, often co-dominating. *E. tetralix* and *Empetrum* are typically very scarce, even in stands in the wetter north of Britain, but both *U. europaeus* and *U. gallii* can occasionally be found and, on the Lizard, *E. vagans* can extend into this kind of heath in considerable abundance. There, too, *Genista pilosa* is quite common.

In general, however, it is in the herbaceous element that the most distinctive features of the vegetation are to be found. All the community constants are well represented and, among these and the preferential species, grasses and sedges are especially prominent. In addition to *Festuca ovina*, which is particularly frequent and plentiful, *Holcus lanatus* and *Danthonia decumbens* are common, and *Agrostis stolonifera*, *A. capillaris* and *Anthoxanthum* occasional. *Koeleria macrantha* can also sometimes be found. Then, *Carex flacca* and *C. caryophyllea* are strongly diagnostic of this kind of *Calluna-Scilla* heath, and the former can be quite abundant.

Among the dicotyledons, this sub-community shares with the last occasional records for *Polygala vulgaris*, *Anthyllis vulneraria*, *Leontodon taraxacoides* and *Daucus carota* ssp. *gummifer*. More strikingly preferential is *Viola riviniana* with, rather less frequent, *Galium verum*, *Achillea millefolium* and, often occurring together in the more southerly part of the range of this heath type, *Serratula tinctoria* and *Stachys betonica*, the latter often as a dwarf ecotype, 5–10 cm tall. *Succisa pratensis* also becomes common here, a floristic feature which continues into the damper heaths of the *Erica* and *Empetrum* sub-communities.

Erica tetralix sub-community. Two features in particular distinguish this kind of *Calluna-Scilla* heath. First, among the sub-shrubs, *E. cinerea* is much reduced and largely replaced as the usual companion to *Calluna* by *E. tetralix*, with *Empetrum nigrum* ssp. *nigrum* and *Salix repens* occasionally present too. The canopy is typically extensive, though again usually of very short stature with the ericoids often very obviously nibbled by herbivores into a tight, densely-branched cover. The impact of grazing is often evident, too, among the herbs and this, together with a shift towards plants of moister and more base-poor soils, constitutes the second diagnostic character. As in the *Viola* sub-community, grasses are prominent but here *F. ovina* is quite often replaced by *F. rubra* as the most abundant species and *Agrostis capillaris*, *Anthoxanthum* and *Danthonia* are also very common and, on occasion, co-dominant in a mixed sward. Sometimes, too, there is locally abundant *Molinia* and *Nardus stricta* can also be found. *Holcus lanatus*, by contrast, is much reduced and *Dactylis* is rare. Among the sedges, *C. flacca* is hardly ever found but *C. nigra* is frequent and *C. panicea* occasional. *Luzula campestris* becomes moderately common as scattered plants.

Among the other species, *Plantago maritima* is often particularly abundant here with *P. lanceolata* also common though not usually so prominent. *Hypochoeris radicata*, however, fares badly on these moist soils and *Thymus praecox* also shows some reduction in frequency and cover. *Viola riviniana* remains quite common but a better preferential among the dicotyledons is *Succisa pratensis*, though it is often nibbled down to a small basal rosette. *Cerastium fontanum*, *Leontodon autumnalis*, *Trifolium repens* and *Prunella vulgaris* can all occur occasionally and sometimes there can be a little *Selaginella selaginoides*.

On the moist soil surface, *Hypnum cupressiforme s.l.* and *Frullania tamarisci* are fairly frequent.

***Empetrum nigrum* ssp. *nigrum* sub-community.** This kind of *Calluna-Scilla* heath shares a number of features with the last. *E. cinerea*, for example, is only very occasionally found, though here its usual replacement is *Empetrum nigrum* ssp. *nigrum* which is often co-dominant with *Calluna* in a low but generally extensive canopy. *E. tetralix* and *Ulex* spp. are rare. Then, among the bushes, *F. ovina*, *F. rubra*, *A. capillaris* and *Anthoxanthum* are all again frequent with *Carex panicea* also common, *C. nigra* more occasional. *Danthonia decumbens*, however, is much scarcer than in the *Erica* sub-community. *Luzula multiflora* joins *L. campestris* as an occasional.

Among the rosette dicotyledons, *Plantago maritima* and *P. lanceolata* are once more very frequent, the former especially being abundant where grazing is heavy, and *Hypochoeris radicata* is again very rare on the moist soils. *Thymus praecox* and *Lotus corniculatus*, though common, both tend to have low covers and *Succisa* is less frequent than in the *Erica* sub-community. *Trifolium repens* and *Rumex acetosa* occur fairly often and *Cerastium fontanum* and *Leontodon autumnalis* are occasional. Among rarer species, *Astragalus danicus* is found in eastern Scotland and, on the north Scottish coast, *Primula scotica* can be locally frequent.

***Calluna vulgaris* sub-community:** Species-poor *Calluno-Scilletum vernae* Malloch 1971. In its general floristics, this type of *Calluna-Scilla* heath resembles impoverished versions of the *Armeria* and *Viola* sub-communities, though it generally has a somewhat taller canopy. *Calluna* is the usual dominant, often forming a wind-waved cover of quite large bushes, though *E. cinerea* is

common and sometimes abundant. All other sub-shrubs are scarce.

F. ovina is the most frequent grass, with *F. rubra* less common and usually less abundant, and *Agrostis capillaris* and *Anthoxanthum* only occasional and generally of low cover. *Hypochoeris radicata* is quite frequent but rosette herbs as a whole are rather poorly represented in the fairly rank herbage with both *Plantago maritima* and *P. lanceolata* reduced in frequency and cover. *Thymus* and *Lotus*, though very common, are often not very abundant. Some stands have a little *Armeria*, *Sedum anglicum*, *Dactylis*, *Daucus* and *Anthyllis* with the vegetation disposed over steeper rocky ground; in others, *Viola riviniana*, *Potentilla erecta* and *Luzula campestris* are the commonest associates. But such enrichment is modest and not consistently maintained throughout.

Habitat

The *Calluna-Scilla* heath occurs over a wide variety of moderately base-poor soils on the less exposed parts of maritime cliffs around much of the coast of Britain. Floristic and structural variation within the community is influenced by climatic and edaphic differences through this considerable geographic range and over particular stretches of cliff. Grazing also affects the composition and appearance of the vegetation and probably contributes to maintaining it against successional change.

The single most distinctive difference between the habitat of this kind of heath and the habitats of other sub-shrub communities is the input of sea-salts from spray, generated by breaking waves and carried inland by the wind. The overall distribution of the community, concentrated in a narrow zone on cliffs around the west coast and then running along the northern and eastern seaboards of Scotland, reflects the general prevalence over the British Isles of south-westerly winds with their long Atlantic fetch and, in the far north-east, the local occurrence of north-westerlies blowing down from the Arctic (Shellard 1976). The *Calluna-Scilla* heath is absent from the more sheltered east coast of England and, where sub-shrub vegetation occurs there close to the sea, it is of an inland heath type. Along much of the south coast, east of Devon, exposure can be high, but many of the otherwise suitable sites for the development of the community are on calcareous rocks, the one kind of hard cliff substrate which cannot support this vegetation.

Of the various chemical constituents of sea-spray, sodium chloride is the most influential on vegetation and the striking effect of heavy deposition is sometimes to be seen here, particularly after strong gales in hot summer weather, in the scorching of the sub-shrubs on their windward side. Compared with almost all other vegetation of maritime cliffs, however, the amount of salt deposition here, measured in terms of soil sodium/loss on ignition (Malloch 1972), is low: on average, about half that recorded in the *Festuca-Holcus* and *Festuca-Plantago* grasslands which characteristically replace the community to seaward. The deposition of spray falls off very rapidly with distance from the point of its production (Malloch 1970, 1972) and, in cliff zonations, the *Calluna-Scilla* heath is thus typically found towards the inner limit of the unenclosed vegetation over relatively gentle slopes. And its abundance, in relation to other more maritime communities, increases with the relative shelter that a local switch to an easterly aspect brings, or with the greater height of the cliffs or where they are bevelled, profiles which set back the upper levels far from the breaking waves.

Apart from observable damage, salt or chloride may influence the water balance of tissues or soils or affect cell metabolism, limiting the vigour of more sensitive species or excluding them altogether, and conferring an absolute or relative advantage on more tolerant ones. In fact, as one might expect from its general position in zonations, the number of strictly maritime species occurring in the community is small. Among the constants, only *Scilla verna* and *Plantago maritima* have general distributions which are largely confined to the coast and such species as *Armeria maritima*, *Plantago coronopus* and *Silene vulgaris* ssp. *maritima* are of limited occurrence. Nonetheless, this small assemblage provides a positive distinction between the *Calluna-Scilla* heath and all other sub-shrub communities. Some of these plants can be found sporadically towards the seaward limit of some other heaths, like the *E. vagans-U. europaeus* and *Calluna-U. gallii* heaths, and *S. verna* in particular can extend a long way inland in some areas, as on the Lizard in Cornwall, but they do not occur consistently throughout these communities or any of their sub-types. Moreover, the gorse spp. and *E. vagans* do not penetrate far into the *Calluna-Scilla* heath because of their sensitivity to salt. In general, therefore, the community occupies the intermediate zone between such inland heaths, or scrub (though these have often been reclaimed, of course) and the limit of final extinction of the most salt-tolerant of the sub-shrubs. And over this zone, the absence of bulkier woody species, and the frequently limited vigour of those which persist in the exposed environment, play an important part in encouraging the species-richness of this kind of heath.

The degree of maritime influence up and along sea-cliffs, interacting with local and regional variations in climate and soils, exerts a strong control on the floristics and distribution of the different kinds of *Calluna-Scilla* heath. In the *Armeria* sub-community, for instance, salt-spray deposition is high and *Calluna*, the most tolerant of the sub-shrubs occurring throughout the whole range of the community, is generally the leading woody plant, though its cover is often discontinuous, and the bushes

wind-waved or windcut and sometimes scorched; and *E. cinerea*, which is more sensitive, is reduced in vigour. Here, too, maritime herbs have their strongest representation.

At this limit of the heath zone on sea-cliffs, maritime influence tends to override smaller differences in substrates and soils, so the *Armeria* sub-community can be found over a wide range of bedrocks, from ultrabasic serpentine to acidic granites and quartzites, though not on limestones which weather to rendzinas inimical to the development of any kind of ericoid heath. In their physical characteristics, however, the profiles are rather uniform, being shallow and rock-dominated, often discontinuously distributed over the steep slopes characteristic here, with numerous rock outcrops whose exact disposition is strongly controlled by the bedding and jointing of the substrates. In summer, particularly towards the sunnier and drier south of Britain, where this kind of *Calluna-Scilla* heath is concentrated, such thin rankers are very susceptible to parching: annual precipitation is generally less than 1000 mm through this region with a clear winter peak (*Climatological Atlas* 1952), the mean annual maximum temperature is usually over 25 °C (Conolly & Dahl 1970) and spring and summer insolation totals are among the highest in the country (Collingbourne 1976). Such parching further checks the vigour of the sub-shrubs, keeping the vegetation open, and favours the abundance of ephemerals and those perennials, like *Sedum anglicum*, which thrive on patches of thin, dry soil. It is these, just as much as the more strictly maritime species, which characterise this sub-community. And, in the warm, oceanic climate of this part of Britain, this kind of *Calluna-Scilla* heath provides a locus for therophytes and more long-lived species with an Oceanic Southern (*Daucus carota* ssp. *gummifer*, *Scilla autumnalis*, *Trifolium bocconei*, *T. subterraneum*) or Oceanic West European distribution (*Bromus hordeaceus* ssp. *ferronii*, *Herniaria ciliolata*).

Both maritime and ephemeral elements extend a little way into the *Calluna* sub-community, together with the general tendency towards heather-dominance. This vegetation can likewise tolerate high levels of salt-deposition but it is characteristic of deeper soils on somewhat gentler slopes than those of the *Armeria* sub-community, still dry, though typical of more uniformly acidic rocks. With a drop in maritime influence and moving on to still shallower slopes, where deeper soils are kept a little moister, and particularly over more base-rich parent materials, the *Viola* sub-community is the usual kind of *Calluna-Scilla* heath. With the greater shelter here, both *Calluna* and *E. cinerea* grow with vigour, and the more sensitive gorse spp. and *E. vagans* can figure occasionally. And, among both constant and preferential species, more mesophytic and base-tolerant plants are well represented, such that the herbaceous element of the vegetation comes to resemble a Mesobromion sward.

In the *Erica* and *Empetrum* sub-communities, low but vigorous sub-shrub growth and a prominent herbaceous component remain characteristic, but there is a shift among both elements which can be related in part to climatic differences in moving to the north of Britain, to where these two types of *Calluna-Scilla* heath are largely confined. Cooler summer temperatures and higher rainfall both play a part in this change. Annual rainfall varies considerably around the northern coasts of Britain, but it is generally over 1000 mm (*Climatological Atlas* 1952), with more than 160 wet days yr^{-1} (Ratcliffe 1968) and high humidity all the year round. In the west, the winters are mild, but mean annual maximum temperature is always less than 24 °C (Conolly & Dahl 1970). Both these factors exert a strong control on the national distribution of *Empetrum nigrum* ssp. *nigrum* (Bell & Tallis 1973), a sub-shrub which is as salt-tolerant as *Calluna* and well able, with its mat-like growth habit, to subsist in very windy situations. This plant becomes very abundant in the *Calluna-Scilla* heath in northern Britain, where the community extends on to deep free-draining acidic soils, particularly on humic rankers over sandstones, which are common on the north and east coasts of Scotland, and on drier peats. The *Empetrum* sub-community is often the most maritime form of this heath in such situations, and can provide a locus for such Continental Northern plants as *Primula scotica* and *Astragalus danicus*.

Along the west Scottish coast, annual rainfall rises to over 2000 mm, with 200 or so wet days yr^{-1} (*Climatological Atlas* 1952, Ratcliffe 1968), so even free-draining mineral soils can be kept very moist and quite deep peats accumulate on steeper slopes down to sea-level. Where such profiles are not subject to strong maritime influence, *E. tetralix* is able to play a prominent role in the *Calluna-Scilla* heath. In the drier south of Britain, this sub-shrub, like *E. cinerea*, can thrive in the more oceanic west but, though the two are roughly similar in their salt-tolerance, *E. tetralix* tends to occur close to cliff tops only where there is drainage impedence, so it is usually confined to very gentle slopes which, set further back from the cliff edge, carry non-maritime heath. In the north, the *Erica* sub-community comes quite close in its floristics to soligenous *Scirpus-Erica* wet heath yet retains many of the typical *Calluna-Scilla* heath features.

However, along with these changes in the sub-shrub canopy of the *Erica* and *Empetrum* sub-communities, and the complementary decline in both of *E. cinerea*, there are also differences in the herbaceous element of these northern kinds of *Calluna-Scilla* heath. The increase in frequency and abundance of *Festuca rubra*, *Agrostis capillaris* and *Anthoxanthum*, together with the

appearance of *Carex nigra* and/or *C. panicea* and weakly preferential occurrence of *Luzula campestris*, *L. multiflora*, *Cerastium fontanum*, *Leontodon autumnalis* and *Trifolium repens*, are all of them fairly gentle shifts but, taken together, they give the grassy matrix of these vegetation types a rather different appearance to that of the *Viola* sub-community, more like a Nardo-Galion sward.

But there is a further environmental feature which plays a part in distinguishing these northern sub-communities, and that is grazing. Many coastal cliffs are open to stock and, except where enclosure has been so assiduous as to take in all the heath zone for improvement, sub-shrub vegetation often constitutes a large proportion of the herbage that is on the more readily accessible gentler slopes. In fact, over much of the southern cliffs of Britain, grazing is now light or non-existent, and from the mid-1950s until recently, there have been few rabbits to compensate for the low stock numbers. Where farm animals do have access, they have generally been beef cattle rather than sheep, and their less close cropping has tended to leave a rougher and uneven sward. Although there are local attempts to re-introduce cliff-top grazing, sometimes using horses or Soay sheep, the general picture is thus one of ungrazed *Calluna-Scilla* heath giving way beyond an artificial boundary to improved pasture.

In the north of Britain, there are areas where grazing of cliff vegetation is light, as around much of the Caithness coast where the quality of the pasture inland is relatively good. Down much of the west coast of Scotland, however, the hinterland is mantled by blanket mire and the cliff tops provide an important source of herbage. Sheep are the usual stock and the effect of their close nibbling is well seen in many stretches of the *Erica* and *Empetrum* sub-communities, where the palatable sub-shrubs, particularly *Calluna*, can be kept down to a very low cover and the rosette hemicryptophytes, particularly *Plantago maritima* and *P. lanceolata*, increase in abundance greatly in relation to the grasses. In more sheltered places, such grazing may be important in maintaining the community as a plagioclimax.

Zonation and succession

The *Calluna-Scilla* heath is characteristically found in zonations developed over maritime cliffs in relation to gradients of salt-spray deposition and soil moisture, themselves complex functions of exposure and cliff physiography. Bedrock type also has an important influence on soil development, the disposition and character of the profiles, and these factors, together with regional variations in climate, affect the vegetation sequences. Grazing can alter zonations and, together with exposure, may prevent successional change. Agricultural improvement, particularly around our southern coasts, has often abruptly truncated the zonations inland, sometimes totally obliterating the heath zone.

Except over limestones, from which the *Calluna-Scilla* heath is usually absent and where ericoid vegetation of any kind is typically limited to drift-contaminated slopes set back some way from the sea, the community forms the inner limit of maritime vegetation. Generally speaking, in the southern part of Britain, where grazing is light or absent, it is replaced to seaward by the *Festuca-Holcus* grassland, which can experience up to twice as much salt-spray deposition. Over basic bedrocks, as on the serpentine, gabbro and hornblende schists of the Lizard and on basalt or calcareous mudstones and shales in Dyfed, Gwynedd, Cumbria and Galloway & Dumfries, the *Calluna-Scilla* heath is typically represented by the *Viola* sub-community over soils which are reasonably deep and moist. To seaward, over more exposed and shallower profiles, this passes to the *Dactylis* sub-community of the *Festuca-Holcus* grassland or, where soil moisture is maintained, as over north-facing slopes for example, to the *Primula* sub-community. The seaward limit of the heath is usually

well marked by the edge of the sub-shrub canopy but, where the wind keeps this very low, the boundary may be scarcely visible from a distance except as a slight tonal difference. And, of course, there is considerable floristic continuity through species like *Festuca rubra*, *Holcus lanatus*, *Plantago lanceolata*, *P. maritima*, *Lotus corniculatus*, *Scilla verna*, *Daucus carota* ssp. *gummifer*, *Hypochoeris radicata*, and *Viola riviniana*.

Within the heath zone itself, there is often a mosaic over more base-rich soils between the *Viola* sub-community and the *Armeria* type according to the distribution of deeper and shallower profiles (Figure 27). And where rock outcrops break the mantle, the zonation may continue to the *Armeria-Cerastium* community, where the sub-shrubs and mesophytic herbs disappear and therophytes and succulents become prominent. These vegetation types can also form a more ordered zonation running down cliffs where the soils become progressively discontinuous but, in complex topography, it may be very difficult to demarcate their limits.

In this southern part of Britain, the sequence of

Figure 27. Zonations on the less exposed tops of maritime cliffs on (*a*) more basic bedrock in the southwest of Britain, with enclosure for agriculture behind, (*b*) more acidic bedrock in southern Britain with light grazing, (*c*) more acidic bedrock in northern Scotland with heavier grazing.

H7a   *Calluna-Scilla* heath, *Armeria* sub-community
H7b   *Calluna-Scilla* heath, *Viola* sub-community
H7c   *Calluna-Scilla* heath, *Erica* sub-community
H7d   *Calluna-Scilla* heath, *Empetrum*
sub-community

H7e   *Calluna-Scilla* heath, *Calluna* sub-community
H8d   *Calluna-Ulex gallii* heath, *Scilla*
sub-community
M15c   *Scirpus-Erica* wet heath, *Cladonia*
sub-community
MC9b   *Festuca-Holcus* grassland, *Dactylis*
sub-community
MC9c   *Festuca-Holcus* grassland, *Achillea*
sub-community
MC10b   *Festuca-Plantago* grassland, *Carex*
sub-community

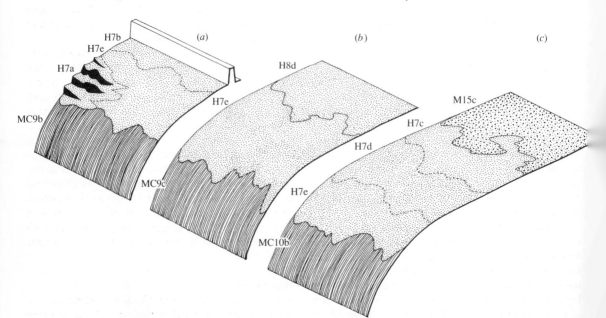

communities is often terminated landwards at the level of the *Calluna-Scilla* heath by the boundary of the improved cliff-top grassland but, in a few places, the zonation continues inland to non-maritime heaths. The most extensive patterns of this kind are to be seen on the Lizard, though they are of a rather unusual type because of the local importance there of *E. vagans*. This, together with occasional *U. europaeus*, can run into the *Viola* sub-community a little way but, on the free-draining, base-rich brown earths of the cliff-top slopes and around the cove valleys, these two sub-shrubs become co-dominant above the *Calluna-Scilla* zone in the *E. vagans-Ulex* heath. Over the Lizard hinterland, this then gives way to the *E. vagans-Schoenus* heath over base-rich gleys or the *U. gallii-Agrostis* heath on base-poor, free-draining soils over acidic superficials.

Essentially similar patterns of *Festuca-Holcus* grassland, *Calluna-Scilla* heath and inland heath can be seen over ungrazed cliffs cut into neutral or acid bedrocks which are of widespread occurrence up the west coast of Britain but particularly important on Land's End, where granite occurs, on the north Cornish and Dyfed coasts, where sandstones and shales crop out, and in Anglesey with its Pre-Cambrian rocks. Here, the *Calluna* sub-community is the usual *Calluna-Scilla* heath of deeper, free-draining soils, passing seawards into the *Achillea* or *Plantago* sub-communities of the *Festuca-Holcus* grassland, the sub-shrubs again fading at the transition but *Festuca rubra*, *Holcus lanatus*, *Plantago* spp., *Scilla verna*, *Rumex acetosa* and *Potentilla erecta* continuing throughout. Again, the *Armeria* sub-community can occur patchily within the *Calluna* type on areas of shallower soil and mark a transition to the *Armeria-Cerastium* therophyte vegetation, or form a seaward transition to this or *Festuca-Armeria* grassland. Landward continuations are once more rare but, in some localities, the sequence passes into the *Scilla* sub-community of the *Calluna-U. gallii* heath.

Where such southern zonations are subject to grazing, the basic pattern of heath and grassland remains, though heavy herbivore pressure may reduce the height of the sub-shrubs further, making the boundary even harder to see, or actually move it inland somewhat where palatable species, notably *Calluna*, are eaten out completely. Also, the *Festuca-Holcus* grassland is often represented just by the *Plantago* sub-community or is totally replaced by the *Festuca-Plantago* grassland, the abundance of *Plantago* spp. running into the heath.

In northern Britain, where grazing is more widespread, this kind of pattern is very common, the *Festuca-Plantago* grassland occupying the bulk of the more accessible cliff slopes with its *Armeria*, *Carex* and *Schoenus* sub-communities distributed according to decreasing maritime influence and increasing soil moisture. Any of these can grade into the *Calluna-Scilla* heath above,

though in this region, of course, the sub-communities tend to be different: on drier, shallower soils the *Calluna* type is characteristic, the *Empetrum* on somewhat deeper mineral profiles and the *Erica* on wet mineral soils or peat. All these kinds of *Calluna-Scilla* heath persist on ungrazed cliffs in the north, though the *Calluna* sub-community seems particularly common there and, with relaxation of herbivore pressure, the *Festuca-Plantago* grassland tends to be replaced by the *Festuca-Holcus* grassland, often in this part of Britain by the *Anthoxanthum* sub-community, an abundance of Nardo-Galion herbs running throughout. Once more, on both grazed and ungrazed cliffs, the *Armeria* sub-community of the *Calluna-Scilla* heath can mark shallower soils under strong maritime influence though it tends to be scarce in the wet climate. And the *Viola* sub-community is likewise rather local, picking out areas of mull soils over more base-rich substrates.

Although, in some parts of Scotland the typical southern pattern of enclosure to the cliff edge is quite common, in the west, the *Calluna-Scilla* heath is often contiguous with inland semi-natural vegetation which, in the wet climate of this part of Britain, is usually Ericion tetralicis wet heath or Erico-Sphagnion blanket mire on peats which thicken up to a continuous deep mantle on inland slopes. Typically, the *Erica* or, on drier ground, the *Empetrum*, sub-community of the heath grades to the *Scirpus-Erica* wet heath or the *Scirpus-Eriophorum* bog, the various sub-shrubs maintaining their cover but the herbaceous component of the heath passing to a *Molinia* or *Scirpus*-dominated suite of associates. On steeper ground where, even in this humid climate, the peat mantle has not been able to accumulate, the *Calluna* sub-community of the heath usually passes to *Festuca-Agrostis-Galium* grassland devoid of sub-shrubs and maritime species.

In sheltered situations on the upper reaches of cliffs, particularly in the south, the *Calluna-Scilla* heath can give way laterally to bracken or scrub, often with a fringe of the *Festuca-Hyacinthoides* grassland or rank *Arrhenatheretum*, and such vegetation often runs inland through the heath zone up head-filled gullies. The main woody colonisers in such situations are *U. europaeus* on drier, more acidic soils and *Prunus spinosa* on moister mulls and, where exposure is not too severe, grazing may play some part in preventing these species gaining a hold in the *Calluna-Scilla* heath. In many cases, however, susceptibility to even small amounts of salt-spray hinders their invasion into the community, so over much of its range this vegetation can be considered a climatic climax.

Distribution

The *Calluna-Scilla* heath occurs all around the coast of the British mainland and offshore islands except to the

east and south between Durham and Dorset. The
Armeria, *Viola* and *Calluna* sub-communities occur
throughout the range, though the first two are better
developed to the south of Galloway, with only local
stations beyond this; the last is also rather rarer to the
north, though it is well represented on Shetland. The
Erica and *Empetrum* sub-communities are both predominantly northern: the former is especially common on
the west coast, particularly in the Hebrides and Sutherland, with scattered occurrences down to Anglesey; the
latter is commonest on the Caithness and Moray Firth
cliffs with outlying stands in Orkney and Shetland and
on the west coast. White & Doyle (1982) have recorded
heath of this general kind along the eastern and northern
coasts of Ireland.

Affinities
Apart from the description of 'Rock Heath' from the
Lizard by Coombe & Frost (1956*a*), the extreme maritime fringe of British sub-shrub vegetation was almost
totally neglected until Malloch (1970, 1972 and unpublished data) began his study of Cornish cliffs, later
extended to cover the entire British coastline. When
variation in this kind of vegetation is seen in such a
national context, the close relationship of the communities to inland heaths is very clear, but, apart from
Malloch's (1972) *U. gallii* sub-community, which is here
transferred to the *Calluna-U. gallii* heath, the vegetation
types coinhere quite well with a unique and strong block
of constants.

Similar communities, considered as true climax
heaths, have since been described by Bridgewater (1970),
Géhu & Géhu (1973) and Géhu (1975) from sea-cliffs in
the Isle of Man, Wales, Cornwall, Brittany and south-
western France, and sometimes placed in a distinct
Dactylo-Ulicion alliance, within which vegetation types
are often distinguished on the basis of the representation
of intra-specific taxa such as *Dactylis glomerata* forma
marina, *Ulex europaeus* forma *maritima*, *U. gallii* forma
humilis, *Genista pilosa* forma *maritima* and *Anthyllis
vulneraria* ssp. *corbierei*. These communities are for the
most part rare and little studied and these taxa not
widespread in Britain and Malloch's (1971) suggestion
that the *Calluna-Scilla* heath be placed in the Ericion
cinereae seems, at the moment, much more sensible.

Floristic table H7

	a	b	c	d	e	7
Calluna vulgaris	V (2–9)	V (4–8)	V (4–9)	V (4–8)	V (4–10)	V (2–10)
Festuca ovina	V (2–8)	V (4–8)	IV (3–7)	IV (2–7)	V (2–8)	V (2–8)
Plantago maritima	IV (2–4)	IV (1–7)	V (3–7)	V (2–5)	III (1–8)	IV (1–8)
Scilla verna	IV (1–4)	IV (2–4)	III (1–4)	IV (1–3)	IV (1–4)	IV (1–4)
Lotus corniculatus	IV (1–5)	V (2–4)	IV (2–4)	IV (2–4)	IV (1–4)	IV (1–5)
Thymus praecox	V (2–5)	V (1–5)	III (2–5)	IV (2–4)	III (1–4)	IV (1–5)
Potentilla erecta	I (2–5)	IV (1–4)	V (2–4)	V (2–4)	III (2–4)	IV (1–5)
Holcus lanatus	V (1–5)	III (1–4)	II (2–4)	IV (2–4)	IV (1–5)	IV (1–5)
Plantago lanceolata	IV (1–4)	V (1–4)	III (1–4)	IV (1–4)	II (1–4)	IV (1–4)
Erica cinerea	IV (2–8)	V (1–8)	II (1–6)	II (4–7)	IV (3–8)	IV (1–8)
Hypochoeris radicata	IV (1–4)	IV (1–4)	II (1–3)	I (1–2)	IV (1–4)	IV (1–4)
Armeria maritima	V (1–4)	I (1–3)		I (3)	II (1–4)	II (1–4)
Sedum anglicum	V (1–4)	I (1–2)	I (2)	I (3)	I (1–4)	II (1–4)
Dactylis glomerata	III (2–7)	II (1–5)	I (2–3)	I (2)	II (1–5)	II (1–7)
Anthyllis vulneraria	III (1–7)	II (1–7)	I (2)	I (2–3)	II (1–6)	I (1–7)
Jasione montana	III (1–3)	I (1–3)		I (1)	I (1–3)	I (1–3)
Aira caryophyllea	II (1–3)	I (2)		I (2)	I (1–3)	I (1–3)
Plantago coronopus	II (1–5)	I (1–3)		I (2)	I (2)	I (1–5)
Centaurium erythraea	II (1–2)	I (1–3)			I (1)	I (1–3)
Silene vulgaris maritima	II (1–4)		I (2–4)	I (2)	I (2–4)	I (1–4)
Cerastium diffusum diffusum	I (1–2)					I (1–2)
Juniperus communis communis	I (3–4)					I (3–4)
Viola riviniana	II (2–3)	IV (1–3)	II (2–3)	II (2–3)	II (1–3)	III (1–3)
Carex flacca	I (2–4)	IV (1–4)	I (1)	I (4)	I (2)	II (1–4)
Polygala vulgaris	II (2–4)	III (1–3)		I (2)	I (1–2)	II (1–4)
Carex caryophyllea	I (2–4)	III (1–4)			I (1)	II (1–4)
Leontodon taraxacoides	II (2–4)	II (2–3)	I (2)			I (2–4)
Koeleria macrantha	I (2–4)	II (2–4)	I (2–4)	I (3)	I (3)	I (2–4)
Agrostis stolonifera		II (2–5)	I (3–5)	I (2–3)	I (2–5)	I (2–5)
Achillea millefolium		II (1–4)		I (3)	I (1)	I (1–4)
Galium verum	I (1–3)	II (1–4)		I (3)	I (1)	I (1–4)

Floristic table H7 *(cont.)*

	a	b	c	d	e	7
Stachys betonica		II (1–4)			I (2)	I (1–4)
Ulex europaeus		II (3–5)		I (5)	I (4)	I (3–5)
Serratula tinctoria		II (1–4)			I (1)	I (1–4)
Erica vagans		II (5–8)				I (5–8)
Leucanthemum vulgare		I (1–3)				I (1–3)
Succisa pratensis	I (3–4)	III (1–4)	IV (1–5)	II (2–4)	I (2–6)	II (1–6)
Danthonia decumbens	I (2–3)	III (1–5)	V (2–5)	I (2–4)	I (2–7)	II (1–7)
Erica tetralix		I (4)	V (4–8)	I (2–4)	I (2)	I (2–8)
Carex nigra		I (3–4)	III (2–4)	II (2–5)	I (1–4)	I (1–5)
Frullania tamarisci	I (4)	I (3)	II (3–4)	I (3–4)	I (2–6)	I (2–6)
Prunella vulgaris		I (1–4)	II (1–4)		I (2–3)	I (1–4)
Selaginella selaginoides			II (2–3)		I (2)	I (2–3)
Molinia caerulea		I (3)	II (2–5)		I (3)	I (2–5)
Salix repens			II (3–7)	I (2)	I (2)	I (2–7)
Nardus stricta			II (2–5)	I (3–5)	I (2–4)	I (2–5)
Carex serotina			I (2–3)			I (2–3)
Dactylorhiza majalis purpurella			I (1–2)			I (1–2)
Agrostis capillaris	II (2–4)	II (2–4)	IV (2–4)	IV (3–4)	II (2–4)	II (2–4)
Anthoxanthum odoratum	II (3–5)	II (2–6)	III (2–5)	IV (3–6)	II (1–7)	II (1–7)
Empetrum nigrum nigrum	I (7)	I (4)	II (2–7)	V (1–8)		II (1–8)
Trifolium repens	I (2–3)	II (1–3)	II (3–4)	III (2–3)	I (2–4)	II (1–4)
Carex panicea	I (3)	I (4)	II (2–4)	III (2–4)	I (2–5)	I (2–5)
Luzula multiflora		I (1)	I (2–4)	II (1–3)		I (1–4)
Festuca rubra	II (1–3)	II (3–7)	III (3–5)	III (2–6)	III (3–8)	III (1–8)
Euphrasia spp.	I (2–3)	II (1–3)	III (1–3)	II (1–3)	II (2–3)	II (1–3)
Hypnum cupressiforme	I (1–3)	II (2–5)	II (3–5)	II (2–4)	II (1–5)	II (1–5)
Luzula campestris	I (2–4)	I (1–3)	II (1–2)	II (1–3)	II (1–3)	I (1–4)
Rumex acetosa	II (1–3)	I (2–3)	I (2)	II (2–3)	II (1–4)	I (1–4)
Daucus carota gummifer	II (1–4)	II (1–3)	I (1)		II (1–4)	I (1–4)
Cerastium fontanum	I (1–3)	I (1–2)	II (1–3)	II (1–2)	I (2)	I (1–3)
Leontodon autumnalis	I (1–3)	I (2)	II (1–3)	II (2–4)	I (1–2)	I (1–4)

	a	b	c	d	e	7
Aira praecox	II (1–4)	I (3)	II (2–3)	I (2–3)	I (2–3)	I (1–4)
Polygala serpyllifolia	I (1–3)	I (1)	I (1–2)	I (1–2)	I (1–3)	I (1–3)
Cladonia rangiformis	I (2–5)	I (3)	I (2–4)	I (2–3)	I (2–3)	I (2–5)
Cladonia chlorophaea	I (2–3)	I (2)		I (2–4)	I (3)	I (2–4)
Carex pulicaris		I (2–4)	I (2)	I (2–3)	I (3)	I (2–4)
Filipendula vulgaris	I (2–4)	I (3–4)				I (2–4)
Genista pilosa	I (3)	I (2–8)				I (2–8)
Astragalus danicus	I (2–3)			I (3)		I (2–3)
Number of samples	60	104	47	40	85	336
Number of species/sample	20 (10–30)	21 (13–31)	19 (10–25)	18 (11–29)	15 (6–20)	19 (6–31)
Vegetation height (cm)	9 (2–50)	9 (2–50)	7 (3–17)	7 (3–20)	12 (2–75)	9 (2–75)
Vegetation cover (%)	96 (70–100)	99 (80–100)	100 (90–100)	99 (90–100)	97 (55–100)	99 (55–100)
Altitude (m)	34 (4–80)	32 (3–70)	17 (7–43)	37 (7–100)	37 (5–65)	32 (3–100)
Slope (°)	19 (0–40)	14 (0–45)	12 (0–40)	12 (0–28)	17 (0–40)	15 (0–45)
Soil pH	4.9 (4.3–5.7)	5.3 (4.3–6.4)	4.7 (4.1–5.4)	4.8 (4.1–5.8)	4.8 (4.3–5.9)	5.0 (4.1–6.4)

a   *Armeria maritima* sub-community

b   *Viola riviniana* sub-community

c   *Erica tetralix* sub-community

d   *Empetrum nigrum* sub-community

e   *Calluna vulgaris* sub-community

7   *Calluna vulgaris-Scilla verna* heath (total)

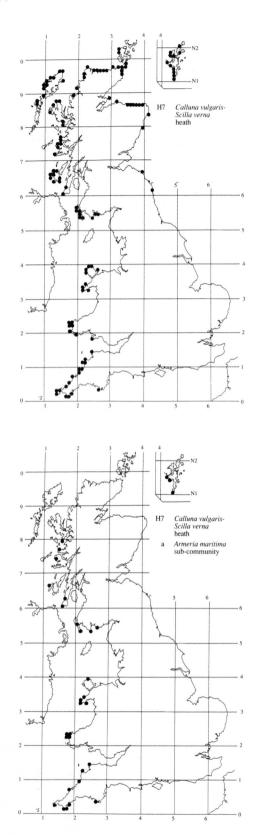

H7   *Calluna vulgaris-*
      Scilla verna
      heath

H7   *Calluna vulgaris-*
      Scilla verna
      heath

a   *Armeria maritima*
      sub-community

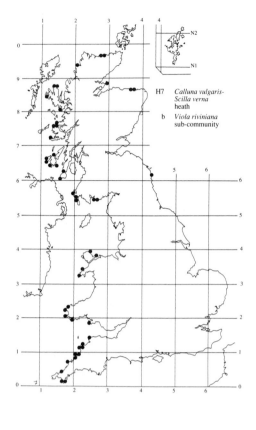

H7   *Calluna vulgaris-*
      Scilla verna
      heath

b   *Viola riviniana*
      sub-community

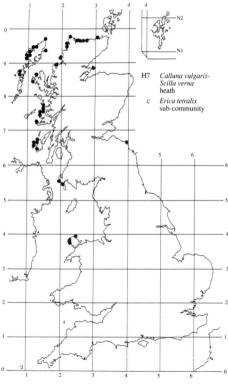

H7   *Calluna vulgaris-*
      Scilla verna
      heath

c   *Erica tetralix*
      sub-community

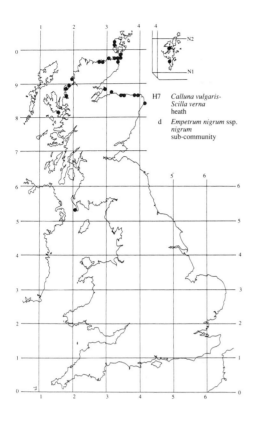

H7   *Calluna vulgaris-*
      Scilla verna
      heath

d   *Empetrum nigrum* ssp.
      nigrum
      sub-community

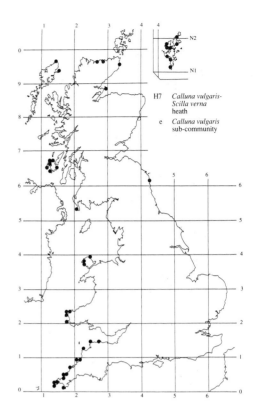

H7   *Calluna vulgaris-*
      Scilla verna
      heath

e   *Calluna vulgaris*
      sub-community

H8
Calluna vulgaris-Ulex gallii heath

Synonymy
Limestone heath   Hope-Simpson & Willis 1955 *p.p.*; *Agrosto-Ulicetum gallii*   Shimwell 1968 *p.p.*; *Ulici-Ericetum cinereae* Bridgewater 1970; *Calluno-Scilletum vernae, Ulex gallii* subassociation   Malloch 1971; *Calluna-Ulex gallii* heaths   Gimingham 1972; Grass-heath Ward *et al.* 1972*a p.p.*

Constant species
Calluna vulgaris, Erica cinerea, Ulex gallii.

Rare species
Agrostis curtisii, Viola lactea.

Physiognomy
The *Calluna vulgaris-Ulex gallii* heath is a floristically rather diverse community with only three constants overall, but its characteristic mixture of *Calluna vulgaris, Ulex gallii* and *Erica cinerea*, typically lacking *E. tetralix, Molinia caerulea* or *Agrostis curtisii*, is quite diagnostic. Often, the three sub-shrubs are co-dominant, though their proportions are quite variable and locally each can be poorly represented, *E. cinerea* in particular tending to become patchy where the community extends on to cooler aspects and to higher altitudes. In the latter situations, too, this kind of heath can have small amounts of *Vaccinium myrtillus*. On disturbed ground, as along pathways or over abandoned fields and settlements, *U. europaeus* may occur with some abundance.

Typically, the sub-shrub canopy is of high cover, sometimes excluding all but a very sparse herbaceous component but, quite often, the bushes are separated by systems of grassy runnels and, where the vegetation has been open to grazing, the structural contrast between these two components can be sharply accentuated. Where the bushes themselves are nibbled, a common occurrence, the canopy can be reduced in height: it is generally less than 30 cm, and a frequent picture is of dense 'hedgehogs' of gorse with short, but rather more untidy, bushes of heather scattered amongst them. In coastal stands, too, exposure to wind helps limit sub-shrub growth and, in exceptional cases, cliff-top heaths which are also heavily grazed can have a very tight cover of woody growth no more than a few centimetres high. Burning can also affect the structure and composition of the vegetation: controlled fires are rare but accidental burns can open up the ground, providing an opportunity for local dominance of *U. gallii* which can sprout vigorously from surviving buried stools, or *E. cinerea*, which can seed prolifically and out-perform *Calluna* in the early years of recovery, particularly on warmer aspects. Such temporary perturbations of dominance may also allow *Pteridium aquilinum* to increase its representation in this kind of heath. Generally speaking, this is an occasional in the community and not abundant, sometimes excluded by the dryness of the soil or by exposure but often limited by the dense sub-shrub cover and thus able to expand where vigorous marginal growth of its rhizomes coincides with opening up of the canopy (e.g. Watt 1955): mosaics of the *Calluna-Ulex* heath and the *Pteridium-Galium* community, presumably mediated at least in part by competition, are common. *Rubus fruticosus* agg, can also occur among the bushes but it is generally infrequent and usually indicative of a local transition to *Ulex-Rubus* scrub over disturbed ground; even there, its abundance can be much reduced by grazing.

Among the herbaceous associates, few species are common throughout and none is consistently abundant, but plants characteristic of Nardo-Galion swards provide the most obvious floristic element, with grasses being especially prominent in many stands, occurring as rough tussocks among the sub-shrubs or forming the bulk of the close-grazed areas of turf between. *Agrostis capillaris* and *Festuca ovina* are the most frequent species overall, with *A. canina* spp. *montana, F. rubra, Anthoxanthum odoratum* and *Danthonia decumbens* occasional to frequent, but rather more unevenly distributed, *Deschampsia flexuosa* and *Nardus stricta* much more patchy in their occurrence. As mentioned above, *Molinia caerulea* is typically scarce in this community but it can occur locally in south-west England and in west Wales, where, together with *E. tetralix*, it transgresses far into drier heath: then, there may be some difficulty in separating this community from the *Ericetum tetralicis* or the *U. gallii-Agrostis* heath. Burning and grazing of stands on somewhat moister soils can also lead to a reduction in sub-shrub cover and the development of vegetation transitional to the *Molinia-Potentilla* grassland. Such problems of diagnosis are a real reflection of the confluence of these communities in the more humid west of Britain.

Among the grassy ground, or growing up as scattered and attenuated plants through the more open bushes, there is often some *Potentilla erecta* and *Galium saxatile*, much more occasionally some *Teucrium scorodonia* and *Polygala serpyllifolia*, but even they can be reduced to very sparse individuals in the densest sub-shrub covers. Then, it may be just scattered patches of scuffed ground, transitions to rocky outcrops or the tops of ant-hills that provide a niche for occasional plants of *Aira praecox*, *Rumex acetosella* and *Jasione montana*. Usually, however, the herbaceous element is a little more extensive than this, though its particular character is strongly dependent upon local edaphic and climatic conditions. By and large, a mesophytic element is poorly represented in this kind of heath, though a slightly preferential occurrence of *Viola riviniana* and *Carex pilulifera*, less so of *C. binervis*, helps define the *Danthonia* sub-community. More striking, on the generally quite sharply draining soils that support this vegetation, is the enrichment of the community calcifuges by more calcicolous plants, or at least by species indicative of a transitional soil base-status, notably *Sanguisorba minor*, *Helianthemum nummularium* and *Carex flacca*, in the *Sanguisorba* sub-community; or by plants typical of maritime heaths, in the *Scilla* sub-community, *S. verna* itself, *Plantago maritima* and *Hypochoeris radicata*. Finally, in the *Vaccinium* sub-community, as well as *V. myrtillus*, there is a preferential occurrence of *Deschampsia flexuosa*, *Nardus stricta* and *Digitalis purpurea*.

In general, the bryophytes and lichens of the community are not very numerous or diverse, nor do they strikingly reflect the floristic trends visible among the vascular plants: quite often, the sub-shrubs are so dense and the grassy herbage so rank or tightly knit as to leave little available ground for their extensive development. However, there is quite often some *Hypnum cupressiforme* and *Dicranum scoparium*, with *Rhytidiadelphus squarrosus* and *Pleurozium schreberi* more occasional, and on moister aspects and at higher altitudes, where the heath is often represented by the *Vaccinium* sub-community, cover of these species can be quite extensive. In more open situations, as on bare ground

exposed by burning or disturbance, such mosses as *Campylopus paradoxus*, *Polytrichum piliferum* or *P. juniperinum* can become abundant, representing early stages in recolonisation. Lichens, too, may make some small contribution in such places or in more open areas of turf, with *Cladonia impexa* and *C. squamosa* occurring occasionally.

Sub-communities

Species-poor sub-community: *Ulici-Ericetum typicum* Bridgewater 1970. Extensive and dense sub-shrub canopies are the rule here, sometimes with one of the woody constants obviously pre-eminent in cover, though often with two or three of them sharing dominance in a patchwork of bushes. The canopy can be quite tall, thicker stands being less penetrable to stock, with the heather and gorse in particular likely to become leggy. The associated flora is characteristically very sparse, with grasses such as *Agrostis capillaris*, *Festuca ovina* and *F. rubra* reduced to small scattered tufts and herbs like *Potentilla erecta* and *Galium saxatile* represented by occasional weak individuals growing through the shorter and more open bushes. In many stands, the only enrichment occurs where there has been some opening of the canopy as a result of burning or disturbance, when *U. europaeus* or *Pteridium* may show a local expansion or, on mobile, sandy substrates, *Carex arenaria*; or where, on exposed and more stable soil surfaces, there can be invasion by plants like *Teucrium scorodonia*, *Aira praecox*, *Rumex acetosella*, *Polytrichum juniperinum*, *P. piliferum* and *Cladonia* spp. Apart from in such places, cryptogams are generally infrequent in this vegetation, with just a few patches of *Hypnum cupressiforme* growing among the bushes.

***Danthonia decumbens* sub-community:** *Ulici-Ericetum caricetosum* Bridgewater 1970. The sub-shrub canopy is well developed in this sub-community, with *U. gallii* and *E. cinerea* especially abundant, but it is not generally so extensive or dense as above and typically there is a system of grassy runnels between the bushes. In these, *Agrostis capillaris* and *Festuca ovina* are quite frequent and occasionally abundant, but much more characteristic is *Danthonia decumbens* with, somewhat less commonly, *Anthoxanthum odoratum*, *F. rubra* and *Agrostis canina* spp. *montana*. Mixtures of these species typically form the bulk of a rough cover growing among the sub-shrubs or, where stands are grazed, the basis of stretches of smooth turf running between them, and then having their taller culms and inflorescences confined to the protected fringes of the bushes. Also very typical of this kind of heath is *Carex pilulifera*, though it seems to be rather patchy in its occurrence and is distinctly more abundant where fires have provided open ground into which it can spread. *Viola lactea* can also show local prominence in such places and disturbance may provide an opportunity for colonisation by *U. europaeus*. Few other floristic features are distinctive though, with *Potentilla erecta* and *Galium saxatile* which become very frequent here, there is occasionally a little *Viola riviniana* and, with *Hypnum cupressiforme* among the grasses, *Pseudoscleropodium purum* is weakly preferential. Locally, in south-west Britain, *Molinia* can become frequent in this vegetation but it is generally of low cover and not usually accompanied by either *Erica tetralix* or *Agrostis curtisii*.

Sanguisorba minor sub-community: Limestone heath Hope-Simpson & Willis 1955 *p.p.*; *Agrosto-Ulicetum gallii* Shimwell 1968a *p.p.* In its floristics, this is the most striking kind of *Calluna-U. gallii* heath. Its sub-shrub cover and the composition of much of the herbaceous component are very similar to the *Danthonia* sub-community, as is the structural variety in the organisation and appearance of these elements in relation to the intensity of grazing. But, here, there is a group of strongly preferential herbs which enrich the vegetation between the bushes. Some of these species are of a more broadly tolerant and mesophytic character, such as *Plantago lanceolata*, *Lotus corniculatus*, *Galium verum*, *Carex caryophyllea*, *Dactylis glomerata*, *Brachypodium sylvaticum* and *Avenula pubescens*, each occurring at generally low covers, though with the grasses showing occasional prominence where they grow tussocky with lack of grazing. Others, notably *Stachys betonica* and *Hypericum pulchrum*, found as scattered individuals, are indicative of some shift in soil base-status while a third, most distinctive, group is more obviously calcicolous, with *Sanguisorba minor*, *Carex flacca*, *Helianthemum nummularium*, *Thymus praecox* and *Linum catharticum*. Where numbers of these are present together in stands which are grazed, a treatment that accentuates the contribution of the hemicryptophytes and chamaephytes to the turf, the vegetation can give the appearance of stretches of Mesobromion grassland growing among islands of the heath sub-shrubs; and, indeed, there is every gradation between this more mixed vegetation and just such mosaics where the *Calluna-U. gallii* heath occurs over limestones (see below).

Scilla verna sub-community: *Calluno-Scilletum vernae*, *Ulex gallii* subassociation Malloch 1971. *Calluna*, *U. gallii* and *E. cinerea* typically occur here as co-dominants in fairly extensive canopies, though ones which, in exposed situations, can be very low, with a tight mat of woody shoots but a few centimetres high. Among the sub-shrubs, there is usually a rather species-poor herbaceous element, with *Festuca ovina* very much the most frequent grass, often the only one, and growing as scattered tussocks. *Potentilla erecta* remains common but other community herbs are scarce, although this vegetation shares with the *Sanguisorba* sub-community occasional records for *Lotus corniculatus* and *Thymus praecox*. More distinctive, however, is the preferential occurrence of *Hypochoeris radicata* and, particularly common, though easily missed when not in its spring flowering period, *Scilla verna*. *Plantago maritima* also occurs quite frequently, and can be abundant in stretches of grazed sward and on cliff-top paths through this vegetation, but this is usually the limit of any maritime element in the flora.

Vaccinium myrtillus sub-community. *Erica cinerea* occurs with reduced frequency in this kind of heath and, though it can remain locally abundant, the most prominent sub-shrub is usually *U. gallii* with smaller amounts of *Calluna* and, strongly diagnostic here, *Vaccinium myrtillus*. Grasses can be of quite high total cover between the bushes with *Agrostis capillaris*, *Festuca ovina* and *Anthoxanthum* all quite frequent, but *Deschampsia flexuosa* is preferential and especially prominent where knolls protrude from among the sub-shrubs. *Nardus stricta* can also occur with local abundance but associated dicotyledonous herbs are few: *Galium saxatile* and *Digitalis purpurea* are occasional. In more open areas, bryophytes can form a fairly lush but patchy cover with *Rhytidiadelphus squarrosus*, *Pleurozium schreberi* and *Dicranum scoparium* supplementing *Hypnum cupressiforme*.

Habitat

The *Calluna-U. gallii* heath is a community of free-draining, generally acid to circumneutral soils in the warm oceanic regions of lowland Britain. Local climatic and edaphic conditions influence floristic variation within this vegetation type and grazing, and sometimes burning, affect its physiognomy and composition. Ultimately, it is these biotic factors, and perhaps, in some situations, exposure to wind, that maintain the community against succession to woodland, and the susceptibility of the soils to improvement for agriculture means that this kind of heath now often survives patchily on marginal grazing land.

The community includes most of the drier, non-maritime heath that is to be found through the British lowlands within the overlapping ranges of *E. cinerea* and *U. gallii*. This region takes in much of the South-West Peninsula, Wales, parts of the north-west of England and the Isle of Man and then swings round through the southern Pennines to take in the seaboard of East Anglia. Climatically, the distinguishing feature of this zone is the equable character of the temperature regime: the summers are generally warm, with mean annual maxima over 25 °C (Conolly & Dahl 1970) but, more importantly for the vigour of oceanic sub-shrubs like *U. gallii* and *E. cinerea*, the winters are mild, February minima being usually over 1 °C (*Climatological Atlas* 1952) and frost days few, particularly close to the coastal fringe where many stands occur.

Across the range of the community, however, certain climatic variables differ considerably and, apart from these two sub-shrubs, this kind of heath has no consistently strong oceanic contingent in its flora. Indeed, summer temperatures themselves, though generally high, differ quite widely across the country from a mean annual maximum close to 30 °C on the east coast to 25 °C on the extremities of Cornwall and Lleyn (Conolly & Dahl 1970). And rainfall varies greatly, from less than 600 mm annually in Suffolk, with less than 120 wet days yr^{-1}, to over 1200 mm on the fringes of Dartmoor, Exmoor, the Welsh uplands and the Pennines, with close to 180 wet days yr^{-1} (*Climatological Atlas* 1952, Ratcliffe 1968). By and large, where the climate becomes cooler and wetter than this, the *Calluna-U. gallii* heath is replaced by other sub-shrub communities and the beginnings of one shift can be seen in the *Vaccinium* sub-community, where *E. cinerea* is partially replaced by bilberry, and where *Deschampsia flexuosa*, *Nardus* and bryophytes increase. Such a floristic change can be seen generally in response to altitude: the mean height of the samples of this kind of *Calluna-U. gallii* heath is about 200 m greater than that of the community as a whole and it is characteristic of upland fringes in Dartmoor and mid-Wales. But it also occurs locally at lower altitudes in response to aspect, with the *Vaccinium* sub-community sometimes figuring on the cooler and moister north faces of hills, as in the Mendips. At the opposite extreme, stands of the species-poor sub-community on the East Anglian coast, with a sparse associated flora of bryophytes and lichens, can resemble the continental *Calluna-Festuca* heath in general appearance.

The other climatic control on the occurrence of the community is maritime influence, more especially the deposition of salt-spray. Quite commonly, it is coastal sites which now offer the most extensive tracts of unimproved soils suitable for the development of this kind of heath and the sub-shrub canopy can persist, as a very low and tight cover, even in very windy situations. But *U. gallii* is intolerant of even moderate amounts of sea-salts, so the community is sharply limited seawards where such winds carry spray on to cliffs. Typically, it is the *Scilla* sub-community that occurs where the *Calluna-U. gallii* heath is represented on more exposed coasts, *S. verna* itself, *Hypochoeris radicata* and *Plantago maritima* all marking a floristic transition to the *Calluna-Scilla* heath, which characteristically replaces the community where sub-shrub vegetation continues seaward. Compared with even the least maritime forms of *Calluna-Scilla* heath, the *Scilla* sub-community

always receives less salt input: measured as soil sodium/ loss on ignition, the latter has been recorded as 87 μmol g^{-1} compared with a mean of 123 μmol g^{-1} for the *Calluna-Scilla* heath (Malloch 1971 and unpublished data).

Between these regional and local climatic extremes, the *Calluna-U. gallii* heath is largely confined to the more free-draining of acid to neutral and generally impoverished soils, which means that the community characteristically marks out exposures of pervious bedrocks or superficials and often occurs on sloping ground, particularly to the west where the higher rainfall can maintain even very immature profiles in a moist state. At one extreme, this kind of heath can occur over excessively-draining sands, as on the East Anglian coast (and perhaps in other stands with the species-poor sub-community); at the other, it can subsist on damp humic soils, as in the west or on the upland fringes (where the *Danthonia* and *Vaccinium* sub-communities predominate). In the latter situations, species such as *Molinia*, *Agrostis curtisii* or *Erica tetralix* may be sparsely represented, but the profiles never show the seasonal water-logging characteristic of wetter heaths.

The particular nature of the parent materials and the profile type are, however, very variable. The *Calluna-U. gallii* heath occurs over a wide range of arenaceous sedimentaries and acidic igneous and metamorphic rocks as well as on silty and sandy superficials like loess and aeolian sands. In the wetter west of Britain such materials show a general tendency to weather to leached profiles with mor but the community can be found on soils of very different degrees of maturity from rankers through brown earths and brown podzolic soils, to podzols proper, even within the single *Danthonia* sub-community. Superficial pH, here and under most other kinds of *Calluna-U. gallii* heath, is usually from 3.5 to 4.5, which accounts for the generally calcifuge character of the associated flora, and the acidifying properties of *Calluna*, and perhaps also of *U. gallii*, help maintain this edaphic environment (Grubb *et al.* 1969, Gimingham 1972).

The predisposition of this kind of heath for free-draining substrates means that it can develop over limestones provided the controlling influence of the calcareous bedrock on pedogenesis is muffled by a mantle of pervious superficials and rainfall is sufficiently high to prevent upward diffusion of lime in the profile. Loess or aeolian sand is an ideal insulation against the renewal of supplies of calcium throughout the profile and over some important exposures of Carboniferous Limestone in the west of Britain, notably in Mendip, on Gower, in Anglesey and on the Great Orme in Gwynedd, the *Calluna-U. gallii* heath is a characteristic feature of deeper deposits of such materials on plateaus and benches (Gittins 1965*a*, Shimwell 1968*a*, Rodwell

1974). On the thick and more acidic mantles, the *Danthonia* sub-community is typical, while on thinner deposits or where the superficials themselves are somewhat calcareous, the mixed vegetation of the *Sanguisorba* sub-community occurs, with its modest contingent of Mesobromion plants.

In all these kinds of *Calluna-U. gallii* heath, the prominence of the associated herbaceous flora, which is largely what distinguishes the different sub-communities, is very variable and much affected by treatments. Where there is no grazing or burning, and where exposure to wind is not such as to keep the sub-shrubs trimmed low, the bushes grow dense and coalesce to produce the kind of extensive, impoverished cover typical of the Species-poor sub-community. With increasing age, the woody cover can become tall and leggy and individual bushes degenerate and die, thus creating a mosaic of more open areas in which a colonising sequence of bryophytes and herbs, appropriate to the edaphic and climatic conditions, can become established as a patch of the *Danthonia*, *Sanguisorba*, *Scilla* or *Vaccinium* sub-community. Where trees do not invade such gaps directly (see below), the sub-shrubs can eventually reassert themselves creating an uneven-aged but again species-poor heath.

Burning, which here seems usually to be an accidental occurrence rather than a deliberate treatment, can produce more extensive areas of regenerating vegetation assignable to the various sub-communities before the sub-shrub canopy again becomes closed in an even-aged woody cover. If grazing by stock is imposed, however, or if rabbits are numerous, then a heathy mosaic may become a more permanent feature, the proportions and arrangement of the bushes and sward reflecting the particular intensity and pattern of grazing. Sheep-grazing is especially likely to produce neat patchworks of close-trimmed sward and low-cropped bushes, the one snaking between islands of the other. Cattle, which are common stock in cliff-top stands in south-west England and in marginal pasture with this heath along the upland fringes, create rougher mosaics, grazing unevenly and not so closely on the sward, leaving ranker herbage around their dung and sometimes trampling open the turf, admitting coarser weedy species like *Teucrium scorodonia* and *Digitalis* or *U. europaeus*.

Zonation and succession

The *Calluna-U. gallii* heath is found in diverse kinds of zonations and mosaics with other vegetation types in relation to differences in the moisture content, base-richness and trophic state of the soils, and to variation in local climatic conditions, most strikingly maritime influence. Burning and grazing can modify these patterns but they also produce effects of their own, being responsible in most situations for the maintenance of the vegetation against succession to woodland. Reclama-

tion of land for improved agriculture has left many stands as fragments in pastoral landscapes.

Towards the south-western part of its range, the community can occur as the dry-heath component in the characteristic sequences of sub-shrub vegetation developed over acidic soils with varying degrees of drainage impedence. Essentially, it replaces the *Calluna-U. minor* heath on the free-draining profiles in such zonations in the region to the west of the Poole basin, which marks the general divide between the ranges of *U. minor* and *U. gallii* (Proctor 1965). In full sequences of this type, the *Calluna-U. gallii* heath gives way, over gleyed podzols and stagnogleys, developed where an argillic B horizon or iron pan impedes drainage, to the *U. gallii-Agrostis* heath. *U. gallii*, *E. cinerea* and *Calluna* all run on, often in considerable abundance, into this community, but the junction between the two is usually marked by the consistent appearance of *E. tetralix*, *Molinia* and *A. curtisii*, which typically share dominance with them. In other cases, where there is a more sudden switch to mineral soils that are strongly seasonally waterlogged, the zonation can pass directly to some kind of *Ericetum tetralicis*. In this kind of wet heath, *E. cinerea* is scarce and *U. gallii* generally confined to sites with some through-put and aeration, so the junction is often quite clearly marked with a shift in dominance to *Calluna*, *E. tetralix* and *Molinia*.

In fact, within the tracts of heaths in south-western Britain, the *Calluna-U. gallii* type occupies a relatively small proportion of the ground, even where more free-draining soils are quite extensive. On the Devonshire Pebble-Bed commons, for example, where gently-dissected Triassic deposits form an undulating dip slope (Ivimey-Cook *et al.* 1975), on the Devon and Gower commons over Carboniferous deposits (NCC Devon Heathland Report 1980, NCC Gower Common unpublished data) and in Pembrokeshire, where suites of sub-shrub communities occur on the plateau cut into Cambrian and Ordovician rocks (NCC Pembrokeshire Heaths Report 1981), this community is rather poorly represented. This is partly because, with the increasing wetness of the climate in moving to the west and into the upland fringes, the damper kinds of heath are able to extend much further on to the free-draining profiles kept continually moist by high precipation:evaporation ratios. Often, then, it is the *U. gallii-Agrostis* heath that occupies the brown podzolic soils or podzols or, beyond its range, some drier kind of *Ericetum tetralicis*, with the *Calluna-U. gallii* heath confined to small stands over knolls and around rock outcrops. Even where tracts of dry heath could be potentially more extensive, the free-draining soils on lowland commons have often been the focus of past disturbance or cultivation, such that they now support *Ulex-Rubus* scrub or *Pteridium-Galium* vegetation.

Even where the effects of human activity have not

been so gross as this, treatments can mask the contribution of the *Calluna-U. gallii* heath to zonations such as these. Where burning occurs, for example, swales often cut across the soil-related boundaries and develop a temporary prominence of regenerating *U. gallii* or *E. cinerea*, or an eventual dominance of *Calluna*, which masks variation in the associated floras of the different constituent heaths. In other cases, burning can be followed by a great expansion in the cover of *A. curtisii* or *Molinia*, centred on the moister soils, but often extending on to the more free-draining profiles: extensive, fire-climax stands of *Molinia-Potentilla* grassland are a marked feature among heath sequences in lowland western Britain. Grazing may contribute to the maintenance of this kind of rank species-poor sward by helping keep the various regenerating sub-shrubs in check but frequently, in zonations of this kind, it produces a different kind of convergence among the associated herbaceous floras of the various heaths, to form extensive stretches of fairly uniform, close-cropped grassland. In the south-west, the most widespread kinds of *Calluna-U. gallii* heath are the *Danthonia* and *Vaccinium* types, the former especially on warmer and drier slopes, the latter in moister, cooler situations. These vegetation types already show a strong measure of floristic continuity in their herbs with some of the damper heaths, particularly the *Festuca* sub-community of the *U. gallii-Agrostis* heath and the *Succisa-Carex* sub-community of the *Ericetum tetralicis*. The effect of grazing over the boundary between the former and the *Calluna-U. gallii* heath is well seen in the sequence of heaths and grass-heaths described from the Dartmoor fringes by Ward *et al.* (1972a) and some perplexing intermediates between the *Calluna-U. gallii* heath and the *Ericetum* are to be seen on certain Pembrokeshire heaths (NCC Pembrokeshire Heaths Report 1981).

The floristic character of the swards derived by grazing the heath vegetation of the *Danthonia* and *Vaccinium* sub-communities places them firmly within the Nardo-Galion: they are characterised by mixtures of such grasses as *Agrostis capillaris*, *A. canina* ssp. *montana*, *Festuca ovina*, *F. rubra*, *Anthoxanthum odoratum*, *Danthonia decumbens*, *Nardus stricta* with *Carex pilulifera*, *Potentilla erecta* and *Galium saxatile*, that is, exactly the herbaceous element of the runnels between the sub-shrubs in the heath itself. Grazing can mediate every gradation between the extremes of dense heath on the one hand and continuous grassy sward on the other, and is the major factor controlling the other common kind of zonation in which the community is found: heath/grassland mosaics over stretches of more or less uniformly free-draining acid soil. These are a particularly characteristic feature of poorer-quality grazing land over acid rocks and superficials around the upland fringes, where the bulk of the sward can usually be referred to some kind of *Festuca-Agrostis-Galium* grass-

land and where the heath forms patches of varying size and organisation over rocky knolls, around field margins and in neglected corners, sometimes even forming hedgerow vegetation on top of inaccessible earth banks. Improvement of the pasture by the application of farmyard manure (or, perhaps, simply a switch from sheep to cattle as the grazing stock) may succeed in converting the sward to a more mesophytic kind, and use of chemical fertilisers, with top-sowing, can readily effect a full transformation to the *Lolio-Cynosuretum*. Towards the limit of in-by land in farms along the margins of Dartmoor and the Welsh upland, such pasture is the frequent context for remaining fragments of the *Calluna-U. gallii* heath.

Variations in base-richness, as well as in the trophic state of the soils, can produce a further element of complexity in these grazing-related sequences over free-draining profiles. As noted earlier, the community can occur over calcareous substrates where these are mantled by a cover of pervious, more base-poor superficials and on sites like the Mendip (Hope-Simpson & Willis 1955), the South Gower coast (NCC South Gower Coast Report 1981) and the Great Orme in Gwynedd (Rodwell 1974), the *Calluna-U. gallii* heath serves as one effective marker of the location of brown earths derived from patches of exotics such as loess or aeolian sand deposited and retained over gentler slopes (Smithson 1953, Perrin 1956, Ball 1960, Findlay 1965; see also Pigott 1962, 1970a). Typically, in such sites, the heath is surrounded by a calcicolous sward, in this case, the *Festuca-Avenula* grassland, developed over thin, dry and highly calcareous rendzina soils on the native limestone, and, where the patches of superficials are sharply defined, there can be an abrupt switch from this to the *Danthonia* sub-community of the heath: this kind of pattern can sometimes be seen where the exotic parent materials have filled up crevices, together with long-weathered and decalcified limestone debris, in the underlying rock surface. Often, however, there is a gradual transition from the rendzinas to the brown earths, with a continuous increase in the soil depth and in the proportion of non-native parent materials, producing some very interesting zonations from the calcicolous sward through to the heath. It is in such situations that the *Sanguisorba* sub-community occurs as an intermediate between the two, sometimes grading directly to the *Festuca-Avenula* grassland, in other cases showing a further transitional zone which can be classified within the *Festuca-Agrostis* grassland. Mixtures of calcicoles and calcifuges are characteristic of this intermediate kind of vegetation, together with some other species such as *Stachys betonica*, *Hypericum pulchrum* and, in Wales and Derbyshire, *Viola lutea* (Balme 1953, 1954, Pigott 1962, Grime 1963a, b). The term 'limestone heath' is often used to describe it, or part of it, but it should be noted that such transitional vegetation can

occur in zonations and mosaics between other constituent grasslands and heaths. The clarity of such patterns, though ultimately dependent upon the edaphic
variation, is also much affected by the grazing intensity:
relaxation allows most of the calcicolous plants to be
overwhelmed by coarse grasses of broad ecological
tolerance and perhaps permits an expansion of the sub-
shrubs on to somewhat more base-rich soils which they
can acidify (e.g. Grubb *et al.* 1969).

One further environmental variable, which can exert a
controlling influence on zonations involving the *Calluna-U. gallii* heath, is salt-spray deposition. Over tracts
of pervious, acidic rocks and superficials, the improvement of deeper profiles has been so extensive that, in
some parts of south-west Britain, the community is now
confined to the unenclosed fringe of land along the cliff
tops, where it is represented by the *Scilla* type. However,
as mentioned earlier, it is only on those parts of cliffs
which are reasonably sheltered from salt-laden winds
that even this kind of sub-shrub vegetation can survive
and quite commonly it represents the limit of the semi-
natural zonation, being terminated abruptly above by a
field boundary with improved pasture or arable beyond.
To seaward, where heath vegetation runs on into the
more maritime zone, there is a transition to the *Calluna-
Scilla* heath: typically, over more acidic soils, this is of
the *Calluna* sub-community, from which *U. gallii* is
absent and where *Armeria maritima* and *Plantago coronopus* occasionally figure, but which otherwise shows
considerable continuity in its floristics, something which
is readily seen where the sub-shrub element is reduced by
exposure to winds to a very short mat through which the
herbs inter-penetrate. Beyond this, the sequence can
continue through maritime grasslands of the *Festuca-
Holcus* and *Festuca-Armeria* types, with the *Festuca-
Plantago* grassland becoming prominent where grazing
is particularly heavy and then to maritime crevice communities. Throughout, the zonation can be interrupted
by patches of the *Armeria-Cerastium* vegetation over
rock outcrops or ant-hills, whose flora can show some
continuity with the heath through species such as
Jasione montana and *Aira praecox*. Such sequences are
especially characteristic of the Cornish granite, the
sandstones and shales of the north Cornish coast and
Dyfed and the Anglesey Pre-Cambrian cliffs.

In certain areas, the *Scilla* sub-community is open to
grazing stock though, within the range of the *Calluna-U.
gallii* heath, it is frequently beef cattle rather than sheep
that are turned on to sea-cliffs and their less assiduous
cropping does not have such a marked effect on the
vegetation. But, even without such treatment, the
invasion of shrubs and trees into this kind of *Calluna-U.
gallii* heath is probably much hindered by exposure to
wind and modest amounts of salt-spray and it can
perhaps be considered a climax vegetation. In more

sheltered situations, episodes of heavy grazing with
scuffing of the ground or fires can both open up the
ground in the short term and allow establishment of
seedlings but, for shrubs and trees to get away, such
treatments must then be withdrawn. Successions have
not been followed in any detail but the most likely
woody invaders on the more base-poor soils are birch
and oak (with a tendency for *Betula pubescens* and
Quercus petraea to be pre-eminent in the areas of wetter
climate), with *Pinus sylvestris* able to seed in from
plantations, and *Sorbus aucuparia* and *Ilex aquifolium*:
such developments would be expected to culminate in
Quercus-Betula-Dicranum woodland. Where the profiles are not so acid, impoverished or excessively draining, it is possible that thorn scrub might develop as a
precursor to more mixed oak-birch woodland of the
Quercus-Pteridium-Rubus or *Quercus-Betula-Oxalis*
types. Along such a successional line, the early establishment of *Ulex-Rubus* or *Pteridium-Rubus* underscrub
may be of considerable importance in breaking the
dominance of the calcifuge sub-shrubs and initiating
some measure of soil enrichment. Disturbance could
provide the initiating step in such a process and the
presence of fragmentary scrubby woodland around
abandoned settlements and wartime emplacements
bears some testimony to this. Where *Pteridium* is locally
established, however, any opening of the heath canopy,
either by disturbance or by degeneration of the bushes,
may allow it to pre-empt the ground forming dense
bracken stands.

Distribution

The *Calluna-U. gallii* heath occurs widely throughout
south-western England and Wales, on the Isle of Man
and, more sporadically, in the southern Pennine fringes
and near the East Anglian coast. The species-poor and
Danthonia sub-communities occur through the whole
range, although there is a tendency for the latter to be
better represented in the wetter west. The *Scilla* sub-
community is confined to the coastal fringe in this more
maritime part of Britain, the *Sanguisorba* sub-community to areas with calcareous bedrock mantled with drift.
The *Vaccinium* sub-community is found mainly at
higher altitudes in the upland fringes.

Affinities

This kind of vegetation has generally attracted detailed
attention in its more unusual forms, like the mixed
calcicole-calcifuge heaths described by Hope-Simpson
& Willis (1955) and the NCC South Gower Coast
Report (1981), and located by Shimwell (1968*a*) in an
Agrosto-Ulicetum; and the maritime heath included by
Malloch (1971) in his *Calluno-Scilletum vernae*. These
assemblages represent marked floristic trends but they
seem best retained along with less peculiar heaths domi-

nated by *Calluna*, *U. gallii* and *E. cinerea* within the ambit of a single community. In general terms, this can be seen as the more oceanic analogue of the *Calluna-U. minor* heath, including most of the sub-shrub vegetation on more free-draining soils in the lowlands west of Poole Harbour.

Within Böcher's (1943) categories, the community falls among the southerly heaths of the Ericion cinereae, what Gimingham (1972) called *Calluna-Ulex gallii* heaths but which in both these schemes also included the vegetation here placed in the *U. gallii-Agrostis* heath. It is certainly true that, in the more humid west of Britain, the floristic distinctions between wet and dry heaths become rather fluid: *E. tetralix* and *Molinia* and, over its smaller range, *A. curtisii* extend far on to less water-logged soils and *U. gallii* and, to a lesser extent, *E. cinerea*, can transgress in the opposite direction, into more enriched forms of the *Ericetum tetralicis*. But, over the region as a whole, there is a strong case for separat-ing off sub-shrub vegetation in which *Calluna*, *U. gallii* and *E. cinerea* are the only overall constants. Similar assemblages have been described from France as the *Ulici-Ericetum cinereae* by Géhu & Géhu (1973) into which Bridgewater (1970) had placed heaths here included in the species-poor and *Danthonia* sub-communities.

The other difficulty of definition relates to the influence of grazing in mediating transitions between the community and calcifuge grasslands of the Nardo-Galion (or, in more specialised habitats, of the Meso-bromion or Silenion maritimae). Herbs of such swards make up the bulk of the associated flora of the *Calluna-U. gallii* heath and, in grazed stands, can compose extensive areas of sward among the bushes. Zonations and mosaics between the heath and the grassland are of diverse form and clarity (e.g. Ward *et al.* 1972*a*), but problems of discerning boundaries are a real reflection of the continuous nature of grazing's impact.

Floristic table H8

	a	b	c	d	e	8
Ulex gallii	V (3–10)	V (5–10)	V (1–9)	V (4–8)	V (5–10)	V (1–10)
Calluna vulgaris	V (3–10)	IV (2–8)	IV (1–6)	V (4–8)	IV (3–5)	V (1–10)
Erica cinerea	V (2–8)	IV (2–8)	IV (1–8)	V (4–8)	II (5–10)	IV (1–10)
Potentilla erecta	I (2–3)	IV (2–5)	III (1–3)	III (1–4)	I (3–5)	III (1–5)
Danthonia decumbens	I (2–3)	IV (2–8)	IV (1–6)		I (3–5)	II (1–8)
Anthoxanthum odoratum	I (5)	III (2–5)	III (1–6)	I (2)	II (3)	II (1–6)
Festuca rubra	I (1–3)	III (2–7)	III (2–5)		I (5)	II (1–7)
Agrostis canina montana		II (2–5)	III (2–6)		II (3)	II (2–6)
Viola riviniana	I (2–7)	II (3)	II (1–3)			I (1–7)
Molinia caerulea	I (1–7)	II (2–3)			II (2–3)	I (1–7)
Carex pilulifera	I (3)	II (2–3)	I (2–3)		I (3)	I (2–3)
Pseudoscleropodium purum		II (3–5)			I (3–5)	I (3–5)
Viola lactea		II (2–3)				I (2–3)
Sanguisorba minor		I (1–3)	IV (1–5)			II (1–5)
Plantago lanceolata		I (2–3)	III (1–5)	I (1–3)		II (1–5)
Helianthemum nummularium			III (1–5)			I (1–5)
Carex flacca			III (1–4)			I (1–4)
Hypericum pulchrum		I (3)	II (1–2)	I (2–3)		I (1–3)
Stachys betonica		I (3)	II (1–2)	I (2–4)		I (1–4)
Galium verum			II (1–3)			I (1–3)
Carex caryophyllea			II (1–3)			I (1–3)
Brachypodium sylvaticum			II (1–6)			I (1–6)
Avenula pubescens			II (1–3)			I (1–3)
Linum catharticum			II (1–2)			I (1–2)
Dactylis glomerata			II (1–5)			I (1–5)
Festuca ovina	I (2–5)	II (3–9)	III (1–6)	V (1–5)	III (5–7)	III (1–9)
Scilla verna		I (3)	I (1–2)	V (1–4)		III (1–4)
Hypochoeris radicata	I (1)	II (2–3)	II (1–3)	IV (1–3)		II (1–3)
Lotus corniculatus	I (3)	I (3)	II (1–3)	III (1–3)		I (1–3)
Thymus praecox		I (3)		III (1–4)		I (1–4)
Plantago maritima		I (2)		III (1–3)		I (1–3)
Deschampsia flexuosa	II (2–7)	II (1–7)	I (1–3)		IV (3–7)	I (1–7)
Vaccinium myrtillus					V (2–5)	I (2–5)

	a	b	c	d	e	8
Nardus stricta		I (2–5)			II (3–5)	I (2–5)
Rhytidiadelphus squarrosus		I (3–5)			II (3–5)	I (3–5)
Pleurozium schreberi		I (3–7)			II (3)	I (3–7)
Agrostis capillaris	II (1–5)	III (2–9)	II (1–4)	I (3)	III (5)	II (1–9)
Galium saxatile	I (1–2)	III (2–5)	II (1–3)	I (2–3)	III (3–7)	II (1–7)
Hypnum cupressiforme	II (1–8)	II (3–7)	II (1–4)		III (3–5)	II (1–8)
Pteridium aquilinum	II (2–5)	I (3)	II (1–7)		II (3)	II (1–7)
Dicranum scoparium	II (1–5)	I (2–5)			II (3–7)	I (1–7)
Teucrium scorodonia	I (2–5)	I (3)	I (1–8)	I (3)	I (5)	I (1–8)
Ulex europaeus	I (2–7)	I (3–5)	I (1–6)	I (1–4)		I (1–7)
Erica tetralix	I (1)	I (3–5)		I (1–5)	I (4)	I (1–5)
Holcus lanatus	I (1)	I (2–3)		I (1)	I (6)	I (1–6)
Carex binervis	I (3–4)	I (2–3)		I (3)	I (3)	I (2–4)
Campylopus paradoxus	I (1–7)	I (2–4)			I (2–3)	I (1–7)
Aira praecox	I (2–5)	I (3–5)			I (3)	I (2–5)
Cladonia impexa	I (1–5)	I (3)			I (2–3)	I (1–5)
Polytrichum piliferum	I (2–5)	I (3)			I (2)	I (2–5)
Rumex acetosella	I (2–3)	I (2)			I (3)	I (2–3)
Cladonia squamosa	I (2)	I (3)			I (3)	I (2–3)
Jasione montana	I (2–4)	I (3)			I (2)	I (2–4)
Pedicularis sylvatica	I (4)	I (3)		I (1–3)		I (1–4)
Agrostis stolonifera	I (3)	I (7)	I (2–4)			I (2–7)
Rubus fruticosus agg.	I (1–2)		I (1–6)		I (4)	I (1–6)
Polytrichum juniperinum	I (1–5)			I (2)	I (2)	I (1–5)
Polygala serpyllifolia		I (3)		I (1–2)	I (3)	I (1–3)
Agrostis curtisii	I (2–4)	I (2–3)				I (2–4)
Number of samples	29	32	30	17	9	117
Number of species/sample	9 (4–25)	12 (8–25)	20 (12–32)	17 (6–24)	15 (4–26)	13 (4–32)
Vegetation height (cm)	36 (20–100)	30 (10–45)	28 (10–80)	15 (3–43)	26 (10–50)	28 (3–100)
Vegetation cover (%)	97 (75–100)	94 (40–100)	97 (70–100)	96 (80–100)	85 (50–100)	96 (40–100)
Altitude (m)	118 (12–350)	201 (90–320)	89 (10–152)	50 (10–110)	328 (100–465)	135 (10–465)
Slope (°)	6 (0–30)	16 (0–35)	11 (0–33)	7 (0–20)	17 (0–50)	15 (0–50)
Soil pH	4.1 (3.3–5.3)	4.0 (3.7–4.5)		4.7 (3.9–5.3)	3.9 (3.6–4.2)	4.1 (3.3–5.3)

a Species-poor sub-community
b *Danthonia decumbens* sub-community
c *Sanguisorba minor* sub-community
d *Scilla verna* sub-community
e *Vaccinium myrtillus* sub-community
8 *Calluna vulgaris-Ulex gallii* heath (total)

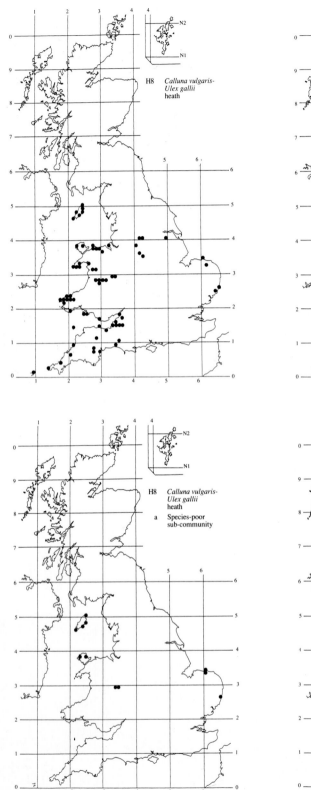
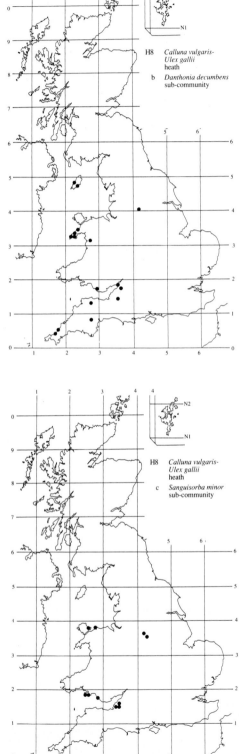

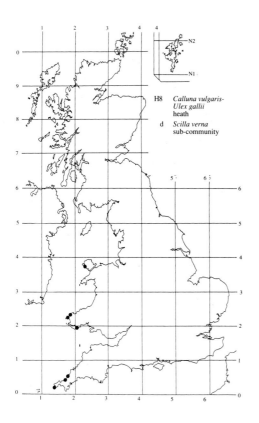

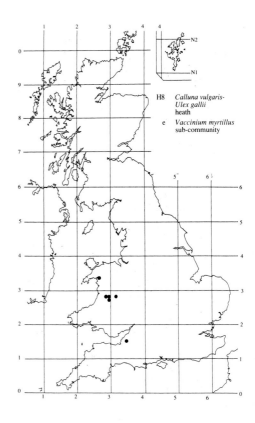

H9
Calluna vulgaris-Deschampsia flexuosa heath

Synonymy
Heather Moor Smith & Moss 1903, Smith & Rankin 1903 *p.p.*; *Callunetum* Lewis & Moss 1911, Elgee 1914, Tansley 1939, Fidler *et al.* 1970 *p.p. Calluno-Ericetum cinereae* Bridgewater 1970 *p.p.*; *Cladonio crispatae-Callunetum* Coppins & Shimwell 1971 *p.p.*; *Pohlio-Callunetum* Shimwell 1975 *p.p.*

Constant species
Calluna vulgaris, Deschampsia flexuosa, Pohlia nutans.

Physiognomy
In the *Calluna vulgaris-Deschampsia flexuosa* heath, *Calluna vulgaris* is almost always the most abundant plant, though it often forms a fairly low, and sometimes quite open, canopy made up of immature individuals: burning is very common on the heathlands where this community occurs, many stands are of a more or less uniform age and relatively few are old enough to have mature or degenerate heather plants. Quite often, too, the cover shows signs of grazing with the heather nibbled close and both these factors have some influence on the contribution of other sub-shrubs. Typically, none of these associates is consistently frequent throughout, though certain species can become quite common and attain a measure of local abundance, if only exceptionally rivalling *Calluna* in their cover. *Vaccinium myrtillus* is the most important of these and, particularly where the community extends to higher altitudes, where the climate is cooler and more humid, its frequency and abundance presage a shift to the *Vaccinium myrtillus-Deschampsia flexuosa* heath. Much more locally, *V. vitis-idaea* and/or the hybrid *V. × intermedium* (Ritchie 1955*b*) can be found here, and *Empetrum nigrum* ssp. *nigrum* has sometimes spread in this kind of vegetation, as on Ilkley Moor in West Yorkshire (Fidler *et al.* 1970, Dalby *et al.* 1971). *Erica cinerea* and *Ulex gallii*, by contrast, are very scarce, although stands of the *Calluna-Ulex gallii* heath, of which both of these species are characteristic, do occur within the range of the *Calluna-*

Deschampsia heath and can be difficult to separate from it. *Erica tetralix* is likewise rare, although it can figure locally in transitions to the *Ericetum tetralicis*, a fairly common feature in the mosaics of dry and wet heaths seen in the North York Moors (e.g. Elgee 1914).

Apart from *Calluna*, the only other vascular constant of the community is *Deschampsia flexuosa* but, even where the cover of the heather is fairly open, this is often represented by fairly sparse tufts and, under very dense canopies, it can be virtually extinguished. Grazed stands, though, can have higher covers of the grass and then the vegetation may grade to the *Deschampsia* grassland, in which the balance has been decisively shifted against heather. Other grasses play a much more limited role even than this: *Molinia caerulea* can become frequent on moister ground, though usually with but little vigour, but apart from this there are just very occasional plants of *Agrostis capillaris, Holcus lanatus, H. mollis* and *Festuca rubra*. Other herbs are likewise few in number and generally of low cover: *Juncus squarrosus* can sometimes be found, *Galium saxatile* and *Potentilla erecta* become frequent in grazed stands and *Pteridium aquilinum* occurs occasionally with moderate abundance. Seedlings of oak, birch and *Pinus sylvestris* are sometimes seen but they rarely survive to the sapling stage with such frequent burning and grazing.

Ground cover in the *Calluna-Deschampsia* heath is rather variable, but bryophytes and lichens are rarely very abundant, even among the more open canopies of heather, and this element of the flora is often poor in species. But its composition is quite characteristic with small acrocarpous mosses, certain leafy hepatics and encrusting lichens predominating over bulky pleuro-carps and fruticose lichens. In contrast to the *Calluna-Vaccinium* and *Vaccinium-Deschampsia* heaths, for example, *Hypnum cupressiforme s.l.* is of restricted occurrence here and species like *Hylocomium splendens, Pleurozium schreberi* and *Ptilidium ciliare* very scarce. *Pohlia nutans*, on the other hand, is very common in all but the most impoverished stands and *Campylopus paradoxus* and *Dicranum scoparium* also occur occasionally, together with the introduced moss *Orthodontium lineare*. This was first recorded in West Yorkshire in 1920 (Watson 1922) but spread rapidly on to bare ground in both heaths and woods through the heavily-polluted south Pennines and is now, indeed, in some areas, the most frequently encountered moss in this community. Then, over patches of exposed mineral soil, where the humic top has been burned or eroded away, there can be locally abundant *Polytrichum juniperinum*, *P. piliferum* and *P. commune*. Among the leafy hepatics, *Gymnocolea inflata* is particularly characteristic, with occasional *Barbilophozia floerkii*, *Cephalozia bicuspidata*, *Cephaloziella divaricata* and *Calypogeia muellerana*.

The commonest lichens of the community are *Cladonia chlorophaea*, *C. floerkeana*, *C. squamosa*, *C. coniocraea* and *C. fimbriata*, forming occasional scattered patches encrusting the surface of the ground, sometimes with *Lecidea granulosa* and *L. uliginosa* over peat and rotting wood and *Hypogymnia physodes* over old *Calluna* branches and stools.

Sub-communities

***Hypnum cupressiforme* sub-community.** Heather is typically strongly dominant here but, rather exceptionally in the *Calluna-Deschampsia* heath, the bushes tend to be large and mature or even degenerate, with their opening centres becoming accessible for recolonisation. *Vaccinium myrtillus* occasionally finds an opportunity to spread in such situations, or *Pteridium aquilinum*, and there are frequent, sometimes dense, tufts of *Deschampsia flexuosa*. But it is among the mosses that the most distinctive features of the vegetation are to be seen, because *Hypnum cupressiforme s.l.* is unusually common and often abundant among the heather stools, with *Dicranum scoparium* also preferential and frequently rivalling *Pohlia nutans* in its cover. Other mosses, though, and leafy hepatics, are sparse and, apart from occasional *Hypogymnia physodes* growing on older *Calluna* branches, lichens are very few.

***Vaccinium myrtillus-Cladonia* spp. sub-community.** This is the richest kind of *Calluna-Deschampsia* heath, characterised by younger canopies of heather, often in a state of obvious recovery from burning. There is frequently, particularly at higher altitudes, a little *Vaccinium myrtillus* and, more locally, other sub-shrubs can figure, with *V. vitis-idaea* and *V. × intermedium* sometimes marking out sites of surface disturbance (Ritchie 1955*b*, Shimwell 1973*b*), *Empetrum nigrum* ssp. *nigrum* spreading where burning combined with grazing has given it an edge over the more palatable species (Fidler *et al.* 1970, Dalby *et al.* 1971). Usually, *D. flexuosa* has rather low cover and, in frequently burned sites, it can be altogether absent,

but over the ground among the sub-shrubs, bryophytes and lichens are more varied than in any other kind of *Calluna-Deschampsia* heath. *Pohlia nutans*, *Campylopus paradoxus* and *Orthodontium lineare* all occur frequently as scattered shoots or in small tufts and there are often patches of the typical leafy hepatics of the community, sometimes further enriched by *Calypogeia fissa*, *Cephalozia connivens* and *Cephaloziella hampeana*. Among the lichens, *Cladonia chlorophaea* and *C. floerkeana* are the most frequent, with occasional records, too, for *C. squamosa*, *C. coccifera* and, more locally on some of the Cheshire and Staffordshire heathlands, *C. crispata*, a fruticose species which Coppins & Shimwell (1971) used to separate off samples into a distinct community (see also Shimwell 1973*b*).

Species-poor sub-community. In this most impoverished form of *Calluna-Deschampsia* heath, *Calluna* and *D. flexuosa* are the only constants and, under very dense or frequently-burned canopies of heather, even the latter can be much attenuated in its cover or totally extinguished. *V. myrtillus* occurs occasionally in small amounts but, even where there are considerable areas of bare ground among the sub-shrubs, the bryophyte and lichen flora is dismally poor. *Pohlia nutans*, *Campylopus paradoxus* and *Orthodontium lineare* all show reduced frequencies compared with the last sub-community and there are just very occasional patches of *Gymnocolea inflata* and *Cladonia squamosa*.

***Galium saxatile* sub-community.** *Calluna* remains constant here but it is quite often rivalled in cover by *D. flexuosa* which can form fairly extensive patches of tussocky turf among the heather, locally enriched by a little *Holcus mollis* or *Festuca rubra*. Then, very commonly, there are scattered plants or locally prominent patches of *Galium saxatile* and occasional *Potentilla erecta* with, on scuffed areas, some *Rumex acetosella*. Oak seedlings can sometimes be found too. Lichens and hepatics are noticeably sparse among the grassy ground and, even among mosses, only *Pohlia nutans* and *Hypnum cupressiforme s.l.* occur more than very occasionally.

***Molinia caerulea* sub-community.** Heather is generally very abundant in this sub-community, but *D. flexuosa* is frequently accompanied by small amounts of *Molinia* which brings a little variegation to the canopy. The ground layer, however, is poorly developed with just very sparse *Pohlia nutans* and *Campylopus paradoxus*.

Habitat

The *Calluna-Deschampsia* heath is the characteristic sub-shrub vegetation of acid and impoverished soils at low to moderate altitudes through the Midlands and northern England. The relatively cool and wet climate of this part of Britain has some influence on the floristics of the community but much of its character derives from a combination of frequent burning and grazing with heavy atmospheric pollution around the industrial conurbations of the region.

Throughout the range of the community, which extends in an arc from south Lancashire, round the southern Pennines, taking in stands in the Cheshire/ Shropshire plain, Staffordshire and Nottinghamshire, up into the North York Moors and Durham, annual rainfall is almost everywhere more than 800 mm with in excess of 140 wet days yr^{-1}. And, towards the centre of its distribution, in the fringes of the Pennine uplands, where this kind of heath frequently extends above 200 m, precipitation rises steeply to 1000 mm and beyond, with more than 160 wet days yr^{-1} (*Climatological Atlas* 1952, Ratcliffe 1968, Atkinson & Smithson 1976). Generally, too, the climate is cool, with, over most of the region, annual accumulated temperatures of less than 1100 day-degrees C (Page 1982). Thus, though the winters are not especially harsh – indeed, at lower altitudes, February minima are often above freezing (Page 1982) – the warm oceanic conditions characteristic of the heathlands of south-western Britain are not found here.

In floristic terms, such climatic differences are seen first in the scarcity in this community of *Ulex gallii*, an Oceanic West European sub-shrub that is often codominant in the *Calluna-U. gallii* dry heath, typical of similar soils to those found here but centred in the more equable western lowlands of Britain. In fact, the geographical boundary between the two communities is not a hard and fast one: *U. gallii* can be found locally right across the central and southern parts of the range of the *Calluna-Deschampsia* heath but, where it marks out definite stands of the *Calluna-U. gallii* heath, the separation between the communities is often an altitudinal one, the latter kind of heath being generally confined to lower levels within the area of overlap. *Erica cinerea*, too, often marks out what is essentially the same geographical and floristic boundary in this part of the country, though it is, of course, much less narrowly oceanic than *U. gallii* (Bannister 1965, Gimingham 1972), remaining an important component of heaths to the far north of Britain outside the more montane and boreal zones.

On the positive side, the cool and wet climatic conditions are reflected in the occurrence here of *Vaccinium-myrtillus* and, more locally, of *V. vitis-idaea*, *V.* × *intermedium* and *Empetrum nigrum* ssp. *nigrum*. The first, though of Continental Northern distribution through Europe as a whole, occurs widely in this country where the annual rainfall is more than the usual minimum for the *Calluna-Deschampsia* heath, 800 mm, so it can be found locally in the more southerly lowlands of Britain

in such sub-shrub communities as the *Calluna-U. minor* heath, which provides a locus in the western Weald, in the *U. gallii-Agrostis* heath around Dartmoor and Exmoor and in the *Calluna-U. gallii* heath along the Welsh Marches. But it is here, particularly on less sunny aspects and, more extensively, at somewhat higher altitues around the Pennine fringes, that it begins to assume the important role that it has in the Myrtillion heaths that are widespread through our uplands. And for *V. vitis-idaea* and *E. nigrum* ssp. *nigrum*, which are Arctic-Alpine in a European context, and much more strongly confined in Britain to the cooler montane zone, the *Calluna-Deschampsia* heath provides an important occasional locus towards the lowland extreme of their ranges. When mixtures of these species occur with heather, the composition of the *Calluna-Deschampsia* heath closely approaches that of the *Vaccinium-Deschampsia* heath, the most widespread sub-shrub vegetation of northern Britain at altitudes over 500 m. Such floristic convergence is best seen in the *Vaccinium-Cladonia* sub-community which, though found throughout most of the range of the *Calluna-Deschampsia* heath, is best developed in the cooler, moister conditions in the Pennine foothills between 200 and 400 m. Even here, however, dominance typically remains with *Calluna*, although various kinds of treatment can give other of the sub-shrubs a temporary or local advantage (see below), thus confusing the natural environmental boundary between the different kinds of heath.

Through the lowland and sub-montane zones of the Midlands and northern England, acidic and impoverished soils suitable for carrying the *Calluna-Deschampsia* heath have developed widely from both pervious drift-free bedrocks and more free-draining superficials. At moderate altitudes in the southern Pennines, for example, down through West and South Yorkshire and on into Derbyshire and Staffordshire, the resistant sandstones of the Millstone Grit provide an important substrate over the prominent scarps and steep, stepped valley sides that mark the first real rise towards the Pennine summit plateaus. Around this region, running into central Lancashire and down the eastern side of the Pennines, south into Nottinghamshire, then reappearing in the West Midlands, there is the more subdued scenery of the Coal Measures, where arenaceous strata cropping out as cuesta scarps provide scattered tracts of suitable ground. Then, through southern Staffordshire, across much of the Cheshire–Shropshire Plain and up through the Nottinghamshire lowlands, Permo-Triassic sandstones and conglomerates underlie the low but resistant flat-topped hills that occur in Cannock Chase, for example, and which form the basis of the gently undulating scenery of Delamere and Sherwood forests. Across these areas, too, certain superficials, notably glacio-fluvial and river terrace drift, together with patches of aeolian sand, underlie this kind of vegetation. Finally, at some remove, in the North York Moors, the *Calluna-Deschampsia* heath can be found over the more pervious strata among the complex sandwich of Jurassic sandstones and shales exposed over the gently-domed surface of the plateau.

Such substrates as these have weathered to a variety of very base-poor soils, with surface acidity generally between pH 3 and 4, highly oligotrophic and at least moderately free-draining, often excessively so. Occasionally, the *Calluna-Deschampsia* heath occurs on steeper slopes and among rock exposures, as around the free faces and weathering detritus of coarser sandstones and grits, and here the profiles may be kept in a more or less permanently immature state, being fragmentary or shallow rankers, often very sandy below. Deeper brown sands, derived from Permo-Triassic sandstones and coarse superficials, and showing some differentiation of horizons, can also be found beneath this kind of vegetation over the gentler ground of the Cheshire–Shropshire plain and around Delamere Forest and Cannock Chase, where the Bridgnorth and Newport series are important, and in Sherwood with its Cuckney profiles (Furness 1978, *Soil Survey* 1983, Ragg *et al.* 1984), though the relative ease with which these soils can be worked and improved for agriculture has left only small fragments with heath in many places. Most often, however, the gentler-sloping surfaces of dips and plateaus of arenaceous rocks, where the community is most widespread and extensive, carry some kind of podzolised profile, like those of the Anglezarke, Belmont and Maw series in the Pennines and North York Moors (Carroll *et al.* 1979, Carroll & Bendelow 1981) and the Goldstone in Delamere and Cannock (Furness 1978, Ragg *et al.* 1984). In such profiles, and particularly those occurring in the rainier environment of the upland fringes, a thick humic top, up to 10 cm or more deep, can develop and be kept quite moist, despite the sharply-draining character of the soil, though frequent burning may repeatedly set back the accumulation of organic detritus and especially intense fires strip the humic horizon right back to the underlying mineral material. In other cases, the development of an iron pan or a bleached hard pan may encourage some drainage impedence in the profile and the *Calluna-Deschampsia* heath can extend some considerable way on to stagnopodzol intergrades to raw peat soils. Typically, however, this is a dry heath community in the sense that conditions are never such as to favour the frequent representation of *Erica tetralix*, although towards the edaphic limits of the *Calluna-Deschampsia* heath on moister ground, burning may play some part in extending the supremacy of heather further than one might expect. Certainly, intimate mosaics of the community with the *Ericetum*

tetralicis wet heath are very common over the more ill-drained parts of the North York Moors and in some places the *Molinia* sub-community can be seen as a transition to such vegetation on wetter soils. Elsewhere, this kind of *Calluna-Deschampsia* heath provides a very local extension of the community on to raised and valley-mire peats scattered through its lowland range, where these have been drained and cut over or otherwise become dry, and in such situations there can be some very modest amelioration of the extremely impoverished soil conditions where the organic matter is disturbed or oxidised.

For the most part, however, the calcifuge and oligotrophic character of the flora of this community accurately reflects the prevailing edaphic conditions and may itself have contributed to the progressive impoverishment of the soils by acidification and promotion of podzolisation following woodland clearance. Some tracts of the *Calluna-Deschampsia* heath may not be very old – in Cannock Chase, for example, tree-felling to supply the local iron industry led to a spread of heath in this Norman forest from the mid-sixteenth century – but much probably dates originally from the combination of worsening climate and deforestation in the Iron Age (e.g. Hicks 1971, Gimingham 1972). And the generally species-poor character of the vegetation bears strong testimony to long histories of burning, to maintain grazing for sheep around the upland fringes and, more recently, as an essential part of the management of moorland for grouse-shooting, particularly important here in the North York Moors. Regular controlled burning, or frequent accidental fires on stretches of heath open for recreation, an important use of stands close to larger centres of population, help maintain the general dominance of *Calluna* in the community and account for the widespread preponderance of more or less permanently immature canopies, sometimes disposed in variegated patterns of burns, or 'swiddens' as they are known in the North York Moors (Elgee 1914), of pioneer and building phases, but not often progressing to the degenerate phase (Watt 1955). Such rare older canopies can be found here, showing the distinctive preferentials of the *Hypnum* sub-community, characteristic of the collapsing centres of the bushes, but these stands are largely confined to fragments of neglected heathland through the Midland plain.

The bulk of younger stands can thus be incorporated in the *Vaccinium-Cladonia* and Species-poor sub-communities, the floristic and structural differences between these two kinds of *Calluna-Deschampsia* heath perhaps reflecting the time since burning and the frequency of fires. Even in the former vegetation, however, the enrichment in bryophytes and lichens beneath the recovering, but still fairly open, heather canopy is of a particular type, and there seems little doubt that, throughout the community, the abundance and diversity of these elements of the flora are further strongly inhibited by atmospheric pollution (Shimwell 1973*b*). Through almost the entire range of the *Calluna-Deschampsia* heath, such pollution, as measured by the sulphur dioxide concentration, is of a high order, between 50 and $100\mu g$ m^{-3} in the data collated by the Warren Spring Laboratory and many stands lie within the immediate ambit of the great conurbations of south Lancashire, Yorkshire and the Midlands, where the effects of industrial and domestic emissions have been strongly felt for well over a century. The preponderance of the acrocarpous mosses *Pohlia nutans* and *Orthodontium lineare*, and the scarcity of pleurocarps, apart from *Hypnum cupressiforme s.l.*, and of Cladina lichens, even in the more mature heaths, can probably be related to this phenomenon (Watson 1922, Hawksworth 1969): these same features can be seen, too, in the impoverished heathlands around the industrial centres of Belgium, The Netherlands and Germany (Barkman in Shimwell 1973*b*). The often rather slimy surfaces of the peats in the *Calluna-Deschampsia* heath also seem to present a congenial habitat for certain leafy hepatics, most distinctively *Gymnocolea inflata*. Taken together with the restricted role that the Vaccinia and *Empetrum nigrum* have here, these characteristics help distinguish the community from the *Calluna-Vaccinium* heath, which is the predominant kind of heather-dominated vegetation through most of the sub-montane zone in northern Britain: essentially, the *Calluna-Deschampsia* heath can be seen as a polluted vicariant of this community, replacing it at the southern lowland extreme of its range.

At such an extreme, it seems that gross disturbance of the heathland habitat provides almost the best opportunity for the local abundance in the *Calluna-Deschampsia* heath of sub-shrubs like *Vaccinium vitis-idaea* and *Empetrum nigrum* ssp. *nigrum* which are very frequent in the *Calluna-Vaccinium* heath. Thus, it is over ground disrupted by such activities as army manoeuvres and quarrying, which destroy the humose topsoil to the profiles, that these species often get a strong hold by rapid rhizomatous spread (Shimwell 1973*b*). Such conditions, too, seem important for the appearance of *V. × intermedium* which also figures very occasionally in such stands of the community. This vigorous hybrid is of very restricted distribution in Britain (Ritchie 1955*b*, Perring 1968), being common only at moderate altitudes in Derbyshire and Staffordshire, despite the frequent occurrence together of *V. myrtillus* and *V. vitis-idaea* through much of upland northern Britain. But what seems particular to this area is the combination of somewhat irregular burning with various kinds of disturbance (e.g. Gourlay 1919) in a zone where there is considerable overlap in the floral phenology of the parents (Ritchie 1955*b*), thus allowing crossing between

plants brought into sudden juxtaposition in an unstable environment.

Combinations of burning with grazing can also help entrench the local abundance of less palatable sub-shrubs like *V. vitis-idaea* (Ritchie 1955*a*) and *Empetrum nigrum* ssp. *nigrum* (Bell & Tallis 1973), and the spread of the latter in the *Calluna-Deschampsia* heath on Ilkley Moor in West Yorkshire, where during and after the Second World War there was heavy stocking with sheep under a common grazing system with no stint, has been very marked (Fidler *et al.* 1970, Dalby *et al.* 1971).

Long-sustained grazing of more typical stands of the community, where heather, perhaps with some bilberry, makes up the canopy, tends to favour elimination of these palatable sub-shrubs with a consequent expansion of *Deschampsia flexuosa*, and the appearance of Nardo-Galion herbs like *Galium saxatile* and *Potentilla erecta*, features which characterise the *Galium* sub-community, a kind of *Calluna-Deschampsia* heath that is especially common on lowland fragments of heathland associated with wood-pasture as in Sherwood.

Zonation and succession

The *Calluna-Deschampsia* heath can be found in zonations and mosaics with wet-heath and mire vegetation where the major influence on floristic variation is natural differences in the moisture content of suites of base-poor and oligotrophic soils. Even in such situations, however, the effects of such factors as burning, grazing and draining complicate the patterns of vegetation types and, in many cases, these are of overriding importance, their effects being especially clearly seen over more uniform tracts of soils, where patchworks of the community, with grasslands and woodlands, represent stages in regeneration and succession. Neglect of traditional treatments has seen extensive invasion by trees at some sites, particularly in the lowlands, though it is there, too, that previously large tracts of heath and woodland have been most assiduously reclaimed for agriculture. In other cases, both in the lowlands and around the upland fringes, the community has been replaced by coniferous plantations.

In a few sites, the *Calluna-Deschampsia* heath forms the dry-heath component in the classic kind of lowland valley-mire zonation that is so characteristic of central southern and south-west Britain, occurring over the most free-draining of the sequence of acidic soils disposed around elongated hollows through which there is concentration of ground waters draining from the surrounds or from particular springs and flushes. Even through the lowland parts of the range of the community, the climate is not sufficiently oceanic to see the transgression of *Erica tetralix* and *Molinia* on to more free-draining soils, so a humid heath zone, such as is typical of these patterns in the more Atlantic parts of the country, is lacking, and there is generally a sharper shift, over ground that shows increasingly strong gleying, to the *Ericetum tetralicis* wet heath, where these two species can have ascendancy over *Calluna*; and then, where there is continuous stagnation more or less to the surface, with accumulation of peat proper, the vegetation changes to the *Narthecio-Sphagnetum* bog. This kind of pattern can still be seen in parts of Cannock Chase, developed around the valleys incised into the Bunter sandstones, pebble-beds and breccias (Ratcliffe 1977), but, though the systems are hydrologically similar to those of, say, the New Forest, the flora is, throughout the sequence, of a more impoverished character. Both the *Ericetum* and *Narthecio-Sphagnetum* are approaching their geographical limits in this part of Britain and, over the drier ground, generations of burning and disturbance have much reduced the *Sphagnum* component of the wet heath and favoured a general abundance of heather throughout, thus blurring the boundaries between the communities. More recently, wholesale lowering of the water-table, probably as a result of coal-mining beneath the Chase, is drying out the mire systems there, eliminating the wetter elements or causing them to migrate downstream.

Essentially the same pattern as this can be seen at somewhat higher altitudes in the North York Moors, where certain of the drainage hollows on the plateau, notably at Fen Bog, have zonations which, though also affected by burning and grazing, still show the controlling influence of a stagnation gradient. In this area, though, and around the upland fringes of the southern Pennines, the *Calluna-Deschampsia* heath is more often found in what is really a reverse sequence of communities, giving way over gentler slopes at higher altitudes to wet heath, then to blanket mire, with the development of increasingly thick ombrogenous peats. In the North York Moors, the *Calluna-Deschampsia* heath and the *Juncus-Dicranum* sub-community of the *Ericetum tetralicis* form the bulk of the vegetation cover over the complex patchworks of Anglezarke humo-ferric podzols, Maw stagnopodzols and Wilcocks stagnohumic gleys, with relatively small areas of *Calluna-Eriophorum* mire marking out Winter Hill peats, almost the whole of these patterns being overlain by systems of swiddens burned for grouse-rearing. On the Pennines, the wet-heath zone is much less extensive and the *Calluna-Deschampsia* heath frequently gives way directly, over what is often a much fretted fringe to the ombrogenous peat, to the blanket mire vegetation, characteristically in this part of Britain of the *Eriophorum* type, in which the effects of pollution, grazing, burning and draining are also all too evident. In such zonations, the increase in the abundance of *Eriophorum vaginatum* among the heather is often the only indicator of the switch in vegetation types, species such as *Deschampsia flexuosa*, the acrocar-

pous mosses and encrusting lichens maintaining their frequency over the dry peat margins. Patches of the *Calluna-Deschampsia* heath can recur at higher levels where the peat blanket is interrupted by grit outcrops but, increasingly in the wetter cooler climate of these high edges, the community is replaced by the *Vaccinium-Deschampsia* heath in which bilberry is the typical dominant.

Rather different kinds of vegetation patterns are to be seen where the *Calluna-Deschampsia* heath occurs on tracts of prevailingly dry, base-poor and oligotrophic soils, such as have developed on sandstone scarps and plateaux and over coarser superficials at lower altitudes through the range of the community. In such situations, it is patchworks of the heath with grasslands and woodlands that make up the bulk of the scenery, so characteristic of places like Delamere and Sherwood Forests and the higher ground in Cannock Chase. Such patterns are largely a reflection of treatment history, with burning and grazing, or their neglect, playing a major part in the disposition of stands of various of the sub-communities of the *Calluna-Deschampsia* heath, of *Deschampsia flexuosa* grassland, which is often a grazed derivative of the *Galium* sub-community within which heather has been substantially reduced, and of the *Quercus-Betula-Deschampsia* woodland, which is the characteristic climax forest of dry, acidic soils in this part of Britain, derived from the heath by invasion of oak and birch. Widespread abandonment of traditional treatments has often favoured the succession to woodland, with the virtual extinction of heath in some sites, though temporary patches of the *Calluna-Deschampsia* community comprise the typical gap flora of the *Quercus-Betula-Deschampsia* woodland. In many places, too, land-use changes have led to an extensive spread of the *Pteridium-Galium* community which can readily replace the *Calluna-Deschampsia* heath on the deeper of the soils that it favours.

Commercial afforestation has also been extensive over many stretches of podzolised soils through the range of the *Calluna-Deschampsia* heath which, with some preparation, can support reasonable crops of *Pinus sylvestris*, *P. nigra* var. *maritima* and larches, all of which, especially the pines, are readily able to seed into remaining tracts of open heathland around. The brown sand soils of the Midland plain too, though droughty, highly impoverished and susceptible to erosion by wind and rain, are readily worked and many areas have been reclaimed and improved for arable and pastoral agriculture, fragmenting the remaining distribution of the community.

Distribution

The *Calluna-Deschampsia* heath is concentrated in the southern Pennines and North York Moors with more local occurrences scattered through the Midland plain. In the former areas, the Species-poor and *Vaccinium-Cladonia* sub-communities are the usual forms, widespread and sometimes extensive over heathlands and moors that are still frequently burned. The *Hypnum*, *Molinia* and *Galium* sub-communities are primarily found on lowland sites where burning has fallen into disuse.

Affinities
The community subsumes much of the heather-dominated vegetation described in the early accounts of the moorland of Northern England (Smith & Moss 1903, Smith & Rankin 1903, Lewis & Moss 1911, Elgee 1914), but its more precise definition essentially follows the proposal of Shimwell (1973*b*) in diagnosing an impoverished heath type distinct from both the lowland *Ulex* heaths of south-west Britain and the Myrtillion communities of the montane zone. In fact, the *Calluna-Deschampsia* heath is somewhat broader than Shimwell's (1973*b*) *Pohlio-Callunetum*, including also some of

what Coppins & Shimwell (1971) grouped in a rather richer *Cladonio crispatae-Callunetum* and it is not so sharply marked off from the upland Myrtillion heaths. Indeed, among the range of other heath types distinguished in Britain, the *Calluna-Deschampsia* heath is best seen as an impoverished replacement for the *Calluna-Vaccinium* heath around its southern limit. From another perspective, the community can be viewed as a degraded northern form of the Continental Genisto-Callunion vegetation included in the *Calluna-Festuca* heath of the eastern lowlands of England, though here the affinities would be based largely on the past distributions of species such as *Genista anglica* and the Cladina lichens, plants which are not in any case restricted to the warmer south-east of the country. In view of its especially impoverished character, Shimwell (1973*b*) proposed erecting a new Pohlio-Callunion alliance to contain this kind of heath and perhaps similar communities described from The Netherlands (Stoutjesdijk 1959, de Smidt 1966, Touw 1969).

Floristic table H9

	a	b	c	d	e	9
Calluna vulgaris	V (8–10)	V (7–10)	V (3–10)	V (4–10)	V (9–10)	V (3–10)
Deschampsia flexuosa	V (3–7)	IV (1–8)	IV (1–8)	V (3–7)	V (3–5)	IV (1–8)
Pohlia nutans	IV (2–5)	IV (2–8)	III (1–6)	III (4–7)	II (2–3)	IV (1–8)
Hypnum cupressiforme	V (3–8)	II (2–5)	I (1)	II (3–5)		II (1–8)
Dicranum scoparium	IV (3–8)	I (1–4)	I (1)	I (4)		II (1–8)
Pteridium aquilinum	II (2–5)	I (1–4)	I (4–7)	I (3–5)		I (1–7)
Hypogymnia physodes	II (1–2)	I (1–2)	I (1)			I (1–2)
Vaccinium myrtillus	II (1–4)	IV (1–7)	II (1–5)	I (4)	I (3)	II (1–5)
Campylopus paradoxus	I (3–5)	III (2–5)	III (1–3)	I (2–3)	II (2–4)	II (1–5)
Cladonia chlorophaea	I (3–5)	III (1–5)	I (2)	I (3)	II (2–4)	II (1–5)
Gymnocolea inflata		III (1–7)	II (1–6)	I (4)	I (2)	II (1–7)
Cladonia floerkeana	I (3)	III (1–4)	I (1–3)			II (1–4)
Orthodontium lineare	I (2–3)	III (1–3)	II (1–6)			II (1–6)
Calypogeia muellerana	I (2)	II (1–3)	I (1–5)		I (2)	I (1–5)
Cladonia squamosa	I (2–5)	II (2–4)	II (1–3)		I (2–3)	I (1–4)
Cephaloziella divaricata	I (3)	II (1–4)	I (3–4)	I (3)	I (2–3)	I (1–4)
Barbilophozia floerkii	I (2–4)	II (1–4)				I (1–4)
Cladonia coccifera	I (3–4)	II (2–7)	I (2)		I (3–4)	I (2–7)
Empetrum nigrum nigrum		I (2–6)				I (2–6)
Vaccinium vitis-idaea		I (5–8)				I (5–8)
Vaccinium × intermedium		I (2–3)				I (2–3)
Cladonia crispata		I (2–6)				I (2–6)
Galium saxatile		I (1–3)		V (2–5)		I (2–5)
Potentilla erecta		I (1)		III (3–4)		I (1–4)
Rumex acetosella				II (3)		I (3)
Festuca rubra		I (2)		II (3–4)		I (2–4)
Holcus mollis				II (3)		I (3)
Quercus spp. seedling				II (2–4)		I (2–4)
Molinia caerulea	I (3–5)	I (4)		I (2)	V (2–5)	I (2–5)

Floristic table H9 *(cont.)*

468   Heaths

	a	b	c	d	e	9
Cladonia coniocraea	I (3)	I (1–3)	I (3)	I (2)	I (2)	I (1–3)
Cladonia fimbriata	I (3)	I (2–3)	I (1)		I (2)	I (1–3)
Agrostis capillaris	I (3)	I (1–4)	I (2–3)	I (3–4)		I (1–4)
Juncus squarrosus	I (3–4)	I (1–5)	I (1)	I (2–4)		I (1–5)
Lepidozia reptans	I (1–3)	I (3)	I (3)			I (1–3)
Holcus lanatus	I (1–3)	I (3)	I (3)			I (1–3)
Pinus sylvestris seedling	I (1)	I (1)			I (1)	I (1)
Ptilidium ciliare	I (2–4)	I (2–3)				I (2–4)
Eriophorum angustifolium	I (3–4)	I (1–4)				I (1–4)
Cladonia uncialis	I (1–4)	I (3–4)				I (1–4)
Cornicularia aculeata	I (3)	I (3)				I (3–5)
Teucrium scorodonia	I (2)	I (5)			I (2)	I (2)
Number of samples	34	96	36	8	8	175
Number of species/sample	7 (4–11)	11 (2–21)	6 (3–12)	9 (5–14)	7 (3–14)	8 (2–15)
Vegetation height (cm)	44 (22–75)	31 (7–100)	41 (6–100)	36 (10–60)	70 (25–100)	39 (6–100)
Shrub/herb cover (%)	97 (80–100)	97 (65–100)	89 (45–100)	93 (60–100)	98 (95–100)	97 (45–100)
Ground layer cover (%)	34 (5–80)	29 (2–85)	9 (0–40)	18 (0–50)	6 (0–40)	25 (0–85)
Altitude (m)	243 (46–594)	234 (5–487)	236 (61–375)	113 (60–229)	169 (15–305)	237 (5–594)
Slope (°)	5 (0–50)	9 (0–45)	6 (0–35)	4 (0–15)	4 (0–30)	7 (0–50)

a *Hypnum cupressiforme* sub-community
b *Vaccinium myrtillus-Cladonia* spp. sub-community
c Species-poor sub-community
d *Galium saxatile* sub-community
e *Molinia caerulea* sub-community
9 *Calluna vulgaris-Deschampsia flexuosa* heath (total)

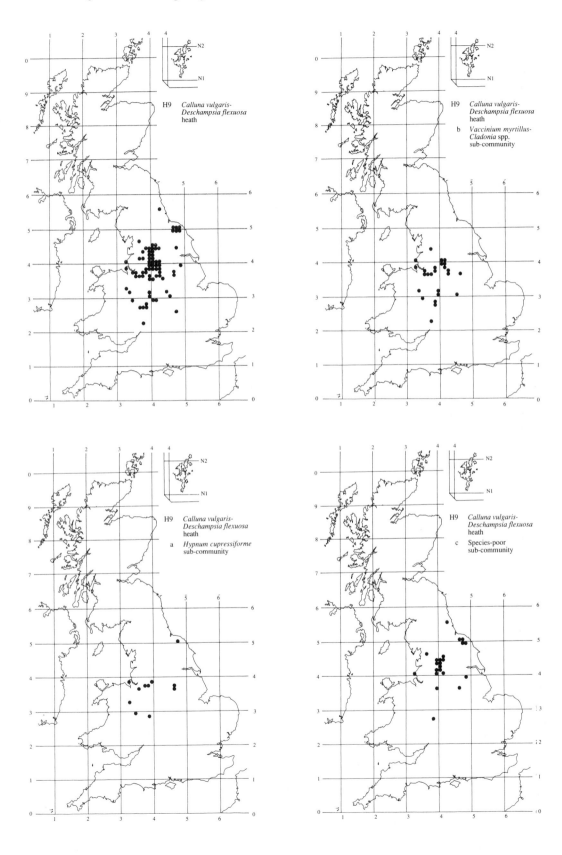

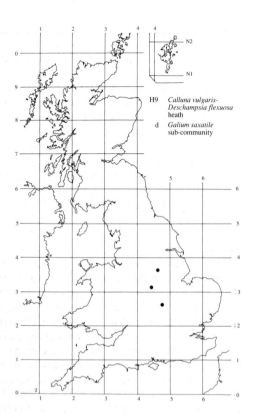

H9 *Calluna vulgaris-Deschampsia flexuosa* heath

d *Galium saxatile* sub-community

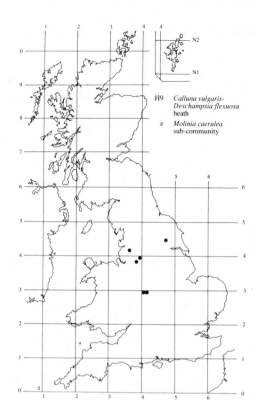

H9 *Calluna vulgaris-Deschampsia flexuosa* heath

e *Molinia caerulea* sub-community

H10
Calluna vulgaris-Erica cinerea heath

Synonymy
Scottish *Calluna* heath Smith 1911, Tansley 1939 *p.p.*; *Callunetum vulgaris* McVean & Ratcliffe 1962 *p.p.*, Birks 1973, Prentice & Prentice 1975, Meek 1975, 1976, Evans *et al.* 1977 *p.p.*, Hill & Evans 1978, Ferreira 1978; *Calluna-Erica cinerea* heaths Muir & Fraser 1940, Gimingham 1964*b*, 1972 *p.p.*; *Calluna vulgaris-Sieglingia decumbens* Association Birks 1973; *Carici binervis-Ericetum cinereae* Br.-Bl. & Tx (1950) 1952 *emend.* Birse 1980 *p.p.*; *Plantago maritima-Erica cinerea* Association Birse 1980.

Constant species
Calluna vulgaris, Erica cinerea, Potentilla erecta.

Rare species
Orobanche alba.

Physiognomy
The *Calluna vulgaris-Erica cinerea* heath is typically dominated by *Calluna vulgaris* but the cover, height and structure of the sub-shrub canopy vary quite markedly according to the incidence of burning and grazing and the degree of exposure. Where the community is more frequently burned, as where it makes some contribution to grouse-moor in parts of southern Scotland and locally around the east-central Highlands, pioneer and building phases of the heather tend to predominate but, with more sporadic firing, there can be patchworks of even-aged stands through to the mature and degenerate. Very commonly, however, there is also some grazing by stock and deer which helps keep the canopy more closely trimmed and, in extreme cases, exposure to wind and sun can reduce the bushes to a tight and even or wind-waved cover but a few centimetres high.

Further modest diversity comes from the frequent but varied contribution of *Erica cinerea*. This is generally subordinate to the heather in abundance though, being more shade-tolerant, it is well able to persist beneath taller *Calluna* canopies as a patchy second tier, the rather straggling branches that it often forms in mixed covers becoming semi-prostrate, rooting adventitiously and spreading laterally (Bannister 1965, Gimingham 1972). And, in certain circumstances, it can become locally abundant in this kind of heath, regenerating well after burning, for example, and, especially on south-facing slopes and where there is little grazing (Gimingham 1949), it can rival *Calluna* in its extent in the middle years of recovery. *Vaccinium myrtillus*, by contrast, is at most occasional here and only rarely of any abundance and *V. vitis-idaea* is decidedly scarce. *Empetrum nigrum* ssp. *nigrum* can occur, but it is very much confined to one particular kind of *Calluna-Erica* heath, and the restricted occurrence of all these sub-shrubs provides a good distinction between the community and the *Calluna-Vaccinium* heath. In this respect, then, the *Calluna-Erica* heath continues northwards the kind of vegetation seen in the *Calluna-Ulex gallii* heath and indeed the community provides a very occasional locus for *U. gallii* at the limit of its range in southern Scotland. Of other woody plants, *Erica tetralix* is sometimes found, particularly where the community forms mosaics with the *Scirpus-Erica* wet heath or its eastern counterpart, the *Ericetum tetralicis*, and rarely there are records for *Arctostaphylos uva-ursi*. Where the vegetation is recovering from burning, scattered plants of *Salix repens* or *S. aurita* can also find a temporary niche.

Despite the characteristic abundance of the constant sub-shrubs, two other floristic features often catch the eye. The first is the frequency of grasses, to a lesser extent sedges and dicotyledons, typical of Nardetalia swards, growing beneath and among the bushes. Of grasses, *Deschampsia flexuosa* is the most consistent throughout, with *Agrostis canina* (probably mostly ssp. *montana*) and *Nardus stricta* occasional to frequent in most kinds of *Calluna-Erica* heath but, in certain of the sub-communities, *Festuca ovina, Anthoxanthum odoratum, Agrostis capillaris* and *Molinia caerulea* also become very common, sometimes too with *Danthonia decumbens, Festuca rubra* or *F. vivipara*. The sedge component is less varied but the Oceanic West European *Carex binervis* is very characteristic of this kind of heath, together with *C. pilulifera*. After burning, mixtures of these plants can become patchily abundant, and *D. flexuosa* and *C. pilulifera* are particularly able to establish themselves as temporary local dominants in such circumstances (Gimingham 1964*a*). In other cases, a more permanent and intimate grass-heath physiognomy, with these species ramifying an often rather short sub-shrub cover, is maintained by grazing. Other monocotyledons which find an occasional place in the community are *Carex panicea* and even *C. pulicaris* where this kind of heath extends some way on to less base-poor soils, *Luzula multiflora*, *L. campestris*, *Juncus squarrosus* and *Scirpus cespitosus*.

Dicotyledons are generally few in number, but *Potentilla erecta* is a constant and *Galium saxatile* fairly common and, in the grassier kinds of *Calluna-Erica* heath, species such as *Lotus corniculatus*, *Plantago lanceolata*, *Campanula rotundifolia*, *Succisa pratensis* and *Hypericum pulchrum* can be found occasionally, sometimes with further enrichment from oceanic or mildly basiphilous elements.

The second striking feature of many stands of the community is the contribution of the ground layer. Early stages of recovery from burning can have a local abundance of *Polytrichum piliferum*, *P. juniperinum* and encrusting *Cladonia* spp., while in exposed stands there is often a patchy carpet of *Racomitrium lanuginosum* and fruticose lichens over the bare ground. But more important than these through the community as a whole are bulky pleurocarpous mosses such as *Hypnum cupressiforme s.l.*, *Pleurozium schreberi* and *Hylocomium splendens*, with *Rhytidiadelphus triquetrus* and *R. loreus* also occurring occasionally. These species, together with *Dicranum scoparium*, typically become abundant with the maturing and opening up of the heather bushes, so frequent burning can repeatedly curtail their development, but they often make a more permanent contribution to grazed stands, where they can form a bulky mat among the more open network of sub-shrub branches and herbs.

Sub-communities

Typical sub-community: *Callunetum vulgaris* McVean & Ratcliffe 1962, Birks 1973, Prentice & Prentice 1975, Meek 1976, Evans *et al.* 1977, Hill & Evans 1978, all *p.p.*; *Calluna-Erica cinerea* heaths Gimingham 1964*b*, 1972 *p.p.*; *Carici binervis-Ericetum cinereae*, Typical subassociation *sensu* Birse 1980. In this, the most generally species-poor kind of *Calluna-Erica* heath, heather is typically dominant and is often overwhelmingly abundant in pioneer or building regrowth after burning. *Erica cinerea* is very frequent and it can show some local

prominence among or beneath the heather or very occasionally replace it as the leading sub-shrub with recovery from fire, but usually it is of sparse cover and sometimes altogether absent. *Vaccinium myrtillus* occurs occasionally, though hardly ever as more than scattered shoots and *Empetrum nigrum* ssp. *nigrum* and *Erica tetralix* are scarce and usually found in stands that are obviously transitional to the *Racomitrium* sub-community.

Monocotyledons are typically few and generally of low cover though *Deschampsia flexuosa*, which is very frequent, sometimes shows local prominence. More striking here, though, is the preferential occurrence of *Molinia caerulea*, usually not more than patchily abundant but, with occasional *Scirpus cespitosus* and *Juncus squarrosus*, often bringing the vegetation close in its composition to degraded forms of *Scirpus-Erica* wet heath, with which this vegetation often forms mosaics. However, one good distinction between the two is provided by *Carex binervis*, better represented here than in any other kind of *Calluna-Erica* heath, though not usually with any abundance, being typically found as scattered shoots or in small clumps. *Agrostis canina*, *A. capillaris* and *Nardus stricta* also occur occasionally at low covers, but *Festuca ovina* and *Anthoxanthum* are characteristically uncommon and the grass-heath physiognomy of some other sub-communities is only rarely developed here. Apart from *Potentilla erecta* and occasional *Galium saxatile*, herbaceous dicotyledons are very sparse but there are sometimes scattered plants of *Blechnum spicant*.

The ground layer is generally also poor in species and of low cover. Younger stands can have patches of *Campylopus paradoxus*, *Polytrichum piliferum* and *P. commune*, together with encrusting lichens such as *Cladonia floerkeana*, growing over the bare peat and small amounts of *Racomitrium lanuginosum* can be found in more exposed situations, but the development of a mat of pleurocarps is usually limited to small wefts of *Hypnum cupressiforme* with occasional *Pleurozium schreberi*, a feature which reflects the scarcity of older, more open canopies of heather in this sub-community. Other particular features are ill-developed, although occasional *Diplophyllum albicans*, *Scapania gracilis* and little tufts of Sphagna can enhance the oceanic character of the vegetation.

***Racomitrium lanuginosum* sub-community:** *Callunetum vulgaris*, *Rhacomitrium* lichen-rich facies Prentice & Prentice 1975; *Callunetum vulgaris* Hill & Evans 1978; *Carici binervis-Ericetum cinereae*, *Rhacomitrium* subassociation *sensu* Birse 1980. *Calluna* is again the usual dominant here but its cover is not so consistently overwhelming as in the Typical sub-community. Indeed, although there is also frequently some *Erica cinerea*, occasional *Vaccinium myrtillus* and, preferential here, quite common *Empetrum nigrum* ssp. *nigrum*, the sub-shrub canopy is altogether more open than usual and often dwarfed, with in extreme cases a patchy cover of wind-pruned, stunted bushes.

Among the sub-shrubs, grasses are especially sparse with just scattered tufts of *Deschampsia flexuosa* and very occasional *Nardus stricta*, *Agrostis canina*, *Festuca ovina* and *F. vivipara*. *Carex binervis* is rather rare, too, its place being often taken here by *C. pilulifera*, sometimes showing modest local abundance, and *C. panicea*. *Scirpus cespitosus* is also frequent and it can be patchily prominent. Of herbaceous dicotyledons, only *Potentilla erecta* shows any frequency, though *Huperzia selago* is preferentially common.

Much more prominent in this sub-community is the ground layer which often occupies substantial areas between the sub-shrubs. Although *Hypnum cupressiforme s.l.* remains frequent, the most important moss here is *Racomitrium lanuginosum* which can become especially abundant among degenerating bushes. Lichens, too, have their best representation in this sub-community with frequent *Cladonia uncialis* and *C. impexa*, and occasional *C. furcata*, *Cornicularia aculeata* and *Cetraria islandica*.

Festuca ovina-Anthoxanthum odoratum sub-community:
Callunetum vulgaris Birks 1973, Prentice & Prentice 1975, Evans *et al.* 1977, Hill & Evans 1978, all *p.p.*; *Carici binervis-Ericetum cinereae*, Typical subassociation *sensu* Birse 1980 *p.p.* *Calluna* is still often abundant in this sub-community, though its cover is quite frequently rivalled by that of *Erica cinerea*. *Vaccinium myrtillus* occurs occasionally, too, though in small amounts, and the sub-shrub canopy can be very extensive in total. But it is usually quite short, sometimes distinctly patchy, and commonly forms a mosaic with, or is thoroughly ramified by, a grassy turf. Most frequent in this element are *Festuca ovina*, *Anthoxanthum* and *Agrostis capillaris*, with occasional *A. canina*, *F. rubra*, *F. vivipara*, *Danthonia decumbens* and *Deschampsia flexuosa*, sometimes growing as tussocky clumps, sometimes as a more even-cropped sward, in which one or more of the grasses can show considerable local abundance or where there is a more balanced mixture of the species. *Carex binervis* and *C. pilulifera* are also both quite common and there are occasional scattered plants of *Luzula multiflora* and *L. campestris*. *Juncus squarrosus* and *Scirpus cespitosus*, by contrast, are both scarce here.

Dicotyledonous herbs, too, are much more numerous than in the above sub-communities. In addition to *Potentilla erecta* and *Galium saxatile*, both of which are very common, there are occasional records for *Campanula rotundifolia*, *Lotus corniculatus*, *Plantago lanceolata*, *P. maritima*, *Polygala serpyllifolia*, *Succisa pratensis*, *Hypericum pulchrum*, *Viola riviniana* and *Thymus praecox*.

Equally distinctive is the consistent role that the bulky pleurocarpous mosses have in this vegetation, with frequent *Hypnum cupressiforme s.l.*, *Pleurozium schreberi* and *Hylocomium splendens*, together with *Dicranum scoparium*, occurring among the more open parts of the canopy and among the grass culms as a patchy carpet.

Thymus praecox-Carex pulicaris sub-community:
Callunetum vulgaris, herb-rich facies McVean & Ratcliffe 1962, Ferreira 1978; *Calluna vulgaris-Sieglingia decumbens* Association Birks 1973; *Carici binervis-Ericetum cinereae*, *Viola riviniana* subassociation *sensu* Birse 1980; *Plantago maritima-Erica cinerea* Association Birse 1980 *p.p.* In its general structure and floristics, this kind of *Calluna-Erica* heath is very similar to the *Festuca-Anthoxanthum* type, with *Calluna* and *E. cinerea* both able to show prominence in the sub-shrub canopy, but with herbs and bryophytes also being of considerable structural importance. Here, however, there are some additional preferentials which make this the most species-rich and distinctive sub-community. Among the grasses, for example, *Danthonia decumbens* is particularly frequent and the commonest sedges are *Carex panicea* and, especially striking here, *C. pulicaris*. Frequent *Thymus praecox* and *Linum catharticum* continue the mildly basiphile aspect of the vegetation, and there are also frequent records for *Viola riviniana*, *Prunella vulgaris* and *Primula vulgaris* with occasional *Solidago virgaurea*, *Antennaria dioica*, *Blechnum spicant*, *Lathyrus montanus* and *Geum rivale*. On Skye, this vegetation provides the main locus for the rare Continental Southern *Orobanche alba* at the northern limit of its British range.

Among the bryophytes, *Dicranum scoparium* and the pleurocarps remain very common, but there is additionally frequent *Rhytidiadelphus triquetrus* and *Breutelia chrysocoma*. In more exposed situations, a local increase in *Racomitrium lanuginosum* and fruticose lichens can produce transitions to the *Racomitrium* sub-community.

Habitat
The *Calluna-Erica* heath is characteristic of acid to circumneutral and generally free-draining soils in the cool oceanic lowlands and upland fringes of northern and western Britain. In more exposed situations, it could perhaps be considered as a climatic or edaphic climax but, very often, grazing and burning play a considerable part in controlling its composition and structure.

The community includes most of the drier, non-maritime heath occurring through the more equable parts of the country beyond the northern limit of *Ulex gallii*. This gorse is found only very sparsely to the north of the Solway, where mean annual maximum temperatures drop below 26 °C (Conolly & Dahl 1970) and beyond which February minima are usually less than the degree or so above freezing which this plant seems to demand (*Climatological Atlas* 1952). The *Calluna-Erica* heath does include stands of dry heath without the gorse at scattered localities down through northern England, Wales and the South-West Peninsula but, for the most part, this is a Scottish community. It is, however, quite strongly confined to the relatively mild regions of northern Britain, becoming distinctly scarce where the mean annual maximum temperature is below 22 °C and where February minima fall more than just below freezing. This, then, is the characteristic 'Atlantic heather moor' of Scotland, as Birse & Robertson (1976) called it, occurring at low to moderate altitudes, almost down to sea-level in some places but only occasionally penetrating over 400 m, through Dumfries & Galloway, the Western Isles, Orkney, Shetland and down the northeast coast to the Moray Firth. Through this region, *Erica cinerea*, which retains a measure of physiological activity through the winter (Bannister 1965, Gimingham 1972), can thrive on more free-draining soils and its vigour here, occurring together with *Molinia* and *Carex binervis*, is a good reflection of the broadly oceanic conditions.

For the Arctic-Alpine sub-shrubs *Vaccinium vitis-idaea* and *Empetrum nigrum*, on the other hand, the climate is not so congenial and even *V. myrtillus*, which becomes generally frequent in Britain where the annual rainfall exceeds 800 mm, is rather scarce here, despite the fact that precipitation is well above this almost everywhere in the range of the community. Its poor showing could be partly attributable to the high incidence of grazing in this vegetation (see below), but the particular combination of temperature and humidity may be less than optimal. Certainly, with the shift to higher altitudes, and particularly in the distinctly cooler and cloudier conditions through the east-central Highlands, these species all become very important in a switch to the *Calluna-Vaccinium* heath, which corresponds largely to what Birse & Robertson (1976) called 'Boreal heather moor'. The two communities do intergrade floristically and their altitudinal and geographical separation are far from absolute but, where they are both found in the same general areas, as in parts of the Grampians and the Southern Uplands, the *Calluna-Erica* heath is typically at lower altitudes than the *Calluna-Vaccinium* heath and/or on warmer slopes. Indeed, almost wherever the community penetrates over 300 m, it is to be found on south- or west-facing aspects, and it is probably the topoclimate on such slopes, locally tilting the balance in favour of *E. cinerea* and against *V. myrtillus*, that gives the *Calluna-Erica* heath its occasional representation at

scattered localities well above 400 m through the Highlands proper.

The other climatic influences on the community are felt through exposure. In very windy situations, over brows and on open plateaus, both at higher altitudes and in the gale-ridden coastal lowlands of the west and in Shetland and Orkney, this kind of heath can have a very stunted and sometimes wind-waved cover of sub-shrubs and develop the extensive mat of *Racomitrium lanuginosum* and fruticose lichens characteristic of the *Racomitrium* sub-community. In very degraded heath of this kind, where frequent and intense burning or heavy grazing have contributed to the run-down of the vegetation, the typical canopy of the community can be very open and the rhizomatous *Empetrum nigrum* ssp. *nigrum* is sometimes able to spread, but *Erica cinerea* maintains its frequency, providing a distinction from the *Calluna-Racomitrium* and *Calluna-Cladonia* heaths of exposed situations through the Highlands.

On exposed coasts, maritime influence is an additional factor in the windy climate that affects the distribution of the community, although along the very rainy western seaboard of Scotland, high precipitation counteracts somewhat the deposition of salt-spray. Nonetheless, although the *Calluna-Erica* heath can be found almost down to sea-level on the tops of coastal cliffs in this part of the country (e.g. Birks 1973, Birse 1980), it is very often replaced in exposed sites, which are otherwise congenial, by the *Calluna-Scilla* heath. Transitional vegetation, in which species such as *Plantago maritima*, *Succisa pratensis* and *Danthonia decumbens* provide a floristic link, can be seen in the *Festuca-Anthoxanthum* and *Thymus-Carex* sub-communities. Over stabilised coastal sands, the community is likewise restricted to the inland edge of dune or machair systems, with the *Calluna-Carex* heath occupying similar soils to seaward.

Throughout its range, there is also a general edaphic limitation to profiles which are moderately to very base-poor and typically free-draining, though of course in the climate of this part of Britain, quite often moist. Surface pH, however, spans quite a broad range, being generally between 3.5 and 6, and in most kinds of *Calluna-Erica* heath, the mean is around 5. On steeper slopes, where the community can extend on to rockier ground around exposures, the soils can be very immature and fragmentary, little more than shallow humic rankers in pockets of rock, but generally they are deeper and better developed than this, sometimes brown earths, more usually brown podzolic profiles or podzols proper. Such soils are widespread over the pervious lime-poor bedrocks that make up so much of northern and western Britain but, within the range of the *Calluna-Erica* heath, particularly important substrates are provided by stretches of arenaceous rocks among the Ordovician and Silurian deposits of the Southern Uplands, the Devonian Old Red Sandstone of north-east Scotland and Orkney, the granite intrusions of Dumfries, Arran and the Grampians, and a variety of other igneous and metamorphic rocks up through the Isles and round into Shetland.

Over such materials, the community is commonest on gentle to moderately steep hill-slopes, though it can be found on level ground provided the drainage remains reasonably free, and the general character of the soils is reflected in the prominence here, among the sub-shrubs, herbs and bryophytes, of calcifuges, many of a rather broad amplitude and some of the most distinctive, like *Carex pilulifera*, quite mesophytic. However, though some of the most frequent plants of the community are tolerant of quite moist soils, this kind of heath extends only a little way on to impeded profiles, such as the gley-podzols and stagnopodzols widely distributed through the range of the *Calluna-Erica* heath on gentle receiving slopes or over impervious bedrock and drift; and species able to stand a measure of stagnation, such as *Scirpus cespitosus*, *Erica tetralix* and *Juncus squarrosus*, find only rather sparse representation towards one edaphic extreme. Similarly, though some of the profiles accumulate a distinct humic top, the community does not normally occur over even the thin ombrogenous peats that, with the increasingly heavy rainfall towards the far north-west, extend on to the moderately steep slopes, even over pervious bedrocks, that further to the south would have the *Calluna-Erica* heath. In both of these situations, the community is typically replaced by the *Scirpus-Erica* wet heath, though the two kinds of vegetation intergrade widely and the boundary between them can be markedly affected by treatment. In extreme cases, the *Calluna-Erica* heath can spread over the dried and fretted margins of quite deep blanket peat that has been drained or cut: such a development appears to be in train on the most severely degraded stretches of blanket mire on Lewis and Harris (Hulme & Blyth 1984) and is well advanced on Shetland (Hulme 1985, Roper-Lindsay & Say 1986). In such transitional habitats, the heath can be of the Typical form or, particularly where there is exposure and wind erosion of the peat mantle down to underlying mineral material, of the *Racomitrium* sub-community.

Another edaphic limit is seen where there is a geological switch to calcareous bedrocks which, within the range of the *Calluna-Erica* heath, occurs most obviously along the Moine Thrust, where the Cambrian Durness Limestone crops out, and over some stretches of Tertiary basalts on Mull and Skye (Birks 1973, Birse 1980). The community can extend some considerable way on to the quite base-rich brown earths that have developed from such substrates, where the pH often rises above 6, and it is characteristically represented there by the *Thymus-Carex* type. The intimate mixtures of calcifuges and calcicoles in this vegetation recall the 'limestone heath' described from more southerly parts of Britain, with the additional interest in this case of a modest floristic element reflecting the oceanic climate, *Blechnum spicant* and *Primula vulgaris* in particular making rather a striking pair. Essentially similar vegetation to this has also been described from serpentine soils on Shetland, particularly on Unst (Birse 1980), where severe exposure makes some stands look transitional to the *Racomitrium* sub-community and encourages the development of a maritime component. And it has been noted, too, at scattered localities over limestones and calcareous meta-sediments through the Highlands (McVean & Ratcliffe 1962) and marking out limy partings in the Silurian shales of the Moorfoot cleughs (Ferreira 1978).

In very exposed situations, it is possible that the *Calluna-Erica* heath represents a fairly natural kind of climax vegetation, from which trees and shrubs are excluded by severe climatic and edaphic conditions, often combined now with a shortage of nearby seed-parents (Birse 1980). More often, though, it is treatments that maintain the community as a plagioclimax, burning and grazing having long been part of the agricultural economy in many parts of its range. In fact, management of the vegetation for grouse-rearing is not so important here as it is in the *Calluna-Vaccinium* heath, which is much more widespread than the community through the major grouse-moor areas of the Grampians and south-east Scotland. But periodic burning to control and regenerate the heather cover in hill-pasture for stock is of major significance in some areas, as in Dumfries & Galloway, and together with the grazing activities of the animals and, in certain regions, of deer too, this has a controlling influence on the floristics and physiognomy of the vegetation.

It is difficult to disentangle the different effects of the two kinds of treatment and, of course, in good pastoral management of the hill-grazings, the point is to use them together to maintain the health of the swards. And, in so far as accounts of burning experiments are ever explicit about the types of heath being treated, which is rarely, these have been mostly conducted on the communities widespread in the east-central Highlands, where the *Calluna-Erica* heath is rather poorly represented. But what we can say, is that, apart from its general effect of superimposing on to the floristic variation in the community the pattern of regeneration of heather universal in burned heaths, fire is particularly important here in providing *E. cinerea* with a repeated opportunity to maintain its representation. Even under the quite dense shade of a maturing *Calluna* canopy, *E. cinerea* can persist as an understorey in this vegetation but, with clearance of the ground in a burn, it can regenerate well from seed, perhaps out-performing *Calluna* on the drier soils here and on warmer slopes, and, maintaining some activity through the winter in the oceanic climate, readily attains co-dominance in the middle years of recovery (Bannister 1964*b*, 1965, Gimingham 1972). Competitive interactions between these two in response to burning may account for much of the canopy diversity that is seen in Typical *Calluna-Erica* heath.

However, grazing of regenerating mixtures of *Calluna* and *E. cinerea*, even light grazing imposed for only a short time, can tip the balance back in favour of the heather which, in response to cropping, seems to adopt a plagiotropic habit, extending its cover laterally, while *E. cinerea* continues to grow erect (Gimingham 1949, 1972). And, of course, very heavy grazing after a burn can severely hinder regrowth of all the more palatable sub-shrubs, perhaps contributing in exposed sites to the kind of structure seen in some stands of the *Racomitrium* sub-community with its open canopy, its local spread of the unpalatable *Empetrum nigrum* ssp. *nigrum* and extensive carpets of *R. lanuginosum* or encouraging in Typical *Calluna-Erica* heath a greater abundance of species such as *Nardus stricta* or *Juncus squarrosus*.

With more judicious grazing applied for considerable periods of time, the major effect is to maintain among a fairly short sub-shrub canopy a continuing contribution from the kinds of grasses and bryophytes that, in ungrazed stands, are able to develop only in a patchy and temporary fashion during the early years of regeneration or among degenerating heather bushes, then to be largely extinguished. Local and somewhat attenuated floras of this kind can sometimes be seen in Typical *Calluna-Erica* heath but it is in the *Festuca-Anthoxanthum* sub-community and, over more base-rich soils, the *Thymus-Carex* type, that such a composition is most characteristic. The former kind of grassy heath is widespread through the range of the community, grazed usually now by sheep, sometimes by cattle and locally by horses, but is particularly important in the hill-pastures of south-west Scotland.

Zonation and succession

The community is typically found with grasslands, other kinds of heath and mires in zonations and mosaics where floristic variation is determined by differences in soils, local climate and treatments. Stable regimes of burning and grazing can hold any successional developments in check but, without some kind of treatment, many stands would probably progress to scrub and woodland.

In the simplest edaphic sequences, the *Calluna-Erica* heath occurs over the driest in series of base-poor soils disposed over the hill-slopes of the upland fringes. In those more oceanic parts of northern Britain where it is concentrated, it typically gives way, over gley-podzols and stagnopodzols which can accumulate a thick humic

top over the periodically waterlogged sub-soil, to the *Scirpus-Erica* wet heath. There *Calluna* retains its high frequency, but *E. cinerea* is usually replaced by *E. tetralix*, and *Molinia* and *Scirpus cespitosus* become very common and often abundant, with more or less total eclipse of the smaller grasses and a replacement of the pleurocarpous mosses by at least some Sphagna. Such transitions are usually seen over ground of decreasing slope, either moving down from hillsides with shedding drainage on to more gentle receiving ground or uphill on to plateaus with intergrades to ombrogenous peat, but they can be very gradual and, where discontinuous mantles of drift result in patchy variations of drainage impedence, intimate mosaics of the two vegetation types can be seen, even over gently-undulating or flat ground. In such patterns as these, Typical *Calluna-Erica* heath is often found grading into Typical *Scirpus-Erica* heath or, in more exposed situations or where the ground has been much degraded, the *Racomitrium* sub-community can be seen juxtaposed with the *Cladonia* sub-community of the *Scirpus-Erica* heath. Quite widely, too, the zonation runs on into the *Scirpus-Eriophorum* blanket mire over thickening ombrogenous peat. The proportion of these communities in the sequences varies with the rainfall because, with the shift into the wetter reaches of northwest Scotland, the wet heath and blanket mire spread on to ground of increasing slope, thus restricting the contribution of the dry heath. Drainage of the peat can reverse this trend and the spread of *E. cinerea* through the wet heath and on into the *Cladonia* sub-community of the blanket mire in the more oceanic parts of Britain, as in the Outer Hebrides (Hulme & Blyth 1984) and Shetland (Hulme 1985, Roper-Lindsay & Say 1986), is a striking indication of the shift of the vegetation towards the *Calluna-Erica* heath where the bogs have been much cut over. Analagous sequences to the above can also be seen where the community extends into eastern Scotland, though here the wet heath is usually represented by the *Ericetum tetralicis* and the blanket bog by the *Calluna-Eriophorum* mire.

More often, though, in these latter zonations, the *Calluna-Erica* heath is itself replaced by the *Calluna-Vaccinium* heath, producing a boreal equivalent to the kinds of patterns normally seen in the oceanic west. However, as explained earlier, the two heath types do show some geographical overlap, when they are sometimes separated altitudinally, the *Calluna-Vaccinium* heath replacing the *Calluna-Erica* type with the move to higher ground, or according to aspect, when the communities can be found in closer proximity, but with the *Calluna-Vaccinium* heath taking over with the shift on to slopes facing north and east. Other local climatic complications can be seen among the dry-heath components of these kinds of sequences where sub-shrub vegetation runs right to the edge of sea-cliffs receiving appreciable amounts of salt-spray. In extreme cases the *Calluna-Erica* heath can be totally replaced by some kind of *Calluna-Scilla* heath, but the two communities sometimes occur together, zoned according to the gradient of spray deposition and showing great continuity in their sub-shrub components, the effect of maritime influence being mainly felt among the associated herb flora, the particular character of which is also influenced by the nature of the soil. Such patterns are well seen on Skye (Birks 1973), along the Caithness cliffs and in Shetland (Birse 1980, Roper-Lindsay & Say 1986), where the *Festuca-Anthoxanthum* sub-community and, on more base-rich soils, the *Thymus-Carex* type, grade to the maritime heath. Analagous patterns occur towards the back of some coastal dune-systems, where the *Calluna-Erica* heath grades to *Calluna-Carex* heath towards the sea.

Zonations of these kinds, influenced primarily by differences in soils or climate or both, are all very often affected by treatments. In particular, burning, with its general tendency to favour vigorous regrowth of heather over the medium term, can blur boundaries between the *Calluna-Erica* heath and either the *Scirpus-Erica* wet heath or other types of drier sub-shrub vegetation, maritime or not, which occur adjacent to it, and it may take a considerable period of time before the floristic differences between the communities grow out to re-assert the zonation that has been cut across by fire. Grazing, too, can confuse patterns by, for example, favouring an abundance of herbaceous associates that are shared by the different kinds of heath: *Molinia* can increase generally across the junction of the *Calluna-Erica* heath and the *Scirpus-Erica* wet heath and, in transitions to maritime sub-shrub vegetation, finer-leaved grasses and small herbs may be encouraged throughout, producing a general grassy appearance that masks the increasing influence of salt-spray.

More generally important, however, is the effect which grazing, either alone or in conjunction with burning, has on mediating the transformation of this kind of heath to Nardo-Galion grasslands. After fire, very heavy grazing can precipitate a run-down of the heath to swards in which *Nardus stricta* or *Juncus squarrosus* play an important part or permit the spread of *Pteridium aquilinum*. But, less catastrophically than this, steady grazing pressure pushes the vegetation towards the *Festuca-Agrostis-Galium* grassland or, over more base-rich soils, the *Festuca-Agrostis-Thymus* grassland. The *Festuca-Anthoxanthum* and *Thymus-Carex* sub-communities are essentially intermediates along this developmental line and are very often found intermixed with the Nardo-Galion swards in mosaics which are determined by differences in grazing intensity. Many of these mixtures are stable and there is some pastoral advantage in having patches of the heath, which can provide a valuable winter bite, among the more generally nutritious grassland, but the long continuance of heavy pasturing, in areas like south-west Scotland, has certainly reduced the proportion of heaths in the mosaics (e.g. Hill & Evans 1978).

Release from grazing and burning, except where the heath occurs in very exposed situations, theoretically permits progression to scrub and woodland though, in many areas, natural seed-parents are now scarce. Over the characteristic soils here, the most likely developments are the *Quercus-Betula-Dicranum* woodland or, where the profiles are not quite so base-poor and oligotrophic, the *Quercus-Betula-Oxalis* woodland, with in both cases birch, mostly *Betula pubescens*, rather than oak being very much the predominant tree. Where the community extends into east-central Scotland, the *Pinus-Hylocomium* woodland is a possible climax of successions involving the *Calluna-Erica* heath, with the *Juniperus-Oxalis* scrub perhaps figuring as a precursor (Birse 1980). In various parts of the range, but particularly in the Dumfries and Galloway hills, ground which would be at least partly occupied by the community has been extensively afforested and conifers from the plantations can seed into the heath.

Distribution

The *Calluna-Erica* heath occurs widely through the more oceanic parts of Scotland with outlying stands in Wales and western England and around the east-central Highlands. It is particularly well seen in south-west Scotland, where the Typical and *Festuca-Anthoxanthum* sub-communities are common, and through some of the Western Isles and on Shetland where the Typical and *Racomitrium* sub-communities are distinctive. The *Thymus-Carex* type is much more local but good stands occur on Skye, Rhum and Uist, and at scattered localities through the Highlands and Southern Uplands.

Affinities

Although the distinctive character of some kinds of the *Calluna-Erica* heath has been recognised in accounts of Scottish sub-shrub communities (e.g. McVean & Ratcliffe 1962, Birks 1973, Ferreira 1978), this vegetation has been mostly subsumed within a broadly-defined *Callunetum* of the British sub-montane, and indeed, with frequent burning, it can be very difficult to separate the community from its boreal counterpart, the *Calluna-Vaccinium* heath. Gimingham (1972), following Muir & Fraser (1940), did distinguish two heath types roughly corresponding with the communities recognised in this scheme, grouping the *Calluna-Erica* heath among the generally oceanic sub-shrub types of Böcher (1943). But it was left to Birse (1980, see also Birse & Robertson 1976) to give a more precise and comprehensive definition of the community, which he saw as equivalent to the

Carici-binervis-Ericetum Braun-Blanquet & Tüxen 1950, 1952. As defined here, the *Calluna-Erica* heath takes in most of Birse's vegetation, but also includes what he saw as some distinctive heath from the Shetland serpentine (see also Roper-Lindsay & Say 1986). With the northern kinds of *Calluna-Scilla* and *Calluna-Carex* heaths, which replace it in more maritime situations, it forms a suite of sub-shrub communities of the cool oceanic parts of northern Britain, giving way south- wards to related vegetation types with *Ulex gallii*. Birse (1980) grouped his emended *Carici-Ericetum* with these warm oceanic communities in the Ulicion gallii. An alternative view, echoing Braun-Blanquet's (1967) suggestion of a major split among Atlantic heaths along the northern boundary of *U. gallii*, would be to place the *Calluna-Erica* heath closer to communities described from Norway (Nordhagen 1928, Böcher 1940) and the Faroes (Böcher 1940).

Floristic table H10

	a	b	c	d	10
Calluna vulgaris	V (1–10)	V (1–8)	V (1–8)	V (1–8)	V (1–10)
Erica cinerea	IV (1–10)	IV (1–6)	V (1–10)	IV (1–6)	IV (1–10)
Potentilla erecta	IV (1–4)	IV (1–4)	V (1–4)	V (1–4)	IV (1–4)
Carex binervis	III (1–4)	I (1–2)	II (1–4)	I (1)	II (1–4)
Molinia caerulea	III (1–5)	I (1–2)	I (1–4)	II (1–4)	II (1–5)
Juncus squarrosus	II (1–4)	I (1–3)	I (1)		I (1–4)
Diplophyllum albicans	II (1–4)	I (1)	I (1–2)	I (1)	I (1–4)
Campylopus paradoxus	II (1–3)	I (1)	I (1–4)		I (1–4)
Rhytidiadelphus loreus	II (1–4)	I (1–3)	I (1)		I (1–4)
Sphagnum capillifolium	I (1–4)				I (1–4)
Cladonia floerkeana	I (1–4)				I (1–4)
Sphagnum recurvum	I (1–8)				I (1–8)
Scapania gracilis	I (1–2)				I (1–2)
Racomitrium lanuginosum	II (1–6)	V (1–6)	I (1–4)	II (1–8)	III (1–6)
Cladonia uncialis	I (1–4)	IV (1–5)	I (1–3)	I (1–4)	II (1–5)
Scirpus cespitosus	II (1–4)	III (1–4)	I (1)	I (1)	II (1–4)
Empetrum nigrum nigrum	I (1–5)	III (1–4)	I (1–8)	I (1–3)	II (1–8)
Cladonia impexa	I (1–4)	III (1–6)	I (1–4)	II (1–4)	II (1–6)
Carex panicea	I (1–3)	III (1–4)	I (1)	II (1–4)	I (1–4)
Huperzia selago	I (1–2)	III (1–3)			I (1–3)
Cornicularia aculeata	I (1–5)	II (1–3)	I (1–2)		I (1–5)
Cladonia furcata	I (1–3)	II (1–3)	I (1)		I (1–3)
Cetraria islandica	I (1–3)	II (1–3)			I (1–3)
Alectoria nigricans	I (1–3)	I (1–3)			I (1–3)
Festuca ovina	I (1–4)	II (1–6)	V (1–6)	V (1–4)	III (1–6)
Anthoxanthum odoratum	I (1–4)	I (1–3)	V (1–6)	V (1–4)	III (1–6)
Galium saxatile	II (1–3)	I (1–3)	III (1–4)	II (1–3)	II (1–4)
Agrostis capillaris	II (1–4)	I (1–3)	III (1–4)	II (1–4)	II (1–4)
Pleurozium schreberi	II (1–4)	I (1–3)	III (1–4)	II (1–3)	II (1–4)
Dicranum scoparium	II (1–4)	I (1–3)	III (1–6)	II (1–3)	II (1–6)
Hylocomium splendens	I (1–6)	I (1–4)	III (1–6)	III (1–6)	II (1–6)

Floristic table H10 (*cont.*)

	a	b	c	d	10
Campanula rotundifolia	I (1–3)	I (1–3)	II (1–3)	I (1)	I (1–3)
Succisa pratensis	I (1–4)	I (1–3)	II (1–4)	II (1–4)	I (1–4)
Hypericum pulchrum	I (1–3)	I (1–3)	II (1–3)	II (1–3)	I (1–3)
Festuca rubra	I (1–4)	I (1)	II (1–6)	II (1–3)	I (1–6)
Lotus corniculatus	I (1–3)	I (1–3)	II (1–4)	II (1–4)	I (1–4)
Luzula multiflora	I (1–3)	I (1–3)	II (1)	I (1)	I (1–3)
Polygala serpyllifolia	I (1–2)	I (1–3)	II (1–3)	II (1–3)	I (1–3)
Luzula campestris	I (1–3)		II (1–3)	II (1–3)	I (1–3)
Plantago maritima			II (1–3)	II (1–2)	I (1–3)
Plantago lanceolata		I (1–3)	II (1–4)	II (1–3)	I (1–4)
Thymus praecox	I (1)	I (1–3)	II (1–4)	V (1–4)	I (1–4)
Danthonia decumbens	I (1–3)	I (1–3)	II (1–4)	V (1–5)	I (1–5)
Viola riviniana	I (1–3)	I (1–3)	II (1–3)	V (1–4)	I (1–4)
Prunella vulgaris			I (1–2)	IV (1–3)	I (1–3)
Carex pulicaris			I (1–3)	IV (1–4)	I (1–4)
Blechnum spicant	II (1–4)	I (1–3)	II (1–4)	III (1–4)	II (1–4)
Rhytidiadelphus triquetrus	I (1–3)	I (1–3)	I (1)	III (1–4)	I (1–4)
Linum catharticum			I (1–3)	III (1–3)	I (1–3)
Primula vulgaris			I (1–2)	III (1–4)	I (1–4)
Breutelia chrysocoma			I (1–4)	III (1–4)	I (1–4)
Solidago virgaurea	I (1–3)	I (1–3)	I (1–3)	II (1–4)	I (1–4)
Deschampsia flexuosa	III (1–4)	III (1–4)	II (1–6)	II (1–3)	III (1–6)
Hypnum cupressiforme	III (1–6)	III (1–4)	II (1–4)	II (1–3)	III (1–6)
Agrostis canina	II (1–4)	II (1–4)	II (1–4)	II (1–4)	II (1–4)
Nardus stricta	II (1–4)	II (1–6)	II (1–3)	I (1)	II (1–6)
Vaccinium myrtillus	II (1–6)	II (1–4)	II (1–4)	I (1)	II (1–6)
Erica tetralix	II (1–4)	II (1–4)	I (1–3)	I (1)	II (1–4)
Carex pilulifera	I (1–4)	III (1–4)	II (1–4)	I (1–3)	II (1–4)
Festuca vivipara	I (1–4)	II (1–6)	II (1–4)	II (1–3)	II (1–6)
Polytrichum commune	I (1–4)	I (1–3)	I (1–3)		I (1–4)
Rhytidiadelphus squarrosus	I (1–3)	I (1–3)	I (1–3)	I (1–3)	I (1–3)

	a	b	c	d	10
Antennaria dioica	I (1–3)				I (1–3)
Polytrichum piliferum	I (1–3)				I (1–3)
Vaccinium vitis-idaea	I (1–4)				I (1–4)
Narthecium ossifragum	I (1–3)		I (1)		I (1–3)
Pteridium aquilinum	I (1–4)		I (1–3)		I (1–4)
Cladonia coccifera	I (1–4)				I (1–4)
Mnium hornum			I (1–3)		I (1–3)
Leontodon autumnalis	I (1–3)		I (1–3)		I (1–3)
Number of samples	84	37	55	13	189
Number of species/sample	16 (5–49)	19 (9–35)	24 (13–58)	36 (12–50)	20 (5–58)
Shrub/herb height (cm)	20 (3–50)	13 (3–31)	15 (3–50)	15 (7–30)	17 (3–50)
Ground layer height (mm)	25 (10–50)	21 (10–50)	15 (10–30)	25 (20–30)	22 (10–50)
Shrub/herb cover (%)	87 (5–100)	60 (15–95)	91 (30–100)	99 (90–100)	89 (5–100)
Ground layer cover (%)	6 (0–70)	36 (25–45)	23 (0–90)	20 (0–70)	18 (0–90)
Altitude (m)	315 (60–780)	403 (120–688)	221 (7–520)	181 (16–390)	295 (7–780)
Slope (°)	15 (0–85)	7 (0–80)	18 (0–80)	21 (5–30)	15 (0–85)
Soil pH	4.8 (3.5–5.7)	5.4 (3.9–6.2)	5.1 (3.4–5.6)	6.4 (6.1–6.6)	5.1 (3.4–6.6)

a Typical sub-community

b *Racomitrium lanuginosum* sub-community

c *Festuca ovina-Anthoxanthum odoratum* sub-community

d *Thymus praecox-Carex pulicaris* sub-community

10 *Calluna vulgaris-Erica cinerea* heath (total)

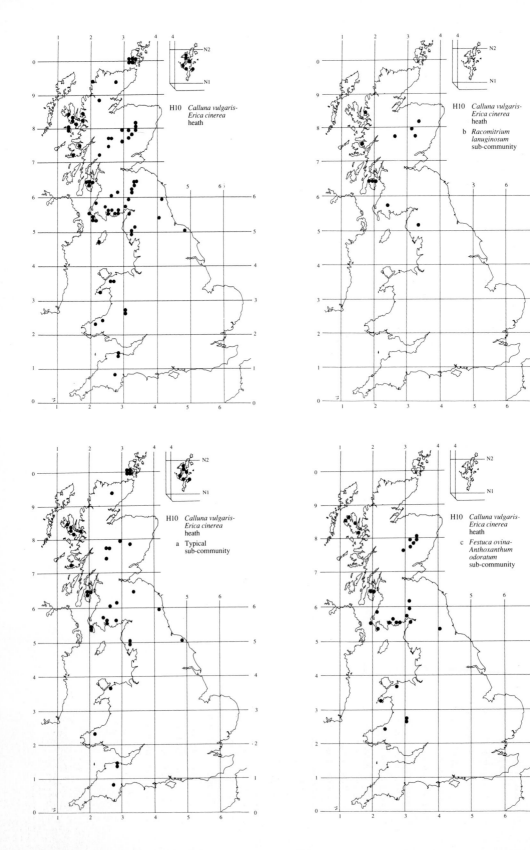

H10 *Calluna vulgaris-Erica cinerea* heath

H10 *Calluna vulgaris-Erica cinerea* heath
b *Racomitrium lanuginosum* sub-community

H10 *Calluna vulgaris-Erica cinerea* heath
a Typical sub-community

H10 *Calluna vulgaris-Erica cinerea* heath
c *Festuca ovina-Anthoxanthum odoratum* sub-community

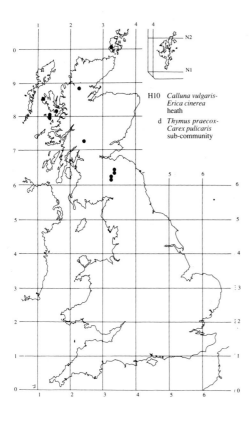

H10 *Calluna vulgaris-
      Erica cinerea*
      heath

d   *Thymus praecox-
      Carex pulicaris*
      sub-community

H11
Calluna vulgaris-Carex arenaria heath

Synonymy
Dune heath   *auct. angl.*, *Calluna-Erica cinerea* heath
Gimingham 1964*b*, 1972 *p.p.*; *Calluna-Empetrum
nigrum* heath   Gimingham 1964*b*, 1972 *p.p.*

Constant species
Calluna vulgaris, Carex arenaria.

Rare species
*Euphorbia portlandica, Trifolium suffocatum, Usnea
flammea.*

Physiognomy
Calluna vulgaris is the only constant sub-shrub throughout the *Calluna-Carex arenaria* heath, and it is often
present in abundance, sometimes as an overwhelming
dominant in rather impoverished vegetation. Generally,
though, the cover of heather is discontinuous, decidedly
patchy in younger or grazed stands, and there is frequently some other sub-shrub present, either intimately
mixed with the *Calluna* or with the bushes of the co-
dominants forming a more distinct mosaic. The most
characteristic of these associates is *Erica cinerea*, with
Empetrum nigrum ssp. *nigrum* figuring in some places,
and each of these can be abundant, locally to the
exclusion of *Calluna* itself. Very occasionally, *Ulex gallii*
can be found and in some stands *Rosa pimpinellifolia* is
plentiful, with *Erica tetralix* and *Salix repens* occurring
in transitions to wetter heath. *Vaccinium myrtillus* is
typically very rare.

Even when some of these other sub-shrubs are present, the community can be distinguished from similar
assemblages by the constancy of *Carex arenaria*, though
this is hardly ever more than moderately abundant and
often distinctly senile, except where the sand substrate
becomes locally mobile, when renewed vigour and cover
in the plant usually presage a transition to patches of
Carex dune vegetation. *Ammophila arenaria* is also
frequent through the community, though it is characteristically sparse and usually decidedly moribund, with
just scattered shoots.

In more species-poor stands, little else than these
plants may be represented, but other tracts of this kind
of heath show considerable enrichment. First, among
grasses, there is often some *Festuca rubra* or *F. ovina*
(inadequately distinguished in the available data), with
Agrostis capillaris and *Anthoxanthum odoratum* also
occurring frequently in some situations and, less commonly, *Poa pratensis* (probably *P. subcaerulea*). The
annuals *Aira praecox* and, more sparsely, *A. caryophyllea* can sometimes be found too. Apart from *C. arenaria*,
sedges are absent though *Luzula campestris* occurs quite
often.

Variation among the dicotyledonous associates is
fairly modest but *Galium verum*, *Lotus corniculatus*,
Viola riviniana and *Thymus praecox* all occur quite
frequently in all but the denser stands of bushes, and
there can also be some *Campanula rotundifolia*, *Galium
saxatile*, *Hypochoeris radicata*, *Jasione montana* and
Sedum anglicum. Often, mixtures of these species,
together with the grasses, form the basis of a sward
running among the sub-shrubs, when there can be
strong floristic continuity with adjacent dune grasslands.

Certain bryophytes, especially hypnoid mosses such
as *Hypnum cupressiforme s.l.*, *Pleurozium schreberi*,
Hylocomium splendens and *Rhytidiadelphus triquetrus*
can also be present among the turf and more open
bushes. In other situations, acrocarps like *Polytrichum
juniperinum*, *P. piliferum* and *Ceratodon purpureus* can
be patchily abundant on areas of bare ground or there
may be extensive carpets of lichens.

Sub-communities

***Erica cinerea* sub-community:** *Calluna-Erica cinerea*
heath, 'dune-heath' variant   Gimingham 1964*b*. In this
sub-community, *Calluna* and *E. cinerea* are generally co-
dominant though one or the other may show local pre-
eminence, with *E. cinerea* sometimes colonising new
sites well in advance of *Calluna*. *Empetrum nigrum* ssp.
nigrum is typically absent but *Ulex gallii* occurs very

occasionally and, at some sites, *Rosa pimpinellifolia* is a distinctive invader.

F. rubra/ovina is very common at low to moderate covers with scattered shoots of *C. arenaria* and *L. campestris* and very sparse *Ammophila*. *Agrostis capillaris* is scarce and *Anthoxanthum* occurs only occasionally but, in more open areas, there is quite often some *Aira praecox* or *A. caryophyllea*. Then, there is occasional *Lotus corniculatus*, *G. verum*, *V. riviniana* and *Thymus* with *H. radicata*, *J. montana* and *Ononis repens* preferential at low frequencies. Patches of *Sedum anglicum* are prominent at some sites, often marking out more pebbly ground, and this kind of *Calluna-Carex* heath also provides a locus for the rare *Euphorbia portlandica* and *Trifolium suffocatum*.

Quite often, however, the really distinctive element among the associated flora here is the cryptogams which, in more attenuated or patchy swards, can occupy the bulk of the ground between the bushes. Apart from occasional plants of *H. cupressiforme* and *P. schreberi*, however, hypnoid mosses are rather sparse and it is usually the acrocarps that are more abundant, with *Dicranum scoparium* an additional preferential. Even more varied and plentiful in many stands, though, are the lichens, particularly *Cladonia* spp., with *C. impexa*, *C. arbuscula*, *C. furcata* all common, *C. floerkeana*, *C. pyxidata*, *C. gracilis*, *C. foliacea* and *C. uncialis* occasional and *C. rangiferina*, *C. rangiformis*, *C. subcervicornis*, *C. bacillaris*, *C. crispata* and *C. coccifera* also recorded infrequently. *Cornicularia aculeata* and *Hypogymnia physodes* are very often found, too, with *C. muricata* and *Peltigera canina* in some stands. A further very distinctive feature at certain localities is the occurrence of ground-growing *Usnea* spp., particularly *U. subfloridana* and *U. articulata*, and sometimes the more local *U. flammea*.

Empetrum nigrum ssp. *nigrum* sub-community: *Calluna-Empetrum nigrum* heath   Gimingham 1964b, 1972 *p.p.* *E. cinerea* can sometimes be found here among the *Calluna*, but *Empetrum nigrum* ssp. *nigrum* is the usual co-dominant and, indeed, it can exceed the heather in cover, its shoots abundant in mixed canopies or the plants forming large circular patches which locally monopolise the ground. *Salix repens* also occurs occasionally in this vegetation with *Erica tetralix* in transitions to wetter heath.

F. rubra/ovina remains common, though usually at low covers, among frequent *C. arenaria* and *L. campestris* and locally tussocky *Ammophila*. In contrast to the *Erica* sub-community, however, *Agrostis capillaris* is very common with occasional *Anthoxanthum* and there is sometimes a little *Poa pratensis*. *L. corniculatus*, *G. verum*, *V. riviniana* and *Thymus* remain quite common, but preferential to this sub-community are *G. saxatile* and *C. rotundifolia*.

Among the cryptogams, acrocarpous mosses remain occasional but more prominent are hypnoid species with *H. cupressiforme*, *Pleurozium* and *R. triquetrus* joined by *Hylocomium splendens* and, less commonly, *Ptilidium ciliare*, each of these sometimes moderately abundant. *Cladonia arbuscula* and *C. impexa* can be patchily plentiful, too, with occasional *Cornicularia aculeata* and *H. physodes*, but there is nothing like the variety and richness of the lichen flora characteristic of the *Erica* sub-community.

Species-poor sub-community. *Calluna* is typically an overwhelming dominant here and often the only sub-shrub, with just occasional bushes of *E. cinerea* or *Empetrum*. *C. arenaria* remains constant but it is usually sparse and *Ammophila* is decidedly infrequent and moribund. *F. rubra/ovina* is generally absent and grasses are sometimes represented just by very puny tufts of *Anthoxanthum* or *Deschampsia flexuosa*. Dicotyledonous associates are likewise often very few indeed and of low cover among the dense heather bushes but *Campanula rotundifolia* and *Galium saxatile* can sometimes be found in more open places. There, too, or over the stools of older, collapsing *Calluna* bushes there can be some *H. cupressiforme*, *Pleurozium*, *D. scoparium*, *R. triquetrus* and very occasional *Cladonia arbuscula*.

Habitat

The *Calluna-Carex* heath is the characteristic sub-shrub vegetation of stabilised base-poor sands on dunes and plains around the coasts of Britain. Unusually among our lowland heaths, the community develops in primary successions by the colonisation of dune grasslands on acidic sands that have become fixed or where, with the passage of time, more lime-rich wind-blown sediments have become leached. However, it is probably dependent now for its establishment on particular episodes of relief from grazing and, though variations in regional climate and substrate affect the character of the vegetation, predation by herbivores continues to influence its composition and structure, and ultimately to control its maintenance against reversion to grassland or progression to scrub and woodland.

The *Calluna-Carex* heath is largely confined to sands with a pH of less than 5, able to develop only on those stretches of coastal dunes and plains which have formed by the accretion of initially quartzitic material or where more calcareous deposits have been eluviated above. Around the seaboard of Britain, more acidic dune sands are actually very local, with sites like South Haven in Dorset and Tentsmuir in Fife providing rather exceptional situations for the development of the community. Elsewhere, the amount of calcium carbonate in the sediment supply, usually derived from comminuted shelly material, can be very considerable and it is only where this has been leached from the upper part of the profile that the surface becomes congenial for the establishment of the sub-shrubs and the more calcifuge herbs and cryptogams. Evidence from a variety of dune systems suggests that, where the initial carbonate content is not more than 5%, most of this will have been lost from the top decimetre of soil within 300 to 400 years (Ranwell 1972, Willis 1985a). Clearly, climate will have some effect on this process, leaching tending to be speedier towards the wetter and cooler north and west of Britain: although dune sands in this part of the country are often initially lime-rich, the *Calluna-Carex* heath is markedly more common there than in the drier south and east. In fact, though, it is possible that only quite superficial or local reduction in the base-richness of the ground is necessary for the sub-shrubs to get an initial hold and, once *Calluna* is established, this may itself speed the shift in conditions by acidifying the substrate.

Even with sands which are initially fairly base-poor or which become quite quickly leached, there is the additional requirement of general surface stability of the sediment before the community can become well established, a condition that is met only in older dunes and on consolidated sand plains. Locally, it appears that *Empetrum nigrum* ssp. *nigrum* can capitalise upon renewed surface disturbance on largely fixed sands in regions of cooler climate pioneering the development of the *Calluna-Carex* heath, a feature familiar from parts of Denmark and Sweden (Landsberg 1955, Gimingham 1964b, 1972) but, in general, this is a vegetation type of the later stages in dune successions, following on from various kinds of grassy sward on stable ground.

Very importantly in such situations, *Ammophila arenaria*, the dominant plant of more mobile base-rich sands all around the British coast, has begun to lose its vigour, proliferative shoot production coming to an end and flowering becoming less free with but sparse regeneration from seed (Gimingham 1964a, Huiskes 1979). The reasons for this decline are unclear, but they are probably connected to the kinds of edaphic changes that follow upon the increased fixity of the sand surface, not simply the eluviation of bases but also the accumulation of organic matter above, the increased capacity for moisture retention and a modest enhancement of the trophic state of the profile (Tansley 1939, Salisbury 1952, Ranwell 1972, Willis 1985a). Such changes favour an expansion of other plants as the competitive edge of the *Ammophila* wanes, its clones becoming reduced to scattered groups of debilitated shoots that are inherited by the vegetation which succeeds the marram-dominated communities. More locally, and especially on acid sands where *Ammophila* performs less well even in the earlier stages of succession, *C. arenaria* assumes this pioneering role, then declining in a similar way as accretion becomes insignificant.

Since water and nutrients often remain strongly limiting to plant growth even on the older and more stable dune surfaces, rather attenuated grassy swards appear to be a fairly natural successor to *Ammophila* or *Carex* vegetation. At this stage, however, the incidence and intensity of grazing probably become of crucial significance to the direction of the sere. The kinds of grassland which develop on more base-poor and impoverished fixed sands are characterised by senile *C. arenaria* and *Ammophila* with *F. rubra*, *F. ovina*, *A. capillaris*, *Anthoxanthum*, *L. campestris*, *G. verum*, *Lotus corniculatus*, *H. radicata*, *D. scoparium*, *Hypnum cupressiforme*, *P. schreberi*, *Hylocomium splendens* and a variety of lichens, particularly *Cornicularia aculeata* and *Cladonia* spp. Clearly, such a flora is virtually identical to the herbaceous and cryptogam elements of the *Calluna-Carex* heath, and what determines whether sub-shrubs establish along with or subsequent to such assemblages seems to be the extent of predation by herbivores. In this kind of habitat, these include rabbits and stock, mostly sheep, though with cattle important in some areas.

In fact, seedlings of the sub-shrubs are extremely rare in what one supposes to be the grassy precursors of the community, or even in transitions to or mosaics with existing stands of the *Calluna-Carex* heath, the occurrence of scattered well-established bushes usually marking the shift on the ground from one vegetation type to the other. It looks, therefore, as if succession to the heath is rarely a steady process even where grazing is slight, but more related to particular perturbations in its intensity (Gimingham 1972). The demise of rabbits with myxomatosis may therefore have been important in the establishment or spread of heath at some sites, though at St Cyrus in Aberdeenshire the epidemic was followed by the appearance of a solitary plant of heather and any subsequent spread was hindered by the recovery of the rabbit population (Gimingham 1972). Perhaps, now, with the widespread reclamation of the dune hinterland, the rarity of seed-parents has become critical to the development of heath. Interestingly, where we have evidence of what happens where stretches of acid dune sands pass in and out of cultivation, as on Tentsmuir, abandonment of arable areas can be marked by a secondary resurgence of the *Calluna-Carex* heath (Leach 1985). As with so many stretches of lowland heath, then, the survival of the community may be strongly dependent upon a distinctive type of sporadic or non-intensive agricultural tradition.

The particular character of the *Calluna-Carex* heath which develops under these conditions depends partly on the nature of the climate experienced by the dune system. Around much of the coast of Britain, and particularly in the west where the temperature regime is more equable, the broadly oceanic *Erica cinerea* can find this kind of habitat very congenial. Indeed, it quite often colonises in advance of *Calluna*, forming substantial bushes before the heather is able to establish (Gimingham 1964*b*). The very free-draining character of the substrate helps maintain the somewhat open, patchy cover of perennial herbs characteristic of the *Erica* subcommunity, with annuals, acrocarps and lichens capitalising on areas of bare ground. This aspect of the vegetation is especially prominent in more southerly stands of the sub-community where rainfall is low. In some sites, as on The Ayres in the Isle of Man, mixtures of sand and consolidated pebbles on the dunes provide an especially distinctive setting for this kind of *Calluna-Carex* heath, supporting a particularly rich lichen flora.

With the shift on to acidic sands in the cooler and wetter north of Britain, the *Erica* sub-community is locally replaced by the *Empetrum* sub-community. *E. nigrum* ssp. *nigrum* is well able to colonise such substrates, particularly where they are kept moist by the damp climate, getting a hold on patches of moss or even, as noted above, on stretches of quite mobile sand. Its long shoots grow prostrate and form roughly circular patches which can help consolidate the sand surface, aiding the subsequent establishment of *Calluna* and, more locally, *Salix repens* (Gimingham 1964*b*, 1972). The frequency of *A. capillaris* and *G. saxatile* among the associates in this kind of *Calluna-Carex* heath perhaps reflects a more base-poor soil surface than in many stands of the *Erica* sub-community, and the prevalence of hypnoid mosses and fruticose lichens, as against acrocarps and encrusting lichens, the less parched and open character of the ground.

In the absence of grazing, the *Calluna-Carex* heath can acquire the dense and impoverished character of the Species-poor sub-community, although renewed predation by stock can reverse such a development in perhaps quite a short time: at Tentsmuir, for example, tall *Calluna* has been virtually eliminated from some areas put under pasturing in the 1950s (Leach 1985). In fact, except where periodic droughts take their toll, the maintenance of shorter and more diverse covers of the *Calluna-Carex* heath is probably dependent on a certain amount of grazing.

Zonation and succession
The *Calluna-Carex* heath is a local element of zonations and mosaics on older stretches of coastal dunes and sand plain, grading to grasslands or various kinds of *C. arenaria* or *Ammophila* vegetation, or to scrub, woodland and bracken, in patterns which reflect the progression of the dune sere, differences in the character of the soils and the influence of past and present treatments. In most cases, the community would probably progress readily to scrub and woodland with relief from grazing, provided seed-parents of suitable shrubs and trees were available.

The most striking complex of vegetation types among which the *Calluna-Carex* heath can be found involves the more calcifuge of the swards characteristic of stable dune sands. Quite often, it occurs within stretches of the *Carex-Festuca-Agrostis* grassland, typically a short turf dominated by mixtures of *C. arenaria*, *F. ovina*, *F. rubra*, *A. capillaris* and *Poa pratensis* with *Anthoxanthum* and *L. campestris* occurring on the more impoverished substrates, and scattered plants of *G. saxatile*, *G. verum* and *L. corniculatus*. Such vegetation can pass imperceptibly into the grassy ground of more open tracts of the *Calluna-Carex* heath, the major difference between the communities being the presence of sub-shrubs. Generally speaking, the distribution of these presents a sharp boundary: occasional bushes are sometimes to be seen isolated among more extensive stretches of grassland, but there is rarely that gentle gradation of size and age among the bushes that denotes a continuous process of invasion. More usually, it looks as if the patterns reflect past colonisation events, with the grassland/heath transitions often subsequently sharpened up by renewed grazing.

Bryophytes can be fairly numerous and abundant in the *Carex-Festuca-Agrostis* grassland, with patches of species like *D. scoparium*, *P. juniperinum*, *C. purpureus*, *Pleurozium schreberi* and *H. splendens* providing further continuity with the heath and, though lichens are generally less varied and abundant in the grassy swards, *Cornicularia aculeata*, *Cladonia arbuscula* and *C. rangiformis* can occur in both vegetation types. This latter element, though, is much more generally obvious in those few localities where the *Calluna-Carex* heath occurs with the *Carex-Cornicularia* community. In that assemblage, *C. arenaria* is the only constant vascular plant in a characteristically very open turf, with just occasional scattered tussocks of *F. ovina* and *F. rubra*, sparse *L. campestris* and only sporadic representation of most of the dicotyledonous associates listed above. However, *H. radicata* is common, together with *Rumex acetosella*, and there are quite often some annuals including *Aira praecox*. Most striking, though, and providing continuity with the lichen assemblage typical of the *Erica* sub-community of the *Calluna-Carex* heath, is *Cornicularia aculeata* and a great variety and abundance of *Cladonia* spp. Again, in such patterns, it is the occurrence of the sub-shrubs which provides the major distinction between the vegetation types.

There is little doubt that the incidence and intensity of grazing, past and present, mediates much of the variation in patterns such as these developed over tracts of fairly uniform base-poor sands. In other situations, where the *Calluna-Carex* heath has managed to become established somewhat earlier in the dune succession on sands that are a little less stable or leached, treatments can interact with edaphic differences to produce other zonations. Then, the community usually passes to some form of *Festuca-Galium* grassland, and where this is represented by the *Luzula* sub-community there can be considerable floristic continuity among the herbaceous element of the vegetation types, with *F. rubra*, *A. capillaris*, *Anthoxanthum*, *L. campestris* and *C. arenaria* contributing throughout to the turf. Quite apart from the disappearance of the sub-shrubs, however, the grassland is usually more varied in its vascular component with frequent records for such plants as *G. verum*, *L. corniculatus*, *Plantago lanceolata*, *Trifolium repens*, *Hieracium pilosella*, *Achillea millefolium*, *Veronica chamaedrys* and *Senecio jacobaea*. Bryophytes and lichens are usually not so plentiful either, with *Brachythecium albicans*, *Homalothecium lutescens*, *Rhytidiadelphus squarrosus* and *Pseudoscleropodium purum* providing the bulk of the cryptogam cover.

More rarely, the *Calluna-Carex* heath can be found in sharper transitions to Typical *Festuca-Galium* grassland in which *Ammophila* may remain locally vigorous with some small measure of continuing accretion, or even to the *Ammophila-Festuca* community which is characteristic of still quite mobile and lime-rich dune sands around our coasts. Generally, though, the passage to this latter vegetation or to the *Carex arenaria* community is indicative of secondary local disturbance of the sand surface in tracts of dune or plain which are largely stable, something which can result from the burrowing or scuffing activities of rabbits or recreational activities of human visitors.

Transitions to moister ground between dune ridges and in the surrounds to slacks can complicate these patterns. Sometimes, in such situations, there is a fairly sharp switch from heath to some fairly damp dune sward such as the *Ranunculus-Bellis* or *Prunella* sub-communities of the *Festuca-Galium* grassland or the *Ononis* sub-community of *Salix-Holcus* slack. In other places, the *Calluna-Carex* heath passes into wet heath or heathy grassland on damp base-poor ground with quite a humic topsoil. There, *C. arenaria*, *F. ovina*, *Anthoxanthum* and *G. saxatile* can maintain some representation, along with certain of the hypnoid mosses, *Calluna* persisting patchily on drier areas, but dominance usually passes to mixtures of *Erica tetralix*, *Nardus stricta*, *Juncus squarrosus* and *Salix repens*. Transitions to such vegetation can be hard to diagnose, but the better-developed assemblages usually seem to be some form of *Nardus-Galium* grassland or run-down Ericion tetralicis wet heath.

Release from grazing can allow stands of the *Erica* and *Empetrum* sub-communities to develop the leggy and impoverished look of the Species-poor sub-community, with the associated sub-shrubs and many of the herbs and cryptogams overwhelmed by the *Calluna*. Where soil impoverishment under long-entrenched heather has not become too severe, and where parching of the surface is relieved somewhat, such vegetation may

progress to woodland provided suitable seed-parents are not beyond the range of dispersal. Where more open stretches of ground remain among the bushes, birch is a very likely early invader, *Betula pendula* predominating on these drier soils, with *B. pubescens* becoming important in transitions to moister ground, where establishment is that much more ready. The appearance of birch, together with *Pinus sylvestris* and other conifers seeding in from the plantations that are an extensive feature of some stretches of fixed dunes, can presage a fairly speedy succession to thickets of the *Quercus-Betula-Deschampsia* woodland, within which the heath flora survives only in more open glades.

In other situations, and particularly where there is some more obvious relief from soil impoverishment after bouts of manuring by stock or infestation with rabbits, or where there is a legacy of disturbance around old enclosures, scrub may supervene in the succession. Often, patches of *Ulex-Rubus* scrub mark out such enriched areas: *Calluna* may persist as occasional bushes in such vegetation, with species such as *F. rubra*, *A. capillaris*, *Anthoxanthum* and *Galium saxatile* forming much of the grassy ground, but the abundance of *U. europaeus* and *R. fruticosus* agg. is very striking and there are often patches of more mesophytic grasses like *Dactylis glomerata* and *Holcus lanatus*. *Pteridium aquilinum* can also occur among such scrub, thickening up where deeper sands are kept a little moister along the foot of dunes or on shady aspects into stands of the *Pteridium-Galium* community.

Distribution

The *Calluna-Carex* heath is a very local community on dunes along the coasts of western England and Wales, becoming commoner in Scotland. Much the most widespread type is the *Erica* sub-community with the *Empetrum* sub-community replacing it locally in north and east Scotland. The Species-poor sub-community can occur throughout the range where conditions are suitable.

Affinities

Early descriptive accounts of British heaths rarely made more than passing reference to the occurrence of sub-shrub vegetation on coastal dune systems, and later studies have usually described the different forms as sub-types of communities with wider, inland distributions. Indeed, the overall similarity of the vegetation included here to these inland heaths is very obvious, with often only *C. arenaria* and *Ammophila* providing any positive characterisation: where the former species occurs in stands of the *Calluna-Festuca* heath in eastern England the separation of the communities can be especially problematical. Generally, however, the *Erica* sub-community has been related to what in this scheme is called the *Calluna-Erica* heath, a vegetation type with a broadly oceanic character and well represented in other sub-maritime habitats around the western seaboard of Britain, as on drying blanket mire close to sea-level and on sea-cliffs outside the influence of salt-spray. Similarly, the *Empetrum* sub-community has been seen as close to what is here termed *Calluna-Vaccinium* heath (Gimingham 1964*b*, 1972). On balance, though, the diagnosis of a distinct community seems a more satisfac-

tory solution, especially in view of the development of this vegetation as part of primary successions on dunes, albeit under the strong influence of biotic, often anthropogenic, factors. As for the phytosociological affinities of the community, the best locus would seem to be in a broadly-defined Ericion alliance in the sense of Böcher (1943).

Floristic table H11

	a	b	c	11
Calluna vulgaris	V (1–9)	V (2–9)	V (7–10)	V (1–10)
Carex arenaria	V (1–5)	V (2–4)	V (2–4)	V (1–5)
Festuca ovina	IV (1–8)	IV (2–5)		III (1–8)
Luzula campestris	III (1–4)	III (2–3)		III (1–4)
Polytrichum piliferum	III (1–6)	III (2–5)		III (1–6)
Cornicularia aculeata	III (1–6)	II (1–5)		II (1–6)
Dicranum scoparium	V (1–8)	II (2)	II (1–4)	III (1–8)
Erica cinerea	V (1–8)	II (2)	I (3–6)	III (1–8)
Cladonia impexa	IV (1–7)	II (2–3)		III (1–7)
Cladonia furcata	II (1–5)	I (2)		II (1–5)
Aira praecox	III (1–5)	I (2–3)		II (1–5)
Cladonia floerkeana	II (1–5)	I (2)		I (1–5)
Cladonia pyxidata	II (1–2)	I (2)		I (1–2)
Cladonia gracilis	II (1–7)			I (1–7)
Cladonia foliacea	II (1–4)			I (1–4)
Aira caryophyllea	II (1–3)			I (1–3)
Rosa pimpinellifolia	II (2–5)			I (2–5)
Hypochoeris radicata	II (1–3)			I (1–3)
Cladonia uncialis	II (1–5)			I (1–5)
Jasione montana	I (1–3)			I (1–3)
Campylopus introflexus	I (1–4)			I (1–4)
Cladonia rangiferina	I (1–5)			I (1–5)
Sedum anglicum	I (1–4)			I (1–4)
Cladonia rangiformis	I (2–4)			I (2–4)
Ulex gallii	I (1–7)			I (1–7)
Ononis repens	I (2–4)			I (2–4)
Usnea articulata	I (1–6)			I (1–6)
Cladonia subcervicornis	I (2–4)			I (2–4)
Cladonia bacillaris	I (1–2)			I (1–2)
Cladonia crispata	I (1–2)			I (1–2)
Cornicularia muricata	I (1–4)			I (1–4)
Empetrum nigrum nigrum		V (2–9)	II (1–7)	II (1–9)
Cladonia arbuscula	III (1–6)	IV (2–9)	I (2)	III (1–9)
Agrostis capillaris	I (2–6)	IV (2–4)		II (2–6)
Galium saxatile	I (2)	III (2–4)	I (2)	II (2–4)
Campanula rotundifolia	I (2)	III (2–4)	I (2–4)	II (2–4)
Hylocomium splendens		III (2–5)	I (2)	II (2–5)
Pohlia nutans	I (3)	II (2)	I (2)	I (2–3)

Floristic table H11 *(cont.)*

	a	b	c	11
Poa pratensis	I (3)	II (2)		I (2–3)
Ptilidium ciliare		II (2)	I (2)	I (2)
Salix repens		II (2–5)		I (2–5)
Vaccinium myrtillus		I (2)		I (2)
Potentilla erecta		I (2)		I (2)
Erica tetralix		I (1–8)		I (1–8)
Hypogymnia physodes	IV (1–8)	II (2)	IV (1–6)	III (1–8)
Hypnum cupressiforme	II (3–6)	III (2–4)	IV (2–6)	III (2–6)
Ammophila arenaria	III (1–4)	III (2–5)	II (1–4)	III (1–5)
Pleurozium schreberi	I (1–6)	III (2–5)	III (1–5)	II (1–6)
Polytrichum juniperinum	III (1–6)	II (2)	II (2)	II (1–6)
Anthoxanthum odoratum	II (1–4)	II (2–4)	I (2)	II (1–4)
Ceratodon purpureus	II (1–3)	II (2–4)	I (2)	II (1–4)
Lotus corniculatus	II (1–2)	II (2)		II (1–2)
Galium verum	II (1–3)	II (2–4)		II (1–4)
Rhytidiadelphus triquetrus		II (2–4)	II (2)	I (2–4)
Peltigera canina	I (1–4)	II (2)	I (1)	I (1–4)
Viola riviniana	I (2–3)	II (2)		I (2–3)
Thymus praecox	I (2–6)	II (2–4)		I (2–6)
Usnea subfloridana	I (2)	I (2)	I (2)	I (2)
Pseudoscleropodium purum	I (4)	I (2)	I (3)	I (2–4)
Deschampsia flexuosa	I (3)	I (2)	I (1–2)	I (1–3)
Lophocolea bidentata s.l.	I (1–2)	I (2)	I (2)	I (1–2)
Cladonia coccifera	I (1–3)	I (2)		I (1–3)
Cerastium fontanum	I (1–2)	I (2)		I (1–2)
Trifolium repens	I (1–3)	I (2)		I (1–3)
Hieracium pilosella	I (1–2)	I (2)		I (1–2)
Rumex acetosella	I (1–3)	I (2)		I (1–3)
Teesdalia nudicaulis	I (3–4)	I (2)		I (2–4)
Racomitrium canescens	I (1–3)	I (2)		I (1–3)
Senecio jacobaea	I (3)	I (2)		I (2–3)
Number of samples	32	25	10	67
Number of species/sample	17 (8–28)	17 (13–26)	10 (3–14)	16 (3–28)
Vegetation height (cm)	14 (3–55)	10 (4–25)	60 (30–100)	18 (3–100)
Shrub/herb cover (%)	56 (10–98)	no data	90 (50–100)	62 (10–100)
Ground layer cover (%)	55 (5–90)	no data	33 (5–60)	50 (5–90)
Altitude (m)	4 (2–8)	no data	2 (0–5)	3 (0–8)
Slope (°)	8 (0–50)	no data	4 (0–25)	6 (0–50)

a   *Erica cinerea* sub-community
b   *Empetrum nigrum nigrum* sub-community
c   Species-poor sub-community
11   *Calluna vulgaris-Carex arenaria* heath (total)

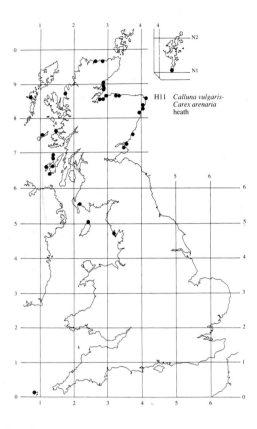

H11   *Calluna vulgaris-*
       Carex arenaria
       heath

H12
Calluna vulgaris-Vaccinium myrtillus heath

Synonymy
Heather Moor   Smith & Rankin 1903, Smith & Moss 1903 *p.p.*; *Callunetum*   Lewis & Moss 1911; *Calluno-Vaccinietum*   Lewis & Moss 1911; *Calluna* heath Ratcliffe 1959 *p.p.*; *Callunetum vulgaris*   McVean & Ratcliffe 1962 *p.p.*; *Calluna*-heath moss sociation   Edgell 1969; *Calluna-Vaccinium* heath   Gimingham 1972 *p.p.*; *Callunetum vulgaris typicum*   Evans *et al.* 1977; *Calluna vulgaris-Vaccinium vitis-idaea* nodum   Huntley & Birks 1979; *Calluna vulgaris-Anemone nemorosa* nodum   Huntley & Birks 1979; *Calluna vulgaris-Deschampsia flexuosa* nodum   Huntley 1979; *Vaccinio-Ericetum cinereae*   Birse 1980 *p.p.*; *Calluna vulgaris-Vaccinium myrtillus* nodum   Hughes & Huntley 1986; *Calluna vulgaris-Hypnum cupressiforme* nodum Hughes & Huntley 1986.

Constant species
Calluna vulgaris, Deschampsia flexuosa, Vaccinium myrtillus, Dicranum scoparium, Hypnum jutlandicum, Pleurozium schreberi.

Rare species
Diphasium × issleri.

Physiognomy
The *Calluna vulgaris-Vaccinium myrtillus* heath is generally dominated by *Calluna vulgaris*, often overwhelmingly so: indeed, this community includes the bulk of the Calluneta so widely described from the less oceanic parts of the sub-montane zone, through which the regular burning of grouse-moor and hill-grazings that encourages a predominance of species-poor building-phase heather is still commonly practised. But older stands, with a more open cover of degenerate *Calluna*, can often be found and there is structural variety, too, in response to differences in grazing intensity and local climate, both of which can affect the height and extent of the sub-shrub canopy. More distinctly, wherever there is some opportunity for a contribution to the cover from other ericoids, capitalising upon the more open ground in the early or late stages of the heather growth cycle, or persisting in usually smaller amounts among the maturing bushes, the potential diversity of this element of the vegetation is quite high.

As in the *Calluna-Erica* heath, *Erica cinerea* can figure among these associates, recovering particularly well from burning on drier slopes and able to persist patchily beneath quite dense heather (Gimingham 1949, 1972, Bannister 1965). But it is much more uneven in its occurrence in this community, rarely of high cover and, in the colder and wetter climatic conditions characteristic of much of the range of this kind of heath, it fares rather badly. *Vaccinium myrtillus*, by contrast, which is only occasional within the *Calluna-Erica* heath, is a constant here and, though generally subordinate to the heather, it can show temporary abundance after fires and being also a shade-tolerant plant, it is able to survive in a sparse lower tier beneath the closing *Calluna* canopy. Out of reach of any grazing animals, and especially in the humic soils that accumulate among boulders or over exposed crags, it can grow with particular vigour and the community includes stands in such situations where *V. myrtillus* is co-dominant with heather in transitions to *Vaccinium-Deschampsia* heath (Smith & Moss 1903, Smith & Rankin 1903, Lewis & Moss 1911, Moss 1913). Very commonly, too, in this community, and especially around the Scottish Highlands where this Arctic-Alpine species has its centre of distribution in Britain, *V. vitis-idaea* can be found, sometimes with local prominence following burning and often persisting best of all the woody associates among denser and taller heather (Ritchie 1955a, Gimingham 1972). Then, *Empetrum nigrum* ssp. *nigrum* is much more frequent through the *Calluna-Vaccinium* heath as a whole than it is in the *Calluna-Erica* heath, although the dense prostrate mats that can develop after fire here are generally overwhelmed fairly quickly by the heather, the *Empetrum* cover being attenuated to far-creeping and sparse-branching shoots which, where they are supported by the *Calluna*, are able to put up scattered sprigs of foliage at the same level.

Among other sub-shrubs, *Empetrum nigrum* ssp. *hermaphroditum* occurs only very occasionally and *Arctostaphylos uva-ursi* is rare, although over less markedly impoverished soils in eastern Scotland, the distinction between this community and the *Calluna-Arctostaphylos uva-ursi* heath, where the bearberry is a constant, is probably at least partly mediated by fire (see below). In such situations around the eastern Highlands, too, small, scattered bushes of *Juniperus communis* ssp. *communis* can sometimes be found here. In other stands, where the *Calluna-Vaccinium* heath extends a little way on to soils with somewhat less free drainage, *Erica tetralix* may occur at low covers but this is an exceptional plant here, and its appearance generally marks a decisive shift to Ericion wet heath (Edgell 1969, Birse 1980).

In very many stands of the community, the contribution of herbs to this vegetation is negligible. Of these, only *Deschampsia flexuosa* is at all frequent throughout and, though it is able to spread quite extensively after burning, particularly where there is also grazing, it is often reduced to sparse scattered tufts beneath dense heather. In other cases, however, and especially where grazing is a more consistent part of the treatment, the herbaceous element is more diverse, with species such as *Festuca ovina*, *Agrostis capillaris*, *A. canina*, *Nardus stricta*, *Potentilla erecta* and *Galium saxatile* becoming preferentially frequent in a distinctly grassier kind of heath. In the less oceanic climate characteristic of the range of this community, however, *Molinia caerulea* and *Carex binervis* hardly ever figure and *Blechnum spicant*, which is quite a frequent plant in some kinds of *Calluna-Erica* heath, is at most occasional. *Juncus squarrosus* can sometimes be found with some frequency and there are scattered records for *Pteridium aquilinum*, *Luzula pilosa*, *Trientalis europaea*, *Listera cordata* and *Lycopodium clavatum*. The rare *Diphasium × issleri*, now reckoned to be a hybrid between *D. alpinum* and *D. complanatum* (the latter not known from Britain) (Jermy *et al.* 1978), has been recorded from this community in the Morrone area of eastern Scotland (Huntley & Birks 1979a).

As in the *Calluna-Erica* heath, the ground layer of the vegetation is often prominent, with distinct patterns of colonisation in relation to the maturity of the heather and other sub-shrubs. Encrusting lichens, *Polytrichum piliferum* and *P. juniperinum* can be locally very abundant in the early years following burning, but again more characteristic of the community as a whole are bulky mosses such as *Dicranum scoparium*, *Pleurozium schreberi*, *Hypnum cupressiforme s.l.* (often obviously *H. jutlandicum*) and *Hylocomium splendens* which, together with larger *Cladonia* spp., make their maximum, and then often very luxuriant, contribution as the heather

canopy opens up with age. Where there is atmospheric pollution, such species can become very sporadic in their occurrence, even where the sub-shrub cover is not dense, and the community includes some stands where there are minor losses of this kind, but, in the southern Pennines, where such pollution is particularly severe, impoverished stands of *Calluna* and the Vaccinia are best placed in the *Calluna-Deschampsia* heath where tolerant acrocarps such as *Pohlia nutans* and *Orthodontium lineare* make up the bulk of an often very sparse ground cover.

Sub-communities

***Calluna vulgaris* sub-community:** *Calluna* heath Ratcliffe 1959; *Callunetum vulgaris* McVean & Ratcliffe 1962 *p.p.*; *Calluna*-heath moss sociation Edgell 1969 *p.p.*; *Callunetum vulgaris typicum* Evans *et al.* 1977 *p.p.*; *Calluna vulgaris-Deschampsia flexuosa* nodum Huntley 1979 *p.p.*; *Vaccinio-Ericetum cinereae*, Typical subassociation Birse 1980 *p.p.*; *Calluna vulgaris-Vaccinium myrtillus* nodum Hughes & Huntley 1986. The vegetation here is typically species-poor, with *Calluna* usually overwhelmingly dominant and often regenerating vigorously after burning. Among such dense, and often quite tall, growth, other sub-shrubs are of low cover. *Vaccinium myrtillus* is very frequent and *Erica cinerea* fairly common, but they generally occur only as scattered plants in a sparse second tier to the canopy, and *Empetrum nigrum* ssp. *nigrum* is usually almost totally overwhelmed. *V. vitis-idaea* is also very scarce but it would probably be more frequent, even if not very abundant, were not so many of the stands from lower altitudes.

Other vascular associates are few. *Deschampsia flexuosa* occurs frequently, though typically as scattered and rather drawn-up individuals, and sparse plants of *Potentilla erecta* and *Pteridium aquilinum* are quite common, but other herbs are very infrequent and the grassier physiognomy found in some stands of *Calluna-Vaccinium* heath is never seen here. The ground cover, too, is usually not very extensive, though such typical community species as *Dicranum scoparium*, *Hypnum jutlandicum* and *Pleurozium schreberi* occur frequently as scattered shoots and there is occasionally some *Hylocomium splendens*, *Rhytidiadelphus loreus*, *Ptilidium ciliare* and *Cladonia impexa*.

***Vaccinium vitis-idaea-Cladonia impexa* sub-community:** *Callunetum vulgaris* McVean & Ratcliffe 1962 *p.p.*; *Calluna vulgaris-Vaccinium vitis-idaea* nodum Huntley & Birks 1979a; *Vaccinio-Ericetum cinereae*, Typical subassociation Birse 1980 *p.p.* This sub-community includes most of the richer stands of *Calluna-Vaccinium* heath in which heather, though still very much the general dominant, is frequently accompanied by *V. myrtillus*, *V. vitis-idaea* and *E. nigrum* ssp. *nigrum*, with occasional *Erica cinerea*. Often, these other sub-shrubs are present in small amounts, with sparse shoots scattered among a fairly uniform canopy of vigorous *Calluna* or with the more shade-sensitive stratified into a second tier, but in recently-burned stands or among degenerate heather which has begun to open up, one or more of them can show local abundance in diverse patchy mosaics of bushes and mats. Very occasionally, too, there can be a little *Arctostaphylos uva-ursi* or *Juniperus*.

As in the *Calluna* sub-community, herbs are usually sparse, with just scattered plants of *Deschampsia flexuosa* and occasional *Potentilla erecta*, *Juncus squarrosus* and *Blechnum spicant*, but bryophytes and lichens are more numerous than there and generally much more abundant. Bulky pleurocarps tend to be especially prominent with *Hylocomium splendens* becoming very common and, along with *Hypnum jutlandicum* and *Pleurozium schreberi*, forming large patches among the sub-shrub stools and over their decumbent branches. *Dicranum scoparium* remains very frequent and there is occasionally some *Rhytidiadelphus loreus*, *R. triquetrus*, *Polytrichum commune*, *Ptilidium ciliare* and *Barbilophozia floerkii*. Among the lichens, clumps of *Cladonia impexa* are particularly distinctive with occasional *C. uncialis*, *C. arbuscula* and *C. crispata* and, among the smaller species, there can sometimes be found *C. pyxidata*, *C. coccifera*, *C. squamosa* and *C. floerkeana*. *Hypogymnia physodes* also occurs quite commonly on old sub-shrub branches.

***Galium saxatile-Festuca ovina* sub-community:** *Callunetum vulgaris typicum* Evans *et al.* 1977 *p.p.*; *Calluna vulgaris-Deschampsia flexuosa* nodum Huntley 1979 *p.p.*; *Calluna vulgaris-Anemone nemorosa* nodum Huntley & Birks 1979; *Vaccinio-Ericetum cinereae*, *Viola riviniana* subassociation Birse 1980. In this sub-community, the dominance of *Calluna* is not nearly so overwhelming as in other kinds of *Calluna-Vaccinium* heath and, though the other sub-shrubs occur at least occasionally in a canopy that can be quite vigorous and tall, the total cover of the bushes is less than usual. Among this more open growth, it is the variety and abundance of herbs that catches the eye. *Deschampsia flexuosa* is frequently joined here by *Festuca ovina*, *Nardus stricta* and *Agrostis capillaris*, with occasional *A. canina*, *F. rubra*, *Anthoxanthum odoratum* and *Danthonia decumbens*, the first two in particular showing locally high cover but more usually occurring in mixtures with one or more of the others, ramifying more diffuse low covers of sub-shrubs or forming stretches of grassy sward between larger bushes. *Galium saxatile* and *Potentilla erecta* occur frequently as scattered plants

among this ground and there is occasional *Carex pilulifera*, *Campanula rotundifolia* and *Polygala serpyllifolia*. Where this kind of heath extends on to soils which are somewhat less base-poor and impoverished this element may be further enriched by such species as *Lotus corniculatus*, *Lathyrus montanus*, *Succisa pratensis*, *Viola riviniana* and *Anemone nemorosa*, the local abundance of one or more of which can create a very distinctive effect (e.g. Huntley & Birks 1979*a*, Birse 1980).

Bryophytes remain quite varied in this sub-community, with *Rhytidiadelphus squarrosus* and *Pseudoscleropodium purum* joining the community species as low-frequency preferentials, but they are generally not so abundant as in the last, occurring often as rather open wefts ramifying among the grass culms. Lichens are also rather few and of low cover, with just occasional patches of *Cladonia impexa* and *C. arbuscula*.

Habitat

The *Calluna-Vaccinium* heath is the typical sub-shrub community of acidic to circumneutral, free-draining mineral soils through the cold and wet sub-montane zone. Climatic and edaphic differences across this range play some part in determining variation within the community but it is generally burning and grazing that exert the major influence on floristics and structure and, in the end, prevent succession to woodland.

This is the commonest type of heather-dominated vegetation occurring at moderate altitudes outside the more oceanic parts of upland Britain. The community can be found, mostly between 200 and 600 m, throughout the western and northern regions of the country wherever the mean annual maximum temperature falls below 26 °C (Conolly & Dahl 1970) and it makes a locally important contribution to moorland vegetation on Dartmoor, through Wales and on the North York Moors. However, within this broadly-defined zone, the *Calluna-Vaccinium* heath is strongly concentrated in areas where the climate, and particularly the winter climate, is more severe, occurring most extensively through the central and north-east Highlands of Scotland, the central reaches of the Southern Uplands and the Northern Pennines, where, for the most part, mean annual maxima are less than 24 °C (Conolly & Dahl 1970), and February minima usually more than half a degree or so below freezing (Page 1982). Throughout these areas, rainfall is not everywhere very high: mostly, there are from 1000 to 1600 mm precipitation annually (*Climatological Atlas* 1952) with usually 160–180 wet days yr⁻¹ (Ratcliffe 1968). But the general conditions are wet and cloudy and, in the harsh winters, there can be more than 40 days with snow lying.

It is such a regime that favours the rise to prominence here, among the *Calluna*, of the broadly-montane *Vaccinium myrtillus* and *Empetrum nigrum* ssp. *nigrum* (Bell &

Tallis 1973) and the more specifically Arctic-Alpine *V. vitis-idaea* (Ritchie 1955*a*), sub-shrubs which are of restricted importance among the heaths of the warmer and less humid lowlands of Britain. In fact, even within the community, there is probably an element of climatically-related variation in the representation of these different species, particularly in the case of the more demanding *V. vitis-idaea*, which becomes, more quickly than the others, rather scarce towards lower altitudes and latitudes (as in many stands of the *Calluna* sub-community) but, by and large, it is in the *Calluna-Vaccinium* heath that this suite of species shows its first consistent rise to prominence.

The same species, too, more particularly the Vaccinia, also help distinguish this community, which falls within what Birse & Robertson (1976) called 'Boreal heather moor', from the *Calluna-Erica* heath, equivalent to their 'Atlantic heather moor', a community which largely replaces it through the more humid but equable lowlands and upland fringes of western Scotland. *V. vitis-idaea* is hardly ever found in that kind of vegetation and, though *Erica cinerea* and *V. myrtillus* occur in both communities, the balance between them shifts decisively from the former, which retains a measure of physiological activity through the winter (Bannister 1965, Gimingham 1972), to the latter with the move to a less oceanic climate. Where the two kinds of heath show some geographical overlap, as through parts of the Southern Uplands and the Grampians, they can often be separated altitudinally or by aspect, the *Calluna-Vaccinium* heath typically occupying the higher ground or the more shady north- and east-facing slopes. The same climatic change plays a part in the scarcity here of *Molinia*, generally a plant of the more equable lowlands, of *Blechnum spicant*, which is broadly Atlantic in its distribution, and of the Oceanic West European *Carex binervis*.

With the generally high rainfall and humidity, the soils beneath the *Calluna-Vaccinium* heath are typically kept moist throughout the year, but they are almost always free-draining, the scarcity of such plants as *Erica tetralix* and *Scirpus cespitosus* serving as a good marker of the but rare occurrences of the community over impeded profiles. Characteristically, this is a vegetation type of more acidic soils developed over shedding slopes, extending on to level ground where the substrate is sharply pervious, but occurring very often over hill slopes and valley sides, screes and crags of gentle to steep angle. Sometimes, the soils are strongly humic and, among boulders or rock crevices, they can amount to little more than fragmentary rankers, though such accumulations often provide very congenial conditions for the growth of *V. myrtillus* in this community. Frequently, however, and particularly on gentler slopes, the profiles are more mature, sometimes brown earths or

brown podzolic soils, though generally humo-ferric podzols. Superficial pH is usually between 3.5 and 4.5 although in the *Galium-Festuca* sub-community in particular, this kind of heath can extend on to more mesotrophic brown soils when there is some small relief from the prevailingly calcifuge character of the flora. Local flushing of podzols by seepage from nearby base-rich rocks can produce a similar effect (Huntley & Birks 1979*a*, Birse 1980).

The range of soils able to support the *Calluna-Vaccinium* heath occurs widely through the cold, wet submontane zone, developing from a variety of more siliceous parent materials such as Silurian sandstones in Wales and the Southern Uplands, the Devonian Old Red Sandstones along the Welsh borders, Carboniferous sandstone up the Pennines and Oolite grits in the North York Moors, and on such rocks the community often marks out scarps and their associated talus slopes or dips free of impervious drift. It can be found, too, over some intrusive igneous rocks like the Lake District Borrowdale Volcanics and the granites of Dartmoor, Galloway and the Grampians. But in the east-central Highlands, where many of the most extensive tracts of this kind of heath occur, it is coarse glacio-fluvial gravels that provide the most distinctive substrate, particularly on the slopes of the terraced moraines in the haughlands above the limits of cultivation through Speyside and Deeside.

And it is in this region, too, that there has been some of the most widespread and continuous use of burning as a treatment for the regeneration of the vegetation. In the uplands, heather has probably long been an important element in the diet of certain cattle breeds but, over the last 200 years or so, it is sheep that have become the more important stock and, throughout its range, the *Calluna-Vaccinium* heath can be found in rough upland grazings, pastured generally in summer but also, in more sheltered situations, in winter too, when the heather and bilberry provide a valuable bite at a time when herbs are in short supply (e.g. MacLeod 1955). And since probably at least 1800, burning has been practised with varying degrees of regularity to curtail any regression to the scrub or woodland cleared for pasturing and to renew the sub-shrub growth (Gimingham 1972). But, increasingly after 1850 or so, efforts were made to capitalise on the rise in the red grouse population that followed the expansion of heathland and, in the North York Moors, parts of the Pennines, Cheviots and Southern Uplands, but especially through the east-central Highlands, this community forms a major element of grouse-moors that are still actively shot over.

Although heather shoots provide the main food of red grouse (Jenkins *et al.* 1963), such that there is a general relationship between bird density and the extent of vigorous *Calluna*, the factors relating the grouse and its environment are complex. Bigger and more successful populations of birds, for example, tend to occur on heaths underlain by less base-poor and oligotrophic substrates (Miller *et al.* 1966, Jenkins *et al.* 1967, Moss 1969) which may, though this is still uncertain, exert an influence by supporting heather of better food value (Moss *et al.* 1975). In fact, on many of the so-called 'rich' moors, it is the *Calluna-Arctostaphylos uva-ursi* heath rather than this community that occupies most of the ground, but among stands of the *Calluna-Vaccinium* heath, it is the *Galium-Festuca* sub-community, with its modest mesophytic element, that extends most often on to these more productive sites (Birse 1980).

However, on 'rich' and 'poor' moors alike, the need for a judicious regime of muirburn, as it is known in Scotland, to maintain high and healthy grouse stocks, was established early (Lovat 1911) and has been since confirmed in the continuing research on the bird and its environment (e.g. Picozzi 1968, Jenkins *et al.* 1970). In Scotland, burning can be carried out between 1 October and 15 April (exceptionally to the end of April in a wet season or to 15 May beyond 457 m) and, though better regeneration has been demonstrated in the north-east after autumn rather than spring burns (Miller & Miles 1970), March and early April have been the traditionally-favoured times in northern Scotland. The usual practice is to burn with the wind, although slower fires can be maintained by back-burning and, of course, the moisture content of the soil and vegetation also affect the intensity of the fire (Gimingham 1972). The size of the burn is of importance too, wider fires tending to be more intense (Hobbs & Gimingham 1984a). The crucial thing seems to be to aim for a ground temperature of less than 200 °C if possible, certainly below 400 °C, even short exposures to temperatures above which can be lethal to the heather stem bases (Whittaker 1960), while at the same time keeping the canopy temperature high enough to burn off the bulk of the above-ground material but not so high as to increase greatly the nutrient losses in smoke (Kenworthy 1964, Allen 1964, Evans & Allen 1971): 500 °C seems to be an optimal canopy temperature to satisfy these requirements (Gimingham 1972).

Where weather conditions remain more or less normal, it is the amount and disposition of the fuel that exert the major control over the temperature trends in a burn (Hobbs & Gimingham 1984a), both the maxima and the duration of high temperatures increasing with the age of the vegetation (Fritsch 1927, Kenworthy 1963). The biomass per unit area in *Calluna* increases at least until stands are about 20 years old and the material becomes increasingly woody with age (Gimingham 1972). But the vegetative regeneration of *Calluna*, which is generally relied upon as the means of replenishing the supply of food, is much impaired in plants which are

more than 15 years old (Lovat 1911, Gimingham 1960, Kayll & Gimingham 1965, Miller & Miles 1970, Mohamed & Gimingham 1970) and there is also the general decline in the density of stems per unit area as stands age (Miller & Miles 1970). Ideally, then, burning is timed towards the close of the building phase in the heather life-cycle, when the regrowth is often 12–15 years old, though longer rotations are necessary in exposed situations or shorter ones in sheltered sites: an average canopy height of 30–38 cm seems to be a good upper limit (Gimingham 1972). And many small burns are better than few large ones, the optimal area being perhaps $\frac{1}{2}$–1 ha, with 2 ha as a maximum, and long thin strips, say 30 m wide, being preferable to rounded or squarish areas (Watson & Miller 1970, Gimingham 1972).

The well-managed grouse-moor thus consists of a mosaic of more or less even-aged stands of heather produced by regular burning to maximise the extent of nutritious pioneer and building regrowth and on many such sites most of the structural and floristic differences among the *Calluna-Vaccinium* heath can be related to this treatment. If vegetative regeneration proceeds well, vigorous sprouting from shoot clusters on the undamaged heather stools (Mohamed & Gimingham 1970) can establish a virtually complete cover by the fourth or fifth season after burning, occasionally even as early as the third (Gimingham 1972) and, as this thickens up to form a dense and even canopy, there is increasingly little opportunity for anything but a very limited contribution from other plants. The *Calluna* sub-community with its patchy lower tier of other shade-tolerant sub-shrubs, puny herbs and rather sparse bryophytes, includes many stands of this kind.

In the early stages of heather regeneration, however, and particularly where this phase is prolonged, the characteristic associates of the community can play a more prominent role. Where older stands have been burned, for example, the more intense heat produced by the combustion of much very woody material, and the inherently poorer sprouting capacity of the aged stools (Mohamed & Gimingham 1970), makes for very slow vegetative recovery. And, though *Calluna* could be said to have fire-adapted sexual regeneration, benefiting from the light and open compacted surface, and perhaps also from the heat-treatment of the seeds previously stored in the congenially humid litter and soil (Gimingham 1972), recolonisation by seeding alone can be very slow (Hobbs & Gimingham 1984b). As in other heath communities, damp peaty surfaces may develop first a covering of algae and then a thin skin of *Lecidea uliginosa* and *L. granulosa*, followed by *Cladonia* spp and locally extensive patches of *Polytrichum piliferum* and *P. juniperinum*. But often very noticeable here is the rapid spread of the *Vaccinia* and *Empetrum*, whose buried rhizomes can largely escape the effects of fire, the intense heat from which may scarcely penetrate the soil (Hobbs & Gimingham 1984a). So, much more variegated covers of sub-shrubs can develop, only slowly becoming stratified and patchy as the heather gradually assumes dominance once more. And, over the accumulating litter, there spread patches of the bulky pleurocarpous mosses and *Cladonia impexa*. It is stands such as this which predominate in the *Vaccinium-Cladonia* sub-community, together with older tracts of *Calluna-Vaccinium* heath where degeneration of the heather bushes has allowed a resurgence of these associates in the gaps.

A different trend of development involves a marked spread of *Deschampsia flexuosa* in the early post-burn succession and, particularly where the humic top-soil is very thin or has been totally burned off, this may be accompanied by other grasses like *Festuca ovina*, *F. rubra*, *Agrostis capillaris*, *A. canina* and *Danthonia*, and herbs such as *Potentilla erecta* and *Galium saxatile*, with the sub-shrubs eventually spreading among them, creating the characteristic flora of the *Galium-Festuca* sub-community, further enrichment occurring where the soils are a little more fertile than usual or flushed.

Very frequently these processes of recovery are subject to the additional and immediate influence of grazing: most burned stands of the *Calluna-Vaccinium* heath are open to stock and wild herbivores and indeed, though moors are often managed primarily for either grouse-rearing or pasturing, the two activities are frequently combined. Even where stock-rearing predominates, regular burning is often practised because although prudent moderate grazing alone ought to be able to maintain a productive cover of heather in this vegetation, this is hardly ever possible to achieve (Gimingham 1972). It was perhaps more likely in former times when small-scale mixed farming predominated in the uplands, with cattle being pastured instead of or along with sheep. But with the widespread shift to the heavy and more selective grazing of hardy sheep breeds, introduced into Scotland from the late 1700s, the productivity of heath vegetation declined and the occasional fires that had probably long been employed to set back tree invasion were replaced in many areas by more regular burning, often on a ten-year rotation (Gimingham 1972).

Provided it is not too heavy, grazing in the early stages of regeneration can speed the development of a closed canopy, though it certainly has marked effects on its proportional composition. Though *Calluna* is palatable, for example, grazing often induces plagiotrophic growth and the adoption of a semi-prostrate habit which may put it at an advantage here over *Vaccinium myrtillus* or the less common *Erica cinerea* (Gimingham 1949). In other situations, preferential grazing may allow the spread of the less palatable *V. vitis-idaea* or *Empetrum*

nigrum, and substantial variations in all these sub-shrubs, developed in response to grazing, can be seen in the *Vaccinium-Cladonia* sub-community.

On less peaty soils, and particularly where these are tending towards podzolised brown earths, grazing often favours the maintenance of the grassy composition typical of the *Galium-Festuca* sub-community and this kind of *Calluna-Vaccinium* heath can also include mosaics of sub-shrubs and sward well on their way towards becoming better quality Nardo-Galion grasslands in a grazing-mediated succession that is generally welcomed by upland farmers. *Calluna-Vaccinium* heaths with quite abundant *Festuca ovina* and *Agrostis canina* can also extend on to podzols, but where heavy grazing follows burning on less fertile soils here, it is *Nardus stricta* or sometimes *Juncus squarrosus* that tend to increase their cover. Stands with moderate amounts of these unpalatable monocotyledons can be grouped in the *Galium-Festuca* sub-community but vegetation in which they become dominant at the expense of the sub-shrubs belongs in the relevant grassland types.

Zonation and succession

The *Calluna-Vaccinium* heath occurs in a wide variety of vegetation patterns with other sub-shrub communities, mires and grasslands, where floristic differences are controlled primarily by variations in soils, climate and treatments. Successional developments are usually held in check by burning and grazing and without these most stands would eventually progress to scrub and woodland, fragments of which can also be found in association with the community at some sites.

The clearest soil-related sequences are seen where the free-draining brown earths or podzols which typically underlie the *Calluna-Vaccinium* heath give way to seasonally-gleyed base-poor soils such as stagnopodzols or gley-podzols. Such zonations sometimes reflect the distribution of impervious drift over otherwise fairly uniform ground but they are often disposed over slopes which are increasingly gentle and, in areas of higher rainfall, this kind of sequence can run on over accumulations of ombrogenous peat. In such situations, the shift on to periodically-waterlogged ground is typically marked by a transition to Ericion tetralicis wet heath. Through the heartland of the *Calluna-Vaccinium* heath, in the east-central Highlands, down through south-east Scotland to the North York Moors, such vegetation is generally represented by the *Juncus-Dicranum* sub-community of the *Ericetum tetralicis*, in which the most characteristic changes are the appearance of *Erica tetralix* and *Scirpus cespitosus* with patchy *Molinia caerulea* and, in more intact stands, some *Sphagnum compactum* and *S. tenellum*. In fact, however, many tracts of this vegetation on grouse-moors and within hill-grazings are also subject to burning and pasturing, which can favour the extension across the soil-related boundaries of fairly uniform canopies of heather, impoverished throughout, or, where the treatments have been especially frequent or severe, a progressive run-down with a spread of *Juncus squarrosus* among a much-reduced cover of sub-shrubs, patterns well seen over some tracts of the North York Moors. This is sometimes a prelude to the development of wet *Juncus-Festuca* grassland though, in the less oceanic uplands, such vegetation tends to be rather local and often related to the occurrence of modest flushing among sequences of dry and wet heaths.

Mosaics of these communities can sometimes be seen over the redistributed peat washed down from the eroding margins of blanket bogs over the high plateaus and watersheds of the uplands or extending on to the shallow intact peat of the bog plane proper where the fringes have become dry. But more complete sequences run on from impeded peaty soils to ombrogenous peats with Erico-Sphagnion vegetation, typically of the *Calluna-Eriophorum* type with its often rather limited *Sphagnum* cover but continuing strong representation of sub-shrubs and bulky pleurocarps over its drier stretches. This kind of pattern can be seen on the grandest scale in the Cairngorms, with the various vegetation types disposed in roughly altitudinal zones over the massively-domed granite mountains (Watt & Jones 1948) but it is visible too in the Moffat and Moorfoot Hills in the Southern Uplands (Ward *et al.* 1972*b*, Ratcliffe 1977, Hill & Evans 1978), the Cheviots and in the unpolluted parts of the Pennines (Tansley 1911, 1939).

Such zonations form a pattern analogous to that seen in the more oceanic parts of western Britain, where the communities are generally replaced by the *Calluna-Erica* dry heath, the *Scirpus-Erica* wet heath and the *Scirpus-Eriophorum* mire. However, the geographical separation of these sequences is not complete and, where there is some overlap in their ranges, as in parts of the Southern Uplands, the drier sub-shrub communities generally partition according to altitude or aspect. The *Calluna-Vaccinium* heath, for example, can be found within the more oceanic zonation in the Moffat Hills and Galloway on higher crags or over slopes with a northerly or easterly aspect; conversely, the *Calluna-Erica* heath sometimes finds a place in the east-central Highlands on south-facing slopes at lower altitudes. Such shifts are to a great extent a reflection of the competitive balance between *Calluna* and *E. cinerea* or the Vaccinia (Figure 28).

Other topoclimatic shifts can be seen in the east-central Highlands too. Where conditions are particularly cold and humid, as around the steep and largely sunless corries on the northern slopes of the Cairngorms, the *Calluna-Vaccinium* heath is itself replaced by the *Vaccinium-Rubus chamaemorus* heath, a widespread community of the north-west and central Highlands but in the latter area often distinctly chionophilous, growing on humic rankers kept wet by long snow-lie. The appearance of *R. chamaemorus* with scattered *Cornus suecica* among the sub-shrubs and of patchy *Sphagnum capillifolium* are good indicators of this transition. Then, in more exposed situations, where rankers and podzols extend over ridges and summits, the *Calluna-Vaccinium* heath gives way to the *Calluna-Cladonia* heath in which the sub-shrub cover is often very dwarfed and where fruticose *Cladonia* spp., of restricted importance in the *Calluna-Vaccinium* heath, together with other large lichens, can be very varied and abundant.

Arctostaphylos uva-ursi sometimes figures locally in such transitions but it is much better represented in zonations and mosaics with *Calluna-Vaccinium* heath at lower altitudes through the east-central Highlands. There, the *Calluna-A. uva-ursi* heath has a very similar altitudinal and geographical range to this community and it is possible that burning has some role in mediating switches from one vegetation type to the other. But the *Calluna-A. uva-ursi* heath is also more consistently associated with less-impoverished soils, typically acidic but often with moder humus rather than mor (McVean & Ratcliffe 1962, Birse 1980), so a combination of treatment and edaphic factors may be responsible for the distribution of the two.

The other widespread interaction between these variables involves transitions to Nardo-Galion grasslands, which are especially likely to develop where there has been grazing in the early stages of post-burn recovery. Then the appearance of the *Galium-Festuca* sub-community may presage a succession to the *Festuca-Agrostis-Galium* grassland, the major plagioclimax pasture of base-poor soils through the upland fringes. Sub-shrubs typically have low frequency and cover in that vegetation but moderately heavy grazing may produce a more or less permanently-maintained mosaic of patches of heath among a grassy ground, or sporadic pasturing sustain a shifting balance of the elements in a more labile pattern with all shades of intermediate vegetation, something well seen in the accounts of Scottish grazings by King (1962) and King & Nicholson (1964). In many sites, however, continuous heavy grazing has favoured a progressive loss of sub-shrub vegetation to grassland, something which, over more fertile brown soils, is usually welcomed by graziers, though not by those with an interest in grouse, the numbers of which decline as the heather cover is reduced (e.g. Yalden 1972, Hewson 1977, Anderson & Yalden 1981). The constant danger in such situations, however, is the spread of *Pteridium aquilinum* which, particularly where the vigour of *Calluna* is reduced by the burning and grazing of *Calluna-Vaccinium* heath over colluvial soils or well-drained drift, is likely to invade vigorously, producing dense

Figure 28. Heath/mire zonations on the drier fringes
of blanket bogs in (a) more atlantic north-west of
Britain and (b) more boreal eastern Highlands of
Scotland, with (c) the fragmentation of heath and
mire in the latter with erosion and grazing.
H10a   *Calluna-Erica* heath, Typical sub-community
H12b   *Calluna-Vaccinium* heath, *Vaccinium-Cladonia*
sub-community
M15c   *Scirpus-Erica* wet heath, *Cladonia*
sub-community

M16d   *Ericetum tetralicis, Juncus-Dicranum*
sub-community
M17b   *Scirpus-Eriophorum* mire, *Cladonia*
sub-community
M19b   *Calluna-Eriophorum* mire, *Empetrum*
sub-community
U4   *Festuca-Agrostis-Galium* grassland
U5   *Nardus-Galium* grassland
U6   *Juncus-Festuca* grassland

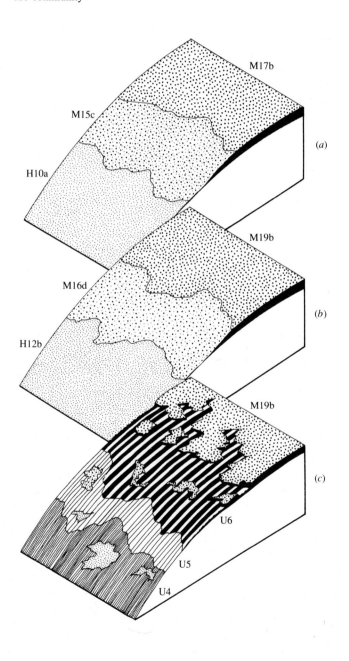

stands of the *Pteridium-Galium* community within which sparse sprigs of the shade-tolerant Vaccinia may represent the limit of sub-shrub survival.

Over less fertile peaty podzols, heavy grazing may result in the no less difficult problem of the expansion of *Nardus stricta* and the development of the *Nardus-Galium* grassland with its poor grazing potential. The kind of pattern seen so well in the map of Moor House in Cumbria (Eddy *et al.* 1969) is now very typical of many upland areas outside the grouse-moors: there, the zonation runs down from blanket bog on the high watershed, impoverished *Eriophorum* mire in many places, through a mosaic of *Nardus-Galium* and *Juncus-Festuca* grasslands, to *Festuca-Agrostis-Galium* swards, with ericoids making only a sparse appearance over inaccessible crags and screes. In extreme situations, heavy grazing may prevent the establishment of any vegetation cover on burned tracts of the *Calluna-Vaccinium* heath and where, by accident or mismanagement, fires have got out of hand and burned deep into the surface humus or even down to the mineral horizons beneath, such a combination of circumstances may be disastrous, triggering erosion by rain run-off or gully-erosion (Gimingham 1972, Maltby 1980, Maltby & Legg 1983).

Burning and grazing are so widespread throughout the range of the *Calluna-Vaccinium* heath that, alone or in combination, they effectively maintain this kind of sub-shrub vegetation as a deflected climax which has little opportunity to progress to woodland, though there is little doubt that, except where it attains higher altitudes or occurs in rather more exposed situations, this would be the natural trend. And, in the centre of its range, through the east-central Highlands, such developments would be likely to involve the invasion of pine and birch, often with juniper as a precursor or coincidental colonist. All these species are well able to take advantage of the open conditions established after fires, or even of scuffed areas in broken stretches of turf among the *Calluna-Vaccinium* heath, but the seedlings are very palatable and, where they are not eaten off, are soon crowded out by the regenerating sub-shrubs or herbs and any further germination, a light-dependent process in these plants, precluded (Miles & Kinnaird 1979, Gilbert 1980). With the widespread clearance of woodland, there is also often the problem of the remoteness of seed-parents, particularly critical in the case of pine. Nonetheless, fragments of woody vegetation survive quite widely on crags or in ravines within tracts of the *Calluna-Vaccinium* heath and over the terraced slopes through Speyside and Deeside there can be found some more extensive mosaics of the community with *Juniperus-Oxalis* scrub and *Pinus-Hylocomium* woodland (McVean & Ratcliffe 1962, Ratcliffe 1974, Huntley & Birks 1979a, b): indeed, it is patchworks of these vegetation types, often with *Calluna-A. uva-ursi* heath,

that comprise the 'Caledonian pine forest' in its broad landscape sense. In such patterns, the floristic similarities of the communities are eminently clear and, in many respects, the heath represents the field layer of more open stretches of scrub and woodland without the shrubs and trees. In this part of Britain at least, the *Calluna-Vaccinium* heath seems to have been derived by clearance of mixed forest of pine, birch and juniper whose elements are now to be seen in largely fragmentary form (McVean & Ratcliffe 1962, O'Sullivan 1977). And the increased and long-continued abundance of *Calluna* in the vegetation which has largely replaced this forest has probably contributed to enhanced podzolisation in the soils (Gimingham 1972, Fitzpatrick 1977), a further factor perhaps in the difficulty of re-establishment of the original kind of climax vegetation.

Elsewhere through the range of the *Calluna-Vaccinium* heath, the community is sometimes found in association with *Juniperus-Oxalis* scrub but, wherever there is any opportunity for seral progression, it is generally birch, usually *Betula pubescens* but locally with *B. pendula*, that is the leading colonist and this can quickly form dense thickets from which the less shade-tolerant heath plants are soon extinguished. Where more mature woodland is able to develop, among which oak can eventually find a place, usually *Quercus petraea* but with *Q. robur* sometimes seeding in from planted stock, it is typically of the *Quercus-Betula-Dicranum* type, with the *Quercus-Betula-Oxalis* woodland locally represented over less infertile soils. Conifers often invade neglected heaths from plantations and can figure in such woodlands and, in places such as the North York Moors, many of the slopes which might carry *Calluna-Vaccinium* heath have been afforested with softwoods or, where unburned, have become covered by scrub in which such trees predominate.

Distribution
The *Calluna-Vaccinium* heath is widely distributed through the less oceanic parts of the sub-montane zone, being particularly extensive in the east-central Highlands but also important in south-east Scotland, the Lake District, parts of Wales and the South-West Peninsula and the North York Moors. It would probably be more extensive in drier areas like the last, and certainly through the Southern Pennines, had not pollution been so severe: here it is largely replaced by the *Calluna-Deschampsia* heath.

Affinities
In many descriptive accounts (e.g. Smith & Rankin 1903, Smith & Moss 1903, Lewis & Moss 1911, Ratcliffe 1959a, McVean & Ratcliffe 1962, Evans *et al.* 1977), this vegetation has been grouped in a compendious *Callunetum*, with often only tentative separation from its oceanic equivalent, the *Calluna-Erica* heath, and of course

with regular burning both communities tend to converge into species-poor stands of heather. But the distinction between these two noted by Muir & Fraser (1940) and Gimingham (1972) seems to be a valid one, both in terms of the floristics of more species-rich stands and their habitat relationships: the *Calluna-Vaccinium* heath belongs fairly clearly among Böcher's (1943) Scano-Danish (Scotch) series and can be included within the 'Boreal heather moor' of Birse (1980).

In Birse's scheme, however, this last vegetation type also takes in some of what is in this scheme separated off into the *Calluna-A. uva-ursi* heath. Certainly, as McVean & Ratcliffe (1962) acknowledged, the relationship between the communities is very close and perhaps partly treatment-mediated. But, on balance, it seems best to retain separate units for these vegetation types while placing them both as close relatives within the Myrtillion boreale alliance.

Floristic table H12

	a	b	c	12
Calluna vulgaris	V (8–10)	V (6–10)	V (4–10)	V (4–10)
Vaccinium myrtillus	V (1–4)	V (1–8)	V (1–6)	V (1–10)
Hypnum jutlandicum	V (1–8)	V (1–8)	V (1–8)	V (1–8)
Dicranum scoparium	V (1–5)	V (1–6)	IV (1–6)	IV (1–6)
Pleurozium schreberi	IV (1–8)	IV (1–8)	IV (1–6)	IV (1–8)
Deschampsia flexuosa	IV (1–4)	IV (1–4)	IV (1–6)	IV (1–6)
Hylocomium splendens	II (1–8)	IV (1–9)	III (1–9)	III (1–9)
Cladonia impexa	II (1–4)	IV (1–8)	II (1–4)	III (1–8)
Vaccinium vitis-idaea	I (1–3)	IV (1–6)	III (1–4)	III (1–6)
Ptilidium ciliare	I (1–3)	II (1–4)	I (1)	II (1–4)
Cladonia pyxidata	I (1–3)	II (1–4)	I (1–3)	II (1–4)
Cladonia uncialis	I (1–3)	II (1–4)	I (1–6)	II (1–6)
Polytrichum commune	I (1–4)	II (1–4)	I (1–4)	II (1–4)
Cladonia coccifera	I (1–3)	II (1–3)	I (1–3)	I (1–3)
Cladonia squamosa	I (1–3)	II (1–4)	I (1–3)	I (1–4)
Juniperus communis		II (1–3)	I (1–3)	I (1–3)
Scirpus cespitosus	I (1–4)	II (1–4)		I (1–4)
Potentilla erecta	III (1–4)	II (1–4)	IV (1–4)	III (1–4)
Galium saxatile	I (1–3)	I (1–3)	IV (1–3)	II (1–3)
Festuca ovina	I (1–3)	I (1–3)	IV (1–6)	II (1–6)
Nardus stricta	I (1–4)	I (1–4)	III (1–4)	II (1–4)
Agrostis capillaris	I (1–4)	I (1–3)	II (1–4)	I (1–4)
Carex pilulifera	I (1–2)	I (1–3)	II (1–4)	I (1–4)
Festuca rubra	I (1–4)	I (1–3)	II (1–6)	I (1–6)
Agrostis canina	I (1–3)	I (1–3)	II (1–4)	I (1–4)
Rhytidiadelphus squarrosus	I (1–4)	I (1–3)	II (1–4)	I (1–4)
Campanula rotundifolia	I (1–3)	I (1–3)	II (1–3)	I (1–3)
Pseudoscleropodium purum	I (1–4)	I (1–3)	II (1–4)	I (1–4)
Danthonia decumbens	I (1–3)		II (1–4)	I (1–4)
Polygala serpyllifolia	I (1–3)		II (1–3)	I (1–3)
Anthoxanthum odoratum	I (1–3)		II (1–6)	I (1–6)
Lotus corniculatus	I (1)		II (1–4)	I (1–4)
Succisa pratensis	I (1–3)		II (1–4)	I (1–4)
Viola riviniana	I (1–3)		II (1–3)	I (1–3)
Lathyrus montanus	I (1)		II (1–4)	I (1–4)
Erica cinerea	III (1–6)	II (1–6)	III (1–6)	III (1–6)
Empetrum nigrum nigrum	I (1–6)	III (1–6)	II (1–4)	III (1–6)

Floristic table H12 *(cont.)*

	a	b	c	12
Juncus squarrosus	I (1–4)	II (1–4)	II (1–4)	II (1–4)
Hypogymnia physodes	I (1–3)	II (1–4)	II (1–6)	II (1–6)
Rhytidiadelphus loreus	II (1–4)	II (1–4)	I (1–4)	II (1–4)
Cladonia arbuscula	I (1–4)	II (1–8)	II (1–4)	II (1–8)
Blechnum spicant	I (1–3)	II (1–3)	II (1–4)	II (1–4)
Pteridium aquilinum	II (1–6)	I (1)	I (1)	I (1–6)
Barbilophozia floerkii	I (1–3)	II (1–4)		I (1–4)
Cladonia crispata	I (1)	I (1–4)	I (1)	I (1–4)
Rhytidiadelphus triquetrus	I (1–3)	I (1–4)	I (1–6)	I (1–4)
Pohlia nutans	I (1–4)	I (1–4)	I (1–3)	I (1–4)
Lophocolea bidentata	I (1–4)	I (1–3)	I (1–3)	I (1–4)
Lycopodium clavatum	I (1–3)	I (1–3)	I (6)	I (1–6)
Luzula pilosa	I (1–3)	I (1–3)	I (1–3)	I (1–3)
Cladonia floerkeana	I (1)	I (1–3)	I (1)	I (1–3)
Trientalis europaea	I (1–3)	I (1–3)	I (1–3)	I (1–3)
Calypogeia muellerana	I (1)	I (1–3)	I (1–3)	I (1–3)
Plagiothecium undulatum	I (1–6)	I (1–4)	I (1–3)	I (1–6)
Lophozia ventricosa	I (1–3)	I (1–3)	I (1)	I (1–3)
Thuidium tamariscinum	I (1–4)	I (1–3)	I (1–4)	I (1–4)
Listera cordata	I (1–3)	I (1–3)	I (1–3)	I (1–3)
Sorbus aucuparia seedling	I (1–3)	I (1–3)	I (1–3)	I (1–3)
Erica tetralix	I (1–4)	I (1–4)	I (1–4)	I (1–4)
Cladonia chlorophaea	I (1–3)	I (1–3)	I (1–3)	I (1–3)
Cladonia gracilis	I (1–3)	I (1–3)	I (1–3)	I (1–3)
Molinia caerulea	I (1–4)	I (1–3)	I (1–6)	I (1–6)
Empetrum nigrum hermaphroditum	I (1–4)	I (1–4)		I (1–4)
Eriophorum vaginatum	I (1)	I (1–4)		I (1–4)
Polytrichum alpestre	I (1–3)	I (1–3)		I (1–3)
Dicranum fuscescens		I (1–4)	I (1–3)	I (1–4)
Cladonia rangiformis		I (1–3)	I (1–3)	I (1–3)
Arctostaphylos uva-ursi		I (1–3)	I (1–3)	I (1–3)
Anemone nemorosa	I (1)		I (1–4)	I (1–4)
Number of samples	56	167	78	311
Number of species/sample	14 (9–28)	21 (4–42)	19 (5–36)	17 (4–42)
Shrub/herb height (cm)	31 (9–75)	23 (8–60)	21 (7–50)	25 (7–75)
Shrub/herb cover (%)	97 (85–100)	92 (70–100)	88 (40–100)	92 (40–100)
Ground layer height (mm)	36 (10–90)	35 (10–100)	39 (10–150)	37 (10–150)
Ground layer cover (%)	18 (0–100)	65 (0–100)	43 (0–100)	60 (0–100)
Altitude (m)	353 (77–724)	494 (274–910)	412 (105–845)	446 (77–910)
Slope (°)	21 (0–45)	13 (0–85)	23 (0–80)	18 (0–85)
Soil pH	4.4 (3.5–6.0)	4.4 (3.3–6.0)	4.5 (3.4–6.1)	4.4 (3.3–6.1)

a   *Calluna vulgaris* sub-community

b   *Vaccinium vitis-idaea-Cladonia impexa* sub-community

c   *Galium saxatile-Festuca ovina* sub-community

12   *Calluna vulgaris-Vaccinium myrtillus* heath (total)

H12 *Calluna vulgaris-*
Vaccinium myrtillus
heath

H12 *Calluna vulgaris-*
Vaccinium myrtillus
heath

b *Vaccinium vitis-idaea-*
Cladonia impexa
sub-community

H12 *Calluna vulgaris-*
Vaccinium myrtillus
heath

a *Calluna vulgaris*
sub-community

H12 *Calluna vulgaris-*
Vaccinium myrtillus
heath

c *Galium saxatile-*
Festuca ovina
sub-community

H13
Calluna vulgaris-Cladonia arbuscula heath

Synonymy
Mountain *Callunetum* Metcalfe 1950 *p.p.*; Lichen heaths Poore & McVean 1957 *p.p.*; *Cladineto-Callunetum* McVean & Ratcliffe 1962, Prentice & Prentice 1975 *p.p.*; *Calluna* heath, dwarf lichen-rich facies Edgell 1969; Dwarf mountain heaths Gimingham 1972 *p.p.*; *Empetrum hermaphroditum-Racomitrium lanuginosum* community Birse & Robertson 1976; *Empetrum hermaphroditum-Cetraria nivalis* nodum Birks & Huntley 1979; *Alectorio-Callunetum vulgaris* (Birse & Robertson 1976) Birse 1980.

Constant species
Calluna vulgaris, Empetrum nigrum hermaphroditum, Racomitrium lanuginosum, Alectoria nigricans, Cetraria islandica, Cladonia arbuscula, C. rangiferina, C. uncialis, Cornicularia aculeata.

Rare species
Arctostaphylos uva-ursi, Loiseleuria procumbens.

Physiognomy
The *Calluna vulgaris-Cladonia arbuscula* heath has a dwarfed mat of sub-shrubs with few vascular associates but with a prominent contribution from lichens, which form an often co-dominant patchwork and which can be so abundant in total as to give stands a grey or yellowish hue even from a distance. The sub-shrub mat is commonly less than 5 cm thick, only rarely more than 8 cm, and it can form a fairly extensive and uniform cover, but it is usually discontinuous, sometimes very fragmentary and occasionally disposed in rather particular patterns related to exposure to wind and frost, characteristically severe in the places where this community occurs. *Calluna vulgaris* is the most frequent component and generally the most abundant, growing in prostrate fashion and often with its flattened branches orientated downwind. These may thus come to overlie older branches of other heather plants, all having their young leafy shoots knotted together into a tight carpet (Metcalfe 1950), or

be arranged in wave-like bands with intervening strips of bare or lichen-covered ground, active shoot growth confined to the sheltered lee faces of the bushes (Crampton 1911, Watt 1947). Over gently-sloping ground, similar vegetated zones may develop on small solifluction terraces, where frost-heave and the filtering downhill of rain water and snow-melt push mineral particles among the distal branches of the heather, building up a riser in which shoots can renew their growth, while behind there forms a bare tread (Metcalfe 1950). Wind can contribute to the formation of this kind of microtopography, blowing often this time against the direction of sub-shrub growth as it whips up and over brows, stunting and trimming the advancing front of the bushes but scouring out the back of each terrace and accentuating the steps (Metcalfe 1950, Prentice & Prentice 1975). Then, over very exposed knolls and summits, wind and frost-heave can reduce the sub-shrub cover to scattered bushes, their contorted wiry stems hugging the ground and being largely bereft of green shoots.

In some stands, *Calluna* is the only sub-shrub but typically other species are present, sometimes finely intermixed with the heather, but often forming a coarse mosaic of patches within the mat. All these associates are rhizomatous or have procumbent rooting branches, so quite far-spread shoots or clumps are frequently found to be joined by straggling stems beneath the heather and these also enable the plants to colonise any sheltered gaps which develop in the mat, often distant from the original point of rooting. Among these other sub-shrubs, *Empetrum nigrum* is the most important overall with ssp. *hermaphroditum* the more characteristic taxon, ssp. *nigrum* partially replacing it towards the lower altitudinal limit of the community, and its shoots can be found intermingled in the mat or aggregated into clumps up to $\frac{1}{2}$ m or so across, often spreading out in the lee of heather bushes on to bare ground.

Of generally similar habit, though of more restricted occurrence in the community, are the Arctic-Alpine rarities *Arctostaphylos uva-ursi* and *Loiseleuria procum-*

bens. A. uva-ursi is at most occasional here and it becomes very scarce where this kind of heath spreads on to the highest mountains but, at lower altitudes, it can be locally abundant, even co-dominant with *Calluna* and *Empetrum*, and is particularly striking where it extends out as a sheltered front to heather waves or on to solifluction treads, where it can rapidly colonise bare ground provided lashing by the wind and erosion of the surface are not too severe, only to be smothered again by the *Calluna* where this can follow (Watt 1947, Metcalfe 1950). However, combinations of *A. uva-ursi, Calluna* and *Empetrum nigrum* ssp. *nigrum* with *Erica cinerea*, such as are characteristic of the *Calluna-Arctostaphylos uva-ursi* heath, are rare here. Vegetation best placed in that community has sometimes been included within the *Calluna-Cladonia* heath (as in Metcalfe 1950) and transitional stands can be found in the *Cladonia* sub-community, but by and large the distinctions between the two kinds of heath are fairly clear and they show a broad altitudinal separation within eastern Scotland (see below). Likewise, the almost total absence of the other rare Arctic-Alpine bearberry, *Arctostaphylos alpinus*, from the *Calluna-Cladonia* heath serves as an effective separation from the *Calluna-Arctostaphylos alpinus* heath, a rather similar vegetation type, though essentially one of the western Highlands.

Loiseleuria procumbens, however, which in the scheme of McVean & Ratcliffe (1962) was very much a characteristic plant of their *Arctoeto-Callunetum*, is again locally quite frequent and abundant within the *Calluna-Cladonia* heath. Indeed, most of the Cairngorm stations of *Loiseleuria* are in this community, a feature well seen in the tables of Birse (1980, where this kind of heath is termed the *Alectoria-Callunetum*), and in the descriptive account of Metcalfe (1950), where *Loiseleuria* can be seen replacing *A. uva-ursi* in some of the higher-altitude stands of the lichen-rich dwarf heaths. It is of much smaller stature than the bearberry, though able in a similar fashion to extend beneath the heather mat and colonise sheltered bare ground beyond, forming quite large cushions.

The other common sub-shrubs of the *Calluna-Cladonia* heath are *Vaccinium myrtillus* and *V. vitis-idaea* and, in the early descriptions of Poore and McVean (1957), lichen-rich heaths in which bilberry was the dominant ericoid in the absence of *Calluna* were included together with those with abundant heather. In this scheme, following McVean & Ratcliffe (1962), a major separation is made between the *Calluna-Cladonia* heath and a *Vaccinium-Cladonia* heath, in which a dwarfed sub-shrub mat with an abundance of lichens remain general structural features, but where *Calluna* is very sparse and where many stands have a distinctly chionophilous character. Here, by contrast, the two Vaccinia, though very frequent, are almost always subordinate in cover:

V. vitis-idaea sometimes plays an important role in stabilising loose mineral material on exposed ground but it is not usually abundant and *V. myrtillus* is generally found as separate shoots growing through the heather mat.

Other vascular associates are few in number and likewise characteristically sparse. The most frequent overall are *Deschampsia flexuosa* and *Carex bigelowii*, both generally occurring as scattered tufts or individual shoots, with some other species such as *Scirpus cespitosus*, *Agrostis canina* and *Molinia caerulea* appearing at lower altitudes, and *Juncus trifidus* becoming occasional towards higher levels. *Huperzia selago* is also frequent in higher-altitude stands with *Diphasium alpinum* sometimes recorded too. Dicotyledonous herbs are very scarce with just occasional records for *Potentilla erecta*, *Galium saxatile* and *Antennaria dioica*.

Much more important structurally than any of these are the lichens which form a variegated and sometimes quite species-rich mat among the sub-shrubs, frequently in discrete patches but often of equal or greater total cover and, with some species particularly, probably of considerable competitive power. Especially common and, in somewhat more sheltered situations, becoming very abundant is *Cladonia arbuscula* with, generally less extensive, though locally co-dominant with it, the similar-looking *C. rangiferina*. *C. uncialis* is also constant at low covers and *C. gracilis* and *C. crispata* become frequent in particular kinds of *Calluna-Cladonia* heath, with smaller species such as *C. squamosa* and *C. coccifera* occasional, though not of any structural significance. *C. impexa*, on the other hand, is rather scarce, which provides a further contrast with many stands of the *Calluna-A. uva-ursi* heath.

Also very common throughout the community is *Cetraria islandica*, the brown thalli of which are especially tolerant of the light snow cover that more sheltered stands can carry. The pale yellow *C. nivalis* can be found, too, though this is very much a plant of the opposite extremes, occurring most often on very exposed brows and summits at higher altitudes (Nordhagen 1943, Poore & McVean 1957). *C. glauca* also occurs but it is very occasional and much more typical of lichen-rich stands of the *Calluna-A. uva-ursi* heath.

Then, there is almost always some *Cornicularia aculeata*, whose little cushions of tangled reddish fronds seem particularly abundant where the large *Cladonia* spp. have become moribund (Metcalfe 1950). Finally, among the constants, is the filamentous *Alectoria nigricans*. *A. ochroleuca* can also sometimes be found, typically in similar exposed situations to *Cetraria nivalis*, but it never attains the luxuriance here that it has among similar Scandinavian heaths (Poore & McVean 1957). Birse (1980) recorded the rare Alpine lichen *A. vexillifera* (= *A. sarmentosa* var. *cincinnata*) in this vegetation.

Other lichens occasionally encountered are the densely-cushioned *Sphaerophorus globosus* and, less commonly, *S. fragilis*, *Thamnolia vermicularis* with its distinctive creeping whitish worm-like thalli and the lumpy grey encrustations of *Ochrolechia frigida*.

Among this mat, bryophytes are generally very few and rarely of any abundance. *Racomitrium lanuginosum* is the commonest and it can form locally conspicuous patches, particularly among the older bare branches of the wind-flattened bushes. But it never has the dominant role among the ground carpet that is typical of the *Calluna-Racomitrium* heath. Occasional associates include *Hypnum jutlandicum*, *Dicranum scoparium* and *Ptilidium ciliare*, with scarce records for *Pleurozium schreberi*, *Dicranum fuscescens*, *Diplophyllum albicans* and *Pohlia nutans*.

Sub-communities

***Cladonia arbuscula-Cladonia rangiferina* sub-community:** Mountain *Callunetum*, *Calluna-Arctostaphylos* facies Metcalfe 1950; *Cladineto-Callunetum sylvaticosum* McVean & Ratcliffe 1962, Prentice & Prentice 1975; *Calluna* heath, dwarf lichen-rich facies Edgell 1969; *Alectoria-Callunetum*, *Agrostis canina* ssp. *montana* subassociation Birse 1980. In this sub-community, larger *Cladonia* spp., particularly *C. arbuscula*, but also on occasion *C. rangiferina*, are especially abundant, often exceeding the sub-shrubs in total cover. *C. impexa* can sometimes be found among the lichen carpet, too, though *C. gracilis* and *C. crispata* are scarce. *Calluna* is usually the most abundant sub-shrub, but *Empetrum nigrum* occurs very commonly, with ssp. *nigrum* often rivalling ssp. *hermaphroditum* in frequency, and each can have quite high cover. *Arctostaphylos uva-ursi* is found in some stands and it can be locally prominent. *Vaccinia* are only occasional and present as sparse shoots but there is sometimes a little *Erica cinerea* and rarely some *E. tetralix*.

Other low-frequency preferentials include *Scirpus cespitosus*, *Molinia*, *Nardus* and *Agrostis canina* (mostly ssp. *montana*), though all these, together with the occasional *Deschampsia flexuosa*, are usually found just as scattered tufts. *Potentilla erecta* and *Huperzia selago* can sometimes be seen but *Carex bigelowii* is rare.

***Empetrum nigrum* ssp. *hermaphroditum-Cetraria nivalis* sub-community:** *Calluna*-lichen heath Poore & McVean 1957; *Cladineto-Callunetum typicum* McVean & Ratcliffe 1962; *Empetrum hermaphroditum-Cetraria nivalis* nodum Huntley & Birks 1979; *Alectorio-Callunetum*, Typical subassociation Birse 1980 *p.p.* Although lichens as a group remain very abundant here, the balance of dominance generally lies with *Calluna* or occasionally *E. nigrum* ssp. *hermaphroditum*,

which is very common and far outweighs ssp. *nigrum* in frequency. Both *V. myrtillus* and *V. vitis-idaea* occur more often here than in the last sub-community, but they usually remain of low cover. Again, some stands have a local abundance of *A. uva-ursi* but here, too, *Loiseleuria* can occasionally be found, though generally in small amounts.

In the lichen mat, *Cladonia arbuscula* is still the most abundant species but, along with *C. rangiferina* and *C. uncialis*, it is frequently joined by *C. gracilis* and more occasionally by *C. crispata* and *C. bellidiflora*. More striking is the common occurrence of *Cetraria nivalis*, sometimes of moderate cover, with scattered records for *Alectoria ochroleuca*.

Mosses make a generally very minor contribution to the ground cover and vascular associates are also usually present as scattered individuals. But *Deschampsia flexuosa* and *Carex bigelowii* are both very common, and preferential at low frequencies are *Diphasium alpinum* and *Juncus trifidus*.

***Cladonia crispata-Loiseleuria procumbens* sub-community:** Mountain *Callunetum*, *Calluna-Loiseleuria* facies Metcalfe 1950; *Alectorio-Callunetum*, Typical subassociation Birse 1980 *p.p.* *Calluna* is again the usual dominant in this sub-community but one or both of the subspecies of *Empetrum nigrum* are commonly present and they can be quite abundant. There is also often some *Loiseleuria*, though *A. uva-ursi* is very rare. The Vaccinia reach the peak of their frequency, but are still usually present just as sparse shoots.

Among the sub-shrub mat, which is often flattened against the ground even more markedly than usual, all the characteristic lichens of the community remain frequent, usually making up a carpet in which no single species is dominant. But there are some clear preferentials to this kind of heath, with *Cladonia crispata* being particularly common, *C. furcata*, *C. coccifera* and *C. squamosa* more occasional, *Ochrolechia frigida* and *Thamnolia vermicularis* frequent. Scattered shoots of *Carex bigelowii*, *Huperzia selago* and *Deschampsia flexuosa* often make up the sparse herbaceous element.

Habitat

The *Calluna-Cladonia* heath is the characteristic sub-shrub vegetation of base-poor soils over exposed ridges and summits of mountains in those parts of Britain with a cold continental climate. Burning and grazing may have curtailed its range in suitable localities in our more southerly uplands but, through its heartland in the eastern Highlands of Scotland, the vegetation seems little affected by treatments and the community can be considered a climax.

It certainly experiences some of the harshest weather conditions of any throughout the country, being typical of what Poore & McVean (1957) called the low-alpine zone, generally between 600 and 900 m, in areas with a mean annual maximum temperature of less than 23 °C (Conolly & Dahl 1970) and where the February minimum falls well below freezing (*Climatological Atlas* 1952). It is thus especially characteristic of the east-central Highlands, with extensions westwards into the central Grampians and up the eastern fringes of the north-west Highlands, and far-flung fragmentary stands in certain parts of the Southern Uplands, the Lake District and north Wales (McVean & Ratcliffe 1962, Edgell 1969), where conditions are locally suitable. Through this region, the kinds of soils favoured by the community, rankers and often shallow podzols, frequently humic above and with a superficial pH between 4 and 5, are very widespread over shedding slopes cut into pervious acidic bedrocks and superficials. In the centre of its range, the *Calluna-Cladonia* heath is especially widespread on such soils over the rounded summits and spurs of the Cairngorms and Lochnagar and the higher slopes above the Spey and the Don, where massive granite intrusions form the basis of many of the mountains, with drift mantling the lower slopes. Other siliceous Dalradian rocks are locally important as parent materials in the Grampians with Moine schists underlying stands on Monadhliath, around Drumochter and on Ben Wyvis and other Moinian metasediments on Beinn Dearg.

In such sites, the *Calluna-Cladonia* heath thus provides an important locus for calcifuges adapted to a climate approaching the Arctic-Alpine in character, for example, *Empetrum nigrum* ssp. *hermaphroditum*, *Vaccinium vitis-idaea*, *Loiseleuria*, *Carex bigelowii* and, to a lesser extent, *Arctostaphylos uva-ursi*. Vascular plants typical of heaths developed in more equable climatic conditions, on the other hand, such as *Erica cinerea* and *Molinia*, tend to be confined to sites with some measure of local shelter or which are at lower altitudes than usual. Such species are best represented here in the *Cladonia* sub-community, together with ssp. *nigrum* as the predominant crowberry taxon, and this vegetation takes the *Calluna-Cladonia* heath to its lowest levels, notably on Orkney where, at well below 600 m, this vegetation is clearly transitional to lichen-rich oceanic *Calluna-Erica* heath (Prentice & Prentice 1975). *Scirpus cespitosus*, though it is a plant of some high-altitude mires, follows the same trend as *E. cinerea* here (Metcalfe 1950, Prentice & Prentice 1975).

Towards higher ground, and perhaps on soils which are more peaty above, offering a congenial surface for its distinctive suite of lichens, the *Cladonia-Ochrolechia* sub-community is characteristic, occurring on some of the bleaker Cairngorm spurs, where *Loiseleuria* is sometimes a prominent associate, but also common over the lower summits through the northern Grampians. Then,

largely confined to the highest situations, with a mean altitude over 150 m more than that of the *Cladonia* sub-community and a mean annual maximum temperature of less than 21 °C (Conolly & Dahl 1970), there is the *Empetrum-Cetraria* sub-community. In the Cairngorms, this kind of vegetation takes heather-dominance to its upper limit, close on 1000 m and though, beyond this zone, dwarf-shrub heaths can run on, it is *E. nigrum* ssp. *hermaphroditum* and *V. myrtillus* that are the predominant woody plants. *V. myrtillus* is always subordinate to *Calluna* here, though the balance between these two species among dwarf-shrub heaths at lower altitudes is also strongly dependent on snow cover.

For the composition and physiognomy of this vegetation, then, it is not just the general character of the climate that is important. Through the low-alpine zone of our more continental mountains, this is inhospitable enough but harsh winds and frosts lend a particular severity to situations that are more exposed. Above all, the *Calluna-Cladonia* heath is a vegetation type of unsheltered slopes in these regions, usually fairly gentle and rounded, on summits and ridges where there is an almost constant blast of strong winds. And this means, too, that the ground is usually blown clear of any substantial accumulations of snow. In fact, though precipitation across this part of Britain is not very high, often below 1600 mm yr^{-1} (*Climatological Atlas* 1952), in the winter months much of this falls as snow, such that there are generally through this region 60–100 or more days annually with snow lying. But it does not settle appreciably on the more wind-lashed slopes, which not only effectively reduces the amount of precipitation locally received by this community but which also exposes the vegetation, and the underlying soils, to the effects of frequent and severe frosts and their subsequent thaw.

Most of the structural features of this kind of heath are attributable to interactions between wind and frost and the different growth responses of the various sub-shrubs to them. There is the generally dwarfed character of the mat and the striking wind-waved physiognomy that sometimes develops, very noticeable in certain of the Cairngorm stands, with the different woody species able to exploit the modicum of shelter provided by the repeated belts of flattened vegetation. Such patterns may show some measure of cyclical alternation of dominance among the sub-shrubs (Watt 1947) but, with freeze-thaw often adding to the erosive effects of the wind on the soil mantle, it is doubtful how far such changes can go undisturbed and the constant check which these processes exert on the vegetation is sufficient to maintain it as a climax and, in extremes, to reduce it to a very sparse cover of plants at their limits of survival.

Provided the ground among the bushes remains

stable, with little or no frost-heaving or wind-blow, the exposure of bare patches of soil and more open areas of decaying sub-shrubs offers an opportunity in the cold continental climate for the local dominance by lichens so characteristic of this vegetation. In fact, in the opinion of Poore & McVean (1957) the maintenance of this typical abundance of lichens is a quite finely balanced matter in the relatively oceanic climate of the British Isles as compared with parts of Scandinavia. The provision of some shelter favours more luxuriant growth of the lichen mat and the preferential prominence of species such as *Cladonia arbuscula*, *C. rangiferina* and *Cetraria islandica* in the *Cladonia* sub-community is frequently related to a locally less exposed physiography than is usual, as where slight hollows perhaps accumulate a little snow, that thus protects the cover from the worst effects of frosts and the biting winds of winter. In such more humid situations, though, there is often a tendency for *Racomitrium lanuginosum* to expand its cover at the expense of the lichens and convert the vegetation to the kind of moss-heath so familiar in the west. At the opposite extreme, in the *Empetrum-Cetraria* sub-community, we can see the best representation of the chionophobous lichens *Cetraria nivalis* and, less commonly, *Alectoria ochroleuca*, over sites which are locally very bleak. But here *Juncus trifidus* begins to exert itself and it is this species which assumes dominance over lichens as altitude and local exposure combine to produce some of the most arctic conditions over the very high Cairngorm summits.

Outside the Scottish Highlands, this kind of heath is rare: vegetation approaching it can be seen on parts of Dollar Law and White Coomb in the Southern Uplands and fragments occur on Skiddaw in Cumbria (McVean & Ratcliffe 1962, Ratcliffe 1977); and some of the more lichen-rich heaths of Cader Idris (Edgell 1969) could perhaps be included in the community. However, even in sites where the topoclimate provides suitable conditions for the local development of this type of vegetation, the virtually ubiquitous treatments of burning and grazing may well have destroyed it. Through most of its present range, however, these seem to be of little importance to the appearance and composition of the community. Muirburns sometimes spread up into the zone in which it is found from the *Calluna-Vaccinium* and *Calluna-A. uva-ursi* heaths but damage does not seem to be widespread. Grazing by sheep and deer is probably high, although ptarmigan, very characteristic and locally numerous birds of these high slopes (Watson 1964, 1966) can take large amounts of heather when it is available, with *E. nigrum* and *V. myrtillus* also figuring in their diet (Gimingham 1972).

Zonation and succession

The *Calluna-Cladonia* heath is typically found with other sub-shrub communities, montane grass- and moss-heaths and snow-bed vegetation in zonations and mosaics which reflect altitudinal and local topographically-determined variations in climate, especially the general temperature trend in moving on to higher ground and exposure. Geological heterogeneity can introduce edaphically-related complications into these patterns and, towards lower altitudes, treatments have often modified the transitions. With the geographical shift to the west, the elements of the vegetation sequence are increasingly replaced by their oceanic counterparts and, in between the climatic extremes, complex intermediate patterns of communities can be seen.

The essential features of the altitudinal zonation characteristic of base-poor soils in our more continental mountains are best seen in the Cairngorms, where there is a strong measure of geological uniformity and a massive scale to the scenery, with the broad granite summits and watersheds and relatively gentle transitions in many areas to the lower slopes (Watt & Jones 1948, Poore & McVean 1957). Moreover, the prevailing infertility of the soils has meant that sheep-pasturing has figured little in the history of land-use there, such that sub-shrub heaths remain extensive throughout their climatic zones (Ratcliffe 1977). Not that there has been no modification to the natural vegetation pattern however: at lower altitudes, the *Pinus-Hylocomium* woodland is probably the climax community and, although stretches of this native pine forest remain, much of the ground below the *Calluna-Cladonia* heath now carries *Calluna-Vaccinium* heath which spread with clearance and which is now often maintained by burning for grouse-rearing. Many of the soils at these levels, though, are derived from superficials, so there is quite frequently some modest variation in soil fertility, with stands of the *Calluna-A. uva-ursi* heath marking out less oligotrophic brown earths, and in drainage impedence, the *Ericetum tetralicis* occupying strongly-gleyed podzols.

In some places, too, the *Juniperus-Oxalis* scrub occurs among mosaics of these heaths, essentially now behaving as a perpetually-renewed seral vegetation. Only on Creag Fhiaclach, at around 640 m on a north-western spur of the Cairngorm range, does this community form a convincing climatically-controlled fringe above the forest zone, grading up to the low-alpine belt, where the *Calluna-Cladonia* heath is typically found (Poore & McVean 1957). Generally, then, the upward transition to the *Calluna-Cladonia* heath is from one or other of the sub-shrub communities mentioned above, such that a predominance of heather is maintained, albeit in a dwarfed form at higher altitudes, right up to about 1000 m in these mountains. The shift from one vegetation type to the other is often gradual and the actual altitude at which the change decisively occurs varies with the degree of local exposure because of the strong dependence, in our prevailingly oceanic climate, on topography to maintain the required element of harshness of climate characteristic of the whole low-alpine zone in more continental parts of northern Europe (Poore & McVean 1957). The stunting and wind-flattening of the sub-shrub mat and the abundance of lichens which are the chief general features of the transition thus appear at lower altitudes and more sharply where there is an early and abrupt move from sheltered slopes on to more open, wind-lashed ground. But there can be a strong continuity in the qualitative composition of the canopy even then and stands of the *Calluna-Vaccinium* and *Calluna-A. uva-ursi* heaths which are not in the dense pioneer or building phase of regeneration from burning sometimes develop lichen floras which approach those of the *Calluna-Cladonia* heath in variety, though not generally in abundance.

To such sub-montane heaths, the *Cladonia* sub-community can be seen as transitional and this characteristically occupies the lower and more sheltered situations that develop a cover of *Calluna-Cladonia* heath, with the *Empetrum-Cetraria* or *Cladonia-Ochrolechia* sub-community ranged above it on the more exposed ridges and summits. Where the zonation continues upwards, on to ground that is too inhospitable for the survival of *Calluna* and where the winds are so strong that even lichens cannot get a hold in any abundance, the *Calluna-Cladonia* heath typically peters out into a zone of the *Juncus-Racomitrium* heath. Again, there can be a measure of floristic continuity among the vegetation types: *J. trifidus* itself makes an occasional appearance in the *Calluna-Cladonia* heath and a number of species remain frequent over the higher ground, such as *Carex bigelowii*, *Deschampsia flexuosa*, *Vaccinium myrtillus* and, with much reduced covers, various of the lichens. But the overall character of the vegetation is very different and, on the extensive tracts of ground above 1000 m through the Cairngorms, the shift to this rush-dominated community, developed as an open network over the loose erosion surfaces of the high plateaus, is very striking (Poore & McVean 1957, Ingram 1958, McVean & Ratcliffe 1962).

The major complications to the higher zones of this kind of transition are related to a decrease in exposure (Poore & McVean 1957). In the low-alpine belt, the sub-communities of the *Calluna-Cladonia* heath themselves can be arranged, not in altitudinal bands, but in patches disposed according to the local variations in the amount of shelter provided by topography. The *Empetrum-Cetraria* and *Cladonia-Ochrolechia* sub-communities, for example, can be found at relatively low altitudes where the exposure is locally extreme, while the *Cladonia* sub-community can attain relatively high levels in hollows and on lee slopes. In this kind of situation, the latter type of *Calluna-Cladonia* heath is essentially transitional to the *Vaccinium-Cladonia* heath, which usually seems to experience longer snow-lie and which in some stands has a marked chionophilous character.

Sometimes, the two sorts of lichen heath form mosaics over gently-undulating slopes, where the hollows catch a little snow and the convexities are largely blown clear, transitions from one community to the other being sometimes a matter of a shift in dominance from heather to bilberry. In other cases, the *Vaccinium-Cladonia* heath marks a clear intermediate zone between the *Calluna-Cladonia* heath and a snow-bed proper with *Nardus-Carex* grass-heath. At lower altitudes, where a shady aspect becomes progressively necessary to maintain long snow-lie, similar zonations to these can be found but with the *Calluna-Cladonia* heath giving way over the wind-blasted lip of the snow-bed, not to the *Vaccinium-Cladonia* heath, but to the *Vaccinium-Rubus* heath. At higher altitudes where, with some shelter, snow cover settles more uniformly, the *Vaccinium-Cladonia* heath may largely replace the *Calluna-Cladonia* heath, maintaining sub-shrub dominance for a further 100 m or so above the heather limit.

Quite often, however, the major variation in zonation at the upper limit of the *Calluna-Cladonia* zone is attributable not so much to increased snow-lie as to the greater constancy of humidity across the cloudy summits away from extreme exposure. Then, there is a progressive tendency for the *Calluna-Cladonia* heath to give way above, not to the *Juncus-Racomitrium* heath, but to the *Carex-Racomitrium* heath, in which dominance typically lies with *R. lanuginosum*. And, in fact, this is essentially the shift from continental to oceanic conditions for, though no other mountain massif in Britain has so much of the *J. trifidus* fell-field vegetation as the Cairngorms, even a slight shift away from the more easterly summits reinforces what is there a local trend. Above Glen Feshie, for example, where there are some particularly fine stands of *Calluna-Cladonia* heath on the south-west facing spurs, the *Carex-Racomitrium* heath predominates above and with the shift to the Moine schist hills around Drumochter and Ben Alder, a little to the west, the *Juncus-Racomitrium* heath can be distinctly patchy above the *Calluna-Cladonia* heath. On Ben Wyvis, a massive bulk of Moine pelitic gneiss some 90 km to the north across the Great Glen, are the biggest tracts of the *Carex-Racomitrium* heath that we have in Britain, disposed over the broad summits and spurs and having some intricate marginal mosaics with the *Calluna-Cladonia* heath, fragments of each community being included within a ground of the other in the boundary zone (McVean & Ratcliffe 1962). Then, moving just 25 km to the west, to Beinn Dearg in Ross-shire, there is a decisive shift to a much more oceanic pattern. This massif is unusual among the mountains of the north-west Highlands in that its Moine granulites and schists have been worn down to the kind of broad domed structure so typical of the Cairngorms but here, on the more exposed spurs and summits, *R. lanuginosum* has a strong edge over the lichens even among the dwarf-shrub heaths. Below the *Carex-Racomitrium* zone, then, the *Calluna-Cladonia* heath is still to be found but it is increasingly replaced through the low-alpine zone in this part of Scotland by the *Calluna-Racomitrium* heath in which it is mosses that are co-dominant among the dwarfed sub-shrub mat. And, with the move downslope into the forest zone, the *Calluna-Vaccinium* heath typical of the eastern part of the country is often replaced by the *Calluna-Vaccinium-Sphagnum* heath or, closer to the coastal fringe, the *Calluna-Erica* heath. As in the more continental sequence, bearberry heath can figure where the soils have moder humus rather than mor, but here it is *Arctostaphylos alpinus* rather than *A. uva-ursi* which is characteristic of the higher-altitude complexes of dwarfed sub-shrub heaths. On Beinn Dearg in Ross and Ben Wyvis, the *Calluna-Cladonia* and *Calluna-Racomitrium* heaths can both be found in patchworks with the *Calluna-A. alpinus* heath, their distributions governed by what are probably quite subtle differences in microclimate and soils.

In the hills fringing the heart of its range, through Speyside and over some of the central Grampians, the characteristic zonations have been more drastically affected below with the sub-montane heath zone often largely occupied by grasslands derived by long-continued pasturing, and the impact of this sometimes penetrates on to the higher exposed ground where the *Calluna-Cladonia* heath is typical. Burns, too, can stray upwards though the diminutive canopy of the community does not readily take fire. For the most part, though, such treatments have little effect and this kind of heath can be considered a climax where progression to scrub and woodland is prevented by locally-severe exposure.

Distribution
The community is most widespread and abundant through the east-central Highlands of Scotland, thinning out westwards into the central Grampians and north-west Highlands where it is progressively replaced by its oceanic counterpart, the *Calluna-Racomitrium* heath, and having but very few fragmentary localities further south into northern England and Wales. The *Cladonia* sub-community extends the distribution to lower altitudes, the *Empetrum-Cetraria* to the higher.

Affinities
Although Smith (1911*a*, *b*) and Crampton (1911) both hinted at the distinctive character of high-altitude heather vegetation in Scotland, it was not until the account of Metcalfe (1950) that the structure and floristics of the *Calluna-Cladonia* heath received detailed attention, continuing upwards the Cairngorm zonation first described by Watt & Jones (1948) and then subsequently modified in the light of further investigations over higher ground still (Burges 1951, Ingram 1958) by Poore & McVean (1957). The diagnoses of both Metcalfe's (1950) Mountain *Callunetum* and, less so, Poore & McVean's (1957) lichen heaths were broader than the *Calluna-Cladonia* heath as defined here, which is essentially an expanded version of McVean & Ratcliffe's (1962) *Cladineto-Callunetum*, distinguished from both the *Calluna-Vaccinium* heaths that extend on to higher ground and from the *Vaccinium-Cladonia* heath, by the balance between the different sub-shrubs and in the representation of lichens. The recent work of Birse (1980; see also Birse & Robertson 1976) has been especially helpful in extending our knowledge of what is here called the *Cladonia-Ochrolechia* sub-community, though Birse's (1980) *Alectoria-Callunetum* takes in a little of what would in this scheme be separated off into the *Calluna-A. alpinus* heath. The distinctions between these various dwarfed lichen-rich heaths of the low-alpine zone are, in fact, sometimes difficult to discern and have to be based on the proportions of the different components in critical cases. But, as a group, they are worthy of separation from the sub-montane and largely anthropogenic heaths of the Calluno-Ulicetalia and represent the British equivalent of the Scandinavian communities placed in the Loisleurieto-Arctostaphylion (Nordhagen 1928, Kalliola 1939) or Arctostaphyleto-Cetrarion (Dahl 1956), alliances of chionophobous dwarf-shrub heaths and grass-heaths usually gathered into the Caricetea curvulae (Braun-Blanquet 1948), though separated off into a Loiseleurio-Vaccinetea by some (see, for example, Birse 1980, 1984). *Cladonia alpestris*, a species perhaps extinct in Britain (Dahl 1968), is very characteristic of the more Continental lichen-rich heaths of this kind in Norway (Dahl 1956), while in the Sylene district, a more oceanic part of Scandinavia, Nordhagen (1928) reported a *Cladonia silvatica-rangiferina*-reiche *Calluna* Assoziation similar to some stands of the *Calluna-Cladonia* heath (McVean & Ratcliffe 1962).

Floristic table H13

	a	b	c	13
Calluna vulgaris	V (4–10)	V (1–10)	V (6–10)	V (1–10)
Cladonia arbuscula	V (1–4)	V (1–4)	V (1–8)	V (1–8)
Cetraria islandica	V (1–4)	V (1–4)	V (1–4)	V (1–4)
Cladonia uncialis	V (1–4)	V (1–4)	V (1–4)	V (1–4)
Cornicularia aculeata	V (1–4)	V (1–6)	IV (1–6)	V (1–6)
Racomitrium lanuginosum	V (1–4)	IV (1–4)	V (1–4)	V (1–4)
Cladonia rangiferina	V (1–4)	V (1–4)	IV (1–4)	V (1–4)
Empetrum nigrum hermaphroditum	III (1–8)	V (1–6)	IV (1–6)	IV (1–8)
Alectoria nigricans	IV (1–4)	III (1–6)	IV (1–6)	IV (1–6)
Scirpus cespitosus	II (1–3)	I (1–3)		I (1–3)
Cladonia impexa	II (1–4)	I (1–4)		I (1–4)
Agrostis canina	II (1–3)	I (1–4)		I (1–4)
Potentilla erecta	II (1–2)			I (1–2)
Erica cinerea	II (1–3)			I (1–3)
Molinia caerulea	II (1–3)			I (1–3)
Nardus stricta	I (1–4)			I (1–4)
Carex panicea	I (1–4)			I (1–4)
Festuca ovina	I (1–3)			I (1–3)
Campylopus paradoxus	I (1–3)			I (1–3)
Erica tetralix	I (1–3)			I (1–3)
Deschampsia flexuosa	II (1–3)	IV (1–4)	II (1–3)	III (1–4)
Cetraria nivalis	I (1–4)	IV (1–4)	II (1–4)	II (1–4)
Cladonia gracilis	I (1–3)	III (1–3)		II (1–3)
Juncus trifidus		II (1–4)	I (1–3)	I (1–4)
Diphasium alpinum	I (1–3)	II (1–3)		I (1–3)
Cladonia bellidiflora	I (1–3)	II (1–3)		I (1–3)
Alchemilla alpina		I (1–6)		I (1–6)
Vaccinium myrtillus	II (1–4)	III (1–4)	IV (1–4)	III (1–4)
Cladonia crispata	I (1–3)	II (1–4)	V (1–4)	II (1–4)
Ochrolechia frigida	II (1–4)	II (1–4)	IV (1–6)	II (1–6)
Loiseleuria procumbens	I (1)	II (1–6)	III (1–4)	II (1–6)
Cladonia coccifera	I (1–3)	I (1–3)	III (1–3)	I (1–3)
Thamnolia vermicularis	I (1–3)		III (1–3)	I (1–3)
Cladonia squamosa	I (1–3)	I (1–3)	II (1–4)	I (1–4)
Cladonia furcata			II (1–3)	I (1–3)
Vaccinium vitis-idaea	II (1–4)	III (1–4)	III (1–4)	III (1–4)
Empetrum nigrum nigrum	III (1–6)	I (3)	III (1–6)	III (1–6)
Carex bigelowii	I (1)	III (1–4)	III (1–3)	III (1–4)
Huperzia selago	II (1–3)	II (1–3)	III (1–4)	II (1–4)
Hypnum jutlandicum	II (1–4)	II (1–3)	II (1–4)	II (1–4)
Dicranum scoparium	II (1–3)	II (1–3)	II (1–3)	II (1–3)
Arctostaphylos uva-ursi	II (1–6)	II (1–6)	I (4)	II (1–6)
Sphaerophorus globosus	II (1–3)	II (1–4)	I (1–3)	II (1–4)
Cetraria glauca	II (1–4)	I (1–4)	II (1–4)	II (1–4)

Floristic table H13 *(cont.)*

	a	b	c	13
Ptilidium ciliare	I (1–3)	II (1–3)	II (1–3)	II (1–3)
Alectoria ochroleuca	I (1–3)	I (1–3)	I (1)	I (1–3)
Pleurozium schreberi	I (1–6)	I (1–4)	I (1)	I (1–6)
Dicranum fuscescens	I (1–3)	I (1–3)	I (1–3)	I (1–3)
Diplophyllum albicans	I (1)	I (1–3)	I (1)	I (1–3)
Pohlia nutans	I (1)	I (1)	I (1–3)	I (1–3)
Hylocomium splendens	I (4)	I (1–3)		I (1–4)
Rhytidiadelphus loreus	I (1)	I (1–3)		I (1–3)
Galium saxatile	I (1–3)	I (1–3)		I (1–3)
Antennaria dioica	I (1)	I (1)		I (1)
Sphaerophorus fragilis	I (1–3)		I (1–3)	I (1–3)
Polytrichum commune	I (1–3)		I (1–4)	I (1–4)
Hypogymnia physodes	I (1–3)		I (1–4)	I (1–4)
Carex pilulifera	I (1–3)		I (1–3)	I (1–3)
Alectoria sarmentosa		I (1)	I (1–3)	I (1–3)
Nardia scalaris		I (1–3)	I (1)	I (1–3)
Polytrichum alpinum		I (1–3)	I (1)	I (1–3)
Salix herbacea		I (1–3)	I (1–4)	I (1–4)
Number of samples	35	52	36	123
Number of species/sample	17 (9–30)	18 (10–28)	17 (12–22)	17 (9–30)
Vegetation height (cm)	5 (2–12)	5 (2–15)	3 (2–12)	4 (2–15)
Shrub/herb layer cover (%)	73 (20–90)	69 (50–90)	71 (60–90)	70 (20–90)
Ground layer cover (%)	64 (40–85)	71 (50–98)	57 (20–85)	63 (20–98)
Altitude (m)	544 (105–885)	805 (640–950)	737 (625–922)	683 (105–950)
Slope (°)	5 (0–18)	9 (0–25)	5 (0–25)	7 (0–25)

a   *Cladonia arbuscula-Cladonia rangiferina* sub-community
b   *Empetrum nigrum hermaphroditum-Cetraria nivalis* sub-community
c   *Cladonia crispata-Loiseleuria procumbens* sub-community
13 *Calluna vulgaris-Cladonia arbuscula* heath (total)

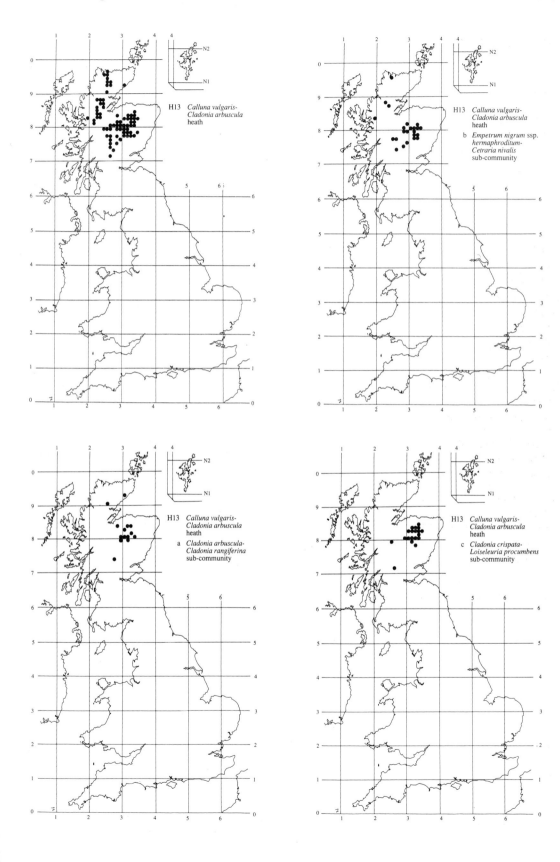

H13 *Calluna vulgaris-Cladonia arbuscula* heath

H13 *Calluna vulgaris-Cladonia arbuscula* heath
b *Empetrum nigrum* ssp. *hermaphroditum-Cetraria nivalis* sub-community

H13 *Calluna vulgaris-Cladonia arbuscula* heath
a *Cladonia arbuscula-Cladonia rangiferina* sub-community

H13 *Calluna vulgaris-Cladonia arbuscula* heath
c *Cladonia crispata-Loiseleuria procumbens* sub-community

H14
Calluna vulgaris-Racomitrium lanuginosum heath

Synonymy

Rhacomitreto-Callunetum McVean & Ratcliffe 1962, Birks 1973; Dwarf mountain heaths Gimingham 1972 *p.p.*; *Calluna vulgaris-Arctostaphylos uva-ursi* nodum Prentice & Prentice 1975 *p.p.*; *Alectorio-Callunetum*, *Agrostis canina* ssp. *montana* subassociation Birse 1980 *p.p.*

Constant species

Calluna vulgaris, Deschampsia flexuosa, Hypnum cupressiforme s.l., Racomitrium lanuginosum, Cladonia arbuscula, C. uncialis.

Rare species

Arctostaphylos alpinus, A. uva-ursi, Loiseleuria procumbens.

Physiognomy

The *Calluna vulgaris-Racomitrium lanuginosum* heath consists essentially of a dwarfed sub-shrub mat, with *Calluna vulgaris* usually predominating as in the *Calluna-Cladonia* heath, but with *Racomitrium lanuginosum* occupying the prominent role which lichens have in that community. As there, the cover of woody plants is typically far from complete, with the heather, severely wind-trimmed to prostrate bushes just a few centimetres thick, often occupying less than half of the ground. Other sub-shrubs generally play a subordinate part, but a number are quite common, sometimes showing local abundance, partly replacing *Calluna* or supplementing its cover, and helping to distinguish the different sub-communities. The most frequent of these woody associates throughout is the Arctic-Alpine *Empetrum nigrum* and, again as in the *Calluna-Cladonia* heath, the two sub-species tend to characterise opposite extremes of the altitudinal range of this community, with ssp. *nigrum* being preferentially common and moderately abundant towards lower levels, ssp. *hermaphroditum* largely confined to the higher. But *Erica cinerea* is frequent too,

much more so than in the continental *Calluna-Cladonia* heath and especially at lower altitudes where, with *Arctostaphylos uva-ursi* and *A. alpinus*, it helps give the *Arctostaphylos* sub-community its distinctive character. Both species of bearberry can show quite high covers here and this kind of heath includes much of the vegetation in north-west Scotland where these two rare Arctic-Alpines occur together. Another rarity occasionally represented in the community is *Loiseleuria procumbens*, and scattered throughout are scarce records for *Juniperus communis*, generally clearly of ssp. *nana. Vaccinium myrtillus* is never more than occasional and typically present as sparse shoots, and *V. vitis-idaea* and *V. uliginosum* are rare.

Among the open patchwork of flattened bushes, other vascular associates are few and far between, with usually scattered and stunted individuals, but this element of the flora can be a little more varied than in the *Calluna-Cladonia* heath. As there, *Deschampsia flexuosa* and *Huperzia selago* can be frequent and *Carex bigelowii* becomes common at higher altitudes, but *Scirpus cespitosus* is found more often and there is frequently some *Carex pilulifera* and *Potentilla erecta*. Then, there can be some *Solidago virgaurea*, in its very diminutive form sometimes called var. *cambrica*, and tiny rosettes of *Succisa pratensis*, and in certain stands grasses can make a modest contribution to the cover.

Much more noticeable than these, however, and helping mark off this community not just from the *Calluna-Cladonia* heath but from other dwarfed sub-shrub mats, is the abundance of *Racomitrium lanuginosum*, typically forming an extensive woolly carpet up to 5–10 cm thick between the sub-shrubs and among their barer branches. Rarely it is the only bryophyte, but *Hypnum cupressiforme s.l.* is also very frequent and it, too, can show some local abundance. Then, in some stands, there can be patches of *Dicranum scoparium, Pleurozium schreberi, Hylocomium splendens* and *Rhytidiadelphus loreus* with, less commonly, scattered plants of *Campylopus paradoxus* and *Sphagnum capillifolium*.

A notable feature of sheltered crevices and lower altitudes is the occasional occurrence of oceanic hepatics such as *Diplophyllum albicans*, *Scapania gracilis* and *Frullania tamarisci*.

Among this mat, lichens are common and sometimes quite varied, but they do not usually occur with any great abundance and species such as *Cetraria nivalis* and *Alectoria ochroleuca*, characteristic of the bleakest ridges and summits in the *Calluna-Cladonia* heath, are characteristically absent. *Cladonia arbuscula* and *C. uncialis* are the most frequent members of this element and each can be found at moderately high cover. *Sphaerophorus globosus* and *Cornicularia aculeata* are also common throughout, with *Cladonia impexa* making a frequent appearance at lower altitudes, *C. gracilis*, *C. bellidiflora*, *Cetraria islandica* and *Ochrolechia frigida* occurring often at higher levels. *Cladonia coccifera* and *C. rangiferina* are very occasional.

Sub-communities

Festuca ovina sub-community: *Rhacomitreto-Callunetum* Birks 1973 *p.p.*; *Alectorio-Callunetum*, *Agrostis canina* ssp. *montana* subassociation Birse 1980 *p.p.* *Calluna* or *R. lanuginosum* or mixtures of the two dominate the vegetation mat here with other plants usually relegated to a minor role. *Empetrum nigrum*, both ssp. *nigrum* and ssp. *hermaphroditum*, and *Erica cinerea* occur occasionally, though hardly ever in any abundance, and in some stands *Loiseleuria* can be quite prominent, though *Arctostaphylos* spp. are normally absent. More striking here is the variety of herbaceous associates. *Deschampsia flexuosa* is actually rather less common than usual but, along with frequent *Carex bigelowii*, there is very often some *C. pilulifera*, *Huperzia selago* and *Potentilla erecta*. More strongly preferential still is *Festuca ovina* (including *F. vivipara*) with *Agrostis canina* (mostly recorded as ssp. *montana*), and *Antennaria dioica* also quite frequent. Less common, but also diagnostic, are *Carex panicea*, *Thymus praecox* and *Salix herbacea* and, on Skye, this vegetation provides a locus for the diminutive Northern Montane orchid *Pseudorchis albida*. *Polytrichum piliferum* occurs occasionally and, again on Skye, Birks (1973) recorded *Racomitrium fasciculare*, *R. heterostichum*, *Andreaea rothii* and *Campylopus atrovirens* in some stands. Lichen cover tends to be lower than in other kinds of *Calluna-Racomitrium* heath but *C. uncialis* is very common and there is occasionally some *Sphaerophorus globosus*, *Cornicularia aculeata* and *Cladonia arbuscula*.

Empetrum nigrum ssp. hermaphroditum sub-community: *Rhacomitreto-Callunetum*, *empetrosum* facies McVean & Ratcliffe 1962. *Calluna* is generally the most abundant plant in this sub-community with *R. lanugino-*

sum sometimes co-dominant but quite often subordinate in cover. *Empetrum nigrum* ssp. *hermaphroditum* is very common, too, and it can be fairly extensive in the mat, occasionally with *Erica cinerea*. The variety of herbs typical of the *Festuca* sub-community is not found here, though the community species *Deschampsia flexuosa* and *Carex bigelowii* are very frequent and *Potentilla erecta*, *Huperzia selago* and *Carex pilulifera* remain quite common. *Nardus stricta* is also often found, though not in any abundance, and *Diphasium alpinum* is strongly preferential. But most distinctive here are the associated cryptogams. In addition to *R. lanuginosum*, there are occasional records for *Dicranum scoparium*, *Pleurozium schreberi*, *Rhytidiadelphus loreus* and *Hylocomium splendens* and the lichen flora typical of the community is supplemented by a variety of species. *Cladonia gracilis* and *C. bellidiflora* occur commonly among the *C. uncialis* and *C. arbuscula* and the combined cover of these bulkier species can be quite considerable. Then, there is very often some *Cetraria islandica* and *Ochrolechia frigida*, with occasional *Cladonia pyxidata*, *C. coccifera* and *Cetraria glauca*.

Arctostaphylos uva-ursi sub-community: *Rhacomitreto-Callunetum*, *arctostaphyletosum* facies McVean & Ratcliffe 1962; *Rhacomitreto-Callunetum* Birks 1973 *p.p.*; *Calluna vulgaris-Arctostaphylos uva-ursi* nodum Prentice & Prentice 1975. *Calluna* and *R. lanuginosum* retain the kind of contribution here that they show in the previous sub-community, with the former generally the more abundant, the latter varying from somewhat sparse to predominant. But the sub-shrub mat is considerably more varied, with *Empetrum nigrum* ssp. *nigrum* and *Erica cinerea* showing their peak of frequency in this kind of *Calluna-Racomitrium* heath with, more strikingly preferential, *Arctostaphylos uva-ursi* and, rather less common but likewise locally abundant, *A. alpinus*. Among the vascular associates, *Scirpus cespitosus*, *Molinia caerulea* and *Carex binervis* become quite common, along with *Deschampsia flexuosa*, *Potentilla erecta* and *Carex pilulifera*, but *C. bigelowii* is very scarce. There are very occasional records for *Dactylorhiza maculata*, *Dryopteris abbreviata* and *Hymenophyllum wilsonii*. In the bryophyte mat, *Dicranum scoparium* and larger pleurocarps such as *Hypnum cupressiforme s.l.* and *Pleurozium schreberi* make their maximum contributions with *Rhytidiadelphus loreus* and *Hylocomium splendens* occasional. Hepatics, too, can be frequent with *Scapania gracilis*, *Diplophyllum albicans* and *Frullania tamarisci* strongly preferential and giving a particularly distinctive feel when occurring together in more sheltered stands included here. The lichen flora has some peculiarities too, with *Cladonia impexa* common among the frequent *C. arbuscula*, *C. uncialis* and *Sphaerophorus globosus*.

Habitat

The *Calluna-Racomitrium* heath is the typical sub-shrub community of base-poor soils on windswept plateaus and ridges at moderate to fairly high altitudes in the cool oceanic climate of the mountains of north-west Scotland. It can be grazed by sheep and deer, but this is probably not ultimately important in maintaining the characteristic composition and physiognomy and this vegetation can be regarded as the natural climax in such exposed situations in this part of Britain.

Like the *Calluna-Cladonia* heath, the counterpart of this community in the eastern Highlands, this is essentially a vegetation type of the low-alpine zone (Poore & McVean 1957), running up to the limit of heather dominance but also including frequent records for a variety of Arctic-Alpine plants tolerant of the generally cold montane climate: *Empetrum nigrum*, *Arctostaphylos uva-ursi*, *A. alpinus*, *Loiseleuria* and *Carex bigelowii*. However, with the shift towards the north-west of Scotland, where summer temperatures at identical altitudes become progressively cooler, the actual upper altitudinal limit of this general kind of vegetation drops (Poore & McVean 1957, McVean & Ratcliffe 1962). Thus, although the *Calluna-Racomitrium* heath can be found up to the same levels as the *Calluna-Cladonia* heath, around 1000 m, this is very exceptional and usually confined to the eastern limits of its range, where the community has a few scattered stations in the central Grampians. For the most part, the upper reaches of this kind of vegetation are around 750 m and, particularly towards the seaward side of the north-west Highlands, within which range of mountains it has its centre of distribution, and on those islands where it is represented, such as Skye, Orkney and Shetland, it can extend down to 250 m or less.

Through this zone in this part of Britain, mean annual maximum temperatures are between 21 and 23 °C (Conolly & Dahl 1970), very much as over the upper reaches of the eastern Highlands. But, particularly towards lower altitudes, the winter temperatures through the north-west Highlands are noticeably milder than to the east, such that the annual fluctuations are considerably reduced and the growing season at equivalent altitudes lengthened. Also very important is the fact that the precipitation is very much higher than in eastern Scotland, with well over 1600 mm annually through much of the range of the community (*Climatological Atlas* 1952) and, more particularly, over 220 wet days yr^{-1} (Ratcliffe 1968). The potential water deficit over most of the region is thus at or very near to zero, with high relative humidity throughout the year and, on the upper slopes, a high percentage of daytime cloudiness (*Climatological Atlas* 1952, Ratcliffe 1968, Chandler & Gregory 1976, Page 1982).

It is this combination of a relatively equable temperature regime and a more or less constantly humid atmosphere that characterises the regional climate of the *Calluna-Racomitrium* heath and determines its general composition, marking out the geographical and floristic boundaries with the *Calluna-Cladonia* heath. By and large, the eastern limits of the former run from Ben Klibrech in Sutherland, down to Ben Wyvis and then to Ben Alder in Inverness, although most of the stands are in fact to the north of the Great Glen (McVean & Ratcliffe 1962). The *Calluna-Cladonia* heath can be found west of this line but, increasingly with the shift into the region of more oceanic climate, the important role of the large *Cladonia* and *Cetraria* spp. among the sub-shrub mat is occluded by the greater competitive power of *R. lanuginosum* and, to a lesser extent, of other bulky bryophytes like *Hypnum cupressiforme s.l.*, *Dicranum scoparium* and *Pleurozium schreberi*, such that the general character of the low-alpine heather-dominated vegetation in this part of Britain becomes that of a 'moss-heath' rather than a 'lichen-heath'. But intermediates between the extreme types are common and, in the area of geographical overlap of their ranges, the large proportion of the flora shared by both kinds of heath is very obvious and separation of stands is sometimes a matter of the pattern of dominance among the cryptogam element of the flora.

Very often, though, the more oceanic character of the *Calluna-Racomitrium* heath is confirmed by the occurrence, too, of *Erica cinerea* and *Carex pilulifera*, plants of very restricted occurrence in the *Calluna-Cladonia* heath. Here, these find their best representation in the *Arctostaphylos* sub-community which takes the *Calluna-Racomitrium* heath to its lowest altitudes and where the most sheltered extremes colonised by the community are reflected in the occasional occurrence of *Molinia*, the Oceanic West European *Carex binervis*, the Atlantic ferns *Hymenophyllum wilsonii* and *Dryopteris abbreviata* (Page 1982), and oceanic hepatics like *Diplophyllum albicans*, *Scapania gracilis*, *Frullania tamarisci* and, more rarely, *Pleurozia purpurea*, *Anastrepta orcadensis* and *Plagiochila spinulosa*. This kind of vegetation brings the community close to some types of *Calluna-Erica* heath, the sub-shrub vegetation which often replaces it in the sub-montane zone.

At the opposite extreme is the *Empetrum* sub-community which extends up to the highest levels occupied by the *Calluna-Racomitrium* heath, with a mean altitude 400 m above that of the *Arctostaphylos* sub-community. Here, ssp. *hermaphroditum* becomes the dominant crowberry taxon, with the preferential occurrence of the montane lichens *Cetraria islandica* and *Ochrolechia frigida* also reflecting the harsher conditions. It is here that the community becomes most similar to the *Calluna-Cladonia* heath, although lichens such as *Cetraria nivalis* and *Alectoria ochroleuca*, which are characteristic of the bleakest situations among the continental heaths, are still absent.

As with the *Calluna-Cladonia* heath, physiography can locally enhance or ameliorate the kind of exposure which favours the optimal development of this vegetation. Generally, it is found over similar gentle to moderately steep slopes as its eastern counterpart, open to the blast of fairly constant and strong winds and blown clear of snow that might provide some shelter in the coldest months. Where spurs or ridges provide locally exposed situations at lower altitudes, where gales can still be frequent and strong in this part of Scotland (Shellard 1976), then the *Calluna-Racomitrium* heath can extend down further than usual whereas, with some degree of shelter higher up within the low-alpine zone, it may be replaced by more chionophilous vegetation where shallow snow can settle.

With the general shift to lower altitudes of the dwarfed sub-shrub heaths in the north-west Highlands, the *Calluna-Racomitrium* heath is excluded from many mountain tops in this part of the country, being more characteristic of plateaus and ridges below the highest summits. In such situations, the moraines which are very widespread over gentler slopes often provide a suitable substrate for the development of the base-poor rankers and podzolic soils typical here, but the community is also frequently found over drift-free granulites and schists of the Moine series, Torridonian sandstone, Lewisian gneiss, the Tertiary basalt of Skye and Devonian Old Red Sandstone on Orkney. The profiles seem generally similar to those of the *Calluna-Cladonia* heath, with a superficial pH of between 4 and 5, but are perhaps more consistently humic above, sometimes amounting to little more than accumulations of raw organic matter over rock fragments.

In the *Festuca* sub-community, with its preferentially frequent *Carex pilulifera*, *Festuca ovina*, *Agrostis canina* ssp. *montana* and occasional *Thymus praecox* and *Carex panicea*, it seems likely that the *Calluna-Racomitrium* heath extends some way on to less humic and perhaps less base-poor soils where there may be some mild influence of calcareous bedrocks, like some mica-schists. But it seems possible, too, that this vegetation is more influenced by grazing than the other sub-communities, although even here the effects are not pronounced. For, in general, this is a climax vegetation type, the essential composition and structure of which are controlled by climate.

Zonation and succession

The *Calluna-Racomitrium* heath characteristically occurs in zonations and mosaics with other sub-shrub vegetation and montane heaths, snow-bed communities and mires where variation is mainly influenced by differences in climate and soils with altitude and local topography. Shifts on to more base-rich profiles derived from less siliceous parent materials can complicate these patterns and treatments have sometimes affected the lower reaches of zonations. The move south-eastwards sees a gradual geographical replacement of the more oceanic vegetation types in these sequences by their continental equivalents and, in the transition zone, intermediate patterns can be found.

The typical altitudinal zonation in which the *Calluna-Racomitrium* heath occurs is well seen in the north-west Highlands over the slopes of Ben More Assynt and Foinaven (McVean & Ratcliffe 1962, Ratcliffe 1977). Here, the community is found on broad, windswept ridges cut into Cambrian quartzite, generally passing downslope into some other kind of sub-shrub vegetation. Locally, in more sheltered situations at lower altitudes, but where the soils remain relatively free-draining, the *Calluna-Racomitrium* heath is replaced by the *Calluna-Erica* heath, the *Arctostaphylos* sub-community of the former and the *Racomitrium* sub-community of the latter representing an almost continuous transition. Generally, however, such a sequence is better seen in the more oceanic conditions found on Skye (Birks 1973) and Orkney (Prentice & Prentice 1975), where the *Calluna-Erica* heath is much more frequent. And, through the north-west Highlands, too, the move downslope often involves a further shift on to less well-drained ground where some sort of wet heath is the more natural development. Usually, in this part of the country, it is the *Scirpus-Erica* heath that occurs in this position in the zonation, where shallow peat is beginning to accumulate over gleyed podzolic soils on the gentle slopes of the low-altitude plateaus. Here, the major floristic change is the marked increase in frequency of *Scirpus cespitosus* and *Molinia* and the appearance of *Erica tetralix* but the pattern of dominance in this kind of vegetation is very varied and, in its *Cladonia* sub-community, the cryptogamic flora is very similar to that of the *Calluna-Racomitrium* heath and, again, the *Arctostaphylos* sub-community can form a gentle transition to it. The usual continuation of this kind of zonation is to *Scirpus-Eriophorum* blanket mire over the deeper and wetter peats of the waterlogged lower ground.

Even where there is some strong element of floristic continuity between the *Calluna-Racomitrium* heath and the vegetation which replaces it downslope, there is usually an obvious physiognomic change from a markedly dwarfed character to a taller, bushier cover in which the vascular associates can make more luxuriant growth. Sometimes, though, as on Foinaven, the community gives way below to prostrate vegetation in which *Juniperus communis* ssp. *nana* becomes very common and often co-dominant with the stunted heather but where such plants as *Erica cinerea*, *Arctosta-phylos uva-ursi*, *A. alpinus* and *Empetrum nigrum* remain quite frequent with usually sparse shoots of *Racomi-trium lanuginosum* and lichens. Such *Calluna-Juniperus* heath is, like the *Calluna-Racomitrium* heath, chiono-phobous but it is really a transitional community on the junction of the low-alpine and sub-alpine zones (Poore & McVean 1957) and it has fairly common *Pleurozia purpurea*, *Frullania tamarisci* and *Diplophyllum albicans*. It can be seen as the north-west Highland equivalent of the *Juniperus-Oxalis* scrub of the east-central High-lands, though there it is ssp. *communis* that colonises up to the limit of the forest zone.

One other community that can be found at levels which are generally below those of the *Calluna-Racomi-trium* heath, but which is also strongly dependent on cool and very humid conditions associated with shel-tered, sunless slopes, is the *Calluna-Vaccinium-Sphag-num* heath. Here, some measure of floristic continuity is provided by *R. lanuginosum* and the bulky pleurocarp associates, but the oceanic hepatic component of the *Calluna-Vaccinium-Sphagnum* heath is so distinctive and its luxuriance so pronounced that there is never any real doubt as to where the transition occurs and, typi-cally, it marks an obvious shift in aspect, from southerly or westerly slopes to those facing east or north. But both this vegetation and the *Calluna-Juniperus* heath can be badly damaged by burning so more diffuse mosaics can sometimes develop.

Moving upslope on to ground that remains equally exposed to high winds as that carrying the *Calluna-Racomitrium* heath, the community is usually replaced by some kind of *Carex-Racomitrium* heath, a more compendious community in this scheme than previously understood and taking in much of the summit moss-heath and some fell-field vegetation of the north-west Highlands. Again, the floristic transition can be a gra-dual one with *R. lanuginosum* in particular retaining high cover, indeed generally increasing its extent to overwhelming dominance of the ground carpet, and *Carex bigelowii*, *Deschampsia flexuosa*, *Diphasium alpi-num* and certain of the lichens continuing in frequency. But *Calluna* is typically unable to survive at the higher altitudes occupied by this vegetation and, in sites like Ben More Assynt and Foinaven, there is often addition-ally a suite of cushion herbs in the *Carex-Racomitrium* heath that gives the vegetation a very distinctive appear-ance. An alternative zonation over higher ground can be seen in parts of Foinaven where the *Calluna-Racomi-trium* heath passes on summit blanket peat to the *Calluna-Eriophorum* mire.

Two other complications to the basic pattern of communities deserve note. First, where there is some increase in shelter within the low-alpine zone and a little above it, such that a modest amount of snow can

accumulate during the winter in shallow hollows, among block scree and over lee slopes, the *Calluna-Racomitrium* heath can give way to the *Vaccinium-Racomitrium* heath. Many species run on into this vegetation, but *Calluna* itself becomes very patchy and dominance generally passes to *Vaccinium myrtillus* or *E. nigrum* ssp. *hermaphroditum* which can thus extend the abundant representation of sub-shrubs to 100 m or more above the upper limit of heather.

In other cases, a zone of the *Vaccinium-Racomitrium* or *Vaccinium-Rubus* heath forms a transitional belt around snow-beds proper, in cold and sheltered hollows among *Calluna-Racomitrium* heath. Such patterns are not well developed on Ben More Assynt and Foinaven, but they can be seen on Beinn Dearg where various types of *Nardus-Carex* vegetation mark out areas of late snow-lie over slopes cut into siliceous granulites, and also on the Lewisian gneiss of the Letterewe hills, though in this range the *Calluna-Racomitrium* heath itself is reduced to a rather fragmentary zone between *Scirpus-Eriophorum* blanket mire and the *Carex-Racomitrium* vegetation of the summits.

The other variation involves the occurrence of dwarfed sub-shrub heath in which *Arctostaphylos alpi-nus* is a consistent component, generally without *A. uva-ursi*, though often with *Loiseleuria* and various other of the *Calluna-Racomitrium* species. This, too, is a chiono-phobous vegetation type, characteristic of situations which are, if anything, even more exposed than those typical here and thus sometimes running on upslope from the *Calluna-Racomitrium* heath a little way over blasted brows. But *A. alpinus* is also associated with soils in which the humus is more like a moder than a mor, so there may be some subtle edaphic differences involved in patchworks of these vegetation types, such as can be seen on Foinaven and Ben Wyvis.

As far east as Ben Wyvis, however, the *Calluna-Racomitrium* heath has almost petered out as a compo-nent of the low-alpine zone which, with increasing continentality of climate, extends to higher altitudes and has the *Calluna-Cladonia* heath as the prevailing dwarfed sub-shrub vegetation. Indeed, this already figures along with the *Calluna-Racomitrium* heath in low-alpine mosaics on Beinn Dearg in Ross and over the Affric-Cannich hills but, beyond the line between Ben Klibreck to the north and Ben Alder in the south, the shift away from the oceanic pattern is visible at all levels in the altitudinal zonation.

Distribution

The *Calluna-Racomitrium* heath is very much a community of the north-west Highlands with scattered occurrences in the central Grampians. In this latter region and at its higher stations through the north-west, the *Empetrum* sub-community is the typical form, with the *Arctostaphylos* and *Festuca* types extending the range to lower levels and on to the western Isles, Orkney and Shetland.

Affinities

Although heath vegetation with sub-shrubs and abundant *R. lanuginosum* figures in some early descriptive accounts of Scottish vegetation (e.g. Smith 1911*a*), such categories were rather compendious and centred mainly on the higher-altitude moss-heath and fell-field vegetation, most of which falls into the *Carex-Racomitrium* heath. Again, it was not until the studies of McVean & Ratcliffe (1962) that this community was clearly distinguished from that kind of heath and from the lichen-rich vegetation of the continental *Calluna-Cladonia* heath. Subsequent work has further clarified their

suggestion of different facies and added what is here defined as the *Festuca* sub-community (Birks 1973; Birse 1980, although he includes this as part of his *Alectorio-Callunetum*, along with much *Calluna-Cladonia* heath).

McVean & Ratcliffe (1962), Birks (1973) and Birse (1980) all agree in including this vegetation in the Arctostaphyleto-Cetrarion or Loiseleurio-Arctostaphylion, although, in the *Arctostaphylos* sub-community, it is difficult to draw a sharp dividing line between this vegetation and the sub-montane Calluno-Ulicetalia community, the *Calluna-Erica* heath. Corresponding associations to the *Calluna-Racomitrium* heath have not been described from the sub-Arctic or Scandinavia and this is very much a vegetation type of the extreme oceanic fringe of the European low-alpine zone.

Floristic table H14

	a	b	c	14
Calluna vulgaris	V (1–8)	V (4–8)	V (4–8)	V (1–8)
Racomitrium lanuginosum	V (1–10)	V (1–6)	V (1–10)	V (1–10)
Cladonia uncialis	V (1–4)	V (1–6)	IV (1–4)	V (1–6)
Hypnum cupressiforme s.l.	IV (1–4)	III (1–4)	IV (1–4)	IV (1–4)
Cladonia arbuscula	II (1–3)	V (1–6)	IV (1–4)	IV (1–6)
Deschampsia flexuosa	II (1–4)	V (1–4)	IV (1–4)	IV (1–4)
Potentilla erecta	V (1–4)	III (1–3)	III (1–2)	III (1–4)
Huperzia selago	IV (1–3)	III (1–3)	II (1–3)	III (1–3)
Carex pilulifera	IV (1–4)	III (1–3)	III (1–3)	III (1–4)
Festuca ovina/vivipara	IV (1–5)	I (1–3)	II (1–4)	II (1–5)
Antennaria dioica	III (1–3)	II (1–3)	II (1–3)	II (1–3)
Agrostis canina	III (1–4)	I (1–3)	II (1–3)	II (1–3)
Alectoria nigricans	II (1–3)	I (1–3)		I (1–3)
Polytrichum piliferum	II (1–3)	I (1–3)		I (1–3)
Carex panicea	II (1–3)			I (1–3)
Euphrasia micrantha	II (1–3)			I (1–3)
Thymus praecox	II (2–3)			I (2–3)
Racomitrium fasciculare	I (1–5)			I (1–5)
Andreaea rothii	I (1–3)			I (1–3)
Campylopus atrovirens	I (1–3)			I (1–3)
Cetraria islandica	II (1–3)	V (1–3)	II (1–3)	III (1–3)
Empetrum nigrum hermaphroditum	II (1–4)	V (1–4)	I (1–3)	III (1–4)
Cladonia gracilis	I (1–3)	IV (1–3)	II (1)	II (1–3)
Ochrolechia frigida	II (1–3)	III (1–3)	I (1)	II (1–3)
Nardus stricta	II (1–4)	III (1–6)	I (1–3)	II (1–6)
Diphasium alpinum	I (1–3)	III (1–4)		II (1–4)
Cladonia bellidiflora	I (1–3)	III (1–3)		II (1–3)
Cladonia pyxidata		II (1–3)	I (1)	I (1–3)
Ptilidium ciliare		II (1–3)		I (1–3)
Cetraria glauca		II (1)		I (1)

Floristic table H14 *(cont.)*

	a	b	c	14
Erica cinerea	II (1–3)	III (1–4)	IV (1–4)	III (1–4)
Arctostaphylos uva-ursi			V (1–6)	III (1–6)
Dicranum scoparium	I (1–3)	II (1–3)	IV (1–3)	III (1–3)
Scirpus cespitosus	II (1–3)	II (1–4)	III (1–3)	III (1–4)
Pleurozium schreberi	I (1–3)	II (1–3)	III (1–4)	III (1–4)
Empetrum nigrum nigrum	II (1–3)	I (1)	III (1–4)	III (1–4)
Diplophyllum albicans	II (1–3)	I (1–3)	III (1–3)	II (1–3)
Cladonia impexa	I (1–3)	I (1–3)	III (1–6)	II (1–6)
Arctostaphylos alpinus	I (1–3)		III (1–7)	II (1–7)
Frullania tamarisci		I (1–3)	III (1–4)	II (1–4)
Scapania gracilis	I (1–3)	I (1)	II (1–3)	I (1–3)
Molinia caerulea	I (1–3)	I (1–3)	II (1–3)	I (1–3)
Carex binervis	I (1)		II (1–4)	I (1–4)
Dactylorhiza maculata		I (1)	II (1–3)	I (1–3)
Hymenophyllum wilsonii			I (1–2)	I (1–2)
Dryopteris abbreviata			I (2)	I (2)
Sphaerophorus globosus	III (1–3)	III (1–3)	III (1–3)	III (1–3)
Cornicularia aculeata	II (1–3)	III (1–3)	II (1–3)	III (1–3)
Carex bigelowii	III (1–4)	III (1–3)	I (1–4)	III (1–4)
Vaccinium myrtillus	II (1–3)	II (1–4)	II (1–4)	II (1–4)
Cladonia coccifera	II (1–3)	II (1–3)	I (1–3)	II (1–3)
Solidago virgaurea	II (1–4)	II (1–3)	I (1–3)	II (1–4)
Rhytidiadelphus loreus	I (1–3)	II (1–3)	II (1–3)	II (1–3)
Hylocomium splendens	I (1–3)	II (1–3)	II (1–3)	II (1–3)
Cladonia rangiferina	I (1–3)	I (1–3)	I (1–3)	I (1–3)
Campylopus paradoxus	I (1–3)	I (1–3)	I (1–3)	I (1–3)
Succisa pratensis	I (1–3)	I (1–2)	I (1–3)	I (1–3)
Juniperus communis	I (1–3)	I (1–3)	I (4–7)	I (1–7)
Alchemilla alpina	I (1–3)	I (1–3)		I (1–3)
Sphagnum capillifolium	I (1–3)	I (1–3)		I (1–3)
Polygala serpyllifolia	I (1–3)	I (1–3)		I (1–3)
Galium saxatile	I (1–3)	I (1–3)		I (1–3)
Agrostis capillaris	I (1–3)	I (1–3)		I (1–3)
Juncus squarrosus	I (1–6)	I (1–3)		I (1–6)
Luzula multiflora	I (1–3)		I (1–3)	I (1–3)
Hypogymnia physodes		I (1–3)	I (1–3)	I (1–3)
Number of samples	21	17	23	61
Number of species/sample	20 (14–30)	21 (13–31)	22 (12–35)	21 (12–35)
Vegetation height (cm)	2 (1–2)	6 (2–20)	8 (3–20)	6 (1–20)
Shrub/herb cover (%)	70 (40–100)	93 (50–100)	88 (60–100)	83 (40–100)
Ground layer cover (%)	30 (20–40)	75	35 (5–40)	44 (5–75)
Altitude (m)	408 (185–615)	670 (457–1067)	259 (31–795)	429 (31–1067)
Slope (°)	5 (1–10)	13 (0–30)	16 (0–45)	13 (0–45)

a   *Festuca ovina* sub-community
b   *Empetrum nigrum hermaphroditum* sub-community
c   *Arctostaphylos uva-ursi* sub-community
14   *Calluna vulgaris-Racomitrium lanuginosum* heath (total)

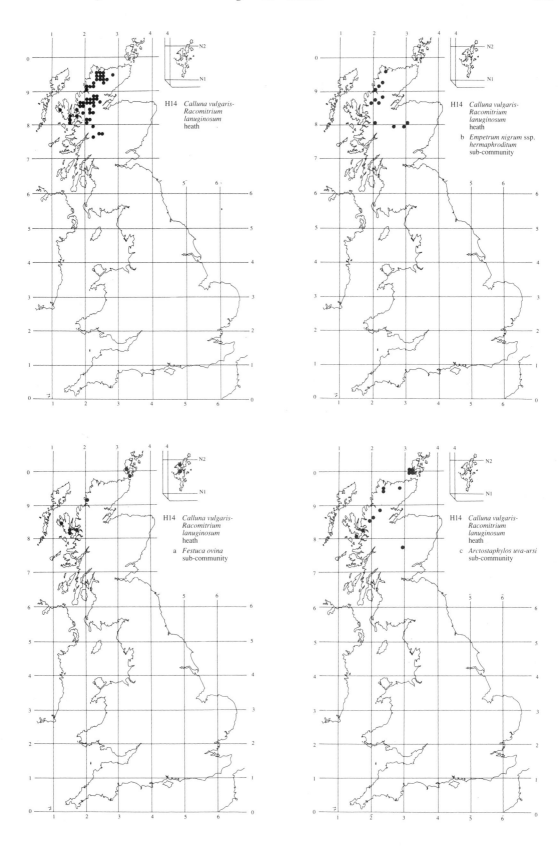

H14 *Calluna vulgaris-Racomitrium lanuginosum* heath

H14 *Calluna vulgaris-Racomitrium lanuginosum* heath

 b *Empetrum nigrum* ssp. *hermaphroditum* sub-community

H14 *Calluna vulgaris-Racomitrium lanuginosum* heath

 a *Festuca ovina* sub-community

H14 *Calluna vulgaris-Racomitrium lanuginosum* heath

 c *Arctostaphylos uva-ursi* sub-community

H15

Calluna vulgaris-Juniperus communis ssp. *nana* heath

Synonymy

Juniperus-Arctostaphylos sociation   Poore & McVean 1957; *Juniperetum nanae*   McVean & Ratcliffe 1962 *p.p.*; *Juniperus nana* nodum   Birks 1975 *p.p.*

Constant species

Calluna vulgaris, Deschampsia flexuosa, Erica cinerea, Juniperus communis ssp. *nana, Potentilla erecta, Scirpus cespitosus, Hypnum cupressiforme s.l., Racomitrium lanuginosum, Pleurozia purpurea, Cladonia uncialis.*

Rare species

Arctostaphylos alpinus, A. uva-ursi, Loiseleuria procumbens, Herbertus borealis, H. stramineus, Plagiochila carringtonii.

Physiognomy

Prostrate juniper referable to *Juniperus communis* ssp. *nana* occurs as an occasional, sometimes with modest local abundance, in a variety of dwarfed sub-shrub heaths in Britain, notably in the *Calluna-Racomitrium* and *Calluna-A. alpinus* types. But in the *Calluna-Juniperus* heath, it is more or less consistently dominant in the sub-shrub mat and accompanied by a small but distinctive element of oceanic hepatics.

The mat is generally less than 10 cm thick and often severely wind-pruned to much less than this, fairly continuous in the best developed stands, though more fragmentary over erosion surfaces and where there has been burning, to which the dwarf juniper is very sensitive (McVean 1961*b*, Poore & McVean 1957). A number of other sub-shrubs are well represented, with *Calluna vulgaris* and *Erica cinerea* being especially frequent, and the former often fairly abundant, though not generally assuming dominance. *Arctostaphylos uva-ursi* and *A. alpinus* are somewhat less common and usually of low cover and *Empetrum nigrum* spp. *hermaphroditum* occurs occasionally. *Vaccinium myrtillus* and *V. uliginosum* are scarce and the community provides an occasional locus for *Loiseleuria procumbens*.

Vascular associates are typically rather few in number and occur as scattered and often stunted individuals among the mat. *Deschampsia flexuosa, Scirpus cespitosus* and *Potentilla erecta* are constant, with *Huperzia selago, Solidago virgaurea* (in its so-called var. *cambrica*), *Dactylorhiza maculata, Polygala serpyllifolia, Succisa pratensis* and *Antennaria dioica* more occasional. *Carex panicea, C. pilulifera, C. bigelowii, Festuca vivipara* and *Nardus stricta* can also sometimes be found and, more unusually, there can be some *Thymus praecox*.

In some stands, the cryptogam flora seems little different from other kinds of dwarfed sub-shrub heath, as in some of the semi-degraded samples of Poore & McVean (1957) and of McVean & Ratcliffe's (1962) *lichenosum* and, indeed, such vegetation where the juniper cover is lower than usual is best shifted into other communities. In typical *Calluna-Juniperus* heath, however, the *Racomitrium lanuginosum, Cladonia uncialis, C. impexa, Sphaerophorus globosus* and *Cornicularia aculeata* common to these communities are frequently accompanied by *Pleurozia purpurea, Frullania tamarisci* and *Diplophyllum albicans*, species which are there scarce and of usually low cover. *P. purpurea* in particular is often quite abundant in this kind of heath and in stands on Beinn Eighe, where the community is very well developed, it can be accompanied by prominent tufts of the very rare *Herbertus borealis*, originally noted by McVean & Ratcliffe (1962) and still recorded only from this site and from western Norway (Ratcliffe 1977). Then, there can be occasional records for *Bazzania tricrenata, Scapania gracilis* and the rare *Plagiochila carringtonii*. *Hylocomium splendens, Dicranum scoparium, Pleurozium schreberi, Rhytidiadelphus loreus* and *Sphagnum capillifolium* also occur at low frequency and, among the lichens, there can be some *Cetraria islandica, Cladonia gracilis, C. arbuscula, C. pyxidata* and *C. leucophaea*. Where the sub-shrub canopy is well developed, however, the total cover of the cryptogams is much less than in the typical moss-heaths of the region.

Habitat

The *Calluna-Juniperus* heath is confined to humic rankers at moderate altitudes in the cool oceanic climate along the western seaboard of the north-west Highlands and some of the western Isles and is best developed on cool, shady slopes which are blown clear of snow. It is readily damaged by burning and its present range may be a remnant of a much wider distribution over suitable substrates in this general part of Scotland. With the geographical shift to the continental climate of the east-central Highlands, the community is replaced at the junction of the sub- and low-alpine zones by the *Juniperus-Oxalis* scrub.

The general climatic conditions favouring the development of the *Calluna-Juniperus* heath are similar to those that pertain over the lower altitudes through the ranges of the *Calluna-Racomitrium* and *Calluna-A. alpinus* heaths, for this too is a vegetation type of those parts of Britain with cool summers, relatively mild winters and very high rainfall. It occurs mostly between 300 and 600 m from just south of Beinn Eighe northwards to Foinaven and Sgribhis Bheinn above Durness, with outlying stands on Skye. Throughout this region, mean annual maxima rarely rise above 22 °C (Conolly & Dahl 1970), but February minima are considerably higher than at equivalent altitudes to the east, so the annual variation in temperature is less than there. Rainfall is also very much higher with often well over 1600 mm yr^{-1} and more than 220 wet days annually (*Climatological Atlas* 1952, Ratcliffe 1968), such that the atmosphere is more or less constantly humid with much low cloud. It is this combination of montane and oceanic conditions that is reflected in the continuing frequency in this community of the mixtures of plants typical of the other sub-shrub vegetation of the region, with Arctic-Alpines like *Arctostaphylos* spp. and *E. nigrum* ssp. *hermaphroditum* and the Northern Montane *Antennaria dioica* growing alongside species such as *Erica cinerea*. This, and the absence of any boreal associates or great abundance of lichens helps mark off the vegetation from the heaths of the more continental parts of Scotland and also from the juniper scrub centred on that part of the country, where the characteristic taxon is ssp. *communis*.

Compared with the habitat of the *Calluna-Racomitrium* and *Calluna-A. alpinus* heaths, however, conditions here are more sheltered. In the first place, as well as being a community that is generally confined to the lower portion of the altitudinal ranges of these other dwarfed sub-shrub heaths, it occurs only along the western seaboard of the region where the moderating effect of the Gulf Stream waters on temperature fluctuations is greatest. And, second, although the vegetation mat here is typically blown clear of any snow, the *Calluna-Juniperus* heath is not usually found in the kind of severely-exposed situations of which the other communities are so characteristic. Most often, it marks out level to gently-sloping sites with a northerly to easterly aspect where the prevailing feature of the topoclimate is its cool, shady nature. Compared with the sub-shrub vegetation of wind-blasted spurs and ridges, then, the woody mat, though still very low, is more extensive and the cryptogam element of the flora different. Plants well-adapted to extreme exposures like *Loiseleuria* are poorly represented, whereas the sheltered atmosphere among the bushes is much more congenial for the oceanic element among the bryophytes even if this is not very species-rich compared with the lush mats of the *Calluna-Vaccinium-Sphagnum* heath, characteristic of very humid, sunless corries at much the same altitudes in the region.

Soil development under the *Calluna-Juniperus* heath is typically rudimentary with just shallow accumulations of decaying juniper and bryophyte litter resting directly on bedrock in humic ranker profiles or with a thin intervening mineral A horizon. Generally, the parent material under remaining stands of the community is Cambrian quartzite, a well-jointed and brittle rock that has weathered to screes and fields of jumbled, frost-shattered debris that catch the eye with their sparkling white colour (Whittow 1979). Over such substrates, the soil and vegetation cover is characteristically discontinuous, like a patchwork of islands with very scanty vegetation on the tracts of bare rock between. On Beinn Eighe and Foinaven, where the *Calluna-Juniperus* heath is now best represented, such mosaics cover many hectares (Poore & McVean 1957, McVean & Ratcliffe 1962).

It seems likely, however, that the community was once more widespread: *J. communis* ssp. *nana* occurs scattered throughout the western Highlands (Perring 1968) and suitable localities where it could attain dominance with the typical assemblage here occur throughout the mountains of Cambrian quartzite and Torridonian sandstone running up the coast of north-west Scotland from Skye to Cape Wrath. Burning may have much reduced its distribution over this area because dwarf juniper and the oceanic hepatics associated with it here are very sensitive to fire: indeed, there seem to be some places where the community has perished after a single burning episode (McVean & Ratcliffe 1962). Durno & McVean (1959) traced the history of vegetation of this kind on Beinn Eighe back to Sub-Atlantic times, concluding that remaining stands were but a relic of a formerly widespread cover.

Zonation and succession

The *Calluna-Juniperus* heath typically occurs at the junction of the sub- and low-alpine zones (Poore & McVean 1957) in sequences of mires, other heaths, chionophilous vegetation and grasslands, where the major lines of floristic variation reflect climatic change with altitude and local exposure, edaphic differences and treatment. The community is a climax but is readily degraded by burning.

Over the long ridge of Foinaven, the characteristic zonation is perhaps better disposed than on the more rugged landscape of the Beinn Eighe massif. Essentially, the *Calluna-Juniperus* heath punctuates the typical regional sequence from mire below to fell-field above where quartzite screes are exposed at moderate altitudes. Immediately beneath it there is generally a mixture of *Calluna-Erica* or *Calluna-Vaccinium-Sphagnum* heaths distributed according to aspect and themselves passing below to *Scirpus-Erica* wet heath and then to *Scirpus-Eriophorum* mire on the low-altitude blanket peats. The downslope boundary is typically marked by a thinning of the cover of juniper among a sub-shrub cover that is generally taller than here, though many of its associates in the *Calluna-Juniperus* heath run on with high frequency and abundance. The same is true above, too, where the community gives way over windswept spurs to the *Calluna-Racomitrium* and/or *Calluna-A. alpinus* heaths or where these vegetation types replace the *Calluna-Juniperus* heath over the more exposed south-western slopes of hills. Indeed, although these are characterised by an increase in the cover of cryptogams among the wind-trimmed mat, juniper itself can continue to have patchy representation in them and, where there is a smear of fine morainic material over the ridges, transitions can be especially gradual. These communities in turn can pass directly to *Carex-Racomitrium* heath or there can be an intervening zone of *Vaccinium-Racomitrium* heath, or a different kind of zonation altogether to high-altitude *Calluna-Eriophorum* mire on the summit blanket peats (McVean & Ratcliffe 1962, Ratcliffe 1977).

Chionophilous vegetation is not well developed among these zonations at Foinaven and, even on Beinn Eighe, there are only moderately late snow-beds but, where north- or east-facing slopes become very shady and cold, the *Calluna-Juniperus* heath passes to *Vaccinium-Rubus* heath or, towards the upper limits of its altitudinal range, *Nardus-Carex* or *Deschampsia-Galium* communities. Similar patterns can be seen in Letterewe Forest, though there the *Calluna-Juniperus* heath itself is rather fragmentary. In some places, too, grassy vegetation is well represented over drier ground above and below the community with *Festuca-Agrostis-Galium* and *Nardus-Galium* grasslands or, where there is some flushing with more base-rich waters, the *Festuca-Agrostis-Thymus* or *Festuca-Agrostis-Alchemilla* swards. At lower altitudes, especially, a variety of sub-shrub vegetation can be converted to such grasslands by grazing, though the rocky ground characterised by the *Calluna-Juniperus* heath is often more difficult of access. Where the community has been burned, however, the vegetation may be degraded to a very fragmentary cover with substantial erosion of the exposed soil mantle and from such a condition recovery may be very slow (Poore & McVean 1957, McVean & Ratcliffe 1962).

Distribution
Although perhaps once more widespread through the north-west Highlands, the community is now of rather patchy occurrence along the western side of the more northerly mountains, with especially good stands on Beinn Eighe and Foinaven.

Affinities
Vegetation with *J. communis* ssp. *nana* received scarcely a mention in early descriptive accounts (Tansley 1939) and only with the surveys of Poore & McVean (1957) and McVean & Ratcliffe (1962) were the first attempts made to distinguish the *Calluna-Juniperus* heath from other sub-shrub communities and from the juniper scrub typical of eastern Scotland. Very few new data have been added to these original samples, though the diagnosis here is somewhat more precise than McVean & Ratcliffe's (1962) *Juniperetum nanae*, transferring some of their *lichenosum* into the *Calluna-Racomitrium* or *Calluna-A. alpinus* heaths where *J. communis* ssp. *nana* is an occasional but from which the oceanic hepatics are missing. It is that distinctive combination that helps define the *Calluna-Juniperus* heath from these closely-related types of the Caricetea curvulae, though there is sufficient similarity for it to be included with them in the Arctostaphyleto-Cetrarion as Birks (1973) proposed.

An alternative view would be to follow McVean & Ratcliffe (1962) in locating the community in the Juniperion nanae, where it might also be possible to place Irish *Juniperus-Arctostaphylos* vegetation. This was seen as part of the Vaccinio-Picetea and such an approach would stress the relationships of the *Calluna-Juniperus* heath with the *Juniperus-Oxalis* scrub whose heartland is in the east-central Highlands and where ssp. *communis* is the juniper taxon. In its typical form, this is a distinctly boreal community with strong seral relationships with native pine forest, though in some situations towards its upper altitudinal limit, it could be seen as itself a climax scrub occupying the equivalent position in the zonation to the *Calluna-Juniperus* heath in the west (Poore & McVean 1957). And, distributed across Scotland, there are stands which are floristically and physiognomically intermediate between the two with junipers that can be difficult to assign between ssp. *communis* and ssp. *nana*.

Floristic table H15

Juniperus communis nana	V (4–8)	*Cetraria islandica*	II (1–2)
Calluna vulgaris	V (3–8)	*Carex panicea*	II (1–3)
Cladonia uncialis	V (1–3)	*Cladonia gracilis*	II (1–2)
Racomitrium lanuginosum	IV (2–7)	*Cladonia arbuscula*	II (1–5)
Scirpus cespitosus	IV (1–3)	*Dicranum scoparium*	I (1–2)
Erica cinerea	IV (1–6)	*Cladonia pyxidata*	I (1)
Deschampsia flexuosa	IV (2–3)	*Pleurozium schreberi*	I (1–3)
Potentilla erecta	IV (1–3)	*Carex pilulifera*	I (1–3)
Pleurozia purpurea	IV (1–4)	*Vaccinium myrtillus*	I (1–3)
Hypnum cupressiforme s.l.	IV (1–4)	*Rhytidiadelphus loreus*	I (1–4)
		Nardus stricta	I (2–3)
Huperzia selago	III (1–2)	*Campylopus paradoxus*	I (1)
Arctostaphylos uva-ursi	III (2–6)	*Herbertus stramineus*	I (1–5)
Cladonia impexa	III (3–5)	*Vaccinium uliginosum*	I (3)
Arctostaphylos alpinus	III (1–4)	*Sphagnum capillifolium*	I (2–3)
Sphaerophorus globosus	III (1–2)	*Cladonia leucophaea*	I (1–2)
Solidago virgaurea	III (1–2)	*Agrostis canina*	I (3)
Empetrum nigrum hermaphroditum	III (1–5)	*Galium saxatile*	I (3)
Frullania tamarisci	III (1–2)	*Viola riviniana*	I (3)
Dactylorhiza maculata	III (1–2)	*Thymus praecox*	I (1–4)
Diplophyllum albicans	III (1–2)		
Cornicularia aculeata	III (1–2)	Number of samples	14
Polygala serpyllifolia	II (1)	Number of species/sample	23 (17–32)
Succisa pratensis	II (1–2)		
Antennaria dioica	II (1–2)	Vegetation height (cm)	3 (2–8)
Molinia caerulea	II (1–3)	Vegetation cover (%)	93 (60–100)
Festuca vivipara	II (1–3)	Altitude (m)	400 (77–610)
Hylocomium splendens	II (1–2)	Slope (°)	6 (2–28)

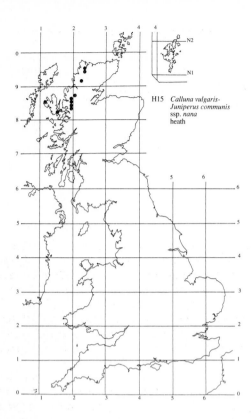

H15 *Calluna vulgaris-*
     Juniperus communis
     ssp. *nana*
     heath

H16
Calluna vulgaris-Arctostaphylos uva-ursi heath

Synonymy

Scottish *Calluna* heath   Smith 1911*b* *p.p.*; *Calluna-Arctostaphylos* heath   Muir & Fraser 1940, Gimingham 1964*a*, 1972; *Arctostaphyleto-Callunetum* McVean & Ratcliffe 1962, Ward 1971; *Vaccinio-Ericetum cinereae* (Birse & Robertson 1976) Birse 1980 *p.p.*; *Arctostaphylos* heath   Urquhart 1986.

Constant species

Arctostaphylos uva-ursi, Calluna vulgaris, Deschampsia flexuosa, Erica cinerea, Vaccinium vitis-idaea, Dicranum scoparium, Hylocomium splendens, Hypnum jutlandicum, Pleurozium schreberi, Cladonia impexa.

Rare species

Arctostaphylos uva-ursi, Lycopodium annotinum, Pyrola media.

Physiognomy

Arctostaphylos uva-ursi can be found in a variety of heath types as an occasional, locally prominent associate but, in the heart of its range, it is most often found in this community which, though it has much in common with other sub-shrub vegetation of the region, frequently has a distinct boreal character. The *Calluna vulgaris-A. uva-ursi* heath, like its closest relative, the *Calluna-Vaccinium* heath, can have a variegated woody cover. *Calluna* is always present and overall it is the most frequent dominant, usually abundant and often overwhelmingly so, in a canopy that usually attains 2–4 dm in height and which can have a substantial total cover. Where, as is often the case, the vegetation is recovering from fairly recent burning, the heather can grow in dense building-phase stands from which many associates are all but excluded, and in such cases the diagnosis of the community is clearly very difficult. Frequently, however, other sub-shrubs make at least a minor contribution to the canopy and are able in various ways to exploit temporal and spatial gaps in the *Calluna* cover, attaining some measure of local abundance early in the heather-

regeneration cycle and sometimes again where the bushes have been allowed to proceed to the degenerate phase. But more precise studies of the effects of fire have been made here than on any other kind of heath (see below) and they show that the patterns of recovery are very varied and often quite persistent (e.g. Hobbs & Gimingham 1984*b*) so, within the general definition, even the proportions of the most frequent contributors to the vegetation can be diverse.

A. uva-ursi, though, is likewise a constant of the community and it can become modestly abundant in gaps within the heather cover but, even there, it typically has a prostrate habit, the branches of the creeping stems forming low mats, often only 5 cm or so thick. Having foliage that is not very shade-tolerant, these patches are readily overwhelmed by the expanding heather cover, but the long monopodial shoots can extend far among the litter and then branch profusely where conditions of better illumination are met (Gimingham 1972). The low growth of the stems and the fact that they are often part-buried among the litter also afford the plant some protection against the effects of burning, and regeneration readily occurs from clusters of dormant buds on older parts of the stems: these may actually increase in numbers with repeated damage although, as in other ericoids, there is a general decline in the vigour of vegetative regeneration in *A. uva-ursi* with age (Mallik & Gimingham 1985).

Erica cinerea is also very common in the community, but it is generally of low cover: it can regrow from sprouting stools or from seed after burning but it rarely gains any ascendancy over *Calluna* in the rather harsh climatic conditions here, even where this vegetation extends on to locally warmer south-facing slopes. In many stands, too, there is some *Vaccinium vitis-idaea* and *V. myrtillus*. These are both rhizomatous so, even after quite severe fires, they have strong regenerative potential right from the start and one or both of them often spreads rapidly in the community in the immediate post-burn phase. They are also shade-tolerant, particularly *V. vitis-idaea*, so they can persist as a patchy second tier beneath the heather as it thickens up. Where they are prominent, and especially where there is also some *Empetrum nigrum* ssp. *nigrum*, not universally well represented here but characteristic of one sub-community, it can become hard to separate this vegetation from the *Calluna-Vaccinium* heath. Then, it may be a matter of whether *A. uva-ursi* is present or not, such that the local extinction of this plant from the canopy can effectively result in a fine mosaic of the two communities.

Quite often, however, there are some other floristic features which help give the *Calluna-A. uva-ursi* heath a distinctive stamp. Quite commonly throughout the community there are small amounts of *Genista anglica*, a rather diffuse and somewhat frail-looking sub-shrub compared with the others frequent here, but actually rather resistant even to quite high fire temperatures, perhaps because of its relatively thick bark (Mallik & Gimingham 1985), and readily able to resprout from the burned stumps. It is a plant with a somewhat odd distribution through Europe, essentially Oceanic West European, but perhaps better described as sub-Atlantic, having a strong centre of its British range located in this heath in eastern Scotland and with some striking outliers towards central Europe (Matthews 1955). Then, more expectedly perhaps, there are some Northern Montane preferentials, such as *Listera cordata*, *Trientalis europaea* and *Antennaria dioica*, and Continental Northern plants, like the nationally rare *Pyrola media*, but these are not very frequent overall and indeed are often distinctly clustered in their occurrence in one particular kind of *Calluna-A. uva-ursi* heath. In fact, except in that sub-community, herbaceous associates are rather few and far between in this vegetation. *Luzula multiflora* and *L. pilosa* are rather more common than in the *Calluna-Vaccinium* heath and *Carex pilulifera* is not so restricted in its occurrence, but otherwise, this element of the flora can be very impoverished.

The bryophyte component is similarly somewhat variable in its composition, and also in its abundance since, as in some other heaths, the bulkier mosses are often strongly associated with particular stages in the heather-regeneration cycle, sometimes spreading among pioneer *Calluna* but frequently making their most prominent contribution among older bushes that are opening up somewhat; and there are also differences with the character of the soil exposed after burning. But *Hypnum jutlandicum*, *Pleurozium schreberi* and *Dicranum scoparium* are very common overall with *Hylocomium splendens* also a constant through much of the community, and *Rhytidiadelphus loreus*, *R. squarrosus* and *Ptilidium ciliare* occasional.

Lichens differ in their representation, too, with only *Cladonia impexa* constant and, in many stands, of but low cover. Fruticose species, such as *C. arbuscula* and *C.*

rangiferina, tend to follow the larger pleurocarps in developing among the shadier and more humid conditions of older heather canopies and, there too, *Hypogymnia physodes* and, less commonly, *Cetraria glauca* can be seen on decaying woody stems. *Cladonia uncialis* also becomes frequent in such vegetation but it is rather more typical of the *Cladonia* sub-community where the lichen contribution tends to be the most extensive and varied found in the *Calluna-A. uva-ursi* heath, with peat-encrusting species especially numerous and abundant.

Sub-communities

***Pyrola media-Lathyrus montanus* sub-community:** *Arctostaphyleto-Callunetum* McVean & Ratcliffe 1962; *Arctostaphyleto-Callunetum*, herb-rich type Ward 1971; *Vaccinio-Ericetum cinereae, Viola riviniana* subassociation Birse 1980 *p.p.*; *Arctostaphylos* heath Urquhart 1986. It is here that *A. uva-ursi* tends to make its most prominent contribution to this community, frequently occupying more than 25% of the ground and occasionally rivalling *Calluna* in abundance and forming a variegated patchwork with it. *Erica cinerea* can have fairly high cover, too, though it is usually present as small, scattered bushes, and the two *Vaccinia* are likewise common but generally sparse. *Genista anglica* is fairly frequent, again at low cover, and very occasionally there can be a little *Juniperus communis* ssp. *communis*. *E. nigrum* ssp. *nigrum* is noticeably sparse.

The most striking feature of this kind of heath, though, is the associated herb flora and it is among this element that the community differs particularly from the *Calluna-Vaccinium* heath. First, grasses are noticeably more common than elsewhere in the *Calluna-A. uva-ursi* heath, with *Festuca ovina* frequently joining *Deschampsia flexuosa* and occasional records for *Anthoxanthum odoratum*, *Agrostis capillaris* and, at some sites, *A. canina* ssp. *montana*. With their clustered tillers protected by insulating old leaf-sheaths, each of these can escape severe damage in fires to attain some degree of abundance in the early years of recovery though they later become sparse. Then, there are some widespread calcifuge plants which can be very frequent in this sub-community, notably *Potentilla erecta* and *Galium saxatile* with, more indicative of the transitional soil base status that seems characteristic here, *Lathyrus montanus*, *Hypericum pulchrum* and, less commonly, *Succisa pratensis*. Third, some distinctly mesophytic species can be found, with *Viola riviniana* and *Lotus corniculatus* occurring frequently, *Campanula rotundifolia* rather less often. Most striking of all, perhaps, is the fairly consistent occurrence of *Pyrola media* and the less common, but still preferential, presence of *Trientalis*, which add a strong boreal feel. *Anemone nemorosa*, too, which is fairly frequent in this sub-community, can be seen as reflecting climatic conditions in its occurrences here outside of woodland and, among the bryophytes, there is a noticeable increase in *Rhytidiadelphus triquetrus*, a feature of a number of different vegetation types in this part of the country. Again, many of the herbs are well adapted vegetatively to survive the effects of burning, having growing points clustered among leaf-bases in rosettes (*Pyrola, Succisa, Viola*) or underground perennating organs which can rapidly produce new shoots (*Potentilla. Lathyrus, Lotus, Campanula, Anemone, Trientalis*) so, in newly-burned tracts, there can sometimes be seen abundant new growth of these plants (Mallik & Gimingham 1985). With age, however, all but the more shade-tolerant, among which *Pyrola* is rather noteworthy, can become very much more thinly distributed and rather patchy from stand to stand (e.g. Hobbs *et al.* 1984).

Overall, with the high total cover of the sub-shrubs and herbs, bryophytes and lichens tend to be less well represented in this sub-community than in other kinds of *Calluna-A. uva-ursi* heath. The community mosses are all very common but neither they, nor the preferential *R. triquetrus*, are very abundant as a rule and, among the lichens, only *Cladonia impexa* occurs frequently and then with low cover.

***Vaccinium myrtillus-Vaccinium vitis-idaea* sub-community.** *A. uva-ursi* remains constant in this sub-community but *Calluna* is usually a fairly uncompromising dominant among the sub-shrub cover with, as a patchy mosaic or a rather sparse understorey, *V. myrtillus* and, usually a little more abundant, *V. vitis-idaea*. *Erica cinerea* and *Genista anglica* are somewhat less common than in the *Pyrola-Lathyrus* sub-community but *E. nigrum* ssp. *nigrum* is much more frequent, occasionally with locally high cover, though generally with fairly sparse shoots intermingled among the other sub-shrubs. More obvious, however, is the scarcity of herbs here because, though *Festuca ovina, Carex pilulifera, Potentilla erecta, Luzula multiflora, L. pilosa* and *Listera cordata* are occasionally recorded, there is nothing like the richness and variety among this element characteristic of the *Pyrola-Lathyrus* type. Conversely, the cryptogams tend to be a little more diverse and considerably more extensive, with the bulky pleurocarpus mosses *Hypnum jutlandicum, Pleurozium schreberi* and *Hylocomium splendens* often abundant beneath the canopy and particularly where the branches are more open. Larger lichens are also more apparent, with *Cladonia arbuscula, C. uncialis* and *C. rangiferina* frequently accompanying *C. impexa*, and *Cetraria glauca* and *Hypogymnia physodes* occasional over older stools.

***Cladonia* spp. sub-community:** *Arctostaphyleto-Callunetum*, herb-poor type Ward 1971. Although *A. uva-ursi* is sometimes quite abundant here, *Calluna* is more often overwhelmingly dominant. Both *Erica cinerea* and

Genista anglica occur frequently, but *V. myrtillus* is absent and *V. vitis-idaea* and *E. nigrum* ssp. *nigrum* very scarce. The herbs of the *Pyrola-Lathyrus* sub-community are hardly ever found but *Carex pilulifera* and *Scirpus cespitosus* are preferentially common and there is occasionally a little *Carex panicea*. More peculiar, however, is the cryptogam element because, in this sub-community, some of the characteristic mosses of the *Calluna-A. uva-ursi* heath, notably *Hylocomium splendens* and *Pleurozium schreberi*, become very patchy in their occurrence. *Hypnum jutlandicum* and *Dicranum scoparium* remain frequent and there is often a little *Pohlia nutans* but it is lichens, especially peat-encrusting species, that are more noticeable. None is ever really abundant but, along with *Cladonia impexa* and *C. uncialis*, there is frequent *C. floerkeana, C. coccifera* and *C. squamosa* and occasional *C. crispata, C. deformis* and *C. cornuta* occurring as scattered patches over the areas of open ground among the bushes.

Habitat

The *Calluna-A. uva-ursi* heath is characteristic of base-poor to circumneutral soils at moderate altitudes in the cold continental climate of the east-central Highlands of Scotland. Edaphic differences play some part in determining floristic variation in the community but their effects are often overlain and modified by the influence of burning which ultimately maintains this vegetation as a plagioclimax.

A. uva-ursi has a wide geographical and altitudinal distribution through the extreme north of Britain where the summers are at their coolest in this country: virtually all its Scottish localities (and those in Eire) fall within the 24 °C mean annual maximum isotherm (Conolly & Dahl 1970). In this community, however, it makes an important contribution to sub-shrub vegetation which is rather strictly confined to those parts of this range with a more continental climate. Thus, while summer maxima experienced by this heath are generally between 21 and 23 °C (Conolly & Dahl 1970) – temperatures which, in eastern Scotland, can prevail at moderately high altitudes – the winter climate is harsh, with low February minima and frequent late frosts (*Climatological Atlas* 1952, Page 1982). Compared with the conditions felt by heaths at identical altitudes to the west, therefore, the annual temperature range is greater. Furthermore, in contrast to the more oceanic zone in Scotland, the rainfall is very much lower. Annual precipitation here is, indeed, often less than 1000 mm and rarely over 1200 mm (*Climatological Atlas* 1952) with only 140–180 wet days yr^{-1} (Ratcliffe 1968). In the winter months, then, snowfall totals are only modest though, with the low spring temperatures, there is usually morning snow-lie on 40 days or more yr^{-1} (Page 1982).

Such conditions as these prevail throughout the foothills of the east-central Highlands where the community can be found, generally between 250 and 600 m, centred around Speyside, but extending eastwards to the Muir of Dinnet and other sites near Morven, west beyond Strathdearn and south, in more fragmentary fashion, through the Forest of Atholl (McVean & Ratcliffe 1962, Ward 1971a, Urquhart 1986). And the distinctive climate of this region is reflected in the vegetation in two ways. First, there is the generally montane character of the sub-shrub canopy with its frequent records for *V. myrtillus* and *E. nigrum* ssp. *nigrum*, and the more strictly Arctic-Alpine *V. vitis-idaea* and *A. uva-ursi*. Winter conditions here are not so blisteringly severe as to exclude *E. cinerea*, a plant which preserves some measure of physiological activity through the coldest months (Bannister 1965) though it tends to be less prominent than in those oceanic heaths, where *A. uva-ursi* can also occasionally figure, along the western seaboard of Scotland. On the other hand, *E. nigrum* ssp. *hermaphroditum* and *Loiseleuria* are very scarce in this vegetation: these species are largely confined to areas within the 22 °C mean annual maximum isotherm (Conolly & Dahl 1970) and tend to take over from *E. nigrum* ssp. *nigrum* and *A. uva-ursi* with the shift to more exposed conditions in eastern Scotland, at altitudes over 600 m and particularly on wind-blasted spurs and summits, where the dwarfed *Calluna-Cladonia* heath replaces the *Calluna-A. uva-ursi* heath and the other sub-shrub community which can be found there at lower altitudes, the *Calluna-Vaccinium* heath.

The second, and more particular, influence of climate is the occurrence of the boreal element in the flora of the *Calluna-A. uva-ursi* heath, though this is only really well seen in the *Pyrola-Lathyrus* sub-community. There the community occasionals *Listera cordata* and *Antennaria dioica* are often joined by *Pyrola media* and *Trientalis europaea* and other less specifically Continental Northern or Northern Montane plants, such as *Anemone nemorosa* and *Rhytidiadelphus triquetrus*, which in other communities in the east-central Highlands, like the *Juniperus-Oxalis* scrub and certain types of *Pinus-Hylocomium* woodland, add a distinctive climatically-related component. These are some of the plants which ultimately help separate the *Calluna-A. uva-ursi* heath from the *Calluna-Vaccinium* heath, which is common (indeed preponderant) through the foothills of this part of Scotland. And, when these species are, for one reason or another, sparse here, as in the *Vaccinium* and *Cladonia* sub-communities, it is that much harder to distinguish these communities along what then becomes a virtual continuum of floristic variation (McVean & Ratcliffe 1962, Ward 1970, 1971a, b).

But there are other peculiarities of the *Pyrola-Lathyrus* sub-community which are related to edaphic variation. By and large, the *Calluna-A. uva-ursi* heath is a vegetation type of more acidic soils developed from lime-poor parent materials, of which there is an abundance through the region, both bedrocks like the granites of the Cairngorms and Monadhliath hills, and superficials which, through Speyside, underlie the terraced haughlands above the limits of cultivation over which this community is particularly well represented. However, the range of profile types which has developed from the pervious rocks and coarse free-draining morainic material is quite broad, including brown earths, brown podzolic soils and humo-ferric podzols. And, among these, there is a clear association between the brown soils and the *Pyrola-Lathyrus* sub-community, many of the preferentials of which reflect more mesophytic edaphic conditions and a base-status transitional between acidic and neutral: superficial pH under this kind of *Calluna-A. uva-ursi* heath is usually from 4.5 to 5.5 and there is a tendency for the formation of moder humus rather than mor, with perhaps a fairly active nutrient turnover (McVean & Ratcliffe 1962, Ward 1971a, Birse 1980). Beneath the more herb-poor sub-communities, on the other hand, podzolic soils tend to predominate.

It is very clear, though, that the full expression of the climatically-related peculiarities of the *Calluna-A. uva-ursi* heath and the development of edaphic trends are strongly influenced by burning: as with the *Calluna-Vaccinium* heath, though not on so extensive a scale, the community forms an important part of grouse-moors in the east-central Highlands, having been burned fairly regularly, if not always judiciously, over at least a century and sometimes nearly two (Gimingham 1972). Indeed, it is in this kind of vegetation that some of the most precise observations on the effects of fire on heaths have been made, notably by Gimingham and his co-workers on the Muir of Dinnet, a site towards the eastern, drier and warmer lowland fringe of the range of the community, but having a fairly typical mix of *Calluna-A. uva-ursi* and *Calluna-Vaccinium* heaths over fluvio-glacial gravels (Ward 1970, Legg 1978, Marren 1979, Hobbs & Gimingham 1984a).

Such studies have shown that most of the associated herbs of the *Calluna-A. uva-ursi* heath reappear in abundance only in those stands which had heather in the pioneer or building phase when burned, but that the variety of such plants afterwards is only partly related to their pre-burn numbers. With pioneer heather stands, virtually all the species appearing after the burn were present before; in building stands, many new species appeared; where older tracts than this were burned, fewer species returned than were originally found (Hobbs & Gimingham 1984b). Such trends reflect differences in the potential for regrowth from either vegetative organs or dormant seed, both of which are influenced by the age of the *Calluna* canopy. In general, the variety and numbers of associated herbs seen in the

Calluna-A. uva-ursi heath decline with the progression from open pioneer to closed building and mature phases, then increase again as the bushes collapse in late maturity and degeneration, so the potential for regrowth from surface rosettes or tillers follows the same trend. The greater diversity of the post-burn herb flora when building stands are fired must therefore originate more from persistent underground organs or stored seed.

Most of the herbs of the community can store seed in the soil (Mallik & Gimingham 1983) and little non-vegetative regeneration in burned stands seems to originate from dispersal from outside (Hobbs *et al.* 1984). But stand-age has a marked effect on local input to the seed-bank because of the demise of fruiting parents with the closure of the heather canopy, and the declining performance of any ageing survivors (Legg 1978, Mallik *et al.* 1984). Furthermore, apart from *Carex pilulifera* (and *Calluna* itself: Hill & Stephens 1981), the soil stores themselves become depleted by the time the mature phase is reached, with some slight increase as the canopy opens again at degeneration (Mallik *et al.* 1984), so re-establishment in stands past their middle years of heather growth is unlikely. Even where soil stores remain appreciable, and well protected from the effects of fire by a mat of pleurocarpous mosses (Hobbs & Gimingham 1984a), conditions for post-burn germination may be poor, perhaps hindered by the same thick bryophyte mats or layers of litter (Miles 1974, Hobbs 1981, Mallik *et al.* 1984).

Thus, although it is generally true that the most important ecological character of the community associates is the ability to regenerate sexually or, even more so, vegetatively in the specialised post-burn environment (Mallik & Gimingham 1985), this may depend on the rather precise coincidence of critical life-history events among both these associates and the dominant sub-shrubs in relation to the burning cycle (Hobbs *et al.* 1984). And, where there is any perturbation to the usual pattern of firing and regrowth, evidence shows that peculiarities of composition can persist long into the post-burn period perpetuating the disturbance (and, incidentally, making it dangerous to devise time-sequences from spatial variation among stands of differing age: Hobbs & Gimingham 1984b).

The survival of the characteristically rich and diverse herb flora of the *Pyrola-Lathyrus* sub-community from one cycle to the next may thus be a rather precarious affair, these plants being ultimately dependent on burning treatments for their renewal, but readily succumbing to local extinction, not only in the short term with canopy closure, but over longer periods with perhaps only slight variation in the fire regime. Few of the species are constant even in the *Pyrola-Lathyrus* sub-community itself, and some appear particularly patchy in their occurrence: *Hypericum pulchrum*, for example, can readily resprout from remnant shoots and seems to have seed whose germination is actually fire-enhanced (Mallik & Gimingham 1985) and it was very characteristic of the more herb-rich stands described by Ward (1970, 1971a), whereas in the samples of very similar vegetation in McVean & Ratcliffe (1962) and Urquhart (1986), it is noticeably sparse.

In the *Vaccinium* and *Cladonia* sub-communities, however, the absence of very nearly all these plants is a much more consistent feature and it is likely that at least some of the stands included here are of vegetation which has lost the *Pyrola-Lathyrus* herbs but which could regain the flora with the re-establishment of a pioneer heather canopy. This is perhaps especially true of the *Vaccinium* sub-community which, with its sprinkling of herbs, abundance of pleurocarps and fruticose lichens and scattered shoots of woody associates among the heather, has much of the character of the late building-phase *Calluna-A. uva-ursi* heath described by Hobbs *et al.* (1984). It does include more stands from higher altitudes than the *Pyrola-Lathyrus* sub-community, with some looking almost transitional to *Calluna-Cladonia* heath, but climatic differences are not such in themselves as to exclude the herbs.

It is difficult to be so precise about interactions between burning and edaphic conditions in the *Calluna-A. uva-ursi* heath but it seems possible that continued exclusion of the *Pyrola-Lathyrus* herbs could sometimes indicate not simply an inimical burning regime but also the kind of soil impoverishment consequent upon long-unrelieved dominance of *Calluna*, such that a more irreversible degradation was initiated. The effects of initially poorer profiles and injudicious management could thus mutually confirm one another and this may have played an important part in the contrast seen today between 'richer' and 'poorer' grouse-moors (Miller *et al.* 1966, Jenkins *et al.* 1967, Moss 1969). On the former, developed over less base-poor substrates and with careful burning to maintain the higher bird numbers, the more species-rich *Calluna-A. uva-ursi* heath and the *Galium-Festuca* sub-community of the *Calluna-Vaccinium* heath predominate; over the latter, which are concentrated on less productive soils, the trend towards the more species-poor kinds of *Calluna-Vaccinium* heath is very marked.

In the *Calluna-A. uva-ursi* heath itself, the floristic impoverishment among the vascular element is most obvious in the *Cladonia* sub-community which, in Ward's (1970, 1971a) studies, was the kind of vegetation associated with podzols with a distinctly humic top-soil. On such substrates, recolonisation after burning was slow such that even after 10–15 years encrusting *Cladonia* spp. were still in occupation of ground that with a good mineral soil would have been long covered by sub-shrubs and herbs. *Carex pilulifera*, the fruits of which are longer lived than most of those of the herbs and which is thus well able to exploit late-developing niches, is best represented here and the high frequency of *Scirpus cespitosus* and occasional occurrence of *Carex panicea* perhaps reflect the surface moisture of such soils where a peaty crust holds rain for an appreciable time. It would seem unlikely that this kind of *Calluna-A. uva-ursi* heath could readily regain the floristic richness of the *Pyrola-Lathyrus* type.

In the more species-poor vegetation included here, it can sometimes be simply the regularity with which *A. uva-ursi* is encountered that helps distinguish this heath from other heather-dominated communities, notably the *Calluna-Vaccinium* heath. The species is certainly a persistent element of regularly-burned sub-shrub vegetation in the region and, though new plants do not attain reproductive maturity until the building phase of the *Calluna* among which they have established, the seeds of *A. uva-ursi* are present in the soil at all stages and vegetative regrowth can also occur immediately after fire. These features, together with the distinctive growth habit of mature plants that enables the far-spreading stems to branch vigorously into local gaps, give the plant advantages which outreach those of its peculiar associates in the species-rich kind of *Calluna-A. uva-ursi* heath.

Stretches of moorland including stands of the community are often open to stock but there is very little information on the impact of their grazing on this vegetation. *A. uva-ursi* figures little in the diet of grouse, but may perhaps provide some bite for sheep among the valuable winter-green elements of the vegetation on these grazings. Urquhart (1986) reported that one of the sites identified by McVean & Ratcliffe (1962) as having this community had become herb-rich sheep pasture and certainly there can be some floristic overlap between the *Pyrola-Lathyrus* sub-community and certain kinds of *Festuca-Agrostis-Galium* and *Nardus-Galium* grasslands. It is fairly easy to see how heavy grazing might lead to an expansion of the grassy element in the richer *Calluna-A. uva-ursi* sward with a reduction in the cover of the sub-shrubs and avoiding such treatment after burning is probably vital for the well-being of the community.

Zonation and succession

With woodland, scrub, other sub-shrub communities and grasslands, the *Calluna-A. uva-ursi* heath comprises a distinctive suite of vegetation types in the sub-montane zone of the east-central Highlands. These are the variously-modified derivatives of the original forest cover, produced by clearance, burning and grazing and, in the case of this community, dependent upon management as grouse-moor or rough pasture to prevent regression to secondary woodland. The *Calluna-A. uva-ursi* heath can be seen in large-scale altitudinal sequences, giving way above to dwarfed heaths and fell-field communities, which are climatic climax vegetation, and there are zonations to wet heath and mire with changes in topography and soils.

Long and quite intensive histories of land use in this part of Scotland (e.g. Birks 1970, O'Sullivan 1977, Carlisle 1977) mean that substantial tracts of more natural vegetation are rare. There is little doubt that, in the east-central Highlands, the climax forest type is similar to the vegetation seen today in the *Pinus-Hylocomium* woodland, but many surviving stands of this have been treated for timber production, sometimes involving planting, and the natural regeneration of the community is often somewhat problematical (Steven & Carlisle 1959, McVean & Ratcliffe 1962, Ratcliffe 1977). Then, although birch and juniper clearly figured in the original forest patchwork, and sometimes remain today in close association with native pine, many tracts of the *Juniperus-Oxalis* scrub behave as remnant understoreys, being perpetuated distant from the kind of woodland one might imagine could succeed them and rarely showing any natural progression. Indeed here, too, there is sometimes only sporadic regeneration of the scrub cover itself (Carlisle & Brown 1968, Carlisle 1977, Miles & Kinnaird 1979). The *Calluna-A. uva-ursi* heath can probably be seen as a further stage in this disruptive reversal of vegetation development, a plagioclimax community that has much in common floristically with the field layers of both these communities, which is sometimes seen among and around them, but which perhaps reverts only with difficulty to its more natural forebears, even when the constraints of regular burning are released.

So, although all three communities are very characteristic of the region and form integral elements in the large-scale landscape of what is understood as 'Caledonian pine forest' (Steven & Carlisle 1959, McVean & Ratcliffe 1962), it is only in a relatively few places that they can be seen in a more intimate juxtaposition and, even then, their relationship is rarely in any sense a dynamic one. The stretch of country between Rothiemurchus and Abernethy, in the middle reaches of the Spey, is one of the best areas where the similarities between the vegetation types can be appreciated. On a general level, there can be frequent records in all three for *Calluna*, *V. myrtillus*, *V. vitis-idaea*, *Deschampsia flexuosa*, *Hypnum jutlandicum*, *Dicranum scoparium*, *Pleurozium schreberi*, *Hylocomium splendens*, such that it often appears as if the heath runs virtually unaltered under the more open stretches of shrubs and trees. More strikingly, although *A. uva-ursi* itself rarely figures in the scrub or woodland, being intolerant of extensive areas of deeper shade from which it cannot extend out, there is quite often some continuity between the different communities in the kind of associated herb flora so distinctive of the *Pyrola-Lathyrus* sub-community of the *Calluna-A. uva-ursi* heath. Where the *Pinus-Hylocomium* woodland and *Juniperus-Oxalis* scrub extend on to the less oligotrophic brown soils, for example, more mesophytic grasses and also herbs such as *Luzula pilosa*, *Viola riviniana* and *Campanula rotundifolia* can be found throughout. Moreover, there are also records in each for certain of the Continental Northern and Northern Montane species which give some stands of the *Calluna-A. uva-ursi* heath a particular phytogeographic stamp: *Listera cordata*, *Trientalis europaea* and *Pyrola media*, for example, provide a common element which confirms the boreal character of these peculiar eastern Scottish vegetation types.

Establishment of juniper and pine in stands of *Calluna-A. uva-ursi* heath seems to be rare: although both can exploit the post-burn habitat (indeed, perhaps depend on it to a great extent), any regeneration is curtailed by repeated burning or overgrowth of heather. Seed-parents are often distant, too, so, if there is any progression to woodland with cessation of treatment, it is birch that generally predominates among the invaders. On the Muir of Dinnet, for example, Urquhart (1986) indicates a very substantial reduction of the area of the community since 1946 as a result of the spread of birch. The likely development over the less-impoverished brown soils would be the *Quercus-Betula-Oxalis* woodland, a common secondary forest type in this region and having there a distinctive *Anemone* sub-community, with, again, such species as *A. nemorosa*, *Trientalis*, *Luzula pilosa*, *Lathyrus montanus* and *Rhytidiadelphus triquetrus* providing strong floristic continuity with the heath.

Where the *Calluna-A. uva-ursi* heath extends on to more base-poor and oligotrophic podzolised profiles, it is increasingly likely to be subordinate in its extent to the *Calluna-Vaccinium* heath and, in fact, through the east-central Highlands as a whole, this latter community is strongly predominant through the sub-montane zone, the former occurring as what are relatively small islands scattered through it (Urquhart 1986). As indicated earlier, treatment for grouse-rearing may have accentuated the partly edaphic contrast between 'rich' and 'poor' moors, and it seems likely that injudicious burning, leading to an overwhelming dominance of *Calluna*, with consequent enhancement of podzolisation, could shift the pattern from an abundance of the richer kind of *Calluna-A. uva-ursi* heath to the poorer *Calluna-Vaccinium* types. Certainly, with a move towards the latter, the peculiar boreal character of eastern Scottish sub-shrub vegetation is lost and, although seral progression

to woodland can occur from such heath, where it is abandoned, the secondary birch-dominated *Quercus-Betula-Dicranum* woodland that develops is likewise much less floristically distinctive than its more mesophytic counterpart.

Mosaics of the two heath communities, with more fragmentary representation of juniper and pine vegetation, are a very characteristic feature of the Grampian foothills throughout Speyside and north of the Dee, and over the lower slopes of the Monadhliath and Cairngorm mountains (Ratcliffe 1977, Urquhart 1986). With the shift to higher altitudes through this region, above the forest zone where they now often comprise the bulk of the cover on more free-draining soils, there is a move to other kinds of sub-shrub vegetation. Over very exposed ridges and spurs, it is the dwarfed *Calluna-Cladonia* heath that is characteristic, replacing the *Calluna-A. uva-ursi* heath at around 670 m in the Cairngorms (McVean & Ratcliffe 1962): *A. uva-ursi* itself can run on in some abundance into the wind-pruned mat of the lichen heath, and more stunted stands of the *Vaccinium* sub-community of the *Calluna-A. uva-ursi* heath can form a transition zone between the two. Beyond, there is then a gradation to *Juncus-Racomitrium* heath. More sheltered situations at intermediate altitudes in such sequences, particularly where a northerly or easterly aspect ensures longer snow-lie, can have stands of the *Vaccinium-Rubus* heath.

Throughout these altitudinal and topographic zonations, a switch to soils with impeded drainage can see transitions to wet heaths and mires. At lower altitudes, as over the Cairngorm slopes, mosaics of *Calluna-A. uva-ursi* and *Calluna-Vaccinium* heaths can pass over gleyed peaty podzols to the *Ericetum tetralicis* in which mixtures of *Calluna* and *Scirpus* can predominate and where *Cladonia impexa*, *C. uncialis* and peat-encrusting lichens can provide additional continuity with the *Cladonia* sub-community of the *Calluna-A. uva-ursi* heath characteristic of surface-damp humic soils. Over higher ground, blanket peats on plateaus can have a cover of the *Calluna-Eriophorum* mire above the sub-montane heaths: a particularly striking sequence of this kind can be seen above the Findhorn River in the northern Grampians, where the *Calluna-A. uva-ursi* heath, including some good *Pyrola-Lathyrus* type, occupies the steeper slopes, passing over the gently-domed summit of Carn Nan Tri-Tighearnan to spectacular lichen-rich blanket bog.

Distribution

All the sub-communities of the *Calluna-A. uva-ursi* heath occur widely but fairly locally through the east-central Highlands, with especially good representation in Speyside. Urquhart (1986) identified a cluster of sites around Craiggowrie, Tulloch Moor, Boat of Garten and Rothiemurchus as having particularly striking examples of the richer *Pyrola-Lathyrus* type with further fine stands between Crathie and Dinnet along the northern side of the Dee.

Affinities

Early descriptive studies regarded *A. uva-ursi* as an occasional, locally prominent plant within a compendious sub-montane *Callunetum* (Smith 1911*b*, Tansley 1939), and, though separate *Calluna-Arctostaphylos* vegetation was later recognised (Muir & Fraser 1940, McVean & Ratcliffe 1962, Gimingham 1964*a*, 1972), the difficulty of distinguishing more species-poor stands from other heaths largely lacking *A. uva-ursi* was acknowledged. The problem is particularly well seen in the investigations of Ward (1970, 1971*a*, *b*) where an *Arctostaphyleto-Callunetum*, as he termed it, following McVean & Ratcliffe (1962), formed part of a floristic continuum among sub-shrub vegetation in the east-central Highlands and, in the scheme of Birse (1980), there is a return to a single *Vaccinio-Ericetum* including virtually all the sub-montane heaths of this region.

Ward (1971b), however, considered it worthwhile to recognise a separate community, though his definition of the *Arctostaphyleto-Callunetum* was somewhat narrower than that of McVean & Ratcliffe (1962), concentrating even more on the richer stands included here in the *Pyrola-Lathyrus* sub-community. Although the pre-

ferentials of this sub-community agree very closely with the revised list of character species which Ward (1971*a*, *b*) proposed for his association, it has been thought better here to return to a somewhat broader diagnostic base for the community, not as all-inclusive as that of Birse (1980), but roughly similar to that of McVean & Ratcliffe (1962). Here, though, the recognition of distinct sub-communities helps clarify transitions to the *Calluna-Vaccinium* heath.

It was Gimingham (1949, 1964*a*) who pointed out the rather unexpected dissimilarity between the kind of vegetation included in the *Calluna-A. uva-ursi* heath and Scandinavian communities with *A. uva-ursi*. In western Norway, for example, the species is found in the kind of oceanic *Calluna-Erica cinerea* heath which, along the northern seaboard of Scotland, can also occasionally have this plant. In south-west Sweden, on the other hand, there is a much more obvious Continental influence with *Genista pilosa* and *Arnica montana* well represented in *A. uva-ursi* heath. Some Danish stands have *A. uva-ursi* with *Empetrum nigrum* ssp. *nigrum* and *Genista anglica*, but the other preferentials of the *Pyrola-Lathyrus* sub-community are generally sparse. Despite these peculiarities among European sub-shrub communities, the *Calluna-A. uva-ursi* heath, together with juniper, pine and birch woods of the same region, form a distinctive boreal component among British vegetation types.

Floristic table H16

	a	b	c	16
Calluna vulgaris	V (5–10)	V (4–10)	V (5–10)	V (4–10)
Arctostaphylos uva-ursi	V (4–8)	V (1–5)	V (1–8)	V (1–8)
Hypnum jutlandicum	IV (1–5)	V (1–10)	V (1–10)	V (1–10)
Cladonia impexa	IV (1–3)	IV (1–6)	V (1–4)	IV (1–6)
Pleurozium schreberi	IV (1–6)	V (1–8)	III (1–3)	IV (1–8)
Erica cinerea	IV (1–6)	III (1–4)	V (1–4)	IV (1–6)
Dicranum scoparium	IV (1–4)	III (1–4)	IV (1–4)	IV (1–4)
Vaccinium vitis-idaea	V (1–4)	IV (1–4)	II (1–6)	IV (1–6)
Deschampsia flexuosa	V (1–4)	IV (1–4)	II (1–4)	IV (1–4)
Hylocomium splendens	IV (1–6)	IV (1–6)	II (1–3)	IV (1–6)
Potentilla erecta	IV (1–4)	II (1–3)	I (1–3)	III (1–4)
Pyrola media	IV (1–3)	I (1–4)		II (1–4)
Lathyrus montanus	IV (1–3)			II (1–3)
Viola riviniana	IV (1–3)			II (1–3)
Festuca ovina	III (1–5)	II (1–3)	I (1–3)	II (1–5)
Rhytidiadelphus triquetrus	III (1–4)	II (1–3)		II (1–4)
Lotus corniculatus	III (1–3)	I (1–3)		II (1–3)
Galium saxatile	III (1–4)			I (1–4)

Hypericum pulchrum	III (1–3)		I (1–3)
Anemone nemorosa	III (1–4)		I (1–4)
Trientalis europaea	II (1–4)	I (1–3)	I (1–4)
Anthoxanthum odoratum	II (1–4)	I (1)	I (1–4)
Agrostis capillaris	II (1–3)	I (1–3)	I (1–3)
Pseudoscleropodium purum	II (1–4)		I (1–4)
Polygala serpyllifolia	II (1–3)		I (1–3)
Campanula rotundifolia	II (1–3)		I (1–3)
Succisa pratensis	II (1–3)		I (1–3)
Luzula campestris	I (1–3)		I (1–3)
Vaccinium myrtillus	III (1–4)	V (1–5)	III (1–5)
Empetrum nigrum nigrum	I (5)	III (1–6)	I (1)	II (1–6)
Cladonia arbuscula		III (1–5)	I (4)	II (1–5)
Cladonia rangiferina		II (1–4)	I (1–4)	I (1–4)
Polytrichum commune	I (1)	II (1–3)		I (1–3)
Cetraria glauca		II (1–4)		I (1–4)
Juncus squarrosus		I (1–4)		I (1–4)
Carex pilulifera	II (1–3)	II (1–3)	IV (1–3)	II (1–3)
Scirpus cespitosus	I (1–3)	I (1–4)	IV (1–5)	II (1–5)
Cladonia uncialis		III (1–3)	IV (1–3)	II (1–3)
Cladonia floerkeana	II (1–3)		III (1–3)	II (1–3)
Carex panicea	I (1)	I (1–3)	III (1–3)	I (1–3)
Pohlia nutans	I (1)	I (1)	III (1–4)	I (1–4)
Cladonia coccifera	I (1)		III (1–3)	I (1–3)
Cladonia squamosa		I (1–3)	III (1–3)	I (1–3)
Cornicularia aculeata		I (1–3)	II (1–3)	I (1–3)
Cladonia crispata			II (1–4)	I (1–4)
Cladonia deformis			II (1–3)	I (1–3)
Cladonia cornuta			II (1–3)	I (1–3)
Genista anglica	III (1–3)	II (1–4)	III (1–3)	III (1–4)
Hypogymnia physodes	II (1–4)	III (1–6)	III (1–4)	II (1–6)
Luzula multiflora	II (1–3)	II (1–3)	I (1)	II (1–3)
Listera cordata	II (1–3)	II (1–3)		II (1–3)
Rhytidiadelphus loreus	II (1–4)	II (1–3)		II (1–4)
Luzula pilosa	II (1–3)	II (1–3)		II (1–3)
Cladonia pyxidata	I (1–3)	I (1–3)	I (1–3)	I (1–3)
Ptilidium ciliare	I (1–4)	I (1–3)	I (1)	I (1–4)
Antennaria dioica	I (1–3)	I (1–3)	I (1–3)	I (1–3)
Rhytidiadelphus squarrosus	I (1–4)	I (1–3)	I (1–3)	I (1–4)
Polytrichum juniperinum	I (1–3)	I (1–3)	I (1–3)	I (1–3)
Erica tetralix	I (1–3)	I (1–4)		I (1–4)
Betula pubescens seedling	I (1–3)	I (1–3)		I (1–3)
Plagiothecium undulatum	I (1–3)	I (1–3)		I (1–3)
Euphrasia officinalis	I (1–3)	I (1)		I (1–3)
Empetrum nigrum hermaphroditum	I (1–3)	I (1–6)		I (1–6)
Danthonia decumbens	I (1–4)		I (1–3)	I (1–4)
Polytrichum piliferum	I (1–3)		I (1)	I (1–3)
Carex bigelowii		I (5)	I (1–3)	I (1–5)
Huperzia selago		I (1–3)	I (1–3)	I (1–3)

Floristic table H16 *(cont.)*

	a	b	c	16
Cladonia furcata		I (1–3)	I (1–3)	I (1–3)
Number of samples	18	25	13	56
Number of species/sample	24 (15–31)	17 (10–27)	17 (8–24)	19 (8–31)
Shrub height (cm)	28 (15–55)	20 (6–40)	24 (8–40)	20 (6–55)
Ground layer height (mm)	35 (30–40)	32 (10–80)	20	31 (10–80)
Shrub/herb cover (%)	96 (70–100)	87 (75–100)	88 (70–100)	90 (70–100)
Ground layer cover (%)	23 (0–80)	66 (40–95)	48 (5–90)	48 (0–95)
Altitude (m)	339 (92–472)	472 (238–750)	398 (251–700)	405 (92–750)
Slope (°)	15 (2–25)	9 (0–38)	15 (0–40)	12 (0–40)

a   *Pyrola media-Lathyrus montanus* sub-community
b   *Vaccinium myrtillus-Vaccinium vitis-idaea* sub-community
c   *Cladonia* spp. sub-community
16 *Calluna vulgaris-Arctostaphylos uva-ursi* heath (total)

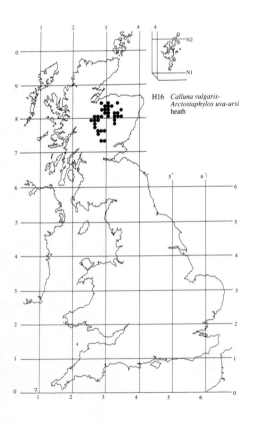

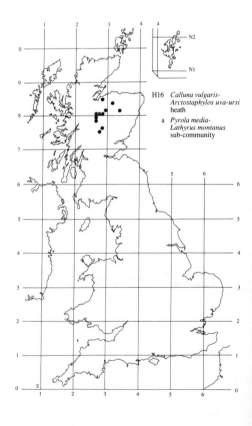

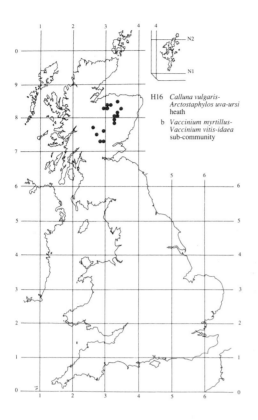

H16 *Calluna vulgaris-*
Arctostaphylos uva-ursi
heath

b *Vaccinium myrtillus-*
Vaccinium vitis-idaea
sub-community

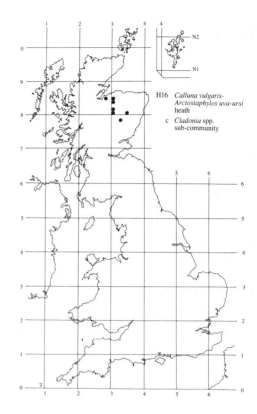

H16 *Calluna vulgaris-*
Arctostaphylos uva-ursi
heath

c *Cladonia* spp.
sub-community

H17
Calluna vulgaris-Arctostaphylos alpinus heath

Synonymy

Arctostaphylos mat   Crampton 1911; *Loiseleuria-Arctous* sociation   Poore & McVean 1957; *Arctoeto-Callunetum*   McVean & Ratcliffe 1962, Prentice & Prentice 1975; *Alectorio-Callunetum vulgaris* (Birse & Robertson 1976) Birse 1980 *p.p.*; *Vaccinio-Ericetum cinereae* (Birse & Robertson 1976) Birse 1980 *p.p.*

Constant species

Arctostaphylos alpinus, Calluna vulgaris, Deschampsia flexuosa, Huperzia selago, Racomitrium lanuginosum, Cladonia arbuscula, C. uncialis.

Rare species

Arctostaphylos alpinus, A. uva-ursi, Loiseleuria procumbens, Cetraria norvegica.

Physiognomy

Arctostaphylos alpinus occurs with some frequency and occasionally with local abundance in various kinds of dwarfed sub-shrub vegetation, but, through the heart of its range, it is most typical of this *Calluna vulgaris-A. alpinus* heath where it is a constant, though generally a subordinate one, in the woody mat. Characteristically here, this is very low, almost always less than a decimetre thick, sometimes just a couple of centimetres and often discontinuous with stretches of bare stones between. It is usually dominated by *Calluna*, the bushes stunted or flattened in the direction of the wind, with the much-contorted branches finely interlocked. *A. alpinus* almost always contributes less than 25% of the cover of the mat, but its prostrate stems can grow beneath the heather and they put up little branched sprigs with rather striking bright green reticulate leaves. *A. uva-ursi*, of generally similar growth habit though with evergreen foliage, also occurs very occasionally in small amounts.

One or other of the crowberries is also very frequent, though of only low to moderate cover, with the creeping stems ramifying among the mat: *Empetrum nigrum* ssp.

hermaphroditum is strongly preferential to the higher altitudes from which this kind of vegetation was first described, and ssp. *nigrum* largely confined to lower situations. *Loiseleuria procumbens*, another creeping evergreen Arctic-Alpine, is also characteristically found with ssp. *hermaphroditum* here, the *Calluna-A. alpinus* heath providing its major locus in the western part of its range. *Erica cinerea*, on the other hand, is a frequent companion of ssp. *nigrum* in lower-altitude stands of the community.

Vaccinium myrtillus occurs commonly throughout, but typically as sparse shoots, and *V. uliginosum* and *V. vitis-idaea* are only very occasionally found. *Salix herbacea* is sometimes seen as small scattered sprigs. *Juniperus communis* ssp. *nana* occurs in some stands, but the generally small contribution of this plant here helps distinguish this vegetation from the *Calluna-Juniperus* heath.

As with other dwarfed sub-shrub communities, few herbs occur frequently throughout the *Calluna-A. alpinus* heath and hardly ever does this element have any abundance. *Huperzia selago* is the commonest of the group and it is often accompanied at higher altitudes by *Diphasium alpinum, Carex bigelowii* and *Antennaria dioica. Deschampsia flexuosa* is also frequent throughout, though more so in lower situations where *Potentilla erecta, Scirpus cespitosus* and *Carex pilulifera* occur most commonly. *Carex panicea, Galium saxatile* and *Solidago virgaurea* can be found very occasionally throughout.

More conspicuous than these are the lichens which form a patchy mosaic growing among and over the sub-shrub mat, sometimes of considerable total cover. *Cladonia arbuscula* and *C. uncialis* are constant, the former often in abundance, and also characteristic, though strongly preferential to higher altitudes, are *Cetraria glauca, C. islandica, Cornicularia aculeata, Alectoria nigricans* and *Sphaerophorus globosus*. The rare *Cetraria norvegica* is a distinctive occasional and there is sometimes a little *Cladonia coccifera* growing on exposed humus with *Ochrolechia frigida* spreading over the vegetation mat. In contrast to the *Calluna-Racomitrium* heath, which has many species in common with this community and a generally similar geographical distribution, mosses are not of great variety or abundance. *Racomitrium lanuginosum* is constant though in small amounts and *Hypnum jutlandicum* becomes frequent at lower altitudes but, apart from these, there is just very occasional *Dicranum scoparium* and *Frullania tamarisci*.

Sub-communities

***Loiseleuria procumbens-Cetraria glauca* sub-community:** *Loiseleuria-Arctous* sociation Poore & McVean 1957; *Arctoeto-Callunetum* McVean & Ratcliffe 1962. In this more distinctive kind of *Calluna-A. alpinus* heath, mixtures of *Calluna* with subordinate *A. alpinus*, *Loiseleuria* and *E. nigrum* ssp. *hermaphroditum* make up the bulk of the mat with scattered shoots of *V. myrtillus*, occasional *V. vitis-idaea*, *Arctostaphylos uva-ursi*, *Juniperus communis* ssp. *nana* and *Salix herbacea*. *Carex bigelowii* and, less commonly, *Deschampsia flexuosa* occur as usually sparsely-distributed individuals with *Antennaria dioica*, *Diphasium alpinum* and, more occasionally, *Euphrasia frigida*, *Lotus corniculatus*, *Festuca vivipara* and *Pinguicula vulgaris* preferential.

Usually small patches of *Racomitrium lanuginosum* are frequent but very much more obvious here is the rich and extensive lichen flora with *Cladonia arbuscula*, *C. uncialis*, *Cetraria glauca*, *C. islandica*, *Alectoria nigricans* and *Sphaerophorus globosus* all very common, *Cladonia coccifera*, *C. pyxidata*, *C. bellidiflora*, *C. gracilis*, *Cetraria norvegica*, *Ochrolechia tartarea* and *O. frigida* occasional.

***Empetrum nigrum* ssp. *nigrum* sub-community:** *Arctoeto-Callunetum* Prentice & Prentice 1975; *Alectorio-Callunetum*, *Agrostis canina* ssp. *montana* subassociation Birse 1980 *p.p.*; *Vaccinio-Ericetum cinereae*, *Cladonia* subassociation Birse 1980 *p.p. Loiseleuria* is found occasionally in this sub-community but there is a noticeable shift away from this plant and *E. nigrum* ssp. *hermaphroditum* as associates in the sub-shrub mat to *E. nigrum* ssp. *nigrum* with *Erica cinerea* also quite common. But *Calluna* still dominates with *A. alpinus* maintaining its constant but usually low cover contribution and, though the mat can be a little thicker, the general appearance is still of very closely wind-pruned vegetation. *Vaccinium myrtillus* again occurs as sparse shoots and *A. uva-ursi* is occasional.

Carex bigelowii is less common here than in the other sub-community but *Deschampsia flexuosa* increases in frequency, *Potentilla erecta* is strongly preferential and

they are often accompanied by small amounts of *Scirpus cespitosus* and *Carex pilulifera* and more occasionally by *Nardus stricta* and *Agrostis canina* (mostly recorded as ssp. *montana*).

Lichens are not so varied or abundant in this subcommunity but *Cladonia arbuscula* and *C. uncialis* remain very common and there is sometimes a little *Cornicularia aculeata* and *Alectoria nigricans* as well as the community occasionals *Ochrolechia frigida* and *Cladonia coccifera*. *Hypnum jutlandicum* frequently joins *Racomitrium lanuginosum* and there are scarce records for such oceanic hepatics as *Diplophyllum albicans* and *Scapania gracilis*.

Habitat

The *Calluna-A. alpinus* heath is the typical climax subshrub vegetation of rather base-poor moder soils over very exposed ridges and crests at moderate to fairly high altitudes in the cold and humid climate of the mountains of north-west Scotland.

Compared with the *Calluna-Racomitrium* heath, the distribution of which is roughly similar to that of this community, the environmental demands of the *Calluna-A. alpinus* heath are a little more tightly drawn. Both are essentially vegetation types of the low-alpine zone (Poore & McVean 1957) in those parts of Scotland beyond the Great Glen, but the *Calluna-A. alpinus* heath is rather more strongly confined to higher ground and hence spread over not so great a geographical area. Thus, it can be found along the north Scottish coast and on Orkney at altitudes down to about 250 m or even less, but generally speaking stands are concentrated between 500 and 750 m, with some exceptional localities up to around 900 m. Absent from many of the lower-level spurs along the seaward edge of the north-west Highlands, the *Calluna-A. alpinus* heath is thus concentrated in the mountains between the Torridan hills to the west and Ben Wyvis in the north with a further cluster of sites through central Sutherland.

Throughout this geographical area, mean annual maxima are for the most part below 21 °C (Conolly & Dahl 1970), so this vegetation often takes *Calluna* to its altitudinal limit and up into the climatic zone where Arctic-Alpines like *A. alpinus*, *E. nigrum* ssp. *hermaphroditum*, *Loiseleuria* and *Carex bigelowii* and other montane plants such as *Antennaria dioica*, *Diphasium alpinum*, *Euphrasia frigida*, *Cetraria islandica*, *C. norvegica*, *Alectoria nigricans* and *Ochrolechia tartarea*, reflect the generally harsh conditions. These are the preferentials of the *Loiseleuria-Cetraria* sub-community which formed the core of early definitions of the *Calluna-A. alpinus* heath (Poore & McVean 1957, McVean & Ratcliffe 1962) and which is the most widespread form of this vegetation towards its upper altitudinal limits.

Compared with similar levels in the eastern High-lands, the winter climate through the range of the community is not so bitterly cold, so annual temperature fluctuations are somewhat reduced and the growing season a little longer. Even in the *Loiseleuria-Cetraria* sub-community, the presence of *Cetraria glauca* as a frequent member of the lichen mat presents a floristic contrast to the *Calluna-Cladonia* heath which is the characteristic dwarfed sub-shrub vegetation at similar altitudes in the more continental east. But, it is in the other sub-community that the influence of the more equable regional climate is best seen. Here are included essentially similar mats of *Calluna* and *A. alpinus* which take the distribution of the community below 500 m, where the mean annual maxima can rise above 21 °C (Conolly & Dahl 1970) and where February minima are not so low as in the mountainous heartland of the north-west of Scotland. In such situations, *E. nigrum* ssp. *nigrum* takes over as the typical crowberry, the more extreme Arctic-Alpines become increasingly sporadic, and such plants as *Erica cinerea*, *Scirpus cespitosus* and *Carex pilulifera* give a more oceanic or sub-montane look to the vegetation. It is here, too, that the bryophyte component makes its maximal contribution with *Hypnum jutlandicum* often joining *Racomitrium lanuginosum*, the community providing an occasional locus, too, for some Atlantic hepatics.

In such situations, where a milder regional climate is combined with very humid atmospheric conditions, the *Calluna-A. alpinus* heath most closely approaches the *Calluna-Racomitrium* heath, where such vascular associates are more uniformly common and where a carpet of bulky pleurocarpous mosses is usually very conspicuous. However, despite the fact that, through the range of the *Calluna-A. alpinus* heath, annual precipitation is usually over 1600 mm (*Climatological Atlas* 1952) with generally more than 180 wet days yr^{-1} (Ratcliffe 1968), the balance of dominance among the cryptogams here is almost always with the lichens, as in the *Calluna-Cladonia* heath. The very marked local exposure to winds, even more severe here than in the *Calluna-Racomitrium* heath, probably plays a part in this, not only producing a stunted and pruned mat of sub-shrubs but also leading to increased evapo-transpiration with no prospects of shelter under winter snow. Even in its low-altitude stations, then, this community is essentially a lichen-heath rather than a moss-heath, characteristic of severely wind-blasted spurs and ridges, flat or gently-sloping, or steeper very exposed crests.

There is perhaps also some slight edaphic difference between the habitats of the *Calluna-A. alpinus* and *Calluna-Racomitrium* heaths. Again, here, the parent materials are generally very lime-poor though lithologically varied, including bedrocks and morainic debris of Torridonian sandstones and Cambrian quartzite to the west, Moine schists through the central and eastern parts of the range, and, on Orkney, Devonian Old Red Sandstone. Typically, these weather to humic rankers, sometimes very fragmentarily disposed over fairly fine detritus with solifluctional disturbance, or occasionally more mature podzolised profiles. Superficial pH is still only 4–5 such that the flora is prevailingly calcifuge but the humus seems distinctly moder-like, so there is perhaps the same amelioration of extreme edaphic impoverishment as is found beneath heaths with *A. uva-ursi* in eastern Scotland.

In general, however, this is a very inhospitable environment and the harsh climatic conditions maintain the vegetation as a climax. The community may provide an occasional bite for sheep and deer, but this probably has little effect on the floristics or physiognomy. Burning, though, is very deleterious and may cause damage from which recovery is extremely slow or perhaps impossible where erosion of the exposed surface is initiated. This kind of heath may have been eliminated from many sites through its range, though it does not seem to have such a relic distribution as the *Calluna-Juniperus* heath.

Zonation and succession

The *Calluna-A. alpinus* heath is typically found with other kinds of sub-shrub vegetation, snow-bed communities, mires and montane sedge-heaths in zonations and mosaics which reflect variations in climate and soils with altitude and local topography and, towards its lower limits, the impact of treatments. With the geographical shift towards the more continental eastern parts of Scotland, there is a replacement of many of the elements in these more oceanic vegetation patterns.

The characteristic position of the *Calluna-A. alpinus* heath, occupying the most exposed situations in the low-alpine zone of the north-west Highlands, is well seen around Beinn Eighe (Ratcliffe 1977). There the community is found on windswept stony moraines on gentle mid-altitude slopes with, below, over the more sheltered drift-covered hillsides, a transition to the *Vaccinium-Rubus* heath or, on steeper sunless slopes with constantly high humidity, the *Calluna-Vaccinium-Sphagnum* heath. *Calluna* usually remains dominant in both these vegetation types and *E. nigrum* ssp. *hermaphroditum* continues as a constant but Vaccinia, particularly *V. myrtillus*, become more prominent and beneath the taller, more luxuriant canopy of sub-shrubs, there is a rich and extensive carpet of pleurocarpous mosses and Atlantic hepatics, especially in the *Calluna-Vaccinium-Sphagnum* heath which gives well-developed stands a strikingly different appearance.

In some places, as above Loch Maree, fragments of the *Pinus-Hylocomium* woodland give a clue as to the original forest cover of this zone of damp heather-moor but usually the vegetational variety towards the submontane slopes is provided by transitions to *Scirpus-Erica* wet heath on moist peaty soils and then to *Scirpus-Eriophorum* mire on the low-level ombrogenous bogs. Direct zonations from the *Calluna-A. alpinus* heath to the *Scirpus-Erica* heath sometimes show considerable floristic continuity where *Calluna, Erica cinerea, Scirpus cespitosus* and *Racomitrium lanuginosum* run on with an abundance of lichens over the eroding margins of a peat mantle on stony morainic material.

In the north-west Highlands, the *Calluna-A. alpinus* heath overlaps altitudinally with the *Calluna-Juniperus* heath and, though this is now a much more local community, the two can sometimes be found in close proximity, as over the north-east slopes of Beinn Eighe, the latter tending to occur a little lower down the mountain sides and best developed on slopes with some shelter and shade. In both composition and physiognomy, the vegetation types are very close with *Calluna, E. nigrum* ssp. *hermaphroditum, A. alpinus* and *A. uva-ursi* all maintaining high frequency in a dwarfed sub-shrub cover, but *Juniperus communis* ssp. *nana* is scarce in the *Calluna-A. alpinus* heath and the distinctive hepatics of the juniper heath, *Pleurozia purpurea, Diplophyllum albicans* and *Frullania tamarisci* hardly ever occur in the more exposed situations at these altitudes.

The *Calluna-Juniperus* heath is a chionophobous community and in hollows among stretches of *Calluna-A. alpinus* heath where snow accumulates, there can be sharp transitions to *Nardus-Carex* vegetation of various kinds. This community can run on up into the mid-alpine zone where shelter is sufficient to allow a snow cover to remain but, very typically through the north-west Highlands, the *Calluna-A. alpinus* heath gives way over exposed rounded high-altitude spurs and summits to the *Carex-Racomitrium* heath, the characteristic moss-heath and fell-field vegetation of our cold oceanic mountains. Again, there can be some floristic continuity through *Racomitrium lanuginosum* and the lichens and *Carex bigelowii, Vaccinium myrtillus* and *Deschampsia flexuosa* and the physiognomic similarity of the communities is well seen where solifluction and wind erosion combine to fragment the cover of the heath into a mosaic of vegetation among stretches of bare morainic gravel. Typically, however, heather does not extend to the higher altitudes and, though *Loiseleuria* can occasionally be found, the summit vegetation is typically moss-dominated and often distributed over striking tracts of open ablation surfaces. In fact, over the rugged peaks of Beinn Eighe, the *Carex-Racomitrium* heath is not nearly so extensive in the zone above the *Calluna-A. alpinus* heath as on Beinn Lair in Letterewe, over the Fannich Hills, on the massive rounded tops of Beinn Dearg and, towards the north-western limit of the mainland range of the community, on Ben Klibreck. On the last site, too,

can be seen an alternative transition over high-altitude plateau to *Calluna-Eriophorum* blanket mire.

Throughout this main part of its range, though again more obviously on some of these other mountains than on Beinn Eighe, the *Calluna-A. alpinus* heath overlaps geographically and altitudinally with the closely-similar *Calluna-Racomitrium* heath. Sometimes, as along the seaboard of the north-west Highlands, where broad exposed spurs extend down to relatively low altitudes, it is this latter community that is predominant, with the *Calluna-A. alpinus* heath prevailing with the shift into the higher, mountainous heartland. But, as in the Letterewe Hills and Beinn Dearg and also on Hoy in Orkney, the two can occur contiguously, the moss-heath perhaps occupying the somewhat less sheltered situations with mor soils but sometimes showing no apparent habitat preference. Towards the eastern margin of the range of the *Calluna-A. alpinus* heath there is a further complication in that the geographical ranges of the western heaths overlap there with that of the *Calluna-Cladonia* heath. On Ben Wyvis and Ben Klibreck, for example, the latter occurs in mosaics with the *Calluna-A. alpinus* heath, the distribution of the alpine bearberry often effectively serving to distinguish the two.

All these three dwarfed sub-shrub heaths are climatic climax communities but they are often contiguous below with vegetation markedly influenced by treatments, particularly the burning and grazing associated with hill-pasturing. The *Calluna-A. alpinus* heath thus sometimes gives way below, or towards its lower altitudinal limits is surrounded by plagioclimax Nardo-Galion swards, such as the *Festuca-Agrostis-Galium* or *Nardus-Galium* grasslands, in which sub-shrubs may survive patchily but where dominance lies with the grasses and dicotyledons which make but an occasional appearance here.

Distribution

The *Calluna-A. alpinus* heath is confined to northern Scotland where the *Loiseleuria-Cetraria* sub-community is widespread and local through the north-west Highlands, with the *Empetrum* sub-community extending the range to lower altitudes along the northern Scottish coast and Orkney.

Affinities

Poore & McVean (1957) provided the first detailed diagnosis of this kind of vegetation, known previously just from Crampton's (1911) account of Caithness: McVean & Ratcliffe's (1962) *Arctoeto-Callunetum* included essentially the same assemblage and is largely preserved here as the more typical and widespread *Loiseleuria-Cetraria* sub-community. But it also seems sensible to include with this the kinds of heather-bearberry mats described from lower altitudes by

Prentice & Prentice (1975) and Birse (1980): this dilutes somewhat the floristic definition provided in the early accounts but illuminates the transitions to the more oceanic *Calluna-Racomitrium* and *Calluno-Erica* heaths.

Both McVean & Ratcliffe (1962) and Prentice & Prentice (1975) saw their *Arctoeto-Callunetum* as belonging to the Caricetea curvulae, rather than the sub-montane heaths, and a variety of dwarfed sub-shrub communities of generally similar type has been des-cribed from Scandinavia: the Dichte *Calluna*-Assozia-tion and *Cetraria nivalis-Alectoria ochroleuca*-reiche *Loiseleuria* Assoziation from Sylene (Nordhagen 1928) and the *Alectorio-Arctostaphyletum uvae-ursi* and *Cetrarietum nivalis typicum* of the Rondane (Dahl 1956).

Floristic table H17

	a	b	17
Calluna vulgaris	V (4–8)	V (5–10)	V (4–10)
Arctostaphylos alpinus	V (1–6)	V (1–6)	V (1–6)
Racomitrium lanuginosum	V (1–6)	V (1–5)	V (1–6)
Cladonia uncialis	V (1–4)	V (1–4)	V (1–4)
Huperzia selago	IV (1–3)	IV (1–3)	IV (1–3)
Cladonia arbuscula	IV (1–6)	IV (1–4)	IV (1–6)
Deschampsia flexuosa	III (1–3)	V (1–4)	IV (1–4)
Loiseleuria procumbens	V (1–6)	II (1–4)	III (1–6)
Cornicularia aculeata	IV (1–3)	II (1–3)	III (1–3)
Alectoria nigricans	IV (1–4)	II (1–4)	III (1–4)
Carex bigelowii	IV (1–4)	II (1–4)	III (1–4)
Cetraria glauca	V (1–4)	I (1)	III (1–4)
Sphaerophorus globosus	V (1–3)	I (1)	III (1–3)
Cetraria islandica	V (1–4)	I (1–3)	III (1–4)
Empetrum nigrum hermaphroditum	IV (1–5)		III (1–5)
Antennaria dioica	III (1–3)	I (1–3)	II (1–3)
Diphasium alpinum	III (1–3)		II (1–3)
Festuca vivipara	II (1–3)	I (1–4)	I (1–4)
Cladonia gracilis	II (1–3)	I (1)	I (1–3)
Ochrolechia tartarea	II (1–3)		I (1–3)
Cladonia pyxidata	II (1–3)		I (1–3)
Euphrasia frigida	II (1–3)		I (1–3)
Cladonia bellidiflora	II (1–3)		I (1–3)
Cetraria norvegica	II (1–3)		I (1–3)
Lotus corniculatus	II (1–3)		I (1–3)
Juniperus communis nana	II (1–4)		I (1–4)
Vaccinium vitis-idaea	II (1–3)		I (1–3)
Pinguicula vulgaris	II (1–3)		I (1–3)
Juncus trifidus	I (1–4)		I (1–4)
Cladonia rangiferina	I (1–4)		I (1–4)
Hypnum jutlandicum	I (1–3)	V (1–4)	III (1–4)
Empetrum nigrum nigrum	I (4)	IV (1–4)	III (1–4)
Potentilla erecta	I (1)	IV (1–3)	III (1–3)
Scirpus cespitosus	II (1–3)	III (1–4)	II (1–4)
Erica cinerea	I (1–3)	III (1–6)	II (1–6)
Carex pilulifera	I (1–3)	III (1–3)	II (1–3)
Nardus stricta	I (1)	II (1–3)	I (1–3)

Agrostis canina	I (1)	II (1–4)	I (1–4)
Diplophyllum albicans	I (1)	II (1–3)	I (1–3)
Scapania gracilis		I (1–3)	I (1–3)
Vaccinium myrtillus	III (1–3)	III (1–4)	III (1–4)
Cladonia coccifera	II (1–3)	II (1–4)	II (1–4)
Salix herbacea	II (1–3)	II (1–4)	II (1–4)
Ochrolechia frigida	II (1–9)	II (1–4)	II (1–9)
Arctostaphylos uva-ursi	II (1–3)	II (1–4)	II (1–4)
Carex panicea	II (1–4)	II (1–4)	II (1–4)
Dicranum scoparium	I (1–3)	I (1–3)	I (1–3)
Hypogymnia physodes	I (1–3)	I (1–3)	I (1–3)
Galium saxatile	I (1–2)	I (1–3)	I (1–3)
Frullania tamarisci	I (1–3)	I (1–3)	I (1–3)
Vaccinium uliginosum	I (1)	I (1–4)	I (1–4)
Festuca ovina	I (1–3)	I (1–3)	I (1–3)
Solidago virgaurea	I (1–3)	I (1–3)	I (1–3)
Number of samples	24	18	42
Number of species/sample	24 (9–31)	18 (10–24)	21 (9–31)
Vegetation height (cm)	3 (1–10)	6 (1–10)	5 (1–10)
Vegetation cover (%)	89 (50–100)	83 (50–90)	86 (50–100)
Altitude (m)	658 (518–890)	331 (185–490)	531 (185–890)
Slope (°)	3 (0–10)	5 (0–32)	5 (0–32)

a *Loiseleuria procumbens-Cetraria glauca* sub-community

b *Empetrum nigrum nigrum* sub-community

17 *Calluna vulgaris-Arctostaphylos alpinus* heath (total)

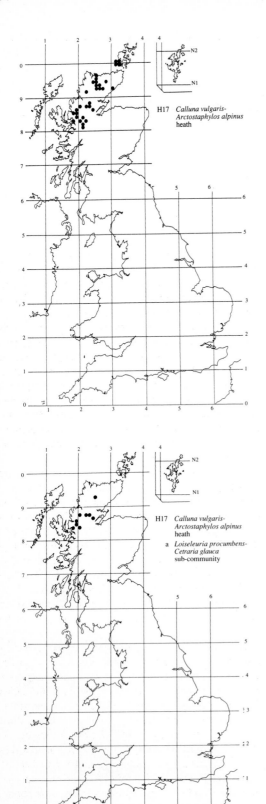

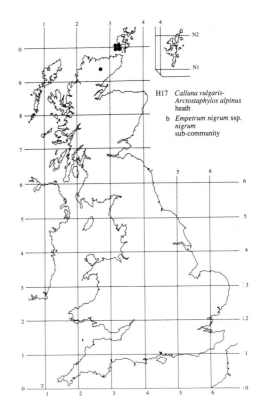

H17   *Calluna vulgaris-*
      Arctostaphylos alpinus
      heath

H17   *Calluna vulgaris-*
      Arctostaphylos alpinus
      heath

b   *Empetrum nigrum* ssp.
     nigrum
     sub-community

H17   *Calluna vulgaris-*
      Arctostaphylos alpinus
      heath

a   *Loiseleuria procumbens-*
     Cetraria glauca
     sub-community

H18
Vaccinium myrtillus-Deschampsia flexuosa heath

Synonymy

Vaccinietum myrtilli Smith 1900, Lewis & Moss 1911, Tansley 1939, Fidler *et al.* 1971; *Vaccinium*-ridge & *Vaccinium*-summit Smith & Moss 1903, Smith & Rankin 1903; *Gramino-Vaccinetum* Smith 1911*b*; *Vaccinium*-edge Moss 1913; *Empetreto-Vaccinetum* Burges 1951 *p.p.*; Lichen-rich *Vaccinium-Festuca* association Poore 1955*b*; *Vaccinium-Chamaepericlymenum* nodum Poore & McVean 1957 *p.p.*; *Vaccineto-Empetretum* McVean & Ratcliffe 1962; *Festuceto-Vaccinetum* McVean & Ratcliffe 1962, Evans *et al.* 1977, Ferreira 1978; *Festuca ovina/Deschampsia flexuosa* grassland King 1962 *p.p.*; *Vaccinium myrtillus*-heath moss sociation Edgell 1969; Mountain *Vaccinium* heaths Gimingham 1972 *p.p.*; *Phyllodoce caerulea* sites Coker & Coker 1973 *p.p.*; *Huperzio-Vaccinetum* Hill & Evans 1978; *Rhytidiadelphus loreus-Vaccinium myrtillus* community Birse 1980.

Constant species

Deschampsia flexuosa, Galium saxatile, Vaccinium myrtillus, Dicranum scoparium, Pleurozium schreberi.

Rare species

Loiseleuria procumbens, Minuartia sedoides, Phyllodoce caerulea, Salix lapponum, Barbilophozia lycopodiodes, Scapania ornithopodiodes.

Physiognomy

The *Vaccinium myrtillus-Deschampsia flexuosa* heath includes a variety of moss-rich and grassy sub-shrub vegetation in which *Vaccinium myrtillus* is the most frequent and generally the most abundant ericoid, with *Calluna vulgaris* typically having a rather inconspicuous role: heather is usually only occasional here and present as scattered plants that are often noticeably lacking in vigour. *V. myrtillus* is not always truly dominant, however, and other sub-shrubs sometimes make a sizeable contribution to the canopy, which is generally 1–2 dm tall, occasionally rather open and, even where extensive,

is without that uniform density of growth found in heaths with much young *Calluna*. Most frequent among the associated sub-shrubs is *Empetrum nigrum*, usually ssp. *hermaphroditum* where it has been possible to distinguish the taxa, though sometimes clearly ssp. *nigrum*, and occasionally both plants growing together. Typically, the procumbent stems of the crowberry penetrate quite far among the bilberry, branching out into patches which can be locally abundant. Some of the vegetation subsumed here has co-dominant *E. nigrum* as in the various kinds of *Vaccinio-Empetretum* described by Burges (1951) and McVean & Ratcliffe (1962), and, of course, being evergreen, the crowberry catches the eye more than the deciduous bilberry in stands without a covering of winter snow.

V. vitis-idaea is about as common as *E. nigrum*, though typically much less abundant. *V. uliginosum*, on the other hand, is rather scarce and preferential for one sub-community but it can have locally high cover there, even attaining co-dominance in some stands (Poore & McVean 1957, McVean & Ratcliffe 1962). Also strongly diagnostic of one particular kind of *Vaccinium-Deschampsia* heath, and occurring occasionally elsewhere, is *Alchemilla alpina*, strictly speaking a woody herb, but able to make quite bushy growth among the ericoids and to contribute substantially to the canopy in certain cases.

The rare *Loiseleuria procumbens*, more a plant of wind-blasted lichen-rich heaths, occurs very occasionally and inaccessible patches of this community sometimes shelter bushes of *Salix lapponum*. The *Vaccinium-Deschampsia* heath is also the locus in this country for the very restricted *Phyllodoce caerulea*, a sub-Arctic ericoid of low, bushy growth, known in Britain only from the Sow of Atholl in Perthshire and the Ben Alder range in Inverness (McBeath 1967, Coker & Coker 1973, Perring & Farrell 1977). Here, its survival looks very precarious and the longest-known colony at the former site has certainly suffered from the predations of collectors and growers, but its close vegetative resemblance to

Empetrum and its rather shy flowering habit afford the plant some protection.

The vascular associates of the sub-shrubs vary considerably in their number and prominence. Constant throughout are *Deschampsia flexuosa* and *Galium saxatile* with *Nardus stricta*, *Agrostis canina* ssp. *montana* and *Potentilla erecta* all very frequent, and in some stands these provide virtually all the herbaceous element and a rather sparse cover among the bushes. *Carex bigelowii* occurs occasionally but its relative infrequency and generally low cover help distinguish the community from the *Vaccinium-Cladonia* heath with which it is sometimes closely associated (e.g. Burges 1951). Then, there is occasionally some *Dryopteris dilatata* and *Cryptogramma crispa* with *Blechnum spicant* becoming very common in one sub-community.

As a group, though, it is grasses which often make the most notable contribution to the herbaceous element of the vegetation, with *Festuca ovina* (much less commonly *F. vivipara* and *F. rubra*), *Agrostis capillaris* and *Anthoxanthum odoratum* occurring at least occasionally and, in some sub-communities, increasing considerably in frequency and abundance, so as to become with *D. flexuosa*, *Nardus* and *A. canina*, a sometimes co-dominant component of the cover, as in the various kinds of grassy-heath subsumed here (Smith 1911*b*, Poore 1955*b*, McVean & Ratcliffe 1962). In such stands, too, small monocotyledons such as *Luzula campestris*, *Carex pilulifera* and *C. binervis* tend to have their best representation in the community.

The other element of the vegetation which is frequently prominent comprises bulky mosses. *Dicranum scoparium*, *Pleurozium schreberi* and *Hypnum cupressiforme s.l.* (often *H. jutlandicum*) are very common throughout, *Hylocomium splendens* is also conspicuous in many stands and, in various kinds of *Vaccinium-Deschampsia* heath, there can be frequent *Rhytidiadelphus loreus*, *R. squarrosus*, *Plagiothecium undulatum*, *Dicranum majus* and *Racomitrium lanuginosum*. *Ptilidium ciliare* and *Polytrichum commune* are also found occasionally throughout and a variety of other species are scarce associates: *Rhytidiadelphus triquetrus*, *Thuidium tamariscinum*, *Polytrichum alpestre*, *P. alpinum* and *Pohlia nutans*.

Some lichens also occur frequently and in all types of *Vaccinium-Deschampsia* heath there can be found stands in which these are rather more varied than usual, but it is in only one of the sub-communites that they become even moderately abundant and there is really never attained here the kind of rich and extensive lichen carpet typical of the *Vaccinium-Cladonia* heath. Among the commonest species are *Cladonia arbuscula*, *C. impexa* and *C. uncialis*, with *C. pyxidata*, *C. rangiferina*, *C. gracilis* and *Cetraria islandica* occurring much less frequently.

Sub-communities

Hylocomium splendens-Rhytidiadelphus loreus sub-community: *Vaccinetum myrtilli* Smith 1900, Lewis & Moss 1911, Tansley 1939; *Vaccinium*-ridge & *Vaccinium*-summit Smith & Moss 1903, Smith & Rankin 1903; *Vaccinium*-edge Moss 1913; *Empetreto-Vaccinetum* mossy facies Burges 1951; *Vaccinium-Chamaepericlymenum* nodum Poore & McVean 1957 *p.p.*; *Vaccineto-Empetretum* McVean & Ratcliffe 1962; *Vaccinium myrtillus*-heath moss sociation Edgell 1969; *Phyllodoce caerulea* sites Coker & Coker 1973 *p.p.*; *Rhytidiadelphus loreus-Vaccinium myrtillus* community Birse 1980. Sub-shrubs are generally clearly dominant here, but with mosses forming an extensive ground carpet. In many stands, *V. myrtillus* is overwhelmingly abundant, but *E. nigrum* is occasionally co-dominant and, in some stands, *V. vitis-idaea* or *V. uliginosum*, this last preferential here, though still not common. *Calluna* is rather more frequent than usual in the community, though still of characteristically low cover and, although *Alchemilla alpina* occurs occasionally this, too, is rarely abundant. The small colonies of *Phyllodoce* on Sgur Iuthurn and Meall an-t-Slugain appear to be in this kind of *Vaccinium-Deschampsia* heath and other stands have been found with bushes of *Salix lapponum*.

Deschampsia flexuosa is very common and sometimes quite abundant and there is occasionally also some *Nardus*, *Agrostis canina* ssp. *montana*, *A. capillaris* and *Anthoxanthum* but, as a rule, grasses are nothing like so consistent in their frequency and total abundance here as in the *Alchemilla-Carex* sub-community. The sedges that are typically found there are also very scarce in this vegetation, though *Carex bigelowii* occurs occasionally and *Luzula sylvatica* is a preferential occasional, sometimes with moderately high cover. A fairly high frequency of *Blechnum spicant* is also a good diagnostic feature and there is occasionally some *Melampyrum pratense*, *Oxalis acetosella* and *Cornus suecica* but, along with *Galium saxatile* and *Potentilla erecta*, more typical of the community as a whole, these are often the only herbs represented.

The bryophyte element, on the other hand, attains its most diverse and extensive cover in this kind of *Vaccinium-Deschampsia* heath, often forming a quite luxuriant carpet among the sub-shrub branches and over their decumbent shoots. *Hylocomium splendens* and *Rhytidiadelphus loreus* join *Dicranum scoparium* and *Pleurozium schreberi* as constants and there are occasional records for *Plagiothecium undulatum*, *Dicranum majus*, *Hylocomium umbratum*, *Sphagnum quinquefarium* and *S. capillifolium*, for *Barbilophozia floerkii*, *Anastrepta orcadensis* and very occasionally for such rare hepatics as

Scapania ornithopodioides and *Barbilophozia lycopo-diodes*. The community associates *Hypnum cupressi-forme s.l.*, *Polytrichum commune* and *Ptilidium ciliare* also remain occasional to frequent.

Lichens are generally much less obvious, although *Cladonia arbuscula* occurs often and *C. uncialis*, *C. gracilis* and *Cetraria islandica* can occasionally be found and when these occur together and with a little *Carex bigelowii*, the vegetation approaches the *Racomitrium-Cladonia* sub-community in its composition.

Alchemilla alpina-Carex pilulifera **sub-community:** *Gra-mino-Vaccinetum* Smith 1911*b*; Lichen-rich *Vacci-nium-Festuca* association Poore 1955*b*; *Festuceto-Vaccinetum* McVean & Ratcliffe 1962, Evans *et al.* 1977, Ferreira 1978; *Festuca ovina/Deschampsia flex-uosa* grassland King 1962; *Phyllodoce caerulea* sites Coker & Coker 1973 *p.p.* *V. myrtillus* can still be quite abundant here but *E. nigrum* is much scarcer than in the first sub-community and *V. vitis-idaea* and *Calluna*, though quite frequent, make little contribution to the cover. Indeed, the ericoids as a whole are often co-dominant with *Alchemilla alpina*, which is strongly preferential here, and/or with grasses which also have their best representation as a group in this sub-commun-ity. Along with *D. flexuosa*, *Festuca ovina* is especially frequent and abundant but *Agrostis capillaris* and *Anthoxanthum* are also common and can have modera-tely high cover and there is occasionally some *Danthonia decumbens*, *Deschampsia cespitosa*, *Festuca rubra*, *F. vivipara*, *Nardus stricta* and *Agrostis canina* ssp. *mon-tana*. *Luzula campestris* and *Carex pilulifera* are also strongly diagnostic, though not usually abundant, and there is occasionally some *C. binervis* and *C. panicea*. It is amongst this kind of vegetation that *Phyllodoce* occurs on the Sow of Atholl (Coker & Coker 1973).

There is also quite often a variety of dicotyledonous herbs. Along with *Galium saxatile*, *Potentilla erecta* attains its highest frequency here and there is occa-sionally some *Campanula rotundifolia*, *Viola riviniana*, *Ranunculus acris*, *Polygala serpyllifolia*, *Anemone nemorosa* and *Veronica officinalis*. Sometimes, bulkier plants such as *Alchemilla glabra* and *Rumex acetosa* can figure and, when *Luzula sylvatica* is also present, the vegetation can approach tall-herb ledge communities in its appearance. More often, however, the herbage is rather short and quite commonly cropped into a heathy sward. Where plants like *Thymus praecox* and *Carex pulicaris* make an occasional appearance in such stands, the *Vaccinium-Deschampsia* heath comes closest to the *Festuca-Agrostis-Alchemilla* grass-heath.

As in the *Hylocomium-Rhytidiadelphus* sub-commun-ity, bulkier mosses can be quite frequent here, though they do not have the same variety and abundance: *Dicranum scoparium*, *Pleurozium schreberi*, *Hyloco-mium splendens* and *Hypnum cupressiforme s.l.* all remain very common and preferentially there is often some *Rhytidiadelphus squarrosus*, but these usually occur as scattered shoots among the turf. In the fairly dense grassy herbage, lichens are sparse with typically just a little *Cladonia arbuscula* and scarce *C. impexa*, *C. uncialis* and *C. gracilis*.

Racomitrium lanuginosum-Cladonia **spp. sub-commun-ity:** *Vaccinetum myrtilli* Fidler *et al.* 1971; *Huperzio-Vaccinetum* Hill & Evans 1978. *V. myrtillus* or mix-tures of this with *E. nigrum* usually dominate in this sub-community, with rather infrequent sparse plants of *Calluna* and *V. vitis-idaea*. *D. flexuosa* and *F. ovina* are both very frequent and each can be abundant but other grasses tend to be rather poorly represented with just occasional plants of *Nardus*, *A. capillaris*, *A. canina* ssp. *montana* and *Anthoxanthum*. *Carex pilulifera* can some-times be found but other preferentials of the *Alchemilla-Carex* sub-community are rare. *Carex bigelowii* is infrequent but locally abundant.

Apart from *Galium saxatile*, herbaceous dicotyledons in general are rather uncommon with just occasional *Potentilla erecta* and the only vascular preferentials are *Diphasium alpinum* and *Huperzia selago*, with even these occurring at low frequency. Except for *Pleurozium schreberi* and *Hypnum cupressiforme s.l.*, the pleuro-carps common elsewhere are rather inconspicuous here, although *Racomitrium lanuginosum* is more frequent than usual and, along with scattered tufts of *Dicranum scoparium*, there is occasionally a little *Campylopus paradoxus*. Lichens, however, tend to have their best representation here, with *Cladonia arbuscula* showing locally high cover and *C. uncialis* and *C. impexa* prefer-ential at low frequency.

Habitat

The *Vaccinium-Deschampsia* heath is typical of moist but free-draining, base-poor to circumneutral soils over steeper slopes at moderate to high altitudes through the uplands of northern Britain. The generally cold and damp character of the climate is often locally enhanced by a sunless aspect and snow-lie in sheltered situations can play some part in determining the floristics and distribution of the community. At higher levels, this kind of vegetation is probably natural but, towards the sub-montane zone, it may have been derived by burning and grazing and, in some places, treatments have preci-pitated its spread on to blanket peats.

In broad terms, this community represents an exten-sion to higher altitudes of the kind of mixed sub-shrub vegetation seen in less assiduously managed stands of the *Calluna-Vaccinium* heath (or, through the southern Pennines, its polluted equivalent, the *Calluna-Des-champsia* heath). The *Vaccinium-Deschampsia* heath has a roughly similar overall geographical range to these two, but they are essentially sub-montane in their distribution, being found mostly between 200 and 600 m and extending into areas where the climate is relatively mild. The *Vaccinium-Deschampsia* heath, by contrast, is largely confined to altitudes above 400 m and it often extends up to 800 m, with a mean height in available samples of around 600 m, and although it can be found over higher ground in Wales, through the Pennines, the Lake District and the Southern Uplands, it is strongly concentrated in northern Scotland, and particularly in the central and eastern Highlands where the climate is distinctly harsh. At these generally higher altitudes the summers are cool, with mean annual maximum temperatures for the most part below 22 °C (Conolly & Dahl 1970) and winters, particularly in the heartlands of its range, are bitter. Such conditions are reflected in the composition of the vegetation in the disappearance of *Erica cinerea*, a rather oceanic plant already at some disadvantage in the *Calluna-Vaccinium* heath, but which here just cannot tolerate the even lower winter temperatures (Bannister 1965); and in the increased frequency of the Arctic-Alpines *V. vitis-idaea* (Ritchie 1955a) and *E. nigrum* ssp. *hermaphroditum* as against the more broadly montane ssp. *nigrum* (Bell & Tallis 1973). The fairly common occurrence of *Carex bigelowii*, a third species whose national distribution pattern roughly matches that of the community, also marks this move into the montane zone, though it is by no means as frequent here as in the more exposed bilberry heaths of these altitudes or of the moss-heaths and fell-field vegetation above.

The second climatic effect relates to precipitation. In fact, this is not especially high through much of the range of the community: conditions are very wet towards the north-west Highlands, where annual precipitation can far exceed 1600 mm with over 200 wet days yr^{-1} but, in many areas, the levels are between 1200 and 1600 mm, with 180–200 wet days yr^{-1} (*Climatological Atlas* 1952, Ratcliffe 1968). This is sufficient, however, to maintain a generally humid atmosphere throughout the year, and to keep the soils moist, particularly where the community extends on to shaded and sheltered northern and eastern slopes, a common occurrence. In such situations, too, the winter snow, which can be frequent and heavy through the range of the community but especially so in the central and eastern Highlands (Manley 1940), is able to persist long. Quite often, then, this is a distinctly chionophilous vegetation type, marking out early snow-beds or the fringes of more long-lasting accumulations or just more sheltered sites over generally wind-lashed slopes. Even where snow-lie is not appreciable, however, the prevailingly damp conditions strongly favour the vigorous growth of *V. myrtillus* (Ritchie 1956) and contribute to the poor performance here of *Calluna*. Most stands of the *Vaccinium-Des-*

champsia heath do, in fact, occur below the altitudinal limit of *Calluna* but, at these heights, heather tends to be better represented over more exposed slopes where, though often reduced to a tight mat of flattened bushes, it can withstand the bitterly cold, but drier, conditions (Watt & Jones 1948). The difference in exposure and humidity between the two kinds of habitat is also seen in the contrasting cryptogam element in the vegetation cover: whereas it is lichens that predominate among the dwarfed sub-shrubs of the *Calluna-Cladonia* heath and the *Vaccinium-Cladonia* heath that replaces it at higher altitudes, mosses are typically much more abundant beneath the taller but often rather open (and in part deciduous) canopy of the *Vaccinium-Deschampsia* heath. *Dicranum scoparium* and pleurocarps such as *Pleurozium schreberi*, *Hypnum cupressiforme*, *Hylocomium splendens* and *Rhytidiadelphus loreus*, which provide the most consistent contribution, are well able to subsist over the fairly loose damp litter that accumulates beneath the sub-shrubs and among the culms in grassier stands.

The combination of cold with some shelter associated with long snow-lie is probably also of prime importance for the survival of *Phyllodoce* in this kind of vegetation. This is a chionophilous plant throughout its range (Polunin 1948, Dahl 1956) and all its Scottish localities have a northerly or easterly aspect with 100 or more days of persistent snow, lasting sometimes into April. Under normal conditions, fresh growth begins under the protective mantle with the shoots expanding fully after the melt in May or early June and, where frosts occur outside the period of snow-lie, damage to growing points can be permanent (Coker & Coker 1973).

Although rainfall and snow-melt help maintain the soil surface beneath the *Vaccinium-Deschampsia* heath in a generally moist state, this is characteristically a community of moderate to steep slopes cut into pervious, drift-free bedrocks, so drainage is free. The tendency to leaching is also strong, such that the community sometimes extends even on to calcareous rocks like the limestones and more lime-rich schists of the Dalradian assemblage in the central Highlands. There, the soils are of a primitive brown podzolic type, micaceous or rich in silt and sand below, and with a superficial pH of as high as 5.5 (McVean & Ratcliffe 1962). Similar profiles can be found beneath the community over less acidic lavas among the Cheviot rocks (King 1962) and the Silurian shales of the Southern Uplands and Wales (Evans *et al.* 1977). Often, though, the soils are more base-poor than this, having developed from quartzites, sandstones or other siliceous rocks that occur widely through the range of the community. Surface pH can then fall as low as 3.5, though the structure of the profiles can vary from very fragmentary rankers over block scree, a very characteristic feature of the '*Vaccinium* edges' of the Pennine grits (Moss 1913), to quite shallow but fully-developed podzols. Typically, however, the soils are strongly organic above, the litter and mor humus providing a very congenial medium among which the bilberry rhizomes can grow.

The floristic differences among the sub-communities can be understood partly in relation to variations in these climatic and edaphic variables. The *Hylocomium-Rhytidiadelphus* type of heath, with its quite luxuriant sub-shrub canopy and well-developed suite of lush bryophytes and preferential records for *Vaccinium uliginosum*, *Blechnum* and *Cornus suecica*, is generally associated with higher altitudes, sunless aspects and sheltered situations over siliceous rocks and, where such conditions coincide, the vegetation has a strongly calcifuge and chionophilous character and provides an occasional niche, even in regions of bitter winter climate, for more oceanic plants which benefit from the locally enhanced humidity and freedom from exposure. This kind of *Vaccinium-Deschampsia* heath is thus typical of large shallow snow-beds and the surrounds of deeper nivation hollows throughout the central and eastern Highlands, where it largely corresponds to McVean & Ratcliffe's (1962) *Vaccinio-Empetretum*. There it comes close floristically to the *Vaccinium-Rubus* heath, a community of similar situations in the sub- and low-alpine zones, where the preferentials noted above become more consistent and are often accompanied by *Rubus chamaemorus*, a plant not generally found here. The *Hylocomium-Rhytidiadelphus* sub-community also extends into the north-west Highlands, where it descends to somewhat lower altitudes and is less tied to sheltered aspects, sometimes having more of the oceanic hepatics associated with the *Vaccinium-Calluna-Sphagnum* heath. It can be found, too, over cold, humid slopes in the Southern Uplands, the Lake District, Wales and down the Pennines, though often with a reduced list of associates, particularly in the last region, where even the higher-altitude heaths have been strongly affected by pollution.

The duration of snow-lie over the *Hylocomium-Rhytidiadelphus* sub-community can be considerable, but it is probably not so long as over the *Nardus-Carex bigelowii* snow-bed vegetation; and there, too, the slopes are generally not so steep, so any melt-water drains away less readily. The contrast in habitats between the two kinds of vegetation is well seen in the terrace profiles described from the Cairngorms by Burges (1951) and in the sketches of snow-beds included in McVean & Ratcliffe (1962) and floristically the change from the one to the other involves a shift in dominance from *V. myrtillus* and pleurocarpous mosses to *Nardus*, *C. bigelowii* and *R. lanuginosum*, with lichens sometimes prominent and *Diphasium alpinum* becoming frequent. The *Racomi-*

trium-Cladonia sub-community includes some stands which can be considered transitional to such vegetation, although even there plants like *Juncus squarrosus* and *Scirpus cespitosus* remain very uncommon. This kind of *Vaccinium-Deschampsia* heath has also been described from wet, humic rankers in the Southern Uplands (Hill & Evans 1978) and it can develop, too, where peaty soils have been burned, either around snow-beds (as in some of the anthropogenic *Vaccinium* heath noted in McVean & Ratcliffe 1962) or on degraded blanket peats at higher altitudes. In the latter kind of situation, *V. myrtillus* and *E. nigrum* are often the sub-shrubs which spread most rapidly over the mire fringes and the *Racomitrium-Cladonia* sub-community sometimes takes in the heathy developments of the kind of retrogressive *Eriophoretum* described from the Pennines (Lewis & Moss 1911, Moss 1913, Fidler *et al.* 1970).

The *Alchemilla-Carex* sub-community represents a different trend of development away from the *Hylocomium-Rhytidiadelphus* type of heath, tending to replace it at somewhat lower altitudes within the overall range of the community and to favour sunnier aspects with less humic and sometimes less base-poor soils. Such slopes are not so cold and humid as those favoured by the *Hylocomium-Rhytidiadelphus* sub-community and they do not accumulate snow for so long, if indeed at all, a shift which encourages the move away from pronounced bilberry dominance towards an abundance of grasses and sedges, with a less luxuriant contribution from the bulkier pleurocarps and even the occasional occurrence of the Oceanic West European *Carex binervis*. Although the bulk of the plants represented are calcifuge to varying degrees, there is often quite a mesophytic character to this kind of *Vaccinium-Deschampsia* heath and the soils, though sometimes fragmentary, often have but a thin layer of mor, with a lithomorphic or brown podzolic structure below. In some cases, too, plants like *Luzula campestris*, *Campanula rotundifolia*, *Viola riviniana*, *Anemone nemorosa* and *Cerastium fontanum*, indicative of more mesotrophic conditions, are joined by *Thymus praecox* and *Carex pulicaris*, a feature particularly well seen over the Dalradian limestones (McVean & Ratcliffe 1962). But the substrates are not always calcareous and, indeed, beneath stands of this sub-community in the Southern Uplands, markedly base-poor soils can occur (King 1962).

In this part of the range of the *Vaccinium-Deschampsia* heath, too, it is very clear that grazing can probably play a major role in favouring the development of the *Alchemilla-Carex* sub-community towards lower altitudes, both by helping tip the balance of dominance away from the palatable *V. myrtillus* towards grasses and by bringing some modest enrichment to the sward through the dunging. Towards the sub-montane zone, therefore, it is possible that the sometimes quite extensive tracts of this vegetation have been biotically derived as a result of woodland clearance and pasturing and, in fact, there is a virtual floristic continuity between the *Alchemilla-Carex* sub-community and, on the one hand, the grazed field layers of our north-western Quercion woods and, on the other, the Nardo-Galion swards of the *Festuca-Agrostis-Alchemilla* grassland. In such situations, then, the *Vaccinium-Deschampsia* heath represents a continuation northwards of the anthropogenic vegetation with bilberry seen in the *Ulex gallii-Agrostis* and *Calluna-U. gallii* heaths. At higher altitudes, the community can probably be seen as a natural climax community, although McVean & Ratcliffe (1962) suggested that marked trampling and fouling might have an effect on montane bilberry vegetation: patches of calcifuge grassland in nivation hollows, for example, appeared to develop where sheep had survived beneath the cover of winter snow.

Zonation and succession
The *Vaccinium-Deschampsia* heath can be seen as a part of altitudinal sequences from sub-montane woodlands, grasslands and sub-shrub vegetation through to high-level moss-heaths and fell-field where transitions reflect increasing harshness of climate above and biotic influences below, with additional zonations to mire communities with edaphic shifts. Within the low-alpine zone, the *Vaccinium-Deschampsia* heath occurs as a climax vegetation type among dwarfed sub-shrub communities and snow-beds, patterns being determined largely by gradients of exposure and snow-lie.

Through much of its range, fragmentation of the forest cover towards the upper limit of the sub-montane zone is such that it is often difficult to see the *Vaccinium-Deschampsia* heath as the high-altitude replacement of woodland or sub-alpine scrub. But the floristic continuity between the different vegetation types is very striking and many of the heath plants form an integral part of the field layers of the Quercion and Dicrano-Pinion woodlands found over more base-poor soils on siliceous bedrocks and drift at lower levels. Towards the west of Scotland and down through northern England and Wales, the *Hylocomium-Rhytidiadelphus* sub-community reaches down to the upper altitudinal limits of the *Quercus-Betula-Dicranum* woodland and, where fragments of this remain in ravines or on screes, they sometimes give way above to a fringe of the bilberry heath (Lewis & Moss 1911, Ferreira 1978). Similar patterns occur in eastern Scotland, too, though here it is the *Pinus-Hylocomium* woodland that probably represents the sub-montane climax vegetation and, around the Cairngorms, the *Hylocomium-Rhytidiadelphus* sub-community replaces it on more sheltered sites above. Here, also, the *Juniperus-Oxalis* scrub can be seen as a convincing intermediate in the sequence, occurring at the natural upper limit of tree growth in some places,

though hardly ever in zonations which run right through from the montane heath above to the woodland below. In other places, it is tall-herb vegetation, generally of the *Luzula-Vaccinium* type, or the *Thelypteris limbosperma* community, that provides floristic continuity between the extremes of the sequence, ferns, tall herbs, shrubs and trees thus representing the structural elements appearing towards lower altitudes.

Analogous patterns to these can be seen where less base-poor profiles developed from calcareous substrates are disposed over slopes of increasing altitude. Here, the *Vaccinium-Deschampsia* heath can maintain its representation on the higher ground, but it occurs generally as the *Alchemilla-Carex* sub-community, a replacement well seen in comparing the zonations of the Cairngorms with those of the Breadalbane-Clova area, or in moving across the boundary of siliceous and calcareous rocks on the summit ridge of Carn a'Chlarsaich near The Cairnwell (McVean & Ratcliffe 1962). And, in the sub-montane, the climax forest is the *Quercus-Betula-Oxalis* woodland or the less markedly calcifuge types of *Pinus-Hylocomium* woodland. Again, floristic similarity is obvious, spatial continuity rare because of forest destruction.

The widespread occurrence of woodland clearance and the prevalence of burning and grazing through the sub-montane zone thus often mean that the *Vaccinium-Deschampsia* heath gives way below to anthropogenic sub-shrub communities maintained for sheep or grouse-rearing. Through much of its range, its replacement at lower altitudes is the *Calluna-Vaccinium* heath, with the *Calluna-A. uva-ursi* heath figuring on less infertile soils in eastern Scotland, and the *Calluna-Deschampsia* heath prevailing in the polluted southern Pennines. Heather tends to be an overwhelming dominant in each of these vegetation types and, even where other sub-shrubs play a part, as in the early stages of regeneration after burning, the Arctic-Alpines are generally less prominent than in the bilberry heath, but the floristic similarities are considerable, both among the vascular plants and the cryptogam element. In some situations, as over the crags and screes of the Pennine grits, which rise to higher ground separated by intervening tracts of blanket mire on the dips, the contrast between the sub-montane *Calluna-Deschampsia* heath and the high-altitude *Vaccinium-Deschampsia* heath can be quite striking (Lewis & Moss 1911, Tansley 1939). In other places, the shift to higher ground is also marked by an increase in slope and, on northern exposures, of shade which favour the *Vaccinium-Deschampsia* heath on cool steep talus and cliffs, well seen in Edgell's (1969) map of Cader Idris. But, often, the transition between the vegetation types is a gradual one, and it can be particularly complicated where shelter below favours bilberry, while the extension of treatments on to higher ground favours heather.

Towards its lower altitudinal limits, and particularly

over warmer south-facing slopes, it seems likely that some stretches of the *Alchemilla-Carex* sub-community have been biotically derived and are now maintained as plagioclimax intermediates between the forest types noted above and Nardo-Galion grasslands. In parts of the Southern Uplands, for example, in the Cheviot (King 1962), Breadalbane (McVean & Ratcliffe 1962) and Caenlochan (Huntley 1979), this kind of *Vaccinium-Deschampsia* heath is commonly found among *Festuca-Agrostis-Thymus* and *Festuca-Agrostis-Alchemilla* grasslands, the disposition of the different elements of the mosaics being a rather complex function of treatments, topoclimate and edaphic factors (King 1962, Huntley 1979). Zonations to the *Festuca-Agrostis-Alchemilla* grassland can be especially gradual, but the difference between the communities is partly one of the proportions of sub-shrubs to grasses and *A. alpina* and partly to do with the more frequent occurrence of mesophytes and mildly calcicolous plants in the grassland.

With increasing altitude, however, and a decisive shift into the low-alpine zone of the higher mountains within the range of the *Vaccinium-Deschampsia* heath, it is natural factors, particularly exposure and snow-lie, which determine the major trends in the vegetation patterns. Thus, wherever there is a move on to slopes which feel the force of strong winds and which are thus blown clear of snow, the community tends to be replaced by dwarfed sub-shrub vegetation in which *V. myrtillus* plays but a small role and where stunted heather is generally abundant. In the central and eastern Highlands, such vegetation is usually of the *Calluna-Cladonia* heath, where the lichens which are generally of small cover in the *Vaccinium-Deschampsia* heath assume a sometimes co-dominant role and where distinctly chionophobous species can be found. In the opposite direction, with an increase in the duration of snow-lie, the *Vaccinium-Deschampsia* heath is replaced by *Nardus-Carex* vegetation, with its shift to abundance of *Nardus* and *C. bigelowii* and preferentially frequent *Scirpus cespitosus* and *Juncus squarrosus*. Zonations between these communities can be found disposed over slopes of differing aspect, around nivation hollows (McVean & Ratcliffe 1962) and over the treads and risers of terraced slopes (Burges 1951), where sometimes quite subtle variations in inclination and shelter are sufficient to influence the balance between the species. A further complication in some sites is the occurrence among these sequences of the *Vaccinium-Rubus* heath, a vegetation type that seems equally chionophilous to the *Vaccinium-Deschampsia* heath, but where *Calluna* can maintain a better representation, with *Rubus chamaemorus* preferentially frequent.

Sometimes, too, the *Vaccinium-Cladonia* heath can be found at similar altitudes to the *Vaccinium-Deschampsia* heath. Like the *Calluna-Cladonia* heath, this is very lichen-rich vegetation in which sub-shrubs are often reduced to a co-dominant role, but it is not a chionophobous community: indeed, it probably experiences similar duration of snow-lie to the *Vaccinium-Deschampsia* heath. *V. myrtillus* maintains its frequency there and both *E. nigrum* ssp. *hermaphroditum* and *V. vitis-idaea* are common, the former often in abundance. But *Cladonia* spp. are much more plentiful along with *Carex bigelowii* and this is a vegetation type which can extend to higher levels than the *Vaccinium-Deschampsia* heath, representing a transition to the *Juncus-Racomitrium* or *Carex-Racomitrium* heath on summit fell-fields. The *Racomitrium-Cladonia* sub-community can sometimes be found as a transition to these low-alpine communities.

With the geographical shift towards the north-west Highlands the elements in these zonations tend to move to somewhat lower altitudes and, in some cases, to be replaced by more oceanic equivalents. The *Vaccinium-Deschampsia* heath can still be found in this part of Scotland, although in this scheme, some of the vegetation which McVean & Ratcliffe (1962) grouped within their *Festuceto-Vaccinetum rhacomitrosum* is transferred to the *Vaccinium-Racomitrium* heath. With its typically western abundance of *R. lanuginosum*, this community, together with the local stands of *Vaccinium-Deschampsia* heath, occupies part of the low-alpine zone, being replaced below by the *Calluna-Vaccinium-Sphagnum* heath, where oceanic hepatics can play a prominent role, and passing at higher altitudes to *Carex-Racomitrium* heath. Sub-shrub vegetation dependent on a humid climate, is, in this part of Britain, less strictly confined to shaded aspects but transitions to *Nardus-Carex* vegetation can be seen where snow persists and the community is also found among stretches of the chionophilous *Deschampsia-Galium* grassland. Over exposed spurs in this part of Scotland, the *Vaccinium-Deschampsia* heath is replaced by the *Calluna-Racomitrium* or *Calluna-A. alpinus* heaths.

Distribution

The community is widespread through the uplands of Britain, but is particularly common in northern Scotland, where the heart of its range occurs in the central and eastern Highlands with more sporadic occurrences to the north-west. All the sub-communities can be found throughout the distribution, but the *Alchemilla-Carex* type is especially characteristic of the Breadalbane-Clova region.

Affinities

Early accounts of this kind of vegetation (Smith 1900, Smith & Moss 1903, Smith & Rankin 1903, Lewis & Moss 1911, Moss 1913, Tansley 1939) tended to concentrate on the dominance of bilberry as opposed to heather as its major distinguishing feature, and indeed there is ecological meaning in the recognition of a *Vaccinetum* alongside a *Callunetum*. Variations within these broad categories were, however, recognised from the start in vegetation types like the *Gramino-Vaccinetum* of Smith (1911*b*), the kind of heath transitional to Nardo-Galion grasslands included in McVean & Ratcliffe's (1962) *Festuceto-Vaccinetum* (see also Poore 1955*b*, King 1962, Evans *et al.* 1977, Ferreira 1978). In this scheme, however, this vegetation, subsumed in the *Alchemilla-Carex* sub-community, is united in the *Vaccinium-Deschampsia* heath with the less grassy bilberry-crowberry stands first described in detail by Burges (1951) and included in the *Vaccineto-Empetretum* of McVean & Ratcliffe (1962: see also Poore & McVean 1957, Birse 1980). This *Hylocomium-Rhytidiadelphus* sub-community represents the core of the revised vegetation type locating it among the mildly chionophilous communities of Nordhagen's (1943) Phyllodoco-Vaccinion myrtilli alliance. Included here would be such Scandinavian relatives of the *Vaccinium-Deschampsia* heath as the *Phyllodoco-Vaccinetum*, an extensive association of the Rondane (Dahl 1956), and the oceanic *Vaccinetum* with *Cornus suecica* from western Norway (Nordhagen 1943). More lichen-rich stands among the *Alchemilla-Carex* and particularly the *Racomitrium-Cladonia* sub-community could then be seen as a link with the fell-field vegetation of the Loiseleurieto-Arctostaphylion.

Floristic table H18

	a	b	c	18
Vaccinium myrtillus	V (1–10)	V (4–8)	V (1–10)	V (1–10)
Deschampsia flexuosa	V (1–6)	V (1–8)	V (1–8)	V (1–8)
Galium saxatile	V (1–4)	V (1–8)	V (1–6)	V (1–8)
Dicranum scoparium	IV (1–3)	IV (1–2)	IV (1–4)	IV (1–4)
Pleurozium schreberi	V (1–8)	V (1–8)	III (1–6)	IV (1–8)
Hylocomium splendens	V (1–10)	IV (1–8)	I (1–4)	III (1–10)
Rhytidiadelphus loreus	IV (1–6)	II (1–4)	II (1–6)	III (1–6)
Blechnum spicant	III (1–4)	II (1–3)		II (1–4)
Plagiothecium undulatum	II (1–4)	I (1–3)	I (1–4)	I (1–4)
Dicranum majus	II (1–9)	I (1–4)	I (1–2)	I (1–9)
Melampyrum pratense	II (1–4)	I (1–3)	I (1–3)	I (1–4)
Barbilophozia floerkii	II (1–3)	I (1–3)	I (1–3)	I (1–3)
Sphagnum capillifolium	II (1–8)	I (1–3)	I (1–4)	I (1–8)
Luzula sylvatica	II (1–6)	I (1–4)	I (1–3)	I (1–6)
Oxalis acetosella	II (1–4)	I (1–3)		I (1–4)
Vaccinium uliginosum	II (1–6)			I (1–6)
Cornus suecica	II (1–4)			I (1–4)
Hylocomium umbratum	I (1–4)			I (1–4)
Sphagnum quinquefarium	I (1–4)			I (1–4)
Ptilium crista-castrensis	I (1–4)			I (1–4)
Anastrepta orcadensis	I (1–4)			I (1–4)
Festuca ovina	II (1–4)	IV (1–8)	IV (1–10)	III (1–10)
Potentilla erecta	III (1–3)	IV (1–4)	II (1–4)	III (1–4)
Agrostis capillaris	II (1–8)	IV (1–6)	II (1–4)	III (1–8)
Anthoxanthum odoratum	II (1–4)	IV (1–6)	II (1–4)	III (1–6)
Rhytidiadelphus squarrosus	II (1–4)	IV (1–8)	II (1–4)	III (1–8)
Alchemilla alpina	II (1–4)	IV (1–6)	I (1–6)	III (1–6)
Carex pilulifera	I (1–3)	IV (1–4)	II (1–3)	II (1–4)
Luzula campestris		IV (1–4)	I (1–3)	II (1–4)
Campanula rotundifolia	I (1–3)	III (1–3)	I (1–3)	II (1–3)
Carex binervis	I (1–3)	II (1–4)	I (1–3)	I (1–4)

Polygala serpyllifolia	I (1–3)	II (1–3)	I (1–3)	I (1–3)
Viola riviniana	I (1–3)	II (1–4)		I (1–4)
Ranunculus acris	I (1–3)	II (1–3)		I (1–3)
Rumex acetosa	I (1–3)	II (1–3)		I (1–3)
Deschampsia cespitosa	I (1–4)	II (1–4)		I (1–4)
Danthonia decumbens		II (1–3)	I (4)	I (1–4)
Anemone nemorosa		II (1–4)	I (1–3)	I (1–4)
Cerastium fontanum		II (1–3)	I (1–3)	I (1–3)
Veronica officinalis		II (1–4)		I (1–4)
Thymus praecox		I (1–4)		I (1–4)
Alchemilla glabra		I (1–4)		I (1–4)
Polygonum viviparum		I (1–3)		I (1–3)
Racomitrium lanuginosum	I (1–6)	II (1–4)	III (1–4)	II (1–6)
Diphasium alpinum	I (1–3)	I (1–3)	II (1–4)	I (1–4)
Cladonia impexa	I (1–3)	I (1–3)	II (1–6)	I (1–6)
Campylopus paradoxus	I (1–3)	I (1–3)	II (1–4)	I (1–4)
Huperzia selago	I (1–3)	I (1–3)	II (1–3)	I (1–3)
Cladonia uncialis	I (1–3)	I (1–3)	II (1–3)	I (1–3)
Hypnum cupressiforme s.l.	III (1–6)	III (1–6)	III (1–8)	III (1–8)
Nardus stricta	III (1–6)	III (1–4)	III (1–4)	III (1–6)
Empetrum nigrum	III (1–10)	II (1–4)	III (1–6)	III (1–10)
Vaccinium vitis-idaea	III (1–6)	III (1–4)	II (1–4)	III (1–6)
Cladonia arbuscula	III (1–6)	II (1–6)	III (1–4)	III (1–6)
Calluna vulgaris	III (1–8)	II (1–6)	II (1–4)	III (1–8)
Polytrichum commune	II (1–6)	II (1–4)	II (1–4)	II (1–6)
Agrostis canina montana	II (1–4)	II (1–6)	II (1–6)	II (1–6)
Carex bigelowii	II (1–4)	II (1–4)	II (1–6)	II (1–6)
Ptilidium ciliare	II (1–4)	II (1–4)	II (1–3)	II (1–4)
Dryopteris dilatata	I (1–4)	I (1–3)	I (1–3)	I (1–4)
Thuidium tamariscinum	I (1–4)	I (1–3)	I (1–3)	I (1–4)
Rhytidiadelphus triquetrus	I (8)	I (1–5)	I (1–3)	I (1–8)
Festuca rubra	I (1–6)	I (1–4)	I (1–4)	I (1–6)
Cryptogramma crispa	I (1–3)	I (1–3)	I (1–4)	I (1–4)
Juncus squarrosus	I (1–3)	I (1–4)	I (1–6)	I (1–6)
Festuca vivipara	I (1–4)	I (1–8)	I (1–6)	I (1–8)
Polytrichum alpestre	I (1–6)	I (1–4)	I (1–3)	I (1–6)
Polytrichum alpinum	I (1–4)	I (1–4)	I (1–4)	I (1–4)
Carex panicea	I (1–3)	I (1–3)	I (1–3)	I (1–3)
Diplophyllum albicans	I (1–3)	I (1–3)	I (1–3)	I (1–3)
Pohlia nutans	I (1–3)	I (1–3)	I (1–3)	I (1–3)
Cladonia pyxidata	I (1–3)	I (1–3)	I (1–3)	I (1–3)
Cetraria islandica	I (1–3)	I (1–3)	I (1–3)	I (1–3)
Cladonia rangiferina	I (1–6)	I (1–4)	I (1–3)	I (1–6)
Cladonia gracilis	I (1–3)	I (1–3)	I (1–3)	I (1–3)
Luzula multiflora	I (1–3)	I (1–3)	I (1–3)	I (1–3)
Solidago virgaurea	I (1–3)	I (1–3)		I (1–3)
Euphrasia officinalis agg.	I (1–3)	I (1–3)		I (1–3)
Number of samples	73	39	58	170
Number of species/sample	20 (9–41)	28 (11–44)	18 (5–33)	21 (5–44)

Floristic table H18 *(cont.)*

	a	b	c	18
Vegetation height (cm)	16 (6–38)	11 (1–40)	12 (2–35)	13 (1–40)
Vegetation cover (%)	96 (10–100)	91 (65–100)	85 (50–100)	92 (10–100)
Altitude (m)	650 (30–910)	558 (198–914)	623 (210–950)	598 (30–950)
Slope (°)	28 (3–70)	29 (2–80)	20 (0–75)	26 (0–80)

a   *Hylocomium splendens-Rhytidiadelphus loreus* sub-community
b   *Alchemilla alpina-Carex pilulifera* sub-community
c   *Racomitrium lanuginosum-Cladonia* spp. sub-community
18   *Vaccinium myrtillus-Deschampsia flexuosa* heath (total)

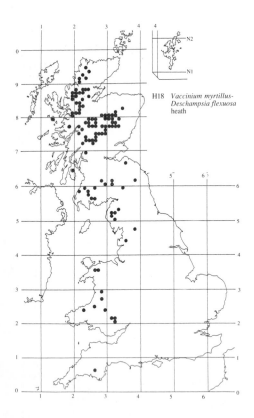

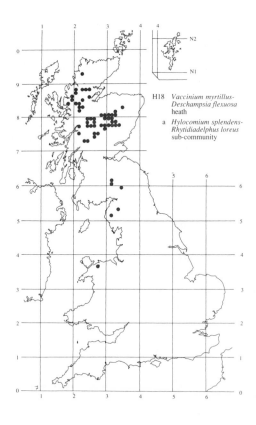

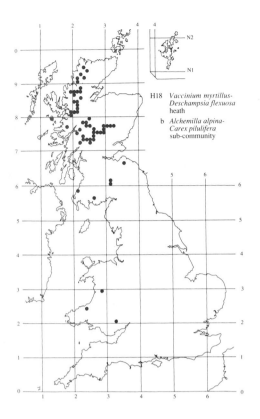

H18   *Vaccinium myrtillus-*
      Deschampsia flexuosa
      heath

   b *Alchemilla alpina-*
      Carex pilulifera
      sub-community

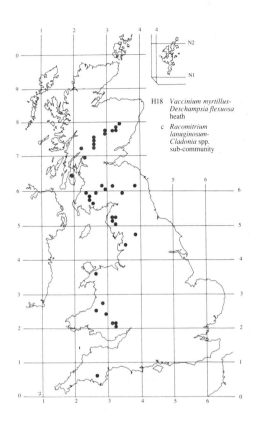

H18   *Vaccinium myrtillus-*
      Deschampsia flexuosa
      heath

   c *Racomitrium*
      lanuginosum-
      Cladonia spp.
      sub-community

H19
Vaccinium myrtillus-Cladonia arbuscula heath

Synonymy

Empetreto-Vaccinetum Burges 1951 *p.p.*; *Vaccinium myrtillus*-lichen heath Poore & McVean 1957; *Empetrum*-lichen heath Poore & McVean 1957; *Cladineto-Vaccinetum* McVean & Ratcliffe 1962; *Festuceto-Vaccinetum rhacomitrosum* McVean & Ratcliffe 1962 *p.p.*; *Vaccinium myrtillus-Empetrum hermaphroditum* nodum Huntley 1979; *Festuceto-Rhacomitrietum lanuginosi* (Birse & Robertson 1976) Birse 1980 *p.p.*; *Carex bigelowii-Festuca vivipara* Association (Birse & Robertson 1976) Birse 1980 *p.p.*

Constant species

Carex bigelowii, Deschampsia flexuosa, Vaccinium myrtillus, V. vitis-idaea, Racomitrium lanuginosum, Cetraria islandica, Cladonia arbuscula, C. uncialis.

Rare species

Loiseleuria procumbens, Kiaeria starkei.

Physiognomy

Like its heather-dominated counterpart, the *Vaccinium myrtillus-Cladonia arbuscula* heath consists essentially of a very low mat of sub-shrubs with an abundance of lichens, which again often mark out stands from a distance with a yellowish or grey-green tinge. Indeed, if anything, the lichens tend to be more extensive throughout this vegetation, and their dominance over the other components is more frequent and more extreme than in the *Calluna-Cladonia* heath. And, among the sub-shrubs, *Calluna* itself is uncommon overall: it increases in frequency in one sub-community but it is typically of low cover even there and of little structural importance, so the kind of wind-waved canopy associated with an abundance of flattened heather bushes is not usual here. The mat is still, however, very short, usually only 5–10 cm thick, but it is *Vaccinium myrtillus* that provides its most consistent element and, although this becomes sparse and noticeably lacking in vigour in more exposed situations, it is quite often abundant and a fairly regular co-dominant in what can be a dense and springy cover. *V. vitis-idaea* is somewhat less common and generally not so extensive and, in certain kinds of *Vaccinium-Cladonia* heath, it becomes distinctly patchy. *V. uliginosum* very occasionally shows local prominence, but overall it is scarce.

The other important sub-shrub in the community is *Empetrum nigrum*, almost always ssp. *hermaphroditum* and, in some of the vegetation included here (what McVean & Ratcliffe (1962) recognised as *Cladineto-Vaccinetum*), it becomes very frequent and often rivals or exceeds the Vaccinia in its cover. *Alchemilla alpina* can also be found, but it is nothing like so common or abundant as among the grassier stands of the *Vaccinium-Deschampsia* heath. The rare Arctic-Alpine *Loiseleuria procumbens* is very occasionally recorded, although neither it, nor the *Arctostaphylos* spp., play the sort of role here that they have in the *Calluna-Cladonia* or *Calluna-A. alpinus* heaths.

As in both those lichen-rich communities, vascular associates of the sub-shrubs in the *Vaccinium-Cladonia* heath are few in number, although they can be somewhat more extensive in their cover. This is especially true of *Carex bigelowii* which is not only more frequent in this community than those heaths but often abundant, indeed, co-dominant with the ericoids and the lichens in many stands, occurring sometimes as numerous small tufts of shoots in intimate mixtures with them, in other cases in large clonal patches forming coarser mosaics. The general prominence of this sedge shifts the composition of the vegetation towards the *Carex-Racomitrium* heath somewhat, but, although transitional stands can be widely found, the general balance of the structural elements is different there, with both sub-shrubs and lichens generally playing a subordinate role.

The other common associate, though one which is usually less abundant, is *Deschampsia flexuosa*. *Festuca ovina* (including many records for *F. vivipara*) is also fairly frequent, particularly in one kind of *Vaccinium-Cladonia* heath, and it can have moderately high cover there, along with *Galium saxatile* and, more occasionally, *Potentilla erecta*. *Carex pilulifera*, *Agrostis canina* (usually recorded as ssp. *montana*) and *A. capillaris* can also be found at low frequencies throughout the community, but the rich and extensive assemblages of Nardo-Galion herbs typical of grassier stands of the *Vaccinium-Deschampsia* heath are not found here. *Nardus stricta* itself is fairly infrequent, too, which is one good feature distinguishing this vegetation from the more chionophilous *Nardus-Carex* community, where *V. myrtillus* and *C. bigelowii* maintain a constant presence, but where *Nardus* is usually strongly dominant. Towards the opposite extreme, in moving towards fell-field vegetation, *Juncus trifidus*, which characteristically predominates over bilberry and the sedge in the *Juncus-Racomitrium* heath, is scarce and of low cover here.

One other feature which helps mark off the *Vaccinium-Cladonia* heath from the vegetation of cloud-ridden summits is the restricted role of *Racomitrium lanuginosum*. This is in fact rather more prominent here than in McVean & Ratcliffe's (1962) original *Cladineto-Vaccinetum*, and the community perhaps takes in some of what these authors would have placed in the *Festuceto-Vaccinetum rhacomitrosum* and the more bilberry-rich stands of the *Festuceto-Rhacomitrietum* of Birse (1980). But only in one of the sub-communities is the moss of any great abundance and in much *Vaccinium-Cladonia* heath it is distinctly sparse. And *Polytrichum alpinum*, which commonly accompanies it in the *Carex-Racomitrium* heath, is only occasional in this community.

Other bryophytes, too, are rather inconsistent in their contribution. *Dicranum fuscescens* is quite frequent and it very occasionally shows local abundance: where this coincides with low covers of the Vaccinia, the community approaches certain kinds of *Carex-Polytrichum* snow-bed vegetation, although McVean & Ratcliffe (1962) reported such transitions as being problematic only in the Clova region. Then, there are some stands in which *Dicranum scoparium*, *Pleurozium schreberi* and *Ptilidium ciliare* become common, but the rich carpets of bulky pleurocarps characteristic of the *Vaccinium-Deschampsia* heath are not usual here. Often, it is just sparse shoots of *Polytrichum alpestre* and *P. piliferum* that are dotted through the mat.

Much more important throughout the community are the lichens, particularly larger fruticose species such as *Cladonia arbuscula* and *C. uncialis* with, less commonly, *C. rangiferina* and *C. gracilis*, mixtures of which can exceed the sub-shrubs in total cover and which can form with them a single layered carpet that can sometimes be peeled away from the substrate intact. *Cetraria islandica* and *Cornicularia aculeata* are also very common, though usually less abundant and there is occasional *Alectoria nigricans*, *Cladonia impexa*, *C. coccifera* and *Thamnolia* *vermicularis*. *Ochrolechia frigida* and *Cetraria nivalis* can also be found, though they are preferential for one particular sub-community.

Sub-communities

***Festuca ovina-Galium saxatile* sub-community:** *Carex bigelowii-Festuca vivipara* Association (Birse & Robertson 1976) Birse 1980 *p.p.* *V. myrtillus* is often abundant here, but it usually shares dominance among the vascular plants with *C. bigelowii* and, strongly preferential to this kind of *Vaccinium-Cladonia* heath, *F. ovina/vivipara*. Among the other sub-shrubs, *E. nigrum* ssp. *nigrum* is sometimes found but ssp. *hermaphroditum* is scarce, while *V. vitis-idaea* attains its maximum frequency and abundance in this sub-community, though it is even then rarely extensive in its cover. *Deschampsia flexuosa* is likewise only moderately abundant. *Galium saxatile* is a good diagnostic species with *Potentilla erecta* also preferential at low frequency and there is occasionally a little *Carex pilulifera* and *Agrostis canina* ssp. *montana*. Where such mixtures occur with a little *Dicranum scoparium*, *Pleurozium schreberi* and *Hypnum jutlandicum*, all of which are occasional here, the vegetation comes close to the *Vaccinium-Deschampsia* heath.

Other stands, with a little less *V. myrtillus* than usual and locally abundant *R. lanuginosum*, are like the grassier forms of the *Carex-Racomitrium* heath but *R. lanuginosum* is usually of low cover and sometimes distinctly patchy and, from both these communities, this vegetation is distinguished by the lichen element. This is not so overwhelmingly extensive as in the *Empetrum-Cladonia* sub-community but *Cladonia arbuscula* is sometimes co-dominant with the vascular plants, *C. uncialis* is moderately abundant throughout and there is commonly a little *Cetraria islandica* and *Cornicularia aculeata* and occasionally some *C. rangiferina*, *C. gracilis*, *C. impexa* and *C. coccifera*.

Racomitrium lanuginosum sub-community: *Festuceto-Vaccinetum rhacomitrosum* McVean & Ratcliffe 1962 *p.p.*; *Festuceto-Rhacomitrietum*, *Cladonia arbuscula* subassociation Birse 1980 *p.p.* In this sub-community, the mat is dominated by various mixtures of *V. myrtillus*, *C. bigelowii*, lichens and, unusually abundant for the *Vaccinium-Cladonia* heath, *R. lanuginosum*. *E. nigrum*, mostly ssp. *hermaphroditum*, is fairly common but *V. vitis-idaea* is very patchy in its occurrence. *F. ovina/vivipara* and *Galium saxatile* can occasionally be found and, where other Nardo-Galion grasses also have sparse representation, the vegetation looks transitional to *Racomitrium*-rich *Vaccinium-Deschampsia* heath, the floristic similarity enhanced by the frequent occurrence in this sub-community of small amounts of *Alchemilla alpina*. Such a trend is exceptional, however, and, though no other species are strongly preferential, the occasional presence of plants like *Salix herbacea* and, among the mosses, the rare *Kiaeria starkei*, both of which can show local abundance, can lend a more pronounced chionophilous look to the vegetation. *Juncus trifidus* also increases in frequency somewhat and there is very occasionally some *Luzula spicata*, *Antennaria dioica*, *Armeria maritima*, *Silene acaulis* and *Sibbaldia procumbens*. In general, though, such plants are still much scarcer than in the *Carex-Racomitrium* heath.

Apart from *R. lanuginosum* and *K. starkei* other mosses are poorly represented, but the lichen cover is quite varied. *C. arbuscula* is sometimes much the most abundant species but it can become very patchy and mixed carpets are more usual, with moderate amounts of *C. uncialis*, *Cetraria islandica* and *Cornicularia aculeata* and, weakly preferential here, *Sphaerophorus globosus*.

Empetrum nigrum ssp. hermaphroditum-Cladonia spp. sub-community:

Vaccinium myrtillus-lichen heath Poore & McVean 1957; *Empetrum*-lichen heath Poore & McVean 1957; *Cladineto-Vaccinetum* McVean & Ratcliffe 1962; *Vaccinium myrtillus-Empetrum hermaphroditum* nodum Huntley 1979. Mixed mats of *V. myrtillus* and *E. nigrum* ssp. *hermaphroditum* provide the bulk of the vascular cover here though their proportions and vigour are very variable and they are quite often exceeded in abundance by the lichens: this sub-community thus encompasses most of the floristic and physiognomic differences used to define the sub-associations of McVean & Ratcliffe's (1962) *Cladineto-Vaccinetum*. Other sub-shrubs are of generally low cover, although *V. vitis-idaea* can be moderately abundant, and there is quite often here a little *Calluna*. *C. bigelowii* remains very frequent but it is usually subordinate and *D. flexuosa*, though very common, is distinctly sparse in cover. Other grasses are very scarce and species like *Carex pilulifera* and *Galium saxatile* are at their most infrequent.

Among the bryophytes, *R. lanuginosum* remains frequent and there is preferentially common *Pleurozium schreberi*, *Dicranum scoparium*, *Ptilidium ciliare* and, more occasionally, *Rhytidiadelphus loreus*, but only in exceptional circumstances is any of these abundant and even their total cover contribution is usually small. Lichens, by contrast, can be overwhelmingly extensive here with *C. arbuscula* especially abundant, *C. uncialis* usually less so, but *C. rangiferina* is also strongly preferential and locally prominent. Frequent *C. gracilis* is also diagnostic though this, together with *Cornicularia aculeata* and *Cetraria islandica*, is usually found at low cover. Then there is occasionally some *Ochrolechia frigida* and *Cetraria nivalis* with sparse *Cladonia pyxidata*, *C. bellidiflora* and *C. leucophaea*.

Habitat

The *Vaccinium-Cladonia* heath is typical of base-poor soils on what are usually moderately sheltered and snow-bound slopes at high altitudes, particularly in the more continental mountains of northern Britain. Floristic variation within the community seems to reflect differences in exposure and soil type, but overall the vegetation can be considered a climatic climax.

Geographically, the community has much the same range as the *Calluna-Cladonia* heath, being strongly concentrated in the central and eastern Highlands of Scotland, though with a somewhat better representation through the mountains of the north-west and at scattered localities in the Southern Uplands and northern England, where the community can be seen in such places as Dollar Law, Skiddaw and Cross Fell (McVean & Ratcliffe 1962, Ratcliffe 1977). And there is considerable altitudinal overlap between the two vegetation types, the *Vaccinium-Cladonia* heath being also characteristic of the low-alpine zone (Poore & McVean 1957), though usually pitched a little higher within it, mostly above 650 m and quite often towards 1000 m or even well beyond. At such levels, the climate is generally harsh, with mean annual maximum temperatures usually less than 21 °C (Conolly & Dahl 1970), and particularly through the heart of its range the winters are very bitter, with February minima well below freezing and frequent late frosts (*Climatological Atlas* 1952, Huntley 1979). Through much of the distribution, precipitation is not especially heavy, often not much more than 1600 mm yr^{-1} (*Climatological Atlas* 1952) with 180–200 wet days yr^{-1} (Ratcliffe 1968), but through the central and eastern Highlands much of the winter share falls as snow and there are up to 100 days or more with morning snow-lie at the altitudes where the community is found.

These general climatic conditions are reflected in the strongly montane character of the vegetation, with Arctic-Alpines such as *C. bigelowii*, *V. vitis-idaea* and *E. nigrum* ssp. *hermaphroditum* all represented with more consistent frequency than in the *Vaccinium-Deschampsia* heath which replaces it at lower altitudes, and plants like *Erica cinerea* and *Molinia caerulea*, which even find a place in the *Calluna-Cladonia* heath at its lower stations, are quite absent. *Alchemilla alpina* adds a further montane element in some stands, and then there is the common occurrence of upland lichens such as *Cetraria islandica* and *Cladonia rangiferina* and more occasional representation of *Juncus trifidus*, *Polytrichum alpinum* and *P. alpestre*.

Just as important, however, to the character of the vegetation is the fact that, compared with most stands of the *Calluna-Cladonia* heath, the local climatic conditions here often seem fairly sheltered. Sometimes the community can be found over more windswept spurs and ridges such as are the typical habitat of that kind of lichen-heath, when it is more difficult to see what particular environmental factors differentiate the two, but often the location of the *Vaccinium-Cladonia* heath is not so exposed to the frequent biting gales typical of these altitudes and, in winter, some snow is able to settle and, over north- and east-facing slopes, persist long over this vegetation, affording protection from frost. The most obvious general effects of this difference in topoclimate are on the composition and relative luxuriance of the sub-shrub canopy compared with the *Calluna-Cladonia* heath, particularly the shift from *Calluna* to *V. myrtillus* as the most prominent ericoid throughout, and on the frequency and abundance of *C. bigelowii*, with a more sporadic occurrence of markedly chionophilous plants.

Details of the snow-lie regime for the community are still not available, but McVean & Ratcliffe (1962) noted that its common position in the low-alpine sequences of the east-central Highlands, between the *Nardus-Carex* community of the early snow-beds and the *Calluna-Cladonia* heath over more exposed slopes, suggested an intermediate duration and thickness of snow-cover: perhaps roughly the same as for the *Vaccinium-Rubus* heath and less grassy stands of the *Vaccinium-Deschampsia* heath, typical of sheltered situations in the sub-alpine zone. In this scheme, McVean & Ratcliffe's (1962) *Cladineto-Vaccinetum* is essentially identical to the *Empetrum-Cladonia* sub-community and within it can be seen those floristic trends in relation to variation in snow-cover which they used to define sub-associations. To one extreme, closest to the *Calluna-Cladonia* heath of more exposed situations, are stands with an abundance of lichens, among which *Cetraria nivalis* figures prominently; to the other, more prominent *R. lanuginosum* and pleurocarps among extensive covers of *V. myrtillus* and *C. bigelowii* mark a transition to early snow-beds proper.

A continuation of the trend towards greater humidity of the environment is reflected in the *Racomitrium* sub-community by the increasing abundance of this moss and of *C. bigelowii* with the local occurrence of *Salix herbacea* and *Kiaeria starkei*, assemblages characteristic of later snow-bed vegetation. Again, actual details of the snow-lie regime are lacking but some of the stands included here are from the shaded northern and eastern aspects in the east-central Highlands that would be expected to retain their snow-cover longer. Others, though, occur outside this region, extending the range of the *Vaccinium-Cladonia* heath into the north-west Highlands, where annual rainfall rises to well over 1600 mm (*Climatological Atlas* 1952) with 200 or more wet days yr^{-1} (Ratcliffe 1968). There, *R. lanuginosum* becomes a much more consistently abundant element of low-alpine bilberry and crowberry heaths, and the *Vaccinium-*

Cladonia heath grades to the *Vaccinium-Racomitrium* heath.

Even in the east of Scotland, however, the rainfall is sufficient to produce marked leaching in the soils beneath the *Vaccinium-Cladonia* heath especially where, as is usually the case with the two sub-communities above, the parent materials are lime-poor, weathering to podzolised profiles, strongly humic above and with a superficial pH usually not much above 4. The *Empetrum-Cladonia* type seems characteristic of the most impoverished soils, derived often from the granites which underlie the Cairngorms, Lochnagar and Monadhliath or the quartzites and quartzose micaschists of the Dalradian assemblage through the Grampians, but the *Racomitrium* sub-community is found widely over the latter too, as well as on soils derived from Lewisian gneiss and Torridonian sandstone in the north-west Highlands.

The distribution of the *Festuca-Galium* sub-community, on the other hand, like the grassier stands of the *Vaccinium-Deschampsia* heath down into the sub-alpine zone, shows a clear correlation with the occurrence of the Dalradian limestone and calcareous mica-schists that run from Breadalbane to Clova and, though the superficial pH does not seem very different from that beneath other kinds of *Vaccinium-Cladonia* heath, the profiles tend towards the brown podzolic type, being less humic and perhaps less oligotrophic. Certainly, the floristic trend among the associates is towards the Nardo-Galion and this sub-community also takes in most of the stands of the *Vaccinium-Cladonia* heath towards the warmer and drier southern fringes of its range, from where McVean & Ratcliffe (1962) noted grassy fragments of their *Cladineto-Vaccinetum*. Even there, however, the continuing prominence of *C. bigelowii* with much *F. vivipara* shows a clear trend from low-alpine heath to fell-field vegetation.

Zonation and succession
The *Vaccinium-Cladonia* heath is a characteristic element in the sequences of sub-shrub communities and snow-bed vegetation of the low-alpine zone of the continental mountains of the east and central Highlands, among which floristic variation is strongly related to exposure and snow-lie. Here, it occurs at the limit of ericoid dominance, marking the altitudinal transition in more sheltered situations from sub-alpine and submontane heaths to summit moss-heath, fell-field or mire. Geology and soils affect the particular kinds of communities represented with the *Vaccinium-Cladonia* heath and, towards lower altitudes, treatments have modified the vegetation cover but these are not important in maintaining the community itself.

The broad altitudinal zonations in which the *Vaccinium-Cladonia* heath plays an integral part are most obvious over the upper slopes of the Cairngorms and on the mountain tops of the Breadalbane-Clova region. On the impoverished granite soils typical of the former area, where there has been relatively little impact of sheep-grazing, sequences of sub-shrub vegetation are extensive, starting with mosaics of the *Calluna-Vaccinium* and *Calluna-A. uva-ursi* heaths with fragments of *Pinus-Hylocomium* woodland and *Juniperus-Oxalis* scrub through the sub-montane and up into the sub-alpine zone. Above these levels, past the present limits of tree growth and beyond the influence of the regular burning employed to regenerate the lower-altitude heaths for grouse-rearing, the particular zonation becomes strongly dependent on exposure. Wind-blasted spurs and ridges have the *Calluna-Cladonia* heath and, though the *Vaccinium-Cladonia* heath finds occasional representation in such situations, it is usually in less exposed places that the community begins to make an obvious contribution to the cover.

Indeed, the effects of local shelter can already be seen at lower altitudes in this region, where the *Hylocomium-Rhytidiadelphus* sub-community of the *Vaccinium-Deschampsia* heath often marks out large shallow snow-beds towards the transition to the sub-alpine zone, with the *Polytrichum-Galium* sub-community of the *Vaccinium-Rubus* heath extending even further below, where easterly or northerly aspects provide shaded and cool conditions. Observation suggests (McVean & Ratcliffe 1962) that each of these vegetation types experiences roughly similar regimes of snow-lie and, among them, the *Empetrum-Cladonia* sub-community of the *Vaccinium-Cladonia* heath can be seen as the high-altitude representative of a mildly chionophilous suite. Quite often, however, mosaics of the *Vaccinium-Cladonia* and *Vaccinium-Deschampsia* heath can be found disposed across hillsides in the low-alpine zone, the enclosed patches of the latter being marked out by a shift from a predominance of lichens among the cryptogamic element to bulky pleurocarps, and by the appearance of species like *Blechnum spicant* and *Cornus suecica*.

Although there are places where the *Vaccinium-Cladonia* heath can be seen in extensive tracts, the more usual picture is for this kind of mosaic to continue, with moves to slopes that are more exposed or less, to the *Calluna-Cladonia* heath on the one hand or to snow-bed vegetation on the other. Transitions to the former are typical of gently undulating hillsides, where the rounded spurs get blown clear of snow, the concavities accumulating a little, and where the move from one community to the other is denoted largely by a replacement of heather by bilberry among a superabundant ground of lichens. Similar transitions can be seen on moving off the exposed tops of the risers on terraced slopes on to the sheltered treads and on shifting round from wind-blasted slopes on to more sheltered aspects and here the

Racomitrium sub-community is sometimes seen as a transition to either *Nardus-Carex* or *Carex-Polytrichum* snow-beds. Moves towards the former involve a reduction in the contribution of *V. myrtillus* and the less snow-tolerant lichens, with an increase in the abundance of *R. lanuginosum* and a marked dominance of *Nardus stricta*. In the latter, *V. myrtillus* and the other sub-shrubs are often absent, but *C. bigelowii* remains prominent with chionophilous bryophytes such as *Polytrichum alpinum* and *Dicranum fuscescens* very common. In such patterns, the patches of the different vegetation types can vary from just a few square metres to hectares in extent.

The greater snow-tolerance of the Vaccinia and *E. nigrum* ssp. *hermaphroditum* over *Calluna* means that the *Vaccinium-Cladonia* heath often runs on above the limit of the *Calluna-Cladonia* heath, maintaining a strong sub-shrub representation for a further hundred metres or more and, in such situations, it can replace it in moderately exposed habitats just within the mid-alpine zone. Over the windy and cloud-ridden summits of the Cairngorms, however, the community typically gives way to the *Juncus-Racomitrium* fell-field vegetation or, more locally, the *Carex-Racomitrium* heath. *V. myrtillus* continues with some frequency into such communities, though at low covers, and dominance is usually shared between such plants as *C. bigelowii*, *R. lanuginosum*, *F. ovina* and *J. trifidus*.

Very similar zonations to these can be seen on Monadhliath and Lochnagar but, with the move on to the Dalradian rocks of the Clova area and down to Breadalbane, where lime-rich exposures begin to figure prominently, there are some differences in the vegetation types represented in the various altitudinal zones and, with many of the mountains having small or sloping summits, the upper end of the sequence is often truncated. The *Empetrum-Cladonia* sub-community of the *Vaccinium-Cladonia* heath can be found over quartzose mica-schists and other siliceous rocks through this region but, quite often, it is the *Festuca-Galium* sub-community that is represented over the more base-rich rocks, and here this continues upwards the mildly chionophilous character of the *Alchemilla-Carex* sub-community of the *Vaccinium-Deschampsia* heath. Towards lower altitudes, this in turn can pass to *Festuca-Agrostis-Alchemilla* grassland where sheep-grazing restricts the contribution from the sub-shrubs. Over the less siliceous rocks, the *Calluna-Cladonia* heath tends to be poorly represented, even where there is quite severe exposure, but some kinds of later snow-beds can be seen over ground where there is a little more shelter than with the *Vaccinium-Cladonia* heath, with transitions to certain kinds of *Nardus-Carex* and *Carex-Polytrichum* vegetation.

Towards the north-west of Scotland, the *Vaccinium-Cladonia* heath retains some representation in sequences that have an altogether more oceanic character, sub-suming some of the vegetation which McVean & Ratcliffe (1962) placed in their *Festuceto-Vaccinetum rhacomitrosum*. In this part of Britain, the effects of the greater humidity of climate are felt to some extent on all aspects and *R. lanuginosum*-rich bilberry heaths occur widely between 400 and 1000 m where there is some measure of shelter from winds. The bulk of them can be grouped within the *Vaccinium-Racomitrium* heath where there is a consistent shift away from lichens as the usual co-dominants in the mat, but the *Racomitrium* sub-community of the *Vaccinium-Cladonia* heath is seen locally as a geographical transition to this more western vegetation type. At lower altitudes the sequences among the heaths pass to *Vaccinium-Rubus* and *Calluna-Vaccinium-Sphagnum* heaths, and upwards there is extensive *Carex-Racomitrium* heath over the broader summits, with *Nardus-Carex* snow-beds in sheltered hollows.

Distribution

The range of the *Vaccinium-Cladonia* heath is strongly centred on the Grampians, where the *Empetrum-Cladonia* sub-community is widespread on the granite and quartzitic mountains, with the *Festuca-Galium* sub-community more characteristic of the lime-rich rocks between Breadalbane and Clova. The *Racomitrium* sub-community, too, can be found scattered through the central and eastern Highlands, but it also extends the distribution of the community into north-west Scotland. Each of the sub-communities has some fragmentary representation in the Southern Uplands and/or northern England.

Affinities

The *Vaccinium-Cladonia* heath is in some senses a parallel community to the *Calluna-Cladonia* heath and, in one of the earliest accounts of this kind of vegetation (Poore & McVean 1957), lichen-heaths rich in either heather or bilberry were grouped together in a single nodum. However, although there are situations within the low-alpine zone where the two vegetation types come very close in their floristics, the difference between them being then largely a matter of the dominance of one sub-shrub rather than the other, the overall characters of the communities present quite a contrast, especially with the somewhat broader definition of the *Vaccinium-Cladonia* heath presented in this scheme as against McVean & Ratcliffe's (1962) understanding of bilberry lichen-heath. Their *Cladineto-Vaccinetum* represents just part of the variation included here, corresponding essentially to the *Empetrum-Cladonia* sub-community, some of the stands of which are from more exposed situations where the floristic and environmental similarities to the *Calluna-Cladonia* heath are most obvious.

In addition to this kind of vegetation, though, the *Vaccinium-Cladonia* heath in this scheme includes many stands which extend the range of the community further towards more sheltered and humid situations. Such a trend is already visible in the *Empetrum-Cladonia* sub-community, the mildly chionophilous character of which was stressed by McVean & Ratcliffe (1962), but it becomes more prominent in the *Racomitrium* sub-community with its local prominence of late snow-bed plants. However, this kind of *Vaccinium-Cladonia* heath also provides greater continuity with the more oceanic bilberry heaths of the low-alpine zone of the north-west Highlands, where it represents a geographical transition to the *Vaccinium-Racomitrium* heath, and perhaps takes in some of the less grassy stands of McVean & Ratcliffe's (1962) *Festuceto-Vaccinetum rhacomitrosum*.

The *Festuca-Galium* sub-community exhibits a different kind of trend towards the vegetation of the *Vaccinium-Deschampsia* heath. Lichen-rich stands of grassy bilberry heaths were noted by both Poore (1955c) and McVean & Ratcliffe (1962), in both cases on the more calcareous rocks of the Breadalbane-Clova region, where this type of *Vaccinium-Cladonia* heath is centred. In both communities, then, there is the same kind of geological and edaphic contrast, the *Festuca-Galium* sub-community being paired with the *Alchemilla-Carex* sub-community of the *Vaccinium-Deschampsia* heath on more base-rich soils, the *Empetrum-Cladonia* sub-community here occurring with the *Hylocomium-Rhytidiadelphus* sub-community of the *Vaccinium-Deschampsia* heath on the more acidic profiles of the granites and quartzites. This latter pairing is the one stressed in Burges's (1951) account of the heaths of the Cairngorm terraces, where the two vegetation types were united as lichen-rich and mossy facies of a single *Empetreto-Vaccinetum*. The former parallel is less well described, but can be seen as continuing the floristic variation included in the early-defined 'sub-alpine grassland' of the Breadalbane area (Smith 1911a), where there is a strong similarity among the sub-montane Nardo-Galion swards and the lichen-rich heaths of the low-alpine zone.

More work is needed to provide some sharper environmental differentiation between these higher-altitude heaths, although it is clear that if the chionophilous character of the *Vaccinium-Cladonia* heath is stressed, then it belongs with the *Vaccinium-Deschampsia* and *Vaccinium-Rubus* heaths in what Nordhagen (1943) termed the Phyllodoco-Vaccinion myrtilli alliance. In this he located a number of vegetation types from the Sylene district of Norway, including *C. arbuscula*-rich heaths with much *V. myrtillus* or *E. nigrum*. These, and the *Empetrum*-lichen heaths described from Middle Sogn by Knaben (1950), bear a strong floristic similarity to the *Vaccinium-Cladonia* heath, more especially to the

Empetrum-Cladonia sub-community, though they generally have a more luxuriant cover of sub-shrubs than their Scottish equivalent. McVean & Ratcliffe's (1962) alternative placing of this kind of vegetation among the lichen-heaths and fell-field communities of the Loiseleurieto-Arctostaphylion (or Arctostaphyleto-Cetrarion) would not be favoured with the broader definition of the *Vaccinium-Cladonia* heath in this scheme.

Floristic table H19

	a	b	c	19
Vaccinium myrtillus	V (1–8)	V (1–8)	V (1–8)	V (1–8)
Carex bigelowii	V (1–10)	V (1–8)	V (1–6)	V (1–10)
Deschampsia flexuosa	V (1–8)	IV (1–8)	V (1–4)	V (1–8)
Cladonia uncialis	IV (1–6)	V (1–5)	V (1–4)	V (1–6)
Racomitrium lanuginosum	IV (1–6)	V (1–10)	IV (1–10)	IV (1–10)
Cetraria islandica	IV (1–4)	IV (1–6)	V (1–6)	IV (1–6)
Cladonia arbuscula	IV (1–8)	III (1–6)	V (1–10)	IV (1–10)
Vaccinium vitis-idaea	IV (1–4)	II (1–4)	IV (1–4)	IV (1–4)
Festuca ovina/vivipara	V (1–8)	III (1–4)	II (1–4)	III (1–8)
Galium saxatile	IV (1–6)	II (1–6)	II (1–3)	III (1–6)
Polytrichum commune	II (1–6)	I (1–3)	I (1–4)	I (1–6)
Potentilla erecta	II (1–4)	I (1–6)		I (1–6)
Hypnum jutlandicum	II (1–8)	I (1–3)	I (1–3)	I (1–8)
Rhytidiadelphus squarrosus	I (1–4)			I (1–4)
Alchemilla alpina	I (1–4)	III (1–8)	I (1–8)	II (1–8)
Kiaeria starkei	I (4)	II (1–8)	I (1–4)	I (1–4)
Huperzia selago	I (1–3)	II (1–3)	I (1–3)	I (1–3)
Salix herbacea	I (1–4)	II (1–6)	I (1–4)	I (1–6)
Sphaerophorus globosus		II (1–3)	I (1–3)	I (1–3)
Armeria maritima		I (1–4)		I (1–4)
Silene acaulis		I (1–3)		I (1–3)
Antennaria dioica		I (1–3)		I (1–3)
Luzula spicata		I (1–3)		I (1–3)
Empetrum nigrum	II (1–6)	III (1–6)	V (1–10)	III (1–10)
Cladonia rangiferina	II (1–4)	II (1–6)	V (1–6)	III (1–6)
Dicranum scoparium	III (1–6)	II (1–3)	IV (1–4)	III (1–6)
Cladonia gracilis	II (1–3)	II (1–4)	IV (1–4)	III (1–4)
Pleurozium schreberi	II (1–8)	I (1–4)	IV (1–6)	III (1–8)
Ptilidium ciliare	II (1–3)	I (1–3)	III (1–3)	II (1–3)
Ochrolechia frigida	I (1–4)	II (1–6)	III (1–4)	II (1–6)
Calluna vulgaris	I (1–4)	I (1–6)	III (1–9)	II (1–9)
Cetraria nivalis	I (1–4)	I (1–3)	II (1–4)	I (1–4)
Cladonia pyxidata	I (1–3)	I (1–3)	II (1–3)	I (1–3)
Rhytidiadelphus loreus	I (1–4)	I (1–3)	II (1–4)	I (1–4)
Cladonia bellidiflora		I (1–3)	II (1–3)	I (1–3)
Cladonia leucophaea			I (1–6)	I (1–6)
Cornicularia aculeata	III (1–6)	III (1–4)	III (1–4)	III (1–6)
Dicranum fuscescens	III (1–6)	II (1–8)	III (1–4)	III (1–8)
Polytrichum alpinum	II (1–4)	II (1–6)	II (1–4)	II (1–6)

Carex pilulifera	II (1–4)	II (1–4)	II (1–3)	II (1–4)
Nardus stricta	II (1–10)	II (1–4)	I (1–3)	II (1–10)
Polytrichum alpestre	II (1–4)	II (1–4)	I (1–3)	II (1–4)
Cladonia coccifera	II (1–3)	II (1–4)	I (1–3)	II (1–4)
Agrostis canina	II (1–6)	II (1–4)	I (1–4)	II (1–6)
Cladonia impexa	II (1–8)	I (1–3)	II (1–6)	II (1–8)
Alectoria nigricans	I (1–4)	II (1–4)	II (1–3)	II (1–4)
Juncus trifidus	I (1–6)	II (1–6)	II (1–6)	II (1–6)
Thamnolia vermicularis	I (1–3)	I (1–4)	I (1–3)	I (1–4)
Cladonia crispata	I (1–3)	I (1–4)	I (1–3)	I (1–4)
Polytrichum piliferum	I (1–4)	I (1–4)	I (1–3)	I (1–4)
Diphasium alpinum	I (1–3)	I (1–6)	I (1–3)	I (1–6)
Agrostis capillaris	I (1–3)	I (1–4)	I (1–3)	I (1–4)
Cladonia squamosa	I (1–3)	I (1–4)	I (1–3)	I (1–4)
Loiseleuria procumbens	I (1–3)	I (1–4)	I (1–8)	I (1–8)
Juncus squarrosus	I (1–4)	I (1–3)	I (1–3)	I (1–4)
Polytrichum longisetum	I (1–4)	I (1–4)		I (1–4)
Diplophyllum albicans	I (1–3)	I (1–3)		I (1–3)
Luzula multiflora	I (1–3)		I (1–3)	I (1–3)
Anthoxanthum odoratum	I (1–4)		I (1–3)	I (1–4)
Lophozia ventricosa	I (1–3)	I (1–3)	I (1–3)	I (1–3)
Number of samples	65	49	85	199
Number of species/sample	18 (13–28)	18 (10–26)	18 (9–30)	18 (9–30)
Shrub/herb height (cm)	7 (3–14)	4 (1–12)	7 (1–35)	7 (1–35)
Shrub/herb cover (%)	73 (15–100)	60 (15–100)	87 (35–100)	75 (15–100)
Ground layer height (mm)	26 (10–50)	23 (10–50)	27 (10–60)	26 (10–60)
Ground layer cover (%)	71 (25–97)	77 (40–100)	59 (1–98)	71 (1–100)
Altitude (m)	764 (50–970)	773 (8–990)	806 (6–1159)	784 (6–1159)
Slope (°)	6 (0–28)	6 (0–28)	9 (0–45)	7 (0–45)

a   *Festuca ovina-Galium saxatile* sub-community

b   *Racomitrium lanuginosum* sub-community

c   *Empetrum nigrum hermaphroditum-Cladonia* spp. sub-community

19   *Vaccinium myrtillus-Cladonia arbuscula* heath (total)

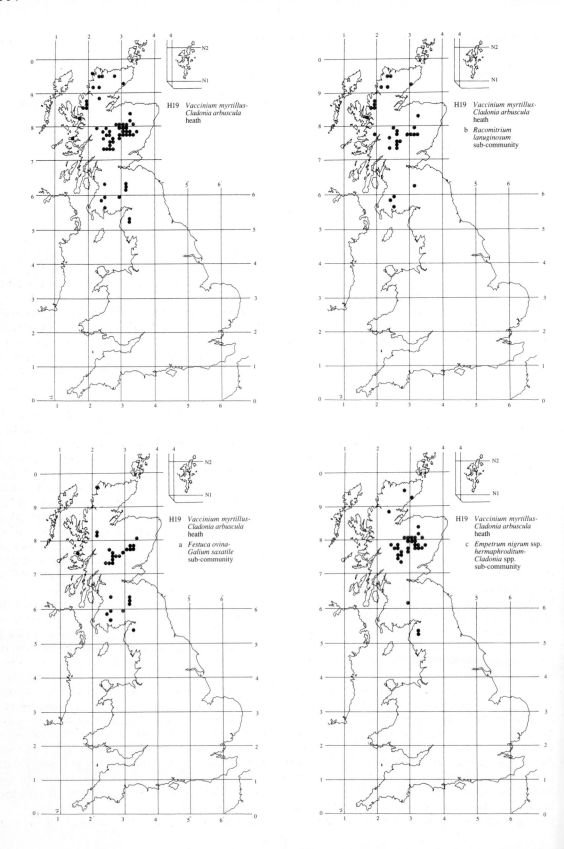

H19 *Vaccinium myrtillus-Cladonia arbuscula* heath

H19 *Vaccinium myrtillus-Cladonia arbuscula* heath
b *Racomitrium lanuginosum* sub-community

H19 *Vaccinium myrtillus-Cladonia arbuscula* heath
a *Festuca ovina-Galium saxatile* sub-community

H19 *Vaccinium myrtillus-Cladonia arbuscula* heath
c *Empetrum nigrum* ssp. *hermaphroditum-Cladonia* spp. sub-community

Vaccinium myrtillus-Racomitrium lanuginosum heath

Synonymy

Rhacomitrio-Vaccinietum Smith 1905 *p.p.*; *Rhacomitrium-Carex bigelowii* nodum, *Empetrum hermaphroditum* facies Poore & McVean 1957; *Empetrum*-hypnoid moss community Poore & McVean 1957, *Rhacomitreto-Empetretum* McVean & Ratcliffe 1962, Birks 1973; *Empetrum*-hypnaceous moss heath McVean & Ratcliffe 1962; *Alchemilla alpina-Vaccinium myrtillus* nodum Birks 1973.

Constant species

Carex bigelowii, Deschampsia flexuosa, Empetrum nigrum ssp. *hermaphroditum, Festuca ovina/vivipara, Galium saxatile, Vaccinium myrtilus, Hylocomium splendens, Pleurozium schreberi, Racomitrium lanuginosum, Rhytidiadelphus loreus, Cladonia uncialis.*

Rare species

Anastrophyllum donianum, A. joergensenii, Bazzania pearsonii, Mastigophora woodsii, Plagiochila carringtonii, Scapania ornithopodioides.

Physiognomy

The *Vaccinium myrtillus-Racomitrum lanuginosum* heath brings together a variety of vegetation types in which *Vaccinium myrtillus* and/or *Empetrum nigrum* ssp. *hermaphroditum*, occasionally with other sub-shrubs, are co-dominant with *Racomitrium lanuginosum* or, in some stands, hypnaceous mosses. *V. myrtillus* is the commonest woody plant overall, although it is quite often subordinate in cover to *E. nigrum* ssp. *hermaphroditum*, the two forming a low mat, generally less than 1 dm thick, which has the appearance of a patchy mosaic of bushes growing among the moss carpet. *V. vitis-idaea* also occurs commonly in most types of *Vaccinium-Racomitrium* heath, though usually at low cover, and there is very occasionally some *V. uliginosum*. *Calluna vulgaris* is typically scarce throughout, though it becomes a little more common at lower altitudes, where the vegetation can approach the *Calluna-Racomitrium*

heath (as on Skye: Birks 1973); *Juniperus communis* ssp. *nana* and *Erica cinerea* can also show local prominence in such stands. In other cases, *Alchemilla alpina*, which is quite a frequent plant through the *Vaccinium-Racomitrium* heath, attains co-dominance among the sub-shrubs. As well as these floristic differences, there is also considerable variation in the total cover of the woody mat, with vegetation at one extreme looking very obviously heathy, other stands resembling *Carex-Racomitrium* moss-heath with a local abundance of sub-shrubs (Poore & McVean 1957).

Floristic continuity in that particular direction is strengthened by the high frequency here of such vascular associates as *Carex bigelowii, Festuca ovina/vivipara, Deschampsia flexuosa* and *Galium saxatile*, although these plants are not generally as abundant as they can be in the *Carex-Racomitrium* heath: *C. bigelowii* sometimes occurs as scattered, quite large, clonal patches and *F. ovina/vivipara* can be sub-dominant but, quite often, they and the others are sparse or distinctly patchy. Also, *Juncus trifidus*, which was locally prominent in some of McVean & Ratcliffe's (1962) *Rhacomitreto-Empetretum*, is of no quantitative importance in this revised *Vaccinium-Racomitrium* heath.

Other common herbs are few, although the grass flora is sometimes a little richer with *Nardus stricta* quite frequent throughout and *Agrostis canina, A. capillaris, Anthoxanthum odoratum* and *Deschampsia cespitosa* occasional, and where these increase their cover somewhat, the vegetation approaches what Poore & McVean (1957) termed a *Nardus*-facies of moss-heath. In other stands, *Huperzia selago* and *Potentilla erecta* become more frequent than usual and, where these are accompanied by such plants as *Thymus praecox, Viola riviniana* and *Carex pilulifera*, of only occasional occurrence overall, a very distinctive composition results. Then, there are sometimes records for *Diphasium alpinum, Luzula sylvatica, L. multiflora, Euphrasia officinalis* agg. and *Armeria maritima*.

But much of the distinctive character of this vegetation and variation within it depends upon the cryptogams, for this is essentially a moss-heath in which *R. lanuginosum* in particular plays a very important role, quite often covering half the ground or more in a rough woolly carpet among the sub-shrubs. Additionally here, though, and this is one good contrast with the *Calluna-Racomitrium* heath, there are frequent records throughout for a variety of other bulky mosses. As well as *Hypnum cupressiforme s.l.* (often *H. jutlandicum*), *Hylocomium splendens*, *Rhytidiadelphus loreus* and *Pleurozium schreberi* are constant in this community and, though of somewhat variable abundance, each can be prominent. Indeed, it is sensible to include here vegetation in which these species locally replace *R. lanuginosum* in an extensive moss mat among *E. nigrum* ssp. *hermaphroditum* and *V. myrtillus*, a feature which both Poore & McVean (1957) and McVean & Ratcliffe (1962) noted at scattered sites through the Scottish Highlands. Some other mosses occur at lower frequencies throughout the *Vaccinium-Racomitrium* heath, *Polytrichum alpinum* and *Dicranum scoparium* being found in small amounts in many stands. Then, among hepatics, *Ptilidium ciliare* and *Diplophyllum albicans* are characteristic of some sub-communities. But greater richness in this element is very much restricted to particularly cool and shady situations, where the *Vaccinium-Racomitrium* heath is one of the vegetation types providing a locus for what Ratcliffe (1968) called the 'mixed northern hepatic mat', a group of Northern Atlantic and more widely-distributed oceanic liverworts which can grow together here in great profusion and luxuriance (see below).

Compared with the *Vaccinium-Cladonia* heath, lichens are much less important in this community, though quite a few are common throughout and locally this element can attain modest abundance. *Cladonia uncialis* and *C. arbuscula* are most frequent overall, with *C. gracilis* and *Cetraria islandica* fairly common but more patchy in occurrence. *Cladonia coccifera*, *C. bellidiflora* and *C. impexa* may show local frequency.

Sub-communities

***Viola riviniana-Thymus praecox* sub-community:** *Rhacomitreto-Empetretum* Birks 1973; *Alchemilla alpina-Vaccinium myrtillus* nodum Birks 1973. This vegetation preserves the general features of the community but the sub-shrub mat is more varied than usual: *E. nigrum* ssp. *hermaphroditum* and *V. myrtillus* can both be abundant but *Alchemilla alpina* is sometimes co-dominant and more locally there can be some *J. communis* ssp. *nana* or *Erica cinerea*, with sparse shoots of *Calluna* also quite common. Community herbs such as *Festuca ovina/vivipara*, *Carex bigelowii*, *Deschampsia flexuosa* and *Galium saxatile* remain very frequent and the first can show modest abundance, but more striking

here is the preferential occurrence among this element of *Huperzia selago* and *Potentilla erecta* with, even more distinctive, *Viola riviniana*, *Thymus praecox* and *Carex pilulifera*. Less common, but still characteristic, are *C. binervis*, *Succisa pratensis*, *Scirpus cespitosus*, *Hypericum pulchrum*, *Antennaria dioica*, *Salix herbacea* and *Selaginella selaginoides*.

The cryptogam flora is poor: *R. lanuginosum* is the typical dominant, but even the other bulky pleurocarps of the community are a little less frequent than usual and *Diplophyllum albicans* is the only common hepatic. *Breutelia chrysocoma* is a quite prominent feature of some stands but no other distinctive characteristics are seen.

***Cetraria islandica* sub-community:** *Rhacomitrium-Carex bigelowii* nodum, *Empetrum hermaphroditum* facies Poore & McVean 1967; *Rhacomitreto-Empetretum* McVean & Ratcliffe 1962 *p.p.* Mixtures of *E. nigrum* ssp. *hermaphroditum* and *V. myrtillus* with abundant *R. lanuginosum* usually dominate here, though some stands with reduced sub-shrub cover make a close approach to the *Carex-Racomitrium* heath. Grasses like *F. ovina/vivipara*, *Deschampsia flexuosa*, *Nardus*, *Agrostis canina* and *Anthoxanthum* tend to be rather more prominent than usual in such vegetation and there can also be a little *Alchemilla alpina*, but the richness of the vascular element typical of the *Viola-Thymus* sub-community is not found.

The bryophytes, too, are not numerous here, although *Rhytidiadelphus loreus*, *Hylocomium splendens* and *Pleurozium schreberi* all occur very commonly together with frequent *Hypnum cupressiforme s.l.* and *Polytrichum alpinum*, and occasional *Ptilidium ciliare*, *Diplophyllum albicans* and *Anastrepta orcadensis*. The lichen flora, though, is a little richer than usual with, in addition to the community species, preferentially frequent *Cladonia gracilis*, *Cetraria islandica* and *Cornicularia aculeata* and, less commonly, *Cladonia leucophaea*, *C. pyxidata*, *Sphaerophorus globosus* and *Alectoria nigricans*. Even then, however, the cover of this element of the vegetation is generally small.

***Bazzania tricrenata-Mylia taylori* sub-community:** *Rhacomitreto-Empetretum* McVean & Ratcliffe 1962 *p.p.*, Birks 1973 *p.p.* In this sub-community, the general floristic features remain very much as above, though *Blechnum spicant* and *Juncus trifidus* figure frequently in the sampled stands. More distinctive, however, is the cryptogam element. *R. lanuginosum* is still the usual dominant, with the other community mosses well represented and, additionally, frequent records for *Dicranum scoparium*, *Plagiothecium undulatum*, *Sphagnum capillifolium* and *Diplophyllum albicans*. But the real richness comes from the occurrence together of the generally oceanic *Pleurozia purpurea*, the Western British *Bazzania tricrenata*, the Sub-Atlantic *Anastrepta orcadensis* and *Scapania gracilis* and the North Atlantic rarities *S. ornithopodioides*, *S. nimbosa*, *Plagiochila carringtonii*, *Bazzania pearsonii*, *Mastigophora woodsii*, *Anastrophyllum donianum* and the very local *A. joergensenii* (Ratcliffe 1968). Mixtures of these species, together with *Mylia taylori*, *Anthelia julacea*, *Tritomaria quinquedentata* and *Dicranodontium uncinatum* can form dense, luxuriant cushions among the sub-shrubs, being especially abundant between the bushes, particular species showing local dominance and imparting a mosaic of yellow, brown and purplish colours to the mat. In such a ground, lichens are sparse, though *Cladonia impexa* shows a notable preference for this kind of *Vaccinium-Racomitrium* heath.

***Rhytidiadelphus loreus-Hylocomium splendens* sub-community:** *Empetrum*-hypnoid moss community Poore & McVean 1957; *Empetrum*-hypnaceous moss heath McVean & Ratcliffe 1962. Although *R. lanuginosum* is much reduced in cover here, the general features of this kind of vegetation accord well with the community as a whole. *E. nigrum* ssp. *hermaphroditum* is usually the dominant sub-shrub with smaller amounts of *V. myrtillus* and, preferentially common here, though usually of low cover, *V. vitis-idaea*. Then, there is frequent *C. bigelowii* and *D. flexuosa*, the latter sometimes quite prominent, with more occasional *Galium saxatile* and *F. ovina/vivipara*.

The really distinctive component, though, is the moss mat where *R. loreus* and *H. splendens*, less abundantly *Pleurozium schreberi*, make up the bulk of the cover. *Hypnum cupressiforme s.l.* is only occasional but *Ptilidium ciliare*, *Dicranum scoparium* and *Anastrepta orcadensis* are all common. *Sphagnum capillifolium* can also be found but the rich assemblages of Atlantic hepatics characteristic of the *Bazzania-Mylia* sub-community do not occur.

Habitat

The *Vaccinium-Racomitrium* heath is characteristic of humic, base-poor soils on fairly exposed slopes and summits at moderate to high altitudes in the cool oceanic mountains of north-west Scotland. Climatic differences and some modest variation in edaphic conditions influence the floristics of the community but this is essentially climax vegetation.

The geographical range of the *Vaccinium-Racomitrium* heath is very similar to that of the *Calluna-Racomitrium* heath, centred firmly on the north-west Highlands of Scotland and extending on to Skye, though with a rather more frequent spread of stands through the Grampians (McVean & Ratcliffe 1962, Birks 1973, Ratcliffe 1977). And the two communities

overlap altitudinally, both being well represented in the low-alpine zone (Poore & McVean 1957), although the *Vaccinium-Racomitrium* heath consistently extends to higher levels than its heather counterpart, occurring mostly on slopes and summits above 600 m and frequently extending over 750 m. Indeed, in its scattered localities south of the Great Glen, it is quite often found up to 1000 m, though along the extreme north-western seaboard and on Skye, its lower limit drops to around 250 m.

The summer climate throughout this zone is cool, with mean annual maxima almost everywhere less than 21 °C and, over the higher peaks of the north-west, below 20 °C (Conolly & Dahl 1970), so *Calluna* itself, *Erica cinerea* and *Carex pilulifera* characteristic of lower-altitude heaths through this part of Scotland, make but a sparse appearance here in the *Viola-Thymus* sub-community which takes the *Vaccinium-Racomitrium* heath to its lowest levels on Skye and around the Forest of Letterewe. More obviously, it is Arctic-Alpines such as *E. nigrum* ssp. *hermaphroditum*, *V. vitis-idaea* and *C. bigelowii*, *Alchemilla alpina*, *Diphasium alpinum* and *Polytrichum alpinum* which give the vegetation its essentially montane character.

Compared with the east-central Highlands, however, where the *Vaccinium-Cladonia* heath is the typical bilberry/crowberry vegetation of the low-alpine zone, the winter climate through much of the range of the community is relatively mild, so the overall temperature regime is fairly equable. Moreover, the climate is very much more humid than through the Grampians, with annual precipitation always over 1600 mm (*Climatological Atlas* 1952) with usually more than 220 wet days yr^{-1} (Ratcliffe 1968) and a high percentage of daytime cloudiness over the higher ground (Chandler & Gregory 1976, Page 1982). And, as in the two corresponding heather communities, the geographical switch to this cool oceanic climate is strongly reflected in the move here from lichens to mosses as the dominant cryptogam element of the vegetation mat. Again, it is *R. lanuginosum* that exerts its formidable competitive power to often exceed the sub-shrubs in cover, though other bulky pleurocarps figure more often here than in the *Calluna-Racomitrium* heath, perhaps reflecting the somewhat more sheltered conditions compared with the typically wind-blasted knolls and summit ridges occupied by that vegetation. The *Rhytidiadelphus-Hylocomium* sub-community, where these other species attain their maximum representation, is not a common vegetation type but at scattered sites through the range of the *Vaccinium-Racomitrium* heath it seems to be associated with slight hollows where modest snow accumulation affords some protection from the worst of the winter cold. In such situations, the shift from *R. lanuginosum* to pleurocarps like *R. loreus*, *H. splendens* and *P. schreberi* can be seen as a floristic transition to the mildly chionophilous *Vaccinium-Deschampsia* heath of more sheltered slopes and snow-bed surrounds.

Topoclimate also plays a part in determining the floristics and distribution of the *Bazzania-Mylia* sub-community. This is confined within the wettest parts of the north-west Highlands, where, with over 220 wet days yr⁻¹, total rainfall approaches 2400 mm yr⁻¹ (Chandler & Gregory 1976) but, even there it is found only locally on ledges and among boulders on north- or east-facing slopes where sunless and especially humid conditions favour the luxuriant development of the 'northern hepatic mat'. This suite of Atlantic liverworts can extend to altitudes below those characteristic of this community: many of the same species figure prominently in the *Calluna-Vaccinium-Sphagnum* heath which occurs commonly down to 300 m or so, though more montane plants such as *Anastrophyllum donianum* and *Scapania nimbosa* peter out there, with *Herbertus aduncus* ssp. *hutchinsiae* becoming correspondingly more prominent. The switch from *Calluna* to *V. myrtillus* and *E. nigrum* ssp. *hermaphroditum* in the heaths of the north-west Highlands does, in fact, show particularly well in the shady situations which encourage the development of these hepatics where, at around 600 m, the generally increasing snow cover tips the balance towards dominance of bilberry and crowberry (Ratcliffe 1968).

Less sheltered situations, on slopes of all aspects and with little snow-lie, favour the occurrence of the central and most widely-distributed type of *Vaccinium-Racomitrium* heath, the *Cetraria* sub-community which is found not only among boulders and on ledges but over open slopes with blocky talus, on summit detritus and on gently-sloping ridges, though even in these latter habitats, a rough rocky surface seems to be characteristic. And, as with the other sub-communities, the soils here are typically strongly humic, sometimes shallow, well-developed peaty podzols but often fragmentary rankers with raw and sometimes rather greasy organic matter resting directly on rock debris. The humidity of the climate keeps the soils moist but they are free-draining.

Almost always, the bedrocks underlying the *Vaccinium-Racomitrium* heath are siliceous in character, with Torridonian sandstone being an especially important substrate in the hills between the Kyle of Lochalsh and An Teallach, Cambrian quartzite and acidic Lewisian gneiss extending north from there to Foinaven, and Moine rocks of various kinds supporting stands south-eastwards into the Grampians. On Skye, intrusive rocks and Tertiary basalts occur beneath the *Vaccinium-Racomitrium* heath, and there, in the *Viola-Thymus* sub-community, there can be slight extension on to mildly basic substrates, particularly where, in that vegetation, *Alchemilla alpina* becomes prominent with *Thymus*

praecox. Even in such circumstances, however, the profiles remain relatively base poor, with a superficial pH of 4–4.8 (Birks 1973).

Zonation and succession

The *Vaccinium-Racomitrium* heath is typically found at the transition between the sub-shrub vegetation of the low-alpine zone and the moss-heaths and fell-field communities of the higher slopes and summits of the mountains of north-west Scotland, the general zonation reflecting the increased harshness of climate with the move to upper slopes. Local variations in exposure over the intermediate slopes commonly result in gradations from the community to wind-blasted heaths and snow-bed vegetation.

The characteristic altitudinal pattern in which the *Vaccinium-Racomitrium* heath occurs is well seen over the upper slopes of mountains like those of the Letterewe Forest, Beinn Dearg and Am Faochagach in Ross and Ben More Assynt and Foinaven in Sutherland, where the community occurs towards the lower limit of the *Carex-Racomitrium* heath. This vegetation is particularly extensive over the more rounded summits in the north-west Highlands, often as the distinctive *Silene* sub-community with its suite of cushion herbs, or in its Typical form with overwhelming dominance of *Racomitrium*, or grading into more open fell-field on active ablation surfaces. Floristic continuity between these kinds of vegetation and the *Vaccinium-Racomitrium* heath is strong, with the *Cetraria* sub-community here looking very much like a sub-shrub facies of Typical *Carex-Racomitrium* heath: indeed in many situations, the former occurs as small patches in hollows or over blocky detritus among the lower reaches of the latter, picked out from a distance by the close canopy of *E. nigrum* ssp. *hermaphroditum* and *V. myrtillus*, but otherwise showing great qualitative similarity among the vascular associates and cryptogams. Usually, the hypnoid mosses are more frequent in the *Vaccinium-Racomitrium* heath and, of course, in the *Rhytidiadelphus-Hylocomium* sub-community, their abundance is very striking. Where this kind of vegetation replaces the *Cetraria* sub-community in mosaics of this sort, as can be seen on Ben Klibreck in Sutherland and locally elsewhere, the contrast between the elements of the pattern is that much greater.

In other situations, the *Vaccinium-Racomitrium* heath forms a more extensive zone below the summit vegetation, becoming especially prominent where the increase in slope is also marked by a shift to rough, bouldery ground on screes. Here, it can be seen as a transition to the *Vaccinium-Deschampsia* heath, the *Hylocomium-Rhytidiadelphus* sub-community of which is very close in its general composition, but where there is a move to more consistent eclipse of *R. lanuginosum* by the hypnaceous mosses with increased shelter and moisture.

Alternatively, decreasing altitude can be marked by a replacement of the *Vaccinium-Racomitrium* heath by sub-shrub vegetation in which *Calluna* plays a more prominent role. This can be seen over both sheltered and exposed slopes. In the former case, where a northerly or easterly aspect results in an especially cool, shady and humid environment, the *Vaccinium-Racomitrium* heath is typically represented by the *Bazzania-Mylia* sub-community and this usually passes to the *Plagiothecium-Anastrepta* sub-community of the *Calluna-Vaccinium-Sphagnum* heath at levels where the snow cover becomes less appreciable: bilberry and crowberry decrease in vigour with some slight shifts in the composition of the luxuriant hepatic mat that characterises both these vegetation types with the move to a less montane environment.

In other places, where a drop in altitude is accompanied by increasing exposure to harsh winds that blow away any snow that does fall, the *Vaccinium-Racomitrium* heath is replaced by the *Calluna-Racomitrium* heath. In general composition and physiognomy, these two are quite similar but the latter usually displays little contribution from *V. myrtillus* and the hypnoid mosses of the former have a limited role there. Nonetheless, the communities can come very close in their suites of vascular associates, especially where both extend to their lowest altitudes along the north-western seaboard of Scotland and on Skye, where the *Viola-Thymus* sub-community here and the *Festuca* sub-community of the *Calluna-Racomitrium* heath show a floristic convergence. At higher altitudes in the hinterland of north-west Scotland, it is usually the *Cetraria* sub-community of the *Vaccinium-Racomitrium* heath and the *Empetrum* sub-community of the *Calluna-Racomitrium* heath that occur together in these altitudinal zonations, sometimes with the *Calluna-A. alpinus* heath, another dwarfed sub-shrub community of wind-blasted spurs, complicating the pattern, as on Foinaven.

In the middle reaches of the low-alpine zone on such mountains, these vegetation types can be disposed not so much in an altitudinal pattern as in patchworks over the slopes reflecting retention of snow in less or more sheltered situations. The *Calluna-Racomitrium* and *Calluna-A. alpinus* heaths represent the most exposed extreme at these levels, with the *Vaccinium-Racomitrium* heath transitional to early snow-bed vegetation. Sometimes, indeed, it can be seen as a surround to more chionophilous communities, though it often forms an element of complex and extensive mosaics with, for example, certain kinds of *Nardus-Carex* vegetation. Patchworks of the *Empetrum-Cetraria* sub-community

and the *Cetraria* sub-community of the *Vaccinium-Racomitrium* heath are especially well seen on the Letterewe Hills, the Affric-Cannich Hills, Foinaven, Beinn Dearg and Ben Klibreck, where sometimes quite gentle transitions between the components depend on the proportions of the sub-shrubs and *Nardus* against a ground of abundant *R. lanuginosum* with frequent hypnoid mosses and lichens. In some sites, other grassy chionophilous vegetation like the *Deschampsia-Galium* community, can also occur in these patterns with *V. myrtillus* and hypnoid mosses running on with some frequency and local abundance. Or there may be sharper transitions to late snow-bed communities.

Distribution
Although stands of the *Vaccinium-Racomitrium* heath, particularly the *Cetraria* sub-community, can be found scattered through the Grampians, the distribution of this kind of vegetation is strongly centred in north-west Scotland. The *Cetraria* sub-community is the most widespread and common type overall, with the *Rhytidiadelphus-Hylocomium* sub-community local throughout the range. The *Bazzania-Mylia* type is strongly confined to the wettest regions and even there is very much restricted to suitably cold and damp aspects. The *Viola-Thymus* sub-community extends the range into the milder foothills of the western seaboard and Skye.

Affinities
The core of the *Vaccinium-Racomitrium* heath as defined here, the *Cetraria* sub-community, is based on the earliest descriptions of this kind of vegetation by Poore & McVean (1957) and McVean & Ratcliffe (1962), distinguishing as a separate *Bazzania-Mylia* sub-community the hepatic-rich stands first noted in detail in Ratcliffe (1968). The stands with hypnaceous mosses, which these authors saw as bearing some relation to their *Empetrum* heaths, can be readily incorporated here and serve to emphasise the relationship between the community and the mildly chionophilous sub-shrub vegetation which extends to lower altitudes. The *Viola-Thymus* sub-community also links the *Vaccinium-Racomitrium* heath with the oceanic heaths of the coastal regions of north-west Scotland. In general, though, the floristic relationships of the community are with the moss-heaths and fell-field vegetation of the high-montane zone. Following McVean & Ratcliffe (1962), it seems best to separate the *Vaccinium-Racomitrium* heath from the *Carex-Racomitrium* heath, though both can be readily located in the Loiseleurieto-Arctostaphylion (Nordhagen 1943) or Arctostaphyleto-Cetrarion (Dahl 1956), where the community can be seen as a bilberry/crowberry analogue to the *Calluna-Racomitrium* heath.

Floristic table H20

	a	b	c	d	20
Vaccinium myrtillus	V (1–6)	V (1–6)	V (1–4)	V (1–8)	V (1–8)
Racomitrium lanuginosum	V (4–10)	V (4–10)	V (4–10)	V (1–4)	V (1–10)
Empetrum nigrum hermaphroditum	IV (1–3)	IV (1–8)	V (4–8)	V (4–10)	V (1–10)
Carex bigelowii	IV (1–4)	IV (1–3)	V (1–3)	V (1–3)	IV (1–4)
Deschampsia flexuosa	IV (1–3)	IV (1–4)	IV (1–3)	V (1–4)	IV (1–4)
Cladonia uncialis	V (1–3)	V (1–4)	V (1–3)	II (1–3)	IV (1–4)
Festuca ovina/vivipara	V (1–4)	IV (1–4)	V (1–4)	II (1–3)	IV (1–4)
Galium saxatile	IV (1–6)	IV (1–4)	IV (1–4)	III (1–3)	IV (1–6)
Rhytidiadelphus loreus	III (1–4)	IV (1–6)	V (1–4)	V (1–8)	IV (1–8)
Hylocomium splendens	III (1–4)	IV (1–4)	V (1–3)	V (1–8)	IV (1–8)
Pleurozium schreberi	III (1–4)	IV (1–4)	III (1–3)	V (1–6)	IV (1–6)
Huperzia selago	IV (1–4)	II (1–3)	IV (1–3)		III (1–4)
Potentilla erecta	V (1–4)	II (1–3)	II (1–3)	I (1)	III (1–4)
Thymus praecox	IV (1–4)	II (1–3)			III (1–4)
Viola riviniana	IV (1–3)		I (1)		II (1–3)
Carex pilulifera	III (1–3)	II (1–3)		I (1–3)	II (1–3)
Calluna vulgaris	III (1–3)	I (1–2)	I (1–3)	I (1–3)	II (1–3)
Succisa pratensis	II (1–3)	I (1)		I (1–3)	II (1–3)
Salix herbacea	II (1–3)				I (1–3)
Scirpus cespitosus	II (1–3)				I (1–3)
Carex binervis	II (1–3)				I (1–3)
Hypericum pulchrum	II (1)				I (1)
Erica cinerea	II (1–4)				I (1–4)
Antennaria dioica	II (1–3)				I (1–3)
Breutelia chrysocoma	II (4–5)				I (4–5)
Selaginella selaginoides	III (1–3)				I (1–3)
Juniperus communis nana	II (1–6)				I (1–6)
Plantago maritima	I (1–3)				I (1–3)
Cetraria islandica	II (1–3)	V (1–4)	III (1–3)	III (1–3)	III (1–4)
Cladonia gracilis	I (1–3)	IV (1–3)	III (1–3)		III (1–3)
Cornicularia aculeata	I (1–3)	III (1–4)		I (1–3)	II (1–4)

Cladonia leucophaea	I (1-3)		I (1-3)	I (1-3)	II (1-3)
Sphaerophorus globosus	II (1-3)				I (1-3)
Cladonia pyxidata	II (1-3)				I (1-3)
Alectoria nigricans	II (1-3)				I (1-3)
Diplophyllum albicans	IV (1-3)	II (1-4)	V (1-4)	I (1-3)	III (1-4)
Anastrepta orcadensis	I (1-3)	II (1-3)	IV (1-4)	IV (1-4)	III (1-4)
Dicranum scoparium	I (1-3)	II (1-3)	IV (1-3)	III (1-3)	III (1-3)
Bazzania tricrenata	I (1-3)	I (1-3)	V (1-3)	I (1)	II (1-3)
Mylia taylori	I (1-3)		V (1-3)		I (1-3)
Sphagnum capillifolium			IV (1-4)		I (1-4)
Pleurozia purpurea	I (1-4)	I (1-3)	IV (3-4)	II (1-4)	I (1-4)
Cladonia impexa		I (1-3)	IV (1-3)		I (1-3)
Scapania ornithopodioides			IV (1-3)		I (1-3)
Plagiochila carringtonii			IV (1-3)		I (1-3)
Blechnum spicant	II (1-4)	I (1-3)	III (1-3)	I (1)	I (1-4)
Scapania gracilis	I (1-3)	II (1-3)	III (1-3)		I (1-3)
Juncus trifidus	I (1-3)	I (1-3)	III (1-3)		I (1-3)
Anthelia julacea	I (1-3)		III (1)		I (1-3)
Tritomaria quinquedentata	I (1-3)	I (1-3)	III (1-3)	I (1)	I (1-3)
Plagiothecium undulatum	I (1-3)	I (1-3)	III (1-3)	I (1-4)	I (1-4)
Anastrophyllum donianum			III (1-4)		I (1-4)
Scapania nimbosa			III (1-4)		I (1-4)
Bazzania pearsonii			III (1-4)		I (1-4)
Dicranodontium uncinatum			III (1-3)		I (1-3)
Hymenophyllum wilsonii	I (1)	II (1)	II (1)		I (1)
Mastigophora woodsii			II (1-3)		I (1-3)
Sphagnum tenellum			II (2)		I (2)
Vaccinium vitis-idaea	I (1-3)	III (1-3)	III (1-3)	IV (1-4)	III (1-4)
Ptilidium ciliare	I (1-3)	III (1-3)	III (1-3)	IV (1-3)	III (1-3)
Oxalis acetosella				I (1-3)	I (1-3)
Hypnum cupressiforme s.l.	III (1-4)	III (1-3)	III (1-3)	II (1-3)	III (1-4)
Cladonia arbuscula	II (1-3)	III (1-6)	III (1-3)	III (1-5)	III (1-6)
Polytrichum alpinum	II (1-3)	III (1-4)	III (1-3)	III (1-3)	III (1-4)
Nardus stricta	III (1-6)	III (1-6)	II (1-3)	II (1-6)	III (1-6)

Floristic table H20 (*cont.*)

	a	b	c	d	20
Alchemilla alpina	III (1–8)	II (1–4)	III (1–4)	II (1–3)	II (1–3)
Agrostis canina	I (4)	III (1–4)	II (1–3)	II (1–3)	II (1–4)
Diphasium alpinum	II (1–3)	II (1–3)	II (1–3)	I (1–3)	II (1–3)
Agrostis capillaris	II (1–4)	II (1–4)	I (1–3)	I (1)	II (1–4)
Vaccinium uliginosum	I (1–3)	II (1–4)	II (1–6)	I (1–3)	II (1–6)
Cladonia coccifera	I (1–3)	II (1–3)	I (1–3)	I (1–3)	I (1–3)
Cladonia bellidiflora	I (1–3)	I (1–3)	II (1–3)	I (1–3)	I (1–3)
Deschampsia cespitosa	I (1–3)	II (1–4)		I (1–3)	I (1–4)
Anthoxanthum odoratum	I (1–3)	II (1–4)		I (1–3)	I (1–4)
Luzula sylvatica	I (1–3)	I (1–4)	II (1–3)		I (1–4)
Armeria maritima	I (1–3)	I (1–3)	II (1–4)		I (1–4)
Dicranum majus	I (1–3)		II (1–4)	I (1–4)	I (1–4)
Barbilophozia floerkii	I (1–3)	I (1–3)	I (1–3)	I (1–3)	I (1–3)
Euphrasia officinalis agg.	I (1–3)	I (1–3)	I (1–3)		I (1–3)
Cladonia rangiferina	I (1–3)	I (1–2)		I (1)	I (1–3)
Luzula multiflora	I (1–3)	I (1–3)		I (1–3)	I (1–3)
Campylopus paradoxus	I (1–3)	I (1–3)			I (1–3)
Thuidium tamariscinum	I (1–3)	I (1–3)			I (1–3)
Polytrichum piliferum	I (1–3)	I (1–4)			I (1–4)
Nardia scalaris	I (1)	I (1–3)			I (1–3)
Agrostis stolonifera	I (1–3)	I (1–3)			I (1–3)
Thelypteris limbosperma	I (1–3)		I (2)		I (1–3)
Tetraplodon mnioides	I (1)		I (1)		I (1)
Thelypteris phegopteris	I (1)		I (2)		I (1–2)
Jungermannia atrovirens	I (1–3)		I (1)		I (1–3)
Plagiochila spinulosa	I (1–3)		I (1–3)		I (1–3)
Pohlia nutans		I (1–3)	I (1–3)		I (1–3)
Sphagnum quinquefarium			I (4)	I (1–3)	I (1–4)
Number of samples	17	14	7	15	53
Number of species/sample	26 (14–35)	22 (12–36)	36 (27–47)	20 (10–29)	24 (10–47)

	a	b	c	d	20
Vegetation height (cm)	5 (1–8)	4 (1–8)		10 (5–27)	7 (1–27)
Vegetation cover (%)	100	89 (30–100)	89 (50–100)	99 (95–100)	95 (30–100)
Altitude (m)	583 (246–823)	754 (560–915)	691 (579–838)	814 (640–1174)	718 (246–1174)
Slope (°)	15 (3–45)	17 (0–40)	17 (0–35)	25 (2–45)	19 (0–45)

a   *Viola riviniana-Thymus praecox* sub-community

b   *Cetraria islandica* sub-community

c   *Bazzania tricrenata-Mylia taylori* sub-community

d   *Rhytidiadelphus loreus-Hylocomium splendens* sub-community

20 *Vaccinium myrtillus-Racomitrium lanuginosum* heath (total)

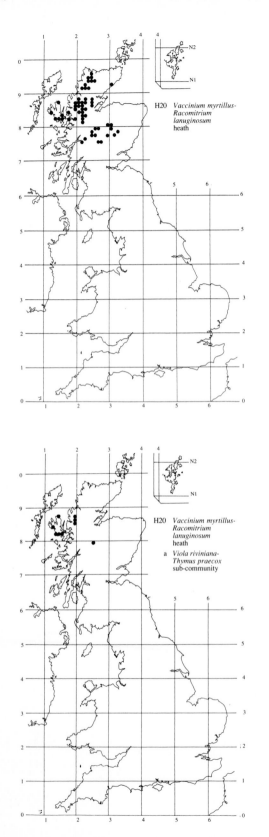

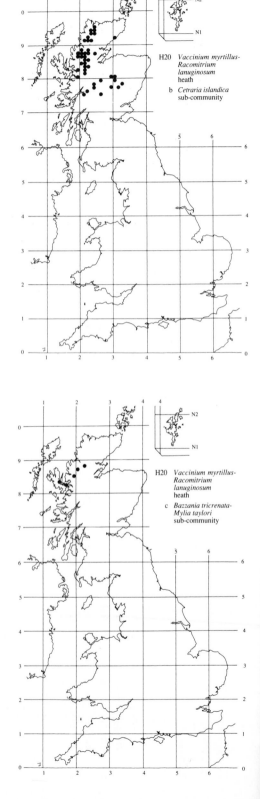

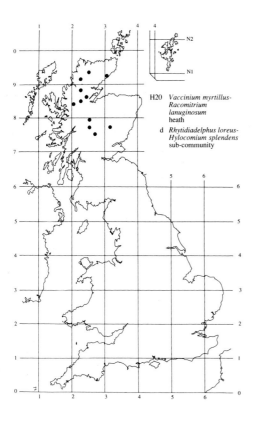

H20   *Vaccinium myrtillus-*
       Racomitrium
       lanuginosum
       heath

d   *Rhytidiadelphus loreus-*
     Hylocomium splendens
     sub-community

H21
Calluna vulgaris-Vaccinium myrtillus-Sphagnum capillifolium heath

Synonymy

Vaccineto-Callunetum hepaticosum McVean & Ratcliffe 1962, Birks 1973; *Vaccinium-Sphagnum-Hymenophyllum wilsonii* nodum Prentice & Prentice 1975; *Herberta adunca-Calluna vulgaris* Association Birse 1984.

Constant species

Blechnum spicant, Calluna vulgaris, Deschampsia flexuosa, Potentilla erecta, Vaccinium myrtillus, Dicranum scoparium, Hylocomium splendens, Hypnum cupressiforme s.l., Plagiothecium undulatum, Pleurozium schreberi, Rhytidiadelphus loreus, Sphagnum capillifolium.

Rare species

Anastrophyllum donianum, A. joergensenii, Bazzania pearsonii, Campylopus setifolius, Colura calyptrifolia, Mastigophora woodsii, Myurium hochstetteri, Plagiochila carringtonii, Scapania ornithopodioides.

Physiognomy

The *Calluna vulgaris-Vaccinium myrtillus-Sphagnum capillifolium* heath has a mixed canopy of sub-shrubs, usually 3–5 dm tall, with a damp layer of luxuriant bryophytes. *Calluna vulgaris* is usually the dominant ericoid, although *Vaccinium myrtillus* is also constant and *Empetrum nigrum*, almost always ssp. *hermaphroditum*, is present in many stands, and each of these can show modest abundance. *Erica cinerea* occurs quite frequently, although it becomes distinctly patchy in the most humid situations occupied by this vegetation, and there is occasionally some *V. vitis-idaea*. In its general composition, then, the canopy is like that of much *Calluna-Vaccinium* heath, although *E. nigrum* ssp. *nigrum* tends to be the predominant crowberry there and, more importantly, regeneration after burning plays a large part in determining the floristics and structure of the sub-shrub element, and the associated flora, at any particular time. With the *Calluna-Vaccinium-Sphagnum* heath, on the other hand, fire is highly inimical to the

distinctive composition of the richest stands and, though this vegetation sometimes experiences light grazing, treatments are not of consequence in maintaining its characteristic features. Other sub-shrubs figure very occasionally, with sometimes a little *Arctostaphylos uva-ursi, A. alpinus* or *Juniperus communis* ssp. *nana*, but the general scarcity of these species helps separate the community from the *Calluna-Juniperus* heath, a vegetation type of rather similar geographical range and particular habitat requirements.

As in the *Calluna-Vaccinium* heath, *Deschampsia flexuosa* and *Potentilla erecta* are both very common among the associates, although they are usually present here as sparse, scattered individuals. More distinctively in this heath, there is constant *Blechnum spicant* and frequent records for *Solidago virgaurea* and *Listera cordata*, the last Northern Montane plant providing one of the floristic links between this community and our native pine and related juniper and birch woods. Other common vascular species are few. *Molinia caerulea* and *Scirpus cespitosus* occur occasionally, giving some continuity with the wet heaths with which this vegetation is often found, and there are scarce records for *Nardus stricta, Agrostis canina, A. capillaris, Festuca vivipara, Carex pilulifera, C. binervis* and *Galium saxatile* but, even in more open parts of the canopy, these never thicken up enough to give the community the look of a grass-heath. In contrast to related vegetation at higher altitudes, notably the *Vaccinium-Racomitrium* heath, *Carex bigelowii* is at most occasional here.

More important than any of these, however, are the bryophytes which form an extensive and lush carpet, becoming especially copious where there is a break in the sub-shrub cover. Constant throughout are bulky hypnaceous mosses such as *Hypnum cupressiforme s.l., Rhytidiadelphus loreus, Pleurozium schreberi* and *Hylocomium splendens*, with *Plagiothecium undulatum, Dicranum scoparium* and *D. majus* also very common, but particularly distinctive here is the high frequency and local abundance of *Sphagnum capillifolium*. Other Acutifolia

Sphagna have also sometimes been recorded, including *S. quinquefarium*, *S. russowii* and *S. girgensohnii*, and *S. tenellum* too can be found occasionally, and together these serve as a good diagnostic feature to separate the community from more bryophyte-rich stands of the *Calluna-Vaccinium* heath.

Racomitrium lanuginosum becomes more frequent at higher altitudes but, even then, it is not generally abundant here, as it is in the *Vaccinium-Racomitrium* heath. Other bryophytes recorded occasionally throughout are *Ptilium crista-castrensis*, another particular link with the woodlands of the region, *Breutelia chrysocoma* and *Thuidium tamariscinum*, *Frullania tamarisci* and *Ptilidium ciliare*. But the most spectacular enrichment of this element of the flora comes from oceanic liverworts, for this community is one of those vegetation types which, in particular habitats, provides a major locus for Ratcliffe's (1968) 'mixed northern hepatic mat'. Species such as *Scapania gracilis*, *Mylia taylori* and *Diplophyllum albicans* can, indeed, be found throughout the *Calluna-Vaccinium-Sphagnum* heath but, in one of the sub-communities, many other more specifically Atlantic species can be found growing together in vegetation which is unique among heather-dominated communities.

Compared with the bryophytes, lichens make a fairly insignificant contribution to the cover. *Cladonia impexa* is the only species which occurs commonly throughout, with *C. gracilis* occasional, and, although *C. uncialis* and *C. arbuscula* become more frequent at higher altitudes, their abundance is generally not very great.

Sub-communities

***Calluna vulgaris-Pteridium aquilinum* sub-community:** *Vaccinium-Sphagnum-Hymenophyllum wilsonii* nodum Prentice & Prentice 1975. *Calluna* is generally a strong dominant in this taller and more species-poor kind of *Calluna-Vaccinium-Sphagnum* heath, in which hepatics in particular are sparsely represented. Among the other sub-shrubs, *V. myrtillus* is very common, with *E. cinerea* and *V. vitis-idaea* also occurring occasionally, but *E. nigrum* ssp. *hermaphroditum* is rather local: it figures among stands on Hoy, for example (Prentice & Prentice 1975), but is quite absent in some other places, particularly where this kind of vegetation extends to its lowest altitudes. Other vascular plants occur sparsely, but distinctive here is the fairly common occurrence of scattered fronds of *Pteridium aquilinum* with occasional *Oxalis acetosella*, *Viola riviniana* and *Luzula sylvatica*.

Bryophytes can have fairly high cover, but the carpet consists almost entirely of the constants of the community, with *Dicranum majus* and, to a lesser extent, *Plagiothecium undulatum* showing preferential frequency. *Sphagnum capillifolium* is sometimes replaced by *S.*

girgensohnii or *S. russowii* and there is occasional *Thuidium tamariscinum*, *Calypogeia muellerana* and *Lophozia ventricosa*. Very occasionally, there can be records for *Hymenophyllum wilsonii* and a few of the oceanic hepatics, but these are no more than fragmentary assemblages compared with the richness of the other sub-community.

***Mastigophora woodsii-Herbertus aduncus* ssp. *hutchinsiae* sub-community:** *Vaccineto-Callunetum hepaticosum* McVean & Ratcliffe 1962, Birks 1973; *Herberta adunca-Calluna vulgaris* Association Birse 1984. In this sub-community, *Calluna* is still usually the most abundant sub-shrub, but the canopy is generally shorter and more mixed than above, with *Empetrum nigrum* ssp. *hermaphroditum* in particular playing a more frequent and sometimes abundant role and *V. uliginosum* figuring occasionally. Other vascular associates are few and even plants such as *Blechnum spicant* and *Listera cordata* can become rather patchy here.

The bryophytes, on the other hand, are extremely well developed with a strikingly consistent enrichment from numerous species. Among the mosses, all the community constants occur frequently, with *Rhytidiadelphus loreus*, *Hylocomium splendens* and *Sphagnum capillifolium* being moderately abundant and, with the shift to higher altitudes here, *R. lanuginosum* becomes more common. But it is the hepatics that catch the eye most obviously, especially where, between the bushes and among the boulders over which this vegetation is often developed, they form soft but bulky patches, each often a pure mass of shoots, variously tinged yellow, orange, red, brown or purple. Some widespread species, like the sub-Atlantic *Scapania gracilis* and *Anastrepta orcadensis*, the western British *Mylia taylori* and *Bazzania tricrenata*, and the North Atlantic *Pleurozia purpurea* and *Diplophyllum albicans*, occur very commonly, together with occasional records for *Dicranodontium uncinatum*, *Tritomaria quinquedentata* and *Plagiochila spinulosa*. But most distinctive are the rarer North Atlantic liverworts *Mastigophora woodsii*, *Plagiochila carringtonii*, *Scapania ornithopodioides*, *Bazzania pearsonii* and, towards higher altitudes, the more montane *Anastrophyllum donianum* (with *A. joergensenii* at a very few sites) and *Scapania nimbosa*. These last two are, in fact, rather more characteristic of the *Vaccinium-Racomitrium* heath, into which many of these same species run on in abundance, whereas here an additional constant member of the hepatic suite is *Herbertus aduncus* ssp. *hutchinsiae*. Indeed, with *Mastigophora woodsii*, this is often the most abundant liverwort forming conspicuous bright orange cushions. On Skye, this kind of vegetation also provides a locus for *Colura calyptrifolia* and the rare mosses *Myurium hochstetteri* and *Campylopus setifolius* (Birks 1973).

Habitat

The *Calluna-Vaccinium-Sphagnum* heath is highly characteristic of fragmentary humic soils developed in situations with a cool but equable climate and consistently shady and extremely humid atmosphere. It is almost wholly confined to low to moderate altitudes through the oceanic mountains of north-west Scotland, and is there very much restricted to steep, sunless slopes of north-west to easterly aspect, often with rock outcrops and blocky talus, among which crevices provide an additional measure of shade. The vegetation is sometimes lightly grazed, but burning is very damaging, and recovery probably extremely slow.

The general floristic features of the community reflect the relatively mild regional climate. For the most part, this is a vegetation type of the sub-alpine zone along the north-western seaboard of Scotland (Poore & McVean 1957, McVean & Ratcliffe 1962), extending down almost to sea-level on Skye (Birks 1973) and not often penetrating much above 500 m. Summers are characteristically cool in this part of Britain, with mean annual maximum temperatures generally less than 22 °C even at lower altitudes (Conolly & Dahl 1970), but the winters are fairly mild, with sea-level February minima above freezing, fairly few frost days and annual accumulated temperatures very much like those throughout the upland fringes of the country (*Climatological Atlas* 1952, Page 1982). These conditions account for the rather muted montane character of the flora here, with Arctic-Alpines like *V. vitis-idaea*, *E. nigrum* ssp. *hermaphroditum* and *C. bigelowii* of only occasional or rather uneven representation throughout, and hypnaceous mosses rather than *R. lanuginosum* providing the most consistent element among the cryptogams.

And, of course, there is the dominance among the sub-shrubs of *Calluna*, rather than *V. myrtillus* and/or *E. nigrum* ssp. *hermaphroditum*: the altitude at which this switch in abundance occurs in more sheltered situations in the north-west Highlands, at around 600 m, corresponds roughly to the upper limit of the *Calluna-Vaccinium-Sphagnum* heath (Ratcliffe 1968). Above this level, increased snow-cover becomes inimical to the vigour of heather (Gimingham 1960) and the community is replaced by the *Vaccinium-Racomitrium* heath. The *Mastigophora-Herbertus* sub-community represents a transition to such low-alpine vegetation, penetrating to the highest levels attained by the *Calluna-Vaccinium-Sphagnum* heath, with a mean altitude in available samples of around 400 m. The *Calluna-Pteridium* sub-community, on the other hand, is characteristic of lower slopes, with a mean altitude some 200 m less and, with its more sporadic Arctic-Alpines, increased frequency of *Blechnum spicant* and *E. cinerea*, and preferential occurrence of *Pteridium aquilinum* it shows floristic continuity with the kind of oceanic sub-shrub vegetation seen in the *Calluna-Erica* heath.

Even in such situations, however, the luxuriance of the carpet of hypnaceous mosses and the frequent presence among them of various Sphagna, help maintain a distinction and is testimony to the high ground and atmospheric humidity here even on the steep slopes characteristically occupied by this community, which are almost always greater than 25° and often up to 45°. The profiles beneath the *Calluna-Vaccinium-Sphagnum* heath are typically fragmentary, sometimes little more than a thin layer of bryophyte remains insulating the vegetation from the mineral substrate, sometimes a thicker accumulation of peat over and between the rock fragments, and, since the bedrocks are characteristically pervious, the soils are free-draining, but the consistently wet climate keeps them, and the atmosphere, permanently moist.

It is this consistency of humidity, rather than the total amount of rainfall, that is of prime importance in determining the distribution and composition of the community. Annual precipitation through the range of the *Calluna-Vaccinium-Sphagnum* heath varies quite considerably, from not much more than 1600 mm through the mountains of Sutherland to over 3200 mm around Beinn Eighe (*Climatological Atlas* 1952), but everywhere in its extent through the north-west Highlands the community experiences more than 220 wet days yr^{-1} (Ratcliffe 1968). Moreover, this kind of vegetation is best developed on slopes which have an aspect between north-west and east, and where the angle of inclination is such as to reduce insolation to a minimum. The characteristic habitat for the *Calluna-Vaccinium-Sphagnum* heath is thus over the lower slopes of the deep, sunless corries that are etched into the northern faces of mountains like Beinn Eighe and Liathach, An Teallach, Foinaven, the peaks of the Monar and Letterewe Forests, the hills of Skye and, on Orkney, the Enegars cliffs (Birks 1973, Prentice & Prentice 1975, Ratcliffe 1977). And the rocks of which many of these mountains are made up, Torridonian sandstone, quartzites or Old Red Sandstone, typically weather to blocky talus, among which the vegetation benefits from additional shade and shelter.

It is in the *Mastigophora-Herbertus* sub-community that the influence of these factors is seen most clearly. This is the kind of *Calluna-Vaccinium-Sphagnum* heath more strictly confined to higher altitudes in the north-west Highlands where summer temperatures are lower and the input from orographic rain much greater, and more closely associated there with the distinctive topography. As Ratcliffe (1968) showed, its most striking feature, the 'mixed northern hepatic mat', is strongly favoured by the particular combination of climatic conditions characteristic of such sites, though we are not really any closer now to understanding which factors are of importance to which members of this distinctive assemblage. Certainly, a very humid climate is a major requirement for the group as a whole, and especially for the North Atlantic species, though, even among them, Ratcliffe (1968) noted some gradation of sensitivity to atmospheric dryness: *M. woodsii*, *H. aduncus* ssp. *hutchinsiae*, *Scapania nimbosa* and *Bazzania pearsonii* seemed to be among the most sensitive, being quickly lost in moving into less favourable habitats in the west and extending least far towards the drier east of the country. And heavy shade may make a vital but indirect contribution to maintaining the required levels of humidity by helping reduce evaporation to negligible levels, although for some species sensitivity to insolation itself might help confine them to the kind of habitat characteristic of this heath. Then, there is the impact of temperature which, for the North Atlantic species, probably exerts its effect through the low summer maxima typical of this part of Britain, even at lower altitudes. Again, however, specific tolerances probably vary. The prominence of *H. aduncus* ssp. *hutchinsiae* in this community, for example, reflects its preference for the warmer conditions typical of moderate elevations: it does not extend much beyond 600 m in the north-west Highlands and is largely absent from the hepatic mat which runs on to higher altitudes in the *Vaccinium-Racomitrium* heath. *Scapania nimbosa*, on the other hand, and *Anastrophyllum donianum* and the very local *A. joergensenii*, are more tolerant of the cooler conditions at altitudes above the limit of the *Calluna-Vaccinium-Sphagnum* heath, although they are often associated there with the shelter provided by fairly lengthy snow cover (Ratcliffe 1968). Such protection does, in fact, allow various members of the hepatic mat to survive in parts of eastern Scotland which would otherwise be too dry in summer and too cold in winter, but such persistence is not in association with this community.

The *Calluna-Pteridium* sub-community can be seen as comprising vegetation occurring in sites which are generally favourable to the development of damp heath rich in hypnaceous mosses and with the characteristic Sphagna of the community, but which, in one particular or another, cannot support the full suite of hepatics. Summer warmth and a drier climate at lower altitudes in the north-west Highlands, on Orkney and in those far-flung localities in Galloway and the Lake District where this vegetation can be found, could play a part here, and there are stands which are not so closely tied to the shadier aspects. It is also possible that this sub-community includes *Calluna-Vaccinium-Sphagnum* heath that has suffered from burning.

McVean & Ratcliffe (1962) noted that even the very precisely-defined habitat demanded by the *Mastigophora-Herbertus* sub-community was more widespread than the vegetation itself, which appeared to have been much reduced by fire and grazing. Even where burned stands had a good regenerating cover of sub-shrubs, the hepatics had seldom reappeared in any abundance, a carpet of mosses developing instead (Ratcliffe 1968), and where grazing hindered the regrowth of the bushes there was the further problem of increased insolation and evaporation.

Zonation and succession

The *Calluna-Vaccinium-Sphagnum* heath is a very local sub-alpine element in the oceanic zonations characteristic of the north-west Highlands, grading above to more montane sub-shrub communities and moss-heaths, and often passing below to wet-heath and mire vegetation. At higher altitudes, variations in exposure and snow-lie affect the patterns of communities, while at lower elevations there is increased influence of burning and grazing. Floristically, the *Calluna-Vaccinium-Sphagnum* heath is very close to certain kinds of woodland and, though there is rarely now any spatial continuity between the vegetation types, it is possible that clearance is responsible for their wide altitudinal separation, and that, in some situations, this is not a climax community.

The usual approach to stands of the *Calluna-Vaccinium-Sphagnum* heath is across tracts of *Scirpus-Eriophorum* bog or, on higher altitude terraces, the *Calluna-Eriophorum* bog, with a zone of *Scirpus-Erica* wet heath marking the transition to the more steeply-sloping ground where the peat cover thins and becomes better drained. The boundary between this last vegetation type and the *Calluna-Vaccinium-Sphagnum* heath running up on to the rocky ground in such zonations, is generally clear, particularly where large amounts of *Molinia* or *Scirpus* figure in the wet heath. In some cases, though, the communities can come quite close: the *Vaccinium* sub-community of the *Scirpus-Erica* wet heath has very frequent and sometimes prominent sub-shrubs and a ground layer in which hypnaceous mosses are as common and abundant as Sphagna, and it can grade fairly imperceptibly into the *Calluna-Pteridium* sub-community here. Alternatively, there may be a transition zone of Typical *Calluna-Erica* heath between the ombrogenous vegetation and the *Calluna-Vaccinium-Sphagnum* heath. Such patterns are well seen around the base of the corries on An Teallach and on Foinaven (Ratcliffe 1977).

In such situations, heather can remain a prominent feature of the vegetation cover to moderately high altitudes. On the shady and damp slopes where the *Calluna-Vaccinium-Sphagnum* heath is found, the community generally marks its upper limit of real vigour, increasing snow-lie over the higher talus and cliff bases being marked by a shift to the *Vaccinium-Racomitrium* heath, the *Bazzania-Mylia* sub-community of which shares many of the same hepatics as the *Mastigophora-Herbertus* sub-community here. Laterally, both vegetation types may fragment over rocky ground, becoming restricted to ledges and crevices, or being replaced there by tall-herb vegetation of the *Luzula-*

Vaccinium or *Luzula-Geum* communities, with the more species-poor examples of which these heaths show considerable floristic continuity, among both the vascular associates and the bryophytes. Liathach and some of the Letterewe crags show this kind of feature.

Towards these higher altitudes, variation in exposure becomes a major factor in determining the kinds of zonation in which the community is found. Over steep banks or in hollows where snow lies longer, the *Calluna-Vaccinium-Sphagnum* heath can be replaced by the *Hylocomium-Rhytidiadelphus* sub-community of the *Vaccinium-Deschampsia* heath or the *Plagiothecium-Anastrepta* sub-community of the *Vaccinium-Rubus* heath, in both of which there is a continuing strong contribution from hypnaceous mosses, with occasional, locally abundant Atlantic hepatics, beneath a bilberry or crowberry canopy. The transitions to the *Vaccinium-Rubus* heath can be especially smooth: indeed, McVean & Ratcliffe (1962) included some of this vegetation in their *Vaccineto-Callunetum* as a *suecicosum* facies. But the constancy of *Cornus suecica* and *Rubus chamaemorus* and occasional occurrence of *Eriophorum vaginatum* in the more chionophilous community will usually serve as a distinction and, though species such as *A. orcadensis*, *Bazzania tricrenata* and *Diplophyllum albicans* remain reasonably frequent, the integrity of the more strictly Atlantic assemblage becomes fragmented. The same is true of the *Vaccinium-Deschampsia* heath, too, although additionally with the passage to this vegetation type, there is an increase in the frequency of grasses, notably *Nardus*, *Agrostis canina*, *A. capillaris* and *F. ovina*. This can presage a shift to the *Nardus-Carex* community of later snow-beds in which the hepatics are very fragmentarily represented. Mosaics of these vegetation types can be seen over the higher slopes of An Teallach and in the hills of the Letterewe and Monar Forests (Ratcliffe 1977).

Some continuity among the hepatics can also be seen in those few places where the *Calluna-Vaccinium-Sphagnum* heath is contiguous with the *Calluna-Juniperus* heath, a community best developed on cool, shady slopes of moderate altitudes where the ground is blown clear of snow in winter. It is especially associated with fields of frost-shattered Cambrian quartzite and is well seen on Foinaven around the steeper ground which has a mosaic of dry and damp heather vegetation (Ratcliffe 1977).

Shifts from the sheltered slopes carrying the *Calluna-Vaccinium-Sphagnum* heath on to ground which is much more exposed to strong winds usually see a sharp replacement of the community by such dwarfed subshrub vegetation as the *Calluna-Racomitrium* or *Calluna-A. alpinus* heaths, in which heather can attain considerably higher altitudes and where *R. lanuginosum* is the predominant component of the often extensive moss carpet. Such transitions can be seen over the windswept lips of corries and cliffs and over lateral transitions to exposed spurs, as on Foinaven (Ratcliffe 1977) and above the Enegars cliffs on Hoy (Prentice & Prentice 1975), where they can mark a shift to summit moss-heath and fell-field communities.

Towards the upper end of its altitudinal limit, the *Calluna-Vaccinium-Sphagnum* heath appears to form a natural component of vegetation patterns controlled largely by variations in local climate and soils. On lower ground, however, it extends into the zone of communities that have often been strongly affected by various kinds of treatment. Where the surrounding heaths have been burned and grazed, for example, the transition to neighbouring communities is often sharpened up by a biotically-related dominance of heather in either the *Calluna-Erica* or *Scirpus-Erica* heaths or, conversely, by a virtual elimination of dwarf-shrubs in Nardo-Galion swards of various kinds, to which the *Vaccinium-Deschampsia* heath can form an intermediate, or *Molinia*-dominated wet heath. Although the damp slopes over which the *Calluna-Vaccinium-Sphagnum* heath occurs are protected somewhat against the ravages of fire by the moistness of the ground and vegetation, it seems certain that the extent of the community has been fragmented and reduced by burning (McVean & Ratcliffe 1962). Ratcliffe (1968) suggested that its original range extended throughout the north-west Highlands, from Ben Hope in the far north to Glencoe and perhaps Ben Cruachan in Argyll, with widespread representation, too, from Jura and Mull right through the Hebrides and round to Orkney and Shetland, in many of which localities it is now extremely scarce or unknown. There is also the possibility that, before the extensive clearance that has characterised the western Highlands, the *Calluna-Vaccinium-Sphagnum* heath could be found among tracts of native pine forest. There is considerable floristic similarity between the community and the field and ground layers of the *Scapania* sub-community of the *Pinus-Hylocomium* woodland, something that is seen very well in the upper reaches of the forest remnants of Coille na Glas-Leitire above Loch Maree, although here, as in other more fragmentary examples, there is now an altitudinal gap between the woodland and the heath in the higher corries above.

Distribution

The community occurs widely but locally through the north-west Highlands and on Skye, with outlying localities on Orkney, in south-west Scotland and the Lake District. The *Mastigophora-Herbertus* sub-community is more restricted in its range, being confined to the more shaded and humid habitats in north-west Scotland.

Affinities

The *Calluna-Vaccinium-Sphagnum* heath takes in McVean & Ratcliffe's (1962) *Vaccineto-Callunetum hepaticosum*, under which head they provided the first description of this kind of vegetation, and unites it with less floristically peculiar stands of bryophyte-rich heath, including samples from Birks (1973) and Prentice & Prentice (1975) from along the fringes of its range. The *Vaccineto-Callunetum suecicosum* of McVean & Ratcliffe (1962) is here transferred to the *Vaccinium-Rubus* heath, where it is united with their *Vaccinetum chionophilum*.

In Ireland, the characteristic assemblage of Atlantic hepatics seen here is found in association with *Calluna*, *V. myrtillus* and *D. flexuosa* in what Mhic Daeid (1976) called the *Herberteto-Polytrichetum alpini*, a community of steep, sheltered slopes in the Kerry and Connemara mountains where, in the warmer climate, the upper altitudinal limit of some of the liverworts is pushed to beyond 1000 m (Ratcliffe 1968). Apart from this parallel, however, heather-dominated vegetation of this extreme oceanic kind is not known from elsewhere. Birks (1973) considered it sufficiently distinct to warrant the erection of a new alliance within the Nardo-Callunetea, though, like Prentice & Prentice (1975), he placed it provisionally in the Myrtillion. To emphasise the close floristic (and perhaps seral) relationship to pine forest, McVean & Ratcliffe (1962) preferred to locate their *Vaccineto-Callunetum hepaticosum* in the Vaccinio-Piceion. On balance, a position among the Myrtillion heaths, or an equivalent alliance, seems better.

Floristic table H21

	a	b	21
Calluna vulgaris	V (4–10)	V (6–10)	V (4–10)
Vaccinium myrtillus	V (1–5)	V (1–4)	V (1–5)
Deschampsia flexuosa	V (1–4)	V (1–3)	V (1–4)
Rhytidiadelphus loreus	IV (1–4)	V (1–4)	V (1–4)
Pleurozium schreberi	IV (1–4)	V (1–4)	V (1–4)
Hylocomium splendens	IV (1–8)	V (1–6)	V (1–8)
Hypnum cupressiforme s.l.	IV (1–6)	V (1–4)	V (1–6)
Dicranum scoparium	IV (1–5)	IV (1–3)	IV (1–5)
Plagiothecium undulatum	V (1–4)	III (1–4)	IV (1–4)
Blechnum spicant	IV (1–3)	III (1–4)	IV (1–4)
Sphagnum capillifolium	III (1–8)	V (1–6)	IV (1–8)
Potentilla erecta	III (1–3)	IV (1–4)	IV (1–4)
Dicranum majus	IV (1–3)	III (1–4)	III (1–4)
Pteridium aquilinum	III (1–3)		II (1–3)
Thuidium tamariscinum	II (1–4)	I (1)	II (1–4)
Oxalis acetosella	II (1–3)		I (1–3)
Luzula sylvatica	II (1–4)		I (1–4)
Viola riviniana	II (1–3)		I (1–3)
Calypogeia muellerana	II (1–3)		I (1–3)
Lophozia ventricosa	II (1–3)		I (1–3)
Sphagnum girgensohnii	I (1–5)		I (1–5)
Sphagnum russowii	I (1–8)		I (1–8)
Scapania gracilis	II (1)	V (1–4)	III (1–4)
Mylia taylori	II (1–4)	V (1–4)	III (1–4)
Racomitrium lanuginosum	I (1–3)	V (1–6)	III (1–6)
Bazzania tricrenata	I (1)	V (1–4)	III (1–4)
Pleurozia purpurea		V (1–4)	II (1–4)
Cladonia uncialis		V (1–3)	II (1–3)
Diplophyllum albicans	II (1–4)	IV (1–4)	II (1–4)
Empetrum nigrum hermaphroditum	II (1–4)	IV (1–6)	II (1–6)
Anastrepta orcadensis	I (1–3)	IV (1–4)	II (1–4)
Mastigophora woodsii	I (1–3)	IV (1–6)	II (1–6)

Floristic table H21 *(cont.)*

	a	b	21
Herbertus aduncus hutchinsiae	I (1–4)	IV (1–5)	II (1–5)
Scapania ornithopodiodes		IV (1–4)	II (1–4)
Cladonia arbuscula	I (1–3)	III (1–4)	II (1–4)
Anastrophyllum donianum		III (1–4)	II (1–4)
Plagiochila carringtonii		III (1–4)	II (1–4)
Dicranodontium uncinatum		III (1–3)	II (1–3)
Tritomaria quinquedentata	I (1–3)	II (1–3)	I (1–3)
Plagiochila spinulosa	I (1–3)	II (1–4)	I (1–4)
Vaccinium uliginosum		II (1–4)	I (1–4)
Bazzania pearsonii		II (1–4)	I (1–4)
Scapania nimbosa		II (1–3)	I (1–3)
Nowellia curvifolia		I (1–3)	I (1–3)
Erica cinerea	III (1–4)	III (1–4)	III (1–4)
Solidago virgaurea	II (1–3)	III (1–3)	III (1–3)
Cladonia impexa	II (1–3)	III (1–4)	III (1–4)
Listera cordata	III (1–3)	II (1–3)	III (1–3)
Vaccinium vitis-idaea	II (1–4)	II (1–3)	II (1–4)
Ptilium crista-castrensis	II (1–4)	II (1–3)	II (1–4)
Sphagnum quinquefarium	II (1–8)	II (1–3)	II (1–8)
Breutelia chrysocoma	II (1–4)	I (1–4)	II (1–4)
Carex bigelowii	I (1–3)	II (1–3)	II (1–3)
Cladonia gracilis	I (1–3)	II (1–3)	II (1–3)
Scirpus cespitosus	I (1–3)	II (1–3)	II (1–3)
Molinia caerulea	I (1–3)	II (1–4)	I (1–4)
Melampyrum pratense	I (1–3)	II (1–3)	I (1–3)
Sphagnum tenellum	I (1–4)	II (1–4)	I (1–4)
Frullania tamarisci	I (1–3)	II (1–3)	I (1–3)
Nardus stricta	I (1–3)	II (1–3)	I (1–3)
Agrostis canina	I (1–3)	I (1–2)	I (1–3)
Carex pilulifera	I (1–3)	I (1–3)	I (1–3)
Agrostis capillaris	I (1–3)	I (1–3)	I (1–3)
Galium saxatile	I (1–3)	I (2)	I (1–3)
Ptilidium ciliare	I (1–3)	I (2–3)	I (1–3)
Festuca vivipara	I (1–3)	I (1–3)	I (1–3)
Carex binervis	I (1–3)	I (1–3)	I (1–3)
Campylopus paradoxus	I (1–3)	I (1)	I (1–3)
Barbilophozia floerkii	I (1–3)	I (1–3)	I (1–3)
Sphaerophorus globosus	I (1–3)	I (1–3)	I (1–3)
Sorbus aucuparia seedling	I (1–3)	I (1–3)	I (1–3)
Dryopteris filix-mas	I (1)	I (1–3)	I (1–3)
Hypericum pulchrum	I (1–3)	I (1–3)	I (1–3)
Succisa pratensis	I (1–3)	I (1–3)	I (1–3)
Arctostaphylos uva-ursi	I (1–3)	I (1)	I (1–3)
Cladonia bellidiflora	I (1–3)	I (1)	I (1–3)
Cladonia pyxidata	I (1–3)	I (1)	I (1–3)
Hylocomium umbratum	I (1–3)	I (1–3)	I (1–3)

Kurzia pauciflora	I (1–3)	I (1–3)	I (1–3)
Hymenophyllum wilsonii	I (1–3)	I (1)	I (1–3)
Kurzia trichoclados	I (1–3)	I (1–2)	I (1–3)
Juniperus communis nana	I (1–3)	I (1–3)	I (1–3)
Dactylorhiza maculata	I (1–3)	I (1)	I (1–3)
Number of samples	28	23	51
Number of species/sample	24 (10–46)	36 (26–42)	29 (10–46)
Vegetation height (cm)	46 (30–75)	32 (6–75)	40 (6–75)
Vegetation cover (%)	95 (30–100)	98 (80–100)	96 (30–100)
Altitude (m)	199 (15–570)	399 (122–640)	289 (15–640)
Slope (°)	38 (3–90)	29 (3–45)	34 (3–90)

a   *Calluna vulgaris-Pteridium aquilinum* sub-community

b   *Mastigophora woodsii-Herbertus aduncus hutchinsiae* sub-community

21   *Calluna vulgaris-Vaccinium myrtillus-Sphagnum capillifolium* heath (total)

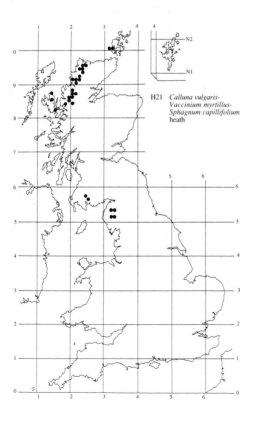

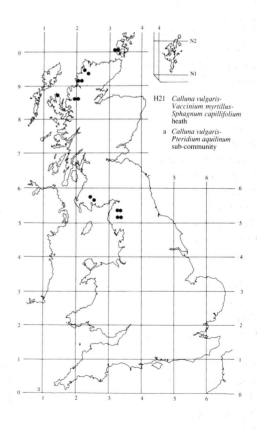

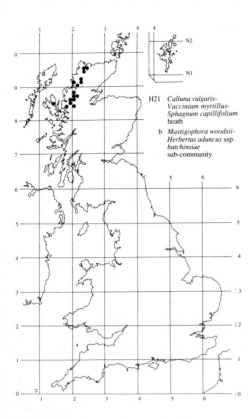

H21 *Calluna vulgaris-*
     Vaccinium myrtillus-
     Sphagnum capillifolium
     heath

  b *Mastigophora woodsii-*
     Herbertus aduncus ssp.
     hutchinsiae
     sub-community

Synonymy

Vaccinio-Callunetum Smith 1905 *p.p.*; *Vaccinium-Chamaepericlymenum* nodum Poore & McVean 1957 *p.p.*; *Vaccineto-Callunetum suecicosum* McVean & Ratcliffe 1962; *Vaccinetum chionophilum* McVean & Ratcliffe 1962 *p.p.*; Mountain *Vaccinium* heaths Gimingham 1972 *p.p.*; *Rubus chamaemorus-Vaccinium myrtillus* nodum Huntley & Birks 1979.

Constant species

Calluna vulgaris, Cornus suecica, Deschampsia flexuosa, Empetrum nigrum ssp. *hermaphroditum, Rubus chamaemorus, Vaccinium myrtillus, V. vitis-idaea, Dicranum scoparium, Hylocomium splendens, Pleurozium schreberi, Rhytidiadelphus loreus, Sphagnum capillifolium, Cladonia arbuscula.*

Rare species

Lycopodium annotinum, Plagiochila carringtonii, Scapania ornithopodioides.

Physiognomy

Like the *Calluna-Vaccinium-Sphagnum* heath, the *Vaccinium myrtillus-Rubus chamaemorus* heath has a mixed cover of sub-shrubs over a moist carpet of bryophytes. However, the canopy here is generally somewhat less tall than in that community, mostly between 1 and 3 dm high, and *Calluna vulgaris* is not invariably the dominant: indeed, in one sub-community, it becomes quite patchy in its abundance, with *Vaccinium myrtillus* usually having greater cover. *Empetrum nigrum* ssp. *hermaphroditum* is also constant and locally sub- or co-dominant, *V. vitis-idaea* rather less frequent and almost always sparse. *V. uliginosum* occurs rarely, though it can be quite extensive. In contrast to the *Calluna-Vaccinium-Sphagnum* heath, *Erica cinerea* is not found in this community.

Other distinctive features can be seen among the vascular associates because, along with constant *Deschampsia flexuosa*, there is very frequently a little *Rubus*

chamaemorus and *Cornus suecica*, species which figure only occasionally in other montane heaths. *Eriophorum vaginatum* can sometimes be found, too, with local abundance in some stands, and there are records for *Potentilla erecta, Melampyrum pratense, Listera cordata, Juncus squarrosus* and *Nardus stricta. Galium saxatile, Carex bigelowii* and *Huperzia selago* are only occasional overall but show preferential frequency in the different sub-communities.

Bryophytes are always a conspicuous feature of the vegetation and in some stands are very abundant. *Dicranum scoparium* and the hypnaceous mosses, *Pleurozium schreberi, Hylocomium splendens* and *Rhytidiadelphus loreus*, provide the most consistent and often the most extensive element, although Sphagna, too, can have high cover, with *Sphagnum capillifolium* especially common, *S. quinquefarium* occasional, *S. subnitens, S. russowii* and *S. fuscum* more scarce, though locally abundant. More uneven in their occurrence are *Polytrichum commune, Plagiothecium undulatum, Hypnum cupressiforme s.l., Ptilidium ciliare* and *Racomitrium lanuginosum* and some broadly Atlantic hepatics such as *Anastrepta orcadensis, Bazzania tricrenata* and *Diplophyllum albicans.*

Lichens are typically less prominent although *Cladonia arbuscula* is constant and can show modest abundance, and *C. bellidiflora, C. uncialis, C. leucophaea* and *C. gracilis* become frequent in one sub-community.

Sub-communities

***Polytrichum commune-Galium saxatile* sub-community:** *Vaccinium-Chamaepericlymenum* nodum Poore & McVean 1957 *p.p.*; *Vaccinetum chionophilum* McVean & Ratcliffe 1962 *p.p. V. myrtillus* is generally dominant in a low sub-shrub canopy, with *Calluna* and/or *E. nigrum* ssp. *hermaphroditum* sub-dominant, *V. vitis-idaea* fairly common but of low cover. *C. suecica* and *R. chamaemorus* are both more consistently frequent here than in the other sub-community though neither is abundant. *D. flexuosa* is rather patchy in its cover, sometimes attaining local prominence, in other stands being very sparse and *Eriophorum vaginatum*, though only occasional, can have fairly high cover. *Galium saxatile* and *Blechnum* are preferentially common, though present just as scattered individuals.

Hypnaceous mosses, particularly *Hylocomium splendens* and *Rhytidiadelphus loreus*, are plentiful with *Sphagnum capillifolium* or occasionally *S. quinquefarium* patchily abundant. Apart from scattered *Cladonia arbuscula*, lichens are rare.

***Plagiothecium undulatum-Anastrepta orcadensis* sub-community:** *Vaccineto-Callunetum suecicosum* McVean and Ratcliffe 1962; *Rubus chamaemorus-Vaccinium myrtillus* nodum   Huntley & Birks 1979. *Calluna* is quite often a strong dominant in a taller canopy here with *E. nigrum* ssp. *hermaphroditum* occasionally abundant, *V. myrtillus* usually of low cover and *V. vitis-idaea* rather uneven in its occurrence. *C. suecica*, *R. chamaemorus* and *D. flexuosa* are all more patchy than in the *Polytrichum-Galium* sub-community and, though *E. vaginatum* can be found quite commonly, it is present as just sparse shoots. *Carex bigelowii* and *Huperzia selago* are preferential at low frequencies.

More striking, though, are the cryptogams. As before, both hypnaceous mosses and Sphagna can be prominent, but the former are enriched by frequent *H. cupressiforme s.l.*, *Ptilidium ciliare*, *Racomitrium lanuginosum* and, particularly common, *Plagiothecium undulatum*, while among the Sphagna there can be found occasional, locally abundant, *S. subnitens*, *S. fuscum* and *S. russowii*. *Dicranum majus* and *Barbilophozia floerkii* are quite frequent and this vegetation provides a locus for a number of Atlantic hepatics, especially those of wider distribution such as *Diplophyllum albicans*, *Anastrepta orcadensis* and *Bazzania tricrenata*, but also some of the rarer species like *Plagiochila carringtonii* and *Scapania ornithopodioides*.

Lichens are more numerous in this sub-community, too, with *Cladonia bellidiflora*, *C. uncialis*, *C. leucophaea*, *C. gracilis* and *C. impexa* occurring along with the constant *C. arbuscula*. Only exceptionally, however, do any of these occur with abundance.

Habitat

The *Vaccinium-Rubus* heath is characteristic of wet, base-poor peats at moderate to high altitudes, where there is some protection against the extremes of dryness and winter cold by virtue of an oceanic climate or locally prolonged snow-lie. It is almost entirely confined to the Scottish Highlands, with a few outlying stands where conditions are locally suitable. It is a climax vegetation, although sometimes affected by grazing and burning where these treatments are applied to surrounding heaths.

Two climatic features above all favour the development of this kind of vegetation. First, this is a community of only moderately cold montane regions. It occurs generally in areas characterised by a cool summer, with almost all the stands falling within the 22 °C mean annual maximum isotherm (Conolly & Dahl 1970), which takes in the bulk of the central and north-west Highlands. Within these areas, however, the *Vaccinium-Rubus* heath is generally confined to the sub-alpine zone (Poore & McVean 1957), most tracts being found between 500 and 800 m, with a mean in available samples of about 650 m and it is best developed where, for one reason or another, there is some shelter from bitter winter cold. Thus, although the Arctic-Alpine sub-shrubs *E. nigrum* ssp. *hermaphroditum* and *V. vitis-idaea* are well represented here and, indeed, the Arctic-Subarctic *R. chamaemorus* and *C. suecica* are particularly distinctive, *Calluna* is still potentially a vigorous plant at these altitudes (Gimingham 1960) and, among the bryophytes, it is hypnaceous mosses rather than *R. lanuginosum* which are generally predominant. This combination of floristic features is what gives the community its stamp.

Towards the north-west part of its range, the required amelioration of the winter conditions is provided by the generally oceanic character of the climate in this part of Scotland. Even at the moderately high levels attained by the *Vaccinium-Rubus* heath at its upper altitudinal limit in this region, winter minima are nothing like so low as in similarly exposed situations in the Central Highlands and frosts not so frequent, nor running so late into the spring (*Climatological Atlas* 1952, Page 1982). With the shift towards the south-eastern area of its distribution, however, the community becomes increasingly associated with the accumulation and persistence of winter snow, and thus more strongly limited to sheltered hollows, the lee sides of crests and ridges and particularly to shady north- and east-facing slopes, in which places it is essentially a vegetation type of early snow-beds (Poore & McVean 1957, McVean & Ratcliffe 1962).

Exactly the same features ensure that the second climatic requirement, for a moist soil and humid atmosphere, is met. In the north-west Highlands, the *Vaccinium-Rubus* heath experiences as much as 3200 mm rain annually (*Climatological Atlas* 1952), with often over 220 wet days yr^{-1} (Ratcliffe 1968), such that the habitat is constantly damp, particularly over the cloudy upper reaches of the mountains, and even over those slopes which do not have a cooler, shadier aspect. In the Central Highlands, the climate is much drier, with sometimes less than half as much total precipitation and less than 180 wet days yr^{-1} in places (Ratcliffe 1968), but a much greater proportion of this falls as snow, so its effect is concentrated, the lie and subsequent melt keeping the soil and vegetation protected from the effects of evaporation and restricting the impact of any drought during the short summer.

The effects of these particular features of the habitat of the *Vaccinium-Rubus* heath throughout its range can be seen in the floristics and distribution of the two sub-communities. To the north-west, it is the *Plagiothecium-Anastrepta* sub-community that is the predominant form, with its less montane and more oceanic character. Here, the continuing predominance of a fairly tall cover of *Calluna* to these quite high altitudes, the great luxuriance of the hypnaceous moss carpet and the patchy occurrence of broadly Atlantic hepatics among some Arctic-Alpines and Arctic-Subarctic plants are clearly related to the humid and more equable conditions. To the south-east, this kind of vegetation can still be found quite widely through the Grampians and, way beyond that, at far-flung localities in the Moffat Hills and in the Hen Hole on The Cheviot, where it provides some of the most southerly stations for *C. suecica*, though, with increasing distance from the north-west Highlands, the sites are more concentrated on to shady, humid slopes. And, with that shift, there is a greater tendency for the *Vaccinium-Rubus* heath to develop the more chionophilous character of the *Polytrichum-Galium* sub-community, *Calluna* declining in vigour with the snow-lie and dominance generally passing to *V. myrtillus*, with a decrease in the diversity of hypnaceous mosses and a loss of the Atlantic hepatics. This kind of *Vaccinium-Rubus* heath is virtually limited to the Grampians with only fragmentary stands occurring north of the Great Glen, apart from on Ben Wyvis (McVean & Ratcliffe 1962).

Although high precipitation or flushing with meltwater maintains the soils beneath the community in a permanently moist condition, the drainage on the moderately steep slopes characteristic here is always free. The profiles are typically poorly developed, often consisting of just a layer of bryophyte or ericoid humus resting directly on blocky talus, derived from a variety of pervious bedrocks through the extensive range of this vegetation. Sometimes, deeper peats occur and there can be underlying pockets of leached mineral soil. But, though the *Vaccinium-Rubus* heath can approach ombrogenous mire vegetation in its composition, with *Eriophorum vaginatum* locally abundant and Sphagna conspicuous among the hypnaceous mosses, conditions are not such as to tip the balance towards the development of the *Calluna-Eriophorum* bog.

Climatic and edaphic factors maintain the *Vaccinium-Rubus* heath as a climax vegetation in most situations although, as Poore & McVean (1957) noted, at its lowest limits it falls within the altitudinal range of historical pine forest. Some stands may be lightly grazed, but burning is deleterious to the floristic richness of the community. Damp and shady conditions can protect against damage by fire and where surrounding drier heaths are burned, it is even possible that something like the *Vaccinium-Rubus* heath can extend its cover. Such anthropogenic bilberry vegetation resulting from this treatment often lacks chionophilous plants like *C. suecica* and *Blechnum*, but is is hard to tell where the transition to it occurs.

Zonation and succession

The *Vaccinium-Rubus* heath is a widespread but local element of the sub-alpine zone in altitudinal transitions throughout the central and north-west Highlands, grading below to sub-montane heaths and mires and passing above to low-alpine sub-shrub vegetation and lichen- or moss-heaths. The amount of local shelter and duration of snow-lie play an important part in determining the patterns at higher levels, and at lower altitudes treatments influence the zonations, though for the most part this vegetation can be considered a climax community.

The general context of the *Vaccinium-Rubus* heath throughout its range is among the middle reaches of the *Vaccinium-Deschampsia* heath, the major sub-shrub community of moist, base-poor soils at moderate to high altitudes through the cold and wet uplands. The two vegetation types come very close floristically and the former can be found as small patches in the latter or in more complex mosaics, the *Vaccinium-Rubus* heath picking out pockets of deeper, wetter peat on shadier slopes or where there is somewhat more prolonged snow-lie among the usually steeper and more rocky ground around (Figure 29). Species such as *Blechnum*, *Cornus suecica*, *Plagiothecium undulatum*, *Sphagnum capillifolium*, *S. quinquefarium* and *Barbilophozia floerkii* occasionally find a place in the *Hylocomium-Rhytidiadelphus* sub-community of the *Vaccinium-Deschampsia* heath but their increase in the shift to the *Vaccinium-Rubus* heath, and the additional appearance of *Rubus chamaemorus*, is usually quite marked. In the north-west Highlands, where the *Plagiothecium-Anastrepta* sub-community is the usual representative of the *Vaccinium-Rubus* heath in such zonations, the pattern may be complicated by the occurrence also of the *Calluna-Vaccinium-Sphagnum* heath. In its general floristics, this community falls somewhere between the other two heath types, having an extensive carpet of hypnaceous mosses and Sphagna, but usually lacking *C. suecica* and *R. chamaemorus*, but additionally, where it occurs over the shadiest and most humid north- or east-facing slopes, it has the best representation of all these vegetation types of Atlantic hepatics which helps give it a very distinctive appearance among these mosaics.

In the north-west Highlands, these sub-alpine heaths often give way over the lower gently-sloping ground of the foothills, where the peats become thicker and less freely-drained, to mire vegetation, characteristically of the *Scirpus-Eriophorum* type at lower altitudes in this oceanic region. Frequently, there is an intervening zone of *Scirpus-Erica* wet heath, the *Vaccinium* sub-community of which, with its increased representation of sub-shrubs and hypnaceous mosses, comes close in composition to the *Vaccinium-Rubus* heath. Patterns of this kind are especially typical of the lower slopes of Foina-ven and Ben More Assynt, An Teallach and the Affric-Cannich Hills (McVean & Ratcliffe 1962).

At higher altitudes in these parts of Scotland, the upward transition from the *Vaccinium-Rubus* heath is generally to sub-shrub vegetation in which *Racomitrium lanuginosum* plays an increasingly important part in the ground carpet. Over slopes that are not too exposed, the community usually passes above to the *Vaccinium-Racomitrium* heath, in some stands of which there is an especially marked continuation of the abundance of hypnaceous mosses characteristic of the *Vaccinium-Rubus* heath (the *Rhytidiadelphus-Hylocomium* sub-community), in others a local richness in Atlantic hepatics, where snow cover accentuates the shelter provided by shady aspect (the *Bazzania-Mylia* sub-community). Often, though, the decisive shift to dominance in the sub-shrub canopy of *V. myrtillus* and *E. nigrum* ssp. *hermaphroditum* in the *Vaccinium-Racomitrium* heath is accompanied by an associated vascular

Figure 29. Transitions to chionophilous and tall-herb vegetation around snow-bed and crags in a heath/mire sequence in the north-west Scottish Highlands.

H10   *Calluna-Erica* heath
H18   *Vaccinium-Deschampsia* heath
H20   *Vaccinium-Racomitrium* heath
H22   *Vaccinium-Rubus* heath
M15   *Scirpus-Erica* wet heath
M17   *Scirpus-Eriophorum* mire
U10   *Carex-Racomitrium* moss-heath
U16   *Luzula-Vaccinium* tall-herb community

and cryptogam flora characteristic of the high montane moss-heaths, so the floristic transition is a clear one, with plants such as *Carex bigelowii*, *Festuca ovina/vivipara*, *Diphasium alpinum* and *Polytrichum alpinum* all becoming very frequent among the *Racomitrium* carpet. Over more windswept brows above patches of *Vaccinium-Rubus* heath, provided these are not at too inhospitably montane an altitude, *Calluna* can continue its dominance in association with such an assemblage as this. Such *Calluna-Racomitrium* heath is typically blown clear of any winter snow and has a dwarfed and wind-trimmed mat of heather, presenting a very different appearance from the vegetation of more sheltered situations below. *Arctostaphylos uva-ursi* and *A. alpinus* sometimes figure in this kind of transition too, with the distinct *Calluna-A. alpinus* heath being separable at some sites. The upper slopes of Foinaven and Ben More Assynt are again classic sites for this kind of zonation, with more fragmentary examples being seen in the Monar Forest, on Beinn Eighe and the Fannich Hills.

With the geographical move towards the south-east part of the Scottish Highlands, where the *Polytrichum-Galium* sub-community becomes the more widespread kind of *Vaccinium-Rubus* heath, the more oceanic kinds of low-alpine moss-heaths are replaced almost entirely by analogous vegetation types in which there is a predominance of lichens associated with abundant heather or bilberry and crowberry. Complex intermediate kinds of zonation are to be seen on Ben Wyvis but, over the slopes of the Cairngorms and around Clova, in the heart of the Grampians, the shift in the vegetation types is complete, stands of *Vaccinium-Rubus* heath giving way to either the *Vaccinium-Cladonia* heath, over moderately sheltered and snow-bound slopes at higher

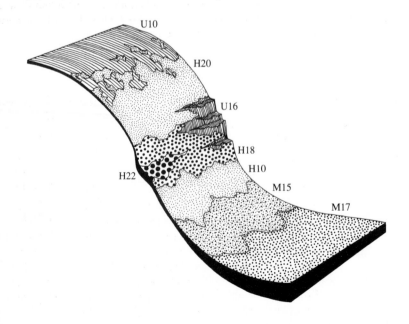

altitudes, with a decisive move to overwhelming lichen-dominance in the mat, or, where there is a stronger influence of bitter winds, to the *Calluna-Cladonia* heath, transitions to which can be a little more gradual. At lower altitudes, too, through the Central Highlands, the zonations tend to be different from those in the north-west, the sub-alpine belt of heaths passing over the foothills to *Calluna-Vaccinium* heath and/or *Calluna-A. uva-ursi* heath, both of these often treated as grouse-moor in this part of Scotland.

The other very important feature of the vegetation patterns in which the *Vaccinium-Rubus* heath is found relates to the influence of longer snow-lie in the sub-alpine zone, because here, and particularly in the Grampians, the community occurs as a mildly chionophilous vegetation type, marking out hollows or sheltered lee slopes or, where there is late persistence of snow, forming a surround to the bed. Typically, in the latter situations, it gives way under the deepest and longest snow cover, to some type of *Nardus-Carex* vegetation, the Typical or *Empetrum-Cetraria* sub-communities of which, with their patchy abundance of hypnaceous mosses beneath a sparse cover of bilberry, show some continuity with the *Vaccinium-Rubus* heath. Generally, however, the great increase in abundance of *Nardus* itself and/or of *C. bigelowii*, with the local prominence of *Scirpus cespitosus* and chionophilous mosses such as *Dicranum fuscescens* and *Kiaeria starkei*, serves to distinguish the central zone. Such patches, fringed above by a narrow arch of *Vaccinium-Deschampsia* heath and then, on the exposed upper lip, by a strip of *Calluna-Cladonia* heath, the whole embedded in a tract of *Calluna-Eriophorum* mire or low-alpine heath, are very characteristic of the Cairngorms (McVean & Ratcliffe 1962).

Even in such situations as these, there can be some influence on the vegetation pattern from grazing and burning which McVean & Ratcliffe (1962) suggested might facilitate the expansion of patches of *Vaccinium-Rubus* heath into a fringe of species-poor anthropogenic bilberry vegetation. And at lower altitudes, certainly, such treatments can greatly modify the context in which the community is found, burning of the drier sub-shrub vegetation tending to encourage the development of a patchwork of heather-dominated and impoverished heath, grazing tending to eliminate ericoids and transform the cover into first, a grassy heath of the *Vaccinium-Deschampsia* type, and then a Nardo-Galion sward.

Distribution

The *Vaccinium-Rubus* heath is almost wholly confined to the central and north-west Highlands, with the *Polytrichum-Galium* sub-community being largely restricted to the former area, the *Plagiothecium-Anastrepta* sub-community very much better developed in the latter.

Affinities

The community as defined in this scheme brings together vegetation types which McVean & Ratcliffe (1962) described as distinct, though closely-related, noda, part of their *Vaccinetum chionophilum* forming the basis of the *Polytrichum-Galium* sub-community, and their *Vaccineto-Callunetum suecicosum* (essentially Poore & McVean's (1957) *Vaccinium-Chamaepericlymenum* nodum) being subsumed by the *Plagiothecium-Anastrepta* sub-community. This solution, which emphasises the presence in both vegetation types of the Arctic-Subarctic *R. chamaemorus* and *C. suecica*, together with occasional *E. vaginatum*, seems the preferable one and helps locate the *Vaccinium-Rubus* heath as a whole among the mildly chionophilous communities of what Nordhagen (1943) termed the Phyllodoco-Vaccinion myrtilli (McVean & Ratcliffe 1962).

Floristic table H22

	a	b	22
Vaccinium myrtillus	V (1–10)	V (1–6)	V (1–10)
Empetrum nigrum hermaphroditum	V (1–8)	V (1–8)	V (1–8)
Sphagnum capillifolium	V (1–4)	V (1–10)	V (1–10)
Calluna vulgaris	V (1–8)	V (6–10)	V (1–10)
Pleurozium schreberi	V (1–6)	V (1–4)	V (1–6)
Hylocomium splendens	V (4–5)	V (1–6)	V (1–6)
Dicranum scoparium	V (1–4)	V (1–4)	V (1–4)
Rhytidiadelphus loreus	V (1–6)	V (1–6)	V (1–6)
Rubus chamaemorus	V (1–3)	IV (1–3)	IV (1–3)
Cornus suecica	V (1–4)	III (1–4)	IV (1–4)
Deschampsia flexuosa	IV (1–6)	IV (1–4)	IV (1–6)

Floristic table H22 *(cont.)*

	a	b	22
Vaccinium vitis-idaea	IV (1–3)	IV (1–4)	IV (1–4)
Cladonia arbuscula	III (1–4)	IV (1–6)	IV (1–6)
Polytrichum commune	V (1–4)	II (1–2)	III (1–4)
Galium saxatile	IV (1–3)	I (1–3)	III (1–3)
Blechnum spicant	III (1–4)	I (1–2)	II (1–4)
Carex echinata	I (1–3)		I (1–3)
Plagiothecium undulatum	I (2–3)	V (1–3)	III (1–3)
Ptilidium ciliare	I (1–2)	IV (2)	III (1–2)
Hypnum cupressiforme s.l.	I (3)	IV (1–3)	III (1–3)
Racomitrium lanuginosum	I (2–3)	IV (2–4)	III (2–4)
Anastrepta orcadensis		IV (1–3)	II (1–3)
Carex bigelowii	I (2–3)	III (1)	II (1–3)
Barbilophozia floerkii	I (1–3)	III (1–3)	II (1–3)
Cladonia bellidiflora	I (1)	III (1–2)	II (1–2)
Cladonia uncialis	I (1)	III (1–2)	II (1–2)
Cladonia leucophaea	I (1)	III (1–3)	II (1–3)
Cladonia impexa	I (1)	III (2–3)	II (1–3)
Cladonia gracilis		III (1–2)	II (1–2)
Huperzia selago	I (1)	II (1)	I (1)
Bazzania tricrenata		II (2–3)	I (2–3)
Dicranum majus		II (2)	I (2)
Diplophyllum albicans		II (1–2)	I (1–2)
Polytrichum alpinum		II (1)	I (1)
Cornicularia aculeata		II (1–3)	I (1–3)
Cladonia coccifera		II (1)	I (1)
Sphagnum russowii		II (2–3)	I (2–3)
Lepidozia pearsonii		II (1–3)	I (1–3)
Eriophorum vaginatum	II (4–6)	III (1–3)	II (1–6)
Sphagnum quinquefarium	II (1–6)	II (3–4)	II (1–6)
Potentilla erecta	II (1–3)	II (1–3)	II (1–3)
Melampyrum pratense	III (1–3)	II (1–2)	II (1–3)
Listera cordata	II (1–3)	I (1–3)	II (1–3)
Juncus squarrosus	II (1)	I (1)	II (1)
Nardus stricta	I (1–3)	II (2)	II (1–3)
Sphagnum subnitens	I (1–3)	I (1)	I (1–3)
Dryopteris dilatata	I (1–3)	I (1)	I (1–3)
Scirpus cespitosus	I (1)	I (1)	I (1)
Vaccinium uliginosum	I (4–7)	I (2–3)	I (2–7)
Cryptogramma crispa	I (1)	I (1)	I (1)
Number of samples	11	12	23
Number of species/sample	19 (17–23)	28 (20–35)	24 (17–35)
Vegetation height (cm)	13 (12–15)	30 (5–75)	26 (5–75)
Vegetation cover (%)	100	94 (60–100)	96 (60–100)

| Altitude (m) | 655 (457–833) | 649 (442–850) | 651 (442–850) |
| Slope (°) | 16 (3–45) | 22 (10–33) | 20 (3–45) |

a   *Polytrichum commune-Galium saxatile* sub-community
b   *Plagiothecium undulatum-Anastrepta orcadensis* sub-community
22   *Vaccinium myrtillus-Rubus chamaemorus* heath (total)

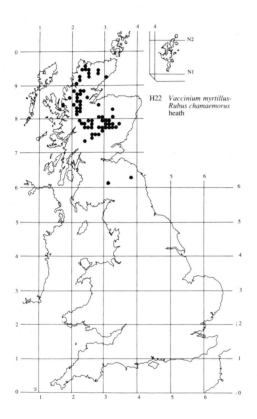

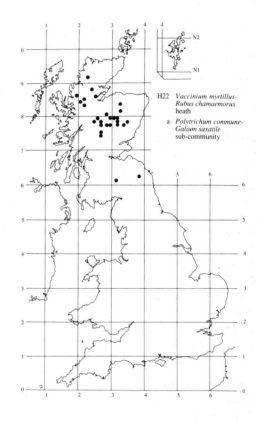

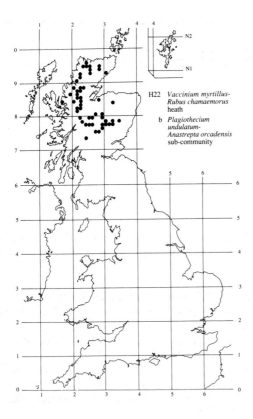

H22 *Vaccinium myrtillus-*
     Rubus chamaemorus
     heath

  b *Plagiothecium*
     undulatum-
     Anastrepta orcadensis
     sub-community

INDEX OF SYNONYMS TO MIRES AND HEATHS

The vegetation types are listed alphabetically, then by date of ascription of the name, with the code number of the equivalent NVC community or communities thereafter. The NVC communities themselves are included in the list with a bold code.

Acrocladio-Caricetum diandrae (Koch 1926) Wheeler 1975   M9

Acrocladio-Caricetum diandrae cicutetosum   Wheeler 1980*b*   M9

Acrocladio-Caricetum diandrae crepetosum   Wheeler 1980*b*   M9

Acrocladio-Caricetum diandrae juncetosum (Koch 1926) Wheeler 1975   M9

Acrocladio-Caricetum diandrae juncetosum subnodulosi   Wheeler 1980*b*   M9

Acrocladio-Caricetum diandrae schoenetosum (Koch 1926) Wheeler 1975   M9

Acrocladio-Caricetum diandrae sphagnetosum (Koch 1926) Wheeler 1975   M9

Acrocladio-Caricetum diandrae typicum   Wheeler 1980*b*   M9

Acrocladium cuspidatum-Carex diandra mire   Ratcliffe & Hattey 1982   M9

Agrostidetum setaceae cornubiense   Coombe & Frost 1956*a*   H4

Agrostidetum setaceae cornubiense   Ivimey-Cook 1959   H4

Agrostidetum setaceae cornubiense   Ivimey-Cook *et al.* 1975   H4

Agrostis curtisii heath   Hopkins 1983   H4

Agrostis setacea 'Short Heath'   Coombe & Frost 1956*a*   H4

Agrostis setacea 'Short Heath'   Marrs & Proctor 1978   H4

Agrostis setacea-Ulex minor heath   Ivimey-Cook 1959   H3

Agrosto setaceae-Ulicetum gallii   Bridgewater 1970   H4

Agrosto setaceae-Ulicetum minoris   Bridgewater 1970   H3

Agrosto-Ulicetum gallii   Shimwell 1968*a*   H8

Alchemilla alpina-Vaccinium myrtillus nodum   Birks 1973   H20

Alectorio-Callunetum vulgaris (Birse & Robertson 1976) Birse 1980   H13,H14,H17

Alectorio-Callunetum, Agrostis canina ssp. *montana* subassociation   Birse 1980   H13,H14,H17

Alectorio-Callunetum, Typical subassociation   Birse 1980   H13

Alpine *Carex-Sphagnum* mire   Ratcliffe 1964   M7

Anagallis tenella-Equisetum variegatum Association   Birse 1980   M10

Anthelia julacea banks   Birks 1973   M31

Anthelia julacea-Deschampsia caespitosa provisional nodum   McVean & Ratcliffe 1962   M31

Anthelia julacea-Sphagnum auriculatum spring   **M31**

Anthyllio corbierei-Ericetum cinereae   Bridgewater 1980   H7

Arctoeto-Callunetum   McVean & Ratcliffe 1962   H17

Arctoeto-Callunetum   Prentice & Prentice 1975   H17

Arctostaphyleto-Callunetum   McVean & Ratcliffe 1962   H16

Arctostaphyleto-Callunetum   Ward 1971   H16

Arctostaphyleto-Callunetum, herb-poor type   Ward 1971   H16

Arctostaphyleto-Callunetum, herb-rich type   Ward 1971   H16

Arctostaphylos heath   Urquhart 1986   H16

Arctostaphylos mat   Crampton 1911   H17

Betula nana bogs   Poore & McVean 1957   M19

Blanket bog   Ward *et al.* 1972   M17

Blanket bogs   Ratcliffe 1959   M19

Blanket bog communities   Adam *et al.* 1977   M15

Bryophyte flushes   Pigott 1956*a*   M38

Calcareous marsh   Sinker 1960   M10

Calliergono sarmentosi-Caricetum saxatilis   Dierssen 1982   M12

Calluna-Arctostaphylos heath   Muir & Fraser 1940   H16

Calluna-Arctostaphylos heath   Gimingham 1964*a*   H16

Calluna-Arctostaphylos heath   Gimingham 1972   H16

Calluna-Cladonia nodum   Bignal & Curtis 1981   M19

Calluna-Empetrum nigrum heath   Gimingham 1964*b*   H11

Calluna-Empetrum nigrum heath   Gimingham 1972   H11

Calluna-Erica cinerea heaths   Muir & Fraser 1940   H10

Calluna-Erica cinerea heath   Gimingham 1964*b*   H10,H11

Calluna-Erica cinerea heaths   Gimingham 1972   H10,H11

Calluna-Erica cinerea heath, 'dune-heath' variant   Gimingham 1964*b*   H11

Calluna-Erica tetralix wet heath   Gimingham 1964   M16

Calluna-Erica tetralix wet heath   Bannister 1966   M16

Calluna-Eriophorum moss   Pearsall 1941   M19

Calluna-Eriophorum-Sphagnetum community   Bignal & Curtis 1981   M18

Calluna heath   Ratcliffe 1959   H12

Calluna heath, dwarf lichen-rich facies   Edgell 1969   H13

Calluna-heath moss sociation   Edgell 1969   H12

Calluna-lichen heath   Poore & McVean 1957   H13

Calluna mire 3.iv   Hulme & Blyth 1984   M17

Calluna moss   Pearsall 1941   M19

Calluna-Pleurozium nodum   Bignal & Curtis 1981   M19

Calluna-Ulex-Erica heath   Fritsch & Parker 1913   H2

Calluna-Ulex-Erica heath   Fritsch & Salisbury 1915   H2

Calluna-Ulex-Erica heath   Haines 1926   H2

Calluna-Ulex gallii heaths   Gimingham 1972   H4,H8

Calluna-Ulex minor heaths   Gimingham 1972   H2,H3

Calluna-Vaccinium heath   Gimingham 1972   H12

Calluna vulgaris-Anemone nemorosa nodum   Huntley & Birks 1979   H12

Calluna vulgaris-Arctostaphylos alpinus heath   **H17**

Calluna vulgaris-Arctostaphylos uva-ursi heath   **H16**

Calluna vulgaris-Arctostaphylos uva-ursi nodum   Prentice & Prentice 1975   H14

Calluna vulgaris-Carex arenaria heath   **H11**

Calluna vulgaris-Cladonia arbuscula heath   **H13**

Calluna vulgaris-Deschampsia flexuosa heath   **H9**

Calluna vulgaris-Deschampsia flexuosa nodum   Huntley 1979   H12

Calluna vulgaris-Erica cinerea heath   **H10**

Calluna vulgaris-Eriophorum vaginatum blanket mire   **M19**

Calluna vulgaris-Eriophorum vaginatum Nodum, Normal Series   Tallis 1973   M18

Calluna vulgaris-Festuca ovina heath   **H1**

Calluna vulgaris Gesellschaft   Dierssen 1982   M20

Calluna vulgaris-Hypnum cupressiforme nodum   Hughes & Huntley 1986   H12

Calluna vulgaris-Juniperus communis ssp. *nana* heath   **H15**

Calluna vulgaris-Racomitrium lanuginosum heath   **H14**

Calluna vulgaris-Scilla verna heath   **H7**

Calluna vulgaris-Sieglingia decumbens Association   Birks 1973   H10

Calluna vulgaris-Ulex gallii heath   **H8**

Calluna vulgaris-Ulex minor heath   **H2**

Calluna vulgaris-Vaccinium myrtillus heath   **H12**

Calluna vulgaris-Vaccinium myrtillus nodum   Hughes & Huntley 1986   H12

Calluna vulgaris-Vaccinium myrtillus-Sphagnum capillifolium heath   **H21**

Calluna vulgaris-Vaccinium vitis-idaea nodum   Huntley & Birks 1979*a*   H12

Calluneto-Eriophoretum   McVean & Ratcliffe 1962   M19

Calluneto-Eriophoretum   Eddy *et al.* 1969   M19

Calluneto-Eriophoretum   Birks 1973   M19

Calluneto-Eriophoretum   Meek 1976   M19

Calluneto-Eriophoretum   Evans *et al.* 1977   M19

Calluneto-Eriophoretum deschampsietosum   Evans *et al.* 1977   M19

Calluneto-Eriophoretum , Empetrum facies   Eddy *et al.* 1969   M19

Calluneto-Eriophoretum, lichen-rich facies   McVean & Ratcliffe 1962   M19

Calluneto-Eriophoretum myrtillosum   Evans *et al.* 1977   M19

Calluneto-Eriophoretum, shrub-rich facies   McVean & Ratcliffe 1962   M19

Calluneto-Eriophoretum, Sphagnum type   McVean & Ratcliffe 1962   M19

Calluneto-Eriophoretum typicum   Evans *et al.* 1977   M19

Callunetum   Lewis & Moss 1911   H9,H12

Callunetum   Elgee 1914   H9

Callunetum   Farrow 1915   H1

Callunetum   Watt 1936   H1

Callunetum   Tansley 1939   H9

Callunetum   Fidler *et al.* 1970   H9

INDEX OF SPECIES IN MIRES AND HEATHS

The species are listed alphabetically, with the code numbers of the NVC communities in which they occur thereafter. Bold codes indicate that a species is constant throughout the community, italic codes that a species is constant in one or more sub-communities.

Achillea millefolium   M38, H7

Achillea ptarmica   M22, M23, M24, M25, M27, M38

Agrostis canina   M25, M35, M37, **M38**, H10, H12, H13, H14, H15, H17, H19, H20, H21

Agrostis canina canina   M2, M3, M4, M5, *M6*, M7, M11, M12, M13, M15, M17, M23, M24, M27, M29, M31, M32

Agrostis canina montana   H2, H4, H5, **H6**, H8, H18

Agrostis capillaris   M6, M8, M12, M15, M23, M25, M32, M33, M37, H1, H4, H6, *H7*, H8, H9, H10, *H11*, H12, H14, H16, *H18*, H19, H20, H21

Agrostis curtisii   M16, M21, M25, H2, **H3**, **H4**, H5, *H6*, H8

Agrostis stolonifera   M4, *M7*, M8, M9, *M10*, M11, M13, M15, *M22*, M23, M24, M25, M26, M27, *M28*, M29, M31, M32, M35, M37, M38, H7, H8, H20

Aira caryophyllea   *H6*, H7, H11

Aira praecox   H7, H8, H11

Ajuga reptans   M22, M27

Alchemilla alpina   M11, M12, H13, H14, *H18*, H20

Alchemilla glabra   M8, M11, M12, M27, M32, M37, M38, H18

Alchemilla filicaulis filicaulis   M11, M12

Alchemilla filicaulis vestita   M8, M38

Alchemilla vulgaris   M8

Alectoria nigricans   H10, **H13**, H14, *H17*, H19, H20

Alectoria ochroleuca   H13

Alectoria sarmentosa   H13

Alnus glutinosa saplings   M24

Alopecurus alpinus   M12, M32

Alopecurus geniculatus   M35, M38

Ammophila arenaria   H11

Anagallis tenella   *M13*, **M14**, M24, M29, **H5**

Anastrepta orcadensis   H18, *H20*, *H21*, *H22*

Anastrophyllum donianum   H20, H21

Andreaea rothii   H14

Andromeda polifolia   *M2*, M18

Anemone nemorosa   M26, H12, H16, H18

Aneura pinguis   M8, M9, **M10**, **M11**, **M12**, **M14**, M15, M21, M29, M31, M32, M37, M38

Angelica sylvestris   M9, M10, M11, *M13*, M22, M23, M24, *M25*, *M26*, *M27*, M28, H5

Antennaria dioica   H10, H13, H14, H15, H16, H17, H19, H20

Anthelia julacea   M12, **M31**, H20

Anthoxanthum odoratum   M4, M6, M7, M8, M10, M11, M13, M15, M17, *M22*, M23, M24, *M25*, *M26*, M27, M28, M32, M38, H4, H6, *H7*, H8, *H10*, H11, H12, H16, *H18*, H20

Anthyllis vulneraria   H6, H7

Apium inundatum   M29

Arctostaphylos alpinus   M19, H14, H15, **H17**

Arctostaphylos uva-ursi   H12, H13, *H14*, H15, **H16**, H17, H21

Armeria maritima   M12, *H7*, H19, H20

Arrhenatherum elatius   M22, M27, M28

Astragalus danicus   H7

Athyrium filix-femina   M23

Atriplex prostrata   *M28*

Aulacomnium palustre   M2, M4, **M5**, M6, M7, **M8**, M9, M13, M15, M16, M17, M18, M19, M21, M23, M24, M25, M26, M29

Avenula pubescens   H8

Baldellia ranunculoides   M29, M30

Barbilophozia floerkii   M19, M20, M31, H12, H18, H20, H21, H22

Barbilophozia lycopodioides   M12

Bartsia alpina   M10

Bazzania pearsonii   H20, H21

Bazzania tricrenata   *H20*, *H21*, H22

BIBLIOGRAPHY

Adam, P. (1976). *Plant sociology and habitat factors in British saltmarshes*. University of Cambridge: PhD thesis.

Adam, P. (1977). On the phytosociological status of *Juncus maritimus* on British saltmarshes. *Vegetatio*, **35**, 81–94.

Adam, P. (1978). Geographical variation in British saltmarsh vegetation. *Journal of Ecology*, **66**, 339–66.

Adam, P. (1981). The vegetation of British saltmarshes. *New Phytologist*, **88**, 143–96.

Adam, P., Birks, H.J.B., Huntley, B. & Prentice, I.C. (1975). Phytosociological studies at Malham Tarn moss and fen, Yorkshire, England. *Vegetatio*, **30**, 117–32.

Adam, P., Birks, H.J.B. & Huntley, B. (1977). Plant communities of the Island of Arran, Scotland. *New Phytologist*, **79**, 689–712.

Adamson, R.S. (1918). On the relationships of some associations of the Southern Pennines. *Journal of Ecology*, **6**, 97–109.

Allen, S.E. (1964). Chemical aspects of heather burning. *Journal of Applied Ecology*, **1**, 347–67.

Allorge, P. (1921–2). Les associations végétales du Vexin français. *Revue générale Botanique*, **33 & 34**.

Anderson, P. & Tallis, J. (1981). The nature and extent of soil and peat erosion in the Peak District. In *Moorland Erosion Study Phase I Report*, ed. J. Phillips, D. Yalden & J. Tallis, pp. 52–64. Bakewell: Peak Park Joint Planning Board.

Anderson, P. & Yalden, D.W. (1981). Increased sheep numbers and loss of heather moorland in the Peak District. *Biological Conservation*, **20**, 195–213.

Atkinson, B.W. & Smithson, P.A. (1976). Precipitation. In *The Climate of the British Isles*, ed. T.J. Chandler & S. Gregory, pp. 129–82. London: Longman.

Avery, B.W. (1980). *Soil Classification for England and Wales (Higher Categories)*. Soil Survey Technical Monograph No. 14. Harpenden: Soil Survey of England and Wales.

Balátová-Tuláčková, E. (1957). Flachmoorwiesen in mittleren und unteren Opava-Tal (Schlesien). *Vegetace ČSSR*, **A4**, 1–201.

Balchin, W.G.V. (1964). The denudation chronology of South-West England. In *Present Views of Some Aspects of the Geology of Cornwall and Devon*, eds. K.F.G. Hosking & G.J. Shrimpton. Penzance: Royal Geographical Society.

Ball, D.F. (1960). *The Soils and Land-Use of the District around Rhyl and Denbigh*. London: HMSO.

Balme, O.E. (1953). Edaphic and vegetational zoning on the Carboniferous Limestone of the Derbyshire Dales. *Journal of Ecology*, **41**, 331–44.

Balme, O.E. (1954). Biological Flora of the British Isles: *Viola lutea* Huds. *Journal of Ecology*, **42**, 234–40.

Bannister, P. (1964a). Stomatal response of heath plants to water deficits. *Journal of Ecology*, **52**, 151–8.

Bannister, P. (1964b). The water relations of certain heath plants with reference to their ecological amplitude. I. Introduction: germination and establishment. *Journal of Ecology*, **52**, 423–32.

Bannister, P. (1964c). The water relations of certain heath plants with reference to their ecological amplitude. II. Field studies. *Journal of Ecology*, **52**, 481–97.

Bannister, P. (1964d). The water relations of certain heath plants with reference to their ecological amplitude. III. Experimental studies: general conclusions. *Journal of Ecology*, **52**, 499–509.

Bannister, P. (1965). Biological Flora of the British Isles: *Erica cinerea* L. *Journal of Ecology*, **53**, 527–42.

Bannister, P. (1966). Biological Flora of the British Isles: *Erica tetralix* L. *Journal of Ecology*, **54**, 795–813.

Barclay-Estrup, P. (1971). The description and interpretation of cyclical processes in a heath community. III. Microclimate in relation to the *Calluna* cycle. *Journal of Ecology*, **59**, 143–66.

Barclay-Estrup, P. & Gimingham, C.H. (1969). The description and interpretation of cyclical processes in a heath community. I. Vegetational change in relation to the *Calluna* cycle. *Journal of Ecology*, **57**, 737–58.

Bell, J.N.B. & Tallis, J.H. (1973). Biological Flora of the British Isles: *Empetrum nigrum* L. *Journal of Ecology*, **61**, 289–305.

Bell, P.R. & Lodge, E. (1963). *Cratoneuron commutatum* as an indicator. *Journal of Ecology*, **51**, 113–22.

Bellamy, D.J. & Rose, F. (1961). The Waveney–Ouse valley fens of the Suffolk-Norfolk border. *Transactions of the Suffolk Naturalists' Society*, **11**, 367–85.

Bellot, F.R. (1966). La vegetacion de Galicia. *Anales del Instituto botànico A.J. Cavanillo*, **1**, 389–444.

Bignal, E.M. & Curtis, D.J. (1981). Peat moss plant communities and site-types of Strathclyde Region, S.W. Scotland. Nature Conservancy Council Report.

Birks, H.J.B. (1969). *The Late-Weichselian and Present Vegetation of the Isle of Skye*. Cambridge University: PhD thesis.

Birks, H.J.B. (1970). The Flandrian forest history of Scotland: a preliminary synthesis. In *British Quaternary Studies, Recent Advances*, ed. F.W. Shotton, pp. 119–35. Oxford: Clarendon Press.

Birks, H.J.B. (1973). *The Past and Present Vegetation of the Isle of Skye: a Palaeoecological Study*. Cambridge: Cambridge University Press.

Birse, E.L. (1980). *Plant Communities of Scotland: A Preliminary Phytocoenonia*. Aberdeen: Macaulay Institute for Soil Research.

Birse, E.L. (1984). *The Phytocoenonia of Scotland: Additions and Revisions*. Aberdeen: Macaulay Institute for Soil Research.

Birse, E.L. & Robertson, J.S. (1967). Vegetation. In *The Soils of the Country around Haddington and Eyemouth*, J.M. Ragg & D.W. Futty. Edinburgh: HMSO.

Birse, E.L. & Robertson, J.S. (1973). Vegetation. In *The Soils of Carrick and the Country around Girvan*, C.J. Brown. Edinburgh: HMSO.

Birse, E.L. & Robertson, J.S. (1976). *Plant Communities and Soils of the Lowland and Southern Upland Regions of Scotland*. Aberdeen: Macaulay Institute for Soil Research.

Boatman, D.J. (1957). An ecological study of two areas of blanket bog on the Galway–Mayo Peninsula, Ireland. *Proceedings of the Royal Irish Academy B*, **59**, 29–42.

Boatman, D.J. (1977). Observations on the growth of *Sphagnum cuspidatum* in a bog pool on the Silver Flowe National Nature Reserve. *Journal of Ecology*, **65**, 119–26.

Boatman, D.J. & Armstrong, W. (1968). A bog type in north-west Sutherland. *Journal of Ecology*, **56**, 129–41.

Boatman, D.J. & Tomlinson, R.W. (1973). The Silver Flowe. I. Some structural and hydrological features of Brishie Bog and their bearing on pool formation. *Journal of Ecology*, **61**, 653–66.

Boatman, D.J. & Tomlinson, R.W. (1977). The Silver Flowe. II. Features of the vegetation and stratigraphy of Brishie Bog, and their bearing on pool formation. *Journal of Ecology*, **65**, 531–46.

Boatman, D.J., Goode, D.A. & Hulme, P.D. (1981). The Silver Flowe. III. Pattern development on Long Loch B and Craigeazle mires. *Journal of Ecology*, **69**, 897–918.

Böcher, T.W. (1940). Studies on the plant geography of the North Atlantic heath formation. I. The heaths of the Faroes. *Kongelige Danske Videnskarbernes Selskabs Skrifter*, **15**, 1–64.

Böcher, T.W. (1943). Studies on the plant-geography of the North Atlantic heath formation. II. Danish dwarf-shrub communities in relation to those of Northern Europe. *Kongelige Norske Videnskabernes Selskabs Skrifter*, **2**, 1–127.

Böcher, T.W. (1954). Oceanic and continental vegetation complexes in south-west Greenland. *Meddelelser Grønland*, **148**, 82–418.

Bond, G. (1951). The fixation of nitrogen associated with the root nodules of *Myrica gale* L. with special reference to its pH relation and ecological significance. *Annals of Botany*, **NS15**, 447–59.

Bond, G. (1967). Fixation of nitrogen by higher plants other than legumes. *Annual Review of Plant Physiology*, **18**, 107–26.

Bower, M.M. (1960). The erosion of blanket peat in the southern Pennines. *East Midland Geographer*, **13**, 22–33.

Bower, M.M. (1961). The distribution of erosion in blanket peat bogs in the Pennines. *Transactions of the Institute of British Geographers*, **29**, 17–30.

Bradshaw, M.E. & Jones, A.V. (1976). *Phytosociology in Upper Teesdale*. Durham: Durham University Department of Extra-Mural Studies.

Braun-Blanquet, J. (1928). *Pflanzensoziologie. Grundzüge der Vegetationskunde*. Berlin: Springer.

Braun-Blanquet, J. (1948). *La végétation alpine des Pyrénées orientales*. Barcelona.

Braun-Blanquet, J. (1967). Vegetationsskizzen aus dem Baskenland mit Ausblicken auf das weitere Ibero-Atlantikum. II Teil. *Vegetatio*, **14**, 1–126.

Braun-Blanquet, J., Pinto Da Silva, A.-R. & Rozeira, A. (1964). Landes à Cistes et Ericacées (Cisto-Lavenduletea et Calluno-Ulicetea). *Agronomia Lusitana*, **1964**, 229–313.

Braun-Blanquet, J. & Tüxen, R. (1952). Irische Pflanzengesellschaften. *Veröffentlichungen des Geobotanischen Institutes Rübel in Zurich*, **25**, 224–415.

Bridgewater, P. (1970). *Phytosociology and community boundaries of the British heath formation*. Durham University: PhD thesis.

Brightmore, D. (1968). Biological Flora of the British Isles: *Lobelia urens* L. *Journal of Ecology*, **56**, 613–20.

British Ecological Society Mires Research Group (unpublished). Notes on the vegetation of Kentra Moss.

Brock, T., Frigge, P. & Van der Ster, H. (1978). *A vegetation study of the pools and surrounding wetlands in the Dooaghtry area, Co. Mayo, Republic of Ireland*. Nijmegen Catholic University: thesis.

Burges, A. (1951). The ecology of the Cairngorms. III. The *Empetrum-Vaccinium* zone. *Journal of Ecology*, **39**, 271–84.

Burtt, B.L. (1950). *Koenigia islandica* in Britain. *Kew Bulletin*, **1950**, 173.

Butler, J.R. (1953). The geochemistry and mineralogy of rock weathering. (1) The Lizard area, Cornwall. *Geochimica et Cosmochimica Acta*, **4**, 157–78.

Carlisle, A. (1977). The impact of man on the native pinewoods of Scotland. In *Native Pinewoods of Scotland*, ed. R.G.H. Bunce & J.N.R. Jeffers, pp. 70–77. Cambridge: Institute of Terrestrial Ecology.

Carlisle, A. & Brown, A.H.F. (1968). Biological Flora of the British Isles: *Pinus sylvestris* L. *Journal of Ecology*, **56**, 269–307.

Carroll, D.M. & Bendelow, V.C. (1981). *Soils of the North York Moors*. Harpenden: Soil Survey of England and Wales.

Carroll, D.M., Hartnup, R. & Jarvis, R.A. (1979). *Soils of South and West Yorkshire*. Harpenden: Soil Survey of England and Wales.

Chandler, T.J. & Gregory, S. (ed.) (1976). *The Climate of the British Isles*. London: Longman.

Chapin, F.S., van Cleve, K. & Chapin, M.C. (1979). Soil temperature and nutrient cycling in the tussock growth form of *Eriophorum vaginatum*. *Journal of Ecology*, **67**, 169–89.

Chapman, S.B. (1975). The distribution and composition of hybrid populations of *Erica ciliaris* L. and *Erica tetralix* L. in Dorset. *Journal of Ecology*, **63**, 809–24.

Clapham, A.R. (1940). The role of bryophytes in the calcareous fens of the Oxford district. *Journal of Ecology*, **38**, 71–80.

Clapham, A.R., Tutin, T.G. & Warburg, E.F. (1962). *Flora of the British Isles*, second edition. Cambridge: Cambridge University Press.

Climatological Atlas of the British Isles (1952). London: Meteorological Office.

Clymo, R.S. (1962). An experimental approach to part of the calcicole problem. *Journal of Ecology*, **50**, 707–31.

Coker, P.D. & Coker, A.M. (1973). Biological Flora of the British Isles: *Phyllodoce caerulea* (L). Bab. *Journal of Ecology*, **61**, 901–13.

Collingbourne, R.H. (1976). Radiation and sunshine. In *The Climate of the British Isles*, ed. T.J. Chandler & S. Gregory, pp. 74–95. London: Longman.

Conolly, A.P. & Dahl, E. (1970). Maximum summer temperature in relation to the modern and Quaternary distributions of certain arctic-montane species in the British Isles. In *Studies in the Vegetational History of the British Isles*, ed. D. Walker and R.G. West, pp. 159–224. Cambridge: Cambridge University Press.

Conway, V.M. (1949). Ringinglow Bog, near Sheffield. II. The present surface. *Journal of Ecology*, **37**, 148–70.

Conway, V.M. (1954). Stratigraphy and pollen analysis of southern Pennine blanket peats. *Journal of Ecology*, **42**, 117–47.

Coombe, D.E. (1961). *Trifolium occidentale*, a new species related to *T. repens* L. *Watsonia*, **5**, 68–87.

Coombe, D.E. & Frost, L.C. (1956*a*). The heaths of the Cornish Serpentine. *Journal of Ecology*, **44**, 226–56.

Coombe, D.E. & Frost, L.C. (1956*b*). The nature and origin of the soils over the Cornish Serpentine. *Journal of Ecology*, **44**, 605–15.

Coombe, D.E. & White, F. (1951). Notes on calcicolous communities and peat formation in Norwegian Lappland. *Journal of Ecology*, **39**, 33–62.

Coppins, B.J. & Shimwell, D.W. (1971). Cryptogam complement and biomass in dry *Calluna* heath of different ages. *Oikos*, **22**, 204–9.

Corbett, W.M. (1973). *Breckland Forest Soils*. Harpenden: Soil Survey of England and Wales.

Corillion, R. (1950). Contribution à l'étude de la répartition de l'*Ulex gallii* Planch., sur le littoral du nord de la Bretagne. *Bulletin de la Société Scientifique de la Bretagne*, **24**. 97–9.

Corillion, R. (1959). Nouvelles précisions sur la répartition de l'*Ulex gallii* Planch. en Bretagne. *Bulletin de la Société Scientifique de la Bretagne*, **34**, 233–6.

Corley, M.F.V. & Hill, M.O. (1981). *Distribution of Bryophytes in the British Isles*. Cardiff: British Bryological Society.

Crampton, C.B. (1911). *The Vegetation of Caithness considered in Relation to the Geology*. Committee for the Survey and Study of British Vegetation.

Curtis, L.F., Courtney, F.M. & Trudgill, S. (1976). *Soils in the British Isles*. London: Longman.

Dahl, E. (1956). *Rondane: Mountain vegetation in south Norway and its relation to the environment*. Oslo: Aschehoug.

Dahl, E. (1968). *Analytical Key to British Macrolichens*, second edition. London: British Lichen Society.

Dahl, E. & Hadač, E. (1941). Strandgesellschaften der Insel Ostoy in Oslofjord. Eine pflanzensoziologische Studie. *Nytt Magasin for Naturvidenskapene B*, **82**, 251–312.

Dahlgren, R. & Lassen, P. (1972). Studies in the flora of northern Morocco. I. Some poor fen communities and notes on a number of Northern and Atlantic plant species. *Botaniska notiser*, **125**, 439.

Dalby, M. (1961). The ecology of crowberry (*Empetrum nigrum*) on Ilkley Moor, 1959–1960. *The Naturalist*, **876**, 37–40.

Dalby, M., Fidler, J.H. & A. & Duncan, J.E. (1971). The vegetative changes on Ilkley Moor. *The Naturalist*, **916**, 49–56.

Daniels, R.E. & Pearson, M.C. (1974). Ecological studies at Roydon Common, Norfolk. *Journal of Ecology*, **62**, 127–50.

des Abbayes, H. & Corillion, R. (1949). Sur la répartition de l'*Ulex gallii* Planch. et d'*Ulex nanus* Sm. dans le Massif Amoricain. *Compte Rendu Sommaire de Séances de la Société de Biogéographie*, **229**, 86–9.

de Smidt, J.T. (1966). The inland-heath communities of the Netherlands. *Wentia*, **15**, 142–62.

Dierssen, K. (1975). Littorelletea uniflorae Br.-Bl. & Tx. 1943. In *Prodomus der Europäischen Pflanzengesellschaften*, ed. R. Tüxen, part 2. Vaduz: Cramer.

Dierssen, K. (1982). *Die wichtigsten Pflanzengesellschaften der Moore NW-Europas*. Genève: Conservatoire et Jardin botaniques.

Dimbleby, G.W. (1962). *The Development of British Heathlands and their Soils*. Oxford University Press.

Dixon, H.N. (1954). *Student's Handbook of British Mosses*, third edition. Eastbourne: Sumfield & Doy.

Dony, J.G. (1967). *Flora of Hertfordshire*. Hitchin: Urban District Council.

Doyle, G.J. (1982). The vegetation, ecology and productivity of atlantic blanket bog in Mayo and Galway, western Ireland, *Journal of Life Sciences of the Royal Dublin Society*, **3**, 147–64.

Doyle, G.J. & Moore, J.J. (1980). Western blanket bog (*Pleurozio purpureae-Ericetum tetralicis*) in Ireland and Great Britain. *Colloques Phytosociologiques*, **7**, 213–23.

Dupont, D. (1973). Synécologie d'une bruyère atlantique: *Erica vagans* L. *Colloques Internationales Phytosociologiques. II. La Végétation des Landes d'Europe Occidentale*, pp. 271–300. Lille.

Durno, S.E. & McVean, D.N. (1959). Forestry history of the Beinn Eighe Nature Reserve. *New Phytologist*, **58**, 228–36.

Dutoit, D. (1924). *Les associations végétales des sous-Alpes de Vevey (Suisse)*. Université de Lausanne: dissertation.

Duvigneaud, P. (1949). Classification phytosociologique des tourbières de l'Europe. *Bulletin de la Société Royale de Botanique de Belgique*, **81**, 59–129.

Duvigneaud, J. (1958). Contribution a l'étude des groupements prairiaux de la plaine alluviale de la Meuse lorraine. *Bulletin de la Société Royale de Botanique de Belgique*. **91**, 7–77.

Duvigneaud, F. (1966). Note sur la biogéochemie des serpentines du sud-ouest de la France. *Bulletin de la Société Royale de Botanique de Belgique*, **99**, 271–329.

Duvigneaud, P. & Vanden Berghen, C. (1945). Associations tourbeuses en Campine occidentale. *Biologisch jaarboek Dodonae*, **12**, 53–90.

Eddy, A., Welch, D. & Rawes, M. (1969). The vegetation of the Moor House National Nature Reserve in the Northern Pennines, England. *Vegetatio*, **16**, 239–84.

Edgell, M.C.R. (1969). Vegetation of an upland ecosystem: Cader Idris, Merionethshire. *Journal of Ecology*, **57**, 335–59.

Elgee, F. (1914). The vegetation of the eastern moorlands of Yorkshire. *Journal of Ecology*, **2**, 1–18.

Ellenberg, H. (1978). *Vegetation Mitteleuropas mit den Alpen*, 2 Auflage. Stuttgart: Ulmer.

Evans, C.C. & Allen, S.E. (1971). Nutrient losses in smoke produced during heather burning. *Oikos*, **22**, 149–54.

Evans, D.F., Hill, M.O. & Ward, S.D. (1977). *A Dichotomous Key to British Submontane Vegetation*. Bangor: Institute of Terrestrial Ecology.

Faegri, K. (1960). *Maps of distribution of Norwegian vascular plants. I. Coast Plants*. Oslo.

Fairburn, W.A. (1968). Climatic zones in the British Isles. *Forestry*, **14**, 117–30.

Farrow, E.P. (1915). On the ecology of the vegetation of Breckland. I. General description of Breckland and its vegetation. *Journal of Ecology*, **3**, 211–28.

Ferguson, N.P. (1979). *The effects of sulphur pollutants in rain on the growth of some species of Sphagnum*. Manchester University: PhD thesis.

Ferguson, N.P. & Lee, J.A. (1983). Past and present sulphur pollution in the southern Pennines. *Atmospheric Environment*, **17**, 1131–7.

Ferreira, R.E.C. (1978). *A Preliminary Vegetation Survey of Selected Cleughs in Western Borders*. Edinburgh: Nature Conservancy Council.

Fidler, J.H., Dalby, M. & Duncan, J.E. (1970). The plant communities of Ilkley Moor, *The Naturalist*, **912**, 41–8.

Findlay, D.C. (1965). *The Soils of the Mendip District of Somerset*. London: HMSO.

Findlay, D.C., Colborne, G.J.N., Cope, D.W., Harrod, T.R., Hogan, D.V. & Staines, S.J. (1984). *Soils and their Use in South West England*. Harpenden: Soil Survey of England and Wales.

Fisher, G.C. (1975a). Terraces, soils and vegetation in the New Forest, Hampshire. *Area*, **7**, 255–61.

Fisher, G.C. (1975b). Some aspects of the phytosociology of heathland and related communities in the New Forest, Hampshire, England. *Journal of Biogeography*, **2**, 103–16.

Fitzpatrick, E.A. (1977). Soils of the native pinewoods of Scotland. In *Native Pinewoods of Scotland*, ed. R.G.H. Bunce & J.N.R. Jeffers, pp. 35–41. Cambridge: Institute of Terrestrial Ecology.

Flett, J.S. (1946). Geology of the Lizard and Meneage. *Memoirs of the Geological Survey, Sheet 359*.

Fransson, S. (1963). Myrvegetation vid Rörvattenan i Nordvästra Jämtland. *Svensk Botanisker Tidskrift*, **57**, 283–332.

Fraser, G.K. (1933). Studies of Scottish moorlands in relation to tree growth. *Forestry Commission Bulletin 15*.

Fritsch, F.E. (1927). The heath association on Hindhead Common, 1911–1926. *Journal of Ecology*, **15**, 344–72.

Fritsch, F.E. & Parker, W.M. (1913). The heath association on Hindhead Common. *New Phytologist*, **12**, 148–63.

Fritsch, F.E. & Salisbury, E.J. (1915). Further observations on the heath association on Hindhead Common. *New Phytologist*, **14**, 116–38.

Furness, R.R. (1978). *Soils of Cheshire*. Harpenden: Soil Survey of England and Wales.

Gadeceau, E. (1903). Essai de géographie botanique sur Belle-Ile-en-Mer. *Mémoires de la Société des Sciences naturelles de Cherbourg*, **33**, 177–368.

Géhu, J.-M. (1961). Les groupements végétaux du Bassin de la Sambre Française. *Vegetatio*, **10**, 69–148, 161–209 & 257–372.

Géhu, J.-M. (1975). Essai systématique et chorologique sur les principales associations végétales du littoral atlantique français. *Anales Real Academia de farmacia*, **47**, 207–27.

Géhu, J.-M. & Géhu, J. (1973). Apport à la connaissance phytosociologique de landes littorales de Bretagne. *Colloques Internationales Phytosociologiques II. La Végétation des Landes d'Europe Occidentale*. Lille.

Gilbert, O.L. (1980). Juniper in Upper Teesdale. *Journal of Ecology*, **68**, 1013–24.

Giller, K.E. (1982). *Aspects of the Ecology of Flood-plain Mire in Broadland*. University of Sheffield: PhD thesis.

Giller, K.E. & Wheeler, B.D. (1986a). Past peat cutting and present vegetation patterns in an undrained fen in the Norfolk broadland. *Journal of Ecology*, **74**, 219–48.

Giller, K.E. & Wheeler, B.D. (1986b). Peat and peat water chemistry of a flood-plain fen in Broadland, Norfolk, U.K. *Freshwater Biology*, **16**, 99–114.

Gillham, M.E. (1957b). Coastal vegetation of Mull and Iona in relation to salinity and soil reaction. *Journal of Ecology*, **45**, 757–78.

Gimingham, C.H. (1949). The effects of grazing on the balance between *Erica cinerea* L. and *Calluna vulgaris* (L.) Hull in upland heath, and their morphological responses. *Journal of Ecology*, **37**, 110–19.

Gimingham, C.H. (1960). Biological Flora of the British Isles: *Calluna vulgaris* (L.) Hull. *Journal of Ecology*, **48**, 455–83.

Gimingham, C.H. (1964a). Maritime and sub-maritime communities. In *The Vegetation of Scotland*, ed. J.H. Burnett, pp. 66–142. Edinburgh: Oliver and Boyd.

Gimingham, C.H. (1964b). Dwarf-shrub heaths. In *The Vegetation of Scotland*, ed. J.H. Burnett, pp. 232–89. Edinburgh: Oliver & Boyd.

Gimingham, C.H. (1972). *Ecology of Heathlands*. London: Chapman & Hall.

Gittins, R.T. (1965a). Multivariate approaches to a limestone grassland community. I. A Stand ordination. *Journal of Ecology*, **53**, 385–401.

Gjaerevøll, O. (1956). The Plant Communities of the Scandinavian Alpine Snow-beds. *Kongelige Norske Videnskabers Selskabs Forhandlinger*, **1**, 1–405.

Godwin, H. (1929). The Sedge and Litter of Wicken Fen. *Journal of Ecology*, **17**, 148–60.

Godwin, H. (1941). Studies in the ecology of Wicken Fen. IV. Crop-taking experiments. *Journal of Ecology*, **29**, 83–106.

Godwin, H. (1975). *The History of the British Flora*, second edition. Cambridge: Cambridge University Press.

Godwin, H. (1978). *Fenland: its Ancient Past and Uncertain*

Future. Cambridge: Cambridge University Press.

Godwin, H. & Conway, V.M. (1939). The ecology of a raised bog near Tregaron, Cardiganshire. *Journal of Ecology*, **27**, 313–63.

Godwin, H. & Tansley, A.G. (1929). The Vegetation of Wicken Fen. In *The Natural History of Wicken Fen. Part V*, pp. 387–446. Cambridge: Bowes & Bowes.

Godwin, H. & Turner, J.S. (1933). Soil acidity in relation to vegetational succession in Calthorpe Broad, Norfolk. *Journal of Ecology*, **21**, 235–62.

Goode, D.A. (1970). *Ecological Studies on the Silver Flowe Nature Reserve*. University of Hull: PhD thesis.

Goode, D.A. & Lindsay, R.A. (1979). The peatland vegetation of Lewis. *Proceedings of the Royal Society of Edinburgh, Series B*, **77**, 279–93.

Goodman, G.T. & Perkins, D.F. (1968). The role of mineral nutrients in *Eriophorum* communities. III. Growth response to added inorganic elements in two *E. vaginatum* communities. *Journal of Ecology*, **56**, 667–83.

Görs, S. (1964). Beiträge zur Kenntnis basiphiler Flachmoorgesellschaften (Tofieldietalia Prsg. apud Oberd.49). 2 Teil: Das Mehlprimel-Kopfbinsen-Moor (Primulo-Schoenetum ferruginei Oberd.(57)62). *Veröffentlichungen des Landesanstalt für Naturschutz und Landschaftspflege*, **32**, 1–42.

Gourlay, W.B. (1919). Notes from Cannock Chase on *Vaccinium intermedium* Ruthe. *Transactions & Proceedings of the Botanical Society of Edinburgh*, **27**, 327–33.

Green, B.H. (1968). Factors influencing the spatial and temporal distribution of *Sphagnum imbricatum* Hornsch. ex Russ. in the British Isles. *Journal of Ecology*, **56**, 47–58.

Green, B.H. & Pearson, M.C. (1968). The ecology of Wybunbury Moss, Cheshire. I. The present vegetation and some physical, chemical and historical factors controlling its nature and distribution. *Journal of Ecology*, **56**, 245–67.

Green, B.H. & Pearson, M.C. (1977). The ecology of Wybunbury Moss, Cheshire. II. Post-Glacial history and the formation of the Cheshire mere and mire landscape. *Journal of Ecology*, **65**, 793–814.

Grime, J.P. (1963a). Factors determining the occurrence of calcifuge species on shallow soils over calcareous substrata. *Journal of Ecology*, **51**, 375–90.

Grime, J.P. (1963b). An ecological investigation at a junction between two plant communities in Coombsdale on the Derbyshire Limestone. *Journal of Ecology*, **51**, 391–402.

Grubb, P.J., Green, H.E. & Merrifield, R.C.J. (1969). The ecology of chalk heath: its relevance to the calcicole-calcifuge and soil acidification problems. *Journal of Ecology*, **57**, 175–212.

Hadač, E. (1971). The vegetation of springs, lakes and 'flags' of Reykjanes Peninsula, S.W. Iceland (Plant Communities of Reykjanes Peninsula, Part 3). *Folia Geobotanica Phytotaxonomica*, **6**, 29–41.

Hallberg, H.P. (1971). Vegetation auf den Schalenablagenungen in Bohuslän, Schweden. *Acta Phytogeographica Suecica*, **56**, 1–136.

Harper, J.L. & Sagar, G.R. (1953). Some aspects of the ecology of buttercups in permanent grassland. *Proceedings of the British Weed Control Conference*, **1**, 256–63.

Harvey, L.A. & St. Leger-Gordon, D. (1974). *Dartmoor*, revised edition. London: Collins.

Haskins, L.E. (1978). *The vegetational history of south-east Dorset*. Southampton University: PhD thesis.

Haslam, S.M. (1965). Ecological studies in Breck fens. I. Vegetation in relation to habitat. *Journal of Ecology*, **53**, 599–619.

Havinden, M. & Wilkinson, F. (1970). Farming. In *Dartmoor: A New Study*, ed. C. Gill. Newton Abbot: David & Charles.

Hawksworth, D.L. (1969). The lichen flora of Derbyshire. *Lichenologist*, **4**, 105–93.

Hegi, G. (1965). *Illustrierte Flora von Mitteleuropa*, **1**, 212–3. Munich: Carl Hanser.

Hewson, R. (1977). The effect on heather *Calluna vulgaris* of excluding sheep from moorland in north-east England. *The Naturalist*, **943**, 133–6.

Hicks, S.P. (1971). Pollen-analytical evidence for the effect of prehistoric agriculture on the vegetation of north Derbyshire. *New Phytologist*, **70**, 647–67.

Hill, M.O. (1979). *TWINSPAN – a FORTRAN program for arranging multivariate data in an ordered two-way table by classification of the individuals and attributes*. New York: Cornell University.

Hill, M.O. & Evans, D.F. (1978). *Types of Vegetation in the Southern Uplands of Scotland*. Bangor: Institute of Terrestrial Ecology.

Hill, M.O. & Stevens, P.A. (1981). Viable seeds in forest soils. *Journal of Ecology*, **69**, 693–709.

Hill, M.O., Bunce, R.G.H. & Shaw, M.W. (1975). Indicator Species Analysis, a divisive polythetic method of classification and its application to a survey of native pinewoods in Scotland. *Journal of Ecology*, **63**, 597–613.

Hilliam, J. (1977). *Phytosociological studies in the Southern Isles of Shetland*. Durham University: PhD thesis.

Hobbs, R.J. (1981). *Post-fire succession in heathland communities*. Aberdeen University: PhD thesis.

Hobbs, R.J. & Gimingham, C.H. (1984a). Studies on fire in Scottish heathland communities. I. Fire characteristics. *Journal of Ecology*, **72**, 223–40.

Hobbs, R.J. & Gimingham, C.H. (1984b). Studies on fire in Scottish heathland communities. II. Post-fire vegetation development. *Journal of Ecology*, **72**, 585–610.

Hobbs, R.J., Mallik, A.U. & Gimingham, C.H. (1984). Studies on fire in Scottish heathland communities. III. Vital attributes of the species. *Journal of Ecology*, **72**, 963–76.

Hodge, C.A.H., Burton, R.G.O., Corbett, W.M., Evans, R. & Seale, R.S. (1984). *Soils and their Use in Eastern England*. Harpenden: Soil Survey of England and Wales.

Holdgate, M.W. (1955a). The vegetation of some springs and wet flushes on Tarn Moor near Orton, Westmorland. *Journal of Ecology*, **43**, 80–9.

Holdgate, M.W. (1955b). The vegetation of some British upland fens. *Journal of Ecology*, **43**, 389–403.

Hope-Simpson, J.F. & Willis, A.J. (1955). Vegetation. In *Bristol and its Adjoining Counties*, ed. C.M. MacInnes & W.F. Whittard, pp. 91–109. Bristol: British Association for the Advancement of Science.

Hopkins, J.J. (1983). *Studies of the Historical Ecology, Vegetation and Flora of the Lizard District, Cornwall, with Particular Reference to Heathland*. Bristol University: PhD thesis.

Huiskes, A.H.L. (1979). Biological Flora of the British Isles: *Ammophila arenaria* (L.) Link. *Journal of Ecology*, **67**, 363–82.

Hulme, P. (1985). The peatland vegetation of the Isle of Lewis and Harris and the Shetland Islands, Scotland. *Aquilo*, **SB21**, 81–8.

Hulme, P.D. & Blyth, A.W. (1984). A classification of the peatland vegetation of the Isle of Lewis and Harris, Scotland. *Proceedings of the 7th International Peat Congress*, **1**, 188–204.

Huntley, B. (1979). The past and present vegetation of the Caenlochan National Nature Reserve, Scotland. I. Present vegetation. *New Phytologist*, **83**, 215–83.

Huntley, B. & Birks, H.J.B. (1979a). The past and present vegetation of the Morrone Birkwoods National Nature Reserve, Scotland. I. A primary phytosociological survey. *Journal of Ecology*, **67**, 419–46.

Huntley, B. & Birks, H.J.B. (1979b). The past and present vegetation of the Morrone Birkwoods National Nature Reserve, Scotland. II. Woodland vegetation and soils. *Journal of Ecology*, **67**, 447–67.

Huntley, B., Huntley, J.P. & Birks, H.J.B. (1981). *PHYTOPAK*: a suite of computer programs designed for the handling and analysis of phytosociological data. *Vegetatio*, **45**, 85–95.

Ingram, M. (1958). The ecology of the Cairngorms. IV. The *Juncus* zone: *Juncus trifidus* communities. *Journal of Ecology*, **46**, 707–37.

Ivimey-Cook, R.B. (1959). Biological Flora of the British Isles: *Agrostis setacea* Curt. *Journal of Ecology*, **47**, 697–706.

Ivimey-Cook, R.B. (1984). *Atlas of the Devon Flora*. Exeter: The Devonshire Association for the Advancement of Science, Literature and Art.

Ivimey-Cook, R.B. & Proctor, M.C.F. (1966b). The plant communities of the Burren, Co. Clare. *Proceedings of the Royal Irish Academy B*, **64**, 211–301.

Ivimey-Cook, R.B. & Proctor, M.C.F. (1967). Factor analysis of data from an east Devon heath: a comparison of principal component and rotated solutions. *Journal of Ecology*, **55**, 405–13.

Ivimey-Cook, R.B., Proctor, M.C.F. & Rowland, D.M. (1975). Analysis of the plant communities of a heathland site: Aylesbeare Common, Devon, England. *Vegetatio*, **31**, 33–45.

Jahns, W. (1962). Zur Kenntnis der Pflanzengesellschaften des Grossen und Wiessen Moores bei Kirchwalsede (Kr. Rotenburg/Hann.). *Mitteilungen der Florist-soziologischen Arbeitsgemeinschaft*, **9**, 88–94.

Jarvis, M.G., Allen, R.H., Fordham, S.J., Hazelden, J., Moffat, A.J. & Sturdy, R.G. (1984). *Soils and their Use in South East England*. Harpenden: Soil Survey of England and Wales.

Jefferies, T.A. (1915). Ecology of the purple heath-grass (*Molinia caerulea*). *Journal of Ecology*, **3**, 93–109.

Jenkins, D., Watson, A. and Miller, G.R. (1963). Population studies on red grouse, *Lagopus lagopus scoticus* (Lath.) in north-east Scotland. *Journal of Animal Ecology*, **32**, 317–76.

Jenkins, D., Watson, A. & Miller, G.R. (1967). Population fluctuations in the red grouse (*Lagopus lagopus scoticus*).

Journal of Animal Ecology, **36**, 97–122.

Jenkins, D., Watson, A. & Miller, G.R. (1970). Practical results of research for the management of Red Grouse. *Biological Conservation*, **4**, 266–72.

Jermy, A.C., Arnold, H.R., Farrell, L. & Perring, F.H. (1978). *Atlas of Ferns of the British Isles*. London: Botanical Society of the British Isles and British Pteridological Society.

Jermy, A.C., Chater, A.O. & David, R.W. (1982). *Sedges of the British Isles*. BSBI Handbook No. 1: a new edition. London: Botanical Society of the British Isles.

Johnson, R.H. (1957). Observations on the stream patterns of some peat moorlands in the southern Pennines. *Memoirs of the Proceedings of the Manchester Literary and Philosophical Society*, **99**, 1–18.

Jonas, F. (1933). Der Hammrich. *Beihefte zum Repertorium specierum novarum regni vegetabilis*, **71**.

Jones, A.V. (1973). *A phytosiological study of Widdybank Fell in Upper Teesdale*. University of Durham: PhD thesis.

Jones, H.E. (1971a). Comparative studies of plant growth and distribution in relation to waterlogging. II. An experimental study of the relationship between transpiration and uptake of iron in *Erica cinerea* L. and *E. tetralix* L. *Journal of Ecology*, **59**, 167–78.

Jones, H.E. (1971b). Comparative studies of plant growth and distribution in relation to waterlogging. III. The response of *Erica cinerea* L. to waterlogging in peat soils of differing iron content. *Journal of Ecology*, **59**, 583–91.

Jones, H.E. & Etherington, J.R. (1970). Comparative studies of plant growth and distribution in relation to waterlogging. I. The survival of *Erica cinerea* L. and *E. tetralix* L. and its apparent relationship to iron and manganese uptake in waterlogged soil. *Journal of Ecology*, **58**, 487–96.

Jones, R.M. (1984). *The Vegetation of Upland Hay Meadows in the North of England with experiments into the Causes of Diversity*. Lancaster University: PhD thesis.

Jovet, P. (1941). Végétation d'une montagne basque siliceuse: La Rhune. *Bulletin de la Société de Botanique de France*, **88**, 69–92.

Kalliola, R. (1939). Pflanzensoziologische untersuchungen in der alpinen Stufe Finnisch-Lapplands. *Annales botanici Societatis zoologicae-botanicae fennicae Vanamo*, **13**.

Kastner, M. & Flössner, W. (1933). Die Pflanzengesellschaften der erzgebirgischen Moore. In *Die Pflanzengesellschaften des westsächsischen Berg- und Hugellandes II*, M. Kastner, W. Flössner & J. Uhlig. Dresden.

Katz, N.J. (1926). *Sphagnum* bogs of Central Russia: phytosociology, ecology and succession. *Journal of Ecology*, **14**, 177–202.

Kayll, A.J. & Gimingham, C.H. (1965). Vegetative regeneration of *Calluna vulgaris* after fire. *Journal of Ecology*, **53**, 729–34.

Kenworthy, J.B. (1963). Temperatures in heather burning. *Nature*, **200**, 1226.

Kenworthy, J.B. (1964). *A study of the changes in plant and soil nutrients associated with moorburning and grazing*. University of St. Andrews: PhD thesis.

King, J. (1962). The *Festuca-Agrostis* grassland complex in south-east Scotland. *Journal of Ecology*, **51**, 321–55.

King, J. & Nicholson, I.A. (1964). Grasslands of the Forest and Sub-Alpine zones. In *The Vegetation of Scotland*, ed. J.H. Burnett, pp. 168–231. Edinburgh: Oliver & Boyd.

Klika, J. (1929). Příspěvek ke geobotanikému prozkumu středního Polabí. *Věstník královské české společnosti náuk*, **2**, 1–25.

Kloss, K. (1965). *Schoenetum, Juncetum subnodulosi* und *Betula pubescens*-Gesellschaften der Kalkreichen Moorniederungen Nordost-Mecklensburgs. *Feddes Repetorium*, **142**, 65–117.

Knaben, G. (1950). Botanical investigations in the Middle District of Western Norway. *Årbok for Universitetet i Bergen*, **8**, 1–117.

Koch, W. (1926). Die Vegetationseinheiten der Linthebene unter Berucksichtigung der Verhaltnisse in der Nordost schweiz. *Jahrbuch der St Gallischen naturwissenschaftlichen Gesellschaft*, **61**, 1–144.

Korneck, D. (1963). Die Pfeifengraswiesen und ihre wichtigsten Kontaktgesellschaften in der nördlichen Oberrheinebene und im Schweinfurther Trockengebiet. III. Kontaktgesellschaften. *Beiträge zur naturkundlichen Forschung in Südwestdeutschland*, **22**, 19–44.

Krausch, H.D. (1964). Die Pflanzengesellschaften des Stechlinsee Gebietes. II. Röchricht und Grossegengesellschaften. *Limnologica*, **2**, 423–82.

Lambert, J.M. (1946). The distribution and status of *Glyceria maxima* (Hartm.) Holmb. in the region of Surlingham and Rockland Broads, Norfolk. *Journal of Ecology*, **33**, 230–67.

Lambert, J.M. (1948). A survey of the Rockland–Claxton level, Norfolk. *Journal of Ecology*, **26**, 120–35.

Lambert, J.M. (1951). Alluvial stratigraphy and vegetational succession in the region of the Bure Valley Broads. III. Classification, status and distribution of communities. *Journal of Ecology*, **39**, 149–70.

Lambert, J.M. (1965). The Vegetation of Broadland. In *The Broads*, ed. E.A. Ellis, pp. 69–92. London: Collins.

Lambert, J.M. & Manners, J.G. (1964). Botany. In *A Survey of Southampton and its Region*, ed. J.F. Monkhouse, pp. 105–17. Southampton: British Association for the Advancement of Science.

Landsberg, S.Y. (1955). *The Morphology and Vegetation of the Sands of Forvie*. Aberdeen University: PhD thesis.

Leach, S.J. (1985). The problem of setting grazing levels on dune-heath, Earlshall Muir, Fife. In *Sand dunes and their management*, ed. P. Doody, pp. 135–43. Peterborough: Nature Conservancy Council.

LeBrun, J., Noirfalise, A., Hienemann, P. & Vanden Berghen, C. (1949). Les Associations végétales de Belgique. *Centre de Recherches écologiques et phytosociologiques de Gembloux, Communication No. 8*, 105–207.

Lee, J. (1981). Atmospheric pollution and the Peak District blanket bogs. In *Moorland Erosion Study, Phase 1 Report*, ed. J. Phillips, D. Yalden & J. Tallis, pp. 104–9. Bakewell: Peak Park Joint Planning Board.

Legg, C.J. (1978). *Succession and homeostasis in heathland vegetation*. Aberdeen University: PhD thesis.

Lemée, G. (1937). Recherches écologiques sur la végétation du Perche. *Revue Générale de Botanique*, **49**.

Lewis, F.J. (1904a). Geographical Distribution of Vegetation of the Basins of the Rivers Eden, Tees, Wear & Tyne. Part I. *Geographical Journal*, **23**, 313–31.

Lewis, F.J. (1904b). Geographical Distribution of Vegetation of the Basins of the Rivers Eden, Tees, Wear & Tyne. Part II. *Geographical Journal*, **24**, 267–285.

Lewis, F.J. & Moss, C.E. (1911). The upland moors of the Pennine chain. In *Types of British Vegetation*, ed. A.G. Tansley, pp. 266–82. Cambridge: Cambridge University Press.

Libbert, W. (1932). Die Vegetationseinheiten der neumärksichen Staubeckenlandschaft unter Berucksichtigung der angrenzenden Landschaften. *Verhandlungen des Botanischen Vereins der Provinz Brandenburg*, **74**, 10–93 & 229–348.

Lid, J. (1959). The vascular plants of Hardangervidda, a mountain plateau of southern Norway. *Nytt magasin for botanikk*, **7**, 61–128.

Lind, E.M. (1949). The history and vegetation of some Cheshire meres. *Memoirs of the Proceedings of the Manchester Literary and Philosophical Society*, **90**, 17–36.

Lindsay, R.A., Riggall, J. & Burd, F.H. (1984). The use of small-scale surface patterns in the classification of British peatlands. *Proceedings of the Field Symposium on Classification of Mire Vegetation*.

Loach, K. (1966). Relations between soil nutrients and vegetation in wet-heaths. I. Soil nutrient content and moisture conditions. *Journal of Ecology*, **54**, 597–608.

Loach, K. (1968a). Relations between soil nutrients and vegetation in wet-heaths. II. Nutrient uptake by the major species in the field and in controlled experiments. *Journal of Ecology*, **56**, 117–27.

Loach, K. (1968b). Seasonal growth and nutrient uptake in a *Molinietum*. *Journal of Ecology*, **56**, 433–44.

Lock, J.M. & Rodwell, J.S. (1981). Observations on the Vegetation of Crag Lough, Northumberland. A report to the National Trust. Unpublished manuscript.

Lovat, Lord (1911). Heather Burning. In *The Grouse in Health and Disease*, ed. A.S. Leslie, pp. 392–412. London.

MacLeod, A.C. (1955). Heather in the seasonal dietary of sheep. *Proceedings of the British Society of Animal Production*, **1955**, 13–17.

Mallik, A.U. & Gimingham, C.H. (1983). Regeneration of heathland plants following burning. *Vegetatio*. **53**, 45–58.

Mallik, A.U. & Gimingham, C.H. (1985). Ecological effects of heather burning. II. Effects on seed germination and vegetative regeneration. *Journal of Ecology*, **73**, 633–44.

Mallik, A.U., Hobbs, R.J. & Legg, C.J. (1984). Seed dynamics in *Calluna-Arctostaphylos* heath in NE Scotland. *Journal of Ecology*, **72**, 855–71.

Malloch, A.J.C. (1970). *Analytical Studies of Cliff-top Vegetation in South-West England*. Cambridge University: PhD thesis.

Malloch, A.J.C. (1971). Vegetation of the maritime cliff-tops of the Lizard and Land's End peninsula, West Cornwall. *New Phytologist*, **70**, 1155–97.

Malloch, A.J.C. (1972). Salt-spray deposition on the maritime cliffs of the Lizard peninsula. *Journal of Ecology*, **60**, 103–12.

Malloch, A.J.C. (1988). VESPAN II. Lancaster: University of Lancaster.

Maltby, E. (1980). The impact of severe fire on *Calluna* moorland in the North Yorks Moors. *Bulletin d'Écologie*, **11**, 683–708.

Maltby, E. & Legg, C.J. (1983). Revegetation of fossil patterned ground exposed by severe fire on the North York Moors. *Permafrost: Fourth International Conference Proceedings*, 792–7.

Manley, G. (1936). The climate of the Northern Pennines. *Quarterly Journal of the Royal Meteorological Society*, **62**, 103–13.

Manley, G. (1940). Snowfall in Britain. *Meteorological Magazine*, **75**, 41ff.

Manley, G. (1942). Meteorological observations on Dun Fell, a mountain station in northern England. *Quarterly Journal of the Royal Meteorological Society*, **68**, 151–62.

Marren, P. (1979). *Muir of Dinnet: Portrait of a National Nature Reserve*. Aberdeen: Nature Conservancy Council.

Marrs, R.H. (1986). The role of catastrophic death of *Calluna* in heathland dynamics. *Vegetatio*, **66**, 109–15.

Marrs, R.H. & Proctor, J. (1978). Chemical and ecological studies of the enclosed heathlands of the Lizard Peninsula, Cornwall, *Vegetatio*, **41**, 121–8.

Matthews, J.R. (1955). *Origin and Distribution of the British Flora*. London: Hutchinson.

Matuszkiewicz, W. (1981). *Przewodnik do oznaczania zbiorowisk roslinnych Polski*. Warzawa: Panstwowe Wydawnictwo Naukowe.

McBeath, R. (1967). *Phyllodoce caerulea* (L.) Bab. on Ben Alder (v.-c.97). *Transactions and Proceedings of the Botanical Society of Edinburgh*, **40**, 335–6.

McVean, D.N. (1961b). Post-glacial history of juniper in Scotland. *Proceedings of the Linnean Society of London*, **172**, 53–5.

McVean, D.N. & Ratcliffe, D.A. (1962). *Plant Communities of the Scottish Highlands*. London: HMSO.

McVicar, S.M. (1912). *The Student's Handbook of British Hepatics*. Eastbourne: Sumfield.

Meade, R. (1981). *A Vegetation Survey of Pembrokeshire Heaths*. Bangor: Nature Conservancy Council.

Meek, V. (1976). *The Vegetation of the Langholm-Newcastleton hills*. Edinburgh: Nature Conservancy Council.

Messenger, G. (1971). *Flora of Rutland*. Leicester: Leicester Museums.

Metcalfe, G. (1950). The ecology of the Cairngorms. Part II. The Mountain Callunetum. *Journal of Ecology*, **38**, 46–74.

Meteorological Office (1977). *Average Rainfall Map for the United Kingdom 1941–71 (South)*. London: HMSO.

Mhic Daeid, C. (1976). *A Phytosociological and Ecological Study of the Vegetation of Peatlands and Heaths in the Killarney Valley*. Trinity College, Dublin: PhD thesis.

Miles, J. (1974). Effects of experimental interference with stand structure on establishment of seedlings in *Callunetum*. *Journal of Ecology*, **62**, 675–87.

Miles, J. & Kinnaird, J.W. (1979). The establishment and regeneration of birch, juniper and scots pine in the Scottish Highlands. *Scottish Forestry*, **33**, 102-19.

Miller, G.R. & Miles, L. (1970). Regeneration of heather (*Calluna vulgaris* (L.) Hull) at different ages and seasons in north-east Scotland. *Journal of Applied Ecology*, **7**, 51–60.

Miller, G.R., Jenkins, D. & Watson, A. (1966). Heather performance and red grouse populations. I. Visual estimates of heather performance. *Journal of Applied Ecology*, **3**, 313–26.

Milton, W.E.J. (1936). Buried viable seeds of enclosed and unenclosed hill land. *Bulletin of the Welsh Plant Breeding Station, Series H.*, **14**, 58–72.

Milton, W.E.J. (1948). Buried viable seed content of upland soils in Montgomery. *Empire Journal of Experimental Agriculture*, **16**, 163–77.

Mohamed, B.F. & Gimingham, C.H. (1970). The morphology of vegetative regeneration in *Calluna vulgaris*. *New Phytologist*, **69**, 743–50.

Moore, D.M. (1958). Biological Flora of the British Isles: *Viola lactea* Sm. *Journal of Ecology*, **46**, 527–35.

Moore, H.I. & Burr, S. (1948). The control of rushes in newly-re-seeded land in Yorkshire. *Journal of the British Grassland Society*, **3**, 283–90.

Moore, J.J. (1962). The Braun-Blanquet System: a reassessment. *Journal of Ecology*, **50**, 761–9.

Moore, J.J. (1968). A Classification of the Bogs and Wet Heaths of Northern Europe. In *Pflanzensoziologische Systematik*, ed. R. Tüxen, pp. 306–320. Den Haag: Junk N.V.

Moore, N.W. (1962). The heaths of Dorset and their conservation. *Journal of Ecology*, **50**, 369–91.

Moore, P.D. (1977). Stratigraphy and pollen analysis of Claish Moss, north-west Scotland: significance of the origin of surface pools and forest history. *Journal of Ecology*, **65**, 375–97.

Moore, P.D. & Bellamy, D.J. (1973). *Peatlands*. London: Elek.

Morrison, M.E.S. (1959). The ecology of a raised bog in Co. Tyrone, Northern Ireland. *Proceedings of the Royal Irish Academy B*, **60**, 291–308.

Moss, C.E. (1911). The vegetation of calcareous soils. A. The sub-formation of the older Limestones. In *Types of British Vegetation*, ed. A.G. Tansley, pp. 146–61. Cambridge: Cambridge University Press.

Moss, C.E. (1913). *Vegetation of the Peak District*. Cambridge: Cambridge University Press.

Moss, R. (1969). A comparison of red grouse (*Lagopus l. scoticus*) stocks with the productivity and nutritive value of heather (*Calluna vulgaris*). *Journal of Animal Ecology*, **38**, 103–22.

Moss, R., Watson, A. & Parr, R. (1975). Maternal nutrition and breeding success in Red Grouse (*Lagopus lagopus scoticus*). *Journal of Animal Ecology*, **44**, 233–44.

Muir, A. & Fraser, G.K. (1940). The soils and vegetation of the Bin and Clashindarroch forests. *Transactions of the Royal Society of Edinburgh*, **60**, 233–341.

NCC Devon Heathland Report (1980). *A Survey of Selected heathlands in Devon*. Banbury: Nature Conservancy Council.

NCC New Forest Bogs Report (1984). *A Survey of Selected New Forest Bogs*. Banbury: Nature Conservancy Council, England Field Unit.

NCC Pembrokeshire Heaths Report (1981). *A vegetation survey of Pembrokeshire heaths*. Bangor: Nature Conservancy Council.

NCC South Gower Coast Report (1981). *South Gower Coast cliff and grassland survey*. Bangor: Nature Conservancy Council.

Newbould, P.J. (1960). The ecology of Cranesmoor, a New Forest valley bog. *Journal of Ecology*, **48**, 361–83.

Newton, A. (1971). *Flora of Cheshire*. Chester: Cheshire Community Council.

Noble, J.C. (1982). Biological Flora of the British Isles: *Carex arenaria* L. *Journal of Ecology*, **70**, 867–86.

Nordhagen, R. (1922). Vegetationsstudien auf der Insel Utsire im westlichen Norwegen. *Bergens Museums Årbok 1920–21*, 1–149.

Nordhagen, R. (1928). *Die Vegetation und Flora des Sylenegebiets*. Oslo.

Nordhagen, R. (1937). Versuch einer neuen Einteilung der sub-alpinen–alpinen Vegetations Norwegens. *Bergens Museums Årbok 1936*, 1–88.

Nordhagen, R. (1940). Stüdien über die maritime Vegetations Norwegens. *Bergens Museums Årbok*, **2**, 1–123.

Nordhagen, R. (1943). *Sikilsdalen og Norges Fjellbeiter*. *Bergens Museums Skrifter 22*. Bergen: Griegs.

Oberdorfer, E. (1957). Süddeutsche Pflanzengesellschaften. *Pflanzensoziologie*, **10**.

Oberdorfer, E. (1977). *Süddeutsche Pflanzengesellschaften*. Teil I. Stuttgart: Fischer.

Oberdorfer, E. (1979). *Pflanzensoziologische Exkursionsflora für Südwestdeutschland, 4 Aufl*. Stuttgart: Ulmer.

Oberdorfer, E. (1983). *Suddeutsche Pflanzengesellschaften*. Teil. III. Stuttgart: Fischer.

Ostenfeld, C.H. (1908). The Land-Vegetation of the Faeröes. In *Botany of the Faeröes*, **3**, 867–1026.

O'Sullivan, A.M. (1968a). Irish Molinietalia communities in relation to those of the Atlantic regions of Europe. In *Pflanzensoziologische Systematik*, ed. R. Tüxen, pp. 273–80. Den Haag: Junk.

O'Sullivan, A.M. (1968b). The lowland grasslands of County Limerick. *An Foras Taluntais, Irish Vegetation Studies*, 2.

O'Sullivan, A.M. (1976). The phytosociology of the Irish wet grasslands belonging to the order Molinietalia. *Colloques Phytosociologiques*, **5**, 259–67.

O'Sullivan, A.M. (1982). The lowland grasslands of Ireland. *Royal Dublin Society Journal of Life Sciences*, **3**, 131–42.

O'Sullivan, P.E. (1977). Vegetation history and the native pinewoods. In *Native Pinewoods of Scotland*, ed. R.G.H. Bunce & J.N.R. Jeffers, pp. 60–9. Cambridge: Institute of Terrestrial Ecology.

Osvald, H. (1923). Die Vegetation des Hochmoores Komosse. *Svenska Växtsociologiska Sällskapet Handlingar*, **1**, 1–436.

Osvald, H. (1949). Notes on the vegetation of British and Irish mosses. *Acta Phytogeographica Suecica*, **26**, 1–62.

Page, C.N. (1982). *The Ferns of Britain and Ireland*. Cambridge: Cambridge University Press.

Page, M.L. (1980). *Phytosociological classification of British neutral grasslands*. Exeter University: PhD thesis.

Pallis, M. (1911). The River Valleys of East Norfolk: their Aquatic and Fen Formations. In *Types of British Vegetation*, ed. A.G. Tansley, pp. 214–45. Cambridge: Cambridge University Press.

Passarge, H. (1964). Pflanzengesellschaften des nordostdeutschen Flachlandes. I. *Pflanzensoziologie (Jena)*, **13**, 1–324.

Pearsall, W.H. (1918). The aquatic and marsh vegetation of Esthwaite Water. *Journal of Ecology*, **5**, 53–74.

Pearsall, W.H. (1938). The soil complex in relation to plant communities. III. Moorland bogs. *Journal of Ecology*, **26**, 298–315.

Pearsall, W.H. (1941). The 'mosses' of the Stainmore district. *Journal of Ecology*, **29**, 161–75.

Pearsall, W.H. (1956). Two blanket bogs in Sutherland. *Journal of Ecology*, **44**, 493–516.

Pearsall, W.H. (1968). *Mountains and Moorlands*. London: Collins.

Pearsall, W.H. & Lind, E.M. (1941). A note on a Connemara bog type. *Journal of Ecology*, **29**, 62–8.

Perrin, R.M.S. (1956). Nature of 'chalk heath' soils. *Nature, London*, **178**, 31–2.

Perring, F.H. (1968). *Critical Supplement to the Atlas of the British Flora*. London: Nelson.

Perring, F.H. & Farrell, L. (1977). *British Red Data Books 1: Vascular Plants*. Nettleham: Society for the Promotion of Nature Conservation.

Perring, F.H. & Walters, S.M. (1962). *Atlas of the British Flora*. London & Edinburgh: Nelson.

Perring, F.H., Sell, P.D., Walters, S.M. & Whitehouse, H.L.K. (1964). *A Flora of Cambridgeshire*. Cambridge: Cambridge University Press.

Persson, Å. (1961). Mire and spring vegetation in an area north of Lake Torneträsk, Torne Lappmark, Sweden. *Opera botanica a Societate botanica ludensi*, **6**, 1–187.

Peterken, G.F. & Tubbs, C.R. (1965). Woodland regeneration in the New Forest, Hampshire, since 1650. *Journal of Applied Ecology*, **2**, 159–70.

Phillips, M.E. (1954). Biological Flora of the British Isles: *Eriophorum angustifolium* Roth. *Journal of Ecology*, **42**, 612–22.

Picozzi, N. (1968). Grouse bags in relation to management and geology of heather moors. *Journal of Applied Ecology*, **5**, 483–8.

Pigott, C.D. (1956a). The vegetation of Upper Teesdale in the North Pennines. *Journal of Ecology*, **44**, 545–86.

Pigott, C.D. (1956b). Vegetation. In *Sheffield and its Region*, ed. D.L. Linton, pp. 78–89. Sheffield: British Association.

Pigott, C.D. (1962). Soil formation and development on the Carboniferous Limestone of Derbyshire. I. Parent materials. *Journal of Ecology*, **50**, 145–56.

Pigott, C.D. (1970a). Soil formation and development on the Carboniferous Limestone of Derbyshire. II. The relation of soil development to vegetation on the plateau near Coombsdale. *Journal of Ecology*, **58**, 529–41.

Pigott, C.D. (1977). The scientific basis of practical conservation: aims and methods of conservation. *Proceedings of the Royal Society of London, Series B*, **197**, 59–68.

Pigott, C.D. (1978a). Climate and Vegetation. In *Upper Teesdale*, ed. A.R. Clapham, pp. 102–21. London: Collins.

Pigott, C.D. (1978b). Soil Development. In *Upper Teesdale*, ed. A.R. Clapham, pp. 129–40. London: Collins.

Pigott, C.D. (1982). The experimental study of vegetation. *New Phytologist*, **90**, 389–404.

Pigott, C.D. (1984). The flora and vegetation of Britain: ecology and conservation. *New Phytologist*, **98**, 119–28.

Pigott, C.D. & Pigott, M.E. (1963). Late-glacial and Post-glacial deposits at Malham, Yorkshire. *New Phytologist*, **62**, 317–34.

Polozova, T.G. (1970). Biological features of *Eriophorum vaginatum* L. as a tussock former (based on observations in tundras of western Taimyr). *Botanicheskii Zhurnal*, **55**, 431–42.

Polunin, N. (1948). Botany of the Canadian Eastern Arctic. Part 2, Ecology and vegetation. *Bulletin of the National Museum of Canada*, **104**.

Poore, M.E.D. (1955a). The use of phytosociological methods in ecological investigations. I. The Braun-Blanquet System. *Journal of Ecology*, **43**, 226–44.

Poore, M.E.D. (1955b). The use of phytosociological methods in ecological investigations. II. Practical issues involved in an attempt to apply the Braun-Blanquet system. *Journal of Ecology*, **43**, 245–69.

Poore, M.E.D. (1955c). The use of phytosociological methods in ecological investigations. III. Practical application. *Journal of Ecology*, **43**, 606–51.

Poore, M.E.D. (1956b). The ecology of Woodwalton Fen. *Journal of Ecology*, **44**, 455–92.

Poore, M.E.D. & McVean, D.N. (1957). A new approach to Scottish mountain vegetation. *Journal of Ecology*, **45**, 401–39.

Poore, M.E.D. & Walker, D. (1959). Wynbury Moss, Cheshire. *Memoirs & Proceedings of the Manchester Literary and Philosophical Society*, **101**, 1–24.

Praeger, R.L. (1934). *The Botanist in Ireland*. Dublin: Hodges, Figgis & Co.

Prentice, H.C. & Prentice, I.C. (1975). The hill vegetation of North Hoy, Orkney. *New Phytologist*, **75**, 313–67.

Proctor, J. (1971). The plant ecology of serpentine. II. Plant response to serpentine soils. *Journal of Ecology*, **59**, 397–410.

Proctor, J. & Woodell, S.R.J. (1971). The plant ecology of serpentine. I. Serpentine vegetation of England and Scotland. *Journal of Ecology*, **59**, 375–95.

Proctor, M.C.F. (1965). The distinguishing characters and geographical distributions of *Ulex minor* and *Ulex gallii*. *Watsonia*, **6**, 177–87.

Proctor, M.C.F. (1974). The vegetation of the Malham Tarn fens. *Field Studies*, **4**, 1–38.

Radley, J. (1962). Peat erosion on the high moors of Derbyshire and West Yorkshire. *East Midland Geographer*, **3**, 40–50.

Ragg, J.M., Beard, G.R., George, H., Heaven, F.W., Hollis, J.M., Jones, R.J.A., Palmer, R.C., Reeve, M.J., Robson, J.D. & Whitfield, W.A.D. (1984). *Soils and their Use in Midland and Western England*. Harpenden: Soil Survey of England and Wales.

Rankin, W.M. (1911a). The Lowland Moors ('Mosses') of Lonsdale (North Lancashire) and their Development from Fens. In *Types of British Vegetation*, ed. A.G. Tansley, pp. 247–59. Cambridge: Cambridge University Press.

Rankin, W.M. (1911b). The Valley Moors of the New Forest. In *Types of British Vegetation*, ed. A.G. Tansley, pp. 259–64. Cambridge: Cambridge University Press.

Ranwell, D.S. (1972). *Ecology of Salt Marshes and Sand Dunes*. London: Chapman & Hall.

Ratcliffe, D. (1959). Biological Flora of the British Isles: *Hornungia petraea* (L.) Rchb. *Journal of Ecology*, **47**, 241–7.

Ratcliffe, D.A. (1959a). The vegetation of the Carnedau, North Wales. I. Grasslands, heaths and bogs. *Journal of Ecology*. **47**, 371–413.

Ratcliffe, D.A. (1959b). The Mountain Plants of the Moffat Hills. *Transactions of the Botanical Society of Edinburgh*, **37**, 257–71.

Ratcliffe, D.A. (1964). Mires and Bogs; Montane Mires and Bogs. In *The Vegetation of Scotland*, ed. J.H. Burnett, pp. 426–78 & 536–58. Edinburgh & London: Oliver & Boyd.

Ratcliffe, D.A. (1965). *A Botanical Survey of the Proposed Cow Green Reservoir Site in Upper Teesdale*. Nature Conservancy manuscript report.

Ratcliffe, D.A. (1968). An ecological account of Atlantic bryophytes in the British Isles. *New Phytologist*, **67**, 365–439.

Ratcliffe, D.A. (1974). The Vegetation. In *The Cairngorms: their Natural History and Scenery*, ed. D. Nethersole-Thompson & A. Watson, pp. 42–76. London: Collins.

Ratcliffe, D.A. ed. (1977). *A Nature Conservation Review*. Cambridge: Cambridge University Press.

Ratcliffe, D.A. (1978). The Plant Communities of Upper Teesdale. In *Upper Teesdale*, ed. A.R. Clapham, pp. 64–87. London: Collins.

Ratcliffe, D.A. & Walker, D. (1958). The Silver Flowe, Galloway. *Journal of Ecology*, **46**, 407–45.

Ratcliffe, J.B. & Hattey, R.P. (1982). Welsh Lowland Peatland Survey. London: Nature Conservancy Council.

Raven, J. & Walters, M. (1956). *Mountain Flowers*. London: Collins.

Rawes, M. (1981). Further results of excluding sheep from high-level grasslands in the North Pennines. *Journal of Ecology*, **69**, 651–69.

Rawes, M. (1983). Changes in two high altitude blanket bogs after the cessation of sheep grazing. *Journal of Ecology*, **71**, 219–35.

Rawes, M. & Heal, O.W. (1978). The blanket bog as part of a Pennine moorland. In *The Ecology of some British Moors and Montane Grasslands*, ed. O.W. Heal & D.F. Perkins, pp. 224–43. Berlin: Springer.

Rawes, M. & Hobbs, R. (1979). Management of semi-natural blanket bog in the northern Pennines. *Journal of Ecology*, **67**, 789–807.

Regnéll, G. (1980). A numerical study of successions in an abandoned damp calcareous meadow in S. Sweden. *Vegetatio*, **43**, 123–30.

Richards, P.W. & Clapham, A.R. (1941a). Biological Flora of the British Isles: *Juncus inflexus* L. *Journal of Ecology*, **29**, 369–74.

Richards, P.W. & Clapham, A.R. (1941b). Biological Flora of the British Isles: *Juncus effusus* L. *Journal of Ecology*, **29**, 375–80.

Richards, P.W. & Clapham, A.R. (1941d). Biological Flora of the British Isles: *Juncus subnodulosus* Schrank. *Journal of Ecology*, **29**, 385–91.

Ritchie, J.C. (1955a). Biological Flora of the British Isles: *Vaccinium vitis-idaea* L. *Journal of Ecology*, **43**, 701–8.

Ritchie, J.C. (1955b). A natural hybrid in *Vaccinium*. *New Phytologist*, **54**, 49–67.

Ritchie, J.C. (1956). Biological Flora of the British Isles: *Vaccinium myrtillus* L. *Journal of Ecology*, **44**, 291–9.

Rodriguez, F.B. (1966). La végétation de Galicia. *Anales del Instituto botánico A.J. Cavanillo*, **24**, 1–306.

Rodwell, J.S. (1974). *The Vegetation of the British Carboniferous Limestone in Relation to Topography and*

Soils. Southampton University: PhD thesis.

Roper-Lindsay, J. & Say, A.M. (1986). Plant communities of the Shetland Isles. *Journal of Ecology*, **74**, 1013–30.

Rose, F. (1950). The East Kent Fens. *Journal of Ecology*, **38**, 292–302.

Rose, F. (1953). A survey of the ecology of British lowland bogs. *Proceedings of the Linnean Society*, **164**, 186–211.

Rübel, E. (1930). *Pflanzengesellschaften der Erde*. Berlin-Bern.

Rutter, A.J. (1955). The composition of wet heath vegetation in relation to the water table. *Journal of Ecology*, **43**, 507–43.

Salisbury, E.J. (1952). *Downs and Dunes*. London: Bell.

Salisbury, E.J. (1964). *Weeds and Aliens*, second edition. London: Collins.

Samuelsson, G. (1934). Die Verbreitung der höheren Wasserpflanzen in Nord Europa. *Acta Phytogeographica Suecica*, **6**, 1–211.

Sayer, J.A. (1969). *Some Aspects of the Management of Grazing Animals on Dartmoor*. University College, London: MSc thesis.

Schoof van Pelt, M.M. (1973). *Littorelletea. A Study of the Vegetation of Some Amphriphytic communities of Western Europe*. Nijmegen: PhD thesis.

Schwickerath, M. (1940). Aufbau und Gliederung der europäischen Hochmoor-gesellschaften. *Botanischer Jahrbücher, Stuttgart*, **71**, 249–66.

Segal, S. (1966). Ecological studies of peat-bog vegetation in the north-western part of the province of Overijsel (The Netherlands). *Wentia*, **15**, 109–41.

Sheail, J. (1979). Documentary evidence of the changes in the use, management and appreciation of the grass-heaths of Breckland. *Journal of Biogeography*, **6**, 277–92.

Sheikh, K.H. (1969a). The effects of competition and nutrition on the inter-relations of some wet-heath plants. *Journal of Ecolgy*, **57**, 87–99.

Sheikh, K.H. (1969b). The responses of *Molinia caerulea* and *Erica tetralix* to soil aeration and related factors. II. Gas concentrations in soil air and soil water. *Journal of Ecology*, **57**, 727–36.

Sheikh, K.H. (1970). The responses of *Molinia caerulea* and *Erica tetralix* to soil aeration and related factors. III. Effects of different gas concentrations on growth in solution culture: and general conclusions. *Journal of Ecology*, **58**, 141–54.

Sheikh, K.H. & Rutter, A.J. (1969). The responses of *Molinia caerulea* and *Erica tetralix* to soil aeration and related factors. I. Root distribution in relation to soil porosity. *Journal of Ecology*, **57**, 713–26.

Shellard, H.C. (1976). Wind. In *The Climate of the British Isles*, ed. T.J. Chandler & S. Gregory, pp. 39–73. London: Longman.

Shimwell, D.W. (1968a). *The Phytosociology of Calcareous Grasslands in the British Isles*. University of Durham: PhD thesis.

Shimwell, D.W. (1972). Anthelion julaceae – A New Alliance of Sub-Alpine Spring Vegetation. *Transactions of the Botanical Society of Edinburgh*, **41**, 445–50.

Shimwell, D.W. (1973b). Man-induced changes in the heathland vegetation of Central England. *Colloques Phytosociologiques*, **2**, 58–74.

Shimwell, D.W. (1974). Sheep grazing intensity in Edale, 1692–1747, and its effect on blanket peat erosion. *Derbyshire Archaeological Journal*, **94**, 35–40.

Shimwell, D.W. (1981). Footpath erosion. In *Moorland Erosion Study. Phase 1 Report*, ed. J. Phillips, D. Yalden & J. Tallis, pp. 160–70. Bakewell: Peak Park Joint Planning Board.

Simmonds, N.W. (1946). Biological Flora of the British Isles: *Gentiana pneumonanthe* L. *Journal of Ecology*, **33**, 295–307.

Sinker, C.A. (1960). The Vegetation of the Malham Tarn Area. *Proceedings of the Leeds Philosophical and Literary Society, Scientific Section*, **8**, 139–75.

Sinker, C.A. (1962). The North Shropshire meres and mosses: a background for ecologists. *Field Studies*, **4**, 101–37.

Sissingh, G. (1946). Rudereto-Secalinetea Br.-Bl. 1936, Klasse der akerondruid-, ruderal-, vloedmerken kaalkapgemeenschappen. In *Overzicht der plantengemenschappen in Nederland*, V. Westhoff, J. Dijk & H. Passchier. Amsterdam: Breughel.

Sjörs, H. (1948). Myrvegetation i Bergslagen. *Acta Phytogeographica Suecica*, **21**, 1–299.

Skipper, E.G. (1922). The ecology of the gorse (*Ulex*) with special reference to the growth forms on Hindhead Common. *Journal of Ecology*, **10**, 24–52.

Skogen, A. (1965). Flora og vegetasjon i Ørland herred, Sør-Trondelag. *Årbok 1965 for det Kongelige Norske videnskabers selskab Museet*, 13–124.

Skogen, A. (1971). Studies in Norwegian maritime heath vegetation. I. The eco-sociological range of *Carex binervis* at its northern distribution limit. *Årbok Universitetet Bergen*, **5**, 1–17.

Skogen, A. (1973). Phytogeographical and ecological studies on *Carex paniculata* L. in Norway. *Årbok Universitetet Bergen*, **3**, 1–12.

Slack, A.A.P. (1970). Flora, In *The Island of Skye*, M. Slesser. Edinburgh: Scottish Mountaineering Trust.

Sledge, W.A. (1949). Distribution and ecology of *Scheuchzeria palustris*. *Watsonia*, **1**, 24–35.

Smith, A.J.E. (1978). *The Moss Flora of Britain and Ireland*. Cambridge: Cambridge University Press.

Smith, L.P. (1976). *The Agricultural Climate of England and Wales*. Ministry of Agriculture, Fisheries and Food Technical Bulletin 35. London: HMSO.

Smith, R. (1900). Botanical Survey of Scotland. *Scottish Geographical Magazine*, **16**.

Smith, W.G. (1911a). Scottish Heaths. In *Types of British Vegetation*, ed. A.G. Tansley, pp. 113–6. Cambridge: Cambridge University Press.

Smith, W.G. (1911b). Arctic-Alpine Vegetation. In *Types of British Vegetation*, ed. A.G. Tansley, pp. 288–329. Cambridge: Cambridge University Press.

Smith, W.G. & Moss, C.E. (1903). Geographical Distribution of Vegetation in Yorkshire. Part I. Leeds and Halifax District. *Geographical Journal*, **21**, 375–401.

Smith, W.G. & Rankin, W.M. (1903). Geographical Distribution of vegetation in Yorkshire. Part II. Harrogate and Skipton District. *Geographical Journal*, **21**.

Smithson, F. (1953). The micro-mineralogy of North Wales soils. *Journal of Soil Science*, **4**, 194–210.

Soil Survey (1983). 1:250,000 Soil Map of England and

Wales: six sheets and legend. Harpenden: Soil Survey of England and Wales.

Sörensen, T. (1942). Untersuchungen über die Therophytengesellschaften auf den isländischen Lehmflächen ('Flags'). *Konigelige Danske Videnskabernes Selskabs Skrifter*, **2**, 1–30.

Sparling, J.H. (1962a). *The Autecology of Schoenus nigricans L.* University of London: PhD thesis.

Sparling, J.H. (1962b). Occurrence of *Schoenus nigricans* L. in the blanket bogs of western Ireland and north-west Scotland. *Nature*, **195**, 723–4.

Sparling, J.H. (1967a). The occurrence of *Schoenus nigricans* L. in blanket bogs. I. Environmental conditions affecting the growth of *S. nigricans* in blanket bog. *Journal of Ecology*, **55**, 1–14.

Sparling, J.H. (1967b). The occurrence of *Schoenus nigricans* L. in blanket bogs. II. Experiments on the growth of *S. nigricans* under controlled conditions. *Journal of Ecology*, **55**, 15–32.

Sparling, J.H. (1968). Biological Flora of the British Isles: *Schoenus nigricans* L. *Journal of Ecology*, **56**, 883–99.

Spence, D.H.N. (1964). The macrophytic vegetation of freshwater lochs, swamps and associated fens. In *The Vegetation of Scotland*, ed. J.H. Burnett, pp. 306–425. Edinburgh: Oliver & Boyd.

Sprent, J.I., Scott, R. & Perry, K.M. (1978). The nitrogen economy of *Myrica gale* in the field. *Journal of Ecology*, **66**, 657–68.

Staines, S.J. (1984). *Soils in Cornwall III: Sheets SW 61, 71 and parts of SW 62, 72, 81 and 82 (The Lizard)*. Harpenden: Soil Survey of England and Wales.

Stapledon, R.G. (1914). *The Sheep Walks of mid-Wales*. Privately printed.

Steindórsson, S. (1963). Um gródur i Papey. *Natturufraedingurinn*, **33**, 214–32.

Steven, H.M. & Carlisle, A. (1959). *The Native Pinewoods of Scotland*. Edinburgh: Oliver & Boyd.

Stoutjesdijk, P. (1959). Heaths and inland dunes of the Veluwe. *Wentia*, **2**, 1–96.

Summerhayes, V.S. & Williams, P.H. (1926). Studies on the ecology of English heaths. II. *Journal of Ecology*, **14**, 203–43.

Svensson, G. (1965). Vegetationsundersökningar på Store mosse. *Botaniker notiser*, **118**, 49–86.

Tallis, J.H. (1961). Some observations on *Sphagnum imbricatum*. *Transactions of the British Bryological Society*, **1**, i–xxi.

Tallis, J.H. (1964b). Studies on southern Pennine peats. II. The pattern of erosion. *Journal of Ecology*, **52**, 333–44.

Tallis, J.H. (1964c). Studies on southern Pennine peats. III. The behaviour of *Sphagnum*. *Journal of Ecology*, **52**, 345–53.

Tallis, J.H. (1965). Studies on southern Pennine peats. IV. Evidence of recent erosion. *Journal of Ecology*, **53**, 509–20.

Tallis, J.H. (1969). The blanket bog vegetation of the Berwyn Mountains, North Wales. *Journal of Ecology*, **57**, 765–87.

Tallis, J.H. (1973a). The terrestrialization of lake basins in north Cheshire with special reference to the development of a 'schwingmoor' structure. *Journal of Ecology*, **61**, 537–67.

Tallis, J.H. (1973b). Studies on southern Pennine peats. V.

Direct observations on peat erosion and peat hydrology at Featherbed Moss, Derbyshire. *Journal of Ecology*, **61**, 1–22.

Tallis, J.H. (1981). Uncontrolled fires. In *Moorland Erosion Study, Phase 1 Report*, ed. J. Phillips, D. Yalden & J. Tallis, pp. 176–82. Bakewell: Peak Park Joint Planning Board.

Tallis, J.H. (1985a). Erosion of blanket peat in the southern Pennines: new light on an old problem. In *The Geomorphology of Northwest England*, ed. R.H. Johnson. Manchester: Manchester University Press.

Tallis, J.H. (1985b). Mass movement and erosion of a southern Pennine blanket peat. *Journal of Ecology*, **73**, 283–315.

Tallis, J.H. & Birks, H.J.B. (1965). The past and present distribution of *Scheuchzeria palustris* in Europe. *Journal of Ecology*, **53**, 287–98.

Tallis, J. & Yalden, D. (1984). *Moorland Restoration Project. Phase 2 Report*. Bakewell: Peak Park Joint Planning Board.

Tansley, A.G. ed. (1911). *Types of British Vegetation*. Cambridge: Cambridge University Press.

Tansley, A.G. (1939). *The British Islands and their Vegetation*. Cambridge: Cambridge University Press.

Taylor, K. (1971). Biological Flora of the British Isles: *Rubus chamaemorus* L. *Journal of Ecology*, **59**, 293–306.

Taylor, K. & Marks, T.C. (1971). The influence of burning and grazing on the growth and development of *Rubus chamaemorus* L. in *Calluna-Eriophorum* bog. In *The Scientific Management of Plant and Animal Communities for Conservation*, ed. E.A.G. Duffey & A.S. Watt, pp. 153–66. Oxford: Blackwell.

Touw, A. (1969). On some liverwort communities in Dutch inland dunes and heaths. *Revue bryologique et lichenologique*, **NS36**, 603–15.

Tubbs, C.R. (1968). *The New Forest: An Ecological History*. Newton Abbot: David & Charles.

Tubbs, C.R. & Dimbleby, G.W. (1965). Early agriculture in the New Forest. *Advances in Science*, **22**, 88–97.

Tubbs, C.R. & Jones, E.L. (1964). The distribution of gorse (*Ulex europaeus* L.) in the New Forest in relation to former land use. *Proceedings of the Hampshire Field Club and Archaeological Society*, **23**, 1–10.

Tutin, T.G., Heywood, V.H., Burges, N.A., Valentine, D.H., Walters, S.M. & Webb, D.A. (1964). *Flora Europaea, Volume 1*. Cambridge: Cambridge University Press.

Tutin, T.G., Heywood, V.H., Burges, N.A., Moore, D.M., Valentine, D.H., Walters, S.M. & Webb, D.A. (1968). *Flora Europaea, Volume 2*. Cambridge: Cambridge University Press.

Tutin, T.G., Heywood, V.H., Burges, N.A., Moore, D.M., Valentine, D.H., Walters, S.M. & Webb, D.A. (1972). *Flora Europaea, Volume 3*. Cambridge: Cambridge University Press.

Tutin, T.G., Heywood, V.H., Burges, N.A., Moore, D.M., Valentine, D.H., Walters, S.M. & Webb, D.A. (1976). *Flora Europaea, Volume 4*. Cambridge: Cambridge University Press.

Tutin, T.G., Heywood, V.H., Burges, N.A., Moore, D.M., Valentine, D.H., Walters, S.M. & Webb, D.A. (1980). *Flora Europaea, Volume 5*. Cambridge: Cambridge

University Press.

Tüxen, R. (1937): Die Pflanzengesellschaften Nordwestdeutschlands. *Mitteilungen der Florist-soziologischen Arbeitsgemeinschaft*, **3**, 1–170.

Tüxen, R. & Oberdorfer, E. (1958). Die Pflanzenwelt Spaniens. II Teil. Eurosibirische Phanerogamen-Gesellschaften Spaniens. *Verföffentlichungen des Geobotanischen Institutes Rübel in Zürich*, **32**, 1–328.

Tüxen, R. & Soyrinki, N. (1958). Die Bullenkuhle bei Bokel. *Abhandlungen hrsg. vom Naturwissenschaftlichen Verein zu Bremen*, **35**, 374–94.

Tveitnes, A. (1945). Fjellbeitne i Hordaland. *Tidsskrift for det norske Landbruk*, **52**, 47–63 & 95–128.

Tyler, C. (1979). Classification of *Schoenus* communities in south and southeast Sweden. *Vegetatio*, **41**, 69–84.

Urquhart, U.H. (1986). *Arctostaphylos Heathland in NE Scotland*. Peterborough: Nature Conservancy Council report.

Urquhart, U.H. & Gimingham, C.H. (1979). *Survey of maritime heath in Northern Scotland*. Edinburgh: Nature Conservancy Council.

Vanden Berghen, C. (1948). La tourbière de Postel. *Biologisch Jaarboek*, **15**, 77–86.

Vanden Berghen, C. (1951). Contribution à l'étude des groupements végétaux des tourbières de Belgique: Landes tourbières bombées à Sphaignes. *Bulletin de la Société Royale de Botanique de Belgique*, **84**, 157–226.

Vanden Berghen, C. (1952). Contribution à l'étude des basmarais de Belgique. *Bulletin du Jardin Botanique de l'État*, **22**, 1–64.

Vanden Berghen, C. (1973). Les Landes à *Erica vagans* de la Haute Soule (Pyrénées-Atlantiques, France). *Colloques Internationales Phytosociologiques II. La Végétation des Landes d'Europe Occidentale*. Lille.

van Groenendael, J.M., Hochstenbach, S.M.H., van Mansfeld, M.J.M. & Roozen, A.J.M. (1979). *The influence of the sea and of parent material on wetlands and blanket bog in west-Connemara, Ireland*. Nijmegen Catholic University: thesis.

Vigerust, Y. (1949). Fjellbeitene i Sikilsdalen. *Utgitt av Selskapet for Norges vel Oslo*, 1–173.

Walker, D. (1970). Direction and rate in some British Post-glacial hydroseres. In *Studies in the Vegetational History of the British Isles*, ed. D. Walker & R.G. West, pp. 117–40. Cambridge: Cambridge University Press.

Walker, D. & Walker, P.M. (1961). Stratigraphic evidence of regeneration in some Irish bogs. *Journal of Ecology*, **49**, 169–85.

Walters, S.M. (1965). Natural History. In *The Cambridge Region*, ed. J.A. Steers, pp. 51–67. Cambridge: British Association for the Advancement of Science.

Ward, S.D. (1970). The phytosociology of *Calluna-Arctostaphylos* heaths in Scotland and Scandinavia. I. Dinnet Moor, Aberdeenshire. *Journal of Ecology*, **58**, 847–64.

Ward, S.D. (1971a). The phytosociology of *Calluna-Arctostaphylos* heaths in Scotland and Scandinavia. II. The north-east Scottish heaths. *Journal of Ecology*, **59**, 679–96.

Ward, S.D. (1971b). The phytosociology of *Calluna-Arctostaphylos* heaths in Scotland and Scandinavia. III. A critical examination of the Arctostaphyleto-Callunetum.

Journal of Ecology, **59**, 697–712.

Ward, S.D., Jones, A.D. & Manton, M. (1972a). The vegetation of Dartmoor. *Field Study*, **3**, 505–34.

Ward, S.D., Evans, D.F. & Millar, R.O. (1972b). *A Vegetation Survey of the Moorfoot Hills Grade 1 Site*. Bangor: Nature Conservancy Montane Grassland Habitat Team.

Watson, A. (1964). The food of ptarmigan (*Lagopus mutus*) in Scotland. *Scottish Naturalist*, **71**, 60–6.

Watson, A. (1966). Hill birds of the Cairngorms. *Scottish Birds*, **4**, 179–203.

Watson, A. & Miller, G.R. (1970). *Grouse Management*. Fordingbridge: The Game Conservancy.

Watson, W. (1922). A new variety of *Orthodontium gracile* Schwegr. *Journal of Botany*, **60**, 139–41.

Watson, W. (1932). The bryophytes and lichens of moorland. *Journal of Ecology*, **20**, 284–313.

Watt, A.S. (1923). On the ecology of British beechwoods with special reference to their regeneration. I. The causes of failure of natural regeneration of the beech (*Fagus sylvatica* L.). *Journal of Ecology*, **11**, 1–48.

Watt, A.S. (1924). On the ecology of British beechwoods with special reference to their regeneration. II. The development and structure of the beech communities on the Sussex Downs. *Journal of Ecology*, **12**, 145–204.

Watt, A.S. (1925). On the ecology of British beechwoods with special reference to their regeneration. II (cont.). The development and structure of the beech communities on the Sussex Downs. *Journal of Ecology*, **13**, 27–73.

Watt, A.S. (1936). Studies in the ecology of Breckland. I. Climate, soils and vegetation. *Journal of Ecology*, **24**, 117–38.

Watt, A.S. (1937). Studies in the ecology of Breckland. II. On the origin and development of blow-outs. *Journal of Ecology*, **25**, 91–112.

Watt, A.S. (1940). Studies in the ecology of Breckland. IV. The grass heath. *Journal of Ecology*, **28**, 42–70.

Watt, A.S. (1947). Pattern and process in the plant community. *Journal of Ecology*, **35**, 1–22.

Watt, A.S. (1955). Bracken versus heather, a study in plant sociology. *Journal of Ecology*, **43**, 490–506.

Watt, A.S. (1971). Factors controlling the floristic composition of some plant communities in Breckland. In *The Scientific Management of Animal and Plant Communities for Conservation*, ed. E.A.G. Duffey & A.S. Watt, pp. 137–52. Oxford: Blackwell.

Watt, A.S. & Jones, E.W. (1948). The ecology of the Cairngorms. I. The environment and the altitudinal zonation of the vegetation. *Journal of Ecology*, **36**, 283–304.

Wattez, J.-R. & Géhu, J.-M. (1972). Documents pour le Caricetum lasiocarpae et le Caricetum diandrae Picards. *Documents Phytosociologiques*, **1**, 47–50.

Webb, D.A. (1956). A new subspecies of *Pedicularis sylvatica* L. *Watsonia*, **3**.

Webb, D.A. (1957). *Hypericum canadense* L., a new American plant in western Ireland. *Irish Naturalists' Journal*, **12**, 113–6.

Webb, D.A. & Halliday, G. (1973). The distribution, habitat and status of *Hypericum canadense* L. in Ireland. *Watsonia*, **9**, 333–44.

Webb, N. (1986). *Heathlands*. London: Collins.
Wein, R.W. (1973). Biological Flora of the British Isles: *Eriophorum vaginatum* L. *Journal of Ecology*, **61**, 601–15.
Welch, D. & Rawes, M. (1966). The intensity of sheep grazing on high-level blanket bog in upper Teesdale. *Irish Journal of Agricultural Research*, **6**, 185–96.
Westhoff, V. & den Held, A.J. (1969). *Plantengemeenschappen in Nederland*. Zutphen: Thieme.
Westhoff, V. & Segal, S. (1961). Cursus Vegetatiekunde 12–17 juni 1961 op Terschelling. Amsterdam: Hugo de Vries Laboratorium.
Wheeler, B.D. (1975). *Phytosociological studies on Rich Fen Systems in England & Wales*. University of Durham: PhD thesis.
Wheeler, B.D. (1978). The wetland plant communities of the River Ant valley, Norfolk. *Transactions of the Norfolk & Norwich Naturalists' Society*, **24**, 153–87.
Wheeler, B.D. (1980a). Plant communities of rich-fen systems in England & Wales. I. Introduction. Tall sedge and reed communities. *Journal of Ecology*, **68**, 368–95.
Wheeler, B.D. (1980b). Plant communities of rich-fen systems in England & Wales. II. Communities of calcareous mires. *Journal of Ecology*, **68**, 405–20.
Wheeler, B.D. (1980c). Plant communities of rich-fen systems in England & Wales. III. Fen meadow, fen grassland and fen woodland communities and contact communities. *Journal of Ecology*, **68**, 761–88.
Wheeler, B.D. (1983). A manuscript copy of a chapter 'British Fens: A Review' now published in *European Mires*, ed. P.D. Moore, pp. 237–81. London: Academic Press (1984).
Wheeler, B.D. (1983b). Vegetation, nutrients and agricultural land-use in a North Buckinghamshire valley fen. *Journal of Ecology*, **71**, 529–44.
Wheeler, B.D. & Giller, K.E. (1982a). Species richness of herbaceous fen vegetation in Broadland, Norfolk, in relation to the quantity of above-ground material. *Journal of Ecology*, **70**, 179–200.
Wheeler, B.D., Al-Farraj, M.M. & Cook, R.E.D. (1985). Iron toxicity to plants in base-rich wetlands: comparative effects on the distribution and growth of *Epilobium hirsutum* L. and *Juncus subnodulosus* Schrank. *New Phytologist*, **100**, 653–69.
Wheeler, B.D., Brookes, B.A. & Smith, R.A.H. (1983). The sites of *Schoenus ferrugineus* L. in Scotland. *Watsonia*, **14**, 249–56.
White, J. & Doyle, G. (1982). The vegetation of Ireland: a catalogue raisonné. *Royal Dublin Society Journal of Life Sciences*, **3**, 289–368.
Whittaker, E. (1960). *Ecological effects of moor burning*. Aberdeen University: PhD thesis.
Whittow, J.B. (1979). *Geology and Scenery in Scotland*. Harmondsworth: Penguin Books.
Wigginton, M.J. & Graham, G.G. (1981). *Guide to the Identification of some Difficult Plant Groups*. Banbury: Nature Conservancy Council, England Field Unit.
Williams, R.B.G. (1964). Fossil patterned ground in eastern England. *Biuletyn peryglacjalny*, **14**, 337–49.
Williams, W.T. & Lambert, J.M. (1959). Multivariate methods in plant ecology. I. Association analysis in plant communities. *Journal of Ecology*, **47**, 83–101.
Williams, W.T. & Lambert, J.M., (1961). Multivariate methods in plant ecology. III. Inverse Association Analysis. *Journal of Ecology*, **49**, 717–29.
Willis, A.J. (1985a). Dune water and nutrient regimes – their ecological relevance. In *Sand Dunes and their Management*, ed. P. Doody, pp. 159–74. Peterborough: Nature Conservancy Council.
Willis, A.J. & Jefferies, R.L. (1959). The plant ecology of the Gordano Valley. *Proceedings of the British Naturalists Society*, **31**, 297–304.
Woodhead, T.W. & Erdtman, O.G.E. (1926). Remains in the peat of the Southern Pennines. *Naturalist (London)*, **835**, 245.
Wooldridge, S.W. & Goldring, F. (1953). *The Weald*. London: Collins.
Yalden, D.W. (1972). The Red Grouse (*Lagopus lagopus scoticus* (Lath.)) in the Peak District. *The Naturalist*, **922**, 89–102.